Reach the Summit!

SUCCESS

Algebra One Interactions–
An Algebra All Can Explore

"With ***Algebra One Interactions****, I discovered just what I was looking for–*
an algebra program that's fun, motivational, and easy to use.
Now all my students have a chance to excel."

HOLT, RINEHART AND WINSTON

Does Algebra One Have to Be Taught IN JUST ONE YEAR?

Consider this. The business world demands more math skills from us than ever before, and this is not going to change. Consider this. Mathematics test scores across the nation show that our students are not keeping pace with the demands for higher math skills in the workplace.

The way we teach is changing, too. Heterogeneous classrooms, students with different learning styles, and fewer remedial math courses challenge us to adopt alternative teaching strategies that will reach a wider range of students.

So does algebra one have to be taught in just one year? Absolutely not.

Enter ***Algebra One Interactions,*** a fun, motivational algebra one course paced over two years. This is an innovative program built for students with varying abilities: students who have difficulty mastering concepts, who need more practice and hands-on experience, or students who are ready for algebra at an earlier age.

With ***Algebra One Interactions,*** algebra is accessible to **all** students. Now all students can enjoy learning algebra and reach the summit of their classroom dreams. Now all students have the opportunity to succeed in a world that demands more mathematics skills of them than ever before.

*"For the first time, I'm really getting into math. It's my subject now. I especially got interested by the **Critical Thinking** feature about extending statistics. Man, just doubling the length and width of a box really changed the volume."*

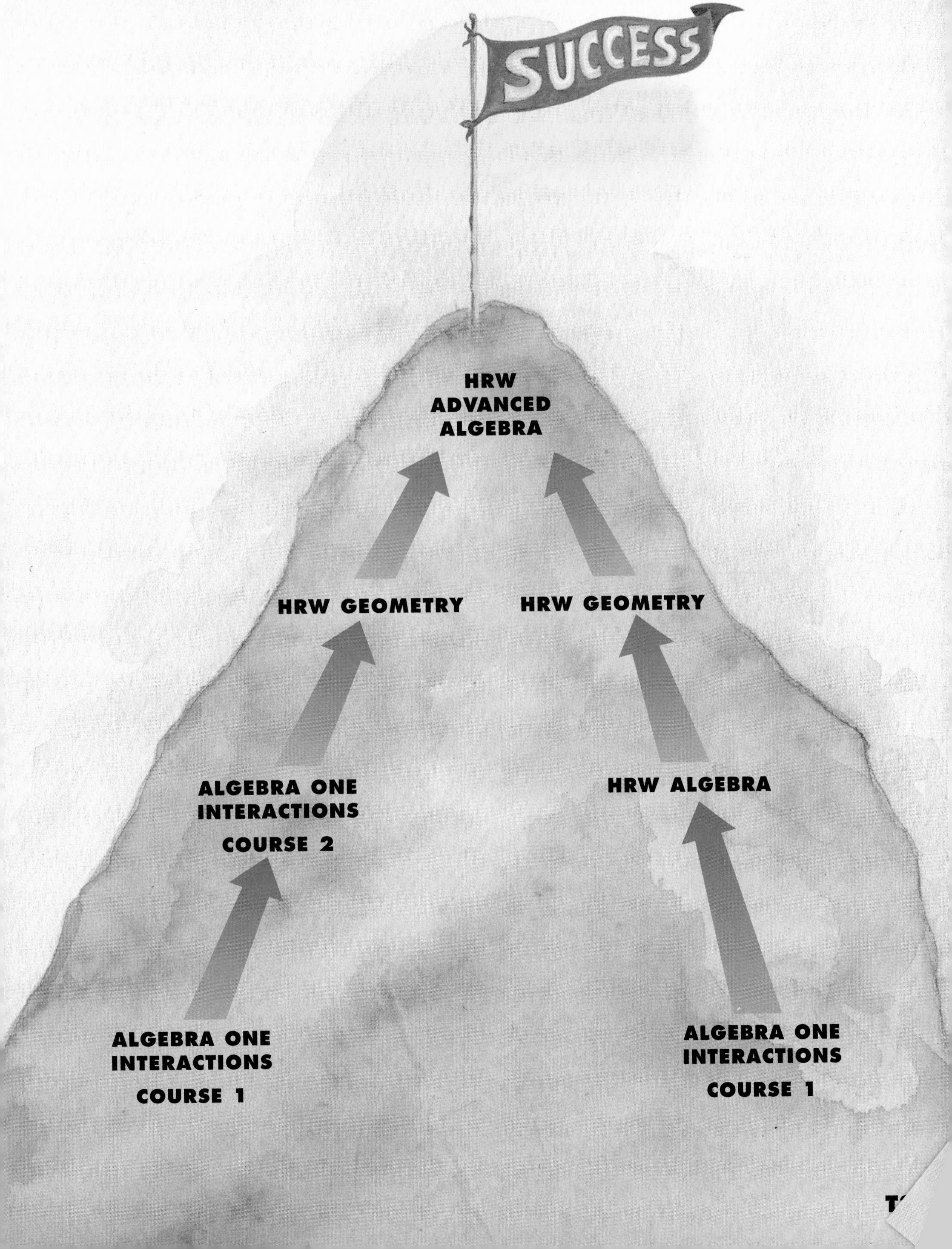

SUCCESS

HRW ADVANCED ALGEBRA

HRW GEOMETRY

HRW GEOMETRY

ALGEBRA ONE INTERACTIONS COURSE 2

HRW ALGEBRA

ALGEBRA ONE INTERACTIONS COURSE 1

ALGEBRA ONE INTERACTIONS COURSE 1

Algebra One Interactions

SETTING A PACE FOR SUCCESS

Establishing a Solid Base Camp

"In ***Algebra One Interactions,*** more than half of *Course 1* is devoted to exploring math concepts that help prepare students for algebra and geometry. By covering integers and data patterns, for example, students learn the basic building blocks of algebra. With paper-folding explorations, they learn geometry. I also like how Chapter 10 serves as a springboard out of linear functions into nonlinear functions in *Course 2*. The overlap really allows students to feel comfortable moving ahead—and believe me, that reassurance never happened in regular algebra."

COURSE 1 TABLE OF CONTENTS

Climbing Higher and Having Fun at the Same Time

"This is definitely heads-up, real-world algebra. ***Algebra One Interactions*** lets my students literally get their hands around symbolic operations and helps them make the jump from the concrete to the abstract. In the *Exploring* lesson in Chapter 7, my class got to use algebra tiles to solve problems with money. You should see the lights go on in their eyes!"

COURSE 2 TABLE OF CONTENTS

"*I never really enjoyed math because I never felt very comfortable with it. But hey, that's history. Now I see how math relates to me. For example, learning about parabolas with a graphics calculator was easy, but I also found out how parabolas can be used in real life.*"

Supplements*

TEAM OF EXPERIENCED GUIDES

"With so many technology tools and teaching resources to support me in the classroom, I know exactly where I stand with my class and exactly where we're going next."

- **Solution Key** Worked–out solutions for exercises in the Pupil's Edition.
- **Practice Workbook** Additional practice for each lesson.
- **Reteaching Masters** Alternative teaching strategies for re-presenting lesson concepts.
- **Technology Masters** Computer and calculator activities that offer additional practice.
- **Technology Handbook** Teacher's guide to using hand-held graphics calculators and computer software (One booklet for both courses).
- **Assessment Resources** A useful collection that includes a Diagnostic Test, Chapter and Mid-Chapter Assessments, Quizzes, and Alternative Assessments that include performance-based assessments and games.
- **Lab Activities and Long-Term Projects** Hands-on activities and projects that engage students inside and outside the classroom and complement regular classroom work.
- **Tech Prep Math Resources** Technical career applications that review the content in each chapter and real-world group projects from various technical fields.
- **Spanish Resources** Translations of key vocabulary and concepts in the Pupil's Edition.
- **Assessment Software** Variety of assessment items and answers delivered on user-friendly software. (Versions available for both Macintosh® and Windows®.)
- **Assessment Software Item Listing** Printout of all the items and answers included on the Assessment Software.
- **Teaching Transparencies** Binder of transparencies that offers over 100 full-color visuals with suggested lesson plans for their use.
- **Teaching Transparencies Directory** Useful guide that makes it easier to review the quality and instructional value of color transparencies and lesson plans.
- **HRW Algebra Tiles and Activities Booklet** Helps students visualize mathematical expressions.
- **Algebra Explorations CD-ROM** Multimedia CD-ROM software program integrated into *Algebra One Interactions*, allowing you to explore the world of functions in an alternative approach to the textbook.
- **Block-Scheduling Handbook** Block-scheduling plans to help you organize indiviual work, group activities, and ongoing assessments.
- **Portfolio Writing Activities** Creative writing activities related to the topics in each chapter.
- **Problem Solving/Critical Thinking Masters** Stimulating problems that take the Pupil's Edition one step further and allow students to extend and enrich their knowledge.
- **Cooperative–Learning Activities** Exploration activities, and games for cooperative learning groups.

****Two booklets, one for each course, except where otherwise noted.***

"The best part was getting to use the computer to explore absolute value functions, and using the ***Algebra Explorations CD-ROM****, we got to help a rescue helicopter find a lost hiker. To me, this kind of math makes sense, this kind of math makes concepts come alive."*

TEACHER'S PLANNING GUIDE

The Trail Is Clearly Marked

"The *Teacher's Planning Guide* is just the thing for taking home and using to plan lessons. It gives me clear, no-nonsense suggestions for teaching in a heterogeneous classroom like mine.

The *Teacher's Answer Edition* is just what I needed, too—all the answers all in one place. It's the perfect reference tool for everyday use in the classroom—easy to use, quick, and I don't have to dig through a lot of teachers' shoptalk."

"I wish I had two teachers' books like these for my other classes. By managing and organizing instruction this way, I save a lot of time. And the flexibility in planning really complements my teaching style. Finally, I can see my way into tomorrow's classes clearly and concisely."

Teacher's Planning Guide—A Superb Compass for Every Lesson

"If other teachers are like me, they know how hard it is to create effective lesson plans—lesson plans that are relevant and also fun for the students. They take time. They take patience. And they also take a lot of experience.

For me, the ***Algebra One Interactions*** *Teacher's Planning Guide* points in exactly the right direction every time and in every situation. For each lesson, it provides two distinct approaches: *Using the Book* if I want my students to follow the material in the Pupil's Edition; and *Using Models* if I want to provide my class with more activity-based instruction. The *Teacher's Planning Guide* even has assessment and enrichment activities that make my job that much easier."

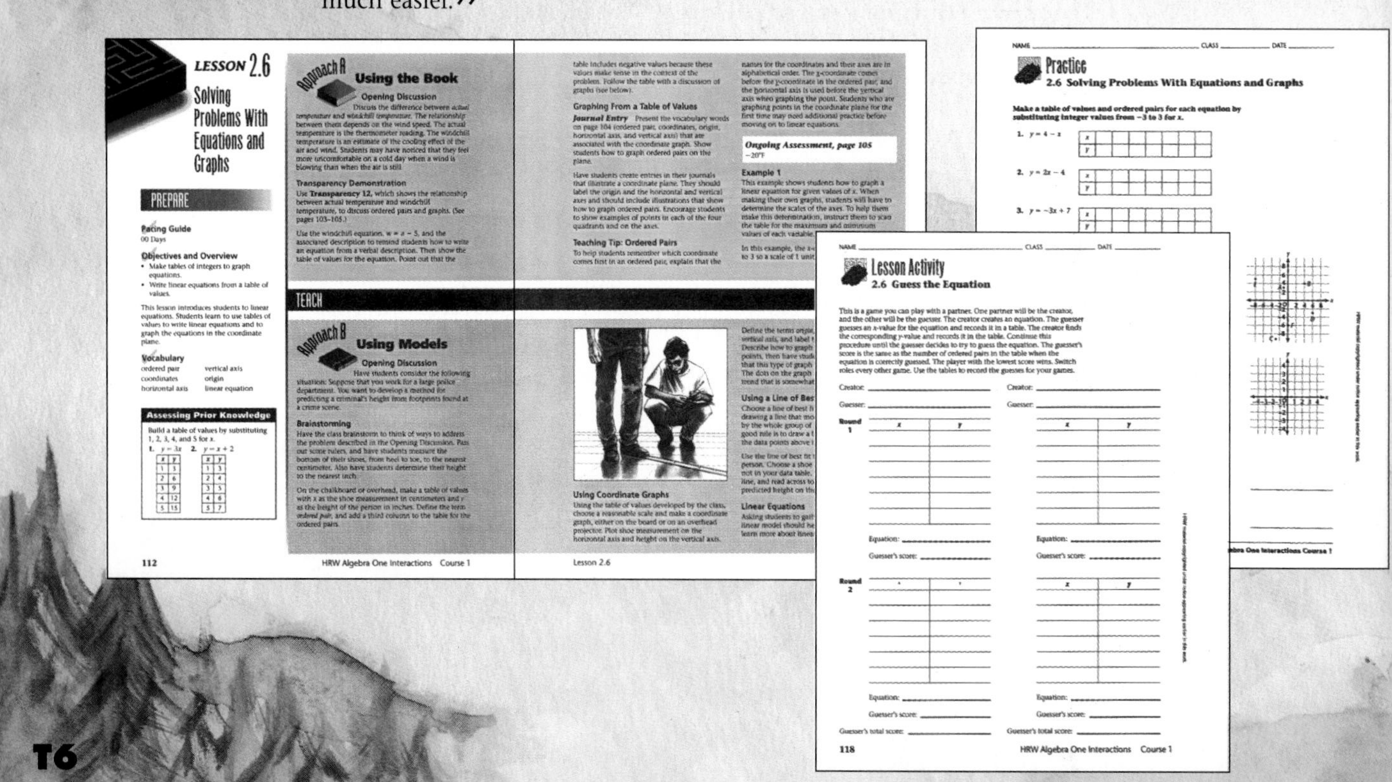

LESSON 2.6
Solving Problems With Equations and Graphs

PREPARE

Pacing Guide
00 Days

Objectives and Overview
- Make tables of integers to graph equations.
- Write linear equations from a table of values.

This lesson introduces students to linear equations. Students learn to use tables of values to write linear equations and to graph the equations in the coordinate plane.

Vocabulary
ordered pair, vertical axis, coordinates, origin, horizontal axis, linear equation

Assessing Prior Knowledge
Build a table of values by substituting 1, 2, 3, 4, and 5 for x.

1. $y = 3x$

x	y
1	3
2	6
3	9
4	12
5	15

2. $y = x + 2$

x	y
1	3
2	4
3	5
4	6
5	7

TEACH

Approach A **Using the Book**

Opening Discussion
Discuss the difference between *actual temperature* and *windchill temperature*. The relationship between them depends on the wind speed. The actual temperature is the thermometer reading. The windchill temperature is an estimate of the cooling effect of the air and wind. Students may have noticed that they feel more uncomfortable on a cold day when a wind is blowing than when the air is still.

Transparency Demonstration
Use **Transparency 12**, which shows the relationship between actual temperature and windchill temperature, to discuss ordered pairs and graphs. (See pages 103–105.)

Use the windchill equation, $w = a - 5$, and the associated description to remind students how to write an equation from a verbal description. Then show the table of values for the equation. Point out that the table includes negative values because these values make sense in the context of the problem. Follow the table with a discussion of graphs (see below).

Graphing From a Table of Values
Journal Entry Present the vocabulary words on page 104 (ordered pair, coordinates, origin, horizontal axis, and vertical axis) that are associated with the coordinate graph. Show students how to graph ordered pairs on the plane.

Have students create entries in their journals that illustrate a coordinate plane. They should label the origin and the horizontal and vertical axes and should include illustrations that show how to graph ordered pairs. Encourage students to show examples of points in each of the four quadrants and on the axes.

Teaching Tip: Ordered Pairs
To help students remember which coordinate comes first in an ordered pair, explain that the names for the coordinates and their axes are in alphabetical order. The x-coordinate comes before the y-coordinate in the ordered pair, and the horizontal axis is used before the vertical axis when graphing the point. Students who are graphing points in the coordinate plane for the first time may need additional practice before moving on to linear equations.

Ongoing Assessment, page 105
−20°F

Example 1
This example shows students how to graph a linear equation for given values of x. When making their own graphs, students will have to determine the scales of the axes. To help them make this determination, instruct them to scan the table for the maximum and minimum values of each variable.

Approach B **Using Models**

Opening Discussion
Have students consider the following situation: Suppose that you work for a large police department. You want to develop a method for predicting a criminal's height from footprints found at a crime scene.

Brainstorming
Have the class brainstorm to think of ways to address the problem described in the Opening Discussion. Pass out some rulers, and have students measure the bottom of their shoes, from heel to toe, to the nearest centimeter. Also have students determine their height to the nearest inch.

On the chalkboard or overhead, make a table of values with x as the shoe measurement in centimeters and y as the height of the person in inches. Define the term *ordered pair*, and add a third column to the table for the ordered pairs.

Using Coordinate Graphs
Using the table of values developed by the class, choose a reasonable scale and make a coordinate graph, either on the board or on an overhead projector. Plot shoe measurement on the horizontal axis and height on the vertical axis.

Using a Line of Best Fit

Linear Equations

112 HRW Algebra One Interactions Course 1 — Lesson 2.6

NAME ______ CLASS ______ DATE ______

Practice
2.6 Solving Problems With Equations and Graphs

Make a table of values and ordered pairs for each equation by substituting integer values from −3 to 3 for x.

1. $y = 4 - x$

x							
y							

2. $y = 2x - 4$

x							
y							

3. $y = -3x + 7$

x							
y							

NAME ______ CLASS ______ DATE ______

Lesson Activity
2.6 Guess the Equation

This is a game you can play with a partner. One partner will be the creator, and the other will be the guesser. The creator creates an equation. The guesser guesses an x-value for the equation and records it in a table. The creator finds the corresponding y-value and records it in the table. Continue this procedure until the guesser decides to try to guess the equation. The guesser's score is the same as the number of ordered pairs in the table when the equation is correctly guessed. The player with the lowest score wins. Switch roles every other game. Use the tables to record the guesses for your games.

Creator: ______ Creator: ______
Guesser: ______ Guesser: ______

Round 1

x	y

Equation: ______
Guesser's score: ______

Round 2

x	y

Equation: ______
Guesser's score: ______

Guesser's total score: ______

118 HRW Algebra One Interactions Course 1

Algebra One Interactions AND TEACHER'S ANSWER EDITION

PREPARE!

"When I opened up the *Teacher's Planning Guide*, I knew I had my hands on an invaluable reference tool I would turn to again and again. The *Prepare* feature took the butterflies out of my stomach because it put the lessons in context. A *Pacing Guide, Objectives and Overview, Vocabulary,* and *Assessing Prior Knowledge* provide crucial information that saves time and establishes goals for both me and my class."

TEACH AND ASSESS!

"Students are sharp—they know if you're prepared; they know if the lesson you're presenting is well thought out. If it's not, their eyes glaze over, their heads start nodding, and then you're doomed.

With *Teach* and *Assess*, that kind of class is not going to happen. You get sound pedagogy, quality teaching strategies that students can relate to."

"Each lesson provides two excellent approaches for teaching: *Using the Book*, if you want to stay close to the text for your instruction; and *Using Models*, for a more kinesthetic, hands-on approach to the content. *Opening Discussion* suggests questions to ask and explanations to give the class to assure that students are keyed into the material. Then, features such as *Journal Entry, Brainstorming, Defining, Cooperative Learning*, and *Explorations*, to name a few, establish definable, tangible goals as well as provide options for teachers to meet individual needs and class styles."

"The *Assignment Guide, Error Analysis*, and *Alternative Assessment* summarize sections and help you evaluate your students progress through the lessons."

RETEACH AND ENRICH

"All students at some point need reinforcement of the skills they're learning. *Reteach* and *Enrich* provide superlative hands-on opportunities for practicing and applying skills. You'll find these features, in particular, help students bridge what to them can seem like an insurmountable gap between concrete and symbolic ways of thinking."

"*Reteach* and *Enrich* exemplify the ways **Algebra One Interactions** puts math back in the real world. *Reteach* in Chapter 1, for example, has students work in groups to resolve how many handshakes occur if ten people in a room shake hands with each person in the room. Talk about a real 'hands on' approach!"

"I like how at the end of each lesson you get three blackline masters right there: *Practice, Lesson Activity*, and *Basic Skills Practice*. These really provide a lot of extra help."

TEACHER'S ANSWER EDITION

Teacher's Answer Edition—Follow the Signs of Progress

"Have you ever felt that teacher's editions were too big and bulky, unmanageable, because they tried to do too many things at once? Well, here's a teacher's edition that simply gives you the answers and that won't feel like you're lugging rocks back and forth to your classroom.

And it dovetails so well with the *Teacher's Planning Guide*. On the one hand, I get a treasure trove of great lesson plans, on the other, a crystal-clear teacher's edition. I couldn't ask for anything more."

"One word about the *Teacher's Answer Edition*: answers, answers, answers. Direct and to the point. No nonsense."

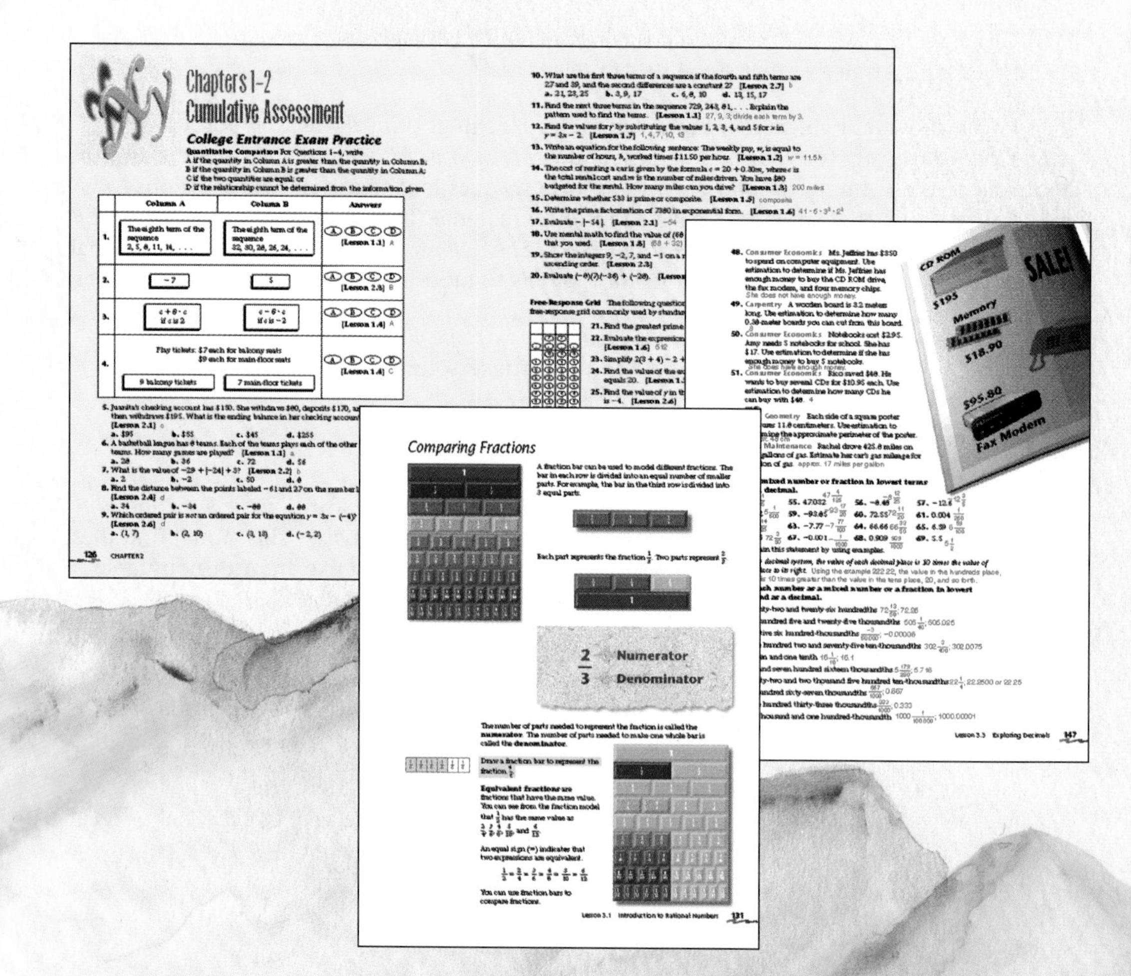

EFFECTIVE INTEGRATION AT EVERY CURVE IN THE TRAIL

Applications that put the fresh air back into algebra

"'Why do I have to learn this?' How many times have we all heard that question? With ***Algebra One Interactions***, the answer is simple. Each lesson opens with *Why*—why, for example, matrices can be used to analyze payroll information for a small company or the results of a swimming competition."

"Algebra doesn't happen out in space somewhere where students can't reach it and apply it to their own lives. ***Algebra One Interactions*** brings math down to earth with all the connections to science and social studies, business and economics, sports and leisure, language arts and life skills, cultures and technology. Then students can decide for themselves what interests them, and how or when they want to travel to the stars."

"The groundspeed of the airplane was only 525.5 miles per hour. Flying into a head wind of 72 miles per hour slowed it down significantly, and by adding the groundspeed to the head wind I figured out the airspeed of the plane. To me, that's what learning math is all about. If I was the pilot, I would've taken the plane to a higher altitude for a faster, smoother ride."

Math connections that let students explore

"Open your eyes and you start to see connections everywhere you look. This is especially true with **Algebra One Interactions**. Geometry, probability and statistics, transformations, and maximum and minimum are all effectively integrated into each bend and curve of the instruction, so students can see how algebra connects with other mathematical disciplines."

"This summer some friends and I are planning a hiking trip up Guadalupe Peak. It won't be that tough though. I figured the slope of the trail with the information I received about the height and length of the trail at certain checkpoints. No problem, the statistical graph showed me the mountain is not as steep as I thought!"

STUDENTS ASCEND TO THE TOP AND COME ALIVE

Practice and assessment built into the instruction—now you're talking!

"Granted, practice and assessment aren't the most important things in life, but they matter in the world of mathematics. And it matters that you have a healthy variety of questions so you can track students progress before, during, and after instruction *and* so you can lead students on different pathways if they have diverse levels of interest and ability."

"***Algebra One Interactions*** won me over with its method of placing ongoing questioning strategies at key points throughout the lessons. Look in the instruction and you'll find interactive, open-ended questions to challenge students understanding. You'll also find *Critical Thinking* and *Communicate* questions that extend concepts and *Try This* exercises that assure students understand what they're learning as they move through the lesson."

"The *Practice & Apply* section is great because it really provides a chance for students to sink their teeth into the concepts. It's not just drill and kill either. Interdisciplinary and cultural connections give the exercises and applications a healthy, real-world focus. When students finish, they move on to *Look Back* and *Look Beyond*. These practice features are superb for reviewing concepts and preparing for new ones. Then, go to *portfolio activities*, long-term *Chapter Projects*, and *Eyewitness Math* activities for alternative practice and assessment to stretch students imaginations."

"*Wow, check it out! I never knew math could be like this. I learned about the stock market, and by the time I got to the* **Practice & Apply** *homework, I knew all the answers.*"

SUCCESS

The heart of algebra is active learning!

"For me, algebra is about getting students involved and having them experience success. I rely on ***Algebra One Interactions*** for this very reason. It gets students involved; it motivates them with kinesthetic learning models, technology, and ongoing questioning strategies that truly work."

"This book is great. For the first time, I understand how math is more than just equations on a page. I can add or subtract integers, for example. I can explore how integers are used in real life, and I can apply what I learn. I wish all my other classes were like this."

TECHNOLOGY—

The Summit is Yours

"The beauty of ***Algebra One Interactions*** is that it's 'technology ready.' When technology becomes available to a teacher, it can be used in the instruction since the option is built right in.

The benefit is obvious: students love using technology and having options. And my class becomes an interactive expedition so students can travel through and explore concepts, not just memorize them and repeat them back on tests."

"Getting to use the graphics calculator was cool! I graphed a translation, a reflection, and a rotation. I learned how the movement of the Earth on its axis is a rotation and a boat gliding across a lake is a translation"

HRW

ALGEBRA ONE INTERACTIONS

COURSE 2

TEACHER'S ANSWER EDITION

Integrating

MATHEMATICS
TECHNOLOGY
EXPLORATIONS
APPLICATIONS
ASSESSMENT

HOLT, RINEHART AND WINSTON

A Harcourt Classroom Education Company

Austin • New York • Orlando • Atlanta • San Francisco • Boston • Dallas • Toronto • London

AUTHORS

Paul A. Kennedy

A professor in the Department of Mathematics at Southwest Texas State University, Dr. Kennedy is a leader in mathematics education reform. His research focuses on developing algebraic thinking by using multiple representations and technology. He has been the author of numerous publications and he is often invited to speak and conduct workshops on the teaching of secondary mathematics.

Diane McGowan

Respected throughout the state of Texas as an educational leader, author, and popular presenter, Ms. McGowan is a past president of the Texas Council of Teachers of Mathematics. She is a mathematics specialist with the Austin Independent School District in Austin, Texas, and has been the recipient of numerous awards for teaching excellence.

James E. Schultz

Dr. Schultz has over 30 years of experience teaching at the high school and college levels and is the Robert L. Morton Professor of Mathematics Education at Ohio University. He helped to establish standards for mathematics instruction as a co-author of the NCTM *Curriculum and Evaluation Standards for School Mathematics* and *A Core Curriculum: Making Mathematics Count for Everyone*.

Kathy Hollowell

Dr. Hollowell is an experienced high school mathematics and computer science teacher who currently serves as Director of the Mathematics & Science Education Resource Center, University of Delaware. Dr. Hollowell is particularly well versed in the special challenge of motivating students and making the classroom a more dynamic place to learn.

Irene "Sam" Jovell

An award-winning teacher at Niskayuna High School, Niskayuna, New York, Ms. Jovell served on the writing team for the New York State Mathematics, Science, and Technology Framework. A popular presenter at state and national conferences, Ms. Jovell's workshops focus on technology-based innovative education.

Portions of this work were published in previous editions.

(Acknowledgments appear on pages 594–595, which are extensions of the copyright page.)

Printed in the United States of America

ISBN: 0-03-055513-2
1 2 3 4 5 6 7 48 04 03 02 01 00 99

TEACHER'S ANSWER EDITION

The most common teacher's edition produced today is a Teacher's Wraparound Edition or a Teacher's Annotated Edition. Its intent is to provide the teacher with helpful suggestions and tips for classroom instruction along with the answers to all the exercises. Although useful in many situations, the standard teacher's edition does not provide the teacher with the assistance needed in the classroom for which *HRW Algebra One Interactions* is written.

Holt, Rinehart and Winston has developed a two-book system for *HRW Algebra One Interactions* that gives teachers the help they need to teach and manage in today's diverse classroom. One of the two teacher's books is this *Teacher's Answer Edition,* which is a duplicate of the *Pupil Edition* with nothing more than overprinted answers. When you need answers, that is exactly what you get.

The other book available to the teacher, is the *Teacher's Planning Guide*. This comprehensive teaching tool provides, in a spacious format, the level of detail that you need. The *Teacher's Planning Guide* provides an extensive presentation of teaching approaches that allows you to create the type of learning environment you want and need for your own classroom.

REVIEWERS

Kenneth L. Gresnick
Sumner High School
St. Louis, Missouri

Judith MacAlpine
Edison High School
Minneapolis, Minnesota

Padma Mani
Andrew Hill High School
San Jose, California

Michael Moloney
Snowden International High School
Boston, Massachusetts

Ida Rheinecker
Orange High School
Orange, California

Evelyn Vieta Robinson
Dade County Public Schools
Miami Beach, Florida

Richard Santoro
Harlan Community Academy
Chicago, Illinois

Susan L. Streeter
South Miami Senior High
Miami, Florida

Pamela Sulzbach
Bell Junior High School
San Diego, California

What Is the Philosophy and Purpose of HRW Algebra One Interactions?

Portfolio Activity

The reform movement in school mathematics has caused a shift not only in the way mathematics is taught, but also in the particular mathematics courses offered. Many of the traditional mathematics courses such as general mathematics and pre-algebra are being eliminated from the high school curriculum. Algebra one, a course at one time reserved for average and above-average students, is becoming a required course for all high school students. However, publishers' algebra textbooks have catered only to the college-bound student. Unable to cope with the abstract presentation of mathematical topics in these textbooks, the remainder of the student population was relegated to a computational skills-based program called general mathematics. Another popular remedial course, pre-algebra, veiled the computational skills practice with the introduction of algebraic symbols and a smattering of geometry topics. This structure in the mathematics curriculum provided students with the necessary number of credits for graduation and, in a time long past, provided students with the minimum mathematics skills needed for non-professional careers.

Prepare Students for the Future

General math and pre-algebra no longer provide a student with the math skills needed to enter into today's technological society. Today, an understanding of algebraic concepts is a minimal requirement for entry into most careers and is a gateway to all forms of higher mathematical study.

The goal is clear and noble and certainly in the best interest of the student. However, the algebra teacher is now faced with a broad base of students of divergent learning levels and learning styles.

Support the Teacher

How can the teacher manage a classroom setting that has suddenly undergone such change? It is obvious that teaching practices and teaching materials must undergo change to accommodate the new algebra classroom. *HRW Algebra One Interactions* is a program designed specifically to accommodate the divergent needs of the students found in this new setting.

Make Algebra Accessible to All

The instructional design of *HRW Algebra One Interactions* is to make algebra accessible to everyone. The lessons in this program are written in such a way that students are presented first with numbers and data, then with tables and graphs, and finally with abstract algebraic concepts. The instruction typically begins with the presentation of real-world data or with the use of manipulatives. In other words, instruction on important mathematical concepts begins with the concrete. As the instruction progresses, the students move to the tabular, then to the graphical,

and then to the symbolic. All levels of student abilities are accounted for in this approach. This is the essence of the instruction found in the student lessons. Labels and separate textbook features that describe students in terms of different ability labels will not be found in the student book. The provisions for different learning levels are embedded in the instructional design of the student lessons. For example, students are often asked to look for patterns by organizing observational data into tables; then they make conjectures from the data. As students generalize or synthesize their observations as conjectures, they progress from the concrete to the abstract. Students who do not possess acute memorization skills will be able to grasp algebraic thinking that was once reserved only for the college-bound student. Students historically labeled as advanced or accelerated will find the instruction more relevant and will gain a deeper understanding of the concepts. They will be more likely to show an interest in mathematically oriented subjects such as chemistry, physics, logic, economics, statistics, and computer science.

Extend Instructional Time

HRW Algebra One Interactions is a two-book algebra one program designed to be taken over a two-year period of time. Twice as much instructional time allows the teacher to take a hands-on, exploratory, and activity-oriented approach to instruction. Manipulatives and calculators will be used extensively, and students will be actively involved in using mathematics in real-world applications. Much group work and many projects will be incorporated into instruction. The program includes all the topics of an algebra one program. It contains a greater coverage of geometry topics and integrates the mathematics of transformations, statistics, probability, and maximum and minimum. *HRW Algebra One Interactions* also includes all of the pre-algebra skills necessary for success in algebra.

Maximize Curriculum Options

Many districts will use this program as a paced or extended algebra one program for students who begin their algebra one study in grade nine and finish at the end of grade ten. Having completed *HRW Algebra One Interactions*, these students will have the prerequisite skills to move into geometry or possibly algebra two. They may or may not take an additional math course before graduation.

Other districts will use *HRW Algebra One Interactions* as an extended algebra one program for students who begin their algebra one study in grade eight. These students will complete the program at the end of grade nine. This allows students to take as many as three other mathematics courses before graduation, much as the traditional curriculum is organized.

And finally, students who are ready to begin a two-year algebra one program in the seventh grade will find *HRW Algebra One Interactions* the best way to complete an algebra one curriculum by the end of grade eight.

ASSIGNMENT GUIDE

Lesson	Exercises for Core Level	Exercises for Core-Plus Level
1.0	1–27	1–27
1.1	1–57	1–10 and 15–59
1.2	1–82	1–35 and 54–73
1.3	1–43	1–8 and 22–45
1.4	1–51	1–9, 18–25, and 28–51
1.5	1–64	1–7, 10–25, and 31–70
1.6	1–75	1–4 and 14–76
2.1	1–61	1–11, 16–21, and 31–65
2.2	1–50	1–5, 7–16, 19–23, and 25–52
2.3	1–41	1–9 and 16–48
2.4	1–39	1–9 and 14–43
2.5	1–44	1–15, 17, and 19–45
2.6	1–44	1–14 and 17–45
2.7	1–36	1–7 and 10–39
2.8	1–24	1–26
3.1	1–33	1–33
3.2	1–40	1–41
3.3	1–51	1–51
3.4	1–41	1–9, 13–29, and 31–41
3.5	1–34	1–34
4.1	1–52	1–28 and 41–52
4.2	1–32	1–14 and 13–33
4.3	1–45	1–18, 21–28, and 33–45
4.4	1–46	1–20 and 31–50
4.5	1–29	1–12 and 21–31
4.6	1–33	1–11 and 23–39
4.7	1–39	1–14 and 22–39
5.1	1–61	1–35, 39–44, and 48–62
5.2	1–44	1–12, 14, 16–18, 20, and 22–47
5.3	1–44	1–44
5.4	1–37	1–37
5.5	1–39	1–41
5.6	1–36	1–36
6.1	1–71	1–71

Lesson	Exercises for Core Level	Exercises for Core-Plus Level
6.2	1–66	1–12, 15–19, 23–31, and 36–70
6.3	1–46	1–48
6.4	1–77	1–7, 12–17, 22–28, and 35–77
6.5	1–43	1–46
6.6	1–43	1–43
7.1	1–39	1–16 and 24–42
7.2	1–53	1–26, 31–37, and 40–54
7.3	1–74	1–14, 24–38, and 47–78
7.4	1–44	1–9 and 17–47
7.5	1–56	1–18 and 26–57
7.6	1–63	1–36 and 42–64
7.7	1–46	1–22 and 30–48
8.1	1–45	1–11, 14–17, 20–26, and 30–46
8.2	1–67	1–27, 38–47, and 49–67
8.3	1–60	1–17, 23–26, and 29–62
8.4	1–63	1–17 and 24–66
8.5	1–59	1–17, 22–32, 37–42, and 46–60
8.6	1–43	1–16, 21–31, and 36–44
9.1	1–56	1–6 and 9–60
9.2	1–66	1–6, 9–18, 21–42, and 45–71
9.3	1–55	1–11, 14–30, and 32–56
9.4	1–38	1–6 and 9–42
9.5	1–44	1–8, 10–14, 16–22, and 24–50
9.6	1–43	1–45
9.7	1–35	1–4, 7–15, and 17–37
9.8	1–66	1–11, 16–33, and 38–68
10.1	1–34	1–8, 11–18, 21–36
10.2	1–30	1–12, 15–26, and 28–33
10.3	1–47	1–12, 15–30, and 35–48
10.4	1–44	1–27 and 33–47
10.5	1–36	1–13, 18–28, and 31–37
10.6	1–42	1–11, 17–37, and 40–43
10.7	1–42	1–16, 19–26, and 30–43

TABLE OF CONTENTS

CHAPTER 3 Matrices 120

CHAPTER 4 Probability and Statistics 164

Math Connections

Applications

CHAPTER 5 Transformations 218

CHAPTER 6 Exponents 264

Math Connections

Applications

CHAPTER 7 Polynomials and Factoring 314

CHAPTER 8 Quadratic Functions 362

Math Connections

Geometry 320, 321, 325, 328, 339, 342, 344, 345, 349, 350, 351, 360, 373, 379, 385, 388, 407 **Coordinate Geometry** 371 **Maximum/Minimum** 365, 376, 397 **Probability** 338, 344, 350, 379, 396 **Statistics** 350, 389

Applications

Science
Physics 367, 373, 378, 379, 389, 401

Language Arts
Communicate 319, 326, 332, 338, 343, 348, 354, 366, 372, 377, 387, 395, 400
Eyewitness Math 380

Business and Economics
Accounting 396
Manufacturing 318
Sales Tax 355
Small Business 337

Sports and Leisure
Sports 406

Other
Band 388
Photography 388

CHAPTER 9 Radicals and Coordinate Geometry 410

CHAPTER 10 Rational Functions 472

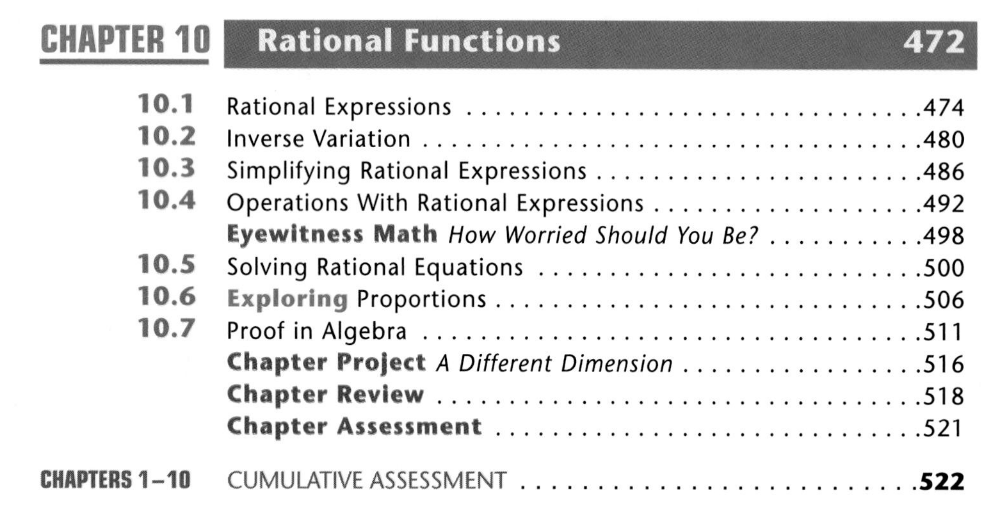

Math Connections

Applications

APPENDIX

Info Bank 524

CHAPTER 1

Functions, Equations, and Inequalities

LESSONS

Equations, graphs, and inequalities can be used to describe and analyze many business situations. In this chapter, you will review how to write an algebraic expression, equation, or inequality to describe relationships such as cost and revenue. You will also review the methods used to solve equations and inequalities.

The ecology club at Hale High School plans to sell T-shirts to raise money for landscaping the central courtyard.

Portfolio Activity

In order for the ecology club to make a profit, the income from selling the T-shirts must be greater than the costs of making the T-shirts. At Ted's Tees, printing costs are 80¢ per shirt, and the cost for each T-shirt is $3. The silk-screen shop charges a $50 set-up fee per order.

In Lessons 1.4 and 1.5, you will be asked to use this information to determine how many shirts need to be sold. You may wish to keep your work for your portfolio.

LESSON 1.0 Exploring Functions

Why *Algebra is the study of relationships between quantities. Many real-world relationships may be modeled by mathematical equations, tables, or graphs.*

The foreign language department at Franklin High School is sponsoring the World Cuisine Festival. The expenses for food, ticket printing, catering supplies, and decorations total $474. Tickets will be sold for $8.00 each. How many tickets do they need to sell to make a profit of $1000?

185

PARTY
PLACE
CATERING SUPPLY REN
Chinese Marke
AL'S
GROCERY STORE
LETTUCE 2.50
5 @ .50
OLIVE OIL 4.89
CHEESE 10.40
4 lb. @ 2.60
FLOUR 3.20
TOMATOES 4.14
6 lb. @ .69
BREAD 1.41
BREAD 1.41
CHICKEN 9.90
5 lb. @ 1.98
SOUR CREAM 4.89
3 @ 1.63
BREAD 1.41
TOTAL 44.15
Expenses:
food $298
catering supplies $57
ticket printing $40
decorations $79
Total expenses: $474

Exploration 1 *Variables and Constants*

1 You can organize the information about the festival in a table. When data is presented in a table format, the patterns are often easier to determine.
Copy and complete the table below to show what the profit (or loss) will be for each number of tickets sold. In the process column, write what you would enter into your calculator to determine the profit.

1.

Process
8(50) − 474
8(100) − 474
8(150) − 474
8(200) − 474
8(250) − 474
8(300) − 474
8(350) − 474

Number of tickets sold	Department expenses	Process	Profit (or loss)
50	$474	8(50) − 474	−$74
100	? $474	?	? $326
150	? $474	?	? $726
200	? $474	?	? $1126
250	? $474	?	? $1526
300	? $474	?	? $1926
350	? $474	?	? $2326

2. The constants in the problem are the department expenses, $474, and the price of a ticket, $8.
4. The amount of profit, p, depends on the number of tickets sold, n. You can tell because when the number of tickets increases, the profit increases.
5. Find the amount of profit by multiplying the number of tickets sold, n, by the ticket price, $8, and then subtract the dept. expenses, $474: $8n - 474 = p$.

2 A **constant** is a number that represents a fixed amount. What are the constants in this problem?

3 A **variable** is a number that represents an amount that is *not* fixed. Its value can *vary,* or change. Describe the two variables in this situation. Use a letter to represent each variable. The variables are number of tickets sold, n, and profit (or loss), p.

4 The amount of profit depends on which variable? How can you tell?

5 Describe the process you used in the table to find the amounts of profit.

6 Since the amount of profit *depends* on another variable in this problem, profit is the **dependent variable**. The dependent variable, profit, depends on an **independent variable**. What is the independent variable? ❖ The number of tickets sold, n

In this example, the profit is a *function* of the number of tickets sold.

A **function** is defined as a set of ordered pairs. In this situation, the ordered pairs consist of the number of tickets sold and the corresponding profit. For example, the ordered pair (50, −74) represents 50 tickets sold and a profit of −\$74, or a loss of \$74. If profit is represented by the variable p, and the number of tickets sold is represented by the variable n, then the ordered pairs are of the form (n, p). The profit, p, depends on the number of tickets sold, n. Thus, p is the dependent variable, and n is the independent variable.

3. Find the amount of profit by multiplying the number of tickets sold by the ticket price, and then subtract the department expenses: $8n - 474 = p$.

Exploration 2 Using an Equation to Represent a Function

1. What is the profit when 275 tickets are sold? when 425 tickets are sold? Explain how you found your answers. \$1726; \$2926

2. How many tickets must be sold to make a profit of \$1126?
200 tickets

3. Describe how to find the amount of profit when you know the number of tickets sold and the expenses.

4. Use your description from Step 3 to write an equation that expresses the amount of profit, y, in terms of the number of tickets sold, x. ❖ $y = 8x - 474$

Since the profit is a function of the number of tickets sold, Tasha and Imelda can count the number of tickets that were sold to determine the amount of profit that should have been made.

Exploration 3 *Using a Graph to Represent a Function*

The equation you wrote for Step 4 of Exploration 2 is graphed here. The graph shows the amount of profit as a function of the number of tickets sold.

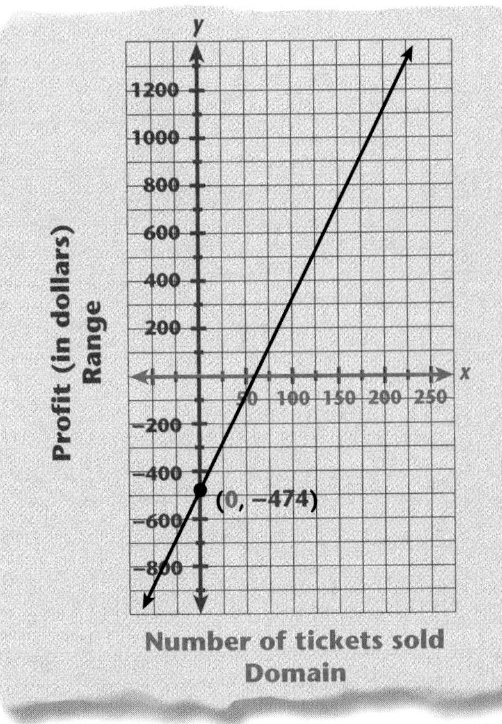

Part I

1. Is the independent variable represented on the horizontal or vertical axis in this graph?
horizontal

2. Is the dependent variable represented on the horizontal or vertical axis in this graph?
vertical

3. How can you use the graph to determine the amount of profit from selling 100 tickets? from selling 200 tickets?

4. Explain how to use the graph to determine how many tickets must be sold to make a profit of about \$800.

Part I 3. To find the amount of profit from selling 100 tickets, find 100 on the horizontal axis. Then move vertically from that point to the graph of the function. The corresponding y-value is the amount of profit (\$326). To find the amount of profit from selling 200 tickets, find 200 on the horizontal axis. Then move vertically from that point to the graph of the function. The corresponding y-value is the amount of profit, (\$1126).

4. To find the number of tickets that must be sold to make a profit of \$800, find 800 on the vertical axis. From that point move horizontally to the graph of the function. The corresponding x-value is the number of tickets (160 tickets).

Part II

1. The values represented on the horizontal axis are called the **domain** of the function. The values represented on the vertical axis are called the **range** of the function.

2. From looking at the graph, what values are in the *domain* of the function? What values are in the *range?* The values of the domain and range are all real numbers.

3. Considering this situation, explain why it is not reasonable to include negative numbers in the domain for this graph. Explain why it is not reasonable to include fractions in the domain for this graph. It is impossible to sell a negative number of tickets. Also, it is impossible to sell a fraction of a ticket.

4. The set of numbers that are reasonable to include in the domain of this graph is called the **reasonable domain** for this situation. What is the reasonable domain for this situation? ❖ All positive integers

1. Describe at least two ways that a real-world problem may be modeled. Answers may vary. Real-world problems may be modeled by equations, tables, or graphs.

Wages Sarah receives a salary of $200 per week plus a commission of 8% of her weekly sales. Copy and complete the table for Exercises 2–7. Then use your table for Exercises 8–11.

	Weekly salary	Weekly sales	Commission	Total income
2.	200	500	40 ?	240 ?
3.	200 ?	1000	80 ?	280 ?
4.	200 ?	1500	120 ?	320 ?
5.	200 ?	2000	160 ?	360 ?
6.	200 ?	2500	200 ?	400 ?
7.	200 ?	3000	240 ?	440 ?

8. What are the constants in this situation? What are the variables? The constant is weekly salary, $200; and the variables are weekly sales, s, commission, c, and total income, i.
9. Choose letters to represent the variables. Write an equation that expresses the total income in terms of the amount of weekly sales and the weekly salary. $i = 200 + 0.08s$
10. What would the income be if the total weekly sales were $5000? $600
11. Use guess-and-check to determine the amount of weekly sales if the income for one week were $640. $5500

Travel Maxwell drives for several minutes at an average speed of 0.8 mile per minute. Copy and complete the table for Exercises 12–16. Then answer Exercises 17 and 18.

	Driving time (in minutes)	Process	Distance traveled (in miles)
12.	1	?	0.8 ?
13.	2	?	1.6 ?
14.	3	?	2.4 ?
15.	4	?	3.2 ?
16.	5	?	4.0 ?

Process
0.8(1)
0.8(2)
0.8(3)
0.8(4)
0.8(5)

17. Write an equation that expresses the distance in miles traveled as a function of time in minutes.
18. Use the equation you wrote in Exercise 17 to find the distance that Maxwell travels if he drives for 1 hour at this average speed.

17. $d = 0.8t$, where d is distance in miles and t is time in minutes.

18. $d = 0.8t$
$d = 0.8(60)$
$d = 48$ miles

Wages The graph shows Lyle's daily wages as a function of the number of hours that he works. Use the graph for Exercises 19–22.

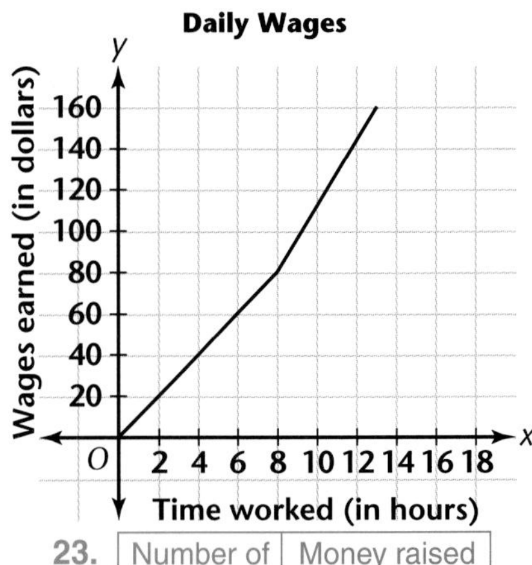

19. What would Lyle's wages be if he worked 7 hours in one day? Explain how you used the graph to find your answer.

20. Suppose that Lyle made $40 in one day. Determine his hourly wage. Explain how you found your answer.

21. What happens to the graph if Lyle works more than 8 hours in one day?

22. What is a reasonable domain for this problem? Explain your reasoning.

Fund-raising The Silver Star dance team is selling homecoming mums. The graph below shows the number of mums sold and the money raised by the team.

23. Construct a table from the graph.

24. Describe the relationship between the number of mums sold and the income displayed in the graph below.

25. Describe how you would use the graph to determine how much money can be raised by selling 10 mums.

26. Explain how to use the graph to determine the number of mums that must be sold to raise $75.

27. What is a reasonable domain for this situation?

23.

Number of mums sold	Money raised (in dollars)
2	$ 30
4	$ 60
6	$ 90
8	$120

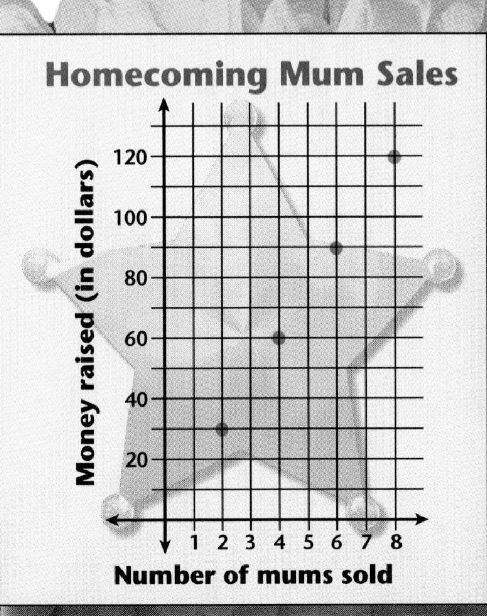

LESSON 1.1 Addition and Subtraction Equations

why *Equations are used in many fields to solve problems. For instance, in business, changes in cost or revenue can be represented by equations.*

Jesse makes clay items to sell. Because the clay items sell quickly, Jesse decides to increase the price per item.

Subtraction Property of Equality

Jesse used to charge \$2 per clay item. He decides to increase the price of each clay item to \$6. This price increase can be represented by the equation $2 + x = 6$, or $x + 2 = 6$, where x represents the amount of increase.

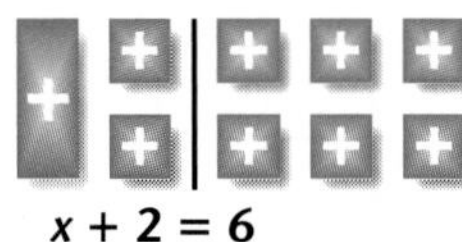

$x + 2 = 6$

The model of the equation $x + 2 = 6$ is shown here. The variable, x, represents the price increase in dollars.

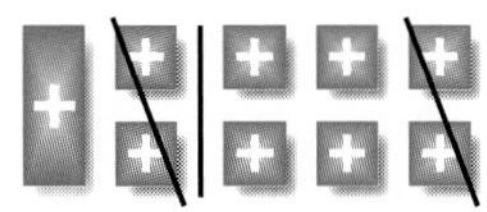

$x + 2 = 6$
$x + 2 - 2 = 6 - 2$

To isolate the x-variable, subtract 2 from each side of the equation.

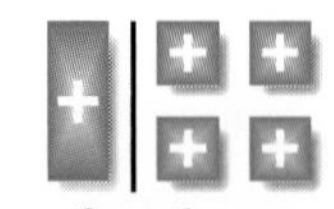

$x + 2 = 6$
$x + 2 - 2 = 6 - 2$
$x = 4$

Simplify.

Jesse increased the price of each clay item by \$4.

Subtracting the same amount from each side of an equation is called the Subtraction Property of Equality.

SUBTRACTION PROPERTY OF EQUALITY

If equal amounts are subtracted from the expressions on each side of an equation, the expressions remain equal.

You can create a graph on a graphics calculator or on graph paper to check the solution of this equation. To find the solution to $x + 2 = 6$, graph the equations $Y_1 = X + 2$ and $Y_2 = 6$. Then find the point of intersection of the two lines. The lines intersect at point (4, 6). The solution to the equation $x + 2 = 6$ is 4.

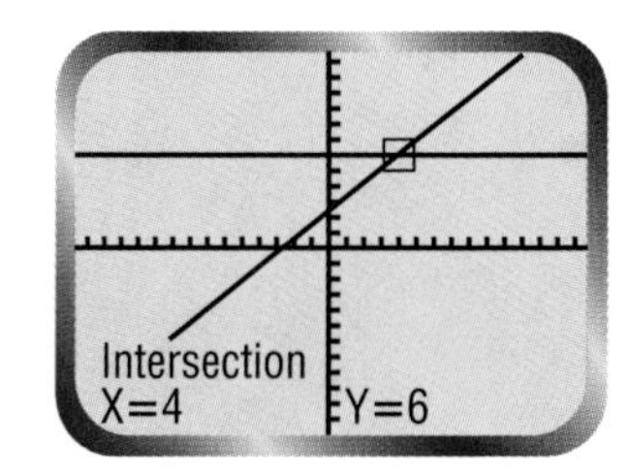

EXAMPLE 1

Solve each equation. Check your solutions.

A $x + 8 = 12$ **B** $t + 9 = 4$

Solution ➤

A To solve the equation, $x + 8 = 12$, subtract 8 from each side of the equation.

$$x + 8 = 12 \quad \text{Given}$$
$$x + 8 - 8 = 12 - 8 \quad \text{Subtraction Property of Equality}$$
$$x = 4 \quad \text{Simplify.}$$

Check by substituting 4 for x in the original equation.

$$x + 8 = 12$$
$$(4) + 8 = 12 \quad \text{True}$$

Since $4 + 8 = 12$, the solution is 4.

B To solve the equation $t + 9 = 4$, subtract 9 from each side of the equation.

$$t + 9 = 4 \quad \text{Given}$$
$$t + 9 - 9 = 4 - 9 \quad \text{Subtraction Property of Equality}$$
$$t = -5 \quad \text{Simplify.}$$

Check by substituting -5 for t in the original equation.

$$t + 9 = 4$$
$$(-5) + 9 = 4 \quad \text{True}$$

Since $-5 + 9 = 4$, the solution is -5.

Try This Solve each equation. Check your solutions. **a.** $y = -15$ **b.** $y = 8$

a. $y + 7 = -8$ **b.** $y - 12 = -4$

Addition Property of Equality

When the same number is added to the expressions on each side of an equation, the Addition Property of Equality is used.

ADDITION PROPERTY OF EQUALITY

If equal amounts are added to the expressions on each side of an equation, the expressions remain equal.

The tile model of the equation $x - 3 = 2$ is shown here.

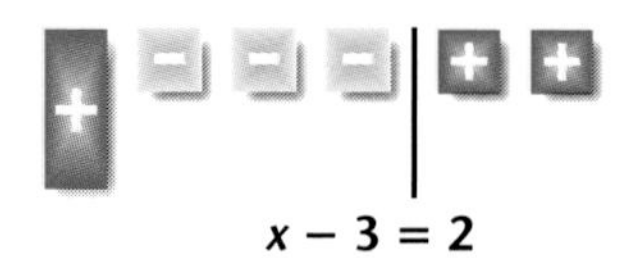

Isolate the x-tile by adding 3 to each side of the equation.

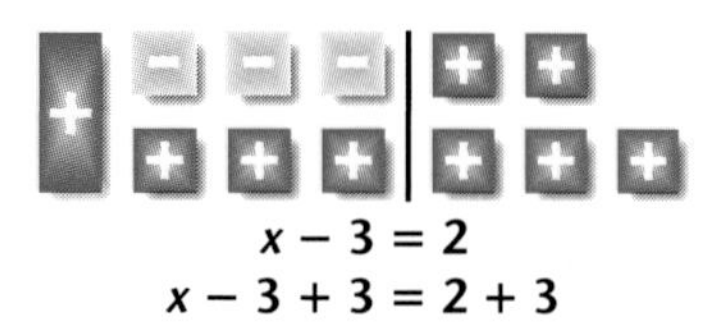

A pair of tiles with opposite signs is called a **neutral pair**. Remove the neutral pairs.

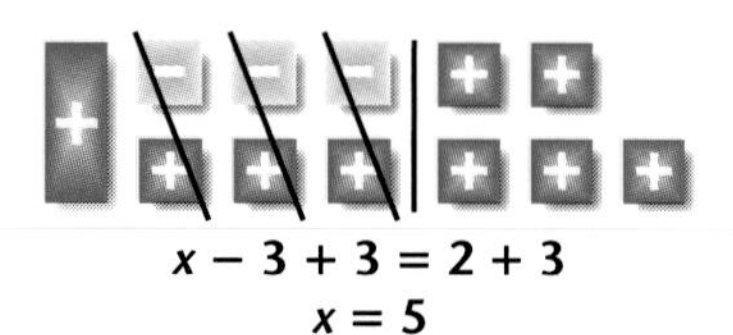

The solution to the equation $x - 3 = 2$ is 5.

Describe how to check this solution. Substitute 5 for the value of x in the equation $x - 3 = 2$: $(5) - 3 = 2$. Since the resulting equation is true, 5 is the correct solution.

EXAMPLE 2

Solve the equation $x - 7 = 2$. Check your solution.

Solution ➤

To solve the equation $x - 7 = 2$, use the Addition Property of Equality. Add 7 to the expressions on each side of the equation.

$x - 7 = 2$	Given
$x - 7 + 7 = 2 + 7$	Addition Property of Equality
$x = 9$	Simplify.

The solution is 9.

Check by substituting 9 for x in the original equation.

$x - 7 = 2$	
$(9) - 7 = 2$	True

Since $9 - 7 = 2$, the solution is 9. ❖

Try This Solve the equation $x - 5 = -4$. Check your solution. $x = 1$

EXAMPLE 3

Sales The *wholesale* price of an item is the amount that a store pays for the item. The *retail* price is the amount that the store charges the customer for the item.

A Write an equation to show that the retail price of this CD player depends on the wholesale price.

B What is the equation if the retail price of the CD player is $130? Solve the equation to determine the wholesale price.

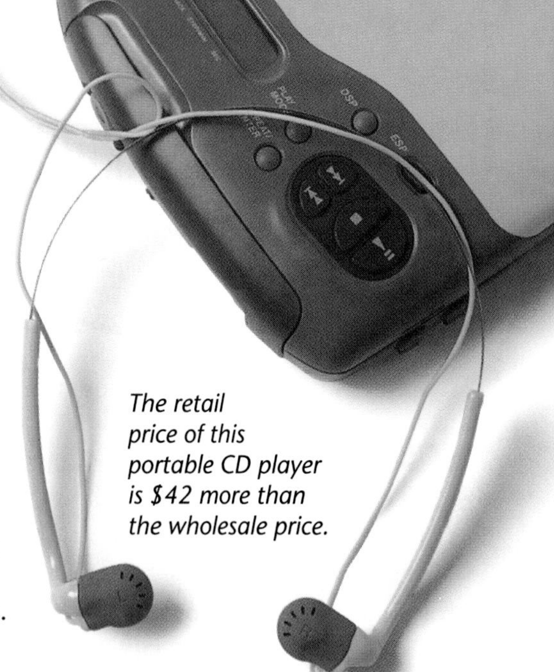

The retail price of this portable CD player is $42 more than the wholesale price.

Solution ➤

A Let w represent the wholesale price, and let r represent the retail price.

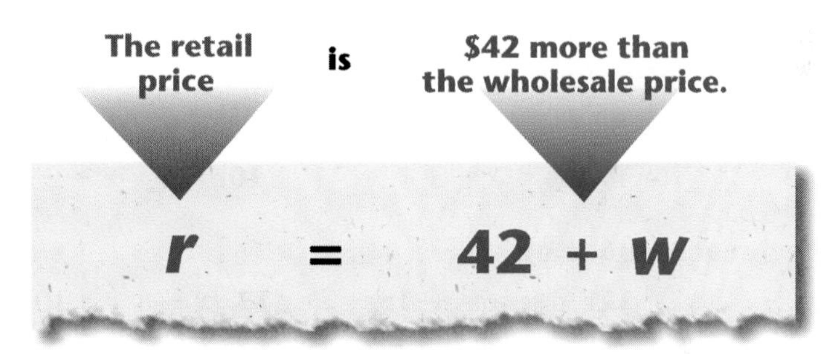
The retail price
is
$42 more than the wholesale price.
r = 42 + w

B The equation for the retail price is $r = 42 + w$.

Since the retail price is $130, substitute 130 for r.
Solve $130 = 42 + w$ for w.

$130 = 42 + w$	Given
$130 - 42 = 42 + w - 42$	Subtraction Property of Equality
$88 = w$	Simplify.

Check by substituting 88 for w in the equation $130 = 42 + w$.

$130 = 42 + w$	
$130 = 42 + (88)$	
$130 = 130$	True

The wholesale price is $88. ❖

a. $s = r - 3$ **b.** $51 = r - 3$; $54 = r$
The retail price is $54.

Try This To determine the sale price of an item, a store manager deducts $3 from the retail price.

a. Write an equation to show that the sale price, s, depends on the retail price, r.

b. What is the equation if the sale price is $51? Solve the equation to determine the retail price.

EXERCISES & PROBLEMS

These students are using algebra tiles to solve the equation $x - 2 = -1$.

Communicate

1. Explain how to solve the equation $x - 2 = -1$ with algebra tiles.

2. Describe how to check your solution to an equation.

3. Explain how to solve an addition or subtraction equation.

4. What is a real-world problem that can be represented by the equation $x + 3 = 12$?

Practice & Apply

Use algebra tiles to solve each equation.

5. $x + 4 = -4$ $x = -8$ **6.** $x - 5 = 8$ $x = 13$ **7.** $x - 2 = -3$ $x = -1$

8. $x + 1 = 6$ $x = 5$ **9.** $-4 = x - 3$ $x = -1$ **10.** $-6 = x + 2$ $x = -8$

Solve and check each equation.

11. $x + 3 = 12$ 9 **12.** $x + 16 = -15$ -31 **13.** $x - 8 = -10$ -2 **14.** $y - 10 = 3$

15. $-14 = x - 14$ 0 **16.** $-5 = x - 12$ 7 **17.** $x - 7.2 = 5.3$ 12.5 **18.** $-1.5 = a + 1.9$

19. $x + 6 = -5$ -11 **20.** $t + 5 = 10$ 5 **21.** $y - 18 = -1$ 17 **22.** $y - 7 = 7$

23. $-1 = x + 11$ -12 **24.** $-15 = r - 2$ -13 **25.** $x - 6.4 = 7.9$ 14.3 **26.** $n + 5 = 11$

27. $-2.5 = x + 4.3$ -6.8 **28.** $b + 6 = -14$ -20 **29.** $y - 9 = -1$ 8 **30.** $h - 20 = 13$

31. $-1 = x + 4$ -5 **32.** $-15 = e - 22$ 7 **33.** $x - 7.5 = 2.3$ 9.8 **34.** $-0.5 = f + 2.8$

14. 13 **18.** -3.4 **22.** 14 **26.** 6 **30.** 33 **34.** -3.3

Geometry Recall that the perimeter of a triangle is the sum of the lengths of all the sides of the triangle.

35. A triangle has perimeter, P, and sides, a, b, and c. Write an equation to describe the perimeter in terms of the sides of the triangle. $P = a + b + c$

36. Rewrite your equation from Exercise 35 to show that the perimeter of the triangle is 12, the length of side a is 4, and the length of side c is 5. Solve the equation for b to find the length of the third side.
$12 = 4 + b + 5; b = 3$

Discounts The sale price of a table is $42 less than the retail price of the table.

37. Write an equation to describe the sale price, s, of the table in terms of the retail price, r. $s = r - 42$

38. Rewrite your equation from Exercise 37 to show that the sale price is $193. Solve the equation to find the retail price. $193 = r - 42; r = 235$

Statistics There were 2500 more residents in Junction City in 1996 than in 1940.

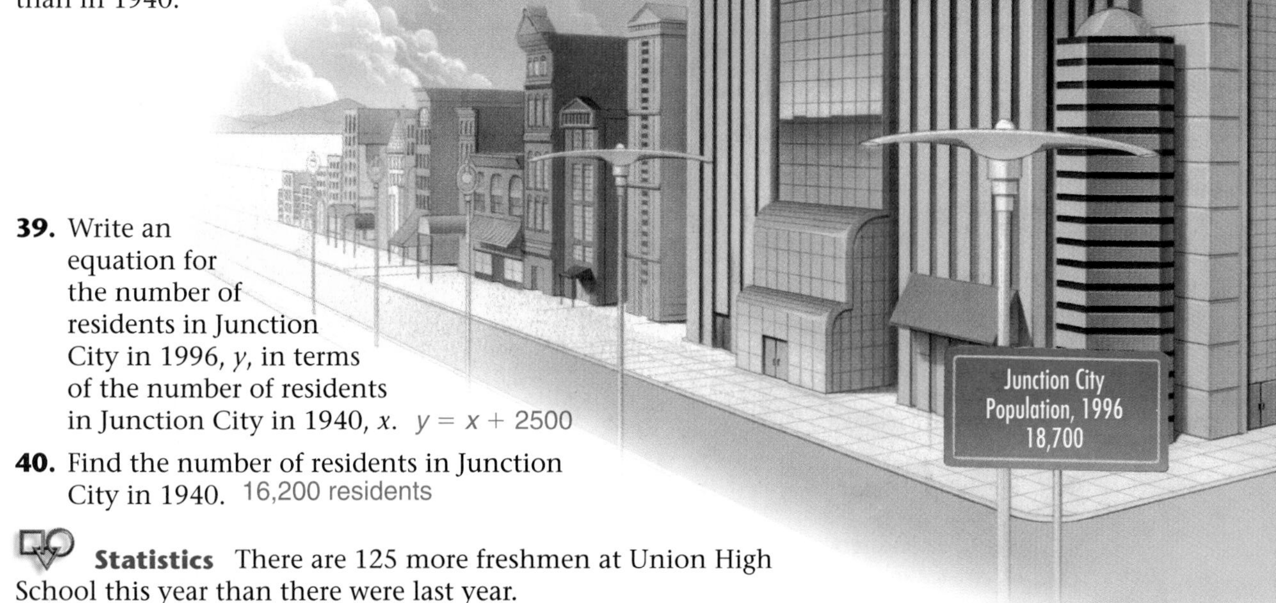

39. Write an equation for the number of residents in Junction City in 1996, y, in terms of the number of residents in Junction City in 1940, x. $y = x + 2500$

40. Find the number of residents in Junction City in 1940. 16,200 residents

Statistics There are 125 more freshmen at Union High School this year than there were last year.

41. Write an equation to describe the number of freshmen this year, t, in terms of the number of freshmen last year, l. $t = l + 125$

42. Rewrite your equation from Exercise 41 to show that the number of freshmen this year is 652. Solve the equation to find the number of freshmen last year. $652 = l + 125$; $l = 527$ freshmen

Look Back

Simplify each expression. [Course 1]

43. $3\frac{1}{2} \cdot 4\frac{1}{3}$ $15\frac{1}{6}$ **44.** $\frac{1}{3} \div \frac{4}{5}$ $\frac{5}{12}$ **45.** $\frac{-4}{7} \cdot \frac{3}{10}$ $-\frac{6}{35}$ **46.** $\frac{2}{3} + \frac{1}{3} + \frac{1}{3}$ $1\frac{1}{3}$

47. $\frac{4}{5} + \frac{1}{10}$ $\frac{9}{10}$ **48.** $\frac{1}{9} + \frac{3}{8}$ $\frac{35}{72}$ **49.** $\frac{2}{3} - \frac{1}{4}$ $\frac{5}{12}$ **50.** $2\frac{1}{3} \div \frac{1}{3}$ 7

Write each percent as a fraction in lowest terms and as a decimal. [Course 1]

51. 25% **52.** 45% **53.** 60% **54.** $33\frac{1}{3}\%$

51. $\frac{1}{4} = 0.25$
52. $\frac{9}{20} = 0.45$
53. $\frac{3}{5} = 0.6$
54. $\frac{1}{3} = 0.\overline{3}$

Use the order of operations to evaluate each expression. [Course 1]

55. $25 + 4 \cdot 2 - 16 \div 2$ 25

56. $32 - 14 \div 2 + 8 \cdot 3$ 49

Look Beyond

Write an equation for each sentence.

57. Five more than a number is 6.

58. Gayle's age is twice Joel's age.

59. The length of the white rope is $\frac{2}{3}$ of the length of the yellow rope.

LESSON 1.2

Multiplication and Division Equations

WHY *Multiplication and division equations are sometimes used to solve problems involving sales and profit.*

The band boosters are selling pizza slices at lunch time. They make a profit of $2 for each whole pizza they sell.

Division Property of Equality

The amount of profit that the band boosters earn depends on the number of pizzas they sell. Let p represent the profit, and let x represent the number of pizzas the band boosters sell. Because they make a profit of $2 for each whole pizza they sell, the equation for this relationship is $p = 2x$.

If the profit is $16, then substitute 16 for p in the equation. Then solve the resulting equation, $16 = 2x$ or $2x = 16$ for x to find the number of pizzas sold.

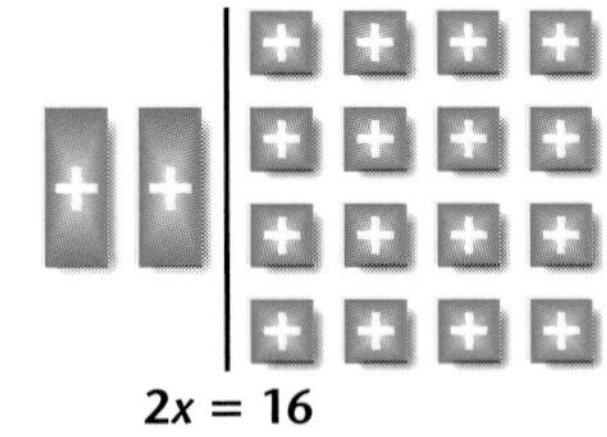

$2x = 16$

To isolate an x-tile, divide each side into two sets.

Since 8 1-tiles are grouped with each x-tile, the solution is 8.

The number of pizzas sold was 8.

$$2x = 16$$
$$\frac{2x}{2} = \frac{16}{2}$$
$$x = 8$$

Dividing the expressions on each side of an equation by the same amount is called the Division Property of Equality.

DIVISION PROPERTY OF EQUALITY

If the expressions on each side of an equation are divided by equal nonzero amounts, the expressions remain equal.

EXAMPLE 1

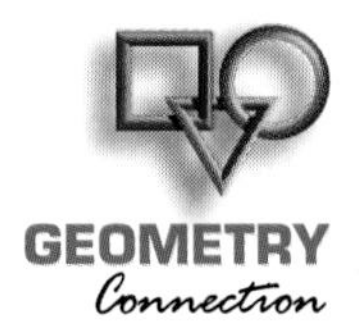

The sum of the angles in a triangle is 180°. Find the measure of each angle in triangle *ABC*.

Solution ➤

The sum of the angles is 180°, so the equation is $x + 2x + 3x = 180°$. Solve this equation for x.

A, x, 3x, 2x, C, B

$$x + 2x + 3x = 180 \quad \text{Given}$$
$$6x = 180 \quad \text{Simplify.}$$
$$\frac{6x}{6} = \frac{180}{6} \quad \text{Division Property of Equality}$$
$$x = 30 \quad \text{Simplify.}$$

Substitute 30 for x in each expression to find the measures of the angles.

$$m\angle A = x = 30° \qquad m\angle B = 2x = 2(30) = 60° \qquad m\angle C = 3x = 3(30) = 90°$$

Check the solution by adding the angle measures.

$$30° + 60° + 90° = 180°$$

Since the sum of these angle measures is 180°, the answer is reasonable. ❖

Multiplication Property of Equality

Each side of an equation can also be multiplied by the same amount. This is called the Multiplication Property of Equality.

MULTIPLICATION PROPERTY OF EQUALITY

If the expressions on each side of an equation are multiplied by equal amounts, the expressions remain equal.

Another important property that can be used to solve equations is the Reciprocal Property.

RECIPROCAL PROPERTY

For any nonzero number, r, there is a number, $\frac{1}{r}$, such that

$$r \cdot \frac{1}{r} = 1.$$

EXAMPLE 2

Solve each equation. **A** $\frac{y}{6} = 5$ **B** $\frac{2}{3}x = 8$

Solution ➤

A To solve $\frac{y}{6} = 5$, multiply the expressions on each side of the equation by 6.

$\frac{y}{6} = 5$	Given
$6\left(\frac{y}{6}\right) = 6 \cdot 5$	Multiplication Property of Equality
$y = 30$	Simplify.

To check, substitute 30 for y in the original equation.

$$\frac{y}{6} = 5$$

$$\frac{30}{6} = 5 \quad \text{True}$$

Since $\frac{30}{6} = 5$, the solution is 30.

B To solve $\frac{2}{3}x = 8$, multiply the expressions on each side of the equation by the reciprocal of $\frac{2}{3}$, which is $\frac{3}{2}$.

$\frac{2}{3}x = 8$	Given
$\frac{3}{2}\left(\frac{2}{3}x\right) = \frac{3}{2}(8)$	Multiplication Property of Equality
$x = \frac{24}{2}$	Simplify.
$x = 12$	Simplify.

To check, substitute 12 for x in the original equation.

$$\frac{2}{3}x = 8$$

$$\frac{2}{3}(12) = 8$$

$$\frac{24}{3} = 8 \quad \text{True}$$

Since $\frac{24}{3} = 8$, the solution is 12. ❖

EXAMPLE 3

Discounts The manager of a book store is having a sale on everything. The sale price is 70% of the regular price.

A Write an equation to describe the sale price, s, in terms of the regular price, r.

B What is the regular price of a book if the sale price is $9.80?

Solution

A Let r represent the regular price and s represent the sale price.

The sale price is 70% of the regular price.

$$s = 0.70 \cdot r$$

CT– If x is the regular price, then the regular price is 100% of x.

100% of x
−30% of x
70% of x

"30% off regular price" means that you subtract 30% from the regular price, 100%.

The equation is $s = 0.70r$.

B Since the sale price of the book is $9.80, substitute 9.80 for s in the equation $s = 0.70r$. Then solve for r to determine the regular price.

$9.80 = 0.70r$ Given

$\frac{9.80}{0.70} = \frac{0.70r}{0.70}$ Division Property of Equality

$14 = r$ Simplify.

The regular price of the book is $14.00. ❖

Explain why the discount "30% off the regular price" is the same as "70% of the regular price."

Try This Helena is a college student who has a job as an engineering intern. She earns $11 per hour.

a. Write an equation to describe the amount, a, that Helena earns in terms of the number of hours, h, that she works. $a = 11h$

b. How many hours will Helena need to work in order to earn $500? About 45.5 hours

Solving Proportions

Suppose that the relationship between the number of males and the number of females in the school choir is 2 to 3. The expression *2 to 3* can also be written as 2:3 or $\frac{2}{3}$. The expression means that for every 2 males in the choir there are 3 females. Expressions that compare two quantities in this way are called **ratios**.

Equivalent ratios are ratios that have the same value. A **proportion** is an equation containing two or more equivalent ratios. The proportions below show some possibilities for the actual number of males and females in the school choir.

$\frac{2}{3} = \frac{6}{9}$ ← males (6), ← females (9) $\frac{2}{3} = \frac{10}{15}$ ← males (10), ← females (15)

EXAMPLE 4

The ratio of white flowers to red flowers in a bouquet is 2 to 3. If there are 30 red flowers, how many white flowers are in the bouquet?

Solution ➤

Let w represent the number of white flowers in the bouquet. Write a proportion.

$$\begin{array}{r}\text{white flowers} \rightarrow \\ \text{red flowers} \rightarrow\end{array} \frac{2}{3} = \frac{w}{30}$$

Solve the proportion to find the number of white flowers in the bouquet.

$\frac{2}{3} = \frac{w}{30}$	Given
$30\left(\frac{2}{3}\right) = 30\left(\frac{w}{30}\right)$	Multiplication Property of Equality
$\frac{60}{3} = w$	Simplify.
$20 = w$	Simplify.

The proportion is $\frac{2}{3} = \frac{20}{30}$. Therefore, there are 20 white flowers and 30 red flowers in the bouquet. ❖

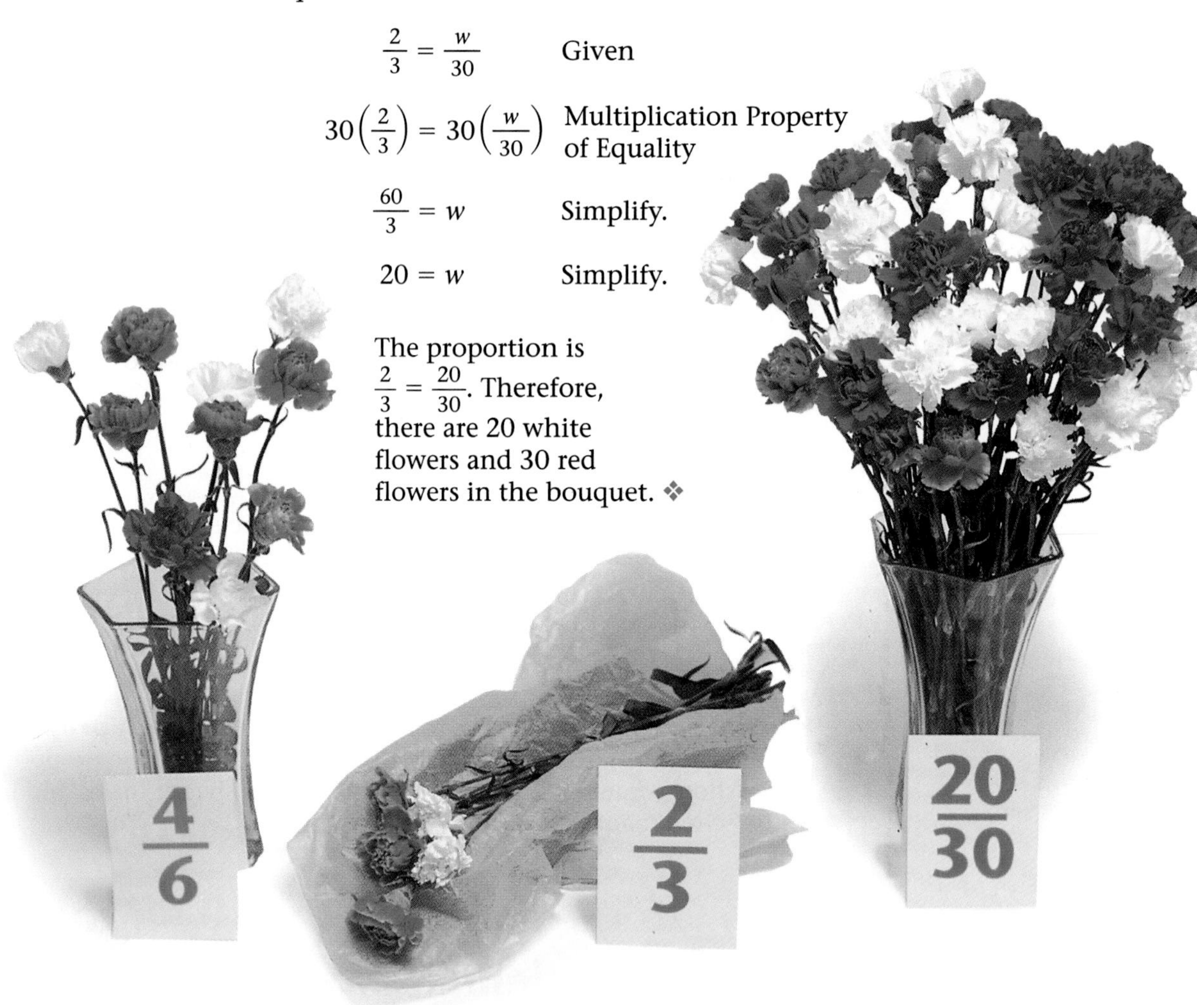

Try This The ratio of freshmen to sophomores at Washington High School is 7 to 5. If there are 343 students in the freshman class, how many students are in the sophomore class? 245 students in the sophomore class

Use the Multiplication Property of Equality to explain why the statement $ad = bc$ is true if $\frac{a}{b} = \frac{c}{d}$ is a true proportion.

Communicate

1. Explain how to solve the equation $\frac{x}{3} = 19$.
2. Write a real-world situation that can be represented by the equation $0.80r = p$.
3. Describe the Reciprocal Property. Give a numerical example.
4. How can you use the Reciprocal Property to solve $\frac{2z}{7} = \frac{4}{9}$?
5. How can you use the Division Property of Equality to solve $5f = 28.5$?

Practice & Apply

Use algebra tiles to solve each equation.

6. $3x = -6$ $x = -2$ **7.** $2x = 8$ $x = 4$ **8.** $5x = 15$ $x = 3$ **9.** $4x = 12$ $x = 3$

Solve and check each equation.

10. $5y = -60$ -12 **11.** $-4w = 3.6$ -0.9 **12.** $-0.9y = -36$ 40 **13.** $-1.56 = 3m$ -0.52

14. $\frac{d}{4} = 15$ 60 **15.** $\frac{g}{-5} = -7$ 35 **16.** $\frac{2}{3}f = 12$ 18 **17.** $\frac{7}{8}a = 14$ 16

18. $6y = -30$ -5 **19.** $-5w = 1.25$ -0.25 **20.** $-0.8y = -48$ 60 **21.** $-1.42 = 4m$ -0.355

22. $\frac{f}{5} = 25$ 125 **23.** $\frac{h}{-3} = 18$ -54 **24.** $\frac{4}{5}t = 16$ 20 **25.** $\frac{3}{7}t = 21$ 49

26. $4y = -20$ -5 **27.** $-2w = 1.6$ -0.8 **28.** $-0.8y = -24$ 30 **29.** $-1.5 = 5m$ -0.3

30. $\frac{w}{3} = -8$ -24 **31.** $\frac{x}{-4} = 27$ -108 **32.** $\frac{3}{4}t = 4$ $5\frac{1}{3}$ **33.** $\frac{5}{9}w = -2$ -3.6

34. $7y = -28$ -4 **35.** $-5.2g = 2.6$ -0.5 **36.** $4y = -4$ -1 **37.** $105 = 50m$ 2.1

38. $\frac{4}{6}x = \frac{1}{2}$ 0.75 **39.** $\frac{r}{5} = \frac{3}{1}$ 15 **40.** $\frac{j}{11} = 12.1$ 133.1 **41.** $13x = 7$ $\frac{7}{13}$

42. $15y = -4.5$ -0.3 **43.** $-3x = 4.8$ -1.6 **44.** $9y = -2.7$ -0.3 **45.** $-4 = 2.4m$ $-1\frac{2}{3}$

Solve each proportion.

46. $\frac{y}{-5} = \frac{4}{15}$ $-1\frac{1}{3}$ **47.** $\frac{24}{64} = \frac{z}{8}$ 3 **48.** $\frac{x}{5} = \frac{-4}{16}$ -1.25 **49.** $\frac{9}{25} = \frac{y}{12}$ 4.32

50. $\frac{y}{-3} = \frac{4}{36}$ $-\frac{1}{3}$ **51.** $\frac{45}{18} = \frac{f}{9}$ 22.5 **52.** $\frac{b}{7} = \frac{14}{32}$ 3.0625 **53.** $\frac{6}{7} = \frac{-h}{21}$ -18

54. $\frac{-p}{17} = \frac{6}{5}$ -20.4 **55.** $\frac{-15}{24} = \frac{-r}{9}$ 5.625 **56.** $\frac{b}{5} = \frac{-3}{20}$ -0.75 **57.** $\frac{-6}{-34} = \frac{m}{17}$ 3

58. $\frac{-6}{r} = \frac{21}{11}$ $-3\frac{1}{7}$ **59.** $\frac{25}{275} = \frac{-n}{15}$ $-1\frac{4}{11}$ **60.** $\frac{15}{34} = \frac{28}{x}$ $63\frac{7}{15}$ **61.** $\frac{-h}{3} = \frac{-9}{270}$ 0.1

62. $\frac{4.5}{180} = \frac{x}{90}$ 2.25 **63.** $\frac{52}{g} = \frac{26}{10}$ 20 **64.** $\frac{30}{7} = \frac{2.1}{y}$ 0.49 **65.** $\frac{2.8}{6.3} = \frac{r}{1.8}$ 0.8

66. $\frac{0.03}{1.6} = \frac{0.8}{n}$ $42\frac{2}{3}$ **67.** $\frac{80}{48} = \frac{0.5}{t}$ 0.3 **68.** $\frac{n}{-7} = \frac{-7}{49}$ 1 **69.** $\frac{3}{5} = \frac{g}{16}$ 9.6

Taxes Property taxes in Randi's neighborhood are 8% of the assessed value of the property.

70. Write an equation to describe the property tax, t, in terms of the assessed value, v. $t = 0.08v$

71. Suppose Randi's property tax is $1200. What is the assessed value of her property? $15,000

Entertainment A music group earns 6% of the total sales of an album it made. Use this information for Exercises 72 and 73.

72. Write an equation to describe the amount of money that the music group earns, a, in terms of the total sales of the album, s. $a = 0.06s$

73. Suppose the total sales of the album is $67,500. How much money does the music group earn from the album? $4050

74. The ratio of seniors to juniors on the student council is 4 to 3. If there are 24 seniors on the student council, how many juniors are on the council? 18 juniors

75. The ratio of cat owners to dog owners in the student population is 5 to 6. If 345 students are cat owners, how many students are dog owners? 414 students own dogs.

A recipe for 5 loaves of harvest grain bread calls for 15 cups of flour.

76. Baking According to the recipe for harvest grain bread described above, how many loaves can be made from 21 cups of flour? 7 loaves

Look Back

Simplify each expression. [Course 1]

77. $3a + 5b + 16a - 9a$ $10a + 5b$

78. $2.5g - 1.4g + 2g$ $3.1g$

79. $5(h + 20)$ $5h + 100$

Graph each equation. [Course 1]

80. $y = x + 6$

81. $y = 2x + 1$

82. $x = -4$

Look Beyond

Determine whether the given value for the variable satisfies the inequality.

83. $3x - 5 < 10; x = -2$ Yes

84. $-2y - 7 > 3; y = -5$ No

85. $2w + 7 \geq 3; w = -2$ Yes

86. $-4z - 5 \leq 3z + 2; z = 2$ Yes

LESSON 1.3 Exploring Two-Step Equations

Why *Equations that can be solved in two steps can be used to solve problems about money and area.*

Holly ordered rolls of color film from a photography supply catalog. Each roll of film costs \$4, and there is a \$7 charge for shipping.

EXAMPLE 1

A Write an equation to describe Holly's cost in terms of the number of rolls of film that she buys.

B How many rolls of film did Holly buy if her total cost was \$55?

Solution ➤

A The total cost depends on the number of rolls of film Holly buys. Let c represent the total cost, and let n represent the number of rolls of film.

The total cost	is	\$4 times the number of rolls of film	plus	\$7.
c	$=$	$4n$	$+$	7

The equation is $c = 4n + 7$.

B Since the total cost is \$55, substitute 55 for c in the equation $c = 4n + 7$. Solve this equation, $55 = 4n + 7$, for n to find the number of rolls of film that Holly bought.

$55 = 4n + 7$	Given
$55 - 7 = 4n + 7 - 7$	Subtraction Property of Equality
$48 = 4n$	Simplify.
$\frac{48}{4} = \frac{4n}{4}$	Division Property of Equality
$12 = n$	Simplify.

If Holly's total cost was \$55, then she bought 12 rolls of film. ❖

Try This Solve each equation.

a. $6x + 7 = 25$ $x = 3$

b. $4.5t - 6.1 = 2.9$ $t = 2$

Geometry Applications

Exploration *Perimeter*

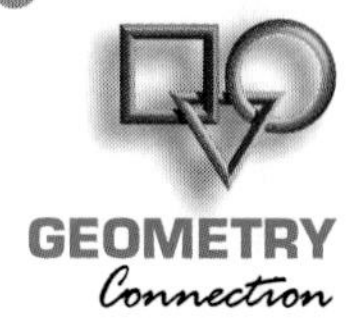
GEOMETRY Connection

1. Let s represent the length of a side of a square, and let p represent the perimeter. Write an equation to describe the perimeter of a square in terms of the length of a side. $p = 4s$

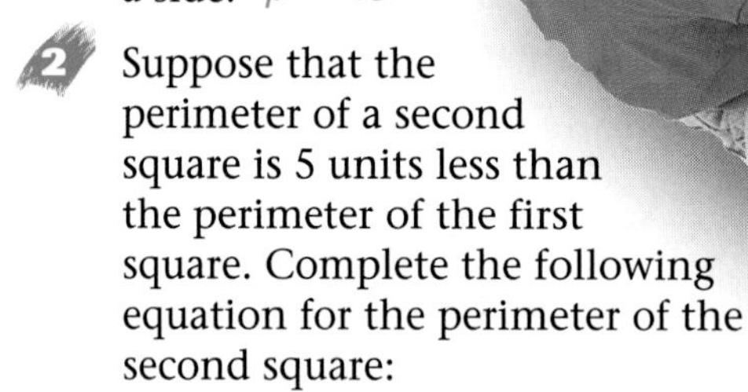

2. Suppose that the perimeter of a second square is 5 units less than the perimeter of the first square. Complete the following equation for the perimeter of the second square:

$$p = 4s \underline{\quad ? \quad} \qquad p = 4s - 5$$

3. Substitute 43 for p in your equation from Step 2 to show that the perimeter of the second square is 43 feet. $43 = 4s - 5$

4. Solve your equation from Step 3 to find the length of each side of the first square, s. Name the property of equality that you used for each step. Check your answer.

5. Suppose that the perimeter of square A is increased by 8 to get the perimeter of square B. The perimeter of square B is 60. Find the length of a side of square A. ❖

EXAMPLE 2

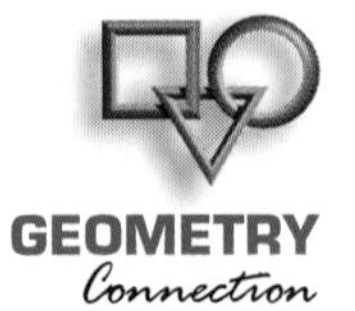

The bases of this rectangular prism are squares with sides that are 5 feet long.

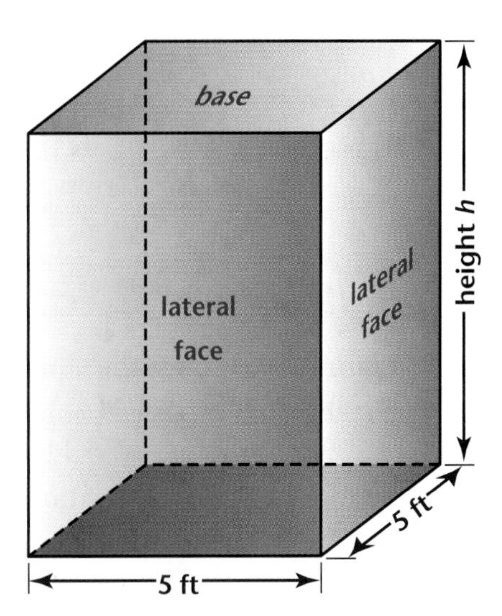

A Write an equation for the surface area, S, of the prism in terms of the height, h.

B What is the height of the prism if the surface area is 200 square feet?

Solution ➤

A The total area of both bases is $2(5 \cdot 5)$, or 50 square feet. The area of each lateral face is $5 \cdot h$, or $5h$. Since there are 4 lateral faces, the lateral surface area is $4 \cdot 5h$, or $20h$.

The equation for the surface area is $S = 20h + 50$.

B If the surface area is 200 square feet, substitute 200 for S.

$$S = 20h + 50$$
$$200 = 20h + 50$$

Solve this equation for h to find the height.

$200 = 20h + 50$	Given
$200 - 50 = 20h + 50 - 50$	Subtraction Property of Equality
$150 = 20h$	Simplify.
$\frac{150}{20} = \frac{20h}{20}$	Division Property of Equality
$7.5 = h$	Simplify.

If the surface area is 200 square feet, then the height of the prism is 7.5 feet. ❖

You can use a graphics calculator to check your solution to the equation $200 = 20h + 50$. Graph $Y_1 = 20X + 50$ and $Y_2 = 200$. The point of intersection is (7.5, 200). The x-coordinate of the point of intersection, 7.5, is the solution.

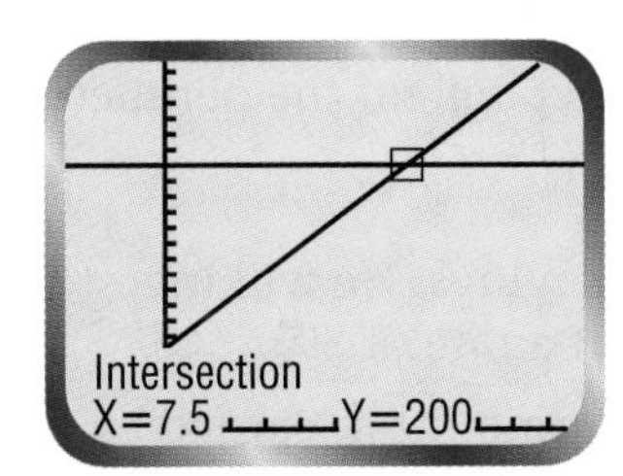

X	Y_1	Y_2
6	170	200
6.5	180	200
7	190	200
7.5	200	200
8	210	200
8.5	220	200
9	230	200

X=7.5

You can also create a table to check your solution. The value of X (height) is 7.5 when both Y_1 and Y_2 are 200.

Communicate

1. Explain how to solve the equation $2k - 6 = 8$.
2. What three methods can you use to determine whether 5 is the solution to the equation $3m - 4 = 11$?
3. Describe a real-world situation that can be represented by the equation $2.50x + 25 = 150$.
4. Explain how to solve the equation $\frac{x-9}{3} = -7$.

Practice & Apply

Use algebra tiles to solve each equation.

5. $2x - 3 = 1$ $x = 2$
6. $3x + 1 = -2$ $x = -1$
7. $2x + 1 = -5$ $x = -3$
8. $4x + 3 = 7$ $x = 1$

Solve and check each equation.

9. $2x + 7 = 13$ 3
10. $-4x - 2 = -22$ 5
11. $5x + 4 = -6$ −2
12. $2r - 1 = 1$ 1
13. $5t + 3 = -6$ −1.8
14. $7m + 11 = -3$ −2
15. $6t + 10 = -3$ $-2\frac{1}{6}$
16. $7g + 3 = -4$ −1
17. $18x - 37 = 1$ $2\frac{1}{9}$
18. $30s + 13 = -2$ −0.5
19. $6p + 14 = -5$ $-3\frac{1}{6}$
20. $4t + 43 = 7$ −9
21. $3p + 5 = -10$ −5
22. $11 = 4t + 3$ 2
23. $6y - 7 = 20$ 4.5
24. $5w - 4 = 17$ 4.2
25. $25z - 18.4 = 26.6$ 1.8
26. $-17y + 8.3 = -49.5$ 3.4
27. $0.8r + 5.7 = 11.46$ 7.2
28. $-11.35 = 1.7a - 15.6$ 2.5
29. $\frac{3}{4}x + 1 = 10$ 12
30. $-\frac{2}{3}y - 9 = 15$ −36
31. $\frac{x}{3} - 42 = 65$ 321
32. $\frac{n}{5.2} + 5.4 = 3.7$ −8.84

Geometry In a rectangle, suppose that the width is 3 feet less than the length.

33. Let l represent the length. Write an expression for the width, w, of the rectangle in terms of the length, l. $w = l - 3$
34. Find the width and the length of the rectangle if the perimeter is 38 feet. $w = 8$ feet, $l = 11$ feet

Investments Darrel decides to invest his savings. He will invest \$2000 at a rate of 7% and the rest of his savings, x, at a rate of 8%.

35. Write an equation to describe the total amount, t, earned by both investments. $t = 0.07(2000) + 0.08x$
36. Determine the amount, x, that Darrel needs to invest at 8% in order to earn a total, t, of \$400 in interest. \$3250

Geometry A right rectangular prism has square bases that have 6-inch sides.

37. Write an equation for the surface area, S, of the prism in terms of the height, h. $S = 24h + 72$

38. What is the height of the prism if the surface area is 228 square inches? 6.5 inches

39. Let y = cost. Let x = number of ceramic items. $y = 10x + 5$

For her monthly supplies, Becky buys $5 worth of paints. Each ceramic piece that she buys to paint costs $10.

Crafts Becky paints ceramic items to sell at 4-H auctions.

39. Write an equation to show that the total monthly cost of Becky's supplies, y, depends on the number of ceramic items that she buys, x.

40. If Becky can spend up to $45 per month, how many ceramic items can she buy? 4 ceramic items

Look Back

Small Business The table at the right lists the charges for a plumber. The total charge is the sum of the service charge and the hourly charge. **[Lessons 1.1, 1.2]**

Service charge	Number of hours	Total charge
$25	1.5	$49
$25	2	$57
$25	2.5	$65
$25	3	$73

41. What would you expect the total charge to be for working 4 hours? $89

42. Write an equation to show that the total charge, c, depends on the number of hours, h. $c = 16h + 25$

43. Write and solve an equation to determine the number of hours that the plumber works if the total charge is $97. $97 = 16h + 25$; $h = 4.5$ hours

Look Beyond

44. Refer to the information above about Becky's ceramic items. If Becky decides to sell the painted ceramic items for $15 each, how many ceramic items must she sell each month to break even? (To break even means that the monthly cost of the supplies equals the monthly income from selling the ceramics.) 1 item

45. How many items must she sell to have a profit of $75? 16 items

LESSON 1.4 Multistep Equations

why *Many real-world problems can be represented with multistep equations. Such equations often have variables on both sides of the equal sign.*

Save Money
Steamy Jack's
Carpet Cleaning
ONLY $30 plus
18¢ per square foot
Courtesy
Carpet Cleaning
$80 plus
10¢ per square foot
Best Value

How can you tell which carpet-cleaning service offers a better deal?

Suppose that you want to have your carpet cleaned and you need to know which company offers the best value. You will solve this problem by solving a multistep equation in Example 1.

Algebra tiles can be used to model the solutions to multistep equations. For example, the equation $3x + 4 = x - 2$ is modeled below with algebra tiles.

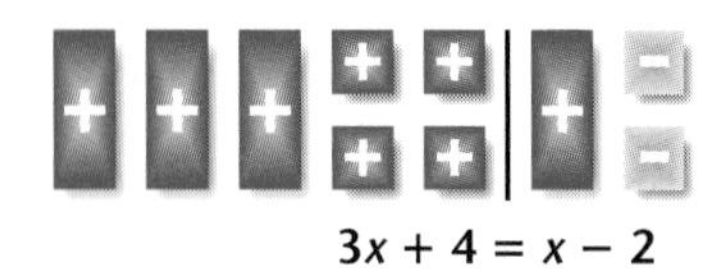

$3x + 4 = x - 2$

First use the Subtraction Property of Equality to get the x-tiles on only one side of the equal sign. Then simplify.

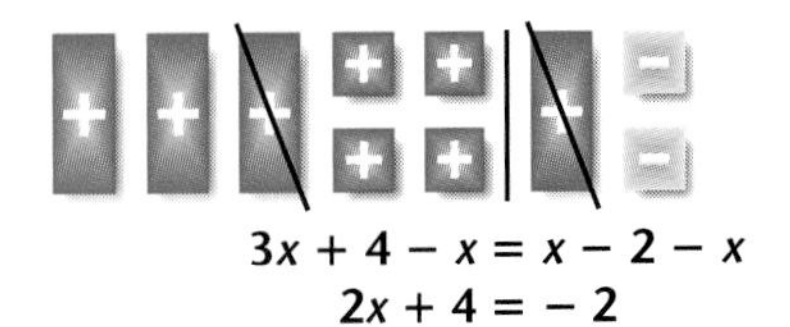

$$3x + 4 - x = x - 2 - x$$
$$2x + 4 = -2$$

To isolate the x-tiles, add 4 negative 1-tiles to each side of the model, and remove the neutral pairs.

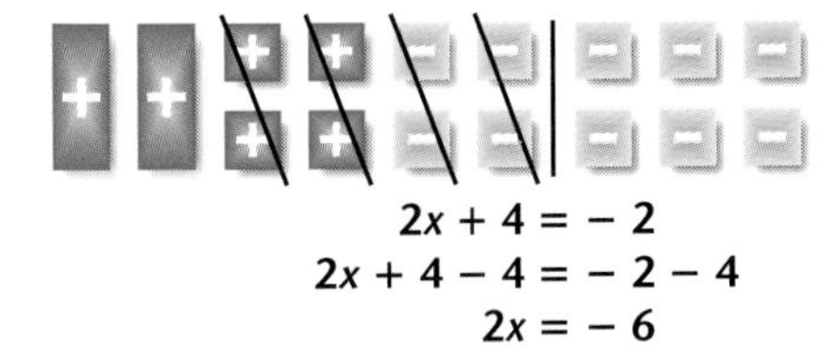

$$2x + 4 = -2$$
$$2x + 4 - 4 = -2 - 4$$
$$2x = -6$$

Use the Division Property of Equality, and simplify.

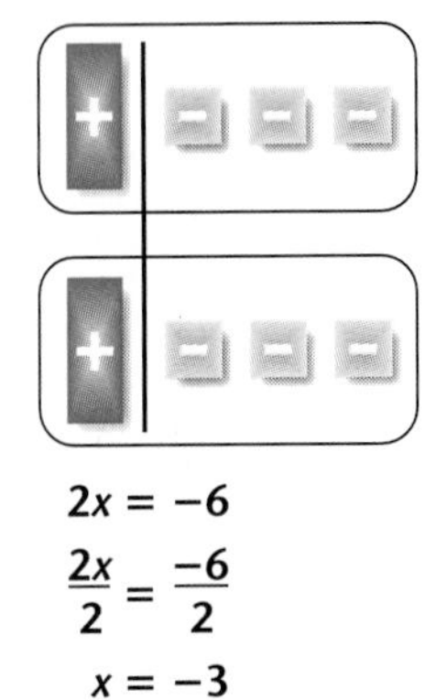

$$2x = -6$$
$$\frac{2x}{2} = \frac{-6}{2}$$
$$x = -3$$

The solution to the equation $3x + 4 = x - 2$ is -3.
Check this solution in the original equation.

$$3x + 4 = x - 2$$
$$3(-3) + 4 = (-3) - 2$$
$$-5 = -5 \text{ True}$$

EXAMPLE 1

Courtesy Carpet Cleaning charges \$80.00 plus \$0.10 per square foot to clean carpet. Steamy Jack's Carpet Cleaning charges \$30.00 plus \$0.18 per square foot.

A Write two equations to show that the total charge for each company depends on the number of square feet of carpet cleaned.

B Write and solve an equation that can be used to find the number of square feet for which the charges of the two companies are equal.

C Use a graphics calculator to check your solution in two different ways.

Solution ➤

A Let x represent the number of square feet cleaned, and let y represent the cost.

Courtesy Carpet Cleaning	$y = 0.10x + 80$
Steamy Jack's Carpet Cleaning	$y = 0.18x + 30$

B To find the number of square feet for which the charges are equal, solve the equation $0.10x + 80 = 0.18x + 30$ for x.

$0.10x + 80 = 0.18x + 30$	Given
$0.10x + 80 - 0.18x = 0.18x + 30 - 0.18x$	Subtraction Property of Equality
$-0.08x + 80 = 30$	Simplify.
$-0.08x + 80 - 80 = 30 - 80$	Subtraction Property of Equality
$-0.08x = -50$	Simplify.
$\frac{-0.08x}{-0.08} = \frac{-50}{-0.08}$	Division Property of Equality
$x = 625$	Simplify.

The charges for both companies are equal when 625 square feet of carpet are cleaned.

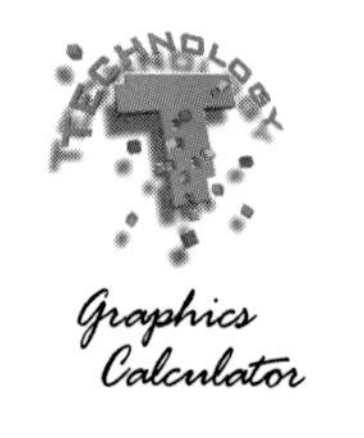

C Graph both equations, $Y_1 = 0.10x + 80$ and $Y_2 = 0.18x + 30$. Find the x-coordinate of the point of intersection, or use a table to find the value of x for which both y-values are equal.

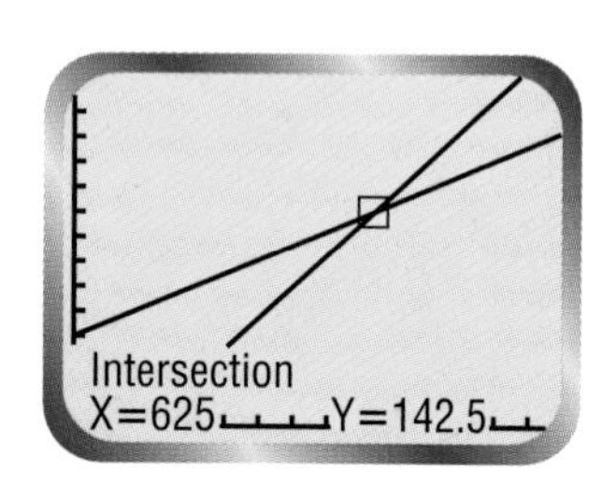

The graph and table show that both companies charge $142.50 when 625 square feet of carpet are cleaned. ❖

X	Y_1	Y_2
622	142.2	141.96
623	142.3	142.14
624	142.4	142.32
625	142.5	142.5
626	142.6	142.68
627	142.7	142.86
628	142.8	143.04

X=625

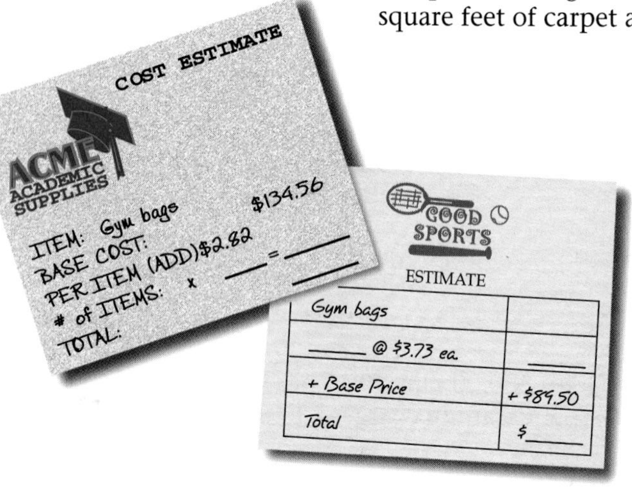

Try This The basketball booster club is selling gym bags that display the school mascot. One company offers to make the bags for $134.56 plus $2.82 per bag. A second company offers to make the bags for $89.50 plus $3.73 per bag. For how many bags are the two companies' charges equal?

$x \approx 49.5$; since you can not sell half a bag, the two companies would be equal in price when about 50 bags are sold.

The Distributive Property

The Distributive Property is often used to solve multistep equations.

> **DISTRIBUTIVE PROPERTY**
>
> For all numbers a, b, and c,
>
> $a(b + c) = ab + ac$ and $(b + c)a = ba + ca$; and
>
> $a(b - c) = ab - ac$ and $(b - c)a = ba - ca$.

EXAMPLE 2

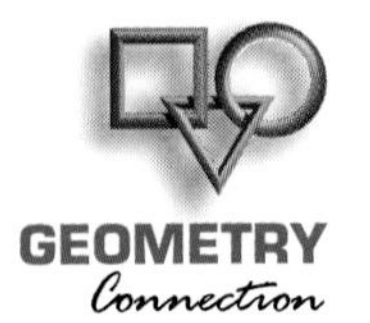

The formula for the area of a trapezoid is $A = \frac{1}{2}h(b_1 + b_2)$. Find the length of b_2 if h is 6 inches, b_1 is 7 inches, and the area, A, is 51 square inches.

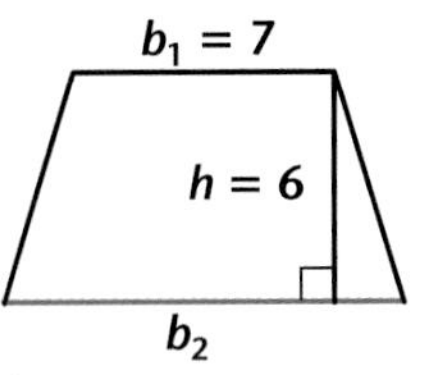

Solution ➤

Substitute the given values into the formula, and solve for b_2.

$A = \frac{1}{2}h(b_1 + b_2)$	Given
$51 = \frac{1}{2}(6)(7 + b_2)$	Substitution
$51 = 3(7 + b_2)$	Simplify.
$51 = 21 + 3b_2$	Distributive Property
$51 - 21 = 21 + 3b_2 - 21$	Subtraction Property of Equality
$30 = 3b_2$	Simplify.
$\frac{30}{3} = \frac{3b_2}{3}$	Division Property of Equality
$10 = b_2$	Simplify.

The length of b_2 is 10 inches. ❖

Unusual Solutions

Some equations do not have a solution. When you try to solve such equations, all variables cancel out, and the resulting statement is false.

EXAMPLE 3

Express Services charges \$4 per ounce plus a \$10 pickup charge for mail delivery. Fast Parcels charges \$4 per ounce plus a \$12 pickup charge. Write and solve an equation to show that the charges for these two companies will never be equal for the same weight.

Solution ➤

Let w represent the weight of a package in ounces , and let c represent the total charge. The equations describing each company's total charge are:

Express Services $c = 4w + 10$ and Fast Parcels $c = 4w + 12$

The charges would be the same when the equation $4w + 10 = 4w + 12$ is true for some value of w.

$4w + 10 = 4w + 12$	Given
$4w + 10 - 4w = 4w + 12 - 4w$	Subtraction Property of Equality
$10 = 12$ False	

This is a false statement. Therefore, the equation has no solution, and the charges for these two companies will never be equal. ❖

You can use a graph to illustrate that the equation $4w + 10 = 4w + 12$ has no solution. Notice that the graphs of the equations $Y_1 = 4X + 10$ and $Y_2 = 4X + 12$ are parallel lines and never intersect. Since the lines have no point in common, there is no solution.

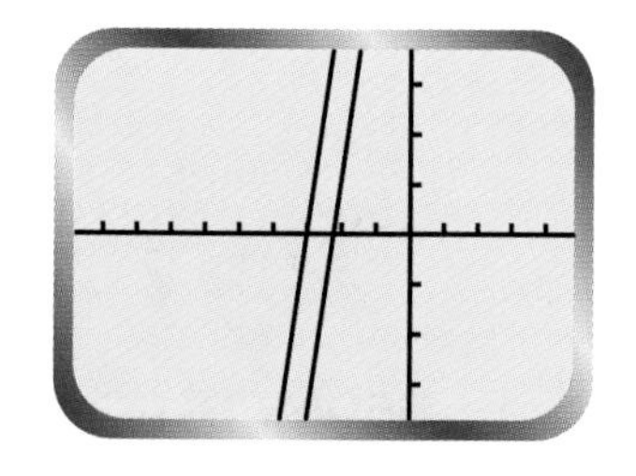

Some equations have an infinite number of solutions. When you try to solve such equations, all of the variables cancel out, and the resulting statement is true.

EXAMPLE 4

Solve the equation $3(x + 4) + 2(x - 5) = 6x - (x - 2)$ for x.

Solution ➤

$3(x + 4) + 2(x - 5) = 6x - (x - 2)$	Given
$3x + 12 + 2x - 10 = 6x - x + 2$	Distributive Property
$5x + 2 = 5x + 2$	Simplify.
$5x + 2 - 5x = 5x + 2 - 5x$	Subtraction Property of Equality
$2 = 2$	Simplify.

This is a true statement regardless of the value of x. Therefore, the solution to this equation is all real numbers. ❖

Describe the graph that would result if you tried to solve the equation in Example 4 by graphing. The two graphs have all points in common and are the same line, $y = 2$.

EXERCISES & PROBLEMS

Communicate

Describe two different ways to solve each equation in Exercises 1–3.

1. $3x + 2 = 4x - 1$
2. $7x - 5 = 2x - 3$
3. $13 + 2x = 5x + 1$
4. Explain how to determine if an equation has no solution.
5. Explain how to determine if the solution to an equation is all real numbers.

Practice & Apply

Construction One home builder charges $45.50 per square foot of the home plus a flat fee of $24,000 for the lot. Another home builder charges a flat fee of $15,000 for a lot of equal size and $51.50 per square foot of the home.

6. Write an equation for the total charge, c, of each home builder in terms of the number of square feet, n. $c = 45.50n + 24{,}000$; $c = 51.50n + 15{,}000$

7. For how many square feet will the charges for the two home builders be equal? 1500 sq ft

8. Which builder would charge less for the house at 206 Linear Lane? The second builder

9. Which builder would charge less for the house at 208 Linear Lane? The first builder

Use algebra tiles to solve each equation.

10. $2x - 2 = 4x + 6$ $x = -4$

11. $3x + 5 = 2x + 2$ $x = -3$

12. $4x + 3 = 5x - 4$ $x = 7$

13. $2x - 5 = 4x - 1$ $x = -2$

Solve and check each equation.

14. $5x + 24 = 2x + 15$ -3

15. $5y - 10 = 14 - 3y$ 3

16. $12 - 6z = 10 - 5z$ 2

17. $5x - 7 = -6x - 29$ -2

18. $-10x + 3 = -3x + 12 - 4x$ -3

19. $6x - 12 = -4x + 18$ 3

20. $3x - 2(x + 6) = 4x - (x - 10)$ -11

21. $8y - 4 + 3(y + 7) = 6y - 3(y - 3)$ -1

22. $(w - 3) - 5(w + 7) = 10(w + 3) - (7w + 5)$ -9

23. $16m - 3(4m + 7) = 3m - (12 - m)$

24. $1.8x - 2.8 = 2.5x + 2.1$ -7

25. $2.6h + 18 = 2.4h + 22$ 20

26. $1.4m - 0.8(m - 2) = 2.4m$ $\frac{8}{9}$

27. $6.3y = 5.2y - 1.1y + 12.1$ 5.5

23. No solution

Technology Use a graphics calculator to solve each equation. Give answers to the nearest tenth.

28. $2.3x - 3.7 = 4x + 6.9$ -6.2

29. $1.2x - 5.6 = 3.1x - 1.4$ -2.2

30. $6.3x + 5 = 5.2x - 2$

30. -6.4

31. Small Business Harry wants to hire a painter to paint his house. Jackson charges \$360 plus \$12 per hour. Davis charges \$279 plus \$15 per hour. When will the two costs be the same? 27 hours

32. **Geometry** For which value of x will the perimeter of the rectangle be equal to the perimeter of the square?

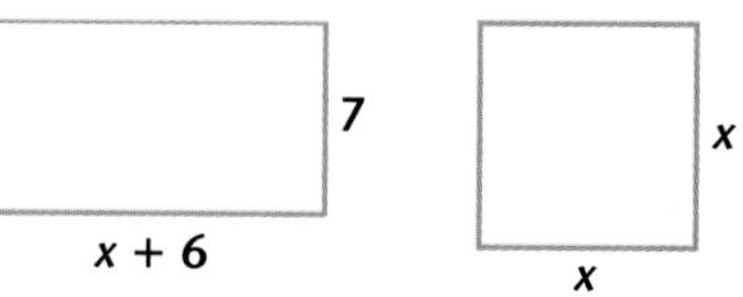

$x = 13$

Chemistry A chemist can make a solution with a given amount of acid by mixing two different pairs of solutions. Pair 1 consists of 12 milliliters of a 2.4% acid solution and an unknown amount of a 1.8% acid solution. Pair 2 consists of 16 milliliters of a 2% acid solution and an unknown amount of a 1.7% acid solution.

33. Write an equation for the amount of acid in the mixture obtained by mixing Pair 1. Let A represent the amount of acid in the final solution, and let x represent the unknown amount of the 1.8% solution. $A = 0.024(12) + 0.018x$

34. Write an equation for the amount of acid in the mixture obtained by mixing Pair 2. Let A represent the amount of acid in the final solution, and let x represent the unknown amount of the 1.7% solution. $A = 0.02(16) + 0.017x$

35. Write an equation to show that the amounts of acid in the mixture from Pair 1 is equal to the amount of acid in the mixture from Pair 2. Solve this equation to find the unknown amount of solution in each pair.
$0.024(12) + 0.018x = 0.02(16) + 0.017x$; $x = 32$ milliliters

36. **Geometry** The sum of the angle measures in a triangle is 180°. In triangle ABC, $m\angle A = 4x + 34$, $m\angle B = 7x - 10$, and $m\angle C = 5x + 12$. Find the measures of angles A, B, and C.
$m\angle A = 70°$, $m\angle B = 53°$, $m\angle C = 57°$

37. 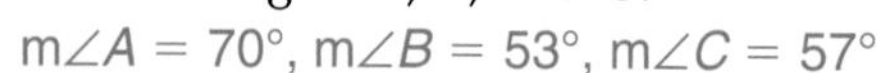 **Portfolio Activity** Refer to the portfolio activity given on page 3. Write and solve an equation to show when the cost of making the T-shirts will equal the income from selling the T-shirts for \$7.50 each.
$7.50x = 0.80x + 3x + 50$; about 14 T-shirts

Look Back

Find the area of each trapezoid described. [Course 1]

38. $b_1 = 6$ cm, $b_2 = 10$ cm, $h = 11$ cm
88 sq cm

39. $b_1 = 27$ ft, $b_2 = 31$ ft, $h = 18$ ft
522 sq ft

Solve each proportion. [Lesson 1.2]

40. $\frac{x}{3} = \frac{2}{6}$ 1

41. $\frac{7}{8} = \frac{y}{-12}$ -10.5

42. $\frac{-2}{7} = \frac{f}{-14}$ 4

43. $\frac{c}{-2} = \frac{14}{15}$ $-1\frac{13}{15}$

44. Describe how to use a graph to solve the equation $2x - 3 = 5$. **[Lesson 1.3]**

44. Graph $y = 2x - 3$ and $y = 5$. The x-coordinate of the point of intersection of the two lines is the solution to the equation; $x = 4$.

Solve and check each equation. [Lesson 1.3]

45. $r - 5 = 5 - 2r$ $3\frac{1}{3}$

46. $5t + 7 = -8$ -3

47. $7m - 18 = -4$ 2

Look Beyond

Solve each inequality.

48. $3x > 4$

49. $-4x < 32$

50. $x - 3 < 5$

51. $2y - 7 > 12$

48. $x > 1\frac{1}{3}$
49. $x > -8$
50. $x < 8$
51. $y > 9.5$

LESSON 1.5 Solving Inequalities

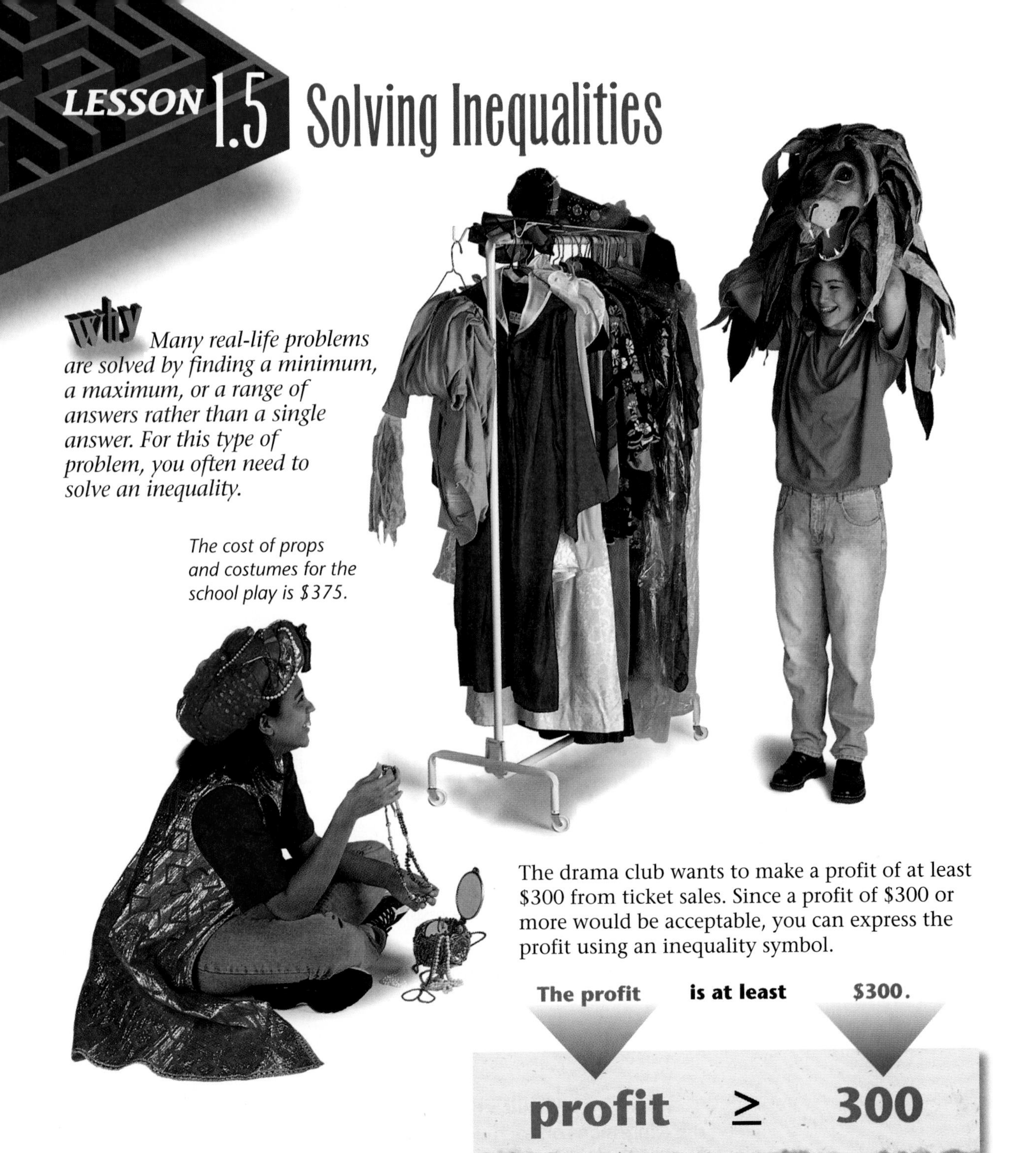

The profit
is at least
$300.
profit ≥ 300

why *Many real-life problems are solved by finding a minimum, a maximum, or a range of answers rather than a single answer. For this type of problem, you often need to solve an inequality.*

The cost of props and costumes for the school play is $375.

The drama club wants to make a profit of at least $300 from ticket sales. Since a profit of $300 or more would be acceptable, you can express the profit using an inequality symbol.

The cost of advertising and theater rental for the school play is $275. The total cost to produce the school play is the sum of the cost of props and costumes and the cost of advertising and theater rental:

$$\$375 + \$275 = \$650$$

Tickets for the play are $4.50 each. In this lesson, you will learn how to find the number of tickets that must be sold to make a profit of *at least* $300.

Let n represent the number of tickets. Since the tickets are \$4.50 each, the expression $4.5n$ represents the amount of money earned from ticket sales. The profit can be expressed as the amount of money earned from ticket sales minus the total cost of \$650.

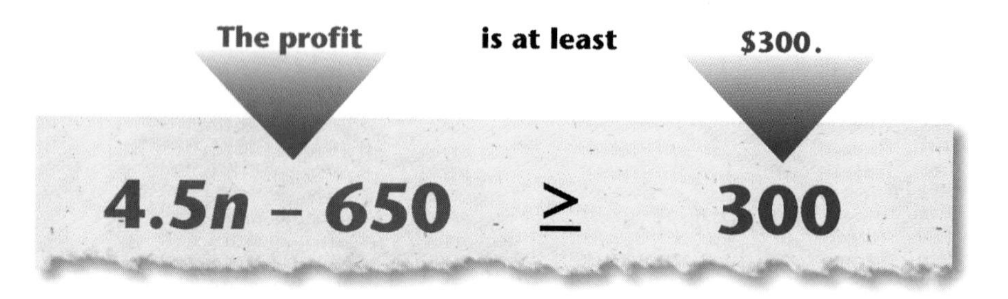

The profit
is at least
$300.
4.5n – 650 ≥ 300

In the same way that you can solve equations by using properties of equality, you can solve inequalities by using *properties of inequality.*

ADDITION PROPERTY OF INEQUALITY

If equal amounts are added to the expressions on each side of an inequality, the resulting inequality is still true.

SUBTRACTION PROPERTY OF INEQUALITY

If equal amounts are subtracted from the expressions on each side of an inequality, the resulting inequality is still true.

MULTIPLICATION PROPERTY OF INEQUALITY

If the expressions on each side of an inequality are multiplied by the same positive number, the resulting inequality is still true.

If the expressions on each side of an inequality are multiplied by the same negative number and the inequality sign is reversed, the resulting inequality is still true.

DIVISION PROPERTY OF INEQUALITY

If the expressions on each side of an inequality are divided by the same positive number, the resulting inequality is still true.

If the expressions on each side of an inequality are divided by the same negative number and the inequality sign is reversed, the resulting inequality is still true.

When multiplying or dividing the expressions on each side of an inequality by a negative number the direction of the inequality sign reverses.

How does multiplying or dividing the expressions on each side of an inequality by a negative number affect the direction of the inequality sign?

EXAMPLE 1

The expression $4.5n - c$, where n is the number of tickets sold and c is the total cost of $650 to produce the play, describes the profit that the drama club earns. The profit needs to be at least $300. Find the number of tickets that the drama club needs to sell by solving the inequality $4.5n - 650 \geq 300$.

Solution ➤

Use the properties of inequality.

$4.5n - 650 \geq 300$	Given
$4.5n - 650 + 650 \geq 300 + 650$	Addition Property of Inequality
$4.5n \geq 950$	Simplify.
$\frac{4.5n}{4.5} \geq \frac{950}{4.5}$	Division Property of Inequality
$n \geq 211.\overline{1}$	Simplify.

The solution is $n \geq 211.\overline{1}$, but only whole tickets can be sold, and 211 tickets is not enough to make a profit of $300. The drama club must sell at least 212 tickets. ❖

Try This A company can spend no more than $40,000 in an advertising campaign. The plan calls for $28,000 to be spent on TV commercials and the remainder on T-shirts that will be given away. The shirts will cost the company $2 each. What is the greatest number of T-shirts that can be given away?

Representing an Inequality on a Number Line

To graph the inequality $n > 5$ on a number line, circle 5 on the number line. Then shade all of the points to the right of 5. The open circle means that 5 is *not* included.

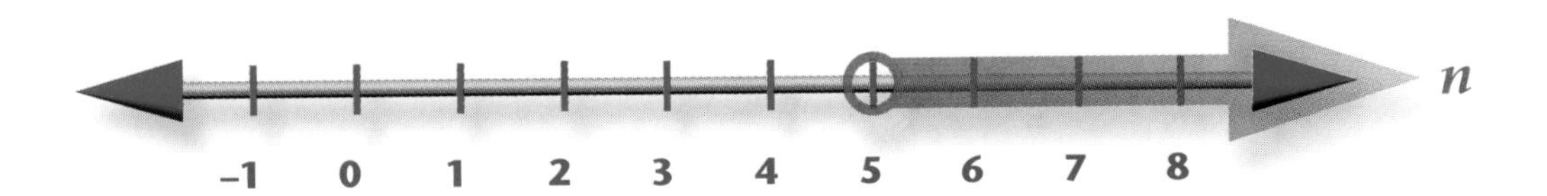

To graph the inequality $t \leq 7$ on a number line, circle 7 and shade in the circle. Then shade all of the points to the left of 7. The shaded circle means that 7 is included.

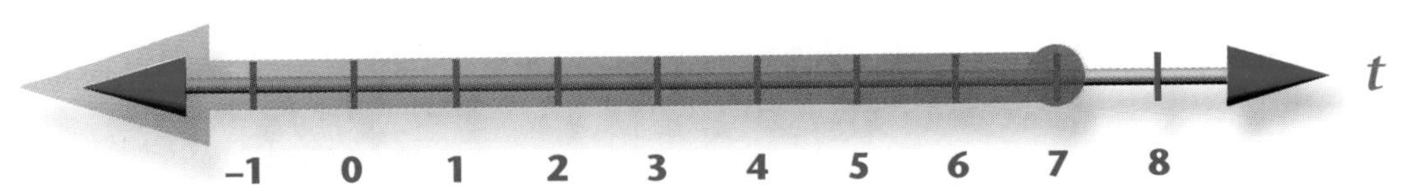

EXAMPLE 2

Solve $-2x + 3 < 15$, and graph the solution on a number line.

Solution ➤

$-2x + 3 < 15$	Given
$-2x + 3 - 3 < 15 - 3$	Subtraction Property of Inequality
$-2x < 12$	Simplify.
$\frac{-2x}{-2} > \frac{12}{-2}$	Division Property of Inequality
$x > -6$	Simplify.

The solution is $x > -6$. Notice that the direction of the inequality sign changes when dividing by a negative number.

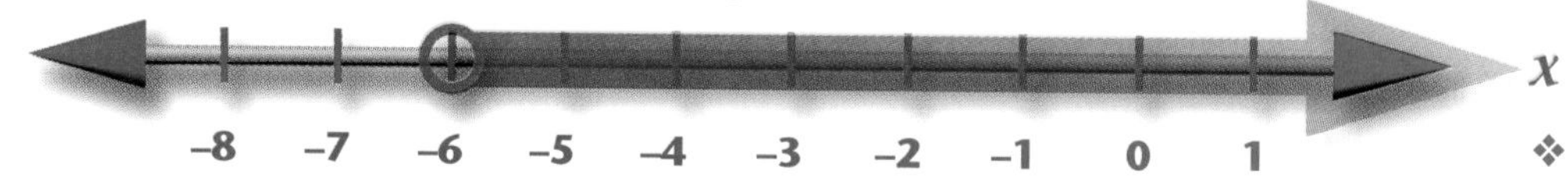

❖

EXERCISES & PROBLEMS

Communicate

1. Explain how to solve and graph the inequality $2x - 5 < 3$.

2. Compare the properties of equality and the properties of inequality. What are the similarities? What are the differences?

3. Why is $x < -2$ *not* the result of solving the inequality $-3x < 6$?

4. Explain how to solve the inequality $-\frac{x}{2} \geq 12$.

Identify the properties used to solve each inequality.

5. $x - 4 \leq 2$ **6.** $-6x > 12$ **7.** $\frac{x}{2} + 3 < 7$ **8.** $x + 4 \geq 4x - 3$

Practice & Apply

Solve each inequality. Graph your solutions on a number line.

9. $x + 1 \leq -4$

10. $-4t > 5$

11. $\frac{r}{-7} \leq 3$

12. $3x - 7 > 18$

13. $-5b + 4 \leq 18$

14. $6y + 7 \geq 2y + 28$

15. $-3y - 7 > 2.3$

16. $1.3x - 9.2 < 7.7$

17. $2.4z + 9.6 \geq 21.3$

18. $10.8 \leq 2x - 11.3$

19. $38.25 > -2q + 29.50$

20. $\frac{x}{4} - 2 \leq 13$

21. $\frac{5}{7}x + 27 \leq 12$

22. $-\frac{7}{8}m - 9 \leq 40$

23. $\frac{-6}{5}a + 18 > 3$

Construction Lyle wants to build a rectangular dog pen with a length that is 12 feet more than the width.

24. Write an equation to express the perimeter, p, in terms of the width, w. $p = 4w + 24$

25. Lyle can afford no more than 560 feet of fence. Write an inequality to show that the perimeter of the dog pen can be no more than 560 feet. $4w + 24 \leq 560$

26. Solve the inequality that you wrote for Exercise 25 to determine the maximum possible width for a 560-foot perimeter. $w \leq 134$
The maximum possible width for a 560-foot perimeter is 134 feet.

Solve each inequality. Graph your solution on a number line.

27. $2x + 5 \leq -3$

28. $-3t > 8$

29. $-\frac{b}{6} \geq \frac{1}{2}$

30. $5x - 9 > 20$

31. $-3b + 2 \leq 16$

32. $4y + 5 \geq y + 26$

33. $-6y - 10 > 3$

34. $3.2x - 8.4 < 10.8$

35. $-5.6z + 8.6 \geq 2.3$

36. $11.7 \leq 2.5x - 1.3$

37. $8.2 > 2.4q - 9.5$

38. $\frac{h}{7} - \frac{1}{14} \leq \frac{1}{2}$

39. $-\frac{2}{3}n + 12 < 21$

40. $\frac{1}{4}d - \frac{3}{8} > 2$

41. $-\frac{2}{5}t - \frac{1}{4}t \geq -2\frac{1}{6}$

Small Business R & G Catering specializes in catering wedding receptions. They charge $550 for setting up the buffet and an additional $6.50 per guest.

42. Mr. and Mrs. Henderson want to spend no more than $1200 on their daughter's wedding reception. Write an inequality in terms of the number of guests, g, that they can invite to the wedding reception. $6.50g + 550 \leq 1200$

43. Solve the inequality that you wrote for Exercise 42. What is the maximum number of guests that they can invite to the reception?

43. $g \leq 100$; They can invite a maximum of 100 guests to the reception.

Entertainment Sharon plans to attend a fair. There is a $4.50 admission fee, and each ride coupon costs $0.25.

44. Write an equation to show the amount, a, that Sharon spends in terms of the number of ride coupons, r, that she buys. $a = 0.25r + 4.50$

45. Sharon can spend no more than $8.50. Write an inequality to represent her spending limit.

46. Solve the inequality that you wrote for Exercise 45 to determine the greatest number of ride coupons that Sharon can buy.

45. $0.25r + 4.50 \leq 8.50$
46. $r \leq 16$; she can buy no more than 16 ride coupons.

Small Business Brian charges $15 plus $5.50 per hour for mowing and trimming yards. Greg charges $12 plus $6.25 per hour.

47. Write two equations to show how much Brian and Greg charge, c, in terms of the number of hours, h, that they work. Brian: $c = 5.50h + 15$ Greg: $c = 6.25h + 12$

48. For how many hours of work is Brian's charge greater than Greg's charge? Less than 4 hours

49. For how many hours of work is Greg's charge greater than Brian's charge? More than 4 hours

50. Fund-raising Members of the Bluebonnet Volunteer Fire Department Auxiliary sponsor an annual chili supper in February. Some materials are donated, but the meat, beans, and bread cost $0.75 per meal. The members also spend $82 on napkins and other materials. They plan to charge $4.50 per meal. How many meals must they serve to make a profit of at least $400? 129 meals

51. Theater Members of the drama club are putting on a play. They need to spend $560 for costumes, sets, and programs. They plan to sell tickets for $7 each. How many tickets must be sold for the drama club to break even? 80 tickets

52. Consumer Economics Anthony is carpeting several rooms in his home. The carpet costs $14.95 per square yard plus $200 for installation. He can afford to spend no more than $3000. How many square yards of carpet can Anthony afford to purchase? 187 sq yd

53. Travel Jeff's car averages 18 miles per gallon of gasoline. What is the greatest number of gallons of gasoline that he will need if he travels no more than 450 miles? 25 gallons

54. **Portfolio Activity** Refer to the portfolio activity given on page 3. Determine the number of T-shirts that must be sold at $7.50 each to make a profit of at least $500. 149 T-shirts

Look Back

Simplify each expression. [Course 1]

55. $\frac{1}{2} + \frac{2}{7}$ $\frac{11}{14}$

56. $\frac{4}{5} - \frac{3}{7}$ $\frac{13}{35}$

57. $-3\frac{2}{5} + 5\frac{1}{5}$ $1\frac{4}{5}$

58. $\frac{4}{5} \cdot \left(-\frac{3}{8}\right)$ $-\frac{3}{10}$

59. $3\frac{1}{4} \div \frac{2}{3}$ $4\frac{7}{8}$

60. $4\frac{5}{8} \div 1\frac{3}{4}$ $2\frac{9}{14}$

Solve each equation. [Lessons 1.3, 1.4]

61. $3x + 1 = 5$ $1\frac{1}{3}$

62. $\frac{r}{6} - 4 = 2$ 36

63. $\frac{3}{4}d + 2 = -10$ -16

64. $5x - 7 = 8x + 15$ $-7\frac{1}{3}$

65. $12 - 3w = 2(w + 8)$ $-\frac{4}{5}$

66. $\frac{2}{3}w - 15 = -5(w + 10)$ $-6\frac{3}{17}$

Look Beyond

Determine whether each given value for the variable satisfies the equation or inequality.

67. $|2x - 3| = 5; x = -1$ Yes

68. $|5x + 7| = 2; x = -2$ No

69. $|3x - 2| > 5; x = -2$ Yes

70. $|2x + 8| < 6; x = -1$ No

LESSON 1.6 Absolute-Value Equations and Inequalities

Absolute-value inequalities are used to describe allowable errors, or tolerance levels, in measurements.

A theme park has a Guess-Your-Weight booth. Houdini, the booth attendant, claims to be able to guess the weight of anyone within 3 pounds. For example, the acceptable range for a 130-pound person is 130 ± 3 pounds. If Houdini does not guess the correct weight within the acceptable range, that person wins a prize. The acceptable range can be expressed as the inequality $|x - 130| \leq 3$, where x is Houdini's guess. You will solve this inequality in Example 2.

A Geometric Interpretation of Absolute Value

The **absolute value** of a number is the distance from 0 to that number on a number line.

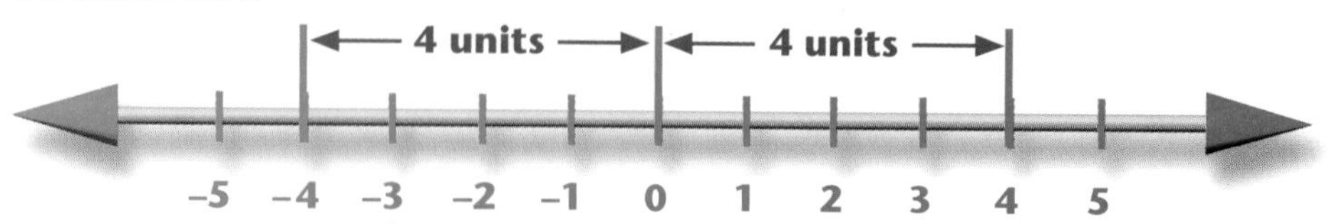

Both 4 and −4 are 4 units from 0, so the absolute value of both −4 and 4 is 4. In mathematics, the symbol | | indicates the absolute value of a number.

$|4| = 4$ *The absolute value of 4 is 4.*

$|-4| = 4$ *The absolute value of −4 is 4.*

The algebraic definition of absolute value follows.

ABSOLUTE VALUE

For any number x,

if x is a positive integer or zero, then the absolute value of x is x, or $|x| = x$; and

if x is a negative integer, then the absolute value of x is the opposite of x, or $|x| = -x$.

The expression ABS is used to tell computers and graphics calculators to compute absolute value. The graphics calculator here shows $|-9|$, $|0|$, and $|15|$.

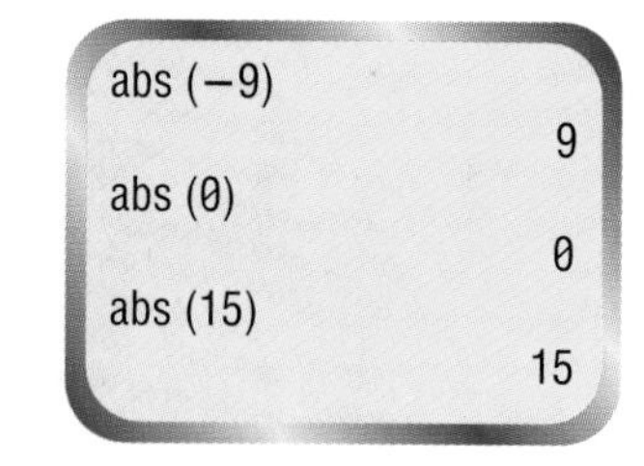

Explain why the absolute value of a number is never negative.

Absolute value can never be negative because absolute value represents the distance that a number is from the origin on a number line, and distance is never negative.

EXAMPLE 1

Solve $|2x - 5| = 6$.

Solution ➤

Since you cannot tell whether the quantity inside the absolute-value symbol is positive or negative, consider both cases.

$$|2x - 5| = 6$$

Case 1	Case 2
If $2x - 5$ is **positive or zero,** then $2x - 5 = 6$.	If $2x - 5$ is **negative,** then $-(2x - 5) = 6$.

Case 1

The quantity inside the absolute-value symbol is positive or zero.

$$\begin{aligned} |2x - 5| &= 6 \\ 2x - 5 &= 6 \\ 2x - 5 + 5 &= 6 + 5 \\ 2x &= 11 \\ \frac{2x}{2} &= \frac{11}{2} \\ x &= 5.5 \end{aligned}$$

Substitute 5.5 for x to check.

$$\begin{aligned} |2x - 5| &= 6 \\ |2(5.5) - 5| &\stackrel{?}{=} 6 \\ |11 - 5| &\stackrel{?}{=} 6 \\ |6| &= 6 \quad \text{True} \end{aligned}$$

Case 2

The quantity inside the absolute-value symbol is negative.

$$\begin{aligned} |2x - 5| &= 6 \\ -(2x - 5) &= 6 \\ -2x + 5 &= 6 \\ -2x + 5 - 5 &= 6 - 5 \\ -2x &= 1 \\ \frac{-2x}{-2} &= \frac{1}{-2} \\ x &= -0.5 \end{aligned}$$

Substitute -0.5 for x to check.

$$\begin{aligned} |2x - 5| &= 6 \\ |2(-0.5) - 5| &\stackrel{?}{=} 6 \\ |-1 - 5| &\stackrel{?}{=} 6 \\ |-6| &= 6 \quad \text{True} \end{aligned}$$

The solution is $x = 5.5$ or $x = -0.5$. ❖

Absolute Value and Error

Absolute value is used to describe acceptable errors in measurement.

EXAMPLE 2

Solve the inequality $|x - 130| \leq 3$ to determine the acceptable range for Houdini's guess for the 130-pound person.

Solution ➤

The variable x represents the weight that Houdini guesses. If Houdini's guess, x, is greater than or equal to 130, then $x - 130$ is positive or zero. If Houdini's guess, x, is less than 130, then $x - 130$ is negative.

Case 1

The quantity inside the absolute-value symbol is positive or zero.

$$\begin{aligned} |x - 130| &\leq 3 \\ x - 130 &\leq 3 \\ x - 130 + 130 &\leq 3 + 130 \\ x &\leq 133 \end{aligned}$$

Case 2

The quantity inside the absolute-value symbol is negative.

$$\begin{aligned} |x - 130| &\leq 3 \\ -(x - 130) &\leq 3 \\ -x + 130 &\leq 3 \\ -x + 130 - 130 &\leq 3 - 130 \\ -x &\leq -127 \\ (-1)(-x) &\geq (-1)(-127) \\ x &\geq 127 \end{aligned}$$

The acceptable guesses that the Houdini can make for the 130-pound person are all of the weights between 127 pounds and 133 pounds, inclusive. This is written as $127 \le x \le 133$ and is represented on a number line as shown below.

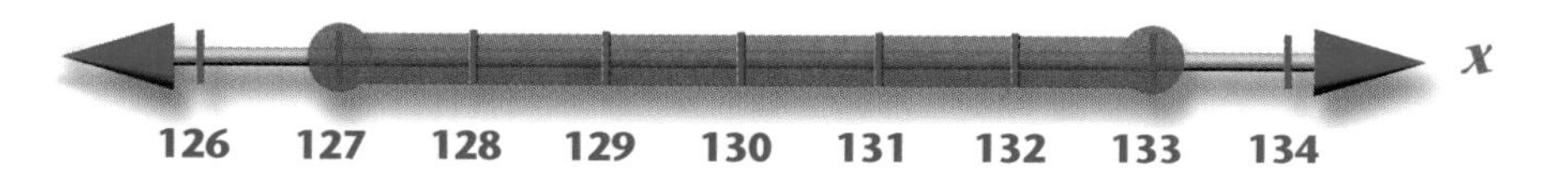

Notice that shaded circles are used at 127 and 133 to indicate that these numbers are included in the solution. ❖

EXAMPLE 3

What values of x are at least 4 units from 2?

Solution ➤

Solve the inequality $|x - 2| \ge 4$.

Case 1

The quantity inside the absolute-value symbol is positive or zero.

$$|x - 2| \ge 4$$
$$x - 2 \ge 4$$
$$x - 2 + 2 \ge 4 + 2$$
$$x \ge 6$$

Case 2

The quantity inside the absolute-value symbol is negative.

$$|x - 2| \ge 4$$
$$-(x - 2) \ge 4$$
$$-x + 2 \ge 4$$
$$-x + 2 - 2 \ge 4 - 2$$
$$-x \ge 2$$
$$(-1)(-x) \le (-1)(2)$$
$$x \le -2$$

The inequality is true when x is greater than or equal to 6. The inequality is also true when x is less than or equal to -2. Therefore, $x \ge 6$ *or* $x \le -2$. Therefore, the solution is shown on the number line below.

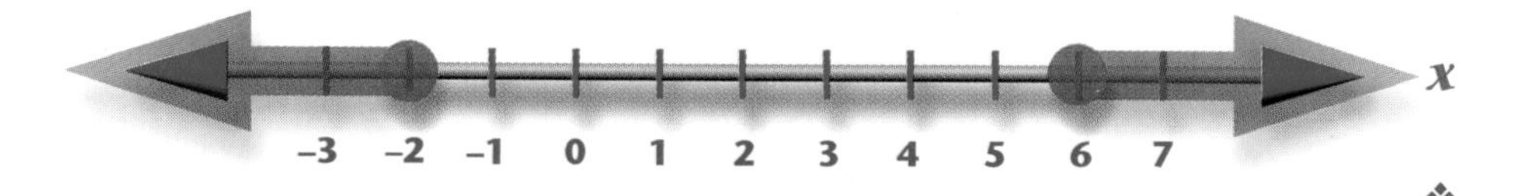

❖

The inequality $|x - 2| \ge 4$ can also be solved by graphing the functions $Y_1 = |x - 2|$ and $Y_2 = 4$. The values of x that satisfy the inequality $|x - 2| \ge 4$ are the x-coordinates of all the points on the graph of $Y_1 = |x - 2|$ that are on or above the graph of $Y_2 = 4$.

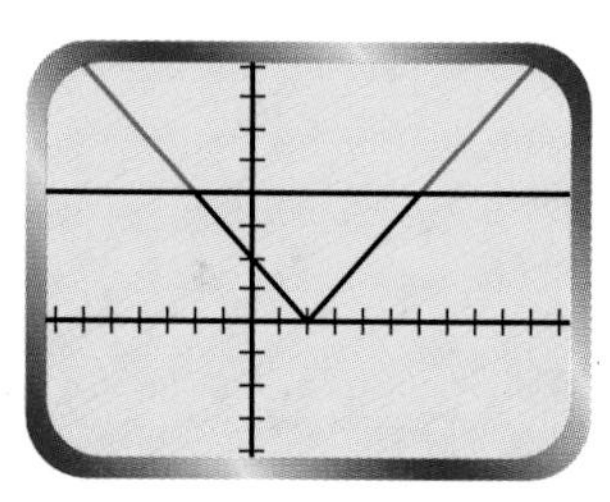

Try This What values of x are no more than 6 units from -4? $-10 \le x \le 2$

EXERCISES & PROBLEMS

Communicate

1. Define the meaning of the absolute value of a number in two ways.
2. Explain why an absolute-value equation may have two solutions.
3. Suppose that the jar shown contains 1500 marbles. Explain how to write an inequality to describe all of the possible guesses, x, of the number of marbles in the jar that are within 10 marbles.
4. Give numerical examples to explain why if x is a negative integer, then $|x| = -x$. (HINT: Remember that $-x$ means *the opposite of x*.

Practice & Apply

Solve and check each equation or inequality.

5. $|x + 5| = 6$ 1, −11
6. $|y - 3| = 3$ 0, 6
7. $|s + 3| = 0$ −3
8. $|3w + 4| = 2$ $-2, -\frac{2}{3}$
9. $|5m - 6| = 14$ 4, −1.6
10. $|8t - 3| = 46$ 6.125, −5.375
11. $|x + 3| = 8$ 5, −11
12. $|y - 7| = 5$ 12, 2
13. $|s + 12| = 10$ −2, −22
14. $|6w + 14| = 21$
15. $|25m - 36| = 4$
16. $|5t - 4| = 6$
17. $|3x - 3| \geq 3$
18. $|2x + 5| \leq 7$
19. $|h + 17| < 20$
20. $|j - 13| > 22$
21. $|27d - 11| > 36$
22. $|5a + 27| > 0$
23. $|34c + 2| \geq 16$
24. $|33y - 99| \geq 11$
25. $|14v - 4| < 6$
26. $|y - 3.5| = 3.4$
27. $|s + 3.7| = 0.9$
28. $|43w + 45| = 28$
29. $|52m - 16| = 4$
30. $|5t - 30| = 45$
31. $|x - 3.6| \geq 4.2$
32. $|x + 3.7| \leq 2$
33. $|h + 3| < 1.02$
34. $|j - 19| > 26$
35. $|7d - 1| > 13$
36. $|55a + 20| < 10$
37. $|4.03c + 2| \geq 2.5$
38. $|3y - 9.3| \geq 3.15$
39. $|x + 5.2| = 6.3$
40. $|46v - 14| < 5.6$

Write an equation or inequality for each situation. Solve the equation or inequality. Graph your solution on a number line.

41. The distance between a number and 5 is equal to 2.
42. The distance between a number and 3 is more than 3.
43. The distance between a number and −2 is less than 4.
44. The distance between a number and −5 is less than 2.3.
45. The distance between a number and 15 is at least 1.5.

Solve and graph each inequality.

46. $|x - 3| \geq 4$ **47.** $|x + 3| \leq 6$ **48.** $|h + 7| < 10$

49. $|j - 9| > 2$ **50.** $|2d - 1| > 3$ **51.** $|5a + 2| < 10$

52. $|4c + 2| \geq 6$ **53.** $|3y - 9| \geq 3$ **54.** $|4v - 4| < 0.6$

Manufacturing A manufacturer needs cardboard boxes to ship merchandise. Each box must be a cube with edges that are 14 centimeters long. The error tolerance for the length of each edge is within 0.2 centimeters, or ± 0.2.

55. Write an absolute-value inequality to show the acceptable lengths of the edges of these boxes. Solve the inequality.

56. Draw a graph on a number line to show the acceptable lengths of the edges of these boxes.

Small Business Rollo's Pizza is located 1.6 miles from City Hall along a highway. Rollo's serves only customers who are within 5 miles of their location and who are located along the highway.

57. Draw a graph on a number line with City Hall located at 0. Show the location of Rollo's and the locations of all the delivery customers.

58. Write an inequality to describe the locations of the delivery customers in terms of their distance from City Hall. Solve the inequality.

59. What does the solution of the inequality represent?

Manufacturing A screw has an ideal length of 1.5 inches. The error tolerance for the length is ±0.03 inches.

60. Write an absolute-value inequality to show the acceptable lengths. Solve the inequality.

61. Draw a graph on a number line to show the acceptable lengths.

Packaging The label on a granola bar states that the weight of the bar is 8 ounces. The actual weight is within 0.1 ounces of the advertised weight.

62. Write an absolute-value inequality to show the acceptable actual weights. Solve the inequality.

63. Draw a graph on a number line to show the acceptable range of actual weights.

Look Back

Solve each equation. [Lessons 1.3, 1.4]

64. $2w - 3 = 4w - 1.5$ -0.75 **65.** $\frac{2}{3}x - 3 = 9$ 18

66. $3y - 7 = 4(y - 2)$ 1 **67.** $6w - 9 = 8(w - 3) - 2(w + 4)$ No solution

68. $5w - 3 = 6w - 7.5$ 4.5 **69.** $\frac{3}{5}b - 7 = 18$ $41\frac{2}{3}$

70. $2y - 8 = 5(2y - 4)$ 1.5 **71.** $7w - 14 = 6(4w - 9) - (3w + 7)$ $3\frac{5}{14}$

Fund-raising Members of the fund-raising committee for the community youth center decide to sell caps with the community center's logo. A graphic artist charges $72.00 to design the logo. Each cap will cost the commitee $3.50. They plan to sell each cap for $6.00. **[Lessons 1.4, 1.5]**

72. Write an equation for the cost of having the caps made, c, in terms of the number of caps, n. $c = 3.50n + 72$

73. Write an equation for the income from the sale of the caps, i, in terms of the number of caps sold, n. $i = 6n$

74. The committee will break even when the cost equals the income. Write an equation to determine when the comittee will break even. Solve the equation. $3.50n + 72 = 6n$ $28.8 = n$ 29 caps must be sold.

75. Profit is the income, i, minus the cost, c. Write an inequality to show a profit of at least $400 from cap sales. Solve the inequality.

$400 \leq 6n - (72 + 3.50n)$

$n \geq 188.8$

189 caps must be sold.

Look Beyond

76. The graph shows that the distance in miles, d, that Marty is from home depends on the number of hours, h, that she drives. How many miles per hour do you think Marty is traveling? Explain how you got your answer.

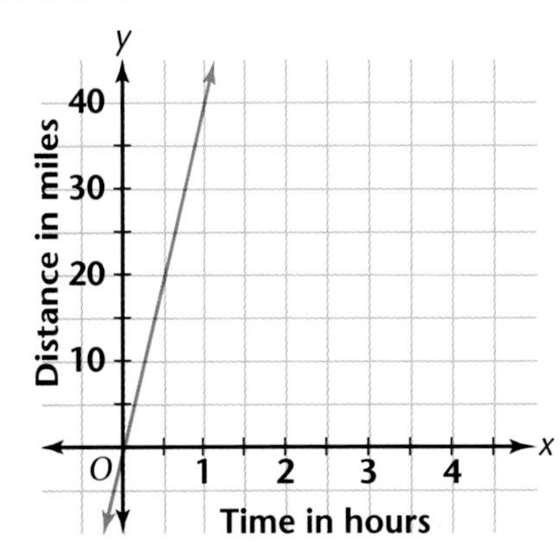

76. 40 miles per hour; the line intercepts the point (1, 40). If Marty has traveled 40 miles in 1 hour, she is traveling 40 mph.

The concepts that you have learned in this chapter can help you design a spreadsheet to solve problems.

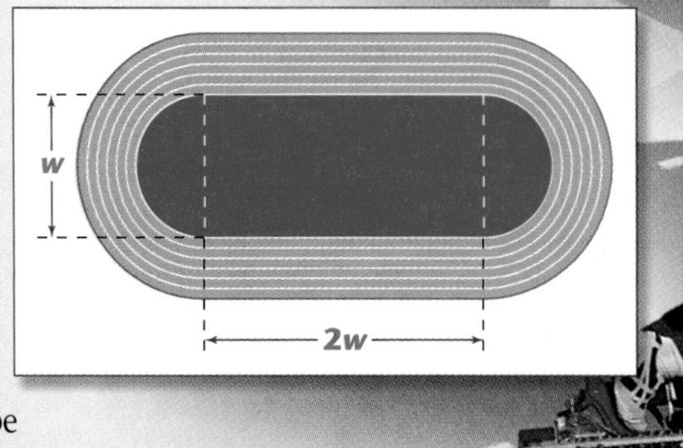

Suppose that the school board has approved the construction of a running track at the high school. Any lane of the track encloses an area that is the shape of a rectangle with a semicircle at each end. The length of the rectangular part of the track should be twice the width.

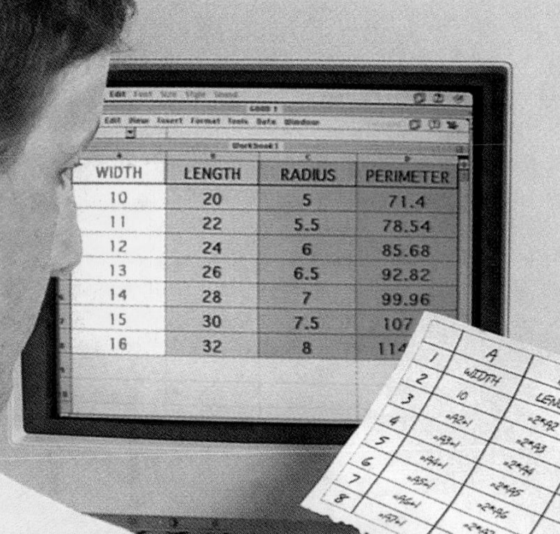

WIDTH LENGTH RADIUS PERIMETER
10 20 5 71.4
11 22 5.5 78.54
12 24 6 85.68
13 26 6.5 92.82
14 28 7 99.96
15 30 7.5 107
16 32 8
A B C D
WIDTH LENGTH RADIUS PERIMETER

You can use a spreadsheet to find dimensions of the track, such as the distance around any lane of the track, depending on the width, w.

Notice that the distance around any lane of the track is the sum of twice the length of the corresponding rectangle and the circumference of the corresponding circle.

$$\text{perimeter} = 2 \cdot 2w + \pi \cdot w$$
$$= 4w + \pi w$$

The spreadsheet shown here begins with a width of 10 meters and increases the width by 1 meter in each new row.

The formulas that were entered into the cells to compute the data are shown below. Notice that 3.14 is used for π.

	A	B	C	D
1	WIDTH	LENGTH	RADIUS	PERIMETER
2	10	=2*A2	=A2/2	=*B2+3.14*A2
3	=A2+1	=2*A3	=A3/2	=*B3+3.14*A3
4	=A3+1	=2*A4	=A4/2	=*B4+3.14*A4
5	=A4+1	=2*A5	=A5/2	=*B5+3.14*A5
6	=A5+1	=2*A6	=A6/2	=*B6+3.14*A6
7	=A6+1	=2*A7	=A7/2	=*B7+3.14*A7
8	=A7+1	=2*A8	=A8/2	=*B8+3.14*A8

ACTIVITY

Answers may vary depending on value of π used. Answers given were calculated using 3.14 for π.

The International Amateur Athletic Federation (IAAF) is the governing body for track and field. The IAAF rules specify that an outdoor running track should measure no less than 400 meters around (perimeter).

Note: Rounding radius <6 places may cause values <400 meters.

1. Find the dimensions of the inside lane of the track that will produce a lap length of no less than 400 meters.
   $w \approx 56.02241$ meters; $2w \approx 112.04482$ meters; $r \approx 28.011205$ meters
2. IAAF rules state that a lane should measure about 1.25 meters in width. Find the radius of the outer edge of the inside lane.
   outer radius of inside lane ≈ 29.261205 meters
3. Suppose that the track is to have 6 lanes. Find the radii of the outer edges of lanes 2–6.
4. Find the distance around the inside of each lane. Assume that the inner edge of the inside lane has a length of 400 meters. Explain how you could stagger the starting point for each lane to account for the differences in the lengths of each lane.

Chapter 1 Review

Vocabulary

absolute value	42	independent variable	5
Addition Property of Equality	12	Multiplication Property of Equality	17
Addition Property of Inequality	36	Multiplication Property of Inequality	36
constant	5	neutral pair	12
dependent variable	5	proportion	19
Distributive Property	30	range	7
Division Property of Equality	17	Reciprocal Property	17
Division Property of Inequality	36	Subtraction Property of Equality	11
domain	7	Subtraction Property of Inequality	36
equivalent ratios	19	variable	5
function	6		

Key Skills & Exercises

Lesson 1.1

➤ Key Skills

Use the Addition and Subtraction Properties of Equality.

Solve the equation $x - 8 = 4$.

$x - 8 = 4$	Given
$x - 8 + 8 = 4 + 8$	Addition Property
$x = 12$	Simplify.

Solve the equation $x + 8 = 4$.

$x + 8 = 4$	Given
$x + 8 - 8 = 4 - 8$	Subtraction Property
$x = -4$	Simplify.

➤ Exercises

Solve and check each equation.

1. $x + 11 = 15$ 4 **2.** $x - 8.6 = 2.5$ 11.1 **3.** $-5 = y - 7$ 2 **4.** $y + 13 = -8$ −21

Lesson 1.2

➤ Key Skills

Use the Multiplication and Division Properties of Equality.

Solve the proportion $\frac{a}{6} = \frac{-7}{15}$.

$\frac{a}{6} = \frac{-7}{15}$	Given
$6\left(\frac{a}{6}\right) = 6\left(\frac{-7}{15}\right)$	Multiplication Property
$a = \frac{-42}{15}$, or -2.8	Simplify.

Solve the equation $-7x = 4.9$.

$-7x = 4.9$	Given
$\frac{-7x}{-7} = \frac{4.9}{-7}$	Division Property
$x = -0.7$	Simplify.

➤ Exercises

Solve and check each equation.

5. $9x = -72$ −8 **6.** $-4.1t = 820$ −200 **7.** $\frac{d}{13} = 26$ 338 **8.** $\frac{12}{35} = \frac{y}{6}$ $y = 2\frac{2}{35}$

Lesson 1.3

➤ *Key Skills*

Use the properties of equality to solve two-step equations.

Solve the equation $8x - 11 = 13$.

$8x - 11 = 13$	Given
$8x - 11 + 11 = 13 + 11$	Addition Property of Equality
$8x = 24$	Simplify.
$\frac{8x}{8} = \frac{24}{8}$	Division Property of Equality
$x = 3$	Simplify.

Check by substituting 3 for x in the original equation.

$$8x - 11 = 13$$
$$8(3) - 11 = 13$$
$$24 - 11 = 13$$
$$13 = 13 \quad \text{True}$$

➤ *Exercises*

Solve and check each equation.

9. $9r - 8 = 19$ 3

10. $33w + 23 = 89$ 2

11. $-8x - 4 = -12$ 1

12. $\frac{m}{6.1} - 6.6 = 3.4$ 61

Lesson 1.4

➤ *Key Skills*

Use the properties of equality and the Distributive Property to solve multistep equations.

Solve the equation $16n + 3(n - 1) = 44.5$.

$16n + 3(n - 1) = 44.5$	Given
$16n + 3n - 3 = 44.5$	Distributive Property
$19n - 3 = 44.5$	Simplify.
$19n - 3 + 3 = 44.5 + 3$	Addition Property of Equality
$19n = 47.5$	Simplify.
$\frac{19n}{19} = \frac{47.5}{19}$	Division Property of Equality
$n = 2.5$	Simplify.

Check by substituting 2.5 for n in the original equation.

$$16n + 3(n - 1) = 44.5$$
$$16(2.5) + 3(2.5 - 1) = 44.5$$
$$40 + 3(1.5) = 44.5$$
$$40 + 4.5 = 44.5$$
$$44.5 = 44.5 \quad \text{True}$$

➤ *Exercises*

Solve and check each equation.

13. $7(x + 4) = 105$ 11

14. $11y - 18 = 5y - 24$ −1

15. $-5z + 11 = 53 - 12z$ 6

16. $9(x + 3) - 5x = 29$ 0.5

17. $3.5b + 2(0.25b - 7) = 2$ 4

18. $4(x + 9) = 9(x + 4)$ 0

Lesson 1.5

➤ Key Skills

Use the Properties of Inequality to solve an inequality, and graph the solution on a number line.

Solve the inequality $-1\frac{1}{3}x - 5 > -3$.

$-1\frac{1}{3}x - 5 > -3$	Given
$-\frac{4}{3}x - 5 > -3$	Change the mixed numeral
$-\frac{4}{3}x - 5 + 5 > -3 + 5$	Addition Property of Equality
$-\frac{4}{3}x > 2$	Simplify.
$-\frac{3}{4}\left(-\frac{4}{3}x\right) < -\frac{3}{4}(2)$	Multiplication Property of Inequality
$x < -\frac{3}{2}$, or -1.5	Simplify.

The graph of $x < -1.5$ is:

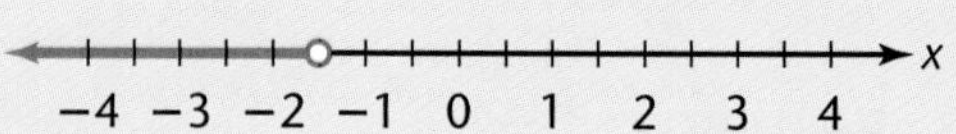

➤ Exercises

Solve and graph each inequality.

19. $-3x + 7 < 31$ **20.** $8x - 3 \geq 13$ **21.** $6.2y - 4.7 > 26.3$

22. $5t \geq 16$ **23.** $\frac{t}{-5} < 0.4$ **24.** $1\frac{3}{8}x + 2 \leq 7\frac{1}{2}$

Lesson 1.6

➤ Key Skills

Solve absolute-value equations and inequalities.

Solve the absolute-value inequality $|4x + 1| \geq 17$.

Case 1: The expression inside the absolute-value symbol is positive or zero.

$$4x + 1 \geq 17$$
$$4x + 1 - 1 \geq 17 - 1$$
$$4x \geq 16$$
$$\frac{4x}{4} \geq \frac{16}{4}$$
$$x \geq 4$$

Case 2: The expression inside the absolute-value symbol is negative.

$$-(4x + 1) \geq 17$$
$$-4x - 1 \geq 17$$
$$-4x - 1 + 1 \geq 17 + 1$$
$$-4x \geq 18$$
$$\frac{-4x}{-4} \leq \frac{18}{-4}$$
$$x \leq -4.5$$

The solution $x \leq -4.5$ *or* $x \geq 4$ is:

x
−5 −4 −3 −2 −1 0 1 2 3 4 5

➤ Exercises

Solve and graph each absolute-value equation or inequality.

25. $|x - 6| = 3$ **26.** $|t + 5| > 6$ **27.** $|7y + 3.5| \geq 24.5$ **28.** $|3z - 6| \leq 21$

Application

$|35 - t| \leq 4$; $31 \leq t \leq 39$; 31 to 39 minutes

29. Travel In order to arrive at school by 7:45 A.M., Cory leaves his home at 7:10 A.M. He always arrives at school within 4 minutes of his goal. Write and solve an absolute-value inequality to find the time, t, that Cory spends traveling to school.

Chapter 1 Assessment

Solve and check each equation.

1. $x + 5 = 2$ -3
2. $17 = y - 4$ 21
3. $7w = -84$ -12
4. $12z - 46 = 2$ 4
5. $\frac{4}{5}m = -20$ -25
6. $13.43 = -28p - 0.57$ -0.5
7. $\frac{1}{3}r + 5 = -7$ -36
8. $7s + 13 = 3s - 17$ -7.5
9. $\frac{a}{2.5} = -100$ -250
10. $150 = -x - 200$ -350
11. $8y - 2(y + 7) = 4$ 3
12. $3t - 4(2t + 3) = 7(5t - 2) - 44t$ -0.5

Consumer Economics The value of a used automobile is \$2300 less than its price when it was new. 13. Let p = new (or original) price, let v = current value

13. Write an equation for the current value, v, of the automobile in terms of the original price, p, of the automobile. $v = p - 2300$
14. Use the equation you wrote for Exercise 13 to write an equation showing that the current value of the automobile is \$9800. Then solve the equation to find the original price of the automobile. $9800 = p - 2300$, $p = 12{,}100$. The original price of the car was \$12,100.

Solve an check each proportion.

15. $\frac{x}{4} = \frac{7}{20}$ 1.4
16. $\frac{13}{27} = \frac{y}{18}$ $8\frac{2}{3}$
17. $\frac{-z}{19} = \frac{4}{5}$ -15.2

Solve and check each inequality. Graph the solution on a number line.

18. $5x + 8 > 43$
19. $-8x - 14 \le x + 91$
20. $-\frac{3}{4}x < 18$

Small Business Ben's housecleaning service charges \$60 for each visit to a client's home. His monthly expenses consist of \$3200 in payroll costs, \$600 in rent, \$150 in utilities, and \$200 in miscellaneous costs.

21. Write an equation for his monthly profit, p, in terms of the number of visits, x, to a client's home. $p = 60x - 4150$
22. Write an inequality to show that Ben's profit must be at least \$3500.
23. How many home visits does Ben's company need to make in order to make the profit of at least \$3500?

22. $60x - 4150 \ge 3500$ 23. Ben's company must make at least 128 visits.

Solve and check each equation or inequality.

24. $|x - 10| = 5$ 15, 5
25. $|t + 6.5| = 3.5$
26. $|c + 8| > 4$
27. $|2d - 7| \le 9$
28. $|3y - 6| < 33$
29. $|6 - z| \ge 14$

25. $-3, -10$
26. $c < -12$ or $c > -4$
27. $-1 \le d \le 8$
28. $-9 < y < 13$
29. $z \le -8$ or $z \ge 20$

Manufacturing A furniture manufacturer assembles bookcases with shelves that must be 35 inches long with an error tolerance of ± 0.2 inches.

30. Write an absolute-value inequality to show the range of acceptable lengths. Then solve the inequality.
31. Draw a graph on a number line to show the acceptable lengths for a shelf.

CHAPTER 2

LESSONS

Linear Functions and Systems

The atmosphere is electric! It's minutes before your favorite rock star begins her huge benefit concert. You can't believe you're here. You had to work hard to make the money and time to attend, but you love her music, and the money goes to charity. Thousands of people like you crowd the hall, eager for the concert to start. A hush passes over the crowd. The music starts. The crowd cheers. It has begun!

Portfolio Activity

You are so excited about the concert that you ask eleven of your friends to join you. The tickets are sold at two different prices, with the center section costing $10 more than the balcony seats. After collecting all the money, you buy 7 center seats and 5 balcony seats for a total of $310. What is the cost of each type of seat?

Use c as the cost of a center-section ticket and b as the cost of a balcony ticket. Write two equations using c and b that show the relationships in this problem. Later you will be asked to solve this problem using various methods. You may wish to use these exercises as part of your portfolio.

LESSON 2.1 Exploring Slope

Why *The slope of a line can have many meanings. It can describe the incline of a hill or a rate of travel. You can find the slope of a line from its graph.*

To gain altitude, hang gliders use updrafts such as slope winds. Slope winds are winds that blow against a hill and are deflected upward. Experienced pilots can use the slope winds of mountain ranges to fly great distances.

SLOPE WIND

The steepness of a hill can be described by the **slope**, which is the ratio of the vertical rise to the horizontal run. The slope of a hill indicates the rate of change in height per horizontal distance.

GEOMETRIC OR GRAPHIC INTERPRETATION OF SLOPE

Slope measures the steepness of a line by the formula

$$\text{slope} = \frac{\text{rise}}{\text{run}}.$$

Positive Slope

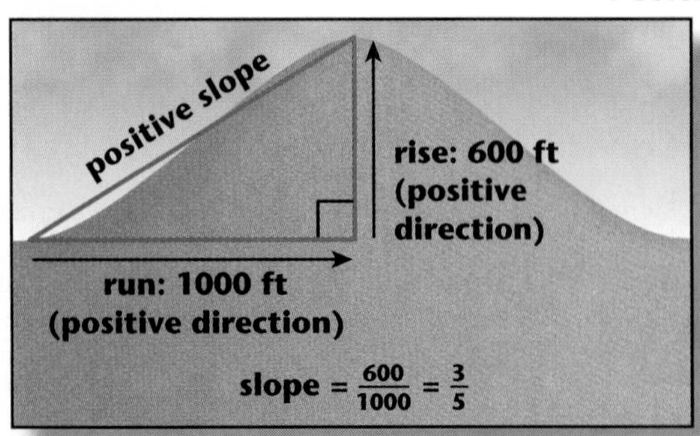

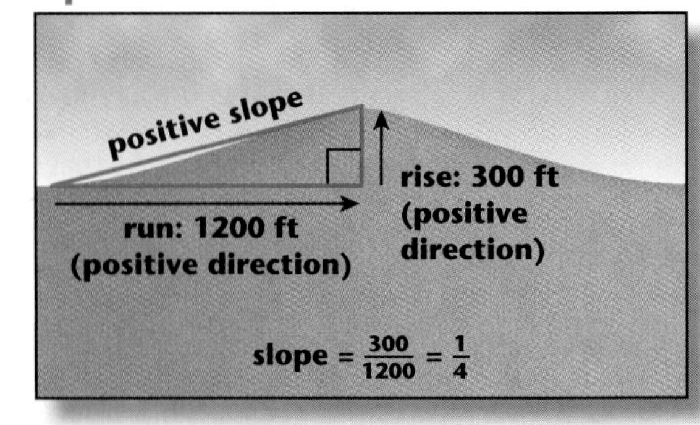

Negative Slope

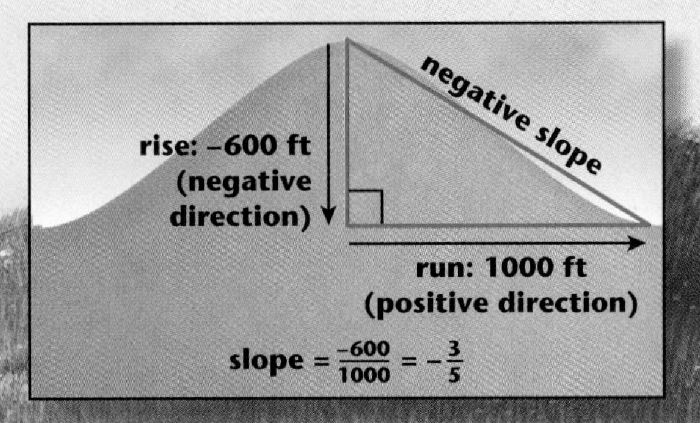

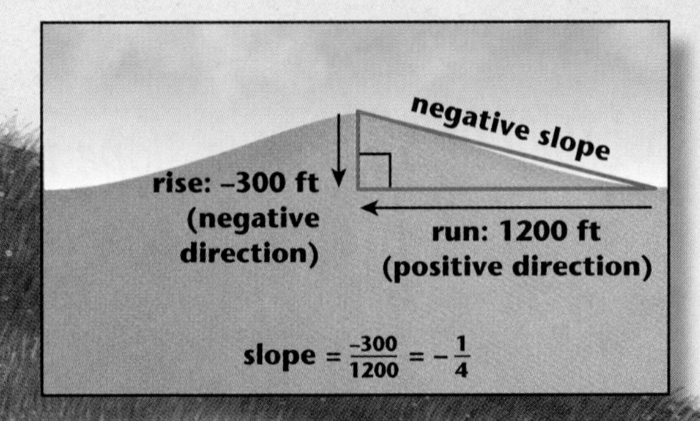

positive slope $= \frac{\text{positive direction}}{\text{positive direction}}$ or $\frac{\text{negative direction}}{\text{negative direction}}$

Positive directions, represented with positive numbers, are *up* and *to the right* on a coordinate plane. Negative directions, represented with negative numbers, are *down* and *to the left* on a coordinate plane.

What combinations of positive and negative directions result in a positive slope? a negative slope? negative slope $= \frac{\text{positive direction}}{\text{negative direction}}$ or $\frac{\text{negative direction}}{\text{positive direction}}$

A line with a positive slope *slants upward* from left to right. A line with negative slope *slants downward* from left to right.

Slope can also be used to describe a rate of speed.

Exploration 1 *Using a Graph*

A hiker leaves a campsite to hike to a lake. The graph of the line shows the relationship between the time since the hiker left the campsite and the distance the hiker is from the campsite.

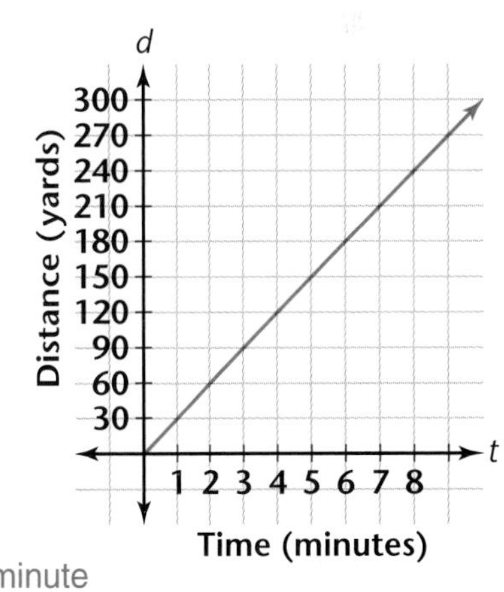

1. How many yards did the hiker travel in 4 minutes? in 8 minutes? in 16 minutes? 120 yards, 240 yards, 480 yards

2. The rate at which the hiker traveled is found by the ratio $\frac{\text{change in distance}}{\text{change in time}}$. What is the hiker's rate? 30 yards per minute

3. Find the slope of the line in the graph. Describe the meaning of the slope in this situation. Slope = 30; the slope represents the hiker's rate.

4. Let d represent the distance and t represent the time. Use d and t to write an equation to show that the distance the hiker traveled depends on the time since he left the campsite. ❖ $d = 30t$

In Exploration 1, you used slope to interpret the graph of a line. You can graph a line if you know the slope and one of the points on the line.

Exploration 2 Using Slope to Graph a Line

Part I: Positive Slope
Suppose a line passes through the point (2, −1) and has a slope of $\frac{2}{3}$.

1. Plot the point (2, −1) on a coordinate grid.
2. A slope of $\frac{2}{3}$ has a rise of _?_ and a run of _?_. 2, 3
3. Find another point on the line by starting at the point (2, −1) and using the slope, $\frac{2}{3}$, to find the next point. Points may vary. (5, 1)
4. Plot the second point as described above. Then draw a line through the two points.
5. Describe how to use a positive slope to graph a line when you know one point on the line.

Part II: Negative Slope
Suppose a line passes through the point (2, −1) and has a slope of $-\frac{2}{3}$.

1. Plot the point (2, −1) on a coordinate grid.
2. Since the slope is negative, you can write it as either $\frac{-2}{3}$ or $\frac{2}{-3}$. So a slope of $-\frac{2}{3}$ has a rise of _?_ and a run of _?_, or a rise of _?_ and a run of _?_. 2, −3, −2, 3
3. Find another point on the line by starting at the point (2, −1) and using the slope, $-\frac{2}{3}$, to find the next point. Points may vary. (−1, 1)
4. Plot the second point as described above. Then draw a line through the two points.
5. Describe how to use a negative slope to graph a line when you know one point on the line. ❖

Horizontal and vertical lines have special slopes.

Exploration 3 Horizontal and Vertical Lines

Part I
The graph of a horizontal line segment is shown here.

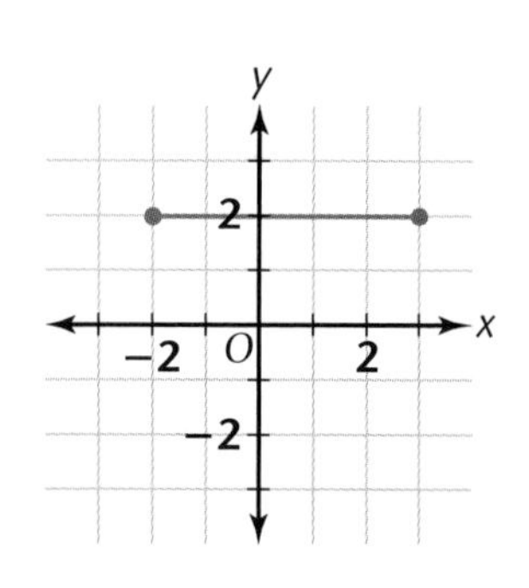

1. What is the rise for the line segment? 0
2. What is the run for the line segment? 5

3. Use $\frac{\text{rise}}{\text{run}}$ to find the slope of the line segment. $\frac{0}{5} = 0$
4. Graph three other horizontal line segments. Find the slope of each. What can you conclude about the slope of any horizontal line?

Part II

The graph of a vertical line segment is shown here.

1. What is the rise for the line segment? 5
2. What is the run for the line segment? 0
3. Use $\frac{\text{rise}}{\text{run}}$ to find the slope of the line segment. $\frac{5}{0}$ = undefined
4. A fraction or ratio with a denominator of 0 is undefined. Complete the following sentence: The slope of the line segment is undefined because the run is ? (zero).
5. Graph three other vertical line segments. Find the slope of each line segment. What can you conclude about the slope of any vertical line? ❖

CT– If the x-coordinates are the same, the line is vertical; if the y-coordinates are the same, the line is horizontal; if neither the x- or y-coordinates are the same, the line is neither vertical nor horizontal.

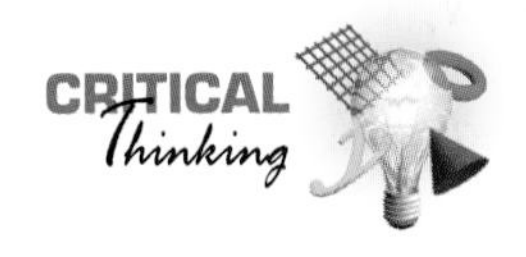

Explain how to determine whether a set of coordinates forms a horizontal line, a vertical line, or neither.

Exploration 4 *Finding Slope From Two Points*

1. Plot the points (3, 5) and (2, 7) on a coordinate plane. Draw a line through the two points.
2. Use your graph to determine the slope of the line. −2
3. Find the change in y-coordinates, or the difference between the y-coordinate of the first point and the y-coordinate of the second point. $5 - 7 = -2$ or $7 - 5 = 2$
4. Find the change in x-coordinates, or the difference between the x-coordinate of the first point and the x-coordinate of the second point. $3 - 2 = 1$ or $2 - 3 = -1$
5. Write the rate of change, or ratio $\frac{\text{difference in } y\text{-coordinates}}{\text{difference in } x\text{-coordinates}}$. How does this rate of change compare with the slope of the line? ❖

   $\frac{-2}{1} = -2$ or $\frac{2}{-1} = -2$ The ratio is the same as the slope of the line.

SLOPE AS A DIFFERENCE RATIO

If the coordinates of two points on a line are given, the slope, m, is the ratio

$$m = \frac{\text{difference in } y\text{-coordinates}}{\text{difference in } x\text{-coordinates}}$$

where the differences are in corresponding order.

EXTENSION

To determine the slope of the line passing through points (2, 5) and (−1, 4), use the definition of slope as a difference ratio.

$$m = \frac{\text{difference in } y\text{-coordinates}}{\text{difference in } x\text{-coordinates}} \qquad m = \frac{\text{difference in } y\text{-coordinates}}{\text{difference in } x\text{-coordinates}}$$

$$= \frac{5-4}{2-(-1)} \qquad \textbf{OR} \qquad = \frac{4-5}{-1-2}$$

$$= \frac{1}{3} \qquad\qquad = \frac{-1}{-3}, \text{ or } \frac{1}{3}$$

The slope is $\frac{1}{3}$. ❖

Direct Variation

Lines that pass through the origin have an equation of the form $y = mx$. In these equations, y varies directly as x is multiplied by the slope, m. The equation $d = rt$, where d is distance, r is rate of speed, and t is time, is an example of a direct variation that can be modeled by a line.

APPLICATION

Kyra walks at a rate of 6 miles per hour. Write a linear equation of direct variation, and draw a graph describing this situation.

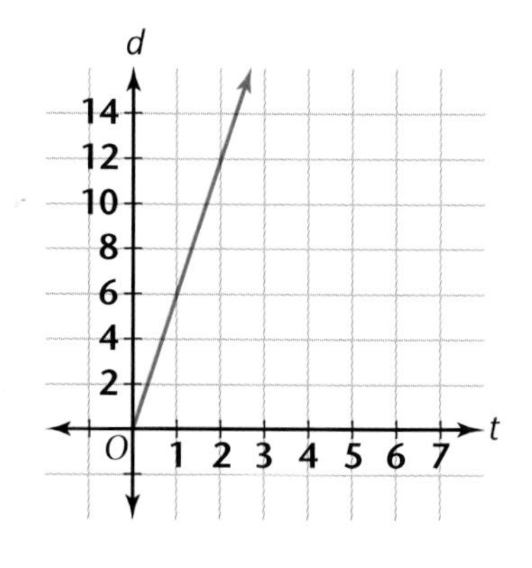

Substitute the rate of speed, 6 miles per hour, for r.

$$d = rt$$
$$d = 6t$$

The graph of $d = 6t$ is a line through the orgin with a slope of 6. ❖

Try This Suppose that hourly wages at a job are \$8. Let x represent the number of hours, and let y represent the total wages earned. Write a linear equation of direct variation, and draw a graph describing this situation.

Let $P_1(x_1, y_1)$ and $P_2(x_2, y_2)$ be two points on a line. Which two of the following ratios give the slope of that line? Which two ratios do not give the slope of that line? Explain.

$$\frac{y_1 - y_2}{x_1 - x_2} \qquad \frac{y_2 - y_1}{x_1 - x_2} \qquad \frac{y_2 - y_1}{x_2 - x_1} \qquad \frac{y_1 - y_2}{x_2 - x_1}$$

EXERCISES & PROBLEMS

2. Choose two points on the line, and calculate the $\frac{\text{rise}}{\text{run}}$ from one point to another.

Communicate

1. What are two different ways to define slope? $\frac{\text{rise}}{\text{run}}$ or $\frac{\text{difference in } y\text{-coordinates}}{\text{difference in } x\text{-coordinates}}$
2. Explain how to find the slope of a line from a graph.
3. A line has a slope of 0 and passes through the point $(-2, 4)$. Describe the line. Horizontal
4. A line has an undefined slope and passes through the point $(4, -9)$. Describe the line. Vertical
5. The graph at the right shows the distance traveled by a bicyclist in terms of the number of minutes she traveled. Describe the trip.
6. Why is the graph at the right shown only in the first quadrant?

Practice & Apply

For Exercises 7–15, determine the slope of each line.

7. −2

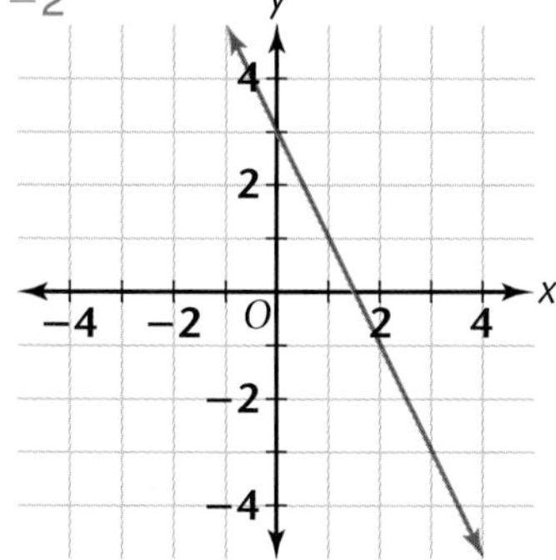

8. $\frac{1}{2}$

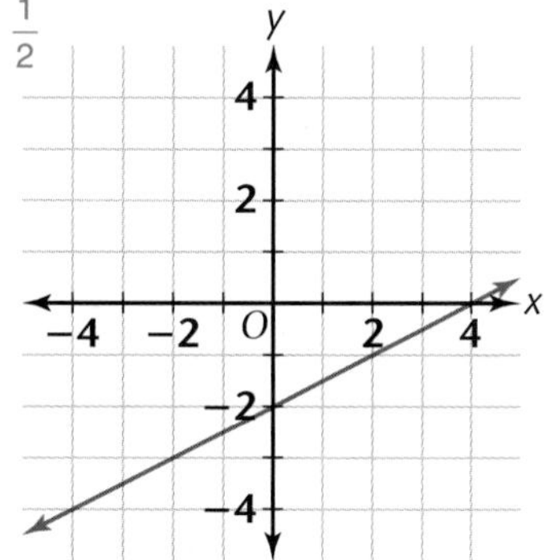

9. 4

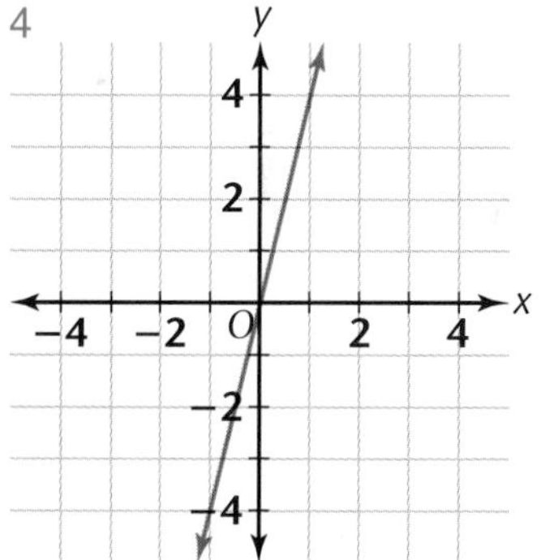

10. Undefined

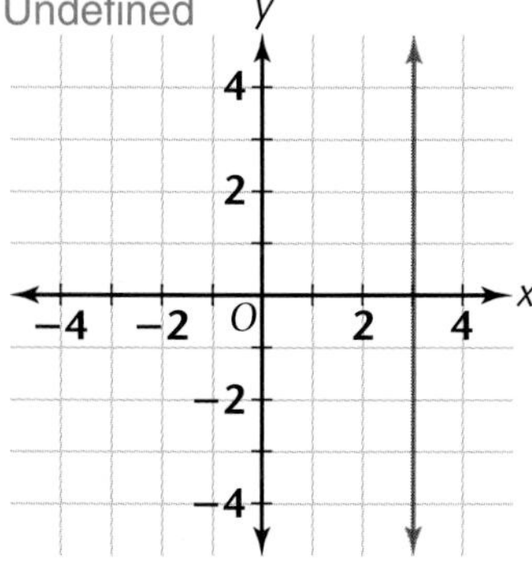

11. $-\frac{1}{3}$

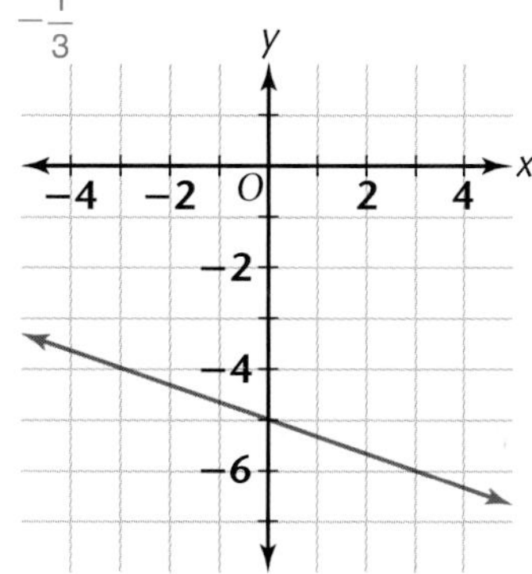

12. 0

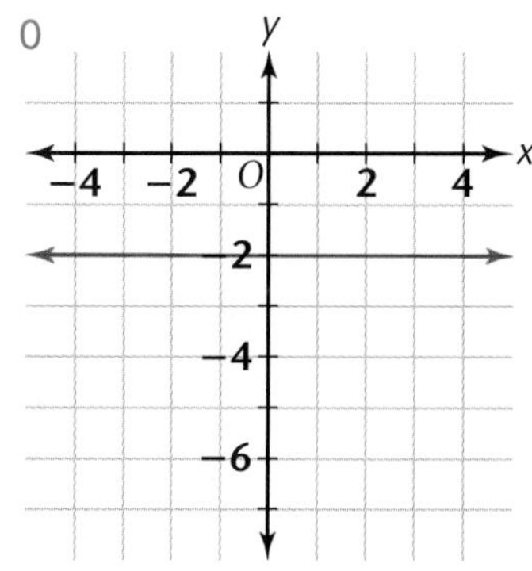

13.

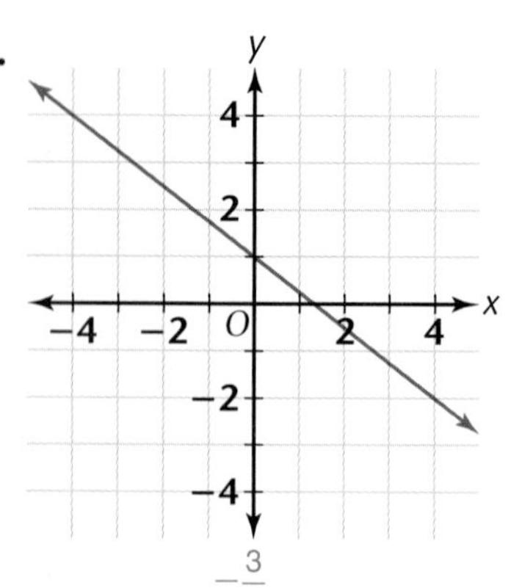

$-\frac{3}{4}$

14.

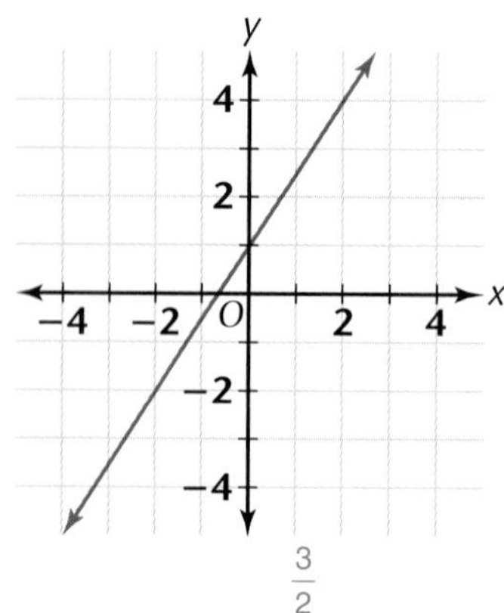

$\frac{3}{2}$

15.

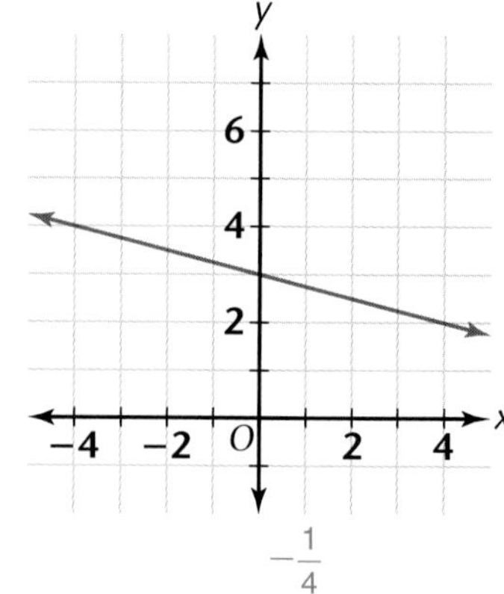

$-\frac{1}{4}$

Determine the slope of the line passing through each pair of points.

16. $(-7, 6)$ and $(5, 2)$ $-\frac{1}{3}$

17. $(5, -2)$ and $(-2, -3)$ $\frac{1}{7}$

18. $(-1, 4)$ and $(-2, 4)$ 0

19. $(-5, -7)$ and $(-3, -4)$ $\frac{3}{2}$

20. $(-4, 2)$ and $(6, 2)$ 0

21. $(4, 5)$ and $(5, 7)$ 2

22. $(-4, 3)$ and $(6, 1)$ $-\frac{1}{5}$

23. $(0, -3)$ and $(-4, -2)$ $-\frac{1}{4}$

24. $(-6, 5)$ and $(-1, 1)$

25. $(-3, 0)$ and $(-3, -6)$ Undef

26. $(-2, 2)$ and $(-2, 7)$ Undef

27. $(1, 5)$ and $(2, 1)$ -4

28. $(-1.5, 2.5)$ and $(0, 3)$ $\frac{1}{3}$

29. $(-8, -10)$ and $(-12, -14)$ 1

30. $(10, 14)$ and $(12, 14)$

24. $-\frac{4}{5}$ **30.** 0

Graph each line described below.

31. Slope of $-\frac{2}{3}$, through the point $(2, -1)$

32. Slope of -1, through the point $(3, -2)$

33. Slope of 3, through the point $(1, 7)$

34. Slope of $\frac{4}{5}$, through the point $(0, -3)$

35. Through point $(1, 4)$ with undefined slope

36. Through the point $(2, -4)$ with slope of 0

37. Slope of $\frac{4}{3}$, through the point $(0, -3)$

38. Slope of 2, through the point $(5, -4)$

39. Slope of 0.5, through the point $(0, 5)$

40. Slope of $-\frac{4}{5}$, through the point $(-1, -2)$

41. Through point $(2, 6)$ with undefined slope

42. Through the point $(-4, -7)$ with slope of 0

Water is being poured at a constant rate into a tank. The table shows the amount of water in the tank after a given number of minutes.

Times (minutes)	2	4	6	8	10	12
Water (gallons)	360	440	520	600	680	760

43. Graph the set of points described in the table.

44. Connect the points. Do they appear to lie on a straight line?

45. Use the points $(2, 360)$ and $(4, 440)$ to determine the slope of the line. Select a different set of points to determine the slope. Is the slope the same regardless of which two points you use? Explain.

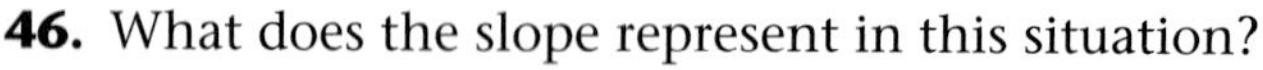

46. What does the slope represent in this situation?

47. Extend the line so that it intersects the y-axis. What does this point represent?

48. Does extending the line to the left of the y-axis make sense for this problem? Explain.

Travel Nick walks from his house to Mike's house. The table shows Nick's distance along the trail as a function of the number of minutes since he began walking.

Time (in minutes)	0	1	3	4	7	8
Distance (in feet)	960	910	810	760	610	560

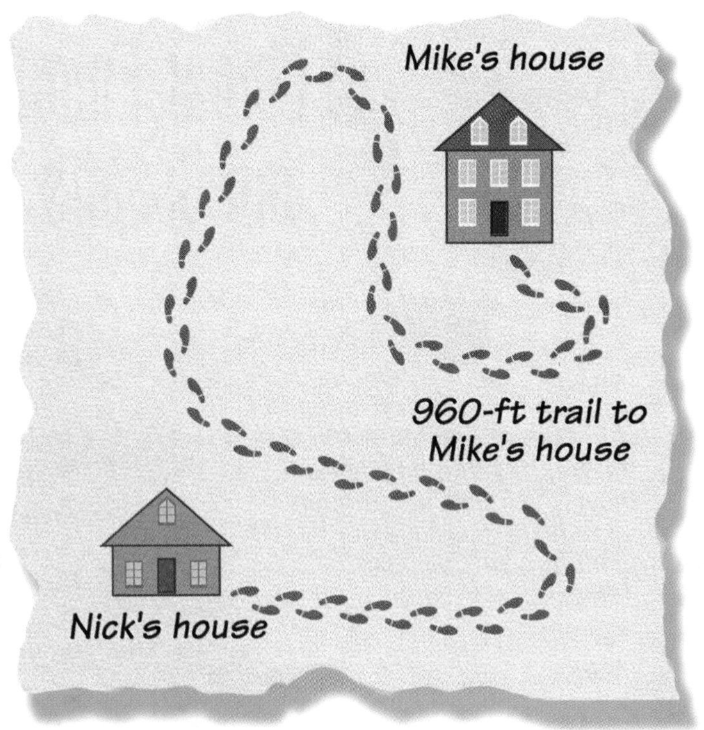

49. Graph the set of points. Explain why these points are in only the first quadrant.

50. Connect the points. Was Nick walking at a constant rate?

51. Determine the slope of the line.

52. Explain why the slope is a negative.

53. An airplane ascends at a rate of 2000 feet per minute. Write a linear equation of direct variation, and draw a graph describing this situation.

Look Back

Business Denise is a salesperson at an appliance store. She earns a weekly salary of $120 plus a commission of 3.2% of her weekly sales. **[Lesson 1.3]**

54. Write an equation to show her weekly income, i, as a function of the amount of weekly sales, s. $i = 0.032s + 120$

55. What must Denise's weekly sales be in order to have a weekly income of $400? $8750

Solve each equation to the nearest tenth. [Lesson 1.4]

56. $1.2h - 4.7 = 5.6h - 18.18$ 3.1

57. $4.9t - 3.2 = 1.04 - 2.5t$ 0.6

58. $13(r + 7) = 2r - 9$ −9.1

59. $-6(b - 3) = 6b + 1$ 1.4

Solve each equation to the nearest tenth. [Lesson 1.4]

60. $5.72x + 24 = 3.6x - 17.3$ −19.5

61. $5p - 7.3 = 4.1p + 16$ 25.9

Look Beyond

Consider each of the following situations, and determine which can be modeled by a linear equation. Explain your reasoning.

62. repair rates: $5 base fee plus $10 per hour

63. The growth of a collection of bacteria which doubles in size each hour

64. rental rates: $6 base fee, plus $8 per hour for the first 3 hours and $6 per hour for the remaining number of hours

65. constant speed of 50 miles per hour

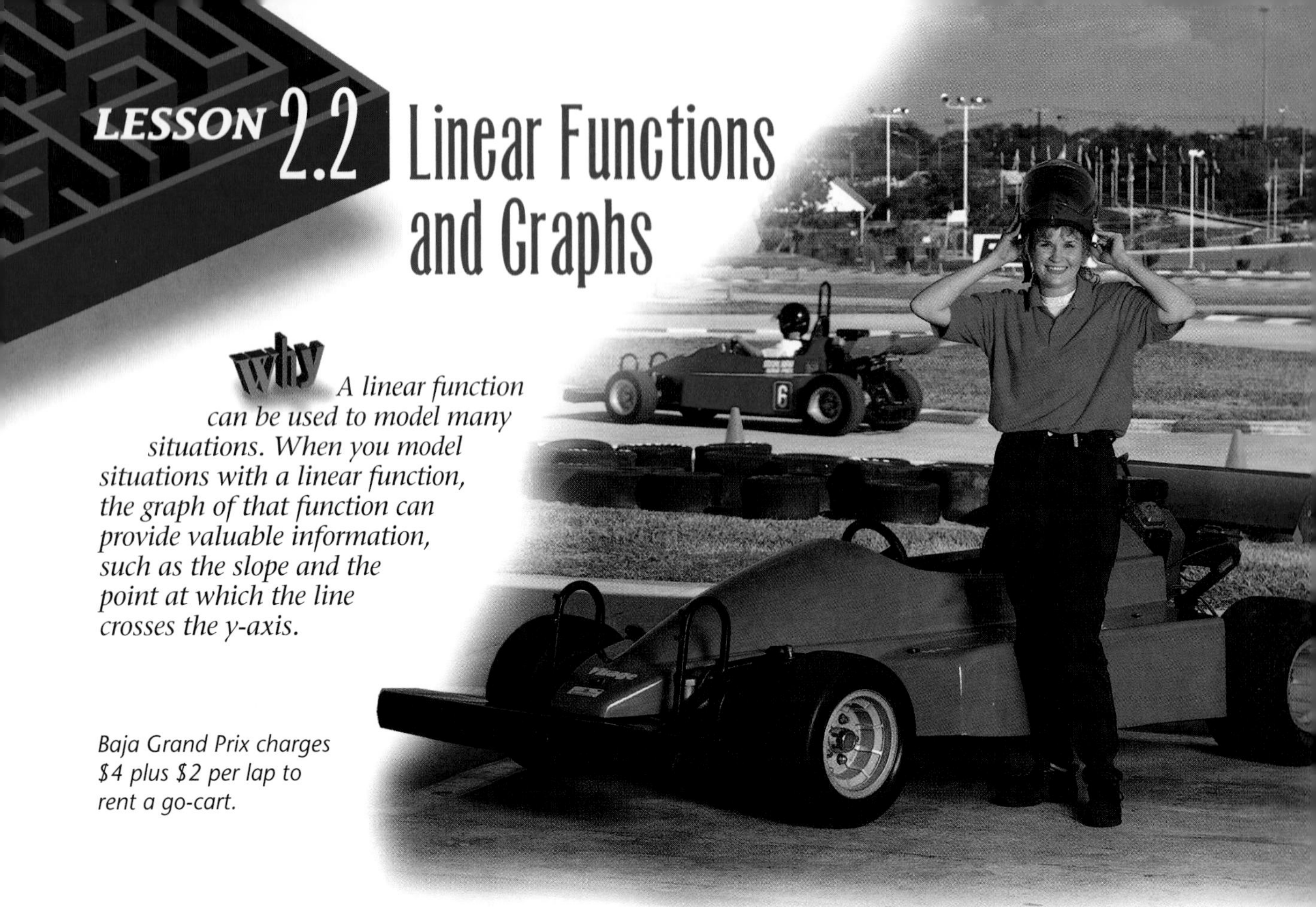

LESSON 2.2 Linear Functions and Graphs

why *A linear function can be used to model many situations. When you model situations with a linear function, the graph of that function can provide valuable information, such as the slope and the point at which the line crosses the y-axis.*

Baja Grand Prix charges $4 plus $2 per lap to rent a go-cart.

The Slope-Intercept Formula

The table below shows the charges for racing a go-cart at Baja Grand Prix.

Number of laps	0	1	2	3	4	5
Cost	$4	$6	$8	$10	$12	$14

The cost of racing a go-cart at Baja Grand Prix can be modeled by a linear function. Let y represent the cost, and let x represent the number of laps.

The cost is $2 per lap plus $4.

$$y = 2x + 4$$

The function is $y = 2x + 4$. The graph of this function is shown at the right.

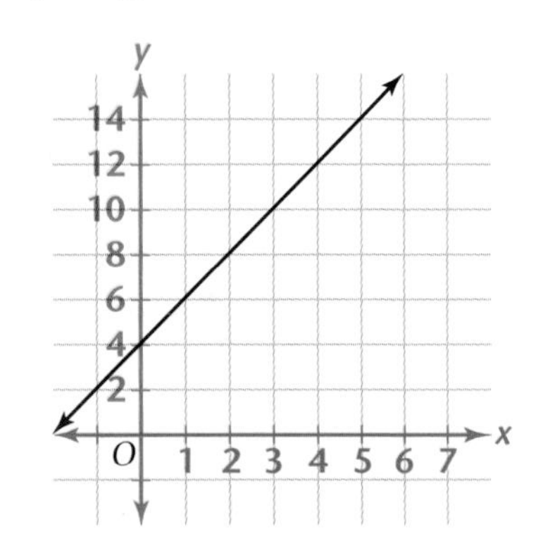

The rate and the slope are the same.

The rate is $2 per lap. How does this rate compare with the slope of the line?

The charge for 0 laps is $4 (if someone rents the go-cart but races less than 1 lap). This is shown on the graph by the **y-intercept**, the point where x is 0 and the line crosses the y-axis. What are the coordinates of the y-intercept?

The y-intercept coordinates are (0, 4).

You can find the slope and the y-intercept of the line by looking at the equation $y = 2x + 4$.

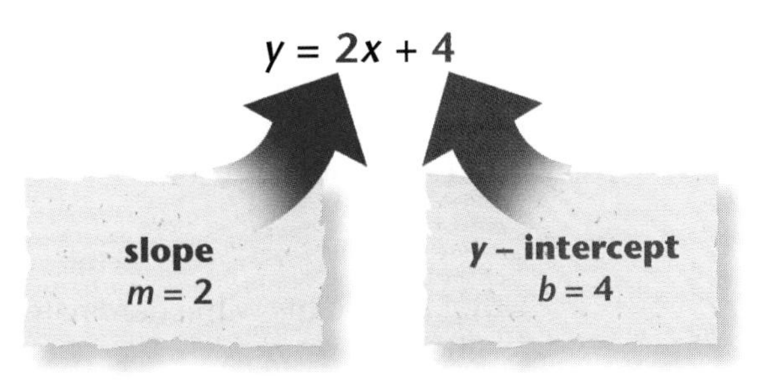

SLOPE-INTERCEPT FORM

The slope-intercept formula or form for a line with slope, m, and y-intercept, b, is

$$y = mx + b.$$

EXAMPLE 1

Graph the line $2y = x - 6$.

***Solution* ➤**

Put the equation in slope-intercept form, $y = mx + b$, by solving for y.

$$2y = x - 6$$
$$\frac{2y}{2} = \frac{x}{2} - \frac{6}{2}$$
$$y = \frac{1}{2}x - 3$$

CT– As the slope increases, the graph becomes steeper in the positive direction. As the slope decreases, the graph flattens out to a horizontal line and then gets steeper in the negative direction. If the y-intercept increases, the graph moves up on the y-axis. If the y-intercept decreases, the graph moves down on the y-axis.

The slope of the line $y = \frac{1}{2}x - 3$ is $\frac{1}{2}$.

The y-intercept of the line $y = \frac{1}{2}x - 3$ is -3, so the line crosses the y-axis at $(0, -3)$.

Plot the point $(0, -3)$, and use the slope to find another point on the line. Then draw the line passing through the two points. ❖

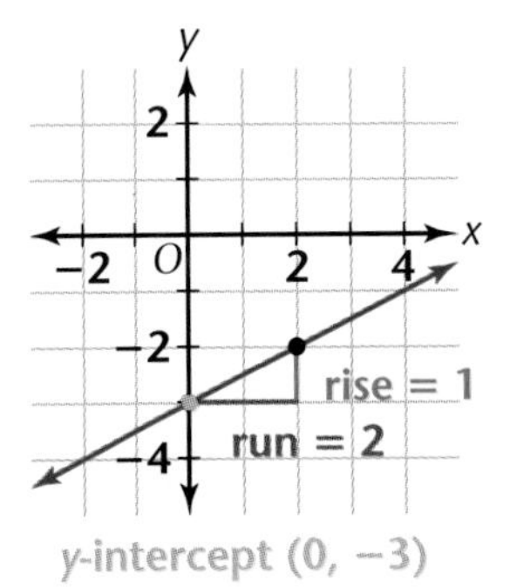

Try This Graph the line $y = -\frac{3}{4}x + 2$.

Describe the effect on the graph in Example 1 if the slope, m, increases or decreases. Describe the effect if the y-intercept, b, increases or decreases.

Writing an Equation From Two Points

If you know the coordinates of two points on a line, you can write an equation of the line. First find the slope of the line. Then use the slope, the coordinates of one of the points, and the slope-intercept formula to find the y-intercept. You can use the slope formula to calculate the slope.

SLOPE FORMULA

Given two points with coordinates (x_1, y_1) and (x_2, y_2), the formula for the slope is

$$m = \frac{\text{change in } y}{\text{change in } x} = \frac{y_2 - y_1}{x_2 - x_1}.$$

EXAMPLE 2

Small Business Mario is an electrician who charges an hourly rate plus a base fee for electrical services. Write an equation to describe Mario's charge in terms of the number of hours he works.

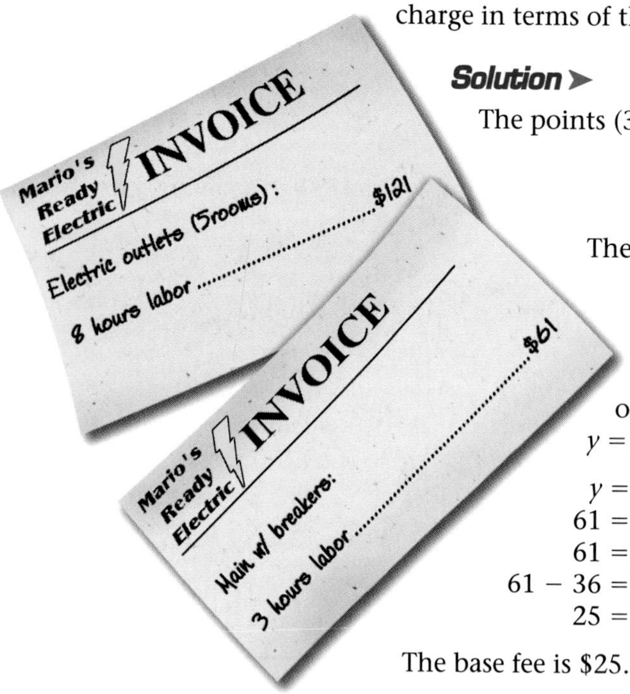

Solution ➤

The points (3, 61) and (8, 121) represent this information.

$$m = \frac{\text{change in } y}{\text{change in } x} = \frac{121 - 61}{8 - 3} = \frac{60}{5} = 12$$

The slope, m, represents Mario's hourly rate of $12.

$$y = 12x + b$$

To determine the base fee that Mario charges, find the value of b. Substitute the coordinates of one of the points, (3, 61) or (8, 121), in $y = 12x + b$, and solve for b.

$y = 12x + b$	Given
$61 = 12(3) + b$	Substitute 61 for y and 3 for x.
$61 = 36 + b$	Simplify.
$61 - 36 = 36 + b - 36$	Subtraction Property of Equality
$25 = b$	Simplify.

The base fee is $25. Therefore, the equation is $y = 12x + 25$. ◆

Standard Form

A linear equation can be written in different forms. One form is called the *standard form* of a linear equation.

STANDARD FORM

An equation of the form **Ax + By = C** is in standard form when **A, B,** and **C** are integers, **A** and **B** are not both zero, and **A** is not negative.

EXAMPLE 3

Crafts Kayla makes wreaths and baskets to sell at a crafts fair. She wants to sell enough wreaths and enough baskets to make a total of \$600.

A Write an equation to model her desired income.

B Graph the equation.

Kayla charges \$4 for each basket and \$12 for each wreath.

Solution ➤

A Let x represent the number of baskets, and let y represent the number of wreaths. Then \$4 times the number of baskets sold plus \$12 times the number of wreaths sold should equal \$600.

$$4x + 12y = 600$$

B To graph the equation, determine the x- and y-intercepts.

When x is 0, the line crosses the y-axis. Substitute 0 for x, and solve for y to find the y-intercept.

$$\begin{aligned} 4x + 12y &= 600 \\ 4(0) + 12y &= 600 \\ 12y &= 600 \\ \frac{12y}{12} &= \frac{600}{12} \\ y &= 50 \end{aligned}$$

The y-intercept is (0, 50).

When y is 0, the line crosses the x-axis. Substitute 0 for y, and solve for x to find the x-intercept.

$$\begin{aligned} 4x + 12y &= 600 \\ 4x + 12(0) &= 600 \\ 4x &= 600 \\ \frac{4x}{4} &= \frac{600}{4} \\ x &= 150 \end{aligned}$$

The x-intercept is (150, 0).

Plot the x- and y-intercepts, and draw the line.

The coordinates (x, y) of any point on the graph of $4x + 12y = 600$ represent the number of baskets and the number of wreaths that Kayla can sell to make a \$600 profit. ❖

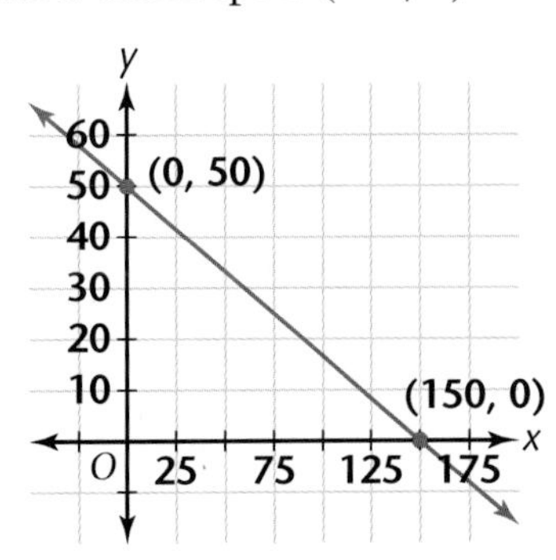

A Line of Best Fit

Real-world data can sometimes be modeled by a linear function. The table below shows the winning times in seconds for the men's 100-meter backstroke in the Summer Olympics.

The winning times for men's 100-meter backstroke appear to be decreasing as the year increases. Notice from the graph that the data points nearly form a line, which means they have a relationship that is nearly linear.

1920

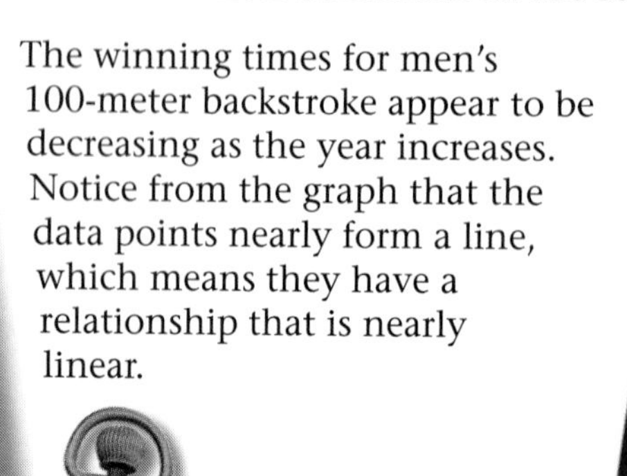

1996

Warren Kealoha, gold-medal winner of the men's 100-meter backstroke in the Summer Olympics of 1920

Jeff Rouse, gold-medal winner of the men's 100-meter backstroke in the Summer Olympics of 1996

Year	Time in seconds
1920	75.2
1924	73.2
1928	68.2
1932	68.6
1936	65.9
1948	66.4
1952	65.4
1956	62.2
1960	61.9
1968	58.7
1972	56.58
1976	55.49
1980	56.53
1984	55.79
1988	55.05
1992	53.98
1996	54.10

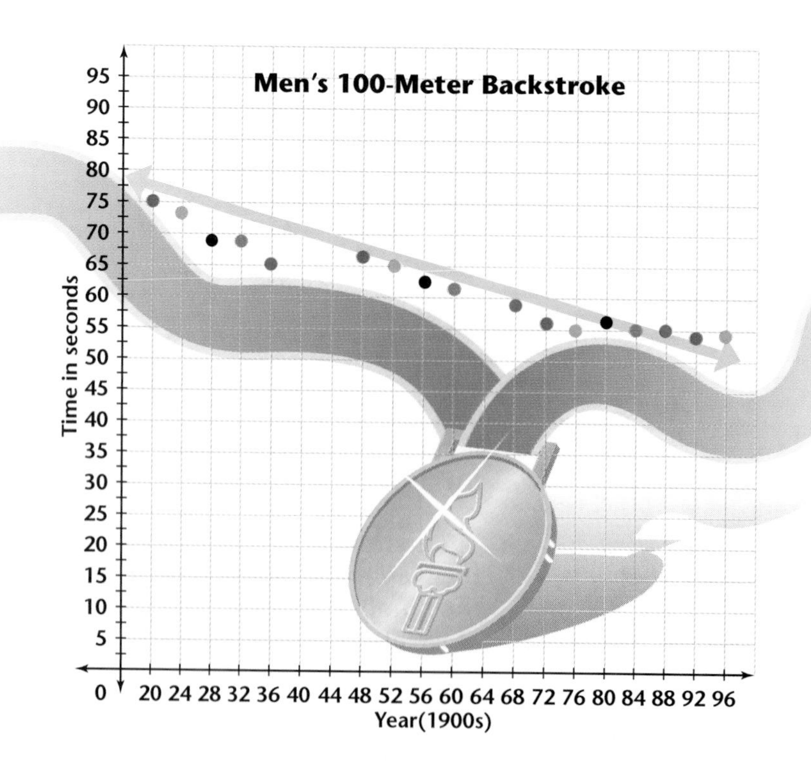

You can use a graphics calculator to determine if the data show a linear relationship. Enter the data into the statistics list of your calculator. Use the last two digits of the year for the first list and the time in seconds for the second list.

Compute the equation of the *regression line*, or the **line of best fit**. The slope is a and the y-intercept is b, so the equation for the regression line is approximately

$$y = -0.267x + 77.83.$$

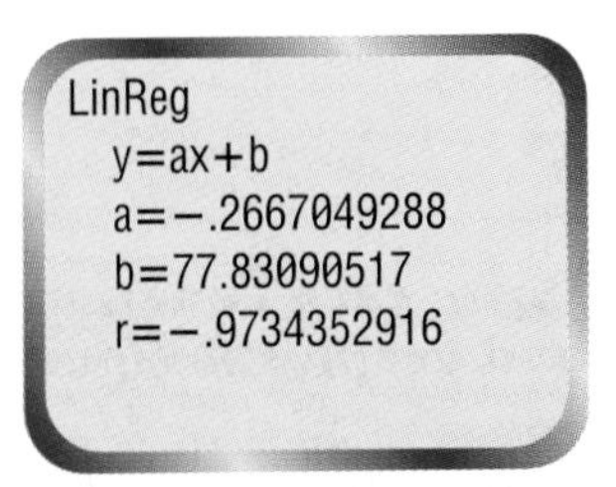

Graph the points and the regression equation. Notice that the line has a *negative slope*. The negative slope indicates a negative correlation. As the years increase, the time decreases.

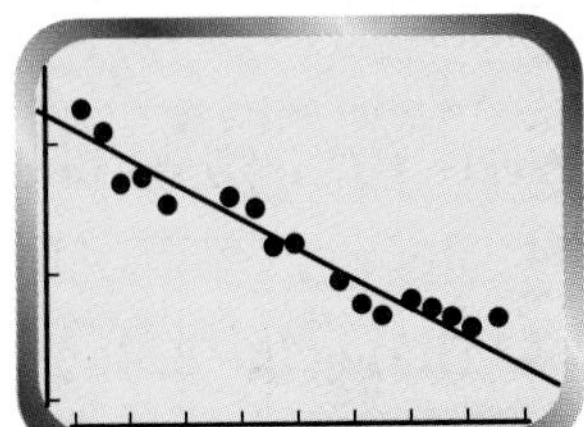

The letter r represents the **correlation coefficient**, which describes how closely the data follow a linear pattern. When r is close to 1 or close to -1, the data very closely follow a linear pattern, and the line is said to be a *good fit*.

In 1964, there was no men's 100-meter backstroke competion. The line of best fit can be used to approximate what the expected winning time for the men's 100-meter backstroke in 1964 would have been. Substitute 64 for x, and solve for y.

$$y = -0.267x + 77.83$$
$$y = -0.267(64) + 77.83$$
$$y \approx 60.7$$

The expected winning time in 1964 would have been approximately 60.7 seconds.

EXERCISES & PROBLEMS

Communicate

1. Explain how to graph the line $y = \frac{3}{4}x + 2$.
2. Describe how to graph $2x - 3y = 12$ using the x- and y-intercepts.
3. How can you tell whether a regression equation is a good fit?
4. Describe how to find the equation of a line if you are given two points on the line.
5. How do you graph the line $3x - 4y = 8$ on a graphics calculator?

Practice & Apply

Use the slope and y-intercept to graph each line.

6. $y = 3x - 2$ **7.** $y = -2x + 4$ **8.** $y = \frac{2}{5}x - 4$

9. $y = \frac{4}{3}x + 5$ **10.** $y = -\frac{1}{2}x - 4$ **11.** $y = -\frac{2}{3}x + 2$

Use the x- and y-intercepts to draw the graph of each line.

12. $3x - 2y = 12$ **13.** $4x + 5y = 10$ **14.** $4x - y = 3$

15. $x + 4y = 8$ **16.** $x - 3y = 6$ **17.** $1.6x + 4y = 12$

Write each equation in slope-intercept form. Then state the slope, and graph the equation.

18. $4x + 5y = 15$ **19.** $3x - y = 24$ **20.** $3.2x + 8y = 30$

21. $5.6x - 4y = 30$ **22.** $7.2x + 3.6y = 480$ **23.** $18.5x + 5y = 26$

Write the equation of each line described below. 26. $y = \frac{5}{3}x + \frac{22}{3}$

24. Slope of $\frac{3}{4}$ with y-intercept -5 $y = \frac{3}{4}x - 5$

25. Slope of -2 with y-intercept 6 $y = -2x + 6$

26. Through $(-2, 4)$ with a slope of $\frac{5}{3}$

27. Through $(-5, -12)$ with a slope of -4 $y = -4x - 32$

28. Through $(-2, 3)$ with a slope of 0 $y = 3$

29. Through $(-2, 3)$ with undefined slope $x = -2$

30. Through $(-5, 6)$ and $(-3, 9)$ $y = \frac{3}{2}x + \frac{27}{2}$

31. Through $(4, 12)$ and $(-2, 56)$ $y = -\frac{22}{3}x + \frac{124}{3}$

32. With x-intercept $(12, 0)$ and y-intercept $(0, 50)$ $y = -\frac{25}{6}x + 50$

33. With x-intercept $(0.5, 0)$ and y-intercept $(0, 3.5)$ $y = -7x + \frac{7}{2}$

34. Graph the lines $3x - y = 6$ and $6x - 2y = 5$ on the same grid. Describe the two graphs.

36. $y = 0.12x + 30$

Write a linear equation to model each situation.

35. The cost, c, to rent a car is \$24 plus 28 cents per mile, m. $c = 0.28m + 24$

36. The cost, y, to paint a house is 12 cents per square foot, x, plus \$30 for supplies.

37. Cashews sell for \$4.25 per pound, and pecans sell for \$3.50 per pound. A combination of cashews, c, and pecans, p, has a value of \$25. $4.25c + 3.5p = 25$

38. A collection of dimes, d, and quarters, q, has a value of \$3.50. $0.10d + 0.25q = 3.50$

39. A solution containing 2.4% acid, x, is mixed with a solution containing 4% acid, y. The mixed solution contains 4.5 milliliters of acid. $0.024x + 0.04y = 4.5$

40. Fertilizer containing 18% phosphorus, x, is mixed with fertilizer containing 10% phosphorus, y. The mixed solution contains 12 pounds of phosphorus. $0.18x + 0.1y = 12$

Fund-raising The drama department has decided to charge \$4 for each student ticket, s, and \$8 for each adult ticket, a, to the spring production. They hope to make \$800 from the sale of tickets.

41. Write an equation to model this situation. $4s + 8a = 800$

42. Graph the equation.

43. Make a table of values. List at least three possible combinations of adult and student tickets that could be sold to make \$800.

The average high and low temperatures for January in 16 selected cities are shown below.

Rio de Janeiro
average high temp. 84°F
average low temp. 73°F

St. Moritz
average high temp. 33°F
average low temp. 29°F

44. Technology Plot the low temperatures on the x-axis and the corresponding high temperatures on the y-axis. Find a line of best fit for all 16 of the selected cities, including Rio de Janeiro and St. Moritz.

45. Suppose the average high temperature in January for Amsterdam is 40°. Use the line of best fit to predict the low temperature in January for Amsterdam.

44. $y \approx 1.05x + 9.84$ **45.** About 29°

CITY	Avg low	Avg high
Vancouver	32°F	41°F
Acapulco	70°F	85°F
Mexico City	42°F	66°F
Sydney	65°F	78°F
Vienna	26°F	34°F
Naussa	65°F	77°F
Hamilton	58°F	68°F
Copenhagen	29°F	36°F
Cairo	47°F	65°F
Paris	32°F	42°F
Athens	42°F	54°F
Hong Kong	56°F	64°F
Tokyo	29°F	47°F
Madrid	33°F	47°F

Look Back

Solve each equation. [Lesson 1.4]

46. $1.3x - 5.6 = 8.4 - 1.2x$ $x = 5.6$

47. $\frac{3x}{5} - \frac{1}{2} = \frac{x}{4} + 2$ $x = \frac{50}{7}$, or $7\frac{1}{7}$

48. What is the slope of all horizontal lines? **[Lesson 2.1]** 0

49. What is the slope of all vertical lines? **[Lesson 2.1]**

50. Determine the slope of the line passing through the points $(-2, 7)$ and $(-2, 12)$. **[Lesson 2.1]**

49. Undefined **50.** Undefined

Look Beyond

51. Is the point $(-3, 5)$ a solution of the equations $2x + 3y = 9$ and $3x - 4y = 29$? No

52. Graph the equations $y = 2x - 4$ and $2x - 4y = 10$ on the same coordinate plane. Find the point of intersection.

LESSON 2.3 Graphing Systems of Equations

why *One of the best ways to present mathematical data is to use a graph. For business decisions and customer choices, graphs can represent costs and show comparisons. When you graph two linear equations on the same coordinate plane, you can see where the lines intersect.*

If the Smiths decide to buy the VCR, they plan to spend $20 per month to rent tapes. How can you determine if it would be less expensive to buy the VCR or to install cable TV?

VCRs on sale NOW
Starting at
$185.00
12:00
12:00

Compare the cost of each purchase over several months. Write an equation to model each purchase.

First, define two variables.

Let c represent the total cost.
Let m represent the number of months.

Next, write an equation to represent each total cost.

Cable TV costs = installation + payments for m months

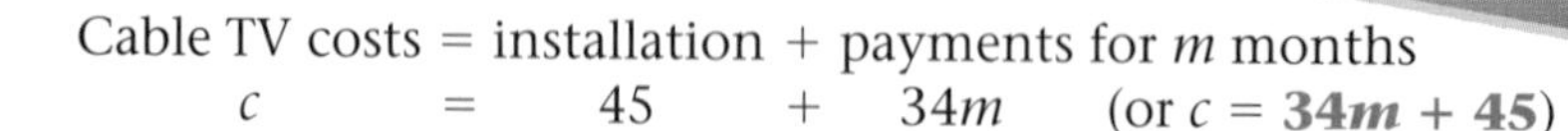

$c = 45 + 34m$ (or $c = \mathbf{34m + 45}$)

VCR and tapes costs = VCR + tape rental for m months

$c = 185 + 20m$ (or $c = \mathbf{20m + 185}$)

Create a table of values from the equations to compare the data. Then graph each equation on the same coordinate plane.

Table of Costs		
m	Cable $34m + 45$	VCR $20m + 185$
1	79	205
2	113	225
3	147	245
4	181	265
5	215	285
6	249	305
7	283	325
8	317	345
9	351	365
10	385	385

Cable TV

$c = 34m + 45$

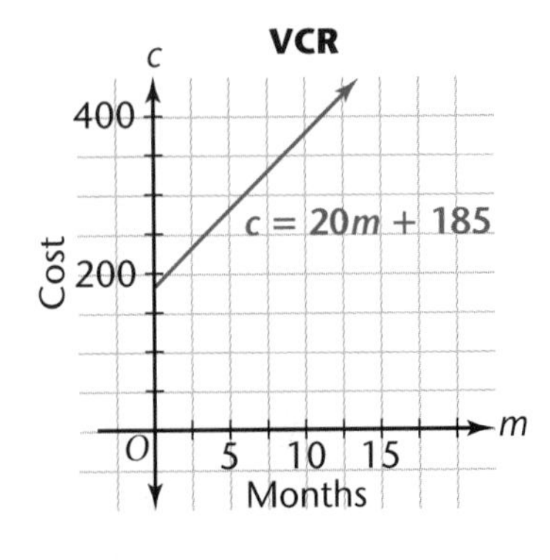

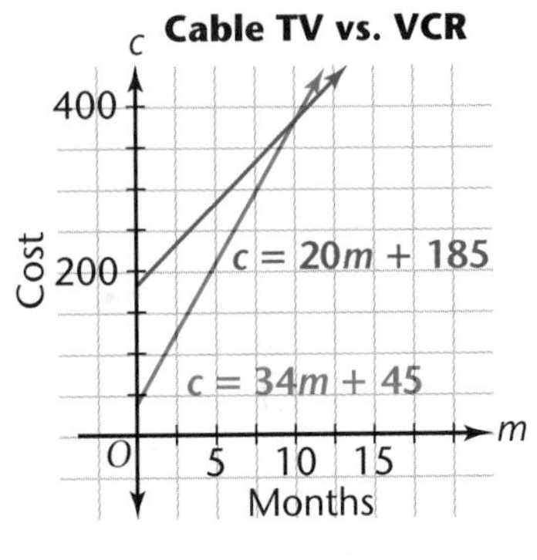

Each cost continues to increase, but one increases at a faster rate than the other. The graphs intersect at (10, 385). This means that at 10 months the cost for both options is \$385. When the x- and y-values are the same for two or more equations, the equations are said to have a **common solution**.

Cable
D: $m \geq 0$; R: $c \geq 45$

VCR
D: $m \geq 0$; R: $c \geq 185$

The equations are graphed only in the first quadrant because cost is a positive value.

What is the reasonable domain and range for these cost equations? Explain why the equations are graphed only in the first quadrant.

Explain which decision is best for the Smith family. Over time it would be better to buy the VCR even though it is more costly in the beginning; after 10 months the cost would be less for the VCR.

A System of Equations

Two or more equations in two or more variables is called a **system of equations**. When a system of two equations in two variables has a single solution, the solution is the ordered pair, (x, y), that satisfies both equations. The solution to the system used in the problem about the Smith family is (10, 385).

EXAMPLE 1

Graphics Calculator

Use a graphics calculator or graph paper to graph the equations $3x + y = 11$ and $x - 2y = 6$. Find the common solution by examining the graph.

Solution ➤

Change each equation to slope-intercept form, $y = mx + b$. Graph the two equations to find a common solution.

$$\begin{aligned} 3x + y &= 11 \\ y &= -3x + 11 \end{aligned} \qquad \begin{aligned} x - 2y &= 6 \\ -2y &= -x + 6 \\ y &= \frac{1}{2}x - 3 \end{aligned}$$

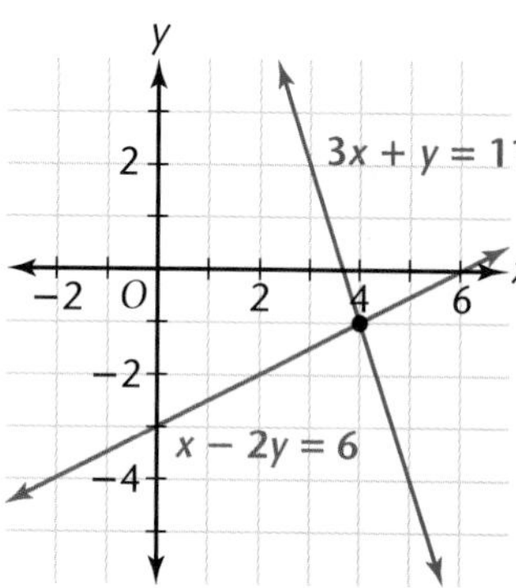

The graphs show that the lines intersect at exactly one point. To find the common solution, find the coordinates of the point of intersection. It appears from the graph that the point of intersection is (4, − 1).

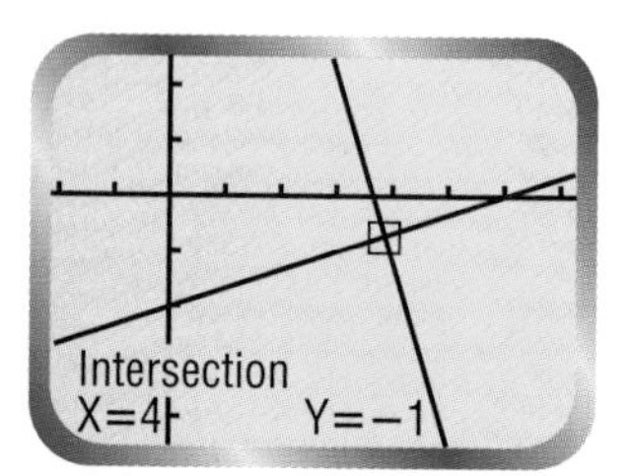

To check the solution, substitute 4 for x and -1 for y in each equation.

$$\begin{aligned} 3x + y &= 11 \\ 3(4) + (-1) &\stackrel{?}{=} 11 \\ 11 &= 11 \quad \text{True} \end{aligned} \qquad \begin{aligned} x - 2y &= 6 \\ 4 - 2(-1) &\stackrel{?}{=} 6 \\ 6 &= 6 \quad \text{True} \end{aligned}$$

Since the substitution is true for each equation, (4, − 1) is the common solution. ❖

Try This Graph the equations $y = 2x + 2$ and $y = -x - 1$. Find a common solution by examining the graph. Check your answer.

EXAMPLE 2

Ms. Alyward gives a science midterm with 200 possible points. There are a total of 38 questions on the test. How many of each type of question are on the test?

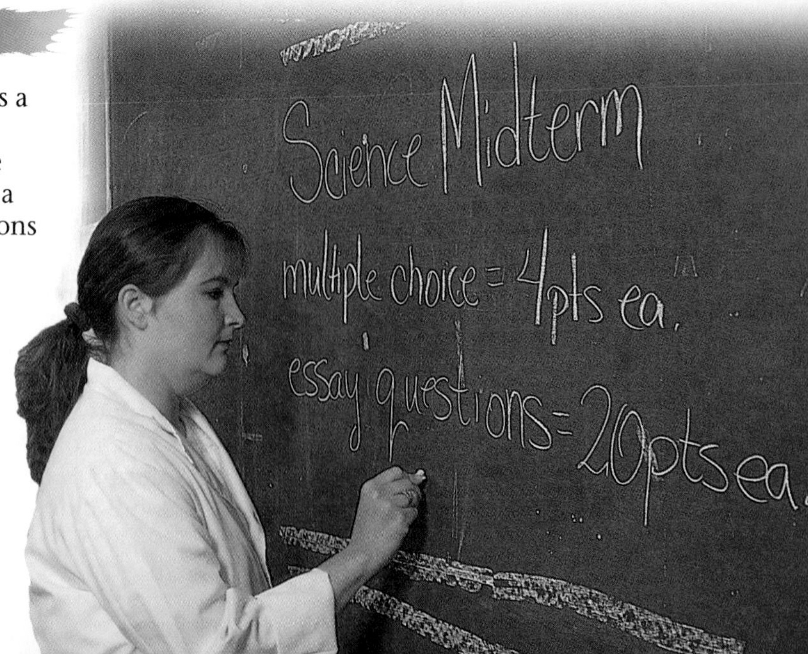

Solution ➤

First define two variables. Then write equations to model the problem.

Organize.

Let x represent the number of 4-point multiple-choice questions.
Let y represent the number of 20-point essay questions.

Write.

number of 4-point items + number of 20-point items = 38 items

$$x + y = 38$$

points for multiple choice + points for essay = 200 points

$$4x + 20y = 200$$

Solve.

To find how many of each type question appear on the test, solve the system of equations for (x, y). Make a table, or graph the two equations to find the point of intersection for a common solution. Change the equations to slope-intercept form to find values for the variables.

$$x + y = 38$$
$$y = -x + 38$$

$$4x + 20y = 200$$
$$20y = -4x + 200$$
$$y = -\frac{1}{5}x + 10$$

Use the table or the graph to find the point of intersection.

Table of Values

Multiple choice x	Essay $-x + 38$	Essay $-\frac{1}{5}x + 10$
10	28	8
15	23	7
20	18	6
25	13	5
30	8	4
✔35	3	3

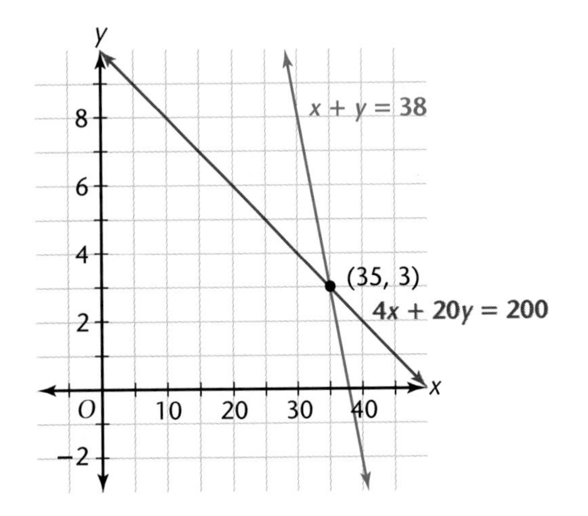

The solution is (35, 3).

There will be 35 multiple-choice questions and 3 essay questions on Ms. Alyward's science midterm.

Check.

$$x + y = 38$$
$$35 + 3 \stackrel{?}{=} 38$$
$$38 = 38 \quad \text{True}$$

$$4x + 20y = 200$$
$$4(35) + 20(3) \stackrel{?}{=} 200$$
$$140 + 60 \stackrel{?}{=} 200$$
$$200 = 200 \quad \text{True}$$ ❖

The slope-intercept form allows you to substitute values for x and find values for y.

Why were the equations written in slope-intercept form before a table of values was made?

Approximate Solution

It is sometimes difficult to find an exact solution from a graph. A reasonable estimate for the coordinates of a point of intersection is an **approximate solution** for a system of equations.

EXAMPLE 3

Solve by graphing. $\begin{cases} 4x + 3y = 6 \\ 2y = x + 2 \end{cases}$

The brace { is used to indicate that the equations form a system.

Solution ➤

Write each equation in slope-intercept form.

$$4x + 3y = 6 \qquad 2y = x + 2$$

$$3y = -4x + 6 \qquad \frac{2y}{2} = \frac{x}{2} + \frac{2}{2}$$

$$y = -\frac{4}{3}x + 2 \qquad y = \frac{1}{2}x + 1$$

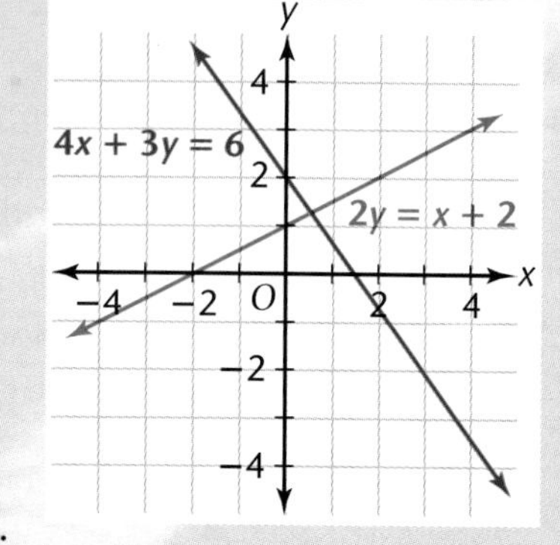

From the graph, locate the point of intersection. Make a reasonable estimate from the graph in order to approximate a common solution. An approximate solution is (0.5, 1.5).

Check your estimated solution by substituting 0.5 for x and 1.5 for y in the original equations.

$$4x + 3y = 6 \qquad 2y = x + 2$$

$$4(0.5) + 3(1.5) \stackrel{?}{=} 6 \qquad 2(1.5) \stackrel{?}{=} (0.5) + 2$$

$$2 + 4.5 \stackrel{?}{=} 6 \qquad 3 \approx 2.5$$

$$6.5 \approx 6$$

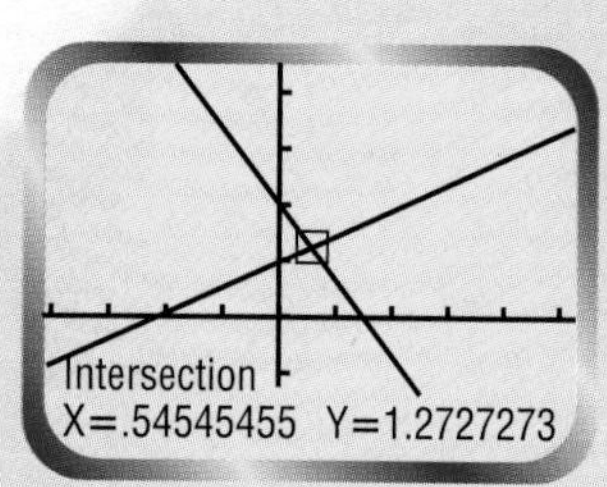

A check shows that the approximate solution (0.5, 1.5) is reasonable. You can also check your solution by graphing on a graphics calculator. ❖

EXERCISES & PROBLEMS

Communicate

For Exercises 1–3, refer to the following system. $\begin{cases} 2x - 3y = 4 \\ x + 4y = -9 \end{cases}$

1. How do you write each equation in slope-intercept form? Explain why this form is used to find values for the variables.

2. Explain how to create a table of values for the system of equations. Discuss how to find the point of intersection from a table.

1. Isolate y on one side of the equal sign. This form of the equation, $y = mx + b$, allows you to substitute values for x and find values for y.
2. Substitute different values for x and then determine the corresponding y-values for each equation. The point of intersection occurs when an x-value generates the same y-values in both equations.

3. Discuss how to graph the system of equations from the table in Exercise 2. Tell how to find a common solution from the graph. Plot the ordered pairs from the table, and connect the points to form two lines. The common solution is the point where the lines intersect, $(-1, -2)$.

4. Discuss how to graph the system at the right. Tell how to find an approximate solution by inspecting the graph. $\begin{cases} x + y = 3 \\ x - y = 4 \end{cases}$

5. Why is it important to check your approximate solution?
A check allows you to see if your estimate is reasonable.

Practice & Apply

6. $(0.8, -0.7)$
7. $(34.3, -12.9)$
8. $(0.5, 0.3)$
9. No solution

Solve by graphing. Round approximate solutions to the nearest tenth. Check algebraically. Approximations may vary.

6. $\begin{cases} 2x + 10y = -5 \\ 6x + 4y = 2 \end{cases}$ 7. $\begin{cases} x - 2y = 60 \\ 3y = 30 - 2x \end{cases}$ 8. $\begin{cases} 3y = 4x - 1 \\ 2x + 3y = 2 \end{cases}$ 9. $\begin{cases} x + 3y = 6 \\ -6y = 2x + 6 \end{cases}$

10. $\begin{cases} 5x + 6y = 14 \\ 3x + 5y = 7 \end{cases}$ 11. $\begin{cases} x = 400 - 2y \\ x - 100 = y \end{cases}$ 12. $\begin{cases} 3x + 5y = 12 \\ 7x - 5y = 18 \end{cases}$ 13. $\begin{cases} x = 2 \\ 2y = 4x + 2 \end{cases}$

10. $(4, -1)$ 11. $(200, 100)$ 12. $(3, 0.6)$ 13. $(2, 5)$

Determine whether the point (2, 10) is an exact solution for each system of equations.

14. $\begin{cases} y = 2x - 4 \\ y = x + 8 \end{cases}$ No 15. $\begin{cases} y = -x + 12 \\ y = -3x + 16 \end{cases}$ Yes 16. $\begin{cases} y = x + 8 \\ y = -3x + 16 \end{cases}$ Yes 17. $\begin{cases} x + 3y = 6 \\ -6y = 2x + 12 \end{cases}$ No

18–21. Graph each system of equations from Exercises 14–17 to check the results.

22. Find the point of intersection of $y = -x + 8$ and $y = -3x + 16$. $(4, 4)$

23. Use a graph to find the approximate point of intersection of $y = -x + 12$ and $y = 2x - 4$. Round your approximation to the nearest tenth. Check your approximation. $(5.3, 6.7)$

24. **Aviation** A plane at 3,800 feet is descending at a rate of 120 feet per minute, and a plane at 520 feet is climbing at a rate of 40 feet per minute. In how many minutes will they be at the same altitude? Write a system of equations. Then graph the system to determine the solution.
$\begin{cases} y = -120x + 3800 \\ y = 40x + 520 \end{cases}$; 20.5 minutes

Refer to points $A(3, 5)$, $B(4, -1)$, and $C(9, 3)$. Lines AB and AC meet at point A.

25. Write an equation for line AB.

26. Write an equation for line AC.

27. Graph your equations from Exercises 25 and 26.

28. Use the coordinates of the point of intersection to check your equations.

Statistics **Use the degree data at the right for Exercises 29–32.**

Degrees Conferred for Five Consecutive Years

Year	1	2	3	4	5
Education degrees	7473	7151	7110	6909	6544
Engineering degrees	2981	3230	3410	3820	4191

29. **Technology** Use a graphics calculator to plot the data as two sets of ordered pairs: (year, education degrees) and (year, engineering degrees). Use your calculator to graph a scatter plot of the data. Sketch the graph.

30. Graph the lines of best fit for both data sets. Sketch these on your graph.

31. Find the point of intersection of the two models. About how many years will it be beyond year 5 before the number of education degrees and the engineering degrees awarded are the same? About 5 years beyond

32. If you had to choose between a career in education or a career in engineering based on this data, which career would you choose and why?

33. Estimate the point of intersection for $y = -\frac{1}{2}x + 5$ and $y = \frac{1}{4}x - 5$.
Answers may vary. Approx $\left(13\frac{1}{3}, -1\frac{2}{3}\right)$

Sports From 1980 to 1990, there was a shift in salaries for basketball players and baseball players. The graph reflects the average salaries for the players.

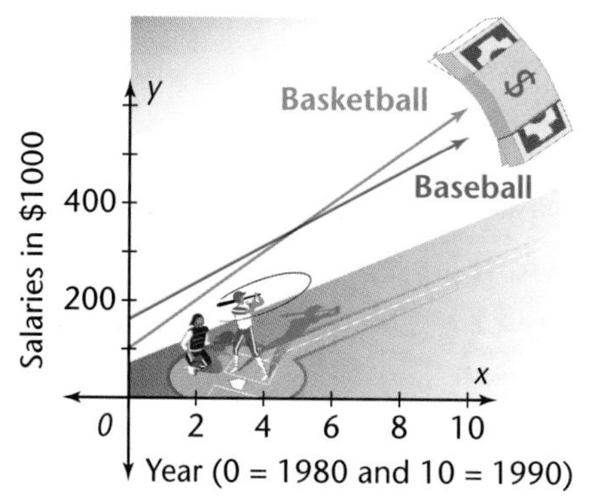

Answers may vary.

34. Estimate the year in which their salaries were equal.

35. Estimate the salary. **34.** About 1985
35. About $360,000

36. **Portfolio Activity** Use graphs to complete the problem in the portfolio activity on page 55.
Floor section: $30 each; Balcony: $20 each

Look Back

37. Place parentheses and brackets to make $[(21 \div 3) - (4 \cdot 0) + 6](-3) = -39$ true. **[Course 1]**

38. If data points on a scatter plot are scattered randomly, what type of correlation exists? **[Course 1]** zero, or none

39. Graph $-3 < x$ *and* $x \leq 5$ on the number line. State the solution. **[Course 1]**

40. Solve $3(5 - 2x) - (8 - 6x) = -9 + 2(3x + 4) - 10$. **[Lesson 1.4]** 3

Find the slope of each line. [Lesson 2.1]

41. $3x + 2y = 12$ $-\frac{3}{2}$

42. $8y = -6x + 12$ $-\frac{3}{4}$

43. $2y = \frac{1}{2}x + 6$ $\frac{1}{4}$

44. $y = -3x - 18$ -3

Look Beyond

Substitute each value of x into the equation $-2x + 4y = 12$, and solve for y. EXAMPLE: Let $x = 2$. $-2(2) + 4y = 12$ $y = 4$

45. Let $x = -6$. 0

46. Let $x = 8y$. -1

47. Let $x = y - 3$. 3

48. Let $x = 3y + 1$. -7

LESSON 2.4 Exploring Substitution Methods

Why *Sometimes graphing can be used to find the solution to a system of equations. However, when an exact solution cannot be determined from a graph, algebraic methods can be used to find the exact solution.*

Sue paid a total of $27 for 3 shares of Allied Sports stock and 4 shares of Best Design stock. What is the price per share of each stock?

Exploration 1 *What Is the Cost?*

You can find an approximate solution by graphing.

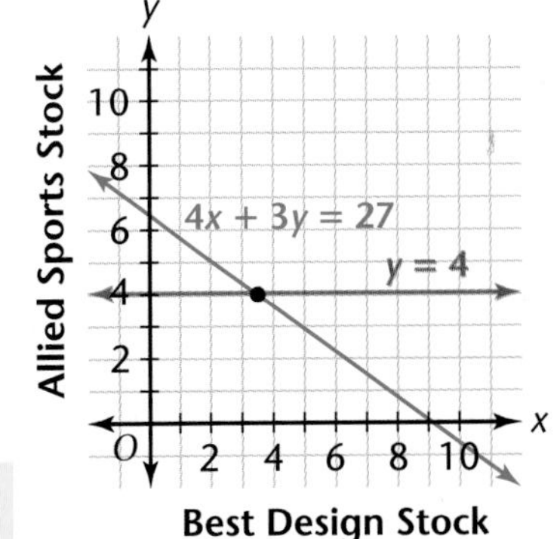

Let x = price of one share of Best Design.
Let y = price of one share of Allied Sports.

Notice at the right that Allied Sports stock costs $4 per share. So, $\begin{cases} 4x + 3y = 27 \\ y = 4 \end{cases}$ is the system that models the problem.

4. Substitute the known value into one of the equations, and solve for the unknown value of the other variable.
5. Substitute 3 for x in the equation $8x + 2y = 19$ to solve for y.

1 Examine the graph. Guess the coordinates of the point of intersection. About (3.5, 4)

2 Explain why your guess is an approximation. The x-coordinate is not a whole number.

3 Substitute 4 for y in $4x + 3y = 27$, and solve the resulting equation for x. $x = 3.75$

4 Use your results from Step 3 to explain how to find the solution to a system if you know the value of one variable.

Allied Sports Stock (y-axis), Best Design Stock (x-axis); $4x + 3y = 27$; $y = 4$

5 How can you use substitution to solve this system? $\begin{cases} 8x + 2y = 19 \\ x = 3 \end{cases}$

6 Does this method result in an exact solution? Solve the system given in Step 5 using substitution. ❖ Yes; (3, −2.5)

Exploration 2 Substituting Expressions

Examine this system. $\begin{cases} 3x + 2y = 4 \\ x + y = 1 \end{cases}$

1. Solve $x + y = 1$ for x. $x = -y + 1$
   a. Since $x = -y + 1$, substitute the expression $(-y + 1)$ for x in $3x + 2y = 4$. $3(-y + 1) + 2y = 4$
   b. Solve the system. Write the solution as an ordered pair. $(2, -1)$

2. Solve $x + y = 1$ for y. $y = -x + 1$
   a. Since $y = -x + 1$, substitute the expression $(-x + 1)$ for y in $3x + 2y = 4$. $3x + 2(-x + 1) = 4$
   b. Solve the system. Write the solution as an ordered pair. $(2, -1)$

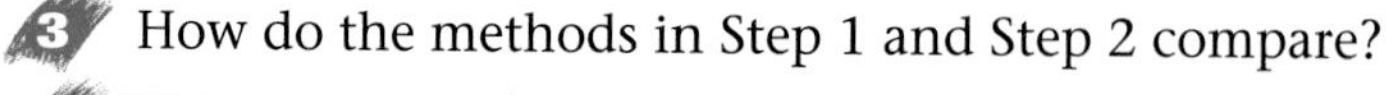

3. How do the methods in Step 1 and Step 2 compare?

4. Describe the substitutions that can be made to solve a system if each coefficient in one of the equations is equal to 1. ❖

Exploration 3 Finding a Value

Examine this system. $\begin{cases} 15x - 5y = 30 \\ -2x + y = -6 \end{cases}$

1. Solve $-2x + y = -6$ for y. $y = 2x - 6$

2. Substitute the expression you found in Step 1 for y in the first equation. $15x - 5(2x - 6) = 30$

3. Solve the system for x. Write the solution as an ordered pair. **3.** $x = 0$; $(0, -6)$

4. Explain how to use substitution to solve a system if at least one variable has a coefficient of 1. ❖

APPLICATION

Investments New Sport Times stock costs $2 per share more than Modern Design stock. Find the price per share of each stock.

In a mock stock market a student purchased 2 shares of Modern Design stock and 3 shares of New Sport Times stock for $96.

Define the variables.

Let m represent the cost of a share of Modern Design stock.

Let n represent the cost of a share of New Sport Times Stock.

Write the equations as a system. $\begin{cases} 2m + 3n = 96 \\ n = m + 2 \end{cases}$

Use the **substitution method** to solve the system of linear equations. Substitute $\boldsymbol{m + 2}$ for $\boldsymbol{n}$ in the equation $2m + 3n = 96$.

$$\begin{aligned} 2m + 3\boldsymbol{n} &= 96 \\ 2m + 3(\boldsymbol{m + 2}) &= 96 \\ 2m + 3m + 6 &= 96 \\ 5m + 6 &= 96 \\ 5m &= 90 \\ m &= 18 \end{aligned}$$

Now substitute 18 for m in the equation $n = m + 2$.

$$\begin{aligned} n &= m + 2 \\ n &= 18 + 2 \\ n &= 20 \end{aligned}$$

The solution is (18, 20).

Check your answer to see that the values satisfy the original equations. The price per share of Modern Design is \$18, and the price per share of New Sport Times is \$20. ❖

APPLICATION

In a mock stock market, Ann purchased 1 share of Disposable Inc. stock and 2 shares of Software Today for a total of \$65. What is the price per share of each stock?

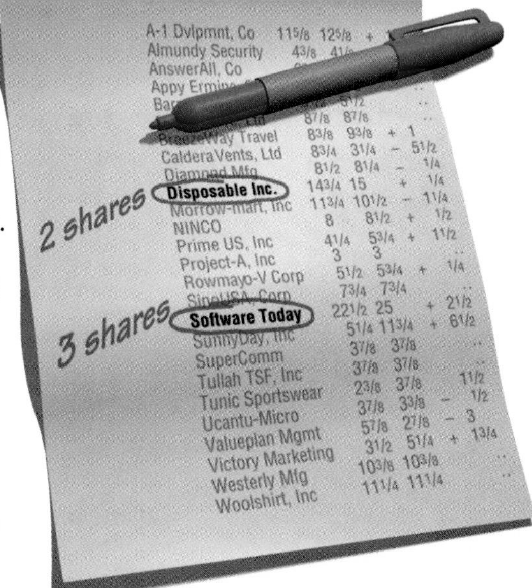

Carl purchased 2 shares of Disposable Inc. and 3 shares of Software Today for a total of \$105.

Define the variables.

Let d represent the cost of each share of Disposable Inc. stock

Let s represent the cost of each share of Software Today stock.

Write the system. $\begin{cases} 2d + 3s = 105 \\ d + 2s = 65 \end{cases}$

To use the substitution method, first solve $d + 2s = 65$ for d.

$$d = -2s + 65$$

Now substitute $\boldsymbol{-2s + 65}$ for $\boldsymbol{d}$ in the equation $2\boldsymbol{d} + 3s = 105$.

1. Solve the equation $2(\boldsymbol{-2s + 65}) + 3s = 105$ for s.
2. Use your value of s to find the value of d in the equation $d = -2s + 65$.
3. Is (15, 25) your solution?

The price per share of Disposable Inc. is \$15. The price per share of Software Today is \$25. ❖

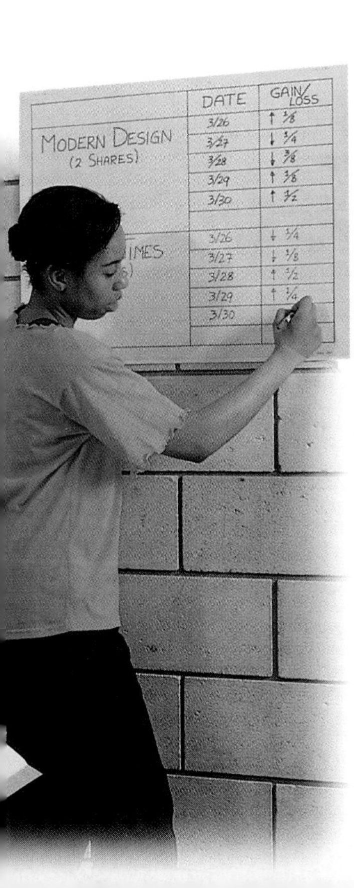

EXERCISES & PROBLEMS

Communicate

1. If you know $y = 42$, explain how to use substitution to solve $y = 2x + 8$.
2. Given the equations $-4x + y = 2$ and $2x + 3y = 34$, explain how to find the expression to substitute and how to find the common solution.
3. Explain how to use substitution to solve the system.

$$\begin{cases} x - 2y = 8 \\ 2x + 3y = 23 \end{cases}$$

Practice & Apply

Solve and check by the substitution method.

7. $(-14, 19)$

4. $\begin{cases} 2x + 8y = 1 \\ x = 2y \end{cases}$ $\left(\frac{1}{6}, \frac{1}{12}\right)$

5. $\begin{cases} x + y = 7 \\ 2x + y = 5 \end{cases}$ $(-2, 9)$

6. $\begin{cases} 3x + y = 5 \\ 2x - y = 10 \end{cases}$ $(3, -4)$

7. $\begin{cases} y = 5 - x \\ 1 = 4x + 3y \end{cases}$

8. $\begin{cases} 2x + y = -92 \\ 2x + 2y = -98 \end{cases}$ $(-43, -6)$

9. $\begin{cases} 4x + 3y = 13 \\ x + y = 4 \end{cases}$ $(1, 3)$

10. $\begin{cases} 6y = x + 18 \\ 2y - x = 6 \end{cases}$ $(0, 3)$

11. $\begin{cases} 2x + y = 1 \\ 10x - 4y = 2 \end{cases}$ $\left(\frac{1}{3}, \frac{1}{3}\right)$

Graph each system of equations and estimate the solution.

12. $\begin{cases} 5x - y = 1 \\ 3x + y = 1 \end{cases}$

13. $\begin{cases} 2x + y = 1 \\ 10x = 4y + 2 \end{cases}$

14. $\begin{cases} 5x = 3y + 12 \\ x = y \end{cases}$

15. $\begin{cases} 3x - 2y = 2 \\ y = 2x + 8 \end{cases}$

16. $\begin{cases} 2x + 3y = 7 \\ x + 4y = 9 \end{cases}$

17. $\begin{cases} 4x - y = 15 \\ -2x + 3y = 12 \end{cases}$

18. $\begin{cases} 2y + x = 4 \\ y - x = -7 \end{cases}$

19. $\begin{cases} 4y - x = 4 \\ y + x = 6 \end{cases}$

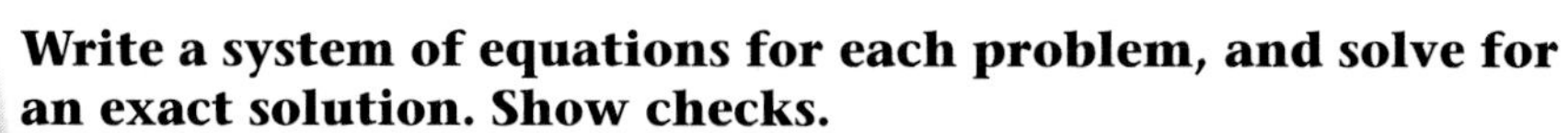

Write a system of equations for each problem, and solve for an exact solution. Show checks.

20. **Aviation** Suppose the hot-air balloon shown at the left is rising at the rate of 4 feet per second. A small aircraft at an altitude of 7452 feet is losing altitude at a rate of 30 feet per second. In how many seconds will both be at the same altitude?

21. The sum of two numbers is 27. The larger number is 3 more than the smaller number. Find the two numbers.

22. One number is 4 less than 3 times a second number. If 3 more than twice the first number is decreased by twice the second, the result is 11. Find both numbers.

23. **Geometry** Find the dimensions of a rectangle with a 210-meter perimeter if its length is twice its width.

Fund-raising At the Boy Scout "all you can eat" spaghetti dinner, the troop served 210 people and raised $935. Complete Exercises 24–27.

Scout Troop 154 Invites you to
SpaghettiFest!
"all you can eat" spaghetti dinner
Adults $6.00
Children $3.50
Westside Community Center
Thursday - 7:00pm

24. Write an equation for the *total amount raised* from adult and child dinners. $6.00a + 3.50c = 935$

25. Write an equation for the *total number* of adult and child dinners served. $a + c = 210$

26. Solve the system of equations from Exercises 24 and 25. How many adult and child dinners were served? 80; 130

27. Name two methods for solving systems of equations that can be used to solve Exercise 26. Why would you choose one method over another?
Graphing; substitution; answers may vary.

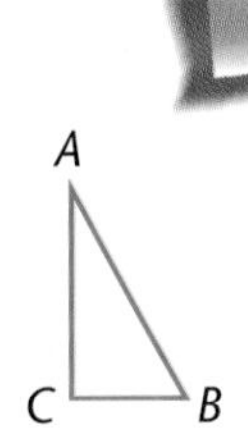

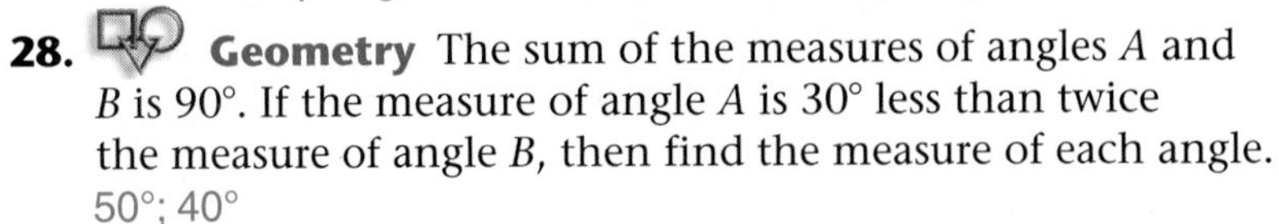

28. **Geometry** The sum of the measures of angles A and B is 90°. If the measure of angle A is 30° less than twice the measure of angle B, then find the measure of each angle.
50°; 40°

29. **Cultural Connection: Asia** A Chinese problem states that several people combined their money to buy a farm tool to share. If each person paid 8 coins, there were 3 coins too many. If each paid 7 coins, there were 4 coins too short. How many people were there, and what was the price of the farm tool? 7 people; 53 coins

30. **Portfolio Activity** Use substitution to complete the problem in the portfolio activity on page 55.
Center: $30 each; Balcony: $20 each

Look Back

31. **Recreation** In a foot race Sam finished 20 feet in front of Joe. Joe was 5 feet behind Mark, and Mark was 10 feet behind Rob. Tom was 15 feet ahead of Rob. In what order did they finish? **[Lesson 1.1]**
Tom; Sam; Rob; Mark; Joe

Solve each equation for x. [Lesson 1.2]

32. $\frac{x}{15} = 3$ 45

33. $\frac{3}{x} = 15$ $\frac{1}{5}$

34. $\frac{15}{x} = 3$ 5

35. $\frac{x}{3} = 15$ 45

36. The number 12.6 is 42% of what number? **[Lesson 1.2]**
30

Determine whether each pair of equations represents lines that are parallel, perpendicular, or neither. [Lesson 2.1]

37. $3y = 2x - 15$
$3x + 2y = 24$
Perpendicular

38. $2y = x - 12$
$2y - x = 12$
Parallel

39. $y = x - 1$
$y = -x + 3$
Perpendicular

Look Beyond

Examine each system of equations. Use substitution to solve for x and y.

40. $\begin{cases} x + 2y + 3z = 8 \\ y + 2z = 3 \\ z = 2 \end{cases}$ $(4, -1)$

41. $\begin{cases} 2x + 3y + 5z = 44 \\ 2y - 6z = 4 \\ z = 4 \end{cases}$ $(-9, 14)$

42. $\begin{cases} x + y = a \\ 2y = 6b \end{cases}$ $(-3b + a, 3b)$

43. $\begin{cases} x - y - d = 0 \\ 6x - 4y = 2d \end{cases}$ $(-d, -2d)$

LESSON 2.5 The Elimination Method

Both the graphing method and the substitution method provide ways to solve systems of equations. Another method used to solve systems is the elimination method.

To find the daily fee and the per-mile cost at Airport Rent-A-Car, you can set up and solve a system of equations by a method called *elimination*. The **elimination method** for solving systems of equations eliminates one variable by adding opposites.

Jason drove 125 miles on the 2-day trip. He drove 350 miles on the 4-day trip.

Exploration Using Opposites

3a. $8a + 4b = 36$

4. Addition of opposites causes one of the variables to be eliminated when using the Addition Property of Equality. This allows the value of the other variable to be determined, which then leads to the determination of the first variable.

1. What is true about the sum of any expression and its opposite? The sum is zero.

2. 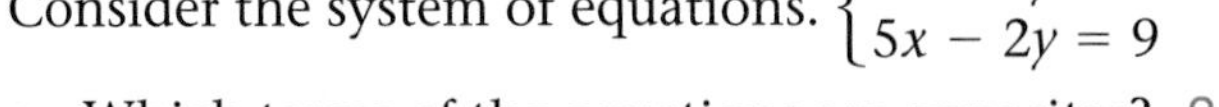
Consider the system of equations. $\begin{cases} 3x + 2y = 7 \\ 5x - 2y = 9 \end{cases}$

   a. Which terms of the equations are opposites? $2y$ and $-2y$

   b. Use the Addition Property of Equality to combine like terms on corresponding sides of the equal signs. $8x = 16$

   c. Solve the resulting equation for x, and then solve for y. $\left(2, \frac{1}{2}\right)$

   d. Check the solution in each equation of the system. $7 = 7$ True, $9 = 9$ True

3. Consider the system of equations. $\begin{cases} 2a + b = 9 & \text{Equation 1} \\ 3a - 4b = 8 & \text{Equation 2} \end{cases}$

   a. Multiply the expressions on each side of Equation 1 by 4.

   b. Use the Addition Property of Equality. Solve for a, and then solve for b. (4, 1)

   c. Check the solution in each equation of the system. $9 = 9$ True, $8 = 8$ True

4. How are opposites used to solve a system of equations? ❖

The method of elimination uses opposites to eliminate one of the variables. This method can be used to solve the rental-car problem on page 84.

Define the variables.

Let x represent the cost per day. Let y represent the cost per mile.

Write the equations in standard form.

$$\begin{cases} 2x + 125y = 95.75 & \leftarrow \text{Bill No. 001} \\ 4x + 350y = 226.50 & \leftarrow \text{Bill No. 002} \end{cases}$$

Multiply each term of the first equation by -2.

$$\begin{aligned} -2(2x) + (-2)(125)y &= -2(95.75) \\ 4x + 350y &= 226.50 \end{aligned}$$

The x-variable can be eliminated with fewer steps than the y-variable; by examining the coefficients, 4 is a multiple of 2, but 350 is not a multiple of 125; choose to eliminate the x.

Explain why x rather than y was chosen to be eliminated.

First simplify by using the Addition Property of Equality.

$$\begin{aligned} -4x + (-250)y &= -191.50 \\ 4x + 350y &= 226.50 \\ \hline 100y &= 35 \\ y &= 0.35 \end{aligned}$$

Then substitute 0.35 in the first equation to find x.

$$\begin{aligned} 2x + 125(0.35) &= 95.75 \\ 2x + 43.75 &= 95.75 \\ 2x &= 52 \\ x &= 26 \end{aligned}$$

The solution is (26, 0.35). A check proves the solution that is correct. Jason rented the car for $26 a day, and the cost per mile was $0.35.

EXAMPLE 1

Solve by elimination. $\begin{cases} \mathbf{2x + 3y = 1} & \textbf{Equation 1} \\ \mathbf{5x + 7y = 3} & \textbf{Equation 2} \end{cases}$

Solution ➤

Multiply **Equation 1** by 5 and **Equation 2** by -2.

$$\begin{cases} (5)\mathbf{2x} + (5)\mathbf{3y} = (5)\mathbf{1} \\ (-2)\mathbf{5x} + (-2)\mathbf{7y} = (-2)\mathbf{3} \end{cases} \longrightarrow \begin{cases} 10x + 15y = 5 \\ -10x - 14y = -6 \end{cases}$$

Add like terms of the two equations.

$$\begin{aligned} 10x + (-10x) + 15y + (-14y) &= 5 + (-6) \\ y &= -1 \end{aligned}$$

Substitute -1 for y, and solve for x.

$$\begin{aligned} 10x + 15y &= 5 \\ 10x + 15(-1) &= 5 \\ 10x - 15 + 15 &= 5 + 15 \\ 10x &= 20 \\ x &= 2 \end{aligned}$$

The solution is $x = 2$ and $y = -1$.
Check by substituting 2 for x and -1 for y in the original equations. ❖

Try This Solve by elimination. $\begin{cases} 3a - 2b = 6 \\ 5a + 7b = 41 \end{cases}$ $a = 4, b = 3$

SUMMARY OF THE ELIMINATION METHOD

1. Write both equations in standard form.
2. Use the Multiplication Property of Equality to write two equations in which corresponding coefficients are opposites.
3. Use the Addition Property of Equality to eliminate one variable, and solve for the remaining variable.
4. Substitute the value of that variable into one of the original equations. Solve for the eliminated variable.
5. State the solution and check.

You now have several methods for solving a system of equations. Graphing is used as a visual model for a problem that involves a system of equations. You can use graphing to check the reliability of a calculated solution or to approximate a solution. The substitution method and the elimination method are used to find an exact solution. Substitution may be the more efficient method if either of the given equations has at least one coefficient of 1 or -1.

EXAMPLE 2

Choose a method for solving the following systems of equations. Explain why you chose each method.

A $\begin{cases} y = 10 - 6x \\ y = 3x - 6 \end{cases}$

B $\begin{cases} 2x - 5y = -20 \\ 4x + 5y = 14 \end{cases}$

C $\begin{cases} 9a - 2b = -11 \\ 8a - 7b = 25 \end{cases}$

D $\begin{cases} 324p + 456t = 225 \\ 178p - 245t = 150 \end{cases}$

Solution ➤

There may be more than one correct answer for each set of equations.

A Use the substitution method. Because each equation is solved for y, the equivalent expression can be substituted for y.

B Use the elimination method because $5y$ and $-5y$ are opposites.

C Use the elimination method. No term has coefficient 1 or -1.

D Use a graphics calculator. Because the coefficients are such large numbers, ordinary algebra and manual graphing techniques may be cumbersome. ❖

CT– AND; the solution must work in the first equation AND the second equation, not just in one or the other.

Try This Choose any method to solve the following system of equations. Explain why you chose that method.

$$\begin{cases} y + 2x = 12 \\ 3y = 5x - 19 \end{cases}$$

Each student's choice of method may vary; (5, 2).

Which word, **AND** or **OR**, connects the equations in a system? Which word helps to explain why a solution to a system of equations must be checked in *both* equations?

EXERCISES & PROBLEMS

Communicate

In the following systems, which terms are opposites? How would you solve each system?

1. $\begin{cases} x + y = 13 \\ x - y = 5 \end{cases}$ **2.** $\begin{cases} 2x - 3y = 8 \\ 5x + 3y = 20 \end{cases}$ **3.** $\begin{cases} 2a + b = 6 \\ -2a - 3b = 8 \end{cases}$

How would you use opposites to solve each system?

4. $\begin{cases} 4x + 3y = 8 \\ x - 2y = 13 \end{cases}$ **5.** $\begin{cases} a - 2b = 7 \\ 4b + 2a = 15 \end{cases}$ **6.** $\begin{cases} 3m - 5n = 11 \\ 2m - 3n = 1 \end{cases}$

Explain how you would use elimination to solve each system.

7. $\begin{cases} 2x + 3y = 9 \\ 3x + 6y = 7 \end{cases}$ **8.** $\begin{cases} 2x - 5y = 1 \\ 3x - 4y = -2 \end{cases}$ **9.** $\begin{cases} 9a + 2b = 2 \\ 21a + 6b = 4 \end{cases}$

10. The sum of two numbers is 156. Their difference is 6. Write a system to model the problem. Which method you would choose to solve the system of equations? Explain why you chose that method.
Answers may vary.
Solve the system $\begin{cases} x + y = 156 \\ x - y = 6 \end{cases}$ by the elimination method. (81, 75)

Practice & Apply

Solve each system of equations by elimination, and check your solution.

11. $\begin{cases} -x + 2y = 12 \\ x + 6y = 20 \end{cases}$ $(-4, 4)$ **12.** $\begin{cases} 2x + 3y = 18 \\ 5x - y = 11 \end{cases}$ $(3, 4)$ **13.** $\begin{cases} -4x + 3y = -1 \\ 8x + 6y = 10 \end{cases}$ $\left(\frac{3}{4}, \frac{2}{3}\right)$

14. $\begin{cases} 2x - 3y = 5 \\ 5x - 3y = 11 \end{cases}$ **15.** $\begin{cases} 6x - 5y = 3 \\ -12x + 8y = 5 \end{cases}$ **16.** $\begin{cases} 4x + 3y = 6 \\ -2x + 6y = 7 \end{cases}$ $\left(\frac{1}{2}, \frac{4}{3}\right)$

$\left(2, -\frac{1}{3}\right)$ **15.** $\left(-\frac{49}{12}, -\frac{11}{2}\right)$

17. Mrs. Jones is celebrating Gauss's birthday by having a pizza party for her two algebra classes. She orders 3 pizzas and 3 bottles of soda for her 2nd period class and 4 pizzas and 6 bottles of soda for her 7th period class. How much does each pizza and bottle of soda cost?

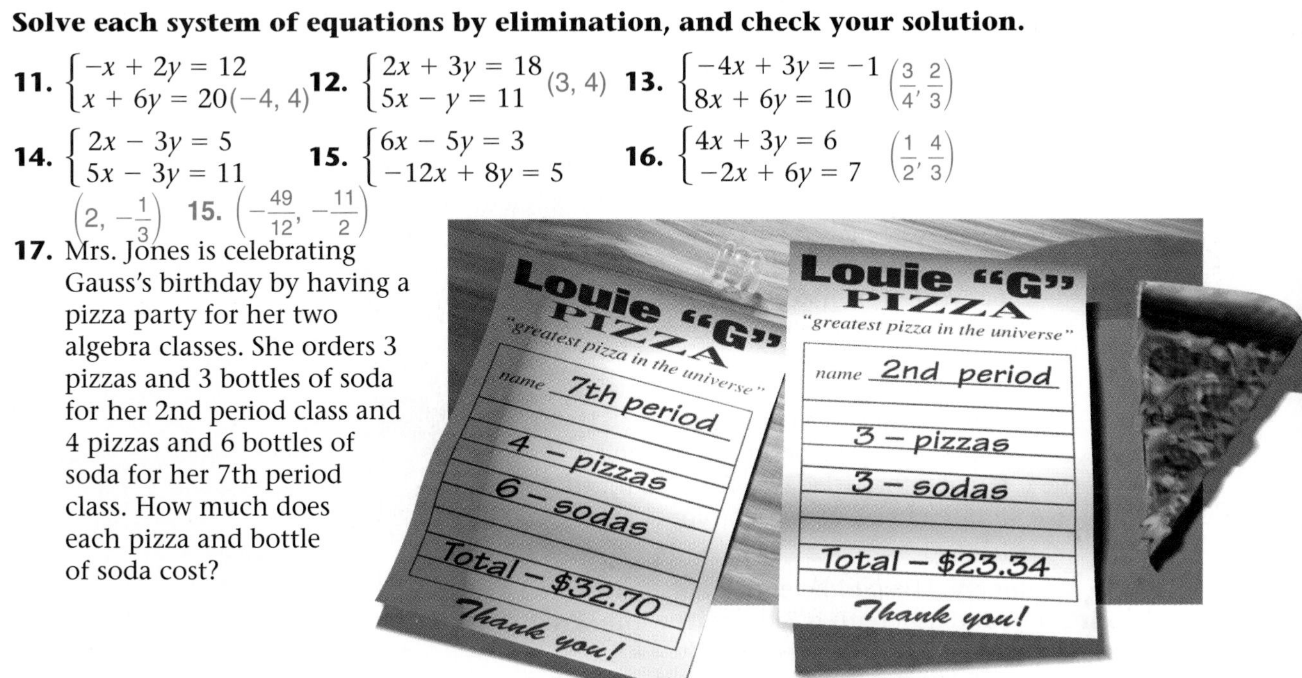

Louie "G" PIZZA
"greatest pizza in the universe"
name 7th period
4 – pizzas
6 – sodas
Total – $32.70
Thank you!
Louie "G" PIZZA
"greatest pizza in the universe"
name 2nd period
3 – pizzas
3 – sodas
Total – $23.34
Thank you!

Solve each system of equations by any method.

18. $\begin{cases} 2m = 2 - 9n \\ 21n = 4 - 6m \end{cases}$ $\left(-\frac{1}{2}, \frac{1}{3}\right)$

19. $\begin{cases} x - 7 = 3y \\ 6y = 2x - 14 \end{cases}$

20. $\begin{cases} 2x = 3 - y \\ y = 3x - 12 \end{cases}$ $(3, -3)$

21. $\begin{cases} 3b = -6a - 3 \\ b = 2a - 1 \end{cases}$ $(0, -1)$

22. $\begin{cases} y = 1.5x + 4 \\ 0.5x + y = -2 \end{cases}$

23. $\begin{cases} 2x = 3y - 12 \\ \frac{1}{3}x = 4y + 5 \end{cases}$ $(-9, -2)$

19. Infinite number of solutions 22. $\left(-3, -\frac{1}{2}\right)$

Write a system of equations for each problem, and choose the best method of solving the system. Solve and show checks. **Complete Exercises 24–30.**

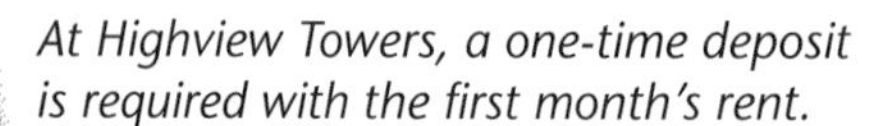

At Highview Towers, a one-time deposit is required with the first month's rent.

24. $\begin{cases} r + d = 900 \\ 12r + d = 6950 \end{cases}$
$d = \$350, r = \$550$

24. Housing At Highview Towers, Roberto paid $900 the first month and a total of $6950 during the first year. Find the monthly rent and the deposit.

25. Wages As a parking attendant at the Saratoga Raceway, Juan earns a fixed salary for the first 15 hours he works each week and then additional pay for any time over 15 hours. During the first week, Juan worked 25 hours and earned $240, and the second week he worked 22.5 hours and earned $213.75. What is Juan's weekly salary for 15 hours and his overtime pay per hour?

26. Shopping at Super Sale Days, Martha buys 3 shirts and 2 pairs of pants for $85.50. She returns during the same sale and buys 4 more shirts and 3 more pairs of pants for $123. What is the sale price of the shirts and pants?

26. $\begin{cases} 3s + 2p = 85.50 \\ 4s + 3p = 123 \end{cases}$
$s = \$10.50, p = \$27$

25. $\begin{cases} s + 10p = 240 \\ s + 7.5p = 213.75 \end{cases}$
$s = \$135, p = \$10.50$ per hour

Mr. Moore sells his tractor for $6000.

27. Investments With the $6000 from the sale of his tractor, Mr. Moore makes two investments, one at a bank paying 5% interest per year and the rest in stocks yielding 9% interest per year. If he makes a total of $380 in interest in the first year, how much is invested at each rate?

28. Sales At a local music store, single-recording tapes sell for $6.99 and concert tapes sell for $10.99. The total number of these tapes sold on Monday was 25. If the total sales for Monday is $230.75 for these two items, find the number of each type sold.

29. Geometry The perimeter of a rectangle is 24 centimeters, and its length is 3 times its width. Find the length and width of this rectangle.

30. Travel At the rates shown in the ad below, what are the daily room rate and the cost of a meal at the Shamrock Inn?

Use the equations $y = x - 2$ and $y = 2x$ for Exercises 31–33.

31. Find the common solution by graphing.

32. Solve the system algebraically. $(-2, -4)$

33. What method did you use to solve the system in Exercise 32? Why?

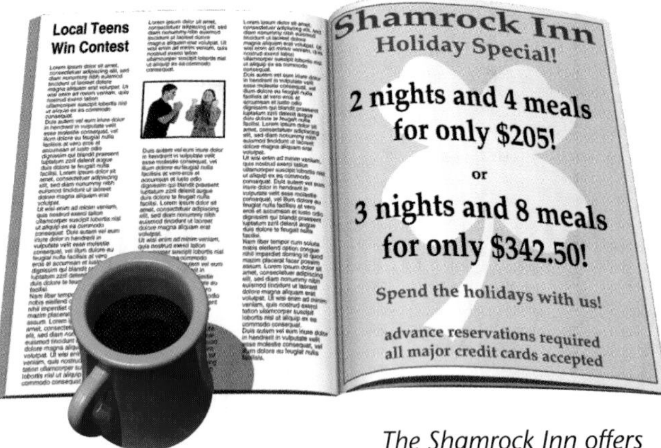

The Shamrock Inn offers two holiday specials.

34. **Portfolio Activity** Complete the problem in the portfolio activity on page 55. Solve the system using the elimination method.
Floor: $30 each; Balcony: $20 each

Look Back

35. Find the next three terms of this sequence.
5, 8, 11, 14, 17, ?, ?, ? **[Course 1]** 20, 23, 26

Evaluate each expression. [Course 1]

36. $\frac{1}{2} - \frac{6}{18}$ $\frac{1}{6}$ **37.** $3.2 + \frac{5}{10}$ 3.7 **38.** $3\frac{5}{9} - 2\frac{7}{8}$ $\frac{49}{72}$

39.

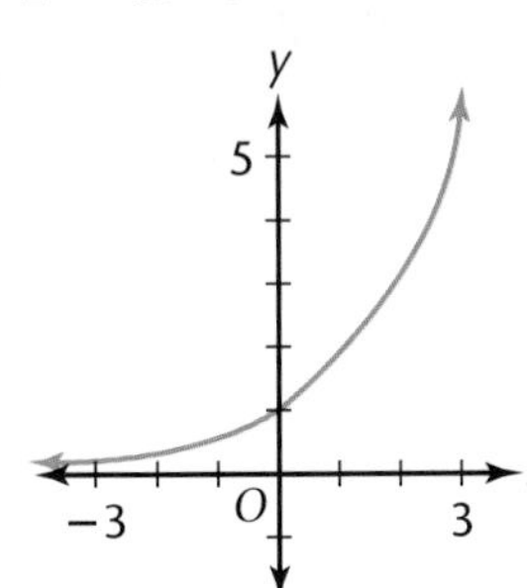

This function can be best described as c.

a. an absolute-value function.
b. a quadratic function.
c. an exponential function.
d. a linear function. **[Course 1]**

40. Cultural Connection: Africa The following problem is in the Rhind papyrus by Ahme: A bag containing equal weights of gold, silver, and lead is purchased for 84 *sha'ty*. What is the weight (in *deben*) of each metal if the price of 1 *deben* is 6 *sha'ty* for silver, 12 *sha'ty* for gold, and 3 *sha'ty* for lead? **[Lesson 1.4]** 4 deben

Find the slope for each line. [Lesson 2.1]

41. $7x - 3y = 22$ $\frac{7}{3}$ **42.** $3x + 2y = 6$ $-\frac{3}{2}$ **43.** $y = 2x$ 2

44. Write an equation for the line passing through $(-3, 8)$ and parallel to $y = 0.8x - 7$. **[Lesson 2.2]** $y = 0.8x + 10.4$

Look Beyond

45. Technology Graph this system of equations, and describe your graph. Use a graphics calculator if you have one. $\begin{cases} 2x - 3y = 6 \\ 4x - 6y = 18 \end{cases}$

LESSON 2.6

Exploring Non-Unique Solutions

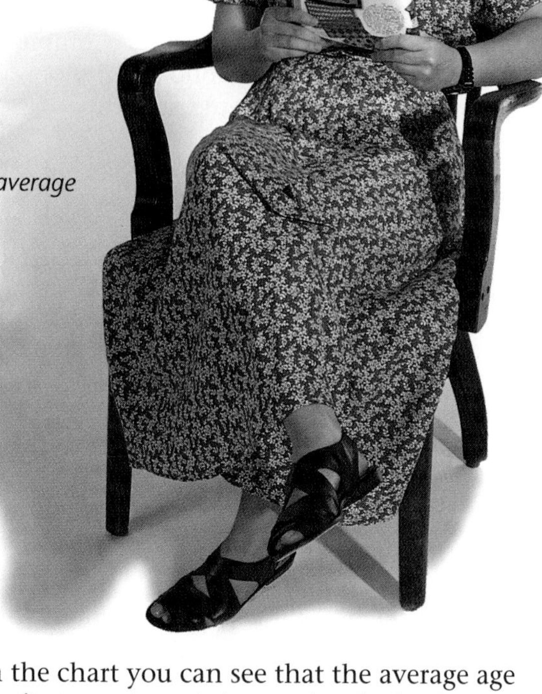

Sara is reading about the average ages for first marriages.

why *You have studied systems of equations that have exactly one solution. There are also systems that have an infinite number of solutions and systems that have no solution at all.*

In the chart you can see that the average age for first marriages is increasing for both men and women. Will the average age for both men and women ever be the same?

Inconsistent Systems

The data in the chart above can be compared by making a graph.

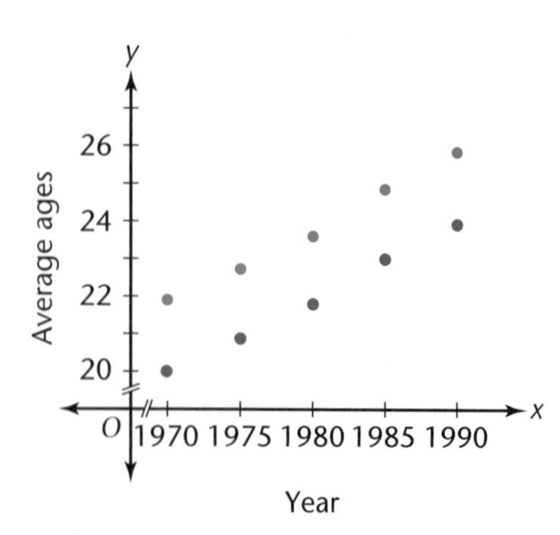

Let x represent the year in 5-year intervals.
Let y represent the average age for first marriages.

On the graph the years begin with 1970 and the ages begin with 20. What pattern do you see? A line of best fit can be used to model each data set.

The data for women can be modeled by $\mathbf{y = 0.2x - 374}$.
The data for men can be modeled by $\mathbf{y = 0.2x - 368}$.

Since the slopes of the equations are equal, the lines are parallel. If the trend continues, the lines never intersect, and the average ages for first marriages will never be equal.

Now consider the system formed by the lines of best fit. $\begin{cases} y = 0.2x - 374 \\ y = 0.2x - 368 \end{cases}$

Solve the system by substituting $0.2x - 368$ for y in the first equation.

$$\begin{aligned} y &= 0.2x - 374 \\ (0.2x - 368) &= 0.2x - 374 \\ -368 &= -374 \qquad \text{False} \end{aligned}$$

Notice that both variables have been eliminated and two unequal numbers remain. This means that there is no possible solution for this system. Since the graphs of the two equations are parallel, they never intersect. This system is called an *inconsistent system*.

An **inconsistent system** has no ordered pair that satisfies both of the original equations. Inconsistent systems have no common solutions. Their graphs are parallel and have the same slope but different y-intercepts.

The graph for mean ages for marriages between 1970 and 1990 is a good example of an inconsistent system. Give another real-life situation that might be modeled by parallel lines.

Answers may vary. One example would be the graphs of functions converting temperatures from Fahrenheit to Celsius and Fahrenheit to Kelvin. The graphs would be parallel.

Consistent Systems

When a system has one or more solutions, the system is **consistent**.

Dependent Systems

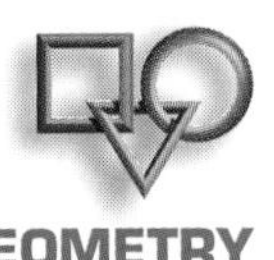

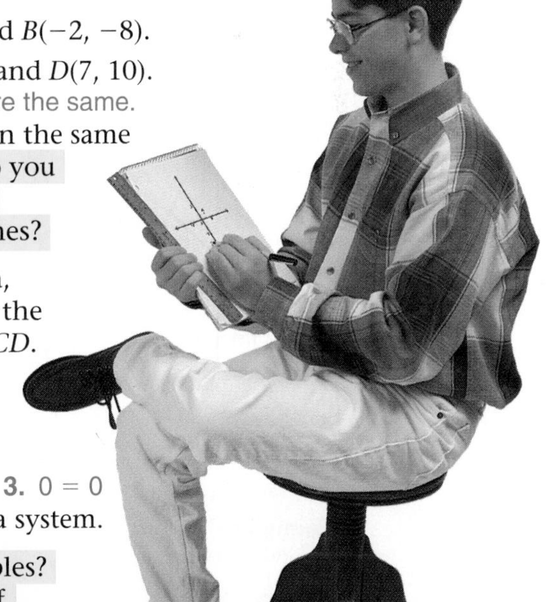

Line AB passes through $A(3, 2)$ and $B(-2, -8)$.
Line CD passes through $C(1, -2)$ and $D(7, 10)$.

1 Draw line AB and line CD on the same coordinate plane. What do you notice about the geometric relationship between the lines?

1. The lines are the same.

2 Use the point-slope formula, $y - y_1 = m(x - x_1)$, to write the equations for lines AB and CD. Compare your equations. What do you notice about the slopes of the two equations?

2. The equation for line AB is $y = 2x - 4$ and the equation for line CD is $y = 2x - 4$; the slopes are the same.

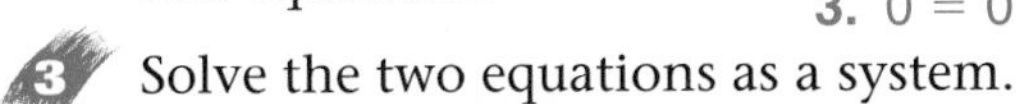

3 Solve the two equations as a system.

3. $0 = 0$

4 What happens to the variables? Is the resulting statement of equality true? Do the two lines have any points in common? Is more than one solution possible? If so, how many? ❖

4. The variables are eliminated; yes; yes; yes; they have an infinite number of points in common.

Exploration 2 Other Systems

Let line 1 have slope m_1 and y-intercept b_1 and be represented by Equation 1. Let line 2 have slope m_2 and y-intercept b_2 and be represented by Equation 2.

Let Equation 1 and Equation 2 form a system.
$$\begin{cases} y = m_1x + b_1 \rightarrow \text{Equation 1} \\ y = m_2x + b_2 \rightarrow \text{Equation 2} \end{cases}$$

1 Suppose line 1 and line 2 intersect in exactly one point. What can you say about

a. m_1 and m_2? The slopes are not equal.

b. b_1 and b_2? The y-intercepts may or may not be equal.

c. the solution to the system? There is only one common solution; consistent.

2 Suppose line 1 and line 2 are parallel. What can you say about

a. m_1 and m_2? The slopes are equal.

b. b_1 and b_2? The y-intercepts are not equal.

c. the solution to the system? There is no common solution; the lines never intersect; inconsistent.

3 Suppose line 1 and line 2 are the same line. What can you say about

a. m_1 and m_2? The slopes are equal.

b. b_1 and b_2? The y-intercepts are equal.

c. the solution to the system? ❖ There are infinitely many solutions because both equations describe the same line; consistent.

Examine the systems shown below.

Independent system

$$\begin{cases} y = x - 3 \\ y = -2x + 3 \end{cases}$$

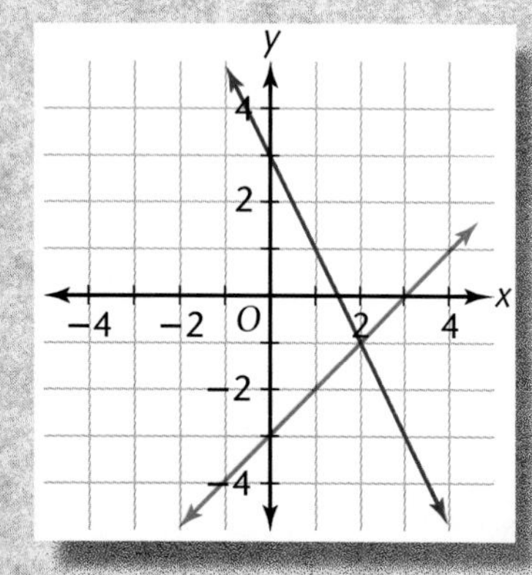

different slope
intersecting lines
one solution
same or different y-intercepts
consistent

Dependent system

$$\begin{cases} y = x - 3 \\ 2y = 2x - 6 \end{cases}$$

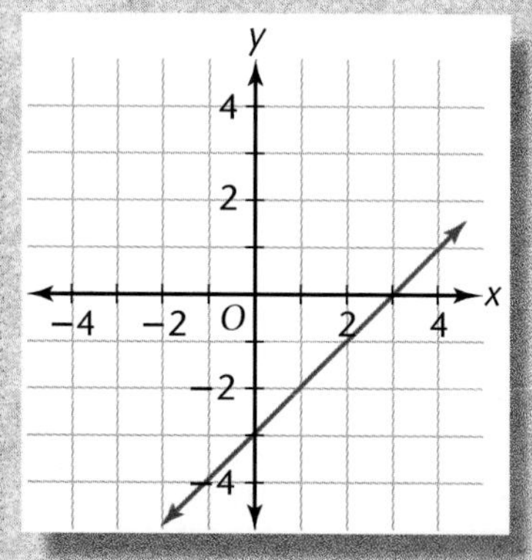

same slope
same line
infinite solutions
same y-intercepts
consistent

Inconsistent system

$$\begin{cases} y = x + 3 \\ y = x - 3 \end{cases}$$

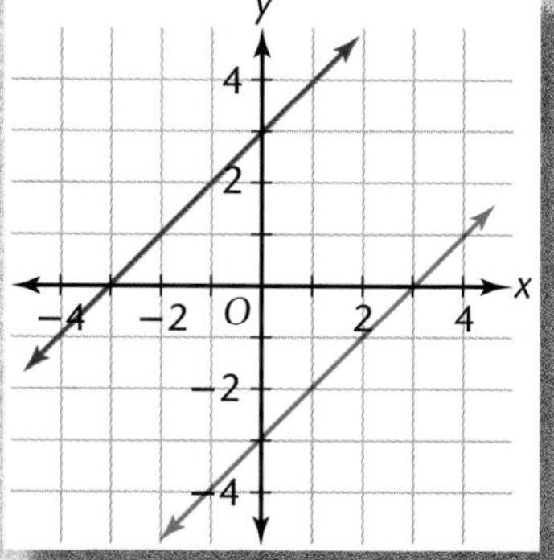

same slope
parallel lines
no solution
different y-intercepts
inconsistent

What characteristics are the same and which are different for each pair of systems?

Cultural Connection: Asia In 213 B.C.E., a Chinese ruler ordered all books to be burned. All of the written knowledge of mathematics was lost. Four hundred fifty years later, Liu Hui came to the rescue. He was able to recover the lost mathematics because people still used rod numerals. The rod numerals were used to represent a decimal system with place value, much like the one we use today. Red rods were used for positive numbers and black rods for negative numbers. Vertical and horizontal numerals were used alternately to keep the place values separate.

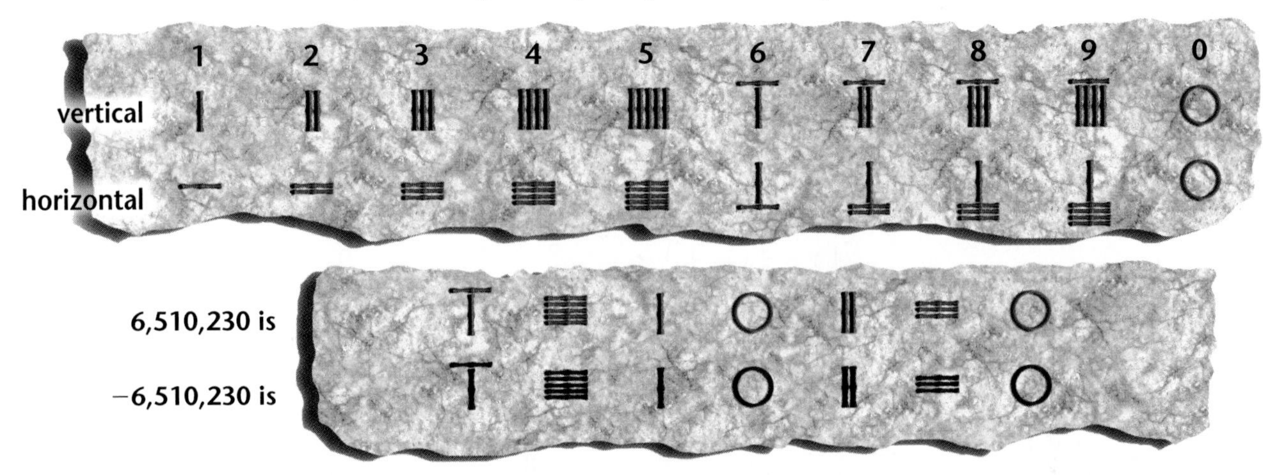

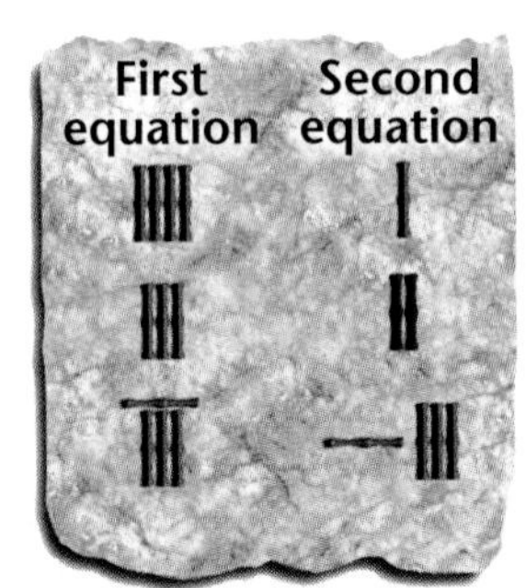

To represent a system of equations, the counting rods were arranged in such a way that one column was assigned to each equation of the system.

The figure at the left represents the system at the right. $\begin{cases} 4x + 3y = 8 \\ x - 2y = 13 \end{cases}$

One row was assigned to the coefficients of each unknown in the equations. The elements of the last row consisted of the entries on the right-hand side of each equation. Red rods were used to represent positive *(cheng)* coefficients and black rods for negative *(fu)* coefficients.

Try This Change the equations in the system to Chinese counting-rod equations.

$$\begin{cases} -23x + 6y = 32 \\ 240x - 56y = 4 \end{cases}$$

Communicate

Explain how to identify each system of equations as inconsistent, dependent, or independent.

1. $\begin{cases} y = 2x - 3 \\ 3y = 6x - 9 \end{cases}$ **2.** $\begin{cases} y = -3x + 2 \\ y = 2x + 2 \end{cases}$ **3.** $\begin{cases} x + y = 4 \\ x + y = 5 \end{cases}$

4. Explain how to write an equation that would form a dependent system with the equation $y = 2x + 3$. Name what parts of the original equation you would use to write the new equation.

5. Discuss how to write another equation that would form an inconsistent system with the equation $y = -3x + 4$.

6. If two equations in a system intersect at exactly one point, describe their slopes and y-intercepts. Explain how you determined your answer.

Practice & Apply

Solve each system algebraically.

7. $\begin{cases} x + y = 7 \\ x + y = -5 \end{cases}$ No solution

8. $\begin{cases} 3y = 3x - 6 \\ y = x - 2 \end{cases}$ Infinite

9. $\begin{cases} 3y = 2x - 24 \\ 4y = 3x - 3 \end{cases}$ $(-87, -66)$

10. $\begin{cases} y = -2x - 4 \\ 2x + y = 6 \end{cases}$ No solution

11. $\begin{cases} 2x + 3y = 11 \\ x - y = -7 \end{cases}$ $(-2, 5)$

12. $\begin{cases} 4x = y + 5 \\ 6x + 4y = -9 \end{cases}$ $\left(\frac{1}{2}, -3\right)$

13. $\begin{cases} 4x + y = 8 \\ y = 4 - 2x \end{cases}$ $(2, 0)$

14. $\begin{cases} y = \frac{3}{2}x + 4 \\ 2y - 8 = 3x \end{cases}$ Infinite

15. $\begin{cases} y = \frac{1}{2}x + 9 \\ 2y - x = 1 \end{cases}$ No solution

Determine whether each system is dependent, independent, or inconsistent. 16. Independent 17. Independent 18. Independent

16. $\begin{cases} 2y + x = 8 \\ y = 2x + 4 \end{cases}$

17. $\begin{cases} 3x + y = 6 \\ y = 3x + 9 \end{cases}$

18. $\begin{cases} y + 2 = 5x \\ y = -3x - 2 \end{cases}$

19. $\begin{cases} y + 6x = 8 \\ y = -6x + 8 \end{cases}$

20. $\begin{cases} x - 5y = 10 \\ -5y = -x + 6 \end{cases}$

21. $\begin{cases} y + 4x = 12 \\ y = 5x + 12 \end{cases}$

19. Dependent 20. Inconsistent 21. Independent

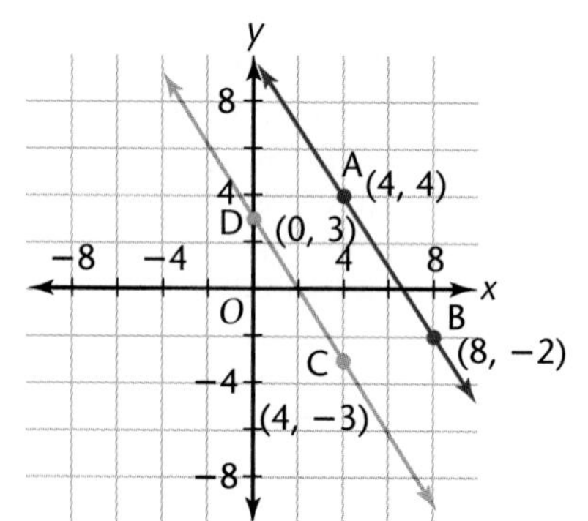

Use the graph on the left to answer Exercises 22–27.

22. Write the equations of lines AB and CD.

23. What kind of system do lines AB and CD represent? Write a complete sentence that states at least two reasons for your answer.

24. Write the equation of a line that would form an independent system with line AB and that contains point A. Graph your line.

25. Write the equation of a line that would form a dependent system with line CD and that contains point D. Graph your line.

26. What is the point of intersection of $y = 6$ and line AB? the point of intersection of $y = 6$ and line CD? $\left(\frac{8}{3}, 6\right)$; $(-2, 6)$

27. Write the equation of a line that would form an inconsistent system with $y = 6$.
Answers may vary. For example, $y = 8$.

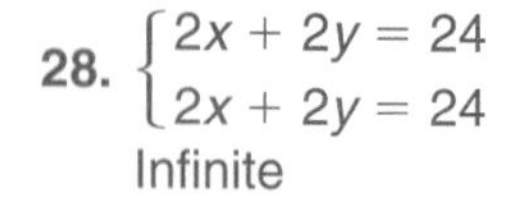

28. $\begin{cases} 2x + 2y = 24 \\ 2x + 2y = 24 \end{cases}$
Infinite

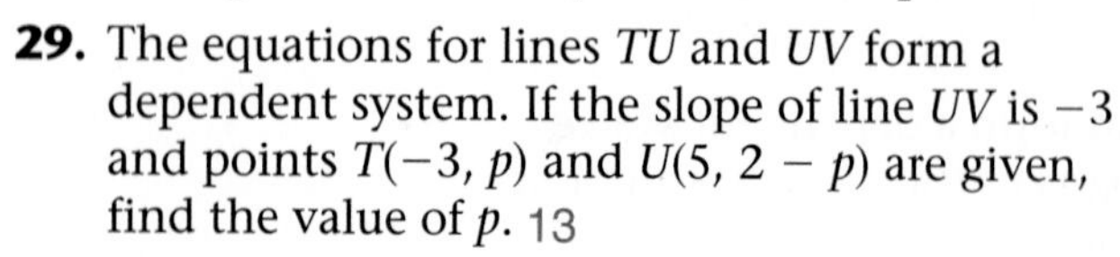

28. **Geometry** Triangle ABC has a perimeter of 18 centimeters, while rectangle $PQRS$ has a perimeter of 24 centimeters. Write a system of equations for the figures. How many solutions are possible?

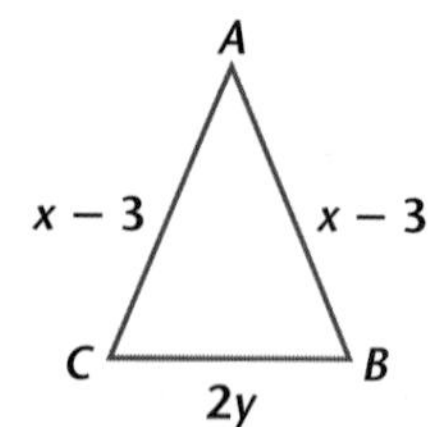

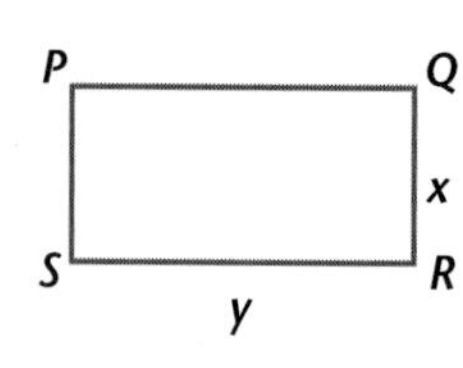

29. The equations for lines TU and UV form a dependent system. If the slope of line UV is -3 and points $T(-3, p)$ and $U(5, 2 - p)$ are given, find the value of p. 13

30. Wages Suppose that after you work for 10 years, you are offered the choice of two different jobs. The first has a starting salary of $2500 per month with an expense account of $750 per month and an annual bonus of $800. The other job pays $3250 per month and has two semi-annual bonuses of $350 each. Explain which of the two jobs you would take and why. Choose the first job. The salary will always be $100 greater than the second job's salary.

31. Geometry The sum of the measures of two angles is 180°. What is the measure of each angle if one angle is 30° more than twice the other? 50° and 130°

32. One number is 24 more than another number. If the sum of the numbers is 260, what are the numbers? 142 and 118

Demography The chart shows the populations of Center City and Bay City from 1970 to 1990. Complete Exercises 33–34.

	Center City	Bay City
1970	40,070	43,750
1975	42,570	46,250
1980	45,070	48,750
1985	47,570	51,250
1990	50,070	53,750

33. Write a system of equations that can be used to predict the population, y, in a given year, x, for each city. Center City: $y = 500x + 40{,}070$ Bay City: $y = 500x + 43{,}750$

34. Will the population of Center City ever equal that of Bay City? If so, when? Never, the system is inconsistent.

First equation	Second equation
I	II
IIII	II
⊥II	–T

35. Cultural Connection: Asia Change these equations from the Chinese counting-rod system to an algebraic system of equations as shown on page 93. Then solve the system.

$\begin{cases} -x + 4y = 7 \\ 2x + 7y = 16 \end{cases}$; (1, 2)

Look Back

Solve for x. [Lesson 1.3]

36. $2x - 5 = 15$ 10 **37.** $-3x + 4 = -14$ 6 **38.** $4x + 7 = -5$ -3 **39.** $ax + b = c$ $\frac{c-b}{a}$

40. Graph the equation $2x + 5y = 7$. **[Lesson 2.2]**

41. Write $3x + 2y = 6x + 9$ in standard form. **[Lesson 2.2]** $3x - 2y = -9$

Solve each system of equations. [Lessons 2.4, 2.5]

42. $\begin{cases} 2x + 3y = 6 \\ x + y = 4 \end{cases}$ (6, −2)

43. $\begin{cases} -x + 3y = 9 \\ x - y = 12 \end{cases}$ (22.5, 10.5)

44. $\begin{cases} -4x - 8y = 24 \\ x - 5y = 2 \end{cases}$ $\left(-3\frac{5}{7}, -1\frac{1}{7}\right)$

Look Beyond

45. To get to work, Charley takes Interstate 87 for 15 miles at 60 miles per hour and then Central Avenue for 12 miles at 40 miles per hour. How many minutes does it take Charley to get to work? 33 minutes

LESSON 2.7 Graphing Linear Inequalities

Why *A cassette player costs more than you have saved. Your grade on the last test was less than you expected. These conditions create inequalities. Sometimes unequal conditions can be reversed. Two inequalities form a system that can be solved by graphing.*

Linear Inequalities

Graphing on a coordinate plane allows you to find solutions to linear inequalities that contain more than one variable.

Exploration *Graphing Linear Inequalities*

1. Graph the line $y = x$ using a graphics calculator or graph paper.
2. Locate and label a point, P, above the line $y = x$. What is the relationship between the x-coordinate and y-coordinate of P? $x < y$
3. Locate and label a point, Q, below the line $y = x$. What is the relationship between the x-coordinate and the y-coordinate of Q? ❖ $x > y$

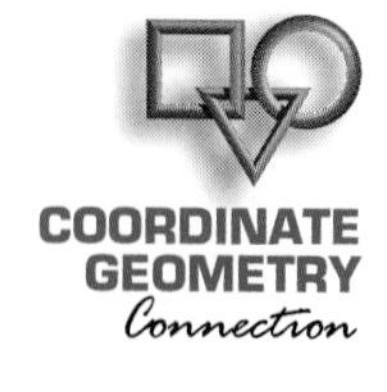

The graph of a linear equation such as $y = x$ divides a coordinate plane into two half-planes. The line graphed for $y = x$ is a **boundary line** for the two half-planes. A solution for the **linear inequality** $y \geq x$ includes *all ordered pairs* that make the inequality true. The shaded area, and the line $y = x$ contain all possible solutions for $y \geq x$. Test the coordinates for points R and S. Which point has coordinates that satisfy the inequality $y \geq x$? Point R

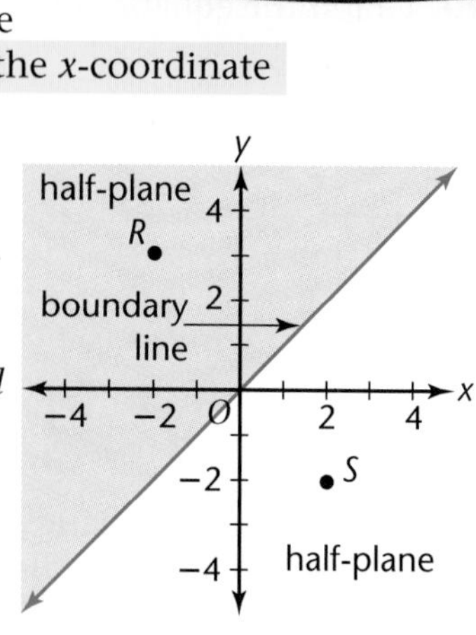

When the inequality symbol ≤ or ≥ is used, the points on the boundary line *are* included. In these cases a solid line is used to represent the boundary. When the inequality symbol < or > is used, the points on the boundary line *are not* included. In these cases a dashed line is used to represent the boundary.

EXAMPLE 1

Graph the inequality to find the solution.

$$x - 2y < 4$$

Solution

Graph the boundary line, $x - 2y = 4$. Find the x- and y-intercepts.

Let $x = 0$.	Let $y = 0$.
$0 - 2y = 4$	$x - 2(0) = 4$
$-2y = 4$	$x - 0 = 4$
$\frac{-2y}{-2} = \frac{4}{-2}$	$x = 4$
$y = -2$	

The points (0, −2) and (4, 0) are on the boundary line. Since the inequality symbol < does not include the equal sign, use a dashed line to connect the points.

Try a point such as (0, 0) to see if it satisfies the inequality. Substitute 0 for x and 0 for y.

$$x - 2y < 4$$
$$0 - 2(0) < 4$$
$$0 < 4 \quad \text{True}$$

Since the coordinates (0, 0) make the inequality true, (0, 0) is a point in the solution. Shade the half-plane that contains that point. ❖

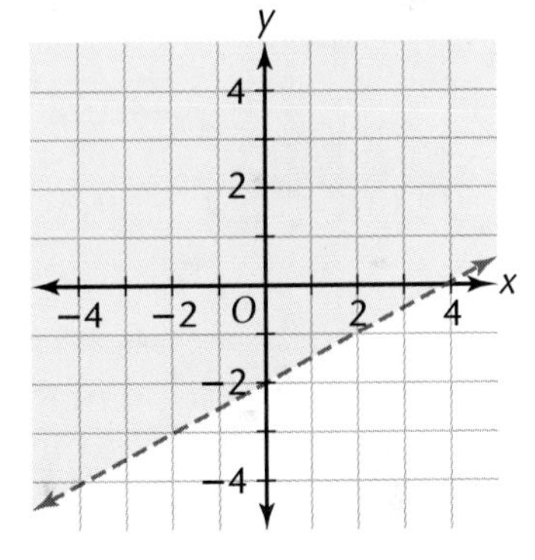

Systems of Linear Inequalities

The solution for a system of linear inequalities is the **intersection** of the solution for each inequality. The intersection on the graph is that portion of the plane where the solutions for each inequality overlap. Every point in the intersection region satisfies the system.

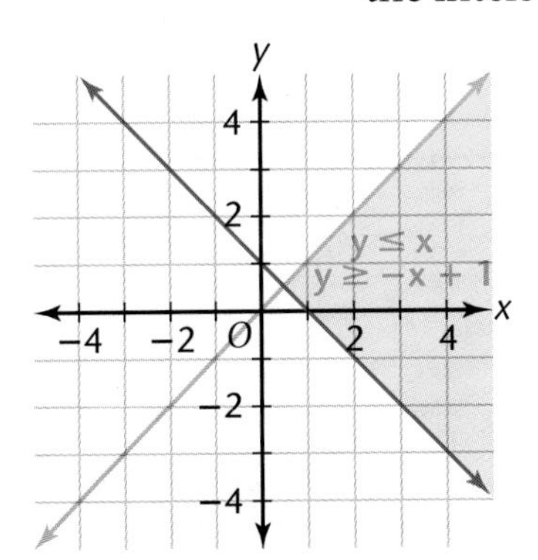

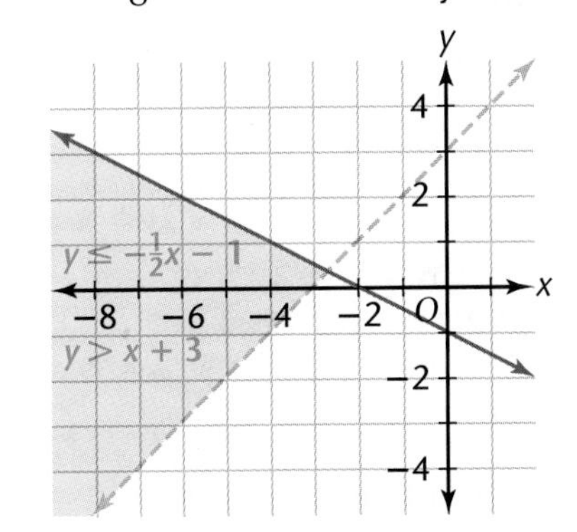

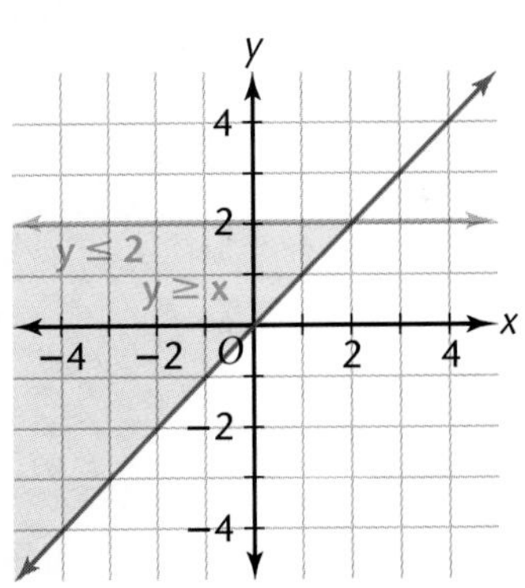

EXAMPLE 2

Solve by graphing. $\begin{cases} x + y \geq -1 \\ -2x + y < 3 \end{cases}$

Solution ➤

Graph the boundary line,
$x + y = -1$.

Let $x = 0$.	Let $y = 0$.
$0 + y = -1$	$x + 0 = -1$
$y = -1$	$x = -1$

The points $(0, -1)$ and $(-1, 0)$ are on the boundary line. Since the inequality symbol $\geq$ includes the equal sign, draw a solid line to connect these points.

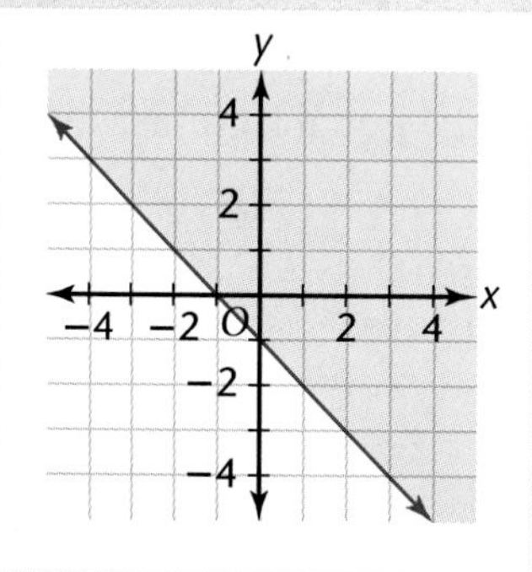

Try the point $(0, 0)$ to see if it satisfies the inequality.

$$0 + 0 \geq -1$$
$$0 \geq -1 \quad \text{True}$$

Since the substitution is true, shade the half-plane that contains $(0, 0)$.

Graph the boundary line,
$-2x + y = 3$.

Let $x = 0$.	Let $y = 0$.
$-2(0) + y = 3$	$-2x + 0 = 3$
$y = 3$	$x = -\frac{3}{2}$

The points $(0, 3)$ and $(-\frac{3}{2}, 0)$ are on the boundary line. Since the inequality symbol $<$ does not include the equal sign, draw a dashed line to connect these points.

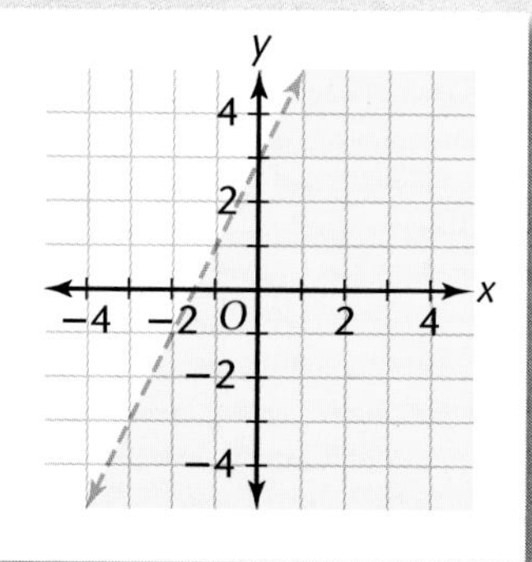

Try the point $(0, 0)$ to see if it satisfies the inequality.

$$-2(0) + 0 < 3$$
$$0 < 3 \quad \text{True}$$

Since the substitution is true, shade the half-plane that contains $(0, 0)$.

The graph shows the solution for the two inequalities. Try a point such as $(1, 1)$ from the intersecting region to test both inequalities.

$$x + y \geq -1$$
$$1 + 1 \geq -1$$
$$2 \geq -1 \quad \text{True}$$

$$-2x + y < 3$$
$$-2(1) + 1 < 3$$
$$-1 < 3 \quad \text{True}$$ ❖

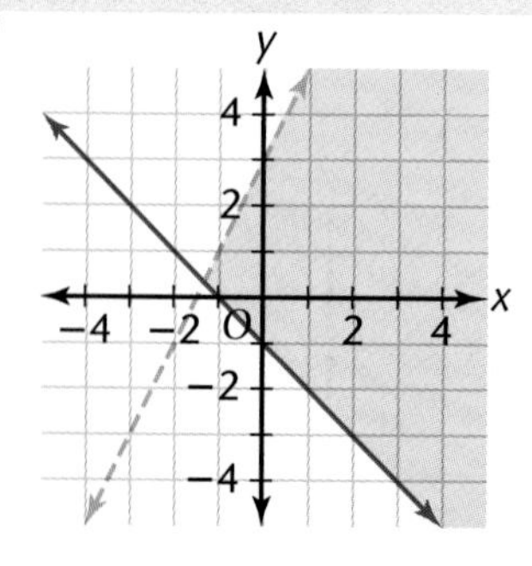

Ongoing Assessment
(1, 1) and (0, 0) are clearly in the overlapping region; and 1 and 0 are small numbers that can be used to quickly evaluate an equation.

Why are (1, 1) and (0, 0) good points to use as checks for inequalities?

EXAMPLE 3

Kwame gets one job mowing lawns at $6 an hour and a second job at the library for $7 an hour.

During the summer, Kwame wants to make at least $126 per week working part-time. He can work no more than 30 hours per week. Write a system of inequalities that represents all of the combinations of hours and jobs that Kwame can work. Graph this solution.

Define the variables.

Let m represent the number of hours mowing.
Let l represent the number of hours at the library.

Write the inequalities that satisfy the conditions in the problem. Use the correct inequality symbols.

Kwame can work *no more* than 30 hours a week.

$$m + l \leq 30$$

Graph $m + l = 30$.

Let $l = 0$.	Let $m = 0$.
$m + 0 = 30$	$0 + l = 30$
$m = 30$	$l = 30$

The points (30, 0) and (0, 30) are on the boundary line. Since the inequality symbol includes the equal sign, use a solid line to connect the points.

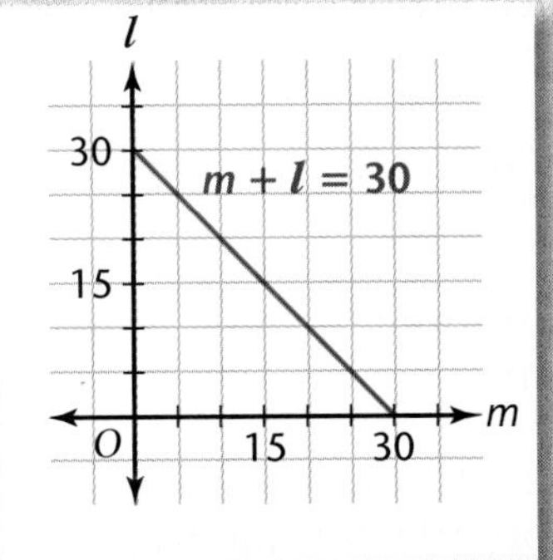

Kwame wants to make *at least* $126 a week.

$$6m + 7l \geq 126$$

Graph $6m + 7l = 126$.

Let $l = 0$.	Let $m = 0$.
$6m + 7(0) = 126$	$6(0) + 7l = 126$
$6m = 126$	$7l = 126$
$m = 21$	$l = 18$

The points (21, 0) and (0, 18) are on the boundary line. Since the inequality symbol includes the equal sign, use a solid line to connect the points.

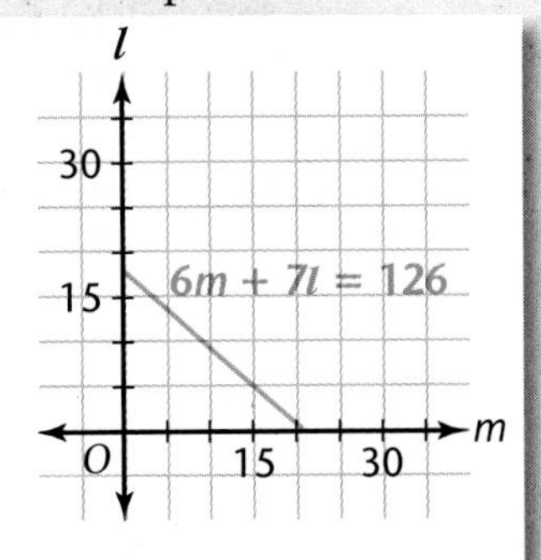

Why do we exclude the points to the left of the l-axis and below the m-axis? Kwame cannot work a negative number of hours.

The point (0, 0) makes the inequality true, so shade below.

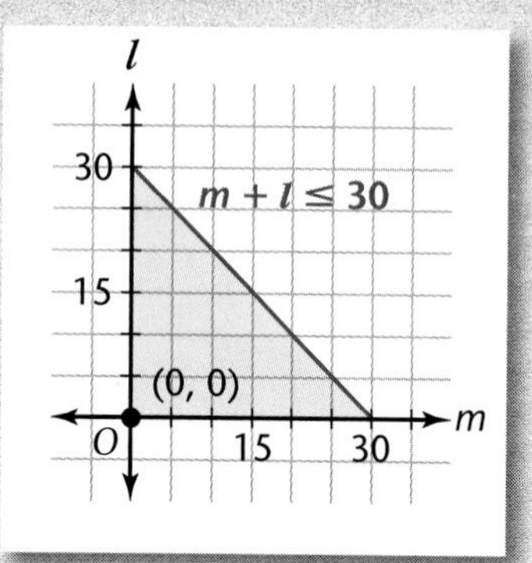

The shaded area shows the greatest number of hours that Kwame can work at each job.

The point (15, 10) makes the inequality true, so shade above.

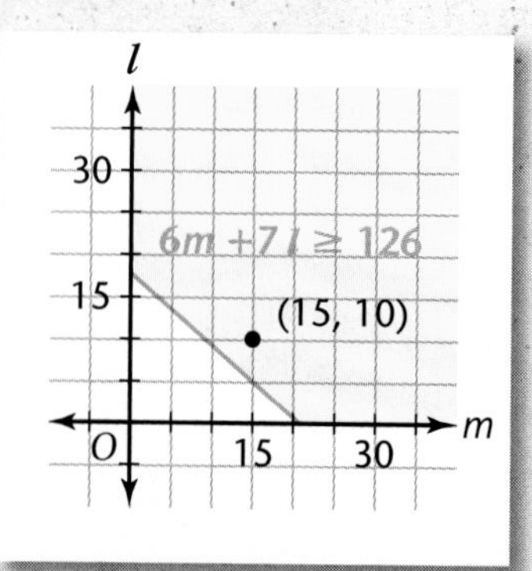

The shaded area shows the least number of hours that Kwame can work to make at least $126.

The graph below represents the solution of the two inequalities.

Any of the points in the intersecting region will satisfy the conditions for both the inequalities.

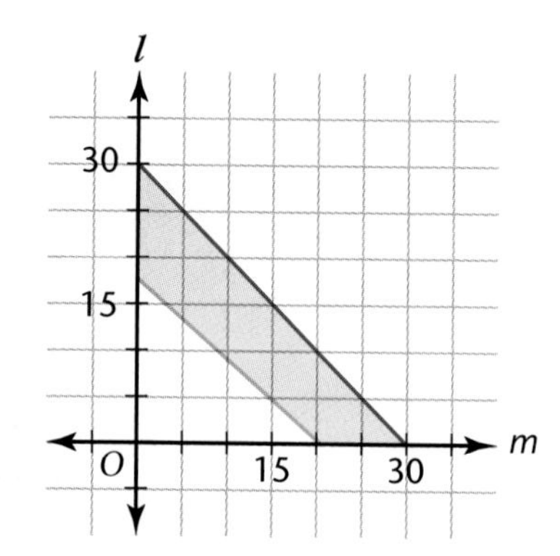

The total hours for each set of points is less than or equal to 30. The total number of dollars Kwame earns for each set of points is $126 or more. ❖

Explain how you would determine how many hours Kwame can work at each job.

To check, try two points from the intersecting region. For example (5, 20) and (10, 15) show that:

- Kwame can work 5 hours mowing and 20 hours at the library.
- Kwame can work 10 hours mowing and 15 hours at the library.

What is the least number of hours Kwame can work to earn at least $126? What is the greatest amount of money he can make and still work 30 hours or less?

EXERCISES & PROBLEMS

Communicate

1. Explain how to decide whether to draw a solid or a dashed line for an inequality.
2. Explain how to choose the area to shade for the inequality $x + 2y < 2$. What is a good point to test in the shaded area?

Explain how to graph the intersection for each system of inequalities.

3. $\begin{cases} y < 1 \\ x < 1 \end{cases}$

4. $\begin{cases} 8x + 4y > 12 \\ y < 3 \end{cases}$

5. $\begin{cases} x \leq 3 \\ x - 2y \geq 2 \end{cases}$

6. $\begin{cases} x + y \leq -2 \\ x + y > -2 \end{cases}$

Explain how to write the system of inequalities to represent the following problem. Describe how to graph the solution for a reasonable domain and range. Show your graph.

7. Each spring Marta and her family add flowers to their garden. They plan on buying no more than 20 new perennials at \$5.00 per pot and a number of annuals at \$1.50 per pot. Marta knows they have budgeted at least \$30 for the plants. What combination of each type of plant can they buy?

Practice & Apply

Graph the common solution for each system of inequalities. Choose a point from the solution. Check both inequalities.

8. $\begin{cases} 2x - 3y > 6 \\ 5x + 4y < 12 \end{cases}$

9. $\begin{cases} x - 4y \leq 12 \\ 4y + x \leq 12 \end{cases}$

10. $\begin{cases} y < x - 5 \\ y \leq 3 \end{cases}$

11. $\begin{cases} 2x + y \leq 4 \\ 2y + x \geq 8 \end{cases}$

12. $\begin{cases} 5y < 2x - 5 \\ 4x + 3y \leq 9 \end{cases}$

13. $\begin{cases} 4y < 3x + 8 \\ y \leq 1 \end{cases}$

14. $\begin{cases} x \geq 4 \\ 0 < y \end{cases}$

15. $\begin{cases} x - 2y < 8 \\ x + y \geq 5 \end{cases}$

16. $\begin{cases} 3x - y > -2 \\ x - y > -1 \end{cases}$

17. **Geometry** Write the system of inequalities defined by each perimeter below, and graph the common solution. Make sure that the solution represents the reasonable domain and range.

The perimeter of rectangle ABCD is at most 30 centimeters, while the perimeter of rectangle PQRS is at least 12 centimeters.

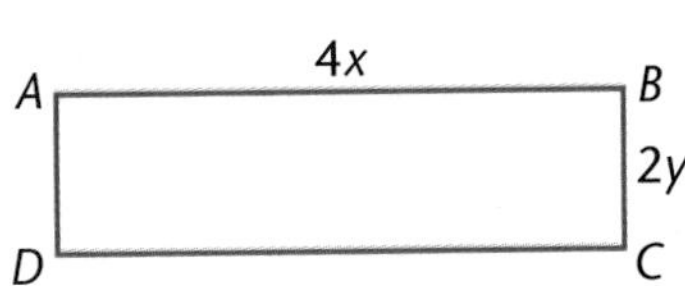

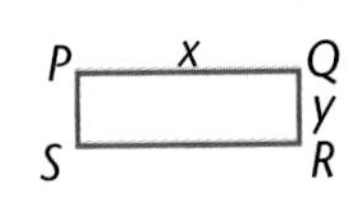

Suppose that you and a friend are going to have lunch at Al's Burgers.

18. You and a friend have \$8.00 total and each of you wants at least one Alburger and at least one order of fries. Is there enough money remaining to buy any more? Yes, one person can have a second order of fries.

Match each set of inequalities to the graph that represents its solution.

19. $y < -2x + 6$ and $y \leq 3x - 4$ b

20. $y \leq -2x + 6$ and $y > 3x - 4$ a

a.

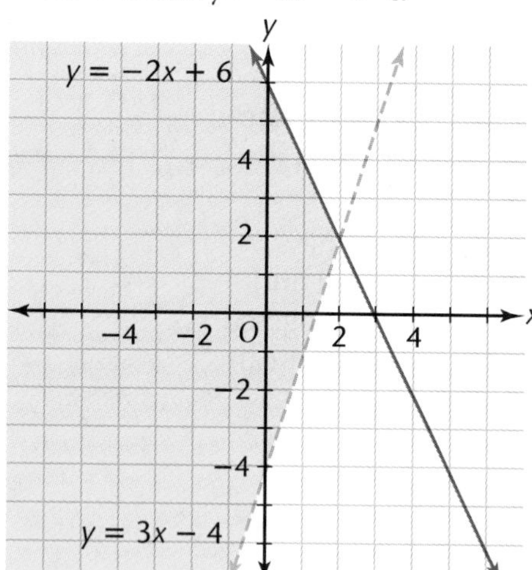

b.

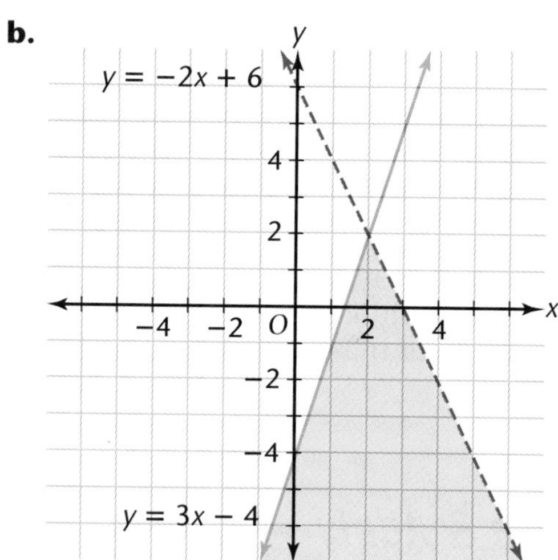

21. Write the system of inequalities that represents the shaded region. The equations of the boundary lines are given.

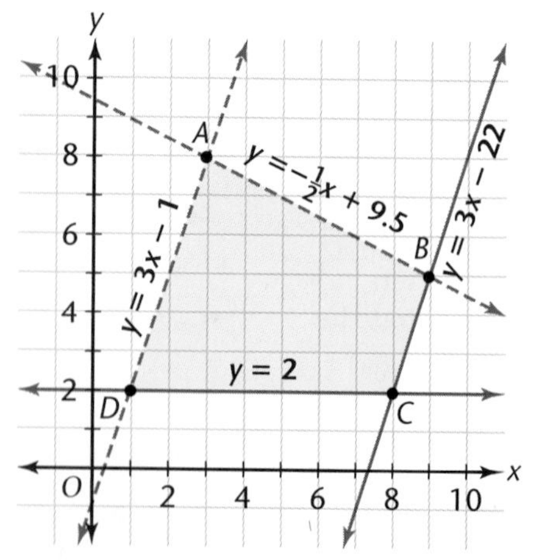

22. Write the system of inequalities that represents the shaded region. Use the points shown to find equations for the boundary lines.

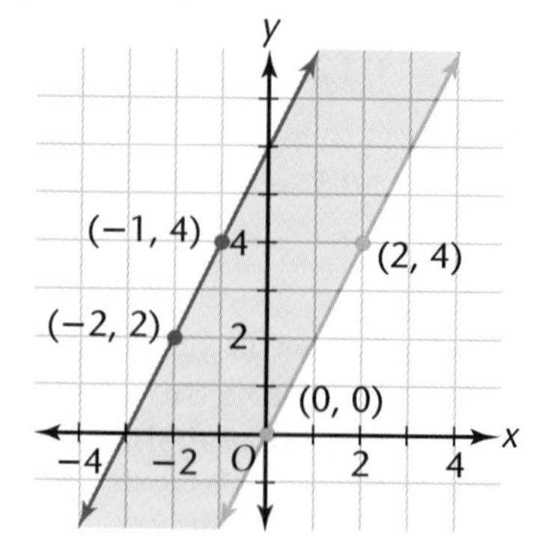

In the Lincoln High basketball game, Todd made only 2-point shots and 1-point free throws.

23. Sports Todd made no more than 16 points in the game. Find the combinations of baskets and free throws Todd could have made.

24. Academics On most days Abel spends no more than 2 hours doing his math and science homework. If math homework always takes about twice as long as science homework, what is the maximum time that Abel usually spends on science homework? $\frac{2}{3}$ hour

Geometry The length of a rectangle is at least 5 units more than its width, while the length of a second rectangle is no more than 5 times its width.

25. Represent the length of the first rectangle as an inequality. $l \geq w + 5$

26. Represent the length of the second rectangle as an inequality. $l \leq 5w$

27. Draw figures to represent the two rectangles, and label the lengths and widths as inequalities.

Look Back

28. Find the next three terms in the sequence. **[Course 1]**

5, 17, 37, 65, 101, . . . 145, 197, 257

29. Simplify the expression $-8 - 2\left[3 + 6\left(5 - \frac{5}{4}\right) + 7\right]$. **[Course 1]** -73

30. Write and solve an equation to find how much each person contributed toward a goal of $1200 if 20 people contributed equally? **[Lesson 1.2]** $20x = 1200$; $60

Solve each equation for x. [Lesson 1.4]

31. $3x + 7 = 2(x - 1)$ -9

32. $4(3x + 9) = -6(3x - 2)$ $-\frac{4}{5}$

Determine whether each pair of lines is parallel, perpendicular, or neither. [Lessons 2.1, 2.2]

33. $x + y = 7$
$x + 3y = 9$
Neither

34. $-2x + y = -5$
$-2x - y = 5$
Neither

35. $-x + 2y = 6$
$-x + 2y = -3$
Parallel

36. $x - 2y = 4$
$2x + y = 1$
Perpendicular

Look Beyond

For $f(x) = 2x + 4$, evaluate the following:

37. $f(3)$ 10

38. $f(-2)$ 0

39. $f\left(-\frac{1}{2}\right)$ 3

LESSON 2.8 Classic Applications

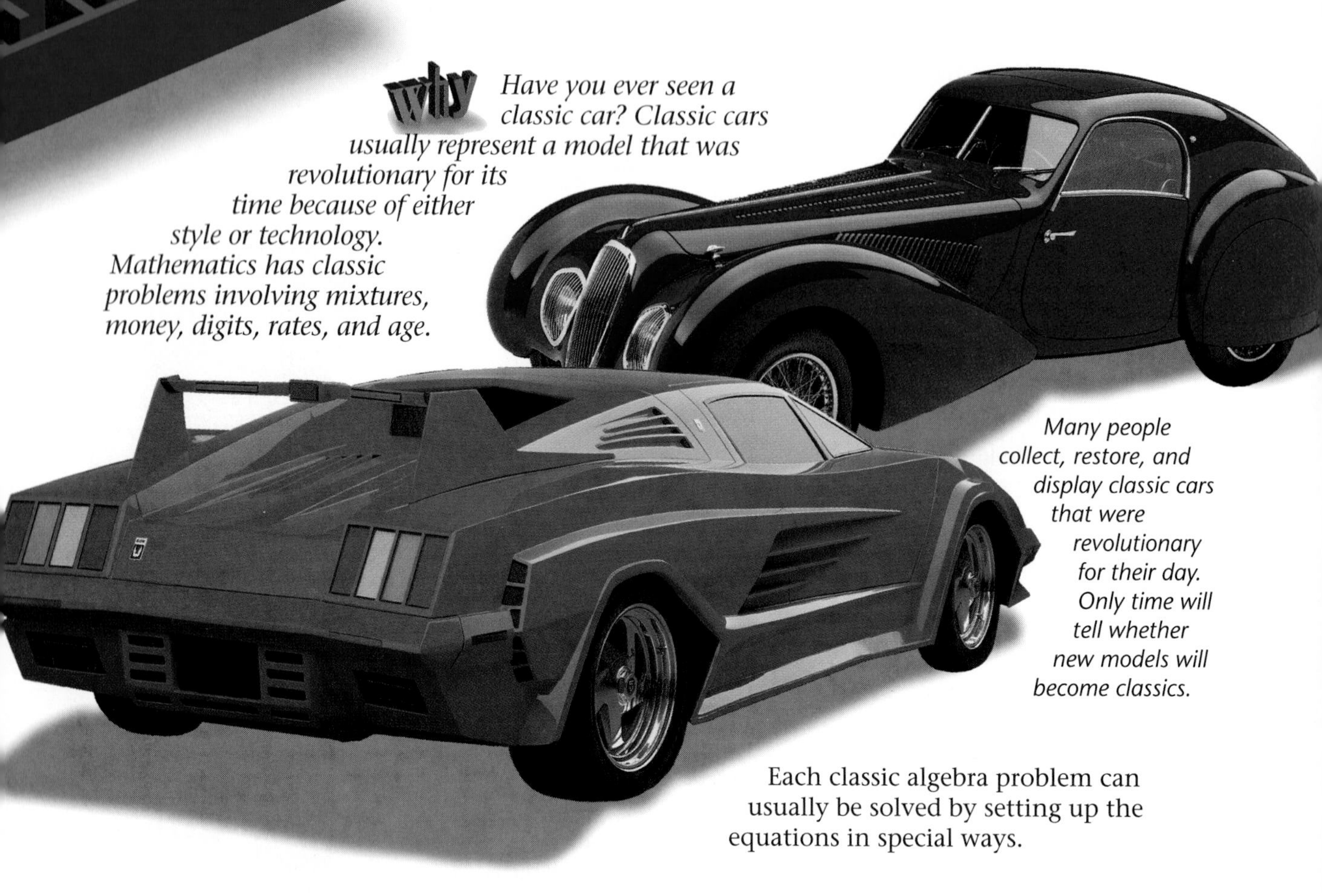

Why Have you ever seen a classic car? Classic cars usually represent a model that was revolutionary for its time because of either style or technology. Mathematics has classic problems involving mixtures, money, digits, rates, and age.

Many people collect, restore, and display classic cars that were revolutionary for their day. Only time will tell whether new models will become classics.

Each classic algebra problem can usually be solved by setting up the equations in special ways.

The classic application in Example 1 uses percent to express how much of a chemical is in a solution.

EXAMPLE 1

Chemistry How many ounces of a 1.5% acid solution should be mixed with a 4% acid solution to produce 60 ounces of a 2.5% acid solution?

Solution

Draw a diagram and make a table to help solve this problem.

Define the variables.

Let x represent the number of ounces of 1.5% acid solution.

Let y represent the number of ounces of 4% acid solution.

Place the information in a chart to help organize the facts.

	Percent acid	Amount of solution	Amount of acid
First solution	1.5%	x	$0.015x$
Second solution	4%	y	$0.04y$
Final solution	2.5%	60	0.025(60)

From the table, write two equations.

$$\begin{cases} x + y = 60 \\ 0.015x + 0.04y = 0.025(60) \end{cases}$$

Solving the system gives the solution (36, 24).

To get the required 60 ounces of the 2.5% acid solution, mix 36 ounces of the 1.5% acid solution with 24 ounces of the 4% acid solution. ❖

In classic coin problems you can usually set up two equations—one involving the number of coins and the other involving the value of the coins.

EXAMPLE 2

In a coin bank there are 250 dimes and quarters worth a total of $39.25. Find how many dimes, d, and how many quarters, q, are in the bank.

Solution ➤

Create a chart to organize the information.

Coin type	Number	Coin value	Value in cents
Quarters	q	25¢	$25q$
Dimes	d	10¢	$10d$
Total	250 coins		3925¢

From the table, write two equations based on the totals.

$$\begin{cases} q + d = 250 \\ 25q + 10d = 3925 \end{cases}$$

By solving the system, you find that q is 95 and d is 155.

The coin bank has 95 quarters and 155 dimes. A check shows that the solution is correct. ❖

Other examples of classic problems are digit problems. In most digit problems the trick is to write the value of a number in expanded form. You can write 52 as $5(10) + 2$. If you reverse the digits in 52, you get 25, and can write it as $2(10) + 5$.

EXAMPLE 3

The sum of the digits of a two-digit number is 7. The original two-digit number is 3 less than 4 times the number with its digits reversed. Find the original two-digit number.

Solution ➤

Let t represent the tens digit of the original number.
Let u represent the units digit of the original number.

The two-digit number expressed in expanded notation is

$$10t + u.$$

The number with its digits reversed in expanded notation is

$$10u + t.$$

The problem says the sum of the digits is 7.

$$t + u = 7$$

The original two-digit number is 3 less than 4 times the number with its digits reversed.

$$10t + u = 4(10u + t) - 3$$

Solving the system $\begin{cases} t + u = 7 \\ 10t + u = 4(10u + t) - 3 \end{cases}$ gives $u = 1$ and $t = 6$.

If t is 6 and u is 1, then the original number is $10(6) + 1$, or 61. The number 61 is a two-digit number which is 3 less than 4 times itself with its digits reversed. Since $4(10(1) + 6) - 3 = 61$, the solution is correct. ❖

Try This A two-digit number whose tens digit is 2 more than the units digit is 3 more than 6 times the sum of its digits. Find the original number. 75

Classic motion problems have some basic elements. Travel by plane, train, or automobile involves speed, distance, and time. Create charts, draw diagrams, and organize the information to solve these types of problems:

EXAMPLE 4

Travel A plane leaves New York City and heads for Chicago, which is 750 miles away. The plane, flying against the wind, takes 2.5 hours to reach Chicago. After refueling the plane returns to New York City, traveling with the wind, in 2 hours. Find the speed of the wind and the speed of the plane with no wind.

Solution ➤

Define the variables. Chart the information.

Let x represent the rate (speed) of the plane without any wind.
Let y represent the rate (speed) of the wind.

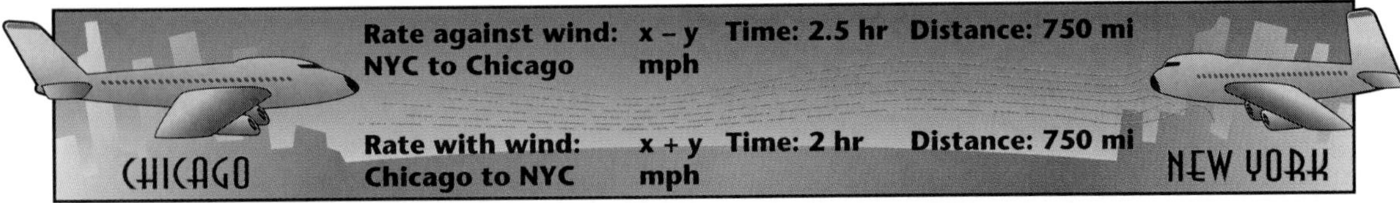

To write the system of equations, recall the relationship between rate, time, and distance.

	Rate	× Time	=	Distance
For Chicago to NYC:	$(x - y)$	(2.5)	=	750
For NYC to Chicago:	$(x + y)$	(2)	=	750

First divide to get simpler equations.

$$(x - y)(2.5) = 750 \qquad (x + y)(2) = 750$$

$$\frac{(x - y)(2.5)}{2.5} = \frac{750}{2.5} \qquad \frac{(x + y)(2)}{2} = \frac{750}{2}$$

$$x - y = 300 \qquad x + y = 375$$

Solving the system gives $x = 337.5$ and $y = 37.5$.

The rate (speed) of the plane is 337.5 miles per hour and the rate (speed) of the wind is 37.5 miles per hour. A check shows that the solution is correct. ❖

How would the speed of the plane be affected if there were no wind?
The speed of the plane would not be affected.

EXAMPLE 5

Age Age problems are classic problems that can usually be solved by writing the two related ages at two points in time. In 4 years the father will be 5 times as old as his son. How old is each now?

A father is 32 years older than his son.

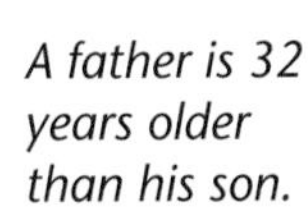

Solution ➤

Write two equations about the age of the father and son at two different times. Define the variables.

Let f represent the father's age. Let s represent the son's age.

In the caption, notice that the father is 32 years older than his son.

$$f = s + 32$$

In 4 years the father will be 5 times the son's age. Add 4 years to each age.

$$f + 4 = 5(s + 4)$$

Solving the system gives $s = 4$ and $f = 36$.

At the present time, the father is 36 years old, and his son is 4 years old. A check shows that the solution is correct. ❖

Exercises & Problems

Communicate

Discuss the problem-solving strategy you would use to solve each problem, and explain how to *set up* the equations.

1. When Jim cleaned out the reflecting pool at the library, he found 20 nickels and quarters. The mixture of nickels and quarters totaled $2.60. How many quarters did Jim find?
2. **Chemistry** A chemistry experiment requires that you mix a 25% glucose solution with a 5% glucose solution. This mixture produces 20 liters of a solution that contains 2.6 liters of glucose. How many liters of the 25% glucose solution do you need?

A bird can fly with the wind 3 times as fast as it can fly against the wind. How can you represent

3. the speed of the bird in still air (no wind) and the speed of the wind?
4. the speed of the bird flying with the wind and against the wind?
5. Discuss how the speed of the bird with the wind compares with the speed of the bird against the wind.

Describe how to set up Exercise 6, discuss how to solve the problem, and tell how to check your answer.

6. The sum of the digits of a two-digit number is 8. If 16 is added to the original number, the result is 3 times the original number with its digits reversed. Find the original number.

8. $18d$ and $108q$

Practice & Apply

7. **Recreation** Ramona paddles upstream, and then turns her canoe around and returns downstream to her cabin in exactly 1 hour. What is the speed of the creek's current and the speed of Ramona's paddling in still water?
   1 mph and 5 mph

Ramona is staying at a vacation resort called Stony Cabins. She rents a canoe and takes 1.5 hours to paddle her canoe upstream.

8. The coin box of a vending machine contains 6 times as many quarters as dimes. If the total amount of money at the end of the day is $28.80, how many of each kind of coin are in the box?

9. Find the two-digit number whose tens digit is 4 less than the units digit. The original number is 2 more than 3 times the sum of the digits. 26

With a tailwind, a jet flew 2000 miles in 8 hours, but the return trip against the same wind required 10 hours.

10. Aviation Find the jet's speed and the wind speed. jet = 225 mph; wind = 25 mph

11. In 15 years, Maya will be twice as old as David is now. In 15 years, David will be the same age that Maya will be 10 years from now. How old are they now? D: 20; M: 25

12. A 4% salt solution is mixed with a 16% salt solution. How many milliliters of each solution are needed to obtain 600 milliliters of a 10% solution? 300 mL each

13. Wymon is 20 years older than Sabrina. In 8 years, Wymon will be twice as old as Sabrina. How old is each today? S: 12; W: 32

14. Cultural Connection: Asia Apply the problem-solving strategy of working backwards (inversion) to a problem supplied by Aryabhata, around 500 C.E.

O beautiful maiden with beaming eyes, tell me, since you understand the method of inversion:

What number multiplied by 3, then increased by $\frac{3}{4}$ of the product, then divided by 7, then diminished by $\frac{1}{3}$ of the result, then multiplied by itself, then diminished by 52, whose square root is then extracted before 8 is added, and then divided by 10, gives the final result of 2? 28

Look Back

Evaluate each expression when x is 2, y is 1, and z is 1. [Course 1]

15. $2x + 4y - 3z$ 5 **16.** $\frac{2x - z}{5}$ $\frac{3}{5}$ **17.** $\frac{x + y}{z}$ 3 **18.** $3xyz$ 6

19. Simplify $(29 + 1) \div 3 \cdot 2 - 4$. **[Course 1]** 16

Simplify. [Course 1]

20. $-[2(a + b)]$ $-2a - 2b$

21. $-[-11(x - y)]$ $11x - 11y$

22. $2(x - y) + 3[-2(x + y)]$ $-4x - 8y$

Find the slope of each equation. [Lesson 2.2]

23. $3x + 4y = 12$ $-\frac{3}{4}$

24. $5y = 10x - 20$ 2

Look Beyond

Find each square root.

25. $\sqrt{\frac{25}{625}}$ $\frac{1}{5}$

26. $\sqrt{a^2}$ For $a \geq 0$, $\sqrt{a^2} = a$; for $a < 0$, $\sqrt{a^2} = -a$

Finding the Right Mixture

A camping supply store wants to make a trail mix of nuts and raisins to provide all of the required nutrients at the lowest cost. The chart shows the nutritional information and cost per ounce.

Backpackers usually have a weight limit for their pack, so each package of trail mix will contain no more than 18 ounces for the mix. The snack should supply at least 1800 calories with no more than 117 grams of fat. How many ounces each of raisins and nuts should each package of trail mix contain to minimize the costs?

	Calories	Fat	Price/ounce
Nuts	150	13 g	\$0.40
Raisins	90	0 g	\$0.10

1 Define the variables for the unknowns.

Let n represent the number of ounces of nuts.

Let r represent the number of ounces of raisins.

2 Write a linear equation, called the optimization equation, to express the relationship between the highest and lowest cost.

Thus, \$0.40n equals the cost of nuts, and \$0.10r equals the cost of raisins. The total cost is \$0.40n + \$0.10r.

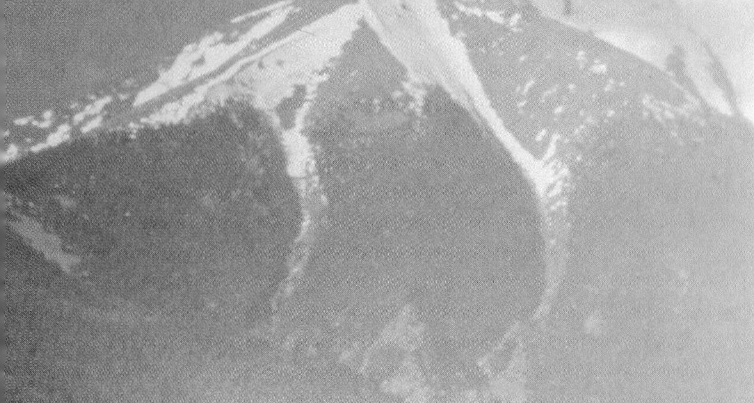

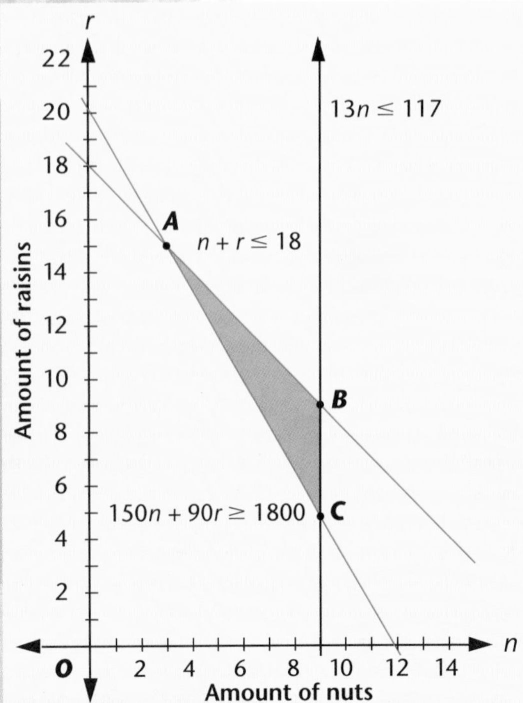

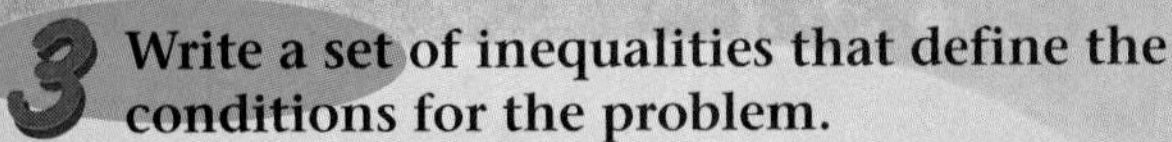

Write a set of inequalities that define the conditions for the problem.

The snack should supply at least 1800 calories.

$150n + 90r \geq 1800$

The snack should supply no more than 117 grams of fat. $13n \leq 117$

Each package should contain no more than 18 ounces of the mix. $n + r \leq 18$

The conditions defined by the inequalities that must be satisfied are called the **constraints.**

Use a graphics calculator or graph paper to graph each of the three inequalities. Shade the region that represents the solution set.

The shaded region, *ABC,* represents the solution set. This region, determined by the linear inequalities and containing the possible solutions, is called the **feasibility region.**

Find the coordinates of the vertices of *ABC*. They are $A(3, 15)$, $B(9, 9)$, and $C(9, 5)$.

Use each vertex of the shaded region to find values for the optimization equation. Then select the equation with the lowest cost.

Substitute $A(3, 15)$.

Total cost = $0.40n + $0.10r

Total cost = $0.40(3) + $0.10(15)

Total cost = $1.20 + $1.50 = $2.70

Substitute $B(9, 9)$.

Total cost = $3.60 + $0.90 = $4.70

Substitute $C(9, 5)$.

Total cost = $3.60 + $0.50 = $4.10

Thus, 3 ounces of nuts and 15 ounces of raisins give a minimum cost of $2.70.

Activity
Solve Using the Method of Linear Programming

The maker of recreational motorcycles lists two models, the Mountain Climber and the Dune Crawler. The chart shows some of the assembly information.

	Labor hours	Maximum produced	Profit
Mountain Crawler	150	60	$120
Dune Crawler	200	45	$180

If the company has no more than 12,000 hours of labor available for production of motorcycles, find the number of each model that should be built to give the maximum profit.

Chapter 2 Review

Vocabulary

common solution	73	inconsistent system	91	slope	56
consistent system	91	independent system	92	standard form	66
dependent system	92	line of best fit	69	substitution method	81
elimination method	84	linear inequality	96	system of equations	73

Key Skills & Exercises

Lesson 2.1

➤ Key Skills

Determine the slope of a line through two given points.

The slope of the line through the points (−5, 6) and (2, −3) is found by using the formula $m = \frac{\text{difference in } y\text{-coordinates}}{\text{difference in } x\text{-coordinates}}$, or $m = \frac{y_2 - y_1}{x_2 - x_1}$.

$$m = \frac{-3 - 6}{2 - (-5)}$$

$$= \frac{-9}{7} \qquad \text{The slope is } -\frac{9}{7}.$$

➤ Exercises

Determine the slope of the line through each pair of points.

1. (7, 9) and (13, 5) $-\frac{2}{3}$

2. (−7, 4) and (−4, −7) $-\frac{11}{3}$

3. (5.5, −2.5) and (−4.4, 10.5) $-\frac{13}{9.9}$ or $-1\frac{31}{99}$

Lesson 2.2

➤ Key Skills

Graph a line for an equation in standard form.

To graph $x - y = 4$, first find the *x*- and *y*-intercepts.

Substitute 0 for *x* to find the *y*-intercept.

$$x - y = 4$$
$$0 - y = 4$$
$$-y = 4$$
$$(-1)(-y) = (-1)(4)$$
$$y = -4$$

The *y*-intercept is (0, −4).

Substitute 0 for *y* to find the *x*-intercept.

$$x - y = 4$$
$$x - 0 = 4$$
$$x = 4$$

The *x*-intercept is (4, 0).

Draw a line through the points (4, 0) and (0, −4) to graph the line.

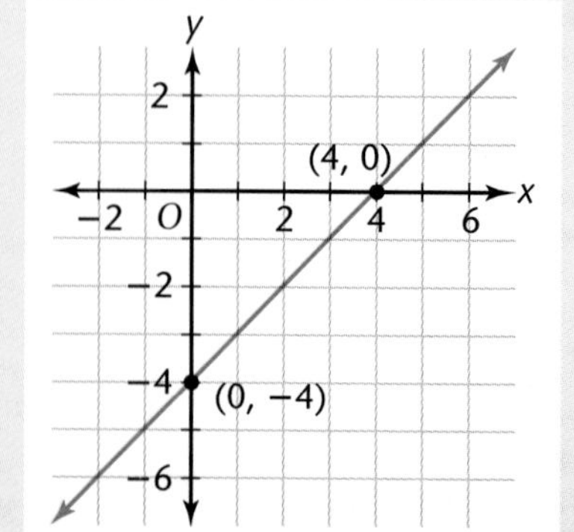

Graph a line for an equation in slope-intercept form.

The slope-intercept form of a line is $y = mx + b$, where m is the slope and $(0, b)$ is the y-intercept.

To graph $y = \frac{1}{2}x + 2$, first plot the y-intercept $(0, 2)$. Then use the slope, $\frac{\text{rise}}{\text{run}} = \frac{1}{2}$, to find a second point.

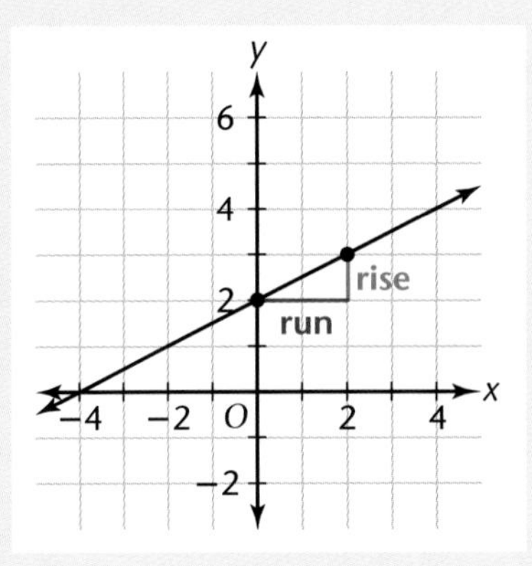

➤ *Exercises*

Use the slope and *y*-intercept or the *x*- and *y*-intercepts to draw a graph of each line.

4. $y = 4x + 2$ **5.** $x + y = 6$ **6.** $3x - 4y = 12$ **7.** $y = -3x + 1$

Lesson 2.3

➤ *Key Skills*

Find the solution to a system of linear equations by graphing the equations and finding the point of intersection.

Solve by graphing. $\begin{cases} x + 3y = 12 \\ x - y = 4 \end{cases}$

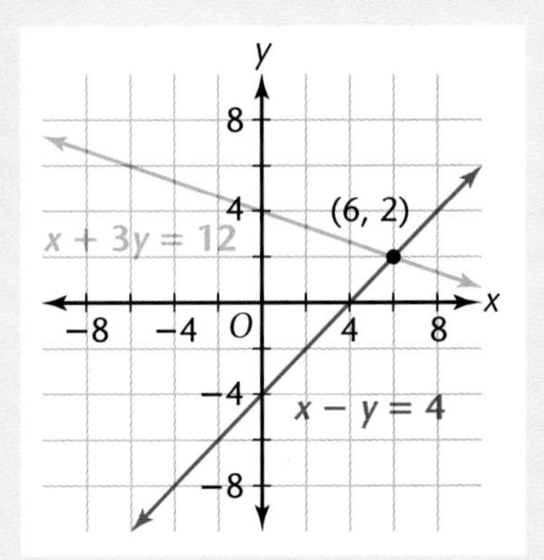

Write the equations in slope-intercept form.

$$\begin{aligned} x + 3y &= 12 \\ 3y &= -x + 12 \\ y &= -\frac{1}{3}x + 4 \end{aligned} \qquad \begin{aligned} x - y &= 4 \\ -y &= -x + 4 \\ y &= x - 4 \end{aligned}$$

Graph the lines with a graphics calculator or using graph paper. Find the point of intersection.

The solution is $(6, 2)$. Check by substituting 6 for x and 2 for y in both of the original equations.

➤ *Exercises*

Solve each system of linear equations by graphing.

8. $\begin{cases} x - y = 6 \\ x + 2y = -9 \end{cases}$ **9.** $\begin{cases} y = -x \\ y = 2x \end{cases}$ **10.** $\begin{cases} x + 6y = 3 \\ 3x + y = -8 \end{cases}$ **11.** $\begin{cases} 3x + y = 6 \\ y + 2 = x \end{cases}$

Lesson 2.4

➤ Key Skills

Find the solution to a system of linear equations by substitution.

Solve. $\begin{cases} x + 4y = 1 \\ 2x + 7y = 6 \end{cases}$

Solve $x + 4y = 1$ for x to get $x = 1 - 4y$. Then substitute the expression $1 - 4y$ for x in the equation $2x + 7y = 6$, and solve for y.

$$\begin{aligned} 2x + 7y &= 6 \\ 2(1 - 4y) + 7y &= 6 \\ 2 - 8y + 7y &= 6 \\ 2 - y &= 6 \\ -y &= 4 \\ y &= -4 \end{aligned}$$

Now substitute -4 for y in one of the original equations, and solve for x.

$$\begin{aligned} x + 4y &= 1 \\ x + 4(-4) &= 1 \\ x - 16 &= 1 \\ x &= 17 \end{aligned}$$

The solution is $(17, -4)$.

➤ Exercises

Solve each system of linear equations by substitution.

12. $\begin{cases} 2x + y = 1 \\ x + y = 2 \end{cases}$ $(-1, 3)$

13. $\begin{cases} x - y = 6 \\ 2x - 4y = 28 \end{cases}$ $(-2, -8)$

14. $\begin{cases} x + 2y = 1 \\ 2x - 8y = -1 \end{cases}$ $\left(\frac{1}{2}, \frac{1}{4}\right)$

15. $\begin{cases} 4x = 3y + 44 \\ x + y = -3 \end{cases}$ $(5, -8)$

Lesson 2.5

➤ Key Skills

Find the solution to a system of linear equations by elimination.

Solve by elimination. $\begin{cases} x + y = -3 \\ 2x = 3y - 11 \end{cases}$

Write the equations in standard form, and compare the coefficients. $\begin{cases} x + y = -3 \\ 2x - 3y = -11 \end{cases}$

Use the Multiplication Property of Equality to get opposite coefficients for one of the variables in the equations.

$$\begin{cases} x + y = -3 \\ 2x - 3y = -11 \end{cases} \rightarrow \begin{cases} 3(x + y) = 3(-3) \\ 2x - 3y = -11 \end{cases} \rightarrow \begin{cases} 3x + 3y = -9 \\ 2x - 3y = -11 \end{cases}$$

Add the expressions on corresponding sides of the equations, and solve for the remaining variable.

$$\begin{aligned} 3x + 3y &= -9 \\ \underline{2x - 3y} &= \underline{-11} \\ 5x \qquad &= -20 \\ x &= -4 \end{aligned}$$

Substitute -4 for x in $x + y = -3$, and solve for y.

$$\begin{aligned} x + y &= -3 \\ -4 + y &= -3 \\ y &= 1 \end{aligned}$$

The solution is $(-4, 1)$.

➤ *Exercises*

Solve each system of linear equations by elimination.

16. $\begin{cases} 4x + 5y = 3 \\ 2x + 5y = -11 \end{cases}$ (7, −5)

17. $\begin{cases} 2x + 3y = -6 \\ -5x - 9y = 14 \end{cases}$ $\left(-4, \frac{2}{3}\right)$

18. $\begin{cases} 0.5x + y = 0 \\ 0.9x - 0.2y = -2 \end{cases}$ (−2, 1)

19. $\begin{cases} -2x + 4y = 12 \\ 3x - 2y = -10 \end{cases}$ (−2, 2)

Lesson 2.6

➤ *Key Skills*

Identify a system of linear equations as independent, dependent, or inconsistent.

To identify the system of linear equations $\begin{cases} 2x + y = 4 \\ 2x + y = 1 \end{cases}$ as independent, dependent, or inconsistent, first solve the system algebraically.

Use the elimination method. $\begin{cases} 2x + y = 4 \\ 2x + y = 1 \end{cases} \rightarrow \begin{cases} 2x + y = 4 \\ -1(2x + y) = -1(1) \end{cases}$

$$\begin{aligned} 2x + y &= 4 \\ \underline{-2x - y} &= \underline{-1} \\ 0 &\neq -3 \end{aligned}$$

All variables are eliminated, and you are left with two unequal numbers. There is no possible solution for this system of linear equations. The system is inconsistent.

➤ *Exercises*

Solve each system of linear equations. Identify the system as independent, dependent, or inconsistent.

20. $\begin{cases} 3x - 2y = 7 \\ 4y = -14 + 6x \end{cases}$

21. $\begin{cases} 4y = 2x + 20 \\ 3x - y = -20 \end{cases}$

22. $\begin{cases} 2x + 5y = 7 \\ 3y = 2x + 17 \end{cases}$

23. $\begin{cases} x + y = 7 \\ 28 - 2y = 2x \end{cases}$

20. Infinite number of solutions; dependent **21.** (−6, 2); independent
22. (−4, 3); independent **23.** No solution; inconsistent

Lesson 2.7

➤ *Key Skills*

Graph the solution for a system of linear inequalities.

Graph the solution set for the system.

$$\begin{cases} y < -x - 1 \\ y \leq x + 1 \end{cases}$$

Graph $y = -x - 1$ as a dashed boundary line and $y = x + 1$ as a solid boundary line. Substitute the coordinates of a point to determine which half-plane to shade for each inequality. The intersection of the solution regions for the two inequalities is the solution to the system of inequalities.

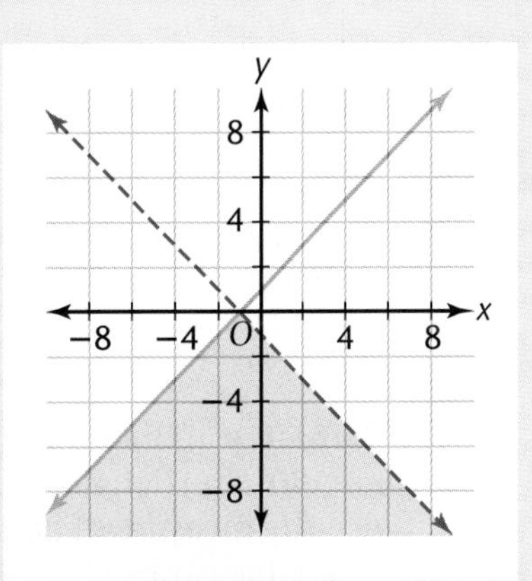

➤ *Exercises*

Graph the solution for each system of inequalities. Use substitution to check.

24. $\begin{cases} y \geq 4 \\ y \leq x + 1 \end{cases}$ **25.** $\begin{cases} 4x - y > 1 \\ 2x + y > -2 \end{cases}$ **26.** $\begin{cases} y - 1 \leq x \\ 4y - 2 \geq 2 \end{cases}$ **27.** $\begin{cases} 2x - y < 3 \\ 2y - 2 \leq 6x \end{cases}$

Lesson 2.8

➤ *Key Skills*

Use a system of equations to solve classic problems.

Investments Mr. Carver invested $5000, part at 5% interest and the rest at 3.9% interest. If he earned $233.50 in interest the first year, how much did he invest at each rate?

Make a table to organize the information.

	Amount	Rate	Interest
	x	5%	$0.05x$
	y	3.9%	$0.039y$
Total:	$x + y = 5000$		233.50

From the table, write two equations.

$$\begin{cases} x + y = 5000 \\ 0.05x + 0.039y = 233.50 \end{cases}$$

The solution to the system is (3500, 1500).

Mr. Carver invested $3500 at 5% and $1500 at 3.9%.

➤ *Exercises*

Use a system of equations to solve each problem.

28. Samantha has a total of 43 quarters and nickels in her wallet. If her change is worth $7.75, how many of each kind of coin does Samantha have? 15 nickels; 28 quarters

29. **Investments** Tina invested $600 of her bonus paycheck at 4.3% interest, and the rest at 9% interest. If she earned $97.80 in interest in one year, how much was her total bonus paycheck? $1400

Applications

30. **Chemistry** How many ounces of a 2.5% acid solution should be mixed with a 5% acid solution to produce 100 ounces of a 4% acid solution? 40 ounces

31. **Geometry** The sum of the measures of all the angles in a triangle is 180°. In triangle *XYZ*, the measure of angle *Y* is 15° more than twice the measure of angle *Z*. The measure of angle *X* is 60°. What are the measures of angles *Y* and *Z*? 85° and 35°

Chapter 2 Assessment

Solve each system by graphing.

1. $\begin{cases} 2x = y + 1 \\ y = 3x + 2 \end{cases}$

2. $\begin{cases} 4y - 2x = -2 \\ 4y - x = 0 \end{cases}$

3. $\begin{cases} 3y - 12x = 18 \\ 3x + y = -1 \end{cases}$

Solve each system by substitution.

4. $\begin{cases} y = x + 3 \\ 2x - 4y = -12 \end{cases}$ (0, 3)

5. $\begin{cases} 2x - 2y = -6 \\ x + 2y = -9 \end{cases}$ (−5, −2)

6. $\begin{cases} 2x - 3y = -2 \\ 4x + y = 1 \end{cases}$ $\left(\frac{1}{14}, \frac{5}{7}\right)$

7. Graph the line that has a slope of $-\frac{3}{4}$ and passes through the point (−3, 5).

Solve each system by elimination.

8. $\begin{cases} 2x + 3y = 3 \\ 4x - 3y = 3 \end{cases}$ $\left(1, \frac{1}{3}\right)$

9. $\begin{cases} 3x + 4y = -2 \\ x - 2y = 6 \end{cases}$ (2, −2)

10. $\begin{cases} 5y = 3x - 18 \\ 2x = 3y + 12 \end{cases}$ (6, 0)

11. The length of a rectangle is 3 inches more than its width. Find the length and width of this rectangle. 7 in. by 4 in.

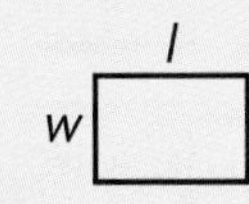

The perimeter is 22 inches.

12. Marla bought 12 books at a garage sale. Some of them were hardback and the rest were paperback. She paid 50¢ for each paperback book and 75¢ for each hardback book. If she spent $6.75, how many of each type of book did she buy? 3 hardback and 9 paperback

Solve each system algebraically. Check your solution.

13. $\begin{cases} y + 5 = x \\ y = x + 2 \end{cases}$ No solution

14. $\begin{cases} 5y - 2x = -15 \\ 4x = 10y + 30 \end{cases}$ Infinite

15. $\begin{cases} 4x + 5y = 14 \\ 5y = 8x + 2 \end{cases}$ (1, 2)

Graph the solution to each system of inequalities. Choose a point from the solution, and check both inequalities.

16. $\begin{cases} y > x \\ y - 2x \leq 2 \end{cases}$

17. $\begin{cases} y \geq 3x - 1 \\ y < 5 \end{cases}$

18. $\begin{cases} 2y + 10x > -12 \\ y + 5 < 4x \end{cases}$

Write each equation in slope-intercept form. Then state the slope, and graph the equation.

19. $x - y = 6$

20. $2x + 6y = -18$

21. $4.2x - 7y = 56$

22. Henry divides Hannah's current age by 2 and then adds that result to her current age. The result is Henry's current age. Henry notices that in 24 years, his age will be 3 times Hannah's current age. How old are Henry and Hannah? Hannah, 16; Henry, 24

23. Tamara drove 810 miles in 14 hours over 2 days. If she drove at an average rate of 55 miles per hour the first day and an average rate of 60 miles per hour the second day, how far and how long did she drive each day? 6 hours; 330 miles and 8 hours; 480 miles

Chapters 1 – 2 Cumulative Assessment

College Entrance Exam Practice

Quantitative Comparison For Questions 1–4, write
A if the quantity in Column A is greater than the quantity in Column B;
B if the quantity in Column B is greater than the quantity in Column A;
C if the two quantities are equal; or
D if the relationship cannot be determined from the information given.

	Column A	Column B	Answers
1. C	The solution to $-13x = 5.21$	The solution to $13x = -5.21$	(A) (B) (C) (D) **[Lesson 1.2]**
2. A	The slope of the line passing through the points (0, 0) and (3, 7)	The slope of the line passing through the points (−3, 5) and (0, −2)	(A) (B) (C) (D) **[Lesson 2.1]**
3. A	$\begin{cases} 3x + 2y = 6 \\ x + y = 0 \end{cases}$ x	y	(A) (B) (C) (D) **[Lesson 2.4]**
4. B	The number of solutions to the system $\begin{cases} 3x - 5y = 2 \\ 9x - 9 = 15y \end{cases}$	The number of solutions to the system $\begin{cases} 3x + 5 = 5y \\ 3x + 5y = 2 \end{cases}$	(A) (B) (C) (D) **[Lesson 2.6]**

5. Which is a solution to the equation $x - 9 = -9$? **[Lesson 1.1]**
a **a.** 0 **b.** 18 **c.** −18 **d.** none of these

6. Which is a solution to the equation $1.8x - 2.7 = 3$? **[Lesson 1.3]**
b **a.** $\frac{50}{3}$ **b.** $\frac{19}{6}$ **c.** $\frac{1}{6}$ **d.** $\frac{5}{3}$

7. Which is equivalent to the inequality $-\frac{3}{4}x + 4 \leq -10$? **[Lesson 1.5]**
d **a.** $x \leq -\frac{56}{3}$ **b.** $x \leq \frac{56}{3}$ **c.** $x \geq -\frac{56}{3}$ **d.** none of these

8. Which is a solution to the system $\begin{cases} 2x + 3y = 9 \\ x - 4y = -23 \end{cases}$? **[Lessons 2.3, 2.4, 2.5]**
b **a.** (3, 5) **b.** (−3, 5) **c.** (3, −5) **d.** (−3, −5)

9. A vending machine accepts only nickels and quarters. When the machine was emptied, 325 coins were counted with a total value of \$76.25. Which system can be used to find the number of nickels and the number of quarters? **[Lesson 2.8]**
b

a. $\begin{cases} n + q = 76.25 \\ 5n + 25q = 325 \end{cases}$ **b.** $\begin{cases} n + q = 325 \\ 5n + 25q = 7625 \end{cases}$

c. $\begin{cases} n + q = 325 \\ 5n + 25q = 76.25 \end{cases}$ **d.** none of these

10. Solve and check $y + 10.6 = 4.4$. **[Lesson 1.1]**
$y = -6.2$

Solve each equation or inequality.

11. $\frac{p}{4.1} + 6 = -8.2$ **[Lesson 1.3]** -58.22

12. $4(x + 7) + 2x = 22$ **[Lesson 1.4]** -1

13. $11.4x + 6.3 > -16.5$ **[Lesson 1.5]** $x > -2$

14. $|2z - 5| \leq 7$ **[Lesson 1.6]** $-1 \leq z \leq 6$

15. Write an equation for the line with a slope of 0 that passes through the point $(-5, 7)$. **[Lesson 2.2]**
$y = 7$

Solve each system. [Lessons 2.3, 2.4, 2.5]

16. $\begin{cases} y = 7x \\ 2x - y = -10 \end{cases}$ $(2, 14)$

17. $\begin{cases} 5x + 2y = 8 \\ x + y = 1 \end{cases}$ $(2, -1)$

18. Identify the system $\begin{cases} 3x - 4 = y \\ 3x - 4y = -20 \end{cases}$ as independent, dependent, or inconsistent. **[Lesson 2.6]** Independent

19. Graph the solution to the system $\begin{cases} 2x - y < 2 \\ y - x \geq 1 \end{cases}$. **[Lesson 2.7]**

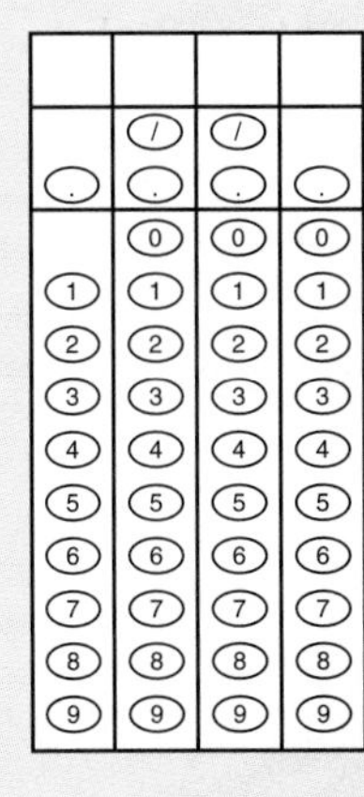

Free-Response Grid The following questions may be answered using a free-response grid commonly used by standardized test services.

20. What is the solution to $\frac{a}{14} = \frac{5}{35}$? **[Lesson 1.2]** 2

21. What is the solution to $-8x + 10 = 4\left(2x - \frac{1}{2}\right)$? **[Lesson 1.4]** $\frac{3}{4}$

22. What is the slope of the line that passes through the points $(-2, -4)$ and $(2, 10)$? **[Lesson 2.1]** $\frac{7}{2}$

23. What is the x-value of the solution to this system?
$\begin{cases} 6x - y = 7 \\ 3x + 4y = 26 \end{cases}$ **[Lesson 2.5]** 2

24. The sum of the digits of a two-digit number is 12. When the digits are reversed and the resulting number is subtracted from the original number, a difference of 18 results. What is the original number? **[Lesson 2.8]** 75

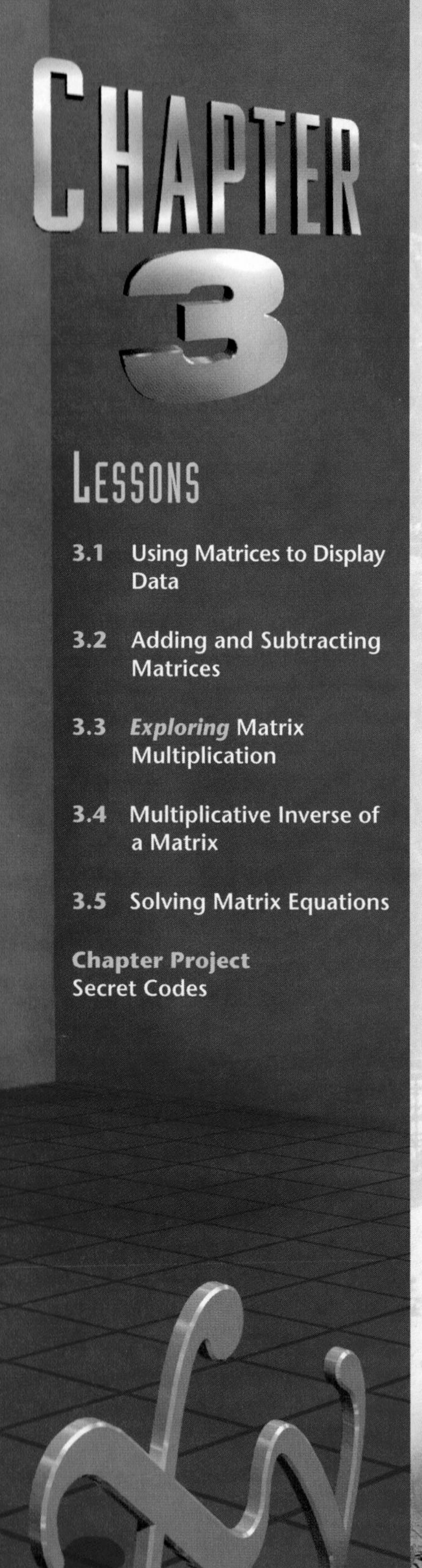

CHAPTER 3

LESSONS

Matrices

Ancient Nubians of Sudan left spectacular pyramids and tombs containing precious artifacts. They left us a puzzle as well. In 1960, the building of the Aswan High Dam on the Nile River meant that the tombs would soon be flooded. Archaeologists from all countries rushed in to excavate the sites and save the artifacts.

Millions of artifacts were recovered, representing civilizations from the Paleolithic to the Middle Ages. But what story do they tell? Researchers used a matrix to organize and classify the treasures. The rows and columns of the matrix quickly filled with numbers. These numbers represented the amount of artifacts with similar characteristics found at burial sites.

There was a problem with the size of the matrix, 512 rows and 171 columns. The matrix did an excellent job of organizing the data, but in the 1960s no computer was powerful enough to do the mathematical analysis. For years the artifacts remained unstudied in museum basements. It was not until the 1990s that a computer was available that could handle the analysis and begin to reveal the secrets of these ancient civilizations.

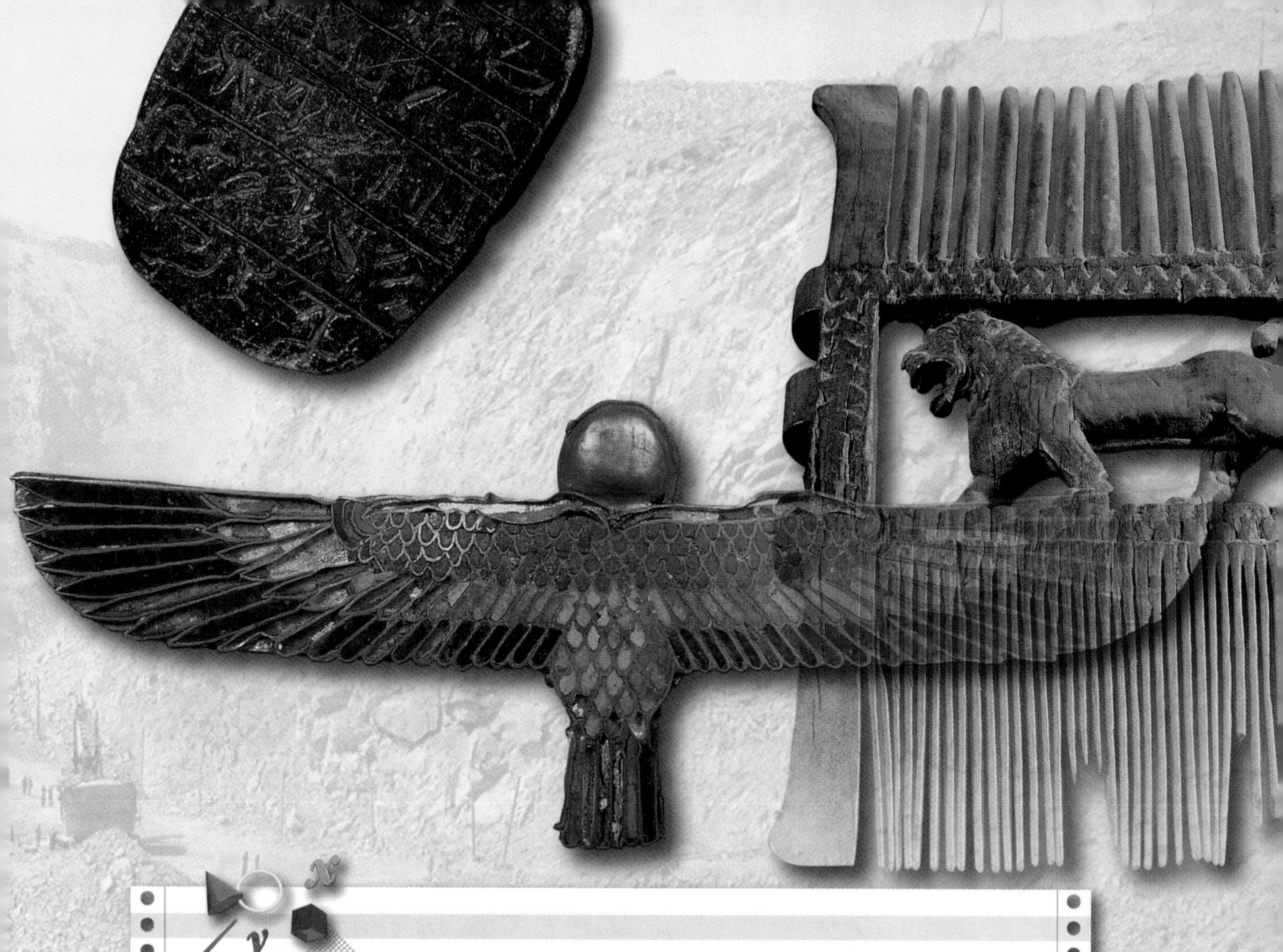

PORTFOLIO ACTIVITY

The many sites on the Aswan High Dam flood plain were divided into three different culture groups: C-Group, Pan-grave, and Transitional. Excitement mounted when excavators realized that some early C-Group sites were royal burial sites. The kings buried at these sites may have been the first pharaohs. Thousands of artifacts needed to be classified. The initial categories were tomb materials, pottery, and nonceramic pieces.

The typical C-Group site might contain 5 pieces of tomb materials, 8 pieces of decorated pottery, and 15 nonceramic artifacts. These nonceramic pieces were often decorative hair clasps, the most identifying feature of the C-Group sites. The average Pan-grave site contained 9 tomb materials, 5 pottery pieces, and 12 nonceramic artifacts. The Pan-grave sites were the only places where nonceramic Kohl pots were found. At the Transitional sites, an average of 18 tomb materials, 3 bits of pottery, and 3 nonceramic artifacts were found. The Transitional sites seem to be characterized by burial rituals.

At the end of this phase of digging, the total number of artifacts in each of the categories was 175 tomb materials, 104 pottery pieces, and 195 nonceramic artifacts.

Organize this information into matrix form, and find how many of each type of site were included in this part of the study.

LESSON 3.1 Using Matrices to Display Data

Why *Writing a research paper requires collecting data. The data that is gathered must be organized. You have already seen how tables of values, graphs, and spreadsheets are used in problem solving. Now you will use a closely related and very important organizational tool called a matrix.*

Organizing Data

Imagine that you are writing a paper on the effect that automobiles have had on society. You want to include the following data as evidence:

Transportation

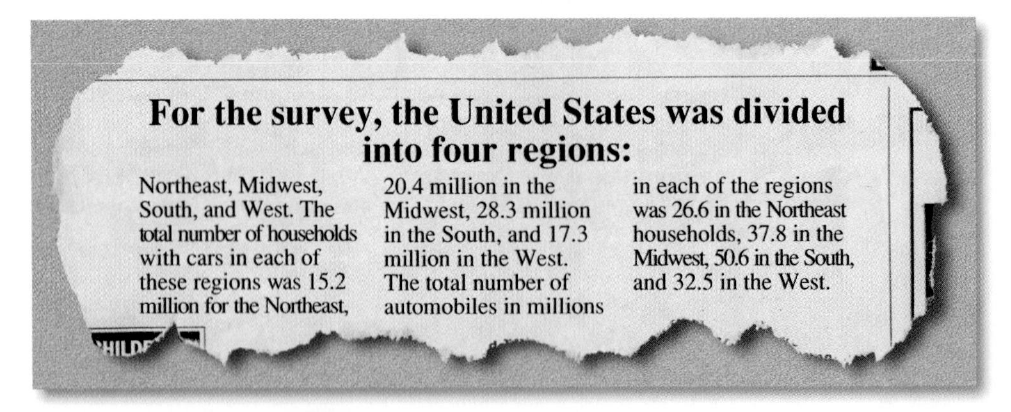

For the survey, the United States was divided into four regions:

Northeast, Midwest, South, and West. The total number of households with cars in each of these regions was 15.2 million for the Northeast, 20.4 million in the Midwest, 28.3 million in the South, and 17.3 million in the West. The total number of automobiles in millions in each of the regions was 26.6 in the Northeast households, 37.8 in the Midwest, 50.6 in the South, and 32.5 in the West.

The amount of information in the paragraph is difficult to interpret. However, when organized in a table, the same information is easier to read and understand.

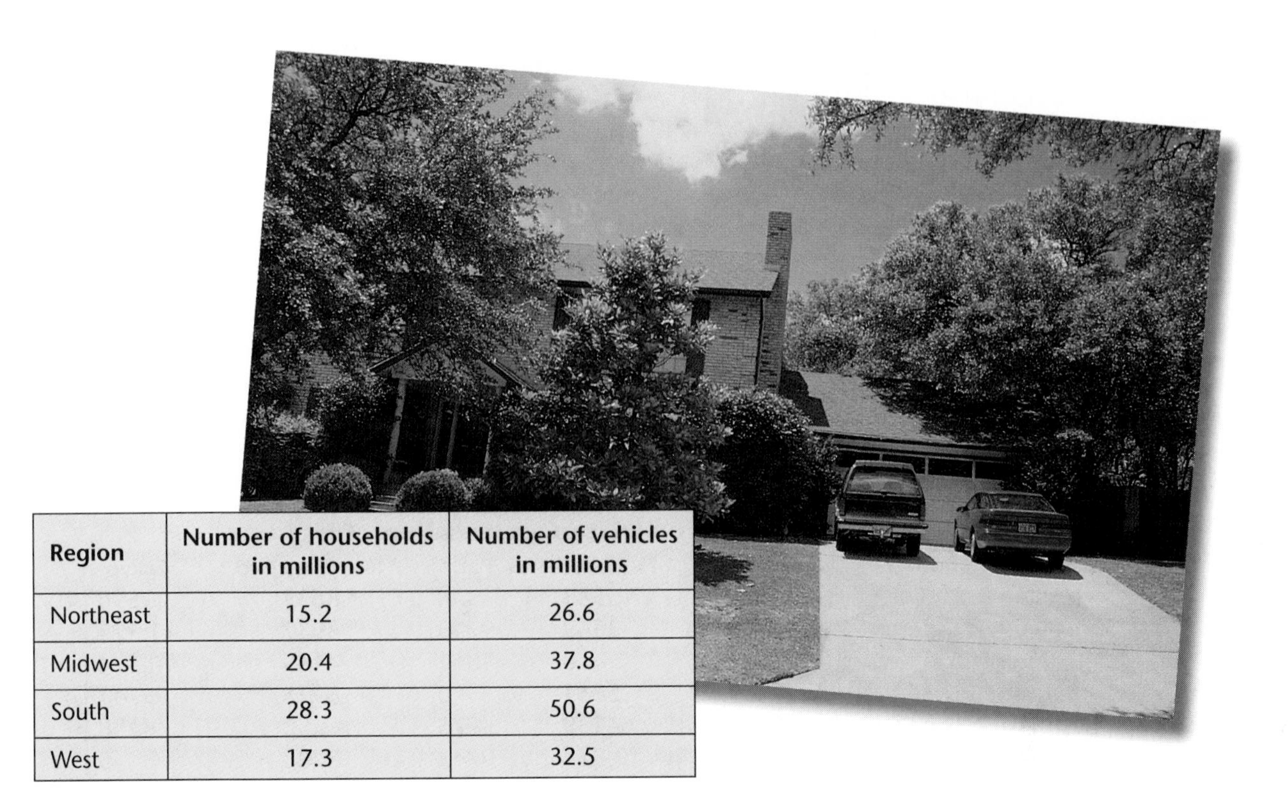

Region	Number of households in millions	Number of vehicles in millions
Northeast	15.2	26.6
Midwest	20.4	37.8
South	28.3	50.6
West	17.3	32.5

$$C = \begin{bmatrix} 15.2 & 26.6 \\ 20.4 & 37.8 \\ 28.3 & 50.6 \\ 17.3 & 32.5 \end{bmatrix}$$

A matrix is usually named with a capital letter.

When data are arranged in a table of rows and columns and enclosed by brackets, [ ], the arrangement is called a **matrix** (plural, *matrices*).

Labels describe the information in the rows and columns but are not part of the matrix.

Matrix C, shown below, has 4 rows and 2 columns. The number of rows by the number of columns describes the *row-by-column* dimensions of the matrix. This is sometimes written as 4×2 or $C_{4\times2}$.

The information that appears in each position of the matrix is called an **entry**. Each entry in the matrix can be located by its matrix **address**.

$$\begin{array}{r c c} & \text{Number of households in millions} & \text{Number of vehicles in millions} \\ \text{Northeast} & \mathbf{15.2} & \mathbf{26.6} \\ \text{Midwest} & \mathbf{20.4} & \mathbf{37.8} \\ \text{South} & \mathbf{28.3} & \mathbf{50.6} \\ \text{West} & \mathbf{17.3} & \mathbf{32.5} \end{array} = C$$

Addresses for matrix C

$$\begin{array}{r c c} & \text{Column } 1 & \text{Column } 2 \\ \text{Row } 1 & c_{11} & c_{12} \\ \text{Row } 2 & c_{21} & c_{22} \\ \text{Row } 3 & c_{31} & c_{32} \\ \text{Row } 4 & c_{41} & c_{42} \end{array}$$

Matrix C has 4 rows and 2 columns. The dimensions are 4 × 2.

The address c_{32} represents the entry in row 3 and column 2 of matrix C. You will find 50.6 at this address. The labels show that c_{32} represents the fact that in the *South* there are *50.6 million vehicles*.

What is the address in matrix C for the entry 37.8? c_{22}

Paris

Mexico City

EXAMPLE 1

Information in an atlas is in table form. This is easily changed to the format of a matrix. Answer the following questions about the location of the information in the matrix.

Airline Distances (in kilometers)

	London	Mexico City	New York	Paris	San Francisco	Tokyo
London	0	8944	5583	344	8637	9590
Mexico City	8944	0	3363	9213	3037	11,321
New York	5583	3363	0	5851	4139	10,874
Paris	344	9213	5851	0	8975	9741
San Francisco	8637	3037	4139	8975	0	8288
Tokyo	9590	11,321	10,874	9741	8288	0

$= D$

A What do the entries in matrix D represent?

B What are the row and column dimensions of matrix D?

C How many kilometers is it from Paris to Mexico City? What is the matrix address of this entry?

D What does the entry d_{36} represent?

E What does the entry d_{22} represent?

Solution ➤

A Each entry in the matrix represents the distance between the cities whose names appear in the corresponding row and column headings.

B Matrix D has dimensions of 6 by 6. A matrix is called a **square matrix** if its row and column dimensions are equal.

C It is 9213 kilometers between Paris and Mexico City. This information is located at d_{42} and d_{24}.

D The entry d_{36} is 10,874. It represents the number of kilometers between New York and Tokyo.

E The entry d_{22} is 0, and it shows that there are 0 kilometers between Mexico City and Mexico City. ❖

Try This What is the entry d_{35}? What are the addresses for the entries with the value 3037? 4139 km; d_{52} or d_{25}

Identify three reasons why matrices are a good way to organize data.

CT– Answers may vary. Data values can be organized and placed in groups and still be viewed as a whole. Specific data values can be easily identified. Relationships between data values are easier to determine because related data items are near to each other.

Matrix Equality

Two matrices are equal when their dimensions are the same and their corresponding entries are equal.

$$\boldsymbol{A} = \begin{bmatrix} 1 & 2 & 3 \\ 4 & 5 & 6 \end{bmatrix} \quad \boldsymbol{B} = \begin{bmatrix} 1 & 3 & 5 \\ 2 & 4 & 6 \end{bmatrix} \quad \boldsymbol{C} = \begin{bmatrix} 1 & 4 \\ 2 & 5 \\ 3 & 6 \end{bmatrix}$$

Matrix A and matrix B have the same dimensions, but unequal entries, so $A \neq B$. Matrix B and matrix C do not have the same dimensions, so $B \neq C$.

EXAMPLE 2

Does $S = T$? Explain why or why not.

$$S = \begin{bmatrix} 2^2 & (12 - 3) \\ \sqrt{49} & 11 \\ 3(-9) & \frac{24}{3} \end{bmatrix} \quad T = \begin{bmatrix} 2(2) & \sqrt{81} \\ 7 & \frac{44}{4} \\ -(30 - 3) & 2^3 \end{bmatrix}$$

Solution ➤

Recall the characteristics that make two matrices equal.

a. The matrices have the same dimensions. In this example, both matrices have dimensions 3 by 2.

b. The matrices have equal entries. Each matrix simplifies to equal the matrix at the right.

$$\begin{bmatrix} 4 & 9 \\ 7 & 11 \\ -27 & 8 \end{bmatrix}$$

Since the dimensions and corresponding entries of the matrices are equal, $S = T$. ❖

EXAMPLE 3

Matrices U and V are equal. Find the values of x, y, and z.

$$U = \begin{bmatrix} 3x + 2 & x + 2 \\ 2xy & 3y \end{bmatrix} \quad V = \begin{bmatrix} 14 & 2y \\ 3z & -4y + 21 \end{bmatrix}$$

Solution ➤

Since $U = V$, the corresponding entries are equal. You can form individual equations from the corresponding entries.

$$3x + 2 = 14 \qquad x + 2 = 2y$$
$$2xy = 3z \qquad 3y = -4y + 21$$

When you solve $3x + 2 = 14$, you find that $\boldsymbol{x} = \mathbf{4}$. You can then substitute **4** for $\boldsymbol{x}$ in any of the other equations.

If you substitute 4 for x in the equation $x + 2 = 2y$ and solve, you get $4 + 2 = 2y$, or $\boldsymbol{y} = \mathbf{3}$. You can then substitute **3** for $\boldsymbol{y}$ in the equation $3y = -4y + 21$ to check the y-value. ❖

How can you find the value for z? Substitute 4 for x and 3 for y into $2xy = 3z$; $z = 8$.

EXERCISES & PROBLEMS

Communicate

1. The paragraph contains information that can be more clearly represented in matrix form. Explain how to create two matrices to organize this data. Place descriptive labels in the left and top margins to identify the rows and columns.

> The retail business relies on data to predict the buying habits of the public. During the years 1985, 1989, and 1993, the Retailers' News records show that the sales of men's suits were 14.6, 10.8, and 11.5 units, respectively. For all apparel, a sales unit represents 1000 items. For the same years, the sales of women's suits were 17.4, 12.3, and 8.6 units. Shirt sales for women were 25.6, 21.3, and 16.2 units, and jeans sales for women were 98.2, 90.1, and 80.3 units. Men's jeans sales were 242.7, 210.5, and 186.9 units. Shirt sales of 16.7, 16.2, and 17.3 units were also posted for men.
>
> ...nd knowing last years numbers. ...xt season.

2. Discuss how to determine the dimensions of each matrix below. Select an entry of your choice from each matrix, and give its address.

$$P = \begin{bmatrix} -11 & 2 \\ -10 & 6 \\ -9 & 4 \\ -8 & 6 \\ -7 & 2 \end{bmatrix} \quad Q = \begin{bmatrix} -5 & 2 \\ -3 & 6 \\ -1 & 2 \\ -2 & 4 \\ -4 & 4 \end{bmatrix} \quad R = \begin{bmatrix} 3 & 2 \\ 3 & 6 \\ 1 & 6 \\ 5 & 6 \end{bmatrix} \quad T = \begin{bmatrix} 7 & 2 \\ 7 & 6 \\ 7 & 4 \\ 9 & 4 \\ 9 & 6 \\ 9 & 2 \end{bmatrix}$$

(columns of each matrix labeled x and y)

2. Answers may vary. Count the number of rows and columns. $P_{5\times 2}$; $Q_{5\times 2}$; $R_{4\times 2}$; $T_{6\times 2}$; $-9 = p_{31}$; $6 = q_{22}$; $5 = r_{41}$; $2 = t_{12} = t_{62}$

3. Discuss how you would tell which matrix or matrices are equal to matrix G.

$$G = \begin{bmatrix} |-5| & \sqrt{4} \\ -(5)^2 & (-2)^3 \end{bmatrix}$$

$$A = \begin{bmatrix} 5 & 2 \\ -25 & -8 \end{bmatrix} \quad B = \begin{bmatrix} -5 & 2 \\ 25 & -8 \end{bmatrix} \quad C = \begin{bmatrix} 8-3 & (\sqrt{2})^2 \\ 20-45 & -20+12 \end{bmatrix} \quad D = \begin{bmatrix} |5| & -2 \\ -\frac{100}{4} & -|8| \end{bmatrix}$$

Simplify entries in each matrix to determine if corresponding addresses contain equal entries. Simplify matrix G. $G = \begin{bmatrix} 5 & 2 \\ -25 & -8 \end{bmatrix}$

Matrix A and matrix C are equal to matrix G.

Practice & Apply

4. Matrices M and N are equal. Find the values of a, b, c, d, g, and k.

$$M = \begin{bmatrix} 2(a+4) & 77 & \frac{1}{3}c \\ -5d-1 & 30 & -\frac{1}{2}k \end{bmatrix} \quad N = \begin{bmatrix} -12 & 11b & 5 \\ -(3-d) & 0.4g & \frac{3}{4}k-3 \end{bmatrix}$$

$a = -10$; $b = 7$; $c = 15$; $d = \frac{1}{3}$; $g = 75$; $k = \frac{12}{5}$ or $2\frac{2}{5}$

Business The Ace Auto Repair shop keeps a record of employee absences. Use this record for Exercises 5–8 on the next page.

	M	T	W	Th	F
Managers	2	1	0	1	4
Mechanics	2	0	2	1	0
Secretaries	3	2	1	0	1

$= A$

5. What is the entry in the address a_{24} of the auto-repair-shop matrix? 1
What does this entry represent? 1 mechanic absent on Thursday

6. On which day were the most employees absent? Monday

7. Which group of employees had the least number of absences? mechanics

8. If there is a total of 30 employees at Ace Auto Repair, what percent of the employees were absent on Friday? 16.7%

9. **Travel** Four cities, A, B, C, and D, and the air routes between them are represented by the diagram. The arrows show the only directions of travel allowed. From the diagram you can see there are 2 routes from A to B. In the matrix there is the number 2 for the entry AB. The same number is shown for BC. Complete the given matrix to show the number of allowable air routes between the cities. Use 0 if there is no direct path between cities.

C

B D

A

To

From	A	B	C	D
A	0	2	?	?
B	?	0	2	?
C	?	?	?	?
D	?	?	?	?

4; 8
4; 4
2; 5; 0; 2
1; 3; 5; 0

Date	Shuttle	Date	Shuttle
9/88	Discovery	10/90	Discovery
12/88	Atlantis	11/90	Atlantis
3/89	Discovery	12/90	Columbia
5/89	Atlantis	3/91	Discovery
8/89	Columbia	4/91	Atlantis
10/89	Atlantis	5/91	Columbia
11/89	Discovery	7/91	Discovery
1/90	Columbia	8/91	Atlantis
2/90	Atlantis	11/91	Discovery
4/90	Discovery	12/91	Atlantis

Space Exploration The table shows the space shuttle names and launch dates for the years 1988–1991.

10. Form matrix L to display the number of shuttle launches for each year. Begin with 1988, and label the rows by year. Label the columns with the shuttle names in alphabetical order.

11. What are the dimensions of your matrix? 4×3

12. What does the matrix address l_{42} represent in matrix L?
1 launch of Columbia in 1991

13. Create another column to represent total launches for the year. In which year did the most launches occur? 1991

14. Over the 4-year period, which year had the least number of launches? 1988
What historical event resulted in so few launches that year? The launching of *Discovery* on Sept. 29, 1988 was the first since the *Challenger* explosion Jan. 28, 1986.

Geography The population of Chula Vista, California, and Abilene, Texas, for the years 1980 and 1990 is represented in the matrix.

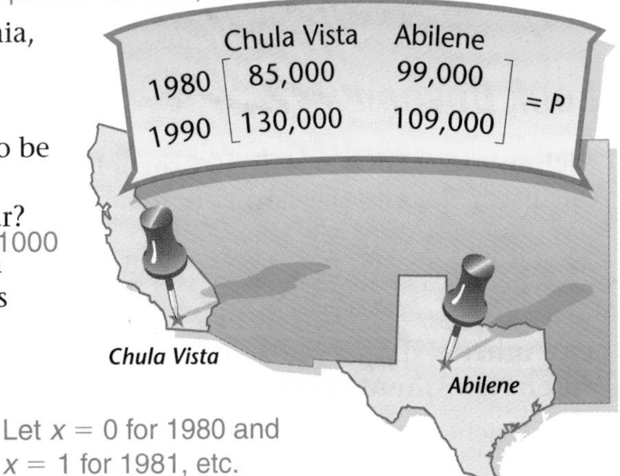

15. If the amount of growth per year was assumed to be constant for both cities over this 10-year period, what was each city's population increase per year?
Chula Vista: 4500; Abilene: 1000

16. Using this assumed population increase for each city, create another matrix that shows each city's population for **each** year from 1980 to 1990.

17. In what year were the populations equal? 1984

18. Write equations that could be used to predict the population of each city. Graphically or algebraically predict the cities' populations in the year 2000.
Let $x = 0$ for 1980 and $x = 1$ for 1981, etc.
Chula Vista: $P = 85{,}000 + 4500x$;
Abilene: $P = 99{,}000 + 1000x$; 175,000; 119,000

Look Back

19. Is the sequence 55, 65, 75, 85, . . . linear or exponential? **[Course 1]**
Linear

Match the name of the property with the example that best illustrates it. [Lessons 1.1–1.4]

	Examples		Properties
20.	$5 + 0 = 5$ c	**a.**	Property of Opposites
21.	$-8 + 8 = 0$ a	**b.**	Associative for Addition
22.	$4 + 9 = 9 + 4$ d	**c.**	Addition Property of Zero
23.	$12 + 18 = 6(2 + 3)$ e	**d.**	Commutative for Addition
24.	$7 + (11 + 4) = (7 + 11) + 4$ b	**e.**	Distributive Property

25. At the end-of-season sale at a department store, Ms. Sanchez bought a suit for 40% off the original price. She paid $96. What was the original price? **[Lesson 1.3]** $160

26. Cultural Connection: Americas The earliest math texts in the Americas were in Spanish. Solve this problem from Guatemala written by D. Juan Joseph de Padilla in 1732. **[Lesson 1.4]**
10; 20; 90

Find 3 numbers whose sum is 120.
The second number is twice the first.
The third number is 3 times the sum
of the other two.

27. Solve each inequality, and graph the solution on a number line. **[Lesson 1.5]**

$$3x - (5 - 2x) \leq x - 21 \quad \textit{and} \quad -7x - 12 \leq 16$$

28. Graph the equation $3x + 4y = 12$. **[Lesson 2.2]**

Look Beyond

Technology Enter and display the following matrices on a graphics calculator. Refer to the calculator manual.

29. $M = \begin{bmatrix} 1 & -5 & 7 \\ 0.5 & 0 & -12 \end{bmatrix}$ **30.** $N = \begin{bmatrix} 4 & 0 \\ 0 & -1 \end{bmatrix}$ **29–30.** Answers may vary.

The matrix displays the number of people who helped with an environmental clean-up project in 3 areas of a city.

	Area 1	Area 2	Area 3
Week 1	25	35	50
Week 2	15	40	32
Week 3	18	42	45
Week 4	21	53	46

31. Which area had the greatest number of helpers in week 1? Area 3

32. How many helpers were in areas 2 and 3 in week 3? 87

33. What was the total number of helpers in week 2? 87

LESSON 3.2 Adding and Subtracting Matrices

Why *A matrix is an effective tool for storing data in a way that makes sense visually. What if the data changes? If you keep getting new data tables, you will need a way to combine the information in the old and new matrices without disturbing the organization. For this you need to know how to add and subtract matrices.*

Your last look at the stock inventory for the Super Sight and Sound Shop was on April 1. Since that time, business has been good, and much of the stock has been sold. The matrices below represent the *stock* available on April 1 and the *sales* for April.

Stock [April 1]:

Videos	DATs
Games	CDs
Laser discs	Tapes

$$A = \begin{bmatrix} 3000 & 1000 \\ 800 & 3000 \\ 1500 & 1000 \end{bmatrix}$$

Sales [April]:

Videos	DATs
Games	CDs
Laser discs	Tapes

$$B = \begin{bmatrix} 2254 & 952 \\ 675 & 1325 \\ 1187 & 548 \end{bmatrix}$$

EXAMPLE 1

Inventory

The sales matrix shows how many of each item were sold in April. Since the original stock has decreased, how would you determine the available stock for May?

Solution ➤

Subtract each entry in B from the corresponding entry in A. The result will be a new matrix, C, that shows the stock available May 1. For example, the video stock available **May 1** is **3000 − 2254**, or **746**. Place **746** in address c_{11}.

$$\begin{array}{ccccc} A & - & B & = & C \\ \textbf{Stock [April 1]} & - & \textbf{Sales [April]} & = & \textbf{Stock [May 1]} \end{array}$$

$$\begin{bmatrix} \mathbf{3000} & 1000 \\ 800 & 3000 \\ 1500 & 1000 \end{bmatrix} - \begin{bmatrix} \mathbf{2254} & 952 \\ 675 & 1325 \\ 1187 & 548 \end{bmatrix} = \begin{bmatrix} \mathbf{746} & ? \\ ? & ? \\ ? & ? \end{bmatrix}$$

Matrix C is completed by subtracting corresponding entries.

Stock [May 1]:

Videos	DATs
Games	CDs
Laser discs	Tapes

$$C = \begin{bmatrix} \mathbf{746} & 48 \\ 125 & 1675 \\ 313 & 452 \end{bmatrix}$$ ❖

EXAMPLE 2

On May 2, a delivery arrives with more stock. The shipment includes an invoice, matrix D, to let the store know what is being delivered.

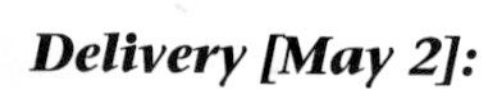

Delivery [May 2]:

Videos	DATs
Games	CDs
Laser discs	Tapes

$$D = \begin{bmatrix} 2500 & 1500 \\ 625 & 775 \\ 1500 & 400 \end{bmatrix}$$

How would you update your stock matrix to show the increase in stock?

Solution ➤

If you add each entry in C to the corresponding entry in D, you will create a new stock matrix, E, for the available stock for May.

$$\begin{array}{ccccc} C & + & D & = & E \\ \textbf{Stock [May 1]} & + & \textbf{Delivery [May 2]} & = & \textbf{Stock [May 2]} \end{array}$$

For example, video stock in the new matrix ***Stock [May 2]*** is 746 + 2500, or 3246, at address e_{11}.

Stock [May 2]:

Videos	DATs
Games	CDs
Laser discs	Tapes

$$E = \begin{bmatrix} 3246 & 1548 \\ 750 & 2450 \\ 1813 & 852 \end{bmatrix}$$ ❖

Try This Let $X = \begin{bmatrix} 5 & -3 \\ 2 & 0 \end{bmatrix}$ and $Y = \begin{bmatrix} -1 & 7 \\ 0 & 3 \end{bmatrix}$. Find $X + Y$. $\begin{bmatrix} 4 & 4 \\ 2 & 3 \end{bmatrix}$

In a large business, how often does this process of updating the stock matrices take place? If there are thousands of different items in the stock of a business, how does a store manage its inventory and update its matrices? Sample answers could be daily or weekly. The manager might group items into categories and keep a matrix for each category.

Exploration Addition and Subtraction Properties

1a,b. $\begin{bmatrix} \frac{27}{4} & -13 \\ 5 & \frac{5}{2} \end{bmatrix}$

1d,e. $\begin{bmatrix} \frac{51}{20} & -12 \\ 5 & -\frac{41}{10} \end{bmatrix}$

Use the following matrices to explore and develop conjectures about the algebra of adding and subtracting matrices. This is a good opportunity to use a calculator or computer.

$$A = \begin{bmatrix} 6 & -3 \\ -4 & \frac{9}{2} \end{bmatrix} \quad B = \begin{bmatrix} -4.2 & 1 \\ 0 & -6.6 \end{bmatrix} \quad C = \begin{bmatrix} \frac{3}{4} & -10 \\ 9 & -2 \end{bmatrix} \quad D = \begin{bmatrix} -2 & 11 & 7 \\ 5 & 9 & -12 \end{bmatrix}$$

1 Add.
a. $A + C$ **b.** $C + A$ **c.** $B + D$
d. $A + (B + C)$ **e.** $(A + B) + C$ **f.** $C + D$

1c,f. no solution

Examine the sums, and describe the procedure for adding matrices. To add matrices, add the corresponding entries of the matrices.

2 Compare and contrast these additions. Are there any restrictions on whether the matrices can be added? Write a conjecture.

2. In order to add matrices, the matrices must have the same dimensions. The additions in **a**, **b**, **d**, and **e** can be performed because the matrices have the same dimensions. The additions in **c** and **f** cannot be performed because the matrices have different dimensions. Addition of matrices is commutative and associative.

3 Subtract.
a. $C - A$ **b.** $B - B$ **c.** $B - A$
d. $A - C$ **e.** $B - C$ **f.** $C - D$

3f. no solution

Examine the differences, and describe the procedure for subtracting matrices. To subtract matrices, subtract the corresponding entries in the matrices.

4 Compare and contrast these subtractions. Are there any restrictions on whether the matrices can be subtracted? Write a conjecture. ❖

3a. $\begin{bmatrix} -\frac{21}{4} & -7 \\ 13 & -\frac{13}{2} \end{bmatrix}$ 3b. $\begin{bmatrix} 0 & 0 \\ 0 & 0 \end{bmatrix}$

3c. $\begin{bmatrix} -\frac{51}{5} & 4 \\ 4 & -\frac{110}{10} \end{bmatrix}$

3d. $\begin{bmatrix} \frac{21}{4} & 7 \\ -13 & \frac{13}{2} \end{bmatrix}$

3e. $\begin{bmatrix} \frac{99}{20} & 11 \\ -9 & -\frac{23}{5} \end{bmatrix}$

The *identity matrix for addition* is a matrix filled with zeros. It is also called the **zero matrix**. The zero matrix for a 2-by-2 matrix is shown here.

$$\begin{bmatrix} 0 & 0 \\ 0 & 0 \end{bmatrix} = Z_{2\times2}$$

Adding the zero matrix to another matrix does not change the original matrix.

$$\begin{bmatrix} 0 & 0 \\ 0 & 0 \end{bmatrix} + \begin{bmatrix} -1 & 5 \\ 7 & 0 \end{bmatrix} = \begin{bmatrix} -1 & 5 \\ 7 & 0 \end{bmatrix}$$

$$\begin{bmatrix} -1 & 5 \\ 7 & 0 \end{bmatrix} + \begin{bmatrix} 0 & 0 \\ 0 & 0 \end{bmatrix} = \begin{bmatrix} -1 & 5 \\ 7 & 0 \end{bmatrix}$$

Consider matrices C and D from the exploration. Form the zero matrix that can be added to C. Then form the zero matrix that can be added to D.

$$? + C = \begin{bmatrix} \frac{3}{4} & -10 \\ 9 & -2 \end{bmatrix} = C + ? \qquad ? + D = \begin{bmatrix} -2 & 11 & 7 \\ 5 & 9 & -12 \end{bmatrix} = D + ?$$

(Answers: $\begin{bmatrix} 0 & 0 \\ 0 & 0 \end{bmatrix}$, $\begin{bmatrix} 0 & 0 \\ 0 & 0 \end{bmatrix}$, $\begin{bmatrix} 0 & 0 & 0 \\ 0 & 0 & 0 \end{bmatrix}$, $\begin{bmatrix} 0 & 0 & 0 \\ 0 & 0 & 0 \end{bmatrix}$)

How are the zero matrices for C and D different?

When you add numbers, you know that certain properties are true.

The *Commutative Property for Addition* states that for numbers a and b, the statement $a + b = b + a$ is true.

The *Associative Property of Addition* states that for numbers a, b, and c, the statement $a + (b + c) = (a + b) + c$ is true.

$L = \begin{bmatrix} 2 & 3 \\ 6 & -1 \end{bmatrix}$

$M = \begin{bmatrix} 4 & 5 \\ -2 & 0 \end{bmatrix}$

$N = \begin{bmatrix} -1 & 4 \\ 7 & -5 \end{bmatrix}$

Use the given matrices L, M, and N at the left to show that matrix addition is also commutative and associative.

$L + M = \begin{bmatrix} 6 & 8 \\ 4 & -1 \end{bmatrix} = M + L \rightarrow L + M = M + L$

$(L + M) + N = \begin{bmatrix} 5 & 12 \\ 11 & -6 \end{bmatrix} = L + (M + N)$
$\rightarrow (L + M) + N = L + (M + N)$

Exercises & Problems

Communicate

Explain how to perform each matrix operation. If an operation is not possible, explain why.

1. $\begin{bmatrix} 10 & 6 \\ 4 & -5 \end{bmatrix} + \begin{bmatrix} -1 & 6 \\ 9 & -7 \end{bmatrix}$

2. $\begin{bmatrix} -11 & 6 \\ 13 & 8 \\ 17 & -9 \end{bmatrix} - \begin{bmatrix} -2 & 10 \\ -16 & 12 \\ 4 & -3 \end{bmatrix}$

3. $\begin{bmatrix} \frac{5}{6} \\ \frac{7}{12} \end{bmatrix} + \begin{bmatrix} \frac{-5}{4} & \frac{2}{3} \end{bmatrix}$ These cannot be added because they have different dimensions.

4. $\begin{bmatrix} \frac{1}{2} & \frac{1}{4} & \frac{-1}{2} \\ \frac{2}{3} & \frac{4}{5} & \frac{1}{4} \end{bmatrix} + \begin{bmatrix} \frac{1}{2} & \frac{1}{4} & \frac{-1}{2} \\ \frac{2}{3} & \frac{4}{5} & \frac{1}{4} \end{bmatrix}$

5. Is matrix subtraction commutative? Explain. No, because it requires individual matrix entries, which are not commutative, to be subtracted.

6. Is matrix subtraction associative? Explain. No, because it requires individual matrix entries, which are not associative, to be subtracted.

7. Is a zero matrix always a square matrix? Explain why or why not. Not necessarily; a zero matrix is simply a matrix in which every element is equal to zero, and this matrix can have any dimensions.

Practice & Apply

Perform each of the matrix operations. If an operation is not possible, explain why.

8. $\begin{bmatrix} 23 & 46 \\ 4 & -35 \end{bmatrix} + \begin{bmatrix} -51 & 16 \\ 29 & -7 \end{bmatrix}$ $\begin{bmatrix} -28 & 62 \\ 33 & -42 \end{bmatrix}$

9. $\begin{bmatrix} 4 & 2 \\ 6 & -7 \\ 3 & 9 \end{bmatrix} + \begin{bmatrix} -5 & 7 \\ -3 & 9 \\ 3 & -6 \end{bmatrix} - \begin{bmatrix} -11 & 3 \\ 8 & -15 \\ -7 & -2 \end{bmatrix}$ $\begin{bmatrix} 10 & 6 \\ -5 & 17 \\ 13 & 5 \end{bmatrix}$

10. $\begin{bmatrix} -6 & 1 \\ -1 & 7 \end{bmatrix} - \begin{bmatrix} -3 & 0 \\ 9 & -2 \end{bmatrix}$ $\begin{bmatrix} -3 & 1 \\ -10 & 9 \end{bmatrix}$

11. $\begin{bmatrix} -1.4 & 4.3 & -9.6 \\ 15.8 & 1 & -3.5 \end{bmatrix} - \begin{bmatrix} 6.2 & 3.2 \\ -9.6 & 7.1 \\ 2.6 & -8.5 \end{bmatrix}$ Not possible, different dimensions

12. $\begin{bmatrix} 3.2 & \frac{3}{5} \\ \frac{-8}{5} & -4.2 \end{bmatrix} - \begin{bmatrix} \frac{-4}{5} & 5.6 \\ -6.7 & \frac{3}{5} \end{bmatrix}$

13. $\begin{bmatrix} 2.3 & -9.3 & 1.8 \\ -8.2 & -4.1 & 10 \\ 2.1 & 6 & -0.16 \end{bmatrix} + \begin{bmatrix} -1.2 & -2.1 & 5.3 \\ 0.2 & -4 & -2.3 \\ -3.4 & 2.4 & 6 \end{bmatrix}$

14. $\begin{bmatrix} \frac{-2}{3} & \frac{4}{5} \\ \frac{1}{15} & \frac{7}{30} \end{bmatrix} - \begin{bmatrix} \frac{9}{15} & \frac{-4}{3} \\ \frac{-2}{5} & \frac{-4}{15} \end{bmatrix}$

15. $\begin{bmatrix} 0.4 & -1.5 & 0.9 \\ 2.6 & 6.9 & 3.7 \end{bmatrix} + \begin{bmatrix} -4.7 & 2.6 & 6.9 \\ -7.3 & 9.8 & -5.5 \end{bmatrix}$

16. $\begin{bmatrix} -3 & 4 \\ -4 & 0 \\ 2 & -5 \end{bmatrix} + \begin{bmatrix} 0 & -1 \\ -6 & 9 \\ -7 & 5 \end{bmatrix}$

17. $\begin{bmatrix} 1 & 6 \\ 3 & -4 \\ -2 & 7 \end{bmatrix} + \begin{bmatrix} 8 & -3 & 11 \\ -9 & 4 & 6 \\ 1 & 0 & -5 \end{bmatrix}$ Not possible, different dimensions

18. $\begin{bmatrix} -25 & 32 & 14 \\ 36 & -42 & -45 \\ -71 & 65 & 29 \end{bmatrix} - \begin{bmatrix} 16 & -34 & -55 \\ 21 & 11 & 22 \\ -43 & -67 & -44 \end{bmatrix} + \begin{bmatrix} 57 & 79 & 64 \\ -38 & -22 & -48 \\ -56 & 88 & 26 \end{bmatrix}$ $\begin{bmatrix} 16 & 145 & 133 \\ -23 & -75 & -115 \\ -84 & 220 & 99 \end{bmatrix}$

19. Inventory At the Super Sight and Sound Shop, the May sales matrix and June 1 delivery invoice have just appeared on your desk. Make a new matrix to update the May 1 stock matrix to reflect the May sales and June 1 delivery. Label your answer matrix Stock [June 1].

Stock [June 1] = $\begin{bmatrix} 3498 & 1967 \\ 798 & 1994 \\ 1934 & 730 \end{bmatrix}$

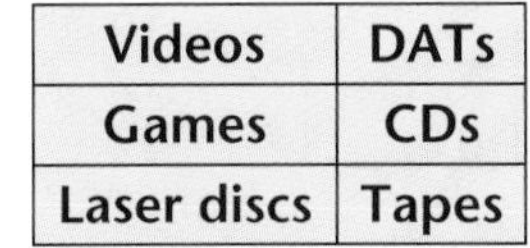

Stock [May 1]

Videos	DATs
Games	CDs
Laser discs	Tapes

$\begin{bmatrix} 3246 & 1548 \\ 750 & 2450 \\ 1813 & 852 \end{bmatrix}$

Sales [May]

Videos	DATs
Games	CDs
Laser discs	Tapes

$\begin{bmatrix} 2748 & 1081 \\ 702 & 1456 \\ 1679 & 622 \end{bmatrix}$

Delivery [June 1]

Videos	DATs
Games	CDs
Laser discs	Tapes

$\begin{bmatrix} 3000 & 1500 \\ 750 & 1000 \\ 1800 & 500 \end{bmatrix}$

Statistics In the Woodlake public school system, there are two high schools, Glenn and Kelly. The enrollment for the electives music (Mu), art (Ar), technology (Te), and health (He) in each school appears in the matrices below.

Glenn H.S.

	Mu	Ar	Te	He
Boys	447	199	514	389
Girls	498	352	432	399

Kelly H.S.

	Mu	Ar	Te	He
Boys	387	276	489	367
Girls	505	392	387	437

20. How many girls are enrolled in art at both schools combined? How many boys are taking health at both schools combined?
744 girls; 756 boys

21. Create a matrix to show the total enrollment in each elective.

22. How many students in both schools are taking technology while in high school? 1822 students

Sports Suppose that you keep the statistics for the girls' basketball team in a matrix, a portion of which is shown below. You record the number of 3-point baskets, 2-point baskets, 1-point free throws (FT), and the total points. The matrices show the statistics for the end of the regular season and the total points scored after the playoff series. Use these matrices to complete Exercises 23–25.

Points scored at end of regular season

	3-pt	2-pt	FT	Total
Waters	3	15	12	51
Riley	1	20	16	59
Sharp	3	17	13	56
Evans	4	6	9	33
Jones	0	14	18	46

Points scored after playoff series

	3-pt	2-pt	FT	Total
Waters	5	25	15	80
Riley	2	22	20	70
Sharp	5	20	15	70
Evans	5	12	12	51
Jones	0	17	22	56

23. Points Scored in Playoff Series

	Total
Waters	29
Riley	11
Sharp	14
Evans	18
Jones	10

23. Create a matrix to show how many points were scored by each player in the playoff series.

24. Which player scored the most points in the playoff series? Waters

25. What is the total number of free throws made during the playoffs? 16

26. Find the values of w, x, y, and z by solving the matrix equation $A + B = C$ for matrices A, B, and C below. $w = -2; x = 6; y = 1; z = -15$

$$A = \begin{bmatrix} -5w + 2 & 2x - 4 \\ 8y - 1 & \frac{1}{5}z \end{bmatrix} \qquad B = \begin{bmatrix} -13 & 5 - x \\ -8 & z + 9 \end{bmatrix} \qquad C = \begin{bmatrix} 2w + 3 & 7 \\ 3y - 4 & -9 \end{bmatrix}$$

27. Use the values you found for w, x, y, and z in Exercise 26 to evaluate each entry in matrices A, B, and C. Solve the matrix equation $A + B = C$ numerically to check your solutions.

27. $A = \begin{bmatrix} 12 & 8 \\ 7 & -3 \end{bmatrix}$ $B = \begin{bmatrix} -13 & -1 \\ -8 & -6 \end{bmatrix}$ $C = \begin{bmatrix} -1 & 7 \\ -1 & -9 \end{bmatrix}$

The symbol that represents the additive inverse of a matrix is $-A$. An additive inverse occurs if $-A + A = Z$ (where Z is the zero matrix, or the identity matrix for addition).

28. Is there one matrix that can be used as the identity matrix for matrices A, B, and C in Exercise 26? Explain.

28. $\begin{bmatrix} 0 & 0 \\ 0 & 0 \end{bmatrix}$ All three matrices are 2×2 matrices.

29. Describe the identity matrix for addition for the matrix below.

$$\begin{bmatrix} 1 & 4 & -3 \\ 7 & 12 & 0 \end{bmatrix}$$

29. $\begin{bmatrix} 0 & 0 & 0 \\ 0 & 0 & 0 \end{bmatrix}$

30. Form matrices $-A$, $-B$, and $-C$ from the given matrices A, B, and C in Exercise 26.

Look Back

31. $x = \frac{3}{2}y - \frac{5}{2}$ 32. $x = -\frac{3}{4}b - \frac{1}{2}$

Solve each equation for x. [Lessons 1.3, 1.4]

31. $-2x + 3y = 5$

32. $-4x - 2b = b + 2$

33. $-x + 3 = 4y$

$x = -4y + 3$

34. $4(-4x - 7) = 8$

$x = -2\frac{1}{4}$

35. **Geometry** You have both a square and a rectangle. The length of the rectangle is 3 times the side of the squares, s, and its width is twice the side of the squares, s. Write expressions for the perimeter and area of the rectangle in terms of s. **[Lesson 1.4]**

$10s$; $6s^2$

36. Graph $-3 < x$ *or* $x \leq -5$ on the number line. State the solution. **[Lesson 1.5]**

37. Write the equation $5x = 2y - 12$ in standard form and in slope-intercept form. **[Lesson 2.2]** $5x - 2y = -12$; $y = \frac{5}{2}x + 6$

38. Graph $y = |x|$ and $y = 9$. Find their common solutions. **[Lesson 2.3]**

Statistics Students collect data by measuring the number of inches around the thickest part of their thumb and then around their wrist. The data for 5 students appear in the matrix below.

	Thumb	Wrist
Samir	3.7	7.3
Betty	2.8	4.7
Ruth	3.1	6.2
Todd	3.6	7.1
Han	3.3	6.6

39. Plot the data above, and describe the correlation between thumb size and wrist size. **[Lesson 2.2]**

40. Solve the system by elimination. $\begin{cases} 3x = 12 + 4y \\ 2y - x = -5 \end{cases}$ **[Lesson 2.5]**

$\left(2, -\frac{3}{2}\right)$

Look Beyond

41. **Geometry** Matrix T represents the coordinates of the vertices of triangle ABC graphed on a coordinate plane. Matrix M represents a translation matrix.

$$T = \begin{bmatrix} 4 & 3 \\ 7 & -2 \\ 10 & 6 \end{bmatrix} \quad M = \begin{bmatrix} -7 & 2 \\ -7 & 2 \\ -7 & 2 \end{bmatrix}$$

(columns of both matrices labeled x, y)

Write the matrix equal to $T + M$. Plot the original points and the translated points on a graph.

EYEWITNESS MATH

Barely Enough GRIZZLIES?

Counting Big Bears

Government Wants Them Off the Endangered List; How Many is Enough?

By Marj Charlier

Staff Reporter of The Wall Street Journal

KALISPELL, Mont.—The solitary monarch of the wild, the grizzly rules the rugged high country. Scientists call it *Ursus arctus horribilus,* and the name suggests the terror it inspires. Fearless and unpredictable, the bear can kill with a single rake of a claw.

Man is the grizzly's only natural enemy, but the clash almost proved fatal to the species. Unrestricted hunting and development of the animal's habitat decimated the bear population over the last century. By 1975 when the grizzly was placed on the federal government's Endangered Species List, only 1,000 were thought to be alive.

Now, federal officials say the grizzly population has rebounded, though they concede there is no hard evidence to support that conclusion, and they are proposing to take the bear off the list.

A Death Sentence?

Many environmentalists, however, believe that even if the number of bears has increased, the grizzly population is still too small to strip the animal of federal protection. Such a plan, they say, would be a death sentence for the species. "If the grizzly bear is delisted, it will be strangled by development and go extinct quite promptly," says Lance Olsen, president of the nonprofit Great Bear Foundation.

Population Unknown

Complicating the issue is the difficulty of determining how many grizzlies roam the Rockies and their foothills. Everyone concedes that the 1,000 figure for 1975 was only a guess. Attempts to take a bear census in the years since then have proved difficult, and no one knows how many grizzlies there are now.

Richard Mace, a wildlife biologist for the state of Montana, knows how tough it is to count grizzlies. Displaying a battered steel box with a jagged rip down one side, he explains that the box once held a tree-mounted camera that was rigged to snap photos when it sensed body heat from a large animal. Biologists had placed some 43 cameras in a section of Montana's Swan Mountains, hoping to get a more accurate bear count.

Apparently, Mr. Mace says, when the camera flashed, the grizzly slashed. Its two-inch long claws ripped through the steel casing, which is a bit thicker than a coffee can. The bear dug out the camera and chewed it to pieces.

Cooperative Learning

Counting animals in the wild is not easy. How do you know if you have counted some more than once? How can you possibly find them all? Wildlife managers sometimes use a technique called *tag and recapture*. To see how the tag-and-recapture method works, you can try it with objects instead of animals.

You will need:

- a batch of several hundred objects (macaroni, beans, crumpled pieces of paper) in a paper bag or other container,
- a marker (or some other way to *tag*, or mark, the objects).

1. Follow steps **a–e** to estimate the number of objects in your batch.

a. Remove a handful of objects from the batch and mark them. Record how many you have marked.

b. Return the marked objects to the batch. Mix thoroughly.

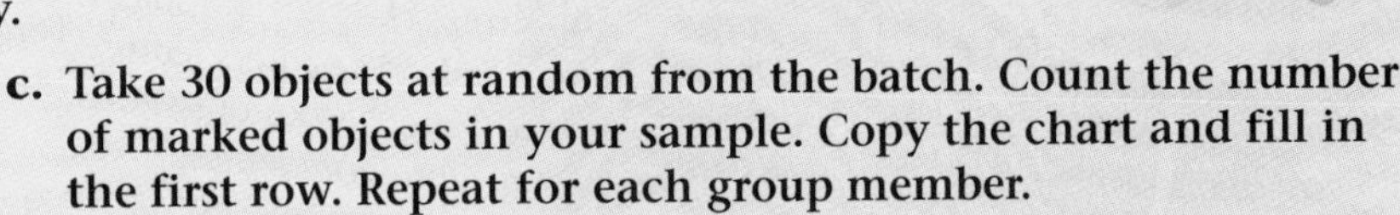

c. Take 30 objects at random from the batch. Count the number of marked objects in your sample. Copy the chart and fill in the first row. Repeat for each group member.

Sample	Total number of objects in sample (N_s)	Number of marked objects in sample (M_s)
1	30	?
2	30	?

d. How do you think the fraction of marked objects in your samples compares with the fraction of marked objects in the whole batch?

e. Use the proportion to estimate the size of the whole batch. (Average your samples for M_s and N_s.)

$$\frac{M_b}{N_b} = \frac{M_s}{N_s}$$

M_b is the number of *marked objects* in the batch.
N_b is the total *number of objects* in the batch.
M_s is the number of *marked objects* in the sample.
N_s is the *total number of objects* in the sample.

f. Check your estimate by counting. Within what percent of the actual total was your estimate?

2. Describe how the tag-and-recapture method could be used to estimate an animal population.

3. Do you think the tag-and-recapture method would work for a grizzly bear population? Why or why not?

LESSON 3.3

Exploring Matrix Multiplication

The Teen Boutique is planning a Back-to-School sale, with cotton sweaters featured. The current stock of sweaters is represented in matrix S.

Why *If the entries in a matrix are doubled or tripled, you can multiply the data in the matrix by a constant. This is called scalar multiplication, and it is related to addition. Another operation, called matrix multiplication, will enable you to find the product of two matrices.*

$$\textbf{\textit{Stock:}}\quad \begin{array}{r} \\ \textbf{Small} \\ \textbf{Medium} \\ \textbf{Large} \end{array} \begin{array}{c} \begin{array}{ccc} \textbf{Blue} & \textbf{Red} & \textbf{Black} \end{array} \\ \begin{bmatrix} 16 & 18 & 15 \\ 14 & 22 & 17 \\ 21 & 20 & 19 \end{bmatrix} \end{array} = S$$

Inventory

In anticipation of the sale, the boutique orders 3 times its present stock. How many of each item are ordered? Can you represent the order in matrix form?

If you want 3 times the stock, find the sum of 3 identical matrices.

$$\begin{array}{ccccccc} S & + & S & + & S & = & T \end{array}$$

$$\begin{bmatrix} 16 & 18 & 15 \\ 14 & 22 & 17 \\ 21 & 20 & 19 \end{bmatrix} + \begin{bmatrix} 16 & 18 & 15 \\ 14 & 22 & 17 \\ 21 & 20 & 19 \end{bmatrix} + \begin{bmatrix} 16 & 18 & 15 \\ 14 & 22 & 17 \\ 21 & 20 & 19 \end{bmatrix} = \begin{bmatrix} 48 & 54 & 45 \\ 42 & 66 & 51 \\ 63 & 60 & 57 \end{bmatrix}$$

Another way to do this is to multiply each entry by 3.

$$3S = T$$

$$3\begin{bmatrix} 16 & 18 & 15 \\ 14 & 22 & 17 \\ 21 & 20 & 19 \end{bmatrix} = \begin{bmatrix} 48 & 54 & 45 \\ 42 & 66 & 51 \\ 63 & 60 & 57 \end{bmatrix}$$

This type of multiplication is called **scalar multiplication**. The **scalar** is the number that multiplies each entry in the matrix. In this case, the scalar is 3.

Multiplying two matrices creates a **product matrix**. This matrix combines part of the information from each of the original matrices. The procedure for multiplying matrices is very different from scalar multiplication.

Exploration 1 Understanding Matrix Multiplication

Manufacturing

Each year Sam and Kim make toy cars and trucks for the fair. The matrices below show how many cars and trucks they plan to make and the number of wheels and nails they will need.

Before you can multiply the two matrices, the *column labels for the left matrix* (**Cars** and **Trucks**) must match the *row labels for the right matrix* (**Cars** and **Trucks**).

$$\begin{array}{c} \\ \text{Sam} \\ \text{Kim} \end{array} \begin{array}{c} \begin{array}{cc} \text{Cars} & \text{Trucks} \end{array} \\ \begin{bmatrix} 8 & 15 \\ 12 & 10 \end{bmatrix} \end{array} \qquad \begin{array}{c} \\ \text{Cars} \\ \text{Trucks} \end{array} \begin{array}{c} \begin{array}{cc} \text{Wheels} & \text{Nails} \end{array} \\ \begin{bmatrix} 4 & 7 \\ 6 & 9 \end{bmatrix} \end{array}$$

Sam builds 8 cars with 4 wheels each. He will need 8(4), or 32 wheels. He is also building 15 trucks with 6 wheels each. He will need 15(6), or 90 more wheels. The total number of wheels will be 8(4) + 15(6), or 122.

Place this information in the Sam-Wheels address for the product matrix.

	Wheels	**Nails**
Sam	8(4) + 15(6) 32 + 90 122	8(7) + 15(9) ? 191
Kim	12(4) + 10(6) ? 108	12(7) + 10(9) ? 174

The matching column and row labels disappear after you multiply. The product matrix will no longer show the labels for **Cars** and **Trucks**. **Sam** and **Kim** and **Wheels** and **Nails** become the labels of the new product matrix.

Answer the following questions to complete the table above.

1. How many wheels will Kim need to complete her work? 108
2. How many nails will Sam need to complete his work? 191
3. How many nails will Kim need to complete her work? 174
4. What matrix represents the product? ❖

Exploration 2 The Procedure for Multiplying Matrices

From Exploration 1, it is possible to create a procedure for multiplying matrices. First, check to see if the number of left matrix columns matches the number of right matrix rows. Then follow the pattern to see how to multiply the matrices.

Let $A = \begin{bmatrix} 3 & 5 \\ 4 & -1 \end{bmatrix}$ and $B = \begin{bmatrix} -2 & -7 \\ 6 & 5 \end{bmatrix}$. Find the product, $P = AB$.

First, find the products of each entry in row 1 of matrix A and each corresponding entry in column 1 of matrix B. Then place the sum of the products at the address p_{11} in the product matrix, P.

Matrix *A* **Matrix *B***

$$\begin{bmatrix} 3 & 5 \\ 4 & -1 \end{bmatrix} \text{ times } \begin{bmatrix} -2 & -7 \\ 6 & 5 \end{bmatrix} \longrightarrow 3(-2) + 5(6) \longrightarrow P = \begin{bmatrix} 24 & \\ & \end{bmatrix}$$

The entry in p_{11} is the sum of [**row 1** entries times **column 1** entries].

Continue this pattern for the remaining entries.

1. What is the entry for address p_{12}? What row and column are multiplied to find p_{12}? 4; row 1, column 2

Matrix *A* **Matrix *B***

$$\begin{bmatrix} 3 & 5 \\ 4 & -1 \end{bmatrix} \text{ times } \begin{bmatrix} -2 & -7 \\ 6 & 5 \end{bmatrix} \longrightarrow 3(\underline{?}) + 5(\underline{?}) \longrightarrow P = \begin{bmatrix} 24 & ? \\ & \end{bmatrix}$$

(answers: −7, 5; 4)

The entry in p_{12} is the sum of [**row** __?__ entries times **column** __?__ entries]. (answers: 1, 2)

2. What is the entry for address p_{21}? What row and column are multiplied to find p_{21}? −14; row 2, column 1

Matrix *A* **Matrix *B***

$$\begin{bmatrix} 3 & 5 \\ 4 & -1 \end{bmatrix} \text{ times } \begin{bmatrix} -2 & -7 \\ 6 & 5 \end{bmatrix} \longrightarrow \underline{?}(-2) + \underline{?}(6) \longrightarrow P = \begin{bmatrix} 24 & p_{12} \\ ? & \end{bmatrix}$$

(answers: 4, −1; −14)

The entry in p_{21} is the sum of [**row** __?__ entries times **column** __?__ entries]. (answers: 2, 1)

3. What is the last entry for matrix P? What row and column are multiplied to find the last entry? What is the address for the last entry in matrix P? −33; row 2, column 2; p_{22}

Matrix *A* **Matrix *B***

$$\begin{bmatrix} 3 & 5 \\ 4 & -1 \end{bmatrix} \text{ times } \begin{bmatrix} -2 & -7 \\ 6 & 5 \end{bmatrix} \longrightarrow \underline{?}(?) + \underline{?}(?) \longrightarrow P = \begin{bmatrix} 24 & p_{12} \\ p_{21} & ? \end{bmatrix}$$

(answers: 4(−7) + −1(5); −33)

The entry in p_{22} is the sum of [**row** __?__ entries times **column** __?__ entries]. (answers: 2, 2)

4. Summarize in your own words the procedure for multiplying matrices. ❖

4. Answers may vary. You can multiply matrices if the number of columns in the first matrix is equal to the number of rows in the second matrix.

To multiply 2-by-2 matrices:

A. Multiply each entry in row 1 times each corresponding entry in column 1, then place the sum of the products at the address p_{11} in a product matrix.
B. Multiply each entry in row 1 times each corresponding entry in column 2, then place the sum of the products at the address p_{12} in a product matrix.
C. Multiply each entry in row 2 times each corresponding entry in column 1, then place the sum of the products at the address p_{21} in a product matrix.
D. Multiply each entry in row 2 times each corresponding entry in column 2, then place the sum of the products at the address p_{22} in a product matrix.

APPLICATION

Mary places two orders for new tableware. Mary's first order is 1 cup, 4 saucers, and 7 plates. Her second order consists of 3 cups, 6 saucers, and 8 plates. What is the cost of each order?

Let matrix C contain the costs, in dollars, for each saucer, cup, and plate. Let matrix D contain the numbers of pieces in each order.

	Cups	Saucers	Plates	
Cost	[2	5	9]	$= C$

	First order	Second order	
Cups	1	3	
Saucers	4	6	$= D$
Plates	7	8	

Check the row and column dimensions of each given matrix to see if multiplication is possible. Then find the product matrix.

Since the **number of columns in *C*** matches the **number of rows in *D***, multiplication is possible.

C (1 by **3**) times D (**3** by 2) $\rightarrow P$ (1 by 2)

Matrix *C* · Matrix *D*

$$\begin{bmatrix} 2 & 5 & 9 \end{bmatrix} \begin{bmatrix} 1 & 3 \\ 4 & 6 \\ 7 & 8 \end{bmatrix} = \mathbf{P}$$

$$CD = [2(1) + 5(4) + 9(7) \quad 2(3) + 5(6) + 9(8)] = [85 \quad 108] = P$$

The cost for the first order is \$85, and the cost for the second order is \$108. ❖

Exploration 3

1a. $\begin{bmatrix} 3 & -7 \\ 16 & 9 \end{bmatrix}$

1b. $\begin{bmatrix} 12 & 9 \\ -7 & -4 \end{bmatrix}$

1c. same as 1b

3. I is an identity matrix. When I is multiplied by each matrix, the product matrix is identical to the original matrix. This is true when the identity matrix is multiplied on either side of a matrix.

4. Matrix multiplication is not commutative.

$LN = \begin{bmatrix} 85 & 55 \\ 129 & 108 \end{bmatrix}$

$NL = \begin{bmatrix} 180 & -3 \\ -85 & 13 \end{bmatrix}$

5. The identity matrix must multiply on either side of a matrix. Thus, the dimensions of the original matrix and its identity matrix must be equal.

Exploration 3 *The Identity Matrix for Multiplication*

1 Let $\mathbf{L} = \begin{bmatrix} 3 & -7 \\ 16 & 9 \end{bmatrix}$, $\mathbf{I} = \begin{bmatrix} 1 & 0 \\ 0 & 1 \end{bmatrix}$, and $\mathbf{N} = \begin{bmatrix} 12 & 9 \\ -7 & -4 \end{bmatrix}$. Multiply.

a. IL **b.** NI **c.** IN

2 What do you notice about the products? They do not contain 0 or 1.

3 Notice that $NI = N$ and $IN = N$. Why is I called an **identity matrix**?

4 Find $L \cdot N$ and $N \cdot L$. Make a conjecture about the Commutative Property and matrix multiplication.

5 Consider the products $IN = N$ and $NI = N$ from Steps 1 and 3. Explain why an identity matrix for multiplication, such as matrix I, must be a square matrix. ❖

Communicate

Explain how to identify the row and column dimensions of the following matrices.

1. $\begin{bmatrix} 1 \\ -5 \\ 3 \end{bmatrix}$ **2.** $\begin{bmatrix} 1 & 3 \\ 0 & 5 \\ 12 & -4 \end{bmatrix}$ **3.** $\begin{bmatrix} 1 & 4 & 3 & 108 \end{bmatrix}$

Explain how to determine whether two matrices can be multiplied. Explain how to perform the multiplication with the given matrices.

4. $3\begin{bmatrix} -6 & 1 \\ 4 & -1 \end{bmatrix}$ **5.** $\begin{bmatrix} 2 & -4 \\ 1 & 0 \end{bmatrix}\begin{bmatrix} 5 & -1 \\ 6 & 5 \end{bmatrix}$ **6.** $\begin{bmatrix} 1 & 2 \\ 1 & -3 \\ 3 & 5 \end{bmatrix}\begin{bmatrix} 4 & 1 & 3 \\ 2 & 1 & 6 \end{bmatrix}$

1. There are 3 numbers going down in the matrix so there are 3 rows. There is 1 number going across the matrix, so there is 1 column.
2. 3 rows and 2 columns
3. 1 row and 4 columns
4. The 3 is a scalar, so it can be multiplied with any matrix. Multiply the scalar by every element in the matrix.

Practice & Apply

If you have technology available to assist you with the matrix operations, you should make use of it as directed by your teacher.

The dimensions of two matrices are given. If the two matrices can be multiplied, indicate the dimensions of the product matrix. If they cannot, write *not possible*.

	A	*B*	*AB*
7.	2 by 4	3 by 5	___
8.	1 by 3	3 by 5	___
9.	4 by 2	2 by 1	___
10.	3 by 3	3 by 2	___
11.	1 by 2	5 by 2	___

7. Not possible
8. 1 by 5
9. 4 by 1
10. 3 by 2
11. Not possible

Use the given matrices to perform each operation. Write *not possible* when appropriate.

$A = \begin{bmatrix} 2 & 0 \\ -1 & -5 \end{bmatrix}$ $B = \begin{bmatrix} 1 & 3 \\ -2 & -3 \\ 4 & 0 \end{bmatrix}$ $C = \begin{bmatrix} 6 & -4 \\ -1 & 12 \\ -2 & 5 \end{bmatrix}$ $D = \begin{bmatrix} -3 & 1 & 4 \\ -2 & 2 & 6 \end{bmatrix}$

12. $B + C$ **13.** $-5D$ **14.** $3B - 4C$

15. AB Not possible **16.** BD **17.** DA Not possible

18. DC **19.** $-BC$ Not possible **20.** CA

12. $\begin{bmatrix} 7 & -1 \\ -3 & 9 \\ 2 & 5 \end{bmatrix}$

13. $\begin{bmatrix} 15 & -5 & -20 \\ 10 & -10 & -30 \end{bmatrix}$

14. $\begin{bmatrix} -21 & 25 \\ -2 & -57 \\ 20 & -20 \end{bmatrix}$

16. $\begin{bmatrix} -9 & 7 & 22 \\ 12 & -8 & -26 \\ -12 & 4 & 16 \end{bmatrix}$

18. $\begin{bmatrix} -27 & 44 \\ -26 & 62 \end{bmatrix}$

20. $\begin{bmatrix} 16 & 20 \\ -14 & -60 \\ -9 & -25 \end{bmatrix}$

For Exercises 21–24, find the missing entries in each product matrix.

21. $-3\begin{bmatrix} -1 & 8 \\ 0 & 3 \end{bmatrix} = \begin{bmatrix} 3 & ? \\ ? & -9 \end{bmatrix}$ −24; 0

22. $5.7\begin{bmatrix} 0.03 & -2.3 \\ 1.8 & 0.4 \end{bmatrix} = \begin{bmatrix} 0.171 & ? \\ 10.26 & ? \end{bmatrix}$ −13.11; 2.28

23. $\begin{bmatrix} 0 & 8 \\ -5 & 1 \\ -2 & 3 \end{bmatrix} \begin{bmatrix} 0 & 2 & -1 \\ 4 & 6 & -3 \end{bmatrix} = \begin{bmatrix} 32 & 48 & ? \\ 4 & ? & 2 \\ ? & 14 & -7 \end{bmatrix}$ −24; −4; 12

24. $\begin{bmatrix} -1 & 8 \\ 0 & 3 \end{bmatrix} \begin{bmatrix} -5 & 6 \\ -9 & 8 \end{bmatrix} = \begin{bmatrix} ? & 58 \\ ? & 24 \end{bmatrix}$ −67; −27

Use matrices *A*, *B*, *C*, and *D* below for Exercises 25–29.

$A = \begin{bmatrix} 0 & 4 \\ -5 & 1 \\ -2 & 3 \end{bmatrix}$ $B = \begin{bmatrix} 2 & -1 \\ 4 & -3 \end{bmatrix}$ $C = \begin{bmatrix} -1 \\ 2 \end{bmatrix}$ $D = \begin{bmatrix} 1 & -2 \\ 3 & -4 \end{bmatrix}$

25. What are the dimensions of matrices *A*, *B*, and *C*? Are the dimensions compatible for multiplication? 3 by 2; 2 by 2; 2 by 1; *AB*, *AC*, and *BC* are possible.

26. Does $A(BC) = (AB)C$? If they are equal, what is the product?

27. What is your conjecture about the Associative Property for Multiplication? Find three more matrices that confirm your conjecture, or find three more matrices that are counterexamples.

26. Yes; $\begin{bmatrix} -40 \\ 10 \\ -22 \end{bmatrix}$

28. What are the row and column dimensions of matrices *B* and *D*? Are these dimensions compatible for multiplication? 2 by 2; 2 by 2; yes

29. Does $BD = DB$? Make a conjecture about the Commutative Property for multiplying matrices. Is this true for all matrices?
No; the Commutative Property of Multiplication does not apply to matrices; no.

30. What characteristics of matrices are necessary to make multiplication possible? The no. of columns in 1st matrix must equal no. of rows in 2nd matrix.

31. If $X = \begin{bmatrix} 1 & 2 & 3 \end{bmatrix}$ and $Y = \begin{bmatrix} 3 \\ 2 \\ 5 \end{bmatrix}$, what are the products *XY* and *YX*?

Wooden Toy Craftsmen
Sam and Kim Grant
764 Harmony Lane
Maplewood, WI 54321
Phone: (123) 456-7891

32. Manufacturing At the fair, toy cars and trucks must be sold at a price that will generate a profit. Use the matrices below to calculate how much it will cost to make a car and how much it will cost to make a truck. Express your answer as a labeled matrix.

	Cost
Cars	2.11
Trucks	3.12

	Wheels	Nails	Blocks
Cars	4	7	2
Trucks	6	9	3

	Cost in cents
Wheels	25
Nails	3
Blocks	45

Cost of supplies:
Wheels $0.25
Nails $0.03
Blocks $0.45

33. Use the matrix at the right to determine how much it will cost Sam and Kim to produce all of their toy cars and trucks. Express your answer as a labeled matrix in dollars.

	Cars	Trucks
Sam	8	15
Kim	12	10

33.

	Cost
Sam	63.68
Kim	56.52

34. Sam and Kim decide to charge $5 per car and $6.50 per truck. Create your selling-price matrix, and determine how much revenue Sam and Kim will collect at the fair if all of the cars and trucks are sold.

34.

	Revenue
Sam	137.50
Kim	125.00

35. Sales Tax If you make a purchase, the total price you pay is the retail price plus sales tax. If sales tax is 6% in Maplewood, create the matrix that represents the amount of sales tax generated by Sam's sales and Kim's sales at the fair.

35.

	Sales Tax
Sam	8.25
Kim	7.50

36. How much profit will Sam and Kim each make after paying the costs of supplies?

36.

	Profit
Sam	65.57
Kim	60.98

Fund-raising The first matrix shows the number of flags and buttons sold by each class. The second gives the purchase cost and selling price of the flags and buttons.

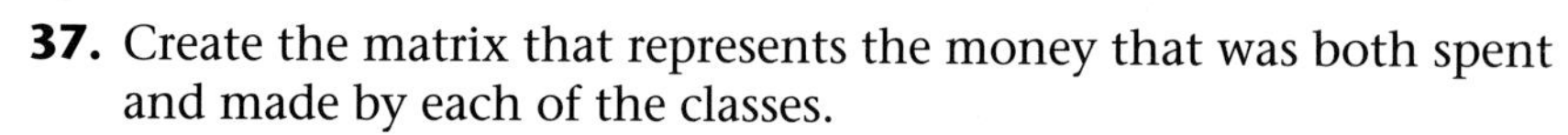

	Flags	Buttons
Freshman	125	200
Sophomore	100	200
Junior	50	150
Senior	200	250

	Purchase cost	Selling price
Flags	0.75	2.00
Buttons	0.35	1.00

At Central City High School, each class raises funds by selling class flags and buttons.

37. Create the matrix that represents the money that was both spent and made by each of the classes.

38. Add a column to the matrix from Exercise 37 to represent the profit made by each class.

39. Why do you think the number of flags and buttons sold decreases each year after the freshman year, but increases again during the senior year? Answers may vary. Seniors have graduation activities and trips that require raising more funds.

Look Back

40. Simplify $5 - 2y\{3 - 2[x + 6y - (3y + x) + 8]\}$. **[Course 1]** $12y^2 + 26y + 5$

41. Solve $\frac{14}{b} = 21$ for b. **[Lesson 1.2]** $\frac{2}{3}$, or approximately 0.667

Solve each equation for *x*. [Lesson 1.3]

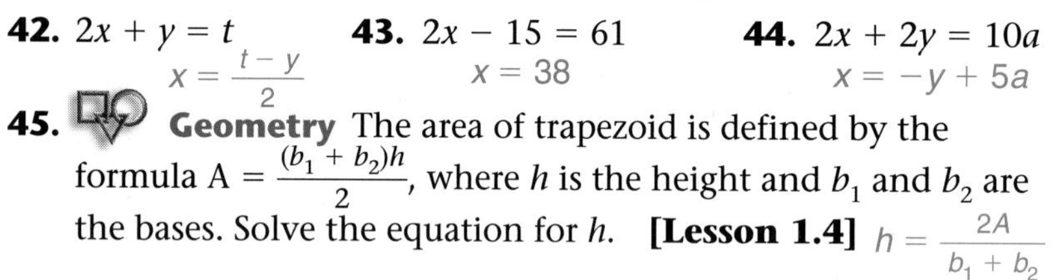

42. $2x + y = t$ $x = \frac{t - y}{2}$

43. $2x - 15 = 61$ $x = 38$

44. $2x + 2y = 10a$ $x = -y + 5a$

45. **Geometry** The area of trapezoid is defined by the formula $A = \frac{(b_1 + b_2)h}{2}$, where h is the height and b_1 and b_2 are the bases. Solve the equation for h. **[Lesson 1.4]** $h = \frac{2A}{b_1 + b_2}$

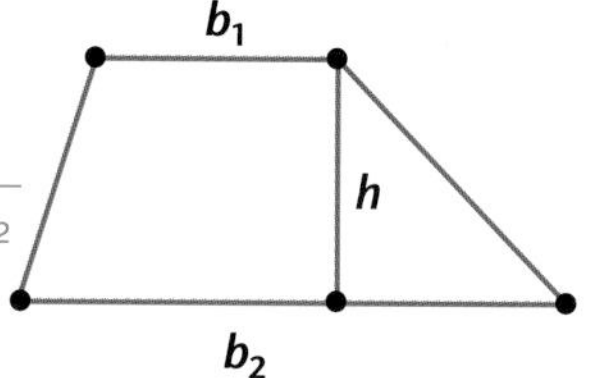

46. Solve $3x + 5(x - 4) = 10x - 25$ for x. **[Lesson 1.4]** $x = 2.5$

47. Graph the equation of the line that is perpendicular to $-3x + 9y = 27$ and contains the point $(-6, 7)$. **[Lessons 2.1, 2.2]**

48. Solve by the elimination method. $\begin{cases} 2x - 5y = 20 \\ 3x - 4y = 37 \end{cases}$ **[Lesson 2.5]** (15, 2)

49. Suppose that you can paddle your canoe 6 miles upstream in 3 hours. It takes you 2 hours to return the same distance downstream. Find the average rate of the current in the stream. **[Lessons 2.4, 2.5, 2.6]**

$\frac{1}{2}$ mph, or 0.5 mph

Look Beyond

Use matrices *A* and *B* to complete Exercises 50 and 51.

$$A = \begin{bmatrix} 2 & -1 \\ 4 & -3 \end{bmatrix} \quad B = \begin{bmatrix} 1.5 & -0.5 \\ 2 & -1 \end{bmatrix}$$

50. Find AB. Label the resulting matrix C.

51. Why is matrix B called the inverse of matrix A?

50. $C = \begin{bmatrix} 1 & 0 \\ 0 & 1 \end{bmatrix}$

51. B is called the inverse of A because the product AB is the identity matrix.

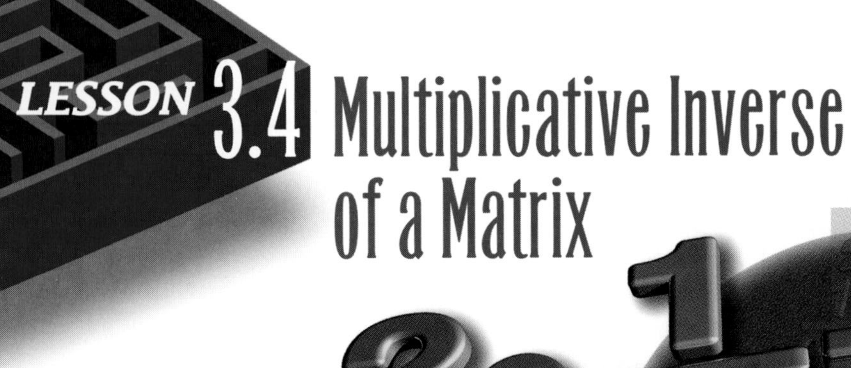

LESSON 3.4 Multiplicative Inverse of a Matrix

why *You have used the Reciprocal Property to solve multiplication equations. When you solve a matrix equation, you use the inverse of a matrix in much the same way.*

You know that $r \cdot \frac{1}{r} = 1$, when $r \neq 0$. Because their product is the multiplicative identity, r and $\frac{1}{r}$ are called reciprocals, or multiplicative inverses. A similar situation occurs with matrices. If the product of two matrices equals an identity matrix, the matrices are **inverse matrices**. The symbol A^{-1} is used to represent the inverse of matrix A.

Examine the matrix multiplication.

$$\underset{A}{\begin{bmatrix} 3 & -1 \\ 4 & 2 \end{bmatrix}} \cdot \underset{B}{\begin{bmatrix} 0.2 & 0.1 \\ -0.4 & 0.3 \end{bmatrix}} = \underset{I}{\begin{bmatrix} 1 & 0 \\ 0 & 1 \end{bmatrix}} = \underset{B}{\begin{bmatrix} 0.2 & 0.1 \\ -0.4 & 0.3 \end{bmatrix}} \cdot \underset{A}{\begin{bmatrix} 3 & -1 \\ 4 & 2 \end{bmatrix}}$$

The matrices $A = \begin{bmatrix} 3 & -1 \\ 4 & 2 \end{bmatrix}$ and $B = \begin{bmatrix} 0.2 & 0.1 \\ -0.4 & 0.3 \end{bmatrix}$ are inverses. The product of A and B is the identity matrix, $\begin{bmatrix} 1 & 0 \\ 0 & 1 \end{bmatrix}$. Thus, $B = A^{-1}$ and $A = B^{-1}$.

INVERSE MATRIX

The matrix A has an inverse matrix, A^{-1}, if

$$AA^{-1} = I = A^{-1}A,$$

where I is the identity matrix for multiplication.

Finding the Inverse Matrix

EXAMPLE 1

If matrix $A = \begin{bmatrix} 3 & 5 \\ 4 & 6 \end{bmatrix}$, find A^{-1}.

Solution

You know from the definition that $AA^{-1} = I = A^{-1}A$. Since A is a 2-by-2 matrix, A^{-1} must be a 2-by-2 matrix. Let $\begin{bmatrix} a & b \\ c & d \end{bmatrix}$ represent A^{-1}, the inverse matrix. Then substitute the matrices in $AA^{-1} = I$.

$$\begin{array}{ccc} A \cdot A^{-1} & = & I \\ \begin{bmatrix} 3 & 5 \\ 4 & 6 \end{bmatrix}\begin{bmatrix} a & b \\ c & d \end{bmatrix} & = & \begin{bmatrix} 1 & 0 \\ 0 & 1 \end{bmatrix} \end{array}$$

Multiply the matrices A and A^{-1}.

$$\begin{array}{ccc} AA^{-1} & = & I \\ \begin{bmatrix} 3a + 5c & 3b + 5d \\ 4a + 6c & 4b + 6d \end{bmatrix} & = & \begin{bmatrix} 1 & 0 \\ 0 & 1 \end{bmatrix} \end{array}$$

The corresponding matrix entries of equal matrices are equal. Write the equations represented by each pair of corresponding entries. Pair the two equations that have the same variables into a system. Then solve each system of equations.

$$\begin{cases} 3a + 5c = 1 \\ 4a + 6c = 0 \end{cases} \qquad \begin{cases} 3b + 5d = 0 \\ 4b + 6d = 1 \end{cases}$$

Use elimination to solve for a and b.

$$\begin{cases} 6(3a + 5c) = 6(1) \\ -5(4a + 6c) = -5(0) \end{cases} \longrightarrow \begin{array}{rl} 18a + 30c &= 6 \\ -20a - 30c &= 0 \\ \hline -2a &= 6 \\ a &= -3 \end{array}$$

$$\begin{cases} 6(3b + 5d) = 6(0) \\ -5(4b + 6d) = -5(1) \end{cases} \longrightarrow \begin{array}{rl} 18b + 30d &= 0 \\ -20b - 30d &= -5 \\ \hline -2b &= -5 \\ b &= 2.5 \end{array}$$

Substitute -3 for a, and solve for c. Substitute 2.5 for b, and solve for d.

$$\begin{array}{rl} 3a + 5c &= 1 \\ 3(-3) + 5c &= 1 \\ -9 + 5c &= 1 \\ 5c &= 10 \\ c &= 2 \end{array} \qquad \begin{array}{rl} 3b + 5d &= 0 \\ 3(2.5) + 5d &= 0 \\ 7.5 + 5d &= 0 \\ 5d &= -7.5 \\ d &= -1.5 \end{array}$$

Replace the variables in $A^{-1} = \begin{bmatrix} a & b \\ c & d \end{bmatrix}$ with the values you found. The inverse matrix A^{-1} is $\begin{bmatrix} -3 & 2.5 \\ 2 & -1.5 \end{bmatrix}$.

Check.

$$\begin{array}{ccccc} A \cdot A^{-1} & = & I & = & A^{-1} \cdot A \\ \begin{bmatrix} 3 & 5 \\ 4 & 6 \end{bmatrix}\begin{bmatrix} -3 & 2.5 \\ 2 & -1.5 \end{bmatrix} & = & \begin{bmatrix} 1 & 0 \\ 0 & 1 \end{bmatrix} & = & \begin{bmatrix} -3 & 2.5 \\ 2 & -1.5 \end{bmatrix}\begin{bmatrix} 3 & 5 \\ 4 & 6 \end{bmatrix} \end{array}$$ ❖

What are the requirements for the row and column dimensions of an inverse matrix? Why? The original matrix and its inverse must have the same row and column dimensions. Since $AA^{-1} = A^{-1}A$, the original matrix and its inverse are commutative.

EXAMPLE 2

Find the inverse of $B = \begin{bmatrix} 4 & 6 \\ 2 & 3 \end{bmatrix}$.

Solution ➤

Let $\begin{bmatrix} a & b \\ c & d \end{bmatrix}$ represent the inverse matrix B^{-1}.

Since $BB^{-1} = I$, write the matrix equation.

$$\begin{matrix} B & \cdot & B^{-1} & = & I \end{matrix}$$

$$\begin{bmatrix} 4 & 6 \\ 2 & 3 \end{bmatrix}\begin{bmatrix} a & b \\ c & d \end{bmatrix} = \begin{bmatrix} 1 & 0 \\ 0 & 1 \end{bmatrix}$$

Show the matrix multiplication.

$$\begin{matrix} BB^{-1} & = & I \end{matrix}$$

$$\begin{bmatrix} 4a + 6c & 4b + 6d \\ 2a + 3c & 2b + 3d \end{bmatrix} = \begin{bmatrix} 1 & 0 \\ 0 & 1 \end{bmatrix}$$

Write the equations from corresponding matrix entries.

$$\begin{cases} 4a + 6c = 1 \\ 2a + 3c = 0 \end{cases} \qquad \begin{cases} 4b + 6d = 0 \\ 2b + 3d = 1 \end{cases}$$

You can use any of the methods you learned in Chapter 2 to solve the systems of linear equations. The elimination method is shown here.

$$\begin{cases} 4a + 6c = 1 \\ -2(2a + 3c) = -2(0) \end{cases} \longrightarrow \begin{array}{r} 4a + 6c = 1 \\ \underline{-4a - 6c = 0} \\ 0 \neq 1 \end{array} \qquad \begin{array}{r} 4b + 6d = 0 \\ \underline{-4b - 6d = -2} \\ 0 \neq -2 \end{array}$$

Ongoing Assessment The elements in row 2 are multiples of the elements of those in row 1. If it is possible to multiply the elements of one row of a matrix by a constant to obtain another row of the matrix, then it has no inverse.

In each system, the variables have been eliminated, and two unequal numbers remain.

There is *no solution* for either system. The matrix $\begin{bmatrix} 4 & 6 \\ 2 & 3 \end{bmatrix}$ *has no inverse.* ❖

Examine the entries in the rows of the matrix in Example 2 that has no inverse. How is one row related to the other? Make a conjecture about a matrix that has no inverse.

a. $\begin{bmatrix} \frac{1}{3} & -\frac{1}{6} \\ 1 & 0 \end{bmatrix}$ **b.** No inverse

Try This Find each inverse, if it exists. **a.** $\begin{bmatrix} 0 & 1 \\ -6 & 2 \end{bmatrix}$ **b.** $\begin{bmatrix} 3 & 9 \\ 4 & 12 \end{bmatrix}$

Matrix Technology

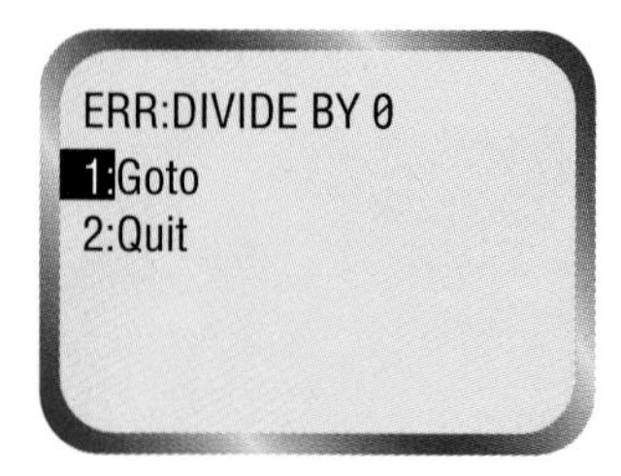

Working with matrices can be tedious because of the many arithmetic calculations. As you continue studying matrices, you may soon encounter larger matrices. Finding an inverse matrix with dimensions beyond 2 by 2 takes many calculations and a lot of time unless you use technology. Computers and calculators can be used to perform these calculations quickly, allowing you to explore the more interesting aspects of matrices.

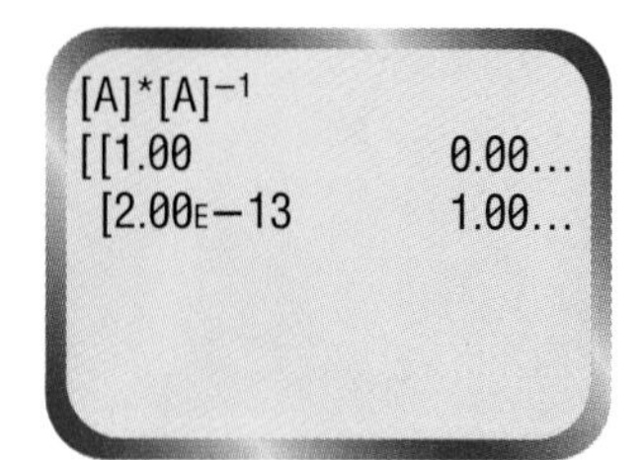

When you use technology, however, you should be aware of unusual situations that cause calculators and computers to display error messages. For example, you can expect an error message if a matrix does not have an inverse or if you try to divide by 0.

Another difficulty could appear when you encounter rounding errors. You might see unusual expressions such as 2.00E − 13 for the number 0.

EXERCISES & PROBLEMS

Communicate

Explain how to find the inverse of each matrix, if it exists.

1. $\begin{bmatrix} -1 & 6 \\ 0 & 3 \end{bmatrix}$

2. $\begin{bmatrix} 1 & 2 \\ 2 & 4 \end{bmatrix}$ Not possible

3. $[2 \quad -5 \quad 7]$ Not possible

4. What conditions cause a matrix not to have an inverse? If the matrix is not square or if its rows are multiples, the matrix does not have an inverse.

5. How is the inverse of a matrix used? It is used to solve matrix equations.

6. What is the product of a matrix multiplied by its inverse? The identity matrix

7. $\begin{bmatrix} 3 & 7 \\ 2 & 5 \end{bmatrix}$ 8. $\begin{bmatrix} -1 & 0.8 \\ 1 & -0.6 \end{bmatrix}$ 10. $\begin{bmatrix} 1 & 0 \\ 0 & -1 \end{bmatrix}$

Practice & Apply

Use the definition of inverse matrix to find the inverse matrix for each of the given matrices.

7. $\begin{bmatrix} 5 & -7 \\ -2 & 3 \end{bmatrix}$

8. $\begin{bmatrix} 3 & 4 \\ 5 & 5 \end{bmatrix}$

9. $\begin{bmatrix} -1 & 0.04 \\ -0.6 & 1 \end{bmatrix}$

9. $\begin{bmatrix} -1.025 & 0.041 \\ -0.615 & 1.025 \end{bmatrix}$ or $\begin{bmatrix} -\frac{125}{122} & \frac{5}{122} \\ -\frac{75}{122} & \frac{125}{122} \end{bmatrix}$

10. $\begin{bmatrix} 1 & 0 \\ 0 & -1 \end{bmatrix}$

11. $\begin{bmatrix} -1 & -2 \\ 2 & 0 \end{bmatrix}$

12. $\begin{bmatrix} -\frac{4}{5} & 1 \\ \frac{9}{5} & -2 \end{bmatrix}$

11. $\begin{bmatrix} 0 & 0.5 \\ -0.5 & -0.25 \end{bmatrix}$ 12. $\begin{bmatrix} 10 & 5 \\ 9 & 4 \end{bmatrix}$

Determine whether each matrix has an inverse.

13. $\begin{bmatrix} 5 & -3 \\ 2 & -4 \end{bmatrix}$ Yes **14.** $\begin{bmatrix} 6 & 8 \\ 3 & 4 \end{bmatrix}$ No **15.** $\begin{bmatrix} -1 & 2 & -3 \\ -4 & 5 & -6 \end{bmatrix}$ No

16. $\begin{bmatrix} \frac{1}{2} & -1 \\ 1 & -2 \end{bmatrix}$ No **17.** $\begin{bmatrix} -0.2 & 1 \\ 0.3 & -1.5 \end{bmatrix}$ No **18.** $\begin{bmatrix} 5 \\ -6 \\ 7 \end{bmatrix}$ No **20.** $\begin{bmatrix} 0.833 & 1 & 0.167 \\ -0.133 & -0.2 & 0.133 \\ 0.083 & 0 & -0.083 \end{bmatrix}$

Technology Use technology to find the inverse of each matrix if it exists.

19. $\begin{bmatrix} 2 & -4 & 6 \\ 5 & 0 & -8 \\ 3 & -6 & 9 \end{bmatrix}$ None **20.** $\begin{bmatrix} 1 & 5 & 10 \\ 0 & -5 & -8 \\ 1 & 5 & -2 \end{bmatrix}$ **21.** $\begin{bmatrix} 6 & 1 & 1 \\ 1 & 6 & 1 \\ 6 & 6 & 1 \end{bmatrix}$ **21.** $\begin{bmatrix} 0 & -0.2 & 0.2 \\ -0.2 & 0 & 0.2 \\ 1.2 & 1.2 & -1.4 \end{bmatrix}$

22. A general 2-by-2 matrix can be written $\begin{bmatrix} a & b \\ c & d \end{bmatrix}$. Calculate the products ad and bc for each matrix $\begin{bmatrix} 10 & 5 \\ 4 & 2 \end{bmatrix}$ and $\begin{bmatrix} 4 & 6 \\ 2 & 3 \end{bmatrix}$. Do either of these matrices have an inverse? Make a conjecture about the products ad and bc in a 2-by-2 matrix that does not have an inverse.

23. Choose at least three of the 2-by-2 matrices that *do not have inverses* from the exercises. Again, for each of these matrices, find the value of the products ad and bc. Do your findings support your conjecture from Exercise 22? Yes, the products ad and bc are equal.

22. ad: 20, bc: 20; ad: 12, bc: 12. Matrices $\begin{bmatrix} 10 & 5 \\ 4 & 2 \end{bmatrix}$ and $\begin{bmatrix} 4 & 6 \\ 2 & 3 \end{bmatrix}$ have no inverse. If the products ad and bc are equal in matrix $\begin{bmatrix} a & b \\ c & d \end{bmatrix}$, then the matrix has no inverse.

For each matrix, find the products *ad* and *bc*. Use your results from Exercises 22 and 23 to write and test your conjecture about the matrices. Use technology to check your conjecture.

24. $\begin{bmatrix} -6 & 60 \\ 0.5 & -5 \end{bmatrix}$ 30, 30; no inverse **25.** $\begin{bmatrix} 7 & -4 \\ 14 & -7 \end{bmatrix}$ −49, −56; has inverse **26.** $\begin{bmatrix} -8 & 3 \\ 7 & 2 \end{bmatrix}$ −16, 21; has inverse

Geometry The matrices represent the endpoints of two line segments AB and CD.

$$\begin{matrix} & x & y \\ A & 0 & 6 \\ B & 4 & 12 \end{matrix} \qquad \begin{matrix} & x & y \\ C & 6 & 0 \\ D & 12 & 4 \end{matrix}$$

27. Graph both line segments. Describe how the slopes of the line segments are related to each other.

28. If $\begin{bmatrix} 0 & 6 \\ 4 & 12 \end{bmatrix} \cdot T = \begin{bmatrix} 6 & 0 \\ 12 & 4 \end{bmatrix}$, find matrix T, where $T = \begin{bmatrix} a & b \\ c & d \end{bmatrix}$.

29. Write a matrix to represent the endpoints of some line segment. Multiply your matrix by matrix T from Exercise 28, and graph the solution. Compare your results with segments AB and CD above.

30. Technology Use technology to find the inverse of matrix A.

$$A = \begin{bmatrix} 1 & 2 & 3 \\ 2 & 4 & 5 \\ 3 & 5 & 6 \end{bmatrix} \qquad \begin{bmatrix} 1 & -3 & 2 \\ -3 & 3 & -1 \\ 2 & -1 & 0 \end{bmatrix}$$

28. $\begin{bmatrix} 0 & 1 \\ 1 & 0 \end{bmatrix}$

29. Answers may vary. $\begin{bmatrix} 4 & 2 \\ 6 & 7 \end{bmatrix} \cdot T = \begin{bmatrix} 2 & 4 \\ 7 & 6 \end{bmatrix}$ The slope of each line segment is the reciprocal of the other.

30. Answers may vary. For example, enter matrix A as a 3×3 matrix. Use the calculator inverse key.

Look Back

31. Life Science Suppose that you want to analyze an experiment. You have collected data that describes the height in centimeters and body temperature in degrees Fahrenheit for 20 people. The data is collected, and a scatter plot is drawn. Describe the function that would best fit this data. **[Lesson 1.1]** Linear

Make a table to evaluate each function when x is -3, -2, -1, 0, 1, 2, and 3. Are any values undefined? [Lesson 1.1]

32. $y = x^2 - 2x + 3$

33. $y = \frac{2}{x - 2}$

34. $y = |x|$

Identify the property of equality that you would use to solve each equation. Then solve each equation. [Lessons 1.2, 1.3]

35. $y - 11 = 67$
Addition, 78

36. $\frac{2}{3}x = 16$
Multiplication, 24

37. $\frac{z}{-3} = 20$
Multiplication, -60

38. Coordinate Geometry A line has slope $\frac{3}{4}$ and passes through the point $P(2, -4)$. Write an equation in standard form for a line that is perpendicular to the original line and that passes through the common point, P. **[Lessons 2.1, 2.2]**
$4x + 3y = -4$

39. Time Management Rick works 2 jobs. He keeps track of his hours using matrices for each day of a five-day workweek. Matrices A and B below represent 2 weeks of work. Create a matrix, C, that shows the total hours worked each day and at each job for the 2 weeks shown. **[Lesson 3.1]**

Hours Worked

Week 1: (M T W Th F)
$$\begin{matrix} \text{Job 1} \\ \text{Job 2} \end{matrix} \begin{bmatrix} 3 & 4 & 2 & 3 & 3 \\ 1 & 6 & 5 & 5 & 4 \end{bmatrix} = A$$

Week 2: (M T W Th F)
$$\begin{matrix} \text{Job 1} \\ \text{Job 2} \end{matrix} \begin{bmatrix} 5 & 0 & 2 & 5 & 3 \\ 3 & 6 & 3 & 5 & 1 \end{bmatrix} = B$$

39. (M T W Th F)
$$C = \begin{matrix} \text{Job 1} \\ \text{Job 2} \end{matrix} \begin{bmatrix} 8 & 4 & 4 & 8 & 6 \\ 4 & 12 & 8 & 10 & 5 \end{bmatrix}$$

32.

x	-3	-2	-1	0	1	2	3
y	18	11	6	3	2	3	6

No

33.

x	-3	-2	-1	0	1	2	3
y	-0.4	-0.5	-0.67	-1	-2	undef.	2

Yes, when $x = 2$

34.

x	-3	-2	-1	0	1	2	3
y	3	2	1	0	1	2	3

No

Look Beyond

The matrix equation below represents a system of equations.

$$\begin{bmatrix} -2 & 5 \\ 4 & -3 \end{bmatrix} \begin{bmatrix} x \\ y \end{bmatrix} = \begin{bmatrix} 10 \\ 1 \end{bmatrix}$$

40. Using matrix multiplication, write the system of equations in standard form.
$\begin{cases} -2x + 5y = 10 \\ 4x - 3y = 1 \end{cases}$

41. Solve the system, and use matrix multiplication to check your solution.
$\left(\frac{5}{2}, 3\right)$

LESSON 3.5 Solving Matrix Equations

NOW SHOWING!
A CARTOON FESTIVAL!
FRIDAY NIGHT IS KID'S NIGHT!
Vintage Movie House
CLASSIC CINEMA

Why *Once you know how to multiply matrices and find inverses, you can solve matrix equations. Matrices can be used to solve systems of equations, even systems with many unknowns. As the size of the matrices increases, technology becomes a more necessary tool.*

The Vintage Movie House shows only classic films. Each Friday is "Kid's Night," when children can buy discounted tickets.

Consumer Economics A group of 3 parents and 5 children went to see a movie on Kid's Night, and the cost was \$34. Another group of 4 parents and 6 children went to *A Cartoon Festival* on Kid's Night, and the cost was \$43. Find the regular cost of a parent's ticket and the discounted cost of a child's ticket.

Let r represent the regular cost of a parent's ticket.
Let d represent the discounted cost of a child's ticket.
Write a system of equations.

3 regular tickets and 5 discount tickets cost \$34
4 regular tickets and 6 discount tickets cost \$43

$$\begin{cases} 3r + 5d = 34 \\ 4r + 6d = 43 \end{cases}$$

This system can be solved by graphing, by elimination, or by substitution. It can also be solved by using matrices.

The Matrix Solution for Systems of Equations

The system of equations $\begin{cases} 3r + 5d = 34 \\ 4r + 6d = 43 \end{cases}$ can be written as a **matrix equation** using the following matrices.

Coefficient matrix	**Variable matrix**	**Constant matrix**
$A = \begin{bmatrix} 3 & 5 \\ 4 & 6 \end{bmatrix}$	$X = \begin{bmatrix} r \\ d \end{bmatrix}$	$B = \begin{bmatrix} 34 \\ 43 \end{bmatrix}$

Notice that together these three matrices form the matrix equation $AX = B$.

$$\begin{matrix} A & \cdot X & = & B \\ \begin{bmatrix} 3 & 5 \\ 4 & 6 \end{bmatrix} & \begin{bmatrix} r \\ d \end{bmatrix} & = & \begin{bmatrix} 34 \\ 43 \end{bmatrix} \end{matrix}$$

The product matrix, AX, is $\begin{bmatrix} 3r + 5d \\ 4r + 6d \end{bmatrix}$. When AX is set equal to the constant matrix, B, the result is the original system of equations.

To solve the matrix equation, $AX = B$, multiply **both sides** of the matrix equation by the inverse A^{-1}. Be sure that you multiply both sides of the matrix equation with A^{-1} *on the left*. Matrix multiplication is not commutative. Recall that $A^{-1}A = I$ and $IX = X$.

$$\begin{aligned} AX &= B \\ A^{-1}AX &= A^{-1}B \\ IX &= A^{-1}B \\ X &= A^{-1}B \end{aligned}$$

Use this information to solve the equation $\begin{matrix} A & \cdot X & = & B \\ \begin{bmatrix} 3 & 5 \\ 4 & 6 \end{bmatrix} & \begin{bmatrix} r \\ d \end{bmatrix} & = & \begin{bmatrix} 34 \\ 43 \end{bmatrix} \end{matrix}$.

In Lesson 3.4, you found that the inverse of $A = \begin{bmatrix} 3 & 5 \\ 4 & 6 \end{bmatrix}$ is $A^{-1} = \begin{bmatrix} -3 & 2.5 \\ 2 & -1.5 \end{bmatrix}$. Multiply each side of $AX = B$ by A^{-1}.

Make sure that both multiplications are on the left of A and B.

$$\begin{matrix} A^{-1} & \cdot & A & \cdot X & = & A^{-1} & \cdot & B \\ \begin{bmatrix} -3 & 2.5 \\ 2 & -1.5 \end{bmatrix} & & \begin{bmatrix} 3 & 5 \\ 4 & 6 \end{bmatrix} & \begin{bmatrix} r \\ d \end{bmatrix} & = & \begin{bmatrix} -3 & 2.5 \\ 2 & -1.5 \end{bmatrix} & & \begin{bmatrix} 34 \\ 43 \end{bmatrix} \end{matrix}$$

$$\begin{matrix} I & \cdot X & = & A^{-1}B \\ \begin{bmatrix} 1 & 0 \\ 0 & 1 \end{bmatrix} & \begin{bmatrix} r \\ d \end{bmatrix} & = & \begin{bmatrix} -3(34) + 2.5(43) \\ 2(34) - 1.5(43) \end{bmatrix} \end{matrix}$$

$$\begin{matrix} X & = & A^{-1}B \\ \begin{bmatrix} r \\ d \end{bmatrix} & = & \begin{bmatrix} 5.50 \\ 3.50 \end{bmatrix} \end{matrix} \quad \text{so} \quad \begin{aligned} r &= 5.50 \\ d &= 3.50 \end{aligned}$$

The regular cost, r, of a parent's ticket is \$5.50, and the discounted cost, d, of a child's ticket is \$3.50.

Why must the inverse matrix be multiplied on the left side of both the coefficient and constant matrices? How does this procedure compare to solving a linear equation using the Multiplication Property of Equality?
The dimensions must match. With linear equations, it does not matter which side you multiply.

Solving Matrix Equations With Technology

Once you know the inverse of matrix A, you can solve the matrix equation $AX = B$. In Lesson 3.4 you used a system of equations to find A^{-1}. If you have a calculator or computer that works with matrices, you can find A^{-1} directly and solve the equation more quickly.

EXAMPLE 1

Graphics Calculator

Use matrices to solve this system of equations. $\begin{cases} 6x - y = 3 \\ 2x - 7y = 1 \end{cases}$

Solution ➤

Write the matrices in the following form.

Coefficient matrix	**Variable matrix**	**Constant matrix**
$A = \begin{bmatrix} 6 & -1 \\ 2 & -7 \end{bmatrix}$	$X = \begin{bmatrix} x \\ y \end{bmatrix}$	$B = \begin{bmatrix} 3 \\ 1 \end{bmatrix}$

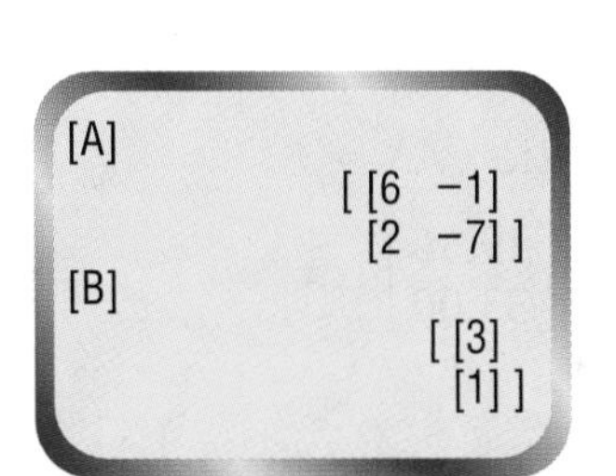

Form the matrix equation $AX = B$.

$$\underset{A}{\begin{bmatrix} 6 & -1 \\ 2 & -7 \end{bmatrix}} \cdot \underset{X}{\begin{bmatrix} x \\ y \end{bmatrix}} = \underset{B}{\begin{bmatrix} 3 \\ 1 \end{bmatrix}}$$

Find the inverse of the coefficient matrix, A. This step can be simplified by using technology. Enter matrix A into the calculator. Place A on the screen, and press the inverse key x^{-1} to get the inverse matrix. You can now solve the matrix equation. Multiply both sides of the matrix equation by the inverse matrix.

[A]$^{-1}$
[[.175 −.025]
[.05 −.15]]
[A]$^{-1}$*[B]
[[.5]
[0]]

$$\underset{A^{-1}}{\begin{bmatrix} 0.175 & -0.025 \\ 0.05 & -0.15 \end{bmatrix}} \cdot \underset{A}{\begin{bmatrix} 6 & -1 \\ 2 & -7 \end{bmatrix}} \cdot \underset{X}{\begin{bmatrix} x \\ y \end{bmatrix}} = \underset{A^{-1}}{\begin{bmatrix} 0.175 & -0.025 \\ 0.05 & -0.15 \end{bmatrix}} \cdot \underset{B}{\begin{bmatrix} 3 \\ 1 \end{bmatrix}}$$

$$\underset{I}{\begin{bmatrix} 1 & 0 \\ 0 & 1 \end{bmatrix}} \cdot \underset{X}{\begin{bmatrix} x \\ y \end{bmatrix}} = \underset{A^{-1}B}{\begin{bmatrix} 0.5 \\ 0 \end{bmatrix}}$$

$$\underset{X}{\begin{bmatrix} x \\ y \end{bmatrix}} = \underset{A^{-1}B}{\begin{bmatrix} 0.5 \\ 0 \end{bmatrix}}$$

The solution to the system is (0.5, 0); x is 0.5, and y is 0. You can check the results by substituting the values for x and y into the original equations. ❖

Try This Use matrices to solve this system. $\begin{cases} -x + y = 1 \\ -2x + 3y = 0 \end{cases}$ (−3, −2)

The Woodwards served 14 hamburgers and 12 hot dogs. The cost was $49.50.

Once you obtain the matrix equation, you can solve the system of equations quickly and easily with technology.

EXAMPLE 2

The Woodwards and the Tates buy food at the same store. They decide to have cookouts on consecutive weekends. On both weekends, food prices were the same. The Tates served 10 hamburgers and 7 hot dogs. The total cost was $33.00. What was the price for each hamburger and each hot dog?

Solution ➤

Organize the information. Define the variables. Let h represent the cost of each hamburger. Let d represent the cost of each hot dog.

Write the equations.

$$14h + 12d = 49.50$$
$$10h + 7d = 33.00$$

Solve the system of equations. Form the matrix equation $AX = B$.

$$A = \begin{bmatrix} 14 & 12 \\ 10 & 7 \end{bmatrix} \qquad X = \begin{bmatrix} h \\ d \end{bmatrix} \qquad B = \begin{bmatrix} 49.50 \\ 33.00 \end{bmatrix}$$

$$\underset{A}{\begin{bmatrix} 14 & 12 \\ 10 & 7 \end{bmatrix}} \cdot \underset{X}{\begin{bmatrix} h \\ d \end{bmatrix}} = \underset{B}{\begin{bmatrix} 49.50 \\ 33.00 \end{bmatrix}}$$

Use a graphics calculator to solve the matrix equation.

Enter the coefficient matrix, $\begin{bmatrix} 14 & 12 \\ 10 & 7 \end{bmatrix}$, for A.

Enter the constant matrix, $\begin{bmatrix} 49.50 \\ 33.00 \end{bmatrix}$, for B.

Multiply A^{-1} by B. This will give you $\begin{bmatrix} 2.25 \\ 1.50 \end{bmatrix}$.

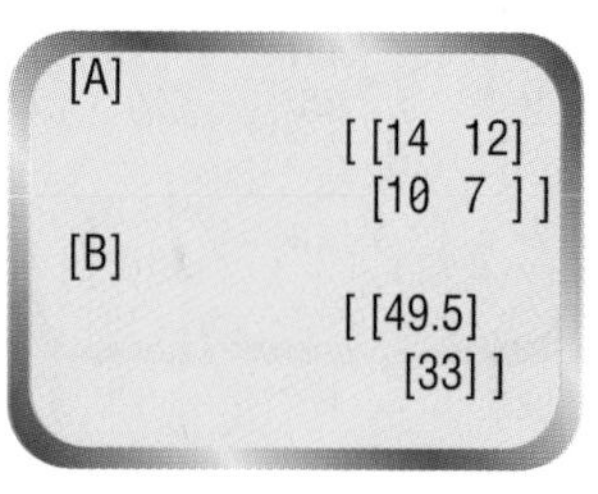

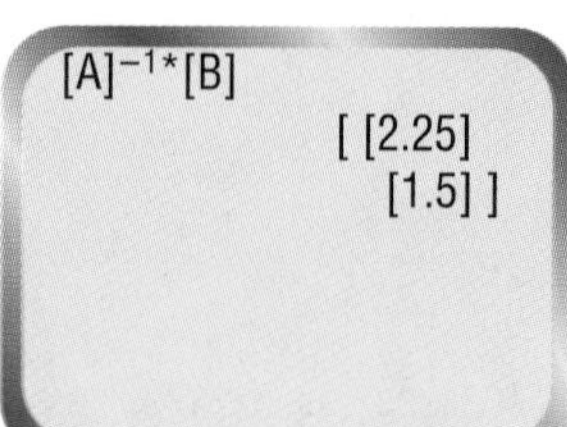

The cost of each hamburger was $2.25, and the cost of each hot dog was $1.50.

Check the answer.

$$14(2.25) + 12(1.50) = 49.50$$
$$10(2.25) + 7(1.50) = 33.00$$ ❖

Ongoing Assessment:
Write the system as a matrix:
$\begin{bmatrix} 3 & 1 & 0 \\ 2 & 3 & 7 \end{bmatrix}$.
Apply scalar multiplication.
$\begin{matrix} 3 \\ 1 \end{matrix}\begin{bmatrix} 3 & 1 & 0 \\ 2 & 3 & 7 \end{bmatrix}$
Simplify.
$\begin{bmatrix} 9 & 3 & 0 \\ 2 & 3 & 7 \end{bmatrix}$
Subtract row 1 from row 2.
$[-7 \quad 0 \quad 7] \rightarrow -7x = 7, x = -1$
Substitute -1 for x into the first equation; $y = 3$.

Cultural Connection: Asia Over 2000 years ago, the Chinese developed a way to solve systems of linear equations written in matrix form.

Given the equations: $\begin{cases} 4x + 5y = 22 \\ 2x + 3y = 13 \end{cases}$ The Chinese would form the matrices: $\begin{matrix} [4 & 5 & 22] \\ [2 & 3 & 13] \end{matrix}$

Apply scalar multiplication, $\begin{matrix} 1[4 & 5 & 22] \\ 2[2 & 3 & 13] \end{matrix}$, to produce $\begin{matrix} [4 & 5 & 22] \\ [4 & 6 & 26] \end{matrix}$.

Subtract row 1 from row 2. $[0 \quad 1 \quad 4] \longrightarrow y = 4$

Finally, substitute 4 for y in the first equation to get $x = \frac{1}{2}$.

Explain how to use the Chinese method to solve the following system.

$$\begin{cases} 3x + y = 0 \\ 2x + 3y = 7 \end{cases}$$

1. Write the equation $AX = B$ and solve: $\begin{bmatrix} 3 & -6 \\ 5 & 2 \end{bmatrix}\begin{bmatrix} a \\ b \end{bmatrix} = \begin{bmatrix} 9 \\ 9 \end{bmatrix}$

Explain how to represent each system in matrix equation form, AX = B.

1. $\begin{cases} 3a - 6b = 9 \\ 5a + 2b = 9 \end{cases}$

2. $\begin{cases} 12x + 5y = 30 \\ 4x - y = 6 \end{cases}$

Explain the method for solving the matrix equation for x and y.

3. $\begin{bmatrix} -2 & -1 \\ 1 & 0 \end{bmatrix}\begin{bmatrix} x \\ y \end{bmatrix} = \begin{bmatrix} -9 \\ -9 \end{bmatrix}$

4. $\begin{bmatrix} 1 & 1 \\ 1 & -1 \end{bmatrix}\begin{bmatrix} x \\ y \end{bmatrix} = \begin{bmatrix} 6 \\ 2 \end{bmatrix}$

2. Write the matrix equation $AX = B$ and solve: $\begin{bmatrix} 12 & 5 \\ 4 & -1 \end{bmatrix}\begin{bmatrix} x \\ y \end{bmatrix} = \begin{bmatrix} 30 \\ 6 \end{bmatrix}$

Practice & Apply

Represent each system of equations as a matrix equation, AX = B.

5. $\begin{cases} 4y = 6x - 12 \\ 5x = 2y + 10 \end{cases}$

6. $\begin{cases} 6a - b - 3c = 2 \\ -3a + b - 3c = 1 \\ -2a + 3b + c = -6 \end{cases}$

7. $\begin{cases} 2x + 3y + 4z = 8 \\ 6x + 12y + 16z = 31 \\ 4x + 9y + 8z = 17 \end{cases}$

Solve each system of equations using matrices, if possible.

8. $\begin{cases} 2x + y = 4 \\ x + y = 3 \end{cases}$ (1, 2)

9. $\begin{cases} 2x = 4 \\ 3x + y = 6 \end{cases}$ (2, 0)

10. $\begin{cases} 2x - 3y = 4 \\ -4x + 6y = 1 \end{cases}$ No solution

11. Which of the following ordered pairs represents the solution to $\begin{bmatrix} 3 & \frac{1}{2} \\ -2 & \frac{1}{3} \end{bmatrix}\begin{bmatrix} x \\ y \end{bmatrix} = \begin{bmatrix} -6 \\ 12 \end{bmatrix}$? c

a. $(-4, 6)$ **b.** $(4, 12)$ **c.** $(-4, 12)$ **d.** $(4, -6)$

Business The potential for growth in a career is an important consideration. The matrix at the right shows the projected job growth in two related fields.

Occupation	U.S. Jobs (in 1000s) 1990	2000
Electrical engineers	440	615
Electrical assemblers	235	135

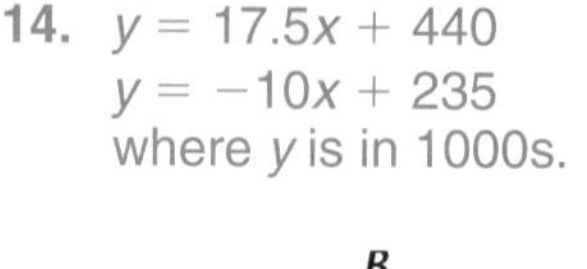

12. What is the projected average increase per year in the engineering jobs? 17,500 per year

13. What is the projected average decline per year in the assembling jobs? 10,000 per year

14. Write the linear equations that best fit this data.

15. In what year were the numbers of these two jobs equal? Use matrices to solve, and check by graphing. 1982

14. $y = 17.5x + 440$
$y = -10x + 235$
where y is in 1000s.

16. **Geometry** The perimeter of rectangle *PQRS* is 30. The perimeter of triangle *ABC* is 40. Find the value of x and y. 3.5, 4

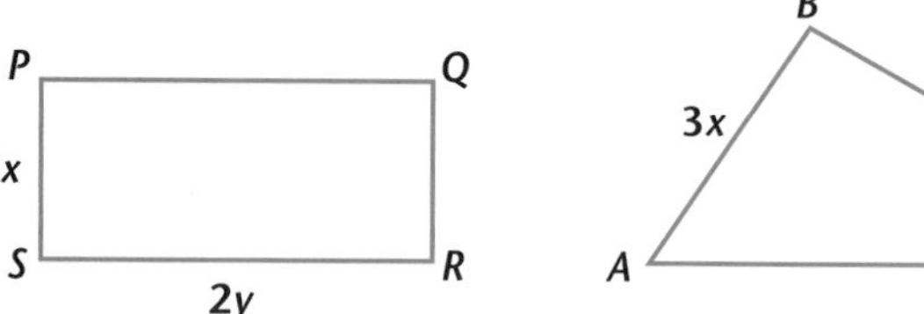

17. At Sid's Burger Barn Kisha and her friend Maya ordered 3 Sid burgers and 2 super shakes. They were joined by Jason and Steve, who ordered 4 Sid burgers and 4 super shakes. When the checks arrived, the girls' check was \$14, and the boys' check was \$21. What was the cost of a Sid burger, and what was the cost of a super shake? \$3.50, \$1.75

Business The matrices below indicate the sweater sales for three days at two branches of a certain department store.

Store #6 Sales

	Cashmere	Cotton	Wool
Day 1	4	8	3
Day 2	3	10	5
Day 3	5	6	8

Store #8 Sales

	Cashmere	Cotton	Wool
Day 1	2	5	3
Day 2	4	7	6
Day 3	3	8	8

The total sales for both stores for each day are

Day 1	\$465
Day 2	\$650
Day 3	\$730

18.

	Cashmere	Cotton	Wool
Day 1	6	13	6
Day 2	7	17	11
Day 3	8	14	16

18. Use matrix addition to find the total number of each type of sweater sold each day.

19. Solve a matrix equation to find the selling price of each type of sweater.
Cashmere: \$25; Cotton: \$15; Wool: \$20

Friday (Concert Ticket Sales)
Adult Tickets 150
Student Tickets 500
Total Tickets 650
Total \$1125

Saturday (Football Game Sales)
Adult Tickets 225
Student Tickets 675
Total Tickets 900
Total \$1575

20. **Entertainment** During Central City High School's spirit weekend, there was a concert on Friday night and an alumni football game on Saturday. For each event, adult and student tickets were sold. The ticket prices were the same for both events. Use the information at the left to find the price of an adult ticket and the price of a student ticket. \$2.50, \$1.50

Wages Kevin and Doug have part time jobs. One week Kevin worked 14 hours, and Doug worked 16 hours. Together they earned $161.50. Another week they each worked 12 hours and had combined earnings of $129.

21. How much does each earn per hour? Kevin $5.25, Doug $5.50

22. How much did Kevin earn during the two weeks? $136.50

23. Cultural Connection: Asia Solve the system using the Chinese method described on page 155. (−1, −4) $\begin{cases} 2x + 5y = -22 \\ 4x - 3y = 8 \end{cases}$

24. What method of solving linear equations does the Chinese method most resemble? Elimination

25. **Portfolio Activity** Complete the problem in the portfolio activity on page 121. There were 8 C-group sites, 5 Pan-grave sites, and 5 Transitional sites. $\begin{bmatrix} 8 \\ 5 \\ 5 \end{bmatrix}$

Look Back

26. If a pair of blue jeans costs a total of $36.70, including 7% sales tax, what was the cost of the jeans before tax? **[Lesson 1.3]** $34.30

27. Chemistry How many ounces of water should be added to a 7.5% acid solution to produce 30 ounces of 5.5% acid solution? **[Lesson 1.4]** 8 oz

Coordinate Geometry Write the equation of the line that passes through the point $(-2, 5)$ and is

28. parallel to the x-axis. $y = 5$

29. perpendicular to the x-axis. **[Lessons 2.1, 2.2]** $x = -2$

Look Beyond

30. A ball is tossed between two players. If x represents the distance from the player tossing the ball, then y represents the height of the ball. Would a third player between 6 and 7 feet from the player tossing the ball be able to intercept the ball? Explain your answer.

31. The given matrix represents 5 ordered pairs of a function. Use differences to show that the function represented by these points is quadratic.

Ordered Pairs

	1	2	3	4	5
x	3	4	5	6	7
y	0	2	6	12	20

Technology On a calculator, graph $y = \left|\frac{4}{3}x + 8\right|$. Find the points where this graph intersects the graph of each function.

32. $y = x$ **33.** $y = x + 5$ **34.** $y = 4$ (−3, 4) and (−9, 4)

32. No intersection points

33. No intersection points

In World War II, the Code Talkers, a group of Navajo marines, developed a code for sending messages in the Navajo language. No country ever broke this code.

SECRET CODES

Cryptography, the science of coding and decoding information, is an interesting application of the matrix techniques that you have learned.

*So, **XRXEPENZ!***

A	B	C	D	E	F	G	H	I	J	K	L	M	N
1	2	3	4	5	6	7	8	9	10	11	12	13	14
O	P	Q	R	S	T	U	V	W	X	Y	Z	*	
15	16	17	18	19	20	21	22	23	24	25	26	27	

Use asterisk * for space.

1 Use this table to switch to and from the number code.

The coding matrix that you will use is $\begin{bmatrix} 1 & -1 \\ -3 & 4 \end{bmatrix}$.

The secret message you want to send is ADIOS AMIGO. Change the letters to numbers.

A D I O S * A M I G O * → 1 4 9 15 19 27 1 13 9 7 15 27

Form a matrix with these numbers. It must be compatible with the 2 by 2 coding matrix so that they can be multiplied. This message matrix is a compatible 2 by 6 matrix with the last position filled with a space, 27.

(Message Matrix) $\begin{bmatrix} 1 & 4 & 9 & 15 & 19 & 27 \\ 1 & 13 & 9 & 7 & 15 & 27 \end{bmatrix}$

2 Multiply the message matrix by the coding matrix.

$$\begin{bmatrix} 1 & -1 \\ -3 & 4 \end{bmatrix}\begin{bmatrix} 1 & 4 & 9 & 15 & 19 & 27 \\ 1 & 13 & 9 & 7 & 15 & 27 \end{bmatrix} = \begin{bmatrix} 0 & -9 & 0 & 8 & 4 & 0 \\ 1 & 40 & 9 & -17 & 3 & 27 \end{bmatrix}$$

3 Some of the numbers in the coded message matrix are not in the alphabet code table because they are either greater than 27 or less than 1. If a number is less than 1, keep adding 27 to that number until the number corresponds to a value on the table. If the number is greater than 27, keep subtracting 27 from that number until the result corresponds to a value in the table.

0	−9	0	8	4	0	1	40	9	−17	3	27
27	18	27	8	4	27	1	13	9	10	3	27

4 Use the alphabet-code table to change the numbers into letters and symbols.

27 18 27 8 4 27 1 13 9 10 3 27 → * R * H D * A M I J C *

5 Send the coded message.

6 Receiver gets **coded message**, and uses the alphabet code table to form the coded-message matrix.

* R * H D * A M I J C * → 27 18 27 8 4 27 1 13 9 10 3 27

$$\begin{bmatrix} 27 & 18 & 27 & 8 & 4 & 27 \\ 1 & 13 & 9 & 10 & 3 & 27 \end{bmatrix}$$

7 Multiply the secret matrix by a **decoding** matrix.

If C (coding) is $\begin{bmatrix} 1 & -1 \\ -3 & 4 \end{bmatrix}$, then C^{-1} is $\begin{bmatrix} 4 & 1 \\ 3 & 1 \end{bmatrix}$.

Use C^{-1} to decode the coded message matrix.

$$\begin{bmatrix} 4 & 1 \\ 3 & 1 \end{bmatrix}\begin{bmatrix} 27 & 18 & 27 & 8 & 4 & 27 \\ 1 & 13 & 9 & 10 & 3 & 27 \end{bmatrix} = \begin{bmatrix} 109 & 85 & 117 & 42 & 19 & 135 \\ 82 & 67 & 90 & 34 & 15 & 108 \end{bmatrix}$$

8 Add or subtract 27 as explained in step 3 to find the numbers that fit the table.

109	85	117	42	19	135	82	67	90	34	15	108
1	4	9	15	19	27	1	13	9	7	15	27

9 Translate from numbers to letters using the alphabet-code table.

A D I O S * A M I G O S *

Activity 1

The message in the introduction of this lesson, XRXEPENZ, was coded with the matrix $\begin{bmatrix} 3 & 7 \\ 2 & 5 \end{bmatrix}$. Decode the message.

Activity 2

Use available technology to code, send, and decode the message BEWARE FEARLESS LEADER. Use the coding matrix $\begin{bmatrix} -1 & 3 \\ -2 & 4 \end{bmatrix}$.

Activity 3

$\begin{bmatrix} 2 & 3 \\ 1 & 4 \end{bmatrix}$ Write your own short message of 15 characters or less. Use the given coding matrix to form the secret message. Give the secret message and the coding matrix to a classmate to decode.

Chapter 3 Review

Vocabulary

Key Skills & Exercises

Lesson 3.1

➤ Key Skills

Determine the dimensions of a matrix and the addresses of its entries.

Let $M = \begin{bmatrix} -5 & 6 \\ 7 & -2 \\ 1 & 0 \end{bmatrix}$ and $N = \begin{bmatrix} -5 & 6 & 7 \\ -2 & 1 & 0 \end{bmatrix}$.

The dimensions of matrix M are 3×2.
The dimensions of matrix N are 2×3.

The entry m_{32}, located in the third row and second column of matrix M, is 0.
The entry n_{21}, located in the second row and first column of matrix N, is -2.
Matrices M and N cannot be equal because their dimensions are not equal.

➤ Exercises

Let $R = \begin{bmatrix} -4 & 1 \\ 2 & -3 \\ 2 & 0 \\ 8 & -5 \end{bmatrix}$.

1. What are the dimensions of matrix R? 4 by 2
2. Find the entry r_{32}. 0
3. Find the entry r_{41}. 8
4. What is the address of -5 in matrix R? r_{42}
5. Give a matrix that is equal to matrix R.

5. Answers may vary.
$\begin{bmatrix} 2(-2) & 3-2 \\ 4(\frac{1}{2}) & 3-6 \\ 16 \div 8 & \frac{0}{6} \\ 1+7 & -25(\frac{1}{5}) \end{bmatrix}$

Lesson 3.2

➤ Key Skills

Add matrices.

Evaluate the matrix operation.

$$\begin{bmatrix} -2.1 & 3.5 & 6.7 \\ 8.1 & -0.4 & -9.9 \\ 2.1 & 5.4 & -4.7 \end{bmatrix} + \begin{bmatrix} -6.4 & 3.2 & 3.5 \\ 6.5 & 7.6 & 1.2 \\ -9.3 & -4.2 & 0.5 \end{bmatrix}$$

Add entries with corresponding addresses.

$$\begin{bmatrix} -2.1+(-6.4) & 3.5+3.2 & 6.7+3.5 \\ 8.1+6.5 & -0.4+7.6 & -9.9+1.2 \\ 2.1+(-9.3) & 5.4+(-4.2) & -4.7+0.5 \end{bmatrix}$$

$$= \begin{bmatrix} -8.5 & 6.7 & 10.2 \\ 14.6 & 7.2 & -8.7 \\ -7.2 & 1.2 & -4.2 \end{bmatrix}$$

Subtract matrices.

Evaluate the matrix operation.

$$\begin{bmatrix} -2.1 & 3.5 & 6.7 \\ 8.1 & -0.4 & -9.9 \\ 2.1 & 5.4 & -4.7 \end{bmatrix} - \begin{bmatrix} -6.4 & 3.2 & 3.5 \\ 6.5 & 7.6 & 1.2 \\ -9.3 & -4.2 & 0.5 \end{bmatrix}$$

Subtract entries with corresponding addresses.

$$\begin{bmatrix} -2.1-(-6.4) & 3.5-3.2 & 6.7-3.5 \\ 8.1-6.5 & -0.4-7.6 & -9.9-1.2 \\ 2.1-(-9.3) & 5.4-(-4.2) & -4.7-0.5 \end{bmatrix}$$

$$= \begin{bmatrix} 4.3 & 0.3 & 3.2 \\ 1.6 & -8.0 & -11.1 \\ 11.4 & 9.6 & -5.2 \end{bmatrix}$$

➤ Exercises

Evaluate each matrix operation.

6. $\begin{bmatrix} 4 & 3 & 5 \\ 2 & 1 & 2 \\ 5 & 3 & 1 \end{bmatrix} + \begin{bmatrix} -3 & 1 & 3 \\ -4 & -3 & 1 \\ -1 & 5 & 9 \end{bmatrix}$

7. $\begin{bmatrix} 5.6 & 6.5 \\ 4.3 & 4.6 \end{bmatrix} + \begin{bmatrix} 7.1 & 3.4 \\ -6.7 & 1.3 \end{bmatrix}$ $\begin{bmatrix} 12.7 & 9.9 \\ -2.4 & 5.9 \end{bmatrix}$

8. $\begin{bmatrix} -6 & 1 \\ -1 & 2 \\ -2 & 7 \end{bmatrix} - \begin{bmatrix} 5 & -7 \\ -5 & 2 \\ -2 & 7 \end{bmatrix}$

9. $\begin{bmatrix} -9.9 & 8.8 \\ -2.3 & 1.2 \end{bmatrix} - \begin{bmatrix} 8.3 & 7.3 \\ -9.6 & -5.5 \end{bmatrix}$ $\begin{bmatrix} -18.2 & 1.5 \\ 7.3 & 6.7 \end{bmatrix}$

6. $\begin{bmatrix} 1 & 4 & 8 \\ -2 & -2 & 3 \\ 4 & 8 & 10 \end{bmatrix}$ 8. $\begin{bmatrix} -11 & 8 \\ 4 & 0 \\ 0 & 0 \end{bmatrix}$

Lesson 3.3

➤ Key Skills

Multiply a matrix by a scalar.

Evaluate the matrix operation.

$$4\begin{bmatrix} -4 & 8 \\ 9 & -5 \end{bmatrix}$$

Multiply each entry of the matrix by 4.

$$\begin{bmatrix} 4 \cdot -4 & 4 \cdot 8 \\ 4 \cdot 9 & 4 \cdot -5 \end{bmatrix} = \begin{bmatrix} -16 & 32 \\ 36 & -20 \end{bmatrix}$$

Multiply two matrices to form a product matrix.

Evaluate the matrix operation.

$$\begin{bmatrix} 4 & 1 & -6 \end{bmatrix} \begin{bmatrix} -1 & 4 \\ 2 & 7 \\ -5 & 3 \end{bmatrix}$$

The product can be found either by the matrix multiplication procedure or by technology. The product is [28 5].

➤ Exercises

Use the given matrices to evaluate the following operations. Write *not possible* when appropriate.

$A = \begin{bmatrix} -2 & -1 \\ 1 & 3 \end{bmatrix}$ $B = \begin{bmatrix} 1 & -3 & -4 \\ 3 & 5 & -1 \end{bmatrix}$ $C = \begin{bmatrix} 5 & 4 & 7 \\ 2 & 1 & 7 \\ 3 & 2 & 6 \end{bmatrix}$ $D = \begin{bmatrix} -1 & -3 \\ 4 & 5 \\ 6 & 1 \end{bmatrix}$

10. $3A$ **11.** $-4D$ **12.** AB **13.** BC **14.** BA **15.** $-DA$

10. $\begin{bmatrix} -6 & -3 \\ 3 & 9 \end{bmatrix}$ 11. $\begin{bmatrix} 4 & 12 \\ -16 & -20 \\ -24 & -4 \end{bmatrix}$ 12. $\begin{bmatrix} -5 & 1 & 9 \\ 10 & 12 & -7 \end{bmatrix}$ 13. $\begin{bmatrix} -13 & -7 & -38 \\ 22 & 15 & 50 \end{bmatrix}$ 14. Not possible 15. $\begin{bmatrix} 1 & 8 \\ 3 & -11 \\ 11 & 3 \end{bmatrix}$

Lesson 3.4

➤ Key Skills

Determine the inverse matrix for multiplication.

Find the inverse of $A = \begin{bmatrix} 4 & 3 \\ -1 & -2 \end{bmatrix}$.

By definition, $AA^{-1} = I$.

$$\begin{bmatrix} 4 & 3 \\ -1 & -2 \end{bmatrix}\begin{bmatrix} a & b \\ c & d \end{bmatrix} = \begin{bmatrix} 1 & 0 \\ 0 & 1 \end{bmatrix}$$

The inverse can be found by solving the following systems.

$$\begin{cases} 4a + 3c = 1 \\ -a - 2c = 0 \end{cases} \quad \begin{cases} 4b + 3d = 0 \\ -b - 2d = 1 \end{cases}$$

The inverse is $\begin{bmatrix} 0.4 & 0.6 \\ -0.2 & -0.8 \end{bmatrix}$.

➤ Exercises

Find the inverse matrix, if it exists, for the given matrices.

16. $\begin{bmatrix} -4 & 1 \\ 3 & -3 \end{bmatrix}$ **17.** $\begin{bmatrix} 0 & -1 \\ -1 & -1 \end{bmatrix}$ **18.** $\begin{bmatrix} 0.1 & 0.2 \\ 1 & 2 \end{bmatrix}$

16. $\begin{bmatrix} -0.33 & -0.11 \\ -0.33 & -0.44 \end{bmatrix}$ or $\begin{bmatrix} -\frac{1}{3} & -\frac{1}{9} \\ -\frac{1}{3} & -\frac{4}{9} \end{bmatrix}$

17. $\begin{bmatrix} 1 & -1 \\ -1 & 0 \end{bmatrix}$

18. Does not exist

Lesson 3.5

➤ *Key Skills*

Solve a system of linear equations using matrices.

Use matrices to solve the system.

$$\begin{cases} 3x + 2y = 16 \\ 2x + 3y = 19 \end{cases}$$

Prepare the matrix equation $AX = B$.

$$\underset{A}{\begin{bmatrix} 3 & 2 \\ 2 & 3 \end{bmatrix}} \underset{X}{\begin{bmatrix} x \\ y \end{bmatrix}} = \underset{B}{\begin{bmatrix} 16 \\ 19 \end{bmatrix}}$$

To solve the matrix equation, multiply each side of the matrix equation by A^{-1} on the left.

$$A^{-1} = \begin{bmatrix} 0.6 & -0.4 \\ -0.4 & 0.6 \end{bmatrix}$$

Since $A^{-1}AX = A^{-1}B$ simplifies to $X = A^{-1}B$, find $A^{-1}B$ to solve the system.

$$\underset{X}{\begin{bmatrix} x \\ y \end{bmatrix}} = \underset{A^{-1}B}{\begin{bmatrix} 2 \\ 5 \end{bmatrix}}$$, so the solution is (2, 5).

➤ *Exercises*

Solve each system of linear equations using matrices, if possible.

19. $\begin{cases} 2x + 5y = 12 \\ x + 3y = 7 \end{cases}$ (1, 2)

20. $\begin{cases} -3x + y = -5 \\ x + 2y = 10 \end{cases}$

21. $\begin{cases} 4x + 3y = 6 \\ x - 2y = -15 \end{cases}$

21. (−3, 6)

22. $\begin{cases} x - 7y = 5 \\ -4x + 5y = -20 \end{cases}$ (5, 0)

23. $\begin{cases} -2x - 6y = 3 \\ 4x + 8y = -5 \end{cases}$ (−0.75, −0.25)

24. $\begin{cases} x + y = 2 \\ 2x - y = 7 \end{cases}$ (3, −1)

20. $\left(\frac{20}{7}, \frac{25}{7}\right)$ or (2.86, 3.57)

Applications

25. Wages Mr. Tunney is giving four of his employees an 8% pay raise. Their current hourly wages are \$6.50, \$6.35, \$7.35, and \$6.75. Form a matrix for these hourly wages. Use scalar multiplication to find the new hourly wage for each employee. \$7.02, \$6.86, \$7.94, \$7.29

26. Stocks The matrix shows the number of shares of three different stocks that are owned by three different people.

	Stock 1	Stock 2	Stock 3
Gain or loss	5.50	−3.00	−4.50

	Julie	Kira	Tony
Stock 1	14	12	10
Stock 2	9	4	8
Stock 3	17	20	2

Julie: −\$26.50
Kira: −\$36.00
Tony: \$22.00

Last Friday, Stock 1 gained \$5.50 per share, Stock 2 lost \$3.00 per share, and Stock 3 lost \$4.50 per share. Use matrix multiplication to determine how much each person gained or lost last Friday?

27. Business Frank's Gym was having a sales promotion for two weeks. During the first week, 4 basic memberships and 2 gold memberships were sold for a total of \$2000. During the second week, 3 basic memberships and 3 gold memberships were sold for a total of \$2250. Write a system of two linear equations, and solve it using matrices to find the cost of a basic membership and the cost of a gold membership. $\begin{cases} 4x + 2y = 2000 \\ 3x + 3y = 2250 \end{cases}$ basic membership: \$250; gold membership: \$500

Chapter 3 Assessment

Use the given matrices to answer Items 1–21. Write *not possible* when appropriate.

$$M = \begin{bmatrix} -2 & -1 \\ 1 & 3 \end{bmatrix} \quad N = \begin{bmatrix} 1 & -3 & -4 \\ 3 & 5 & -1 \end{bmatrix} \quad P = \begin{bmatrix} 5 & 4 & 7 \\ 2 & 1 & 7 \end{bmatrix} \quad Q = \begin{bmatrix} -1 & -3 \\ 4 & 5 \\ 6 & 1 \end{bmatrix}$$

1. Give the dimensions of each matrix. 2×2, 2×3, 2×3, 3×2

2. Find the entry n_{23} in matrix N. -1

3. Find the entry q_{12} in matrix Q. -3

4. What is the address of 4 in matrix P? p_{12}

5. What is the address of 3 in matrix M? m_{22}

6. Explain how to determine whether a matrix is equal to matrix N.

7. Explain how to determine whether two matrices can be multiplied.

6. A matrix equal to N must have 2 rows and 3 col. and have entries that are equal to each corresponding entry in N.

7. Two matrices can be multiplied if the number of col. in the 1st matrix is equal to the number of rows in the 2nd matrix.

Evaluate each matrix operation.

8. $N + P$ **9.** $Q - P$ Not possible **10.** $P - N$

11. $2M$ **12.** $-4N$ **13.** MN

14. QP **15.** $-PQ$ **16.** NP Not possible

17. M^{-1} **18.** MM^{-1} **19.** P^{-1} Not possible

20. Let I be the identity matrix for addition. Evaluate $N + I$.

21. Let I be the identity matrix for multiplication. Evaluate MI.

8. $\begin{bmatrix} 6 & 1 & 3 \\ 5 & 6 & 6 \end{bmatrix}$ 10. $\begin{bmatrix} 4 & 7 & 11 \\ -1 & -4 & 8 \end{bmatrix}$ 11. $\begin{bmatrix} -4 & -2 \\ 2 & 6 \end{bmatrix}$

12. $\begin{bmatrix} -4 & 12 & 16 \\ -12 & -20 & 4 \end{bmatrix}$ 13. $\begin{bmatrix} -5 & 1 & 9 \\ 10 & 12 & -7 \end{bmatrix}$

14. $\begin{bmatrix} -11 & -7 & -28 \\ 30 & 21 & 63 \\ 32 & 25 & 49 \end{bmatrix}$ 15. $\begin{bmatrix} -53 & -12 \\ -44 & -6 \end{bmatrix}$

17. $\begin{bmatrix} -0.6 & -0.2 \\ 0.2 & 0.4 \end{bmatrix}$

18. $\begin{bmatrix} 1 & 0 \\ 0 & 1 \end{bmatrix}$

20. $\begin{bmatrix} 1 & -3 & -4 \\ 3 & 5 & -1 \end{bmatrix}$

21. $\begin{bmatrix} -2 & -1 \\ 1 & 3 \end{bmatrix}$

22. If it is possible to multiply matrix A times matrix B, is it also possible to multiply matrix B times matrix A? Explain your answer. Sometimes, if A and B are square matrices.

23. **Business** Kyle and Kodie both work at the Ice Cream Shop and at the Burger Palace. Kyle makes \$6.50 per hour at the Ice Cream Shop and \$6.00 per hour at the Burger Palace. Kodie makes \$6.75 per hour at the Ice Cream Shop and \$6.50 per hour at the Burger Palace. Last week, they each worked 15 hours at the Ice Cream Shop and 12 hours at the Burger Palace. Form a matrix containing hourly wages and a matrix containing hours worked. Use matrix multiplication to find the amount that each earned last week. \$169.50; \$179.25

Solve each system of linear equations using matrices, if possible.

24. $\begin{cases} -2x - 3y = 5 \\ x + 2y = -2 \end{cases}$ **25.** $\begin{cases} -5x + 2y = -3 \\ 10x - 4y = 6 \end{cases}$ Infinite no. of solutions **26.** $\begin{cases} -2x + 7y = 0 \\ x + 3y = 1 \end{cases}$

24. $\begin{bmatrix} -4 \\ 1 \end{bmatrix}$

26. $\begin{bmatrix} \frac{7}{13} \\ \frac{2}{13} \end{bmatrix}$

27. **Geometry** Two supplementary angles have measures whose sum is 180°. Two complementary angles have measures whose sum is 90°. Two supplementary angles have measures of $3x$ and $3y$, and two complementary angles have measures of $4x$ and y. Write a system of two linear equations. Use matrices to solve the system, and find the measures of the four angles. 30°, 150°, 40°, 50°

CHAPTER 4

Probability and Statistics

LESSONS

Probability touches your life in many ways. Games, insurance, health and diet, and countless other situations involve an element of chance. Statistics makes extensive use of probability to study, predict, and draw conclusions from data. Sometimes probabilities are found experimentally by studying data. Other times, probabilities are found theoretically by considering all the possibilities. In this chapter, you will learn about both kinds of probability and explore examples of probability in action.

Portfolio Activity

Suppose that on a true-false quiz, a student guesses at all five answers. Use experimental or theoretical methods to find the probability that the student gets at least 80% of the answers correct.

LESSON 4.1 Exploring Experimental Probability

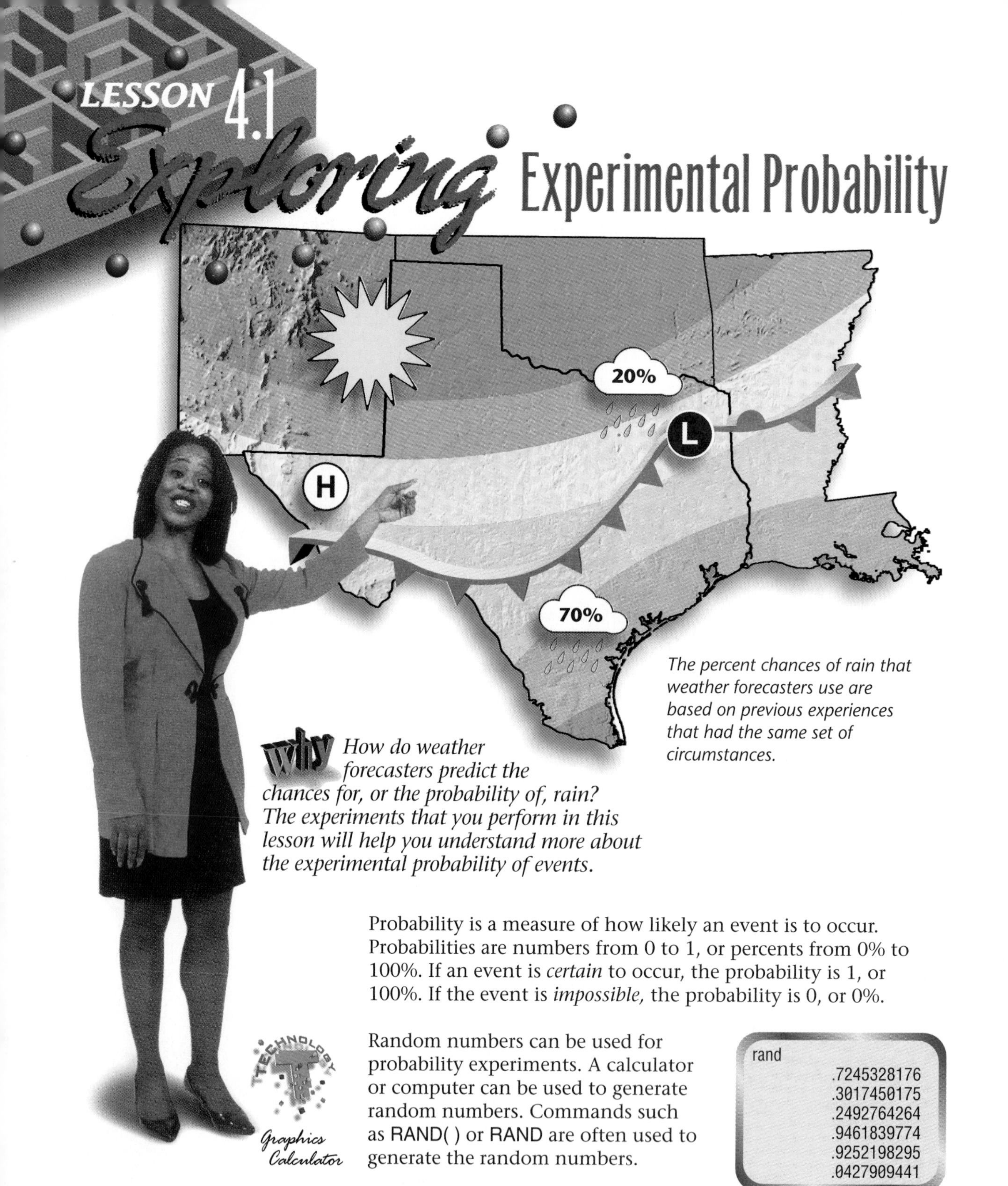

The percent chances of rain that weather forecasters use are based on previous experiences that had the same set of circumstances.

why *How do weather forecasters predict the chances for, or the probability of, rain? The experiments that you perform in this lesson will help you understand more about the experimental probability of events.*

Probability is a measure of how likely an event is to occur. Probabilities are numbers from 0 to 1, or percents from 0% to 100%. If an event is *certain* to occur, the probability is 1, or 100%. If the event is *impossible,* the probability is 0, or 0%.

TECHNOLOGY Graphics Calculator

Random numbers can be used for probability experiments. A calculator or computer can be used to generate random numbers. Commands such as RAND( ) or RAND are often used to generate the random numbers.

```
rand
       .7245328176
       .3017450175
       .2492764264
       .9461839774
       .9252198295
       .0427909441
```

To simulate a coin toss, you can use your calculator or computer to randomly generate the numbers 1 (for *heads*) and 2 (for *tails*). For example, suppose that you want to simulate tossing a coin 6 times and find the number of times that heads occur.

INT(RAND*2) + 1

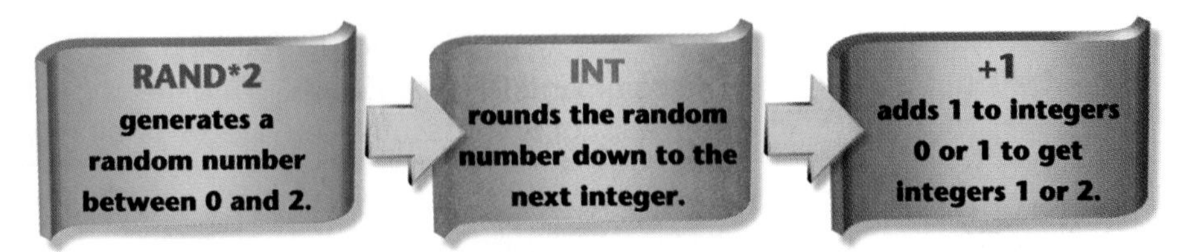
RAND*2
generates a random number between 0 and 2.
INT
rounds the random number down to the next integer.
+1
adds 1 to integers 0 or 1 to get integers 1 or 2.

Use the RAND function multiplied by 2 to generate numbers from 0 to 2, not inclusive. Then use the integer function INT to round the random number *down* to the next integer. Since this will produce the numbers 0 and 1, add 1 to adjust the numbers up to 1 and 2.

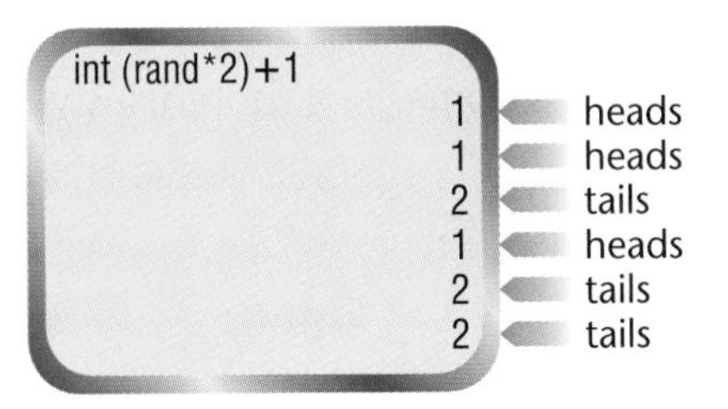
int (rand*2)+1
1 heads
1 heads
2 tails
1 heads
2 tails
2 tails

In this experiment, the number of **trials** is 6 (6 coin tosses), and the number of times that the **successful event** (heads) occurred is 3. So based on this experiment, the *experimental probability* of getting heads on a coin toss is

$$P(\text{heads}) = \frac{\text{number of heads}}{\text{number of trials}} = \frac{3}{6} = \frac{1}{2} = 50\%.$$

EXPERIMENTAL PROBABILITY

Let t be the number of trials in the experiment. Let f be the number of times that a successful event occurs.

The **experimental probability,** P, of the event is given by

$$P = \frac{f}{t}.$$

Exploration 1

Experiments With Coin Tosses

You can use a random-number generator to simulate tossing a coin in this exploration.

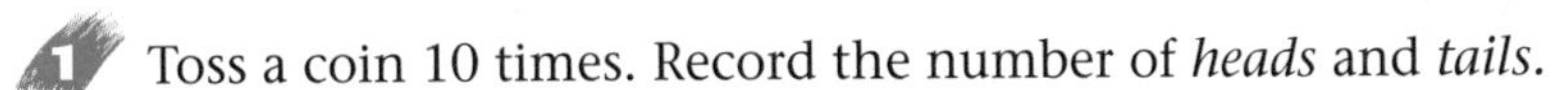

1. Toss a coin 10 times. Record the number of *heads* and *tails*.
2. Find the experimental probability of *heads*. Find the experimental probability of *tails*.
3. Combine all of the data from your class. Find the experimental probability of *heads* from the class data. Find the experimental probability of *tails* from the class data.
4. Find the sum of probabilities for heads and for tails from Step 2. Find the sum of probabilities for heads and for tails from Step 3. Are the two sums equal? Explain why or why not.
5. How does the experimental probability that you found in Step 2 compare with the experimental probability from the class data that you found in Step 3? ❖

CRITICAL Thinking

How can you use your experimental probabilities from Exploration 1 to predict the number of times you will get heads in 20 tosses of a coin?

Experimental probability may vary when an experiment is conducted more than one time.

GENERATING RANDOM INTEGERS

The command INT(RAND * K) + A generates random integers from a list of K consecutive integers beginning with A.

EXTENSION

What are the possible numbers that can appear at random when you use the following command?

INT(RAND * 4) + 3

The command **RAND * 4** results in a decimal value from 0 to 4, not inclusive.

The command **INT(RAND * 4)** generates integers 0, 1, 2, and 3.

The command **INT(RAND * 4) + 3** generates integers 3, 4, 5, and 6. ❖

Exploration 2 *Probability From a Number Cube*

You can use a random-number generator to simulate rolling an ordinary 6-sided number cube in this exploration.

1. The faces on an ordinary 6-sided number cube are numbered 1 to 6. If the cube is rolled 12 times, how many times would you *expect* the number 4 to appear on the top of the cube?

2. Roll an ordinary 6-sided number cube 12 times. Explain how to use a random-number generator to simulate this experiment. Count the number of times you get a 4.

3. In this experiment, what is a successful event? What is a trial?

3. A successful event is rolling a 4; a trial is rolling the cube.

4. What is your experimental probability of getting a 4 in 12 rolls of an ordinary 6-sided number cube?

5. How can you use the experimental probability that you found in Step 4 to predict the number of times you would expect to get a 4 in 24 rolls of an ordinary 6-sided number cube? ❖

$24 \cdot \frac{1}{12} = 2$ times

or

$\frac{1}{12} = \frac{f}{24}$; $f = 2$ times

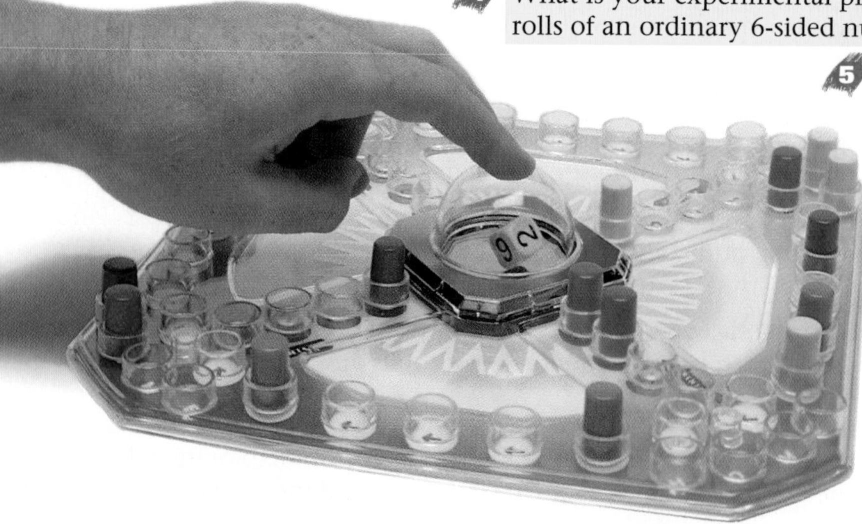

Communicate

1. Explain what experimental probability is. Give two examples.
2. In an experiment in which you toss a coin, explain how you can determine the experimental probability of *heads* if you know the experimental probability of *tails*.
3. To determine an experimental probability, Janis conducted 10 trials and Juanita conducted 12 trials. Is it possible for both Janis and Juanita to get the same experimental probability? Explain.
4. Explain how to use a random-number generator to simulate 8 tosses of a coin. Use the random-number generator INT(RAND*2), where 0 indicates heads and 1 indicates tails. Use this random number generator 8 times.

Practice & Apply

The results of spinning a spinner with regions I, II, and III 100 times are given in the table below. Use this table for Exercises 5–9.

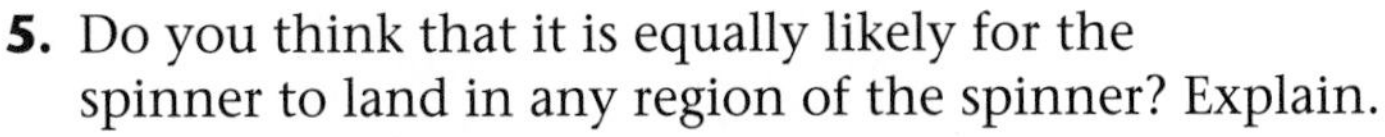

Region	I	II	III
Frequency	21	26	53

Notice that regions I, II, and III are not equal in size.

5. Do you think that it is equally likely for the spinner to land in any region of the spinner? Explain.
6. Find the experimental probability for landing in region I. 21%
7. Find the experimental probability for landing in region II. 26%
8. Find the experimental probability for landing in region III. 53%
9. Find the sum of the experimental probabilities for landing in region I and for landing in region II from Exercises 6 and 7. Is this close to the experimental probability for landing in region III from Exercise 8? Why or why not?

The results of 5 groups each tossing a coin 10 times and counting the number of *heads* is shown.

10. What is each group's experimental probability of getting heads?
11. Find the overall experimental probability of getting heads. 44%
12. What is each group's experimental probability of getting tails?
13. Find the overall experimental probability of getting tails. 56%
14. For each group, what is the sum of the answers to Exercises 10 and 12? Why?
15. What is the sum of the answers to Exercises 11 and 13? Why?

The chances of rain are based on experimental probabilities

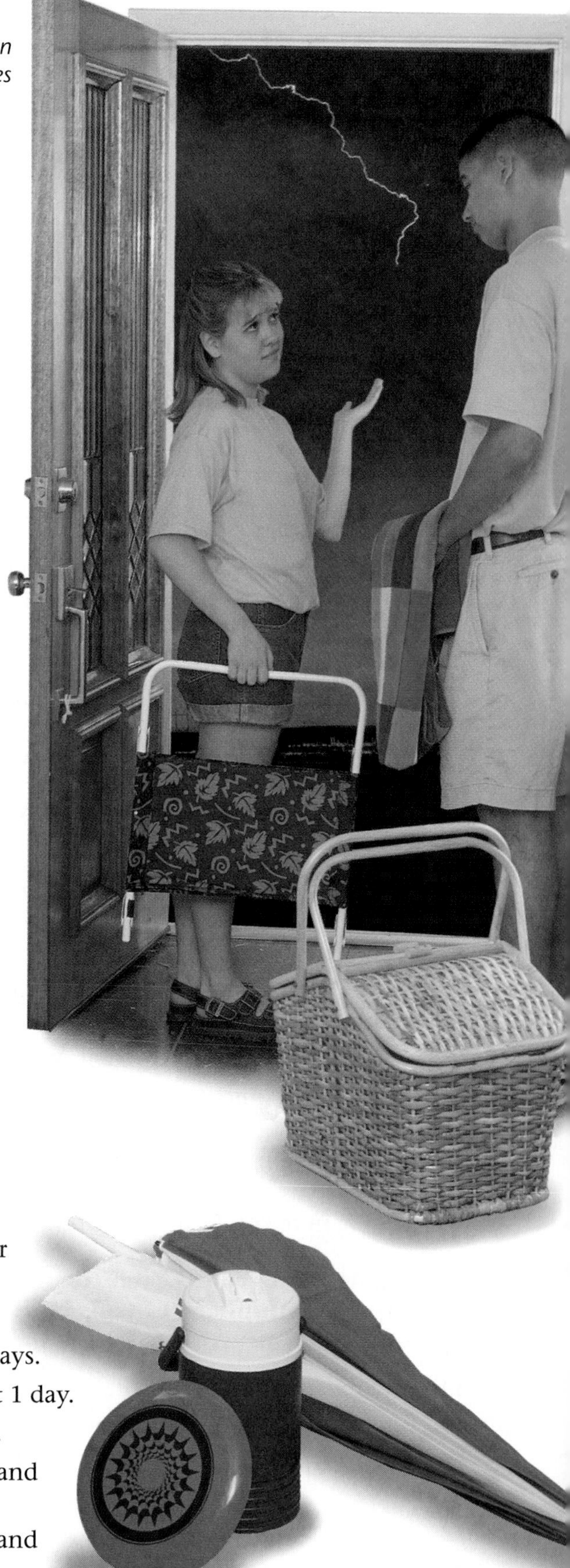

The results of 20 trials of tossing a coin 2 times is shown below. The experiment simulates a 50% probability of rain on 2 consecutive days. In the experiment, H represents rain and T represents no rain. For example, HT represents rain on the first day and no rain on the second day.

TT	TT	TH	HT	HT
HT	HH	HT	HH	HT
HH	HT	HH	TT	TH
HT	TH	HH	TH	TH

Use the results above to find the experimental probability of each event.

16. rain on both days 25%

17. rain on at least one day 85%

18. rain on the first day 60%

19. rain on the second day 50%

20. no rain on either day 15%

21. Find the sum of the experimental probabilities from Exercises 16 and 20. Is the sum 1? Explain.

22. Find the sum of the experimental probabilities from Exercises 17 and 20. Is the sum 1? Explain.

Technology Describe the output of each command.

23. INT(RAND)

24. INT(RAND*5)

25. INT(RAND*10)

26. INT(RAND*10) + 1

27. INT(RAND*12) + 1

28. INT(RAND*365) + 1

Technology The probability of rain is 40% on 2 consecutive days. Simulate the 2 consecutive days 10 times. Use integer values from 1 to 4 for rain and integer values from 5 to 10 for no rain.

29. Write the command used, and record your results.

30. Find the experimental probability of rain on both days.

31. Find the experimental probability of rain on at least 1 day.

32. Find the experimental probability of rain on 0 days.

33. Find the sum of the probabilities from Exercises 31 and 32. Is the sum 1? Why or why not?

34. Find the sum of the probabilities from Exercises 30 and 32. Is the sum 1? Why or why not?

Technology The probability of rain is 60% on each of 2 consecutive days. Simulate the 2 consecutive days 10 times. Use integer values from 1 to 6 for rain and integer values from 7 to 10 for no rain.

35. Write the calculator command used, and record your results.

36. Find the experimental probability of rain on both days.

37. Find the experimental probability of rain on at least 1 day.

38. Find the experimental probability of rain on 0 days.

39. Find the sum of the probabilities from Exercises 36 and 38. Is the sum 1? Why or why not?

40. Find the sum of the probabilities from Exercises 37 and 38. Is the sum 1? Why or why not? 1 or 100%; Yes; All possible events are included in this sum.

Technology Write commands to generate the random numbers described. If possible, check your results with a graphics calculator or a spreadsheet.

41. integer values from 0 to 3
INT(RAND*4)

42. integer values from 0 to 7
INT(RAND*8)

43. integer values from 1 to 7
INT(RAND*7) + 1

44. integer values from 4 to 9
INT(RAND*6) + 4

Look Back

Find the slope of the line through each pair of points. [Lesson 2.1]

45. $(-2, 4)$ and $(-2, -5)$
Undefined

46. $(6, 1)$ and $(-9, 4)$ $-\frac{1}{5}$

47. Recreation Jenny pays a membership fee of \$47 plus \$4 per hour at a tennis club. Carrie pays a flat fee of \$275. How many hours can Jenny play tennis before her cost equals Carrie's cost? **[Lesson 2.4]** 57 hours

48. Solve this system of equations. **[Lessons 2.4, 2.5]**

$$\begin{cases} 3x + 5y = 2 \\ x - 3y = -4 \end{cases} \qquad (-1, 1)$$

49. Technology Write the system of equations below as a matrix equation, and solve by using matrix methods. **[Lesson 3.5]**

$$\begin{bmatrix} 4 & 1 \\ 1 & -2 \end{bmatrix}\begin{bmatrix} x \\ y \end{bmatrix} = \begin{bmatrix} 3 \\ -5 \end{bmatrix} \qquad \begin{cases} 4x + y = 3 \\ x - 2y = -5 \end{cases}$$

Approx $(0.11, 2.56)$ or $\left(\frac{1}{9}, 2\frac{5}{9}\right)$

Look Beyond

Probability Fred conducts 20 trials of an experiment in which he tosses a coin 4 times. He uses the results to find the experimental probability that a family with 4 children has 4 boys and 0 girls. He finds that 4 boys occur in 3 of the 20 trials.

50. What is the experimental probability that a family with 4 children has 4 boys? $\frac{3}{20}$ or 15%

51. What are all of the possible outcomes for the 4 children?

52. Out of 20 families with 4 children, how many families would you *expect* to have 4 boys and 0 girls? 1 or 2 families

LESSON 4.2

Exploring Simulations

Amy makes 50% of her field-goal attempts.

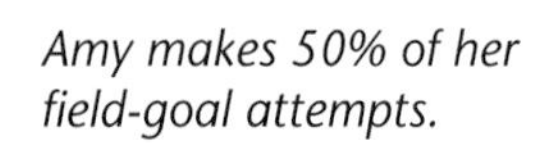

Probabilities derived from coin, number cube, or technology experiments can be used to simulate outcomes when you want to find experimental probabilities. Simulations are used in studying games, sports, weather, political elections, and many other situations.

Amy made four consecutive successful shots against Johnson High School. How often do you think Amy makes four successful shots in a row? If you had the actual shot-by-shot records of Amy's performance, you could see how often this happens. Even without the records, you can find the experimental probability by using a *simulation.* A **simulation** is an experiment with mathematical characteristics that are similar to the actual event.

Suppose Amy makes 50% of the shots she takes in a game. If Amy takes 20 shots, a simulation can be designed to find the experimental probability that she will make four in a row.

A spreadsheet can be used to generate random numbers.

1. Let 1 represent each shot made, and let 0 represent each shot missed.
2. Each row represents a trial. Generate 20 random numbers for each row. Thus, each row will have 20 columns.
3. As many rows as needed can be generated. For this simulation examine the first 10 rows or trials, as shown on the next page. Count the number of trials in which a sequence of at least four consecutive 1s appear.

B	C	D	E
Results			
1	1	1	0

The beginning of the spreadsheet shows that in the first trial Amy made her first three shots and missed the next one.

At the end of the first trial, Amy made 4 in a row.

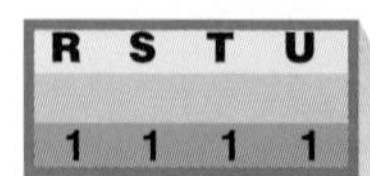

=INT(RAND()*2)

	A	B	C	D	E	F	G	H	I	J	K	L	M	N	O	P	Q	R	S	T	U
1	Trial	Results																			
2	1	1	1	1	0	1	1	1	0	1	0	0	0	0	1	1	0	1	1	1	1
3	2	1	0	0	0	1	0	0	0	0	0	0	0	0	1	1	1	0	1	1	0
4	3	0	0	1	1	1	1	1	0	1	1	1	1	0	1	0	0	0	0	0	1
5	4	1	1	1	0	1	0	0	1	0	1	1	0	0	0	1	1	1	0	0	0
6	5	1	1	1	0	1	1	0	1	1	0	1	1	0	0	0	1	0	1	1	1
7	6	0	0	1	0	1	0	0	1	0	0	1	1	1	1	0	1	1	1	1	0
8	7	1	0	0	0	1	0	0	0	1	1	1	1	0	1	0	0	0	0	0	0
9	8	1	0	1	0	1	0	1	1	1	0	1	1	1	0	0	1	0	1	0	0
10	9	1	0	1	0	1	1	1	1	1	0	0	1	0	0	1	1	0	0	0	0
11	10	1	1	1	1	1	1	0	1	1	1	0	1	1	0	0	0	1	0	1	1

Of the 10 trials in rows 2 through 11, there were at least four consecutive 1s recorded in 6 of the trials. The experimental probability that Amy will make 4 shots in a row is $\frac{6}{10}$. ❖

1–4. Answers will depend on the number of heads.

$$P = \frac{\text{at least 4 heads in a row}}{20}$$

Exploration 1 *Basketball Simulation*

Toss a coin to simulate a 50% chance of making a shot.

1. Toss a coin 20 times, and record the sequence of heads and tails.
2. Does the sequence have at least four consecutive heads?
3. Combine the results of the entire class.
4. Give the experimental probability based on the entire class results. ❖

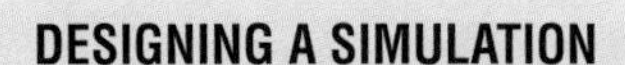

DESIGNING A SIMULATION

1. Choose a random number generator such as a coin toss, number-cube roll, calculator, computer, or other method that can be used to simulate a situation. Describe what each result represents.
2. Plan how to perform the experiment to simulate one trial.
3. Perform a large number of trials, and record the results.

Exploration 2 *Weather Simulation*

Perform a simulation to find the experimental probability that it rains at least one of the two days. **1–3.** Answers may vary depending on the results of the experiment.

The weather report predicts that the probability of rain is 40% on Saturday and 40% on Sunday.

Simulate a 40% chance of rain. If you use technology, generate a sequence of random integers from 1 to 100. A number less than or equal to 40 means that it will rain. A number greater than 40 means that it will not rain.

You can also use slips of paper or chips. Make 4 slips that show *rain* and 6 that show *no rain*. Then draw at random from a bag.

To simulate the weather for two days, generate two numbers. This represents one trial.

Perform a large number of trials, and record the results.

The spreadsheet shows the simulation for 10 trials.

=INT(RAND()*100)+1				
	A	**B**	**C**	
1	Trial	1st Number	2nd Number	*At least one number less than or equal to 40?*
2	1	98	68	*no*
3	2	33	82	*yes*
4	3	21	17	*yes*
5	4	94	14	*yes*
6	5	87	36	*yes*
7	6	73	83	*no*
8	7	56	73	*no*
9	8	65	12	*yes*
10	9	18	58	*yes*
11	10	99	**18**	*yes*

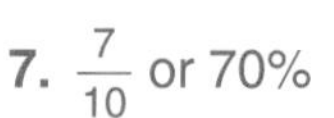

What do the numbers 98 and 68 in Trial 1 indicate about rain on those days? It will not rain on Sat. or Sun.

What do the numbers 33 and 82 in Trial 2 indicate about rain on those days? It will rain on Sat. but not on Sun.

According to the data, how many weekends had rain at least one day in the 10 trials? 7 weekends

What is the experimental probability of having rain on at least one of the two days? ❖

APPLICATION

The probability that Allessandro shows up on time is 60%. The probability that Zita shows up on time is 80%.

Allessandro and Zita plan to meet for lunch. A simulation can be designed to find the experimental probability that both Allessandro and Zita show up on time for lunch.

Design 1: Using number cubes

1. Roll two number cubes, a red one and a green one, for example. Let a 1, 2, or 3 on the red cube mean that Allessandro is on time, and let a 4 or 5 mean that he is late. Roll again if you get a 6. Note that 60% of the rolls that count mean that he is on time, and let 40% mean that he is late. Let a 1, 2, 3, or 4 on the green cube mean that Zita is on time, and let a 5 mean that she is late. Roll again if you get a 6.
2. Roll the two number cubes to represent one trial, which represents one meeting.
3. Roll the cubes 10 times. Disregard any rolls that include a 6. Determine how many times, out of the 10 attempts, they meet on time for lunch. How many times does the red cube show a 1, 2, or 3 and the green one show a 1, 2, 3, or 4?
4. Express the experimental probability as a fraction with *the number of times the friends are on time* over *the total number of trials*. This represents the experimental probability that both friends are on time for lunch.

Spreadsheet

Design 2: Using technology

1. Use a calculator or computer to generate a pair of random integers from 1 to 100. If the first number is less than or equal to 60, it means that Allessandro is on time. If the second number is less than or equal to 80, it means that Zita is on time.
2. Each pair of numbers represents one trial, or meeting.
3. Generate 10 pairs. Determine how many times out of 10 that the friends meet on time.
4. Write the fraction with the number of successful trials over the total number of trials. This represents the experimental probability that both friends are on time for lunch. ❖

CT– Situations will vary. Sample answers: chance of a baby being a boy or a girl; chances of it raining, based on a weather forecast.

Make a list of real-life situations that can be simulated by a coin toss and by a roll of a number cube. Compare the characteristics of the different types of situations.

EXERCISES & PROBLEMS

Communicate

1. What are the three steps for designing a simulation?
2. Assume that the chance of making a shot is 50%. Give three different ways to simulate making or missing a shot.
3. Suppose you want to simulate a day when there is a 30% chance of rain. Tell how to do this by using random numbers from 1 to 10.
4. Name three things that could be simulated by tossing a coin one time.
5. Tell how you might simulate the situation of correctly guessing a multiple-choice item with 5 possible responses.
6. Suppose 90% of the flights for an airline are on time. Explain how to design a simulation to find the experimental probability that three consecutive flights are on time.

Practice & Apply

Sports Look at the data from the basketball simulation on page 173.

7. How many shots did Amy make in the first trial? 13
8. In how many trials did Amy make more than 50% of her shots? 6
9. Give the value in cell Q4. What formula was used to generate the value? 0; INT(RAND*2)
10. What does the value in cell Q4 tell you? Amy did not make her 16th shot in trial 3.

Meteorology Suppose you want to simulate two days during which there is an 80% chance of rain in order to find out if it rains on at least one of the two days. Explain how you would perform the simulation by

11. generating random numbers from 1 to 10.
12. rolling a six-sided number cube.

Describe one way to simulate selecting at random

13. a day of the week.
14. a day of the year.
15. 1 student out of 24 students.
16. which of 8 team captains gets to choose first.

Describe a simulation for Exercises 17–22.

17. You are given two random numbers between 1 and 100. Find the experimental probability that both numbers are less than or equal to 20.
18. There are 10 ways to make a choice. There are 3 possible winning choices. Find the experimental probability that a person will win on 2 of 5 choices.

19. Find the experimental probability that a family with 4 children includes 2 boys and 2 girls. Assume that the births of boys and girls are equally likely.

20. Find the experimental probability that a student guesses correctly on all 3 questions of a true-false quiz.

21. Find the experimental probability of guessing the correct answer at least 70% of the time on a 4-question multiple choice quiz, with each question having 5 possible responses.

22. Repeat Exercise 21 for a 6-question quiz.

Look Back

23. Draw examples, and explain the difference in the graphs of a quadratic function and an exponential function. **[Course 1]**

24. If the variable n represents the number of the term, write an expression based on n that will generate any term of the following sequence: 4, 8, 12, 16, 20, 24, . . . **[Lesson 1.1]** The nth term is equal to $4n$.

25. Write an expression to represent the length, l, of a piece of fabric that is 74 centimeters longer than the width, w. **[Lesson 1.2]** $l = w + 74$

26. Solve Exercise 25 for the width, w, if the length, l, is 86 centimeters. **[Lesson 1.2]** 12 cm

27. Solve $ac = p$ for c. **[Lesson 1.2]** $c = \frac{p}{a}$

28. Graph the solution to $5n + 7 < 8$ on a number line. **[Lesson 1.5]**

29. **Coordinate Geometry** A line passes through the origin and the point $A(3, -2)$. Write the equation for the line that passes through point A and is perpendicular to the original line. **[Lesson 2.2]** $3x - 2y = 13$

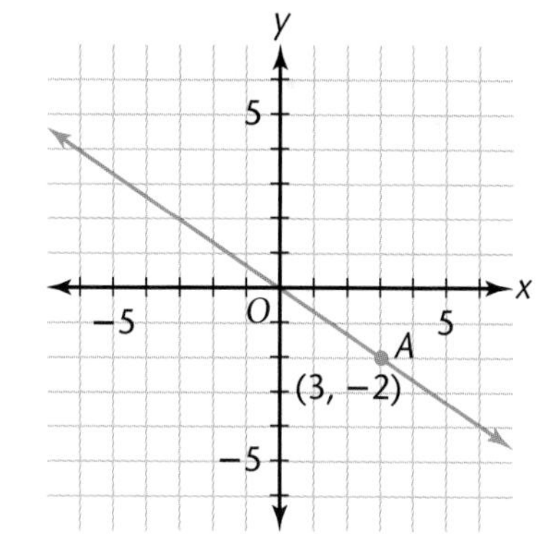

30. Multiply the two pairs of matrices. Are the results the same for each pair? **[Lesson 3.3]** no

$$\begin{bmatrix} -1 & 3 \\ 5 & 2 \end{bmatrix}\begin{bmatrix} 2 & -4 \\ 1 & -3 \end{bmatrix} \quad \text{and} \quad \begin{bmatrix} 2 & -4 \\ 1 & -3 \end{bmatrix}\begin{bmatrix} -1 & 3 \\ 5 & 2 \end{bmatrix}$$

31. Multiply the two pairs of matrices. Are the results the same for each pair? **[Lesson 3.3]** Yes

$$\begin{bmatrix} -1 & 3 \\ 3 & -1 \end{bmatrix}\begin{bmatrix} 2 & 1 \\ 1 & 2 \end{bmatrix} \quad \text{and} \quad \begin{bmatrix} 2 & 1 \\ 1 & 2 \end{bmatrix}\begin{bmatrix} -1 & 3 \\ 3 & -1 \end{bmatrix}$$

32. Is matrix multiplication commutative? Explain using examples. **[Lesson 3.3]** No; although the two matrices in Exercise 31 are commutative, the matrices in Exercise 30 are not.

Look Beyond

$\frac{1}{8}$ or 12.5%

33. Find the *theoretical* probability that a student guesses correctly on all 3 questions of a true-false quiz.

LESSON 4.3 Statistics and Probability

You already may be familiar with several topics from statistics, such as mean, median, mode, range, scatter plots, correlation, and lines of best fit. Descriptive statistics are often used to make predictions based on experimental probabilities.

Batting Averages

National League

Year	Name	Team	Avg.
1989	Tony Gwynn	Pittsburgh	.336
1990	Willie McGee	St. Louis	.335
1991	Terry Pendleton	Atlanta	.319
1992	Gary Sheffield	San Diego	.330
1993	Andres Galarraga	Colorado	.370

American League

Year	Name	Team	Avg.
1989	Kirby Puckett	Minnesota	.339
1990	George Brett	Kansas City	.329
1991	Julio Franco	Texas	.342
1992	Edgar Martinez	Seattle	.343
1993	John Olerud	Toronto	.363

Descriptive Statistics

There are many ways statistics can help you describe data. For example, *mean, median,* and *mode* indicate the central tendency of data. The *range* provides information about the spread of the data.

EXAMPLE 1

Give the mean, median, mode, and range for the batting averages in each league. Use these statistics to write a summary statement.

Solution ➤

Sports

To find the **mean**, add the batting averages for each player in each league. Then divide each sum by 5. The means are .338 and .3432, respectively. The .3432 is customarily rounded to .343.

To find the **median**, arrange the batting averages in order, and then choose the middle number. If the number of batting averages is even, the median is the mean of the two middle numbers. The National League batting averages in order from least to greatest are .319, .330, .335, .336, and .370. The median is .335. The median for the American League is .342.

To find the **mode**, choose the batting average that occurs the most. Since no number appears more than once in either set of data, there is no mode.

To find the **range**, determine the difference between the highest and lowest batting averages. The range of the National League is .370 − .319 = .051. The range of the American League is .363 − .329 = .034.

In summary, the mean and median batting averages were higher in the American League. The batting averages in the National League had a greater spread, or range. ❖

Try This Find the mean, median, and mode of the number of fans with season tickets that entered the stadium during a 15-game study.

Number of fans: 24, 38, 26, 31, 31, 30, 29, 29, 37, 32, 39, 40, 29, 35, 45

mean = 33, median = 31, mode = 29

Recall from Course 1 that a **scatter plot** is a display of data that has been organized into ordered pairs and graphed on the coordinate plane. A **line of best fit** can be used to model data that have an approximately linear relationship. A **correlation coefficient,** ranging from −1 to 1, measures how closely the data points fall along the line of best fit.

EXAMPLE 2

Graphics Calculator

Make a scatter plot of the batting averages for each year for the National League. Find the correlation coefficient and the line of best fit.

Solution ➤

Using the data from page 178, plot the data points for a scatter plot and find the line of best fit.

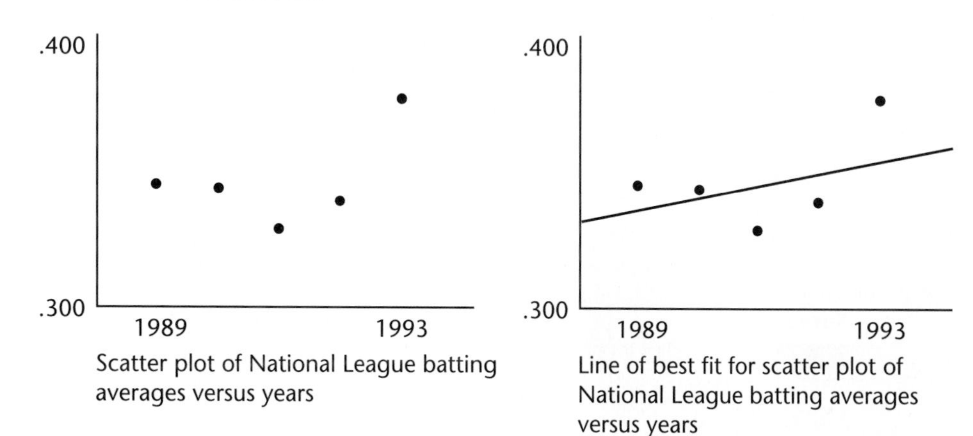

Scatter plot of National League batting averages versus years

Line of best fit for scatter plot of National League batting averages versus years

The variables show a moderate relationship. The correlation is about 0.52. What does this trend suggest about the batting averages over the five-year period? ❖ This trend suggests that the batting averages increased over time.

Using Survey Information

You have probably heard or read the results of many different *surveys*. **Surveys** give people information about entire groups, or **populations.**

Mr. Phillips's class created a survey to find out students' musical tastes. The survey asked students to rate different kinds of music on a scale of 1 to 5. The results for country music were tallied in a **frequency table,** shown below.

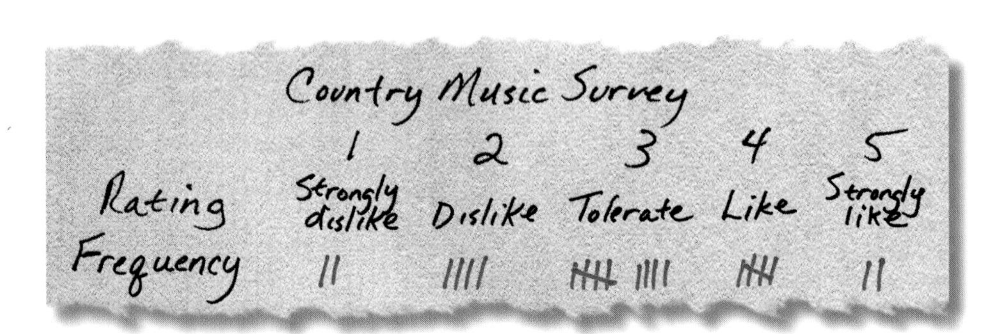

Country Music Survey

	1	2	3	4	5
Rating	Strongly dislike	Dislike	Tolerate	Like	Strongly like
Frequency	II	IIII	~~IIII~~ IIII	~~IIII~~	II

You can use a frequency table to construct a bar graph of the data.

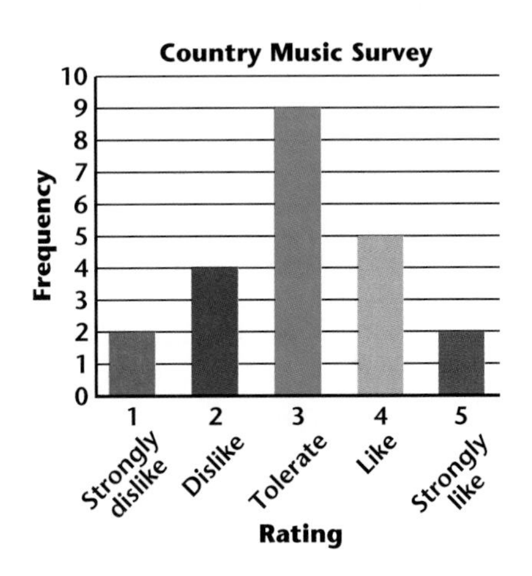

Since the ratings can be expressed as numbers on a scale of 1 to 5, you can find any of the measures of central tendency for these ratings. For example, a mean rating of 3.4 indicates a rating between "tolerate" and "like."

Comparing Surveys

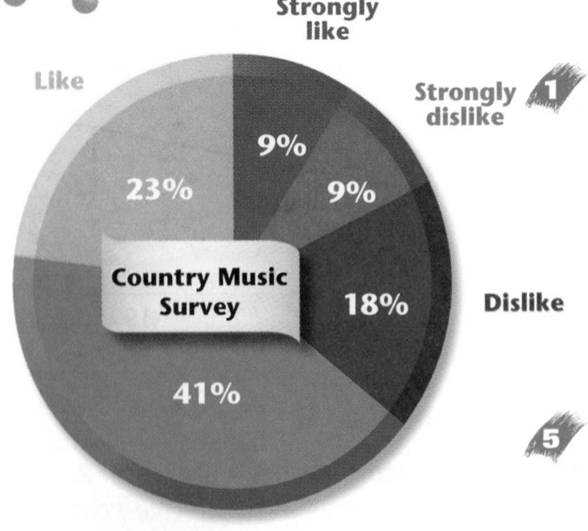

Part I

1. Compare the frequency table with the bar graph. How can you use a frequency table to construct a bar graph?
2. What is the mode? Is the mode a useful measure for this data set? Explain your reasoning.
3. How can you tell what the median is from the bar graph? What is the median?
4. List all of the 22 numerical ratings from the Country Music Survey. Find the mean rating.
5. What is the experimental probability that a student chosen at random from Mr. Phillips's class tolerates country music? either strongly dislikes or strongly likes country music?

Part II

The survey results for rap music are shown in the frequency tables below.

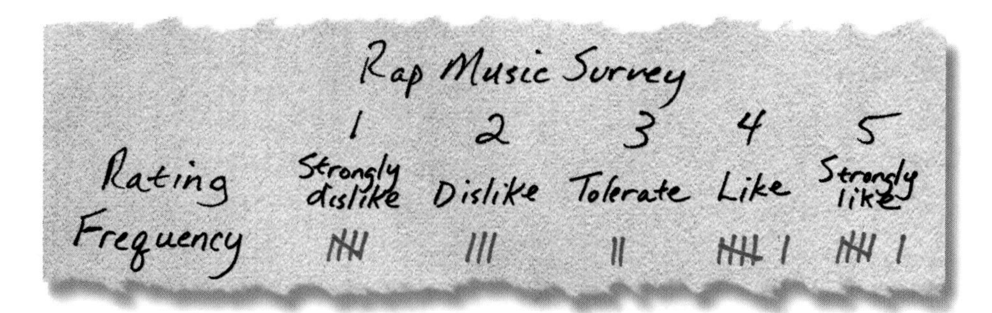

Rap Music Survey

Rating	1 Strongly dislike	2 Dislike	3 Tolerate	4 Like	5 Strongly like
Frequency	卌	|||	||	卌 |	卌 |

1. Find the median rating for the rap music survey. How does this median compare with the median rating for the country music survey?

2. Find the mode rating for the rap music survey. How does this mode compare with the mode rating for the country music survey?

3. Find the mean rating the rap music survey. How does this mean compare with the mean rating for the country music survey?

4. Construct a bar graph to display the results of the rap music survey. How does the bar graph for the rap music survey compare with the bar graph for the country music survey.

5. What is the experimental probability that a student chosen at random from Mr. Philips's class tolerates rap music? What is the probability that a student chosen at random from Mr. Philips's class either strongly dislikes or strongly likes rap music?

6. How are the survey results for the country music survey and the rap music survey alike? How are they different? Is there a measure of central tendency that you think is best for comparing these two surveys? Explain. ❖

EXERCISES & PROBLEMS

Communicate

1. Compare and contrast mean, median, and mode.
2. Explain how the median was found for the American League in Example 1 on page 178.
3. One student thought that the median for the data 10, 20, 8, 5, and 17 was 8. Was this student right or wrong? Explain.
4. Explain how to find the median for a set of 6 of data values.

Rating	Frequency
1-Always	15
2-Often	41
3-Sometimes	34
4-Rarely	7
5-Never	3

For Exercises 5–7, use the survey results at the left to the question, "Do you wear green on St. Patrick's Day?"

5. Explain how to build a bar graph for the survey results.

6. How can you find the mean, median, and mode rating for this survey?

7. Explain how to use this information to find the experimental probability that a person chosen at random from this group would say that he or she never wears green on St. Patrick's Day.
There are 3 "never" ratings, so divide by 100: $\frac{3}{100}$ or 3%.

Practice & Apply

Given the data 24, 26, 19, 20, and 33, find

8. the mean. 24.4 **9.** the median. 24 **10.** the mode. None **11.** the range. 14

Given the data 12.4, 14.1, 14.1, 14.6, 15.0, and 15.3, find

12. the mean 14.25 **13.** the median. 14.35 **14.** the mode. 14.1 **15.** the range. 2.9

16. Tell how to find the median for a set of 9 data values.
Arrange in order; then find the middle value.

17. Tell how to find the median for a set of 10 data values.

18. If the lowest score is 62, the median score is 76, and the highest score is 98, find the range. 36

19. If the range of scores is 32, the median score is 80, and the highest score is 100, find the lowest score. 68

20. Construct a set of 5 pieces of data with mode 18, median 20, and mean 21, or explain why this is impossible. Answers may vary.
One possibility: 18, 18, 20, 24, 25

21. Construct a set of 5 pieces of data with mode 18, median 20, and mean 19, or explain why this is impossible.

Meteorology The monthly normal mean temperatures for three cities are shown below.

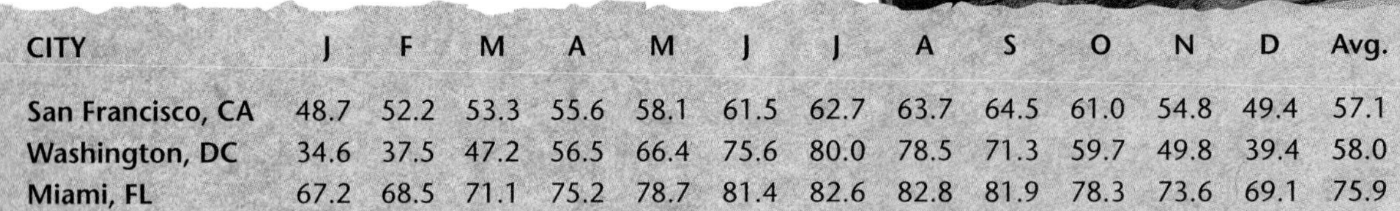

CITY	J	F	M	A	M	J	J	A	S	O	N	D	Avg.
San Francisco, CA	48.7	52.2	53.3	55.6	58.1	61.5	62.7	63.7	64.5	61.0	54.8	49.4	57.1
Washington, DC	34.6	37.5	47.2	56.5	66.4	75.6	80.0	78.5	71.3	59.7	49.8	39.4	58.0
Miami, FL	67.2	68.5	71.1	75.2	78.7	81.4	82.6	82.8	81.9	78.3	73.6	69.1	75.9

22. Compute the range of temperatures for each city. Compare the temperatures of these cities without using numbers.

23. Which city stands out because of its average temperature? Miami, it is significantly higher.

24. Which city stands out because of its temperature range? Washington, it is significantly larger.

National League

Year	Name	Team	Avg.
1989	Tony Gwynn	Pittsburgh	.336
1990	Willie McGee	St. Louis	.335
1991	Terry Pendleton	Atlanta	.319
1992	Gary Sheffield	San Diego	.330
1993	Andres Galarraga	Colorado	.370

American League

Year	Name	Team	Avg.
1989	Kirby Puckett	Minnesota	.339
1990	George Brett	Kansas City	.329
1991	Julio Franco	Texas	.342
1992	Edgar Martinez	Seattle	.343
1993	John Olerud	Toronto	.363

25. **Sports** Make a scatter plot of the batting averages versus years for the American League.

26. Are the batting averages for the American League getting better or worse? How much is the change per year?

27. **Technology** Find the correlation coefficient for the data in Exercise 25. $r \approx 0.79$

28. Compare the correlation coefficient in Exercise 27 with the correlation coefficient that you found for the National League in Example 2 on page 179. Is the relationship stronger or weaker for the American League?

29. Use an almanac or another source to find data for more recent National League batting champions. Analyze the data, and compare the results with the data given in Example 1 on pages 178 and 179. Answers may vary.

30. Repeat Exercise 29 for the American League. Answers may vary.

31. **Advertising** Suppose a lawn-care company used an ad to show that its products were preferred 3 to 1 by showing that its product covered a lawn 3 times as long and 3 times as wide as the lawn of its competition. This would result in an area how many times as large? 9 times as large

Mr. Phillips's class student survey data on a scale of 1 to 5 (1 meaning strongly dislike and 5 meaning strongly like) for jazz music and classical music are shown here as a bar graph.

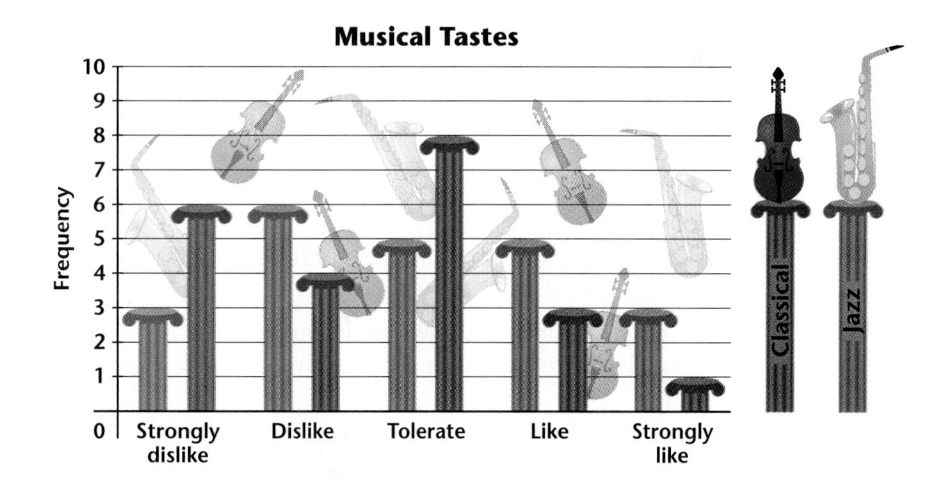

32. Use the numerical ratings to find the mean rating for classical music. Use the bar graph to find the mean rating for jazz music. Classical: 2.5; Jazz: ≈ 2.95

33. Find the experimental probability that a student chosen at random from Mr. Phillips's class only tolerates jazz music. $\frac{5}{22}$ or approx 22.7%

34. Find the experimental probability that a student chosen at random from Mr. Phillips's class either likes or strongly likes classical music. $\frac{2}{11}$ or approx 18.2%

35. Find three examples of graphs and statistics presented in newspapers, magazines, or other textbooks. For each graph, interpret the information presented, and determine whether the graph is misleading in any way. Explain. Answers may vary.

Look Back

36. The sequence 1, 6, 13, 22, 33, 46, . . . can be represented by what kind of function? How do you know? **[Course 1]** Quadratic; second differences are constant.

37. **Meteorology** The lowest temperature on record for three cities is shown in the following table. What are all the differences in temperature? **[Course 1]**

CITY	RECORD LOW
Juneau, AK	−22
Duluth, MN	−39
Houston, TX	−7

38. A radio is priced \$63. What is the total price of the radio if you also pay a sales tax of 6%? **[Lesson 1.2]** \$66.78

39. An automatic coin counter counts Denise's collection of dimes and nickels and gives a total of \$36.45. Earlier, she counted 231 nickels. Write and solve an equation to find the number of dimes. **[Lessons 1.4]** $(0.05)(231) + 0.10D = 36.45$; $D = 249$

40. **Coordinate Geometry** What figure is shown when you connect the points $A(-1, -2)$, $B(1, -1)$, $C(3, 0)$, $D(1, 2)$, $E(0, 0)$, and then A again? What is the equation for a line containing points A and D? **[Lesson 2.2]**

41. In Exercise 40 what is the slope of a line parallel to a line containing points D and C? **[Lesson 2.2]**

42. Solve the matrix equation. Then write the matrix equation as a system of linear equations, and solve the system. Are your answers the same? **[Lesson 3.5]**

$$\begin{cases} 9x - 4y = 61 \\ -3x + 2y = -23 \end{cases} \qquad \begin{bmatrix} 9 & -4 \\ -3 & 2 \end{bmatrix}\begin{bmatrix} x \\ y \end{bmatrix} = \begin{bmatrix} 61 \\ -23 \end{bmatrix}$$

The answers are the same. (5, −4)

Look Beyond

43. How many ways can the letters in the word *and* be arranged? 6

44. How many ways can the letters in the word *sand* be arranged? 24

45. Compare the answers from Exercises 43 and 44. How does adding one letter affect the number of possible arrangements?

LESSON 4.4

Exploring the Addition Principle of Counting

why *The process of counting entries in lists such as databases can sometimes be tedious. Once you learn the shortcuts to counting, you can save time and effort. You can also discover many useful and interesting mathematical properties.*

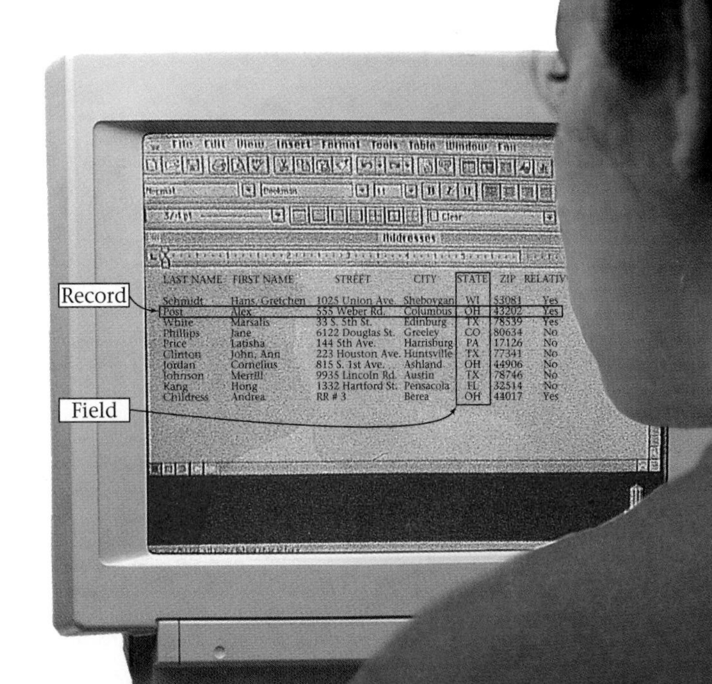

Databases display records that have a selected entry in one or more fields. For example, you can print the records of all the people who live in Texas in the database at the right.

White	Marsalis	TX	Yes
Clinton	John	TX	No
Johnson	Merrill	TX	No

Similarly, if you print the records of all the people in this database who are relatives, you will get the following list.

Schmidt	Hans	WI	Yes
Post	Alex	OH	Yes
White	Marsalis	TX	Yes
Childress	Andrea	OH	Yes

Meg keeps a database of addresses of her friends and relatives. Each row, such as the information about Alex Post, is a *record*. Each column, such as state, is a *field*.

Databases can also combine information from two different fields. When you request combined information, you will need to use the instructions AND and OR. The words **AND** and **OR** have special meaning when working with databases. For example, if you print the records of all the people who are living in Texas OR who are relatives, you will get the following list:

Schmidt	Hans	WI	Yes
Post	Alex	OH	Yes
White	Marsalis	TX	Yes
Clinton	John	TX	No
Johnson	Merrill	TX	No
Childress	Andrea	OH	Yes

If you print the records of all the people who are living in Texas AND who are relatives, you will get the following list:

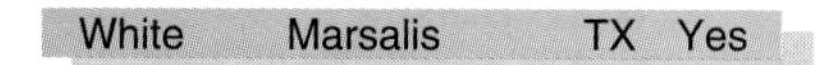

White	Marsalis	TX	Yes

The relationship between AND and OR can be illustrated with a **Venn diagram**.

Name and Address Database

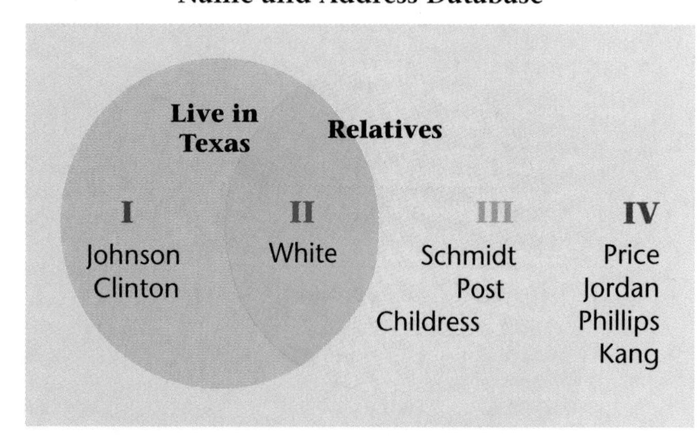

Region I *contains people who live in Texas and who are not relatives.*

Region II *contains people who live in Texas and are relatives.*

Region III *contains people who are relatives and who do not live in Texas.*

Region IV *contains people who do not live in Texas and who are not relatives.*

Region II represents the overlapping regions, or **intersection**, of the people who are in the first set **AND** the second set. Regions I, II, and III together represent the combined regions, or **union**, of the two sets of people. These people are in the first set **OR** the second set.

When two sets have no members in common, the sets are called **disjoint**. For example, the set of people who live in Ohio and the set of people who live in Texas are disjoint.

intersection

union

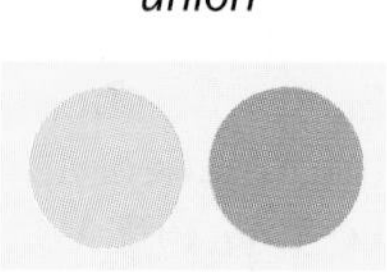

disjoint

CT– If A is contained in B and B is contained in A, then A and B are the same sets and contain all the same elements.

What is the relationship between set A and set B if A is contained in B and B is contained in A?

Original Database

Last name	First name	Street	City	State	Zip	Relative?
Schmidt	Hans	1025 Union Ave.	Sheboygan	WI	53081	Yes
Post	Alex	555 Weber Rd.	Columbus	OH	43202	Yes
White	Marsalis	33 S. 5th St.	Edinburg	TX	78539	Yes
Phillips	Jane	6122 Douglas St.	Greeley	CO	80634	No
Price	Latisha	144 5th Ave.	Harrisburg	PA	17126	No
Clinton	John	223 Houston Ave.	Huntsville	TX	77341	No
Jordan	Cornelius	815 S. 1st Ave.	Ashland	OH	44906	No
Johnson	Merrill	9935 Lincoln Rd.	Austin	TX	78746	No
Kang	Hong	1332 Hartford St.	Pensacola	FL	32514	No
Childress	Andrea	RR # 3	Berea	OH	44017	Yes

Exploration 1 *Interpreting a Venn Diagram*

For each step, use the original database above that shows all 10 records.

1. Draw a Venn diagram using the last names of people living in Ohio and people living in Texas as your sets.
2. How many of the people live in Ohio? 3
3. How many of the people live in Texas? 3
4. How many of the people live in either Ohio OR Texas? 6
5. How many of the people live in both Ohio AND Texas? 0
6. Recall the number that you gave as an answer in Step 4. Then find the sum of the numbers that you gave as answers in Steps 2 and 3. What do you notice when you compare these two amounts? ❖

   They are both 6.

Exploration 2 *Counting From a Venn Diagram*

For each step, use the original database above that shows all 10 records.

1. Draw a Venn diagram that represents the people who live in Ohio and who are relatives.
2. How many of the people live in Ohio? 3
3. How many of the people are relatives? 4
4. How many of the people either live in Ohio OR are relatives? 5
5. How many of the people live in Ohio AND are relatives? 2
6. Does your answer to Step 4 equal the sum of the answers to Steps 2 and 3? Explain. No, two people have been counted twice.
7. Show how to produce the answer to Step 4 based on the numbers from Steps 2, 3, and 5. ❖

   Ohio + Relatives − Relatives in Ohio = Total $(3 + 4 - 2 = 5)$

The application on page 188 illustrates the use of a counting technique in statistics.

APPLICATION

	Favor rule	Oppose rule	Total
Boys	4	9	13
Girls	7	10	17
Total	11	19	30

Mark took a class survey to get student opinions about a new rule concerning students driving to school.

STATISTICS *Connection*

After reviewing the data, the following questions were asked. Notice that you need to know what is meant by AND and OR.

Q: How many students surveyed are boys?
A: The total for the first row is 13, the total number of boys.

Q: How many students favor the rule?
A: The total for the first column is 11, the total number who favor the rule.

Q: How many students are boys AND favor the rule?
A: If you look under *Favor rule* and across from *Boys,* the intersection shows the number 4. This is the number of students who favor the rule and are boys.

Q: How many students are boys OR favor the rule?
A: The answer is the number of boys plus the number favoring the rule, except that some have been counted twice. Subtract the number who are boys AND favor the rule. If you use the information from the responses to the previous questions, the answer is $13 + 11 - 4$, or 20. ❖

The response to these questions leads to an important principle used in counting.

ADDITION PRINCIPLE OF COUNTING

Suppose there are m ways to make a first choice, n ways to make a second choice, and t ways that have been counted twice. Then there are $m + n - t$ ways to make the first choice OR the second choice.

APPLICATION

Games In a card game called *Crazy Eights,* if the queen of hearts is showing, you can play either a queen, a heart, or an eight. How many different cards can be played? Keep in mind that an ordinary deck of cards has 4 queens, 13 hearts, and 4 eights.

Do not count the queen of hearts, since it has already been played. There are 3 other queens, 12 other hearts, and 4 eights, for a total of 19 cards. The eight of hearts has been counted twice, so subtract 1. Thus, $3 + 12 + 4 - 1$ totals 18 cards that can be played. ❖

EXERCISES & PROBLEMS

Communicate

1. Explain the difference between a field and a record in a database. A field is a category; a record is each line of info.
2. In the original database for this lesson shown on page 187, name another entry other than Schmidt that is in the intersection of the set of relatives and the set of people who do not live in Texas. Post or Childress
3. In the game of *Crazy Eights* described on page 188, explain how the 8 of hearts is counted twice. It counts once as an 8, and once as a heart.
4. What do you know about two sets if, in the union of the sets, you do not have to subtract? Intersection is empty.

Statistics A survey of participation in school music programs has the results shown in the table at the right. Explain how to find each of the following.

5. How many students are girls? $49 + 57 = 106$
6. How many students participated? $43 + 49 = 92$
7. How many students are girls AND participated? 49
8. How many students are girls OR participated? $49 + 57 + 43 = 149$

	Participated	Did not participate
Boys	43	59
Girls	49	57

Practice & Apply

List the integers from 1 to 10, inclusive, that are

9. even. 2, 4, 6, 8, 10
10. multiples of 3. 3, 6, 9
11. even AND multiples of 3. 6
12. even OR multiples of 3. 2, 3, 4, 6, 8, 9, 10

Games Tell how many of each of the following are in an ordinary deck of 52 cards.

13. aces 4
14. red cards 26
15. red cards OR aces 28

In the first application on page 188, how many students

16. are girls? 17
17. oppose the rule? 19
18. are girls AND oppose the rule? 10
19. are girls OR oppose the rule? 26

20. When does the number of members in the union of two sets equal the sum of the number of members in each set? When they do not intersect

List the integers from 1 to 20, inclusive, that are multiples

21. of 5. 5, 10, 15, 20

22. of 3. 3, 6, 9, 12, 15, 18

23. of 5 AND multiples of 3. 15

24. of 5 OR multiples of 3. 3, 5, 6, 9, 10, 12, 15, 18, 20

25. People waiting in line for a concert were given numbers. At one point they let people with tickets numbered 100 to 150 enter. How many people entered? 51

26. If it takes 15 minutes to saw a log into 3 pieces, how long does it take to saw it into 4 pieces? HINT: Three pieces require two cuts. 22.5 min

27. A spreadsheet has 25 rows of data, beginning in row 5. What row number is the last row of data? 29

Statistics In a class of 28 students, 17 have brown eyes and 13 have black hair. This includes 10 who have both.

28. Show this in a Venn diagram. Include all 28 students.

29. How many students have either brown eyes OR black hair? 20

30. In a poll, 13 students said they had taken music lessons, and 5 said they had taken dance lessons. Assume that it is possible for some of these students to have taken both music and dance lessons. What is the greatest possible number of students polled? the least? The most is 18, the least is 13.

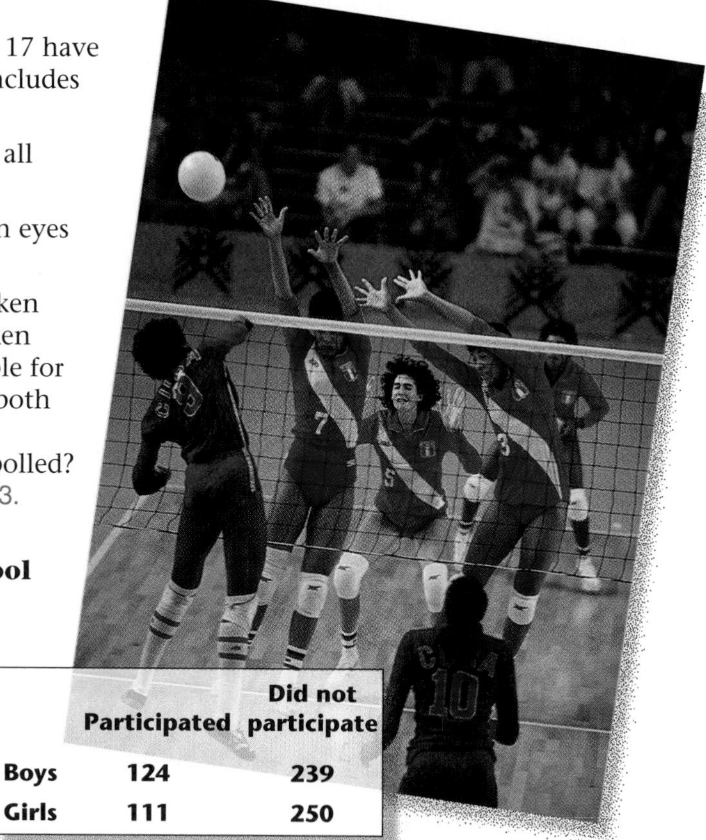

A survey about participation in school sports had the results shown in the table at the right.

	Participated	Did not participate
Boys	124	239
Girls	111	250

31. How many students are girls? 361

32. How many students participated? 235

33. How many students are girls AND participated? 111

34. How many students are girls OR participated? 485

Student Government The survey in the first application on page 188 had the following results for the entire school:

Grade	Favor rule	Oppose rule	Total
9	2	97	99
10	34	61	95
11	57	40	97
12	81	7	88

35. How many 9th graders opposed the rule? 97

36. How many students were either in 9th or 10th grade AND opposed the rule? 158

37. How many students were either in 9th or 10th grade OR opposed the rule? 241

38. How many non-seniors favored the rule? 93

39. How many non-seniors opposed the rule? 198

Look Back

40. Savings Chou deposits \$150 in a bank that pays 3% interest per year. How much money will she have in this account after 5 years? **[Lesson 1.3]** \$173.89

Solve. [Lesson 1.4]

41. $-(-3x + 7) = 23 - 2x$ $x = 6$ **42.** $4a + 2 = -5(2 - 3a)$ $a = \frac{12}{11}$

43. Geometry The volume of a cone is $V = \frac{1}{3}\pi r^2 h$. Solve the equation for h. **[Lesson 1.4]** $h = \frac{3V}{\pi r^2}$

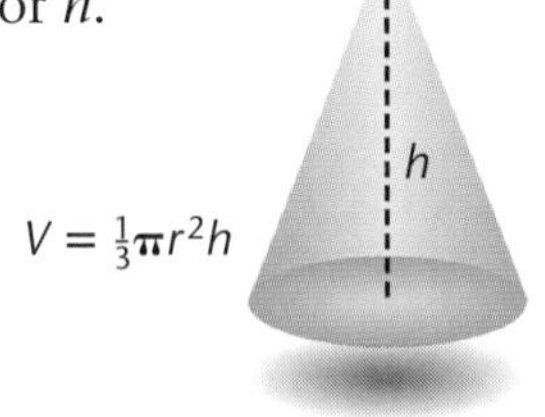

Solve. [Lesson 1.4]

44. $3(x + 5) - 23 = 2x - 47$ -39

45. $\frac{x}{3} = -3x + 5$ $\frac{3}{2}$ or 1.5

46. Sports A major league baseball player has a .300 batting average. Use a simulation to estimate the probability that the batter gets at least 2 hits in his next 4 times at bat. Describe your process. **[Lesson 4.2]**

Look Beyond

Of the 28 equally likely ways of choosing a student, there are 17 ways of choosing a student with brown eyes.

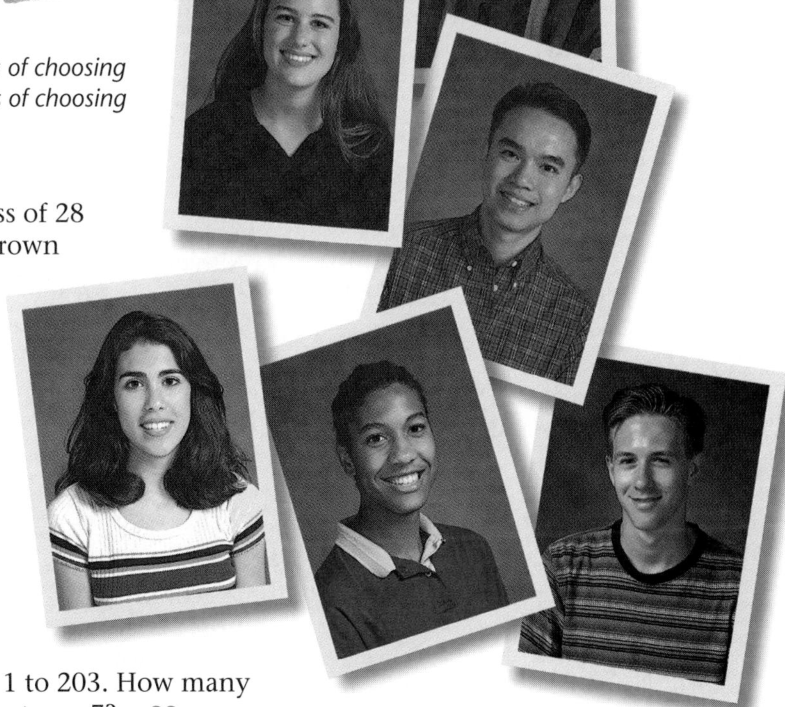

In Exercises 28 and 29, there was a class of 28 students described. Of those, 17 had brown eyes and 13 had black hair. There were 10 who had both. If a student is chosen at random from this class, the theoretical probability that the student has brown eyes is $\frac{17}{28}$. Find the theoretical probability that a student chosen at random from this class has

47. brown eyes AND black hair. $\frac{5}{14}$

48. black hair. $\frac{13}{28}$

49. brown eyes OR black hair. $\frac{5}{7}$

50. A book has pages numbered from 1 to 203. How many of the page numbers contain at least one 7? 38

LESSON 4.5

Multiplication Principle of Counting

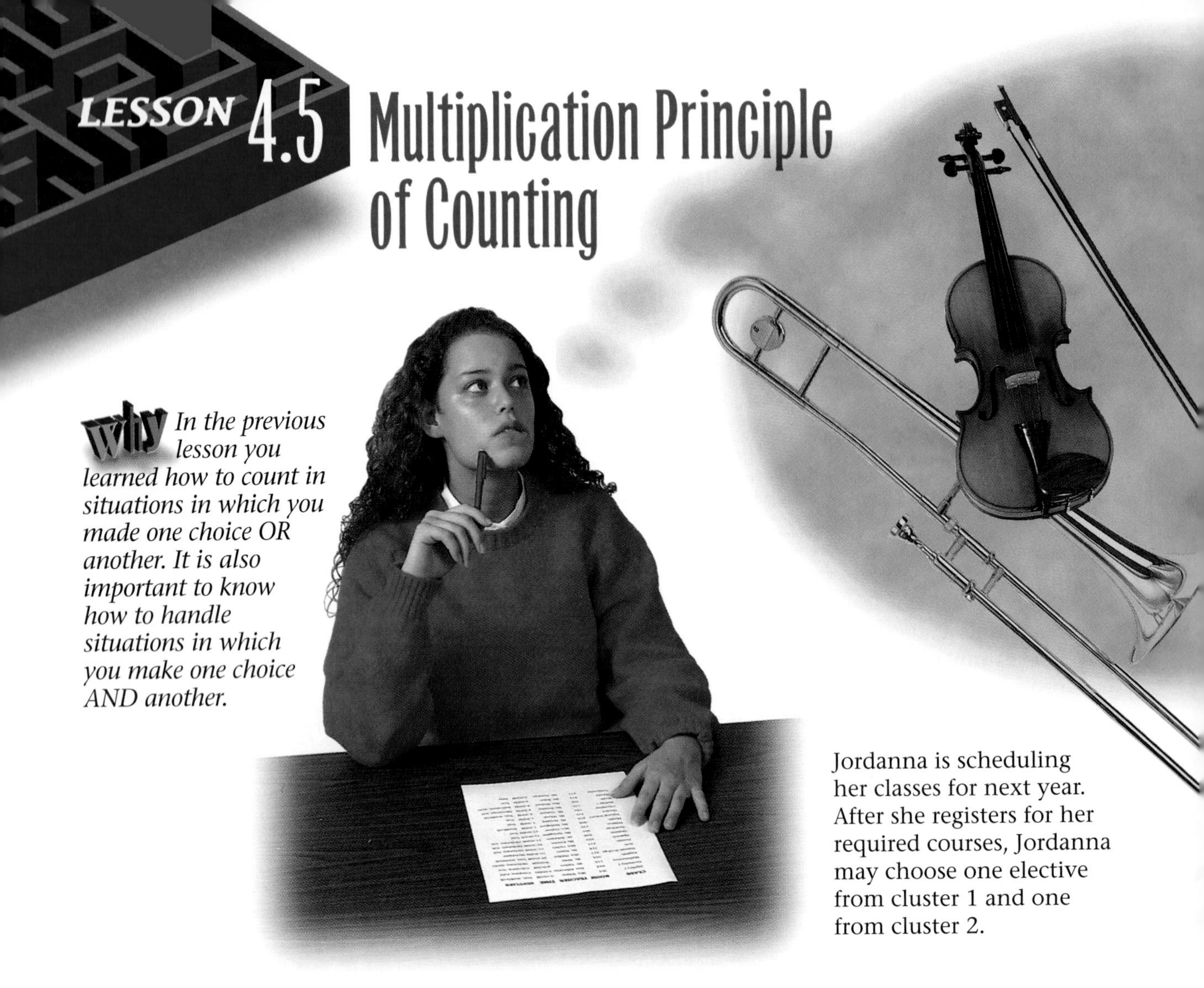

why *In the previous lesson you learned how to count in situations in which you made one choice OR another. It is also important to know how to handle situations in which you make one choice AND another.*

Jordanna is scheduling her classes for next year. After she registers for her required courses, Jordanna may choose one elective from cluster 1 and one from cluster 2.

Academics One way to find all the possible selections is to make a **tree diagram.**

The tree diagram shows that for each of the 3 ways to choose a Cluster-1 elective, there are 2 ways to choose a Cluster-2 elective.

You can see by following along the branches of the tree diagram that there are $3 \cdot 2$, or 6, possible choices.

Cluster 1	Cluster 2
band	home economics
	woodworking
orchestra	home economics
	woodworking
chorus	home economics
	woodworking

- band and home economics
- orchestra and home economics
- chorus and home economics
- band and woodworking
- orchestra and woodworking
- chorus and woodworking

There are 6 possible selections.

EXAMPLE 1

Sports There are 2 volleyball teams, each with 6 members. One player is to be selected at random. Flip a coin to select the team, then roll a number cube to select a player from that team. How many pairings of outcomes are possible?

Solution ➤

Make a tree diagram.

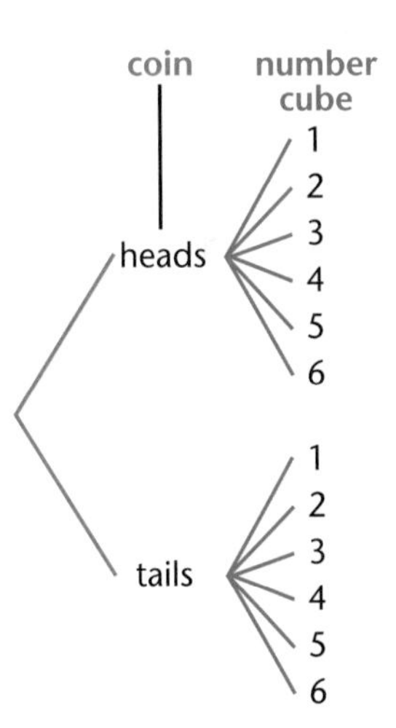

For each of the 2 outcomes from the coin, there are 6 outcomes from the number cube. That is, $2 \cdot 6$, or 12 pairings. ❖

What if there were 3 teams instead of 2? How could you find the number of possible outcomes? Multiply 3 times 6 for 18 outcomes.

These examples lead to another useful principle used in counting.

> **MULTIPLICATION PRINCIPLE OF COUNTING**
>
> If there are m ways to make a first choice and n ways to make a second choice, then there are $m \cdot n$ ways to make a first choice AND a second choice.

The Multiplication Principle of Counting is often used instead of a tree diagram, especially when the numbers in counting problems are large. Even when the original numbers are small, some multiplications can produce numbers that are too inconvenient for a tree diagram.

EXAMPLE 2

Inventory

20 styles!
Each in 8 Sizes
and 3 Colors!

A shoe store stocks 20 styles of shoes. Each style has 8 sizes and 3 colors. How many pairs of shoes must the store carry to have one pair of each combination?

Solution ➤

For each of the 20 styles there are 8 sizes. This means there are 20 • 8, or 160, possible combinations of styles and sizes. For each of these 160 combinations there are 3 colors, so there are 160 • 3, or 480, combinations in all. You can also perform the multiplications together.

$$20 \cdot 8 \cdot 3 = 480$$

As you can see, the Multiplication Principle of Counting can be extended beyond two choices. ❖

Try This

The tree diagram should show the number of choices: 2, 4, and 2. The Multiplication Principle of Counting says that there are 2 • 4 • 2, or 16 possible outcomes.

Tammy is selecting colored pens. The first characteristic is the nib—felt tip or ballpoint. The second characteristic is the color—red, blue, green, or black. The third characteristic is the brand—Able or Blakely. Use a tree diagram to show the possible selections for a pen. Explain how this relates to the Multiplication Principle of Counting.

The next example shows how counting principles are used in geometry.

EXAMPLE 3

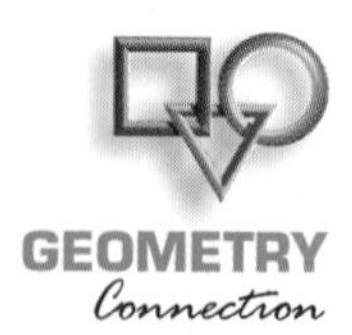

Points *A*, *B*, and *C* lie on the same line. How many ways are there to name the line with two of the letters?

Solution A ➤

Make a tree diagram.

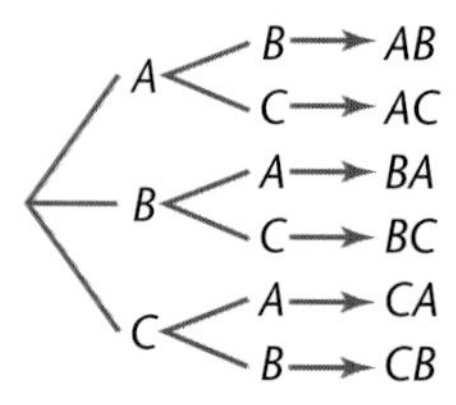

There are 6 ways to name the line.

$\overleftrightarrow{AB}$, $\overleftrightarrow{AC}$, $\overleftrightarrow{BA}$, $\overleftrightarrow{BC}$, $\overleftrightarrow{CA}$, and $\overleftrightarrow{CB}$

Solution B ➤

Use the Multiplication Principle of Counting. There are 3 ways to choose the first letter. Once this is done, there are 2 ways remaining to choose the second letter. There are $3 \cdot 2$, or 6 ways. ❖

What are the advantages of using a tree diagram, and what are the advantages of using the Multiplication Principle of Counting?

The Addition Principle of Counting and the Multiplication Principle of Counting are very useful methods to simplify counting. Nevertheless, you must be careful to understand when and where to use each principle.

EXAMPLE 4

Games A regular deck of 52 playing cards contains 26 red cards and 26 black cards. The deck contains 4 aces—2 black aces and 2 red aces.

A One card is drawn. How many ways are there to draw a black ace?

B One card is drawn. How many ways are there to draw a black card or an ace?

C Two cards are drawn. The first card is replaced before drawing the second card. How many ways are there to draw a black card followed by an ace?

Solution ➤

A Two aces are black, the ace of spades and the ace of clubs. Thus, there are 2 ways to draw a black ace.

B There are 26 black cards and 4 aces, but the 2 black aces are counted twice. According to the Addition Principle of Counting, there are $26 + 4 - 2$, or 28, ways to draw a black card OR an ace.

C There are 26 ways to draw a black card. No matter which card is drawn first, there are 4 ways to draw an ace as the second card. Using the Multiplication Principle of Counting, there are $26 \cdot 4$, or 104, ways to draw a black card AND an ace in two draws if the first card is replaced. ❖

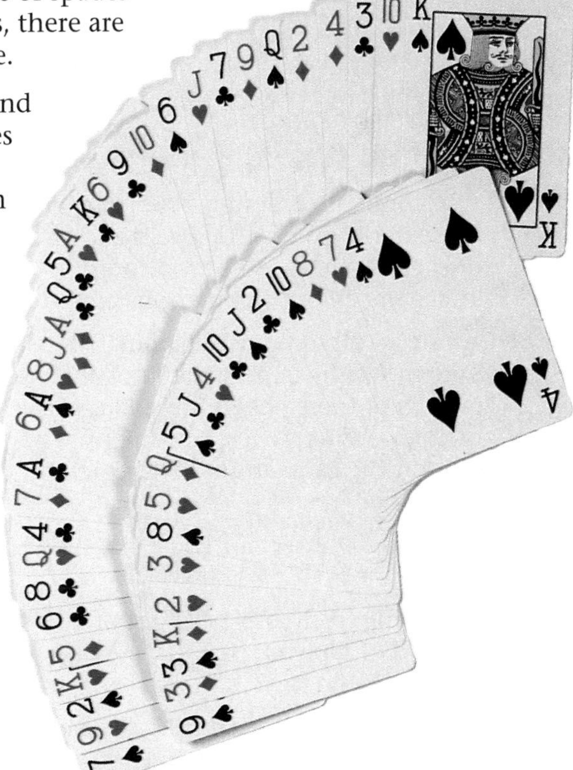

EXERCISES & PROBLEMS

Communicate

1. Explain the difference between these two sentences and explain the special meaning of AND and OR.
"I am going to the mall AND I am going to study."
"I am going to the mall OR I am going to study."
AND: you do both things; OR: you do one or the other.

Suppose in the scheduling choices described on page 192, there were 4 Cluster-1 electives instead of 3. Explain how to determine the number of possible choices if Jordanna were to select

2. one Cluster-1 elective OR one Cluster-2 elective.
4 ways for 1, 2 ways for 2, total is 4 + 2 or 6

3. one Cluster-1 elective AND one Cluster-2 elective.
4 ways for 1, 2 ways for 2, total is 4 × 2 or 8

4. Suppose that there are 2 volleyball teams, and each team has 8 players. One player is to be selected at random. Explain how to find the number of branches that the corresponding tree diagram has.

5. Two cards are drawn from an ordinary deck of 52 cards. The first card is replaced before drawing the second card. How can you find the number of possible ways to draw a 3 followed by a red card?

6. **Geometry** To name a ray, always start with the endpoint. In this case it is A. How many ways are there to use two of the letters to name this ray?
3 ways; $\overrightarrow{AB}$, $\overrightarrow{AC}$, $\overrightarrow{AD}$

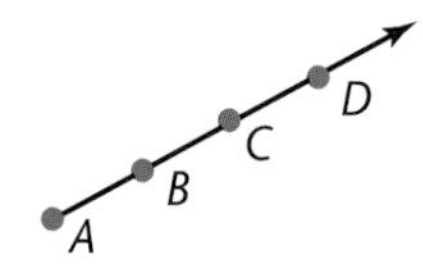

Practice & Apply

7. The Raholls are ordering a new car. There are 9 exterior colors from which to choose. For each of these, there are 4 interior colors. How many color combinations are possible?
36

8. **Games** At the start of a game of chess, white moves first and has 20 possible moves. For each of these possibilities, black has 20 possible countermoves. How many ways are there for the game to start with each player making one move? 400

9. In how many ways can Tricia buy 2 out of 4 CDs AND 3 out of 5 tapes? 60

10. Ed is trying to choose 1 CD out of 2 that he likes AND 1 tape out of 3 that he likes. In how many ways can he do this? 6

11. A menu contains 4 appetizers, 3 salads, 5 main courses, and 6 desserts. How many ways are there to choose one of each? 360

Language Arts A word that is obtained by rearranging the letters of another word is called an *anagram*. For example, *art* is an anagram of *rat.*

12. Find another anagram of *rat.* tar

13. How many ways are there to arrange the letters of the word *rat,* using each letter exactly once? 6

14. How many ways are there to arrange the 5 letters of the word *stake?* HINT: There are 5 ways to choose the first letter. For each of these 5 ways, there are 4 ways to choose the second letter, and so on. 120

15. Write three anagrams of *stake.* Answers may vary.

skate, takes, steak, teaks

16. How many ways are there to arrange the 6 letters of the word *foster?* 720

17. Find two anagrams of *foster.* softer, fortes

18. How many different 7-digit phone numbers are possible if the first 3 digits are 555? Assume 555-0000 is allowed. $10^4 = 10{,}000$

19. How many different 7-digit phone numbers are possible? Assume phone numbers cannot start with zero.

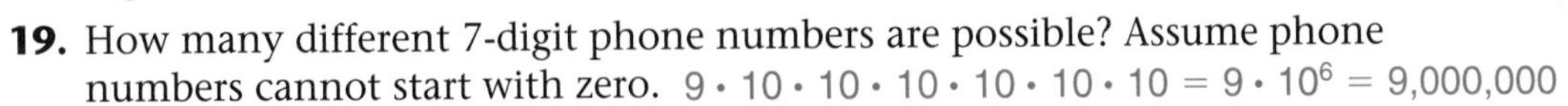
$9 \cdot 10 \cdot 10 \cdot 10 \cdot 10 \cdot 10 \cdot 10 = 9 \cdot 10^6 = 9{,}000{,}000$

20. **Geometry** How many ways are there to name the triangle at the right using the letters *R*, *S*, and *T*? 6
$\triangle RST, \triangle RTS, \triangle STR, \triangle SRT, \triangle TSR, \triangle TRS$

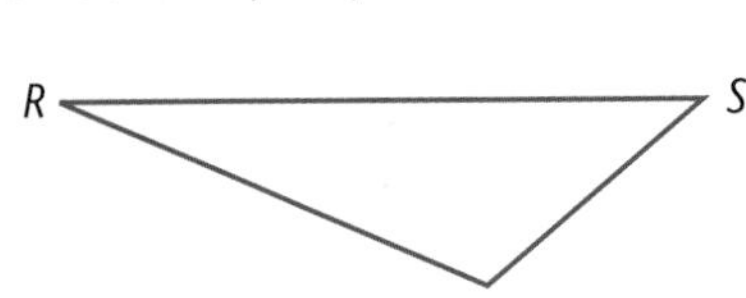

Look Back

21. Hexagonal numbers are in a sequence 1, 6, 15, 28, 45, 66, 91, . . . Find the next two hexagonal numbers. **[Course 1]** 120, 153

Solve. [Lesson 1.2]

22. $\frac{3}{2}x = 6$ 4 **23.** $\frac{x}{-4} = -\frac{5}{8}$ 2.5 **24.** $\frac{x}{5} = 10$ 50

25. **Coordinate Geometry** The points $P(7, -4)$ and $Q(-2, 5)$ are on a line. What is the slope of that line? **[Lesson 2.1]** -1

26. If lines on a graph are parallel, how are their slopes related? **[Lesson 2.1]** Their slopes are equal.

27. **Chemistry** How many milliliters of a 9% acid solution are needed to mix with 450 milliliters of a 1.6% acid solution to produce a 5% acid solution? **[Lessons 2.4, 2.5]** 382.5 mL

Look Beyond

Academics Assuming that Jordanna on page 192 likes all of the electives equally, the probability that she chooses band and home economics is $\frac{1}{6}$ because that choice is 1 of 6 equally likely possibilities. Find the following probabilities:

28. She chooses band for her Cluster-1 elective. $\frac{1}{3}$

29. She chooses woodworking for her Cluster-2 elective. $\frac{1}{2}$

30. She chooses either band or woodworking (or both). $\frac{2}{3}$

31. She chooses both band and woodworking. $\frac{1}{6}$

LESSON 4.6 Theoretical Probability

why

In Lesson 4.1 you explored experimental probability. Another way to find the numerical measure of chance is to use theoretical probability. Many probabilities can be found more exactly with theoretical probability.

Mrs. Miller is going to choose 3 students from her first period class of 20 students to collect data. She will also choose 4 students from her second period class of 28 students. In which class is the chance of being selected greater?

More students will be chosen from Mrs. Miller's second period class, but this does not mean that the chances of being selected are greater. There are also more students from which to choose. To make it easier to compare, find the fraction that represents the number to be chosen out of the total number for each class.

First period (3 out of 20) The chances are $\frac{3}{20}$, or 15%.

Second period (4 out of 28) The chances are $\frac{4}{28}$, or about 14%.

The chances are slightly greater that a certain person will be chosen in the first period class even though fewer students will be chosen from that class.

The *event* is *being chosen to collect data*. Since each of the students has an equal chance of being chosen, the number of *possible outcomes* is the number of students in the class. The number of *successful outcomes* in the event is the number of students to be chosen.

THEORETICAL PROBABILITY

Let n be the number of *equally likely* outcomes in the event. Let s be the number of *successful outcomes* in the event. Then the **theoretical probability** that an event will occur is $P = \frac{s}{n}$.

The condition that the outcomes are **equally likely** means that each outcome has the same probability of happening.

Suppose that getting a 5 on a toss of an ordinary 6-sided number cube is considered a success. You might think that the probability is $\frac{1}{2}$ since only two things can happen—either you will roll a 5 or you will not. Experimentally, however, 5 does not usually come up half the time, so you know that $\frac{1}{2}$ is not correct.

The fact is that the two possibilities are *not equally likely.* The theoretical probability must take into account all 6 *equally likely outcomes.* Notice that you can roll a 1, 2, 3, 4, 5, or 6, but only one of these is a 5. Each of the numbers has 1 chance in 6 that it will appear. The probability of rolling a 5 is $\frac{1}{6}$, not $\frac{1}{2}$.

Always be sure to consider *equally likely outcomes* when determining the theoretical probability. Note that when you are asked to find probability, it will be the theoretical probability unless otherwise specified.

EXAMPLE 1

Find the probability that a name selected at random from the database shown below is from Ohio (OH).

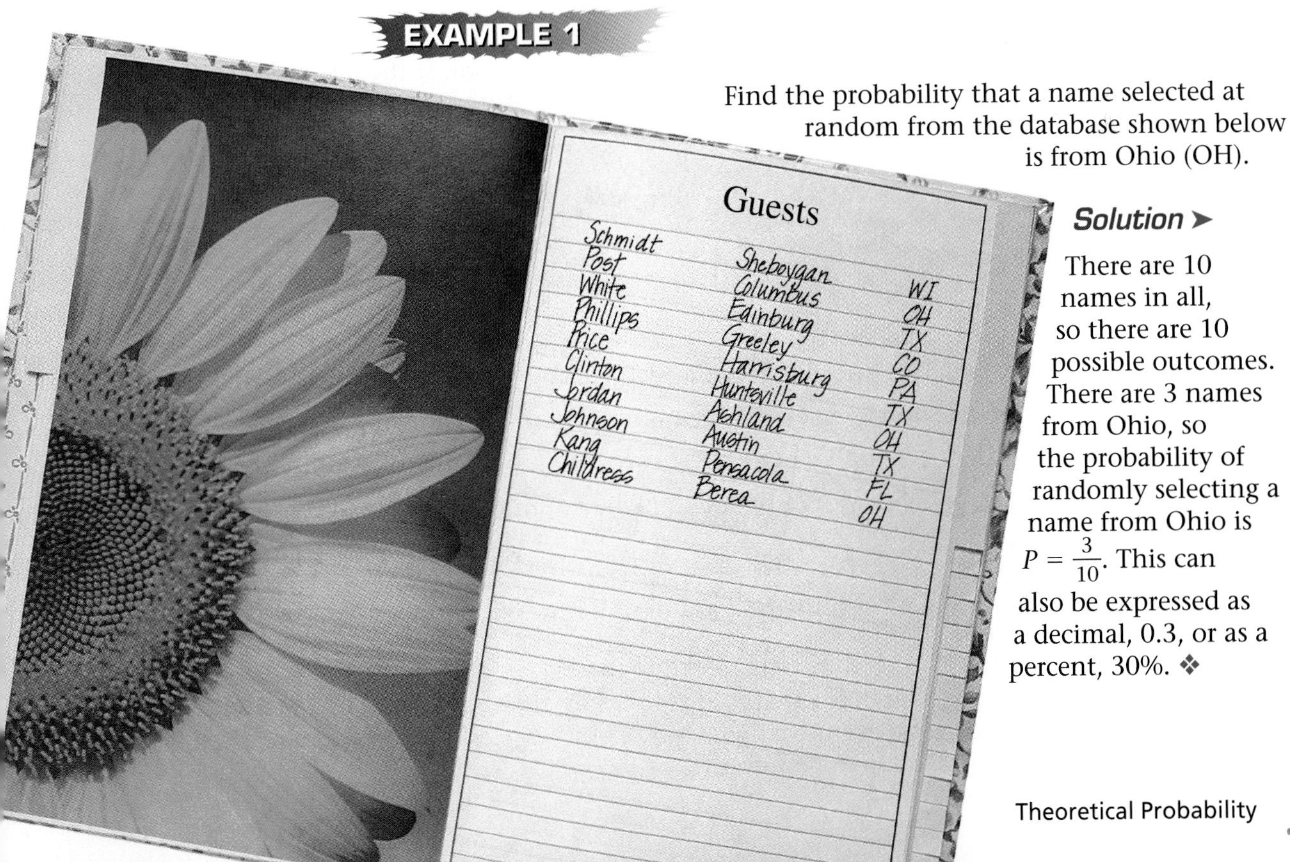

Guests
Schmidt Sheboygan WI
Post Columbus OH
White Edinburg TX
Phillips Greeley CO
Price Harrisburg PA
Clinton Huntsville TX
Jordan Ashland OH
Johnson Austin TX
Kang Pensacola FL
Childress Berea OH

Solution ➤

There are 10 names in all, so there are 10 possible outcomes. There are 3 names from Ohio, so the probability of randomly selecting a name from Ohio is $P = \frac{3}{10}$. This can also be expressed as a decimal, 0.3, or as a percent, 30%. ❖

EXAMPLE 2

Suppose that you roll a number cube once. What is the probability that the result is less than 3?

Solution ➤

The equally likely outcomes are 1, 2, 3, 4, 5, or 6. Two of these outcomes (the event 1 appears and the event 2 appears) are successes, so the probability is $\frac{2}{6}$, or $\frac{1}{3}$. ❖

EXAMPLE 3

Suppose that you roll two number cubes one time each. What is the probability that the sum is 5?

Solution ➤

There are 6 equally likely outcomes for the first number cube. For each of these, there are 6 equally likely outcomes for the second number cube. By the Multiplication Principle of Counting, there are 6 • 6, or 36, ways to get a number on the first cube AND a number on the second cube. These results are summarized in the table.

Second number cube (columns); **First number cube** (rows)

	1	2	3	4	5	6
1	2	3	4	5	6	7
2	3	4	5	6	7	8
3	4	5	6	7	8	9
4	5	6	7	8	9	10
5	6	7	8	9	10	11
6	7	8	9	10	11	12

Look at the table. How many sums are there in all? How many sums of 5 are there? The probability of rolling a sum of 5 is $\frac{4}{36}$, or $\frac{1}{9}$. ❖ There are 36 sums in all; 4 of the 36 have a sum of 5.

Notice that there are 11 possible sums in Example 3: 2, 3, 4, 5, 6, 7, 8, 9, 10, 11, and 12. But the probability of getting a sum of 7 is not $\frac{1}{11}$. These sums are *not* equally likely. For example, a sum of 7 can happen 6 ways, while a sum of 12 can happen only 1 way. The probability of a sum of 7 is $\frac{6}{36}$, or $\frac{1}{6}$. The possibility of a sum of 12 is $\frac{1}{36}$. A sum of 7 is more likely to occur than a sum of 12.

Try This Find the probability that a sum of 9 is tossed on one roll of two number cubes. The probability of getting a sum of 9 is $\frac{4}{36}$, or $\frac{1}{9}$.

Describe an event that results in a sum that has a probability of 0. Then describe an event that has a probability of 1. What is the range of values for the probability of an event?

EXAMPLE 4

The table below shows the results of a class survey about a new rule at the school. A student who participated in this survey is chosen at random. What is the probability that the student

	Favor rule	Oppose rule	Total
Boys	4	9	13
Girls	7	10	17
Total	11	19	30

A is a boy?

B favors the rule?

C is a boy AND favors the rule?

D is a boy OR favors the rule?

Solution ➤

The denominator of the fraction is the number of *equally-likely* possibilities. In this case, it is the number of students in the survey, 30. You can use counting principles to find the values for the numerator.

A 13 students are boys. The probability is $\frac{13}{30}$.

B 11 students favor the rule. The probability is $\frac{11}{30}$.

C 4 students are boys who favor the rule. The probability is $\frac{4}{30}$, or $\frac{2}{15}$.

D 13 students are boys and 11 students favor the rule, but this counts the 4 boys who favor the rule twice. By the Additional Principle of Counting, there are $13 + 11 - 4$, or 20, students who are boys OR who favor the rule. The probability is $\frac{20}{30}$, or $\frac{2}{3}$. ❖

EXERCISES & PROBLEMS

Communicate

1. Explain why Example 4 above is considered experimental probability instead of theoretical probability.
2. John said he had a 50-50 chance of getting all 10 questions right on a quiz because either he would or he would not get each one right. Is he correct? Explain.
3. Based on his record so far this season, the probability that Nat will make a successful free throw is 40%. Is this theoretical probability or experimental probability? Explain.
4. In a two-player game, 4 coins are flipped. The first player wins if the coins come up all heads or all tails. Otherwise the second player wins. Is this game fair? Explain.
5. When finding theoretical probabilities, why must all of the possibilities be equally likely?

Practice & Apply

An integer between 1 and 25, inclusive, is drawn at random. Find the probability that the integer is

6. even. $\frac{12}{25}$ **7.** odd. $\frac{13}{25}$ **8.** a multiple of 3. $\frac{8}{25}$

9. a prime. $\frac{9}{25}$ **10.** even AND a multiple of 3. $\frac{4}{25}$ **11.** even OR a multiple of 3. $\frac{16}{25}$

Language Arts Suppose that you select a letter of the English alphabet at random. Find the probability it is

12. an *r*. **13.** a vowel (*a*, *e*, *i*, *o*, *u*, or *y*). **14.** a consonant.

15. one of the letters of the word *mathematics*.

12. $\frac{1}{26}$ 13. $\frac{3}{13}$ 14. $\frac{10}{13}$ 15. $\frac{4}{13}$

16. Geography Find the probability that the name of a state chosen at random from the United States begins with the letter *M*. HINT: The states starting with *M* are Maine, Maryland, Massachusetts, Michigan, Minnesota, Mississippi, Missouri, and Montana. $\frac{4}{25}$

17. Astronomy Find the probability that a planet in our solar system chosen at random is closer to the Sun than Earth is. HINT: Of the 9 planets, only Mercury and Venus are closer to the Sun than Earth is. $\frac{2}{9}$

Travel According to the National Transportation Safety Board, there were about 0.243 fatal crashes for every 100,000 departures of commuter planes in 1992.

18. $\frac{1}{411,523}$

18. Express this ratio as a fraction with a numerator of 1.

19. Is this experimental probability or theoretical probability? Experimental

20. For larger planes (carrying over 30 passengers), the number of fatal crashes was 0.05 per 100,000. Express this ratio as a fraction with a numerator of 1. $\frac{1}{2,000,000}$

21. According to these figures, a commuter plane is how many times more likely to have a fatal crash than a larger plane? About 5 times

22. Language Arts If the letters of the word *tap* are rearranged, find the probability that an anagram is formed. $\frac{1}{2}$ with *tap*

23. **Geometry** A dart is thrown at a circular target. Assuming that it is equally likely to land anywhere in the target, what is the probability that it hits the bull's-eye, the red region in the center? $\frac{1}{16}$

24. John is trying to solve the equation $2x - 1 = 15$ on a quiz in algebra. He doesn't remember how to do it, but he hopes that the answer is a positive one-digit number and makes a guess. Find the probability that he guesses the correct solution. $\frac{1}{9}$

25. Repeat Exercise 24 for the equation $1 - 2x = 15$. 0

Look Back

26. An item that costs $53 totals $56.18 with the sales tax. Demonstrate the proportion method for finding the rate of the sales tax. **[Lesson 1.2]**

27. Solve the equation $y = mx + b$ for m. **[Lesson 1.3]**

27. $m = \frac{y - b}{x}$

28. If $-6 \leq n - 4 < 10$, show the solution to the inequality on a number line when the values of n are integers. **[Lesson 1.5]**

29. Draw the graph of the function $y = \frac{1}{x}$ for the values $0 < x \leq 4$. What is the range? **[Lesson 2.2]**

30. Suppose that a coin is improperly balanced, and the probability of *heads* is 0.64. If Suzanne tosses the coin 50 times, how many times would you expect *heads* to appear? **[Lesson 4.1]** 32

31. Technology Make a scatter plot for the points (2, 3), (3, 4), (5, 3), (5, 5), (7, 5), and (8, 7), and find the line of best fit. **[Lesson 4.3]**

Karl's typing speeds on 5 tests were 19, 21, 26, 21, and 27 words per minute.

32. What is the mode for Karl's typing speeds? **[Lesson 4.3]** 21

33. What is the median for the typing speeds? **[Lesson 4.3]** 21

Look Beyond

The expression 3! means 3 *factorial*, or the product of the whole numbers from 3 to 1, inclusive. Thus, $3! = 3 \cdot 2 \cdot 1 = 6$.

Find each value.

34. 4! 24 **35.** 5! 120 **36.** 6! 720 **37.** $\frac{3!}{2!}$ 3 **38.** $\frac{7!}{5!}$ 42 **39.** $\frac{10!}{6!}$ 5040

LESSON 4.7 Independent Events

Why *Many experiments involve more than one event. Sometimes the occurrence of a previous event will affect the probability of the events that follow. As probability becomes more difficult to determine, geometry can provide assistance.*

In Lesson 4.5, Jordanna selects one elective from Cluster 1 (band, orchestra, or chorus) and one elective from Cluster 2 (home economics or woodworking). If all of the choices are equally likely, how can you find find the probability that she chooses band and woodworking?

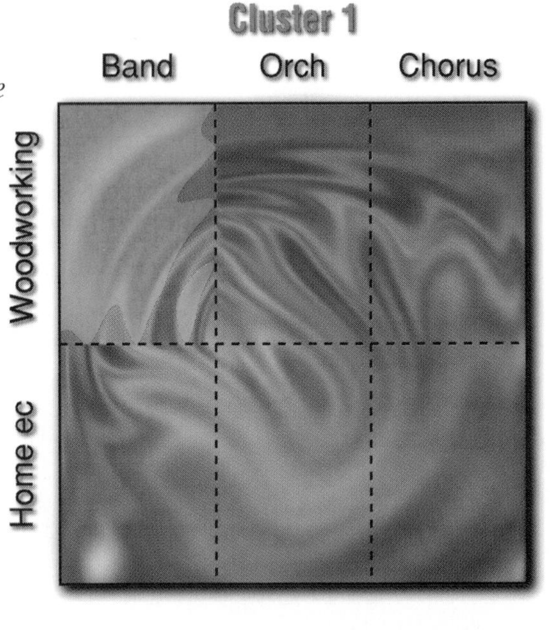

GEOMETRY *Connection*

Use two vertical lines to divide a square into three equal regions, one for each Cluster-1 elective. Then divide it again with a horizontal line, so that there is one region for each Cluster-2 elective. The six regions of the square represent the six equally-likely possibilities or outcomes. One region represents the pair, *band* and *woodworking*. The probability for this choice is $\frac{1}{6}$.

EXAMPLE 1

Recall from page 175 that Allessandro and Zita plan to meet for lunch. Use an area model and the bar graph to find the probability that both Allessandro and Zita show up on time.

100%
90%
80%
70%
60%
50%
40%
30%
20%
10%
0%
Allessandro
Zita

Zita
Allessandro

Solution ➤

On graph paper, divide a 10 by 10 square into two parts with a vertical line, as shown in the first figure. This shows a 60% probability that Allessandro is on time. Divide the square into two parts with a horizontal line, as shown in the second figure. This shows an 80% probability that Zita is on time.

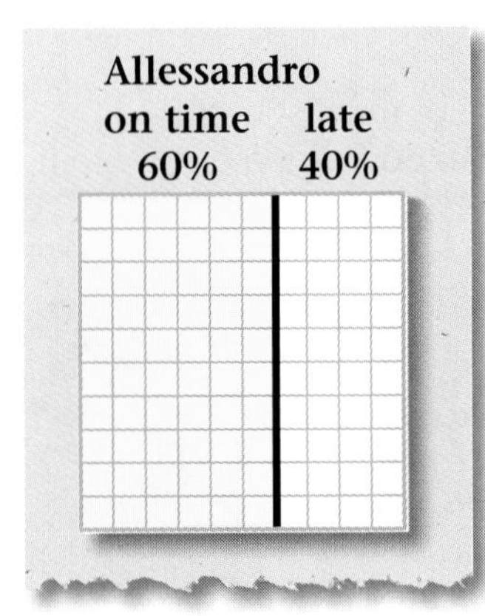

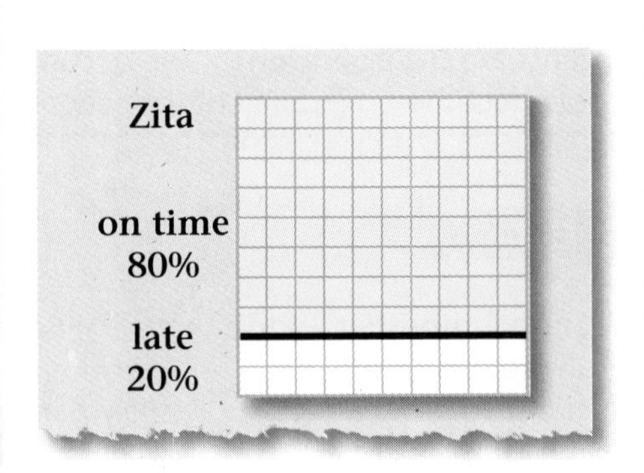

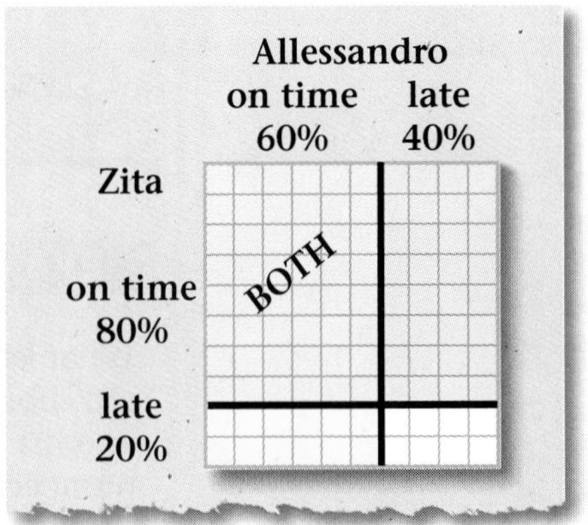

When the squares for Allessandro and Zita are combined, they divide the square into four regions. The region marked **BOTH** represents the probability that both are on time. It contains 8 • 6, or 48, out of 100 possible squares. The probability that both are on time is $\frac{48}{100}$ = or 48%. ❖

Try This Use the combined figure to determine the probability that Allessandro is on time and Zita is not. The probability is $\frac{12}{100}$ or 12%.

Sometimes you can solve a problem without using a model.

EXAMPLE 2

Games Sarah draws a card and *without* replacing it, draws another from a regular 52-card deck. Find the probability that the 2 cards she draws are black. In a regular deck, half of the cards are black.

Solution ➤

There are 52 ways of drawing the first card. After Sarah draws the first card, there are 51 ways of drawing the second card. Since she draws a first card and a second card, use the Multiplication Principle of Counting. There are 52 • 51, or 2652, equally likely ways of drawing the two cards when the first card is not replaced.

A success occurs if Sarah draws two black cards. There are 26 ways to draw the first black card. After the first black card is drawn, there are 25 ways to draw the second black card. Apply the Multiplication Principle of Counting. There are 26 • 25, or 650, ways to draw two black cards.

The probability of drawing two black cards is $\frac{650}{2652}$, or approximately 24.5%. ❖

Probability of Independent Events

Independence has a special significance in probability. In Jordanna's problem on page 192, she first selects an elective from Cluster 1 and then selects an elective from Cluster 2. The outcome of her first selection has no effect on her second selection. The two events are independent.

INDEPENDENT EVENTS

Two events are **independent** if the occurrence of the first event does *not* affect the probability of the second event occurring.

EXAMPLE 3

What is the probability that Todd's red cube shows an odd number and his green cube shows a number greater than or equal to 5? Are the events independent?

Todd rolls a red number cube and records the number shown. He then rolls a green number cube and records the number.

Solution

Apply the Multiplication Principle of Counting. There are $6 \cdot 6$, or 36, equally likely ways that two numbers can appear in two rolls. On the first roll, 3 of the 6 numbers are odd. On the second roll, 2 of the 6 numbers are greater than or equal to 5. There are $3 \cdot 2$, or 6, ways of rolling successful events on both rolls. The probability of success is $\frac{6}{36}$, or $\frac{1}{6}$. Because the result of the first roll does not affect the second roll, the events are independent. ❖

Examine the probabilities of success on the first roll, the second roll, and both rolls.

The probability of rolling an odd number on the red cube is $\frac{1}{2}$.

The probability of rolling a 5 or greater on the green cube is $\frac{1}{3}$.

The probability of rolling both an odd number and a 5 or greater is $\frac{1}{2} \cdot \frac{1}{3}$, or $\frac{1}{6}$.

What pattern do you see between the probability of each independent event and the probability of both independent events together?

The probability of both occurring is the product of the probabilities of each occurring.

Finding Probabilities With the Complement

In Lesson 4.6, you determined a theoretical probability in three steps.

1. Find the denominator, or the number of equally likely possibilities.
2. Find the numerator, or the number of successes.
3. Express the probability as a fraction or percent.

Another useful technique in solving probability problems is to count the ways that something *does not* happen. In some cases this provides an easier way to solve the problem.

EXAMPLE 4

Games Two different cards are drawn from an ordinary deck of 52 cards. Find the probability that at least one of them is red.

Solution ➤

The event *at least one card is red* includes every event that can happen when two cards are drawn *except* an event of getting two black cards. Recall from Example 2 on page 205 that there are 650 out of 2652 ways of getting two black cards. Therefore, a successful event includes all of the possible outcomes except the 650 ways of getting two black cards. This is $2652 - 650$, or 2002, ways. The probability of getting at least one red card is $\frac{2002}{2652}$, or approximately 75.5%. ❖

When you find the number of possible ways that an event occurs by considering the number of ways that it does *not* occur, you are using the **complement**.

Try This Use the complement to determine the following theoretical probability: When two number cubes are tossed, what is the probability that the sum is greater than or equal to 3? HINT: Subtract from 1 the probability that the sum is less than 3.

A box contains 7 quarters, 6 dimes, 5 nickels, and 8 pennies. One coin is drawn from the box. Explain how to find the probability that the coin is worth at least 10 cents. There are 26 coins; 13 of these coins are worth 10¢ or more. The probability is $\frac{13}{26}$ or $\frac{1}{2}$.

Communicate

1. Use a square divided into regions for each event; use a tree diagram.
2. The outcome of one event does not affect the outcome of the other.

1. Name two ways to represent independent events with a diagram.
2. Explain what it means for two events to be independent.
3. When you find probabilities of independent events, do you usually use the Addition Principle of Counting or the Multiplication Principle of Counting? Multiplication Principle of Counting
4. How many paths are there for this tree diagram? 8
5. Explain how to find a probability by finding the complement.

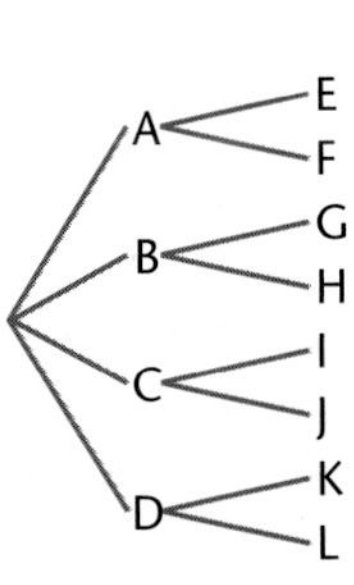

In this area model, the small squares to the left of the vertical red line represent the occurrence of event A; the small squares above the horizontal blue line represent the occurrence of event C, and so on. Tell how to find the following probabilities.

6. A occurs.

7. C occurs.

8. A occurs and C occurs.

9. A occurs or C occurs.

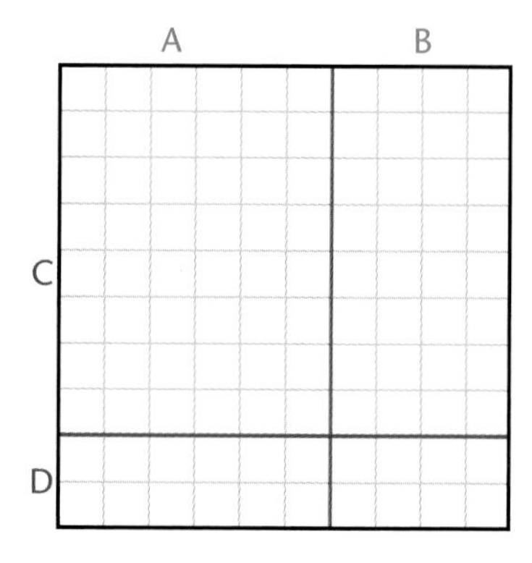

Practice & Apply

Travel John randomly selects which trails to use while hiking from Wapitu Falls to Songbird Lake.

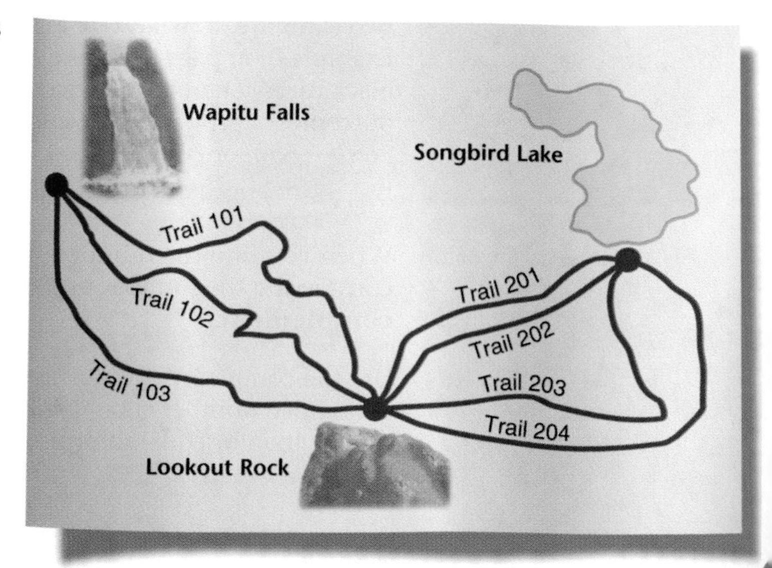

10. Find the probability that John takes Trail 101 from Wapitu Falls to Lookout Rock. $\frac{1}{3}$

11. Find the probability that John takes Trail 201 from Lookout Rock to Songbird Lake. $\frac{1}{4}$

12. Find the probability that John takes Trail 101 from Wapitu Falls to Lookout Rock AND takes Trail 201 from Lookout Rock to Songbird Lake. $\frac{1}{12}$

13. Find the probability that John takes Trail 101 from Wapitu Falls to Lookout Rock OR takes Trail 201 from Lookout Rock to Songbird Lake. $\frac{1}{2}$

Five chips numbered 1 through 5 are placed in a bag. A chip is drawn and replaced. Then a second chip is drawn. Find the probability that

14. both chips are even. $\frac{4}{25}$

15. both chips are odd. $\frac{9}{25}$

16. the first chip is even and the second chip is odd. $\frac{6}{25}$

17. one chip is even and the other chip is odd. $\frac{12}{25}$

A fair coin is tossed, the result is recorded, and then the coin is tossed again. Are the events below dependent or independent? Explain why.

18. The second toss is heads.

19. The second toss is tails.

20. Two heads occur in a row.

One number is selected from the list 1, 3, 5, 7. Another is selected from the list 5, 6, and 7. Find the probability that

21. both are even. 0

22. they are the same. $\frac{1}{6}$

23. the sum is even. $\frac{2}{3}$

24. the number from the second list is greater than the number from the first list. $\frac{2}{3}$

Use the grid at the right to find the probability that

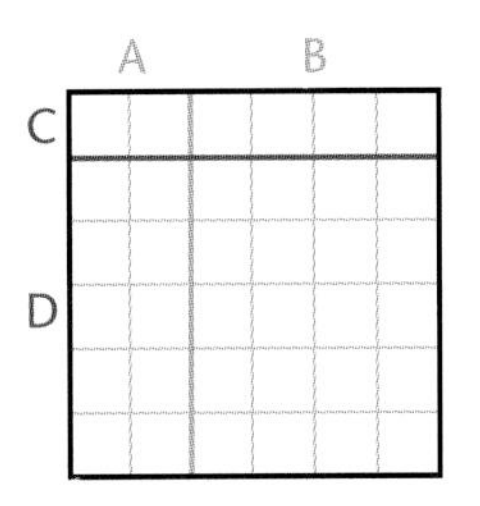

25. event A occurs. $\frac{1}{3}$

26. event C occurs. $\frac{1}{6}$

27. event A occurs AND event C occurs. $\frac{1}{18}$

28. event A occurs OR event C occurs. $\frac{4}{9}$

29. **Portfolio Activity** Complete the problem in the portfolio activity on page 165.

Look Back

Find the *y*-values of the following functions when *x* is −2, −1, 0, 1, and 2. Then identify the type of function. [Course 1]

30. $y = \frac{4}{x}$

31. $y = 4x^2$

32. $y = |4x|$

33. $y = -4x$

34. Find the product $-6(2)\left(\frac{5}{-3}\right)$. **[Course 1]** 20

35. Tim leaves home on his bike and rides at a rate of 12 miles per hour. Two hours after he leaves, Tim's mother remembers that Tim has a dentist's appointment. She sets out after him in her car at 40 miles per hour. How long does it take her to catch up to him? **[Lesson 2.8]** $\frac{6}{7}$ hour

36. Academics At one school, 48 students signed up to take Spanish, and 23 signed up to take French. If there were 12 that signed up for both, how many students total signed up for these foreign languages? **[Lesson 4.4]** 59 students

37. If Rich has 5 books, in how many ways can he arrange them on a shelf of his bookcase? **[Lesson 4.5]** 120

38. Travel There are 3 highways from Birchville to Pine Springs. From Pine Springs to Clearwater there are 5 roads. How many possible ways are there to travel from Birchville to Clearwater? **[Lesson 4.5]** 15

Look Beyond

39. Two radio stations are each having a contest. At the first station, the prize is \$1000, and your probability of winning is $\frac{1}{100,000}$. At the second station, the prize is \$25, and your probability of winning is $\frac{1}{2000}$. Which provides a better return for the listener's time? Compare your chances of winning the two contests. HINT: What would you expect to happen if you played each contest 100,000 times?

WINNING WAYS

Only 4 out of 10 students will be selected to compete against an academic team from another school. If the order in which the students are chosen to appear matters, how many ways can the 4 students be selected?

If you use the Multiplication Principle of Counting, there are 10 ways to make the first choice. For each of these choices, there are 9 ways to make the second choice.

First choice	Second choice	Third choice	Fourth choice
10	**9**	**?**	**?**

That means that there are 10 • 9 ways to make the first two choices. If you continue in the same way, there are 10 • 9 • 8 • 7, or 5040, ways to make the 4 choices.

First choice	Second choice	Third choice	Fourth choice
10	**9**	**8**	**7**

When r objects are chosen and the order matters, the result is called a permutation on n things taken r at a time. The number of possible **permutations** is written ${}_nP_r$. For example, ${}_{10}P_4 = 10 \bullet 9 \bullet 8 \bullet 7$, or 5040.

Other examples include the following:

a. The permutations of 5 letters taken 2 at a time is ${}_5P_2 = 5 \bullet 4$, or 20.

b. The permutation of 7 digits taken 3 at a time is ${}_7P_3 = 7 \bullet 6 \bullet 5$, or 210.

Of the 5040 permutations in the first example, some of the choices contain the same students. For example, the choice Andy, Bob, Cardi, and Donna is the same as Cardi, Bob, Donna, and Andy when order does not matter. There are 4 • 3 • 2 • 1, or 24, permutations for each group of 4 students. Thus, if the total number of permutations, 5040, is divided by 24, the result is 210.

Each of these 210 sets of 4 students is called a **combination**. Therefore, while there are 5040 permutations of 10 objects taken 4 at a time, there are only 210 combinations of 10 objects taken 4 at a time.

Thus, when r objects are chosen from n objects and the order does not matter, it is called the **combination** of n things taken r at a time. This is written as ${}_nC_r$. For the example above, ${}_{10}C_4 = \frac{10 \cdot 9 \cdot 8 \cdot 7}{4 \cdot 3 \cdot 2 \cdot 1} = \frac{5040}{24}$, or 210.

Other examples include the following:

a. The combination of 5 letters taken 2 at a time is ${}_5C_2 = \frac{5 \cdot 4}{2 \cdot 1} = \frac{20}{2}$, or 10.

b. The combination of 7 digits taken 3 at a time is ${}_7C_3 = \frac{7 \cdot 6 \cdot 5}{3 \cdot 2 \cdot 1} = \frac{210}{6}$, or 35.

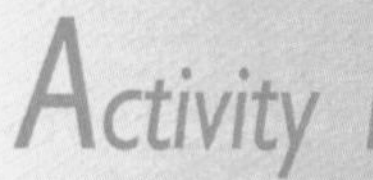

Activity 1

Use a tree diagram to show all of the permutations of the letters A, B, C, and D taken 2 at a time. Explain how the result can be used to find the number of combinations of 4 objects taken 2 at a time.

Activity 2

Find the number of arrangements consisting of 1, 2, 3, 4, and 5 letters from the word *point*. What is the sum of all possible arrangements?

Show all of the 3-letter arrangements, and determine which ones form English words. How many English words did you find? What is the probability that a 3-letter word can be formed from the letters in the word *point*?

Chapter 4 Review

Vocabulary

Addition Principle of Counting	188	frequency table	180	successful event	167
complement	207	independent events	206	survey	180
correlation coefficient	179	intersection	186	theoretical probability	199
disjoint sets	186	Multiplication Principle of Counting	193	tree diagram	192
equally likely outcomes	199	range	179	trial	167
experimental probability	167	simulation	172	union	186
				Venn diagram	186

Key Skills & Exercises

Lesson 4.1

➤ Key Skills

Calculate experimental probability.

Two coins are tossed 5 times with the following results:

Trial	1	2	3	4	5
Coin 1	H	T	T	T	H
Coin 2	T	T	H	H	H

The experimental probability that both coins are alike is $\frac{2}{5}$.

Use technology to generate random numbers.

RAND generates a number from 0 to 1, not inclusive.

The output of the command INT(RAND * 4) is 0, 1, 2, or 3.

The output of the command INT(RAND * 4) + 2 is 2, 3, 4, or 5.

In general, the function INT(RAND * K) + A generates random integers from a list of K consecutive integers beginning with A.

➤ Exercises

Use the data above to find each experimental probability.

1. At least one coin is heads.
2. Both coins are tails.

1. $\frac{4}{5}$ or 80%
2. $\frac{1}{5}$ or 20%

Write the command to generate random numbers from each list.

3. 4, 5, 6, 7, 8
   INT(RAND*5) + 4
4. 0, 1, 2, 3, 4, 5, 6, 7, 8, 9, 10
   INT(RAND*11)

Lesson 4.2

➤ Key Skills

Design a simulation to determine experimental probability.

Rolling a number cube can be used to simulate guessing the correct response to a multiple-choice question. Let 1 through 5 represent the 5 possible choices. Roll again if you get a 6, and do not count it as a trial. Let a 1 represent a correct response. The results are shown below.

Trial	1	2	3	4	5	6	7	8	9	10
Roll	1	4	2	1	3	2	1	4	3	5

➤ Exercises

Suppose you want to simulate guessing all of the answers on a 10-question true-false quiz. Design a simulation using

5. random numbers. **6.** coins.

Lesson 4.3

➤ Key Skills

Use statistics and probability to evaluate survey information.

Each member of a focus group of 23 people was asked to rate his or her support for Mr. Smith as a candidate for mayor on a scale from 1 (strongly oppose) to 5 (strongly support). You can see from the bar graph that the mean score is:

$$(1 \cdot 1 + 3 \cdot 2 + 9 \cdot 3 + 7 \cdot 4 + 3 \cdot 5) \div 23, \text{ or } 3.35.$$

➤ Exercises

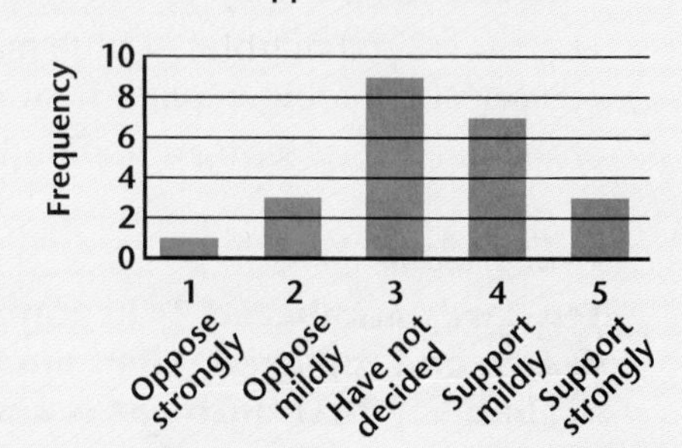

Refer to the graph of the focus group's ratings of support.

7. What is the median score of the group's ratings? How does it compare with the mean score?

8. What is the probability that a person in the focus group will vote for Mr. Smith for mayor?

8. $\frac{10}{23}$, or approx 43%

Lesson 4.4

➤ Key Skills

Use the Addition Principle of Counting to determine the number of elements in the union of sets.

In a music class, 11 students play the clarinet, 15 play the flute, and 8 play both. You can use the Addition Principle of Counting to determine the number of students in the music class. There are 11 students who play the clarinet, 15 who play the flute, and 8 who have been counted twice. So, there are 11 + 15 − 8, or 18, students.

➤ Exercises

Consider an ordinary deck of 52 playing cards.

9. Draw a Venn diagram showing the cards that are black OR face cards.

10. How many cards are black OR face cards? red OR numbered cards? 32; 44

Lesson 4.5

➤ Key Skills

Use the Multiplication Principle of Counting to find the number of possible ways to make multiple choices.

A craft store stocks 5 styles of T-shirts in 4 sizes and 10 different colors. The craft store stocks 5 • 4 • 10, or 200 kinds of T-shirts.

➤ Exercises

11. How many ways are there to arrange the letters in the word *math*? 24

12. A dress shop stocks 6 styles of blouses. Each of these styles comes in 7 sizes and 5 colors. How many different selections are possible? 210

Lesson 4.6

➤ Key Skills

Find the theoretical probability.

Suppose you roll an ordinary 6-sided number cube once. The probability that you will roll a number greater than 4 is $\frac{2}{6}$, or $\frac{1}{3}$.

➤ Exercises

An integer between −10 and 10 is drawn from a bag containing numbered chips. Find the probability that the integer is

13. negative. **14.** zero. **15.** even.

13. $\frac{10}{21}$ or approx 48% 14. $\frac{1}{21}$ or approx 5% 15. $\frac{11}{21}$ or approx 52%

Lesson 4.7

➤ Key Skills

Find the probability of independent events.

There are 2 red marbles, 4 blue marbles, and 3 white marbles in a bag. If you select a marble, replace it, and then select another marble, the probability of selecting a red marble *and* a blue marble is $\frac{2}{9} \cdot \frac{4}{9}$, or $\frac{8}{81}$.

➤ Exercises

Ten cards numbered 1 through 10 are placed in a box. One card is drawn and replaced. Then another card is drawn. Find the probability that both cards are

16. less than 5. $\frac{4}{25}$ or 16%

17. multiples of 3. $\frac{9}{100}$ or 9%

18. the same. $\frac{1}{10}$ or 10%

Applications

Business The Hamburger Hut is giving away free food during its grand opening. It is distributing 1,000 coupons, of which 100 are for a free hamburger, 150 are for free fries, and 200 are for a free soft drink. What is the probability of winning

19. a hamburger? **20.** fries? **21.** a soft drink?

19. $\frac{1}{10}$ or 10%
20. $\frac{3}{20}$ or 15%
21. $\frac{1}{5}$ or 20%

22. The Hamburger Hut advertises 15 different ways to order its hamburgers, 2 different types of fries, and 8 different kinds of soft drinks. How many ways are there to choose one of each? 240

23. The probability that a household in the United States will have some type of pet is about $\frac{3}{5}$. The probability that a household will have at least one child is $\frac{1}{3}$. What is the probability that a household in the United States has a pet AND a child? $\frac{1}{5}$ or 20%

24. Tim, Mary, and Paul have applied for a loan to buy a used car. The probability that Tim's application is approved is 75%. The probability that Mary's application is approved is 80%. The probability that Paul's application is approved is 62.5%. What is the probability that all three loans will be approved? 37.5%

25. In Exercise 24, what is the probability that only Mary's application will be approved? 7.5%

Chapter 4 Assessment

If RAND generates a random number between 0 and 1 (including 0, but not including 1), then describe the output of the following commands:

1. RAND * 5 **2.** INT(RAND * 5) **3.** INT(RAND * 5) + 4

4. Write a command with RAND to generate random numbers from the list 1, 2, 3, 4, 5, 6, 7, and 8. INT(RAND*8) + 1

5. Show how to simulate randomly selecting a phone number from a list of 6 phone numbers.

6. If the range for a set of scores is 45, the mean is 75, and the highest score is 98, find the lowest score. 53

The frequency table below shows the range of opinion scores on the food at the school cafeteria from 1 (very poor) to 5 (very good).

Scores	1 (Very poor)	2 (Poor)	3 (Fair)	4 (Good)	5 (Very good)
Frequency	/	////	~~////~~ /	~~////~~ ////	~~////~~ //

7. What is the median score? What is the mean score? 4; 4

8. What is the experimental probability that a student surveyed thinks the school food is very good? $\frac{7}{27}$ or approx 26%

For an ordinary deck of 52 cards, tell how many of the cards are

9. numbered. 36

10. red. 26

11. red AND numbered. 18

12. red OR numbered. 44

13. There are 16 girls on a softball team and 11 girls on a track team. If there are 5 girls on both teams, how many girls total are on the softball team or the track team? 22

14. What does it mean to say that outcomes are equally likely? Why is this important?

Two ordinary 6-sided number cubes are rolled. Find the probability that the sum is

15. 10. $\frac{1}{12}$ or approx 8.3%

16. less than 5. $\frac{1}{6}$ or approx 16.7%

17. a multiple of 2 AND 3.

18. a multiple of 2 OR 3. $\frac{2}{3}$ or approx 66.7%

17. $\frac{1}{6}$ or approx 16.7%

An integer from 1 to 10 is selected. Then a letter from the alphabet is selected. Find the probability that

19. the integer is odd AND the letter is a vowel (*a*, *e*, *i*, *o*, *u*, or *y*). $\frac{3}{26}$ or approx 11.5%

20. the integer is a multiple of 5 AND the letter is a consonant. $\frac{2}{13}$ or approx 15.4%

21. the integer is a multiple of 3 AND the letter selected is *t*. $\frac{3}{260}$ or approx 1.2%

A box contains 3 blue stickers, 4 red stickers, and 11 yellow stickers. Maria chooses one sticker and then another. Find the probability that she chooses

22. a blue sticker, puts it back, and then chooses a yellow sticker. $\frac{11}{108}$ or approx 10.2%

23. a blue sticker, does *not* put it back, and then chooses a yellow sticker. $\frac{11}{102}$ or approx 10.8%

24. two red stickers without putting the first one back. $\frac{2}{51}$ or approx 3.9%

25. Explain the difference between independent and dependent events.

Chapters 1 – 4 Cumulative Assessment

College Entrance Exam Practice

Quantitative Comparison For Questions 1–4, write
A if the quantity in Column A is greater than the quantity in Column B;
B if the quantity in Column B is greater than the quantity in Column A;
C if the two quantities are equal; or
D if the relationship cannot be determined from the information given.

	Column A	Column B	Answers
1.	The greater of the two solutions of $\lvert x - 4 \rvert = 2$	The greater of the two solutions of $\lvert x + 4 \rvert = 2$	A (A) (B) (C) (D) **[Lesson 1.6]**
2.	$[3 \quad 5]\begin{bmatrix} 2 & -1 \\ 4 & 6 \end{bmatrix} = [a \quad b]$ a	b	B (A) (B) (C) (D) **[Lesson 3.3]**
3.	The mode of the data: 9, 11, 13, 18, 13	The mean of the data: 9, 11, 13, 18, 13	A (A) (B) (C) (D) **[Lesson 4.3]**
4.	The probability of two heads on two successive coin tosses	The probability of heads followed by tails on two successive coin tosses	C (A) (B) (C) (D) **[Lesson 4.6]**

5. Which two properties can be used to solve $3x + 2 = 5$? **[Lessons 1.3, 1.4]** c
a. the Addition and Subtraction Properties of Equality
b. the Division and Multiplication Properties of Equality
c. the Subtraction and Division Properties of Equality
d. none of these

6. Let $A = \begin{bmatrix} -1 & 0 \\ 2 & 3 \end{bmatrix}$ and $B = \begin{bmatrix} 1 & 3 \\ 2 & 0 \end{bmatrix}$. Which of the matrix expressions below is equivalent to $\begin{bmatrix} 2 & 3 \\ 0 & -3 \end{bmatrix}$? **[Lesson 3.2]** c

a. $A + B$ **b.** $A - B$ **c.** $B - A$ **d.** none of these

7. Let A be a matrix that has an inverse A^{-1}. What is the product $A \cdot A^{-1}$? **[Lesson 3.4]** c

a. $\begin{bmatrix} 0 & 1 \\ 1 & 0 \end{bmatrix}$ **b.** $\begin{bmatrix} 1 & 1 \\ 1 & 1 \end{bmatrix}$ **c.** $\begin{bmatrix} 1 & 0 \\ 0 & 1 \end{bmatrix}$ **d.** $\begin{bmatrix} 1 & 0 \\ 0 & 0 \end{bmatrix}$

8. What are the values of x and y in the matrix equation

$\begin{bmatrix} 3 & 2 \\ 2 & 1 \end{bmatrix}\begin{bmatrix} x \\ y \end{bmatrix} = \begin{bmatrix} 2 \\ -2 \end{bmatrix}$? **[Lesson 3.5]** d

a. $x = 6, y = -10$ **b.** $x = 6, y = 10$
c. $x = 10, y = -6$ **d.** none of these

9. Which command generates a random number from the list 3, 4, and 5? **[Lesson 4.1]** d

a. INT(RAND * 5) + 5 **b.** INT(RAND * 3) + 5
c. INT(RAND * 5) + 3 **d.** INT(RAND * 3) + 3

10. In a poll, 15 students said that they like basketball, and 8 said that they like baseball. What was the number of students polled? **[Lesson 4.4]** d

a. 23 **b.** 15 **c.** 7 **d.** not enough information to tell

11. Solve the inequality $-5x - 7 < 13$, and graph the solution on a number line. **[Lesson 1.5]**

12. Write the equation $-4x + 5 = 10y$ in slope-intercept form. Then state the slope, and graph the equation. **[Lesson 2.2]**

13. Solve the system $\begin{cases} y = 2x - 7 \\ 3x - 1.5y = 10.5 \end{cases}$. State the solution, and identify the system as independent, dependent, or inconsistent. **[Lesson 2.6]**
Infinite solutions; dependent system

For Exercises 14–17, let $X = \begin{bmatrix} 2 & -1 & 5 \\ 1 & 0 & -2 \end{bmatrix}$ and $Y = \begin{bmatrix} 4 & 9 & -6 \\ -1 & 4 & 3 \end{bmatrix}$.

14. What are the dimensions of matrix X? **[Lesson 3.1]** 2×3

15. Find the entry x_{23}. **[Lesson 3.1]** -2

16. Find $X + Y$. **[Lesson 3.2]**

17. Find $3X$. **[Lesson 3.2]**

16. $\begin{bmatrix} 6 & 8 & -1 \\ 0 & 4 & 1 \end{bmatrix}$ 17. $\begin{bmatrix} 6 & -3 & 15 \\ 3 & 0 & -6 \end{bmatrix}$

18. Find the inverse of $\begin{bmatrix} -2 & 3 \\ 4 & -1 \end{bmatrix}$, if the inverse exists. **[Lesson 3.4]** $\begin{bmatrix} 0.1 & 0.3 \\ 0.4 & 0.2 \end{bmatrix}$

19. There are 48 students in the choir and 50 students in the orchestra. If 15 students are in both the choir and the orchestra, what is the total number of students in the choir and orchestra? **[Lesson 4.4]** 83

20. Mark has 3 nickels, 5 dimes, and 7 quarters in his pocket. He chooses one coin, replaces it, and then chooses another. What is the probability that he chose a dime first and then a quarter? **[Lesson 4.7]** $\frac{5}{15} \cdot \frac{7}{15} \approx 0.16$, or approx 16%

Free-Response Grid The following questions may be answered using a free-response grid commonly used by standardized test services.

21. Jason invested part of his $10,500 inheritance in a money-market fund that paid an annual rate of 4%. He then invested the rest in a high-yield bond that had an annual return of 10%. The annual return from both investments was $660. How much was invested at 4%? **[Lesson 2.8]** $6500 at 4%

22. How many ways are there to arrange 2 different letters of the alphabet? **[Lesson 4.5]** 650 ways

23. Suppose that you roll an ordinary 6-sided number cube. What is the probability that you will roll a number that is greater than 4? **[Lesson 4.6]** $\frac{1}{3}$ or approx 33.3%

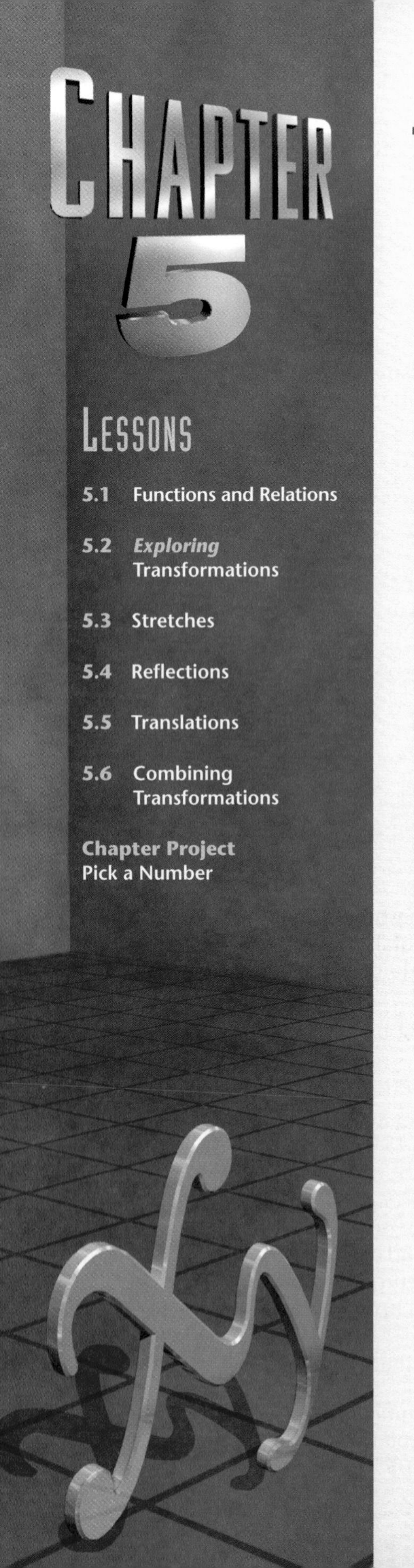

Transformations

LESSONS

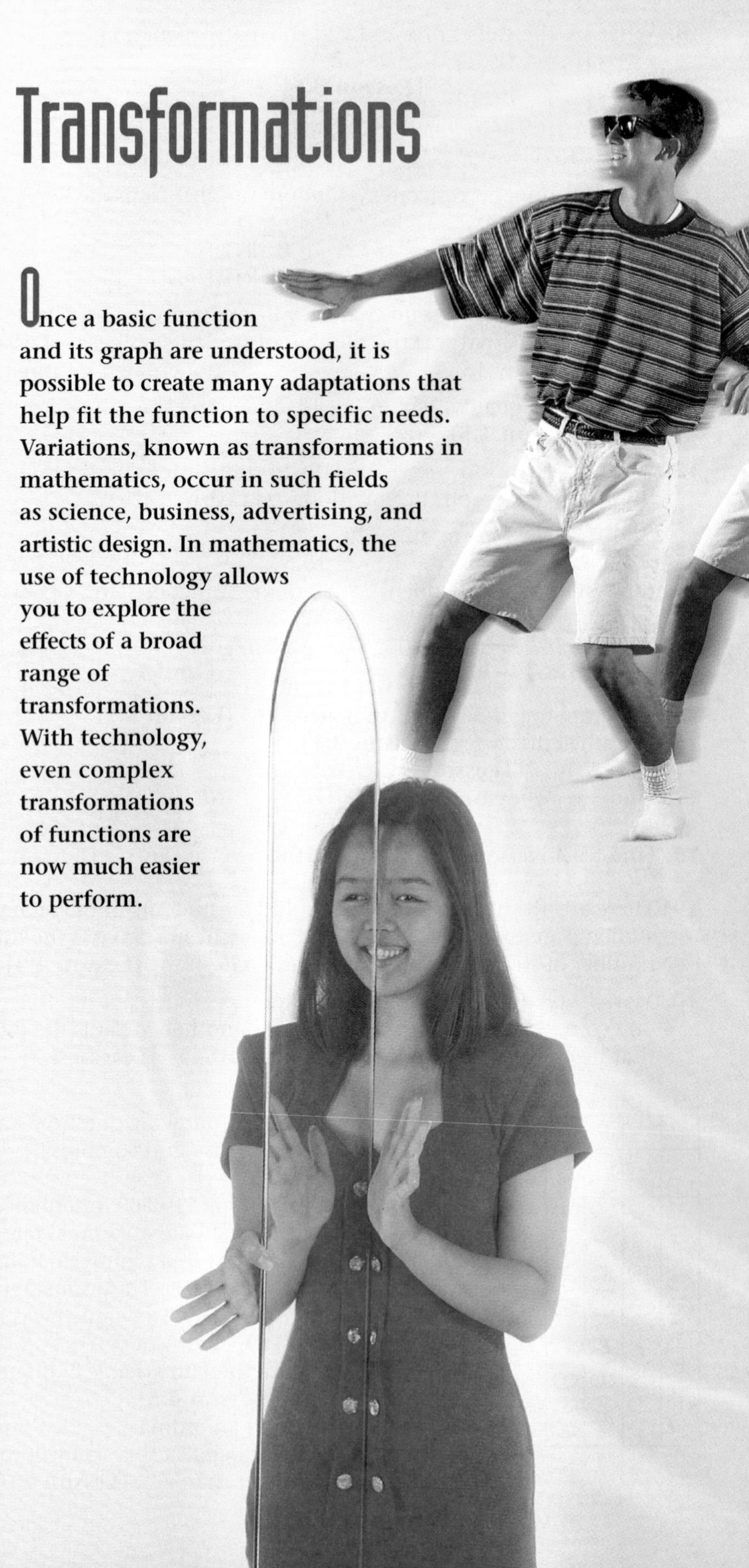

Once a basic function and its graph are understood, it is possible to create many adaptations that help fit the function to specific needs. Variations, known as transformations in mathematics, occur in such fields as science, business, advertising, and artistic design. In mathematics, the use of technology allows you to explore the effects of a broad range of transformations. With technology, even complex transformations of functions are now much easier to perform.

Portfolio Activity

Draw a coordinate plane on a piece of graph paper. Put the *y*-axis close to the left edge of the paper. Then design a figure in the first quadrant like the one shown below. Reflect the figure through the *x*-axis. Translate the reflection to the right far enough so that the translation does not overlap the reflection. Then reflect the translation through the *x*-axis. Continue the procedure to create a repeating pattern.

Draw a new figure, and perform several different reflections and translations to create other repeating patterns.

Identify one point on your original figure by its coordinates. List the coordinates of that point after each successive transformation.

LESSON 5.1 Functions and Relations

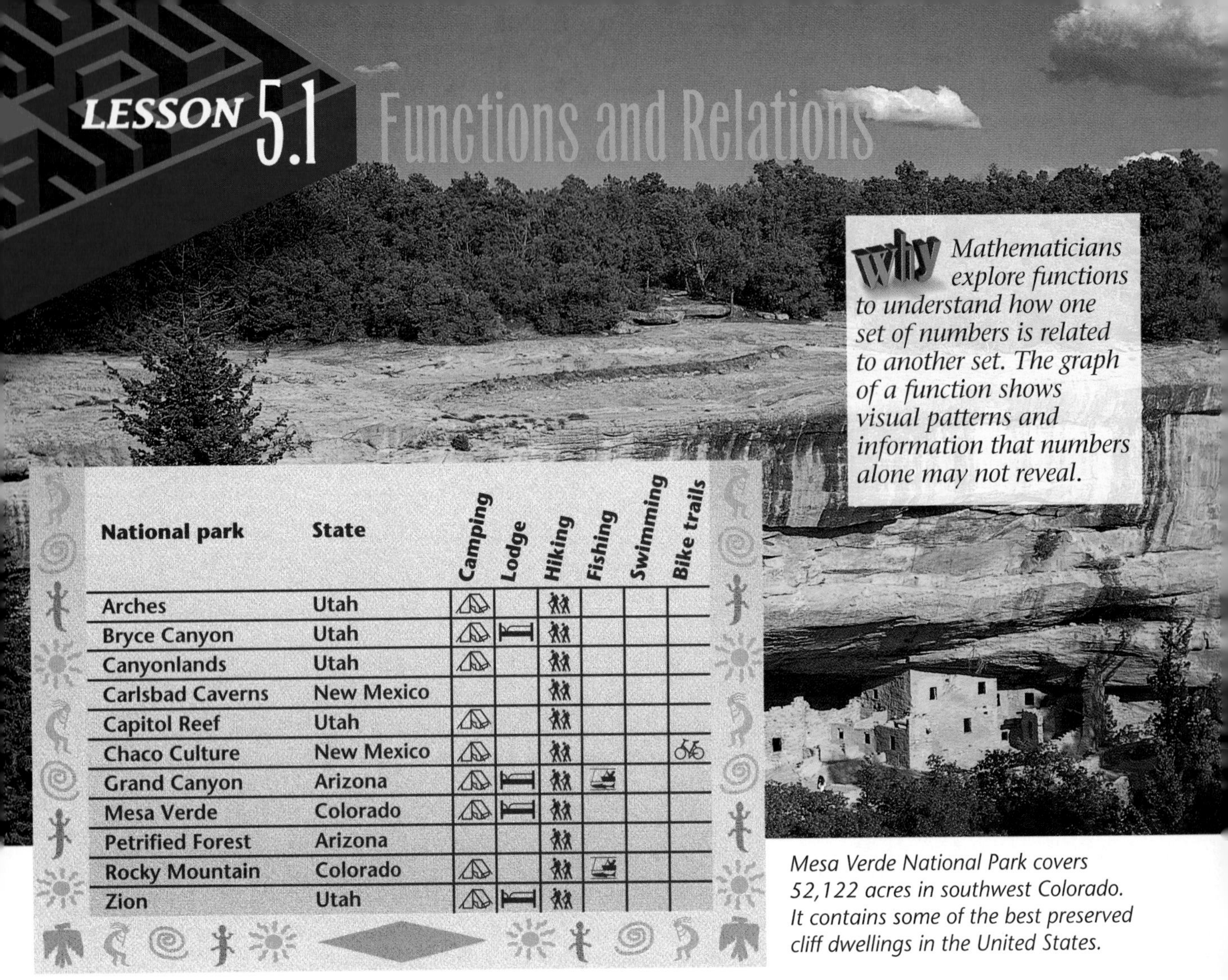
WHY Mathematicians explore functions to understand how one set of numbers is related to another set. The graph of a function shows visual patterns and information that numbers alone may not reveal.

National park	State	Camping	Lodge	Hiking	Fishing	Swimming	Bike trails
Arches	Utah	✓		✓			
Bryce Canyon	Utah	✓	✓	✓			
Canyonlands	Utah	✓		✓			
Carlsbad Caverns	New Mexico			✓			
Capitol Reef	Utah	✓		✓			
Chaco Culture	New Mexico	✓		✓			✓
Grand Canyon	Arizona	✓	✓	✓	✓		
Mesa Verde	Colorado	✓	✓	✓			
Petrified Forest	Arizona			✓			
Rocky Mountain	Colorado	✓		✓	✓		
Zion	Utah	✓	✓	✓			

Mesa Verde National Park covers 52,122 acres in southwest Colorado. It contains some of the best preserved cliff dwellings in the United States.

The Alvarez family is deciding which national parks to visit on their summer vacation. Their road atlas contains a chart that lists the camping features of each park.

Defining Relation and Function

Two different pairings appear in the table. The first pairs a park with the features available at that park. For example, Mesa Verde is paired with camping, a lodge, and hiking trails. This pairing is an example of a *relation*.

Mesa Verde — **Camping**, **Lodge**, **Hiking trails**

A **relation** pairs elements from one set with elements of another. In other words, a relation is a set of ordered pairs.

The second pairs each park with *exactly one* state. For example, Mesa Verde is paired only with Colorado. This pairing is an example of a *function*. A function is a special relation.

Mesa Verde — **Colorado**

FUNCTION

A **function** is a set of ordered pairs for which there is exactly one second coordinate for each first coordinate.

In Course 1 you were introduced to several elementary functions and examined their graphs by plotting ordered pairs. Recall that to graph a function, you use the coordinates of ordered pairs, such as (x, y), to locate a set of points on a coordinate plane.

CT– Answers may vary. For example, they began the trip in a residential area and had to stop for stoplights, stop signs, or traffic. They may have stopped to buy gas 6 minutes from home. Their speed levels off at 55 mph after about 20 minutes because they reached a highway.

The graph below shows speed as a function of time. The points on the graph form a set of ordered pairs (time, speed). There is exactly one speed for any moment in time.

The graph shows the relationship between

- the speed in miles per hour at which the Alvarez's car travels, and
- each moment of time during the first 26 minutes of their trip.

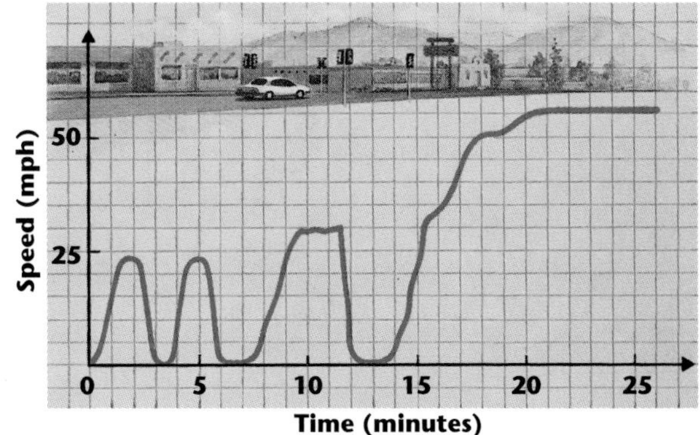

Use the graph to determine what might have happened during the first 26 minutes of the trip to cause the changes in the graph.

Testing for a Function

Once a relation is graphed, a **vertical-line test** will show whether the relation is a function. A relation is a function *if any vertical line intersects the graph of the relation no more than once.*

EXAMPLE 1

Which of the following graphs represent functions?

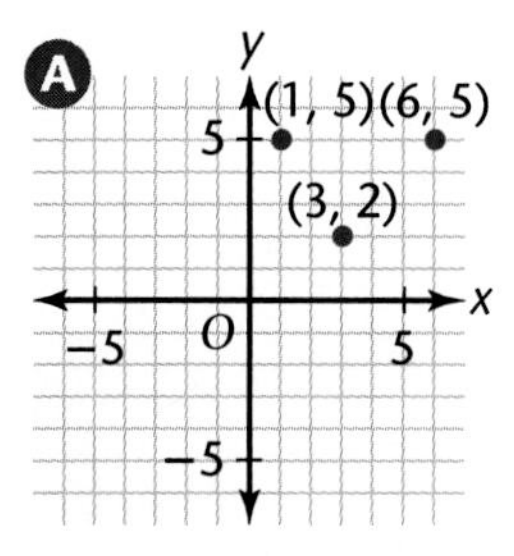

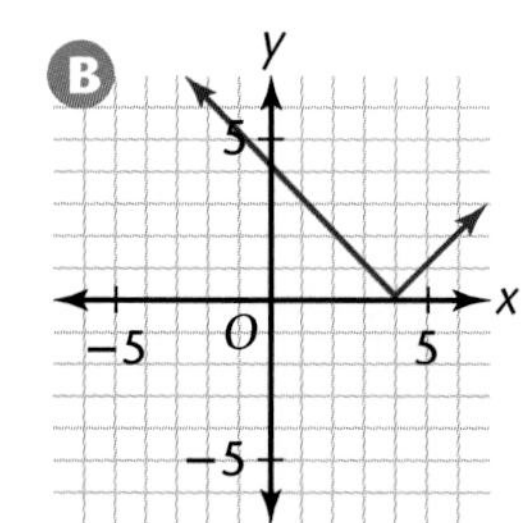

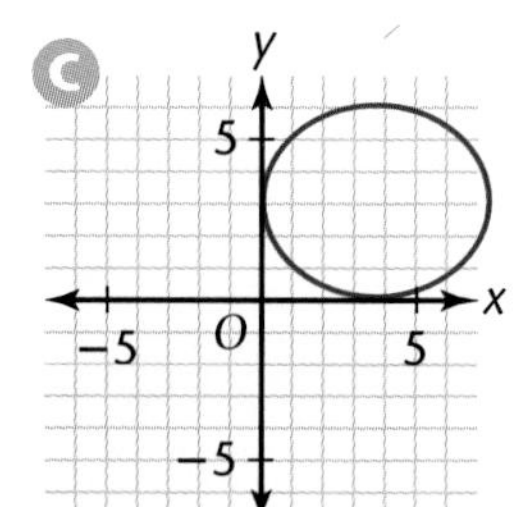

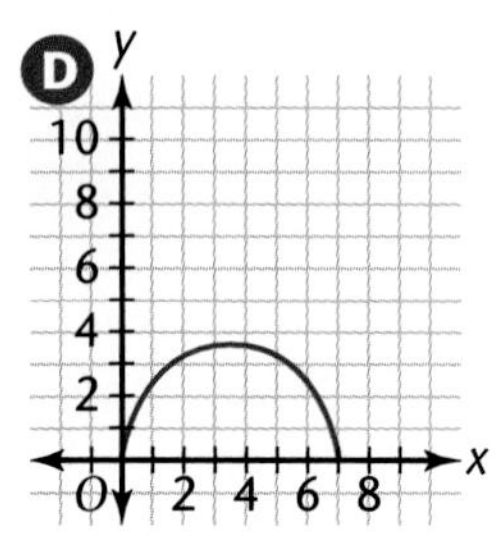

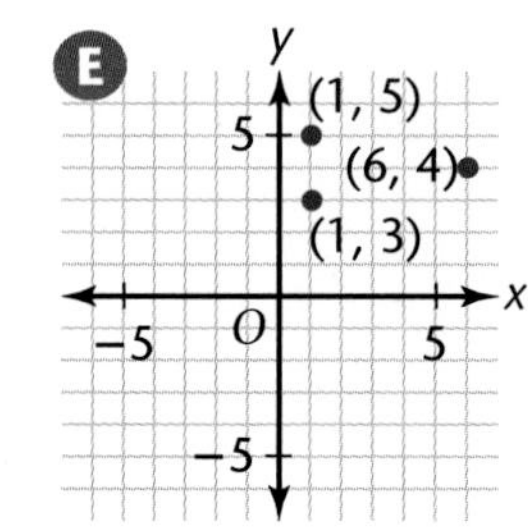

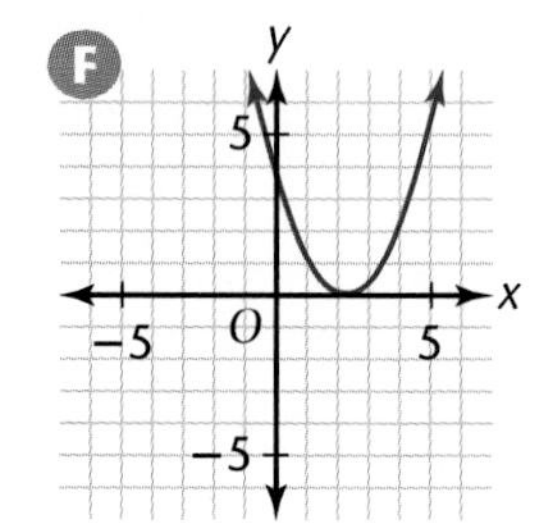

Solution ➤

Any vertical line will intersect graphs A, B, D, and F *no more than once,* so these graphs represent functions. The vertical lines on graphs C and E intersect the graph of the relation twice. These graphs do *not* represent functions.

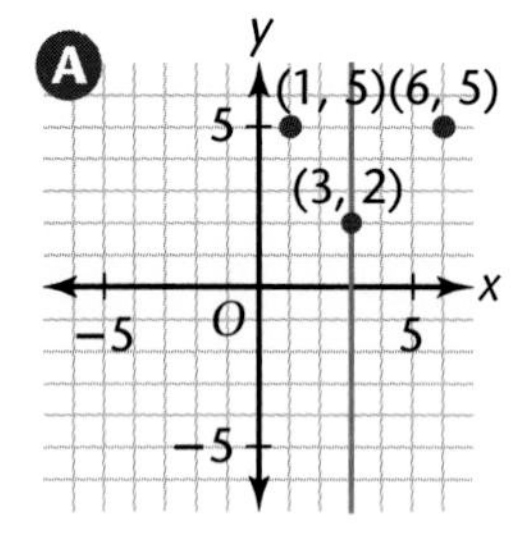

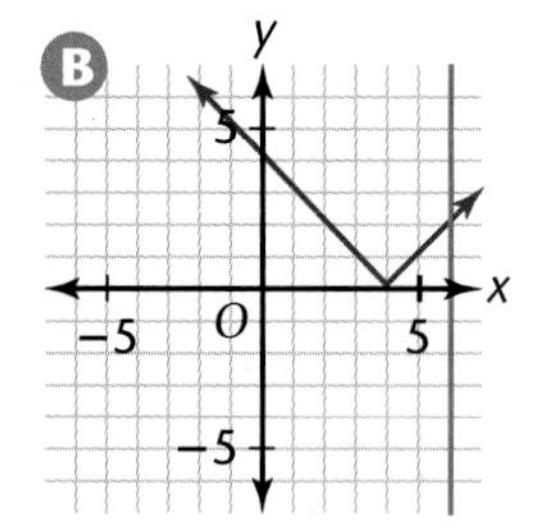

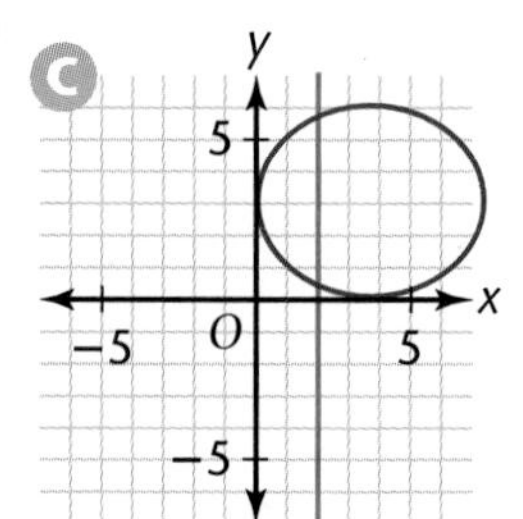

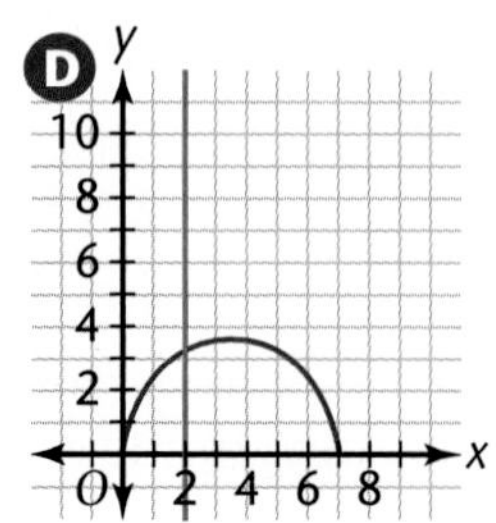

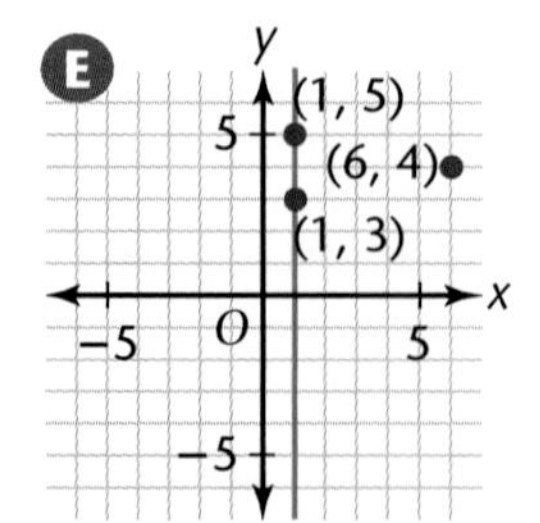

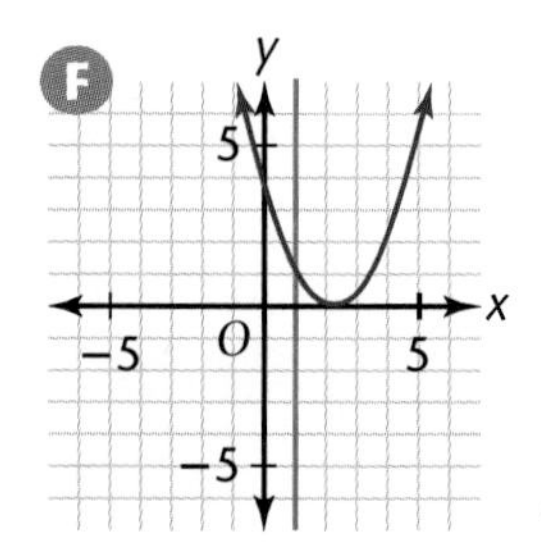

For any function, the set of x-values is the **domain**, and the set of y-values is the **range**.

The domain contains only the numbers that can be used in the function. The range contains only the numbers that the function can produce. For example, division by 0 is not defined, so 0 cannot be in the domain of $y = \frac{1}{x}$. Explain why there are no negative numbers in the range of the function $y = |x|$.

Examine the graphs in Example 1. Can you identify the domain and range of functions A, B, D, and F?

Example	Domain (x-values)	Range (y-values)
A	1, 3, and 6	2 and 5
B	all numbers	all numbers greater than or equal to zero
D	$0 \leq x \leq 7$	$0 \leq y \leq 3.5$
F	all numbers	all numbers greater than or equal to zero

Function Notation

The equation $y = x^2$ represents a quadratic function. Another way to express $y = x^2$ is to use *function notation*. **Function notation** uses a symbol such as $f(x)$ to write $y = x^2$ as $f(x) = x^2$.

$f(x) = x^2$ is read as "f of x equals x squared." The variable in parentheses, x, is the **independent variable**. The expression on the right side of the equal sign is the *function rule*. The value of the function rule is represented by the **dependent variable**, $f(x)$. To evaluate a function, replace the independent variable, x, with a value from the domain. Then perform the operations to find the corresponding value of the dependent variable.

For example, to evaluate the function $f(x) = x^2$ for $x = 3$, substitute 3 for x.

$$f(3) = 3^2 = 9$$

The domain of a function is sometimes called the input values. The range is sometimes called the output values.

input values ⟶ function rule ⟶ output values
(domain) (range)

EXAMPLE 2

Suppose h is the function $h(x) = x^2 - 2x + 1$. Because the independent variable, x, can be any real number, the domain of the function is *all real numbers*. Evaluate $h(4)$.

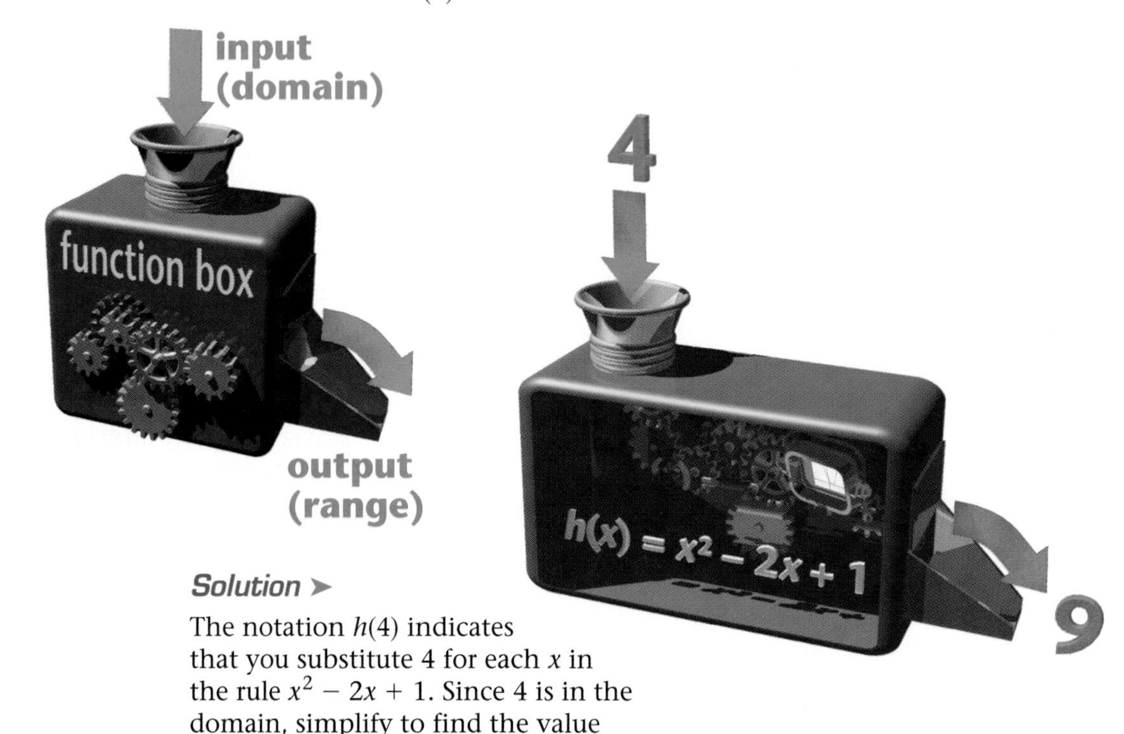

Solution ➤

The notation $h(4)$ indicates that you substitute 4 for each x in the rule $x^2 - 2x + 1$. Since 4 is in the domain, simplify to find the value of the function when x is 4.

$$h(x) = x^2 - 2x + 1$$
$$h(4) = 4^2 - 2(4) + 1$$
$$h(4) = 9$$ ❖

The h identifies or names the function. The input values for x in Example 2 are all real numbers. The function rule is the expression $x^2 - 2x + 1$. The values that you get after substituting each input value and simplifying the expression are the output values.

The values shown in $h(4) = 9$ represent the ordered pair (4, 9). Notice that 4 is an element of the domain and 9 is an element of the range. If you graph this relation, you can use the vertical-line test to show that it is a function.

Try This For the function $f(x) = 10x^2$, evaluate $f(-3)$. $f(-3) = 90$

The distance between two points on a number line is always a nonnegative number. This makes the absolute-value function a good representation for distance on the number line. For example, the absolute-value function can represent the distance between a given number, such as 5, and some other number, x. This is shown in Example 3.

EXAMPLE 3

Write the function that represents the distance between any number and the fixed point 5. Find the distance between -4 and 5 and between 0 and 5.

Solution ➤

The function $g(x) = |x - 5|$, where the domain is all numbers, gives the distance between any real number, x, and 5.

To find the distance between -4 and 5, evaluate $g(-4)$.

The distance between x and 5 when x is -4 is $g(-4) = |-4 - 5| = 9$.

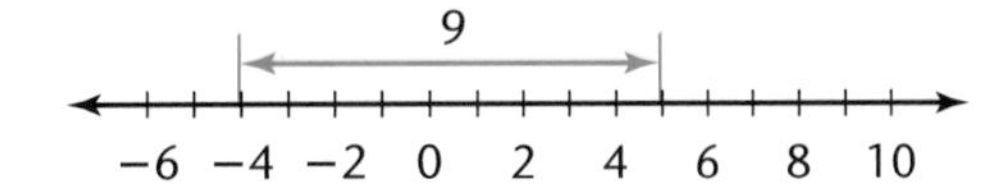

To find the distance between 0 and 5, evaluate $g(0)$.

The distance between x and 5 when x is 0 is $g(0) = |0 - 5| = 5$.

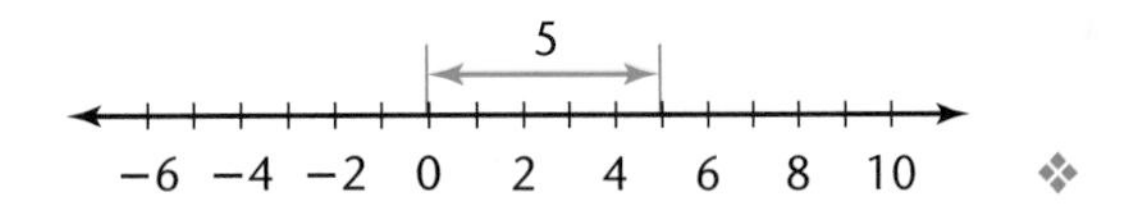

❖

EXERCISES & PROBLEMS

Communicate

1. What is the difference between a function and a relation?
2. Explain how the independent and dependent variables are related to the domain and range.
3. Describe the relationship between the definition of a function and the vertical-line test.
4. Explain how to find the range of the function $f(x) = 2x^2 - 5x$ if the domain is restricted to integers from -2 to 2, inclusive.
5. Suppose that you are given the ordered pair $(6, -46)$ from the function $h(x)$. How do you know which element is from the domain and which is from the range?
6. Explain how to represent the information in the function notation $h(-24) = 16$ as an ordered pair.

Practice & Apply

10. No; there is an x-value that has many y-values.

Which of the following relations are functions? Explain.

7. {(3, 4), (4, 4), (5, 4)}

8. {(1, 1)}

9.

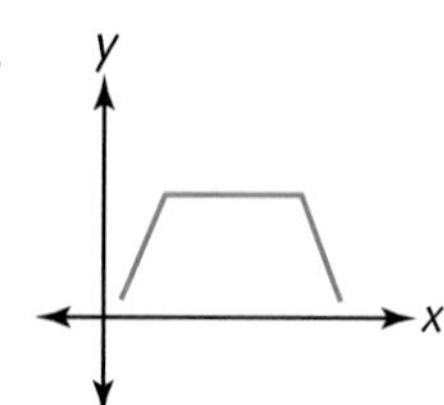

10.

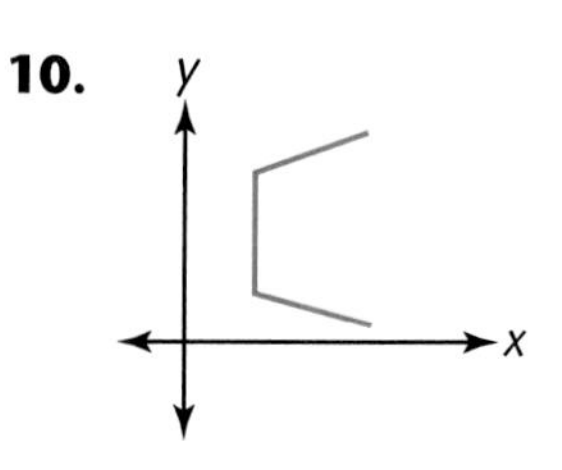

7. Yes; there is one y-coordinate for each x-coordinate.
8. Yes; there is one y-coordinate for each x-coordinate.
9. Yes; there is one y-coordinate for each x-coordinate.

Find the domain and range of each function.

11. {(3, 2), (4, 3), (5, 2)}

12.

13.

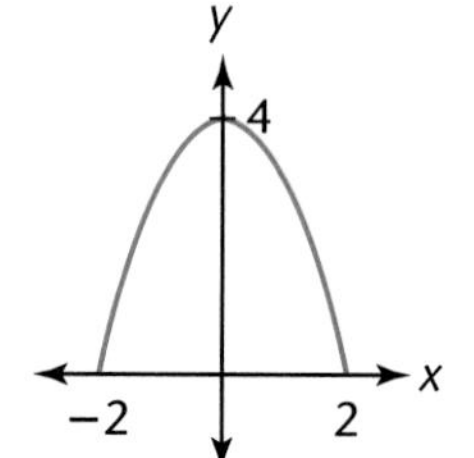

11. D: {3, 4, 5}; R: {2, 3}

12. D: {all real numbers}; R: {all real numbers}

13. D: $\{-2 \leq x \leq 2\}$; R: $\{0 \leq y \leq 4\}$

14. Given $y = \frac{1}{x}$, write the ordered pair (x, y) when x is 3. $\left(3, \frac{1}{3}\right)$

15. If $h(x) = x + 3$, find $h(7)$. $h(7) = 10$

Refer to the national parks table on page 220.

16. Which parks have fishing? Grand Canyon, Rocky Mt.

17. What are the features at Canyonlands? Camping and hiking

18. Which parks have bike trails? Chaco Culture

19. Which have swimming? None

20. Which have a lodge or fishing? Bryce Canyon, Grand Canyon, Mesa Verde, Rocky Mt., Zion

21. Which have a lodge and fishing? Grand Canyon

Technology Enter the national parks information in a database. Sort the database to list the parks that

22. are in Utah.

23. have camping.

24. do *not* have a lodge.

25. have a lodge and fishing. Grand Canyon

Which of the following relations are functions? Explain.

26. {(9, 5), (9, −5)}

27. {(5, 9), (−5, 9)}

28.

29.

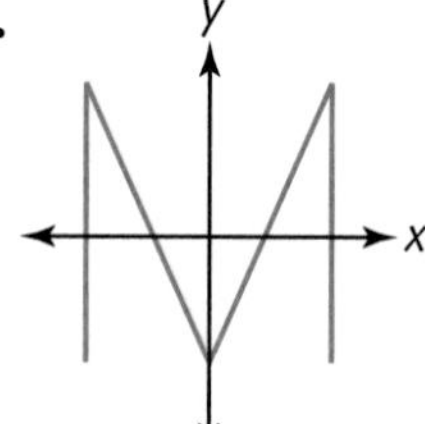

30. D: {1, 2, 3, 4}; R: $\left\{\frac{1}{2}, \frac{1}{3}, \frac{1}{4}\right\}$

31. D: {0.1, 0.2, 0.3}; R: {1, 2, 3}

Find the domain and range of each function.

30. $\left\{\left(1, \frac{1}{2}\right), \left(2, \frac{1}{2}\right), \left(3, \frac{1}{3}\right), \left(4, \frac{1}{4}\right)\right\}$

31. {(0.1, 1), (0.2, 2), (0.3, 3)}

32.

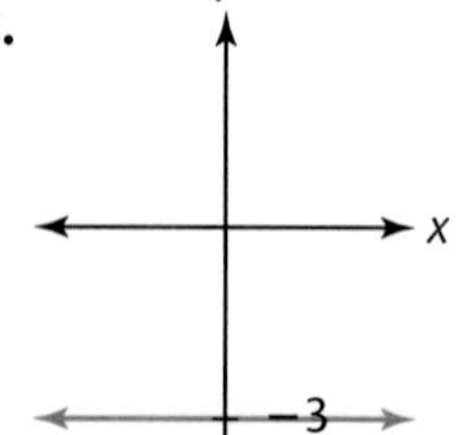

33.

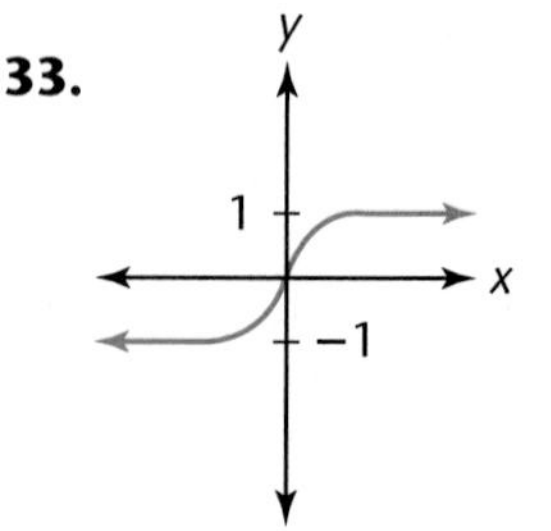

32. D: {all real numbers}; R: {−3}

33. D: {all real numbers}; R: $\{-1 \leq y \leq 1\}$

Let $f(x) = 5x$. Evaluate.

34. $f(3)$ 15 **35.** $f(0)$ 0 **36.** $f(-2)$ -10 **37.** $f(-6)$ -30

Evaluate each function when x is 3.

38. $g(x) = x^2$ 9 **39.** $f(x) = 2^x$ 8 **40.** $h(x) = |x|$ 3 **41.** $k(x) = \frac{1}{x}$ $\frac{1}{3}$

For each function, identify the independent and dependent variable. Then describe the domain and range.

42. $f(x) = x - 5$ **43.** $f(x) = x^2$ **44.** $f(x) = (x + 4)^2$

45. $f(x) = (x - 7)^2$ **46.** $f(x) = -3x^2$ **47.** $f(x) = -|x|$

48. $f(x) = |x + 2|$ **49.** $f(x) = |x - 5|$

50. **Temperature** If you know the Celsius temperature, do you necessarily know the Fahrenheit temperature? Is Fahrenheit temperature a function of Celsius temperature? Explain. HINT: $F = \frac{9}{5}C + 32$

51. If you know the Fahrenheit temperature, do you necessarily know the Celsius temperature? Is Celsius temperature a function of Fahrenheit temperature? Explain. HINT: $C = \frac{5}{9}(F - 32)$

Look Back

Solve each equation. [Lessons 1.3, 1.4]

52. $3r + 4 = -2 + 6r$ 2 **53.** $8(x - 7) = 3x + 4$ 12

54. Solve $A = \frac{h}{2}(B + b)$ for B. **[Lesson 1.3]** $B = \frac{2A}{h} - b$

55. Graph the linear equation $4x - 2y = 24$. **[Lesson 2.2]**

56. A line crosses the y-axis at 5 and is perpendicular to $y = -\frac{3}{5}x$. Write the equation, in standard form, for this line. **[Lesson 2.2]** $5x - 3y = -15$

57. Find the product $\begin{bmatrix} 1 & 3 \\ -5 & 7 \end{bmatrix}\begin{bmatrix} 6 & 0 \\ 1 & -2 \end{bmatrix}$. **[Lesson 3.3]** $\begin{bmatrix} 9 & -6 \\ -23 & -14 \end{bmatrix}$

58. Find the inverse of $A = \begin{bmatrix} 3 & -5 \\ -1 & 2 \end{bmatrix}$. **[Lesson 3.4]** $\begin{bmatrix} 2 & 5 \\ 1 & 3 \end{bmatrix}$

59. There are 5 friends that play tennis. How many games must be played for each player to play all the others one time? **[Lesson 4.5]** 10

Probability There are 8 red, 4 green, 9 yellow, and 3 white tickets in a jar. The tickets are thoroughly mixed. **[Lesson 4.6]**

60. Jacob draws one ticket from the jar without looking. What is the probability that he will draw a red ticket? $\frac{1}{3}$

61. Jacob puts the ticket back and draws another without looking. What is the probability that he will now draw a yellow or a green ticket? $\frac{13}{24}$

Look Beyond

Graph each on a graphics calculator. HINT: Graph the top half using $Y_1 =$ and the bottom half using $Y_2 =$.

62. $x = y^2$ **63.** $x^2 + y^2 = 25$

LESSON 5.2 Exploring Transformations

Why *Computers and graphics calculators can draw several graphs on the same coordinate plane. Using technology or sketching the graphs can help you explore the relationship between a change in the function rule and a change in its graph.*

In Course 1 you learned how to draw the graphs of several functions. One of those functions is the absolute-value function, $f(x) = |x|$ or $y = |x|$. The **vertex** of this graph is where the V comes to a point. This vertex has its lowest, or **minimum**, point at the origin (0, 0).

This graph is also **symmetric** with respect to the y-axis. The axis that passes vertically through the vertex of this function is the **axis of symmetry**. If you fold the graph along this axis, the left and right halves of the graph will match exactly.

x	$\lvert x \rvert$
−3	3
−2	2
−1	1
0	0
1	1
2	2
3	3

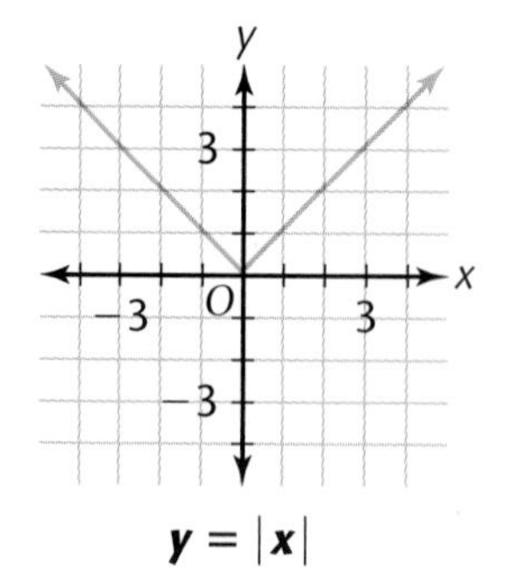

$y = |x|$

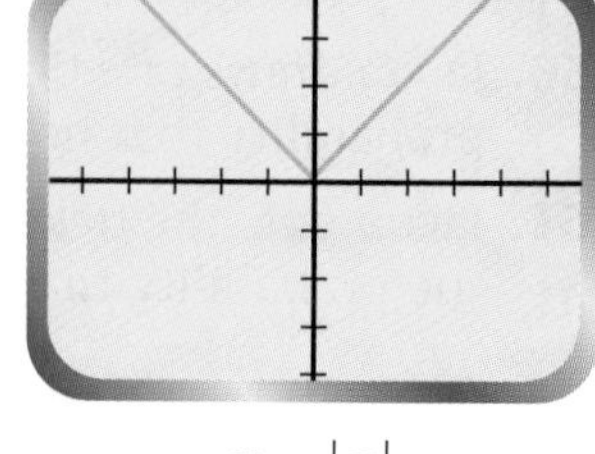

$y = |x|$

The original function $y = |x|$ is an example of a *parent function*. A **parent function** is the most basic of a family of functions. A variation such as a stretch, reflection, or shift is a **transformation** of the parent function.

There are various ways to stretch, reflect, and shift the graph of $y = |x|$. Examine the graphs in **I–VI**. The parent function $y = |x|$ has been

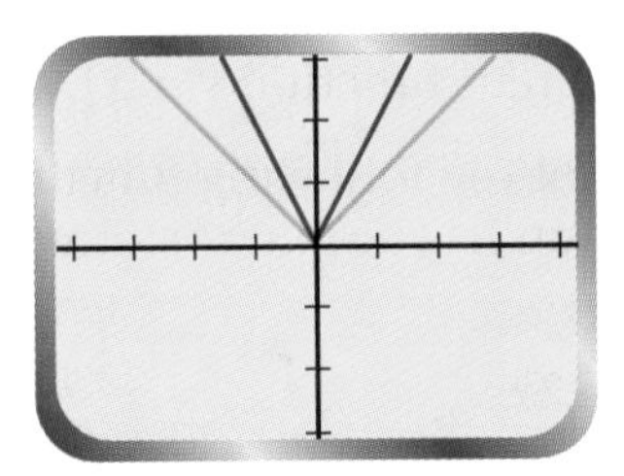

I. *stretched by 2.*

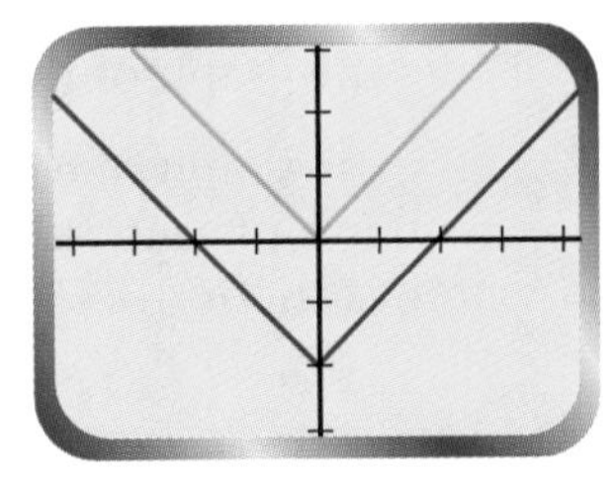

II. *shifted vertically by −2.*

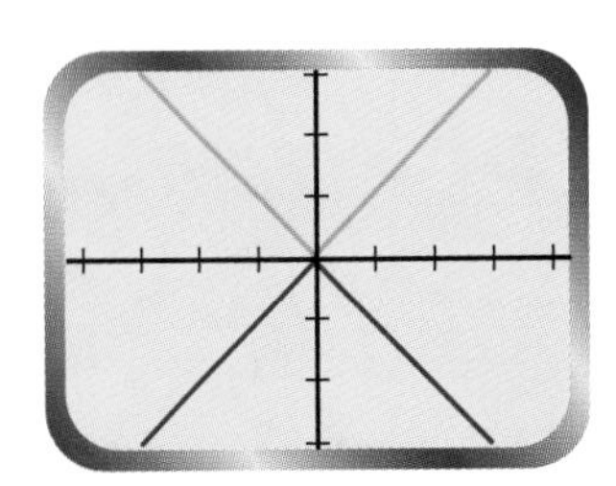

III. *reflected through the x-axis.*

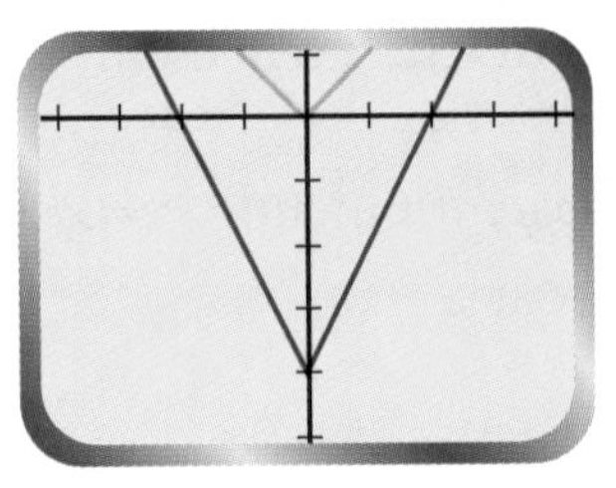

IV. *stretched by 2 and shifted vertically by −4.*

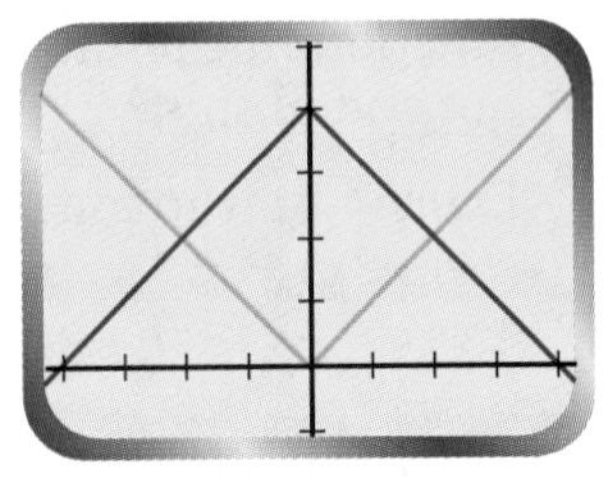

V. *reflected through the x-axis and shifted vertically by 4.*

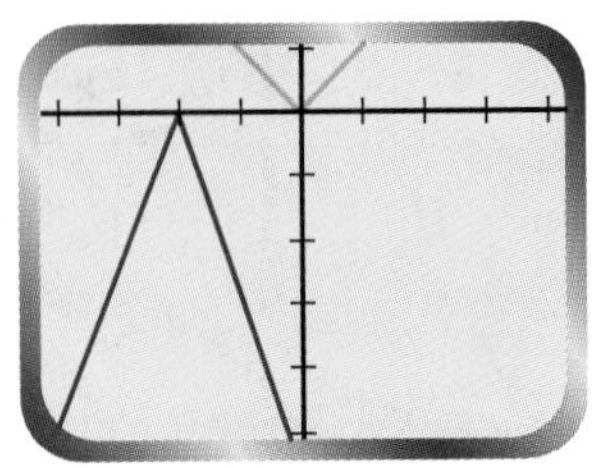

VI. *stretched by 3, reflected through the x-axis, and shifted horizontally by −2.*

Exploration 1 *Transformations of Absolute-Value Functions*

Graphics Calculator

1 Draw the graph of the parent function $y = |x|$. If you have a graphics calculator, enter ABS X or ABS(X) to select the function.

2 Graph each function below. Describe how the graph has been transformed from the parent function.

a. $y = -|x|$ **b.** $y = 2|x|$ **c.** $y = |x| - 2$

d. $y = -|x| + 4$ **e.** $y = 2|x| - 4$ **f.** $y = -3|x + 2|$

3 Match your graphs **a–f** from Step 2 with the graphs **I–VI** at the top of the page.

4 What must you do to the parent function $y = |x|$ to make the graph of the new function result in a stretch? a reflection? a vertical shift? a horizontal shift?

5 How can you tell if the graph of the absolute-value function has been stretched, reflected, or shifted?

6 Explain how to use the parent function $y = |x|$ to graph the following:

a. $y = 3|x|$ **b.** $y = -5|x|$ **c.** $y = |x| + 2$

7 Make a conjecture about $y = |x + 1|$ as a transformation of $y = |x|$.

8 Test your conjecture by graphing the following:

a. $y = |x + 3|$ **b.** $y = |x - 3|$ ❖

Exploration 2 Transformations of Quadratic Functions

1. Draw the graph of the parent function $y = x^2$.
2. Draw the graph of each of the following functions. Describe how the graph of each function has been transformed from the parent function.

   a. $y = -x^2$ **b.** $y = 2x^2$ **c.** $y = x^2 - 2$ **d.** $y = (x + 4)^2$

3. Compare your graphs in Step 2 above with the graphs in Step 2 of Exploration 1. How are the transformations alike?
4. What must you do to the function rule for $y = x^2$ to make the new function rule result in a stretch? a reflection? a vertical shift? a horizontal shift?
5. Explain how to use the parent function $y = x^2$ to graph the function $y = -2x^2 + 3$. ❖

APPLICATION

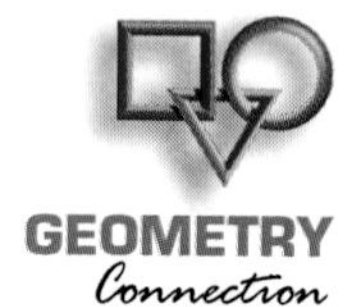

Suppose a photo has a border that adds a total of 2 inches to each dimension of the photo. The area of a photo including its border is a function of the length of a side of the photo. How can you model this application with a function?

Make a table. Write the lengths of the sides of the photo as they increase by 1 inch. Then find the corresponding areas.

Length of photo	x	0	1	2	3	4	5	6	x
Area with border	A(x)	4	9	16	25	36	49	64	$(x+2)^2$

To get the area, $A(x)$, 2 is added to x and that sum is squared. The function is $A(x) = (x + 2)^2$.

To model this function, graph the ordered pairs from the table, and draw a smooth curve through the points.

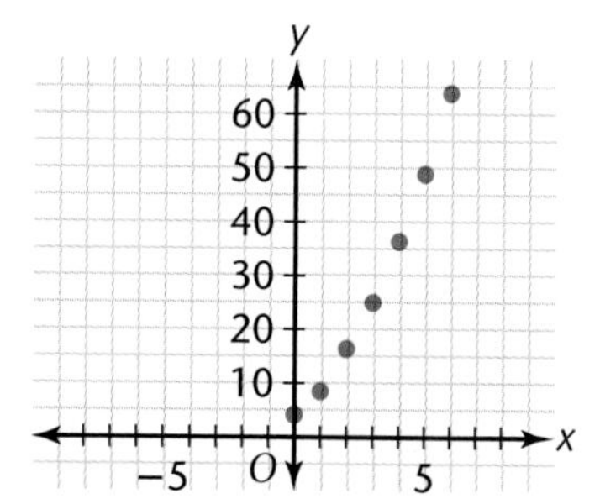

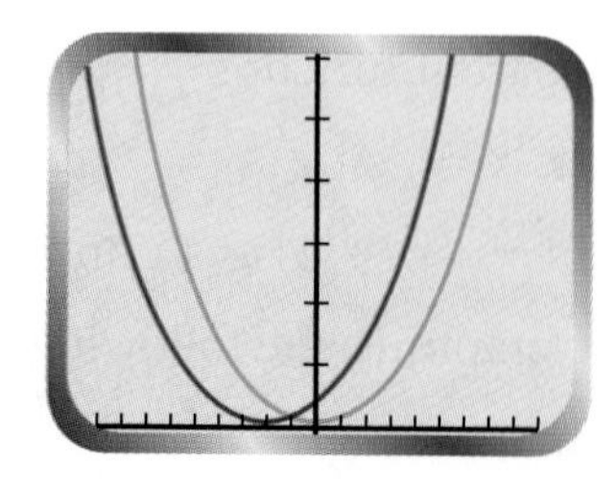

The entire parabola is shown on a graphics calculator. However, since length and area cannot be negative, the reasonable domain and range are restricted to non-negative values. Therefore, the graph only makes sense in the first quadrant.

Compare the graph with the parent function $y = x^2$. Try various transformations until one fits. Notice that the $y = x^2$ function is shifted 2 units to the left. Thus, the function rule for the application is $(x + 2)^2$ and $A(x) = (x + 2)^2$ is the function that models the problem. ❖

The following is a summary of some parent functions:

SUMMARY OF PARENT FUNCTIONS

FUNCTION	NAME
$y = x$	Linear
$y = 2^x$ and $y = 10^x$	Exponential
$y = x^2$	Quadratic (parabolic)
$y = \frac{1}{x}$	Reciprocal
$y = \|x\|$	Absolute value

In later lessons you will examine some transformations of these parent functions.

1. The vertex is the point (0, 0). The line of symmetry is the vertical line through the vertex; that is, the y-axis, or $x = 0$.
2. All of the y-values are increased by a factor of the stretch, thereby causing the graph to be narrower vertically when compared with the parent function.
3. All of the y-values are negated (opposite), thereby causing the graph to be upside-down when compared with the parent function.

Exercises & Problems

4. The graph moves either up or down for a vertical shift. All points of the graph shift by the same amount and in the same direction.

Communicate

1. Explain how to identify the vertex and axis of symmetry for the graph of the function $y = x^2$.
2. Describe how the graph changes when the parent function is stretched.
3. Describe how the graph changes when the parent function is reflected.
4. Describe how the graph changes when the parent function is shifted vertically.
5. Explain how to use the parent function $y = |x|$ to graph $y = 2|x| + 2$.
6. Explain how to use the parent function $y = x^2$ to graph $y = -(x - 2)^2$.

5. Stretch the graph by a factor of 2, and then shift vertically by 2 units.
6. Shift the graph horizontally to the right by 2 units, and then reflect this graph with respect to the x-axis.

Practice & Apply

Give the parent function for each graph. Then describe the transformation using the terms *stretch, reflect,* and *shift.*

7. $y = x^2$; reflected through the x-axis

8.

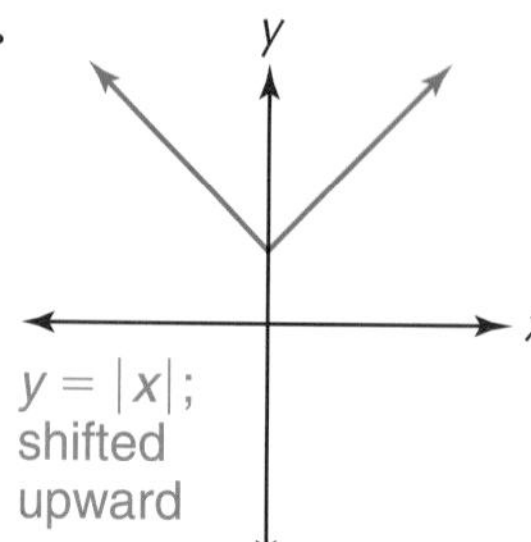

9.

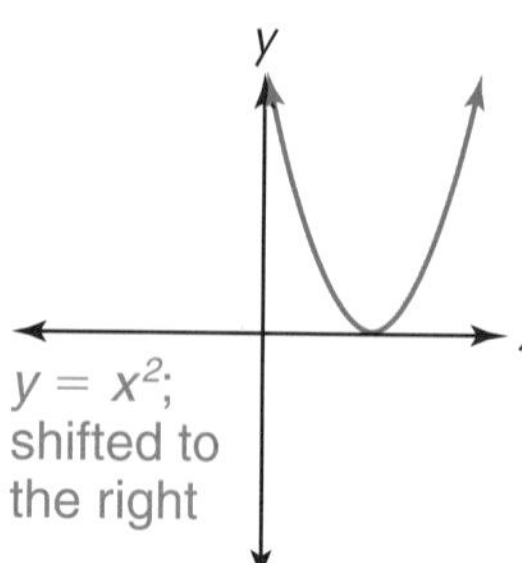

Graph each function below. Then tell whether the graph of $y = |x|$ has been stretched, reflected, or shifted.

10. $y = 4|x|$ **11.** $y = 4 + |x|$ **12.** $y = -|x|$

13. $y = 0.5|x|$ **14.** $y = |x - 5|$ **15.** $y = |x| - 3$

Graph each function below. Then tell whether the graph of $y = x^2$ has been stretched, reflected, or shifted.

16. $y = x^2 + 5$ **17.** $y = 4x^2$ **18.** $y = -x^2$

19. $y = x^2 + \frac{1}{3}$ **20.** $y = (x + 2)^2$ **21.** $y = x^2 - 3$

Identify the parent function. Tell what transformation has been applied to each, and then graph the function.

22. $y = 5x^2$ **23.** $y = 10x$ **24.** $y = \frac{1}{x + 2}$ **25.** $y = |x - 2|$

A square picture is to be framed with a 2-inch mat around it. The area of the picture without the mat is $A = x^2$, where each side of the picture is x inches. The area of the picture including the mat is $A = (x + 4)^2$.

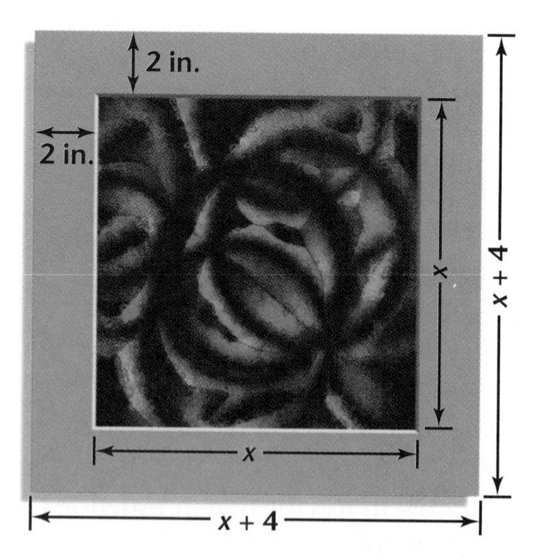

26. Make a table of values for $A = x^2$.

27. Make a table of values for $A = (x + 4)^2$.

28. Graph the two functions on the same coordinate plane.

29. Use transformations to tell how the two graphs are related.
$A = (x + 4)^2$ is the graph of $A = x^2$ shifted 4 units to the left.

30. A group of students are weighed at the beginning of the year. What transformation of the data takes place if each student gains 3 pounds?
It is shifted up 3 pounds.

31. Suppose that a teacher decides to double each student's score on a quiz. What kind of transformation is this? Vertical stretch by a factor of 2

Geometry Tell what kind of transformation moves $\triangle ABC$ to $\triangle A'B'C'$, which is read as "*A* prime, *B* prime, *C* prime."

32.

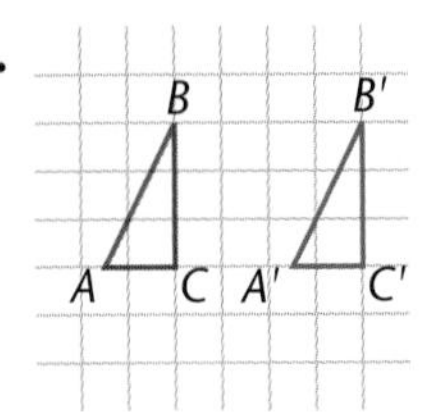

Horizontal shift of 4 units to the right

33.

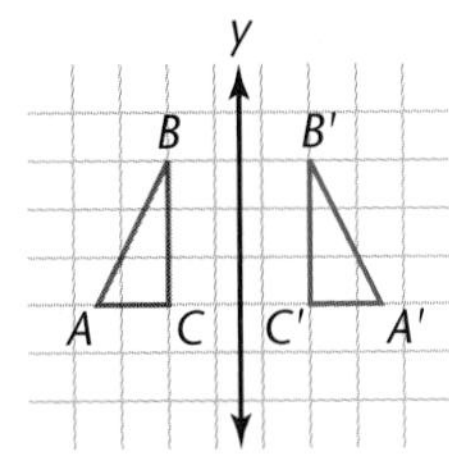

Reflection with respect to the *y*-axis

34. Find the equation of the axis of symmetry for this graph.
$x = 3$

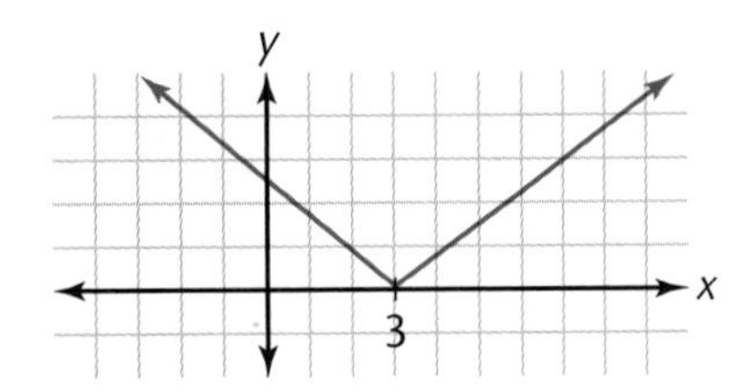

Look Back

35. What are the missing numbers in the sequence? **[Course 1]**
3, 6, 11, 18, 27, _?_, _?_, 66, _?_ 38, 51, 83

36. Evaluate $6 + 36 \div 3 - 1$. **[Course 1]** 17

37. The cost of lunch was \$6.09. The total amount for the food was \$5.80. What percent of the total was the sales tax? **[Lesson 1.2]** 5%

Graph the solution to each inequality on a number line. [Lesson 1.5]

38. $3x - 5 < 5x - 17$

39. $7x + 4 \geq -10$ and $2x + 3 < 13$

40. Write the equation of the line that contains the points $(3, -7)$ and $(-4, 3)$. **[Lesson 2.2]** $y = -\frac{10}{7}x - \frac{19}{7}$, or $10x + 7y = -19$

Solve each system of equations, if possible. If the system has no solution, explain why. [Lessons 2.4, 2.5]

41. $\begin{cases} 8x - 3y = -1 \\ 2x + y = 5 \end{cases}$ (1, 3)

42. $\begin{cases} 4x - 3y = 6 \\ x = 12 \end{cases}$ (12, 14)

43. $\begin{cases} 4x - 6y = 3 \\ 2x - 3y = 4 \end{cases}$ No solution; parallel lines

44. Find the matrix products for both *YA* and *XA*. **[Lesson 3.4]**

$$A = \begin{bmatrix} -6 & 3 \\ -8 & 7 \end{bmatrix} \qquad Y = \begin{bmatrix} -1 & 0 \\ 0 & 1 \end{bmatrix} \qquad X = \begin{bmatrix} 1 & 0 \\ 0 & -1 \end{bmatrix}$$

$$YA = \begin{bmatrix} 6 & -3 \\ -8 & 7 \end{bmatrix} \qquad XA = \begin{bmatrix} -6 & 3 \\ 8 & -7 \end{bmatrix}$$

Look Beyond

Statistics Tell what happens to the mean of a set of data if each data point is

45. increased by 5. Increases by 5

46. doubled. Doubles

47. replaced by its opposite. Becomes the opposite

LESSON 5.3 Stretches

why *When you evaluate a function and plot the points, the function will have a distinctive shape. Changes in the scale will change the shape of that graph. One such change is a vertical stretch. Once you know the effects of a change in scale on a parent function, it is easier to sketch or visualize the transformed graph.*

The height of a waterfall can be approximated if you know the time it takes the water to fall. This relationship can be modeled by the function $D = 16t^2$, where D is the distance in feet and t is the time in seconds. For example, if it takes the water 3 seconds to fall, the height is about $16(3)^2$, or 144 feet. What is the height if the time is 4 seconds? 256 feet

The table shows the values of the function $D = 16t^2$ for times from 0 to 3 seconds.

How do these values compare with the values of the parent function $D = t^2$? That is, compare the functions $f(x) = x^2$ and $g(x) = 16x^2$.

$D = 16t^2$		
t	$16t^2$	D
0	$16(0)^2$	0
1	$16(1)^2$	16
2	$16(2)^2$	64
3	$16(3)^2$	144

What is the value of the parent function $f(x) = x^2$ when x is 5?

If x is 5, then $f(5) = 5^2 = 25$.

What is the value of the transformed function $g(x) = 16x^2$ when x is 5?

If x is 5, then $g(5) = 16 \cdot 5^2 = 400$.

The values of the function $y = 16x^2$ are 16 times greater than the corresponding values of the parent function $y = x^2$. The coefficient, 16, is called the **scale factor** of the function.

What does $g(5)$ represent in the waterfall problem? $g(5)$ represents the height of the waterfall if it takes the water 5 seconds to fall from the top to the bottom.

Can you visualize the graph of $y = 16x^2$? Can you sketch it without plotting the points? To do this you must understand how the scale factor affects the parent function.

Compare the graphs.

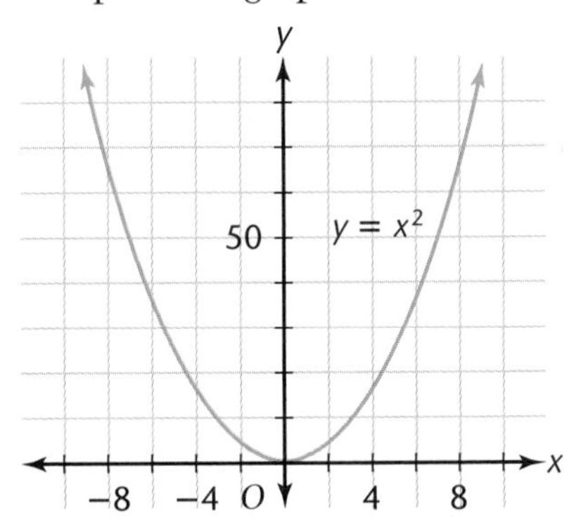

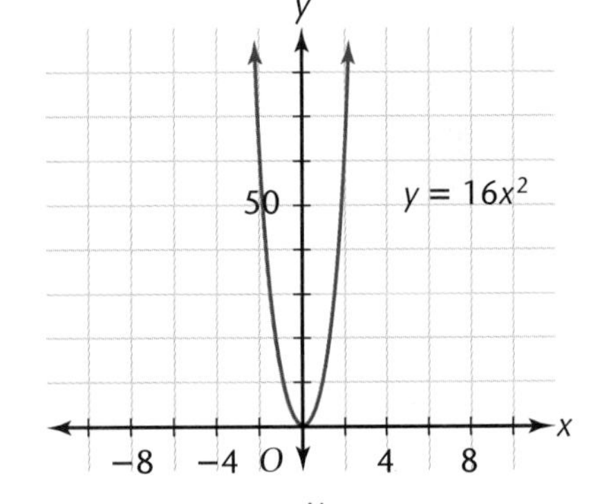

The graph of $y = x^2$ is **stretched** vertically to become the graph $y = 16x^2$.

When you apply the scale factor of 16, the point (2, 4) on the parent graph becomes the point (2, 64) on the transformed graph.

$$16 \cdot 2^2 = 64$$

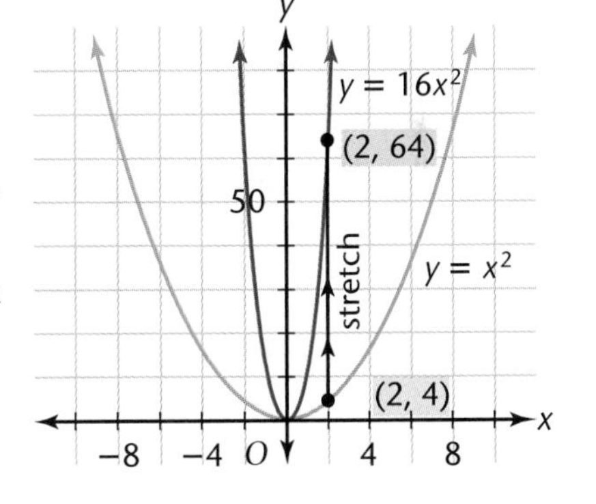

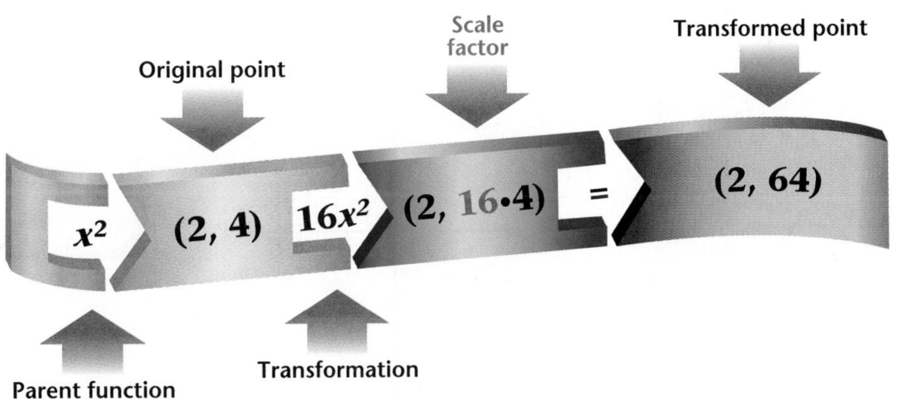

EXAMPLE 1

Use the parent function to sketch the graph of $y = \frac{|x|}{2}$.

Solution ➤

The function $y = \frac{|x|}{2}$ can be written as $y = \frac{1}{2}|x|$. The parent function is $y = |x|$. The transformation has a scale factor of $\frac{1}{2}$. Each of the y-values is $\frac{1}{2}$ the corresponding value of the parent function. The graph will be stretched vertically to only $\frac{1}{2}$ the height of $y = |x|$.

x	$\frac{1}{2}\|x\|$	y
-2	$\frac{1}{2}\|-2\|$	1
-1	$\frac{1}{2}\|-1\|$	$\frac{1}{2}$
0	$\frac{1}{2}\|0\|$	0
1	$\frac{1}{2}\|1\|$	$\frac{1}{2}$
2	$\frac{1}{2}\|2\|$	1

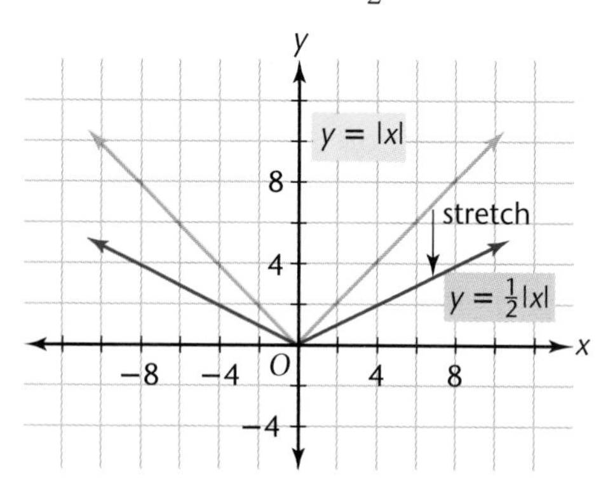

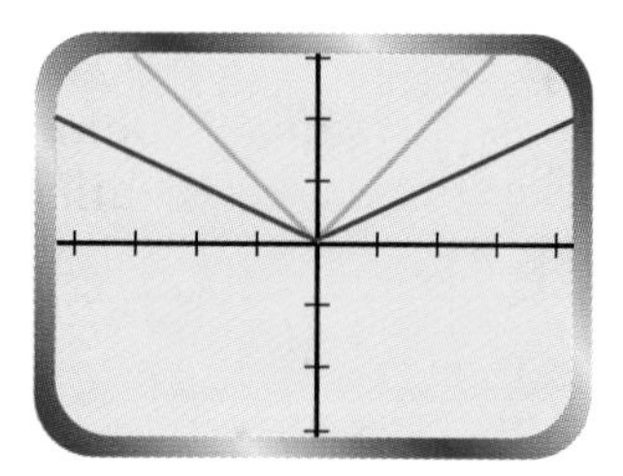

❖

The graph of $y = 16x^2$ and $y = \frac{1}{2}|x|$ have been stretched from the graph of their parent functions. When $a > 0$, the graph of $y = a \cdot f(x)$ is stretched vertically from the graph of its parent function, $y = f(x)$, by the scale factor a.

Try This Sketch the functions $f(x) = 4x^2$ and $g(x) = \frac{|x|}{4}$.

EXAMPLE 2

Sketch the graph of $y = \frac{6}{x}$. How does it compare with its parent function?

Solution ➤

The function $y = \frac{6}{x}$ can be written as $y = 6\left(\frac{1}{x}\right)$. The parent function is the reciprocal function, $y = \frac{1}{x}$. Each y-value of the parent graph is stretched vertically by a scale factor of 6.

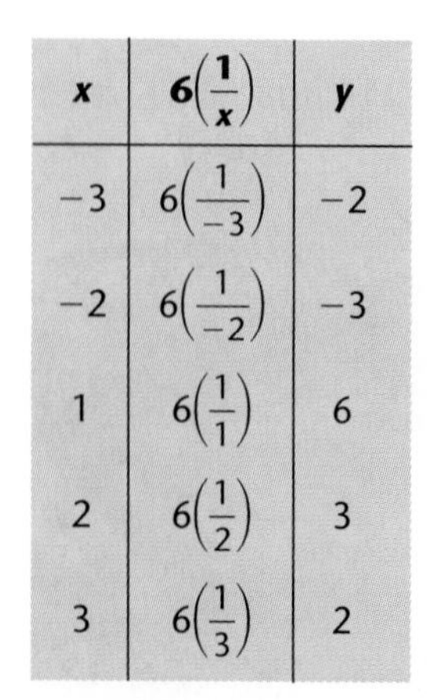

x	$6\left(\frac{1}{x}\right)$	y
-3	$6\left(\frac{1}{-3}\right)$	-2
-2	$6\left(\frac{1}{-2}\right)$	-3
1	$6\left(\frac{1}{1}\right)$	6
2	$6\left(\frac{1}{2}\right)$	3
3	$6\left(\frac{1}{3}\right)$	2

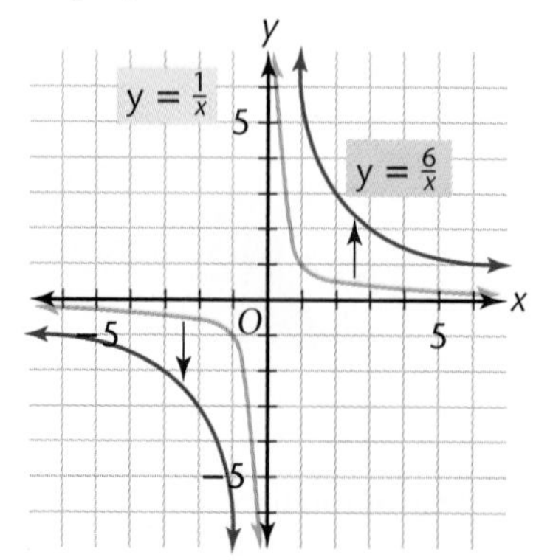

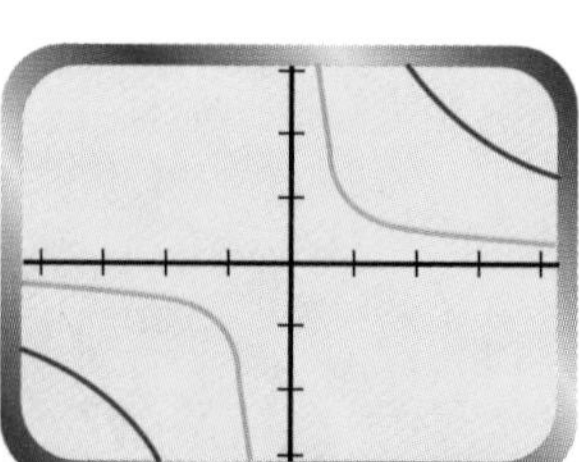

❖

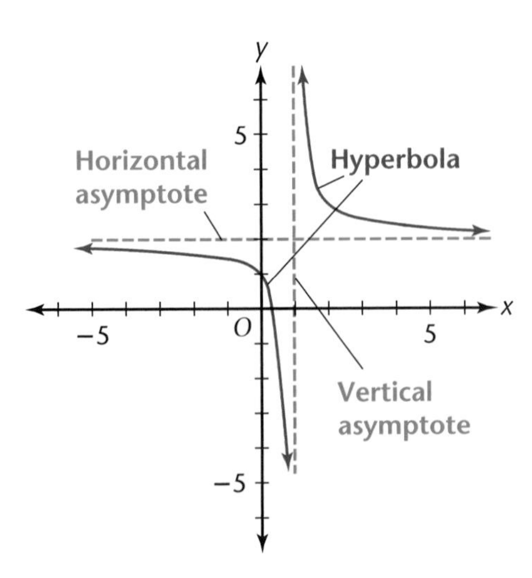

The graph of the reciprocal function forms a curve called a **hyperbola**. Notice that the graph of $y = \frac{1}{x}$ gets closer and closer to the x- and y-axes but never touches them. The lines that a hyperbola approaches, but does not touch, are called **asymptotes**. The intersection of the asymptotes provides a reference point when transforming reciprocal functions. At what point do the asymptotes intersect in the graph at the left?

At the point (1, 2)

Compare the changes in the graphs of $y = ax^2$ and $y = a\left(\frac{1}{x}\right)$ as the scale factor a increases when $a > 0$. Assuming $a > 0$, in $y = ax^2$, as a increases, the graph becomes narrower, or moves *closer* to the positive y-axis. In $y = a\left(\frac{1}{x}\right)$, as a increases, the graph moves away from the origin, or moves *away from* both the x- and y-axes.

EXAMPLE 3

Which functions result in vertical stretches of the parent graph?

A $y = \frac{x^2}{3}$ **B** $y = x^2 + \frac{3}{4}$ **C** $y = 5|x|$

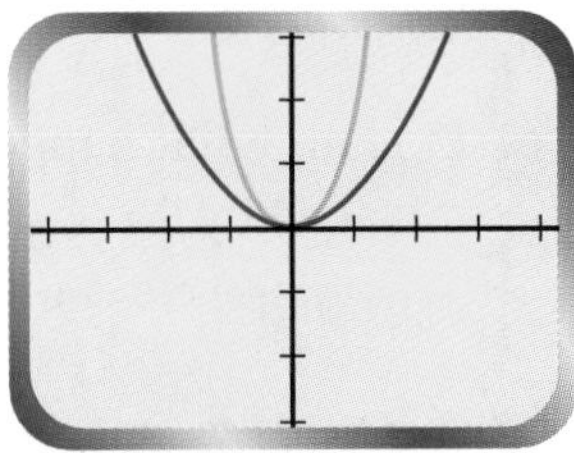

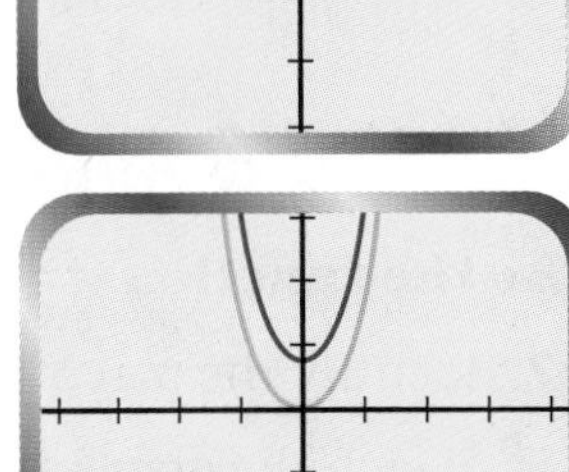

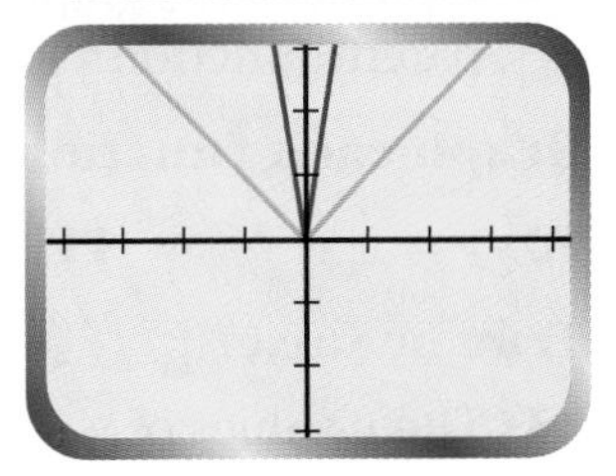

Solution

A Since x^2 is multiplied by $\frac{1}{3}$, this is a vertical stretch by a scale factor of $\frac{1}{3}$. The y-values are only $\frac{1}{3}$ of the y-values of the parent function.

B Since $\frac{3}{4}$ is *added* to x^2, this is *not* a vertical stretch; it is a vertical shift. The graph of the function is shifted upward.

C Since $|x|$ is multiplied by a scale factor of 5, each y-value of the parent function stretches vertically by 5. ❖

SUMMARY

To graph a vertical stretch, you should

1. identify the parent function, and
2. determine the scale factor.

The graph of the parent function will be stretched vertically by the amount of the scale factor.

EXERCISES & PROBLEMS

The formula for the average velocity of this car is $v = \frac{4}{t}$, where 4 is the length of the ramp, in feet, and t is the time it takes the car to go down the ramp.

Communicate

1. Explain the pattern of changes in the graph of $y = ax^2$ for different values of a from 1 to 4.
2. Explain the pattern of changes in the graph of $y = ax^2$ for different values of a from 0 to 1.
3. Describe what is meant by vertical stretch.
4. Explain the pattern of changes in the graph of $y = a\left(\frac{1}{x}\right)$ for different values of a from 1 to 4.
5. Explain the pattern of changes in the graph of $y = a\left(\frac{1}{x}\right)$ for different values of a, such as $a = \frac{1}{2}, \frac{1}{3}$, and $\frac{1}{4}$.
6. Explain what happens to the graph of $v = \frac{4}{t}$ when 4 is replaced by 2.
   The graph would move closer to the origin.

Practice & Apply

Use the function $y = 5x^2$ for Exercises 7–9.

7. Identify the parent function. $y = x^2$
8. Sketch the graphs of the function and the parent function on the same coordinate plane.
9. How does the graph of $y = 5x^2$ compare with the graph of the parent function? The graph of $y = 5x^2$ is a vertical stretch by a factor 5 of the graph of the parent function, $y = x^2$.

Graph each function.

10. $y = 2|x|$
11. $y = 0.5x^2$
12. $y = 3x^2$

True or false?

13. The y-values of $y = 5|x|$ are 5 times the y-values of $y = |x|$. True
14. The x-values of $y = 5|x|$ must be $\frac{1}{5}$ of the x-values of $y = |x|$ in order to get the same y-values. True

Use the function $y = \frac{1}{2x}$ for Exercises 15–17.

15. Identify the parent function. $y = \frac{1}{x}$
16. Sketch the graphs of the function and the parent function on the same coordinate plane.
17. Explain how the graph of the parent function was changed by the 2.

Which functions result in vertical stretches of the parent function?

18. $y = 3x^2$ Stretch by a factor of 3

19. $y = |x| + 1$ No stretch

20. $y = \frac{x^2}{5}$ Stretch by a factor of $\frac{1}{5}$

21. $y = \frac{5}{x}$ Stretch by a factor of 5

Write an equation for each function.

22.

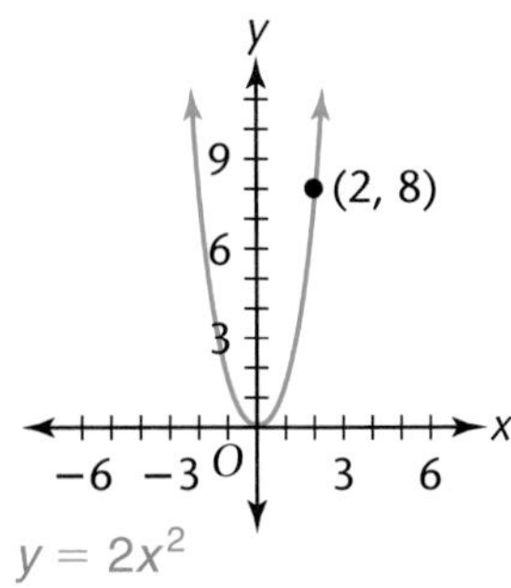

$y = 2x^2$

23.

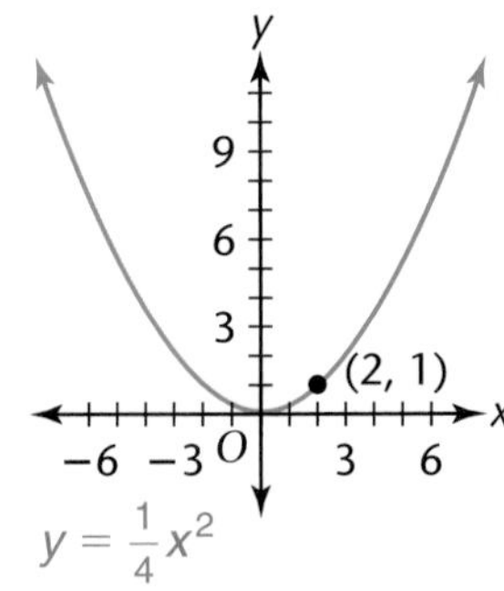

$y = \frac{1}{4}x^2$

24. Art Pictures can be enlarged by using a grid. Cut a cartoon or other figure from a newspaper or magazine. Draw a grid on your picture. Make a larger grid on another sheet of paper, and use it to make an enlargement of the picture. Answers may vary.

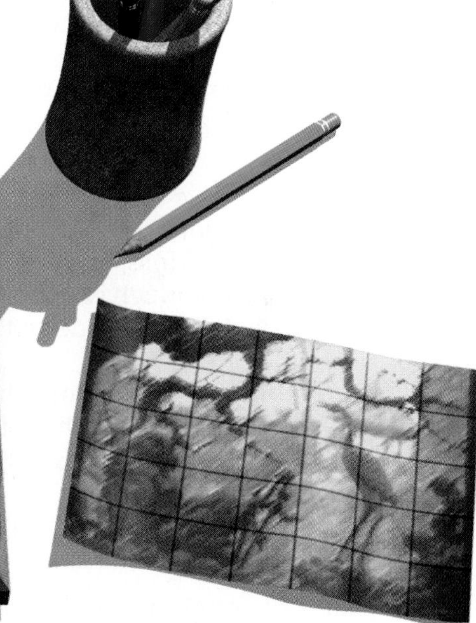

25. The workers in an electronics plant got an "across the board" raise of 5%. Find the scale factor, which is the amount that their salaries were "stretched." HINT: The scale factor is *not* 0.05. 1.05

For each function, identify the parent function, draw the graph, and tell how the graph of the parent function is transformed by the 3.

26. $y = \text{ABS}(3x)$

27. $y = x^2 + 3$

28. $y = \left(\frac{x}{3}\right)^2$

29. **Statistics** Draw a bar graph for this frequency distribution of test scores.

Score	50	55	60	65	70	75	80	85	90	95	100
Frequency	1	0	2	5	10	6	2	0	0	0	0

30. Suppose that the teacher adds 20 bonus points to every student's score. Draw the new bar graph, and explain how it compares with the first graph.

31. Suppose that the teacher multiplies every student's original score by $\frac{5}{4}$. Draw a new bar graph, and explain how it compares with the graph you drew in Exercise 30.

Look Back

32. Solve the equation $4x + 3 = \frac{2x}{7}$. **[Lesson 1.3]** $x = -\frac{21}{26}$

33. The cost for two assistants, each of whom work 12 hours part-time, was $129.60. One was paid $0.30 more per hour because of experience. What is the hourly pay for each of the workers? **[Lesson 1.4]** $5.25; $5.55

34. Solve the equation $5x - (7x + 4) = 3(x + 5) + 4(5 - 2x)$. **[Lesson 1.4]** 13

Solve each system of linear equations. [Lessons 2.4, 2.5]

35. $\begin{cases} 3x + 7y = -6 \\ x - 2y = 11 \end{cases}$ (5, −3)

36. $\begin{cases} 5y - 3x = -31 \\ 4y = 16 \end{cases}$ (17, 4)

37. $\begin{cases} 1.25x + 2y = 5 \\ 3.75x + 6y = 15 \end{cases}$ Same line, infinite solutions

38. **Probability** The menu of a restaurant has 6 different sandwiches, 4 different soft drinks, and 2 different desserts. How many different combinations of lunch can you select if each lunch consists of a sandwich, a drink, and a dessert? **[Lesson 4.5]** 48

Probability A spinner on a hexagonal base has 1 through 6 as equally likely numbers. After the spinner determines the first number, a coin is tossed. Complete Exercises 39 and 40.

39. What are the possible combinations of numbers and heads or tails that could appear? **[Lesson 4.5]** 1H, 1T, 2H, 2T, 3H, 3T, 4H, 4T, 5H, 5T, 6H, 6T

40. What is the probability of selecting any one of the pairs? **[Lesson 4.6]** 1/12, or about 8.3%

41. Two sets of numbers form the ordered pairs (3, 7), (5, −1), (1, 0), (2, 5), (16, −6), (0, 15), (1, −1), and (−2, 5). Is this relation a function? Explain. **[Lesson 5.1]** No, 1 is paired with both 0 and −1.

Look Beyond

Match each equation with graph A, B, or C.

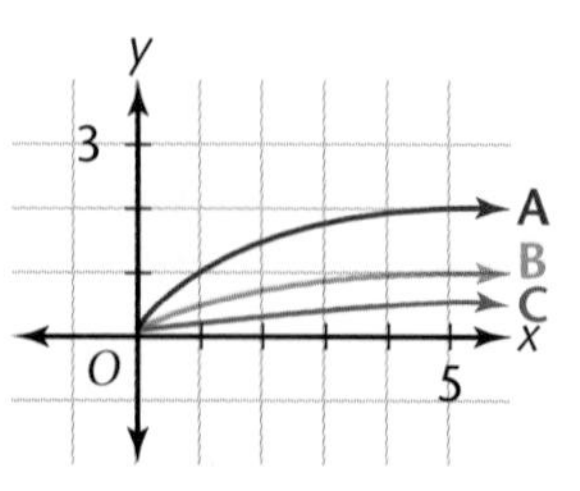

42. $y = \sqrt{x}$ A

43. $y = \frac{\sqrt{x}}{4}$ C

44. $y = \sqrt{\frac{x}{4}}$ B

LESSON 5.4 Reflections

why *A mirror reflects a figure as its image on the other side. The graph of a function can be reflected through an axis or a given line. This reflection can be represented by a function rule.*

Exploration A Reflection

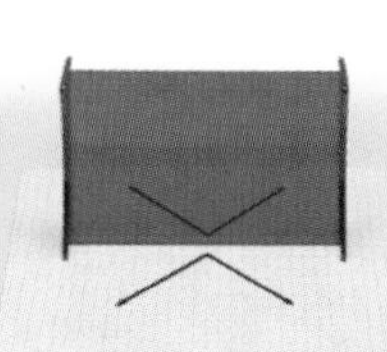

This reflection tool is being used to reflect an absolute-value function through the x-axis.

1. Use the values −2, −1, 0, 1, and 2 for x in this exploration. Make a table of values, and graph the parent function $f(x) = x^2$.
2. Reflect f through the x-axis. That is, create a mirror image of f as if the mirror were placed on the x-axis. Let g represent the function for this reflected image.
3. Make a table of values for the reflected function g. Use the same x-values that you used to graph f.
4. Compare the tables for the two graphs. The point (2, 4) is on the parent graph. What is the value of y on the reflected graph when x is 2? If (−3, 9) is a point on the parent graph, what is the corresponding y-value on the reflected graph? −4; −9
5. Evaluate $g(x)$ for the given values of x.

   a. $g(3) = \underline{?}$ −9 **b.** $g(0) = \underline{?}$ 0 **c.** $g(-3) = \underline{?}$ −9 **d.** $g(a) = \underline{?}$ $-a^2$
6. Follow Steps 1–3 for the following functions:

   a. $f(x) = |x|$

   b. $f(x) = \frac{2}{x}$

   c. $f(x) = -2x^2$

   7. $y = -f(x)$

   Examine the graphs, and describe in general the relationship between the graph of the function and the graph of its reflection.
7. Suppose that the graph of $y = f(x)$ is reflected through the x-axis. What is the function for the reflected graph? ❖

Compare the graphs of $y = x$ and $y = -x$. Describe the transformation. How do the graphs of $y = x$ and $y = -x$ compare with $y = |x|$? The transformation is a reflection through the x- or y-axis. The combined graphs look the same as the graph of $y = |x|$ and its reflection through the x-axis.

EXAMPLE 1

Describe each graph. **A** $y = -|x|$ **B** $y = -3x^2$

Solution ➤

A The parent function is $y = |x|$. The y-value of every point on the graph of the parent function changes to its opposite. Thus, the graph of $y = -|x|$ is a reflection of the graph of $y = |x|$ through the x-axis.

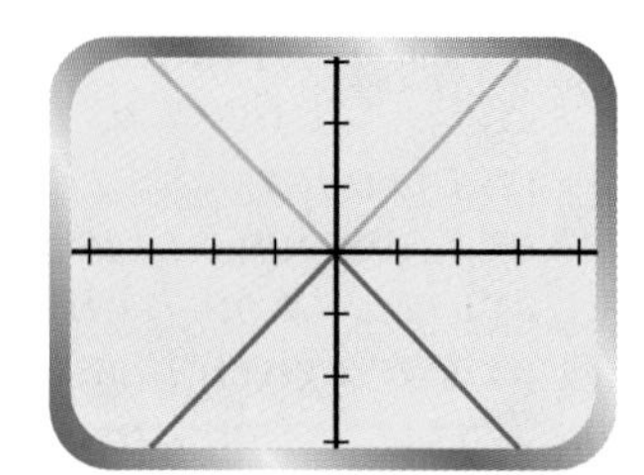

Notice that the vertex of the graph of $y = -|x|$ is now the maximum point on the graph.

B The parent function is $y = x^2$. First, every point on the graph of the parent function is stretched vertically by a scale factor of 3. Then the negative sign in $y = -3x^2$ indicates a reflection of the graph $y = 3x^2$ through the x-axis.

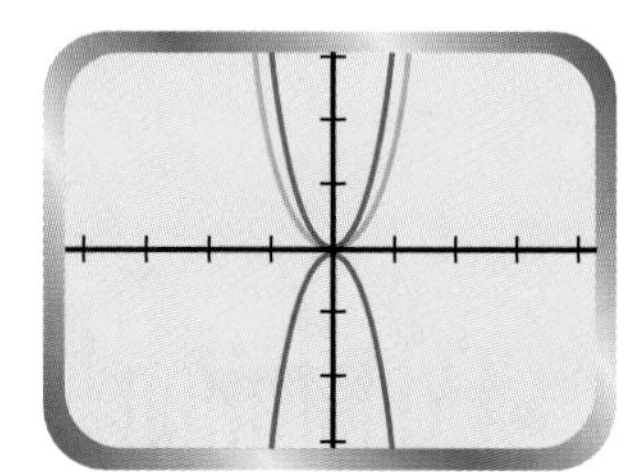

The parabola for $y = -3x^2$ opens downward, so the vertex of the parabola is the maximum point on the graph. ❖

A **vertical reflection** through the x-axis results when each point (a, b) of the original graph is replaced by the point $(a, -b)$. Thus, in a vertical reflection the function $y = f(x)$ is transformed to $y = -f(x)$.

Let $f(x) = x + 2$. If a mirror image of f is created as if the mirror were placed on the y-axis, g would represent the function for the reflected image.

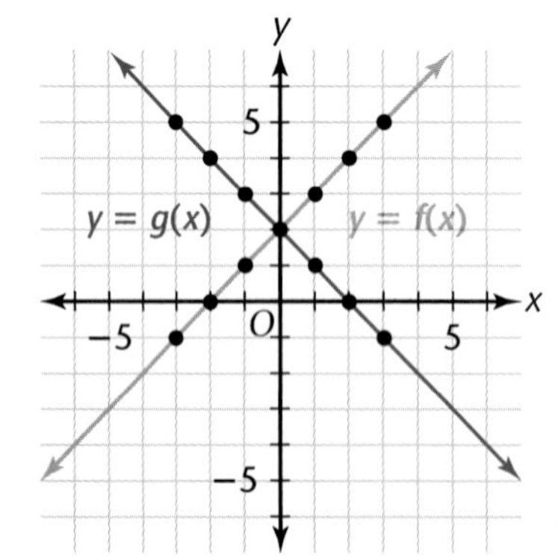

x	$y = f(x)$
3	5
2	4
1	3
0	2
−1	1
−2	0
−3	−1

x	$y = g(x)$
−3	5
−2	4
−1	3
0	2
1	1
2	0
3	−1

Since $(3, 5)$ is replaced by $(-3, 5)$, this is a reflection through the y-axis, a *horizontal reflection.*

A **horizontal reflection** through the y-axis results when each point (a, b) of the original graph is replaced by the point $(-a, b)$. Thus, in a horizontal reflection the function $y = f(x)$ is transformed to $y = f(-x)$.

EXAMPLE 2

Identify which graphs are vertical reflections through the x-axis of the graphs of parent functions.

A $y = -2^x$ **B** $y = x^2 - 2$ **C** $y = -\frac{1}{x}$

Solution ➤

A The y-values of $y = -2^x$ are the opposites of the y-values of the function $y = 2^x$. This is a vertical reflection of $y = 2^x$ through the x-axis.

B This is not a reflection because -2 is *added* to x^2.

C The y-values of $y = -\frac{1}{x}$ are the opposites of the y-values of the function $y = \frac{1}{x}$. This is a vertical reflection of $y = \frac{1}{x}$ through the x-axis. ❖

Remember, a vertical reflection occurs when the y-values of the function are replaced by their opposites. For this reason, look for vertical reflections when negative signs affect the y-values of functions. Plot a few points to check.

SUMMARY

To graph a *vertical* reflection,

1. identify the function that is reflected, and
2. replace the rule with its opposite.

The graph of the function will be a vertical reflection through the x-axis.

EXERCISES & PROBLEMS

Communicate

1. Explain the effect of the negative sign in the function $y = -ax^2$. The negative sign will reflect the graph of $y = ax$ through the x-axis.
2. Describe how to write the reflection of the function $h(x) = a\frac{1}{x}$ through the x-axis.
3. Describe how to graph the reflection of the function $f(x) = 2x$ through the x-axis.
4. How does a reflection of a function through the x-axis change the function rule?
5. What is the effect of reflecting a function through the y-axis when the function is symmetric about the y-axis? Give an example.
6. Identify the function $y = -(x)^2$ and the function $y = (-x)^2$ as a vertical or horizontal reflection of the parent function $y = x^2$. Describe the effect of the negative sign in each function.

Practice & Apply

7. A vertical reflection is a reflection through which axis?

8. A horizontal reflection is a reflection through which axis?

9. If you see the reflection of a mountain in a lake, is this a vertical reflection or a horizontal reflection?

10. If the right half of a fir tree is the mirror image of the left half, is this a vertical reflection or a horizontal reflection?

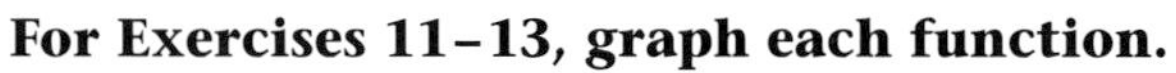

For Exercises 11–13, graph each function.

11. $y = -x^2$ **12.** $y = -|x|$ **13.** $y = -\frac{1}{x}$

14. Compare the graphs of $y = -\frac{x^2}{2}$ and $y = \frac{x^2}{-2}$. Explain how they are the same or different.

15. Graph the function $y = 2^x$ reflected through the x-axis and reflected through the y-axis.

16. Graph $f(x) = 2x + 1$. **17.** For $f(x) = 2x + 1$, find $f(3)$. 7

7. x-axis
8. y-axis
9. Vertical
10. Horizontal

For Exercises 18 and 19, find $f(3)$, and use this information to tell which kind of reflection (horizontal or vertical) is applied to the function $y = 2x + 1$.

18. $y = -(2x + 1)$
−7; vertical

19. $y = 2(-x) + 1$
−5; horizontal

20. Show the result when $y = 2x + 1$ is reflected vertically, then horizontally.

21. How are the graphs of the functions in Exercises 18 and 19 alike? How are they different?

22. Draw a sketch that reflects (2, 5) into (5, 2) and (10, 3) into (3, 10). What happens to (a, b) as a result of the reflection?
transformation $(a,b) \rightarrow (b,a)$

Stone figure in Kong Woods, the family cemetery of Confucius, a famous Chinese philosopher

23. Cultural Connection: Asia

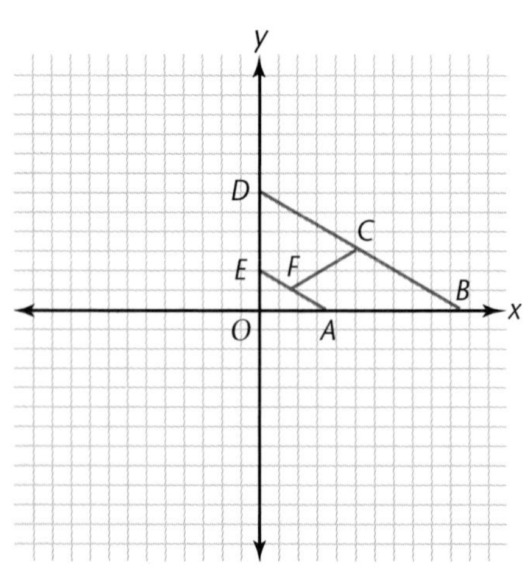

On a piece of graph paper, construct the line segments shown. Point E is (0, 2), and D is (0, 6). Segment EA is twice the length of segment OE, and segment DB is twice segment OD. Segment FC connects the midpoints of segments EA and DB. Reflect these segments through both the x- and y-axes into all 4 quadrants to complete the pattern. This basic pattern appears in a carving found in a tomb of the Han dynasty in China (202 B.C.E. to 220 C.E.). When the segments are reflected through both axes the shape is said to have *bilateral symmetry*.

24. Explain how the symmetry of a parabola can help you graph $y = x^2$.

Look Back

25. **Physics** Seven seconds after a flash of lightning, Mark heard thunder. If the speed of sound through air is approximately 330 meters per second, how far away did the lightning strike? **[Lesson 1.2]** 2310 meters

26. **Geometry** The length of a rectangular garden needs to be 3 times the width. If the perimeter of the rectangular garden is 72 meters, what is the area? **[Lesson 1.4]** 243 square meters

Solve. [Lesson 1.4]

27. $6(3 - 2r) = -(5r + 3)$ 3 **28.** $\frac{x}{2} + 5 = \frac{3x - 2}{4}$ 22 **29.** $3x = -9(x - 4)$ 3

30. Graph the solution set for $x \geq 3$ *or* $x < -1$ on a number line. **[Lesson 1.5]**

31. **Statistics** The quality control department measures the lengths of 5 random samples of glass rods in a 15-minute production run. What is the mean length of the glass rods tested in this run based on the following sample measurements? **[Lesson 4.3]** 23.74 cm

Sample number	1	2	3	4	5
Length (in cm)	23.5	23.9	23.6	24.0	23.7

32. **Probability** A number cube and a number tetrahedron (4 faces) are used for a probability experiment. How many possible outcomes are there when both are rolled once? **[Lesson 4.6]** 24

Look Beyond

Match each function with the appropriate graph.

33. $y = \sqrt{x}$ b **34.** $y = \sqrt{-x}$ a **35.** $y = -\sqrt{x}$ d **36.** $y = -\sqrt{-x}$ c

a.

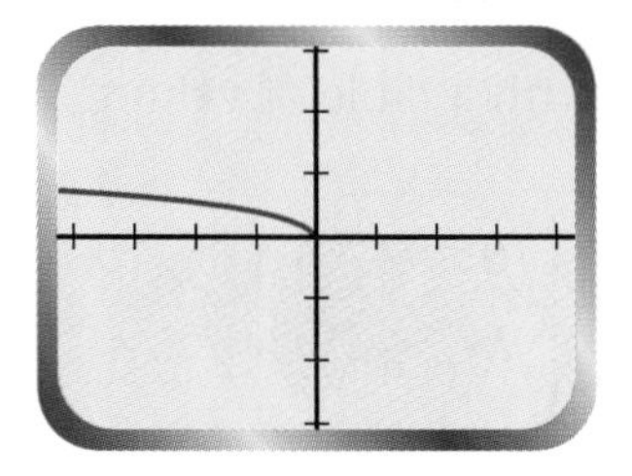

b.

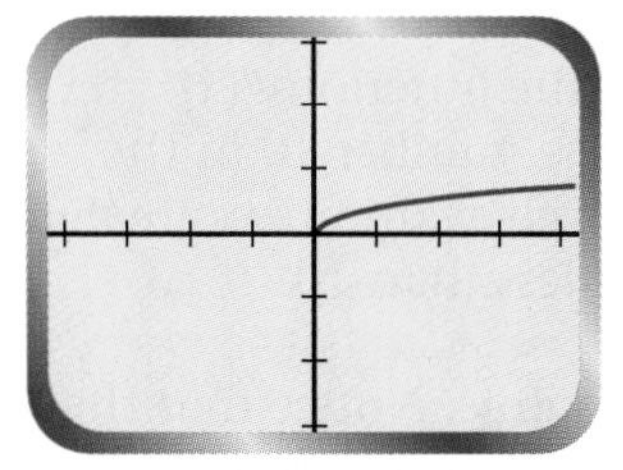

c.

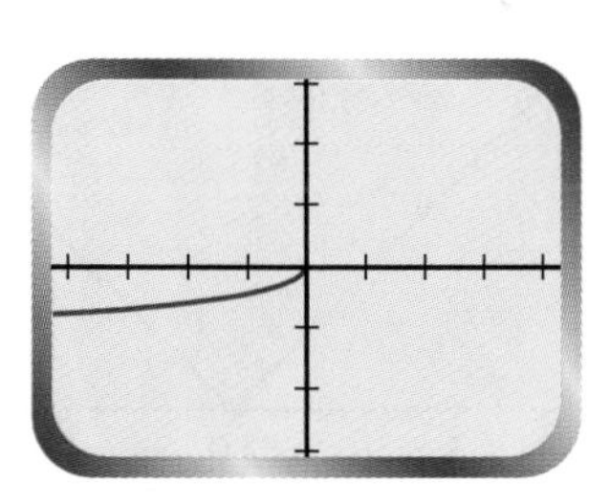

d.

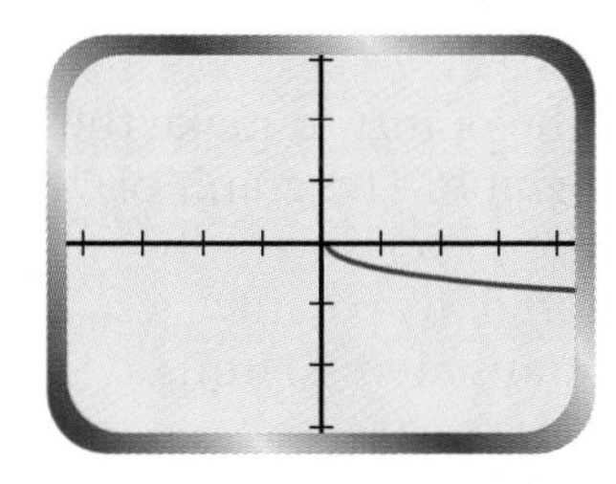

37. Graph $y = 5 - x^2$.

LESSON 5.5 Translations

In 1960 Escher designed this tiled facade for the hall of a school in The Hague, Holland. It was executed in concrete in two colors.

why *Sometimes a graph looks like a parent function that has been shifted vertically or horizontally. When this occurs, there is a simple relationship between the rules for each function.*

A transformation that shifts the graph of a function horizontally or vertically is called a **translation**. In the Escher facade above, point A is translated to point B by shifting point A to the right.

EXAMPLE 1

Compare the graph of $E(x) = |x - 10|$ with the graph of its parent function.

Solution ➤

To graph the function $E(x) = |x - 10|$, start with a table of values. Include the number 10 in the table, as well as numbers greater than and less than 10.

Plot the points. Notice that the function is a variation of the absolute-value function, $y = |x|$.

Connect the points to make the V-shaped graph. The graph of the function, $E(x) = |x - 10|$, is the graph of its parent function translated 10 units to the right. ❖

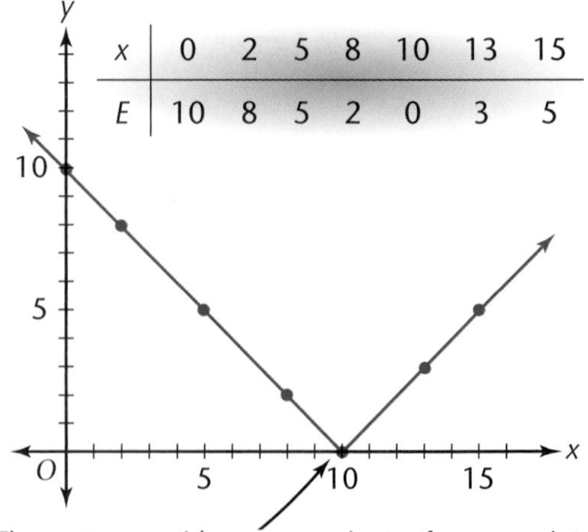

x	0	2	5	8	10	13	15
E	10	8	5	2	0	3	5

The vertex provides a convenient reference point.

Vertical Translation

A transformation that shifts the graph of a function up (positive direction) or down (negative direction) is a **vertical translation**.

EXAMPLE 2

Compare the graph $y = x^2 - 3$ with the graph of the parent function $y = x^2$. In what direction does the translation shift the vertex? How far?

Solution ➤

Evaluate $h(x) = x^2 - 3$ for several values of x, and plot the graph.

$h(x) = x^2 - 3$		
x	$x^2 - 3$	$h(x)$
−2	$(-2)^2 - 3$	1
−1	$(-1)^2 - 3$	−2
0	$(0)^2 - 3$	−3
1	$(1)^2 - 3$	−2
2	$(2)^2 - 3$	1

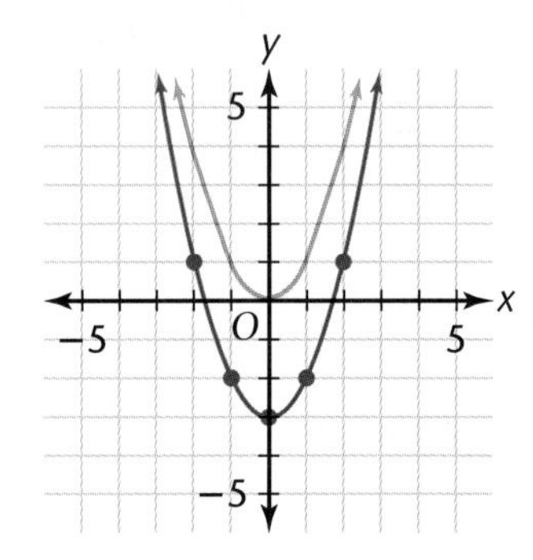

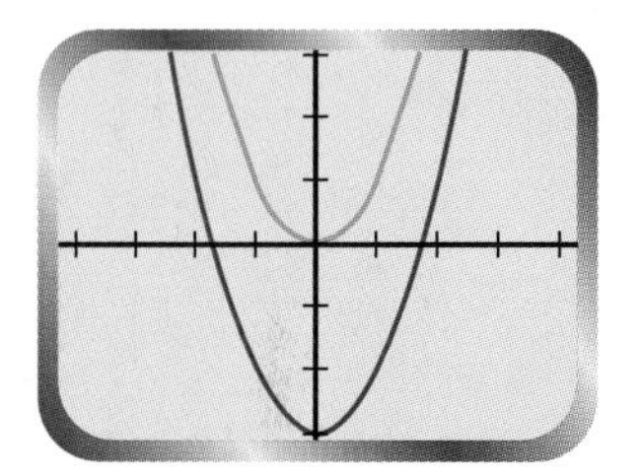

The vertex is vertically translated 3 units downward. ❖

Try This Sketch the graph of the function $y = |x| + 4$. Describe the translation.

VERTICAL TRANSLATION

The graph of $y = f(x) + \boldsymbol{k}$ is translated vertically $\boldsymbol{k}$ units from the graph of $y = f(x)$.

The following translations show how each formula applies to basic parent functions. The graph of a function shifts vertically when you add a positive or negative constant to the function.

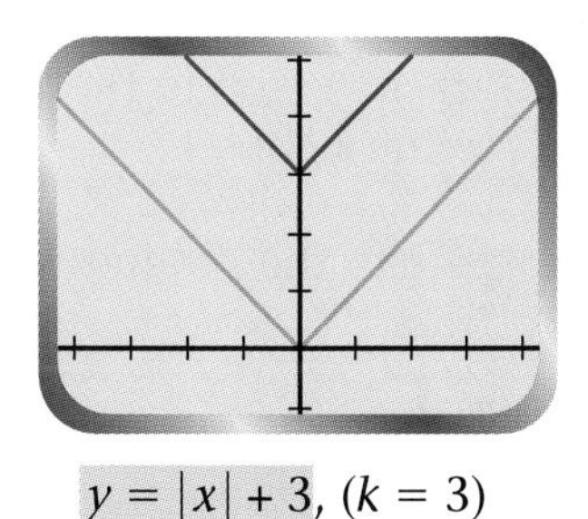

$y = |x| + 3$, $(k = 3)$

The graph is shifted up 3 units from the graph of $y = |x|$.

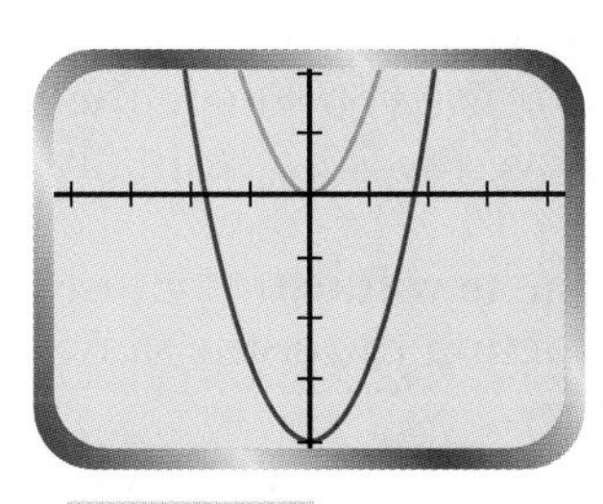

$y = x^2 - 4$, $(k = -4)$

The graph is shifted down 4 units from the graph of $y = x^2$.

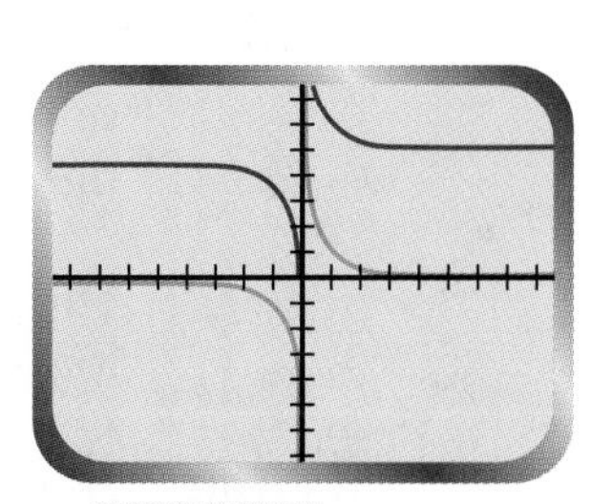

$y = \frac{1}{x} + 5$, $(k = 5)$

The graph is shifted up 5 units from the graph of $y = \frac{1}{x}$.

Horizontal Translation

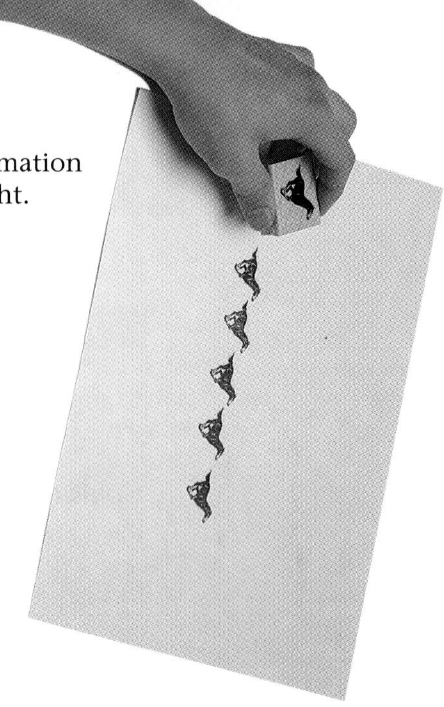

A **horizontal translation** is a transformation that shifts the function left or right. Remember, the positive direction is to the right and the negative direction is to the left.

EXAMPLE 3

Compare the graph of $y = (x - 3)^2$ with the graph of its parent function $y = x^2$. In which direction does the translation shift the vertex? How much?

Solution ➤

Graph the parent function $y = x^2$. The transformed function, $y = (x - 3)^2$, can be interpreted as "Subtract 3 from x, and then square the difference." The table shows some of the values of this function. Sketch this function and the parent function on the same coordinate plane.

$y = (x - 3)^2$

x	$(x - 3)^2$	y
-2	$(-2 - 3)^2$	25
-1	$(-1 - 3)^2$	16
0	$(0 - 3)^2$	9
1	$(1 - 3)^2$	4
2	$(2 - 3)^2$	1
3	$(3 - 3)^2$	0
4	$(4 - 3)^2$	1

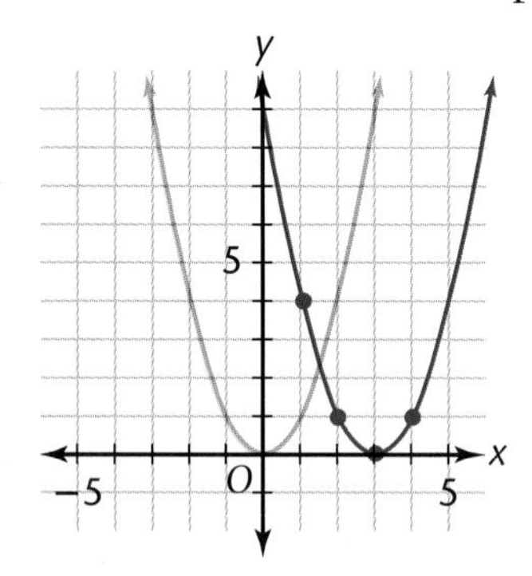

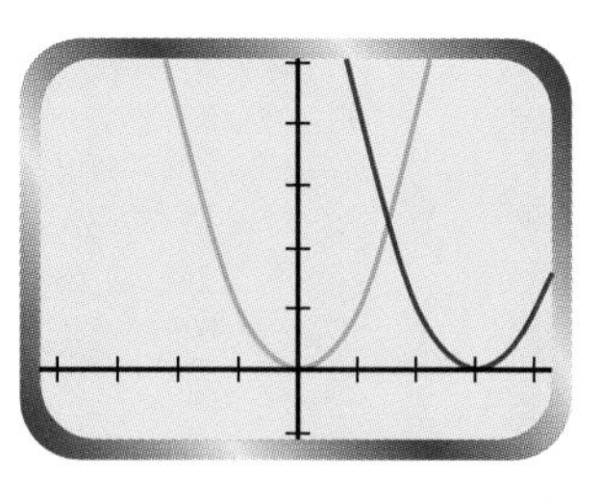

Notice that a translation does not change the shape of the parabola. If you subtract 3 from x *before* you square the quantity, you shift the graph to the right 3 units. The vertex is horizontally shifted 3 units to the right.

The x-values in the new function must be 3 units greater than the x-values for the parent function to get the same y-value. ❖

Try This Sketch the graph of $y = |x - 6|$. Describe the translation.

> **HORIZONTAL TRANSLATION**
>
> The graph of $y = f(x - \boldsymbol{h})$ is translated horizontally $\boldsymbol{h}$ units from the graph of $y = f(x)$.

The following examples show how the graphs and the formulas represent the horizontal translation of various parent functions.

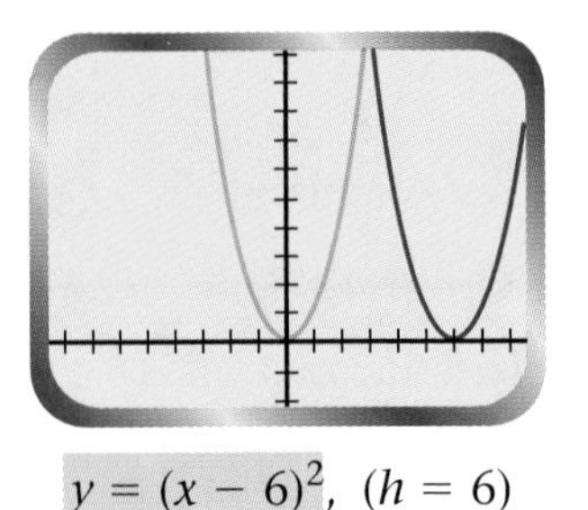

$y = (x - 6)^2$, $(h = 6)$

The graph shifts 6 units to the right from the graph of $y = x^2$.

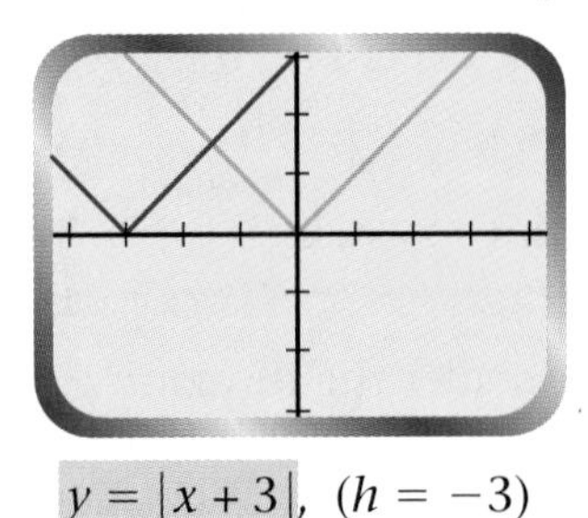

$y = |x + 3|$, $(h = -3)$

The graph shifts 3 units to the left from the graph of $y = |x|$.

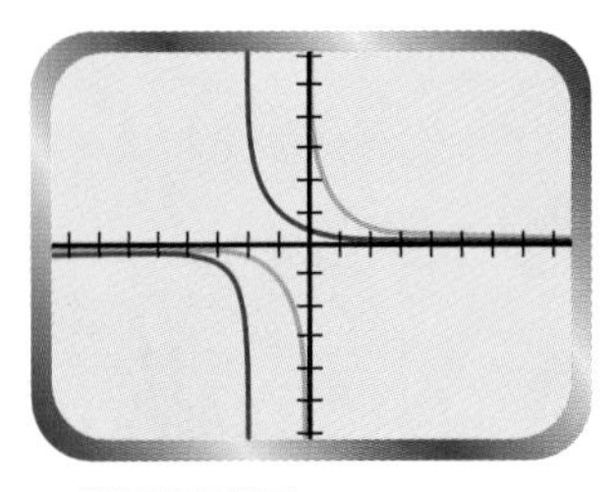

$y = \frac{1}{x + 2}$, $(h = -2)$

The graph shifts 2 units to the left from the graph of $y = \frac{1}{x}$.

EXAMPLE 4

Which of the functions represent translations of the parent graph?

A $y = x^2 + 3$ **B** $y = \frac{3}{4} + 2^x$ **C** $y = \frac{|x|}{9}$ **D** $y = \frac{1}{x - 10}$

Solution ➤

A Since 3 is added to x^2, this is a vertical translation of the parent function $y = x^2$ by 3 units upward.

B Since $\frac{3}{4}$ is added to 2^x, the parent function $y = 2x$ is translated upward by $\frac{3}{4}$ of a unit.

C Since $|x|$ is multiplied by $\frac{1}{9}$, this is a stretch, *not* a translation.

D Since 10 is subtracted from x, this is a horizontal translation of the parent function $y = \frac{1}{x}$ by 10 units to the right. ❖

EXAMPLE 5

Three salespeople have salaries of \$21,000, \$24,000, and \$30,000. What happens to the average salary if each gets a \$3000 raise?

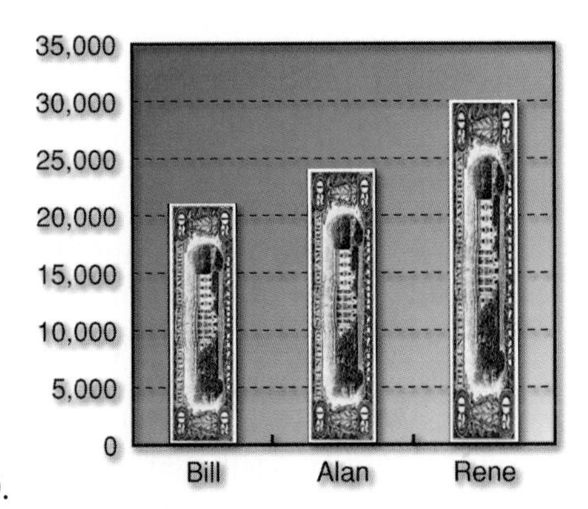

Solution ➤

The average salary before the raise is

$$\frac{21,000 + 24,000 + 30,000}{3} = 25,000.$$

If each gets a \$3000 raise, the average will be $\frac{24,000 + 27,000 + 33,000}{3} = 28,000$.

If each salary increases by \$3000, the average will increase by \$3000. In general, if each salary is translated by a number, k, the average is translated by the same number, k. ❖

SUMMARY

A translation of the parent function $y = f(x)$ occurs if a constant is added or subtracted from

1. the variable x. $y = f(x - h)$
2. the function value $f(x)$. $y = f(x) + k$

For example, consider the function $y = (x - 5)^2$. The parent function is $y = x^2$. When x is 5, $(x - 5)^2$ is 0. Thus, the vertex (0, 0) of $y = x^2$ has been shifted to the right 5 units. What happens to the vertex of $y = x^2$ in the translation $y = x^2 - 5$? The vertex moves down by 5 units.

Cultural Connection: Africa The designs of African art for pottery and fabric show the use of translation and reflection for decoration. Such examples show that geometric transformations have existed for centuries in this as well as in many other cultures. Algebra has created a way to represent these geometric features by using symbols and expressions.

The earthenware vessel in the background dates from the early twentieth century, and the other one is from the thirteenth century.

EXERCISES & PROBLEMS

Communicate

1. What is meant by a *translation* in mathematics?
2. A transformation of the parent quadratic function is $f(x) = (x - 3)^2 + 5$. What value tells you the amount and direction of the *horizontal* translation? -3
3. A transformation of the parent absolute-value function is $g(x) = |x - 3| + 5$. What value tells you the amount and direction of the *vertical* translation? $+5$
4. Explain why the graph of $f(x) = (x - h)^2$ is translated horizontally when a constant, h, is subtracted from x before the quantity is squared.
5. Explain why the graph of $f(x) = x^2 + k$ is translated vertically when a constant, k, is added after the variable is squared.
6. How is the graph of $y = x^2$ changed by the 4 and the 3 in the graph of $y = (x - 4)^2 + 3$? What is the location of the vertex of the graph for the equation?

Practice & Apply

Identify the parent function. Tell what type of transformation is applied to each, and then draw the graph. HINT: Remember to plot a few points to see what happens to the parent function.

7. $y = -x^2$ **8.** $y = (-x)^2$ **9.** $y = x^2 + 3$ **10.** $y = (x + 3)^2$

11. $y = (x - 6)^2 - 1$ **12.** $y = -x^2 - 6$ **13.** $y = \text{ABS}(x) - 1$ **14.** $y = \text{ABS}(x - 1)$

Consider the parent function $f(x) = x^2$. Note that $f(3) = 9$, so the point (3, 9) is on the graph. In each exercise below, use the given fact to tell how the parent function is translated.

15. $f(x) = x^2 + 5$ contains the point (3, 14). Translated 5 units up

16. $f(x) = x^2 - 5$ contains the point (3, 4). Translated 5 units down

17. $f(x) = (x + 5)^2$ contains the point (−2, 9). Translated 5 units left

18. $f(x) = (x - 5)^2$ contains the point (8, 9). Translated 5 units right

The point (5, 8) is on the graph of $f(x)$. Tell what happens to (5, 8) when each transformation is applied to the function.

19. vertical translation by 6 (5, 14)

20. vertical translation by −2 (5, 6)

21. vertical translation by −10 (5, −2)

22. vertical stretch by 3 (5, 24)

23. horizontal translation by 3 (8, 8)

24. horizontal translation by −1 (4, 8)

25. horizontal translation by −12 (−7, 8)

26. vertical stretch by 10 (5, 80)

Tell whether the graph of each function is stretched or translated from the graph of the parent function, $y = |x|$.

27. $y = \frac{2}{3}|x|$ Stretched

28. $y = \frac{2}{3} + |x|$ Translated

29. What happens to the average salary of the workers in Example 5 on page 249 if all of the salaries are increased by 10%? Average salary increases by 10% to $27,500.

30. Suppose that one ball is thrown from level ground in a parabolic path as shown. A second ball is thrown in an identical path, but from a point on top of a cliff to a point on another cliff. What transformation relates the higher path to the lower one?
Vertical translation

Look Back

31. Simplify the expression $-3(8 - 4x) - 5(x + 2)$. **[Lesson 1.4]** $7x - 34$

32. **Geometry** The area of a trapezoid can be determined by the formula $A = \frac{(b_1 + b_2)h}{2}$. The area, A, of a given trapezoid is 60. The altitude, h, is 5, and the larger base, b_2, is twice the smaller base, b_1. What is the length of each base? **[Lesson 1.4]** 8, 16

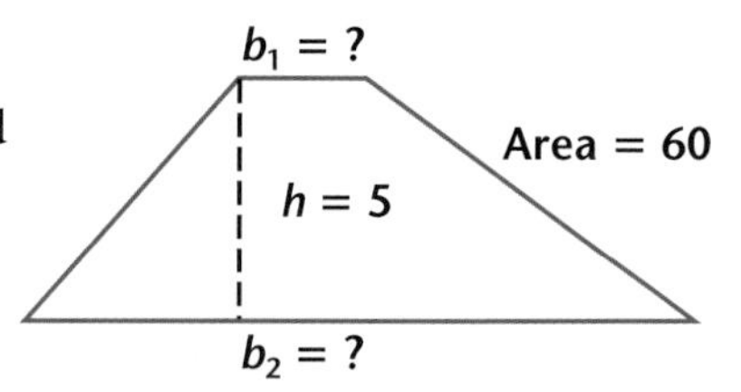

33. Solve $7x - (24 + 3x) = 0$ for x. **[Lesson 1.4]** 6

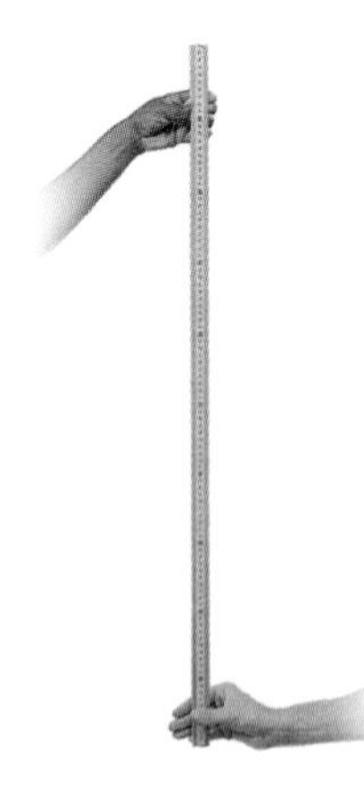

Statistics Investigate differences in reaction time. Place the zero point of a ruler between a person's thumb and index finger. Release the ruler. The person should catch the ruler as quickly as possible. Determine the point at which the ruler was caught. **[Lesson 4.3]**

34. Find the average catch distance along the ruler.
Answers may vary.

35. Compare the average catch distance along the ruler for two arbitrary groups. Which group showed the shortest average catch distance? What does this indicate about the reaction time of each group?
Answers may vary.

36. Sketch the graph of $y = -\frac{10}{x}$. **[Lesson 5.3]**

37. Determine a function rule for $f(x)$ that will model the data below. Give the steps that you used to find the formula. **[Lesson 5.3]**

There should be a 4-centimeter distance between thumb and index finger, with the ruler centered between them.

x	0	1	2	3	4	5	6	7
$f(x)$	0	3	12	27	48	75	108	147

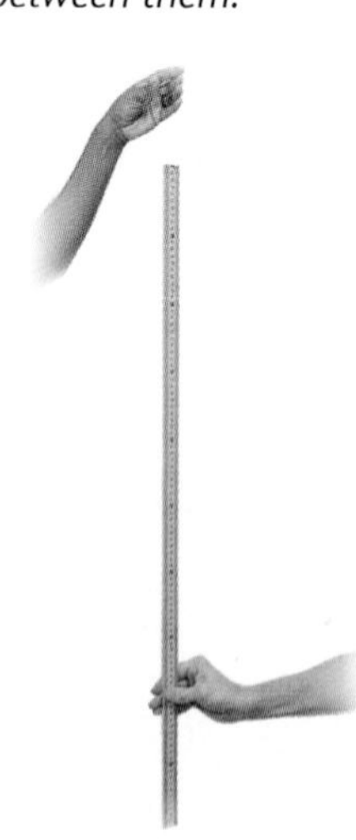

Geometry For Exercises 38 and 39, use the triangle on the coordinate plane. Its vertices are $A(2, 1)$, $B(4, 4)$, and $C(7, 2)$. **[Lesson 5.4]**

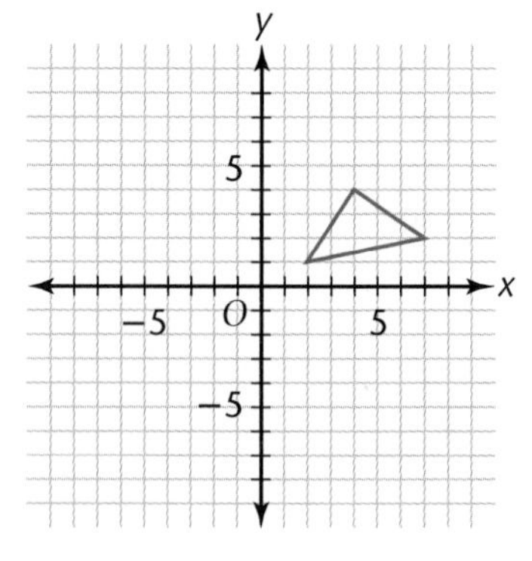

38. Reflect the triangle through the x-axis. What are the coordinates for the vertices of the transformed triangle?

39. Reflect the original triangle through the y-axis, and give the new coordinates.

Look Beyond

Graph the parent function $y = x^3$ and each transformation. How is each transformed function different from the parent function?

40. $y = -x^3$ **41.** $y = (-x)^3$

LESSON 5.6 Combining Transformations

Bob and Daniel allow a total of 2 hours for rest breaks.

WHY *Once a function is recognized as being related to a known parent function, its graph can be sketched by identifying and performing several transformations.*

Travel Bob and Daniel are planning a 100-mile bike ride. How fast must they ride to complete the whole trip in 6 hours? 7 hours? h hours?

To finish the trip in 6 hours, they expect to ride 4 hours and rest 2 hours $(6 - 2 = 4)$. That means they would have to ride 100 miles in 4 hours.

$$\text{rate} = \frac{100 \text{ miles}}{4 \text{ hours}} = \frac{25 \text{ miles}}{1 \text{ hour}}, \text{ or } 25 \text{ mph}$$

To finish the trip in 7 hours, they would need to ride 5 hours and rest 2 hours. They would have to ride 100 miles in 5 hours.

$$\text{rate} = \frac{100 \text{ miles}}{5 \text{ hours}} = \frac{20 \text{ miles}}{1 \text{ hour}}, \text{ or } 20 \text{ mph}$$

To finish the trip in h hours, they would need to ride $h - 2$ hours. They would have to ride 100 miles in $h - 2$ hours.

$$\text{rate} = \frac{100 \text{ miles}}{h - 2 \text{ hours}}, \text{ or } \frac{100}{h - 2} \text{ mph}$$

If R is the rate and h is the total number of hours for the trip, then the rate that they would have to travel is expressed as

$$R = \frac{100}{h - 2}.$$

To sketch the graph of a transformed function, begin with the parent function, and then perform the individual transformations one at a time.

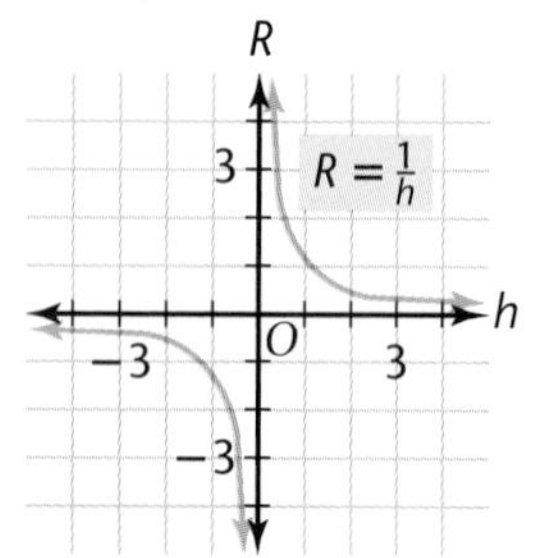

Sketch the graph of $R = \frac{100}{h-2}$.

First determine the parent function. Since h is in the denominator, you should first consider the parent reciprocal function $R = \frac{1}{h}$. Start with $R = \frac{1}{h}$ and find the transformations needed to end with $R = \frac{100}{h-2}$.

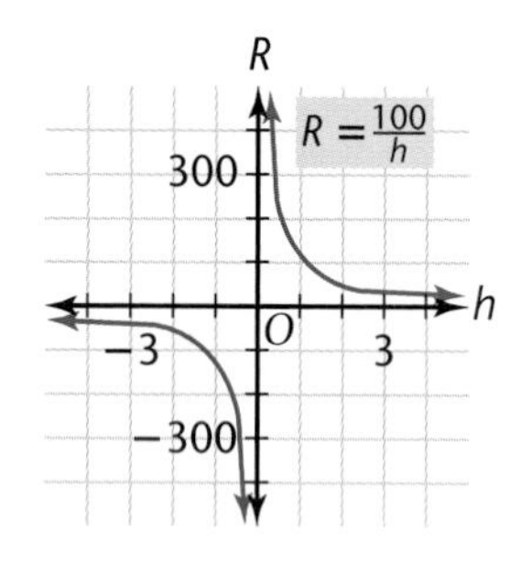

Next determine whether the function has been stretched or reflected. A graph is stretched when its parent function is multiplied by a nonzero constant. If the constant is negative, the graph is also reflected.

Examine $R = \frac{100}{h}$. Since $R = \frac{100}{h}$ can be written as $100\left(\frac{1}{h}\right)$, the parent function is stretched vertically by 100, and there is no reflection. Notice on the blue graph that the R-axis is labeled in hundreds. The graphs of $R = \frac{1}{h}$ and $R = \frac{100}{h}$ look alike because the scale on the R-axis is changed. Each R-value in the graph of $R = \frac{100}{h}$ is 100 times the amount of the R-value in the graph of $R = \frac{1}{h}$.

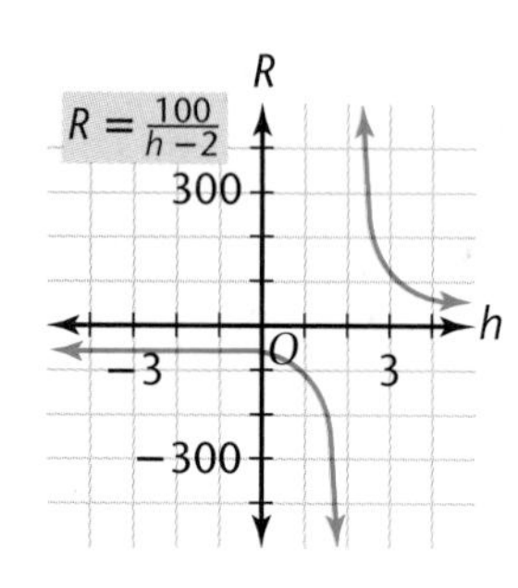

Then determine whether the function has been translated. A graph is translated when you add or subtract a constant. Since 2 is subtracted from h in the function $R = \frac{100}{h-2}$, the h coordinate needs to be 2 units greater to get the same value as $R = \frac{100}{h}$. This indicates a horizontal translation of 2 units to the right.

To summarize, the transformation from $R = \frac{1}{h}$ to $R = \frac{100}{h-2}$ includes a vertical stretch of 100, followed by a translation of 2 units to the right. The point $\left(3, \frac{1}{3}\right)$ on the original graph, is transformed to the point $\left(5, \frac{100}{3}\right)$.

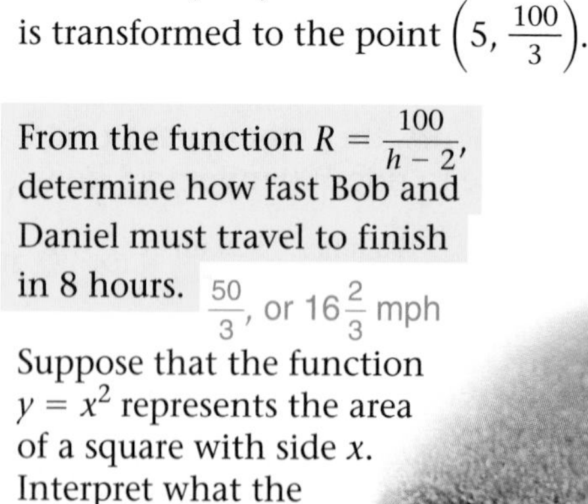

From the function $R = \frac{100}{h-2}$, determine how fast Bob and Daniel must travel to finish in 8 hours. $\frac{50}{3}$, or $16\frac{2}{3}$ mph

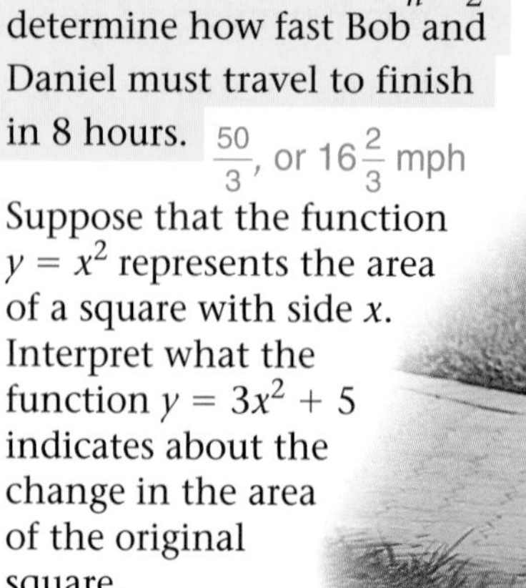

CRITICAL *Thinking*

Suppose that the function $y = x^2$ represents the area of a square with side x. Interpret what the function $y = 3x^2 + 5$ indicates about the change in the area of the original square.

The area of the original square is multiplied by 3, then that product is increased by 5.

EXAMPLE 1

Sketch the graph of $y = 2|x + 4|$.

Solution ➤

1. The parent function is $y = |x|$.
2. Since the scale factor is 2, the y-values will be twice as great as those of the parent function. This is a vertical stretch of 2 units.
3. Because 4 is added to the variable, the x-values in the transformed function need to be 4 units less to produce the same y-values as the stretched parent function produces. The result is a translation of 4 units to the left.

The graph of $y = |x|$ is stretched vertically by 2 units and translated 4 units to the left.

The vertex is at -4 on the x-axis. Since $y = 2|0| = 0$, the new position of the vertex is $(-4, 0)$. ❖

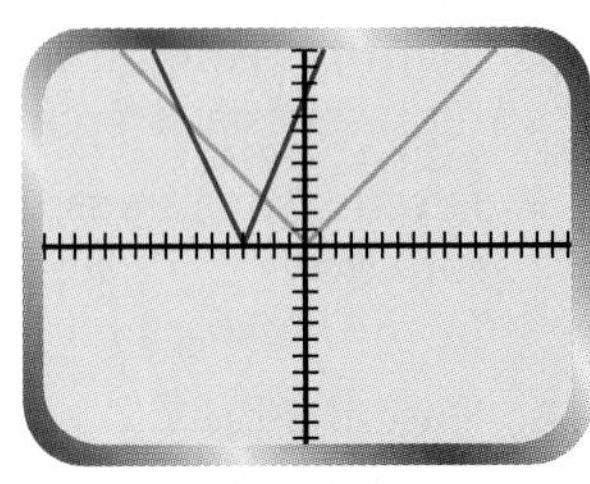

A graphics calculator can plot the parent function and transformation quickly.

Try This Check several points for $y = 2|x + 4|$ to see how steep the V of the graph is. Is the graph symmetric? If so, what is the axis of symmetry?

The graph is symmetric with respect to the line $x = -4$.

SUMMARY

For many transformation functions, the graph can be sketched directly from the information that appears in the function rule.

$$y = -a \cdot f(x - h) + k$$

Reflection ($-$), Vertical stretch (a), Horizontal translation (h), Vertical translation (k)

EXAMPLE 2

Use the information from the summary above to sketch the graph of the function $y = -(x + 3)^2 - 2$.

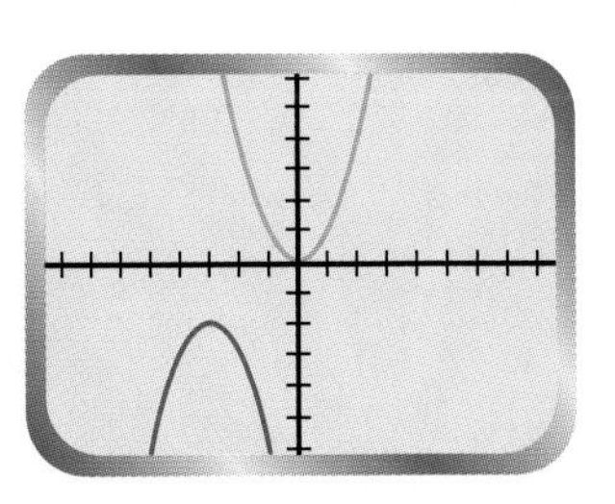

Solution ➤

The parent function is $y = x^2$, so the graph is a parabola. Because the sign is negative, the parabola is reflected through the x-axis. There is no constant following the negative sign. Thus, there is no vertical stretch. The value of h must be -3, since $x - (-3) = x + 3$. This indicates a horizontal translation of 3 spaces to the left. Finally, k is -2, so the parabola is translated vertically 2 units downward. ❖

EXERCISES & PROBLEMS

Communicate

1. In the 100-mile bike ride problem on page 253, explain how to identify which number in the function represents the translation.
2. In the bike ride problem, explain how to identify which number in the function represents a stretch of the graph.
3. Discuss the characteristics of various parent functions that can help you to identify them from transformed functions.
4. Describe how to write the function for a translation of 4 units to the left of the parent function $f(x) = \frac{1}{x}$.
5. Given the function $y = -a \cdot f(x - h) + k$, identify and explain the effect of each transformation, using the letters and symbols in the formula.
6. Explain the steps you would perform to sketch the graph of the function $y = |x + 7| - 3$.

Practice & Apply

Sketch the graph of each function.

7. $j(x) = 2(x + 5)^2$
8. $m(x) = 2x^2 + 5$
9. $t(x) = |x + 3| - 4$
10. $v(x) = 10|x - 4|$
11. $f(x) = \frac{3}{x - 2}$
12. $g(x) = \frac{3}{x} - 2$
13. $h(x) = 0.5(2^x) - 1$
14. $z(x) = \frac{3}{4} + \frac{2}{x}$
15. $p(x) = \frac{3}{4} + \frac{2}{4 + x}$

For Exercises 16–20, tell what happens to the point (4, 10) on the graph of $y = \frac{5}{2}x$ when each of the transformations are applied in the order given. HINT: If you are not sure, draw a sketch.

16. a vertical stretch of 2, followed by a vertical translation of 3 (4, 23)
17. a vertical reflection, followed by a vertical stretch of 2 (4, −20)
18. a horizontal reflection, followed by a vertical translation of −3 (−4, 7)
19. a vertical stretch of −3, followed by a horizontal translation of 5 (9, −30)
20. a vertical reflection, then a horizontal translation of −5, followed by a vertical stretch of $\frac{1}{2}$ (−1, −5)
21. Simplify the formula $y = 3(x^2 + 5)$ so that it can be sketched from the clues in the function rule. $3x^2 + 15$

22. In the figure at the right, a triangle is shown in the first quadrant. Its reflection and translation are also shown. If the coordinates of A are (5, 3), find the coordinates of A' and A''. Do the other vertices change according to the same pattern?

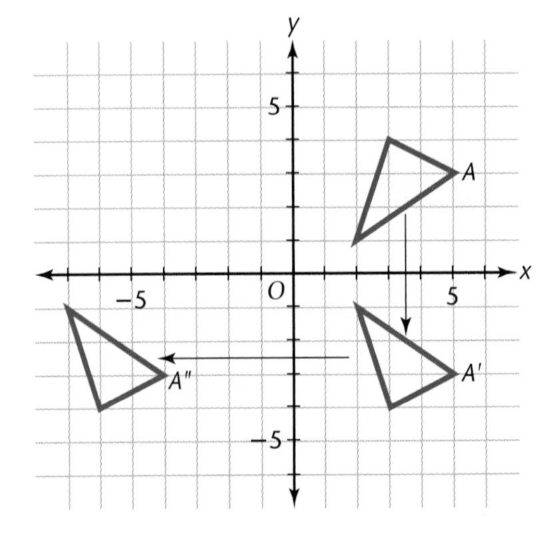

23. Use the information in the table to determine the transformation of a parent function that fits the data.

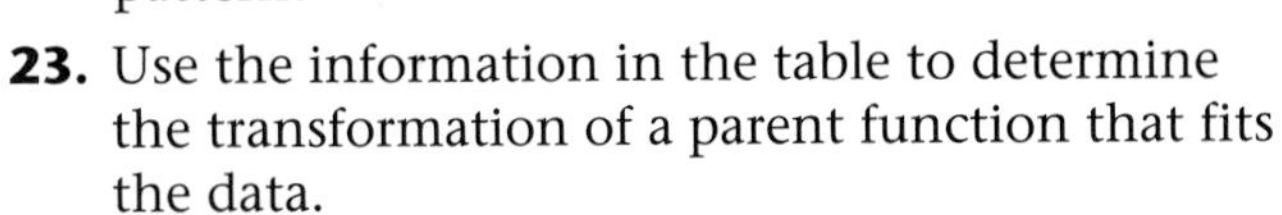

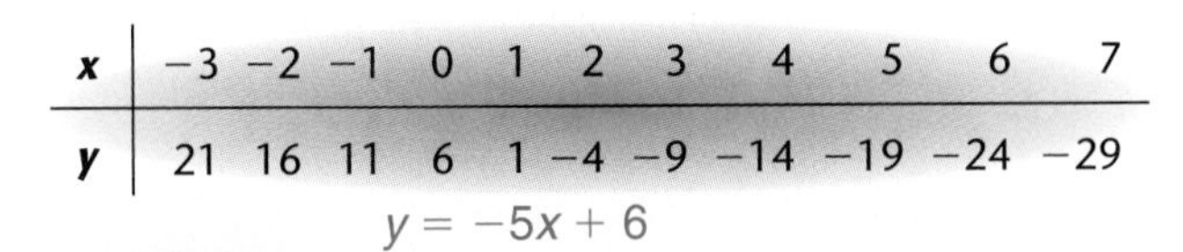

x	−3	−2	−1	0	1	2	3	4	5	6	7
y	21	16	11	6	1	−4	−9	−14	−19	−24	−29

$y = -5x + 6$

24. **Portfolio Activity** Complete the problem in the portfolio activity on page 219. Answers may vary since designs and transformations may vary.

Look Back

25. Manufacturing A roll of tape contains 32 meters of tape. Each box requires 75 centimeters of tape to seal it for delivery. If 35 centimeters of tape must be left for sealing other items, write and solve an equation that shows how many boxes can be sealed with one roll of tape. **[Lesson 1.4]** $32 = 0.75b + 0.35$; $b = 42$ boxes

26. Solve the inequality $|x - 5| \leq 17$, and graph the solution. **[Lesson 1.6]**

27. Graph the function $4x - 5y = 3$. Describe the procedure for graphing this function. **[Lesson 2.2]**

28. If the graph of a line contains the points $A(-3, 7)$ and $B(2, -9)$, what is the equation for the line perpendicular to the given line and passing through point A? **[Lessons 2.1, 2.2]** $5x - 16y = -127$

Solve this system of equations by each method below.

$$\begin{cases} 2x - 3y = 8 \\ 6x + 5y = -4 \end{cases}$$

29. graphing **[Lesson 2.3]** (1, −2)

30. elimination **[Lesson 2.5]** (1, −2)

31. matrices **[Lesson 3.5]** (1, −2)

32. **Probability** What is the theoretical probability of rolling a 7 on an ordinary 6-sided number cube? **[Lesson 4.6]** 0

Look Beyond

33. Graph the function $y = -16(x - 7)^2 + 784$.

34. Describe each transformation of the parent function in Exercise 33.

Graph each equation.

35. $y = 3 + \sqrt{x + 4}$

36. $y = -7 + 10\sqrt{x}$

Pick a Number

Try this with a classmate. Ask your classmate to

pick a number from 1 to 9, triple the number, add 6, and then divide the result by 3.

Ask for the number. If you mentally subtract 2, you can tell your classmate the original number.

Try other numbers. What do you notice?

Algebra can be used to see why this happens. Let x be the number chosen.

	Arithmetic	Algebra
Pick a number.	5	x
Triple the number and add 6.	$15 + 6 = 21$	$3x + 6$
Divide by 3.	$21 \div 3 = 7$	$x + 2$

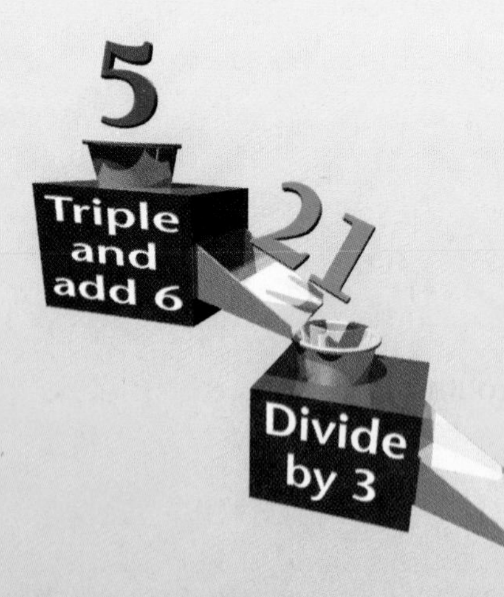

The algebra shows you how to go directly to the result. This is an example of finding a function of a function. In terms of a function machine, the output from the first machine is used as the input for the second machine.

When you find a function of a function, you are forming a **composite function**. If f is one function and g is another function, write $g(f(x))$ to indicate the value of the composite function for any x value.

$$\text{If } f(x) = 3x + 6 \text{ and } g(x) = \frac{x}{3}, \text{ then}$$

$$g(f(x)) = \frac{f(x)}{3} = \frac{3x + 6}{3} = x + 2.$$

The composition of two functions produces a new function with a new graph.

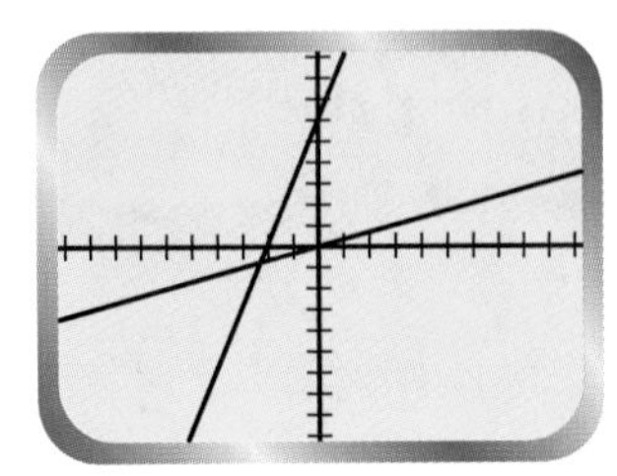

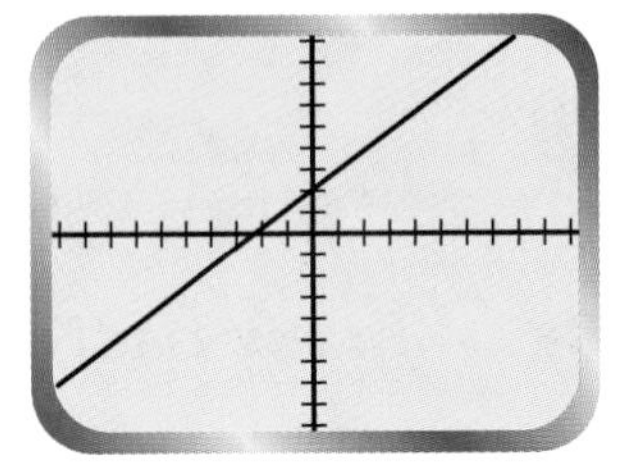

Composition of functions is not a commutative operation. Thus $g(f(x)) = x + 2$, but

$$f(g(x)) = 3(g(x)) + 6 = 3\left(\frac{x}{3}\right) + 6 = x + 6.$$

Activity 1

Make up a two-step computation game like the one in the opener of the project on page 258. Show how the domain and range are related through the steps. Then show the simplified composition of the functions that allows you to perform the calculation easily. Try the game with your friends.

Activity 2

Use what you have learned about composing functions to graph the composition of two functions. Begin by selecting a parent function from those you studied in this chapter. Replace the x in the function rule with a function, and graph the composition.

a. Identify the domains and ranges of the first function and the composite function.

b. Make a table of the input values, the intermediate values for input in the second step, and the final composite values.

c. Use a graphics calculator or draw the graph of the new composite function, and comment on the graph.

Chapter 5 Review

Vocabulary

asymptotes	**237**	**parent function**	**228**	**stretch**	**235**
axis of symmetry	**228**	**range**	**223**	**symmetric**	**228**
domain	**223**	**reflection**	**242**	**transformation**	**228**
function	**221**	**relation**	**220**	**translation**	**246**
hyperbola	**237**	**scale factor**	**235**	**vertex**	**228**

Key Skills & Exercises

Lesson 5.1

➤ Key Skills

Test a graph to identify a function by using the vertical-line test.

Any vertical line will intersect the graph at the right no more than once, so the graph is a function.

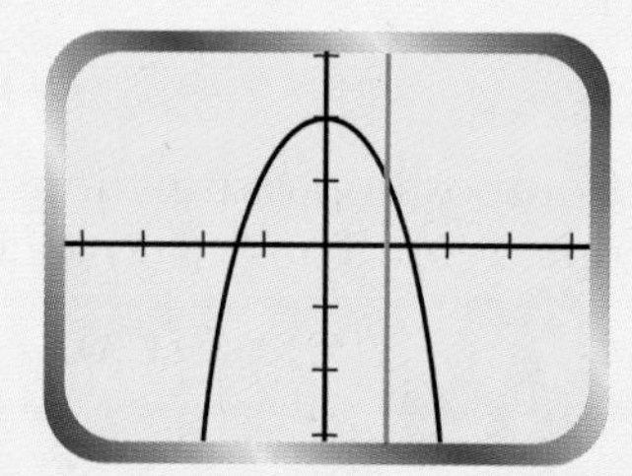

Evaluate a function.

To evaluate the function $f(x) = x^2 + 1$ for $f(-3)$, substitute -3 for x, and simplify.

$$f(x) = x^2 + 1$$
$$f(-3) = (-3)^2 + 1$$
$$f(-3) = 10$$

➤ Exercises

Which of the following are graphs of functions?

1.

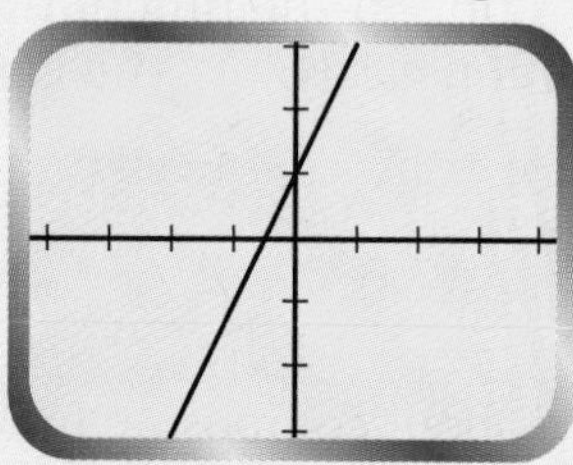

Yes

2.

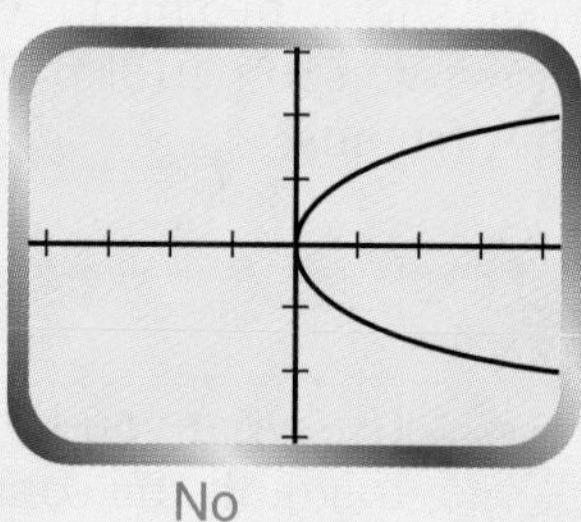

No

3.

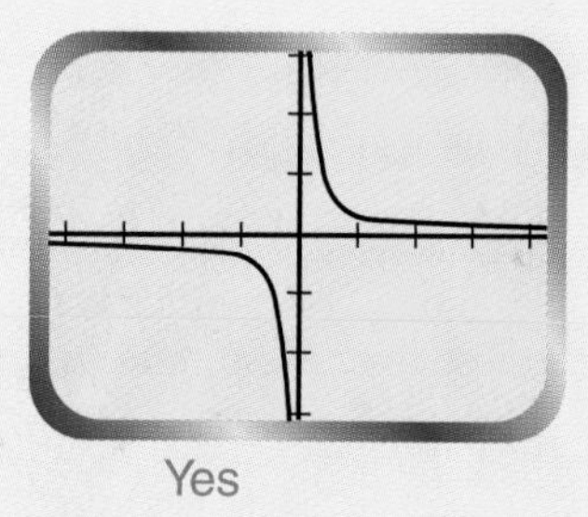

Yes

Evaluate each function when *x* is 4.

4. $f(x) = x - 8$ -4 **5.** $g(x) = |x + 2|$ 6 **6.** $h(x) = 3^x$ 81

Lesson 5.2

➤ Key Skills

Identify which transformation—stretch, reflection, or translation—was applied to a parent function.

The parent function for the graph at the right is $y = x^2$. It has been reflected through the x-axis and shifted vertically.

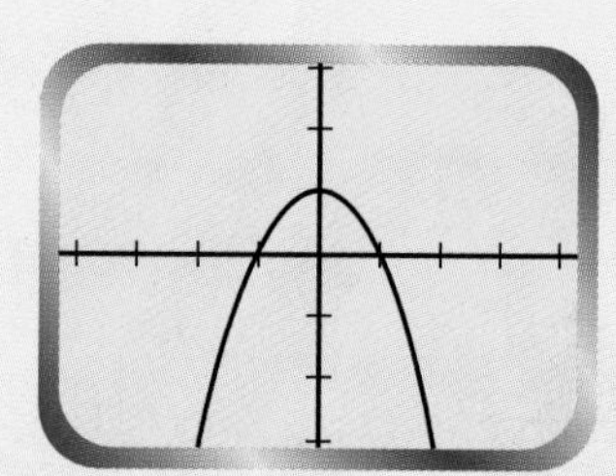

➤ Exercises

Give the parent function for each graph. Then tell what kind of transformation was applied.

7.

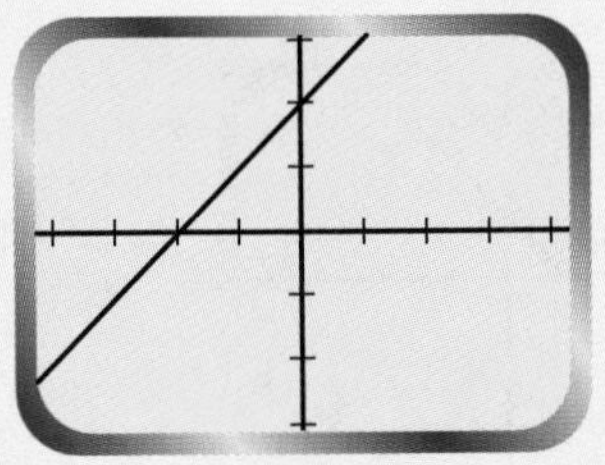

$y = x$; vertical shift by 2 units

8.

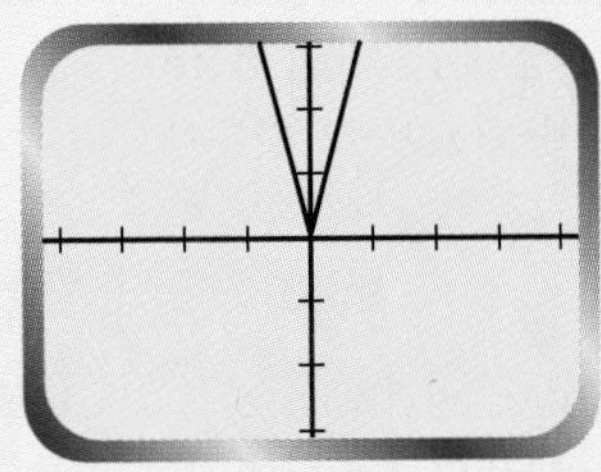

$y = |x|$; stretch by a factor of 3

9.

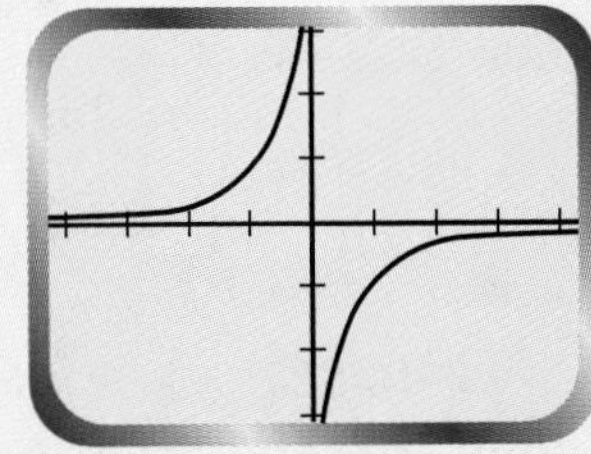

$y = \frac{1}{x}$; reflection through the x-axis

Lesson 5.3

➤ Key Skills

Describe the effect of a stretch on the graph of a function.

The parent function of $y = 3|x|$ is $y = |x|$. Each of the y-values is 3 times the corresponding y-value of the parent function.

x	$3\|x\|$	y
-2	$3\|-2\|$	6
-1	$3\|-1\|$	3
0	$3\|0\|$	0
1	$3\|1\|$	3
2	$3\|2\|$	6

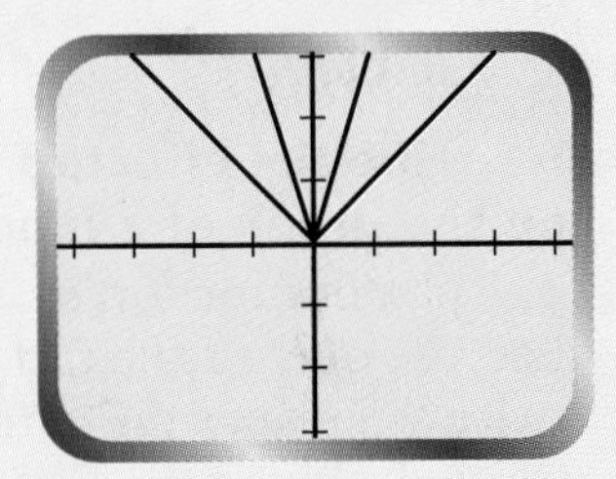

➤ Exercises

Graph each function.

10. $y = 4|x|$ **11.** $y = 3x^2$ **12.** $y = \frac{1}{4x}$

13. $y = \frac{x^2}{2}$ **14.** $y = \frac{|x|}{5}$ **15.** $y = \frac{3}{x}$

Lesson 5.4

➤ Key Skills

Describe the effect of a reflection on the graph of the parent function.

The parent function of $y = -|x|$ is $y = |x|$. The values of $-|x|$ are the opposites of the values of $|x|$. This is a vertical reflection through the x-axis of the parent graph, $y = |x|$.

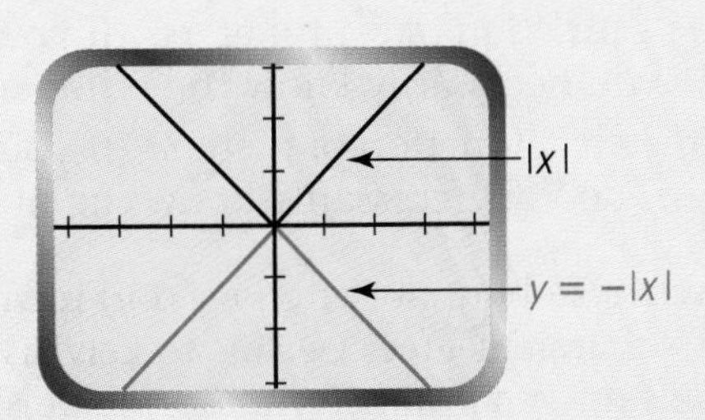

➤ Exercises

Graph each function.

16. $y = -x$ **17.** $y = -\frac{1}{x}$ **18.** $y = -3|x|$ **19.** $y = -3x^2$

Lesson 5.5

➤ Key Skills

Describe the effect of a translation on the graph of a function.

The parent function of $y = |x| + 2$ is $y = |x|$. The graph of $y = |x| + 2$ is shifted 2 units up from the graph of $y = |x|$. The graph of $y = |x + 1|$ is shifted to the left 1 unit from the graph of $y = |x|$.

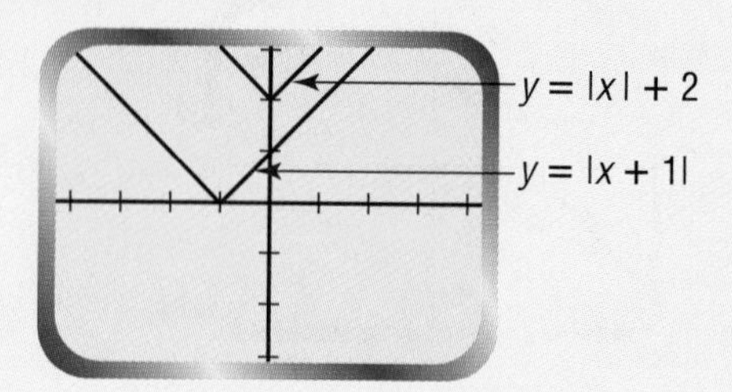

➤ Exercises

Identify the parent function. Describe the translation applied to each, and then draw the graph.

20. $y = x^2 + 1$ **21.** $y = (x + 1)^2$ **22.** $y = |x| - 4$ **23.** $y = |x - 4|$

Lesson 5.6

➤ Key Skills

Describe the effect of a combination of transformations on the graph of a function.

The parent function of $y = -2|x + 2|$ is $y = |x|$. Because of the scale factor -2, the y-values will be twice as great as the parent function and reflected through the x-axis. This is a vertical stretch of 2 and a reflection. Because 2 is added to x (or -2 is subtracted from x), this is a horizontal translation of 2 units to the left.

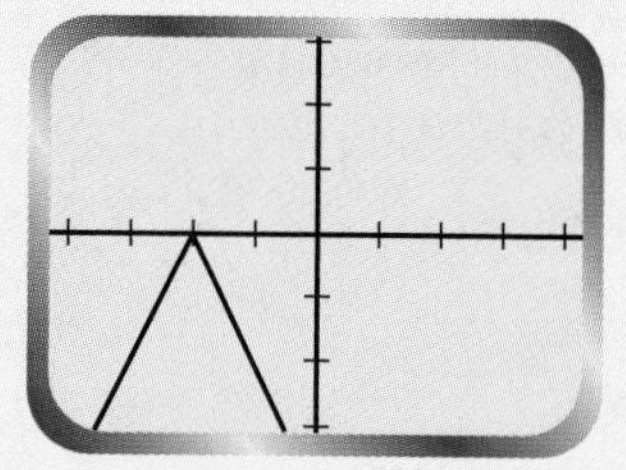

➤ Exercises

Graph each function.

24. $f(x) = 3x^2 + 1$ **25.** $g(x) = |x - 6| + 2$ **26.** $h(x) = -\frac{2}{x} + 1$

Applications

27. Are meters a function of centimeters? (If you know the number of meters, do you necessarily know the number of centimeters?) Yes

28. Are centimeters a function of meters? (If you know the number of centimeters, do you necessarily know the number of meters?) Yes

29. Temperature Kelvins are often used to measure temperature in science. To change degrees Celsius to kelvins, add 273.16 to the degrees Celsius. What transformation relates degrees Celsius to kelvins? Shift up by 273.16 units

30. Discounts Tomas decides to have a sale at his art store. He decides to reduce the price of everything in the store by 20%. Find the scale factor for the amount by which the cost of the items were "stretched." 0.80 or $\frac{4}{5}$

31. Belinda reflected a line segment with the endpoints (2, 4) and (6, 6) through the x-axis, and then shifted it to the left 5 units. What were the endpoints of the new line segment? (−3, −4) and (1, −6)

Chapter 5 Assessment

1. When is a relation a function? When no two ordered pairs have the same first coordinate

2. Explain how to use the vertical-line test to identify the graph of a function. Use a pencil or some other straight item. If the pencil held vertically can intersect the graph in more than one point, the relation is not a function.

Evaluate each function when x is 3.

3. $f(x) = -2^x$ -8 **4.** $g(x) = -x^2$ -9 **5.** $h(x) = 2|x|$ 6

Graph each function. Then tell whether the graph of $y = |x|$ is stretched, reflected, or shifted.

6. $y = 3|x|$ **7.** $y = |x| - 5$ **8.** $y = -|x|$ **9.** $y = |x + 4|$

Use the function $y = \frac{|x|}{2}$, for Items 10–12.

10. Identify the parent function. $y = |x|$

11. Sketch the graphs of the function and the parent function on the same axes.

12. Explain how the graph of the parent function is changed by the 2.
The parent function, $y = |x|$, is stretched by a factor of $\frac{1}{2}$.

Tell what happens to the point (3, 2) when each transformation is applied.

13. vertical translation of 5 (3, 7) **14.** horizontal translation of -3 (0, 2)

15. vertical translation of -6 (3, -4) **16.** vertical stretch of 4 (3, 8)

17. What happens to the average hourly wage of the workers at a restaurant if all hourly wages are increased by 5%? Increases by 5%

18. Explain how you know when the graph of a quadratic function is a parabola that opens downward.
In the function $y = ax^2 + bx + c$, when $a < 0$, the parabola opens downward.

Graph each function.

19. $f(x) = (x - 3)^2$ **20.** $h(x) = 5|x + 3|$ **21.** $t(x) = \frac{-2}{x + 2}$

22. Discounts A local department store is having a sale. They marked down their clearance items by 25%, and for one day only they are offering an additional 30% off each clearance item. Miki decided to leave the original prices on the price tags and make a sign that shows 55% off. Is Miki's sign correct? Why or why not? No; The reduction is off the sale price, not the original price; $x - 0.55x \neq (x - 0.25x) - 0.30(x - 0.25x)$

23. Explain how to determine the minimum point of the graph of the quadratic function $y = 2(x + 4)^2 - 1$.

Explain how to use a graphics calculator to solve each equation for x.

24. $\frac{4}{x} = 3 + x$ **25.** $|1 - 3x| = |x| + 5$

CHAPTER 6

Exponents

LESSONS

Exponential functions model a variety of important real-world activities. The growth of living organisms, population growth and decline, radioactive dating of fossils, and the spread of certain diseases are all examples of exponential growth or decay. This chapter will give you the opportunity to explore exponential growth and decay.

One of the areas in which exponential functions play an important role is finance. The portfolio activity gives you an example of interest on savings and loans that typically behaves exponentially.

A credit card company charges 1.5% interest on the unpaid balance and requires a minimum payment of $10 or 3% of the unpaid balance, whichever is greater. Suppose you borrow $1000 and pay it back by making the minimum payment each month. How long will it take to pay it back? What is the total amount that you will pay back?

LESSON 6.1

Exploring Exponents

why *Numbers with many digits can be expressed in a simpler way by using exponents.*

By using an exponent 2•2•2•2 is represented as 2^4.

Base — Exponent

$$2 \cdot 2 \cdot 2 \cdot 2 = 2^4 = 16$$

The fourth power of 2 is 16.

The expression 2^{1000} represents 2 to the thousandth power and is the product of a thousand 2s.

Some powers have special names. The first power of 2 is 2^1. It is read *two to the first power.* The second power of 2 is 2^2. It is read *two squared* or *two to the second power.* The third power of 2 is 2^3. It is read *two cubed* or *two to the third power.* The *n*th power of 2 is 2^n. It is read *two to the* n*th power.*

PRINTOUT OF 2^1000

10715086071862673209484250490600018105
61404811705533607443750388303510511249
36122493198378815695858127594672917553
14682518714528569231404359845775746985
74813934567774824230985421074605062371
14187795418215304647498358194126739876
75591655439460770629145711964776865421
67660429837652624386837205668069376

EXPONENTS AND POWERS

If x is any number and a is an integer greater than 1, then

$$x^a = \underbrace{x \cdot x \cdot x \cdot \cdots \cdot x}_{a \text{ factors}}$$

If a is 1, then $x^a = x^1 = x$.

Our number system uses 10 as its base. A number written in the form **2315** is in **customary notation**, or decimal notation. It can be expressed in **expanded notation** as

$$\mathbf{2} \cdot 1000 + \mathbf{3} \cdot 100 + \mathbf{1} \cdot 10 + \mathbf{5}$$

or in **exponential notation** as

$$\mathbf{2} \cdot 10^3 + \mathbf{3} \cdot 10^2 + \mathbf{1} \cdot 10^1 + \mathbf{5}.$$

How would you write 15,208 in exponential notation?
$1 \cdot 10^4 + 5 \cdot 10^3 + 2 \cdot 10^2 + 8$

Exploration 1 *Multiplying Powers*

1 Examine the steps in the example below.

Product form → $10^3 \cdot 10^1$ → $1000 \cdot 10$ → $10{,}000$ → 10^4 ← Simplified form

Complete the steps as shown in Step 1.

a. $10^3 \cdot 10^2 = 1000 \cdot 100 = \underline{?}$ (100,000) $= \underline{?}$ (10^5)

b. $10^3 \cdot 10^3 = (\underline{?})(\underline{?})$ ($1000 \cdot 1000$) $= \underline{?}$ (1,000,000) $= \underline{?}$ (10^6)

2 Complete the table for $10^4 \cdot 10^2$. Write the missing product as a single power of 10. This product is now in simplest form.

Decimal form	$10{,}000 \cdot 100 = 1{,}000{,}000$
Exponential form	$10^4 \cdot 10^2 = \underline{?}$ 10^6

3 Make a table similar to the one above for each of the following:

a. $10^3 \cdot 10^6$ **b.** $10^1 \cdot 10^3$ **c.** $2^2 \cdot 2^4$ **d.** $3^2 \cdot 3^2$

4 How would you simplify $a^m \cdot a^n$ using exponents? Make a table to check your guess.

5 For any positive number a and any positive integers m and n, what is the simplified form for $a^m \cdot a^n$? ❖ $a^m \cdot a^n = a^{m+n}$

The answer to Step 5 suggests that multiplying powers *with the same base* involves the addition of exponents. Next you will explore what happens to the exponents when you divide powers with the same base.

Exploration 2 *Dividing Powers*

1 Use the table to write the quotient $\frac{10^6}{10^2}$ as a power of 10.

	Decimal	Exponent
Numerator	1,000,000	10^6
Denominator	100	10^2
Quotient	10,000	10^4 ?

2 Make a table similar to the one above for each of the following:

a. $\frac{10^5}{10^2}$ **b.** $\frac{2^6}{2^2}$ **c.** $\frac{3^4}{3^3}$

3 Guess the simplified form in exponential notation for $\frac{10^5}{10^1}$. Make a table to check your guess.

4 For any positive number a and any positive integers m and n, with $m > n$, what do you think is the simplified form for $\frac{a^m}{a^n}$? $\frac{a^m}{a^n} = a^{m-n}$

5 Explain how you would simplify $\frac{2^5}{2^3}$.
Subtract the exponents: $\frac{2^5}{2^3} = 2^{5-3} = 2^2$.

6 Can you use this procedure to simplify $\frac{5^2}{3^2}$? Make a conjecture about the bases of exponents when using this procedure. ❖ No, because the bases are different. If the bases are the same, then you can divide powers by subtracting the exponents.

Explain why $(ab)^n = a^nb^n$ and $\left(\frac{a}{b}\right)^n = \frac{a^n}{b^n}$ when n is any positive integer.
$(ab)^n$ means that ab is a factor n times. By rearranging the factors, a is a factor n times and b is a factor n times. So $(ab)^n = a^nb^n$. Similarly $\left(\frac{a}{b}\right)^n$ means $\frac{a}{b}$ is a factor n times. So, a is a factor in the numerator n times and b is a factor in the denominator n times. So $\left(\frac{a}{b}\right)^n = \frac{a^n}{b^n}$.

Exploration 3 *The Power of a Power*

1 Write both 10^5 and 10^6 in customary notation. 100,000; 1,000,000

2 Since $(10^3)^2$ means $10^3 \cdot 10^3$, change 10^3 to customary notation and square the number. $1000 \cdot 1000 = 1{,}000{,}000$

3 Change the resulting number from customary notation back to exponential notation. Does $(10^3)^2$ equal 10^5 or 10^6? 10^6; 10^6

4 Does $(5^2)^4$ equal 5^6 or 5^8? What was done with the exponents in $(5^2)^4$?
5^8; The exponents were multiplied.

5 For any positive number a and any positive integers m and n, what is the equivalent expression for $(a^m)^n$? ❖ $(a^m)^n = a^{mn}$

APPLICATION

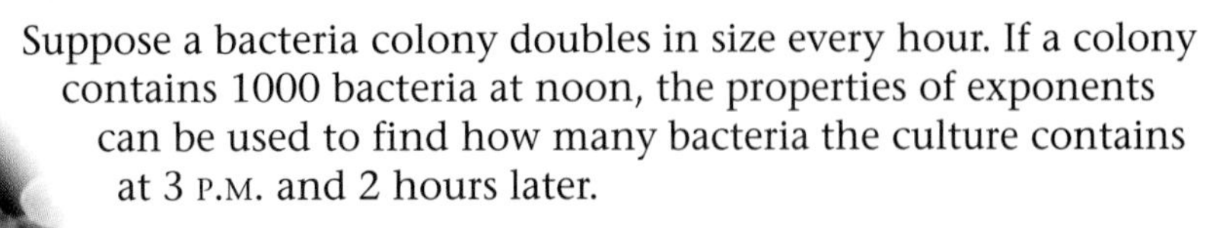

Suppose a bacteria colony doubles in size every hour. If a colony contains 1000 bacteria at noon, the properties of exponents can be used to find how many bacteria the culture contains at 3 P.M. and 2 hours later.

At 3 P.M. there will be $1000 \cdot 2^3$ or 8000 bacteria. Two hours later the bacteria will double two more times. There will be $(1000 \cdot 2^3) \cdot 2^2 = 1000 \cdot 2^{3+2} = 1000 \cdot 2^5$ or 32,000 bacteria in the culture. ❖

Once you are familiar with the properties of exponents, you can simplify many exponential expressions.

a. $3^2 \cdot 3^3 = 3^5$
$9 \cdot 27 = 243$

b. $5^4 \div 5^1 = 5^3$
$625 \div 5 = 125$

c. $(2^2)^3 \div (2^3)^2 = 2^6 \div 2^6$
$64 \div 64 = 1$

EXTENSION

Most computations with exponents can be performed easily on a calculator. Check the calculator you use to find out how it performs these computations.

Most calculators will use the y^x or ^ key to indicate an exponent. Parentheses are used for more complicated numerators and denominators.

For example, to simplify $\frac{10^3 + 10^4}{10^2}$, the numerator contains addition of the powers, so it should be included in parentheses. On many calculators, you can use

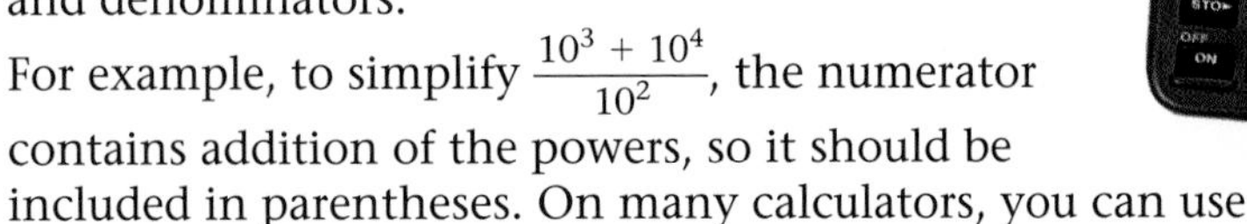

(10 y^x 3 + 10 y^x 4) ÷ 10 y^x 2 = or

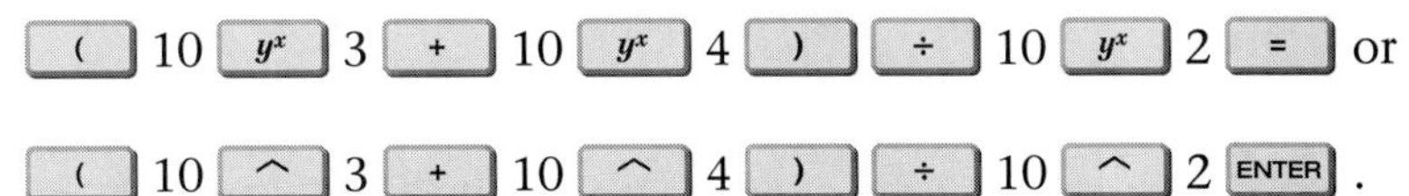

CT– Exponents can only be added when the powers are being multiplied and when the bases are the same.

Which quotient below is correct? Explain.

a. $\frac{10^3 + 10^4}{10^2} = \frac{10^3(1 + 10)}{10^2} = 10^1(11) = 110$ True

b. $\frac{10^3 + 10^4}{10^2} = \frac{10^7}{10^2} = 10^5 = 100{,}000$ ❖ False; $10^3 + 10^4 \neq 10^7$

Compare the way that the exponents are treated when the powers are added with the way that the exponents are treated when the powers are multiplied.

EXERCISES & PROBLEMS

Communicate

1. Explain the relationship between the exponent in a power of 10 and the number of zeros in a number written in customary notation. Use 10^{12} as an example.
2. How can you tell how many times the base is used as a factor in 3^6?
3. When you multiply powers, what operation is applied to the exponents of the common base? Explain why.
4. When you divide powers, what operation is applied to the exponents of the common base? Explain why.
5. When you raise a power to a power, what operation is applied to the exponents of the common base? Explain why.
6. Explain how you can determine, without computing anything, which value is greatest: 5^{17}, 8^{16}, or 8^{17}.

Killer whale

Explain why each equation is true.

7. $a^r \cdot a^s = a^{r+s}$ **8.** $\frac{a^p}{a^q} = a^{p-q}$ **9.** $(a^s)^t = a^{st}$

Practice & Apply

Make a table like the one in Exploration 1 for each product.

10. $10^6 \cdot 10^4$ **11.** $10^3 \cdot 10^5$ **12.** $2^3 \cdot 2^4$ **13.** $3^4 \cdot 3^2$

Make a table like the one in Exploration 2 for each quotient.

14. $\frac{10^3}{10^2}$ **15.** $\frac{10^6}{10^3}$ **16.** $\frac{2^6}{2^3}$ **17.** $\frac{3^5}{3^4}$

Change each to customary notation.

18. 2^5 32 **19.** 2^2 4 **20.** 10^5 100,000

21. Evaluate 2^5 and 2^2 to find $2^5 \cdot 2^2$. 128

22. Express the product $2^5 \cdot 2^2$ as a single power of 2. 2^7

23. Evaluate 2^5 and 2^2 to find $\frac{2^5}{2^2}$. 8

24. Express the quotient $\frac{2^5}{2^2}$ as a single power of 2. 2^3

25. Evaluate 2^5 to find $(2^5)^2$. 1024

26. Express $(2^5)^2$ as a single power of 2. 2^{10}

squid

Evaluate each expression by using the rules for exponents.

27. $(6^2)^3$ 46,656 **28.** $\frac{4^3}{4^2}$ 4 **29.** $\frac{10^2 \cdot 10^5}{10^3}$ 10,000 **30.** $\frac{10^2 + 10^4}{10^2}$ 101

Evaluate each expression.

31. $(x^2)^3$ x^6 **32.** $\frac{c^4}{c^1}$ c^3 **33.** $\frac{r^3(r^4)}{r^2}$ r^5 **34.** $\frac{2^3 + 2^4}{2^2}$ 6

Write each expression as a single power of 10, if possible. Then evaluate each expression.

35. $10^4 \cdot 10^2$ $10^6 = 1{,}000{,}000$ **36.** $10^4 + 10^2$ 10,100 **37.** $10^4 \div 10^2$ $10^2 = 100$

38. $10^4 - 10^2$ 9900 **39.** $(10^2)^3$ $10^6 = 1{,}000{,}000$ **40.** 10^{2+3} $10^5 = 100{,}000$

Technology For exponents, 2^3 is entered as 2 [^] 3 on some calculators and as 2 [y^x] 3 on others. Evaluate 2^3. Familiarize yourself with the way that the calculator you are using handles exponents. Then use the calculator to evaluate each expression.

41. 2^{20} 1,048,576 **42.** 3^{10} 59,049 **43.** 5^{10} 9,765,625 **44.** 0.5^4 0.0625

The table shows how to read certain large numbers.

Customary	Exponential	Word
1,000,000	10^6	million
1,000,000,000	10^9	billion
1,000,000,000,000	10^{12}	trillion

For example, the number 2,450,000,000 is read "two billion, four hundred fifty million," and 3×10^6 is read "three million." Write out each number in words.

45. 7,000,000,000,000 Seven trillion

46. 2,030,400,000,000 **46.** Two trillion, thirty billion, four hundred million

47. 5×10^9 Five billion

48. 3×10^{12} Three trillion

49. A newspaper article refers to a debt of 4.2 trillion dollars. Write this in customary notation.
4,200,000,000,000

Algae

Yellowtail fish

Krill (zooplankton)

Biology The food chain explains why even small organisms affect much larger ones. It works in the following way:

- On average, a killer whale may eat about 10 pounds of seal per day.
- Each pound of seals requires about 10 pounds of fish or squid per day.
- Each pound of fish or squid requires about 10 pounds of zooplankton per day.
- Each pound of zooplankton requires about 10 pounds of algae per day.

Express each number of pounds per day as a single power of 10.

50. seals needed to sustain a killer whale 10^1

51. fish or squid needed to sustain a killer whale 10^2

52. zooplankton needed to sustain a killer whale 10^3

53. algae needed to sustain a killer whale 10^4

54. **Statistics** Make a bar graph from a frequency distribution of the digits in 2^{1000} given at the beginning of this lesson. Count how many 0s, 1s, 2s, . . . , 9s there are. Would you say that the distribution is about the same for each digit, or does it favor some particular digits?

55. Write 2^{10} in customary notation. 1024

56. **Technology** A calculator shows the value 20^{10}. Compare this answer with the answer in Exercise 55. What do you think this notation means? Write out the answer in customary notation.
1.024×10^{13}; 10,240,000,000,000

Look Back

Solve for *x*. [Lesson 1.6]

57. $|x - 3| \leq 7$ **58.** $|2x + 4| > 8$

57. $-4 \leq x \leq 10$
58. $x < -6$ or $x > 2$

59. **Coordinate Geometry** A line passes through the points $P(14, -3)$ and $Q(-5, 27)$. Write an equation for the line that passes through point P and is perpendicular to line PQ. **[Lessons 2.1, 2.2]** $y = \frac{19}{30}x - \frac{178}{15}$

60. Find the inverse of $A = \begin{bmatrix} 3 & -1 \\ -2 & 1 \end{bmatrix}$. What is the product of $A \cdot A^{-1}$? **[Lesson 3.4]**

60. $A^{-1} = \begin{bmatrix} 1 & 1 \\ 2 & 3 \end{bmatrix}$; $A \cdot A^{-1} = \begin{bmatrix} 1 & 0 \\ 0 & 1 \end{bmatrix}$

61. **Probability** Joshua records the results of several weeks of his daily basketball free throw practice and decides that he will usually make about 100 shots out of 150 attempts. What is the probability that at any given time Joshua will make a successful free throw shot? Is this experimental probability or theoretical probability? Explain. **[Lessons 4.1, 4.6]**

62. **Probability** From the menu at the right, how many different ways can you order an ice cream cone with one topping? **[Lesson 4.5]** 252

Super Scooper

flavors

1. vanilla
2. chocolate
3. strawberry
4. banana
5. rocky road
6. pralines & cream
7. chocolate chip
8. cookies & cream
9. pineapple
10. lime sherbet
11. orange sherbet
12. raspberry sherbet

toppings
• sprinkles • shaved chocolate
• malt • pecans • peanuts
• gummy bits • marshmallows

cones
plain
sugar
waffle

63. $A(x) = 2(x + 3)^2$

63. **Geometry** The area in square centimeters of a square piece of cardboard is determined from the length of its sides. Write the function for the area if 3 centimeters are added to each side of the original square and that area is doubled. **[Lesson 5.6]**

64. Use your understanding of transformations to describe the graph of the function in Exercise 63 without drawing it. **[Lesson 5.6]** $y = 2(x + 3)^2$ is a parabola shifted 3 units left and stretched by a factor of 2.

Look Beyond

65. Look at the pattern in the table for Exploration 2. Based on this table, 10^n can be written as a 1 followed by how many zeros?
n zeros

66. What does Exercise 65 suggest for the value of 10^0?
$10^0 = 1$

Use a calculator to evaluate each expression.

67. $4^{0.5}$ 2 **68.** $9^{0.5}$ 3 **69.** $81^{0.5}$ 9 **70.** $100^{0.5}$ 10

71. What do you think is the value of $b^{0.5}$, for any positive number b? $\sqrt{b}$

LESSON 6.2 Multiplying and Dividing Monomials

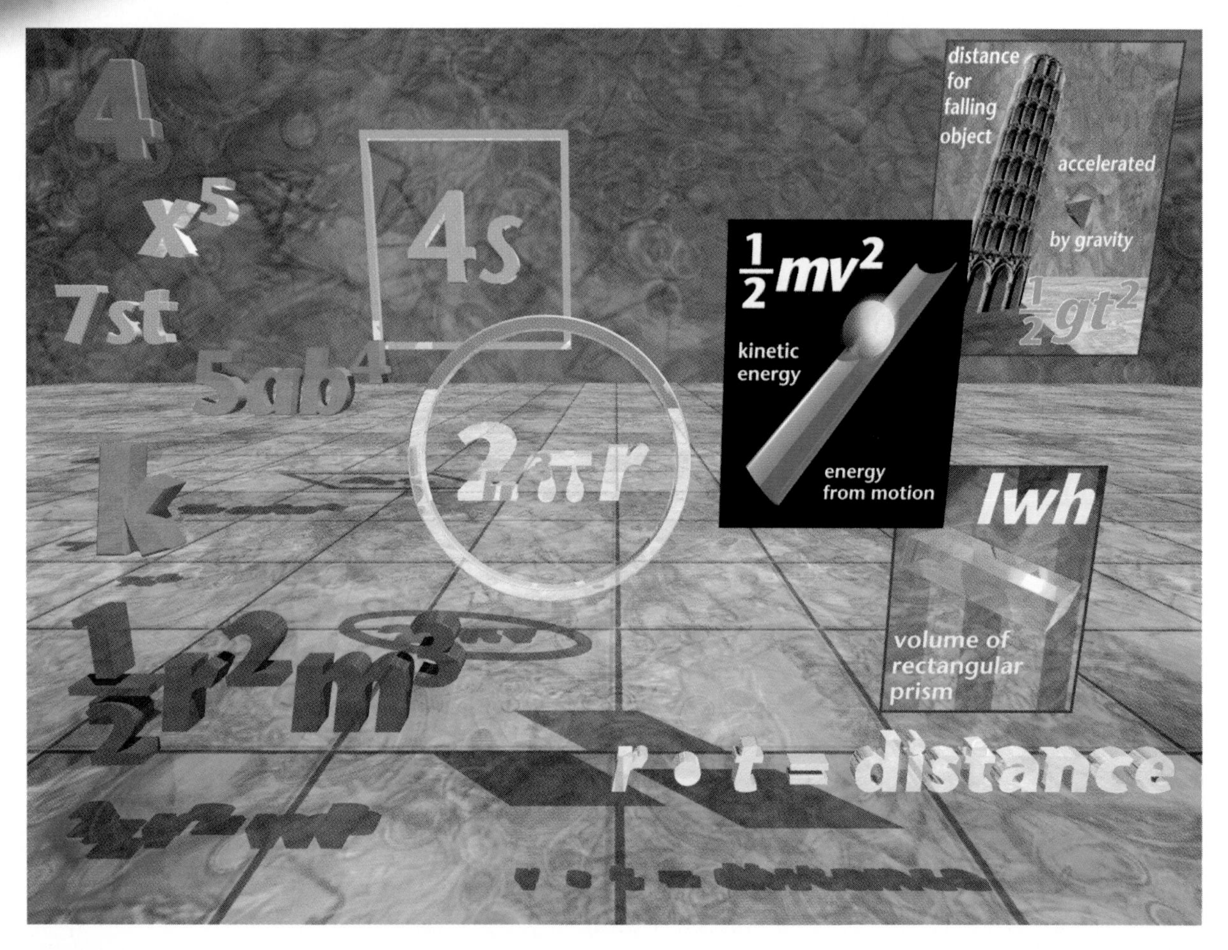

Computation with powers of variables can be simplified by using the same properties as those used for numerical powers.

In the previous lesson you discovered a property that lets you simplify the product of two powers with the same base. The property tells you to keep the base and add the exponents. For example,

$$x^3 \cdot x^5 = x^{3+5} = x^8.$$

PRODUCT-OF-POWERS PROPERTY

If x is any number and a and b are any positive integers, then

$$x^a \cdot x^b = x^{a+b}.$$

The Product-of-Powers Property can be used to find the product of more complex expressions such as $5a^2b$ and $-2ab^3$. Expressions such as $5a^2b$ and $-2ab^3$ are called *monomials*.

A **monomial** is an algebraic expression that is either a constant, a variable, or a product of a constant and one or more variables. A monomial can also contain powers of variables with positive integer exponents.

$(5a^2b^2)(-2ab^3) = -10(a^2 \cdot a^1)(b^2 \cdot b^3)$	Multiply the coefficients. Then group the powers of like bases.
$= -10a^3b^5$	Product-of-Powers Property

EXAMPLE 1

Find each product.

A $(6st)(-2s^2t)$

B $(-4a^2b)(-ac^2)(3b^2c^2)$

Solution ➤

Multiply the coefficients and group powers of the same base. Then use the Product-of-Powers Property.

A $(6st)(-2s^2t) = -12s^3t^2$

B $(-4a^2b)(-ac^2)(3b^2c^2) = 12a^3b^3c^4$ ❖

Explain how to use the Product-of-Powers Property to simplify each expression.

a. $x^{3c} \cdot x^{4c}$ **b.** $(x + c)^d \cdot (x + c)^{2d}$

a. $x^{3c} \cdot x^{4c} = x^{(3c+4c)} = x^{7c}$
b. $(x + c)^d \cdot (x + c)^{2d} = (x + c)^{(d+2d)} = (x + c)^{3d}$

The Product-of-Powers Property can be used to simplify $(a^2)^3$.

$$(a^2)^3 = a^2 \cdot a^2 \cdot a^2 = a^{2+2+2} = a^6$$

Notice that $(a^2)^3 = a^{2 \cdot 3}$. To raise a power to a power, multiply the exponents.

POWER-OF-A-POWER PROPERTY

If x is any number and a and b are any positive integers, then

$$(x^a)^b = x^{ab}.$$

EXAMPLE 2

Simplify.

A $(3^2)^4$

B $(p^2)^5$

Solution ➤

Use the Power-of-a-Power Property to multiply the exponents.

A $(3^2)^4 = 3^{2 \cdot 4} = 3^8$

B $(p^2)^5 = p^{2 \cdot 5} = p^{10}$ ❖

Sometimes an exponential expression has a monomial for a base.

EXAMPLE 3

Simplify $(xy^2)^3$.

***Solution* ➤**

The exponent 3 outside the parentheses indicates that the monomial xy^2 is used as a factor 3 times. Simplify by regrouping and multiplying.

$$\begin{aligned}(xy^2)^3 &= (xy^2)(xy^2)(xy^2)\\ &= (x \cdot x \cdot x)(y^2 \cdot y^2 \cdot y^2)\\ &= x^3y^6 \end{aligned}$$ ❖

POWER-OF-A-PRODUCT PROPERTY

If x and y are any numbers and n is a positive integer, then

$$(xy)^n = x^ny^n.$$

If a is a numerical coefficient, then $(axy)^n = a^nx^ny^n$. For example,

$$(3xy)^2 = (3xy)(3xy) = (3^2)(x^2)(y^2) = 9x^2y^2.$$

Remember to apply the exponent that is outside of the parentheses to *each* factor of the monomial that is inside the parentheses. This includes the coefficient, 3. Thus, $(3xy)^2 = 9x^2y^2$.

EXAMPLE 4

CT– For $(-x)^m$, if x is positive and m is even, the value is positive. If x is positive and m is odd, the value is negative. Note that if x is negative, the value is positive, whether m is even or odd. Exponents of a power are added if powers with the same base are multiplied. The exponents of a power are multiplied to find a power of a power.

Simplify each expression.

A $(3x^2y^3)^3$ **B** $(-t)^5$ **C** $(-t)^6$ **D** $-t^4$ **E** $(-5x)^3$

***Solution* ➤**

A $(3x^2y^3)^3 = (3)^3(x^2)^3(y^3)^3 = (3^3)(x^{2 \cdot 3})(y^{3 \cdot 3}) = 27x^6y^9$

B $(-t)^5 = (-1 \cdot t)^5 = (-1)^5 \cdot t^5 = -1 \cdot t^5 = -t^5$

C $(-t)^6 = (-1 \cdot t)^6 = (-1)^6 \cdot t^6 = 1 \cdot t^6 = t^6$

D $-t^4$ is in simplest form. $(-1) \cdot t^4 = -t^4$
The exponent applies only to t and not to -1.

E $(-5x)^3 = (-5 \cdot x)^3 = (-5)^3 \cdot x^3 = -125x^3$ ❖

Explain whether $(-x)^m$ is positive or negative when x is positive and m is even. Do the same when m is odd. When are the exponents of a power added, and when are they multiplied?

Earlier you found that if you simplify quotients such as $\frac{10^6}{10^2}$, the quotient can be written without a denominator. A similar property holds for variables and monomials.

QUOTIENT-OF-POWERS PROPERTY

If x is any number except 0 and a and b are any positive integers, with $a > b$, then

$$\frac{x^a}{x^b} = x^{a-b}.$$

EXAMPLE 5

Simplify. **A** $\frac{-4x^2y^5}{2xy^3}$ **B** $\frac{c^4b}{c^2a}$

Solution

Simplify any numerical coefficients. Then subtract the exponent in the denominator from the exponent in the numerator with the same base.

A $\frac{-4x^2y^5}{2xy^3} = \left(\frac{-4}{2}\right)(x^{2-1})(y^{5-3}) = -2xy^2$

B $\frac{c^4b}{c^2a} = (c^{4-2})\left(\frac{b}{a}\right) = \frac{c^2b}{a}$ ❖

Try This Simplify each expression.

a. $\frac{6ab^2}{-2b}$ $-3ab$ **b.** $\frac{(25^2)(-81t^3)}{45^3t}$ $-\frac{5}{9}t^2$

EXERCISES & PROBLEMS

Communicate

1. Explain why $3x^2$ is a monomial but $3x^{-2}$ is not.
2. Compare the Product-of-Powers Property with the Quotient-of-Powers Property.
3. Explain why $(x^4y^3)^6 = x^{24}y^{18}$ is true.
4. Explain why $(x^a)^b = (x^b)^a$ is true.
5. Explain why $y^2 \cdot y^4 \neq y^8$ is true.

Practice & Apply

Simplify each expression.

6. $(3x)^2$ $9x^2$ **7.** $\left(\frac{a}{b}\right)^3$ $\frac{a^3}{b^3}$ **8.** $\left(\frac{10x}{y^3}\right)^2$ $\frac{100x^2}{y^6}$ **9.** $-3t^4$ $-3t^4$

Find each product.

10. $(8r^2)(4r^3)$ $32r^5$ **11.** $(70x^4)(7x^3)$ $490x^7$ **12.** $(-2a^5)(4a^2)$ $-8a^7$

13. $(-p^2)(10p^3)$ $-10p^5$ **14.** $(-2a^2)(-5a^4)$ $10a^6$ **15.** $(-x^2y^5)(-x^3y^2)$ x^5y^7

16. $(48a^4b^2)(-0.2ac^5)$ $-9.6a^5b^2c^5$ **17.** $(4.3d^2n^{10}k)(0.1n^2k^3)$ $0.43d^2n^{12}k^4$

Find each quotient. 24. $-\frac{40a^3b^5}{c^3}$

18. $\frac{8r^3}{4r^2}$ $2r$ **19.** $\frac{70x^4}{7x^3}$ $10x$ **20.** $\frac{-2a^5}{4a^2}$ $-\frac{a^3}{2}$ **21.** $\frac{-p^4}{10p^3}$ $-\frac{p}{10}$ **22.** $\frac{-2a^6b^7}{-5a^4b}$ $\frac{2a^2b^6}{5}$

23. $\frac{-x^4y^5}{-x^3y^2}$ xy^3 **24.** $\frac{48a^4b^5}{-1.2ac^3}$ **25.** $\frac{0.8r^6}{0.004r^3}$ $200r^3$ **26.** $\frac{4u^2v^{10}w^4}{0.1w^2v^3}$ $40u^2v^7w^2$ **27.** $\frac{(5u^2)^3}{(-5u)^2}$ $5u^4$

Simplify each expression. 32. $625j^8k^{12}$ 33. $9x^2y^8z^{10}$

28. $(2x^4)^3$ $8x^{12}$ **29.** $(3b^2)^5$ $243b^{10}$ **30.** $(-2r^3)^2$ $4r^6$ **31.** $(-10m^4)^3$ $-1000m^{12}$

32. $(5j^2k^3)^4$ **33.** $(3xy^4z^5)^2$ **34.** $2(3a^2)^3$ $54a^6$ **35.** $10(-5b^5)^2$ $250b^{10}$

36. $(8n^2p)^3$ $512n^6p^3$ **37.** $(7j^2)^3$ $343j^6$ **38.** $(ab^5)(a^3)^2$ a^7b^5 **39.** $(v^4w^3)^2(v^3)^4$ $v^{20}w^6$

40. Evaluate the two monomials $2A^3$ and $(2A)^3$ for $A = 10$. 2000; 8000

41. Evaluate $(-1)^1$, $(-1)^2$, $(-1)^3$, $(-1)^4$, and $(-1)^5$. $-1, 1, -1, 1, -1$

42. Use the pattern in Exercise 41 to find the value of $(-1)^{100}$. Explain what you notice about $(-1)^n$ for various values of n.
$(-1)^{100} = 1$; $(-1)^n = 1$ if n is even, $(-1)^n = -1$ if n is odd.

43. **Geometry** Find the area of a square if each side is twice as long as a side of the square shown. $4s^2$

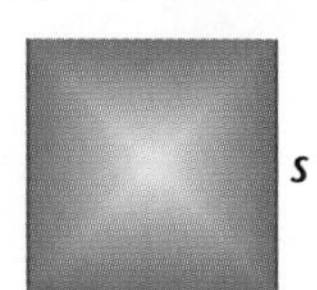

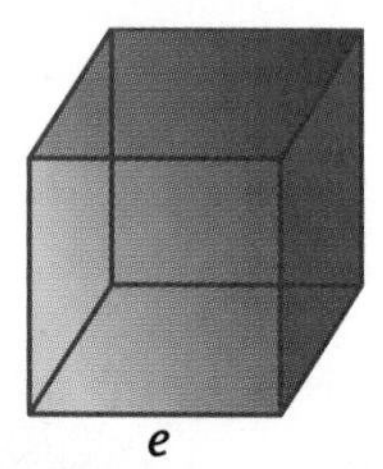

44. **Geometry** Find the volume of a cube if each edge is double each edge of the cube shown. $8e^3$

Technology Calculators can be used to evaluate expressions involving monomials. Store 10 for A and 2 for B in the memory of a graphics calculator. Use a calculator to evaluate each expression.

45. A^2 100 **46.** B^3 8 **47.** $A^2 + B^3$ 108 **48.** A^2B^3 800 **49.** $\frac{A^2}{B^3}$ 12.5 **50.** $\frac{1}{A^2 + B^3}$ 0.0093

Evaluate each expression for $A = 9.8$ and $B = 2.1$.

51. A^2 96.04 **52.** B^3 9.261 **53.** $A^2 + B^3$ 105.301 **54.** A^2B^3 889.42644 **55.** $\frac{A^2}{B^3}$ 10.37037 **56.** $\frac{1}{A^2 + B^3}$ 0.0095

57. Does $(10^2)^3$ equal 10^8, 10^5, or 100^3? Explain your reasoning.

Look Back

Solve each equation for *n*. [Lessons 1.3, 1.4]

58. $(4n - 20) - n = n + 12$ $n = 16$

59. $9(n + k) = 5n + 17k + 12$
$n = 2k + 3$

60. Travel Distance, d, equals rate of speed, r, times time, t. If a car travels 195 miles in 3 hours, determine the rate of speed. What distance will the car travel in 4 hours? **[Lesson 1.3]**
rate = 65 mph; 260 miles

Solve each system. [Lessons 2.4, 2.5]

61. $\begin{cases} 5x - 3y = 32 \\ 2x = 12.8 \end{cases}$
(6.4, 0)

62. $\begin{cases} 7x - 3y = 2 \\ 2x - y = -5 \end{cases}$
(17, 39)

When traveling in foreign countries, distances are usually given in kilometers.

Statistics
Use the data at the left to find

63. the median. 3

64. the mean. 4

65. the mode. **[Lesson 4.3]** 3

66. A math class has 28 students. Of those students, 16 are girls. What is the probability that a random selection of one student will be a boy? **[Lesson 4.6]** $\frac{3}{7}$

Look Beyond

Use Exercises 67–70 below to discover a property of sums of squares.

67. Show that $1^2 + 8^2 = 4^2 + 7^2$. Show that $14^2 + 87^2 = 41^2 + 78^2$. How were the numbers 14, 87, 41, and 78 chosen?

68. Does $17^2 + 84^2 = 71^2 + 48^2$? How were the numbers 17, 84, 71, and 48 chosen?

69. Show that $0^2 + 5^2 = 3^2 + 4^2$. Find new numbers x and y so that $(03)^2 + 54^2 = x^2 + y^2$.

70. Find new numbers x and y so that $(04)^2 + 53^2 = x^2 + y^2$. Show that $4^2 + 6^2 = 3^2 + 7^2$ is *false*. Does $43^2 + 67^2 = 34^2 + 76^2$?

LESSON 6.3 Negative and Zero Exponents

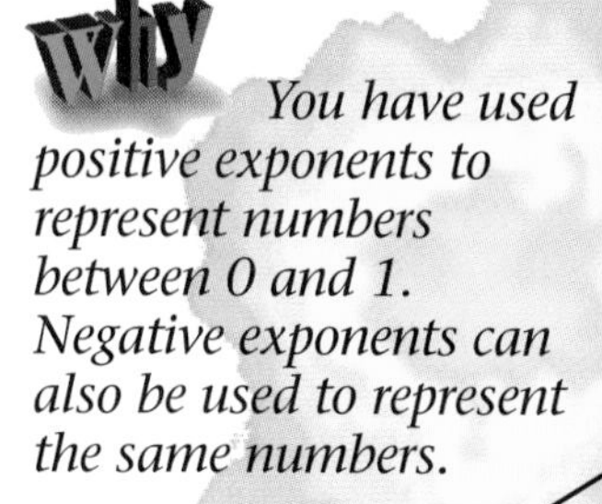

You have used positive exponents to represent numbers between 0 and 1. Negative exponents can also be used to represent the same numbers.

In chemistry, the approximate mass of a neutron or proton of an atom is 1.660×10^{-24} grams.

The expression 10^1 is defined to be 10. How should the expression 10^{-1} be defined?

Calculator

When the Quotient-of-Powers Property was defined, the exponent in the numerator was greater than the exponent in the denominator. For example, $\frac{10^4}{10^3} = 10^{4-3} = 10^1$. If the Quotient-of-Powers Property is applied when the exponent in the numerator is less than the exponent in the denominator, the result is $\frac{10^3}{10^4} = 10^{3-4} = 10^{-1}$.

If you write the powers as factors, the result is

$$\frac{10^3}{10^4} = \frac{10 \cdot 10 \cdot 10}{10 \cdot 10 \cdot 10 \cdot 10} = \frac{1}{10}.$$

The results above indicate that 10^{-1} should be defined as $\frac{1}{10}$. How can you enter 10^{-1} on your calculator?

NEGATIVE EXPONENT

If x is any number except zero and n is any integer, then

$$x^{-n} = \frac{1}{x^n}.$$

EXAMPLE 1

Simplify each expression.

A $2^{-3} \cdot 2^2$ **B** $\frac{10^3}{10^{-1}}$ **C** $10^2 \cdot 3^2$

Solution

A Use the Product-of-Powers Property.

$$2^{-3} \cdot 2^2 = 2^{-3+2} = 2^{-1} = \frac{1}{2^1} = \frac{1}{2}$$

B Use the Quotient-of-Powers Property. Subtract the exponents, and simplify the expression.

$$\frac{10^3}{10^{-1}} = 10^{3-(-1)} = 10^4 = 10{,}000$$

C Since 10^2 and 3^2 do not have the same base, $10^2 \cdot 3^2$ cannot be simplified using the properties of exponents.

$$10^2 \cdot 3^2 = 100 \cdot 9 = 900$$ ❖

Compare the expressions $(-2)^3$ and $(-2)^{-3}$. Simplify each expression. How are the results related? $(-2)^3 = -8$; $(-2)^{-3} = -\frac{1}{8}$; the numbers are reciprocals of each other.

You have simplified expressions containing positive and negative exponents. Next, you will explore the significance of an expression that contains a **zero exponent**.

Exploration *Defining x^0*

1 Simplify each expression.

a. $\frac{10^5}{10^5}$ 1 **b.** $\frac{10^3}{10^3}$ 1 **c.** $\frac{10^1}{10^1}$ 1

2 Use the Quotient-of-Powers Property to simplify each expression in Step 1. **2a.** 10^0 **2b.** 10^0 **2c.** 10^0

3 What is the simplified numerical value of 10^0? $10^0 = 1$

4 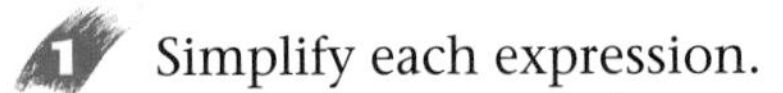 How would you define x^0 for any nonzero number x? ❖

$x^0 = 1$ for any nonzero number x.

Expressions containing variables with negative and zero exponents can be simplified by using the properties of powers.

EXAMPLE 2

Simplify each expression.

A $c^{-4} \cdot c^4$ **B** $-3y^{-2}$ **C** $\frac{m^2}{n^{-3}}$

Solution ➤

A $c^{-4} \cdot c^4 = c^{-4+4} = c^0 = 1$
Does this agree with the definition you wrote in Step 4 of the exploration on page 280?

B $-3y^{-2} = (-3)(y^{-2}) = \frac{-3}{y^2}$

C $\frac{m^2}{n^{-3}} = m^2 \cdot n^{-(-3)} = m^2n^3$

Notice that $\frac{m^2}{n^{-3}} = \frac{m^2}{\frac{1}{n^3}} = m^2 \div \frac{1}{n^3} = m^2 \cdot \frac{n^3}{1} = m^2n^3$. ❖

SUMMARY OF POWER PROPERTIES

Let *a* and *b* be any numbers with integer exponents *m* and *n*.

Product of Powers	Quotient of Powers ($b \neq 0$)	Power of a Power	Power of a Product
$b^m \cdot b^n = b^{m+n}$	$\frac{b^m}{b^n} = b^{m-n}$	$(b^m)^n = b^{mn}$	$(ab)^m = a^mb^m$

If *a* and *b* are non-zero numbers and *n* is any integer, explain why the following rule is true.

$$\left(\frac{a}{b}\right)^{-n} = \left(\frac{b}{a}\right)^n$$

CT– $\left(\frac{a}{b}\right)^{-n} = \frac{a^{-n}}{b^{-n}} = \frac{\frac{1}{a^n}}{\frac{1}{b^n}} = \frac{b^n}{a^n} = \left(\frac{b}{a}\right)^n$

Communicate

1. What is the meaning of a negative exponent?
2. Explain why $5a^{-2}$ does not equal $\frac{1}{5a^2}$.
3. Use $\frac{2^5}{2^5}$ to explain why $2^0 = 1$. For what other bases does this apply?
4. Can $5^2 \cdot 4^{-3}$ be simplified using the properties of exponents? Explain.
5. Explain why $(3a)^{-2}$ does not equal $\frac{3}{a^2}$.
6. Explain why the reciprocal of $\frac{3}{5}$ can be written as $\left(\frac{3}{5}\right)^{-1}$.

1. A negative exponent is equal to 1 over the number raised to the opposite of the negative exponent: $a^{-n} = \frac{1}{a^n}$, for any nonzero number, *a*.

6. $\left(\frac{3}{5}\right)^{-1} = \frac{3^{-1}}{5^{-1}} = \frac{\frac{1}{3^1}}{\frac{1}{5^1}} = \frac{5^1}{3^1} = \frac{5}{3}$

Practice & Apply

Evaluate each expression.

7. -3^{-2} $-\frac{1}{9}$ **8.** $(-3)^2$ 9 **9.** 3^{-2} $\frac{1}{9}$ **10.** -3^2 -9

11. Copy and complete the table by continuing the pattern.

Decimal form	10,000	? 1000	100	10	1	? 0.1	0.01
Exponential form	10^4	10^3	? 10^2	10^1	? 10^0	10^{-1}	? 10^{-2}

12. Write 10^{-8} in customary notation. $\frac{1}{100{,}000{,}000}$

Write each expression without a negative or zero exponent.

13. 2^{-3} $\frac{1}{2^3}$ **14.** 10^{-5} $\frac{1}{10^5}$ **15.** a^3b^{-2} $\frac{a^3}{b^2}$ **16.** $c^{-4}d^3$ $\frac{d^3}{c^4}$

17. $v^0w^2y^{-1}$ $\frac{w^2}{y}$ **18.** $(a^2b^{-7})^0$ 1 **19.** r^6r^{-2} r^4 **20.** $-t^{-1}t^{-2}$ $-\frac{1}{t^3}$

21. $\frac{m^2}{m^{-3}}$ m^5 **22.** $\frac{2a^{-5}}{a^{-6}}$ $2a$ **23.** $\frac{(2a^3)(10a^5)}{4a^{-1}}$ $5a^9$ **24.** $\frac{b^{-2}b^4}{b^{-3}b^4}$ b

Probability The probability p of getting a multiple-choice question with four choices for each item correct by guessing is $\frac{1}{4}$, and the probability q of getting it wrong is $\frac{3}{4}$. On a test with 5 items, a formula for finding the probability of getting a certain number of items correct is shown in the chart below.

25. Find the remaining values in the chart to the nearest thousandth.

No. correct	Probability	Value
0	p^0q^5	$\left(\frac{1}{4}\right)^0\left(\frac{3}{4}\right)^5 \approx 0.237$
1	$5p^1q^4$	? 0.396
2	$10p^2q^3$	? 0.264
3	$10p^3q^2$	? 0.088
4	$5p^4q^1$	? 0.015
5	p^5q^0	? 0.001

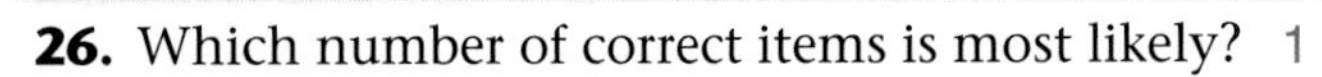

26. Which number of correct items is most likely? 1

27. Which number of correct items is least likely? 5

28. Find the sum of all six probabilities in the chart. 1

Technology Tell whether each expression is easier to evaluate on a calculator or mentally. Explain why you think so. Then simplify.

29. $\frac{2.56^7}{2.56^6}$ **30.** $\frac{2.56^6}{2.56^7}$ **31.** 0^7

32. 7^0 **33.** $(2.992 \times 9.554)^0$ **34.** $(19.43 \times 0)^{18}$

35. **Technology** Try to compute 2^{236} on a calculator. What happens?

36. Try to compute 0^0 on a calculator. What happens?

37. Explain why 0^0 does not equal 1 and is undefined.

38. **Technology** Make a table using $Y_1 = 0^x$ and $Y_2 = x^0$. What happens when x is 0? Why?

Look Back

39. What are the next three terms of the sequence 8, 11, 16, 23, 32, . . . ? **[Course 1]** 43, 56, 71

Solve each equation for y. How are the equations related? [Lesson 1.3]

40. $5(3y) = 3y - 24$ -2 41. $aby = by - c$

42. On a number line, graph the solution to the inequality $3x - 6 > -4x + 12$. Describe the solution to this inequality in words. **[Lesson 1.5]**

43. Describe the slopes of vertical lines and the slopes of horizontal lines. **[Lesson 2.1]** Vertical: undefined slope; horizontal: 0 slope

44. The graph of the quadratic parent function is stretched vertically by a scale factor of 2. It is then translated -3 units horizontally and 4 units vertically. Finally, it is reflected through the x-axis. Draw a graph and write a formula for this transformation. **[Lesson 5.6]**

45. Let $A = 3 \cdot 3 \cdot 3 \cdot 3 \cdot 3 \cdot x \cdot x \cdot x \cdot x \cdot x \cdot x \cdot y \cdot y \cdot y \cdot y$ and $B = 2 \cdot 2 \cdot 2 \cdot 2 \cdot 3 \cdot 3 \cdot 3 \cdot x \cdot x \cdot x \cdot x \cdot y$. Simplify each expression. **[Lesson 6.2]** $A = 243x^6y^4$; $B = 432x^4y$

46. Write the product $A \cdot B$ from Exercise 45 in exponential form. **[Lesson 6.2]** $2^4 \cdot 3^8x^{10}y^5$ or $104{,}976x^{10}y^5$

Look Beyond

Cultural Connection: Europe Pierre de Fermat (1601–1665) was a French mathematician who helped develop the concepts of number theory and probability. He studied numbers of the form $2^{2^n} + 1$, which are now called Fermat numbers.

n	Fermat number
0	$2^{2^0} + 1 = 2^1 + 1 = 3$
1	$2^{2^1} + 1 = 2^2 + 1 = 5$
2	$2^{2^2} + 1 = 2^4 + 1 = ?$ 17
3	$2^{2^3} + 1 = 2^8 + 1 = ?$ 257
4	$2^{2^4} + 1 = ? = ?$ $2^{16} + 1$; 65,537
5	$2^{2^5} + 1 = ? = ?$ $2^{32} + 1$; 4,294,967,297

47. Copy and complete the table of Fermat numbers.

48. Fermat observed that each of the numbers was prime. He made the conjecture that all such numbers were prime. Though an unusually gifted mathematician, he was wrong in thinking that the sixth number, $2^{2^5} + 1$, was prime; it has a factor of 641. Use a calculator to find the other factor. Later research suggested that *no* more Fermat numbers are prime, so even great mathematicians can be wrong. 6,700,417

EYEWITNESS MATH

All Mixed Up?

You are playing a card game and it's your turn to deal. How many times do you shuffle the deck? Is that enough to be sure the cards are thoroughly mixed? Is there such a thing as shuffling too much? Read the article below to learn how two mathematicians, one of them a magician as well, found some answers that surprised shufflers around the world.

In Shuffling Cards, 7 Is Winning Number

By Gina Kolata

It takes just seven ordinary, imperfect shuffles to mix a deck of cards thoroughly, researchers have found. Fewer are not enough and more do not significantly improve the mixing.

The mathematical proof, discovered after studies of results from elaborate computer calculations and careful observation of card games, confirms the intuition of many players that most shuffling is inadequate.

The finding has implications for everyone who plays cards and everyone who has a stake in knowing whether a shuffle is random . . .

No one expected that the shuffling problem would have a simple answer, said Dr. Dave Bayer, a mathematician and computer scientist at Columbia who is coauthor of the recent discovery. Other problems in statistics, like analyzing speech patterns to identify speakers, might be amenable to similar approaches, he said . . .

Dr. Persi Diaconis, a mathematician and statistician at Harvard University who is another author of the discovery, said the methods used are already helping mathematicians analyze problems in abstract mathematics that have nothing to do with shuffling or with any known real-world phenomena.

Dr. Diaconis has been carefully watching card players shuffle for the past 20 years. The researchers studied the dovetail or riffle shuffle, in which the deck of cards is cut in half and the two halves are riffled together. They said this was the most commonly used method of shuffling cards. But, Dr. Diaconis said it produces a card order that is "far from random." . . .

Dr. Diaconis began working with Dr. Jim Reeds at Bell Laboratories and showed that a deck is perfectly mixed if it is shuffled between 5 and 20 times.

Next, Dr. Diaconis worked with Dr. Aldous and showed that it takes 5 to 12 shuffles to perfectly mix a deck.

In the meantime, he also worked on "perfect shuffles," those that exactly interlace the cards. He derived a mathematical proof showing that if a deck is perfectly shuffled eight times, the cards will be in the same order as they were before the shuffling.

To find out how many ordinary shuffles were necessary to mix a deck, Dr. Diaconis and Dr. Bayer watched players shuffle. He also watched Las Vegas dealers to see how perfectly they would interlace the cards they shuffled . . .

The researchers did extensive simulations of shuffling on a computer. To get the proof, the researchers looked at a lot of shuffles, guessed that the answer was seven, and finally proved it by finding an abstract way to describe what happens when cards are shuffled.

"When you take an honest description of something realistic and try to write it out in mathematics, usually it's a mess," Dr. Diaconis said. "We were lucky that the formula fit the real problem. That is just miraculous, somehow."

Getting Lost in the Shuffle

Even after a deck of cards is cut and shuffled, fragments of the original arrangement remain. In this example, the hearts are cut into two sequences: ace through six, and seven through king. Then the sequences are shuffled.

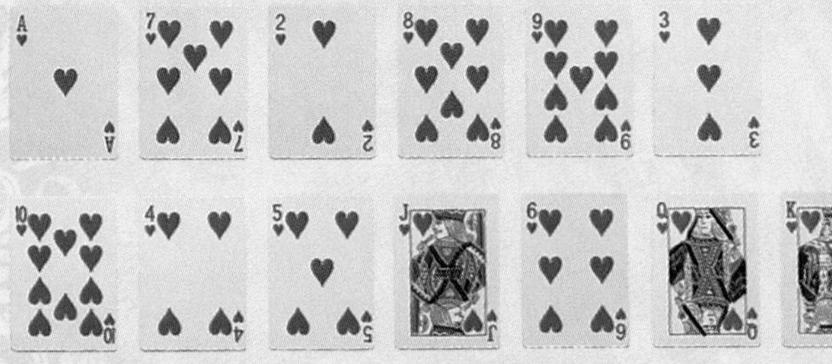

The cards are no longer ordered, but the sequence of ace through six remains, with cards from the other fragment interspersed, also in sequence. The larger the number of cards in play, the more shuffles required to mix them.

Curved lines show the odds that the first card in a deck will occupy any other position in the deck after one to seven shuffles. After one shuffle, for example, the first card is very likely to be one of the first few cards in the deck and very unlikely to be even five or six cards back. After four shuffles, it is still far more likely to be at the beginning of the deck than at the end. Only after seven shuffles does the card have about the same odds of being in any given position.

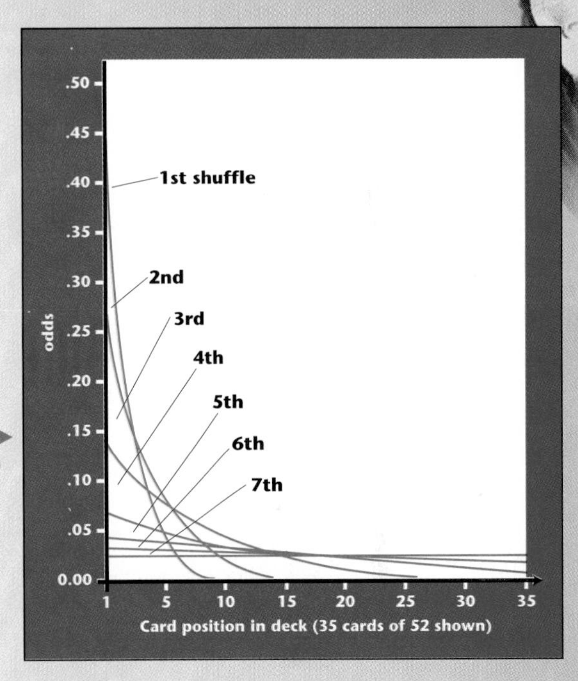

Cooperative Learning

1. Based on the graph in the article, how many times more likely is the first card to be in the first position than in the 35th position after 4 shuffles? Explain.
2. **a.** What is the difference between a perfect shuffle and an ordinary shuffle?
   b. Is the shuffle in the diagram above perfect? Explain.
3. Use a diagram or model to find out how many perfect shuffles it takes for a deck to return to its original order if the deck consists of
   a. 2^2 cards. **b.** 2^3 cards. **c.** 2^4 cards.
4. **a.** How many perfect shuffles do you think it would take for a deck of 32 cards to return to its original order? Why?
   b. Use a diagram or model to check your prediction. What did you find?
5. Would a machine that always made perfect shuffles be useful when you play a card game? Why or why not?

LESSON 6.4 Scientific Notation

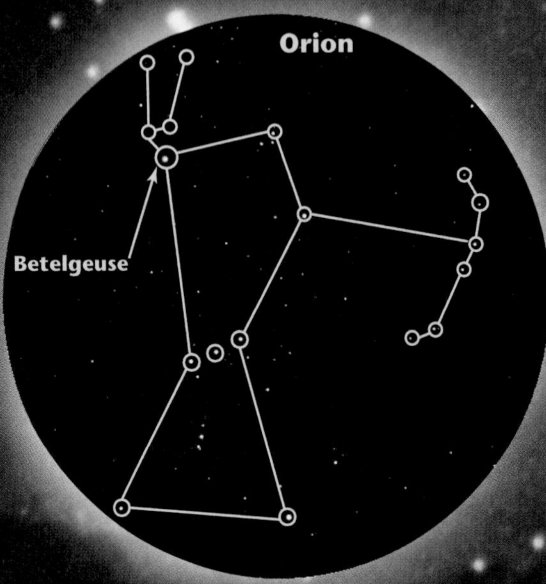

why *The study of the universe requires measurements of very large and very small quantities. The distance from Earth to one of the stars, for instance, is about 6,000,000,000,000,000 miles. The wavelength of an X-ray is approximately 0.0000000001 meters.*

These numbers, with all their zeros, are hard to read and to use in calculations. They do not even fit on a calculator display. A special way to express these large and small quantities in a compact form is *scientific notation.*

Astronomy Betelgeuse is one of the stars in the constellation Orion, named for a great hunter in Greek and Roman mythology. The approximate distance in miles from Earth to Betelgeuse is written as a 6 followed by 15 zeros. In scientific notation this distance is written 6×10^{15}.

A number written in **scientific notation** is written with two factors, a number from 1 to 10, including 1 but not including 10, and a power of 10.

6×10^{15}

First factor Number from 1 to 10

Second factor Power of 10

EXAMPLE 1

The distance from Earth to Proxima Centauri, the nearest star other than the Sun, is about 24,000,000,000,000 miles. It is read as 24 trillion miles. Write this number in scientific notation.

***Solution* ➤**

To express the first factor, place the decimal point after the 2, since the first factor, 2.4, must be a number between 1 and 10. This moves the decimal point 13 places to the left. To compensate for moving the decimal point, multiply the expression 2.4 by 10^{13}. This moves the decimal point 13 places to the right again. The two representations are now equivalent.

Customary Notation	Scientific Notation
24,000,000,000,000	2.4×10^{13} ❖

Try This Write 875,000 in scientific notation. 8.75×10^5

EXAMPLE 2

Astronomy Recall from page 286 that Betelgeuse is 6×10^{15} miles from Earth. If you travel at a speed of 1000 miles per hour, how long will it take to get to the star Betelgeuse?

***Solution A* ➤**

Find $6 \times 10^{15} \div 1000$. Using long division,

$$1000\overline{)6{,}000{,}000{,}000{,}000{,}000}\ \text{quotient } 6{,}000{,}000{,}000{,}000 = 6 \times 10^{12}.$$

The flight will take about 6 trillion hours! How many years is this?

About 685 million

***Solution B* ➤**

You can also use the properties of exponents.

$$6 \times 10^{15} \div 1000 = \frac{6 \times 10^{15}}{10^3} = 6 \times 10^{15-3} = 6 \times 10^{12} \text{ ❖}$$

Try This Find $3 \times 10^{28} \div 10^4$. 3×10^{24}

Numbers for small quantities can also be written in scientific notation. Recall that a decimal such as 0.01 can be written with a negative exponent, 10^{-2}.

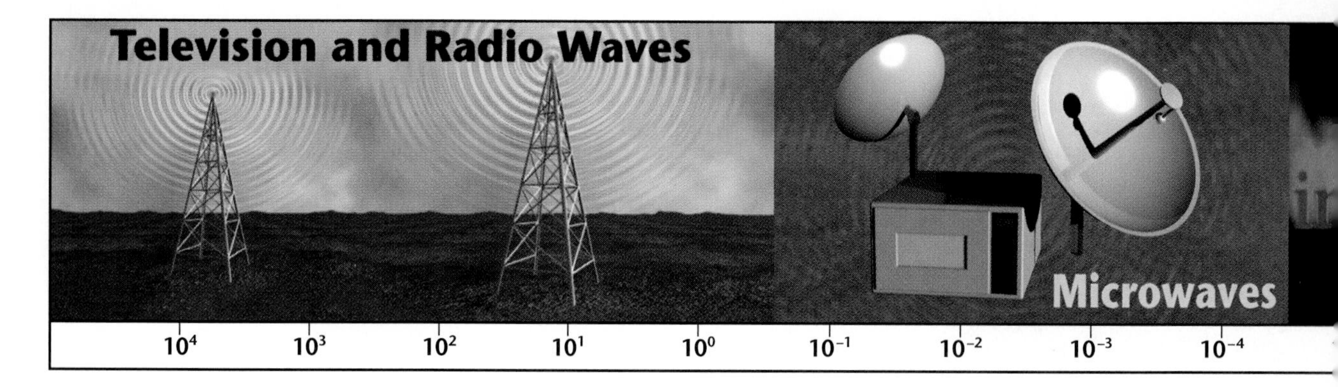

EXAMPLE 3

A typical X-ray has a wavelength of about 0.000000000125 meters. Write this number in scientific notation.

Solution ➤

Since the first factor must be a number between 1 and 10 (including 1), place the decimal after the 1. This, in effect, moves the decimal point 10 places to the right.

To compensate for moving the decimal point, multiply 1.25 by 10^{-10}. This is equivalent to dividing by 10^{10}, which moves the decimal point 10 places to the left, where it belongs.

Customary Notation		**Scientific Notation**
Thus, 0.000000000125	$=$	1.25×10^{-10} ❖

Try This Write 0.0000000402 in scientific notation. 4.02×10^{-8}

Cultural Connection: Europe About 300 years ago, the English mathematician, Sir Isaac Newton, passed a ray of sunlight through a prism. Newton observed the colors of the rainbow. Today we know that visible light is only part of a larger set of waves that make up the electromagnetic spectrum.

EXAMPLE 4

Physics How many times longer than an X-ray of 10^{-10} meters is a microwave of 10^{-2} meters?

Solution ➤

Divide the length of the microwave by the length of the X-ray.

$$\frac{10^{-2}}{10^{-10}} = 10^{-2-(-10)} = 10^{-2+10} = 10^8$$

The microwave is 10^8, or 100 million, times longer than the X-ray. ❖

How can you change 12×10^{-2} and 0.12×10^{-2} to scientific notation?

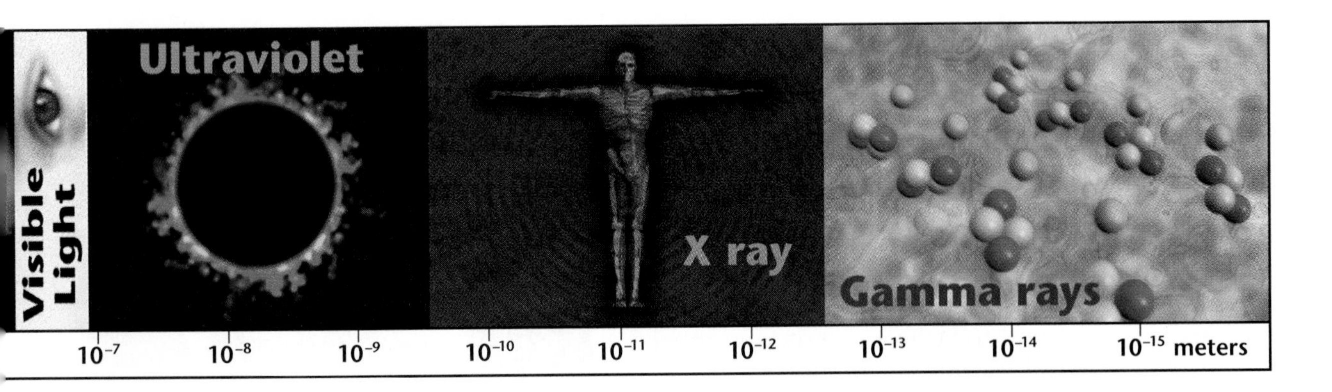

When you multiply or divide two numbers written in scientific notation, you are working with pairs of numbers that are each represented by two factors. Rearrange the factors to simplify the expression.

EXAMPLE 5

Express the product and quotient in scientific notation.

A $(3 \times 10^3)(4 \times 10^{-5})$ **B** $\dfrac{2.5 \times 10^6}{5 \times 10^2}$

Solution ➤

A $(3 \times 10^3)(4 \times 10^{-5}) = (3 \times 4)(10^3 \times 10^{-5})$
$= 12 \times 10^{-2}$
$= 1.2 \times 10^{-1}$

B $\dfrac{2.5 \times 10^6}{5 \times 10^2} = \left(\dfrac{2.5}{5}\right)(10^6 \times 10^{-2})$
$= 0.5 \times 10^4$
$= 5 \times 10^3$ ❖

Most computations with scientific notation can be done on a calculator. Check to see if the calculator you use has an EE key or EXP key.

EXAMPLE 6

Calculator

The speed of light is about 2.98×10^5 kilometers per second. If the Sun is about 1.49×10^8 kilometers from Earth, how long does it take sunlight to reach Earth?

Solution ➤

Divide 1.49×10^8 by 2.98×10^5. Depending on your calculator, use the EE key or EXP key.

1.49 EE 8 ÷ 2.98 EE 5 ENTER → 500

1.49 EXP 8 ÷ 2.98 EXP 5 = → 500

It takes about 500 seconds, or $8\frac{1}{3}$ minutes, for sunlight to reach Earth. ❖

EXERCISES & PROBLEMS

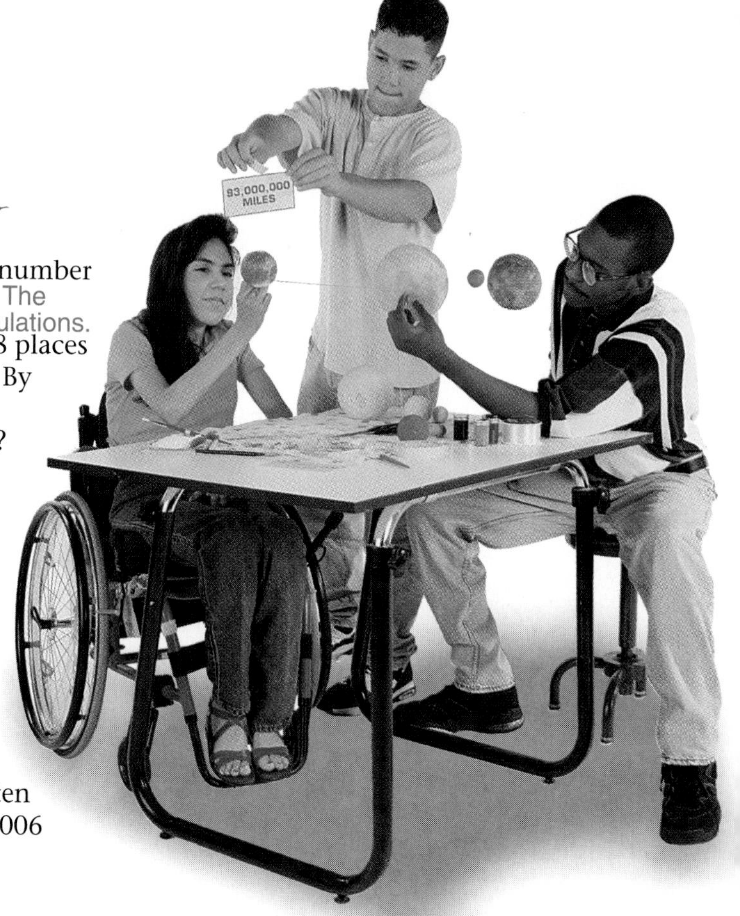

Communicate

1. Why is it important to be able to write a number with many zeros in scientific notation? The numbers are easier to read and use in calculations.
2. Explain what moving the decimal point 8 places to the left does to the value of a number. By what power of 10 must you multiply to compensate for that decimal point move?
3. Explain what moving the decimal point 12 places to the right does to the value of a number. By what power of 10 must you multiply to compensate for that decimal point move?
4. Explain how to write the distance in miles from Earth to the Sun (about 93,000,000 miles) in scientific notation.
5. Explain how to use the properties of exponents to multiply two numbers written in scientific notation. Use 240,000 and 0.006 in your explanation.

Practice & Apply

Write each number in scientific notation.

6. 2,000,000 2×10^6
7. 8,000,000,000 8×10^9
8. 340,000 3.4×10^5
9. 58,000 5.8×10^4
10. 0.00008 8×10^{-5}
11. 0.0000005 5×10^{-7}
12. 0.000234 2.34×10^{-4}
13. 0.000000082 8.2×10^{-8}

Write each number in customary notation.

14. 3×10^4 30,000
15. 4×10^8 400,000,000
16. 6.7×10^{10} 67,000,000,000
17. 9.01×10^5 901,000
18. 4×10^{-7} 0.0000004
19. 5×10^{-9} 0.000000005
20. 8.8×10^{-12}
21. 7.2×10^{-10}

20. 0.0000000000088
21. 0.00000000072

Social Studies The following information gives examples of large numbers that are used when studying different aspects of American culture. Write each number in scientific notation.

22. 125,000: the number of passengers bumped from United States airlines flights each year due to overbooking 1.25×10^5
23. 5 trillion: the total value of outstanding stock in the stock market 5×10^{12}
24. 5.4 million: the number of American businesses owned by women 5.4×10^6
25. 3.7 million: the number of square feet in the Pentagon, the world's largest office building 3.7×10^6

Evaluate each expression. Write your answers in scientific notation. 31. 1.96×10^{17}

26. $(9 \times 10^6)(7 \times 10^6)$ 6.3×10^{13}
27. $(9 \times 10^6) + (7 \times 10^6)$ 1.6×10^7
28. $(9 \times 10^6) - (7 \times 10^6)$ 2×10^6
29. $(2 \times 10^4)(3 \times 10^5)$ 6×10^9
30. $(6.34 \times 10^8)(1.1 \times 10^6)$ 6.974×10^{14}
31. $(8 \times 10^6)(2.45 \times 10^{10})$
32. $(8.2 \times 10^6)(3.1 \times 10^6)$ 2.542×10^{13}
33. $(1.9 \times 10^8)(2 \times 10^{10})$ 3.8×10^{18}
34. $\frac{8 \times 10^6}{2 \times 10^2}$ 4×10^4
35. $\frac{9 \times 10^8}{2 \times 10^4}$ 4.5×10^4
36. $\frac{3.6 \times 10^{10}}{6 \times 10^4}$ 6×10^5
37. $\frac{2 \times 10^6}{5.6 \times 10^5}$ ≈ 3.57 or $\approx 3.57 \times 10^0$
38. $\frac{(2 \times 10^6)(9 \times 10^4)}{6 \times 10^4}$ 3×10^6
39. $(2 \times 10^{10})(6 \times 10^7)(4 \times 10^{-8})$ 4.8×10^{10}

Economics The national debt for the United States in 1992 was listed as 4064.6 billion dollars. Use this information for Exercises 40–42.

40. Write this number in scientific notation. 4.0646×10^{12}
41. Express this number in customary notation. 4,064,600,000,000
42. Assuming that the population of the United States was about 250 million in 1992, how much debt was this per person? About \$16,000

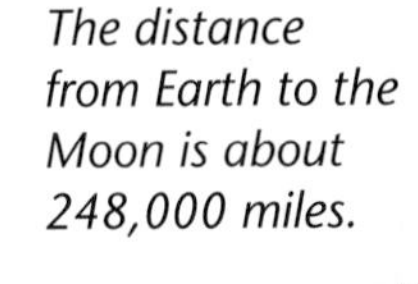
The distance from Earth to the Moon is about 248,000 miles.

43. Write your age at your last birthday in *seconds*. Use scientific notation. Answers may vary; 1 year = 3.1536×10^7 seconds.
44. **Astronomy** Write the distance in miles from Earth to the Moon (given at the right) in scientific notation. 2.48×10^5 miles

Astronomy The distance that light travels in one year, 5.87×10^{12} miles, is called a light-year.

45. Write this number in customary notation. 5,870,000,000,000 miles
46. If the distance from Betelgeuse to Earth is 6×10^{15} miles, how long does it take light to travel to Earth from that star? About 1.022×10^3 years
47. If the distance from Proxima Centauri to Earth is 2.4×10^{13} miles, how long does it take light to travel to Earth from Proxima Centauri? About 4.089 years

48. **Astronomy** According to Nobel Prize-winning physicist Leon Lederman, the universe is 10^{18} seconds old. How many years is this? Write your answer in words. 31.7 billion

Significant Digits Sometimes scientific notation is used to indicate a level of accuracy. For example, an attendance figure for a rock concert, given as 1.2×10^4, (or 12,000), would be assumed to be accurate to the nearest *thousand*, while a figure of 1.20×10^4 (also 12,000) would be assumed to be accurate to the nearest hundred. This is because the appearance of the 0 in 1.20 indicates that it is significant. Similarly, 1.2000×10^4 indicates that all the zeros are significant, so it is accurate to the nearest *one*. Write each in scientific notation to indicate which digits are significant. 55. 2×10^{-3} 56. 2.0×10^{-3}

49. 76,000 to the nearest thousand 7.6×10^4
50. 76,000 to the nearest hundred 7.60×10^4
51. 8,000,000 to the nearest million 8×10^6
52. 8,000,000 to the nearest thousand 8.000×10^6
53. 8,000,000 to the nearest hundred 8.0000×10^6
54. 8,000,000 to the nearest ten 8.00000×10^6
55. 0.002 to the nearest thousandth
56. 0.0020 to the nearest ten-thousandth

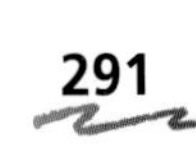

57. A **googol** is the number formed by writing 1 followed by 100 zeros. Write this number in scientific notation. 1×10^{100}

Technology Write the following numbers in customary notation.

58. 2.3 E 04 23,000 **59.** 5.6 E 03 5600 **60.** 7.22 E −03 0.00722

61. 1.01 E −04 0.000101 **62.** −2.8 E 02 −280 **63.** −9.303 E −04 −0.0009303

64. **Technology** What is the largest number that can be shown on your calculator without using scientific notation?

65. What is the largest number that can be shown on your calculator using scientific notation?

66. What is the smallest number that can be shown on your calculator without using scientific notation?

67. What is the smallest number that can be shown on your calculator using scientific notation?

64.–67. Answers may vary depending on the calculator. Check your calculator manual.

Look Back

68. Susan made the following test grades: 89, 97, 78, 86, 90. What grade does she need to make on her sixth test so that her test average is at least 90? **[Lesson 1.5]** 100

Solve each inequality and graph the solution. [Lessons 1.5, 1.6]

69. $5r + 7 < 6r - 9$ **70.** $7(g + 2) \leq -6g + 1$

71. $|8x + 1| \geq 17$ **72.** $|3t + 9| > -3$

73. **Coordinate Geometry** Given points $A(3, 7)$ and $B(-3, 12)$ on the coordinate plane, find the slope of the line containing the two points. **[Lesson 2.1]** $-\frac{5}{6}$

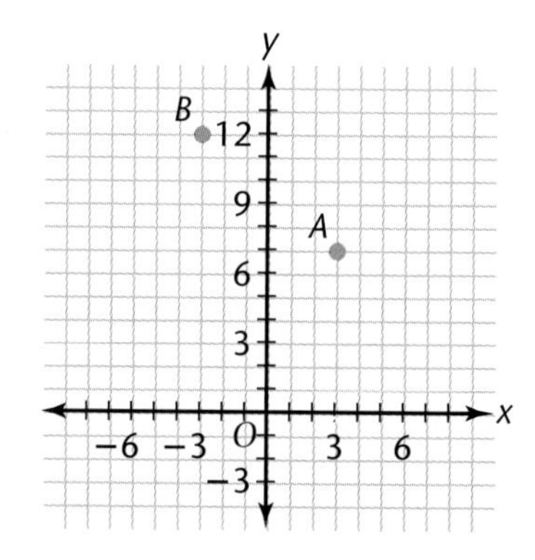

74. Graph this system of inequalities, and describe the solution region. **[Lesson 2.7]**

$$\begin{cases} x - y > 5 \\ -3x + y \leq 9 \end{cases}$$

Look Beyond

Technology Use the constant feature of your calculator to write the following sequences in customary notation.

75. $0.9^1, 0.9^2, 0.9^3, 0.9^4, 0.9^5, \ldots$ 0.9, 0.81, 0.729, 0.6561, 0.59049, . . .

76. $1.1^1, 1.1^2, 1.1^3, 1.1^4, 1.1^5, \ldots$ 1.1, 1.21, 1.331, 1.4641, 1.61051, . . .

77. Compare what happens when these numbers are raised to higher and higher powers. 0.9^n gets smaller, approaching 0; 1.1^n gets larger.

LESSON 6.5 Exponential Functions

Jan,1,1991 **The News** Weather Cold & Snow

World Population Hits 5.5 Billion-- Growing at a Rate of 1.7% per year

why *An exponential function can be used to model problems of growth and decline in population, finance, and science.*

Social Studies To better understand the effects of population growth, it is helpful to understand the underlying mathematics. This involves a look back to the exponential functions.

According to the headline above, how many people are added to the population in one year? How many are added in one minute?

$$\begin{aligned} 1.7\% \times 5.5 \text{ billion} &= 0.017 \times 5{,}500{,}000{,}000 \\ &= 93{,}500{,}000 \text{ people per year,} \\ &\text{or about } 256{,}000 \text{ people per day,} \\ &\text{or about } 10{,}700 \text{ people per hour,} \\ &\text{or about } 178 \text{ people per minute.} \end{aligned}$$

In the time it took you to read this page, the world population probably increased by several dozen people.

How was the increase for one day obtained from the increase for one year?

The figure for one year was divided by 365 and rounded to the nearest thousandth.

Population in Billions

2

1

Year 0 C.E. 250 C.E. 500 C.E.

EXAMPLE 1

A Estimate the world's population in the year 2000.

B If population growth continues at the same rate, how many people will be added to the population each minute in the year 2000?

Solution ➤

B3	=B2*1.017	
	A	B
1	Year	Population
2	1990	5.50
3	1991	5.59
4	1992	5.69
5	1993	5.79
6	1994	5.88
7	1995	5.98
8	1996	6.09
9	1997	6.19
10	1998	6.29
11	1999	6.40
12	2000	6.50

A Start with the 1990 population of 5.5 billion. Since the population increases by 1.7% each year, the new population will be 101.7%, or 1.017, times the old population. Use 1.017 as a multiplier to get the population for 1991, 1992, 1993, and so on.

If you round to the nearest 0.1 billion, the population in the year 2000 will be about 6.5 billion.

B If this new population of 6.5 billion increases by 1.7%, the number of people added in the year 2000 is calculated as follows:

$$1.7\% \times 6.5 \text{ billion} = 0.017 \times 6{,}500{,}000{,}000$$
$$= 110{,}500{,}000 \text{ people per year,}$$

or about 303,000 people per day,
or about 12,600 people per hour,
or about 210 people per minute.

This means that while the population was increasing by 178 people per minute in 1990, it will be increasing by 210 people per minute in 2000. ❖

What is the formula for finding the population if you know its rate of growth? Examine the pattern.

Years	Population (billions)	Exponential notation
0	5.5	$5.5(1.017)^0$
1	5.5(1.017)	$5.5(1.017)^1$
2	5.5(1.017)(1.017)	$5.5(1.017)^2$
3	5.5(1.017)(1.017)(1.017)	$5.5(1.017)^3$

Each year the world population increases by another factor of 1.017. The population, P in billions, after t years can be represented by

$$P = 5.5(1.017)^t.$$

The formula $P = 5.5(1.017)^t$ shows the population, P, after t years, for an original population is 5.5 (billion) and a rate of growth of 1.7% or 0.017. This leads to a general formula for yearly growth. This formula works for population, and many other situations in which growth takes place exponentially.

GENERAL GROWTH FORMULA

Let P be the amount after t years at a yearly growth rate of r, expressed as a decimal. If the original amount is A, then

$$P = A(1 + r)^t.$$

EXAMPLE 2

Social Studies India had an estimated population of about 886 million in 1992 and was growing at a yearly rate of 1.9%. How much will the population increase in 10 years?

Solution ➤

Use the growth formula. If you have a calculator, use the y^x or ^ key. You may also use a constant multiplier of 1.019 in your computation.

$$\begin{aligned} P &= A(1 + r)^t \\ &= 886(1.019)^{10} \\ &\approx 1069 \end{aligned}$$

The population increase in 10 years is $1069 - 886 = 183$, or about 183 million people. ❖

Try This Suppose that the population of a country was 200,000,000 inhabitants at the last census. If the population grows at an annual rate of 1.5%, by how many people will the population increase in the next 6 years? about 19 million people

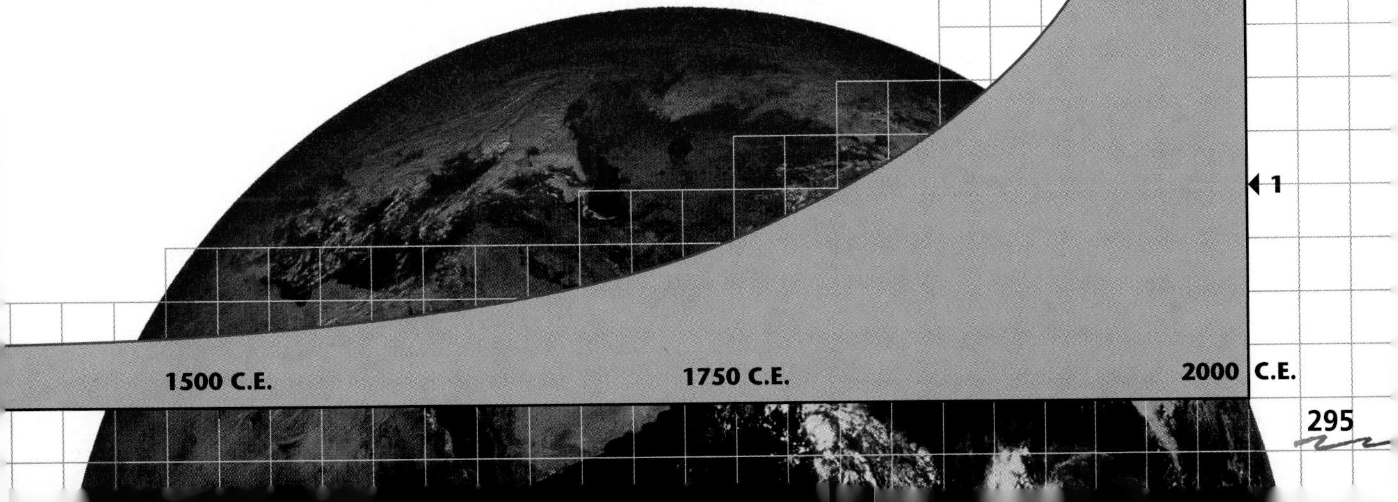

Population in Billions

8
7
6
5
4
3
2
1

1500 C.E.
1750 C.E.
2000 C.E.

EXAMPLE 3

Make a table of values. Graph and describe the function $f(x) = 2^x$. Show why it is a function.

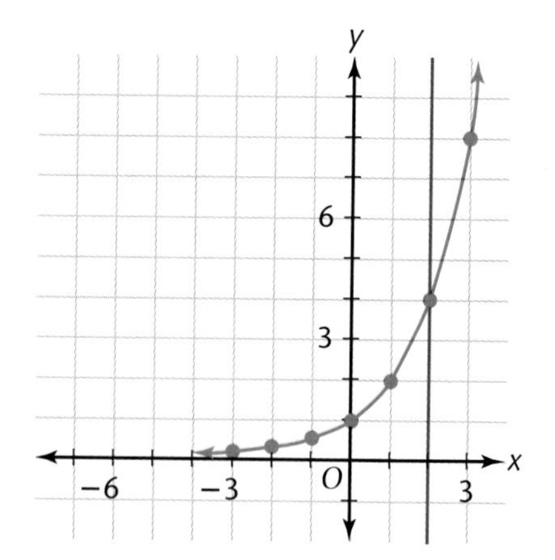

Solution ➤

x	−3	−2	−1	0	1	2	3
2^x	0.125	0.25	0.5	1	2	4	8

The function $f(x) = 2^x$ shows the characteristics of an exponential function. Since the graph remains above the x-axis for all values of x, the range is all y-values greater than zero. When x is 0 the function equals 1. After crossing the y-axis, the function increases rapidly.

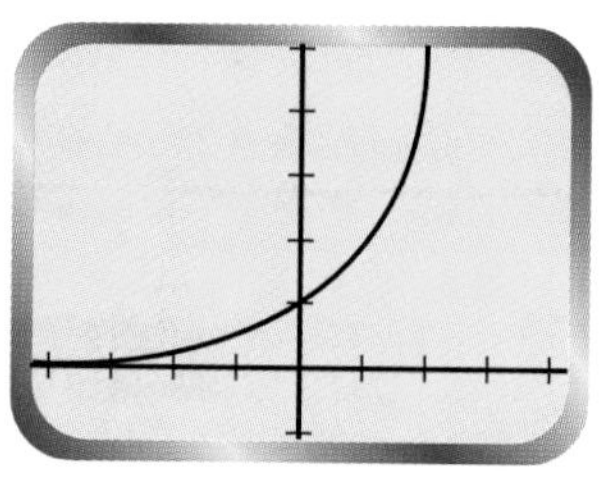

The vertical-line test shows that $f(x) = 2^x$ is a function.

A calculator produces the same distinctive curve. ❖

EXAMPLE 4

Graph $y = 3^x$ and $y = 4^x$ on the same axes. Explain how they are similar and how are they different.

Technology: Graphics Calculator

Solution ➤

You can graph each function on a graphics calculator, or you can make a table of values and graph them on paper. Let the domain be all integers.

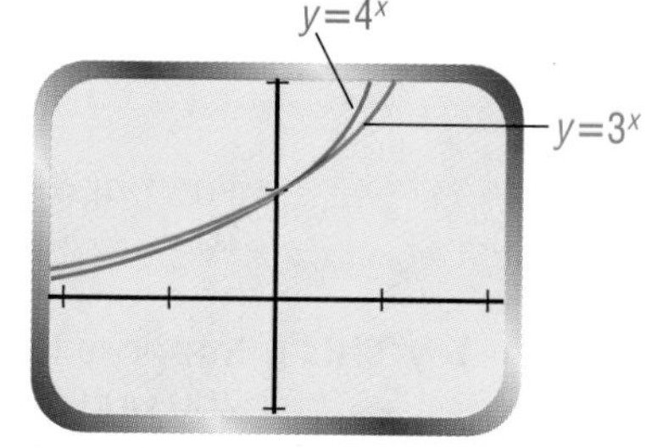

x	−3	−2	−1	0	1	2	3
3^x	0.037	0.11	0.33	1	3	9	27
4^x	0.0156	0.0625	0.25	1	4	16	64

The graphs of the functions are similar in that each of them rises from left to right. Each crosses the y-axis at $y = 1$. Each gets closer and closer to the x-axis as x decreases.

They are different in that the graph of $y = 4^x$ rises faster than the graph of $y = 3^x$ for values of x greater than 0. ❖

For the graphs of $f(x) = a^x$, for $a > 0$, as a increases, the graph rises faster.

Continue graphing on the same coordinate plane some powers with other bases, such as 5^x, 6^x, and so on. What do you notice about the pattern?

EXERCISES & PROBLEMS

Communicate

1. The population of Zimbabwe in 1995 was about 11 million and was growing at a rate of about 1.9% per year. Explain how to find the constant multiplier.
2. Explain how a graph can give you a clearer understanding of the effects of population growth.
3. Identify and explain each of the variables in the growth formula, $P = A(1 + r)^t$.
4. Describe how an exponential function changes as its base increases.
   As the base gets larger, the amount grows more quickly.
5. Describe the relationship between a constant multiplier on a calculator and an exponent in mathematics.
   The constant multiplier is the base to which the exponent is applied.
6. Explain why population growth and money that earns interest can be represented by the same formula.
   Both are increasing exponentially from an original amount at a given rate over a specified number of years.

Practice & Apply

Graph each exponential function.

7. $y = 2^x$ 8. $y = 0.4^x$ 9. $y = 5^x$ 10. $y = 10^x$ 11. $y = 0.5^x$ 12. $y = 0.1^x$

For Exercises 13–14, examine your graphs from Exercises 7–12.

13. For what values of b does the graph of $y = b^x$ rise from left to right? $b > 1$
14. For what values of b does the graph of $y = b^x$ fall from left to right? $0 < b < 1$

Graph each function below. In each case, describe the effect of the 3 in the transformation of the parent function $y = 2^x$.

15. $y = 2^x + 3$ 16. $y = 2^{x+3}$ 17. $y = 3 \cdot 2^x$ 18. $y = 2^{3x}$

19. Is the graph of $y = x^4$ an exponential function? Explain. No, the variable is not in the exponent.
20. Is the graph of $y = 4^x$ an exponential function? Explain. Yes, the variable is in the exponent.

Let $f(x) = 2^x$. Evaluate.

21. $f(3)$ 8 22. $f(0)$ 1 23. $f(-1)$ $\frac{1}{2}$ 24. $f(-4)$ $\frac{1}{16}$

Demographics The population of Spain was about 39 million in 1992 and was growing at a rate of about 0.3% per year.

25. What multiplier is used to find the new population each year? 1.003
26. Use this information to estimate the population in 1993. 39.117 million
27. Estimate the population in the year 2000. About 39.95 million

Demographics The population of Italy was about 58 million in 1995 and was growing at a rate of about 0.1% per year. The population of France was about 57.8 million in 1995 and was growing at a rate of about 0.4% per year. Use this information to estimate each population.

28. Italy in 1996 About 58.058 million

29. Italy in 1998 About 58.174 million

30. Italy in 2000 About 58.291 million

31. France in 1996 About 58.031 million

32. France in 1998 About 58.496 million

33. France in 2000 About 58.965 million

34. In what year might the population of France be greater than the population of Italy? 1997

35. **Portfolio Activity** Complete the problem in the portfolio activity on page 265.

The first payment of 3% on an amount of $1015 is $30.45. After 12 payments, $842.01 is still owed. In approximately 9 years and 10 months, the amount paid back is $1779 (rounded to the nearest dollar).

Look Back

Refer to matrix *A*. $A = \begin{bmatrix} 1 & -3 \\ 5 & 2 \end{bmatrix}$ **[Lesson 3.3]**

36. What is A^2? **37.** What is A^3?

36. $\begin{bmatrix} -14 & -9 \\ 15 & -11 \end{bmatrix}$ 37. $\begin{bmatrix} -59 & 24 \\ -40 & -67 \end{bmatrix}$

38. Explain the characteristic that a relation must have in order to be a function. **[Lesson 5.1]** For each x-value, there is only one y-value.

39. What must happen to the parent function for its graph to be vertically reflected through the x-axis? **[Lesson 5.4]**
$-f(x)$ is a reflection of $f(x)$ through the x-axis; the y-values become the opposite (multiplied by -1).

Evaluate each expression. [Lesson 6.3]

40. 10^{-2} $\frac{1}{100}$

41. -10^2 -100

42. $(-10)^2$ 100

43. $(-10)^{-2}$ $\frac{1}{100}$

Look Beyond

Cultural Connection: Europe Maria Gaetana Agnesi was born in Italy in 1718. She is considered to be the first woman in modern times to achieve a reputation for mathematical work. She wrote her best-known work in 1748; in this work she first discussed a curve that is now known as the **Agnesi curve**.

This curve is represented by the function $f(x) = \frac{a^3}{x^2 + a^2}$. Let the domain of f be $\{-3, -2, -1, 0, 1, 2, 3\}$.

44. Graph the function when a is 1.

45. Graph the function when a is 5.

46. How do the graphs compare?

46. Both graphs have a maximum value when $x = 0$ and are symmetric about the y-axis.

LESSON 6.6 Applications of Exponential Functions

why *How old is a dinosaur bone? How much money should be invested to reach a savings target? These are a few of the many questions that can be answered with the use of exponential functions.*

Carbon-14 dating

Exponential functions can be used to determine the age of fossils and archeological finds. This method is based on the fact that the amount of radioactivity in an object decreases, or decays, over time.

Paleontology Carbon-14 dating is a reliable method of determining the age of objects up to 40,000 years old. Carbon-14 is a radioactive form of carbon that has a half-life of 5700 years. This means that half of the substance is converted to nonradioactive carbon every 5700 years. For example, it can be concluded that an object with half as much carbon-14 as its living counterpart died 5700 years ago. Every 5700 years, half of the remaining amount of carbon-14 decays.

Years	Fraction of carbon-14 remaining
0	$\left(\frac{1}{2}\right)^0 = 1$
5700	$\left(\frac{1}{2}\right)^1 = \frac{1}{2}$
11,400	$\left(\frac{1}{2}\right)^2 = \frac{1}{4}$
17,100	$\left(\frac{1}{2}\right)^3 = \frac{1}{8}$
22,800	$\left(\frac{1}{2}\right)^4 = \frac{1}{16}$
28,500	$\left(\frac{1}{2}\right)^5 = \frac{1}{32}$
34,200	$\left(\frac{1}{2}\right)^6 = \frac{1}{64}$

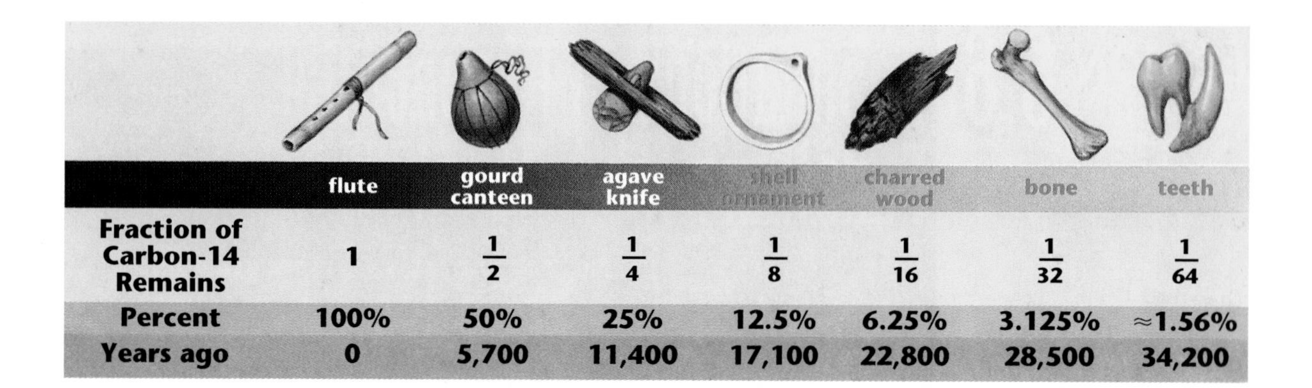

	flute	gourd canteen	agave knife	shell ornament	charred wood	bone	teeth
Fraction of Carbon-14 Remains	1	$\frac{1}{2}$	$\frac{1}{4}$	$\frac{1}{8}$	$\frac{1}{16}$	$\frac{1}{32}$	$\frac{1}{64}$
Percent	100%	50%	25%	12.5%	6.25%	3.125%	≈1.56%
Years ago	0	5,700	11,400	17,100	22,800	28,500	34,200

EXAMPLE 1

Use the table to estimate the age of an object that has 10% of its original carbon-14.

Solution ➤

Since 10%, or $\frac{1}{10}$, is between $\frac{1}{8}$ and $\frac{1}{16}$, the age is between 17,100 and 22,800 years. A reasonable estimate is about 19,000 years old. ❖

EXAMPLE 2

Savings Suppose that you wish to save $1000 for a down payment on a car and that you want to buy the car when you graduate in 3 years. The current interest rate for savings is 5% compounded annually. How much will you need to invest in a savings account now to buy the car in 3 years?

Solution ➤

If A represents the original amount, then $P = A(1 + r)^t$ is the formula for compound interest. Let P represent the amount of money after t years. Let r represent the yearly rate of growth (expressed as a decimal) compounded once each year. Substitute the values in the formula, and solve for A.

$$\begin{aligned} P &= A(1 + r)^t \\ 1000 &= A(1.05)^3 \\ 1000 &= A(1.157625) \\ \frac{1000}{1.157625} &= A \\ 863.84 &\approx A \end{aligned}$$

You will need to invest about $863.84. ❖

The example can also be solved by working backward. Think of the $1000 as being invested 3 years ago, and use −3 as the number of years. Substitute the values into the formula, and then solve for P.

$$\begin{aligned} P &= A(1 + r)^t \\ P &= 1000(1.05)^{-3} \\ P &\approx 863.84 \end{aligned}$$

EXAMPLE 3

Investments If the sculpture is now worth $14,586.08, what was its value 4 years ago?

The value of a sculpture has been growing at a rate of 5% per year for 4 years.

Solution ➤

Use the formula. Since you are trying to find the value 4 years ago, substitute −4 for t.

$$\begin{aligned} P &= A(1 + r)^t \\ &= 14{,}586.08(1.05)^{-4} \\ &\approx 14{,}586.08(0.8227) \\ &\approx 12{,}000 \end{aligned}$$

The value was about $12,000. ❖

Try This An art museum recently sold a painting for $4,500,000. The directors claim that its value has been growing at a rate of 7% per year for the last 50 years. What did the museum pay for the painting 50 years ago? about $152,765

EXAMPLE 4

Ginny has a 90% probability of getting a given step correct in an algebra problem. She does 100 problems, each with 2 steps.

A In how many problems can you expect her to get both steps correct?

B Calculate the probability if the problems have 3 steps.

C Calculate the probability if the problems have 10 steps.

Solution ➤

A In 100 problems you can expect her to get the first step correct 90 times. Of these 90 times, she would get the second step correct 90% of the time. Since 90% of 90 is 81, she would get both steps correct 81 times out of 100. Notice that the answer is $100(0.90)^2$.

B From the solution in Part A, you know that in 100 problems she would have the first 2 steps correct 81 times. She would also get the third step correct 90% of these times, or about 73 times. Notice that the answer is $100(0.90)^3$.

C By the same reasoning, the answer is $100(0.90)^{10}$. If you use the exponent key on a calculator, you get approximately 35. Thus, she gets only about 35% correct when doing 10-step problems, even though she has a 90% chance of getting any particular step correct. ❖

How many steps would a problem have to have for Ginny to have less than a 20% probability of getting the problem correct? over 15 steps

In this lesson you have seen several examples of *exponential growth* and *exponential decay*. The general formula for exponential growth gives the amount, P, when A is the initial amount, r is the percent of increase expressed in decimal form, and t is the time expressed in years, or

$$P = A(1 + r)^t$$

For **exponential growth**, r is positive, but when **exponential decay** or decrease takes place, the value of r is negative.

EXAMPLE 5

Demographics

The population of a city in the United States was 152,494 in 1990 and was *decreasing* at a rate of about 1% per year. When will the city's population fall below 140,000? Solve the problem using the given method.

A with a graph **B** with a table

Solution ➤

Each year the population will be 99% of what it was the previous year, so the multiplier is 1.00 − 0.01, or 0.99. The population x years after 1990 is given by $y = 152{,}494(0.99)^x$.

A Graph $Y_1 = 152{,}494(0.99)^x$ and $Y_2 = 140{,}000$, and find the point of intersection. The y-value of 140,000 occurs when x is approximately 8.5. Therefore, the city's population will fall below 140,000 after about $8\frac{1}{2}$ years.

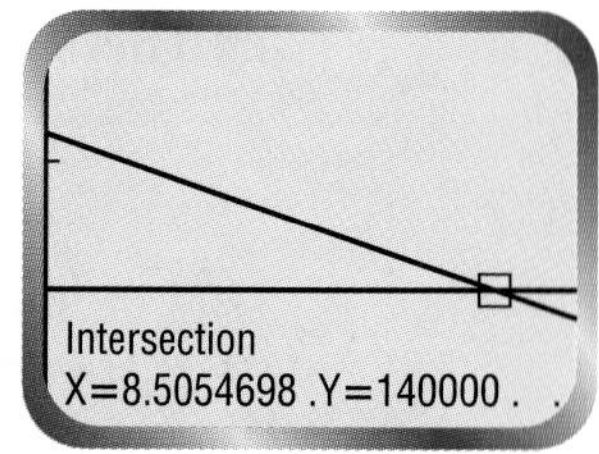

B The population decreases by a factor of 0.99 each year. Use the constant feature on a calculator to generate a table of values for the population. Continue the table until you reach the year in which the value falls below 140,000.

x	$y = 152{,}494(0.99)^x$ Population	x	$y = 152{,}494(0.99)^x$ Population
0	152,494	5	145,020
1	150,969	6	143,570
2	149,459	7	142,134
3	147,965	8	140,713
4	146,485	9	139,306

According to the table, the population falls below 140,000 between 8 and 9 years. ❖

In Chapter 5 you learned how to use transformations to translate the graph of a parent function. Graphs that represent the exponential function $y = a^x$ can be translated in the same way.

EXAMPLE 6

Compare the graphs of the transformations with the graph of the parent function $y = 2^x$.

A $y = 2^x - 3$ **B** $y = 2^{x-3}$

Solution ➤

Graph each function.

Parent function
$y = 2^x$

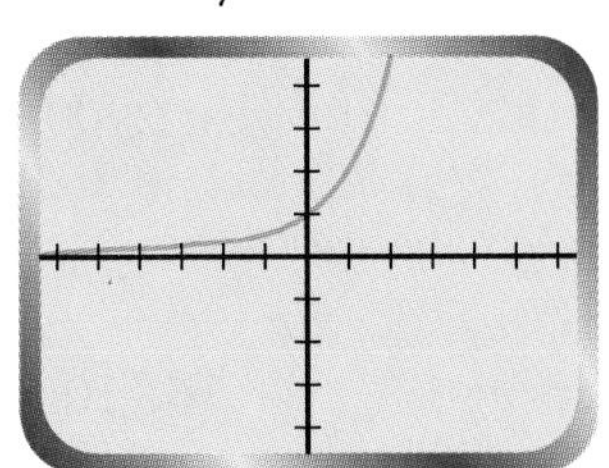

A $y = 2^x - 3$

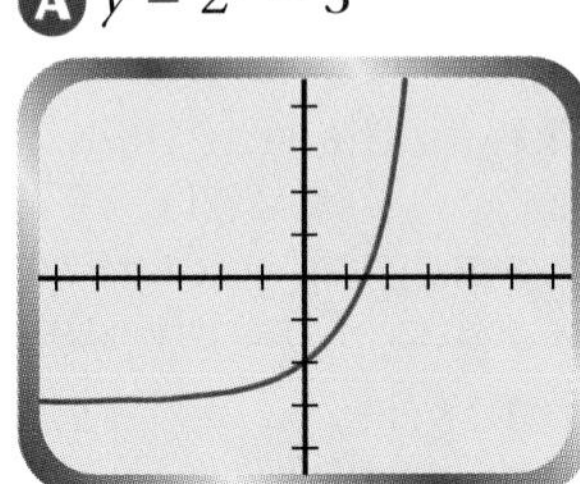

B $y = 2^{x-3}$

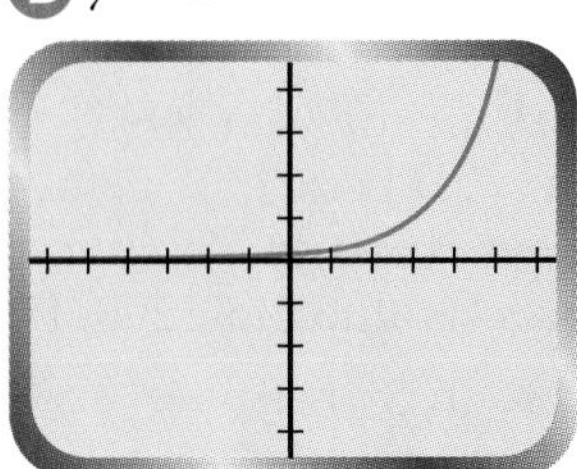

Note that all of the graphs have the same shape.

A The graph of $y = 2^x - 3$ is shifted *down* 3 units from the graph of $y = 2^x$. Each y-value of $y = 2^x - 3$ is 3 units less than the corresponding y-value of $y = 2^x$.

B The graph of $y = 2^{x-3}$ is shifted 3 units *to the right* of the graph of $y = 2^x$. ❖

EXERCISES & PROBLEMS

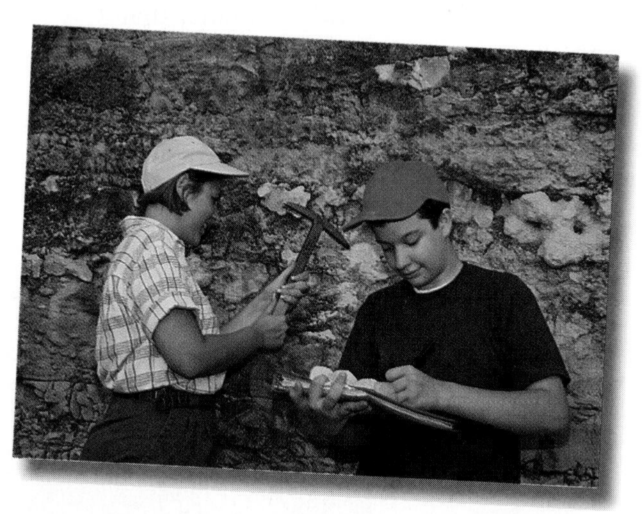

Communicate

1. List some real-life situations that are described by exponential functions.
2. Describe the formula for exponential decay.
3. Explain why r is negative in the formula for exponential decay.
4. Tell how a horizontal translation is shown in the function rule for exponential functions.
5. Tell how a vertical translation is shown in the function rule for exponential functions.
6. Choose a base, and form an exponential sequence of five or six terms. Describe what you find when you examine the differences.

Practice & Apply

Chemistry Use the method of carbon-14 dating to estimate the age of an object that has

7. 25% of its original carbon-14 remaining. About 11,400 years

8. 1% of its original carbon-14 remaining. About 37,000 years

9. **Technology** The population of the metropolitan area of a city was 361,280 in 1993 and was *decreasing* at a rate of about 0.1% per year. Draw a graph of the population after x years.

Demography Use your graph from Exercise 9 to estimate the population of the city

10. in the year 1995. About 360,558 people

11. in the year 1998. About 359,477 people

12. in the year 2000. About 358,759 people

Investments An investment is growing at a rate of 8% per year and now has a value of $8200. Find the value of the investment

13. in 5 years. About $12,048

14. in 10 years. About $17,703

15. 5 years ago. About $5581

16. 10 years ago. About $3798

Investments An investment is losing value at a rate of 2% per year and now has a value of $94,000. Find the value of the investment

17. in 5 years. About $84,969

18. in 10 years. About $76,805

19. 5 years ago. About $103,991

20. 10 years ago. About $115,045

21. **Probability** If Ginny, from Example 4 on page 301, does 100 4-step problems, in how many problems will she get *all* of the steps correct? About 66 questions

22. **Probability** If the x-value in 2^x is chosen randomly from the list 1, 2, 3, 4, and 5, find the probability that 2^x is greater than 10. $\frac{2}{5}$

23. Repeat Exercise 22 if the x-value is chosen from the list 1, 2, 3, . . . , 100. $\frac{97}{100}$

The digits of a repeating decimal such as $\frac{1}{3} = 0.33333\ldots$ can be described in terms of an exponential function. Evaluate each expression.

24. $3\left(\frac{1}{10}\right)^1$ 0.3

25. $3\left(\frac{1}{10}\right)^2$ 0.03

26. $3\left(\frac{1}{10}\right)^3$ 0.003

27. $3\left(\frac{1}{10}\right)^4$ 0.0003

Write the digit in the sixth decimal place of

28. $\frac{1}{3}$. 3 **29.** $\frac{2}{3}$. 6 **30.** $\frac{4}{9}$. 4

Technology Graph each function, and describe the effect of 5 in the transformation of the parent function $y = 10^x$.

31. $y = 10^x + 5$ **32.** $y = 10^{x+5}$ **33.** $y = 5 \cdot 10^x$ **34.** $y = 10^{5x}$

Look Back

35. Jamie bought tapes and cassettes and paid the cashier $42.75. If the cost of the items before tax was $39.33, what was the tax rate? **[Lesson 1.2]** About 8.7%

36. Solve $6(2x - 9) = -3(x + 6)$. **[Lesson 1.4]** $x = 2\frac{2}{5}$

37. **Probability** What is the probability of rolling an ordinary 6-sided number cube once and getting a prime number? **[Lesson 4.1]** $\frac{1}{2}$

38. **Probability** What is the probability of rolling an even number on a number cube and then getting heads on a coin toss? **[Lesson 4.1]** $\frac{1}{4}$

39. There are 44 students who signed up for band or art. Of those students, 21 chose band, and 29 chose art. How many students are taking both band and art? **[Lesson 4.4]** 6 students

40. Tamitra flips a coin, records the result, and then flips it again. She wants to know the probability of heads on both tosses. Are the events independent? How do you know? **[Lesson 4.7]**

41. Describe the differences between the graph of $y = 2^x$ and the graph of $y = 2^{-x}$. **[Lesson 6.3]** If $x > 0$, then $y = 2^x$ increases from left to right; $y = 2^{-x}$ decreases from left to right.

Look Beyond

42. **Technology** A number that is often used to describe growth of natural phenomena and money is e, which is about 2.718. Find e^x on your calculator. Write the first 8 digits of the decimal expansion. Use the e^x key and 1 for x. Then draw the graph of $y = e^x$.

43. Compare.

a. The value of $1000 invested at 8% interest compounded *once a year* is given by $y = 1000(1.08)^x$, where x is the number of years. Find the value after 10 years. About $2159

b. The value of $1000 invested at 8% interest compounded *continuously* is given by $y = 1000e^{0.08x}$. Find the value after 10 years. About $2226; this method earns more interest.

Please Don't Sneeze

Suppose that one person in your class gets the flu. While contagious, this person makes contact with two other people so that two more people are exposed to the virus. If these two have not yet had the flu, assume that each of them will now get it and make contact with two more people. If these people have already had contact, they are immune and will not pass it on. Use a simulation to find out how many people in the class will get the flu before the epidemic is over.

To do this, assign each person in your class a number. Then use random numbers from a calculator or a computer spreadsheet or random numbers drawn from a hat to see who is exposed to the flu virus according to the conditions given above.

The formula INT(RAND*30) + 1 will generate a random number from 1 to 30, where RAND is a random number from 0 to 1, not including 0 or 1. (If your class has 28 students, replace the 30 in the formula by 28, etc.) Summarize the results in a tree diagram.

In the example, the following 13 random numbers were initially generated. 10, 23, 12, 25, 29, 25, 4, 5, 29, 23, 9, 3, 5 . . .

The first random number was 10, so student 10 was the first to get the flu.

The next two random numbers were 23 and 12. This means that while contagious, student 10 had contact with students 23 and 12, who also got the flu.

The next two random numbers were 25 and 29, indicating that student 23 passed the flu on to students 25 and 29.

The next two random numbers were 25 and 4, indicating that student 12 had contact with 25 and 4. Since student 25 already had the virus, this branch ends.

The next random numbers generated were 5, 29, 23, 9, 3, and 5, so the tree diagram so far looks like this. Remember that a circle around a number indicates that the branch ends.

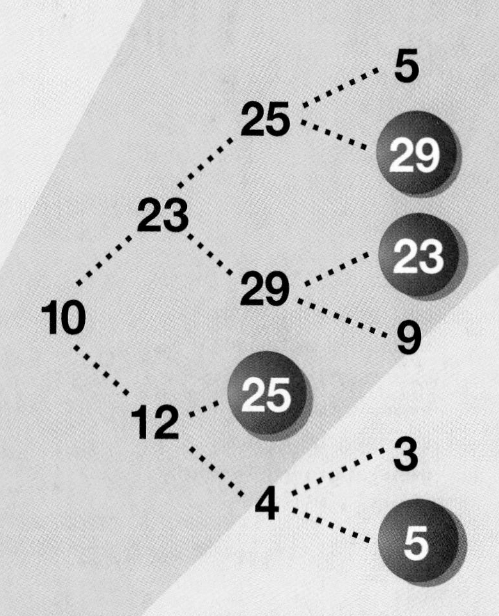

When the above simulation was continued, all students except 1, 16, 17, 18, 22, 27, 28, and 30 got the virus before all the branches ended. Thus, 22 of the 30 students got the flu.

Activity

Conduct a simulation of an epidemic. Then pool your data for the class, and summarize the results. What is the average number of people who are infected before the simulated epidemic comes to an end?

Chapter 6 Review

Vocabulary

Key Skills & Exercises

Lesson 6.1

➤ Key Skills

Use the definition of exponent to simplify expressions.

$3^5 = 3 \cdot 3 \cdot 3 \cdot 3 \cdot 3 = 243$

$4^3 = 4 \cdot 4 \cdot 4 = 64$

$5^4 = 5 \cdot 5 \cdot 5 \cdot 5 = 625$

The exponent in a power of 10 tells you the number of zeros to use when writing the number in decimal notation.

$10^4 = 10 \cdot 10 \cdot 10 \cdot 10 = 10{,}000$

➤ Exercises

Simplify each expression.

1. 6^3 216 **2.** 5^5 3125 **3.** 10^8 100,000,000 **4.** 10^6 1,000,000

Lesson 6.2

➤ Key Skills

Use the Product-of-Powers Property or Quotient-of-Powers Property to simplify expressions.

Product of Powers

$$(6x^2y^3)(-3x^4y) = (6 \cdot -3)(x^{2+4})(y^{3+1}) = -18x^6y^4$$

Quotient of Powers

$$\frac{15x^4y^3z}{3xy^2} = (5)(x^{4-1})(y^{3-2})(z) = 5x^3yz$$

Use Power-of-a-Power Property or Power-of-a-Product Property to simplify expressions.

Power of a Power

$$(x^5)^3 = x^5 \cdot x^5 \cdot x^5 = x^{5 \cdot 3} = x^{15}$$

Power of a Product

$$(4x^4y^3)^2 = (4^2)(x^{4 \cdot 2})(y^{3 \cdot 2}) = 16x^8y^6$$

➤ *Exercises*

Simplify each expression.

5. $(2t^4)(3t^2)$ $6t^6$ **6.** $(-a^2b^3)(4ab^2)$ $-4a^3b^5$ **7.** $(-0.1w^2x^2y)(-0.5w^5)$ $0.05w^7x^2y$

8. $\frac{10w^5}{2w^2}$ $5w^3$ **9.** $\frac{-21wx^2y^4}{-3xy^3}$ $7wxy$ **10.** $(r^3)^4$ r^{12}

11. $(-3c^2)^3$ $-27c^6$ **12.** $(2p^2q^3)^2$ $4p^4q^6$ **13.** $(m^4n^2)^2(m^3)^4$ $m^{20}n^4$

Lesson 6.3

➤ *Key Skills*

Simplify expressions containing negative exponents.

$$5x^{-3}y^2 = \frac{5y^2}{x^3}$$

$$\frac{m^{-3}p^2}{n^{-2}p^2} = \frac{n^2}{m^3}$$

Simplify expressions containing zero exponents.

$$(3x)^0 = 1$$

$$7a^0b^2 = 7b^2$$

➤ *Exercises*

Write each expression without a negative or zero exponent.

14. 3^{-2} $\frac{1}{9}$ **15.** a^2b^{-3} $\frac{a^2}{b^3}$ **16.** $(-3)^0$ 1 **17.** $a^0b^{-1}c^2$ $\frac{c^2}{b}$

18. $(2c^2d^{-1})^0$ 1 **19.** $-q^{-3}q^2$ $-\frac{1}{q}$ **20.** $\frac{b^{-2}}{b^3}$ $\frac{1}{b^5}$ **21.** $\frac{t^{-2}u}{t^{-4}u^2}$ $\frac{t^2}{u}$

Lesson 6.4

➤ *Key Skills*

Write numbers in scientific notation.

Customary notation

a. 32,000,000,000

b. 0.000000784

Scientific notation

a. 3.2×10^{10}

b. 7.84×10^{-7}

Perform computations with numbers written in scientific notation.

$$\begin{aligned}(3 \times 10^{-3})(4 \times 10^8) &= (3 \cdot 4)(10^{-3} \times 10^8)\\ &= 12 \times 10^5\\ &= 1.2 \times 10^6\end{aligned}$$

$$\frac{4 \times 10^6}{2 \times 10^3} = \left(\frac{4}{2}\right)(10^{6-3}) = 2 \times 10^3$$

➤ *Exercises*

Write each number in scientific notation.

22. 5,900,000 5.9×10^6 **23.** 368,000,000,000 3.68×10^{11} **24.** 0.0000075 7.5×10^{-6}

Write each number in customary notation.

25. 2×10^3 2000 **26.** 7.9×10^{-5} 0.000079 **27.** 8.34×10^9 8,340,000,000

Evaluate each expression. Write your answer in scientific notation. **28.** 1.5×10^8

28. $(3 \times 10^2)(5 \times 10^5)$ **29.** $(2.1 \times 10^5)(4 \times 10^{-3})$ 8.4×10^2 **30.** 1×10^3

30. $(8 \times 10^2) + (2 \times 10^2)$ **31.** $(9 \times 10^5) - (3 \times 10^5)$ 6×10^5

32. $\frac{9 \times 10^7}{3 \times 10^2}$ 3×10^5 **33.** $\frac{8 \times 10^4}{2 \times 10^{-2}}$ 4×10^6

Lesson 6.5

➤ *Key Skills*

Use exponential functions to estimate population growth.

The population of a country was estimated to be about 750 million in 1990 and was growing at a yearly rate of about 1.7%. At this rate of growth, the population of this country in 10 years will be $P = 750(1 + 0.017)^{10} \approx 888{,}000{,}000$, or about 888 million people.

➤ *Exercises*

The population of New Zealand was about 3.3 million in 1991 and was growing at a rate of about 0.8% per year. Estimate the population

34. in the year 1998.
About 3.49 million

35. in the year 2000.
About 3.55 million

Lesson 6.6

➤ *Key Skills*

Use exponential functions to estimate growth or decay.

An investment is losing value at a rate of 3% per year and now has a value of $6500. The value of the investment in 5 years will be $y = 6500(1 - 0.03)^5 \approx \$5581.77$. The value of the investment 5 years ago was $y = 6500(1 - 0.03)^{-5} \approx \$7569.28$.

➤ *Exercises*

An investment is losing value at a rate of 5.5% per year and now has a value of $6300. Find the value of the investment at the given times.

36. in 5 years $4747.88

37. in 10 years $3578.15

38. 5 years ago $8359.52

39. 10 years ago $11,092.32

Applications

40. **Geometry** Suppose that the radius of a circle is length r. Then the area of the circle is πr^2. Find the area of another circle with a radius 5 times as long as the radius of the original circle.
$25\pi r^2$

41. **Astronomy** The Sun has a diameter of about 864,000 miles. Write this number in scientific notation.
8.64×10^5

42. **Astronomy** The interior temperature of the Sun is about 3.5×10^7°F. Write this number in decimal notation.
35,000,000°F

43. **Time** The calendar year is increasing at a rate of about 1.7×10^{-6} minutes each year. How many seconds is this? Write your answer in scientific notation.
About 1.02×10^{-4}

Probability Suppose that a number x is chosen randomly from the list −1, −2, −3, −4, and −5. Find the probability that

44. 2^x is less than 0.5. 80%

45. 3^x is greater than $\frac{1}{27}$. 40%

46. 10^x is greater than 0.001. 40%

47. $\frac{1}{2^x}$ is less than 16. 60%

Chapter 6 Assessment

1. What is a monomial? A monomial is a constant, a variable, or the product of a constant and one or more variables.
2. The product of two powers can be found by adding the exponents. What must be true about these powers in order to use this rule? The powers must have the same base.

Simplify each expression.

3. $(2y)^3$ $8y^3$
4. $(2f^3)(3f^4)$ $6f^7$
5. $(-x^3y)(5x^2y^2z)$ $-5x^5y^3z$
6. $\frac{36t^5}{6t^2}$ $6t^3$
7. $\frac{-2w^3z^4}{-3wz^5}$ $\frac{2w^2}{3z}$
8. $\frac{ab^4}{-2b^3}$ $-\frac{ab}{2}$
9. $(g^3)^4$ g^{12}
10. $(4c^2)^5$ $1024c^{10}$
11. $6(2cd^3)^2$ $24c^2d^6$

12. What is the difference between an exponent of 0 and an exponent of 1?
Exponent 0: the value of the power is 1. Exponent 1: the value of the power is the base.

Write each expression without a negative or zero exponent.

13. 2^{-4} $\frac{1}{2^4}$ or $\frac{1}{16}$
14. $a^{-3}b^2$ $\frac{b^2}{a^3}$
15. $m^{-4}n^0p^4$ $\frac{p^4}{m^4}$
16. $-2t^{-2}$ $-\frac{2}{t^2}$
17. y^5y^{-3} y^2
18. $\frac{3h^{-2}}{h^{-1}}$ $\frac{3}{h}$

19. What is the meaning of a negative exponent in a number that is written in scientific notation? The value of the number is less than 1.

Taxes In 1970, the IRS collected about 196 million dollars in taxes. In 1992, about 1.1 billion dollars was collected.

20. Write 196 million in scientific notation. 1.96×10^8
21. Write 1.1 billion in scientific notation. 1.1×10^9
22. How much more did the IRS collect in 1992 than in 1970? Write your answer in scientific notation and in customary notation.
9.04×10^8; $904,000,000

Evaluate each expression. Write your answer in scientific notation.

23. $(3 \times 10^{-2})(5 \times 10^5)$
1.5×10^4
24. $\frac{3 \times 10^6}{5 \times 10^4}$
6×10^1

Demographics The population of Peru in 1992 was about 23 million and was growing at a rate of 2% per year.

25. What is the multiplier to get the new population each year? 1.02
26. Estimate the population for the year 1997. 25.4 million
27. Estimate the population for the year 2000. About 26.9 million

Investments An investment is losing money at a rate of 0.5% per year and 5 years ago had a value of $8500. Find the value of the investment

28. 10 years ago. About $8716
29. now. $8290
30. in 5 years. About $8084
31. in 10 years. About $7884

Chapters 1 – 6 Cumulative Assessment

College Entrance Exam Practice

Quantitative Comparison For Questions 1–4, write
A if the quantity in Column A is greater than the quantity in Column B;
B if the quantity in Column B is greater than the quantity in Column A;
C if the two quantities are equal; or
D if the relationship cannot be determined from the information given.

	Column A	Column B	Answers
1.	The solution to $-5(x-2) = 10x - 8$	The solution to $6y + 2(2y-1) = 9$	A Ⓐ Ⓑ Ⓒ Ⓓ **[Lesson 1.4]**
2.	$\begin{bmatrix} 6 & -5 \\ 2 & 3 \end{bmatrix}\begin{bmatrix} x \\ y \end{bmatrix} = \begin{bmatrix} 9 \\ 17 \end{bmatrix}$ x	y	A Ⓐ Ⓑ Ⓒ Ⓓ **[Lesson 3.5]**
3.	2^8	8^2	A Ⓐ Ⓑ Ⓒ Ⓓ **[Lesson 6.1]**
4.	5×10^{-3}	0.005	C Ⓐ Ⓑ Ⓒ Ⓓ **[Lesson 6.4]**

5. Which is the solution of $|2x - 3| < 4$? **[Lesson 1.6]** c
- **a.** $x < -0.5$ *or* $x > 3.5$
- **b.** $x > -0.5$ *or* $x > 3.5$
- **c.** $-0.5 < x < 3.5$
- **d.** $0.5 < x < 3.5$

6. Which are the coordinates for two points on the line with a slope of $-\frac{2}{3}$? **[Lesson 2.1]** c
- **a.** (0, 3) and (2, 0)
- **b.** (3, 0) and (0, −2)
- **c.** (4, 7) and (7, 5)
- **d.** (5, 8) and (2, 6)

7. Which is the graph of a function? **[Lesson 5.1]** c

a.

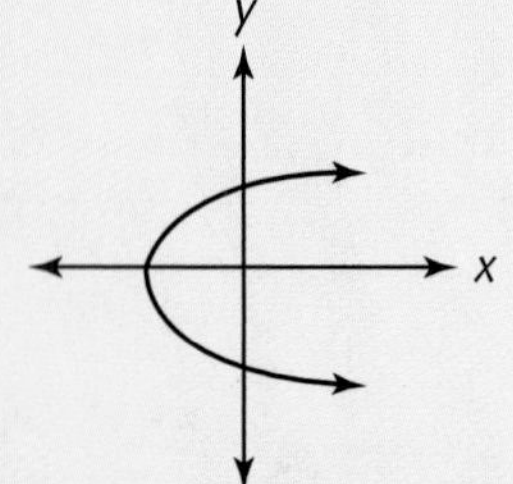

b.

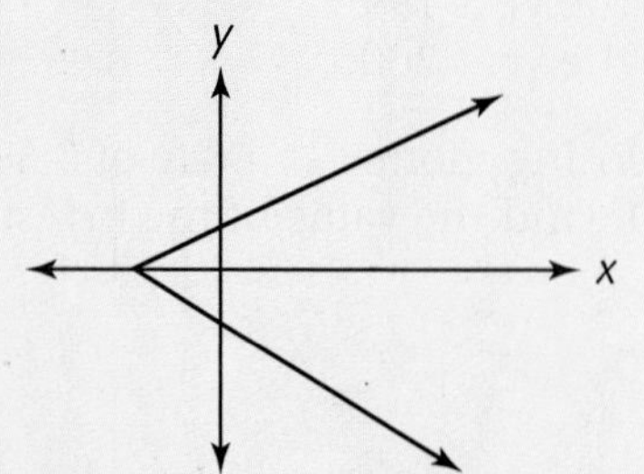

c.

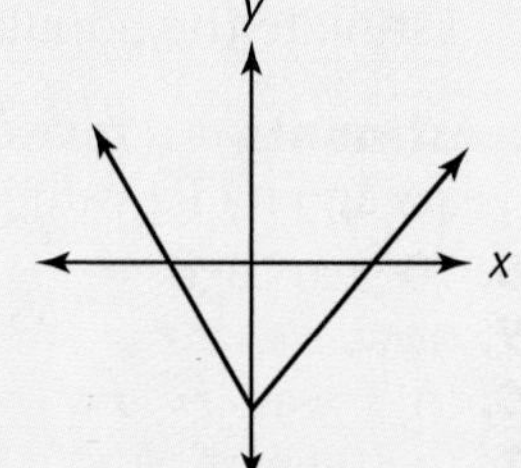

8. Which is a *stretch* of the function $y = |x|$? **[Lesson 5.2]** c

a. $y = -|x|$ b. $y = |x - 3|$ c. $y = 3|x|$ d. $y = |x| + 3$

9. Which is a vertical stretch of the graph of $f(x) = x^2$ followed by a horizontal translation? **[Lesson 5.6]** b

a. $f(x) = 2x^2 + 4$
b. $f(x) = 2(x + 4)^2$
c. $f(x) = (x + 4)^2 + 2$
d. none of these

10. Which is a simplified form of $(-3p^2)^3$? **[Lesson 6.2]** c

a. $-27p^5$ b. $-729p^6$ c. $-27p^6$ d. $-3p^6$

11. Cathy has $17.40 in dimes and quarters, for a total of 81 coins. How many of each type of coin does she have? **[Lessons 2.4, 2.5]** 19*d*; 62*q*

12. If $A = \begin{bmatrix} 4 & -2 \\ 5 & -3 \end{bmatrix}$ and $B = \begin{bmatrix} -2 \\ 6 \end{bmatrix}$, find AB. **[Lesson 3.3]** $AB = \begin{bmatrix} -20 \\ -28 \end{bmatrix}$

13. **Probability** A bag contains 3 blue chips, 7 red chips, and 2 yellow chips. If you select a chip, replace it, and select another chip, what is the probability of selecting a red chip and then a yellow chip? **[Lesson 4.7]** $\frac{7}{72}$ or 9.7%

14. Identify the parent function for $y = -2|x|$. Tell what type of transformation is applied, and then draw a carefully labeled graph. **[Lesson 5.3]**

For Items 15–17, use the parent function $y = x^2$ to graph each function. [Lessons 5.4, 5.5, 5.6]

15. $y = -x^2$ 16. $y = x^2 + 1$ 17. $y = -(x - 1)^2 - 2$

18. Simplify $\frac{-100a^8b^{10}}{-20a^5b^9}$. **[Lesson 6.2]** $5a^3b$

19. Simplify $(3x^2y^{-3})^0$. **[Lesson 6.3]** 1, if $x \neq 0$ and $y \neq 0$

20. After a few years of ownership, the value of an automobile depreciates at a rate of about 6% a year. At this rate, what will be the value in 5 years of a car that is currently worth $10,000? **[Lesson 6.6]** $7339.04

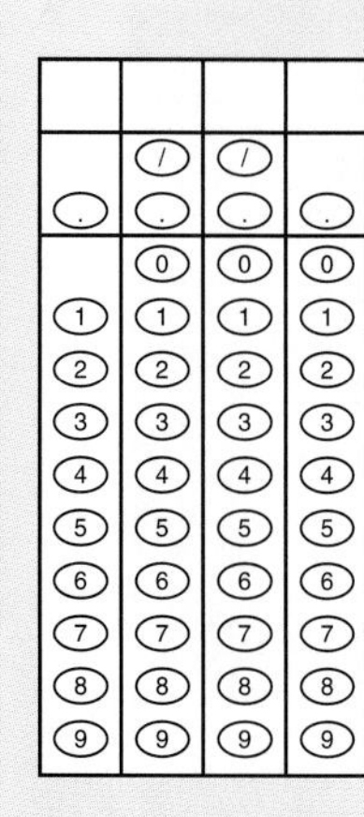

Free-Response Grid The following questions may be answered using a free-response grid commonly used by standardized test services.

21. The two matrices $\begin{bmatrix} a & 4(x-2) & c \\ d & e & f \end{bmatrix}$ and $\begin{bmatrix} g & 32 & i \\ j & k & l \end{bmatrix}$ are equal. Find the value of x. **[Lesson 3.1]** $x = 10$

22. How many 3-digit numbers can be formed from the digits 0, 3, 4, 5, 6, and 7? **[Lesson 4.5]** 180

23. **Probability** An ordinary 6-sided number cube is rolled. What is the probability that the number rolled is an even number that is less than 6? **[Lesson 4.6]** $\frac{1}{3}$

24. A country with a population of 2 million people is growing at the rate of 1.5 percent per year. By how many thousands of people will the population increase in 10 years? Round your answer to the nearest thousand. **[Lesson 6.5]** 321,000 people

25. Find, to the nearest whole number, the value of a $100 investment in 21 years if the annual rate of growth of the investment is 10%. **[Lesson 6.6]** $740

CHAPTER 7

Polynomials and Factoring

LESSONS

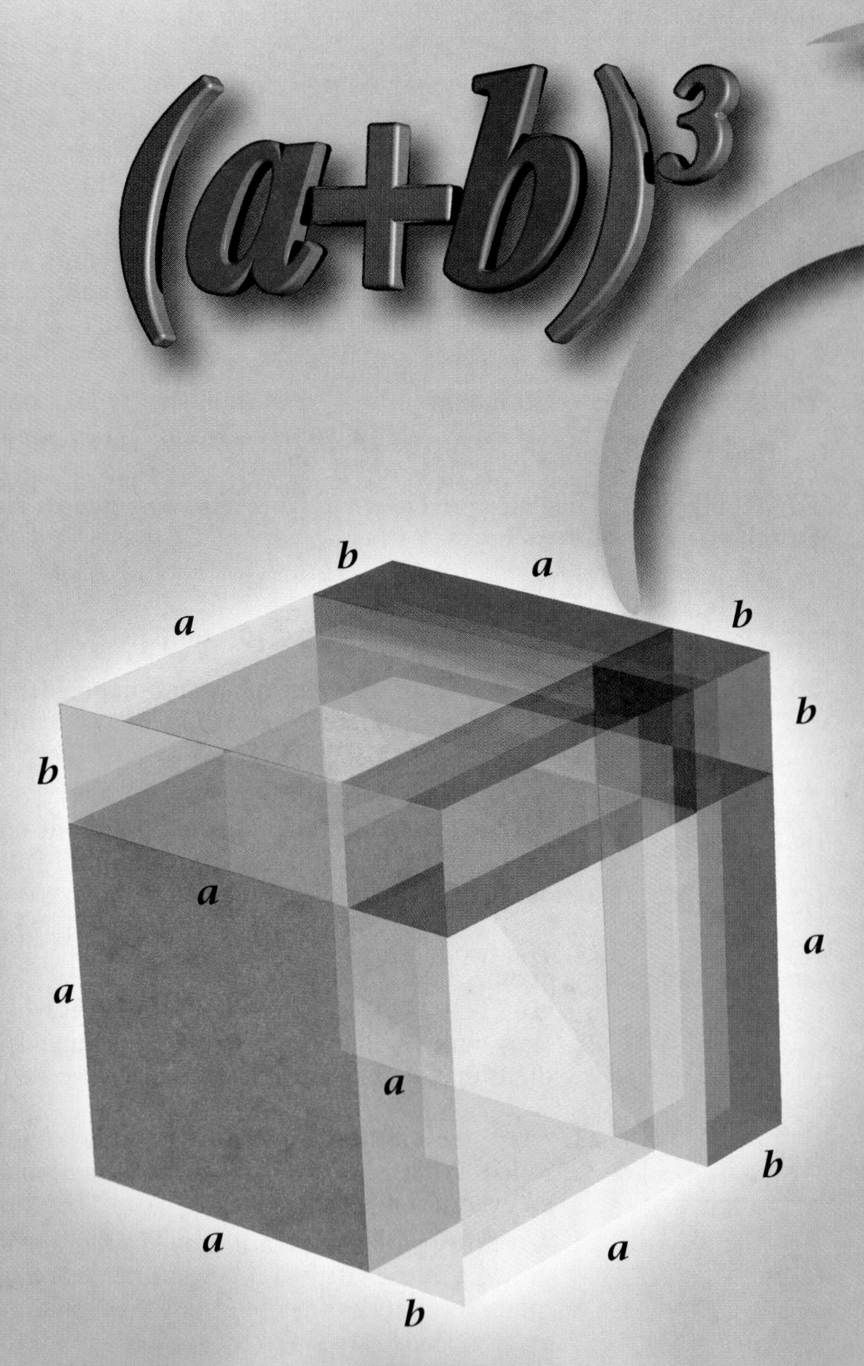

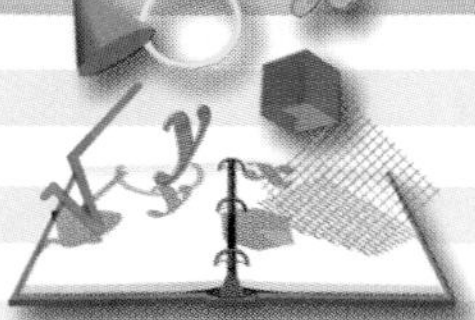

Portfolio Activity

Punnett squares are used in genetics to determine the possible characteristic of offspring. Assume that each parent has both the recessive and the dominant form of the gene for height. Show how to determine the probability that the offspring from a cross between these two parents will be tall if tall is dominant and short is recessive. How is this related to polynomials?

LESSON 7.1

Exploring Polynomial Functions

why *Surface area and volume formulas are polynomial functions that can be used in manufacturing.*

The production manager wants to change the size of the Super Suds box. Currently, the length of the base is 20 centimeters, the width of the base is 10 centimeters, and the height is 30 centimeters.

Power Wash Incorporated manufactures a variety of powder laundry soaps. The production manager notices that production costs of Super Suds laundry soap are rising. Instead of raising prices, the production manager investigates lowering production costs by changing the size of the package.

Exploration 1 *Volume and Surface-Area Functions*

Part I: Volume

You can use the formula $V = Bh$, where B is the base area and h is the height, to find the volume, V, of the Super Suds box.

3. Since $V = lwh$, you can substitute $2x$ for l, x for w, and 30 for h, to get $V(x) = 2x \cdot x \cdot 30$.

1. What is the volume of the Super Suds box? 6000 cubic cm

2. The production manager wants to reduce the volume by 10%. What would the new volume be? 5400 cubic centimeters

3. The production manager decides to reduce the size of the rectangular base, keeping the same height and keeping the length twice the width. Explain why you can use the following function, where x is the width of the base, to represent the volume of the Super Suds box:

$$V(x) = 2x \cdot x \cdot 30$$
$$= 60x^2$$

4 For the function $V(x) = 60x^2$, find $V(10)$ to compute the volume of the original Super Suds box. Is this the same as your answer to Step 1? 6000 cu cm; yes

5 Examine the graphics calculator table. What width gives the volume of the original box? What size base can be used to reduce the original volume by approximately 10%? 10; 9.5

Width Volume

X	Y_1
9.4	5301.6
9.5	5415
9.6	5529.6
9.7	5645.4
9.8	5762.4
9.9	5880.6
10	6000

$X = 60X^2$

Part II: Surface Area

The *surface area* of the Super Suds box is the sum of the areas of the six *lateral faces* of the box. In other words, the surface area is the area of the material used to make the box.

1 Find the surface area of the original Super Suds box. 2200 sq cm

2 If the production manager wants to reduce the surface area by 10%, what should the new surface area be? 1980 sq cm

3 The production manager decides to reduce the size of the rectangular base, keeping the same height and keeping the length twice the width. Explain why you can use the following function, where x is the width of the base, to represent the surface area of the Super Suds box:

$$\begin{aligned} S(x) &= 2(2x \cdot x) + 2(2x \cdot 30) + 2(x \cdot 30) \\ &= 4x^2 + 120x + 60x \\ &= 4x^2 + 180x \end{aligned}$$

3. Since $S = 2(l \cdot w) + 2(l \cdot h) + 2(w \cdot h)$, you can substitute $2x$ for l, x for w, and 30 for h, to get $S(x) = 2(2x \cdot x) + 2(2x \cdot 30) + 2(x \cdot 30)$.

4 For the function $S(x) = 4x^2 + 180x$, find $S(10)$ to compute the surface area of the original box. 2200 square centimeters

5 Examine Table 1 at the left. What width, X, gives the surface area, Y_1, of the original Super Suds box? 10

6 Examine Table 2 at the left. What width, X, can be used to reduce the original surface area of the Super Suds box by approximately 10%?

7 Suppose the production manager decides to reduce the width of the base to 9.3 centimeters. What would be the volume and surface area of the new Super Suds box? What percent reduction in volume is this? What percent reduction in surface area is this? ❖

Table 1

X	Y_1
9.7	2122.4
9.8	2148.2
9.9	2174
10	2200
10.1	2226
10.2	2252.2
10.3	2278.4

$X = 4X^2+180X$

Table 2

X	Y_1
8.8	1893.8
8.9	1918.8
9	1944
9.1	1969.2
9.2	1994.6
9.3	2020
9.4	2045.4

$X = 4X^2+180X$

6. Between 9.1 and 9.2; about 9.15

7. $V = 5189.4$ cubic cm; $S = 2019.96$ sq cm; 14%; about 8%

APPLICATION

Manufacturing

Quick-Quench, a thirst-quenching drink mix, comes in a cylindrical container. To increase sales, manufacturers plan to increase the volume by 20% by changing the height of the can. Write a function to represent the volume in terms of the height of the can. Then find the new height that would increase the volume of the can by 20% of the original height.

The formula for the volume of a cylinder is $V = Bh$, or $V = \pi r^2 h$. Find the volume of the original can.

$$\begin{aligned} V &= \pi r^2 h \\ &= \pi \cdot 6^2 \cdot 10 \\ &= 360\pi, \text{ or approximately } 1131 \end{aligned}$$

The volume of the original can is approximately 1131 cubic centimeters. Find the new volume after a 20% increase in volume.

$$\begin{aligned} \text{New volume} &= 1131 + 20\% \cdot 1131 \\ &\approx 1131 + 226.2 \\ &\approx 1357.2 \text{ cubic centimeters} \end{aligned}$$

Use a graphics calculator to make a table for the function $V(h) = 36\pi h$, where h is the height of the can. Substitute X for h, and enter $Y = 36\pi X$ into your graphics calculator.

X	Y_1
11.7	1323.2
11.8	1334.5
11.9	1345.9
12	1357.2
12.1	1368.5
12.2	1379.8
12.3	1391.1

X=12

The height that increases the volume of the can by approximately 20% is 12 centimeters. ❖

The functions $V(x) = 60x^2$ and $S(x) = 4x^2 + 180x$ from the exploration are called *polynomial functions*.

POLYNOMIAL FUNCTION

A **polynomial function** is a function that consists of one or more *monomials*. A **monomial** is an algebraic expression that consists of either a constant, a variable, or a product of a constant and one or more variables.

EXTENSION

Graphics Calculator

Show that the functions $y = (x + 2)^3$ and $y = x^3 + 6x^2 + 12x + 8$ are equivalent for integer values from -3 to 3. That is, show that $(x + 2)^3 = x^3 + 6x^2 + 12x + 8$.

Build a table of values for the equations. Enter Y_1 = (X + 2)^3 and Y_2 = X^3 + 6X^2 + 12X + 8, and compare the y-values in the table.

X	Y_1	Y_2
−3	−1	−1
−2	0	0
−1	1	1
0	8	8
1	27	27
2	64	64
3	125	125

Since the columns for Y_1 and Y_2 have the same values, the functions are equivalent for integer values from -3 to 3. ❖

Try This Use a graphics calculator to determine whether the functions $y = 2\pi x^2 + 20\pi x$ and $y = 2\pi x(x + 10)$ are equivalent for integer values from -3 to 3.

X	Y_1	Y_2
−3	−131.9	−131.9
−2	−100.5	−100.5
−1	−56.55	−56.55
0	0	0
1	69.115	69.115
2	150.8	150.8
3	245.04	245.04

EXERCISES & PROBLEMS

Communicate

1. Define polynomial function. Give two examples.
2. Write functions for the volume and surface area of a cube with edge x.
3. Explain how you can use a graphics calculator table to show that $x^2 - 4 = (x + 2)(x - 2)$ is true for integer values from -3 to 3.

X	Y_1	Y_2
−1	−3	−3
0	−4	−4
1	−3	−3
2	0	0
3	5	5
4	12	12
5	21	21

Practice & Apply

Geometry Give the indicated functions for each geometric solid.

4. The volume of a cube with an edge that is x centimeters long $V(x) = x^3$ cu cm

5. The surface area of a cube that is x inches long $S(x) = 6x^2$ sq in.

6. The surface area of a rectangular solid with a base length of 7 inches, a base width of 3 inches, and a height of x inches $S(x) = 20x + 42$ sq in.

7. The volume of a cylinder with height of 5 meters and a radius of x meters $V(x) = 5\pi x^2$ cu m

Show that each equation is true by substituting x-values -1, 0, and 1.

8. $x^2 + 5x + 6 = (x + 2)(x + 3)$

9. $(x - 4)(x - 2) = x^2 - 6x + 8$

10. $x^2 - 5x + 6 = (x - 2)(x - 3)$

11. $(x + 7)(x - 1) = x^2 + 6x - 7$

12. $x^2 - 9 = (x - 3)(x + 3)$

13. $(x - 6)(x + 4) = x^2 - 2x - 24$

14. $x^3 - 8 = (x - 2)(x^2 + 2x + 4)$

15. $(x + 5)(x + 3) = x^2 + 8x + 15$

16. $x^3 + 9x^2 + 27x + 27 = (x + 3)^3$

17. $(x - 6)(x - 7) = x^2 - 13x + 42$

18. $(x - 3)^2 = x^2 - 6x + 9$

19. $(x + 5)^2 = x^2 + 10x + 25$

Geometry Sports balls come in all different sizes. Complete the table for each radius given. Round your answers to the nearest tenth.

	Radius	Circumference	Surface area	Volume
	r	$C(r) = 2\pi r$	$S(r) = 4\pi r^2$	$V(r) = \frac{4\pi r^3}{3}$
20.	18 inches	?	?	?
21.	8 inches	?	?	?
22.	5 inches	?	?	?
23.	2 inches	?	?	?
24.	1 inch	?	?	?
25.	0.5 inch	?	?	?

Geometry A liquid-propane tank is made of a 20-foot long cylinder with two hemispheres on each end (a hemisphere is half of a sphere). A manufacturer investigates several tanks with different radii.

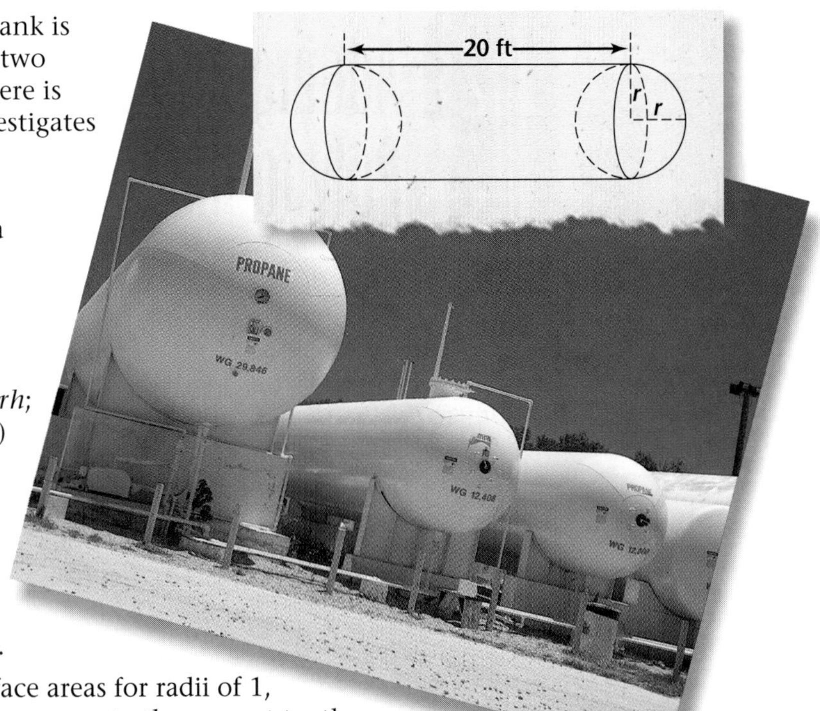

26. If the radius is 5 feet, what is the volume of the tank? (volume of a cylinder: $V = \pi r^2 h$; volume of a sphere: $V = \frac{4}{3}\pi r^3$)

27. If the radius is 5 feet, what is the surface area of the tank? (lateral surface area of a cylinder: $L = 2\pi r h$; surface area of a sphere: $S = 4\pi r^2$)

28. Define a function $V(x)$ for the volume of the tank, where x is the radius of the hemisphere.

29. Define a function $S(x)$ for the surface area of the tank, where x is the radius of the hemisphere.

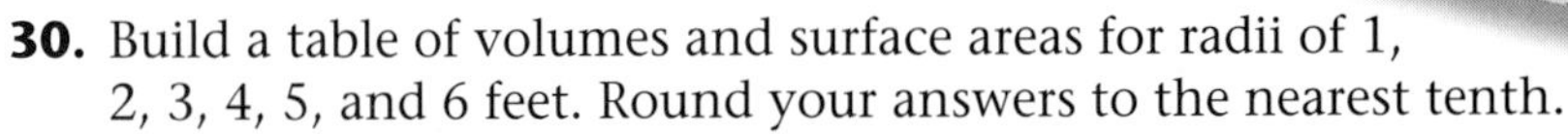

30. Build a table of volumes and surface areas for radii of 1, 2, 3, 4, 5, and 6 feet. Round your answers to the nearest tenth.

Look Back

Write an equation for the line containing each pair of points. [Lesson 2.2]

31. (2, 3) (−2, 3) $y = 3$ **32.** (0, −1) (4, 3) $y = x - 1$ **33.** (−1, 1) (−2, 2) $y = -x$

A department consists of 10 women and 15 men. Committee work is assigned randomly. [Lesson 4.6]

34. How many combinations of 3-person committees are possible? 2300

35. What is the probability that a three-person committee will consist of 3 women? 5%

36. What is the probability that a three-person committee will consist of 3 men? About 20%

Describe each transformation of the parent function $f(x) = x^2$. Check your answer with a graphics calculator. [Lesson 5.6]

37. $y = x^2 - 1$ **38.** $y = (x + 2)^2$ **39.** $y = -(x + 2)^2 - 1$

37. Vertical shift down by 1 unit **38.** Horizontal shift left by 2 units
39. Vertical reflection, horizontal shift left by 2, and vertical shift down by 1.

Look Beyond

Technology Graph each set of polynomial functions on the same coordinate plane. Describe the similarities and differences.

40. $y = x, y = x^3, y = x^5$ **41.** $y = x^2, y = x^4, y = x^6$

42. $y = x^2 - 2, y = x^3 - 2x, y = x^4 - 2x^2$

LESSON 7.2 Adding and Subtracting Polynomials

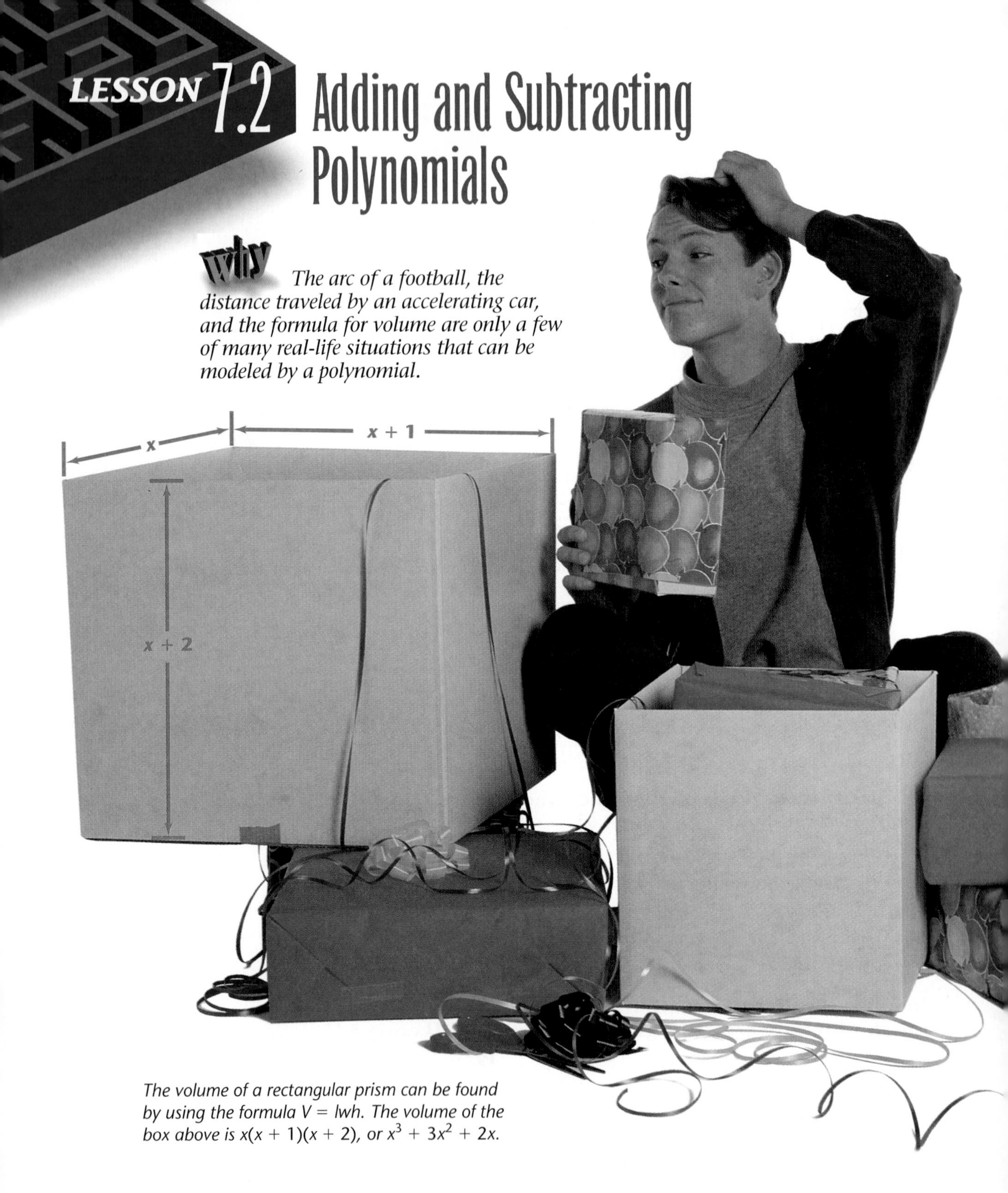

why
The arc of a football, the distance traveled by an accelerating car, and the formula for volume are only a few of many real-life situations that can be modeled by a polynomial.
x
x + 1
x + 2

The volume of a rectangular prism can be found by using the formula $V = lwh$. The volume of the box above is $x(x + 1)(x + 2)$, or $x^3 + 3x^2 + 2x$.

A **polynomial** is a monomial or the sum or difference of two or more monomials. The volume of the box above, $x^3 + 3x^2 + 2x$, is an example of a polynomial in one variable.

The **degree of a polynomial** in one variable is the *exponent with the greatest value of any of the polynomial's terms*. The degree of $9 - 4x^2$ is 2. What is the degree of $2a^2 + 5a^4 - 3a^3 + 1$? 4

The terms of a polynomial may appear in any order. However, in **standard form**, the terms of a polynomial are ordered from left to right, from the greatest to the least degree of the variable. This is called **descending order**. When the terms are ordered from least to greatest, this is called **ascending order**.

The polynomial $3x^2 - 4x + 6$ is in standard form. The degrees of the x-terms are in order from greatest to least. The polynomial $9 - 4x^2$ can be written in standard form as $-4x^2 + 9$. Write $5x^2 + x^3$ in standard form.

$x^3 + 5x^2$

Some polynomial expressions have special names that are dependent on either their *degree* or their *number of terms*, as illustrated in the table.

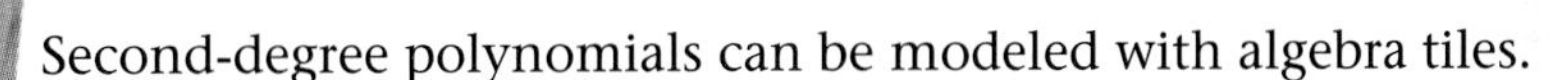

Polynomial	# of terms	Name by # of terms	Degree	Name by degree
12	1	monomial	0	constant
$8x$	1	monomial	1	linear
$9 - 4x^2$	2	binomial	2	quadratic
$5x^2 + x^3$	2	binomial	3	cubic
$3x^2 - 4x + 6$	3	trinomial	2	quadratic
$3x^4 - 4x^3 + 6x^2 - 7$	4	polynomial	4	quartic

Second-degree polynomials can be modeled with algebra tiles.

x^2	$-x^2$	x	$-x$	1	-1

1.     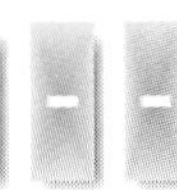  represents $3x^2 - 4x + 6$

2.  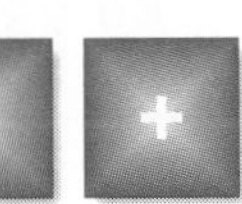    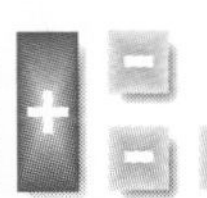 represents $5x^2 + x - 3$

3.      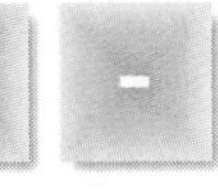 represents $9 - 4x^2$

Adding Polynomials

Algebra tiles can be used to model polynomial addition. To add $x^2 + x + 1$ and $2x^2 + 3x + 2$, model each polynomial and group like tiles.

Polynomials can be added in vertical or horizontal form. In vertical form align the like terms and add.

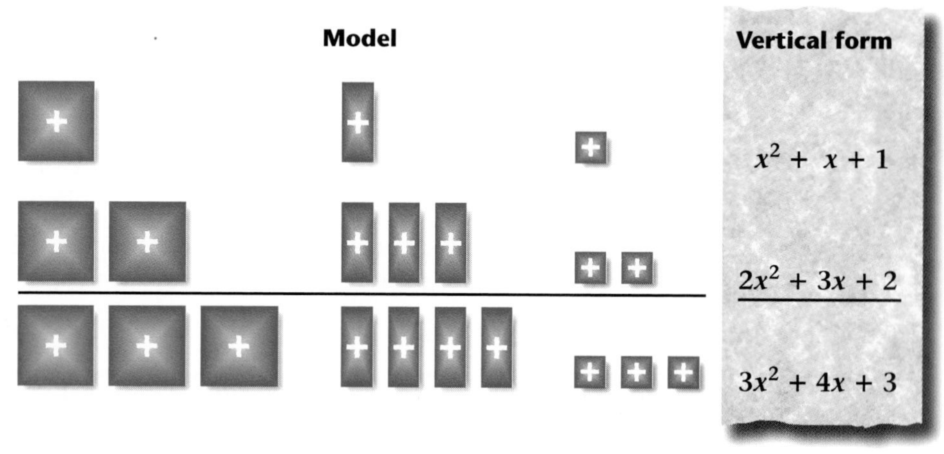

In horizontal form, write $(x^2 + x + 1) + (2x^2 + 3x + 2)$ and add like terms.

EXAMPLE 1

Use algebra tiles to model $(2x^2 - 3x + 5) + (4x^2 + 7x - 2)$. Find the sum.

Solution A ➤

Model each polynomial, and group like tiles. Notice that when an equal number of negative and positive tiles of the same type are put together, the combination equals zero.

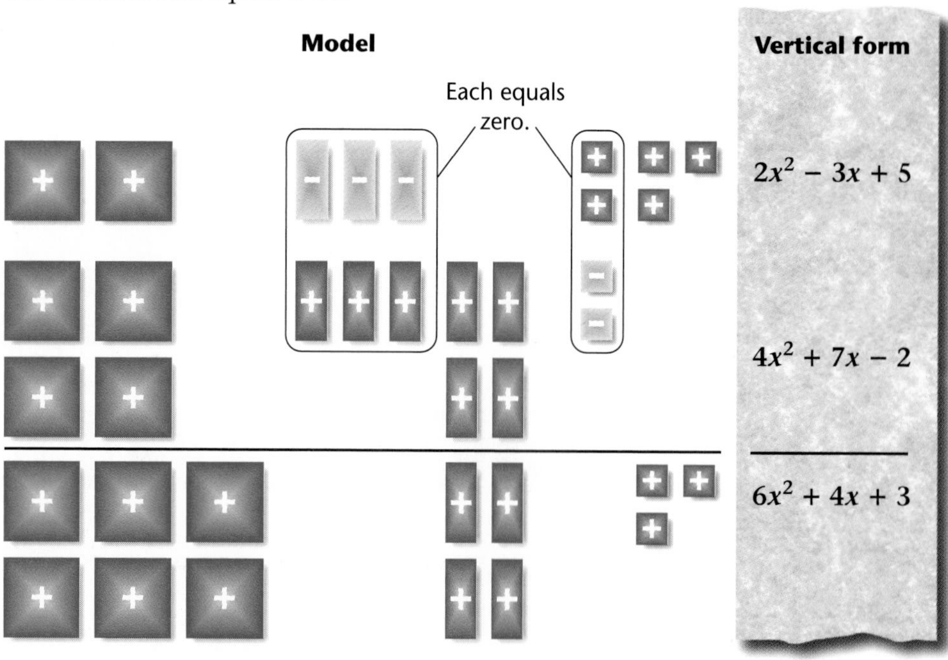

Solution B ➤

In horizontal form, add like terms.

$$(2x^2 - 3x + 5) + (4x^2 + 7x - 2)$$
$$= (2x^2 + 4x^2) + (-3x + 7x) + (5 - 2)$$
$$= 6x^2 + 4x + 3$$ ❖

Try This Use algebra tiles to model $(3x^2 + 5x) + (4 - 6x - 2x^2)$. Then find the sum in the horizontal form. $x^2 - x + 4$

EXAMPLE 2

Find the perimeter of each polygon. Check the addition by substituting a value for each variable.

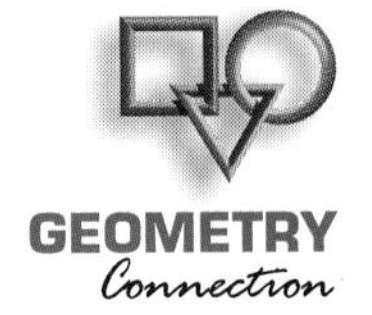

A

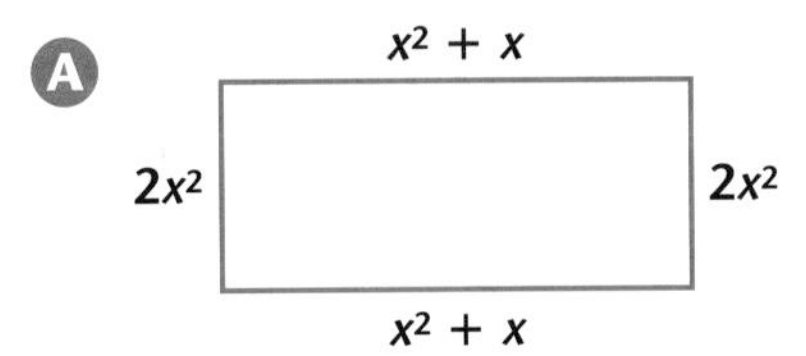

B

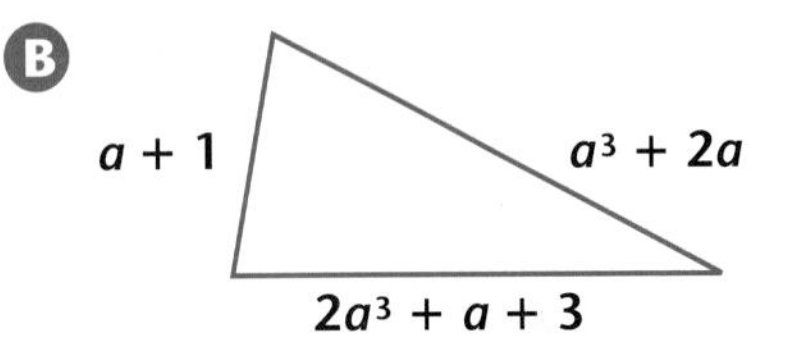

Solution ➤

In vertical form, add like terms. Substitute 10 for each variable to check.

A

Vertical form

$$\begin{array}{l} x^2 + x \\ 2x^2 \\ 2x^2 \\ \underline{x^2 + x} \\ 6x^2 + 2x \end{array}$$

Check: $x = 10$

$$\begin{array}{rcl} 10^2 + 10 &=& 110 \\ 2 \cdot 10^2 &=& 200 \\ 2 \cdot 10^2 &=& 200 \\ \underline{10^2 + 10} &=& \underline{110} \\ 6 \cdot 10^2 + 2 \cdot 10 &=& 620 \end{array}$$

B

Vertical form

$$\begin{array}{l} a^3 + 2a \\ \quad a + 1 \\ \underline{2a^3 + a + 3} \\ 3a^3 + 4a + 4 \end{array}$$

Check: $x = 10$

$$\begin{array}{rcl} 10^3 + 2 \cdot 10 &=& 1020 \\ 10 + 1 &=& 11 \\ \underline{2 \cdot 10^3 + 10 + 3} &=& \underline{2013} \\ 3 \cdot 10^3 + 4 \cdot 10 + 4 &=& 3044 \end{array}$$

CT– By using the Commutative and Associative properties, terms are rearranged so like terms are together, making mental math easier.

The perimeter can also be found in horizontal form.

A $(x^2 + x) + 2x^2 + 2x^2 + (x^2 + x) = 6x^2 + 2x$

B $(a^3 + 2a) + (a + 1) + (2a^3 + a + 3) = 3a^3 + 4a + 4$ ❖

CRITICAL *Thinking* Explain how using the horizontal form allows you to add polynomials mentally.

Subtracting Polynomials

Recall that the definition of subtraction is to add the opposite.

EXAMPLE 3

CT– Our number system has a base of 10, so $1 = 10^0$, $100 = 10^2$, and $1000 = 10^3$. Thus, our numbers can be represented as polynomials by letting $x = 10$ when numbers are written in expanded form.

Subtract $x^2 - 4$ from $3x^2 - 2x + 8$.

Solution

The binomial $x^2 - 4$ can be written as $x^2 + 0x - 4$, where $0x$ is a "placeholder" for the missing x term. Remember to change signs when subtracting.

Vertical form

$$\begin{array}{r} 3x^2 - 2x + 8 \\ -\ (x^2 + 0x - 4) \\ \hline 2x^2 - 2x + 12 \end{array}$$

Horizontal form

$$\begin{aligned} 3x^2 - 2x + 8 - (x^2 - 4) &= 3x^2 - 2x + 8 - x^2 + 4 \\ &= 2x^2 - 2x + 12 \end{aligned}$$

❖

Study the two examples below. How is the addition of whole numbers related to the addition of polynomials?

1.
$$\begin{array}{rl} 485 &= 4 \cdot 10^2 + 8 \cdot 10 + 5 \\ 113 &= 1 \cdot 10^2 + 1 \cdot 10 + 3 \\ \hline 598 &= 5 \cdot 10^2 + 9 \cdot 10 + 8 \end{array}$$

2.
$$\begin{array}{r} 4x^2 + 8x + 5 \\ x^2 + x + 3 \\ \hline 5x^2 + 9x + 8 \end{array}$$

Exercises & Problems

Communicate

1. Explain how to use algebra tiles to represent $5x^2 - 2x + 3$.
2. How do you determine the degree of a polynomial?
3. Give two different names for the polynomial $3x^3 + 4x^2 - 7$.
4. Draw an algebra-tile model of $(2x^2 + 3) + (x^2 - 1)$. Find the sum.
5. Explain how to find $(3b^3 - 2b + 1) - (b^3 + b - 3)$.
6. Explain how to write a polynomial in standard form. Use $5x^2 - 2x + 3x^4 - 6$ as a model.

Practice & Apply

7. Use algebra tiles to represent $x^2 + 4x + 4$.
8. Draw an algebra-tile model of $(3x^2 + 6) + (2x^2 - 1)$.

Rewrite each polynomial in standard form.

9. $6 + c + c^3$ $c^3 + c + 6$
10. $5x^3 - 1 + 5x^4 + 5x^2$ $5x^4 + 5x^3 + 5x^2 - 1$
11. $10 + p^7$ $p^7 + 10$

Write the degree of each polynomial.

12. $4r + 1$ 1
13. $x^3 + x^4 + x - 1$ 4
14. $y + y^3$ 3

Name each polynomial by the number of terms and by degree.

15. $3x + 1$ binomial; linear
16. $8x^2 - 1$ binomial; quadratic
17. $8x^2 - 2x + 3$ trinomial; quadratic

Give an example of a polynomial that is a

18. quadratic trinomial.
19. linear binomial.
20. cubic monomial.

Answers may vary for **18–20.** **18.** $6x^2 - 5x + 9$ **19.** $10n + 3$ **20.** $8d^3$

Use vertical form to add.

21. $3x^2 + 4x^4 - x + 1$ and $3x^4 + x^2 - 6$
22. $2y^3 + y^2 + 1$ and $3y^3 - y^2 + 2$
23. $4r^4 + r^3 - 6$ and $r^3 + r^2$ $4r^4 + 2r^3 + r^2 - 6$
24. $2c - 3$ and $c^2 + c + 4$ $c^2 + 3c + 1$

21. $7x^4 + 4x^2 - x - 5$ **22.** $5y^3 + 3$

Use horizontal form to add.

25. $y^3 - 4$ and $y^2 - 2$ $y^3 + y^2 - 6$
26. $x^3 + 2x - 1$ and $3x^2 + 4$

26. $x^3 + 3x^2 + 2x + 3$

27. $3s^2 + 7s - 6$ and $s^3 + s^2 - s - 1$ $s^3 + 4s^2 + 6s - 7$
28. $w^3 + w - 2$ and $4w^3 - 7w + 2$ $5w^3 - 6w$

Use vertical form to subtract.

29. $x^2 + x$ from $x^3 + x^2 + 7$ $x^3 - x + 7$
30. $3y^2 - 4$ from $4y^2 - y + 6$

30. $y^2 - y + 10$

31. $4c^3 - c^2 - 1$ from $5c^3 + 10c + 5$ $c^3 + c^2 + 10c + 6$
32. $8x^3$ from $x^3 - x + 4$ $-7x^3 - x + 4$

Use horizontal form to subtract.

33. $y^2 + 3y + 2 - (3y - 2)$ $y^2 + 4$
34. $3x^2 - 2x + 10 - (2x^2 + 4x - 6)$

34. $x^2 - 6x + 16$

35. $3x^2 - 5x + 3 - (2x^2 - x - 4)$ $x^2 - 4x + 7$
36. $2x^2 + 5x - (x^2 - 3)$ $x^2 + 5x + 3$

Simplify. Express all answers in standard form.

37. $(1 - 4x - x^4) + (x - 3x^2 + 9)$ $-x^4 - 3x^2 - 3x + 10$
38. $5 - 3x - 1.4x^2 - (13.7x - 62 + 5.6x^2)$ $-7x^2 - 16.7x + 67$
39. Subtract $x^5 - 2x^2 + 3x + 5$ from $4x^5 - 3x^3 - 3x - 5$. $3x^5 - 3x^3 + 2x^2 - 6x - 10$

$34{,}276 = 3(10^4) + 4(10^3) + 2(10^2) + 7(10^1) + 6$

40. **Geometry** Find the perimeter of the isosceles triangle. $8y^2 + y - 1$

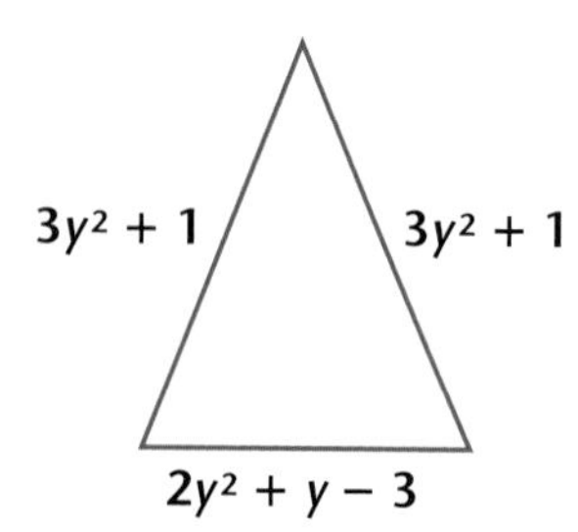

Geometry Exercises 41–43 refer to the dimensions of the figure at the right.

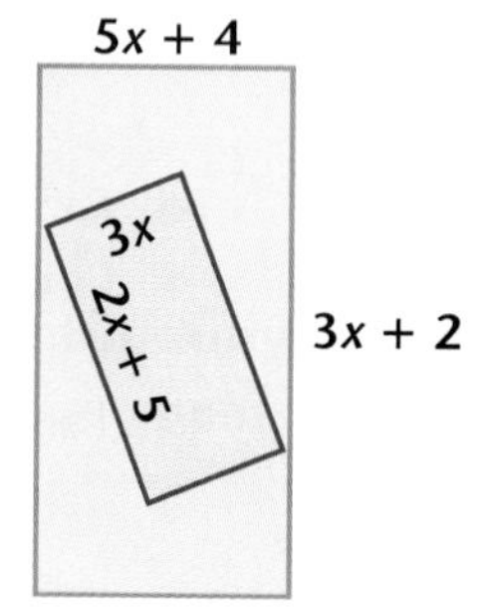

41. Find the polynomial expression that represents the perimeter of the large rectangle. $16x + 12$

42. Find the polynomial expression that represents the perimeter of the smaller rectangle. $10x + 10$

43. Find the difference between the perimeters of the two rectangles. $6x + 2$

Look Back

Mult. previous term by 3.

44. State the pattern, and find the next three terms of the sequence. 1, 3, 9, 27, 81, ?, ?, ? **[Course 1]** 243, 729, 2187

45. Solve $\frac{6x - 12x + 18}{3} = 1$ for x. **[Lesson 1.4]** $2\frac{1}{2}$ or 2.5

46. Solve $-2(a + 3) = 5 - 6(2a - 7)$ for a. **[Lesson 1.4]** $5\frac{3}{10}$ or 5.3

47. Solve this system of equations. $\begin{cases} 6x = 4 - 2y \\ 12x - 4y = 16 \end{cases}$ **[Lesson 2.5]** $(1, -1)$

48. Without repeating any digits, how many different 3-digit codes can be made with the digits 1–9? **[Lesson 4.5]** 504

Multiply. 49. $6x^3 - 24x^2$ 50. $8m^5 + 48m^4$

49. $6x(x^2 - 4x)$ **50.** $8m^3(m^2 + 6m)$ **[Lesson 6.2]**

51. Simplify $(3x^5)(-2x^3)(-x)$. $6x^9$

Write each number in scientific notation.

52. 7,100,000 7.1×10^6 **53.** 8,900,000,000 **[Lesson 6.4]** 8.9×10^9

Look Beyond

54. Use algebra tiles to model a square whose area can be represented as $x^2 + 6x + 9$. What is the length of a side of the square? $x + 3$

LESSON 7.3

Exploring Multiplication Models

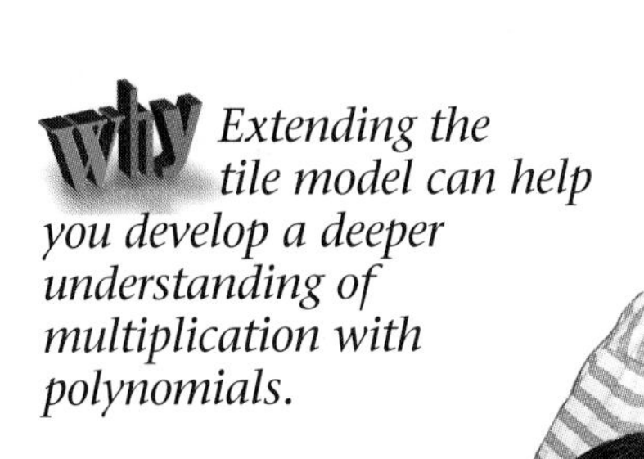

why *Extending the tile model can help you develop a deeper understanding of multiplication with polynomials.*

These students are exploring multiplication models to find that multiplication of polynomials can be represented by a geometric model.

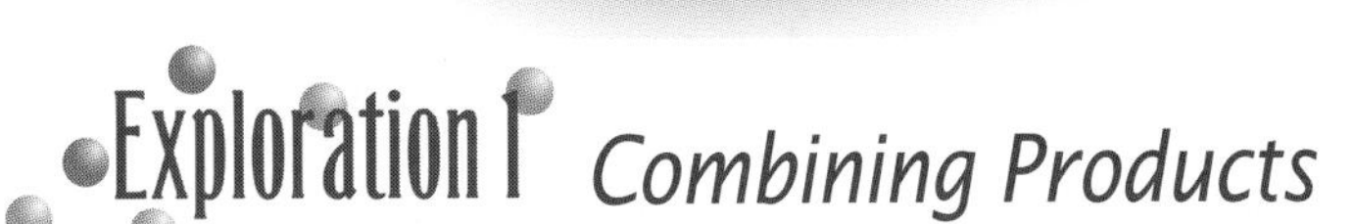

Exploration 1 *Combining Products*

1. The model shows that $(x + 1)$ can be multiplied by 2 to get $(2x + 2)$. This is an example of which property? Distributive Property

$x + 1$

$2(x + 1)$ → 2 { } $2x + 2$ → $2(x + 1) = 2(x) + 2(1)$
$= 2x + 2$

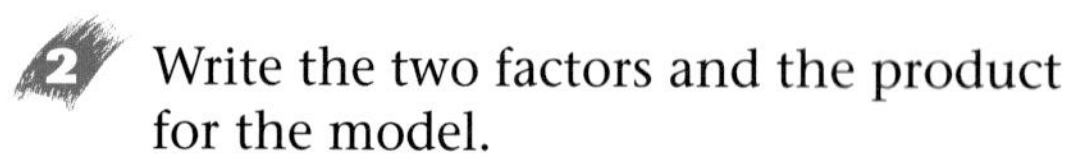

2. Write the two factors and the product for the model.
$x(x + 1) = x^2 + x$

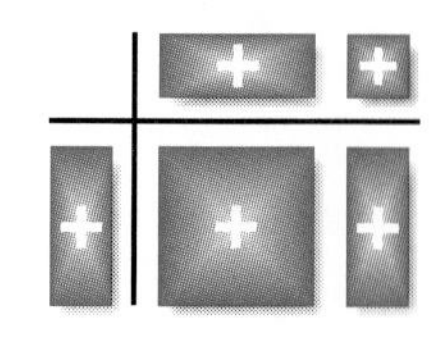

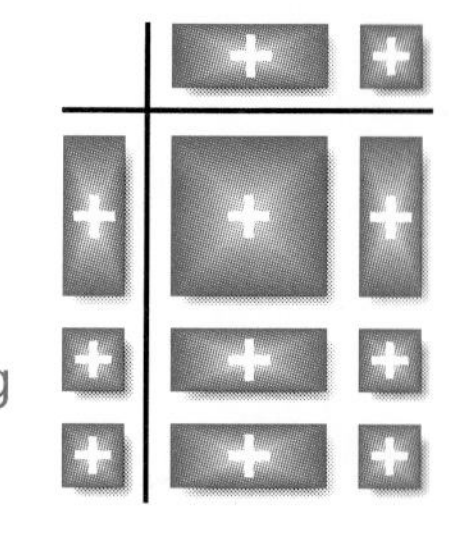

3. Explain how you can combine the products in Steps 1 and 2 to represent the product of the binomials $(x + 2)$ and $(x + 1)$. What is the product of $(x + 2)(x + 1)$? By combining the product in Step 2 with the product from Step 1 and arranging the tiles so that $x(x + 1) = x^2 + x$ and $2(x + 1) = 2x + 2$, the new product is $x^2 + 3x + 2$.

4. How can you check your multiplication of $(x + 2)(x + 1)$?
Substitute a value for x in $(x + 2)(x + 1) = x^2 + 3x + 2$ to see if the equation is true; let $x = 1$; $(1 + 2)(1 + 1) = 1^2 + 3 + 2$, $6 = 6$, True.

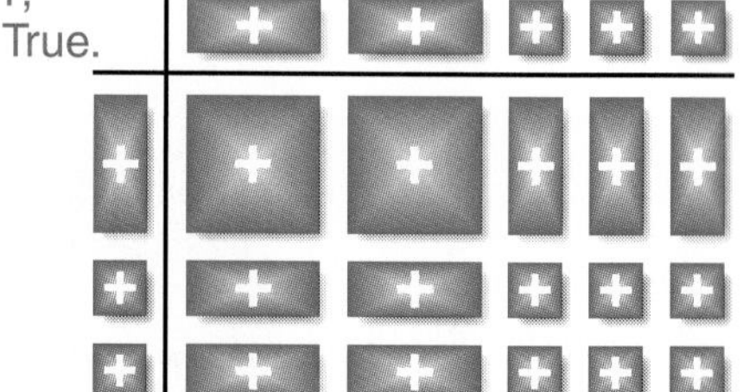

5. What two factors are multiplied in the model at the right? What is the product?
$(x + 2)$ and $(2x + 3)$; $2x^2 + 7x + 6$

6. Check your multiplication in Step 5 by substituting 10 for x.

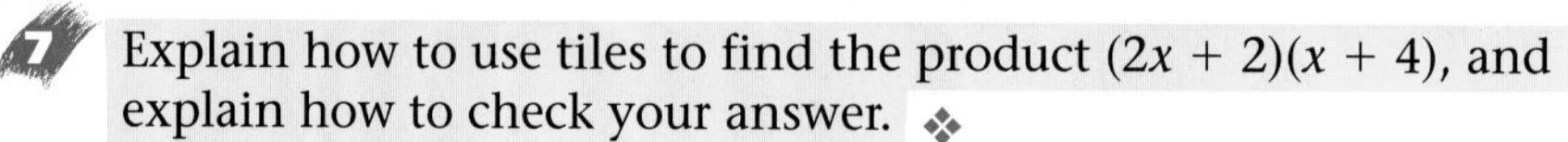

7. Explain how to use tiles to find the product $(2x + 2)(x + 4)$, and explain how to check your answer. ❖

Exploration 2 *Multiplying With Negatives*

1. Describe the algebra-tile model for the expression $x + 2$.
One positive x-tile and 2 positive 1-tiles

2. Describe the algebra-tile model for the expression $x - 3$.
One positive x-tile and 3 negative 1-tiles

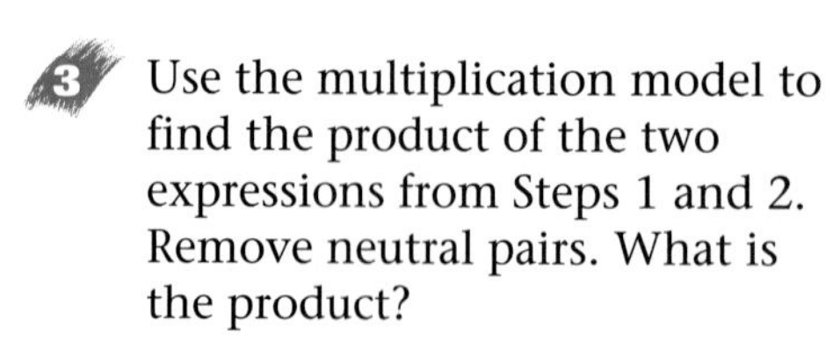

3. Use the multiplication model to find the product of the two expressions from Steps 1 and 2. Remove neutral pairs. What is the product?

$(x + 2)(x - 3) = \underline{\ ?\ }$ $x^2 - x - 6$

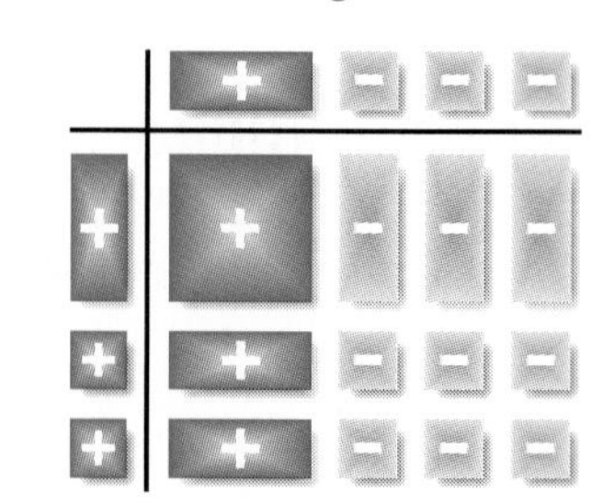

4. Check the multiplication by substituting any value, such as 10, for x.

5. Use tiles to find the product $(x - 2)(x - 4)$. Check by substituting 10 for x.

6 What is the product when you substitute 2 for x in $(x - 2)(x - 4)$? What is the product when you substitute 4 for x in $(x - 2)(x - 4)$? For the function $y = (x - 2)(x - 4)$, 2 and 4 are called **zeros**. Why do you think they are called this?

7 What do you think are the zeros for $y = (x - 5)(x + 3)$? Explain why you chose those numbers. ❖ 5; −3; When 5 or −3 is substituted for the x-value, one of the factors becomes zero.

EXTENSION

X	Y_1	
−6	4	
−5	0	
−4	−2	
−3	−2	
−2	0	
−1	4	
0	10	
Y_1=0		

Use your graphics calculator to find the zeros of the function $y = (x + 5)(x + 2)$.

Build a table of values for the function, and look for the x-values that give a y-value of zero.

X	Y_1	
−6	4	
−5	0	
−4	−2	
−3	−2	
−2	0	
−1	4	
0	10	
Y_1=0		

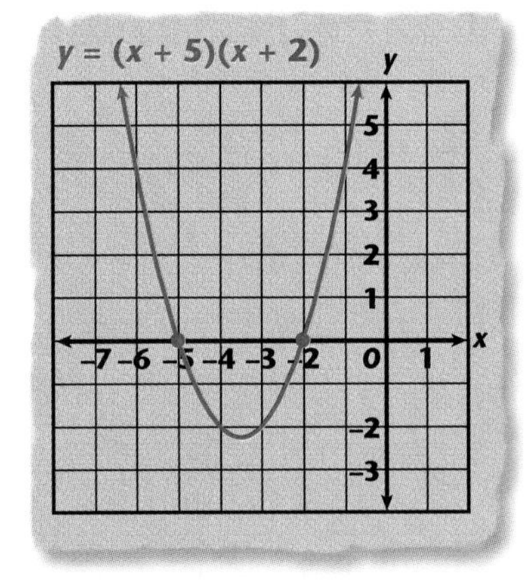

You can see that the zeros of $y = (x + 5)(x + 2)$ are at x-values of −5 and −2.

On the graph, you can see that the graph crosses at (−5, 0) and (−2, 0). Therefore, the zeros of $y = (x + 5)(x + 2)$ are −5 and −2.

Exploration 3 *Special Products*

Part I

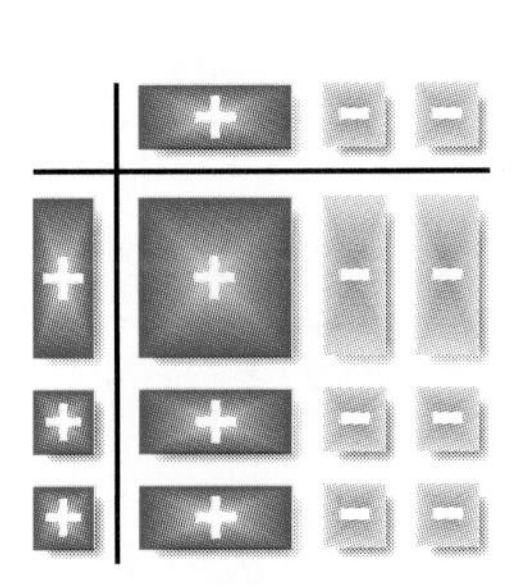

1 Model the product $(x + 2)(x - 2)$. At the right

2 Remove neutral pairs. What happens to the x-terms in the product? The x-terms cancel.

3 Use a tile model to represent the product $(x + 3)(x - 3)$. Remove neutral pairs. What happens to the x-terms in the product?

4 How can you find the product $(x + 5)(x - 5)$ without tiles? What is the product?

Part II

1. $x^2 + 4x + 4$

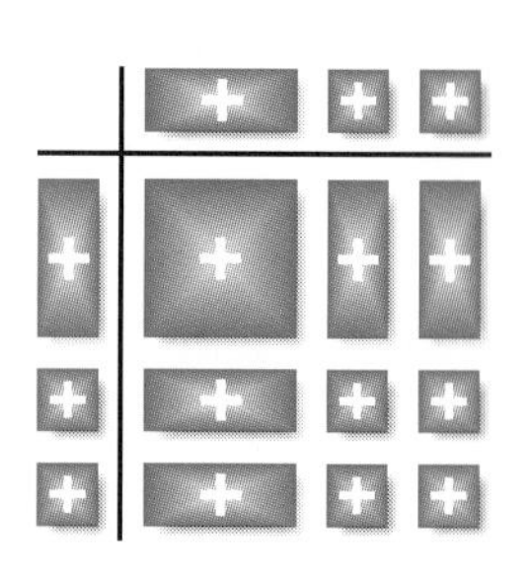

1. Model $(x + 2)(x + 2)$. Write the product.
2. Use a tile model to find the product $(x + 3)(x + 3)$.
3. Explain how the product $(x + 2)(x + 2)$ is similar to the product $(x + 3)(x + 3)$. Compare x-terms.
4. How can you find the product $(x + 5)(x + 5)$ without tiles? What is the product? ❖

EXERCISES & PROBLEMS

Communicate

1. Explain how you can use algebra tiles to find each product.

   $(x + 2)(x + 3)$ $(x - 4)(x - 4)$ $(x + 6)(x + 6)$
2. Give an example of a product in which you remove neutral pairs.
3. Explain how you can multiply $(x + 7)(x - 7)$ mentally.
4. Use a model to show why $(x + 4)^2$ does not equal $x^2 + 16$.

Practice & Apply

Write two factors and a product for each model. Check each product by substituting any number for *x*.

5.

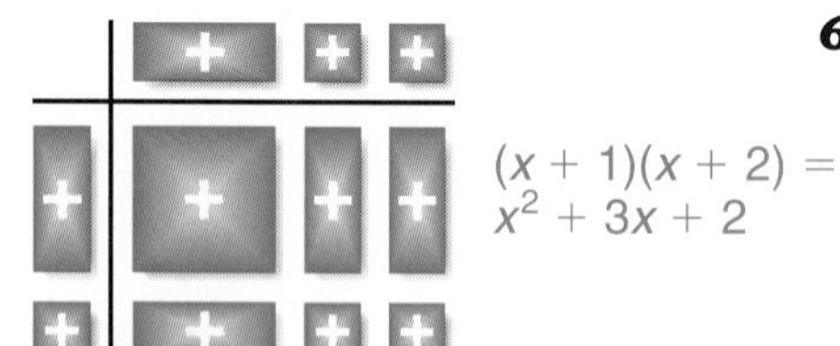

$(x + 1)(x + 2) = x^2 + 3x + 2$

6.

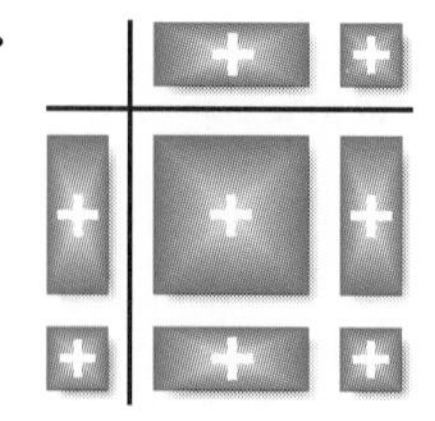

$(x + 1)(x + 1) = x^2 + 2x + 1$

7.

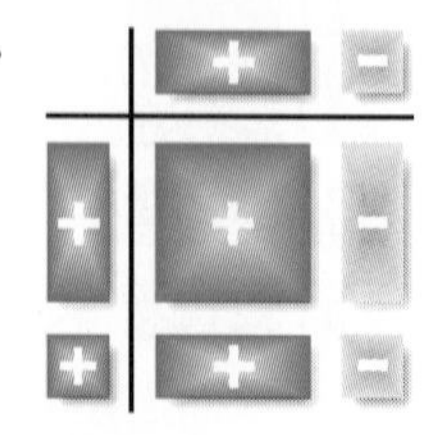

$(x + 1)(x - 1) = x^2 - 1$

8.

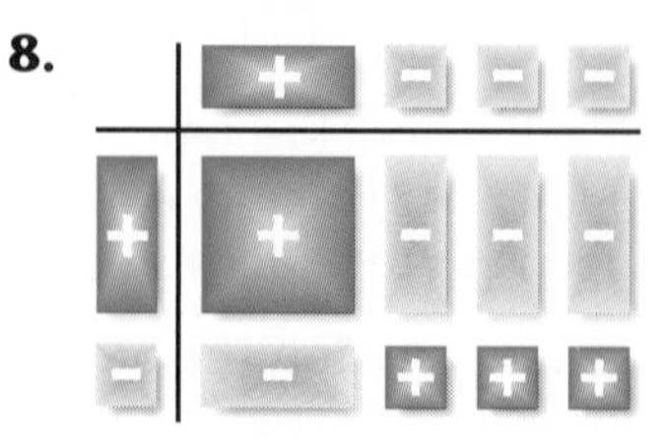

$(x - 1)(x - 3) = x^2 - 4x + 3$

9.

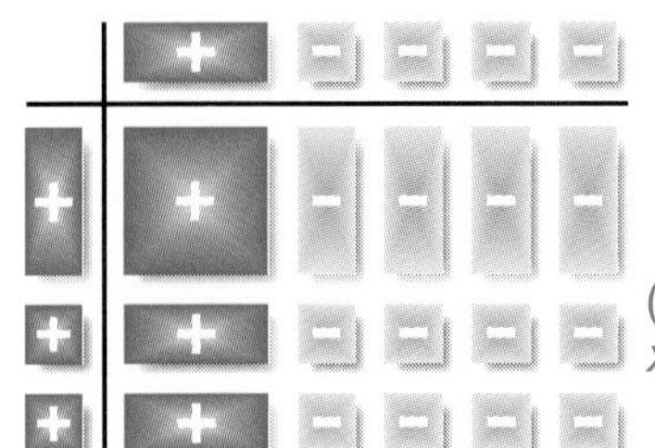

$(x + 2)(x - 4) = x^2 - 2x - 8$

10.

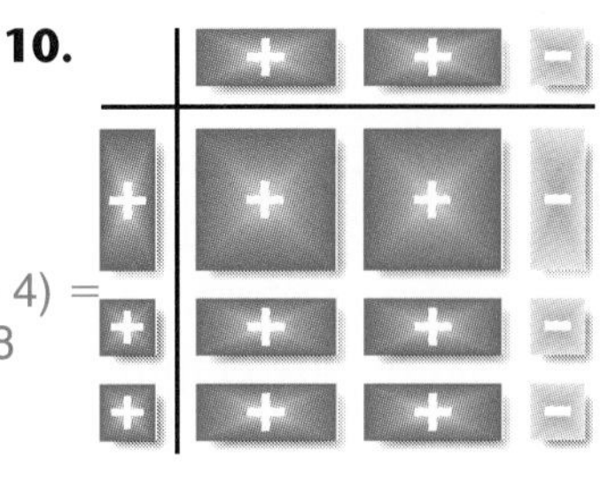

$(x + 2)(2x - 1) = 2x^2 + 3x - 2$

Determine whether each statement is true by substituting 10 for x.

11. $x(2x) = 2x^2$ True

12. $x(2x + 1) = 2x^2 + 2$ Not true

13. $x(2x - 1) = 2x^2 - x$ True

14. $(x + 3)(x + 1) = 2x^2 + 3x + 3$

15. $(x - 4)(x + 3) = x^2 - x - 12$ True

16. $(x - 2)(x + 7) = x^2 + 5x - 14$

14. Not true 16. True

Use the Distributive Property to find each product.

17. $4(x + 2)$ $4x + 8$

18. $6(2x + 7)$ $12x + 42$

19. $5(x + 10)$ $5x + 50$

20. $4(x - 4)$ $4x - 16$

21. $3(x + 8)$ $3x + 24$

22. $-2(x - 3)$ $-2x + 6$

23. $8(3x - 4)$ $24x - 32$

24. $-3(x + 4)$ $-3x - 12$

25. $x(1 - x)$ $x - x^2$

26. $-x(2 - x)$ $-2x + x^2$

27. $x(-x + 5)$ $-x^2 + 5x$

28. $4x(3 - x)$ $12x - 4x^2$

Find each product. You may use tiles to check.

29. $(x + 5)(x + 2)$

30. $(x + 2)(x - 1)$

31. $(2x + 1)(x + 1)$

32. $(x + 5)(x - 2)$

33. $(2x + 2)(2x - 2)$

34. $(x + 2)(x + 1)$

35. $(3x + 5)(x - 7)$

36. $(5x + 1)(5x - 1)$

37. $(x + 2)(x - 2)$

38. $(x + 1)(x + 1)$

39. $(x - 2)(4x - 3)$

40. $(x + 6)(x + 6)$

41. $(x + 4)(x - 4)$

42. $(x + 5)(x - 7)$

43. $(x - 6)(x - 6)$

Identify the zeros of each function.

44. $y = (x + 5)(x + 2)$

45. $y = (x + 2)(x - 10)$

46. $y = (x + 11)(x + 13)$

47. $y = (x - 2)(x + 1)$

48. $y = (x + 5)(x - 2)$

49. $y = (x + 15)(x - 7)$

50. $y = (x + 6)(x + 6)$

51. $y = (x + 4)(x - 3)$

52. $y = (x + 10)(x - 3)$

53. $y = (2x + 2)(2x - 2)$

54. $y = \left(x - \frac{1}{2}\right)\left(x + \frac{1}{2}\right)$

55. $y = \left(x + \frac{2}{3}\right)(x - 6)$

53. $-1, 1$

54. $\frac{1}{2}, -\frac{1}{2}$

55. $-\frac{2}{3}, 6$

44. $-5, -2$
45. $-2, 10$
46. $-11, -13$
47. $2, -1$
48. $-5, 2$
49. $-15, 7$
50. -6
51. $-4, 3$
52. $-10, 3$

56. Explain how the diagram at the right shows that $6 \cdot 23 = 138$.

	20	3
6	120	18

57. Explain how the diagram at the right shows that $14 \cdot 12 = 168$.

	10	2
10	100	20
4	40	8

58. Explain how the diagram at the right shows that $(x + 1)(x - 1) = x^2 - 1$.

	x	-1
x	x^2	$-x$
1	x	-1

Look Back

Solve each equation or inequality. [Lesson 1.6]

59. $|x - 2| = 5$
$x = 7$ or $x = -3$

60. $|2x - 1| = 1$
$x = 0$ or $x = 1$

61. $|x| \leq 5$
$-5 \leq x \leq 5$

62. Solve the system of linear equations by graphing. **[Lesson 2.3]** $\begin{cases} 6y = 8x - 2 \\ 2x + 3y = 2 \end{cases}$

63. Solve the system of linear equations by substitution. **[Lesson 2.4]** $(-2, 9)$ $\begin{cases} x + y = 7 \\ 4x + 2y = 10 \end{cases}$

64. Solve the system of linear equations by elimination. **[Lesson 2.5]** $\left(\frac{3}{4}, \frac{2}{3}\right)$ $\begin{cases} -4x + 3y = -1 \\ -8x - 6y = -10 \end{cases}$

In Exercises 65–66, multiply both pairs of matrices, and determine whether products are the same. [Lesson 3.3]

65. $\begin{bmatrix} 2 & 1 \\ -1 & 3 \end{bmatrix}\begin{bmatrix} 4 & 0 \\ -2 & 1 \end{bmatrix}$ and $\begin{bmatrix} 4 & 0 \\ -2 & 1 \end{bmatrix}\begin{bmatrix} 2 & 1 \\ -1 & 3 \end{bmatrix}$ Not the same

66. $\begin{bmatrix} 1 & 6 \\ 2 & 4 \end{bmatrix}\begin{bmatrix} 0 & 1 \\ 4 & -3 \end{bmatrix}$ and $\begin{bmatrix} 6 & 4 \\ 1 & 2 \end{bmatrix}\begin{bmatrix} 1 & -3 \\ 0 & 4 \end{bmatrix}$ Not the same

67. How many different committees of three can be formed with 5 people? **[Lesson 4.4]** 10 committees

68. What is the theoretical probability that you will roll a number greater than 3 in one roll of an ordinary six-sided number cube? **[Lesson 4.6]** $\frac{1}{2}$ or 50%

69. Suppose Carl draws one card and then a second card from a deck of cards. Explain how these events could be independent and how these events could be dependent. **[Lesson 4.7]**

70. Graph the function $y = |x| - 4$. Is the graph of $y = |x|$ stretched, reflected, or shifted in the graph of $y = |x| - 4$? **[Lesson 5.2]**

71. Graph the function $y = \frac{1}{x - 4}$. Is the graph of $y = \frac{1}{x}$ stretched, reflected, or shifted in the graph of $y = \frac{1}{x - 4}$? What happens at $x = 4$? **[Lesson 5.2]**

Simplify. [Lesson 7.2]

72. $(x^2 + 2x) + (x + 1)$
$x^2 + 3x + 1$

73. $(x^2 + 2x) - (x + 1)$
$x^2 + x - 1$

74. $(2x^2 + x) - (2x^2 - x)$
$2x$

Look Beyond

The *zeros* of the polynomial function $y = x^2 - 4$ are 2 and -2 because when substituted for x, they result in the value 0 for y. Find the zeros of each function below.

75. $y = x^2 - 9$ 3, -3

76. $y = x^2 - 16$ 4, -4

77. $y = x^2 - 25$ 5, -5

78. Technology Graph the functions in Exercises 75–77. How can you identify the zeros on a graph? Explain your reasoning.

LESSON 7.4

Exploring Multiplication of Binomials

The Distributive Property can be used to multiply a monomial and a binomial. It can also be used to multiply two binomials.

Exploration 1 *Multiplying With Tiles*

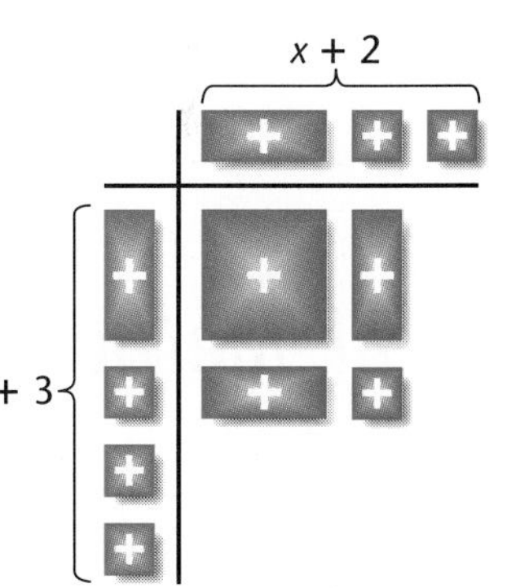

1. Complete the algebra-tile model.
2. What two factors are multiplied? $x + 3$ and $x + 2$
3. What product is represented by the completed model? $x^2 + 5x + 6$
4. If a rectangle has sides of $x + 3$ and $x + 2$, what is the area? $x^2 + 5x + 6$
5. Explain how to use an algebra-tile model to show that $(x + 4)(x + 5) = x^2 + 9x + 20$. ❖

5. One factor is 1 positive x-tile and 4 positive 1-tiles. The other factor is 1 positive x-tile and 5 positive 1-tiles. The product rectangle will have 1 positive x^2-tile, 9 positive x-tiles, and 20 positive 1-tiles.

The Distributive Property can be used to show that $(x + 3)(x + 2) = x^2 + 5x + 6$.

$(x + 3)(x + 2)$	Given
$= x(x + 3) + 2(x + 3)$	Distribute the first binomial to *each term of the second binomial*.
$= x^2 + 3x + 2x + 6$	Distribute *each term* of the second binomial *to each term* of the first binomial.
$= x^2 + \quad 5x \quad + 6$	Add like terms.

Use the Distributive Property to show that $(x + 4)(x + 5) = x^2 + 9x + 20$.

Exploration 2 *Multiplying Binomials*

$$(x + 3)(x + 5) = x^2 + 8x + 15$$

$$(x + 3)(x + 5) = x^2 + 5x + 3x + 15$$

$$(x + 3)(x + 5) = x^2 + \quad 8x \quad + 15$$

$$(x + 3)(x + 5) = x^2 + 8x + 15$$

1. What is the relationship between the first term of each binomial factor and the first term of the trinomial? The product of the first terms of each factor is the first term of the trinomial.

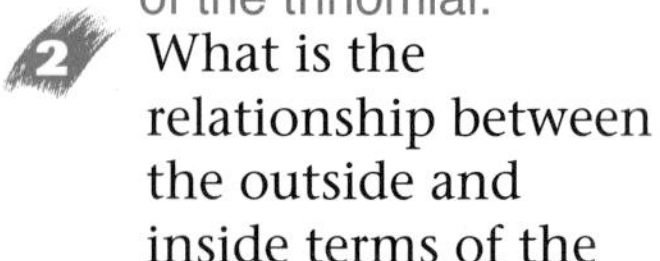

2. What is the relationship between the outside and inside terms of the binomial factors and the middle term of the trinomial? The sum of the products of the outside terms and inside terms is the middle term of the trinomial.

3. What is the relationship between the last term of each binomial factor and the last term of the trinomial? The product of the last terms of each factor is the last term of the trinomial.

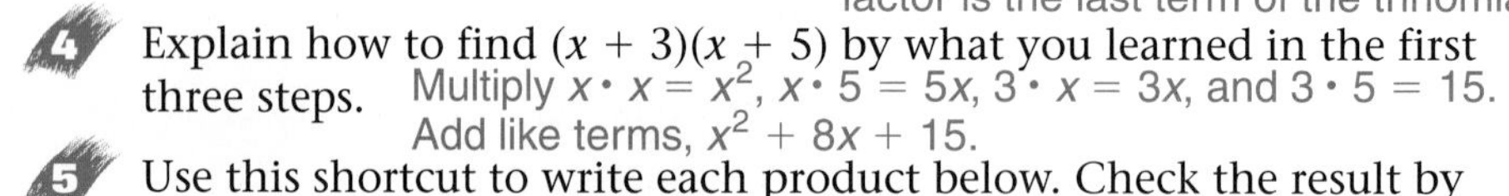

4. Explain how to find $(x + 3)(x + 5)$ by what you learned in the first three steps. Multiply $x \cdot x = x^2$, $x \cdot 5 = 5x$, $3 \cdot x = 3x$, and $3 \cdot 5 = 15$. Add like terms, $x^2 + 8x + 15$.

5. Use this shortcut to write each product below. Check the result by using a tile model or the Distributive Property.

   a. $(x + 3)(x - 5)$ **b.** $(x - 3)(x + 5)$ **c.** $(x - 3)(x - 5)$

5a. $x^2 - 2x - 15$
5b. $x^2 + 2x - 15$
5c. $x^2 - 8x + 15$

6. Show how you would use this shortcut to complete the following statement: $(x + a)(x + b) = \underline{?}$ ❖

   $x^2 + xb + ax + ab$

FOIL Method

In Exploration 2 you used a method for multiplying two binomials that is usually referred to as **FOIL**.

- Multiply **F**irst terms.
- Multiply **O**utside terms. Multiply **I**nside terms. Add outside and inside products.
- Multiply **L**ast terms.

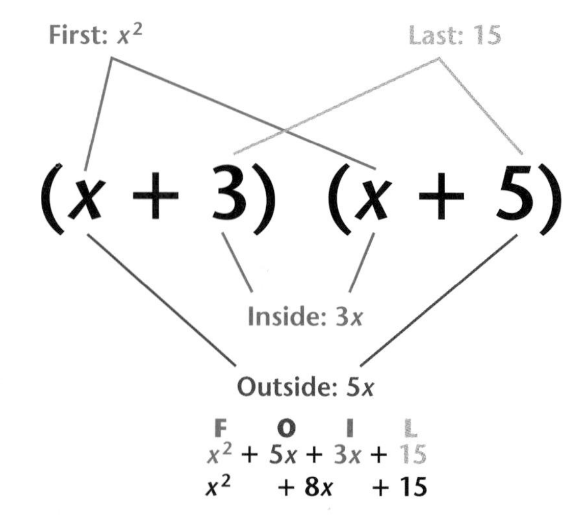

EXAMPLE 1

Use the FOIL method to find each product.

A $(3x + 1)(3x - 2)$ **B** $(2a - 1)(5a + 3)$ **C** $(5m - 2)(5m + 2)$

Solution ➤

Multiply to find each term, and then simplify.

A $(3x + 1)(3x - 2) = 9x^2 - 6x + 3x - 2 = 9x^2 - 3x - 2$

B $(2a - 1)(5a + 3) = 10a^2 + 6a - 5a - 3 = 10a^2 + a - 3$

C $(5m - 2)(5m + 2) = 25m^2 + 10m - 10m - 4 = 25m^2 - 4$ ❖

Draw an algebra-tile diagram to show that $(2x + 3)(x + 1) = 2x^2 + 5x + 3$. Explain how you can check your diagram.

EXAMPLE 2

Small Business Cruz's Frame Shop makes a mat by cutting out the inside of a rectangular board. Find the length and width of the original board if the area of the mat is 148 square inches.

$2x - 1$

$2x - 5$

$x + 6$

$x + 2$

Solution ➤

Find the areas of the inside and outside rectangles formed by the mat.

Area of the outside rectangle	Area of the inside rectangle
$(2x - 1)(x + 6)$	$(x + 2)(2x - 5)$
$2x^2 + 12x - x - 6$	$2x^2 - 5x + 4x - 10$
$2x^2 + 11x - 6$	$2x^2 - x - 10$

Substitute each area into the formula, and then solve for x.

Area outside rectangle	−	Area inside rectangle	=	Area of the mat
$2x^2 + 11x - 6$	−	$(2x^2 - x - 10)$	=	148
$2x^2 + 11x - 6$	−	$2x^2 + x + 10$	=	148
		$12x + 4$	=	148
		$12x$	=	144
		x	=	**12**

The length of the original board is $2x - 1$. Substitute 12 for x.

$$2x - 1 = 2(\mathbf{12}) - 1 = 23$$

The width is $x + 6$. Substitute 12 for x.

$$x + 6 = \mathbf{12} + 6 = 18$$

The original board was 23 inches by 18 inches. ❖

The Punnett square shows the possible results of crossing two flowers, each containing a purple gene and a white gene. Since the purple gene is dominant, only flowers with a *ww* genetic makeup are white.

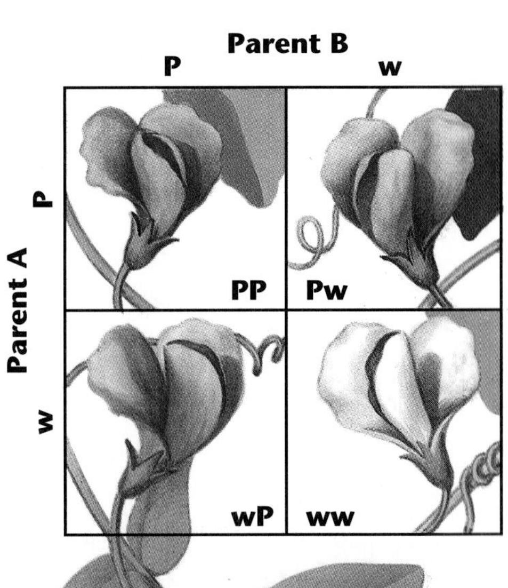

EXAMPLE 3

Show how squaring a binomial models the probability that crossing two hybrid purple (*Pw*) flowers results in a flower that has a 25% probability of being white.

Solution

Square the binomial $P + w$.

$(P + w)^2 = P^2 + 2Pw + w^2 = PP + Pw + Pw + ww$

Since there are four possibilities and one is w^2 or *ww*, the probability that the offspring will be white is $\frac{1}{4}$ or 25%. ❖

EXERCISES & PROBLEMS

Communicate

1. Show how to model each product.

a. $(x + 1)(x + 2)$

b. $(x - 1)(x - 2)$

2. Explain how to use the Distributive Property to find each product.

a. $(x + 1)(x + 2)$

b. $(x - 1)(x - 2)$

3. Describe how to find $(2x + 3)(x - 4)$ by using the FOIL method.

4. Use algebra tiles to model the areas $(x + 2)^2$ and $x^2 + 2^2$. How are the two models different?

5. $x^2 + 7x + 10$ 6. $a^2 + a - 12$ 7. $b^2 - 4b + 3$
8. $y^2 + y - 6$ 9. $c^2 + 10c + 25$ 10. $d^2 - 25$

Practice & Apply

Use the Distributive Property to find each product.

5. $(x + 2)(x + 5)$ **6.** $(a - 3)(a + 4)$ **7.** $(b - 1)(b - 3)$

8. $(y + 3)(y - 2)$ **9.** $(c + 5)(c + 5)$ **10.** $(d + 5)(d - 5)$

Use the FOIL method to find each product.

11. $(y + 5)(y + 3)$
12. $(w + 9)(w + 1)$
13. $(b - 7)(b + 3)$
14. $(3y - 2)(y - 1)$
15. $(5p + 3)(p + 1)$
16. $(2q - 1)(2q + 1)$
17. $(2x + 5)(2x - 3)$
18. $(4m + 1)(5m - 3)$
19. $(2w - 9)(3w - 8)$
20. $(3x - 5)(3x - 5)$
21. $(7s + 2)(2s - 3)$
22. $\left(y - \frac{1}{2}\right)\left(y + \frac{1}{2}\right)$
23. $\left(y - \frac{1}{3}\right)\left(y + \frac{5}{9}\right)$
24. $(c^2 + 1)(c^2 + 2)$
25. $(2a^2 + 3)(2a^2 + 3)$
26. $(a + c)(a + 2c)$
27. $(p + q)(p + q)$
28. $(a^2 + b)(a^2 - b)$
29. $(x^2 + y)(x + y)$
30. $(c + d)(2c + d)$
31. $(1.2m + 5)(0.8m - 4)$

Geometry A rectangular garden has a length of $x + 8$ units and a width of $(x - 4)$ units.

32. Draw a diagram, and label the dimensions.

33. Find the area. $x^2 + 4x - 32$ sq units

34. **Geometry** Find the area of a square rug that is $(y + 6)$ units on a side. $y^2 + 12y + 36$

35. **Geometry** Which has the greater area, a square with sides $(x + 1)$ units long or a rectangle with length $(x + 2)$ units and width x units? How much greater? square; 1 unit

36. **Geometry** Which has a greater area, a square with sides $(x - 1)$ units long or a rectangle with length x units and width $(x - 2)$ units? How much greater? square; 1 unit

Look Back

Solve. [Lesson 1.6]

37. $|5 - 3x| = 9$ $\quad -1\frac{1}{3}, 4\frac{2}{3}$
38. $|16x + 2| \leq 4$ $\quad x \leq \frac{1}{8}$ or $x \geq \frac{3}{8}$
39. $|7x - 9| > -11$ $\quad -\frac{2}{7} < x < 2\frac{6}{7}$

Find the slope of each line. [Lesson 2.1]

40. $4x + 3y = 12$ $\quad -\frac{4}{3}$
41. $y = 4x$ $\quad 4$

42. Write the equation of the line through $(-2, -1)$ and parallel to $y = -2x - 4$. **[Lesson 2.2]** $y = -2x - 5$

Solve each system of equations by graphing. Check your answers by substitution. [Lesson 2.3]

43. $\begin{cases} 2x + y = 9 \\ -\frac{1}{2}x + y = -1 \end{cases}$
44. $\begin{cases} x + y = 4 \\ x + y = 10 \end{cases}$

Look Beyond

Where does the graph of each function cross the x-axis?

45. $f(x) = (x + 3)(x - 2)$ $\quad -3, 2$
46. $g(x) = (x + 2)^2$ $\quad -2$
47. $h(x) = x^2 - 4$ $\quad -2, 2$

LESSON 7.5 Common Factors

The product of two factors is found by multiplication. Sometimes the product and one of the two factors are known and it is necessary to find the second factor.

Two unit squares can be used to form a rectangle with a height of 1 and a length of 2. A rectangle with a height of 2 and a length of 1 can also be formed.

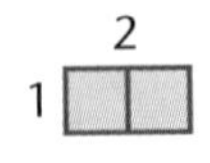

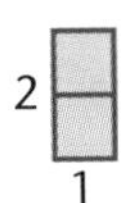

However, the two rectangles will be considered the same because the second rectangle is a rotation of the first.

Four unit squares form two different rectangles, a 2-by-2 and a 1-by-4. A 4-by-1 rectangle is the same as a 1-by-4 rectangle.

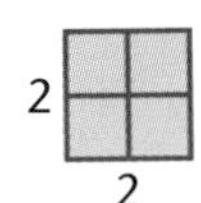

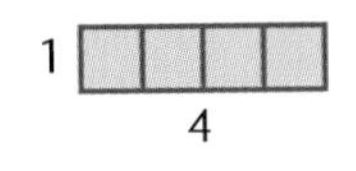

Exploration *Prime Numbers*

1. d. 1
   e. 2
   f. 1
   g. 2
   h. 2
   i. 2
   j. 1
   k. 3
   l. 1

1 Use unit squares to form rectangles. Record the number of different rectangles that you can make.

	Number of unit squares	Number of rectangles		Number of unit squares	Number of rectangles
a.	2	1	g.	8	?
b.	3	1	h.	9	?
c.	4	2	i.	10	?
d.	5	?	j.	11	?
e.	6	?	k.	12	?
f.	7	?	l.	13	?

2. 2, 3, 5, 7, 11, 13; 2, 3, 5, 7, 11, 13, 17, 19, 23, 29
3. 4, 6, 8, 9, 10, 12, 14, 15, 16, 18, 20, 21, 22, 24, 25, 26, 27, 28

2 Which numbers in Step 1 form exactly 1 rectangle? These numbers are called **prime numbers**. List the prime numbers less than 30.

3 The numbers in the list that are not prime are called **composite numbers**. List the composite numbers less than 30.

4 The number 1 is not considered to be prime or composite. Why do you think this is so?

5 Give a definition of a prime number that includes the word *factor*. ❖

When two numbers are multiplied, they form a product. Each number is called a **factor** of that product. Thus, 3 and 8 are factors of 24 because $3 \cdot 8 = 24$, and one way to express 24 in factored form is $3 \cdot 8$. Sometimes individual factors can be factored further. The process ends when all of the factors are prime. Name the prime factors of 24. $24 = 2 \cdot 2 \cdot 2 \cdot 3 = 2^3 \cdot 3$

When you are asked to factor $3x^2 + 12x$, the first step is to examine the terms for a common monomial factor. Since $\mathbf{3x}$ is common to both terms, write

$$3x^2 + 12x = \mathbf{3x}(x + 4).$$

↑
common monomial factor

The polynomial $3x^2 + 12x$ has been factored over the integers because each term in $3x(x + 4)$ has an integer coefficient. When a polynomial has no polynomial factors with integral coefficients except itself and 1, it is a **prime polynomial** with respect to the integers. In this text, to factor means to factor over the integers.

EXAMPLE 1

Factor each polynomial.

A $5am - 5an$ **B** $5x^3 - 3y^2$ **C** $2c^4 - 4c^3 + 6c^2$

Solution ➤

Factor the greatest common factor, or GCF, from each term.

A The GCF is $\mathbf{5a}$. $5am - 5an = \mathbf{5a}(m - n)$

B The GCF is 1. $5x^3 - 3y^2$ is prime.

C The GCF is $\mathbf{2c^2}$. $2c^4 - 4c^3 + 6c^2 = \mathbf{2c^2}(c^2 - 2c + 3)$

To check each factorization, multiply the two factors to find the original expression. ❖

CT– The only common factor of 2 and 3 is 1.

Substitute numbers for the variables in the previous example and simplify. Compare the results. What do you find? The results are the same.

CRITICAL *Thinking*

Even though the expression $2x + 3y$ can be written as the product of a monomial and a binomial, $6\left(\frac{1}{3}x + \frac{1}{2}y\right)$, the binomial $2x + 3y$ is considered prime. Explain why.

Sometimes a polynomial has a common factor that contains two terms. This expression is called a **common binomial factor**.

EXAMPLE 2

Factor $r(t + 1) + s(t + 1)$.

Solution ➤

Notice that $r(t + 1) + s(t + 1)$ contains the common binomial factor $t + 1$. Think of $t + 1$ as c.

$r(t + 1) + s(t + 1) = rc + sc$	Substitute c for $(t + 1)$.
$= (r + s)c$	Factor.
$= (r + s)(t + 1)$	Replace c with $(t + 1)$.

The expression can also be factored mentally if you apply the Distributive Property.

$$r(t + 1) + s(t + 1) = (r + s)(t + 1)$$ ❖

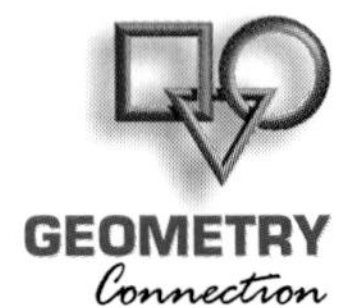

You can visualize this factoring procedure with an area model. It shows the same total area in different arrangements.

$t + 1$ $t + 1$ $t + 1$

r [ ] + s [ ] = $r + s$ [ ]

$$r(t + 1) + s(t + 1) = (r + s)(t + 1)$$

Try This Factor $a(x - 3) + 6(x - 3)$. $(a + 6)(x - 3)$

When a polynomial has terms with more than one common factor, the expression can sometimes be factored by grouping terms.

EXAMPLE 3

Factor $x^2 + x + 2x + 2$ by grouping.

Solution ➤

There are several ways to factor this expression by grouping. One way is to group the terms that have a common coefficient or variable. Treat $x^2 + x$ as one expression, and treat $2x + 2$ as another expression.

$x^2 + x + 2x + 2 = (x^2 + x) + (2x + 2)$	Group terms.
$= x(x + 1) + 2(x + 1)$	Factor the GCF from each group.
$= (x + 2)(x + 1)$	Write the expression as the product of two binomials.

To check, multiply the factors. ❖

Try This Factor $ax + bx + ay + by$ by grouping. $(a + b)(x + y)$

EXERCISES & PROBLEMS

Communicate

1. Define a *prime* polynomial.
2. What property of mathematics is used when you factor by removing a GCF?
3. Once you have found the GCF, how do you find the remaining factor of the polynomial?

Each of the given pairs contains a GCF. Find the GCF, and explain how you found it.

4. 60 and 150
5. x^3y^5 and x^5y^2
6. $25(x + y)$ and $39(x + y)$
7. Explain how to group the terms to factor $y^2 + 2y + 3y + 6$.

Practice & Apply

Identify each polynomial as prime or composite.

8. $4x^2 - 16$ composite
9. $r^2 + 10$ prime
10. $n^2 + 4$ prime

Factor each polynomial by removing the GCF.

11. $2x^2 - 4$ $2(x^2 - 2)$
12. $5n^2 - 10$ $5(n^2 - 2)$
13. $3x^2 + 6x$ $3x(x + 2)$
14. $x^9 - x^2$ $x^2(x^7 - 1)$
15. $k^5 + k^2$ $k^2(k^3 + 1)$
16. $4a^8 - 20a^6 + 8a^4$
17. $4x^2 + 2x - 6$ $2(2x^2 + x - 3)$
18. $7x^2 - 28x - 14$ $7(x^2 - 4x - 2)$
19. $27y^3 + 18y^2 - 81y$
20. $3m^3 - 9m^2 + 3m$
21. $90 + 15a^5 - 45a$
22. $2x^3y - 18x^2y^2 + 17xy^3$

16. $4a^4(a^4 - 5a^2 + 2)$ 19. $9y(3y^2 + 2y - 9)$ 20. $3m(m^2 - 3m + 1)$ 21. $15(6 + a^5 - 3a)$ 22. $xy(2x^2 - 18xy + 17y^2)$

Write each as the product of two binomials.

23. $x(x + 1) + 2(x + 1)$
24. $5(y + 3) - x(y + 3)$
25. $a(x + y) + b(x + y)$
26. $(4 + p)3q - 4(4 + p)$
27. $x(x - 1) + 2(x - 1)$
28. $r(x - 4) + t(x - 4)$
29. $5a(a - 3) + 4(a - 3)$
30. $2w(w + 4) - 3(w + 4)$
31. $2(x - 2) + x(2 - x)$
32. $8(y - 1) - x(y - 1)$
33. $2r(r - s)^2 - 3(r - s)^2$
34. $ax(u - v)^n + bz(u - v)^n$

Factor.

35. $2x + 2y + ax + ay$
36. $nu + nv + 3u + 3v$
37. $12ab - 15a - 8b + 10$
38. $ax + ay + 12x + 12y$
39. $3(x + y) + 12(x + y)$
40. $x^2 + 3x + 4x + 12$
41. $2n^2 - 6n + 14n - 42$
42. $6pq + 12p^2 - 8qp + 2p^2$
43. $5x(2d + 3)^3 - 10(2d + 3)^3$

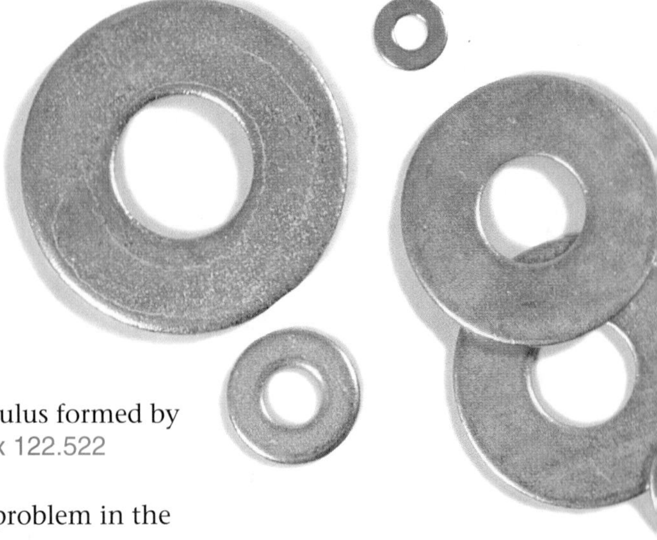

An annulus is an object shaped like a washer.

Geometry The shaded area between the two concentric circles is called an **annulus**. The formula for the area of the *annulus* is $\pi R^2 - \pi r^2$, where the radius of the larger circle is R and the radius of the smaller circle is r.

44. Write the formula for the area of an annulus in factored form. $\pi(R^2 - r^2)$

45. Use the factored form to find the area of an annulus formed by concentric circles with radii of 8 and 5. Approx 122.522

46. **Portfolio Activity** Complete the problem in the portfolio activity on page 315. $\frac{3}{4}$

47. No, the first differences are not constant.

Look Back

47. Is the sequence 3, 6, 12, 24, 48 representative of a linear relationship? How do you know? **[Course 1]**

48. Evaluate $8 - 5(2 - 6) - 3^2$. **[Course 1]** 19

49. Solve the formula $A = \frac{1}{2}h(b_1 + b_2)$ for h. **[Lesson 1.4]** $\frac{2A}{b_1 + b_2}$

50. Solve $3(x - 5) - 2(x + 8) = 7x + 3$. **[Lesson 1.4]** $-\frac{17}{3}$, or $-5\frac{2}{3}$

51. Graph $x - 4y = 8$. **[Lesson 2.2]**

52. Solve this system. $\begin{cases} 4x = 11 + 15y \\ 6x + 5y = 0 \end{cases}$ **[Lesson 2.5]** $\left(\frac{1}{2}, -\frac{3}{5}\right)$

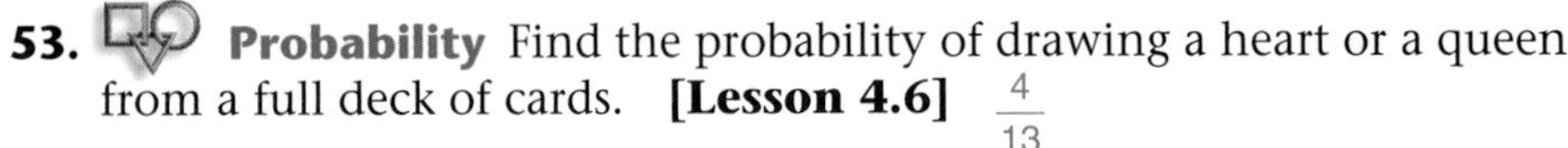

53. **Probability** Find the probability of drawing a heart or a queen from a full deck of cards. **[Lesson 4.6]** $\frac{4}{13}$

Given $y = 3x - 5$ as the original function, write the equation that represents each transformation.

54. a reflection across the x-axis **[Lesson 5.4]** $y = -3x + 5$

55. a translation of 4 units to the right and 2 units down **[Lesson 5.5]** $y = 3x - 19$

56. Evaluate $\frac{2^2 \cdot 2^4}{2^5} - 2^0$. **[Lesson 6.3]** 1

Look Beyond

57. **Physics** When a projectile is fired vertically into the air, its motion can be modeled by the equation $h = -16t^2 + 320t$, where h is the height in feet at time t in seconds. From the graph of the equation, find the time when $h = 1200$.

LESSON 7.6 Factoring Special Polynomials

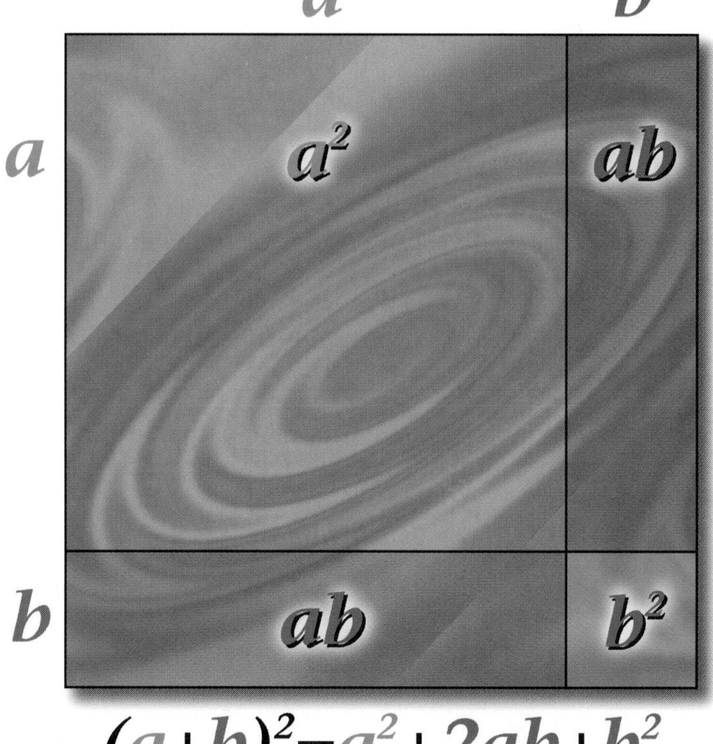

$$(a+b)^2=a^2+2ab+b^2$$

why *If the length of one side of a square is **a**, the expression for its area is **a**^{2}. If the length of the side is increased, then the new expression for the area exhibits a simple pattern.*

Perfect-Square Trinomial

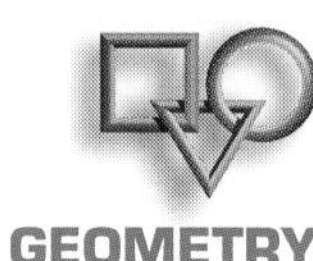

GEOMETRY *Connection*

You can see from the geometry of the figure that the area of the large square is $(a + b)^2$, which is $a^2 + 2ab + b^2$. This shows why the product $a^2 + 2ab + b^2$ is called a **perfect-square trinomial**.

The same pattern is used to square the expression $a - b$. Square the first term, subtract twice the product of the binomial terms, and add the square of the last term.

$$(a - b)^2 = a^2 - 2ab + b^2$$

Explain how to find the perfect-square trinomials for $(2x + 6)^2$ and $(2x - 6)^2$. How do their products compare?

To see if a trinomial such as $4x^2 + 12x + 9$ is a perfect square, use this pattern.

- The first term, $4x^2$, is a perfect square, $(\mathbf{2x})^2$.
- The last term, 9, is a perfect square, $(\mathbf{3})^2$.
- The middle term, $12x$, is $2(\mathbf{2x})(\mathbf{3})$.

Thus, $4x^2 + 12x + 9$ is a perfect square trinomial.

EXAMPLE 1

Examine each expression. Is the expression a perfect-square trinomial? Explain.

A $x^2 + 8x + 16$ **B** $3x^2 + 16x + 16$ **C** $4x^2 - 3xy - y^2$

D $4a^2 + 12ab + 9b^2$ **E** $9y^2 + 6xy + x^2$ **F** $16x^2 + 12xy + y^2$

Solution ➤

A Yes; $(x + 4)^2$ **B** No; $3x^2$ is not a perfect square.

C No; $4x^2$ and y^2 are perfect squares, but the sign in front of y^2 is negative, not positive. Also the middle term does not equal $2(2x)(y)$.

D Yes; $(2a + 3b)^2$ **E** Yes; $(3y + x)^2$

F No, $16x^2$ and y^2 are perfect squares, but the middle term is not $2(4x)(y)$. ❖

The perfect-square trinomial pattern can be used to factor expressions in the form of $a^2 + 2ab + b^2$ or $a^2 - 2ab + b^2$.

For example, $9m^2 + 12mn + 4n^2$ fits the pattern.
The first term is a perfect square, $\mathbf{(3m)^2}$.
The last term is a perfect square, $\mathbf{(2n)^2}$.
The middle term, $\mathbf{12mn}$, is $2(3m)(2n)$.
Thus, $\mathbf{9m^2} + \mathbf{12mn} + \mathbf{4n^2} = (3m + 2n)^2$.

FACTORING A PERFECT-SQUARE TRINOMIAL

For all numbers a and b,

$$a^2 + 2ab + b^2 = (a + b)(a + b) = (a + b)^2, \text{ and}$$
$$a^2 - 2ab + b^2 = (a - b)(a - b) = (a - b)^2.$$

EXAMPLE 2

Factor each expression.

A $x^2 - 10x + 25$ **B** $9s^2 + 24s + 16$

C $64a^2 - 16ab + b^2$ **D** $49y^4 + 14y^2 + 1$

Solution ➤

A $x^2 - 10x + 25 = (x - 5)^2$ **B** $9s^2 + 24s + 16 = (3s + 4)^2$

C $64a^2 - 16ab + b^2 = (8a - b)^2$ **D** $49y^4 + 14y^2 + 1 = (7y^2 + 1)^2$

To check, substitute a number for the variable and evaluate. ❖

$x^2 + 6x + 9 = (x + 3)^2$ and $x^2 + 6x + 9 = 5^2$. So, $x + 3 = 5 \Rightarrow x = 2$.

The area of a square is $x^2 + 6x + 9$ square units. If the length of one side is 5 units, find the value of x by factoring.

Difference of Two Squares

When you multiply $(a + b)(a - b)$, the product is $a^2 - b^2$. This product is called the **difference of two squares**.

$$\begin{aligned}(\mathbf{a} + \mathbf{b})(\mathbf{a} - \mathbf{b}) &= a(a - b) + b(a - b) \\ &= a^2 - ab + ab - b^2 \\ &= \mathbf{a^2} \quad - \quad \mathbf{b^2}\end{aligned}$$

EXAMPLE 3

Examine each expression. Is the expression a difference of two squares? Explain.

A $4a^2 - 25$ **B** $9x^2 - 15$ **C** $b^2 + 49$
D $c^2 - 4d^2$ **E** $a^3 - 9$

Solution ➤

A Yes; $(2a)^2 - 5^2$

B No; 15 is not a perfect square.

C No; $b^2 + 49$ is not a difference; it is a sum.

D Yes; $c^2 - (2d)^2$

E No; a^3 is not a perfect square. ❖

The difference-of-two-squares pattern can be used to factor expressions in the form $a^2 - b^2$.

For example, $4c^2 - 81d^2$ fits the pattern.
The first term is a perfect square, $(\mathbf{2c})^2$.
The second term is a perfect square, $(\mathbf{9d})^2$.
The terms are subtracted.

Thus, $\mathbf{4c^2} - \mathbf{81d^2} = (2c + 9d)(2c - 9d)$.

FACTORING A DIFFERENCE OF TWO SQUARES
For all numbers a and b, $a^2 - b^2 = (a + b)(a - b)$.

EXAMPLE 4

Factor each expression.

A $x^2 - 4$ **B** $36a^2 - 49b^2$ **C** $16x^2 - 25$ **D** $m^4 - n^4$

Solution ➤

A $x^2 - 4 = (x + 2)(x - 2)$

B $36a^2 - 49b^2 = (6a + 7b)(6a - 7b)$

C $16x^2 - 25 = (4x + 5)(4x - 5)$

D $m^4 - n^4 = (m^2 + n^2)(m^2 - n^2)$
$= (m^2 + n^2)(m + n)(m - n)$ ❖

Try This Factor $25w^2 - 81$. $(5w + 9)(5w - 9)$

Explain why $a^8 - b^8$ has more than two factors. What are they?

CT– When $a^8 - b^8$ is factored, you get $(a^4 + b^4)(a^4 - b^4)$. Notice that $a^4 - b^4$ can also be factored: $(a^4 - b^4) = (a^2 + b^2)(a^2 - b^2)$. Finally, the difference of two squares can be factored: $(a^2 - b^2) = (a + b)(a - b)$. So $a^8 - b^8 = (a^4 + b^4)(a^2 + b^2)(a + b)(a - b)$.

EXAMPLE 5

Find each product by using the difference of two squares.

A $31 \cdot 29$ **B** $17 \cdot 13$ **C** $34 \cdot 26$

Solution ➤

A Think of $31 \cdot 29$ as $(30 + 1)(30 - 1)$.
The product is $30^2 - 1^2 = 900 - 1 = 899$.

B Think of $17 \cdot 13$ as $(15 + 2)(15 - 2)$.
The product is $15^2 - 2^2 = 225 - 4 = 221$.

C Think of $34 \cdot 26$ as $(30 + 4)(30 - 4)$.
The product is $30^2 - 4^2 = 900 - 16 = 884$. ❖

Graph $y = x^2 - 10x + 25$. Explain how the graph of a perfect-square trinomial can give you the factors of the expression. Use your graphics calculator to factor $x^2 - 16x + 64$.

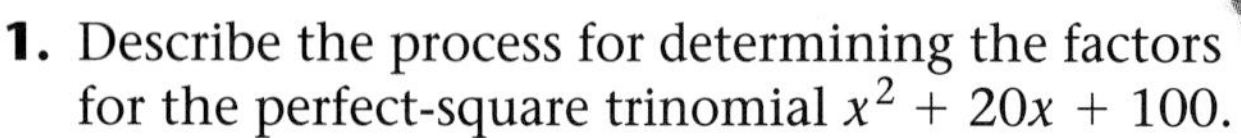

Communicate

1. Describe the process for determining the factors for the perfect-square trinomial $x^2 + 20x + 100$.
2. Explain how to factor $4x^2 - 12x + 9$.
3. Explain how to factor the difference of two squares, $p^2 - 121$.

Practice & Apply

Use the generalization of a perfect-square trinomial or the difference of two squares to find each product.

4. $(p + 3)^2$ $p^2 + 6p + 9$
5. $(2x - 1)^2$ $4x^2 - 4x + 1$
6. $(a - 4)(a + 4)$ $a^2 - 16$
7. $(7y - 9)(7y + 9)$ $49y^2 - 81$
8. $(8x - 3y)(8x + 3y)$ $64x^2 - 9y^2$
9. $(5z - 12)^2$ $25z^2 - 120z + 144$

Find the missing term in each perfect-square trinomial.

10. $x^2 - 14x + \underline{?}$ 49
11. $16y^2 + \underline{?} + 9$ $24y$
12. $25a^2 + 60a + \underline{?}$ 36
13. $9x^2 + \underline{?} + 25$ $30x$
14. $x^2 - 12x + \underline{?}$ 36
15. $\underline{?} - 36y + 81$ $4y^2$

Factor each polynomial completely. 26. $4(5 + 3q)(5 - 3q)$ 29. $(3c + 2d)(3c - 2d)$

16. $x^2 - 4$ $(x + 2)(x - 2)$
17. $x^2 + 4x + 4$ $(x + 2)^2$
18. $y^2 - 100$ $(y + 10)(y - 10)$
19. $y^2 + 8y + 16$ $(y + 4)^2$
20. $16c^2 - 25$ $(4c + 5)(4c - 5)$
21. $4t^2 - 1$ $(2t + 1)(2t - 1)$
22. $81 - 4m^2$ $(9 + 2m)(9 - 2m)$
23. $25x^2 - 9$ $(5x + 3)(5x - 3)$
24. $r^2 - 18r + 81$ $(r - 9)^2$
25. $4x^2 - 20x + 25$ $(2x - 5)^2$
26. $100 - 36q^2$
27. $36d^2 + 12d + 1$ $(6d + 1)^2$
28. $p^2 - q^2$ $(p + q)(p - q)$
29. $9c^2 - 4d^2$
30. $16x^2 + 72xy + 81y^2$
31. $9a^2 - 12a + 4$ $(3a - 2)^2$
32. $49x^2 - 42xy + 9y^2$
33. $a^2x^2 + 2axb + b^2$ $(ax + b)^2$
34. $4m^2 + 4mn + n^2$ $(2m + n)^2$
35. $81a^4 - 9b^2$
36. $x^4 - y^4$
37. $x^2(25 - x^2) - 4(25 - x^2)$

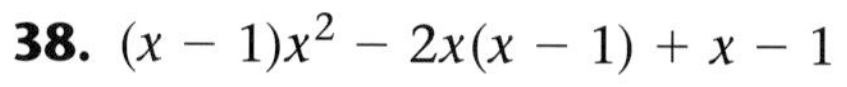

38. $(x - 1)x^2 - 2x(x - 1) + x - 1$
39. $(3x + 5)(x^2 - 3) - (3x + 5)$
40. $(x^2 - y^2)(x^2 + 2xy) + (x^2 - y^2)(y^2)$

30. $(4x + 9y)^2$ 32. $(7x - 3y)^2$ 35. $9(3a^2 + b)(3a^2 - b)$ 36. $(x^2 + y^2)(x + y)(x - y)$

41. Cultural Connection: Africa Abu Kamil, known as the "Egyptian Calculator," used geometric models around the year 900 C.E. to solve problems. You can make his models with algebra tiles or paper rectangles that you can cut.

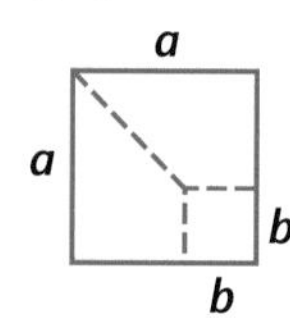

Use a sheet of paper to create a model of $a^2 - b^2$ as shown. Remove the b^2 section from the corner. Next cut along the dotted line, and move the pieces to form a rectangle. Explain how this shows the factorization of $a^2 - b^2$.

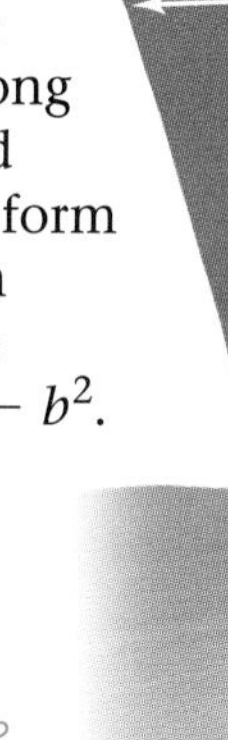

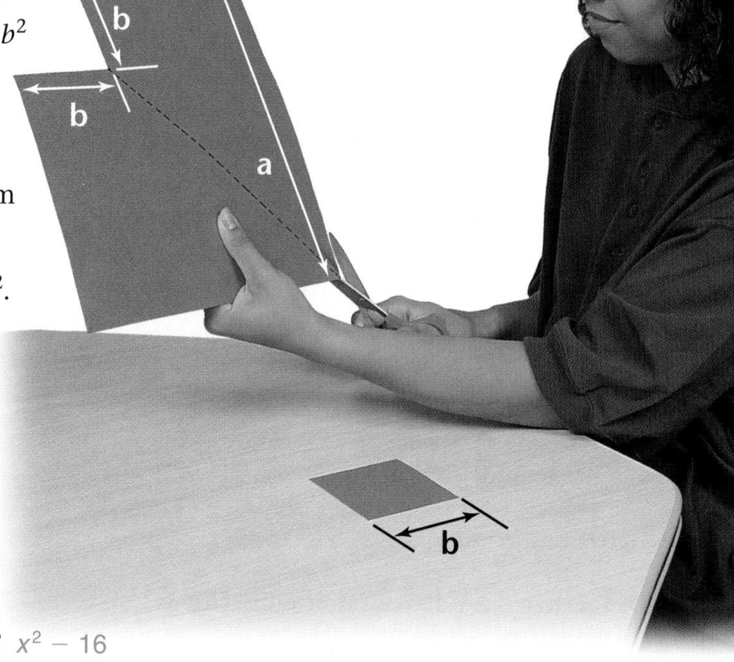

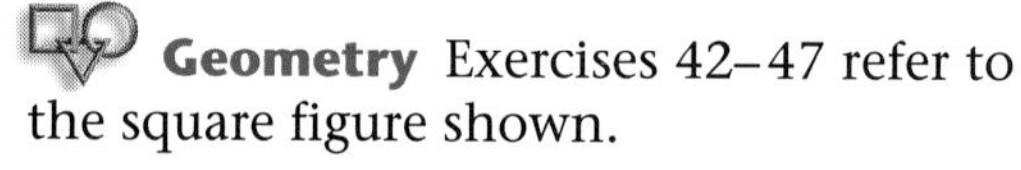

Geometry Exercises 42–47 refer to the square figure shown.

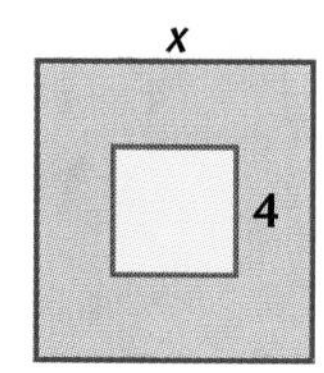

42. What is the area of the large square? x^2

43. What is the area of the small square? 16

44. If the small square is removed, what is the area of the blue region? $x^2 - 16$

45. Factor the polynomial that represents the area of the blue region. $(x + 4)(x - 4)$

46. Draw a rectangle whose dimensions are the factors you just found.

47. Show that the area of this rectangle equals the area of the blue region.

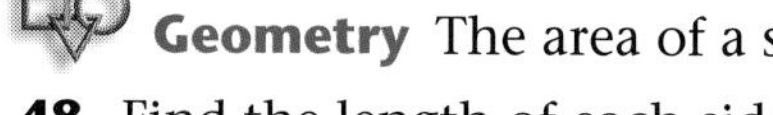

Geometry The area of a square is represented by $n^2 - 12n + 36$.

48. Find the length of each side. $n - 6$

49. Find the perimeter of the square. $4n - 24$

50. a^2; 9; $a^2 - 9$; $(a + 3)(a - 3)$

50. **Geometry** Within a large square whose side is a units is a small square whose side is 3 units. What is the area of the large square? What is the area of the small square? What is the area of the yellow surface between the two squares? Factor the expression for the yellow area.

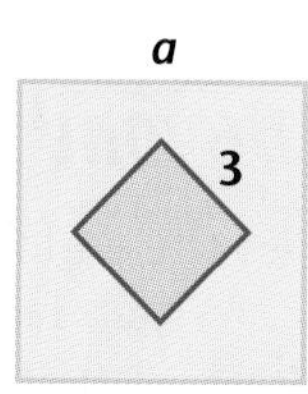

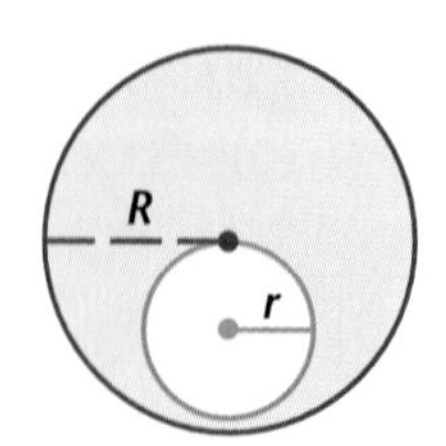

51. **Geometry** A small circle with a radius of r units is drawn within a large circle with a radius of R units. What is the area of the large circle? What is the area of the shaded part? Factor the expression for the shaded area. HINT: First, factor out the common monomial factor, π. Then factor the binomial.

πR^2; $\pi R^2 - \pi r^2$; $\pi(R + r)(R - r)$

Look Back

52. There is exactly one line that connects any two points. Use the table and the number pattern to find how many lines can be drawn that connect two points at a time for 10 points. **[Course 1]**
45

Points	2	3	4	5	6	. . .	10
Lines	1	3	6	?	?	. . .	?
				10	15		45

53. **Statistics** Mark has scores of 78, 83, and 92 on his first 3 tests. What must his score be on the next test to have an average of 87? **[Lesson 1.4]** 95

54. Graph the solution of $|x| < 4$ on a number line. **[Lesson 1.6]**

A line is represented by the equation $3y = 5 - 4x$.

55. Find the slope of a line parallel to the given line. **[Lesson 2.1]** $-\frac{4}{3}$

56. Find the slope of a line perpendicular to the given line. **[Lesson 2.1]** $\frac{3}{4}$

57. Graph this system of inequalities. $\begin{cases} y < x - 5 \\ y \geq 3 \end{cases}$ **[Lesson 2.7]**

58. Find the inverse matrix of $\begin{bmatrix} 1 & 2 \\ -2 & 1 \end{bmatrix}$. **[Lesson 3.4]** $\begin{bmatrix} \frac{1}{5} & -\frac{2}{5} \\ \frac{2}{5} & \frac{1}{5} \end{bmatrix}$

59. **Probability** Two coins are tossed. Find the probability that at least one coin is heads. **[Lesson 4.6]** $\frac{3}{4}$ or 75%

Determine whether each relation is a function. [Lesson 5.1]

60. $\{(1, 2), (2, 2), (3, 2), (4, 2)\}$ Yes

61. $\{(2, 7), (3, 5), (-2, 4), (3, -2)\}$ No

62. What is the name for the point where the graph of an absolute-value function changes direction? **[Lesson 5.2]** Vertex

63. What is the graph of the equation $y = (x - 5)^2$ called? **[Lesson 5.5]** Parabola

Look Beyond

64. Given the trinomial $x^2 + x - 42$, what are possible pairs of factors for -42? Which factors result in the sum of 1? What does the sign before the last term determine?

LESSON 7.7 Factoring Trinomials

The area of a rectangular garden can be found by using the formula $A = lw$.

Factoring can be thought of as undoing multiplication. If you examine certain products carefully, you will discover patterns that enable you to recognize the factors.

Factoring With Tiles

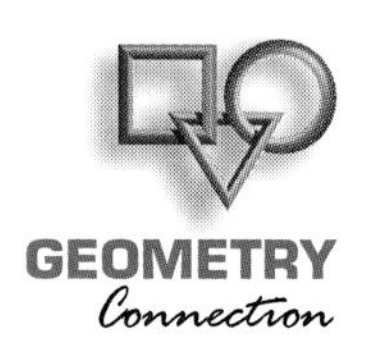

GEOMETRY Connection

Suppose you know that the area of a rectangle is represented by $x^2 + 6x + 8$. How can you find the representation of its length and width? One way to find the length and width is to use algebra tiles. If the tiles can be arranged to form a rectangle, the length and width can be determined.

Start with the tiles that model $x^2 + 6x + 8$.

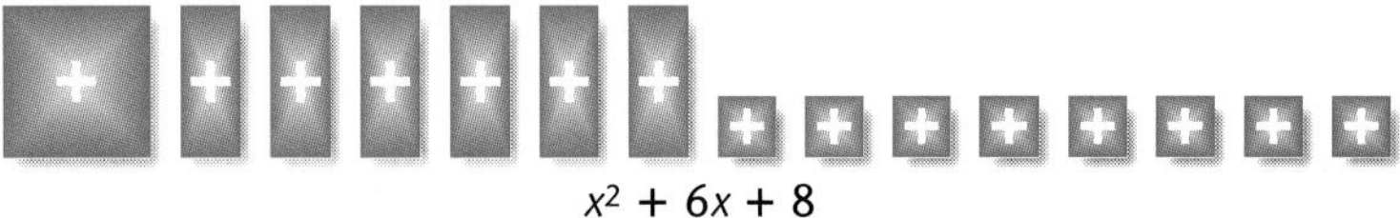

Arrange the tiles in a rectangle.

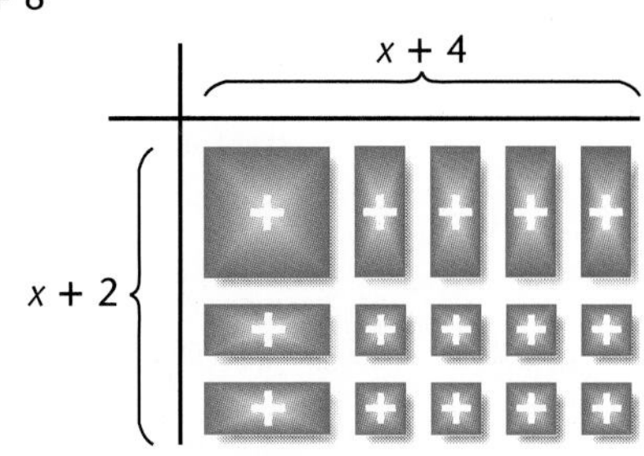

The length is represented by $x + 4$, and the width is represented by $x + 2$. To check, multiply the factors.

$$(x + 2)(x + 4) = x^2 + 2x + 4x + 8$$
$$= x^2 + 6x + 8$$

Sometimes using tiles may take more than one try. Examine the tiles for $x^2 + 2x - 8$.

Place the x^2-tile. Try a 1-by-8 arrangement of the negative 1-tiles. It is impossible to complete a rectangle.

Try a 2-by-4 arrangement of the negative 1-tiles.

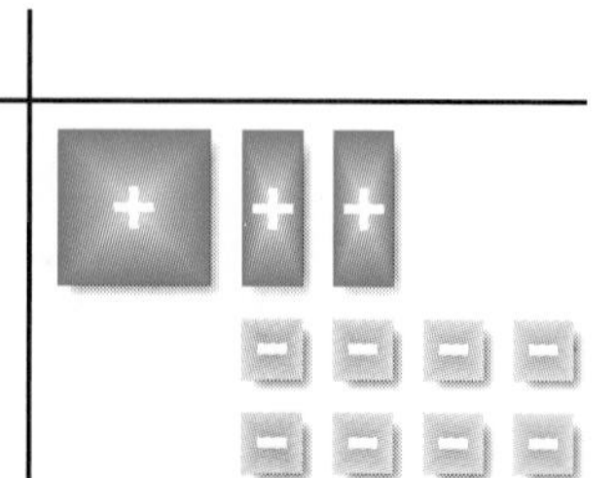

Complete a rectangle by adding 2 positive and 2 negative x-tiles.

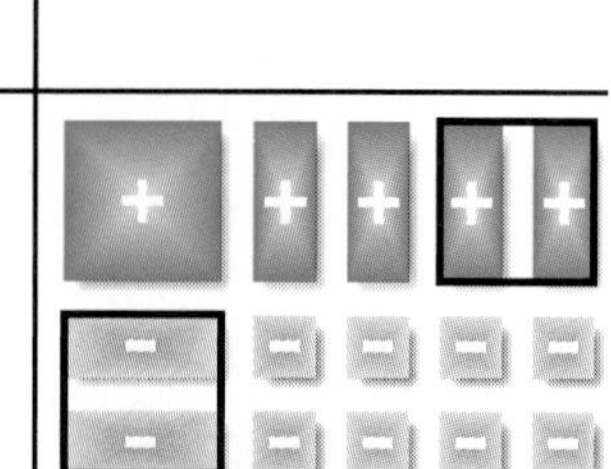

Finally, determine the factor tiles.

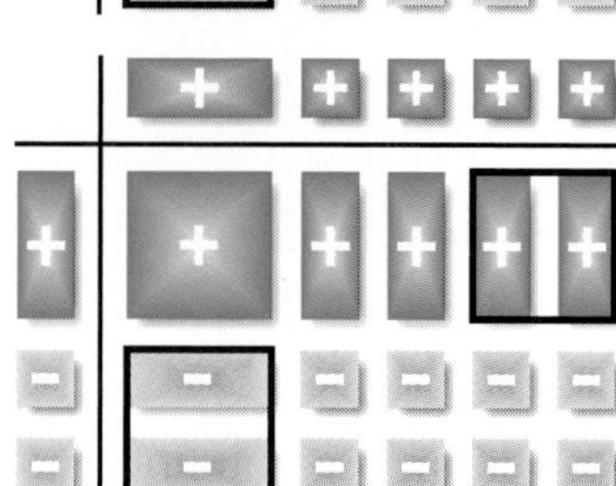

Thus, $x^2 + 2x - 8 = (x - 2)(x + 4)$. To check, multiply the factors.

$$\begin{aligned}(x - 2)(x + 4) &= x^2 - 2x + 4x - 8 \\ &= x^2 + 2x - 8\end{aligned}$$

Try This **a.** Explain how to use tiles to factor $x^2 - 2x - 8$. Check by multiplication.

b. Explain how to use tiles to factor $x^2 - 6x + 8$. Check by multiplication.

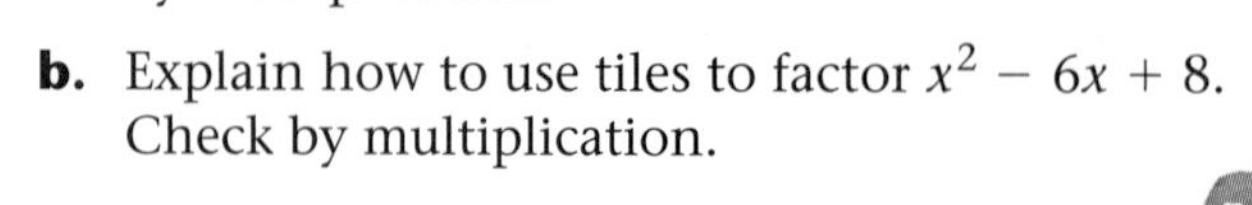

Factoring by Guess-and-Check

Examine the following trinomial patterns and their factors:

a. $x^2 + 6x + 8 = (x + 2)(x + 4)$

b. $x^2 - 6x + 8 = (x - 2)(x - 4)$

c. $x^2 + 2x - 8 = (x - 2)(x + 4)$

d. $x^2 - 2x - 8 = (x + 2)(x - 4)$

In each trinomial, the coefficient of x^2 is 1, and the last term is either 8 or -8. Since factoring is related to multiplication, a trinomial can be factored by working backward with the FOIL method. You can use guess-and-check to write the correct factors and signs in the FOIL model.

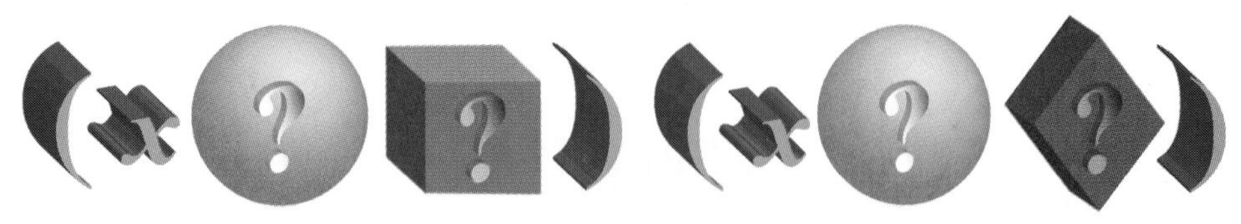

Follow the steps to factor $x^2 + x - 12$.

1. Examine the last term of the trinomial. Since there is a negative sign before the 12 and the coefficient of x is positive, $x^2 + x - 12$ is an example like trinomial **c** above. The appropriate signs in the FOIL model are $-$ and $+$.

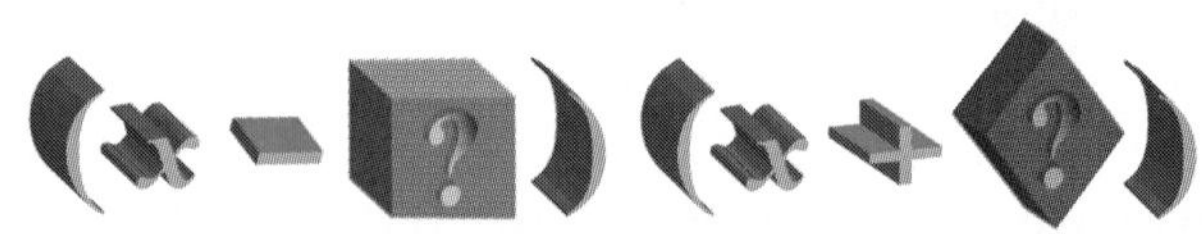

2. To find the values of ■ and ◆, remember that their product must be 12. Possible factor pairs of 12 are $1 \cdot 12$, $2 \cdot 6$, and $3 \cdot 4$. Choose the factors 3 and 4, because the sign for the middle term of the trinomial is positive and the coefficient is 1. Let 3 replace ■ and 4 replace ◆, so that $-3x + 4x$ will equal x.

Thus, the result of factoring $x^2 + x - 12$ is $(x - 3)(x + 4)$.

3. Explain why the other factor pairs do not work.

4. Check the factors. $(x - 3)(x + 4) = x^2 + 4x - 3x - 12$
$= x^2 + x - 12$

EXAMPLE

Factor each trinomial.

A $x^2 - x - 20$

B $x^2 - 10x + 16$

C $x^2 + 4x - 21$

D $x^2 + 9x + 18$

Solution ➤

A Use trinomial pattern **d**.
$x^2 - x - 20 = (x + 4)(x - 5)$

B Use trinomial pattern **b**.
$x^2 - 10x + 16 = (x - 2)(x - 8)$

C Use trinomial pattern **c**.
$x^2 + 4x - 21 = (x - 3)(x + 7)$

D Use trinomial pattern **a**.
$x^2 + 9x + 18 = (x + 6)(x + 3)$ ❖

Sometimes a trinomial has a common monomial factor. Explain why $2x^3 + 16x^2 + 24x$ is written in factored form as $2x(x + 2)(x + 6)$.

CT– Common monomial factors should always be factored first.

There are some trinomials such as $x^2 - 6x - 8$ that cannot be factored by any method. For the middle term to equal $-6x$, the factors for the constant term, 8, would have to be -2 and -4. The constant term would then equal $+8$. There are no signs for the factor pair that fit the pattern of signs for the trinomial. The polynomial $x^2 - 6x - 8$ is prime or *irreducible* over the integers.

Communicate

Explain how to use algebra tiles to factor each polynomial.

1. $x^2 - 5x + 4$ **2.** $x^2 - 4x - 12$ **3.** $x^2 + 6x + 9$

Write each trinomial in factored form. Tell why the signs of the factors are the same or opposite.

4. $x^2 + x - 6$ **5.** $x^2 - 7x + 10$ **6.** $x^2 + 2x - 15$

7. Explain how to use the guess-and-check method to factor $x^2 - 5x - 24$.

8. If the third term of a trinomial is 36, write the possible factor pairs.

1, 36; 2, 18; 3, 12; 4, 9; 6, 6; −1, −36; −2, −18; −3, −12; −4, −9; −6, −6

Practice & Apply

9. Use algebra tiles to factor $x^2 + 5x + 6$.

Write each trinomial in factored form. Are the signs of the factors the same or opposite?

10. $x^2 - 13x + 36$ **11.** $x^2 + 11x + 24$

12. $x^2 + 10x - 24$ **13.** $x^2 - 35x - 36$

10. $(x - 4)(x - 9)$; same
11. $(x + 3)(x + 8)$; same
12. $(x + 12)(x - 2)$; opposite
13. $(x + 1)(x - 36)$; opposite

For each polynomial, write all the factor pairs of the third term, and then circle the pair that would successfully factor the polynomial.

For example, $x^2 - 3x - 4$ has factor pairs of $1 \cdot -4$ (circled), $-1 \cdot 4$, $2 \cdot -2$, and $-2 \cdot 2$.

14. $y^2 - 9y - 36$ **15.** $x^2 - 21x + 54$

16. $x^2 + 19x + 48$ **17.** $z^2 + 10z - 144$

Write each trinomial as a product of its factors. Use factoring patterns or algebra tiles to assist you in your work.

18. $a^2 - 2a - 35$ **19.** $p^2 + 4p - 12$ **20.** $y^2 - 5y + 6$ **21.** $b^2 - 5b - 24$

22. $n^2 - 11n + 18$ **23.** $z^2 + z - 20$ **24.** $x^2 - 3x - 28$ **25.** $s^2 - 24s + 63$

This set of exercises includes all factoring patterns used throughout the chapter. Write each polynomial as a product of its factors.

26. $4x^3y - 20x^2y + 16xy$ **27.** $6y^3 - 18y^2 + 12y$ **28.** $x^2 - 18x + 81$

29. $(a + 3)(a^2 + 5a) - 6(a + 3)$ **30.** $5x^3 - 50x^2 + 45x$ **31.** $x^3 + 2x^2 - 36x - 72$

32. $-x^4 + 2x^2 + 8$ **33.** $64p^4 - 16$ **34.** $z^2 - 5z - 36$

35. $x^2 - 2x + 1$ **36.** $125x^2y - 5x^4y$ **37.** $2ax + ay + 2bx + by$

Look Back

38. Write the next three terms in this sequence.
3, 9, 19, 33, . . . **[Course 1]** 51, 73, 99

Solve. [Lesson 1.2]

39. $\frac{3}{4}a = 363$ 484 **40.** $\frac{z}{-8} = \frac{9}{12}$ -6 **41.** $w - \frac{7}{9} = 23$ $23\frac{7}{9}$

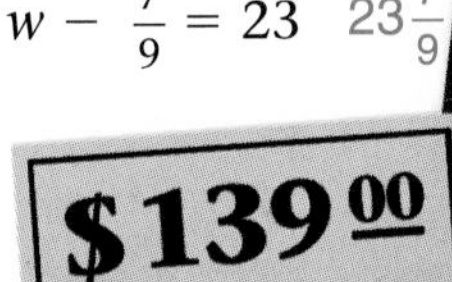

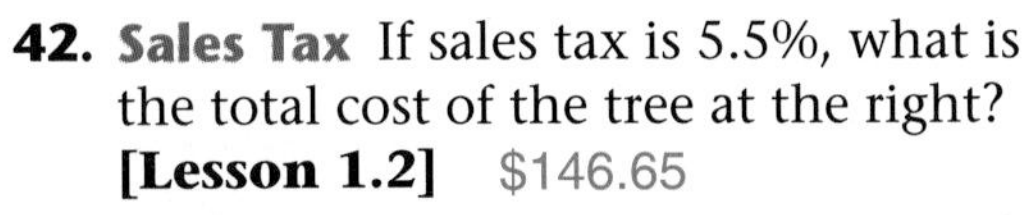

42. **Sales Tax** If sales tax is 5.5%, what is the total cost of the tree at the right? **[Lesson 1.2]** $146.65

43. A line with slope -2 passes through point $A(4, -1)$. Write the equation of the line that passes through the point and is perpendicular to the original line. **[Lesson 2.2]** $y = \frac{1}{2}x - 3$

A freshman class buys a tree to donate to the childrens' hospital.

44. Graph the inequality $-4 < x < 5$ on the number line. **[Lesson 2.7]**

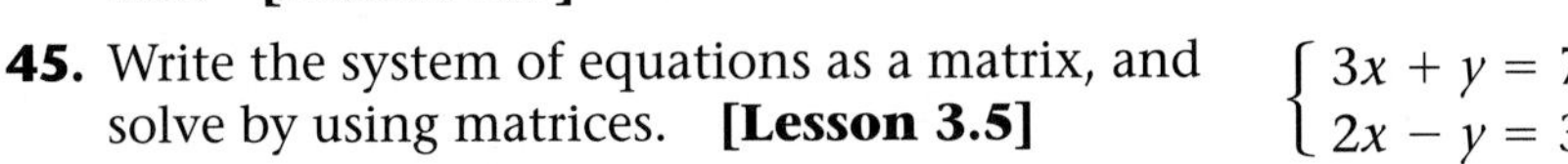

45. Write the system of equations as a matrix, and solve by using matrices. **[Lesson 3.5]**
$$\begin{cases} 3x + y = 7 \\ 2x - y = 3 \end{cases}$$

46. List the integers from 1 to 25, inclusive, that are multiples of 2. **[Lesson 4.4]** 2, 4, 6, 8, 10, 12, 14, 16, 18, 20, 22, 24

Look Beyond

Cultural Connection: Americas De Padilla, a Guatemalan mathematician, wrote this problem 250 years ago.

47. Find two numbers, given that the second is triple the first. If you multiply the first number by the second number and the second number by 4, the sum of the products is 420. 10, 30; -14, -42

48. Graph $y = |x|$ and $y = x^2$. If these graphs are "folded" together along a vertical line, what is the equation of the line that divides them both equally?

CHAPTER 7 PROJECT

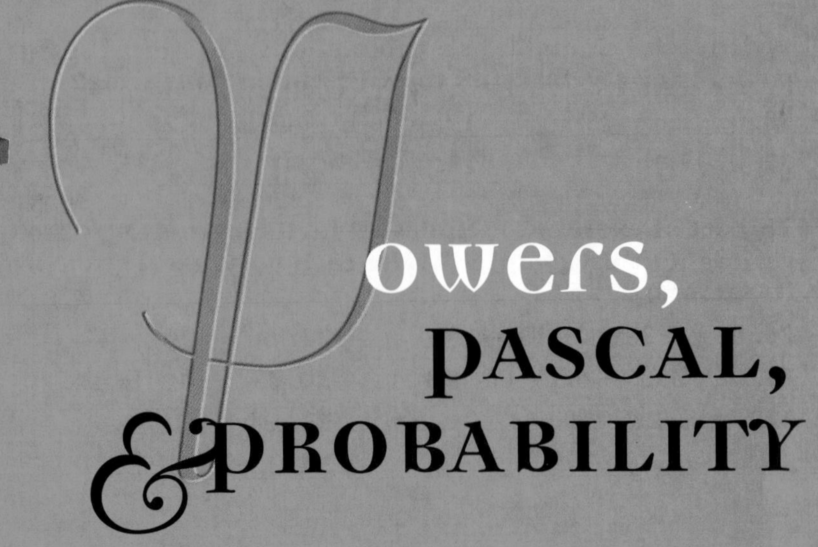

Powers, Pascal, & Probability

Cultural Connection: ASIA If a great mathematical discovery were made today, the discoverer and his work would be on the evening news. This has not been the case throughout history. Pascal's triangle, named after the French mathematician Blaise Pascal, appeared in work published in Europe in 1665. However, historical writings show that mathematicians from the ancient Islamic civilization were already using the remarkable properties of Pascal's triangle in the tenth century.

ACTIVITY

Look at the different powers of the binomial $(a + b)$, and compare the expanded form with the Pascal triangle.

POWER	EXPANDED FORM	PASCAL'S TRIANGLE
$(a+b)^0 =$	1	1
$(a+b)^1 =$	$1a+1b$	1 1
$(a+b)^2 =$	$1a^2+2ab+1b^2$	1 2 1
$(a+b)^3 =$	$1a^3+3a^2b+3ab^2+1b^3$	1 3 3 1
$(a+b)^4 =$	$1a^4+4a^3b+6a^2b^2+4ab^3+1b^4$	1 4 6 4 1
$(a+b)^5 =$	$1a^5+5a^4b+10a^3b^2+10a^2b^3+5ab^4+1b^5$	1 5 10 10 5 1

1. Start with the expanded form of $(a + b)^2$, and multiply it by $a + b$. What do you get?
2. Do the same with the expanded forms of $(a + b)^3$ and $(a + b)^4$. What do you get?
3. How does the number of terms in each row of the expanded form compare with the exponent of the binomial?
4. How many terms would you expect in the expansion of $(a + b)^7$?
5. Recalling that the triangle's elements are created by adding pairs of elements from a previous row, extend the rows of the triangle through row 10.
6. Write the coefficients for each term in the expansion of $(a + b)^7$.
7. Look at the algebraic terms of the expanded form. Describe the pattern in the exponents as you read them from left to right. What happens to the exponents of a? What happens to the exponents of b?
8. Write the complete algebraic expansion of $(a + b)^7$ without multiplying.

A Chinese version of the triangle appeared in 1303 C.E.

	3 heads	2 heads 1 tails	1 heads 2 tails	3 tails
Tally marks				
Totals				
Totals ÷ 4				

ACTIVITY 2

1. Copy the table.
2. Flip three coins a total of 32 times. Record the results in the table.
3. Divide each total by 4, and round the answer to the nearest whole number. What are the results for each column?

ACTIVITY 3

1. Look at the tree diagram for the coin-tossing activity.
2. How many ways are there to toss 3 heads?
3. How many ways are there to toss 3 tails?
4. How many ways are there to toss 2 heads and 1 tails?
5. How many ways are there to toss 1 heads and 2 tails?
6. What is the probability of each outcome?

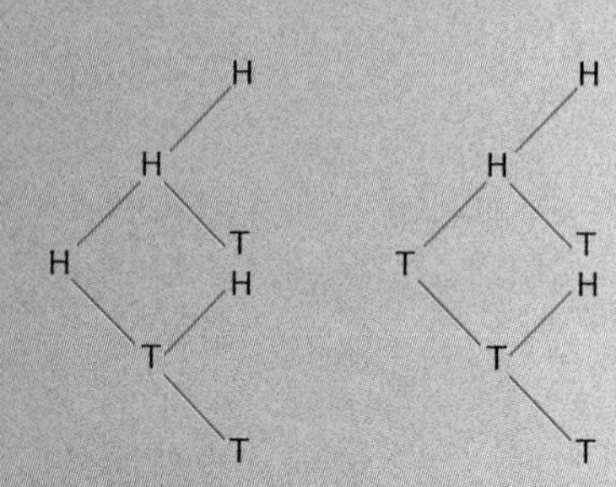

ACTIVITY 4

1. Write the expanded form of $(h + t)^3$.
2. The probability of getting heads when tossing a coin is $\frac{1}{2}$. The probability of getting tails when tossing a coin is $\frac{1}{2}$. Let h and t each equal $\frac{1}{2}$, and find the value of *each term* in the expanded form of $(h + t)^3$.
3. How are the values for each term related to
   a. the probabilities that you calculated in Activity 3?
   b. the results that you obtained in Activity 2?
4. If you tossed 5 coins at a time, what would be the probability of getting 5 heads? 4 heads and 1 tails?

Chapter 7 Review

Vocabulary

Key Skills & Exercises

Lesson 7.1

➤ Key Skills

Use polynomial functions to express the area and volume of solid figures.

The volume of a cylinder that is 10 inches high is $10B$, where B is the area of the circular base. The area of a circle with radius, r, is πr^2, so the volume of the cylinder is expressed by the monomial $10\pi r^2$.

Express a polynomial as the product of other polynomials.

The equation $(x - 1)^2 = x^2 - 2x + 1$ is an identity (true for all values of x). For example if $x = -3$, then the left side, $(-3 - 1)^2$, is $(-4)^2$, or 16. The right side, $(-3^2) - 2(-3) + 1$, is $9 + 6 + 1$, or 16.

➤ Exercises

1. The formula for the volume of a cone is $V = \frac{1}{3}Bh$, where B is the area of the circular base. Write a monomial that represents the volume of a cone with a height of 9 feet and a base that has a radius r. $V = \frac{1}{3}\pi r^2 \cdot 9 = 3\pi r^2$

2. The equation $(x + 2)(x - 2) = x^2 - 4$ is an identity. Show that each side of the equation has the same value when x is replaced by each of the following numbers: $-3, -2, -1, 0, 1, 2$, and 3.

Lesson 7.2

➤ Key Skills

Use vertical form to add and subtract polynomials.

Subtract $5m^2 - 4$ from $m^2 - 5m - 10$.

$$\begin{array}{r} m^2 - 5m - 10 \\ \underline{-(5m^2 \qquad\quad - 4)} \\ -4m^2 - 5m - 6 \end{array}$$

Use horizontal form to add and subtract polynomials.

Add $5y^2 - 3y + 8$ and $y^2 + y - 9$.

$(5y^2 - 3y + 8) + (y^2 + y - 9) = 6y^2 - 2y - 1$

➤ Exercises

Simplify. Express all answers in standard form.

3. $(3x^2 - 4x + 2) + (2x^2 + 3x - 2)$ **4.** $(c^3 + 4c^2 + 6) + (c^2 + 3c - 5)$

5. $(8d^2 - d) - (2d^2 + 4d - 5)$ **6.** $(w^3 - 3w + 9) - (8w^3)$

3. $5x^2 - x$ 4. $c^3 + 5c^2 + 3c + 1$ 5. $6d^2 - 5d + 5$ 6. $-7w^3 - 3w + 9$

Lesson 7.3

➤ Key Skills

Use the Distributive Property to find products.

$3x(2x - 3) = 3x(2x) - 3x(3) = 6x^2 - 9x$

➤ Exercises

Use the Distributive Property to find each product.

7. $5(x - 5)$ $5x - 25$ **8.** $y(y + 4)$ $y^2 + 4y$ **9.** $4t(t^2 + 7)$ $4t^3 + 28t$ **10.** $2r^2(r^2 - 3r)$ $2r^4 - 6r^3$

11. $b(12b^2 + 11b)$ $12b^3 + 11b^2$ **12.** $4y(y + 5)$ $4y^2 + 20y$ **13.** $5x^2(2x^2 - x)$ $10x^4 - 5x^3$ **14.** $6d^2(d^2 - 1)$ $6d^4 - 6d^2$

Lesson 7.4

➤ Key Skills

Use the Distributive Property to find a product of two binomials.

$$(x + 6)(x + 2)$$
$$x(x + 6) + 2(x + 6)$$
$$x^2 + 6x + 2x + 12$$
$$x^2 + 8x + 12$$

Use the FOIL method to multiply binomials.

$$(2x + 3)(x - 1)$$

$$(2x + 3)(x - 1) = 2x^2 - 2x + 3x - 3$$
$$2x^2 + x - 3$$

15. $x^2 + 5x - 6$ 16. $y^2 + 7y - 18$ 17. $z^2 - 9z + 18$ 18. $3m^2 + 20m + 25$ 19. $2p^2 + p - 45$

➤ Exercises

Use the Distributive Property to find a product of two binomials.

15. $(x + 6)(x - 1)$ **16.** $(y + 9)(y - 2)$ **17.** $(z - 3)(z - 6)$

18. $(3m + 5)(m + 5)$ **19.** $(2p - 9)(p + 5)$ **20.** $(2d + 7)(2d - 6)$

20. $4d^2 + 2d - 42$
21. $x^2 - x - 12$

Use the FOIL method to find each product.

21. $(x + 3)(x - 4)$ **22.** $(5d - 8)(d - 1)$ **23.** $(4w + 3z)(w + z)$

24. $(y + 4)(y + 5)$ **25.** $(x + 2)(x - 4)$ **26.** $(3z + 1)(4z - 1)$

22. $5d^2 - 13d + 8$ 23. $4w^2 + 7wz + 3z^2$ 24. $y^2 + 9y + 20$
25. $x^2 - 2x - 8$ 26. $12z^2 + z - 1$

Lesson 7.5

➤ Key Skills

Factor a polynomial by removing the GCF.

$$8m^6 + 4m^4 - 2m^2$$
$$= 2m^2(4m^4 + 2m^2 - 1)$$

Find a binomial as the greatest common factor.

$$2x^2 + 2 + x^3 + x$$
$$= (2x^2 + 2) + (x^3 + x)$$
$$= 2(x^2 + 1) + x(x^2 + 1)$$
$$= (2 + x)(x^2 + 1)$$

➤ Exercises

Factor each polynomial by removing the GCF.

27. $16x^3 + 8x^2$ **28.** $9y^7 + 6y^3 + 3y$ **29.** $b^6 + 15b^3 - 30b^2$

30. $24m^9 - 16m^4 + 8m^3$ **31.** $60a^4 + 20a^3 + 10a^2$ **32.** $100p^8 - 50p^6 - 25p$

Factor the common binomial factor from each polynomial.

33. $d(f + 1) + h(f + 1)$ **34.** $3y(y - 3) - 4(y - 3)$ **35.** $5x - 5y + x^2 - xy$

36. $x(z - 4) + y(z - 4)$ **37.** $10 - 5t - 2t + t^2$ **38.** $6c^2 + 6 + c^3 + c$

Lesson 7.6

➤ Key Skills

Use a generalization of special polynomials to find each product.

$(x + 7)^2 = (x + 7)(x + 7) = x^2 + 14x + 49$

$(p + 8)(p - 8) = p^2 - 64$

Factor perfect-square trinomials.

$x^2 - 18x + 81 = (x - 9)^2$

Factor the difference of two squares.

$4x^2 - 9 = (2x + 3)(2x - 3)$

➤ Exercises

39. $c^2 - 18c + 81$ 40. $b^2 - 100$ 41. $25a^2 - 4$

Use a generalization of special polynomials to find each product.

39. $(c - 9)^2$ **40.** $(b - 10)(b + 10)$ **41.** $(5a - 2)(5a + 2)$

Factor each special polynomial. 42. $(a + 3)^2$ 43. $(w - 8)^2$ 44. $(6p + 1)^2$

42. $a^2 + 6a + 9$ **43.** $w^2 - 16w + 64$ **44.** $36p^2 + 12p + 1$

45. $y^2 - 81$ **46.** $16r^2 - 25$ **47.** $9z^2 - 1$

45. $(y + 9)(y - 9)$ 46. $(4r + 5)(4r - 5)$ 47. $(3z + 1)(3z - 1)$

Lesson 7.7

➤ Key Skills

Use sign and number patterns to determine the factors of a trinomial.

$x^2 - x - 12 = (x + \underline{?})(x - \underline{?})$ Factor pairs: $1 \cdot 12$, $2 \cdot 6$, $3 \cdot 4$

The factors are $(x + 3)(x - 4)$.

48. $(n - 6)(n + 4)$ 49. $(h + 10)(h + 4)$ 50. $(y + 6)(y - 3)$ 51. $(a - 4)(a - 3)$

52. $(x - 4)(x - 1)$ 53. $2(a + 1)(a + 3)$

➤ Exercises

Write each polynomial as a product of its factors.

48. $n^2 - 2n - 24$ **49.** $h^2 + 14h + 40$ **50.** $y^2 + 3y - 18$

51. $a^2 - 7a + 12$ **52.** $x^2 - 5x + 4$ **53.** $2a^2 + 8a + 6$

Applications

Geometry Given the rectangle with sides as shown, solve the following.

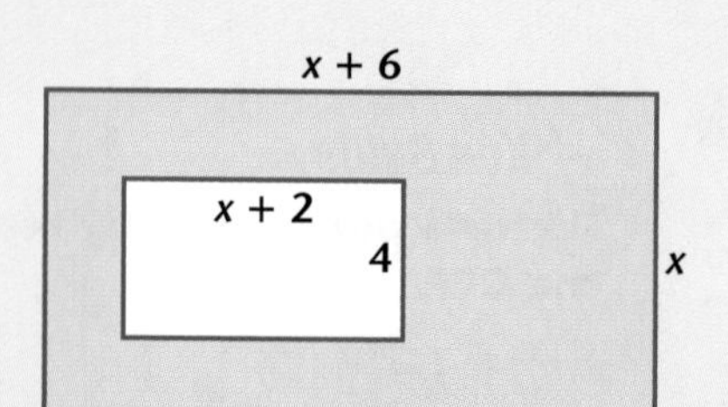

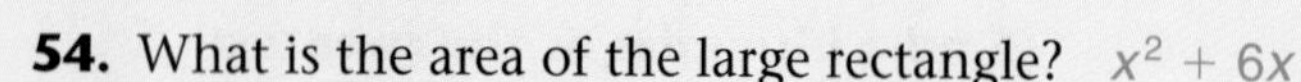

54. What is the area of the large rectangle? $x^2 + 6x$

55. What is the area of the small rectangle? $4x + 8$

56. If the smaller region is removed, what is the area of the colored region? $x^2 + 2x - 8$ square units

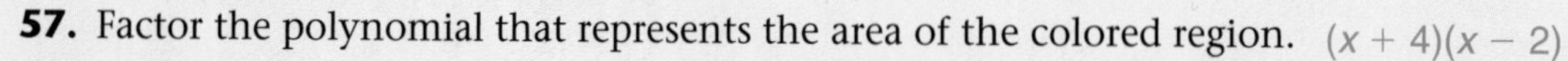

57. Factor the polynomial that represents the area of the colored region. $(x + 4)(x - 2)$

Chapter 7 Assessment

1. The dimensions of the rectangular base of a box are x and $3x$. The height of the box is 10 inches. Write a polynomial function for the surface area of the box. $6x^2 + 80x$

Simplify. Express all answers in standard form.

2. $(2 + 4x + 2x^2) + (4x^2 - 6)$ $6x^2 + 4x - 4$

3. $9 - 5v^2 + 5v^4 - (10v^2 + 7v - 11)$ $5v^4 - 15v^2 - 7v + 20$

Use the Distributive Property to find each product.

4. $8(x - 7)$ $8x - 56$

5. $r(r - 6)$ $r^2 - 6r$

6. $3a^2(a - 1)$ $3a^3 - 3a^2$

7. $q(2q^2 + q)$ $2q^3 + q^2$

Write each polynomial in factored form.

8. $6x^6 + 3x^4$ $3x^4(2x^2 + 1)$

9. $25a^8 - 15a^4$ $5a^4(5a^4 - 3)$

10. $8x^5 - 4x^4 + x^3$ $x^3(8x^2 - 4x + 1)$

11. $25y^9 + 15y^7 - 5y^2$ $5y^2(5y^7 + 3y^5 - 1)$

12. Describe how to find $(x + 4)(x - 8)$ using the FOIL method.

13. Write the degree of the polynomial $6y^2 + 3y^4 - 1$. 4

Factor the common binomial factor from each polynomial.

14. $5(r + 1) - t(r + 1)$ $(r + 1)(5 - t)$

15. $p^2 + 2p + 3p + 6$ $(p + 2)(p + 3)$

Use the Distributive Property or the FOIL method to find each product.

16. $y^2 + 5y + 6$ 17. $c^2 - c - 12$ 18. $\frac{1}{2}a^2 + \frac{1}{2}a - 1$

16. $(y + 3)(y + 2)$

17. $(c + 3)(c - 4)$

18. $\left(\frac{1}{2}a + 1\right)(a - 1)$

19. $(4x - y)(3x + 2y)$

20. $(w - 9)(w - 6)$

21. $(2n + 1)(n + 1)$

19. $12x^2 + 5xy - 2y^2$ 20. $w^2 - 15w + 54$ 21. $2n^2 + 3n + 1$ 23. $p^2 - 144$

Use a generalization of special polynomials to find each product.

22. $(x + 11)^2$ $x^2 + 22x + 121$

23. $(p + 12)(p - 12)$

24. $(m - 9)^2$ $m^2 - 18m + 81$

25. Find the area of the floor of a square garage that is $t + 5$ units on a side. $t^2 + 10t + 25$ sq units

26. Can $7c^2 + 45c - 28$ be factored? Explain why or why not. Yes; there is no GCF, but by factoring by guess-and-check, it can be factored into $(c + 7)(7c - 4)$.

Use any method to factor each polynomial completely.

27. $4r + 4s + rt + st$ $(r + s)(4 + t)$

28. $14x^3 + 7x - 8x^2 - 4$ $(7x - 4)(2x^2 + 1)$

29. $4v^2 - 144$ $4(v + 6)(v - 6)$

30. $a^4 - 81$ $(a^2 + 9)(a + 3)(a - 3)$

31. $n^2 - 12n + 36$ $(n - 6)^2$

32. $15z^2 - z - 2$ $(5z - 2)(3z + 1)$

33. $w^4 - 7w^3 - 18w^2$ $w^2(w - 9)(w + 2)$

34. $d^4 + 5d^2 + 6$ $(d^2 + 2)(d^2 + 3)$

Find the greatest common factor.

35. $y^3 - y^2 + 4y$ y

36. $6w^3 + 3w^2 + 3$ 3

37. $7m^3 - m^2$ m^2

38. $x^2 + 20x + 96$ sq units 39. $x^2 + 6x$ sq units

Within a large rectangle there is a smaller rectangle.

38. What is the area in square units of the large rectangle?

39. What is the area in square units of the smaller rectangle?

40. What is the area in square units of the colored area between the two figures? $14x + 96$ square units

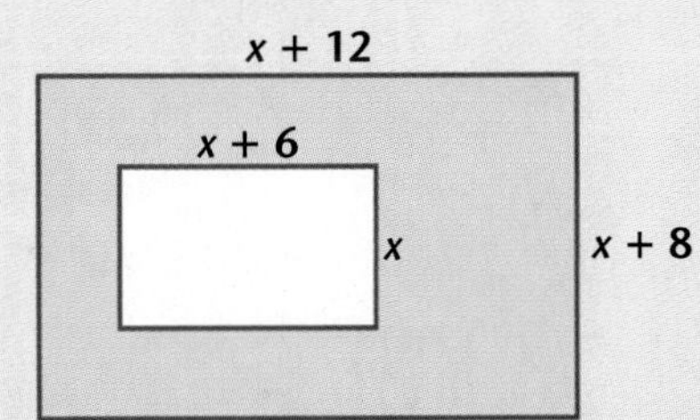

CHAPTER 8

Quadratic Functions

LESSONS

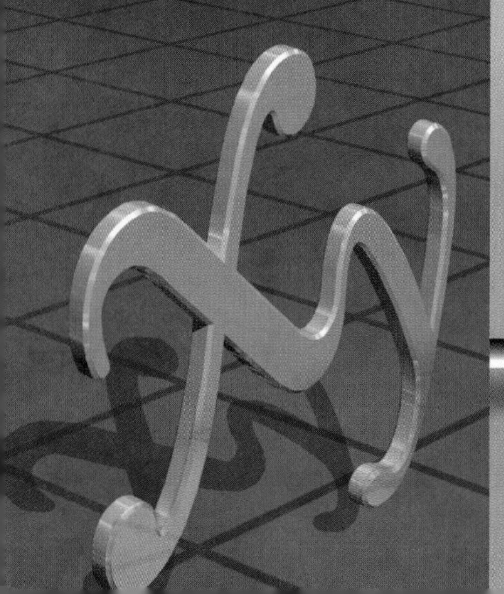

You have seen that the height of a projectile can be modeled by a quadratic function. Often the trajectory of a projectile will resemble a parabola, the graph of a quadratic function. The cost of purchasing an automobile can also be modeled by a quadratic function. In this chapter you will learn different methods of analyzing quadratic functions and quadratic equations.

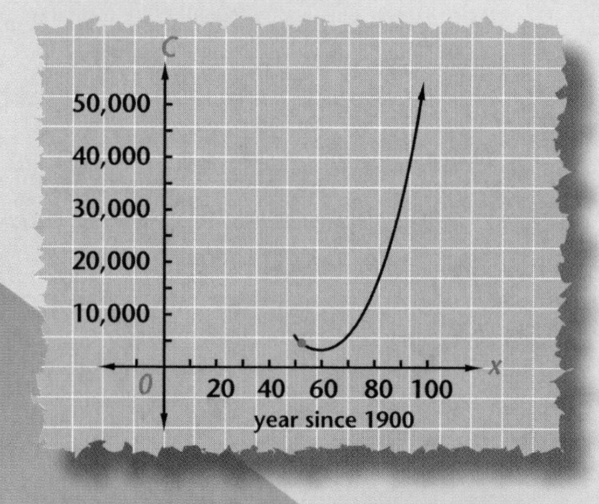

Sometimes the cost of a product can be approximated by a quadratic function.

1953

1962

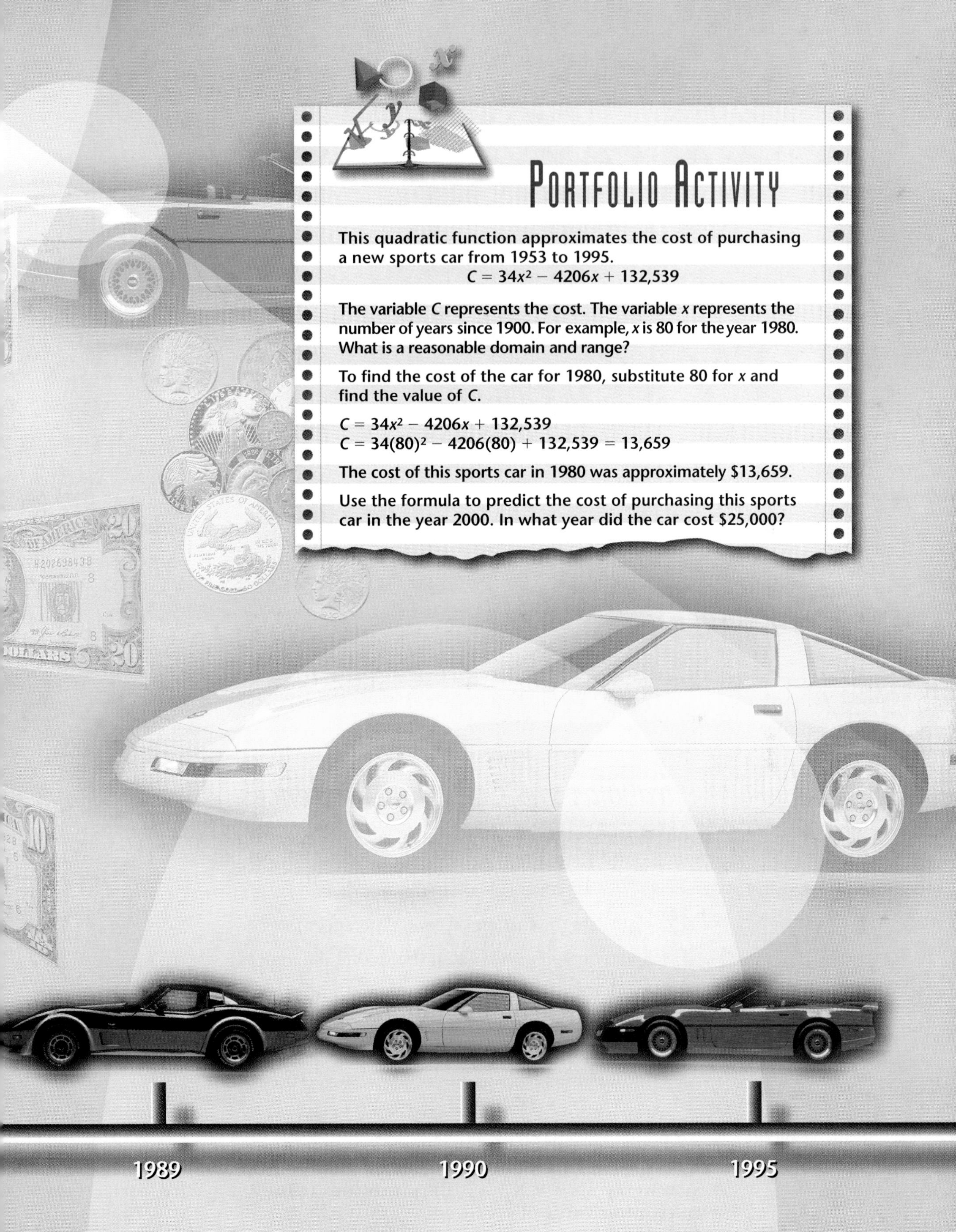

Portfolio Activity

This quadratic function approximates the cost of purchasing a new sports car from 1953 to 1995.

$$C = 34x^2 - 4206x + 132,539$$

The variable C represents the cost. The variable x represents the number of years since 1900. For example, x is 80 for the year 1980. What is a reasonable domain and range?

To find the cost of the car for 1980, substitute 80 for x and find the value of C.

$C = 34x^2 - 4206x + 132,539$
$C = 34(80)^2 - 4206(80) + 132,539 = 13,659$

The cost of this sports car in 1980 was approximately $13,659.

Use the formula to predict the cost of purchasing this sports car in the year 2000. In what year did the car cost $25,000?

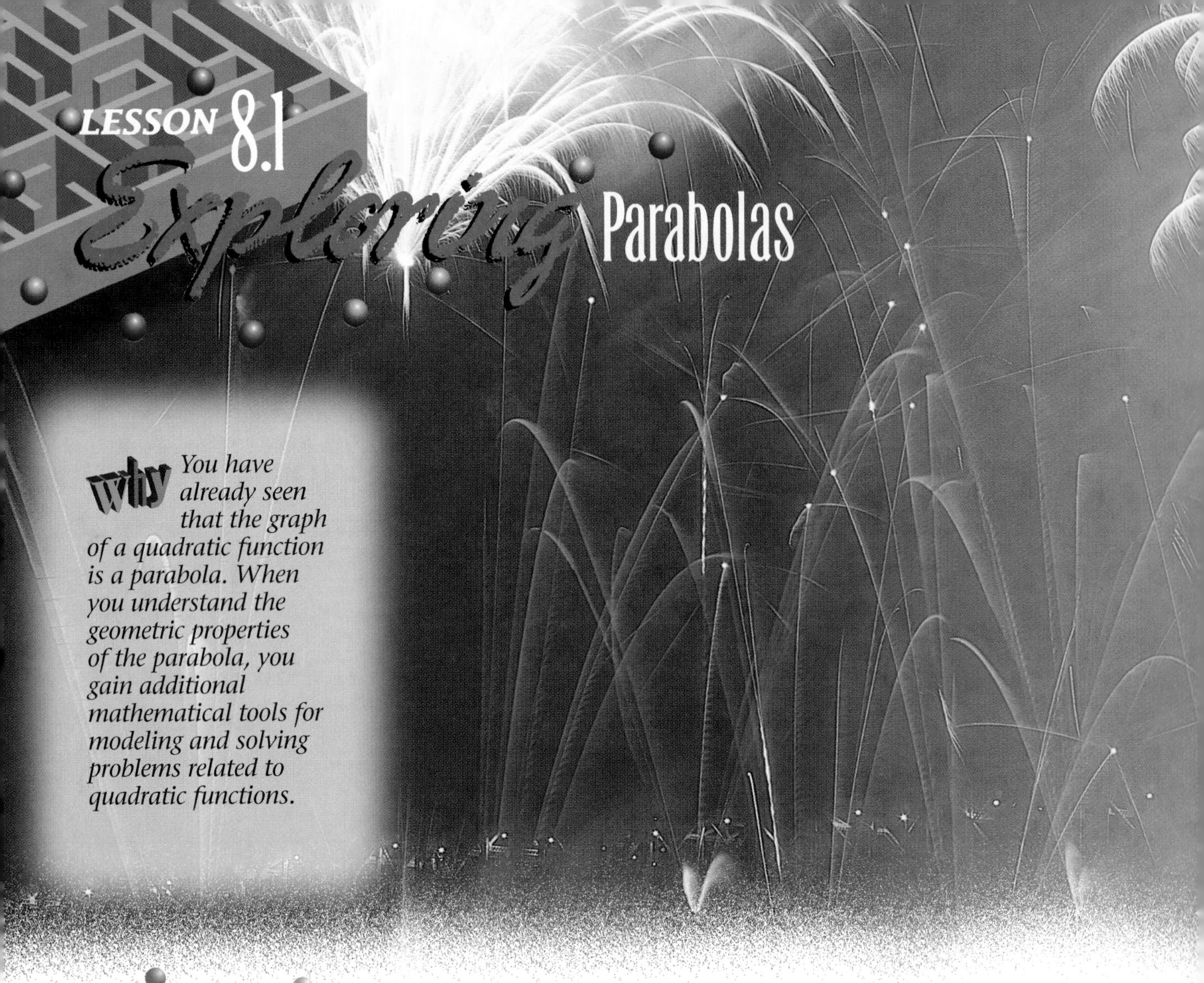

Why You have already seen that the graph of a quadratic function is a parabola. When you understand the geometric properties of the parabola, you gain additional mathematical tools for modeling and solving problems related to quadratic functions.

Exploration 1 *Parabolas and Constant Differences*

The table shows some points on the graph of the function $y = f(x)$.

x	0	1	2	3	4	5	6
$f(x)$	10	0	−6	−8	−6	?	?

1. Calculate the first and second differences for $f(x)$.
   1st differences: −10, −6, −2, 2; 2nd differences: 4
2. What do you notice about the second differences?
   They are constant.
3. Work backward by using the pattern, and predict $f(5)$ and $f(6)$.
   $f(5) = 0$; $f(6) = 10$
4. Plot $y = f(x)$. Connect the points with a smooth curve.
5. What kind of curve did you draw? Explain the relationship between second differences and the curve that you drew. ❖

5. The graph is a parabola. If the 2nd differences are constant, then the function is a quadratic function.

In Chapter 5, you learned that the function $g(x) = a(x - h)^2 + k$ transforms the parent quadratic function $f(x) = x^2$ by stretching the parent function by a factor of a and moving its **vertex** from (0, 0) to (h, k). The **axis of symmetry** is $x = h$. If $a > 0$, the **minimum value** of g is k. If $a < 0$ the **maximum value** of g is k.

Exploration 2 *Parabolas and Transformations*

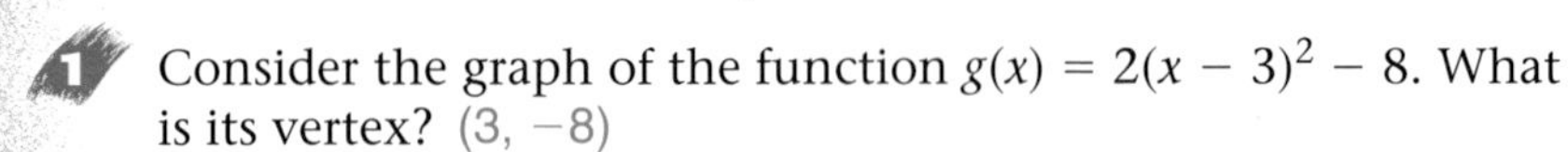

1. Consider the graph of the function $g(x) = 2(x - 3)^2 - 8$. What is its vertex? (3, −8)
2. Find four more points on the graph. (0, 10), (1, 0), (2, −6), (5, 0)
3. Plot the points and sketch the graph.
4. What kind of curve did you draw? How does your graph compare with the graph you drew in Exploration 1? ❖
   Parabola; same graph as in Exploration 1

The graph of a quadratic function is a parabola. The x-value of the point where the parabola crosses the x-axis is called a **zero of the function**. How many zeros can a quadratic function possibly have? Explain your answer. There are *no zeros* if the parabola does not cross the x-axis, *one zero* if it touches the x-axis at one point, and *two zeros* if it intersects the x-axis at two points.

Exploration 3 *Parabolas and Polynomials*

MAXIMUM MINIMUM *Connection*

1. Graph the function $t(x) = 2x^2 - 12x + 10$.
2. Find the zeros of the function. 5 and 1
3. Find the average, h, of the two zeros. 3
4. Find $t(h)$. −8
5. What is the vertex of the graph? (3, −8)
6. What is the axis of symmetry of the graph? $x = 3$
7. What is the minimum value of the function? −8
8. Compare the graphs for Exploration 2 and 3. Same graph
9. Simplify $g(x) = 2(x - 3)^2 - 8$. Compare $g(x)$ with $t(x)$. ❖
   Same value

APPLICATION

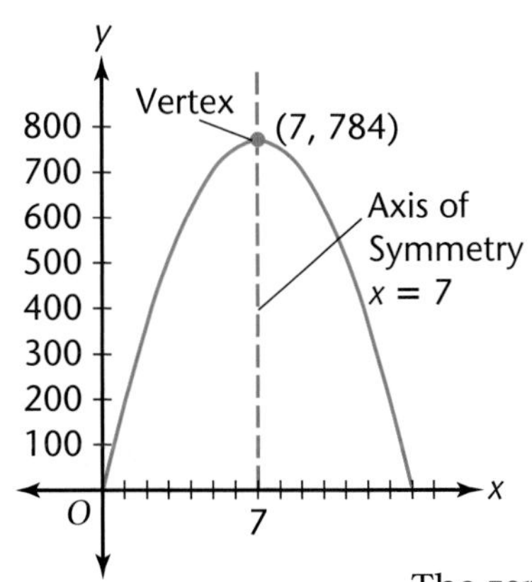

In Course 1, you used a table and differences to find the maximum height of the flight of a small rocket. The flight can be modeled with the function $f(x) = -16(x - 7)^2 + 784$. In this form, the vertex, the axis of symmetry, and the maximum value can be read directly from the equation. The zeros can be read from the graph.

function	$f(x) = -16(x - 7)^2 + 784$
vertex	(7, 784)
axis of symmetry	$x = 7$
maximum value	784
zeros	0, 14

The zeros of f are 0 and 14. Check by substitution. ❖

CRITICAL *Thinking*

How can the zeros of $f(x) = -16(x - 7)^2 + 784$ be interpreted in terms of the flight of the rocket?

EXERCISES & PROBLEMS

Communicate

1. Explain how you can tell from second differences that the values of a function represent a quadratic.

x	−4	−3	−2	−1	0
$f(x)$	0	−1	0	3	8

2. Explain how the graph of the quadratic function $g(x) = 2(x - 3)^2 - 8$ differs from the graph of the parent function $y = x^2$.

Refer to the function $g(x) = a(x - h)^2 + k$ to complete Exercises 3–4.

3. Explain how to determine the vertex of $g(x) = (x - 3)^2 - 4$.

4. Explain how to determine the axis of symmetry.

5. Explain how to determine the zeros of $g(x) = x^2 - 8x + 16$.

6. Explain how to determine the axis of symmetry for $g(x) = x^2 - 8x + 16$ from the zeros of the polynomial.

Practice & Apply

7. Do the values represent a quadratic function? Graph to check.

x	0	1	2	3	4	5	6
$f(x)$	−5	0	3	4	3	0	−5

Compare the graph of each function with the graph of $y = x^2$. Describe the transformation on the parent function $y = x^2$.

8. $y = (x - 2)^2 + 3$

9. $y = 3(x - 5)^2 - 2$

10. $y = -(x - 2)^2 + 1$

11. $y = (x + 1)^2$

12. $y = -3(x - 2)^2 + 1$

13. $y = \frac{1}{2}(x - 2)^2 + 3$

Determine the vertex and axis of symmetry for each function.

14. $y = -2(x + 4)^2 - 3$

15. $y = \frac{1}{2}(x - 2)^2 + 3$

16. $y = (x - 3)^2 - 7$

17. $y = -3(x - 5)^2 + 2$

18. $y = -(x + 3)^2 - 2$

19. $y = 2(x + 5)^2 + 7$

Find the zeros of each function by graphing.

20. $f(x) = x^2 + 8x - 9$ 1, −9
21. $f(x) = x^2 - 20x + 100$ 10
22. $f(x) = x^2 - x - 72$ 9, −8
23. $f(x) = x^2 + 6x - 7$ 1, −7
24. $f(x) = x^2 + 4x - 5$ 1, −5
25. $f(x) = x^2 + 2x - 24$ 4, −6
26. $f(x) = 2x^2 - 2x - 144$ 9, −8
27. $f(x) = 3x^2 + 9x - 12$ 1, −4
28. $f(x) = 5x^2 + 15x - 20$ 1, −4

First graph each function. Then find the zeros, the axis of symmetry, and the minimum value of each function.

29. $f(x) = x^2 + 18x + 81$
30. $f(x) = x^2 + 6x - 7$
31. $f(x) = x^2 - 5x + 6$

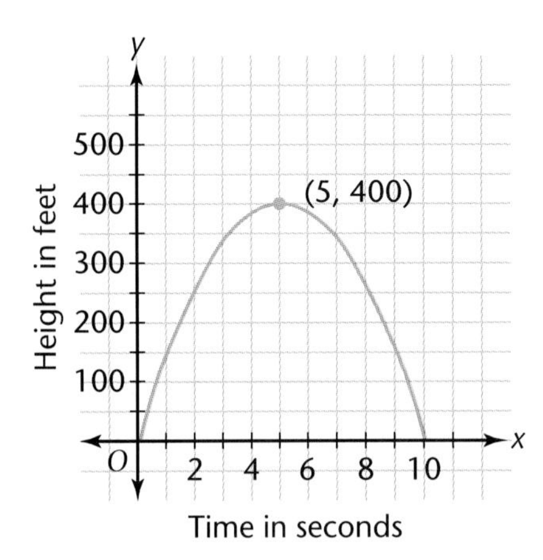

Physics The graph represents a relationship between the time when a projectile is propelled vertically into the air and the height that it reaches.

32. What is the maximum height reached by the projectile? 400 feet

33. How long does it take for the projectile to reach its maximum height? 5 seconds

34. How long does it take for the projectile to return to Earth?

35. What is the equation of the axis of symmetry of the graph?

36. Which equation was used to graph the parabola? c

a. $y = -16(x + 5)^2 + 400$ **b.** $y = -16(x + 5)^2 - 400$

c. $y = -16(x - 5)^2 + 400$ **d.** $y = -16(x - 5)^2 - 400$

34. 10 seconds 35. $x = 5$

37. Physics Use the formula in the caption to find the height in meters after 5 seconds. 76 m

The formula $h = 40t - 5t^2 + 1$ can be used to find the height in meters of the arrow after t seconds.

Look Back

Solve each system of inequalities. [Lesson 2.7]

38. $\begin{cases} y \le -2x + 1 \\ y < 2x - 3 \end{cases}$

39. $\begin{cases} 3x + y < -3 \\ x - y \ge 2 \end{cases}$

Perform each matrix operation. [Lesson 3.2]

40. $\begin{bmatrix} 6 & -2 \\ 8 & 3 \end{bmatrix} + \begin{bmatrix} -2 & 4 \\ 3 & 6 \end{bmatrix}$

41. $\begin{bmatrix} -4 & 5 \\ 6 & 2 \end{bmatrix} - \begin{bmatrix} 3 & -1 \\ -5 & 6 \end{bmatrix}$

42. $\begin{bmatrix} -3 & 7 \\ 4 & -5 \end{bmatrix} + \begin{bmatrix} 6 & 8 \\ -1 & -3 \end{bmatrix}$

Factor. [Lesson 7.7]

43. $x^2 + 10x + 25$ $(x + 5)^2$
44. $x^2 + 14x + 49$ $(x + 7)^2$
45. $x^2 - 18x + 81$ $(x - 9)^2$

40. $\begin{bmatrix} 4 & 2 \\ 11 & 9 \end{bmatrix}$ 41. $\begin{bmatrix} -7 & 6 \\ 11 & -4 \end{bmatrix}$ 42. $\begin{bmatrix} 3 & 15 \\ 3 & -8 \end{bmatrix}$

Look Beyond

46. Several students are able to collect 100,000 cans to recycle. What is the minimum space the cans will take? Assume that a can has a circumference of 21 centimeters and a height of 12.5 centimeters.

LESSON 8.2

Solving Equations of the Form $x^2 = k$

Galileo Galilei, born in Pisa in 1564, proved theoretically that objects fall according to the law of uniformly accelerated motion.

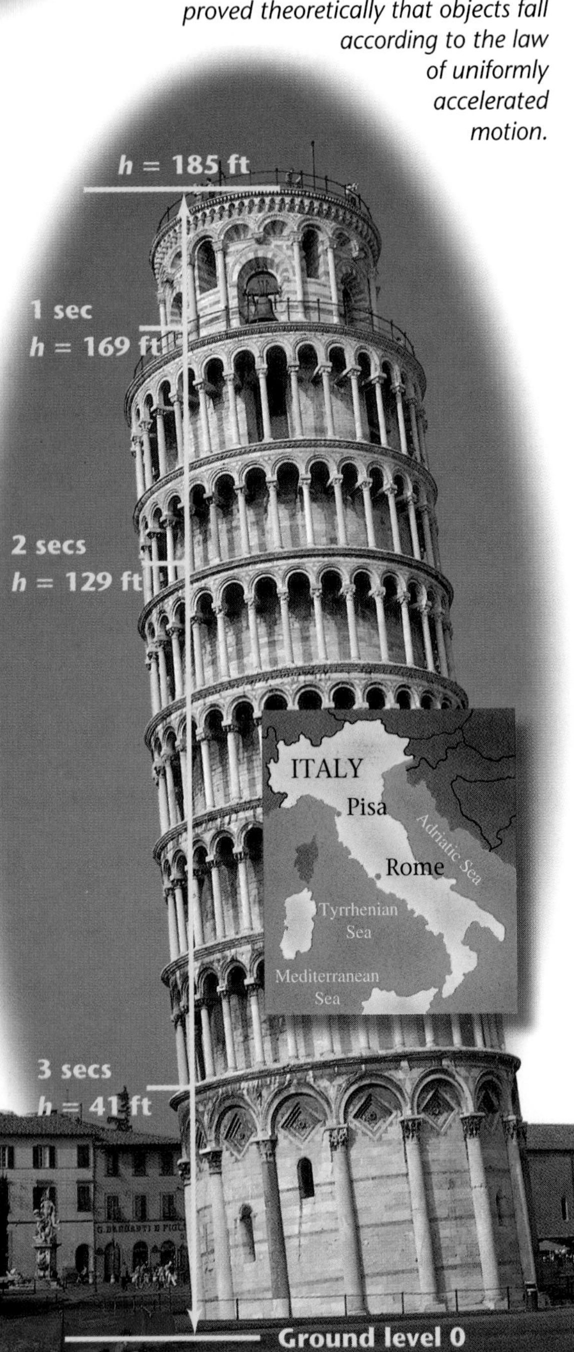

Why *You already know how to find approximate solutions to a quadratic equation by using technology or graphing. Algebraic techniques can be used to find exact solutions.*

An object falls from the top of the Leaning Tower of Pisa, 185 feet above ground level. Its height after t seconds is given by the function $h(t) = -16t^2 + 185$. How long will it take the object to reach the ground?

When the object reaches the ground, the height is 0 feet. Thus, we substitute 0 for $h(t)$ in the given quadratic function to obtain the quadratic equation $0 = -16t^2 + 185$.

Tabular Method

Begin by making a table of values for the function $h(t) = -16t^2 + 185$. Substitute integer values of t into $h(t) = -16t^2 + 185$ to find values of $h(t)$. According to the table, $h(t)$ will be 0 somewhere between $t = 3$ and $t = 4$ seconds. A new table can be constructed in which t changes by tenths.

Time (t)	Height (h)
0	185
1	169
2	121
3	41
4	−71 ← **Stop here.**

Time (t)	Height (h)
3.0	41.00
3.1	31.24
3.2	21.16
3.3	10.76
3.4	0.04
3.5	−11.00 ← **Stop here.**

Closer investigation using the table shows that 3.4 is an approximate solution to the nearest tenth for the equation $0 = -16t^2 + 185$.

How can you use a table to get a solution to the nearest hundredth for the equation $0 = -16t^2 + 185$?

Graphing Method

Graph the function $h(t) = -16t^2 + 185$ with t on the x-axis and $h(t)$ on the y-axis. Find the point where $y \approx 0$. According to the graph, $x \approx 3.4$. Therefore, the object hits the ground after about 3.4 seconds.

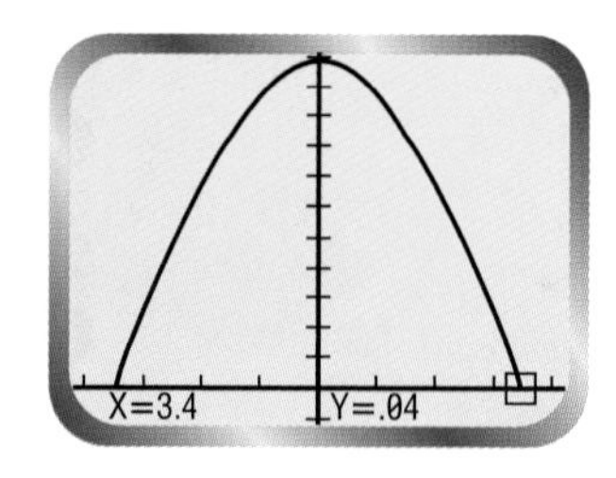

In fact, the quadratic equation $0 = -16t^2 + 185$ has two solutions. A more complete picture of the graph of $0 = -16t^2 + 185$ suggests that the solutions are approximately 3.4 or -3.4. Since the falling-object problem is about height and time, only the positive values are reasonable for the domain and range.

Finding Square Roots

Finding the exact solutions for $0 = -16t^2 + 185$ involves finding **square roots**. Every positive number has a positive and a negative square root.

The positive square root of 9 is 3. $\qquad \sqrt{9} = 3$

The negative square root of 9 is -3. $\qquad -\sqrt{9} = -3$

When solving a quadratic equation, you can use the symbol $\pm$, which is read as "plus or minus."

Therefore, to solve the equation $x^2 = 9$, take the square root of the expression on each side of the equal sign.

$$x^2 = 9$$
$$x = \pm\sqrt{9}$$
$$x = 3 \text{ or } x = -3$$

Remember to consider both the positive and negative square root of 9.

EXAMPLE 1

Show that $\sqrt{\frac{4}{9}} = \frac{\sqrt{4}}{\sqrt{9}}$.

Solution ➤

Simplify the square root under the radical signs.

$$\sqrt{\frac{4}{9}} = \frac{2}{3} \text{ because } \frac{2}{3} \cdot \frac{2}{3} = \frac{4}{9}$$

$$\frac{\sqrt{4}}{\sqrt{9}} = \frac{2}{3} \text{ because } \frac{2 \cdot 2}{3 \cdot 3} = \frac{4}{9}$$

Since $\sqrt{\frac{4}{9}} = \frac{2}{3}$ and $\frac{\sqrt{4}}{\sqrt{9}} = \frac{2}{3}$, $\sqrt{\frac{4}{9}} = \frac{\sqrt{4}}{\sqrt{9}}$ by substitution. ❖

EXAMPLE 2

Solve each equation.

A $x^2 = \frac{4}{9}$ **B** $x^2 = 1.44$ **C** $x^2 = 10$

Solution ➤

A There are two solutions: $\sqrt{\frac{4}{9}} = \frac{2}{3}$ and $-\sqrt{\frac{4}{9}} = -\frac{2}{3}$

B There are two solutions: $\sqrt{1.44} = 1.2$ and $-\sqrt{1.44} = -1.2$

C The solutions are $\sqrt{10}$ and $-\sqrt{10}$. There is no rational number answer. An approximation can be found by using the [√] key on your calculator.

[√] 10 [=] 3.16227766

The approximate solutions are 3.16 and −3.16.

The results of Example 2 lead to a generalization for solving a quadratic equation of the form $x^2 = k$. ❖

SOLVING $x^2 = k$ WHEN $k \geq 0$

If $x^2 = k$, and $k \geq 0$, then

1. $x = \pm\sqrt{k}$ and

2. the solutions are $\sqrt{k}$ and $-\sqrt{k}$.

For example, if $x^2 = 16$, then $x = \pm\sqrt{16}$, or ± 4. The solutions are 4 and −4.

Algebraic Method

Now the *exact* time that it takes for the object to fall from the top of the Tower of Pisa to the ground can be found.

$$-16t^2 + 185 = 0$$
$$-16t^2 = -185$$
$$t^2 \approx 11.56$$
$$t \approx \pm\sqrt{11.56}$$

The time is approximately $\sqrt{11.56}$ seconds.

CT– The zeros of $h(t)$ are the values of t that make $h(t) = 0$ true. Since $h(t) = -16t^2 + 185$, solving $0 = -16t^2 + 185$ gives the zeros of h.

Why is $-\sqrt{11.56}$ not a solution to this problem?

Time cannot be negative.

A calculator gives the solution 3.4, which agrees with the tabular and graphing methods.

Explain why the algebraic method for solving the quadratic equation $-16t^2 + 185 = 0$ gives another way to find the zeros for the quadratic function $h(t) = -16t^2 + 185$.

EXAMPLE 3

Solve each equation.

A $(a - 2)^2 - 9 = 0$ **B** $(x - 2)^2 = 11$

Solution ➤

A In this equation, the expression $a - 2$ plays the role of x in the statement $x^2 = k$.

$$(a - 2)^2 - 9 = 0$$
$$(a - 2)^2 = 9$$
$$a - 2 = \pm\sqrt{9}$$
$$a - 2 = \pm 3$$
$$a = 2 \pm 3$$

The solutions are $2 + 3$, or 5, and $2 - 3$, or -1. Check each solution in the original equation.

B $(x - 2)^2 = 11$

$$x - 2 = \pm\sqrt{11}$$
$$x = 2 \pm \sqrt{11}$$

The approximate solutions are $2 + \sqrt{11}$, or 5.32, and $2 - \sqrt{11}$, or -1.32. Check each solution in the original equation. ❖

The solution to Example 3, Part **A** will help you sketch the graph of the function $f(x) = (x - 2)^2 - 9$.

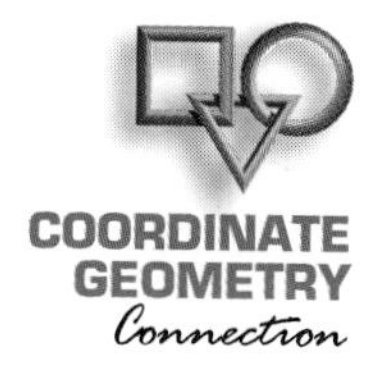

In the form $f(x) = (x - 2)^2 - 9$, the vertex is $(2, -9)$ and the axis of symmetry is $x = 2$. Since the coefficient of the quadratic term is positive, the parabola opens upward and has a minimum value. The solutions to $(x - 2)^2 - 9 = 0$ are 5 and -1. Thus, the graph is a parabola that crosses the x-axis at $(5, 0)$ and $(-1, 0)$.

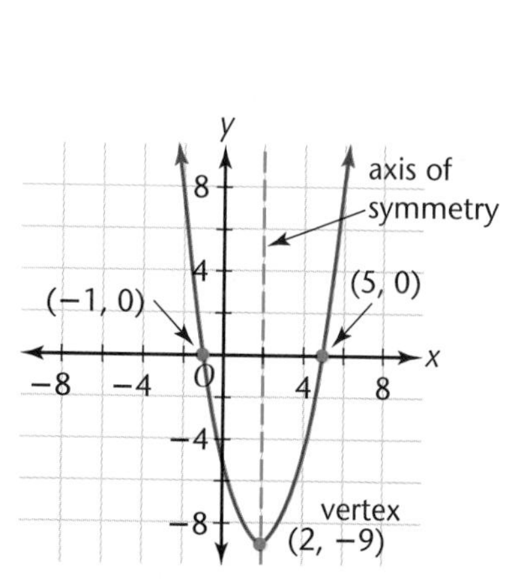

Communicate

1. Explain how to make a table of values for the function $y = -12x^2 + 300$ in order to determine where the function crosses the x-axis. Why would it be important to know when the values of the function are no longer positive?
2. Explain how to find the square root of 100. What is meant by the sign $\pm$?

Explain how to solve each equation.

3. $x^2 = 64$ 4. $x^2 = 8$ 5. $x^2 = \frac{16}{100}$

Explain how to solve each equation by using square roots.

6. $(x + 3)^2 - 25 = 0$ 7. $(x - 8)^2 = 2$
8. Explain how to sketch a graph from the vertex, axis of symmetry, and zeros of $f(x) = (x + 4)^2 - 5$.

21. ± 25 33. ± 7.94 39. 4.41, 1.59
25. ± 7.35 37. 4.45, -0.45 41. 14, -10
29. $\pm\frac{1}{5}$ 38. -9.24, -4.76

Practice & Apply

9. Make a table of values for the function $f(t) = -14t^2 + 300$. Where does the function cross the t-axis?

Find each square root. Round when necessary to the nearest hundredth.

10. $\sqrt{121}$ 11 11. $\sqrt{144}$ 12 12. $\sqrt{625}$ 25 13. $\sqrt{36}$ 6
14. $\sqrt{44}$ 6.63 15. $\sqrt{90}$ 9.49 16. $\sqrt{88}$ 9.38 17. $\sqrt{19}$ 4.36

Solve each equation. Round when necessary to the nearest hundredth.

18. $x^2 = 25$ ± 5 19. $x^2 = 169$ ± 13 20. $x^2 = 81$ ± 9 21. $x^2 = 625$
22. $x^2 = 12$ ± 3.46 23. $x^2 = 24$ ± 4.90 24. $x^2 = 18$ ± 4.24 25. $x^2 = 54$
26. $x^2 = \frac{25}{81}$ $\pm\frac{5}{9}$ 27. $x^2 = \frac{49}{121}$ $\pm\frac{7}{11}$ 28. $x^2 = \frac{36}{49}$ $\pm\frac{6}{7}$ 29. $x^2 = \frac{4}{100}$
30. $x^2 = 32$ ± 5.66 31. $x^2 = 45$ ± 6.71 32. $x^2 = 28$ ± 5.29 33. $x^2 = 63$
34. $(x + 4)^2 - 25 = 0$ 1, -9 35. $(x - 5)^2 - 9 = 0$ 8, 2 36. $(x + 1)^2 - 1 = 0$ 0, -2 37. $(x - 2)^2 - 6 = 0$
38. $(x + 7)^2 - 5 = 0$ 39. $(x - 3)^2 - 2 = 0$ 40. $(x + 3)^2 = 36$ 3, -9 41. $(x - 2)^2 = 144$
42. $(x - 8)^2 = 81$ 17, -1 43. $(x - 1)^2 = 11$ 4.32, -2.32 44. $(x + 5)^2 = 10$ -1.84, -8.16 45. $(x + 6)^2 = 15$ -2.13, -9.87

Find the vertex, axis of symmetry, and zeros of each function.

46. $f(x) = (x - 4)^2 - 9$ 47. $g(x) = (x + 2)^2 - 1$ 48. $h(x) = (x - 4)^2 - 3$

Refer to the graph of $f(x) = (x + 4)^2 - 4$ for Exercises 49–53.

49. Find the vertex. $(-4, -4)$

50. Find the axis of symmetry. $x = -4$

51. Find the zeros. $-2, -6$

52. Sketch a graph.

53. Check Exercises 49–52 by graphing $f(x) = (x + 4)^2 - 4$.

54. **Geometry** The area of a square garden is 169 square feet. Find the length of each side. $(A = s^2)$
13 feet

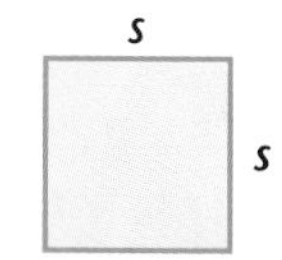

55. **Geometry** Use the formula $S = 4\pi r^2$ to find the radius of a sphere with a surface area of 90 square meters. (Use $\pi = 3.14$.)
About 2.68 meters

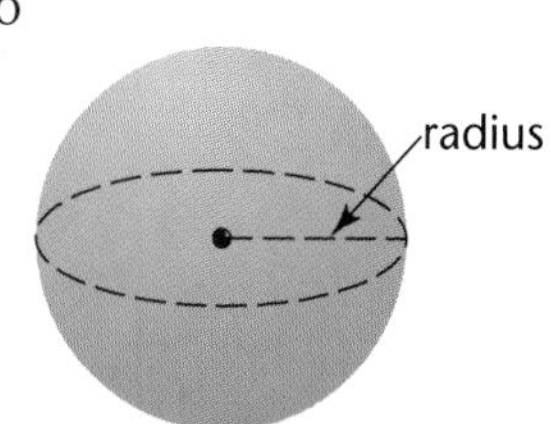

700 ft

The height of the Met Life Building in New York City is 700 feet.

Physics If an object falls from the Met Life Building, and its height after t seconds is given by the function $h(t) = -16t^2 + 700$, find

56. how long it will take for the object to reach the ground. About 6.61 seconds

57. how long it will take for the object to reach a height of 100 feet. About 6.12 seconds

58. $\begin{bmatrix} 23 & -14 \\ 24 & -18 \end{bmatrix}$ 59. $\begin{bmatrix} 62 & 70 \\ 20 & 28 \end{bmatrix}$

Look Back

Find each product matrix. [Lesson 3.3]

58. $\begin{bmatrix} 4 & -1 \\ 6 & 0 \end{bmatrix}\begin{bmatrix} 4 & -3 \\ -7 & 2 \end{bmatrix}$

59. $\begin{bmatrix} 9 & -4 \\ 2 & -4 \end{bmatrix}\begin{bmatrix} 6 & 6 \\ -2 & -4 \end{bmatrix}$

Simplify each expression. [Lesson 6.2]

60. $\frac{-6x^2y^2}{2x}$ $-3xy^2$

61. $\frac{b^3c^4}{bc}$ b^2c^3

62. $(p^2)^4(2a^2b^3)^2$ $4a^4b^6p^8$

Add or subtract each polynomial. [Lesson 7.2]

63. $(3x^2 - 2x + 1) + (2x^2 + 4x + 6)$ $5x^2 + 2x + 7$

64. $x^2 + 2x - 1 - (2x^2 - 5x + 7)$ $-x^2 + 7x - 8$

65. Find the factors of $a^2 - b^2$, $a^4 - b^4$, and $a^8 - b^8$. How many factors does $a^{64} - b^{64}$ have? Explain. **[Lesson 7.6]**

66. About 4.12 67. 0

Look Beyond

66. Find $\sqrt{5^2 - 4 \cdot 1 \cdot 2}$.

67. Find $\sqrt{6^2 - 4(1)(9)}$.

LESSON 8.3 Completing the Square

Why *If you rewrite the quadratic function $y = x^2 + bx + c$ in the form $y = (x - h)^2 + k$, you can find the vertex and axis of symmetry of the parabola. A technique called completing the square will enable you to do this.*

In the previous chapter you used algebra tiles to factor perfect-square trinomials such as $x^2 + 6x + 9$. How can a square be constructed from the following tiles?

To make a square, start with the x^2-tile and arrange the six x-tiles into 2 groups of 3.

$x^2 + 6x$

Count the number of 1-tiles that it takes to fill in the corner.

$x^2 + 6x + 9$

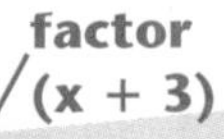

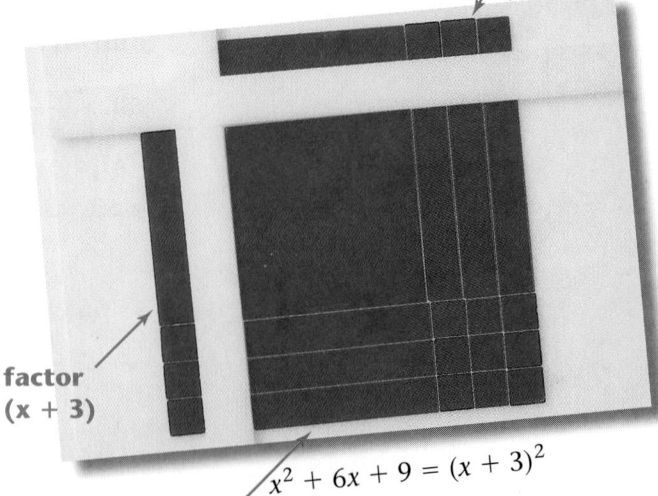

$x^2 + 6x + 9 = (x + 3)^2$

completed square $x^2 + 6x + 9$

Why do you think the process of adding the nine 1-tiles is called **completing the square**? When the nine 1-tiles are added, the figure is a square.

Exploration *How Many Ones?*

Use algebra tiles to complete the square for $x^2 + 10x$.

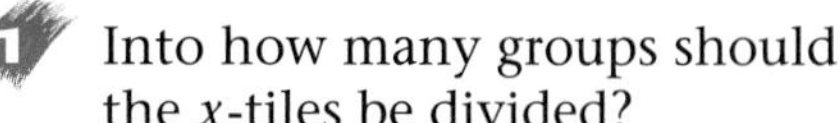

1. Into how many groups should the x-tiles be divided?
   2 groups with 5 x-tiles per group

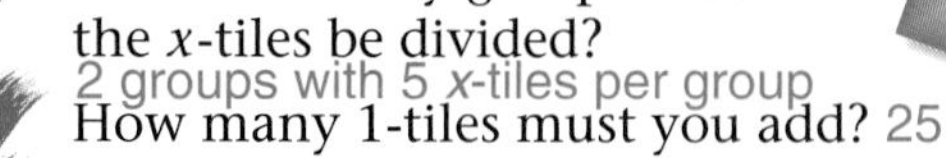

2. How many 1-tiles must you add? 25

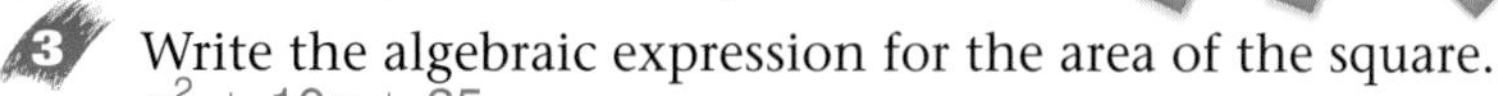

3. Write the algebraic expression for the area of the square.
   $x^2 + 10x + 25$

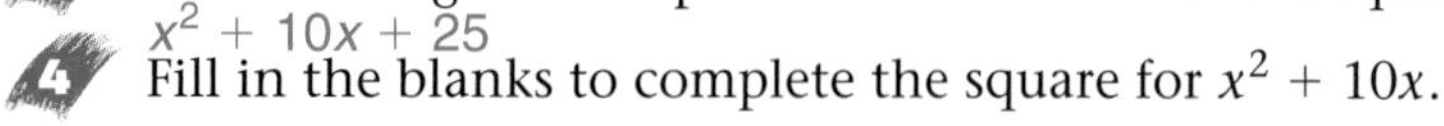

4. Fill in the blanks to complete the square for $x^2 + 10x$.
   a. Divide the x-tiles into 2 sets of __?__ x-tiles. 5
   b. Add __?__ 1-tiles. 25
   c. $x^2 + 10x + \underline{?} = (x + 5)^2$ 25

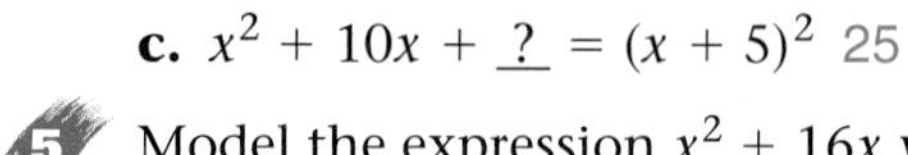

5. $x + 8$; the constant that is added is the square of half the coefficient of x: $\left(\frac{1}{2} \cdot 16\right)^2 = 8^2 = 64.$

5. Model the expression $x^2 + 16x$ with algebra tiles. What is the length of the side of your new square? How does the constant that is added to the expression compare with the number 16?
   a. Write the area of your new square in the form $x^2 + bx + c$. $x^2 + 16x + 64$
   b. Write the area of your new square in the form $(x + \underline{?})^2$. $(x + 8)^2$

6. Suppose you have 1 x^2-tile and 74 x-tiles. How many 1-tiles do you need to complete the square? Explain how you got your answer. ❖

6. 1369 1-tiles: $\left(\frac{1}{2} \cdot 74\right)^2 = 1369$

In the exploration above, the coefficient of x is an even number. It is easy to divide an even number of x-tiles into 2 sets. What happens if the coefficient of x is odd? What happens if the coefficient of x is negative? This is shown in Examples 1 and 2.

EXAMPLE 1

Complete the square for $x^2 + 5x$.

Solution ➤

It is not possible to use actual tiles to complete the square. However, a picture will serve as a good model in this case.

Split the 5 x-tiles into 2 groups, each with $2\frac{1}{2}$ x-tiles. Then add 1-tiles to complete the square.

You need $\left(\frac{5}{2}\right)^2$, or $\frac{25}{4}$, 1-tiles.

By arranging $6\frac{1}{4}$ 1-tiles in a certain way, the result is a square whose area is $x^2 + 5x + 6\frac{1}{4}$, or $\left(x + \frac{5}{2}\right)^2$. ❖

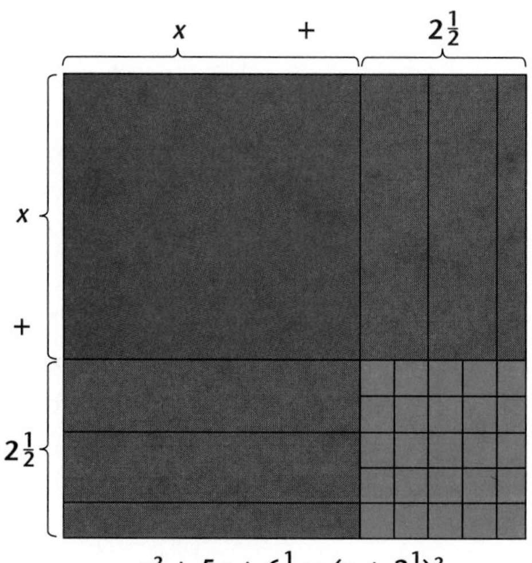

$x^2 + 5x + 6\frac{1}{4} = (x + 2\frac{1}{2})^2$

EXAMPLE 2

Complete the square for $x^2 - 6x$.

Solution ➤

One-half the coefficient of x is -3.
Add $(-3)^2$, or 9, to form a perfect square.
The perfect square is $x^2 - 6x + 9 = x^2 - 6x + (-3)^2 = (x - 3)^2$. ❖

How do you decide what to add to $x^2 + 6x$ and $x^2 - 6x$ to form a perfect square? Divide the coefficient of x by 2, and square the result.

Completing the square gives a technique for rewriting the quadratic function $y = x^2 + 6x$ in the form $y = (x - h)^2 + k$.

Start with the quadratic function.	$y = x^2 + 6x$
Find what you need to add to complete the square.	$\frac{1}{2} \cdot 6 = 3; \quad 3^2 = 9$
Add and subtract the square from the quadratic.	$y = x^2 + 6x + 9 - 9$
Group the terms.	$y = (x^2 + 6x + 9) - 9$
Write in the form $y = (x - h)^2 + k$.	$y = (x + 3)^2 - 9$

Why can you add $9 - 9$ to $x^2 + 6x$ without changing the value of the expression? It is equivalent to adding zero.

The minimum value or the maximum value of a quadratic function, $y = (x - h)^2 + k$, is the y-value of the vertex.

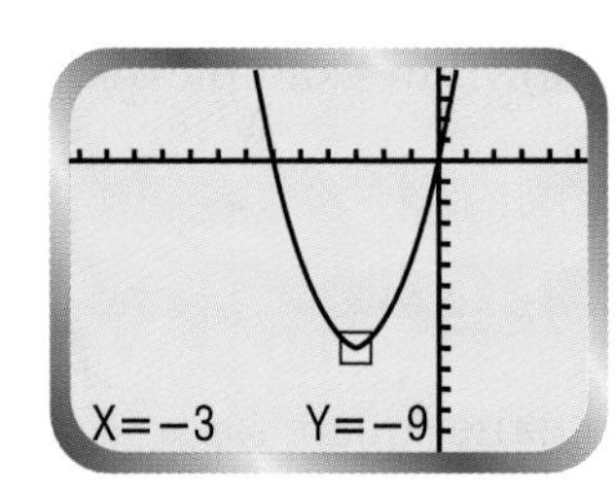

The vertex of the graph of $y = (x + 3)^2 - 9$ is $(-3, -9)$. Thus, the minimum value of $y = x^2 + 6x$ is -9.

EXAMPLE 3

Find the minimum value for the function $f(x) = x^2 - 3x$.

Solution ➤

The vertex of the parabola can be found by completing the square. The coefficient of x is -3. Add and subtract the square of $-\frac{3}{2}$ to the equation.

$$\begin{aligned} f(x) &= x^2 - 3x + \left(-\frac{3}{2}\right)^2 - \left(-\frac{3}{2}\right)^2 \\ &= \left(x - \frac{3}{2}\right)^2 - \frac{9}{4} \end{aligned}$$

The graph of $f(x) = \left(x - \frac{3}{2}\right)^2 - \frac{9}{4}$ is a parabola with the vertex at $\left(\frac{3}{2}, -\frac{9}{4}\right)$.

The graph opens upward, so the minimum value is $-\frac{9}{4}$. ❖

You can now complete the square for a quadratic of the form $y = x^2 + bx$. The next step is to complete the square for a quadratic of the form $y = x^2 + bx + c$.

EXAMPLE 4

A Rewrite $y = x^2 - 8x + 7$ in the form $y = (x - h)^2 + k$.

B Find the vertex of the parabola.

Solution ➤

A The new feature in this problem is the constant 7.

Original function	$y = x^2 - 8x + 7$
Group the x^2 and x terms.	$y = (x^2 - 8x) + 7$

First complete the square of $x^2 - 8x$.

$\frac{1}{2}(-8) = -4$ and $(-4)^2 = \mathbf{16}$

Now add and subtract **16**.	$y = (x^2 - 8x + \mathbf{16}) + 7 - \mathbf{16}$
Write in the form $y = (x - h)^2 + k$.	$y = (x - 4)^2 - 9$

B The vertex is $(4, -9)$. Check by graphing. ❖

Try This Find the vertex of $y = x^2 - 6x + 11$. (3, 2)

EXERCISES & PROBLEMS

Communicate

1. Describe how to use algebra tiles to complete the square for $x^2 + 4x$.
2. Discuss how to complete the square algebraically for $x^2 - 7x$.
3. How do you rewrite $y = x^2 + 10x + 25 - 25$ in the form $y = (x - h)^2 + k$?

Describe how to find the minimum value of each quadratic function.

4. $y = x^2 + 2$
5. $y = x^2 - x$
6. $y = x^2 - 3$

7. Explain how to rewrite $y = x^2 - 10x + 11$ in the form $y = (x - h)^2 + k$. How do you find the vertex from this form of the equation?

Practice & Apply

Use algebra tiles to complete the square for each expression. Write your answer in the form $x^2 + bx + c$ and in the form $(x + \underline{?})^2$.

8. $x^2 + 14x$ **9.** $x^2 - 14x$ **10.** $x^2 + 8x$ **11.** $x^2 - 8x$

Complete the square. Write your answer in the form $x^2 + bx + c$ and in the form $(x + \underline{?})^2$.

12. $x^2 + 6x$ **13.** $x^2 - 2x$ **14.** $x^2 + 12x$ **15.** $x^2 - 12x$

16. $x^2 + 7x$ **17.** $x^2 - 10x$ **18.** $x^2 + 15x$ **19.** $x^2 - 5x$

20. $x^2 + 16x$ **21.** $x^2 + 20x$ **22.** $x^2 - 9x$ **23.** $x^2 + 40x$

Rewrite each function in the form $y = (x - h)^2 + k$.

24. $y = x^2 + 8x + 16 - 16$ **25.** $y = x^2 - 4x + 4 - 4$

26. $y = x^2 - 10x + 25 - 25$ **27.** $y = x^2 + 14x + 49 - 49$

28. $y = x^2 - 16x + 64 - 64$ **29.** $y = x^2 + 20x + 100 - 100$

30. $y = (x + 5)^2 - 25$

Refer to the equation $y = x^2 + 10x$ for Exercises 30–32.

30. Complete the square and rewrite it in the form $y = (x - h)^2 + k$.

31. Find the vertex. $(-5, -25)$

32. Find the maximum or minimum value.
Minimum: -25

Find the vertex of each quadratic function.

33. $f(x) = x^2 + 3$ $(0, 3)$ **34.** $f(x) = x^2 - 4$ $(0, -4)$ **35.** $f(x) = x^2 + 2$ $(0, 2)$

36. $f(x) = x^2 - 1$ $(0, -1)$ **37.** $f(x) = x^2 + 6$ $(0, 6)$ **38.** $f(x) = x^2 - 5$ $(0, -5)$

39. $y = x^2 + 4x - 1$ **40.** $y = x^2 - 2x - 3$ $(1, -4)$ **41.** $y = x^2 + 10x - 12$

42. $y = x^2 - x + 2$ **43.** $y = x^2 - 6x - 5$ **44.** $y = x^2 + \frac{1}{3}x - 3$

39. $(-2, -5)$ 41. $(-5, -37)$ 42. $(0.5, 1.75)$ 43. $(3, -14)$

44. $\left(-\frac{1}{6}, -3\frac{1}{36}\right)$

A water rocket is shot vertically into the air with an initial velocity of 192 feet per second.

Physics The relationship between time, t, and height, h, is given by the formula $h = -16t^2 + 192t$.

45. Graph the function. (If you use graphing technology, you will need to use x in place of t and y in place of h.)

46. Complete the square. Find the vertex of the parabola. $(6, 576)$

47. What is the maximum height reached by the projectile? 576 feet

48. How long does it take for the projectile to reach its maximum height? 6 seconds

A tennis pro throws a ball vertically into the air and watches as it begins to descend. He extends his racket to make contact with the ball.

Physics The height of the ball can be modeled by the function $h(t) = -16(t - 0.6)^2 + 11$, where $h(t)$ is the height of the ball in feet after t seconds have elapsed.

49. What is the maximum height reached by the tennis ball? 11 feet

50. How long did it take the tennis ball to reach its maximum height? About $\frac{1}{2}$ second

51. The player can make contact with the ball at a height of about $8\frac{1}{2}$ feet. About how much time will elapse before the player hits the tennis ball? About 1 second

52. If the tennis ball is allowed to fall to the ground, how much time will elapse? About $1\frac{1}{2}$ seconds

Look Back

Solve each system by graphing. [Lesson 2.3]

53. $\begin{cases} x - 3y = 3 \\ 2x - y = -4 \end{cases}$ **54.** $\begin{cases} x - 2y = 0 \\ x + y = 3 \end{cases}$

55. Can you find the inverse of $\begin{bmatrix} 4 & 6 \\ 2 & 3 \end{bmatrix}$? Explain. **[Lesson 3.4]** No, the matrix has no inverse.

56. **Probability** Find the probability that an even number will be drawn from a bag containing the whole numbers from 1 to 20 inclusive. **[Lesson 4.6]** $\frac{1}{2}$ or 50%

Write each expression without a negative exponent. [Lesson 6.3]

57. a^2b^{-3} $\frac{a^2}{b^3}$ **58.** $\frac{m^5}{m^{-2}}$ m^7 **59.** $\frac{2n^{-6}}{n^{-4}}$ $\frac{2}{n^2}$

60. **Geometry** Find the area of the square rug at the right. **[Lesson 7.4]** $x^2 + 8x + 16$ square units

Look Beyond

The equation for the height in feet of a model rocket after t seconds is given by $h(t) = -16t^2 + 200t$. Use the indicated method to find t when the rocket has a height of 400 feet.

61. graphing

62. algebra 2.5 seconds, 10 seconds

EYEWITNESS MATH

Rescue at 2000 Feet

A Miraculous Sky Rescue

The jump began as a routine skydiving exercise, part of a convention of 420 parachutists sponsored by Skydive, Arizona, but it quickly turned into a test of nerve, instinct, and courage . . .

Moments after he went out the open hatch of a four-engine DC-4 airplane near Coolidge, Ariz., sky diver Gregory Robertson, 35, could see that Debbie Williams, 31, a fellow parachutist with a modest 50 jumps to her credit, was in big trouble. Instead of "floating" in the proper stretched-out position parallel to the earth, Williams was tumbling like a rag doll. In attempting to join three other divers in a hand-holding ring formation, she had slammed into the backpack of another chutist and was knocked unconscious.

From his instructor's position above the other divers, Robertson reacted with instincts that had been honed by 1700 jumps during time away from his job as an AT&T engineer in Phoenix. He straightened into a vertical dart, arms pinned to his body, ankles crossed, head aimed at the ground in what chutists call a "no-lift" dive, and plummeted toward Williams . . . At 3500 ft., about 10 seconds before impact, Robertson caught up with Williams, almost hitting her but slowing his own descent by assuming the open-body froglike position. He angled the unconscious sky diver so her chute could open readily and at 2000 ft., with some 6 seconds left, yanked the ripcord on her emergency chute then pulled his own ripcord. The two sky divers floated to the ground. Williams, a fifth-grade teacher from Post, Texas, landed on her back, suffering a skull fracture and a perforated kidney—but alive. In the history of recreational skydiving, there has never been such a daring rescue in anyone's recollection.

Does the article give you a complete picture of the rescue? How can you tell if the numbers in the article are accurate?

You can use what you know about distance, time, and speed to check the facts in the article and to get a fuller sense of what took place during this amazing feat.

First, study these diagrams to get an idea of some of the forces that affect sky divers in free fall.

Robertson explains how he saved a life at 2000 ft.

When a parachutist jumps out of a plane, the force of gravity causes the diver to accelerate downward.

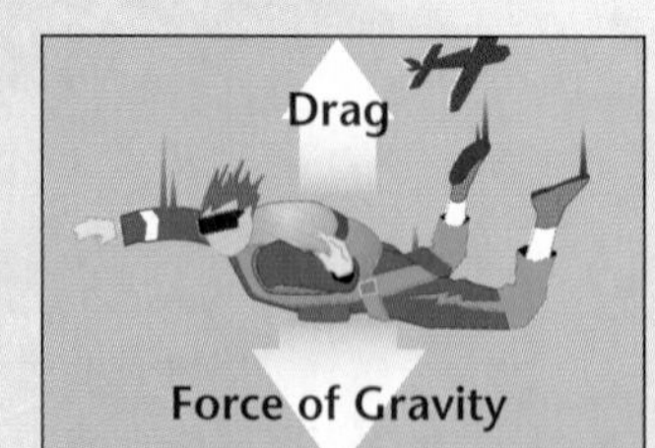

As the sky diver's speed increases, the force of air resistance (or drag) plays a larger and larger role.

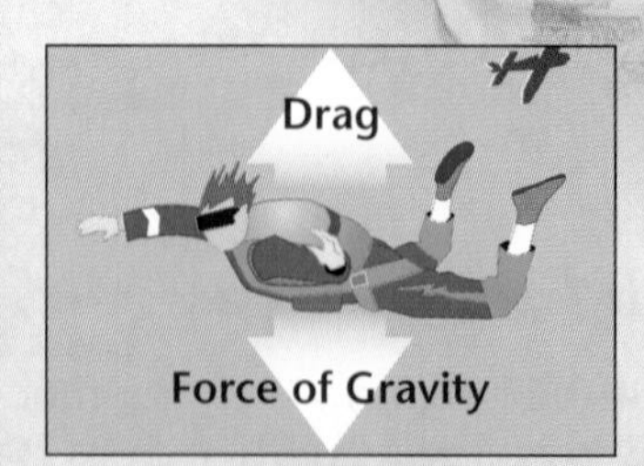

At some point, the two forces balance and the sky diver no longer accelerates, but falls at a steady rate, called the terminal velocity.

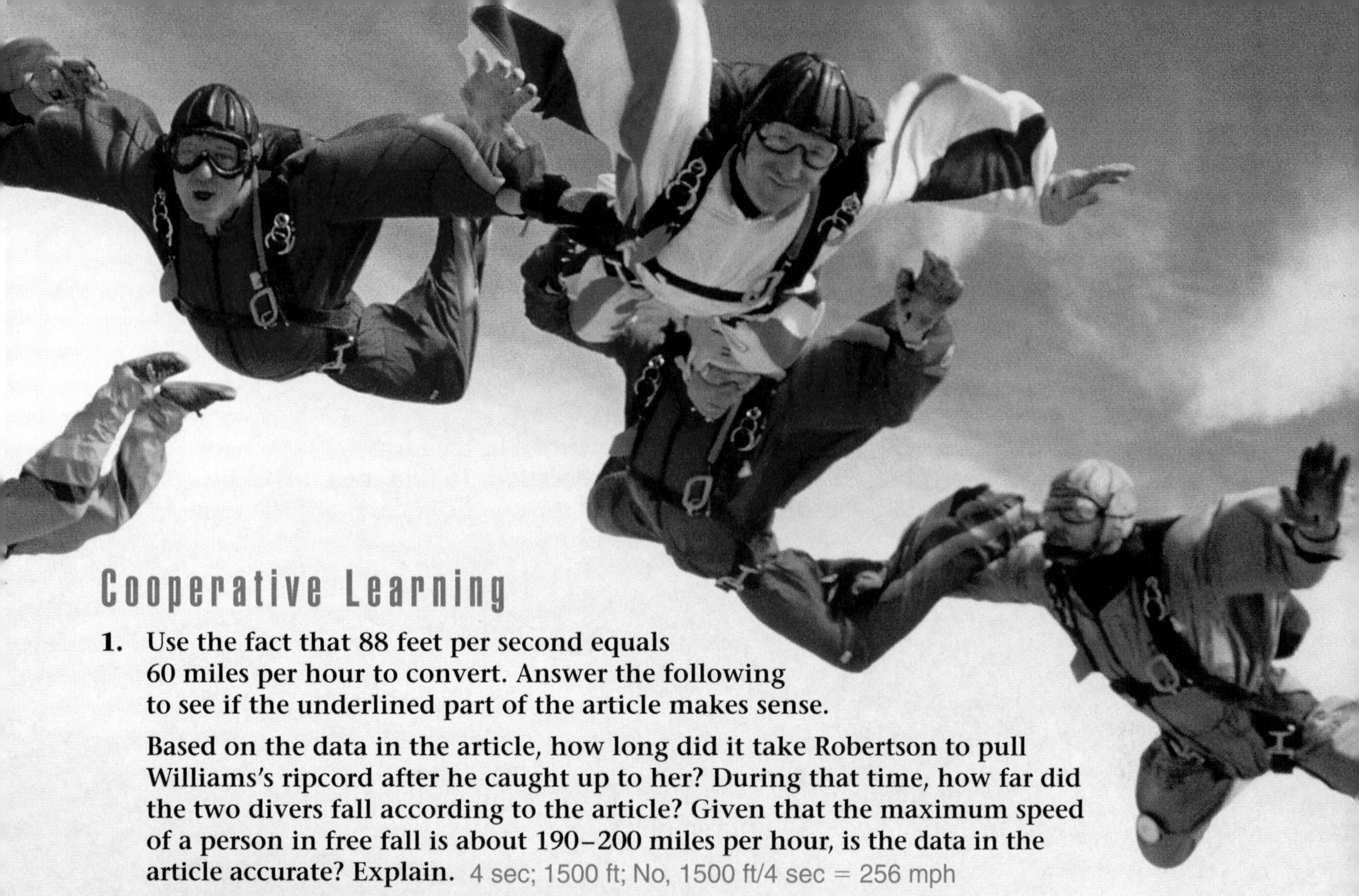

Cooperative Learning

1. Use the fact that 88 feet per second equals 60 miles per hour to convert. Answer the following to see if the underlined part of the article makes sense.

   Based on the data in the article, how long did it take Robertson to pull Williams's ripcord after he caught up to her? During that time, how far did the two divers fall according to the article? Given that the maximum speed of a person in free fall is about 190–200 miles per hour, is the data in the article accurate? Explain. 4 sec; 1500 ft; No, 1500 ft/4 sec = 256 mph

2. After Robertson caught up with Williams, the two divers were probably falling at a speed between 125 and 150 miles per hour. Suppose they fell at 135 miles per hour. You can use that value to find more reasonable time estimates for the rescue. 135 mph = 198 ft/sec

   About how many seconds from impact would the two sky divers be at 3500 feet? at 2000 feet? Suppose Robertson did catch up to Williams at 3500 feet and did pull her ripcord at 2000 feet. About how much time would have elapsed between those two events if the divers were falling at 135 miles per hour? 17.7 sec; 10.1 sec; about 7 to 8 sec

3. While Robertson was catching up to Williams, he changed his body position to speed up and then slow down. Such motion is difficult to calculate exactly, but you can make estimates for this part of the rescue by using a single average velocity for Robertson. Assume that Robertson was at 8500 feet and Williams was at 8300 feet when Robertson started trying to catch up.

   Write an equation for t in terms of v_r and v_w, where

   v_r is Robertson's average velocity in feet per second while catching up to Williams,

   v_w is Williams' velocity in feet per second, and $t = \frac{200}{v_r - v_w}$

   t is the time in seconds from when Robertson starts to go after Williams to when he reaches her.

4. Use your equation to find t if $v_r = 206$ (about 140 miles per hour) and $v_w = 198$ (about 135 miles per hour). Based on your results, at what altitude would Robertson catch up to Williams? $t = 25$ sec; 3350 ft

5. Imagine that you are a journalist writing about the rescue. Use your data from Steps 2, 3, and 4 to create a paragraph that will give your readers a reasonably accurate sense of the event. Answers may vary.

LESSON 8.4

Solving Equations of the Form $x^2 + bx + c = 0$

Why *You have already solved equations of the form $x^2 = k$. Equations of the form $x^2 + bx + c = 0$ can be solved by completing the square.*

In Lesson 8.3 you found the vertex of $f(x) = x^2 - 8x + 7$ by completing the square. When this function is rewritten in the form $f(x) = (x - 4)^2 - 9$, the vertex $(4, -9)$ can be noted by inspection. To find the zeros of this function, substitute 0 for $f(x)$, and rewrite the equation in the form $x^2 = k$.

$$(x - 4)^2 - 9 = 0$$
$$(x - 4)^2 = 9$$
$$x - 4 = \pm 3$$

$$x - 4 = 3 \quad \text{or} \quad x - 4 = -3$$
$$x = 7 \quad \text{or} \quad x = 1$$

The solutions are 7 and 1. These two solutions indicate what the x-intercepts in the graph of $y = x^2 - 8x + 7$ are. Substitute 7 and 1 for x.

$$y = 7^2 - 8(7) + 7 = 49 - 56 + 7 = 0$$

$$y = 1^2 - 8(1) + 7 = 1 - 8 + 7 = 0$$

Thus, the parabola intersects the x-axis at (7, 0) and (1, 0).

Why are the zeros of the function $f(x) = x^2 - 8x + 6$ and the solutions to the equation $x^2 - 8x + 6 = 0$ the same? When $f(x) = 0$, the graph crosses the x-axis, which is the same as the roots or solutions of the equation.

EXAMPLE 1

A Solve $x^2 - 6x + 8 = 0$.

B Find the zeros of $h(x) = x^2 - 6x + 7$.

Solution ➤

For each problem, write the expression in the form $(x - h)^2 + k$ by completing the square. Then rewrite the equation in $x^2 = k$ form to find the solution. Check by substitution.

A

$$\begin{aligned} x^2 - 6x + 8 &= 0 \\ x^2 - 6x + 9 - 9 + 8 &= 0 \\ (x^2 - 6x + 9) - 1 &= 0 \\ (x - 3)^2 - 1 &= 0 \\ (x - 3)^2 &= 1 \end{aligned}$$

$$x - 3 = \sqrt{1} \quad \text{or} \quad x - 3 = -\sqrt{1}$$

$$x = 4 \quad \text{or} \quad x = 2$$

The solutions are 4 and 2.

B

$$\begin{aligned} x^2 - 6x + 7 &= 0 \\ x^2 - 6x + 9 - 9 + 7 &= 0 \\ (x^2 - 6x + 9) - 2 &= 0 \\ (x - 3)^2 - 2 &= 0 \\ (x - 3)^2 &= 2 \end{aligned}$$

$$x - 3 = \sqrt{2} \quad \text{or} \quad x - 3 = -\sqrt{2}$$

$$x = 3 + \sqrt{2} \quad \text{or} \quad x = 3 - \sqrt{2}$$

The zeros are $3 + \sqrt{2}$ and $3 - \sqrt{2}$. ❖

Try This Solve $x^2 - 7x + 10 = 0$. $x = 5$ or $x = 2$

When a quadratic can be factored mentally, you can quickly solve the quadratic equation. For example, factor $x^2 - 6x + 8$ into $(x - 4)(x - 2)$.

Thus, $x^2 - 6x + 8 = 0$ is the same as $(x - 4)(x - 2) = 0$.

If the product of two factors is equal to zero, then at least one of the factors must be zero. This generalization is called the *Zero Product Property.*

ZERO PRODUCT PROPERTY

If a and b are real numbers such that $ab = 0$, then $a = 0$ or $b = 0$.

The Zero Product Property can be used to solve $(x - 4)(x - 2) = 0$.

$$x - 4 = 0 \quad \text{or} \quad x - 2 = 0$$

$$x = 4 \qquad\qquad x = 2$$

The solutions are 4 and 2.

Check by substituting 4 and 2 for x in the original equation.

$$x^2 - 6x + 8 = (4)^2 - 6(4) + 8 = 16 - 24 + 8 = 0 \quad \text{True}$$

$$x^2 - 6x + 8 = (2)^2 - 6(2) + 8 = 4 - 12 + 8 = 0 \quad \text{True}$$

EXAMPLE 2

Solve $x^2 + x = 2$ by the indicated method.

A Completing the square **B** Factoring

Solution ➤

A

$$x^2 + x = 2$$
$$x^2 + x + \left(\frac{1}{2}\right)^2 = 2 + \frac{1}{4}$$
$$\left(x + \frac{1}{2}\right)^2 = \frac{9}{4}$$
$$x + \frac{1}{2} = \frac{3}{2} \quad \text{or} \quad x + \frac{1}{2} = -\frac{3}{2}$$
$$x = \frac{3}{2} - \frac{1}{2} \quad \text{or} \quad x = -\frac{3}{2} - \frac{1}{2}$$
$$x = 1 \quad \text{or} \quad x = -2$$

B

$$x^2 + x = 2$$
$$x^2 + x - 2 = 0$$
$$(x + 2)(x - 1) = 0$$
$$x + 2 = 0 \quad \text{or} \quad x - 1 = 0$$
$$x = -2 \quad \text{or} \quad x = 1$$

The solutions are 1 and −2. ❖

Sometimes either method can be used to solve a quadratic in the form $x^2 + bx + c = 0$. How would you decide which method to use? Most students will choose the method that makes the most sense to them. Either method can be used.

EXAMPLE 3

Let $f(x) = x^2 + 3x - 15$. Find the value of x when $f(x)$ is -5.

Solution A ➤

Graph $y = x^2 + 3x - 15$ and $y = -5$ on the same set of axes. Use a graphics calculator if you have one. Find the point(s) of intersection. You see that when $x = 2$ or $x = -5$, then $f(x) = -5$.

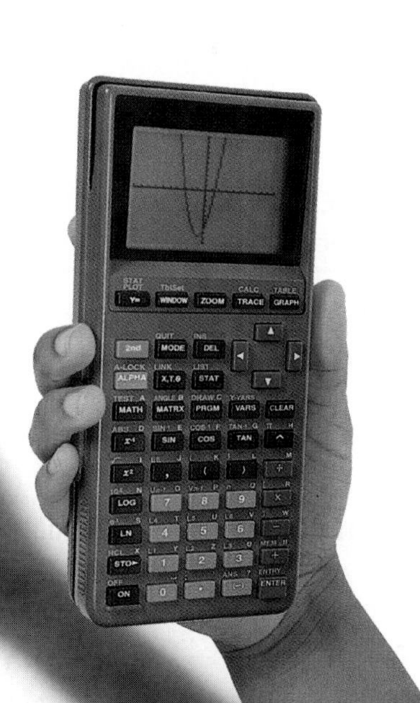

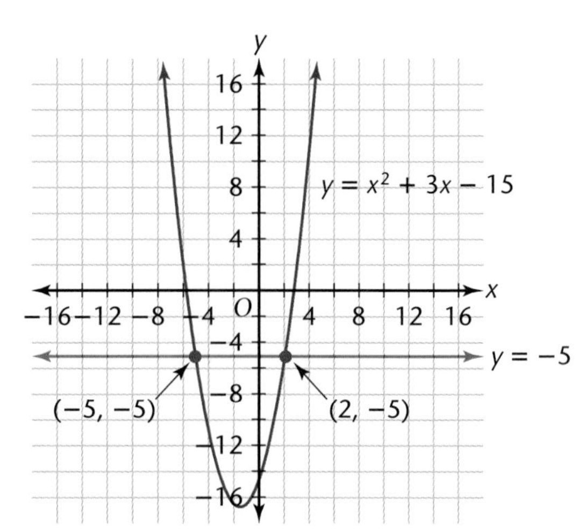

Solution B ➤

Solve $x^2 + 3x - 15 = -5$. Add 5 to each side of the equation to get $x^2 + 3x - 10 = 0$. Factor, and use the Zero Product Property to find the possible values of x.

$$(x - 2)(x + 5) = 0$$

$$x - 2 = 0 \quad \text{or} \quad x + 5 = 0$$

$$x = 2 \quad \text{or} \quad x = -5$$

Check by substitution. ❖

Try This Solve $x^2 - 8x + 22 = 10$. $x = 2$ or $x = 6$

EXAMPLE 4

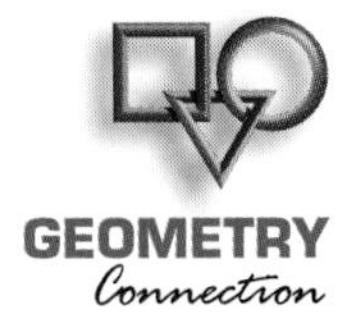

GEOMETRY *Connection*

A box with a square base and no top is to be made by cutting out 2-inch squares from each corner of a square piece of cardboard and folding up the sides. The volume of the box is to be 8 cubic inches. What size should the piece of cardboard be?

Solution ➤

Let x be the length and width of the piece of cardboard. When the corner squares are cut out the length and width of the box are both $(x - 4)$.

$V = 8$ cubic inches

The volume of a rectangular box is $V = \text{length} \cdot \text{width} \cdot \text{height}$.

$8 = (x - 4)(x - 4)(2)$	$V = lwh$
$4 = (x - 4)(x - 4)$	Divide both sides by 2.
$4 = x^2 - 8x + 16$	Simplify.
$0 = x^2 - 8x + 12$	Subtract 4 from both sides.
$0 = (x - 6)(x - 2)$	Factor.
$x - 6 = 0$ or $x - 2 = 0$	Zero Product Property
$x = 6$ or $x = 2$	Addition Property of Equality

Because you cannot cut 2-inch squares from each corner of a 2-inch square, the value of x cannot equal 2. Thus, the piece of cardboard should be 6 inches by 6 inches. ❖

EXAMPLE 5

Find the points where the graph of $y = x - 3$ intersects the graph of $y = x^2 - 10x + 21$.

Solution A ➤

Graph the equations, and find the points of intersection.

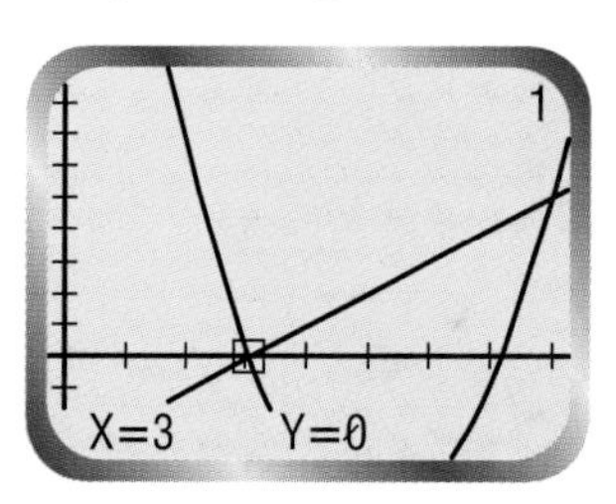

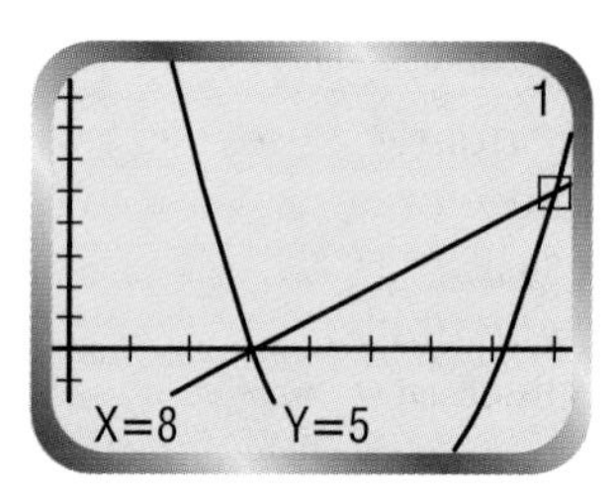

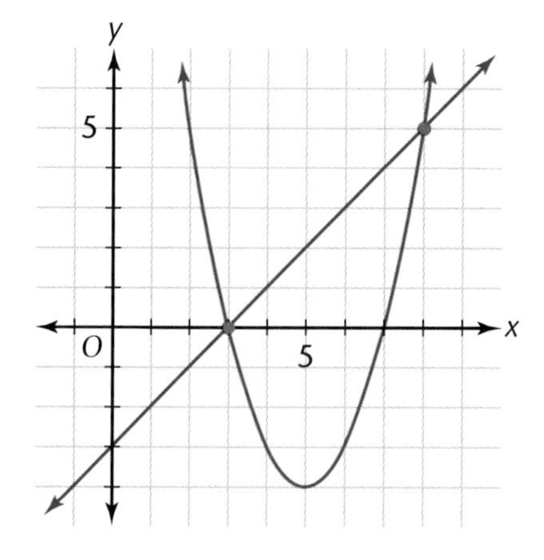

The graphs intersect at (8, 5) and (3, 0).

Solution B ➤

Solve the system. $\begin{cases} x^2 - 10x + 21 = y \\ y = x - 3 \end{cases}$

First, substitute $x - 3$ (from the second equation) for y (in the first equation) and simplify.

$$x^2 - 10x + 21 = x - 3$$
$$x^2 - 10x + 21 - x + 3 = x - 3 - x + 3$$
$$x^2 - 11x + 24 = 0$$

Then solve the new quadratic by factoring.

$$(x - 8)(x - 3) = 0$$
$$x - 8 = 0 \quad \text{or} \quad x - 3 = 0$$
$$x = 8 \quad \text{or} \quad x = 3$$

Substitute 8 and 3 for x into the equation $y = x - 3$.

$$\begin{array}{ll} y = x - 3 & y = x - 3 \\ y = 8 - 3 & y = 3 - 3 \\ y = 5 & y = 0 \end{array}$$

The graphs intersect at (8, 5) and (3, 0). ❖

Try This Find the points where the graph of $y = -2x + 4$ intersects the graph of $y = x^2 - 4x + 1$. (3, −2) and (−1, 6)

Exercises & Problems

Communicate

1. How are x-intercepts related to the zeros of a function?
2. Explain how to find the zeros of $f(x) = x^2 - 6x + 8$ by completing the square.
3. Describe how to solve $x^2 - 2x = 15$ by completing the square.
4. Discuss how to solve $x^2 + 10x = 24$ by factoring.

Which method would you choose to solve each problem? Explain.

5. $x^2 + 12x + 36 = 0$
6. $x^2 - 8x - 9 = 0$
7. Graph $f(x) = x^2 - 7x + 12$. Explain how to find the value of x when $f(x)$ is 2.
8. Describe how to find where the graph of $y = x - 1$ intersects the graph of $y = x^2 - 3x + 2$.

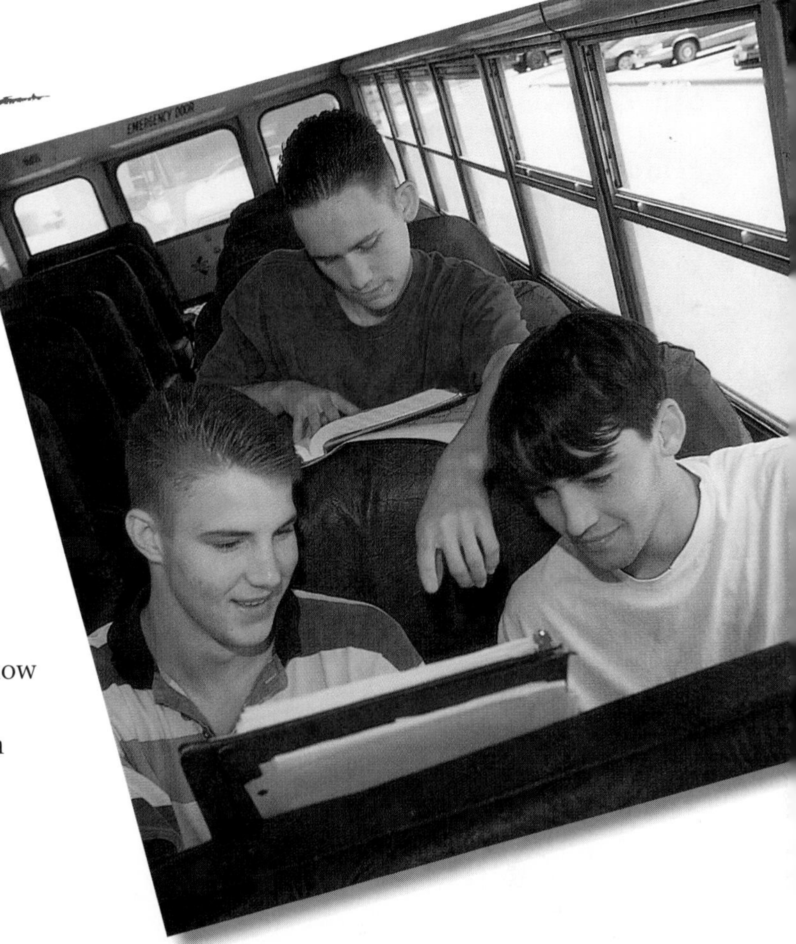

Practice & Apply

Find the x-intercepts, or zeros, of each function. Graph to check.

9. $y = x^2 - 2x - 8$ 4, −2
10. $y = x^2 + 6x + 5$ −1, −5
11. $y = x^2 - 4x + 4$ 2

Solve each equation by factoring.

12. $x^2 - 2x - 3 = 0$ −1, 3
13. $x^2 + 4x - 5 = 0$ 1, −5
14. $x^2 + 7x + 12 = 0$ −3, −4
15. $x^2 - 10x + 24 = 0$ 4, 6
16. $x^2 - 3x - 10 = 0$ 5, −2
17. $x^2 - 8x + 15 = 0$ 3, 5
18. $x^2 - 6x + 9 = 0$ 3
19. $x^2 + 10x + 25 = 0$ −5
20. $x^2 - 2x + 1 = 0$ 1

27. ≈1.45, ≈ −3.45 28. ≈0.24, ≈−4.24 29. ≈−2.27, ≈−5.73

Solve by completing the square. Round to the nearest hundredth when necessary.

21. $x^2 - 2x - 15 = 0$ 5, −3
22. $x^2 + 4x - 5 = 0$ 1, −5
23. $x^2 - x - 20 = 0$ 5, −4
24. $x^2 + 2x - 8 = 0$ 2, −4
25. $x^2 - 4x - 12 = 0$ 6, −2
26. $x^2 + x - 6 = 0$ 2, −3
27. $x^2 + 2x - 5 = 0$
28. $x^2 + 4x - 1 = 0$
29. $x^2 + 8x + 13 = 0$

Solve each equation by factoring or completing the square.

30. $b^2 + 10b = 0$ **31.** $r^2 - 10r + 24 = 0$ **32.** $x^2 - x - 3 = 0$ **33.** $x^2 + 4x - 12 = 0$

34. $x^2 + 3x - 5 = 0$ **35.** $x^2 - 3x - 1 = 0$ **36.** $x^2 - 5x - 3 = 0$ **37.** $x^2 - x = 0$

Refer to the equation $f(x) = x^2 - 4x + 3$ for Exercises 38–41. Graph the equation to find a value for x when $f(x)$ is the indicated value. Check by substituting the indicated value of $f(x)$.

38. $f(x) = 3$ 0, 4 **39.** $f(x) = 8$ 5, −1 **40.** $f(x) = -1$ 2 **41.** $f(x) = 15$ 6, −2

Find the point or points where the graphs intersect. Graph to check.

42. $\begin{cases} y = 9 \\ y = x^2 \end{cases}$ (−3, 9) and (3, 9)

43. $\begin{cases} y = 4 \\ y = x^2 - 2x + 1 \end{cases}$ (3, 4) and (−1, 4)

44. $\begin{cases} y = x - 1 \\ y = x^2 - 3x + 3 \end{cases}$ (2, 1)

45. $\begin{cases} y = x + 3 \\ y = x^2 - 4x + 3 \end{cases}$ (0, 3) and (5, 8)

46. Find two consecutive even integers whose product is 224.
14 and 16; −14 and −16

47. **Geometry** The length of the rectangle on the right is 4 yards longer than the width. Find the length and the width.
10 yards by 6 yards

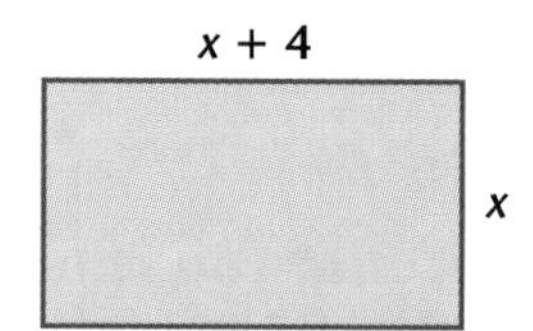

A = 60 square yards

48. **Photography** The perimeter of a photograph is 80 centimeters. Find the dimensions of the photo if the area is 396 square centimeters.
22 cm by 18 cm

Last school year the marching band used a rectangular formation with 8 columns and 10 rows.

49. **Band** An equal number of columns and rows were added to the formation when 40 new members joined the marching band at the start of a new school year. How many rows and columns were added?
2 rows, 2 columns

50. The difference of two numbers is 3. What are the numbers if the sum of their squares is 117?
9 and 6; -9 and -6

51. Physics A boat fires an emergency flare that travels upward at an initial velocity of 25 meters per second. To find the height in meters, h, of the flare at t seconds, use the formula $h = 25t - 5t^2$. After how many seconds will the flare be at a height of 10 meters? Round your answers to the nearest tenth.
0.4 second and 4.6 seconds

Look Back

Write each system in its matrix equation form. [Lesson 3.5]

52. $\begin{cases} x - 2y = 1 \\ 4x + 2y = -1 \end{cases}$

53. $\begin{cases} 5x - 2y = 11 \\ 3x + 5y = 19 \end{cases}$

Statistics Find the mean and the mode for each data set. **[Lesson 4.3]**

54. 10, 15, 20, 36, 10, 25 ≈ 19.3; 10

55. 1, 3, 6, 8, 2, 3, 6 ≈ 4.1; 3 and 6

56. 8, 10.5, 30, 15, 12, 10.5 ≈ 14.3; 10.5

57. 89, 71, 96, 82, 100 87.6; no mode

Write each number in customary notation. [Lesson 6.4]

58. 4×10^4 40,000

59. 6.5×10^7 65,000,000

60. 9.6×10^{-5} 0.000096

Write each polynomial in factored form. [Lessons 7.5, 7.6]

61. $4b^2 + 6b - 36$ $2(2b^2 + 3b - 18)$

62. $6w(w^2 - w)$ $6w^2(w - 1)$

63. $8p^4 - 16p$ $8p(p^3 - 2)$

Look Beyond

64. Factor $2x^2 + x - 6$. $(2x - 3)(x + 2)$

65. Use numerical examples and a calculator to decide if the following statements are true.

a. $\sqrt{a + b} = \sqrt{a} + \sqrt{b}$ False

b. $\sqrt{ab} = \sqrt{a} \cdot \sqrt{b}$ True

66. Imagine that someone invented a number for $\sqrt{-1}$ and called it i. Consider i^0 to be 1. If $i = \sqrt{-1}$, then what is i^2 and i^3? What is i^4? i^5? i^6? Can you find a pattern that will let you predict the value of i^{92}?

LESSON 8.5 The Quadratic Formula

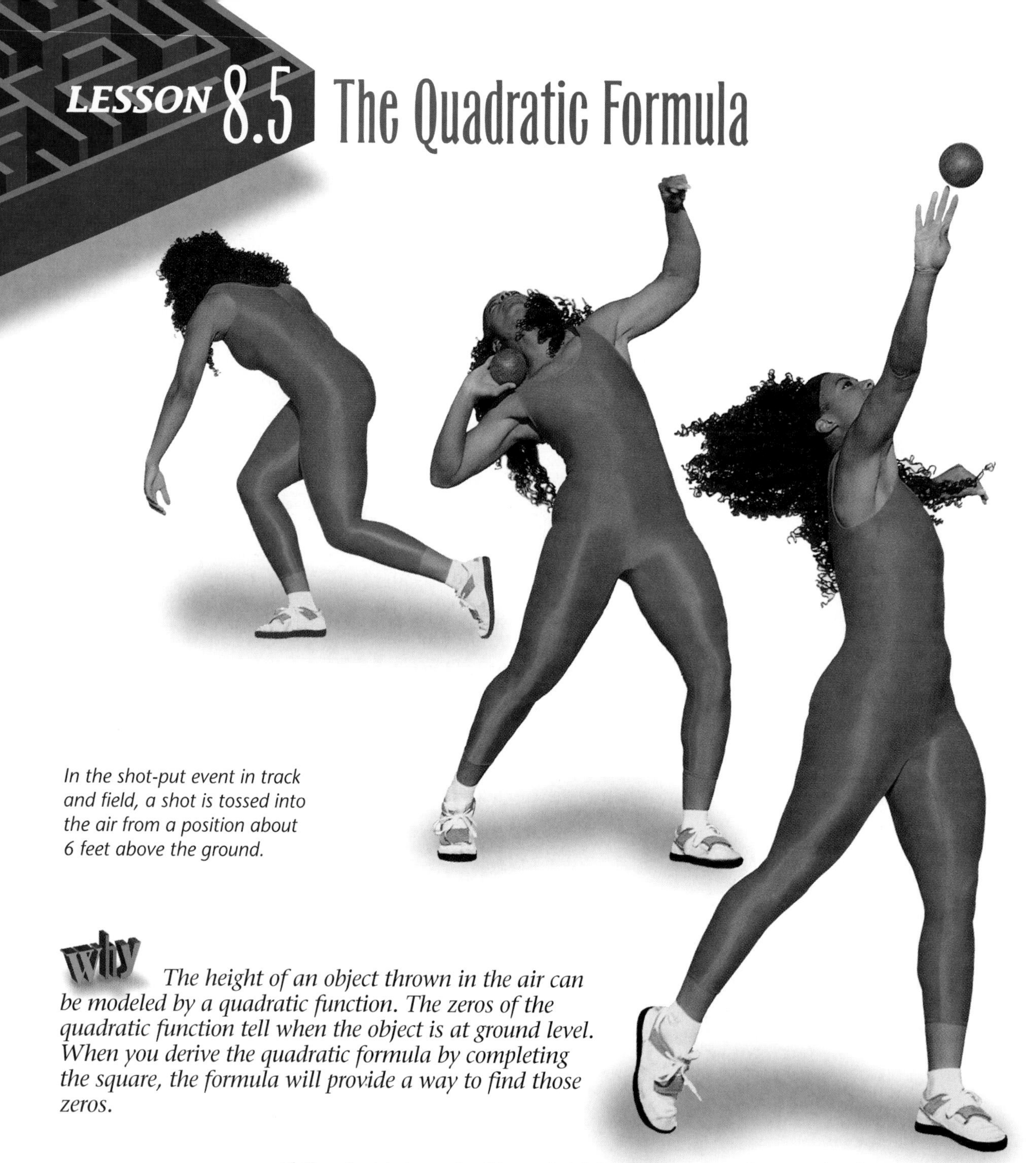

In the shot-put event in track and field, a shot is tossed into the air from a position about 6 feet above the ground.

Why *The height of an object thrown in the air can be modeled by a quadratic function. The zeros of the quadratic function tell when the object is at ground level. When you derive the quadratic formula by completing the square, the formula will provide a way to find those zeros.*

If the shot is tossed with an initial horizontal and vertical velocity of 31 feet per second, the time that the shot travels can be found by solving the quadratic equation $-16t^2 + 31t + 6.375 = 0$. You can then use t to solve for the horizontal distance, d, in the formula $d = 31t$.

Cultural Connection: Asia Since early Babylonian times scholars have known how to solve the standard quadratic equation $ax^2 + bx + c = 0$ by using a formula. To see how this quadratic formula can be derived, follow the steps as the equations $2x^2 + 5x + 1 = 0$ and $ax^2 + bx + c = 0$ are solved.

Divide both sides of each equation by the coefficient of x^2.

$$2x^2 + 5x + 1 = 0 \qquad ax^2 + bx + c = 0$$

$$x^2 + \frac{5}{2}x + \frac{1}{2} = 0 \qquad x^2 + \frac{b}{a}x + \frac{c}{a} = 0$$

Subtract the constant term from both sides of each equation.

$$x^2 + \frac{5}{2}x = -\frac{1}{2} \qquad x^2 + \frac{b}{a}x = -\frac{c}{a}$$

Complete the square.

$$x^2 + \frac{5}{2}x + \left(\frac{5}{4}\right)^2 = -\frac{1}{2} + \left(\frac{5}{4}\right)^2 \qquad x^2 + \frac{b}{a}x + \left(\frac{b}{2a}\right)^2 = -\frac{c}{a} + \left(\frac{b}{2a}\right)^2$$

$$\left(x + \frac{5}{4}\right)^2 = -\frac{1}{2} + \frac{25}{16} \qquad \left(x + \frac{b}{2a}\right)^2 = -\frac{c}{d} + \frac{b^2}{4a^2}$$

Work with the right side of each equation to simplify.

$$\left(x + \frac{5}{4}\right)^2 = -\frac{8 \cdot 1}{8 \cdot 2} + \frac{25}{16} \qquad \left(x + \frac{b}{2a}\right)^2 = -\frac{4a \cdot c}{4a \cdot a} + \frac{b^2}{4a^2}$$

$$\left(x + \frac{5}{4}\right)^2 = \frac{25 - 8}{16} \text{ or } \frac{17}{16} \qquad \left(x + \frac{b}{2a}\right)^2 = \frac{b^2 - 4ac}{4a^2}$$

Recall that if $x^2 = k$, then $x = \pm\sqrt{k}$.

$$x + \frac{5}{4} = \pm\sqrt{\frac{17}{16}} \qquad x + \frac{b}{2a} = \sqrt{\frac{b^2 - 4ac}{4a^2}}$$

$$x + \frac{5}{4} = \pm\frac{\sqrt{17}}{\sqrt{16}} \qquad x + \frac{b}{2a} = \pm\frac{\sqrt{b^2 - 4ac}}{\sqrt{4a^2}}$$

$$x = -\frac{5}{4} \pm \frac{\sqrt{17}}{4} \qquad x = -\frac{b}{2a} \pm \frac{\sqrt{b^2 - 4ac}}{2a}$$

$$x = \frac{-5 \pm \sqrt{17}}{4} \qquad x = \frac{-b \pm \sqrt{b^2 - 4ac}}{2a}$$

You have derived the quadratic formula.

THE QUADRATIC FORMULA

The solutions of the quadratic equation $ax^2 + bx + c = 0$ are

$$\frac{-b \pm \sqrt{b^2 - 4ac}}{2a}.$$

Notice that if you substitute 2 for a, 5 for b, and 1 for c from $2x^2 + 5x + 1 = 0$ into the formula $x = \frac{-b \pm \sqrt{b^2 - 4ac}}{2a}$, you get the following result:

$$x = \frac{-5 \pm \sqrt{25 - (4 \cdot 2 \cdot 1)}}{2 \cdot 2} = \frac{-5 \pm \sqrt{17}}{4}$$

EXAMPLE 1

Use the quadratic formula to solve $x^2 - 10 + 3x = 0$.

Solution ➤

Rewrite $x^2 - 10 + 3x = 0$ in the form $x^2 + 3x - 10 = 0$. Notice that $a = 1$, $b = 3$, and $c = -10$. Substitute and simplify.

$$x = \frac{-3 \pm \sqrt{(3)^2 - 4(1)(-10)}}{2(1)}$$

$$x = \frac{-3 \pm \sqrt{9 + 40}}{2}$$

$$x = \frac{-3 + 7}{2} \quad \text{or} \quad x = \frac{-3 - 7}{2}$$

$$x = \frac{4}{2} = 2 \quad \text{or} \quad x = \frac{-10}{2} = -5$$

The solutions are 2 and −5. ❖

Try This Use the quadratic formula to solve $x^2 + x - 12 = 0$.
$x = -4$ or $x = 3$

You now have three different methods for solving a quadratic equation.

1. factoring
2. completing the square
3. the quadratic formula

Use all three methods to solve $2x^2 - 7x - 4 = 0$. Which method do you think is best for solving real-world applications? Is factoring always a possible way to solve a quadratic equation? $x = 4$ or $x = -\frac{1}{2}$; answers may vary, but most students will choose the quadratic equation; no, sometimes a quadratic equation cannot be factored easily.

EXAMPLE 2

Find the zeros of $h(x) = 3x^2 + 2x - 4$.

Solution ➤

Solve $0 = 3x^2 + 2x - 4$. Use the quadratic formula, where a is 3, b is 2, and c is −4.

$$x = \frac{-2 \pm \sqrt{4 - (-48)}}{6}$$

$$x = \frac{-2 \pm \sqrt{52}}{6}$$

A calculator shows that $x \approx 0.87$ or $x \approx -1.54$. Check by substitution. ❖

Ongoing Assessment
Write the equation as $3n^2 - 2n - 2 = 0$. Substitute 3 for a, −2 for b, and −2 for c in the quadratic formula.

How would you use the quadratic formula to solve $3n^2 - 2n = 2$?

You can now solve the quadratic equation $-16t^2 + 31t + 6.375 = 0$ that models the shot-put problem.

Let $a = -16$, $b = 31$, and $c = 6.375$.

$$t = \frac{-31 \pm \sqrt{(31)^2 - (4)(-16)(6.375)}}{2(-16)} = \frac{-31 \pm \sqrt{961 + 408}}{-32}$$

$t = -0.1875$ or $t = 2.125$ and $d = 31t$, so $d = 31(2.125) = 65.875$

The shot travels a horizontal distance of about 66 feet.

Exploration *The Discriminant*

Examine each graph.

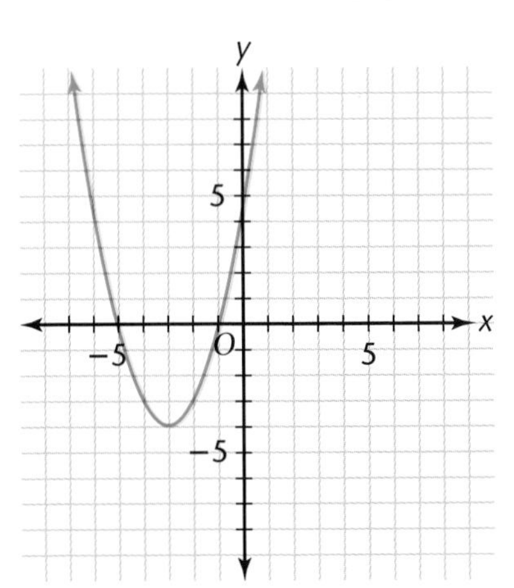

$f(x) = x^2 + 6x + 5$

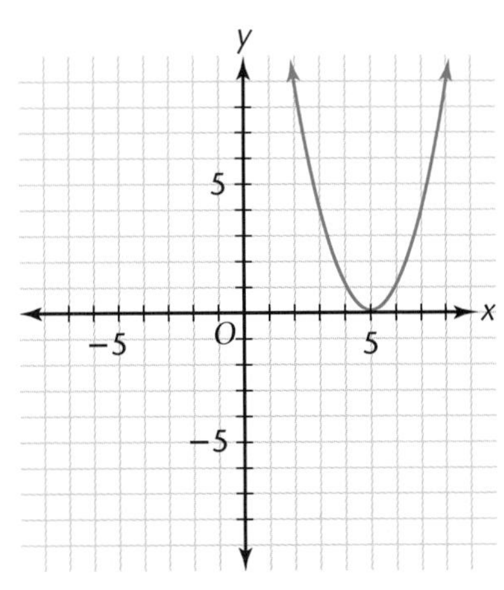

$g(x) = x^2 - 10x + 25$

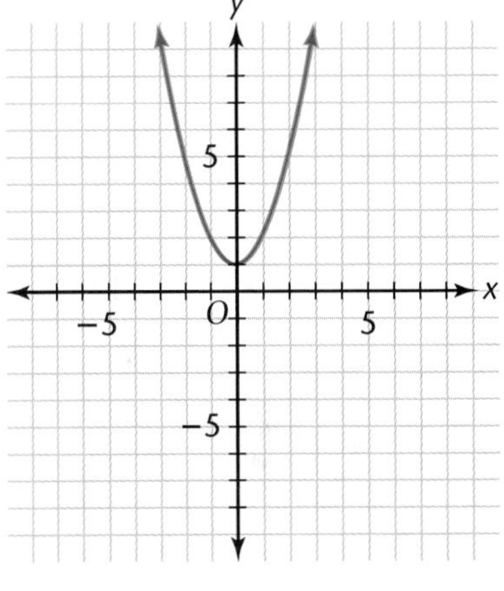

$h(x) = x^2 + 1$

Each expression is written in the form $ax^2 + bx + c$.

Find the values of a, b, and c, and complete the chart. The expression $b^2 - 4ac$ is called the **discriminant** of the quadratic formula.

Function	a	b	c	$b^2 - 4ac$	Number of x-intercepts
f	1 ?	?	?	16 ?	2 ?
g	1 ?	?	?	0 ?	1 ?
h	1 ?	?	?	−4 ?	none ?

b	c
6	5
−10	25
0	1

How do you think the value of the discriminant can help you determine the number of solutions to a quadratic equation? ❖

EXAMPLE 3

What does the discriminant tell you about the solution to each of the following equations?

A $3x^2 - 2x + 1 = 0$ **B** $4x = 4x^2 + 1$ **C** $2x^3 + 3x + 1 = 0$

Solution ➤

A Find $b^2 - 4ac$ where $a = 3$, $b = -2$, and $c = 1$.
The discriminant is $(-2)^2 - 4 \cdot 3 \cdot 1 = 4 - 12 = -8$. Since the discriminant is negative, $f(x) = 3x^2 - 2x + 1$ has no x-intercepts, and $3x^2 - 2x + 1 = 0$ has no real solutions.

B Rewrite the equation in standard form to find a, b, and c. If $4x^2 - 4x + 1 = 0$, then $a = 4$, $b = -4$, and $c = 1$. The discriminant is $(-4)^2 - 4 \cdot 4 \cdot 1 = 16 - 16 = 0$. Since the discriminant equals 0, $f(x) = 4x^2 - 4x + 1$ has exactly one x-intercept, and $4x = 4x^2 + 1$ has one real solution.

C The equation is not a quadratic. There is no discriminant. ❖

When the discriminant is a perfect square, the quadratic equation can be factored. You can use the quadratic formula to find the factors.

EXAMPLE 4

Factor $6x^2 + 23x + 20$.

Solution ➤

Evaluate the discriminant.

$$b^2 - 4ac = (23)^2 - 4(6)(20) = 49, \text{ which is a perfect square}$$

Find the solution to $6x^2 + 23x + 20 = 0$.

$$x = \frac{-b \pm \sqrt{b^2 - 4ac}}{2a} = \frac{-23 \pm \sqrt{49}}{12} = \frac{-23 \pm 7}{12}$$

Simplify.

$$x = -\frac{30}{12} = -\frac{5}{2} \quad \text{or} \quad x = -\frac{16}{12} = -\frac{4}{3}$$

Work backward to find the factors.

$x = -\frac{5}{2}$	or	$x = -\frac{4}{3}$	Given
$2x = -5$	or	$3x = -4$	Multiplication Property of Equality
$2x + 5 = 0$	or	$3x + 4 = 0$	Addition Property of Equality

The factors of $6x^2 + 23x + 20$ are $(2x + 5)$ and $(3x + 4)$. Check by multiplying the factors. ❖

EXERCISES & PROBLEMS

Communicate

1. Discuss how to identify the coefficients a, b, and c in the equation $2n^2 + 3n = 7$ in order to use the quadratic formula. Why is the standard form of the equation important?
2. How do you find the discriminant for $2t^2 + 3t - 7 = 0$?

Discuss how to use the discriminant to determine the number of solutions to each equation.

3. $2x^2 - x - 2 = 0$
4. $x^2 - 2x = -7$
5. Describe how to use the quadratic formula to find the solution to $4y^2 + 12y = 7$.
6. If the solutions to $4x^2 + 12x + 5 = 0$ are $-\frac{1}{2}$ and $-\frac{5}{2}$, discuss how to work backward to find the factors of $4x^2 + 12x + 5$.

Practice & Apply

Identify *a*, *b*, and *c* for each quadratic equation.

7. $x^2 - 4x - 5 = 0$
8. $r^2 - 5r - 4 = 0$
9. $5n^2 - 2n - 1 = 0$
10. $-p^2 - 3p + 7 = 0$
11. $3t^2 - 45 = 0$
12. $-3m + 7 - m^2 = 0$

Find the value of the discriminant, and determine the number of solutions for each equation.

13. $x^2 - x - 3 = 0$
14. $x^2 + 2x - 8 = 0$
15. $x^2 + 8x + 13 = 0$
16. $z^2 + 4z - 21 = 0$
17. $2y^2 - 8y - 8 = 0$
18. $8y^2 - 2 = 0$
19. $4k^2 - 3k = 5$
20. $3x^2 - x + 4 = 0$
21. $4x^2 - 12x + 9 = 0$

33. $-0.55, -5.45$ **35.** $1.90, -1.23$ **36.** $0.23, -1.10$

Use the quadratic formula to solve each equation. Give answers to the nearest hundredth when necessary. Check by substitution.

22. $a^2 - 4a - 21 = 0$ 7, −3
23. $t^2 + 6t - 16 = 0$ 2, −8
24. $m^2 + 4m - 5 = 0$ 1, −5
25. $w^2 - 4w = 0$ 0, 4
26. $x^2 + 9x = 0$ 0, −9
27. $x^2 - 9 = 0$ 3, −3
28. $x^2 + 2x = 0$ 0, −2
29. $3m^2 = 2m + 1$ 1, −0.33
30. $-x^2 + 6x - 9 = 0$ 3
31. $2r^2 - r - 3 = 0$ 1.5, −1
32. $2w^2 - 5w + 3 = 0$ 1, 1.5
33. $x^2 + 6x + 3 = 0$
34. $10y^2 + 7y = 12$ 0.8, −1.5
35. $3x^2 - 2x - 7 = 0$
36. $8y^2 + 7y - 2 = 0$

Choose any method to solve each quadratic equation. Give answers to the nearest hundredth when necessary.

37. $x^2 + 4x - 12 = 0$ 2, −6
38. $x^2 - 13 = 0$ 3.61, −3.61
39. $2x^2 + 5x - 3 = 0$ 0.5, −3
40. $x^2 + 7x + 10 = 0$ −2, −5
41. $2a^2 - 5a - 12 = 0$
42. $3x^2 - 2x - 2 = 0$ 1.22, −0.55
43. $3x^2 - 7x - 3 = 0$
44. $8x^2 + 10x + 3 = 0$
45. $3x^2 + 10x - 5 = 0$ 0.44, −3.77

41. 4, −1.5 43. 2.70, −0.37 44. −0.5, −0.75

46. Accounting To approximate the profit per day for her business, Mrs. Howe uses the formula $p = -x^2 + 50x - 350$. The profit, p, depends on the number of cases of decorator napkins that are sold. How many cases of napkins must she sell to break even ($p = 0$)? to make a maximum profit? Find the maximum profit.
9 or 42 to break even; 25 for a maximum profit of $275

47. The seats in a theater are arranged in parallel rows forming a rectangular seating region. The number of seats in each row of the theater is 16 fewer than the number of rows. How many seats are in each row of the 1161-seat theater? 27 seats

48. **Portfolio Activity** Complete the problem in the portfolio activity on page 363. Domain: $x > 53$, range: $y > 2463$; cost is $51,939 in the year 2000; cost was $25,000 in 1987; use 87.6 for x for a cost of $25,001.24.

Look Back

49. **Probability** Find the probability that a letter chosen at random from the English alphabet would be one of the letters of the word *probability*. **[Lesson 4.6]** $\frac{9}{26}$

Write each of the following numbers in scientific notation: [Lesson 6.4]

50. 32,000,000 3.2×10^7
51. 67,000 6.7×10^4
52. 0.00654 6.54×10^{-3}
53. 0.00000091 9.1×10^{-7}

Graph. [Lesson 6.5]

54. $y = 2^x + 4$
55. $y = 2 \cdot 3^x$
56. $y = 2^{x-1}$

57. **Geometry** Write the expression for the perimeter of the rectangle. **[Lesson 7.2]** $4x + 28$

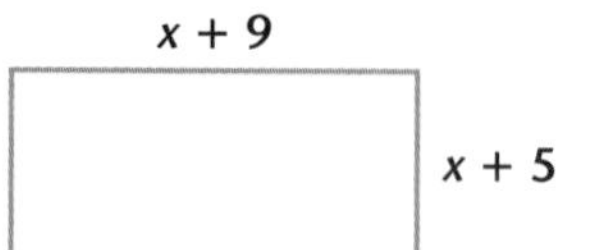

Multiply. [Lesson 7.4]

58. $(x - 5)(x + 3)$ $x^2 - 2x - 15$
59. $(2x + 4)(2x - 2)$ $4x^2 + 4x - 8$

Look Beyond

60. From what you know about graphing linear inequalities, describe what you think the graph of $y < x^2 + 3x + 2$ looks like.

LESSON 8.6 Graphing Quadratic Inequalities

Why *Not all problems involving quadratics are modeled by equations. Some are modeled by inequalities. Solving a quadratic inequality includes techniques that you already know plus one important step.*

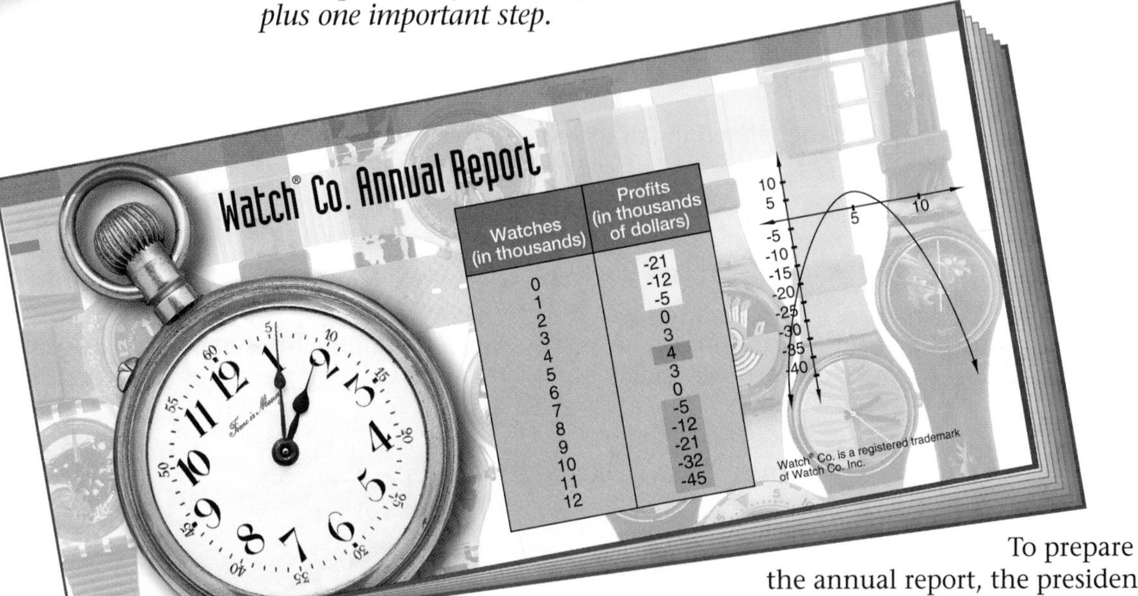

Watch® Co. Annual Report
Watch® Co. is a registered trademark of Watch Co. Inc.

Watches (in thousands)	Profits (in thousands of dollars)
0	-21
1	-12
2	-5
3	0
4	3
5	4
6	3
7	0
8	-5
9	-12
10	-21
11	-32
12	-45

To prepare the annual report, the president of a watch company gathers data about the number of watches produced per week and the corresponding profit. He uses the model $p = -x^2 + 10x - 21$, in which p represents profit and x represents the number of watches produced per week.

Exploration Production and Profit

MAXIMUM MINIMUM *Connection*

Use the graph of this profit function and the table of values to answer each question.

1. At what levels of production is the profit equal to 0?
   3000 and 7000 watches per week
2. At what levels of production is the profit greater than 0?
   When more than 3000 but less than 7000 are produced per week
3. At what levels of production is the profit less than 0?
   When less than 3000 or more than 7000 are produced per week
4. How many watches should be produced to maximize profit?
   5000 watches per week
5. According to the profit model, what is the maximum profit that the company can make? ❖ $4000 per week

In the exploration above, you solved a quadratic equation and two **quadratic inequalities**.

$-x^2 + 10x - 21 = 0$	$-x^2 + 10x - 21 > 0$	$-x^2 + 10x - 21 < 0$
Quadratic equation	**Quadratic inequality**	**Quadratic inequality**

EXAMPLE 1

Solve $x^2 - 2x - 15 > 0$.

Solution A ➤

Graph $y = x^2 - 2x - 15$, and find the values of x for which values of y are greater than 0.

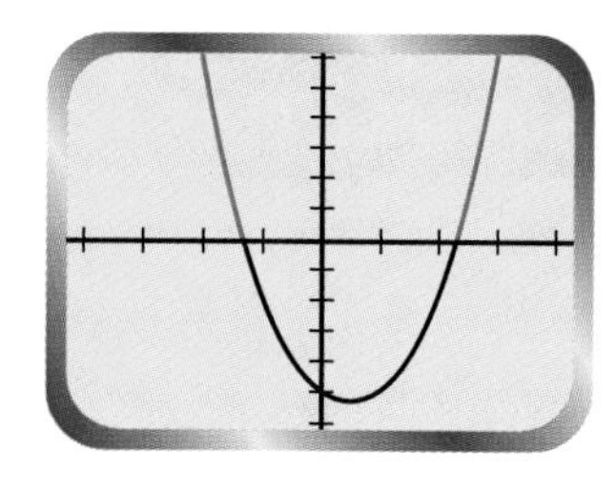

Solution B ➤

Use the x-axis as a number line.

Solve the related quadratic equation, $x^2 - 2x - 15 = 0$.

$$x^2 - 2x - 15 = 0$$
$$(x - 5)(x + 3) = 0$$
$$x - 5 = 0 \quad \text{or} \quad x + 3 = 0$$
$$x = 5 \quad \text{or} \quad x = -3$$

Plot the solutions on an x-axis. Since the inequality symbol is $>$, use an open circle at -3 and 5.

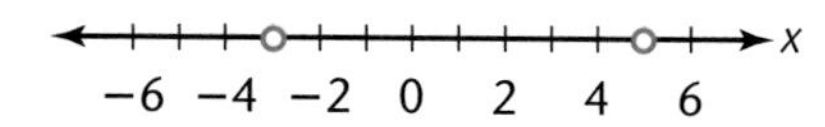

The two solutions divide the x-axis into three intervals, to the left, between, and to the right of the two points. Test each interval. Choose a number from each interval, and substitute it for x in the original inequality.

$x^2 - 2x - 15 > 0$	$x^2 - 2x - 15 > 0$	$x^2 - 2x - 15 > 0$
$(-5)^2 - 2(-5) - 15 \ ? \ 0$	$(0)^2 - 2(0) - 15 \ ? \ 0$	$(6)^2 - 2(6) - 15 \ ? \ 0$
$25 + 10 - 15 \ ? \ 0$	$0 - 0 - 15 \ ? \ 0$	$36 - 12 - 15 \ ? \ 0$
$20 > 0$ **True**	$-15 > 0$ **False**	$9 > 0$ **True**
Solution interval	**No**	**Solution interval**

CT– Test the interval between the two boundary points. If the number satisfies the inequality, shade that area; if not, shade outside the boundary points.

Graph the solutions.

x
−3 0 5

$x < -3$ or $x > 5$ ❖

Try This Solve $x^2 + x - 12 \geq 0$. $x \leq -4$ or $x \geq 3$

Explain how you can determine which intervals to shade without testing numbers from all of the intervals.

EXAMPLE 2

Graph $y > -x^2 + 1$.

Solution

Graph $y = -x^2 + 1$. Draw the parabola with a dashed curve to show that the solutions of $y = x^2 + 1$ are *not* on the graph.

To graph $y > -x^2 + 1$, shade the points above the parabola.

As a check, test points inside and outside the parabola.

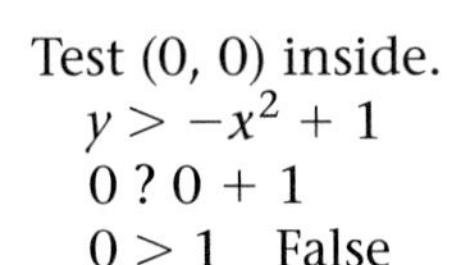

Test (0, 0) inside.	Test (0, 3) outside.
$y > -x^2 + 1$	$y > -x^2 + 1$
$0 \mathbin{?} 0 + 1$	$3 \mathbin{?} 0 + 1$
$0 > 1$ False	$3 > 1$ True

Shade the region outside the parabola.

The graph of $y > -x^2 + 1$ is the set of points outside the parabola. ❖

You can test points inside and outside the graph of $y = -x^2 + 1$ to show that the shaded region on each graph below represents the given set of points.

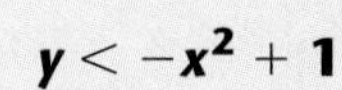

$y < -x^2 + 1$

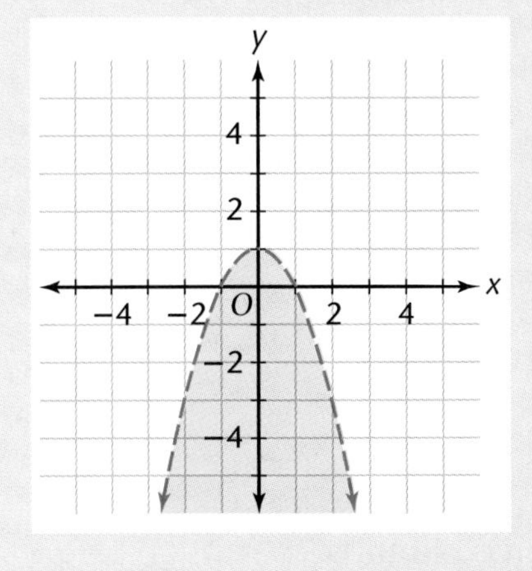

$y \geq -x^2 + 1$

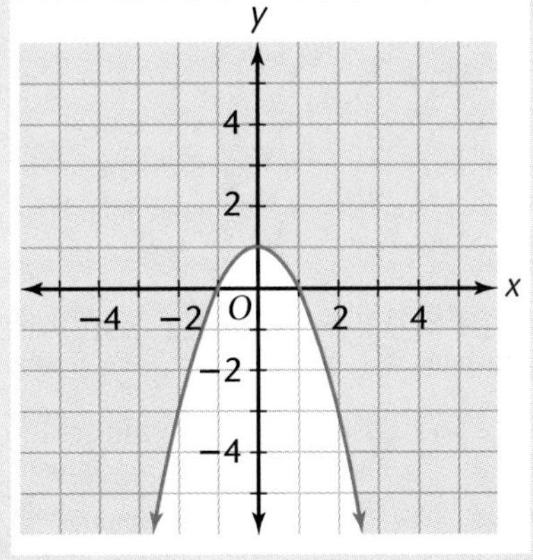

$y \leq -x^2 + 1$

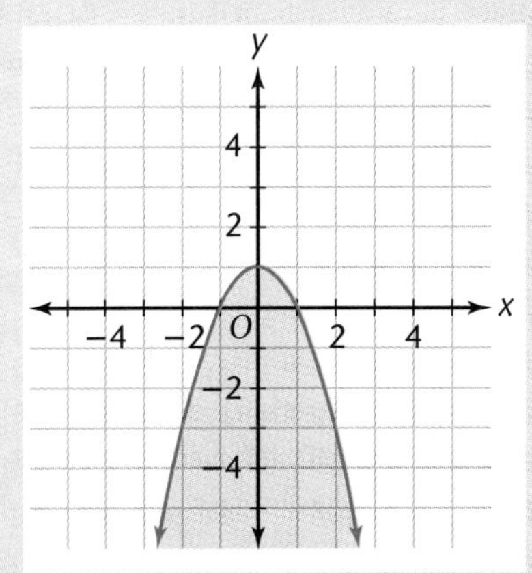

EXAMPLE 3

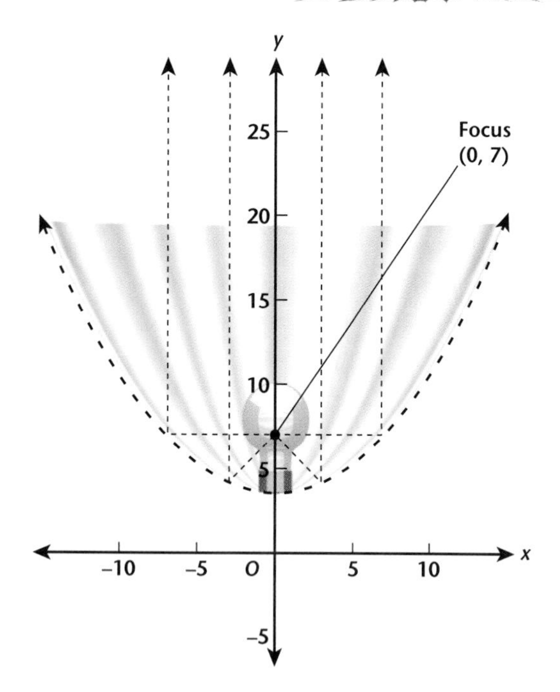

A cross section of the parabolic light reflector on the left is described by the equation $y = \frac{1}{14}x^2 + \frac{7}{2}$. The bulb inside the reflector is located at the point (0, 7). Light bounces off the reflector in parallel rays. This allows a flashlight to direct light in a narrow beam. Determine which of the inequalities indicate the region in which the light bulb is located.

A $y > \frac{1}{14}x^2 + \frac{7}{2}$ **B** $y < \frac{1}{14}x^2 + \frac{7}{2}$

Solution

A $y > \frac{1}{14}x^2 + \frac{7}{2}$

Test (0, 7).

$7 > \left(\frac{1}{14}\right)(0) + \frac{7}{2}$

$7 > \frac{7}{2}$ True

B $y < \frac{1}{14}x^2 + \frac{7}{2}$

Test (0, 7).

$7 < \left(\frac{1}{14}\right)(0) + \frac{7}{2}$

$7 < \frac{7}{2}$ False

The inequality $y > \frac{1}{14}x^2 + \frac{7}{2}$ contains the point of focus inside the parabola. ❖

EXERCISES & PROBLEMS

Communicate

1. Explain how to solve $x^2 - 9x + 18 > 0$ by using the Zero Product Property.

2. Describe how to graph and check the solution intervals to Exercise 1 on a number line.

Explain how to determine if the boundary line is solid or dashed for the graph of each inequality.

3. $y < x^2 - 3$ **4.** $y \leq x^2 - 3$ **5.** $y \geq x^2 - 3$

Discuss how to graph each inequality. Explain how to decide whether to shade inside or outside the parabola for each solution.

6. $y < x^2 - 3$ **7.** $y \leq x^2 - 3$ **8.** $y \geq x^2 - 3$

Practice & Apply

Solve each quadratic inequality by using the Zero Product Property. Graph the solution on a number line.

9. $x^2 - 6x + 8 > 0$
10. $x^2 + 3x + 2 < 0$
11. $x^2 - 9x + 14 \le 0$
12. $x^2 + 2x - 15 > 0$
13. $x^2 - 7x + 6 < 0$
14. $x^2 - 8x - 20 < 0$
15. $x^2 - 8x + 15 \le 0$
16. $x^2 - 3x - 18 \le 0$
17. $x^2 - 11x + 30 \ge 0$
18. $x^2 - 4x - 12 \ge 0$
19. $x^2 + 5x - 14 > 0$
20. $x^2 + 10x + 24 \ge 0$

Graph each quadratic inequality. Shade the solution region.

21. $y \ge x^2$
22. $y \ge x^2 + 4$
23. $y < -x^2$
24. $y > x^2$
25. $y > x^2 - 2x$
26. $y \ge 1 - x^2$
27. $y > -2x^2$
28. $y \le x^2 - 3x$
29. $y \le x^2 + 2x - 8$
30. $y \ge x^2 - 2x + 1$
31. $y \le x^2 - 4x + 4$
32. $y \le x^2 - 5x - 6$

A projectile is fired vertically into the air. Its motion is described by $h = -16t^2 + 320t$, where h is its height (in feet) after t seconds.

Physics Use a quadratic inequality to answer each question.

33. During what time intervals will the height of the projectile be below 1024 feet?

34. During what time intervals will the height of the projectile be above 1024 feet?

35. Technology Use graphing technology to graph $h = -16t^2 + 320t$, and check your answers to Exercises 33 and 34.

Look Back

Find each product. [Lesson 6.2]

36. $(-ab^2c)(a^2bc^3)$
$-a^3b^3c^4$

37. $(3p^2q^3r^4)(-2pqr)$
$-6p^3q^4r^5$

Write each number in scientific notation. [Lesson 6.4]

38. 0.825
8.25×10^{-1}

39. 0.000001
1×10^{-6}

40. 0.0000074
7.4×10^{-6}

Factor each polynomial. Begin by removing the greatest common factor. [Lesson 7.5]

41. $12y^2 - 2y$
$2y(6y - 1)$

42. $2ax + 6x + ab + 3b$
$(a + 3)(2x + b)$

43. $4x^2 - 24x + 32$
$4(x - 4)(x - 2)$

Look Beyond

44. Three identical rectangles with side lengths that are positive integers, each have a perimeter of 24 inches. Find the area of the square formed when the rectangles are joined. 81 square inches

WHAT'S THE DIFFERENCE?

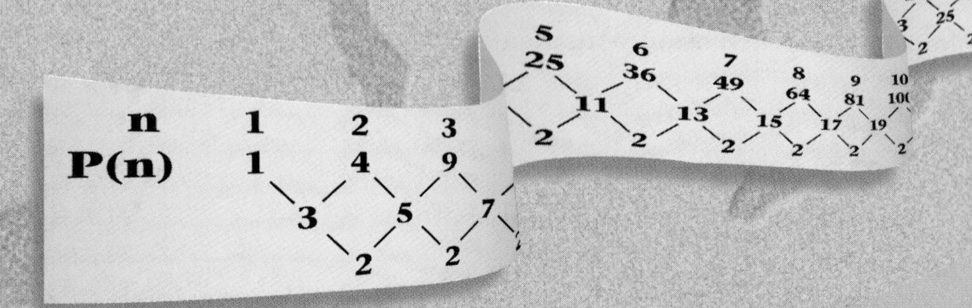

If n represents the dots on a side, the sequence shows the total number of dots in a square pattern.

A function whose domain is consecutive positive integers can produce a sequence. Many sequences have a function rule that you can discover.

A function that will produce this sequence is $A(n) = n^2$. The variable n is usually used when working with sequences, and it represents the number of the terms in the sequence.

How do you determine the function rule? Sometimes the pattern in the sequence is clear enough to use guess-and-check to find the rule. When the sequence is more complicated, other methods are needed. If a sequence eventually produces a constant difference, one way to find the general rule for that sequence is to use **finite differences**.

Examine the sequence 0, 5, 12, 21, 32, 45, 60. You want to find a general rule that will generate any term of this sequence. Begin by finding differences until they are constant.

Table 1

n	1	2	3	4
P(n) (1st Term)	0	5	12	21
1st Difference		5	7	9
2nd Difference			2	2

$$P(n) = an^2 + bn + c$$
$$P(1) = a(1)^2 + b(1) + c = a + b + c$$
$$P(2) = a(2)^2 + b(2) + c = 4a + 2b + c$$
$$P(3) = a(3)^2 + b(3) + c = 9a + 3b + c$$
$$P(4) = a(4)^2 + b(4) + c = 16a + 4b + c$$

If the sequence produces constant second differences, you can expect a quadratic equation. Write the quadratic equation in the form of a polynomial function, $P(n) = an^2 + bn + c$. Evaluate the function for n-values from 1 to at least 3. Next, make a table of the terms and differences, using the expressions you get by evaluating the general quadratic polynomial.

The sequence in Table 1 is a special case of the corresponding general expressions in Table 2. Match the first expression in each row of Table 2 with the corresponding expression in Table 1.

Table 2		Table 1
$a + b + c$	$=$	0
$3a + b$	$=$	5
$2a$	$=$	2

Use *substitution* to find the values of a, b, and c.

Use $2a = 2$ to solve for a.

$a = 1$

Use $3a + b = 5$ and the value for a to solve for b.

$3 + b = 5$, so $b = 2$

Use $a + b + c = 0$ and the values for a and b to solve for c.

$1 + 2 + c = 0$, so $c = -3$

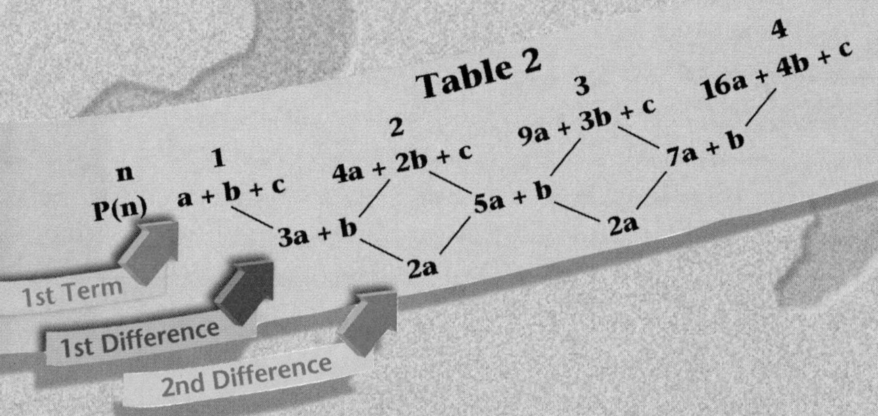

Finally, replace the letters in the quadratic polynomial, $P(n) = an^2 + bn + c$ with the values $a = 1$, $b = 2$, and $c = -3$.

$$P(n) = n^2 + 2n - 3$$

Check your rule by substitution to see if it generates the appropriate sequence, 0, 5, 12, 21, 32, 45, 60, . . .

Thus, $P(n) = n^2 + 2n - 3$ is a rule that generates the sequence.

Activity

Find a Rule

In groups of 2 to 5 students, the leader begins by choosing a quadratic equation with integer coefficients and generating a sequence of 5 numbers. The leader then shows only the sequence to the other people in the group. The group tries to find the function rule that generated the sequence. Finally, the group uses the rule they found to generate any 2 terms between 6 and 20. The leader confirms the results by showing the original function rule and sequence. Let group members take turns being the leader.

Chapter 8 Review

Vocabulary

axis of symmetry	364	minimum value	364	square root	369
completing the square	374	quadratic formula	391	vertex	364
discriminant	393	quadratic inequality	397	Zero Product Property	383
maximum value	364				

Key Skills & Exercises

Lesson 8.1

➤ Key Skills

Examine quadratic functions to find the vertex, axis of symmetry, and zeros of the quadratic function in order to sketch a graph.

$$g(x) = (x - 1)^2 - 4$$

The vertex is $(1, -4)$, and the axis of symmetry is $x = 1$.

$$\begin{aligned} g(x) &= (x - 1)^2 - 4 \\ &= x^2 - 2x - 3 \\ &= (x - 3)(x + 1) \end{aligned}$$

The zeros of $g(x)$ are 3 and -1.

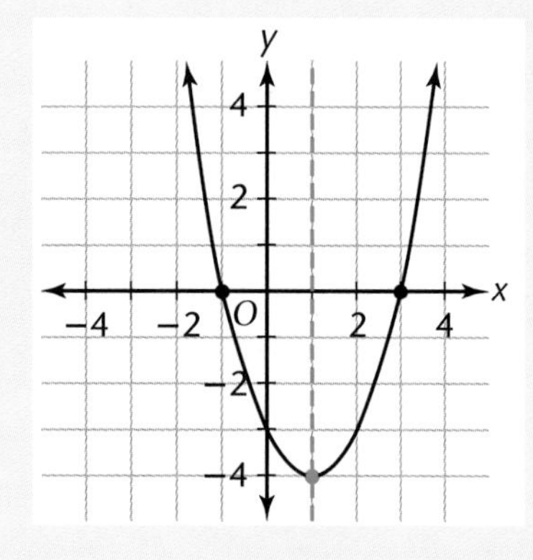

➤ Exercises

Determine the vertex, axis of symmetry, and zeros for each function. Then sketch a graph from the information.

1. $y = (x + 1)^2 - 4$ **2.** $y = (x - 5)^2 - 1$ **3.** $y = 2(x - 3)^2 - 2$

Lesson 8.2

➤ Key Skills

Use square roots to solve quadratic equations.

$x^2 = 36 \rightarrow x = \pm\sqrt{36} \rightarrow x = 6 \text{ or } -6$

$(x - 3)^2 = 15 \rightarrow x - 3 = \pm\sqrt{15} \rightarrow x = 3 + \sqrt{15} \approx 6.87 \text{ or } x = 3 - \sqrt{15} \approx -0.87$

➤ Exercises

Solve each equation. Give answers to the nearest hundredth when necessary.

4. $x^2 = 225$ ± 15 **5.** $x^2 = \frac{9}{144}$ $\pm\frac{1}{4}$ **6.** $x^2 = 50$ ± 7.07

7. $(x + 1)^2 - 4 = 0$ 1, −3 **8.** $(x + 7)^2 - 81 = 0$ 2, −16 **9.** $(x - 3)^2 - 3 = 0$ 4.73, 1.27

Lesson 8.3

➤ Key Skills

Complete the square.

To complete the square for $x^2 - 6x$, find one-half of the coefficient of $-6x$, which is -3. Square -3, and add it to $x^2 - 6x$.

$$x^2 - 6x + (-3)^2 = x^2 - 6x + 9$$

Write quadratic equations in the form $y = (x - h)^2 + k$. Find the vertex.

To find the vertex of $y = x^2 - 6x + 2$, complete the square. Add and subtract 9.

$$y = (x^2 - 6x + 9) + 2 - 9 = (x - 3)^2 - 7$$

The vertex is $(3, -7)$.

➤ Exercises

Complete the square.

10. $x^2 + 4x$ $+4$ **11.** $x^2 - 16x$ $+64$ **12.** $x^2 - x$ $+\frac{1}{4}$ **13.** $x^2 + 3x$ $+\frac{9}{4}$

Rewrite each function in the form $y = (x - h)^2 + k$. Find each vertex.

14. $y = x^2 + 4x + 3$
$y = (x + 2)^2 - 1; (-2, -1)$

15. $y = x^2 - 8x + 18$
$y = (x - 4)^2 + 2; (4, 2)$

16. $y = x^2 - 12x + 32$
$y = (x - 6)^2 - 4; (6, -4)$

Lesson 8.4

➤ Key Skills

Solve quadratic equations by completing the square.

$$x^2 + 2x = 24$$
$$x^2 + 2x + 1 = 24 + 1$$
$$(x + 1)^2 = 25$$
$$x + 1 = 5 \quad \text{or} \quad x + 1 = -5$$
$$x = 4 \quad \text{or} \quad x = -6$$

Solve quadratic equations by factoring.

$$x^2 + 2x = 24$$
$$x^2 + 2x - 24 = 0$$
$$(x + 6)(x - 4) = 0$$
$$x + 6 = 0 \quad \text{or} \quad x - 4 = 0$$
$$x = -6 \quad \text{or} \quad x = 4$$

➤ Exercises

Solve each equation by factoring or by completing the square.

17. $x^2 + 3x - 10 = 0$
2, −5

18. $x^2 + 5x + 6 = 0$
−2, −3

19. $x^2 - 8x + 16 = 0$
4

Lesson 8.5

➤ Key Skills

Evaluate the discriminant. Determine the number of solutions.

$$3x^2 + 4x + 7 = 0$$

Substitute $a = 3$, $b = 4$, and $c = 7$ in $b^2 - 4ac$. Simplify. If the discriminant is negative, the equation has no real solutions. If the discriminant is 0, there is exactly one real solution. If the discriminant is positive, the equation has two real solutions.

$$b^2 - 4ac = (4)^2 - (4 \cdot 3 \cdot 7) = -68$$

The equation has no real solutions.

Use the quadratic formula to find solutions to quadratic equations.

For the equation $x^2 + 3x - 14 = 0$, substitute 1 for a, 3 for b, and -14 for c in the quadratic formula, and simplify.

$$x = \frac{-b \pm \sqrt{b^2 - 4ac}}{2a}$$
$$x = \frac{-3 \pm \sqrt{9 - (4 \cdot 1 \cdot -14)}}{2 \cdot 1}$$
$$x = \frac{-3 \pm \sqrt{65}}{2}$$
$$x \approx 2.53 \text{ or } x \approx -5.53$$

➤ Exercises

Evaluate the discriminant to determine the number of real solutions. Then solve. Give answers to the nearest hundredth.

20. $x^2 - 2x + 9 = 0$ **21.** $4x^2 - 5x - 4 = 0$ **22.** $4x^2 - 4x + 1 = 0$

23. $x^2 - 9x + 18 = 0$ **24.** $6x^2 - 7x - 3 = 0$ **25.** $8x^2 + 2x - 1 = 0$

Lesson 8.6

➤ Key Skills

Solve and graph quadratic inequalities.

Use factoring and the Zero Product Property to solve $x^2 + 3x - 4 > 0$.

$$x^2 + 3x - 4 = 0$$
$$(x + 4)(x - 1) = 0$$
$$x = -4 \text{ or } x = 1$$

Test numbers in the inequality to determine the solutions. Since numbers less than −4 or numbers greater than 1 produce a true inequality, the solution is $x < -4$ or $x > 1$.

−6 −4 −2 0 2 4 6

To graph $y \le x^2 + 2$, graph $y = x^2 + 2$. Draw a solid parabola to show that the graph of $y = x^2 + 2$ is included in the solution. Since $y \le x^2 + 2$, shade the points below the curve.

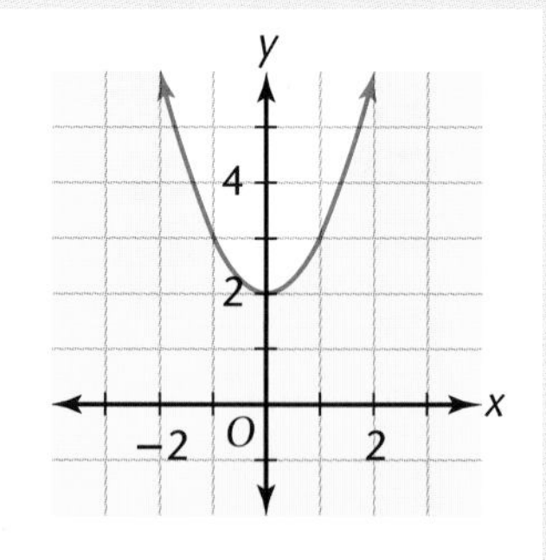

➤ Exercises

Solve each quadratic inequality by using the Zero Product Property. Graph the solution on a number line.

26. $x^2 + 6x + 8 \ge 0$ **27.** $x^2 - 3x - 10 < 0$ **28.** $x^2 + 9x + 18 > 0$

Graph each inequality. Shade the solution region.

29. $y \le x^2$ **30.** $y > x^2 - 4x$ **31.** $y \ge x^2 - x - 10$

Applications

32. About 12 feet

Sports The graph represents a relationship between the number of yards a football is thrown and its height above the ground.

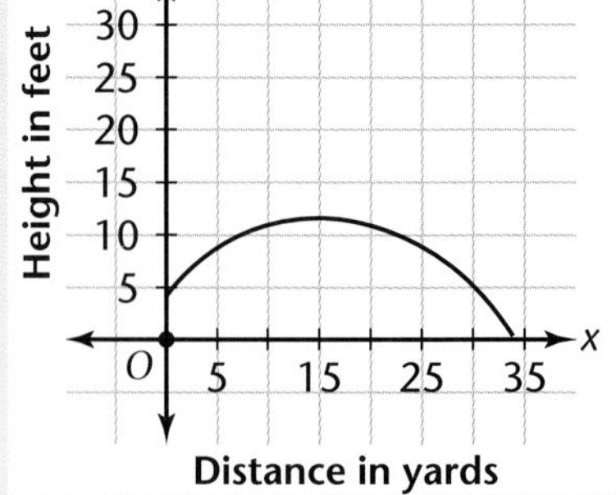

32. What is the maximum height reached by the football?

33. How far does the football travel before it reaches its maximum height? 15 yards

34. If a receiver catches the ball 30 yards from where the ball was thrown, what was the height of the ball? About 5 ft

35. Which equation was used to graph the parabola? a

a. $y = -\frac{1}{30}(x - 15)^2 + 12$ **b.** $y = -\frac{1}{30}(x + 15)^2 + 12$

c. $y = -\frac{1}{30}(x - 15)^2 - 12$ **d.** $y = -\frac{1}{30}(x + 15)^2 - 12$

36. Find two numbers whose sum is 55 and whose product is 600. 40 and 15

37. **Geometry** A piece of wrapping paper has an area of 480 square inches. If Orin cuts a square piece from the paper and the new piece of paper has an area of 455 square inches, what is the length of each side of the square? 5 inches

Chapter 8 Assessment

1. Compare the graph of $y = (x + 2)^2 - 4$ with the graph of $y = x^2$. Describe the transformation on the parent function, $y = x^2$.
The graph moves left 2 units and down 4 units.

Determine the vertex, axis of symmetry, and zeros for each function. Then sketch a graph from the information.

2. $y = (x - 6)^2 - 4$ **3.** $y = (x + 3)^2 - 1$

4. Why does an equation of the form $x^2 = k$, when $k > 0$, have two solutions? Because $\sqrt{k} \cdot \sqrt{k} = k$ and $-\sqrt{k} \cdot -\sqrt{k} = k$, so $x^2 = k \Rightarrow x = \pm\sqrt{k}$

Solve each equation. Give answers to the nearest hundredth.

5. $x^2 = 900$ ± 30 **6.** $x^2 = 12$ ± 3.46 **7.** $(x + 5)^2 - 16 = 0$ $-1, -9$

8. **Geometry** Use the formula $A = \pi r^2$ to find the radius of a circle with an area of 100 square feet. (Use 3.14 for π.) About 5.64 feet

9. An object is dropped, and its height, h, after t seconds is given by the function $h(t) = -16t^2 + 550$. About how long will it take for the object to reach the ground? About 5.86 seconds

Complete the square.

10. $x^2 + 8x$
$x^2 + 8x + 16$

11. $x^2 - 30x$
$x^2 - 30x + 225$

12. $x^2 + 11x$
$x^2 + 11x + \frac{121}{4}$

Rewrite each function in the form $y = (x - h)^2 + k$. Find each vertex.

13. $y = x^2 + 4x + 1$
$y = (x + 2)^2 - 3$; $(-2, -3)$

14. $y = x^2 - 14x + 50$
$y = (x - 7)^2 + 1$; $(7, 1)$

Solve by factoring or by completing the square. Round to the nearest hundredth when necessary.

15. $x^2 - 35 - 2x = 0$ $7, -5$ **16.** $x^2 + 6x - 12 = 0$ $\approx 1.58, \approx -7.58$

17. Why does a quadratic equation need to be written in standard form in order to use the quadratic formula to solve it? In standard form, you can determine the values of a, b, and c to substitute in the quadratic formula.

Find the value of the discriminant for each quadratic equation. Find the number of real solutions.

18. $x^2 + 10x - 13 = 0$
152; 2

19. $5x^2 - 4x + 2 = 0$
-24; none

Use the quadratic formula to solve each equation. Give answers to the nearest hundredth when necessary.

20. $x^2 - 4x - 60 = 0$
$10, -6$

21. $2x^2 + 3x - 4 = 0$
$0.85, -2.35$

Solve each quadratic inequality by using the Zero Product Property. Graph the solution on a number line.

22. $x^2 + 4x - 5 > 0$ **23.** $x^2 + 10x + 21 \le 0$

Graph each quadratic inequality. Shade the solution region.

24. $y \le -x^2 + 2$ **25.** $y > x^2 + 5x + 6$

Chapters 1–8 Cumulative Assessment

College Entrance Exam Practice

Quantitative Comparison For Questions 1–4, write
A if the quantity in Column A is greater than the quantity in Column B;
B if the quantity in Column B is greater than the quantity in Column A;
C if the two quantities are equal; or
D if the relationship cannot be determined from the information given.

	Column A	Column B	Answers
1. B	The solution to $12x = -108$	The solution to $\frac{32}{25} = \frac{x}{-6}$	(A) (B) (C) (D) **[Lesson 1.2]**
2. A	The number of ways to arrange 3 different vowels in a row	The number of ways to choose a meal from 2 entrees, 3 beverages, and 3 desserts	(A) (B) (C) (D) **[Lesson 4.5]**
3. A	7^0	7^{-1}	(A) (B) (C) (D) **[Lesson 6.3]**
4. B	The discriminant of $2x^2 - 5x + 3 = 0$	The discriminant of $35x - 21 = 14x^2$	(A) (B) (C) (D) **[Lesson 8.5]**

5. Which is a step used to solve the system by substitution?
b $\begin{cases} x + 3y = 18 \\ 5x - 15y = 30 \end{cases}$ **[Lesson 2.4]**

a. $5x - 15(18 - x) = 30$ **b.** $5(18 - 3y) - 15y = 30$
c. $5(x + 3y) = 5(18)$ **d.** $x + 3(5x - 30) = 18$

6. Which function is *not* a translation of the parent function $y = 2^x$?
b **[Lesson 5.5]**

a. $y = 2^x + 3$ **b.** $y = \frac{2x}{3}$
c. $y = 2^x - 3$ **d.** $y = 2^{x-3}$

7. Which is the result of simplifying $(-2x^3)^2(x^4)$? **[Lesson 6.2]**
a **a.** $4x^{10}$ **b.** $4x^9$ **c.** $-4x^{10}$ **d.** $4x^{20}$

8. Which is the factored form of $b^2 - 8b + 16$? **[Lesson7.6]**
b **a.** $(b - 4)(b + 4)$ **b.** $(b - 4)^2$
c. $(b + 4)^2$ **d.** $(b - 8)^2$

9. Which is the number of real solutions of the equation
b $3x^2 + 12x + 12 = 0$? **[Lesson 8.4]**
a. 0 **b.** 1
c. 2 **d.** an infinite number

10. Does the matrix $\begin{bmatrix} -2 & 2 \\ 1 & -1 \end{bmatrix}$ have an inverse? If so, find it. **[Lesson 3.4]**
No

11. Suppose that you choose a card from a deck of 52 cards. What is the probability that you will choose an ace or an eight? **[Lesson 4.7]** $\frac{2}{13}$

12. Write 93,000,000 in scientific notation. **[Lesson 6.4]** 9.3×10^7

13. Simplify the expression $(z^3 - 8z^2 + 4z + 7) - (2z^2 + 5z - 8)$. Write your answer in standard form. **[Lesson 7.2]** $z^3 - 10z^2 - z + 15$

14. Explain how to use the FOIL method to multiply $(x + 4)$ and $(2x - 5)$. Then find the product. **[Lesson 7.4]**

15. Factor $3a(2c + d)^2 - 5(2c + d)^2$. **[Lesson 7.5]** $(2c + d)^2(3a - 5)$

16. Write as a product of binominal factors: $2b^2 - 3b - 20$. **[Lesson 7.6]**
$(2b - 5)(b + 4)$

17. Rewrite $y = x^2 - 6x + 4$ in the form $y = (x - h)^2 + k$. Find the vertex of the corresponding parabola. **[Lesson 8.3]** $y = (x - 3)^2 - 5$; $(3, -5)$

18. Solve by factoring or completing the square: $x^2 + 6x - 55 = 0$. **[Lessons 8.3, 8.4]** $-11, 5$

19. Graph the inequality $y \geq x^2 + 2x - 4$. Shade the solution region. **[Lesson 8.6]**

Free-Response Grid The following questions may be answered using a free-response grid commonly used by standardized test services.

20. Determine the slope of the line through the points (2, 5) and (4, 10). **[Lesson 2.1]** $\frac{5}{2}$

21. Suppose that you roll an ordinary 6-sided number cube. What is the probability that you will roll an even number? **[Lessons 4.6, 4.7]** $\frac{1}{2}$

22. Change 3^4 to customary notation. **[Lesson 6.1]** 81

23. Write the *positive* solution of $(x - 1)^2 - 15 = 0$ to the nearest hundredth. **[Lesson 8.2]** 4.87

24. Calvin has the following test grades: 87, 92, 78, and 81. What does he need to make on his fifth test in order to have an average of at least 85? 87 **[Lessons 1.5, 4.3]**

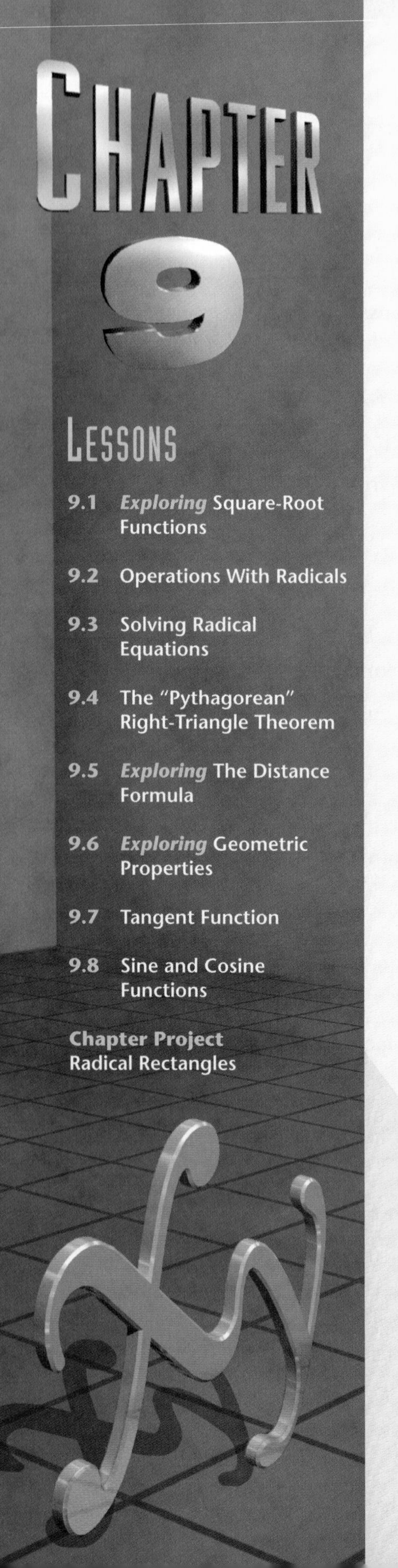

CHAPTER 9

Radicals and Coordinate Geometry

LESSONS

In this chapter you can examine square roots and their relationship to the process of squaring. Square roots are used whenever the quadratic formula or the distance formula are used. For this reason, they play a central and useful role in both algebra and geometry.

Square roots appear in many applications. For example,

- **You know the length of a pendulum and want to know how long it will take for one complete swing.**

$$t = 2\pi\sqrt{\frac{l}{g}}$$

- **You know the braking distance of a car and want to know the speed of the car.**

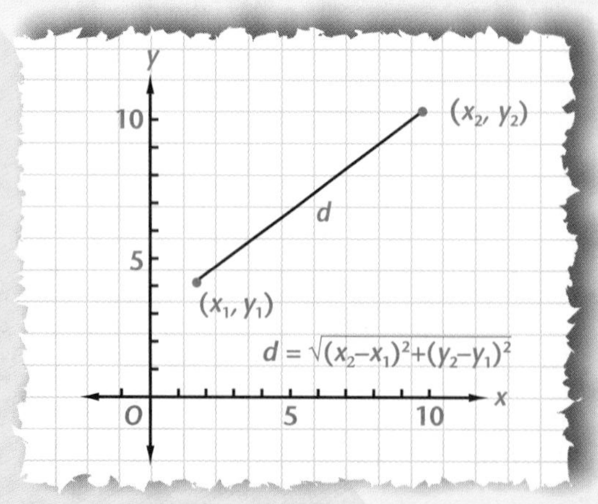

$r = \sqrt{\frac{A}{\pi}}$

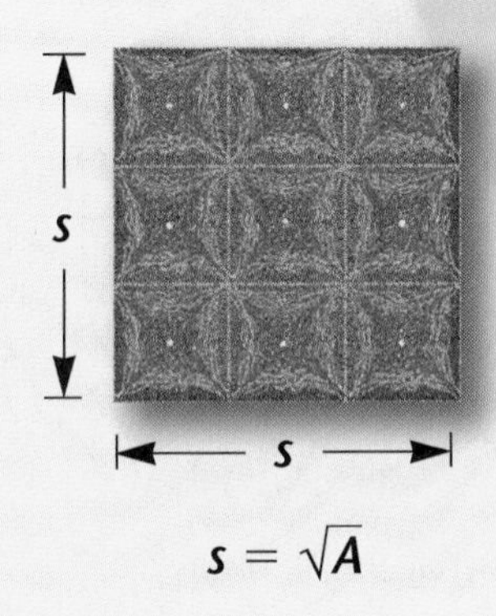

$s = \sqrt{A}$

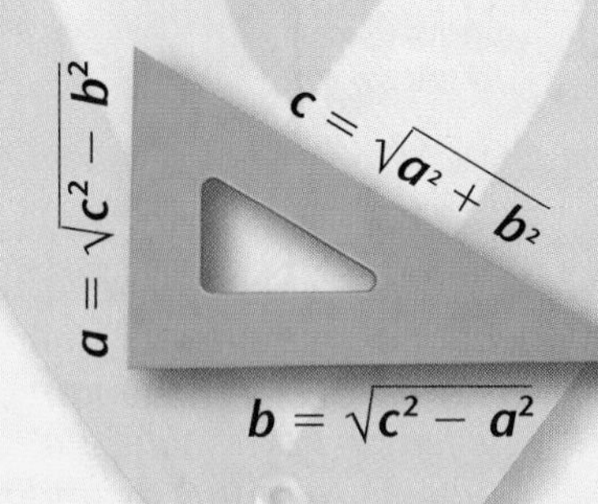

$b = \sqrt{c^2 - a^2}$

Portfolio Activity

The data in the table below comes from a driver's manual.

Braking distance on dry concrete, d, (in feet)	Speed, s, (in miles per hour)
22	20
50	30
88	40
138	50
198	60
270	70

a. Use the data from the table to create a graph that models speed as a function of braking distance.

b. Use your graph to answer the following questions:

1. If the braking distance is 100 feet, what is the speed?
2. If the speed is 55 miles per hour, what is the braking distance?

c. A skid mark at the scene of an accident that occurred on dry concrete is 36 feet long. Assume that this number represents the braking distance. Note that 36 is midway between 22 and 50 feet, the braking distances for 20 and 30 miles per hour, respectively.

$s = a\sqrt{d}$

The speed limit at the location of the accident is 25 miles per hour.

The driver says, "I was going below the speed limit."
A passenger says, " We were going at exactly the speed limit."
A policeman says, "You were going above the speed limit."

You be the judge. Decide who is correct and present a mathematical argument to support your conclusion.

LESSON 9.1

Exploring Square-Root Functions

Why *How can you find the length of one side of a square if you know its area? Finding the length of the side of a square is the same as finding the positive square root of a number.*

An artist is working on a square tabletop mosaic covered by 36 square tiles. What is the length of one side of that table?

Since the mosaic is square, count the tiles on any side to find the length. The length of each side is 6 tiles.

area = 36 square tiles
side = 6 tile lengths

You can check your answer by multiplying the length of the side by itself.

$6 \cdot 6 = 6^2$, or 36

Since the 36 tiles can be arranged in a square with the measure of each side equal to the same integer, the number 36 is called a **perfect square**.

List the first 10 perfect-square numbers.

1, 4, 9, 16, 25, 36, 49, 64, 81, 100

Exploration 1 *Estimating Square Roots*

How can you determine the length of the side of the blue square if its area is 12 square centimeters?

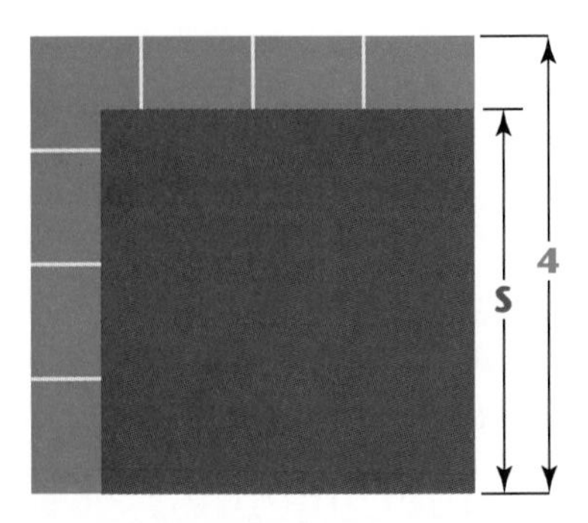

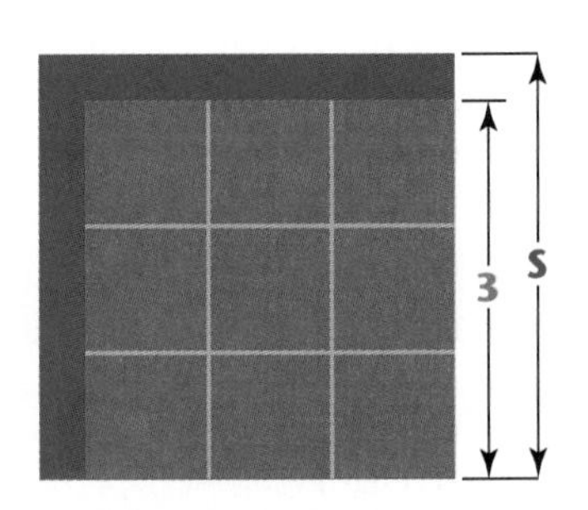

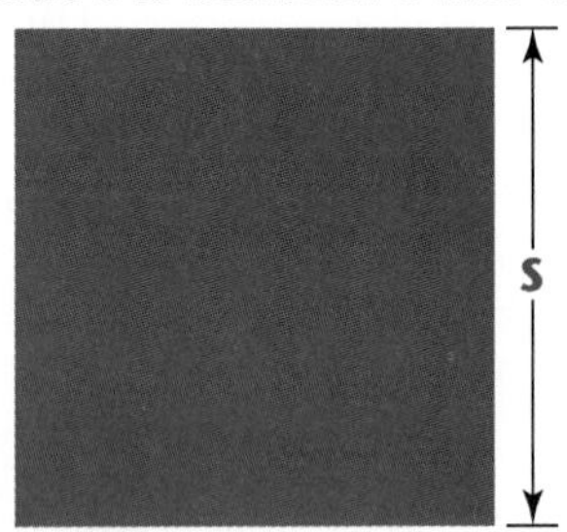

1. What is the area of the large green square?
16 sq cm

2. What is the area of the large red square?
9 sq cm

3. Estimate the length of a side of the blue square.
Answers may vary; 3.4 cm

4. Multiply the length you find in Step 3 by itself.
Answers may vary; 11.56 sq cm

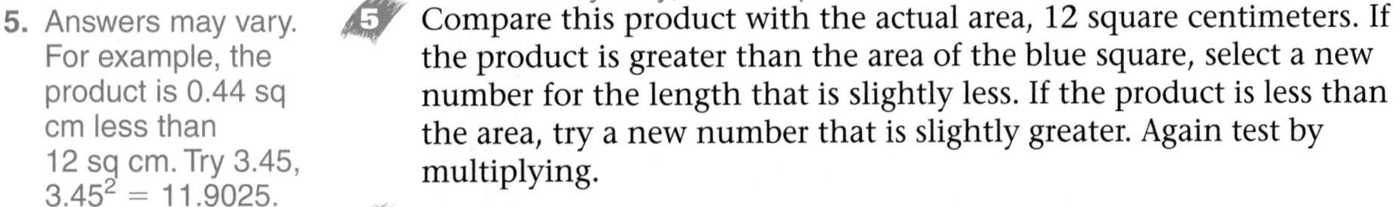
5. Compare this product with the actual area, 12 square centimeters. If the product is greater than the area of the blue square, select a new number for the length that is slightly less. If the product is less than the area, try a new number that is slightly greater. Again test by multiplying.

6. Continue to guess-and-check until you have an estimate to 2 decimal places. You have estimated $\sqrt{12}$.

7. Use this procedure to estimate the decimal value of the length of a side of a square with an area of 20 square centimeters. Describe how to find $\sqrt{20}$ to the nearest hundredth. ❖

5. Answers may vary. For example, the product is 0.44 sq cm less than 12 sq cm. Try 3.45, $3.45^2 = 11.9025$.
6. Answers may vary. An estimate of 3.46 or 3.47 is good since $\sqrt{12}$ is between 3.46 and 3.47.

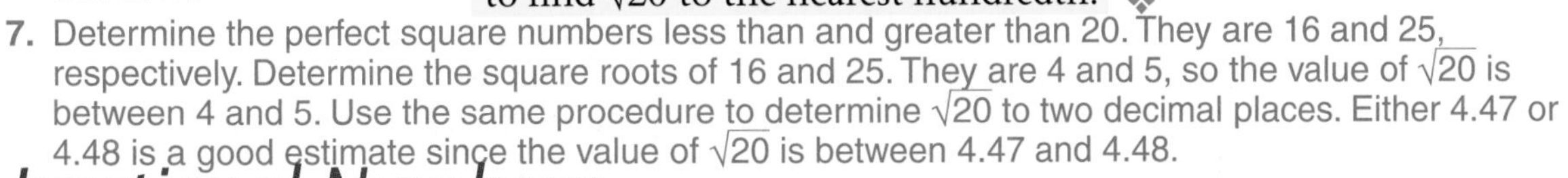
7. Determine the perfect square numbers less than and greater than 20. They are 16 and 25, respectively. Determine the square roots of 16 and 25. They are 4 and 5, so the value of $\sqrt{20}$ is between 4 and 5. Use the same procedure to determine $\sqrt{20}$ to two decimal places. Either 4.47 or 4.48 is a good estimate since the value of $\sqrt{20}$ is between 4.47 and 4.48.

Irrational Numbers

Every **rational number** can be represented by a terminating or repeating decimal. A repeating decimal is written with a bar over the numbers that repeat. On the other hand, the decimal part of $\sqrt{12}$ will never terminate or repeat a pattern. Numbers such as $\sqrt{12}$ or π are called **irrational numbers**.

Rational Numbers	Irrational Numbers	
$\frac{5}{8} = 0.\mathbf{625}$	$\sqrt{2} = 1.\mathbf{4142}\ldots$	$\pi = 3.1415926\ldots$
$\frac{2}{3} = 0.\overline{6}$ (6 repeats forever.)	$\sqrt{12} = 3.\mathbf{4641}\ldots$	$a = 0.121221222\ldots$

To save time and effort, many people use a calculator to find square roots. The square root key on most calculators is identified by the **radical sign**, $\sqrt{\ }$. When the $\boxed{\sqrt{\ }}$ key is pressed, the square root is given as an exact number or as an approximation. On some calculators, $\sqrt{2}$ is represented by 1.414213562. This, however, is only an approximation. When a calculator displays an irrational number, it cannot keep producing digits indefinitely, so it rounds the value.

Between any two rational numbers there is another rational number. However, between any two integers, say 1 and 2, there may not be an integer.

The *real numbers* include all the rational and irrational numbers. The real numbers form a **dense set**. That is, between any two real numbers there is another real number. Explain why the rationals are dense but the integers are not.

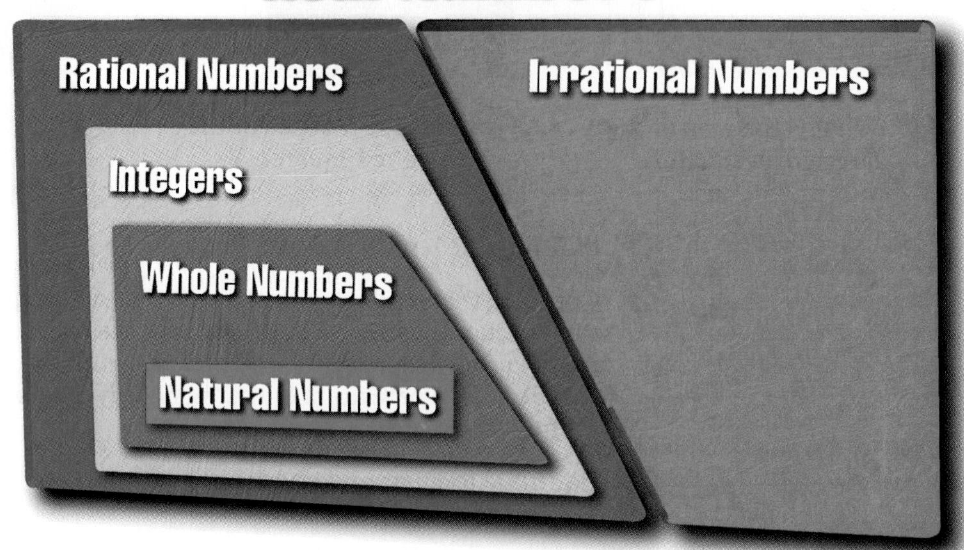

The Principal-Square-Root Function

In Lesson 8.3 you found the positive and negative square roots of a positive number. For example, 2 and -2 are square roots of 4 because $(-2)^2 = 4$ and $2^2 = 4$. The positive square root, sometimes called the **principal square root**, is indicated by the radical sign $\sqrt{\ }$.

Thus, $\sqrt{4} = 2$.

The function $f(x) = \sqrt{x}$ represents the parent function of the principal square root.

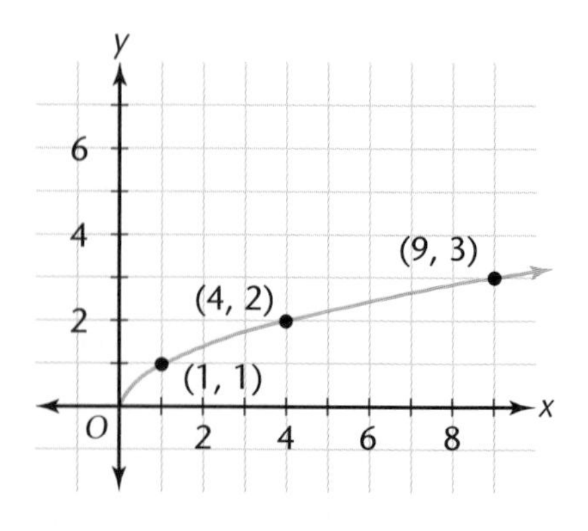

What are the domain and range of $f(x) = \sqrt{x}$? The domain and range of $f(x)$ are all non-negative real numbers.

CT– The positive part of $y = x^2$ corresponds to the positive part of $y = \sqrt{x}$, but it has been reflected through the line $y = x$.

The squaring function and the principal-square-root function are related. Plot $f(x) = \sqrt{x}$ and $g(x) = x^2$ on the same coordinate plane. Compare the two graphs. What is the relationship between these two functions?

Exploration 2 Transforming the Square-Root Function

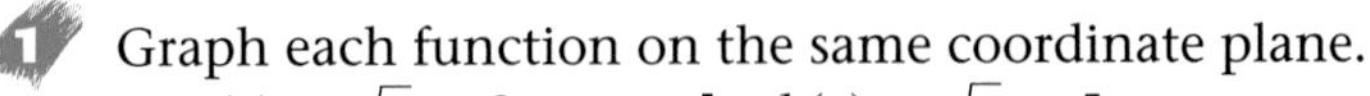

1 Graph each function on the same coordinate plane.

a. $g(x) = \sqrt{x} + 2$ **b.** $h(x) = \sqrt{x} + 5$ **c.** $p(x) = \sqrt{x} - 3$

2 Compare each function in Step 1 with $f(x) = \sqrt{x}$. What transformation results when $f(x) = \sqrt{x}$ is changed to $q(x) = \sqrt{x} + k$?

3 Graph each function on the same coordinate plane.

a. $g(x) = \sqrt{x + 2}$ **b.** $h(x) = \sqrt{x + 5}$ **c.** $p(x) = \sqrt{x - 3}$

4 Compare each function in Step 3 with $f(x) = \sqrt{x}$. What transformation results when $f(x) = \sqrt{x}$ is changed to $q(x) = \sqrt{x - h}$?

5 Graph each function on the same coordinate plane.

a. $g(x) = 2\sqrt{x}$ **b.** $h(x) = 3\sqrt{x}$ **c.** $p(x) = 5\sqrt{x}$

6 Compare each function in Step 5 with $f(x) = \sqrt{x}$. What transformation results when $f(x) = \sqrt{x}$ is changed to $g(x) = a\sqrt{x}$, where $a > 0$?

7 Graph each function on the same coordinate plane.

a. $g(x) = -2\sqrt{x}$ **b.** $h(x) = -3\sqrt{x}$ **c.** $p(x) = -5\sqrt{x}$

8 Compare each function in Step 7 with $f(x) = \sqrt{x}$. What transformation results when $f(x) = \sqrt{x}$ is changed to $g(x) = a\sqrt{x}$, where $a < 0$?

9 Summarize your findings by explaining the results of Steps 2, 4, 6, and 8. ❖

Compare the graph of $y = \sqrt{x}$ with the graph of $y = -\sqrt{x}$. The graph of $y = -\sqrt{x}$ shows that it is also a function. However, the relation $y = \pm\sqrt{x}$ is not a function.

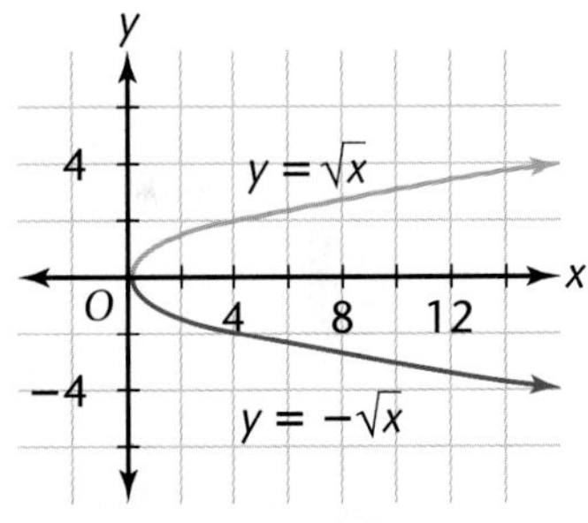

Explain the transformations that were applied to $f(x) = \sqrt{x}$ to produce g and h, which are graphed below.

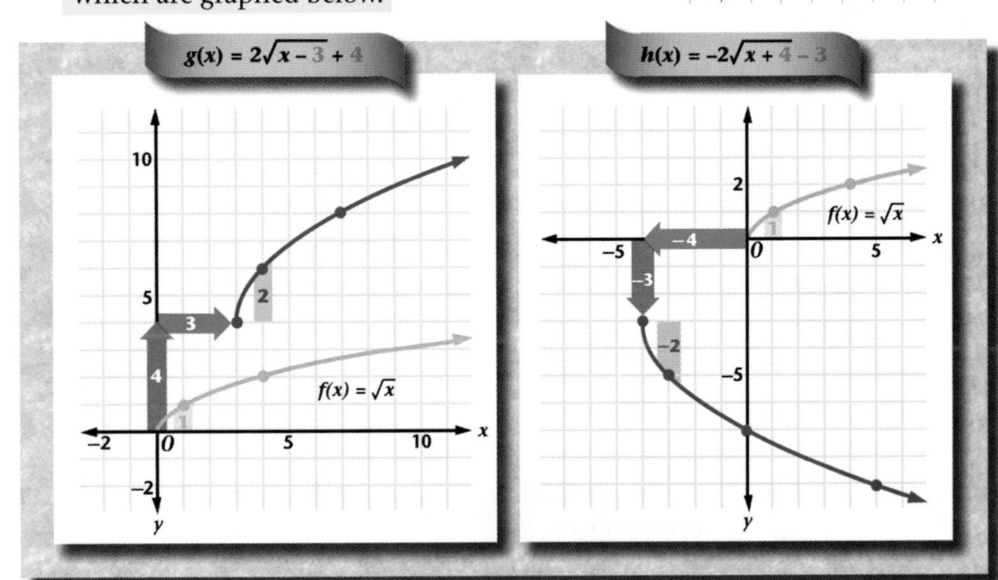

EXTENSION

The transformations in Exploration 2 can be combined to graph functions such as $f(x) = 3 - \sqrt{x + 2}$. First write the function in the form $f(x) = -\sqrt{x + 2} + 3$.

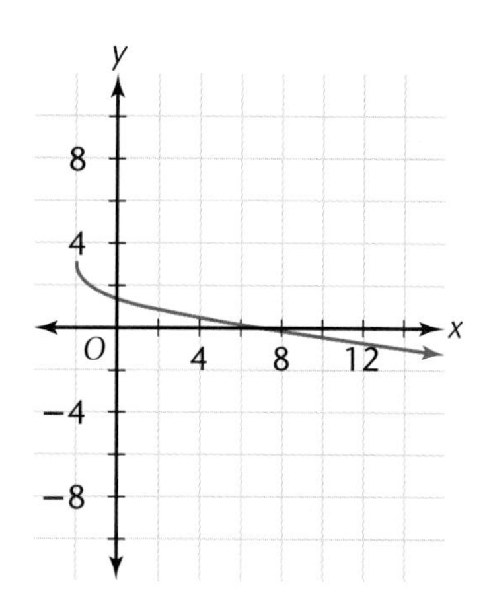

The transformation $\sqrt{x + 2}$ shifts the graph of $\sqrt{x}$ left 2 units. Then $-\sqrt{x + 2}$ reflects the graph through the x-axis. Finally, $-\sqrt{x + 2} + 3$ shifts the graph up 3 units.

Thus, the transformation $f(x) = -\sqrt{x + 2} + 3$ shifts the graph to the left 2 units, reflects it through the x-axis, and then shifts it up 3 units. ❖

Janice Brown made the first solar-powered flight in the Solar Challenger on December 3, 1980.

APPLICATION

The approximate distance to the horizon (in kilometers) is given by the formula $D = 3.56\sqrt{x}$. D represents the distance to the horizon in kilometers, and x is the altitude in meters. Find the approximate distance in kilometers to the horizon. Use the formula $3.56\sqrt{x}$ with $x = 20$.

$$D = 3.56\sqrt{20}$$

$$D \approx 3.56 \cdot 4.472$$

$$D \approx 15.92$$

The distance to the horizon is approximately 15.92 kilometers. ❖

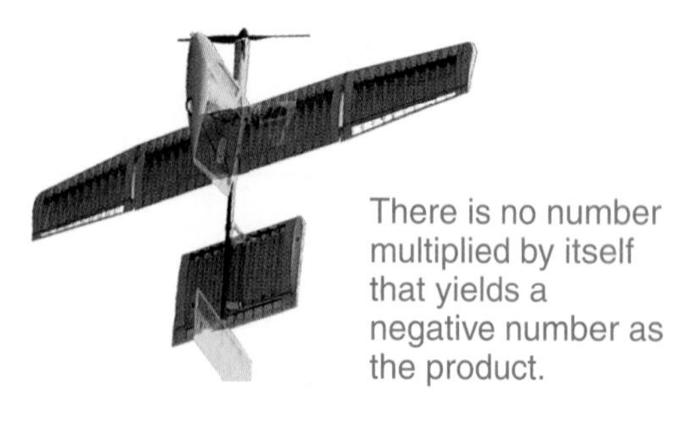

There is no number multiplied by itself that yields a negative number as the product.

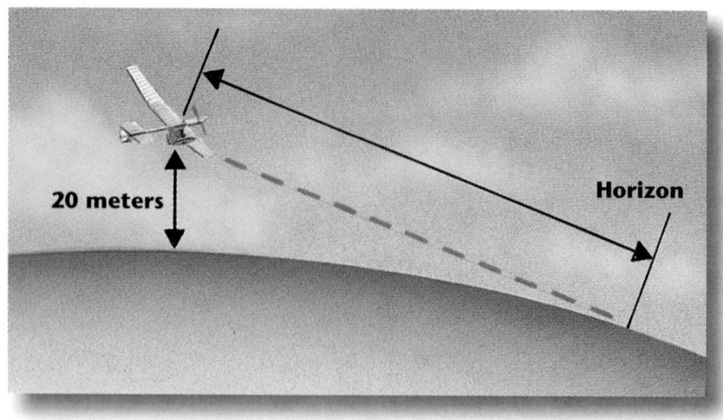

Why do you think that the square root of a negative number is not a real number?

Communicate

1. Describe how you can determine the square root of a perfect square with graph paper.
2. How can you estimate the square root of a number that is not a perfect square with graph paper?
3. What are the characteristics of an irrational number? Give three examples.
4. Explain how to estimate $\sqrt{7}$.
5. Explain how to write the transformation that moves the graph of the square-root function up 3 and right 2.
6. Explain why every positive number has two square roots.

Practice & Apply

Estimate each square root. If the square root is irrational, find the value to the nearest hundredth.

7. $\sqrt{225}$ 15 **8.** $-\sqrt{169}$ −13 **9.** $\sqrt{11}$ 3.32 **10.** $\sqrt{\frac{4}{9}}$ $\frac{2}{3}$ or ≈0.67 **11.** $-\sqrt{40}$ −6.32

12. $-\sqrt{27}$ −5.2 **13.** $\sqrt{1000}$ 31.62 **14.** $\sqrt{10{,}000}$ 100 **15.** $-\sqrt{0.04}$ −0.2 **16.** $\sqrt{0.059}$ 0.24

Geometry Find the length of the side of a square with each area.

17. 250 square meters About 15.81 m **18.** 144 square centimeters 12 cm **19.** 28 square miles About 5.29 miles

20. A yard is in the shape of a square. If the area of the yard is 676 square feet, what is the length of each side? 26 feet

21. Use the formula $D = 1.22\sqrt{x}$, where x is the altitude and D is the distance in miles, to find the distance to the horizon if an airplane is flying at an altitude of 30,000 feet. ≈211 miles

22. **Technology** Choose five positive numbers greater than 50. Find the square of each positive number. Then find the square root of the result. Record your results in a table with the headings shown below. One example has been started for you.

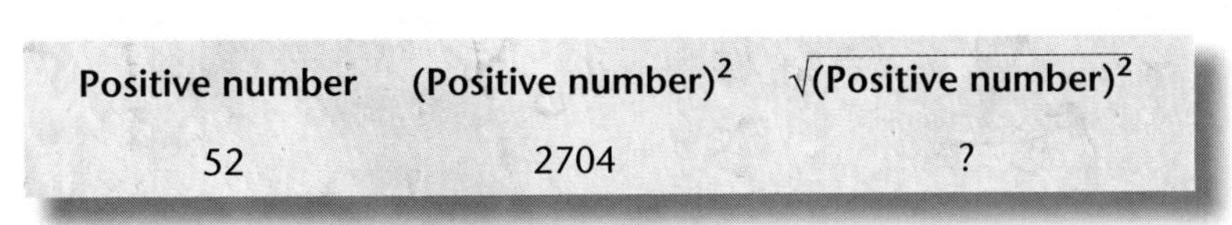

Positive number	(Positive number)^2	$\sqrt{(\text{Positive number})^2}$
52	2704	?

Make a conjecture based on the data in your table.

23. Technology Choose five negative numbers. Find the square of each negative number, and then find the square root of the result. Record your results in a table with the following headings:

Negative number	(Negative number)^2	$\sqrt{(\text{Negative number})^2}$
?	?	?

Make a conjecture based on the data in your table.

Graph each function, and describe it as a transformation of $y = \sqrt{x}$.

24. $y = \sqrt{x} + 6$ **25.** $y = 4\sqrt{x}$ **26.** $y = \sqrt{x - 2}$

27. $y = -\sqrt{x} + 1$ **28.** $y = 2\sqrt{x + 1} - 3$ **29.** $y = -2\sqrt{x - 1} + 6$

30. Write equations for two radical functions whose graphs create a parabola with the vertex at (3, 0). Answers may vary. For example, $y = \sqrt{x - 3}$ and $y = -\sqrt{x - 3}$.

31. Find the vertex and three additional ordered pairs for each pair of functions. Plot the ordered pairs on the same coordinate plane. State the domain and range of each function.

a. $y = \sqrt{4 - x}$ and $y = -\sqrt{4 - x}$
V: (4, 0); D: $x \leq 4$; R: all numbers y

b. $y = \sqrt{-x} + 3$ and $y = -\sqrt{-x} + 3$
V: (0, 3); D: $x \leq 0$; R: all numbers y

32. Test each statement with numbers. Then determine which statements are always true, and explain your reasoning. Give a counterexample if the statement is false.

a. $\sqrt{k^2} = k$, for any k F, $k < 0$ **b.** $\sqrt{k^2} = |k|$, for any k T

c. $\sqrt{k^2} = \pm k$ T

33. For what values of x is $\sqrt{x}$ defined? $x \geq 0$

34. For what values of x is $\sqrt{x + 5}$ defined? $x \geq -5$

35. What is the domain of the function $y = \sqrt{x + 5}$? $x \geq -5$

Describe each transformation, and describe the effect on the parent function, $y = \sqrt{x}$.

36. $y = -\sqrt{x - 4}$ **37.** $y = 5\sqrt{x + 3} - 4$ **38.** $y = 2.00 - 3.50\sqrt{x + 3.75}$

Write the equation of each square-root function graphed.

39. $y = \sqrt{x + 3}$

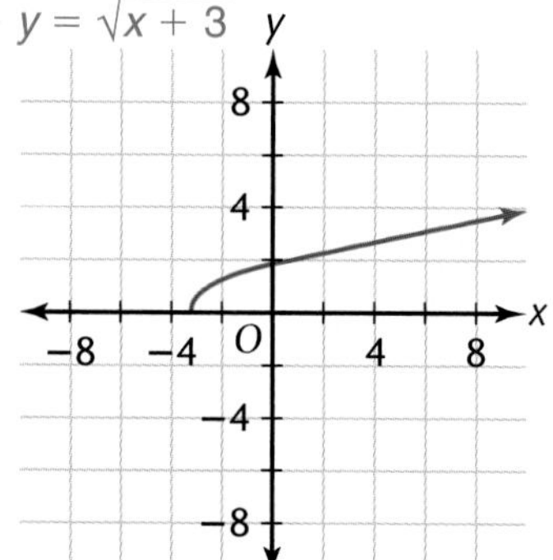

40. $y = -\sqrt{x + 4}$

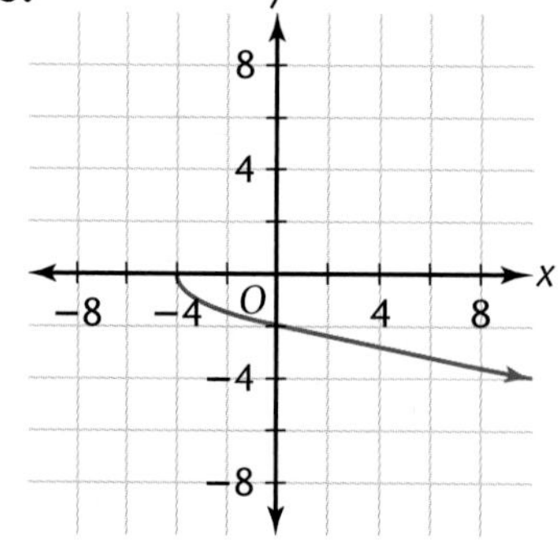

41. $y = \sqrt{x - 2} + 6$

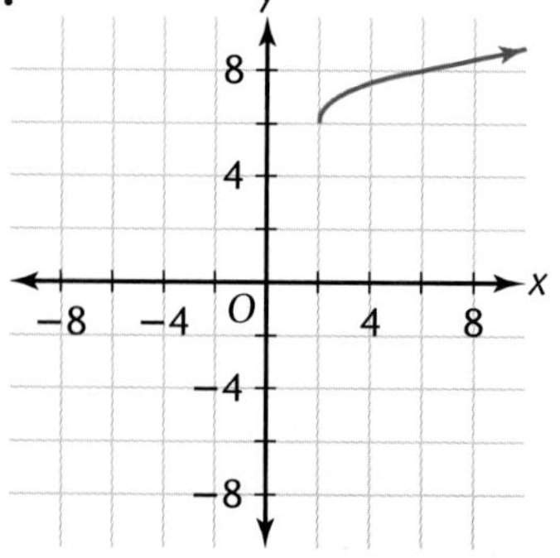

42. **Technology** Copy and complete the table below. Use a calculator to obtain decimal approximations to the nearest tenth for the square roots.

x	0	1	2	3	4	5	6	7	8	9	10
$\sqrt{x}$	?	?	?	?	?	?	?	?	?	?	?

43. Plot the data in your table on graph paper with x on the horizontal axis and $\sqrt{x}$ on the vertical axis.

Is it possible to have a point on the graph in Exercise 43 in the

44. second quadrant? No **45.** third quadrant? No **46.** fourth quadrant? No

47. **Technology** Use graphing technology to graph $y = \sqrt{x}$. Trace the function. Do the values in your table agree with the trace values on the calculator?

Look Back

Without using graphing technology, sketch the graph from what you know about transformations. [Lessons 5.2–5.6]

48. $y = 3x + 2$ **49.** $y = x^2 - 4$ **50.** $y = 3 - 2^x$ **51.** $y = -x^2 + 6$

Find the vertex of each absolute-value function. Use graphing technology to check your answer. [Lesson 5.6]

52. $y = |x - 2|$ (2, 0) **53.** $y = |x| + 5$ (0, 5)

54. $y = 2|x + 3| - 4$ (−3, −4) **55.** $y = -|x - 1| + 3$ (1, 3)

56. Solve the equation $x^2 - 5x - 7 = 0$. **[Lesson 8.3]** $\frac{5 \pm \sqrt{53}}{2}$, or ≈ 6.14 and ≈ -1.14

Look Beyond

Decide whether each statement is true or false. If you believe that a statement is false, give a counterexample.

57. If $a^2 = b^2$, then $a = b$. F, let $a = -1$, $b = 1$.

58. If $a = b$, then $a^2 = b^2$. T

59. If $a^2 = b^2$ and both a and b have the same sign, then $a = b$. T

60. Caitlin; Barbara; David; Andrew

60. In a long-distance race of 4 runners, Caitlin finished ahead of Andrew. David finished with a time 0.1 second more than Barbara, who finished 1 second after the winner. David was faster than only 1 other runner. Who won the race, and what was the order of finish?

LESSON 9.2 Operations With Radicals

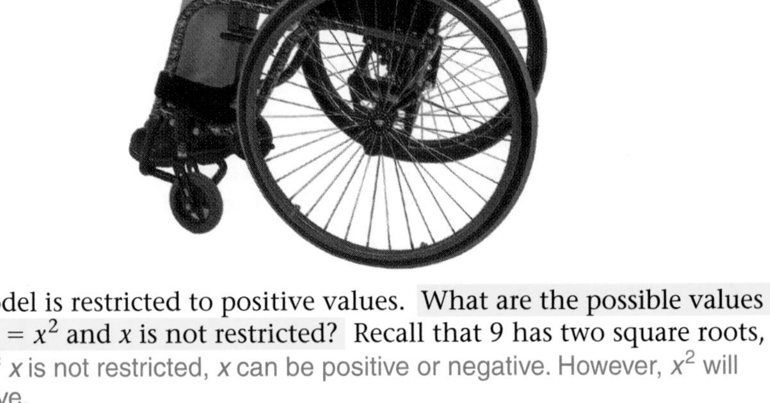

Why *Radicals often appear when computing statistics or evaluating formulas in physics. Radicals are sometimes used with numbers, but they are also used in algebraic expressions. Several methods are used to simplify radicals.*

The area model is restricted to positive values. What are the possible values of x when $A = x^2$ and x is not restricted? Recall that 9 has two square roots, 3 and -3. If x is not restricted, x can be positive or negative. However, x^2 will still be positive.

DEFINITION OF SQUARE ROOT

If a is a number greater than or equal to zero, $\sqrt{a}$ and $-\sqrt{a}$ represent the square roots of a. Each square root of a has the following properties:

$$\sqrt{a} \cdot \sqrt{a} = a \qquad (-\sqrt{a})(-\sqrt{a}) = a$$

When you work with radicals, it is helpful to know the name for each part.

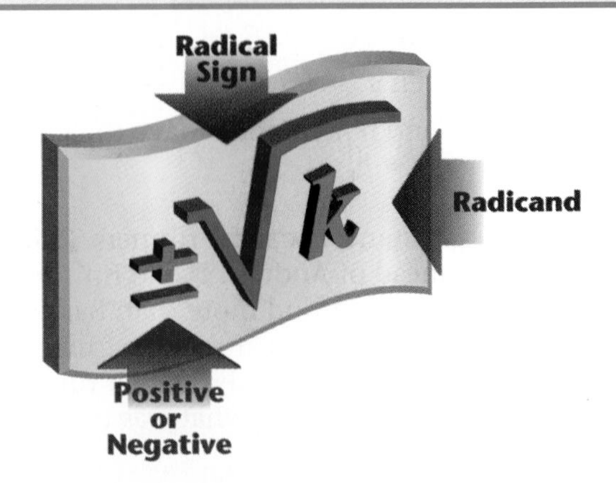

EXAMPLE 1

Find each square root. **A** $\sqrt{25}$ **B** $-\sqrt{144}$ **C** $\sqrt{7}$

Solution ➤

A $\sqrt{25} = 5$ **B** $-\sqrt{144} = -12$

C $\sqrt{7}$ is not a perfect square. It is approximately 2.646. ❖

Try This Find each square root. **a.** $\sqrt{196}$ 14 **b.** $-\sqrt{225}$ −15

Simplifying Radicals and Radical Expressions

Radical expressions that are simplified allow you to manipulate algebraic expressions more easily. In some cases the calculations can be done without technology. A radical expression is in **simplest radical form** if the expression within the radical sign

1. contains no perfect squares greater than 1,
2. contains no fractions, and
3. is not in the denominator of a fraction.

EXAMPLE 2

Simplify.

A $\sqrt{20}$ **B** $\frac{\sqrt{12}}{4}$ **C** $\frac{\sqrt{18}}{\sqrt{2}}$

Solution ➤

A $\sqrt{20}$ is not a perfect square, but it can be factored, so the radicand contains a perfect square. Simplify further.

$\sqrt{20} = \sqrt{2^2 \cdot 5} = \sqrt{2^2}\sqrt{5} = 2\sqrt{5}$

B $\frac{\sqrt{12}}{4} \cdot \frac{\sqrt{2^2 \cdot 3}}{4} = \frac{2\sqrt{3}}{4} = \frac{\sqrt{3}}{2}$

C $\frac{\sqrt{18}}{\sqrt{2}} = \sqrt{\frac{18}{2}} = \sqrt{9} = 3$ ❖

When the radicand of a radical expression contains variables such as a^6 or b^5, you can rewrite the expression using powers of 2. Then simplify.

$$\begin{aligned}\sqrt{a^6 \cdot b^5} &= \sqrt{a^2 \cdot a^2 \cdot a^2 \cdot b^2 \cdot b^2 \cdot b} \\ &= \sqrt{a^2} \cdot \sqrt{a^2} \cdot \sqrt{a^2} \cdot \sqrt{b^2} \cdot \sqrt{b^2} \cdot \sqrt{b} \\ &= a \cdot a \cdot a \cdot b \cdot b \cdot \sqrt{b} \\ &= a^3b^2\sqrt{b}\end{aligned}$$

The procedures used in Example 2 involve two important properties of square roots.

MULTIPLICATION PROPERTY OF SQUARE ROOTS

For all numbers a and b, such that a and $b \geq 0$,

$$\sqrt{ab} = \sqrt{a}\sqrt{b}.$$

EXAMPLE 3

Simplify.

A $\sqrt{12}$ **B** $\sqrt{400}$ **C** $\sqrt{a^5b^{12}}$, where a and $b \geq 0$

Solution ➤

Look for perfect square factors, and apply the Multiplication Property of Square Roots. Then find the square roots of the perfect squares. Leave the factor that is not a perfect square inside the radical symbol.

A $\sqrt{12} = \sqrt{4 \cdot 3} = \sqrt{4} \cdot \sqrt{3} = 2\sqrt{3}$

B $\sqrt{400} = \sqrt{4 \cdot 100} = \sqrt{4} \cdot \sqrt{100} = 2 \cdot 10 = 20$

C $\sqrt{a^5b^{12}} = \sqrt{(a^2)^2 \cdot a \cdot (b^6)^2} = a^2b^6\sqrt{a}$ ❖

Try This Simplify $\sqrt{72m^2n^5}$, where m and $n \geq 0$. $6mn^2\sqrt{2n}$

DIVISION PROPERTY OF SQUARE ROOTS

For all numbers $a \geq 0$ *and* $b > 0$,

$$\sqrt{\frac{a}{b}} = \frac{\sqrt{a}}{\sqrt{b}}.$$

EXAMPLE 4

Simplify.

A $\sqrt{\frac{16}{25}}$ **B** $\sqrt{\frac{7}{16}}$ **C** $\sqrt{\frac{9}{5}}$ **D** $\sqrt{\frac{a^2b^3}{c^2}}$

Solution ➤

Rewrite the square root using the Division Property of Square Roots. Then simplify the numerator and denominator separately.

A $\sqrt{\frac{16}{25}} = \frac{\sqrt{16}}{\sqrt{25}} = \frac{4}{5}$

B $\sqrt{\frac{7}{16}} = \frac{\sqrt{7}}{\sqrt{16}} = \frac{\sqrt{7}}{4}$

C $\sqrt{\frac{9}{5}} = \frac{\sqrt{9}}{\sqrt{5}} = \frac{3}{\sqrt{5}}$ You can can simplify this expression by multiplying the fraction by $\frac{\sqrt{5}}{\sqrt{5}}$, or 1. $\frac{3}{\sqrt{5}} \cdot \frac{\sqrt{5}}{\sqrt{5}} = \frac{3\sqrt{5}}{\sqrt{5}\sqrt{5}} = \frac{3\sqrt{5}}{\sqrt{25}} = \frac{3\sqrt{5}}{5}$

This is known as **rationalizing the denominator**.

D For a and $b \geq 0$ and $c > 0$, $\sqrt{\frac{a^2b^3}{c^2}} = \frac{\sqrt{a^2b^2b}}{\sqrt{c^2}} = \frac{ab\sqrt{b}}{c}$. ❖

Operations and Radical Expressions

You know that $\sqrt{16 \cdot 9} = \sqrt{16} \cdot \sqrt{9}$. Is it true that $\sqrt{16 + 9} = \sqrt{16} + \sqrt{9}$? Examine the computations below to find the answer.

$$\sqrt{16 + 9} = \sqrt{25} \text{ or } 5$$
$$\sqrt{16} + \sqrt{9} = 4 + 3 \text{ or } 7$$

Since 5 does not equal 7, $\sqrt{16 + 9}$ does not equal $\sqrt{16} + \sqrt{9}$. In general,

$\sqrt{a + b} \neq \sqrt{a} + \sqrt{b}$.

In order to add expressions that contain radicals, the radicands must be the same. Like terms that contain $\sqrt{5}$, for example, can be added by the Distributive Property.

$2\sqrt{5} + 4\sqrt{5} = (2 + 4)\sqrt{5} = 6\sqrt{5}$

EXAMPLE 5

Simplify.

A $5\sqrt{6} - 2\sqrt{6}$

B $5 + 6\sqrt{7} - 2\sqrt{7} - 3$

C $8\sqrt{3} + 6\sqrt{2} - \sqrt{3} + 2\sqrt{2}$

D $a\sqrt{x} + b\sqrt{x}$

Solution ➤

A Use the Distributive Property to combine like terms. In this case, $\sqrt{6}$ is the common factor. Then simplify.

$$5\sqrt{6} - 2\sqrt{6} = (5 - 2)\sqrt{6} = 3\sqrt{6}$$

B Rearrange and combine like terms.

$$5 + 6\sqrt{7} - 2\sqrt{7} - 3 = (5 - 3) + (6 - 2)\sqrt{7} = 2 + 4\sqrt{7}$$

C Rearrange and combine like terms.

$$8\sqrt{3} + 6\sqrt{2} - \sqrt{3} + 2\sqrt{2} = (8 - 1)\sqrt{3} + (6 + 2)\sqrt{2} = 7\sqrt{3} + 8\sqrt{2}$$

D Literal expressions are treated in a similar way.

$$a\sqrt{x} + b\sqrt{x} = (a + b)\sqrt{x}$$ ❖

Does $\sqrt{a} - \sqrt{b}$ equal $\sqrt{a - b}$ for any a and b? Substitute values to test the idea. Explain the results of the test. Are there any values of a and b that allow the statement to be true? No. Explanations may vary. The only conditions that allow the statement to be true is if b is 0, both a and b are 0, or if $a = b$.

There are several ways that radicals may appear in multiplication.

EXAMPLE 6

Simplify.

A $(5\sqrt{3})^2$

B $\sqrt{3}\sqrt{6}$

C $\sqrt{2}(6 + \sqrt{12})$

D $(3 - \sqrt{2})(4 + \sqrt{2})$

Solution ➤

A Recall from the properties of exponents that the second power means the product of two identical factors. Rearrange the factors and multiply.

$$(5\sqrt{3})^2 = (5\sqrt{3})(5\sqrt{3}) = (5 \cdot 5)(\sqrt{3}\sqrt{3}) = 25 \cdot 3 = 75$$

B The Multiplication Property of Square Roots allows you to multiply separate radicals. You can then factor the new product in a different way and simplify.

$$\sqrt{3}\sqrt{6} = \sqrt{3 \cdot 6} = \sqrt{18} = \sqrt{9 \cdot 2} = \sqrt{9}\sqrt{2} = 3\sqrt{2}$$

C Use the Distributive Property to multiply, then factor again, and simplify the result. Notice that the radical is usually written last in each term.

$$\begin{aligned} \sqrt{2}(6 + \sqrt{12}) &= \sqrt{2} \cdot 6 + \sqrt{2} \cdot \sqrt{12} \\ &= 6\sqrt{2} + \sqrt{2 \cdot 12} \\ &= 6\sqrt{2} + \sqrt{24} \\ &= 6\sqrt{2} + \sqrt{2 \cdot 2 \cdot 2 \cdot 3} \\ &= 6\sqrt{2} + 2\sqrt{6} \end{aligned}$$

D To multiply differences and sums, multiply the two binomials. The FOIL method is used here.

$$(3 - \sqrt{2})(4 + \sqrt{2}) = 12 + 3\sqrt{2} - 4\sqrt{2} - 2 = 10 - \sqrt{2}$$ ❖

When you enter more complicated expressions into a calculator, you may need to use parentheses. To find the square root of $3^2 + 4^2$, enter $\sqrt{(3^2 + 4^2)}$ into the calculator. You must put parentheses around the entire expression under the radical sign.

What do you think the calculator will display as an answer if the parentheses are not used? 19

EXERCISES & PROBLEMS

Communicate

1. Explain when $\sqrt{a^2} = a$.
2. How is factoring used to simplify expressions that contain radicals? What is simplest radical form?
3. For what values of a can you find $\sqrt{-a + 3}$?
4. Explain in your own words what the Multiplication and Division Properties of Square Roots allow you to do.
5. When is it possible to add two radical expressions?
6. Describe a procedure for multiplying $(\sqrt{3} + 2)(\sqrt{2} + 3)$.

Practice & Apply

Simplify each radical expression by factoring.

7. $\sqrt{49}$ 7
8. $\sqrt{196}$ 14
9. $\sqrt{576}$ 24
10. $\sqrt{3600}$ 60
11. $\sqrt{75}$ $5\sqrt{3}$
12. $\sqrt{98}$ $7\sqrt{2}$
13. $\sqrt{1620}$ $18\sqrt{5}$
14. $\sqrt{264}$ $2\sqrt{66}$

Decide whether the given statement is true or false for all values of *a* and *b*. Assume that *a* and *b* ≥ 0.

15. $\sqrt{a + b} = \sqrt{a} + \sqrt{b}$ F
16. $\sqrt{a - b} = \sqrt{a} - \sqrt{b}$ F
17. $\sqrt{ab} = \sqrt{a}\sqrt{b}$ T
18. $\sqrt{\frac{a}{b}} = \frac{\sqrt{a}}{\sqrt{b}}, b \neq 0$ T

Express each radical expression in simplest radical form.

19. $\sqrt{3}\sqrt{12}$ 6
20. $\sqrt{8}\sqrt{18}$ 12
21. $\sqrt{48}\sqrt{3}$ 12
22. $\sqrt{54}\sqrt{6}$ 18
23. $\sqrt{\frac{64}{16}}$ 2
24. $\sqrt{\frac{96}{2}}$ $4\sqrt{3}$
25. $\frac{\sqrt{50}}{\sqrt{25}}$ $\sqrt{2}$
26. $\frac{\sqrt{150}}{\sqrt{6}}$ 5
27. $\sqrt{5}\sqrt{15}$ $5\sqrt{3}$
28. $\sqrt{98}\sqrt{14}$ $14\sqrt{7}$
29. $\sqrt{\frac{56}{8}}$ $\sqrt{7}$
30. $\frac{\sqrt{96}}{\sqrt{8}}$ $2\sqrt{3}$

Simplify each radical expression. Assume that all variables are non-negative and that all denominators are non-zero.

31. $\sqrt{a^4b^6}$ a^2b^3
32. $\sqrt{x^8y^9}$ $x^4y^4\sqrt{y}$
33. $\sqrt{\frac{p^9}{q^{10}}}$ $\frac{p^4\sqrt{p}}{q^5}$
34. $\sqrt{\frac{x^3}{y^6}}$ $\frac{x\sqrt{x}}{y^3}$

If possible, perform each indicated operation, and simplify your answer.

35. $3\sqrt{5} + 4\sqrt{5}$ $7\sqrt{5}$

36. $3\sqrt{2} + 4\sqrt{8} - 3\sqrt{18}$ $2\sqrt{2}$

37. $\sqrt{7} + \sqrt{29}$ simplified

38. $4\sqrt{5} + 2\sqrt{5} - 5\sqrt{5}$ $\sqrt{5}$

39. $\sqrt{6} + 2\sqrt{3} - \sqrt{6}$ $2\sqrt{3}$

40. $(4 + \sqrt{3}) + (1 - \sqrt{2})$ $5 + \sqrt{3} - \sqrt{2}$

41. $\frac{6 + \sqrt{18}}{3}$ $2 + \sqrt{2}$

42. $\frac{\sqrt{15} + \sqrt{10}}{\sqrt{5}}$ $\sqrt{3} + \sqrt{2}$

Simplify each radical expression.

43. $(3\sqrt{5})^2$ 45

44. $(4\sqrt{25})^2$ 400

45. $\sqrt{12}\sqrt{6}$ $6\sqrt{2}$

46. $\sqrt{72}\sqrt{32}$ 48

47 $3(\sqrt{5} + 9)$ $3\sqrt{5} + 27$

48. $\sqrt{5}(6 - \sqrt{15})$ $6\sqrt{5} - 5\sqrt{3}$

49. $\sqrt{6}(6 + \sqrt{18})$ $6\sqrt{6} + 6\sqrt{3}$

50. $(\sqrt{5} - 2)(\sqrt{5} + 2)$ 1

51. $(\sqrt{3} - 4)(\sqrt{3} + 2)$ $-5 - 2\sqrt{3}$

52. $(\sqrt{5} + 7)(\sqrt{2} - 8)$ $\sqrt{10} - 8\sqrt{5} + 7\sqrt{2} - 56$

53. $(\sqrt{6} + 2)^2$ $10 + 4\sqrt{6}$

54. $\sqrt{3}(\sqrt{3} + 2)^2$ $12 + 7\sqrt{3}$

55. **Geometry** Copy and complete the table. Each entry should be simplified.

Area of square	6	12	24	48	96
Length of side	?	?	?	?	?

$\sqrt{6}$ $2\sqrt{3}$ $2\sqrt{6}$ $4\sqrt{3}$ $4\sqrt{6}$

56. Add three more columns to your table according to the number pattern established in the Area-of-square column.

Area of square	192	384	768
Length of side	$8\sqrt{3}$	$8\sqrt{6}$	$16\sqrt{3}$

Look Back

57. Find the solution to this system of equations. **[Lessons 2.4, 2.5]** (7, 13)

$$\begin{cases} y = -x + 20 \\ y = 2x - 1 \end{cases}$$

Simplify each expression. [Lesson 6.2]

58. $(-a^2b^2)^3(a^4b)^2$ $-a^{14}b^8$

59. $\frac{x^5y^7}{x^2y^3}, x \neq 0, y \neq 0$ x^3y^4

60. $\left(\frac{20x^3}{-4x^2}\right)^3$ $-125x^3$

Find each product. [Lesson 7.4]

61. $(2x - 4)(2x - 4)$ $4x^2 - 16x + 16$

62. $(3a + 5)(2a - 6)$ $6a^2 - 8a - 30$

63. $(6b + 1)(3b - 1)$ $18b^2 - 3b - 1$

Solve each equation by completing the square. [Lessons 8.3, 8.4]

64. $y^2 - 8y + 12 = 0$ 2, 6

65. $x^2 + 14x - 15 = 0$ 1, −15

66. $r^2 - 24r + 63 = 0$ 21, 3

Look Beyond

67. Notice that $2 \cdot 2 \cdot 2 = 2^3 = 8$. Then $\sqrt[3]{8} = 2$ (read "the cube root of 8 equals 2"). Copy and complete the following table.
64; 125; 216; 343; 512; 729; 1000

x	1	2	3	4	5	6	7	8	9	10
x^3	1	8	27	?	?	?	?	?	?	?

Based on the data in the table, answer Exercises 68–71.

68. $\sqrt[3]{125} = \underline{?}$ 5

69. $\sqrt[3]{1000} = \underline{?}$ 10

70. $\sqrt[3]{100}$ is between what two whole numbers? 4 and 5

71. **Technology** Use a calculator to approximate $\sqrt[3]{100}$ to the nearest tenth. ≈4.6

LESSON 9.3 Solving Radical Equations

Why *Equations that contain radicals are often used in science to model natural phenomena. Scientists also use radicals to solve equations that contain squares of numbers and variables. The mathematics of radicals provides a necessary tool to solve many problems that occur in science and other areas.*

The relationship of the length of a pendulum to the number of swings per minute is one example of an equation that contains a radical.

Physics In a physical science class, six groups are given six strings of different sizes, each tied to a washer. They are asked to time one complete swing of each pendulum. Then they are asked to put all their data together, make a table, and draw a graph of the data. The point (0, 0) represents the time for a string of zero length.

Examine the data in the table and on the graph.

Let l represent the length of the string in centimeters and t represent the time of one swing in seconds.

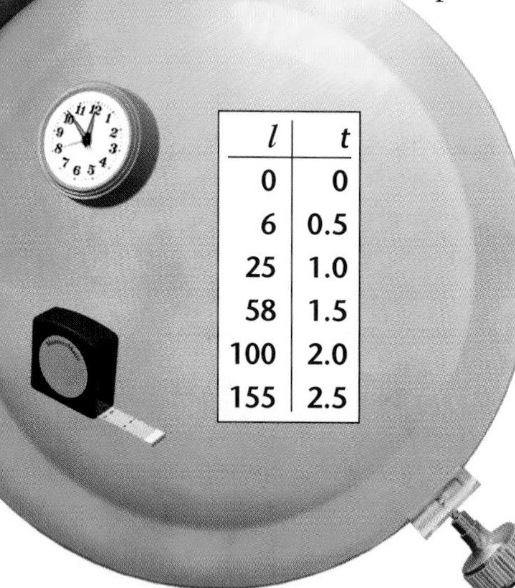

l	t
0	0
6	0.5
25	1.0
58	1.5
100	2.0
155	2.5

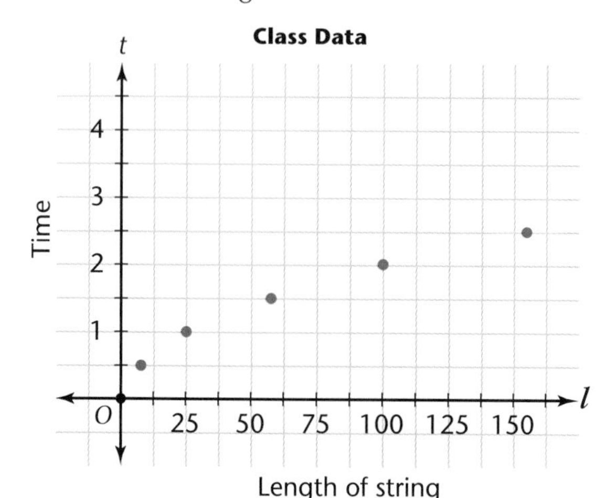

What conclusions can be drawn from the information? How can you use the graph to predict the time that it will take for a string of any length to make a complete swing?

First, notice that the curve resembles a transformation of the parent function $y = \sqrt{x}$. In this case, the parent function is written $t = \sqrt{l}$. The transformation for a stretch of the function is $t = a\sqrt{l}$. The physical science class found that $a = \frac{1}{5}$ worked for their data. Check to see if the class was right. Use the function $t = \frac{1}{5}\sqrt{l}$ to predict the time it will take for a string that is 225 centimeters long to make a complete swing.

3 seconds

The equation that models the motion of a pendulum is $t = 2\pi\sqrt{\frac{l}{g}}$. The variable l is the length in centimeters, t is the time in seconds, and g is acceleration due to gravity (980 centimeters per second per second). This formula can be used to see how close the physical science class's data is to the theoretical value determined by the formula $t = 2\pi\sqrt{\frac{l}{980}}$.

$$\begin{aligned} t &= 2\pi\sqrt{\frac{l}{980}} \\ &= \frac{2\pi \cdot \sqrt{l}}{\sqrt{980}} \\ &= \frac{2\pi}{\sqrt{980}} \cdot \sqrt{l} \\ &\approx 0.20\sqrt{l} \end{aligned}$$

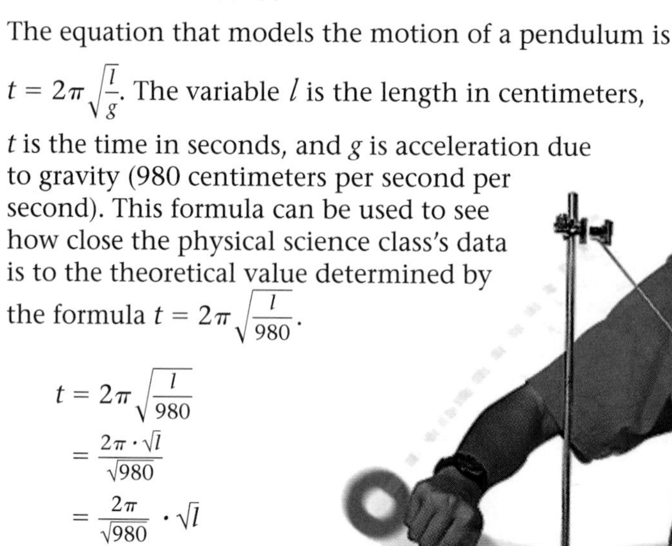

EXAMPLE 1

A Find the time that it takes a 75-centimeter pendulum to make a complete swing.

B Find the length of a pendulum that takes 3 seconds to make a complete swing.

Solution ➤

A Substitute 75 for l into the pendulum formula.

$$t = 2\pi\sqrt{\frac{l}{g}} \longrightarrow t = 2\pi\sqrt{\frac{75}{980}}$$

Use a calculator. $\quad t \approx 1.74$

One swing takes about 1.7 seconds.

B Substitute the known information into the formula, and solve for length, l.

$3 = 2\pi\frac{\sqrt{l}}{\sqrt{980}}$	Given
$3\sqrt{980} = 2\pi\sqrt{l}$	Multiply both sides by $\sqrt{980}$.
$\frac{3\sqrt{980}}{2\pi} = \sqrt{l}$	Divide both sides by 2π.
$\left(\frac{3\sqrt{980}}{2\pi}\right)^2 = l$	Use the definition of square root.
$223.4 \approx l$	Use a calculator.

The length of the pendulum is about 223.4 centimeters. ❖

Solving Equations Containing Radicals

When you square both sides of an equation, you may obtain **extraneous roots**. For example, $-5 = 5$ is false, but $(-5)^2 = 5^2$ is true. Therefore, you *must* check your solution in the original equation.

EXAMPLE 2

Solve $\sqrt{x + 2} = 3$ **A** graphically. **B** algebraically.

Solution ➤

Graphics Calculator

A Enter the left side of the equation in the graphics calculator as $Y_1 = \sqrt{x + 2}$. Then enter the right side of the equation as $Y_2 = 3$. Find the point(s) of intersection. Since the point of intersection is (7, 3), the solution is $x = 7$.

X=7 Y=3

B The square root and the squaring operations are inverses and can be used to solve equations.

$\sqrt{x+2} = 3$	Given
$(\sqrt{x+2})^2 = 3^2$	Square both sides.
$x + 2 = 9$	Simplify.
$x = 7$	Subtraction Property of Equality

You *must* check by substituting 7 for x in the original equation. The solution is 7. ❖

Try This Solve $\sqrt{2x - 3} = 4$. $x = 9\frac{1}{2}$

What is the domain of the function $y = \sqrt{x+2}$? Explain.
$x \geq -2$. The radicand must be positive.

EXAMPLE 3

Solve $\sqrt{x+6} = x$.

Solution ➤

Notice that the variable appears on both sides of the equation.

$\sqrt{x+6} = x$	Given
$(\sqrt{x+6})^2 = x^2$	Square both sides of the equation.
$x + 6 = x^2$	Simplify.
$0 = x^2 - x - 6$	Form a quadratic equal to 0.
$0 = (x-3)(x+2)$	Factor.
$x - 3 = 0$ or $x + 2 = 0$	Set each factor equal to zero.
$x = 3$ or $x = -2$	Simplify.

The possible solutions are 3 and -2. You *must* check each possible solution by substituting it into the original equation to determine whether it is, in fact, a solution.

For $x = 3$, $\sqrt{3+6} = \sqrt{9} = 3$, so 3 is a solution.

For $x = -2$, $\sqrt{-2+6} = \sqrt{4}$. Since $\sqrt{4} \neq -2$, -2 is *not* a solution. ❖

The graphs of $y = \sqrt{x+6}$ and $y = x$ show that the equation $\sqrt{x+6} = x$ has only one solution. How do you know this? The graphs intersect at only one point.

Solving Equations by Using Radicals

To solve $x^2 = 225$, recall that if $x^2 = k$, where $k \geq 0$, then $x = \pm\sqrt{k}$. Recall that there are two solutions, because the value of x can be positive or negative.

$$x^2 = 225$$
$$x = \pm\sqrt{225}$$
$$x = \pm 15$$

EXAMPLE 4

Solve each equation for x.

A $x^2 = 150$ **B** $x^2 = 4^2 + 3^2$ **C** $x^2 = y^2 + z^2$

Solution ➤

A $x^2 = 150$

$$x = \pm\sqrt{150} = \pm\sqrt{25 \cdot 6} = \pm 5\sqrt{6}$$

Thus, $x = 5\sqrt{6}$ or $x = -5\sqrt{6}$.

Check.
$$5\sqrt{6} \cdot 5\sqrt{6} = 25 \cdot 6 = 150$$
$$(-5\sqrt{6}) \cdot (-5\sqrt{6}) = 25 \cdot 6 = 150$$

B $x^2 = 4^2 + 3^2$

$$x = \pm\sqrt{4^2 + 3^2} = \pm\sqrt{25} = \pm 5$$

C For a literal equation, follow the same procedure that you would use for numbers. Remember to include everything on the right side of the equation under one radical. Account for positive and negative values as possible solutions.

$$x^2 = y^2 + z^2$$
$$x = \pm\sqrt{y^2 + z^2}$$
$$x = \sqrt{y^2 + z^2} \text{ or } -\sqrt{y^2 + z^2}$$ ❖

Problems that have two possible answers must also be considered in context. For example, if a geometric problem relates to distance, the answer must be positive or zero. In such a case, the negative value should be disregarded.

EXAMPLE 5

The area of the circular flower garden is 23 square yards. Find the radius.

Solution ➤

Substitute 23 for A, and solve for r.
Since $\pi r^2 = 23$ and $r^2 = \frac{23}{\pi}$,

$$r = \pm\sqrt{\frac{23}{\pi}} \approx \sqrt{\frac{23}{3.14}} \approx \pm\sqrt{7.325} \approx \pm 2.706.$$

Only the positive square root can be used. The length of the radius is approximately 2.7 yards. ❖

EXAMPLE 6

Solve the equation $x^2 - 8x + 16 = 36$ by using radicals.

Solution ➤

Recall the methods from the previous two chapters. The trinomial $x^2 - 8x + 16$ is a perfect square. When factored, it becomes $(x - 4)^2$. Write the equation in this form, and solve.

Let $x = 10$.
$10^2 - 8(10) + 16 \overset{?}{=} 36;$
$36 = 36$
True

$$(x - 4)^2 = 36$$
$$x - 4 = \pm\sqrt{36}$$
$$x - 4 = \pm 6$$
$$x = 4 \pm 6$$
$$x = 10 \quad \text{or} \quad x = -2$$

Let $x = -2$.
$(-2)^2 - 8(-2) + 16 \overset{?}{=} 36;$
$36 = 36$
True

Since x can be any number, you must consider both positive and negative values. ❖

Check to see if the values actually satisfy the original equation.

Solve the equation $x^2 - 8x + 16 = 36$ by graphing $y = x^2 - 8x + 16$ and $y = 36$. How does the graph justify two solutions for x?
The graphs of $y = x^2 - 8x + 16$ and $y = 36$ intersect at $(-2, 36)$ and $(10, 36)$.

EXERCISES & PROBLEMS

Communicate

1. Explain why you must consider both positive and negative values if you take the square root of a number.
2. Describe a step-by-step method for finding the solution to an equation of the form $\sqrt{x + 7} = 114$.
3. Does $\sqrt{-x} = -3$ have a solution? Explain.
4. Explain the difference between solving $a^2 = b^2 + c^2$ for a and solving $a^2 = (b + c)^2$ for a.
5. Give a step-by-step procedure for finding the solution to the equation $x^2 + 2bx + b^2 = c^2$ for x. Then test the procedure by substituting numbers for b and c.
6. Find the explanation of the quadratic formula in Lesson 8.5, and identify how radicals are used to develop the formula.

Practice & Apply

Solve each equation algebraically. Be sure to check your solution.

7. $\sqrt{x - 5} = 2$ 9
8. $\sqrt{x + 7} = 5$ 18
9. $\sqrt{2x} = 6$ 18
10. $\sqrt{10 - x} = 3$ 1
11. $\sqrt{2x + 9} = 7$ 20
12. $\sqrt{2x - 1} = 4$ $\frac{17}{2}$, or 8.5
13. $\sqrt{x + 2} = x$ 2
14. $\sqrt{6 - x} = x$ 2
15. $\sqrt{5x - 6} = x$ 2, 3
16. $\sqrt{x - 1} = x - 7$ 10
17. $\sqrt{x + 3} = x + 1$ 1
18. $\sqrt{2x + 6} = x - 1$ 5
19. $\sqrt{x^2 + 3x - 6} = x$ 2
20. $\sqrt{x - 1} = x - 1$ 1, 2
21. $\sqrt{x^2 + 5x + 11} = x + 3$ 2

Use graphing technology or sketch the graphs to solve each equation that has a solution.

22. $\sqrt{x + 12} = x$
23. $\sqrt{x - 2} = x$
24. $\sqrt{x} = \frac{1}{3}x + \frac{2}{3}$

Physics The motion of a pendulum can be modeled by $t = 2\pi\sqrt{\frac{l}{32}}$, where l is the length of the pendulum in feet and t is the number of seconds required for one complete swing.

25. A grandfather clock is based on the motion of a pendulum. About how long should a pendulum be so that the time required for 1 complete swing is 1 second? About 0.81 feet

26. If the time required for 1 complete swing is doubled, by what number is the length multiplied? 4

27. If the time is multiplied by 3, by what number is the length multiplied? 9

28. The time required for 1 complete swing of a pendulum is multiplied by c. By what number will the corresponding length be multiplied? Generalize. c^2

Sports The formula for kinetic energy is $E = \frac{1}{2}mv^2$, where E is the kinetic energy in joules, m is the mass of the object in kilograms, and v is the velocity of a moving object in meters per second.

29. Solve the formula for v. $v = \frac{\sqrt{2Em}}{m}$

30. If a baseball is thrown with 50 joules of energy and has a mass of 0.14 kilograms, what is its velocity? About 26.73 meters per second

Solve each equation, and simplify the solution.

31. $x^2 = 90$ $3\sqrt{10}, -3\sqrt{10}$
32. $4x^2 = 7$ $\frac{\sqrt{7}}{2}, -\frac{\sqrt{7}}{2}$
33. $x^2 + 9 = 6x$ 3
34. $2x^2 = 48$ $2\sqrt{6}, -2\sqrt{6}$
35. $x^2 - 8x + 16 = 0$ 4
36. $x^2 - 12x + 36 = 0$ 6

Solve each equation if possible. If not possible, explain why.

37. $x^2 = 9$ 3, −3 **38.** $\sqrt{x} = 9$ 81 **39.** $|x| = 9$ 9, −9

40. $x^2 = -9$ **41.** $\sqrt{x} = -9$ **42.** $|x| = -9$

Physics A beam is balanced when the product of the weight and the distance from the fulcrum on one side is equal to the product of the weight and the distance from the fulcrum on the other side.

W_l W_r d_l d_r

$$W_l d_l = W_r d_r, \text{ or } \frac{W_l}{W_r} = \frac{d_r}{d_l}$$

To find the weight of the metal cube, W, first place W on the right, and balance it with weight W_l. Then switch sides, and balance W with weight W_r.

43. The ratios $\frac{W_l}{W} = \frac{W}{W_r}$ are equal. So $W_l W_r = W^2$. Express the value of W in terms of W_l and W_r. $W = \sqrt{W_l W_r}$

44. A **geometric mean**, g, of two factors, f_1 and f_2, is the square root of the product of the two factors: $g = \sqrt{f_1 \cdot f_2}$. Is W the geometric mean of W_l and W_r? Explain. Yes, because $W = \sqrt{W_l W_r}$.

45. If W_l is 9 pounds and W_r is 16 pounds, what is the weight of W? 12

Look Back

46. Solve $-2x + 3 < 11$. **[Lesson 1.5]** $x > -4$

47. Find the inverse of the matrix $A = \begin{bmatrix} 6 & -1 \\ 2 & -7 \end{bmatrix}$. **[Lesson 3.4]**

47. $\begin{bmatrix} \frac{7}{40} & -\frac{1}{40} \\ \frac{1}{20} & -\frac{3}{20} \end{bmatrix}$ or $\begin{bmatrix} 0.175 & -0.025 \\ 0.05 & -0.15 \end{bmatrix}$

Graph each pair of functions on the same coordinate plane. How are the graphs alike and how are they different? [Lesson 5.6]

48. $y = x + 3$
$y = (x - 2) + 3$

49. $y = x^2 + 3$
$y = (x - 2)^2 + 3$

50. $y = |x| + 3$
$y = |x - 2| + 3$

51. Factor $4x^4 - 16y^4$ completely. **[Lesson 7.6]** $4(x^2 - 2y^2)(x^2 + 2y^2)$

52. What conditions of the discriminant indicate that a quadratic has no real solution? Give an example. **[Lesson 8.5]**
Less than zero; $y = x^2 + 2x + 2$; discriminant $= -4$

Solve and graph each inequality. [Lesson 8.6]

53. $x^2 - x - 12 > 0$ **54.** $y^2 + 5y + 6 < 0$ **55.** $a^2 - 6a - 16 < 0$

Look Beyond

56. Solve $x^2 + y^2 = 4$ for y, and graph the two functions on the same coordinate plane to form a circle. What is the center of the circle?

LESSON 9.4 The "Pythagorean" Right-Triangle Theorem

Why *The "Pythagorean" Right-Triangle theorem is one of the most familiar and significant theorems in mathematics. It makes extensive use of squares and square roots. You can use this theorem to find the length of one side of a right triangle if you know the lengths of the other two sides.*

Copy and complete the table for the given sides of a right triangle.

Side a	Side b	Side c	a^2	b^2	c^2	$a^2 + b^2$
3	4	5				
5	12	13				
16	30	34				
13	84	85				

Compare the last two columns of the table. What do you notice about the relationship between c^2 and $a^2 + b^2$?

Cultural Connection: Asia The relationships of the sides of right triangles have been studied in many parts of the world throughout history. The ancient Chinese, Babylonians, and Hindus had all discovered the right-triangle relationship. Throughout history, many people have discovered clever proofs of the relationship.

European culture provided the name we currently associate with this right-triangle relationship. It is usually referred to as the "Pythagorean" Right-Triangle Theorem, although Babylonians knew of the relationship over 1000 years earlier than Pythagoras's discovery.

"PYTHAGOREAN" RIGHT-TRIANGLE THEOREM

Given a right triangle with **legs** of length a and b and **hypotenuse** of length c, then $c^2 = a^2 + b^2$ or $a^2 + b^2 = c^2$.

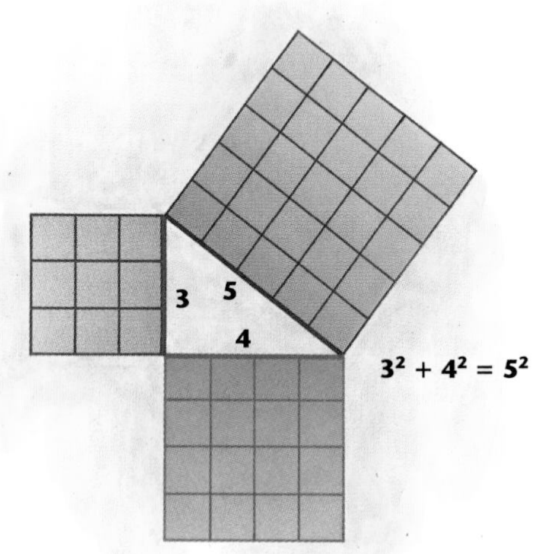

Over the years this illustration has been used to show that the sum of the squares of the legs of a right triangle is equal to the square of the hypotenuse. Notice that the combined areas of the red square and the blue square is equal to the area of the purple square.

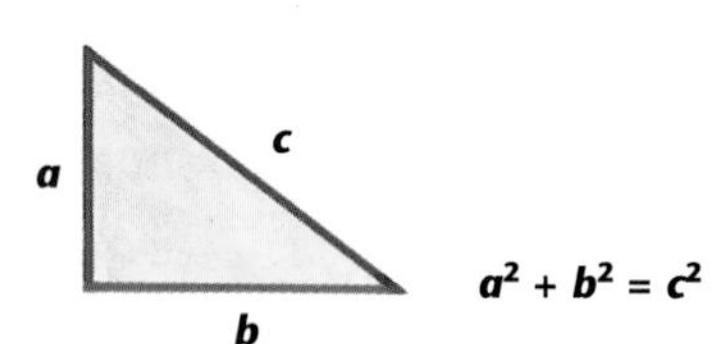

EXAMPLE 1

Find the length of x, y, and z.

A

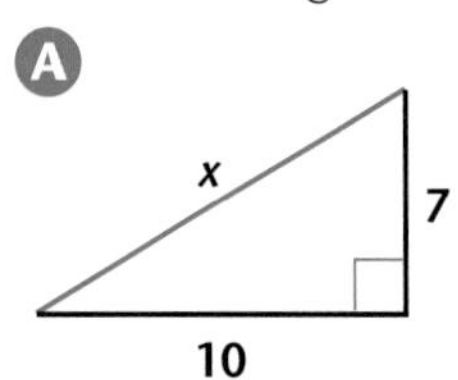

B

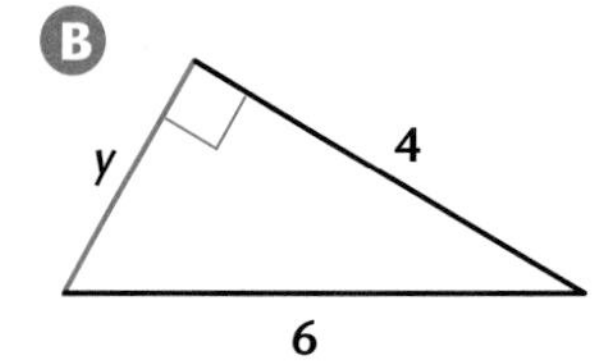

C

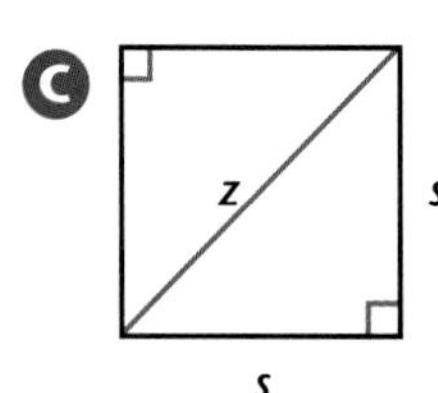

Solution ➤

A Since x is opposite the right angle, x represents the length of the hypotenuse.

$$x^2 = 7^2 + 10^2$$
$$x^2 = 149$$
$$x = \pm\sqrt{149} \approx \pm 12.21$$

Since x represents the length of a segment, it must be positive. Therefore, $x = \sqrt{149} \approx 12.21$.

B The length of the hypotenuse is 6.

$$y^2 + 4^2 = 6^2$$
$$y^2 + 16 = 36$$
$$y^2 = 20$$
$$y = \pm\sqrt{20} \text{ or } \pm 2\sqrt{5}$$
$$y \approx 4.47 \qquad \text{The negative root is not a solution.}$$

C The diagonal, z, of a square is the hypotenuse. Each leg is s.

$$z^2 = s^2 + s^2$$
$$z^2 = 2s^2$$
$$z = \sqrt{2s^2}$$
$$z = s\sqrt{2} \qquad \text{Only positive numbers represent a length.}$$

Thus, the formula for the diagonal of a square with sides s is $s\sqrt{2}$. ❖

EXAMPLE 2

Travel Your boat is traveling due north at 20 miles per hour. Your friend's boat left at the same time from the same location headed due west at 15 miles per hour. After an hour you get a call from your friend that he has engine trouble. How far must you travel to reach your friend?

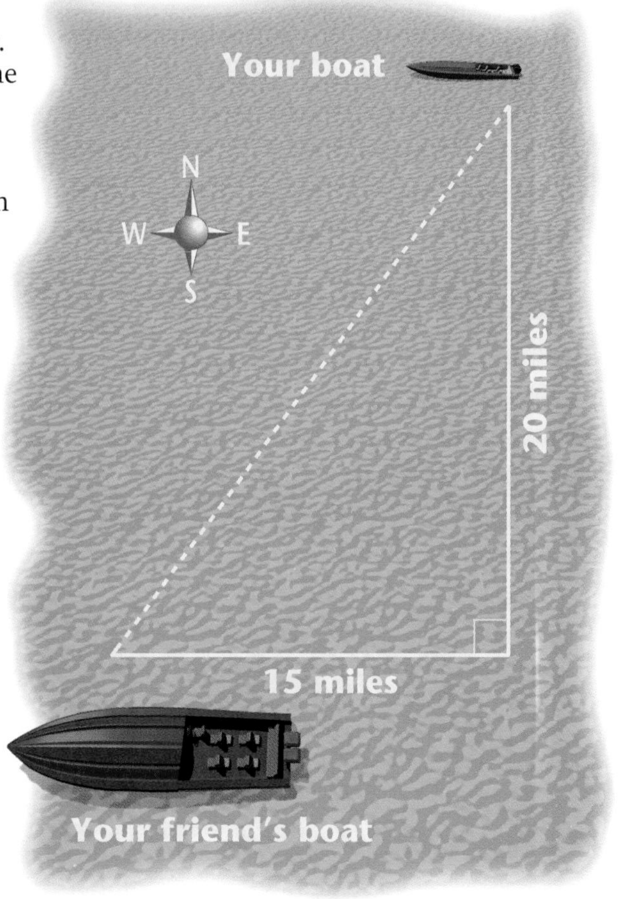

Solution ➤

The legs will be 20 miles and 15 miles. The distance you must travel is the length of the hypotenuse. Use the "Pythagorean" Right-Triangle Theorem to find the distance. Add, and use the definition of square root to solve.

$$c^2 = a^2 + b^2$$
$$c^2 = 20^2 + 15^2$$
$$c^2 = 625$$

Since distance is involved, only positive values are used.

$$c = \sqrt{625} \rightarrow c = 25 \text{ miles}$$ ❖

Try This Find the hypotenuse of a right triangle if the legs are 36 and 48. 60

What do you notice about multiples of the sides of a 3-by-4-by-5 right triangle? If you do not know, make a table of multiples of the legs and see if you can guess the hypotenuse. Then check your results.

Use algebra to solve the equation $a^2 + b^2 = c^2$ for a or b. Then you can use the new equation to determine the unknown side of the right triangle. All you need to know is the hypotenuse and the other side.

EXAMPLE 3

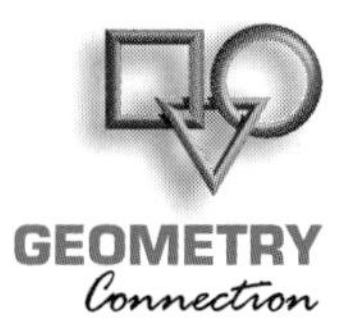

The height of a pyramid with a square base can be determined from measurements of the base and the slant height. The measurement of the base is 40 meters, and the slant height is 52 meters. How high is the pyramid?

Solution ➤

Use half the base of the pyramid, 20 meters, as one side of the triangle. Use the slant height, 52 meters, as the hypotenuse.

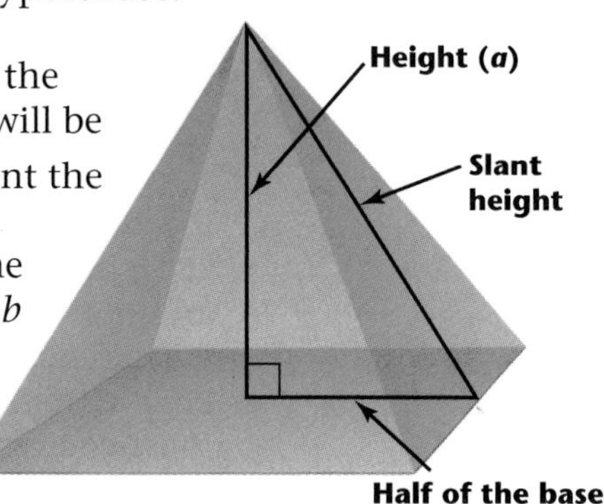

If you solve for side a, the formula from the "Pythagorean" Right-Triangle Theorem will be $a = \pm\sqrt{c^2 - b^2}$. Let the height, a, represent the leg whose length we do not know. Since distance is positive, you can disregard the negative value. Substitute the values for b and c to find the height of the pyramid.

$$a = \sqrt{52^2 - 20^2}$$
$$a = \sqrt{2304} = 48$$

The height of the pyramid is 48 meters. ❖

An important theorem in geometry states that for all positive numbers a and b, $\sqrt{a^2 + b^2}$ is less than $a + b$. Use this theorem to explain why the path along the hypotenuse is always shorter than the path along the two legs.

The path along the legs has a distance of $a + b$, and by the "Pythagorean" Right-Triangle Theorem, the path along the hypotenuse has a distance of $\sqrt{a^2 + b^2}$. The path along the hypotenuse is always shorter than the path along the two legs.

Communicate

1. Name the sides of a right triangle.
2. State the "Pythagorean" Right-Triangle Theorem in your own words.
3. Explain how to find the hypotenuse of a right triangle if you know the lengths of the legs.
4. Describe how to change the basic equation of the "Pythagorean" Right-Triangle Theorem to find the length of one of the legs.
5. How would you show someone that the "Pythagorean" Right-Triangle Theorem is true?
6. Explain how to use the "Pythagorean" Right-Triangle Theorem to find the length of a diagonal of a rectangle.

Practice & Apply

Technology Copy and complete the table. Use a calculator and round the answer to the nearest tenth.

	Leg	Leg	Hypotenuse		Leg	Leg	Hypotenuse
7.	24	45	? 51	**8.**	10	?	26 24
9.	?	8	17 15	**10.**	5	9	? 10.3
11.	12	?	25 21.9	**12.**	?	6	25 24.3
13.	30	40	? 50	**14.**	0.75	?	1.25 1

Solve for *x*.

15. 10

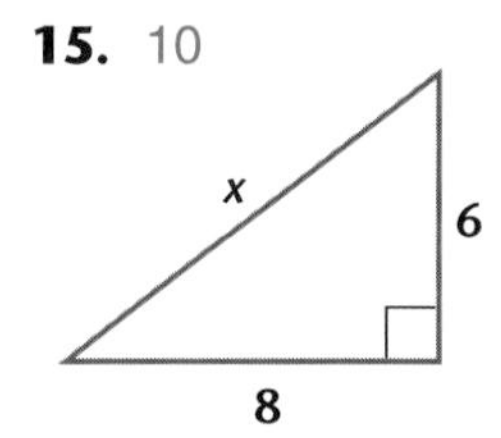

16.

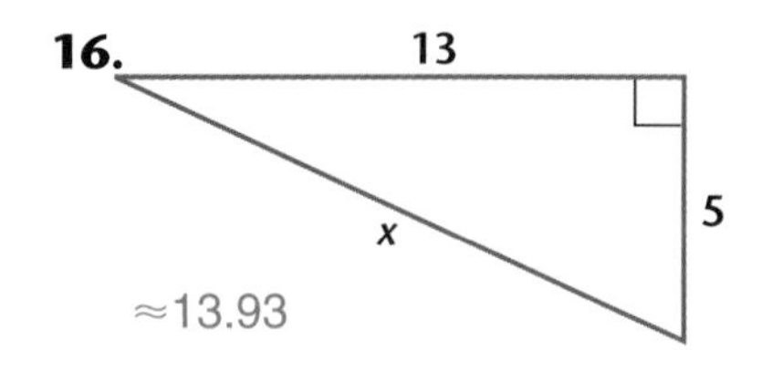

≈13.93

17. ≈2.45

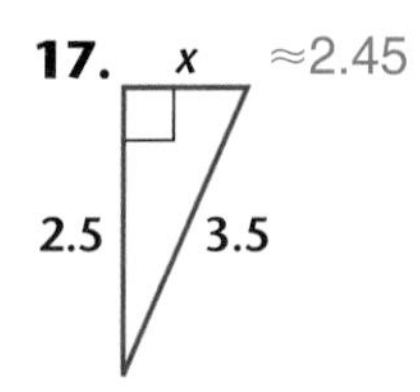

Landscaping A diagram for a garden in the shape of a right triangle is shown at the right.

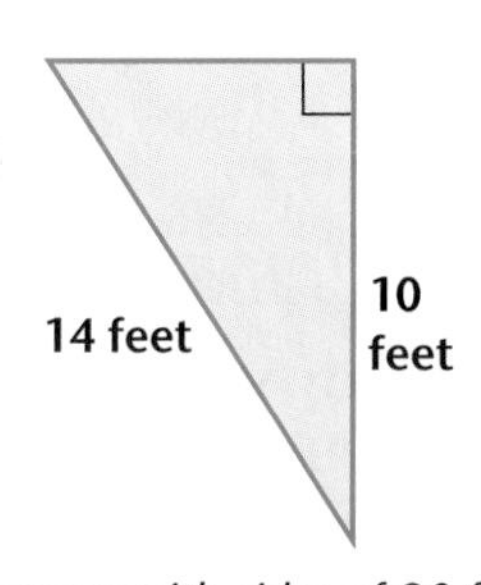

18. How many feet of fencing must be bought to enclose the garden? Assume that fencing is sold by the foot. 34 feet

19. If the cost of fencing is $4.98 per foot, how much will it cost to enclose the garden? $169.32

20. **Sports** To the nearest foot, how long is a throw from third base to first base? About 127 feet

21. Find the diagonal of a square that is 5 inches on each side. About 7.07 in.

22. A hiker leaves camp and walks 5 miles east. He then walks 10 miles south. How far from camp is the hiker? About 11.18 miles

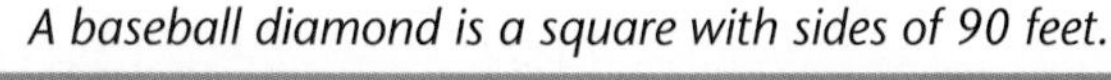
A baseball diamond is a square with sides of 90 feet.

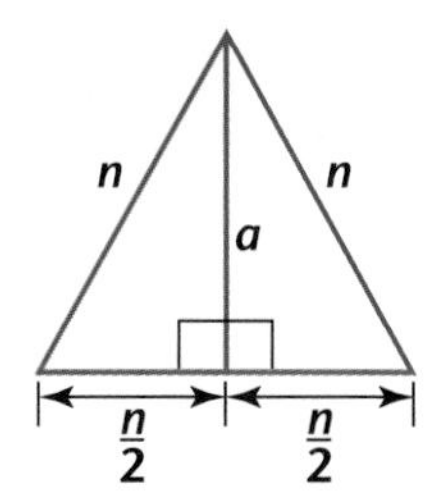

Geometry An equilateral triangle has an axis of symmetry from each vertex that is perpendicular to the opposite side at its midpoint. The axis of symmetry divides the equilateral triangle into two right triangles that have the same size and shape. A representative equilateral triangle is shown where a is the altitude of the triangle.

23. Copy and complete the table. Simplify each answer.

Equilateral Triangles

Side	4	6	8	10	12	20	n	?
Half of a side	2	?	?	?	?	?	?	?
Altitude	?	?	?	?	?	?	?	$17\sqrt{3}$

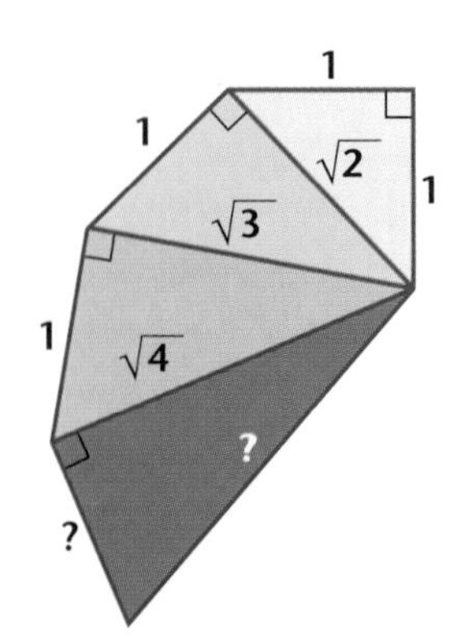

Cultural Connection: Europe The given figure is known as the wheel of Theodorus, a Greek mathematician who lived in the mid-fifth century B.C.E.

24. Continue the spiral by extending the length of the hypotenuse of each new right triangle. Check the numbers with the "Pythagorean"Right-Triangle Theorem. Make a table and extend the number to 9 right triangles.

Right triangle number	1	2	3	4	5	6	7	8	9
Hypotenuse	$\sqrt{2}$	?	?	?	?	?	?	?	?

$\sqrt{3}$; $\sqrt{4}$ or 2; $\sqrt{5}$; $\sqrt{6}$; $\sqrt{7}$; $\sqrt{8}$ or $2\sqrt{2}$; $\sqrt{9}$ or 3; $\sqrt{10}$

25. Predict the length of the hypotenuse for the 10th right triangle. $\sqrt{11}$

26. Describe the length of the hypotenuse for the *n*th right triangle in terms of *n*. $\sqrt{n+1}$

27. The hypotenuse of a right triangle is 8 centimeters. Find the lengths of the two legs if one leg is 1 centimeter longer than the other. Give both an exact and an approximate answer rounded to the nearest hundredth.
$\frac{-2 \pm \sqrt{508}}{4}$, or $\frac{-1 \pm \sqrt{127}}{2}$; ≈ 5.13, ≈ -6.13

Look Back

Darnell is paid a weekly salary of $20 plus $0.50 for every T-shirt that he sells. Copy and complete the table. [Lesson 1.2]

T-shirts sold	0	1	2	3	4
Total pay	**28.** ? 20.00	**29.** ? 20.50	**30.** ? 21.00	**31.** ? 21.50	**32.** ? 22.00

33. Write an equation that describes his weekly salary, *w*, in terms of the number of T-shirts sold, *t*. $w = 20.00 + 0.50t$

34. If Darnell sells 29 T-shirts in a given week, what will his salary be? $34.50

35. If Darnell needs to earn $49 in a given week in order to buy a particular sweater, how many T-shirts must he sell? 58

Decide whether the lines are parallel, perpendicular, or neither. [Lessons 2.1, 2.2]

36. $y = \frac{1}{2}x + 3$
$y = 4x + 3$
neither

37. $y = 3x - 4$
$y = -\frac{1}{3}x + 2$
perpendicular

38. $-2x + y = 8$
$-6x + 3y = 15$
parallel

Look Beyond

Plot each pair of points on graph paper, and find the distance between them.

39. (3, 2), (10, 2) **40.** (−7, −1), (5, −1) **41.** (4, 1), (4, 8) **42.** (−6, −4), (−6, −1)

LESSON 9.5

Exploring the Distance Formula

Why *The ability to find the distance between two points is of critical importance to sailors, surveyors, mathematicians, architects, astronomers, and draftsmen. Countless others, taxi drivers, carpenters, and emergency medical technicians, must possess the ability to measure indirect distance.*

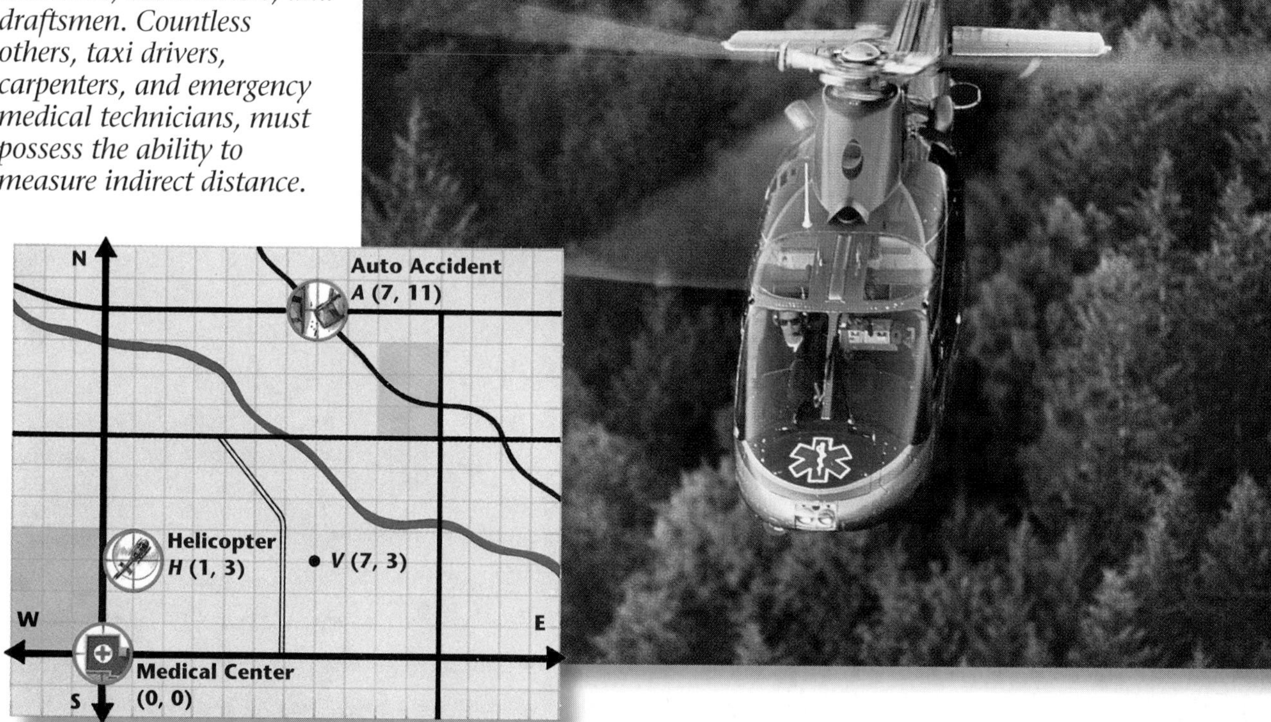

Rescue Service A medical center learns of an auto accident located 7 miles east and 11 miles north of its location. If the medical center's helicopter is now located 1 mile east and 3 miles north of the medical center, what distance must the helicopter travel to get to the scene of the accident?

Visualize a segment drawn from H to A. This segment represents the distance the helicopter must travel. Now visualize a line through A perpendicular to the west-east axis and a line through H, perpendicular to the north-south axis. The intersection of these two lines meets at point V (7, 3). The 3 lines form the right triangle AVH with the right angle at V. How can you find the length of HA?

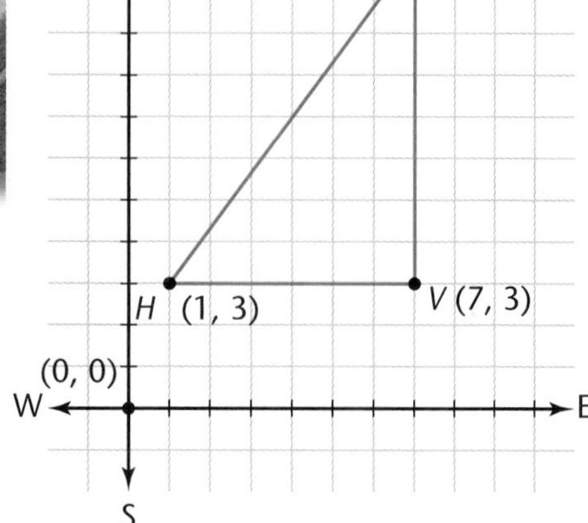

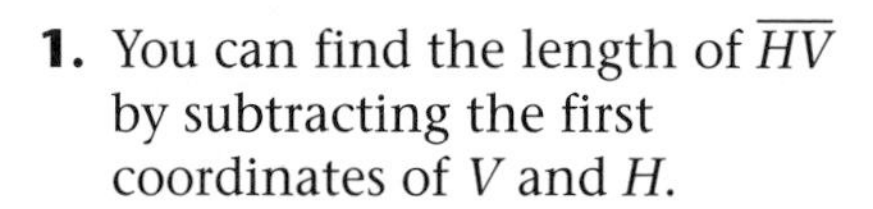

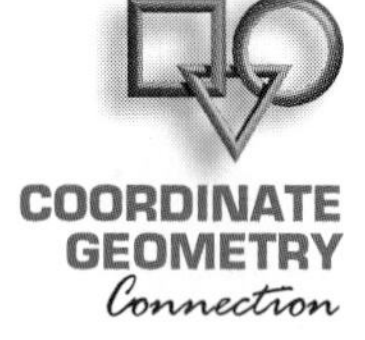

1. You can find the length of $\overline{HV}$ by subtracting the first coordinates of V and H.

   length of $\overline{HV} = 7 - 1 = 6$

2. You can find the length of $\overline{AV}$ by subtracting the second coordinates of A and V.

   length of $\overline{AV} = 11 - 3 = 8$

3. Now use the "Pythagorean" Right-Triangle Theorem.

$$c^2 = a^2 + b^2$$

$$(\text{length of } \overline{AH})^2 = 6^2 + 8^2 = 100$$

$$\text{length of } \overline{AH} = \sqrt{100} = 10$$

The medical center's helicopter must travel 10 miles.

Exploration 1 Finding the Distance Formula

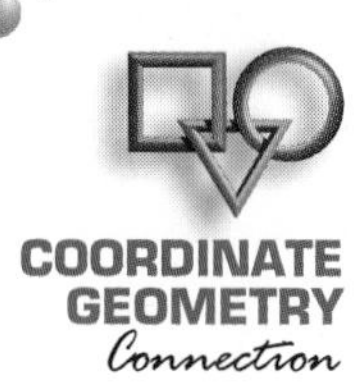

Use graph paper and the method illustrated in the medical center problem to find the distance between each pair of given points.

a. $K(11, 2)$, $L(14, 6)$ 5 **b.** $M(2, 3)$, $N(7, 15)$ 13

c. $R(-4, 3)$, $S(5, 8)$ ≈10.3 **d.** $T(-2, 3)$, $U(-6, -2)$ ≈6.4

The **distance** between two points A and B is the **length** of $\overline{AB}$. This is the segment with A and B as endpoints.

Follow the same procedure as before to find the distance from A to B, but use letters for the coordinates of the points, $A(x_1, y_1)$ and $B(x_2, y_2)$, instead of numbers. Compare your results with the distance formula. ❖ The results are the same.

DISTANCE FORMULA

The distance, d, between two points $A(x_1, y_1)$ and $B(x_2, y_2)$ is given by

$$d = \sqrt{(x_2 - x_1)^2 + (y_2 - y_1)^2}.$$

The Converse of the Theorem

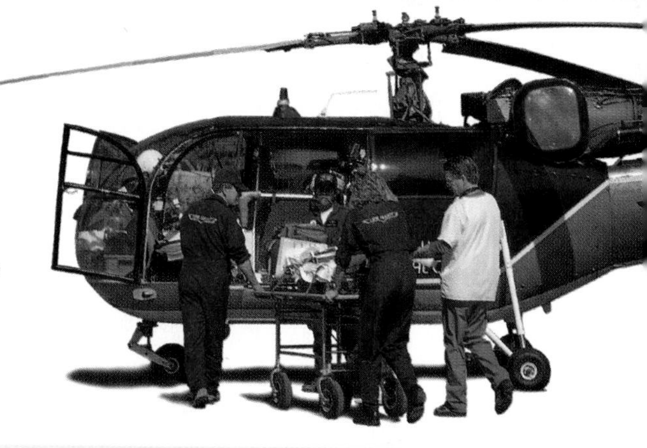

When the *if* and *then* portions of a theorem are interchanged, the new statement is called a **converse**.

CT– No. Counterexamples will vary. For instance, "If you are in Chicago, then you are in Illinois," is true. The converse "If you are in Illinois, then you are in Chicago," is not true.

CONVERSE OF THE "PYTHAGOREAN" RIGHT-TRIANGLE THEOREM

If a, b, and c are the lengths of the sides of a triangle with $a^2 + b^2 = c^2$, then the triangle is a right triangle with hypotenuse c.

Is the *converse* of an if-then statement always true? If not, give an example.

EXAMPLE 1

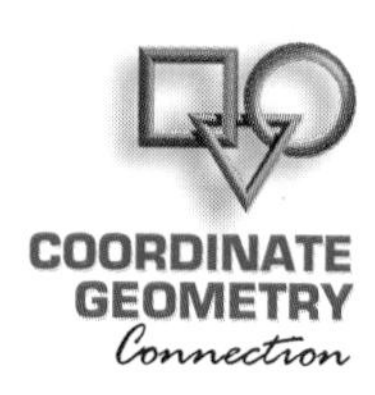

Given points $P(1, 2)$, $Q(3, -1)$, and $R(-5, -2)$, determine whether triangle PQR is a right triangle.

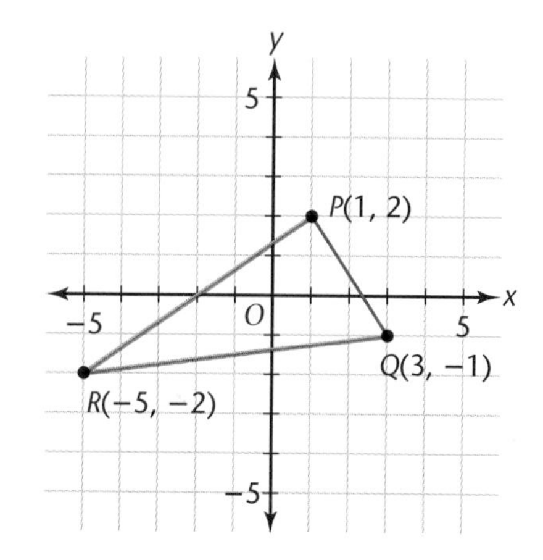

Solution ➤

Draw triangle PQR on graph paper.

Use the distance formula to find the lengths of each side.

$$\text{Length of } \overline{PR} = \sqrt{(-5-1)^2 + (-2-2)^2} = \sqrt{(-6)^2 + (-4)^2} = \sqrt{52}$$

$$\text{Length of } \overline{PQ} = \sqrt{(3-1)^2 + (-1-2)^2} = \sqrt{2^2 + (-3)^2} = \sqrt{13}$$

$$\text{Length of } \overline{RQ} = \sqrt{[3-(-5)]^2 + [-1-(-2)]^2} = \sqrt{8^2 + 1^2} = \sqrt{65}$$

CT– If the slopes of the two legs are negative reciprocals of each other, then the lines are perpendicular. In other words, if $m_1 \cdot m_2 = -1$, then the lines are perpendicular.

Now use the converse of the "Pythagorean" Right-Triangle Theorem to test whether the three lengths can be the sides of a right triangle.

Does $(\sqrt{52})^2 + (\sqrt{13})^2 = (\sqrt{65})^2$?

Since $52 + 13 = 65$, the triangle is a right triangle. ❖

Try This The sides of a triangle are 60, 90, and 109 centimeters. Is it a right triangle? Is a triangle with sides of 54, 71, 90 a right triangle? No. They are close to right triangles, but they are not right triangles.

How can you use the idea of slope to show that the triangle PQR in Example 1 is a right triangle?

The Midpoint Formula

Exploration 2 Finding the Midpoint Formula

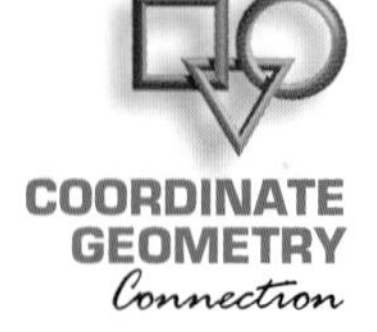

1. Locate the points on graph paper. Guess the midpoint of $\overline{PQ}$.
   a. $P(1, 6)$, $Q(7, 10)$ **b.** $P(-2, 3)$, $Q(6, 7)$ **c.** $P(-4, -1)$, $Q(8, 7)$
2. Use the distance formula to verify that each of your proposed midpoints is the midpoint of the given segment.
3. What rule can be applied to the x-coordinates of the endpoints of a segment to obtain the x-coordinate of the midpoint of the segment?
4. What rule can be applied to the y-coordinates of the endpoints of a segment to obtain the y-coordinate of the midpoint of the segment?
5. Apply your rule from Steps 3 and 4 to find the coordinates of the midpoint of $\overline{AB}$, for $A(12, 63)$ and $B(43, 20)$.
6. Use the distance formula to verify your answer. ❖

MIDPOINT FORMULA

Given: $\overline{PQ}$ with $P(x_1, y_1)$ and $Q(x_2, y_2)$

The coordinates of the midpoint, M, of $\overline{PQ}$, are $\left(\frac{x_1 + x_2}{2}, \frac{y_1 + y_2}{2}\right)$.

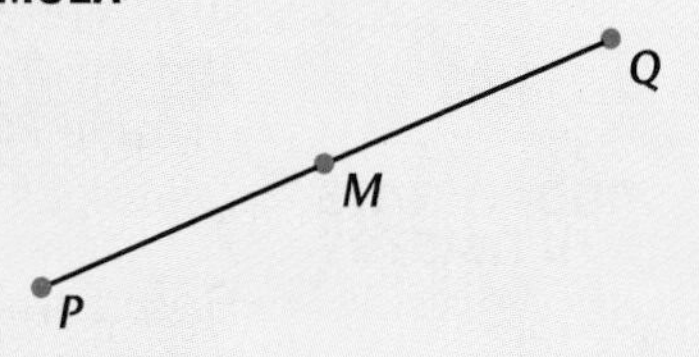

EXAMPLE 2

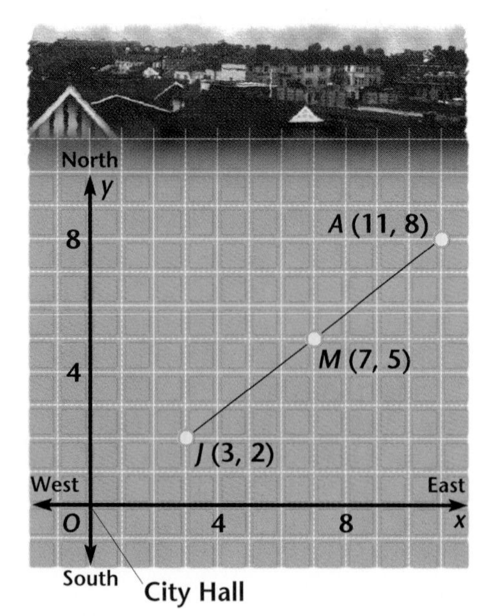

The streets of a city are laid out like a coordinate plane with the origin at City Hall. Jacques lives 3 blocks east and 2 blocks north of City Hall. His friend Alise lives 11 blocks east and 8 blocks north of City Hall. They want to meet at the point midway between their two locations. At what point should they meet?

Solution ➤

Plot $J(3, 2)$ and $A(11, 8)$ on a coordinate plane as shown in the figure. Find the midpoint, $M(x, y)$, of $\overline{JA}$ by using the midpoint formula.

$$x = \frac{x_1 + x_2}{2} = \frac{3 + 11}{2} = \frac{14}{2} = 7$$

$$y = \frac{y_1 + y_2}{2} = \frac{2 + 8}{2} = \frac{10}{2} = 5$$

The midpoint $M(7, 5)$ is where they should meet. ❖

EXAMPLE 3

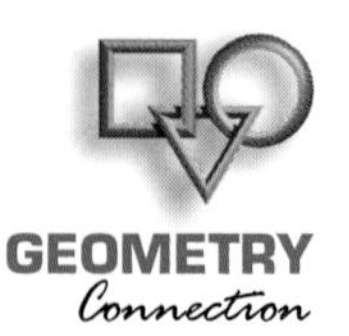

The midpoint of a diameter of a circle is $M(3, 4)$. If one endpoint of a diameter is $A(-3, 6)$, what are the coordinates of the other endpoint, $B(x_2, y_2)$?

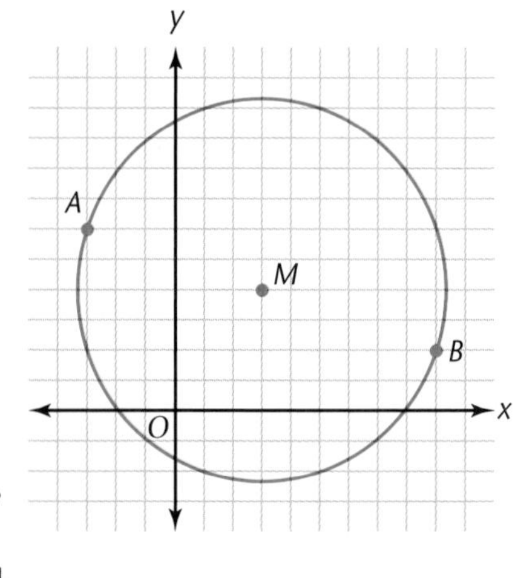

Solution ➤

Substitute the information into the midpoint formula, and solve.

$$\frac{x_1 + x_2}{2} = x \qquad \frac{y_1 + y_2}{2} = y$$

$$\frac{-3 + x_2}{2} = 3 \qquad \frac{6 + y_2}{2} = 4$$

$$2\left(\frac{-3 + x_2}{2}\right) = 2(3) \qquad 2\left(\frac{6 + y_2}{2}\right) = 2(4)$$

$$-3 + x_2 = 6 \qquad 6 + y_2 = 8$$

$$x_2 = 9 \qquad y_2 = 2$$

The other endpoint is $B(9, 2)$. ❖

EXERCISES & PROBLEMS

Communicate

1. What is the difference between a point and a coordinate?
2. What is the relationship between the "Pythagorean" Right-Triangle Theorem and the distance formula?
3. Explain how to find the distance between points on a coordinate plane.
4. Describe a way to find the midpoint of a segment on a coordinate plane.
5. Explain the *converse* of an if-then statement.

Practice & Apply

For each figure, find the coordinates of *Q* and the length of the hypotenuse of each triangle.

6.

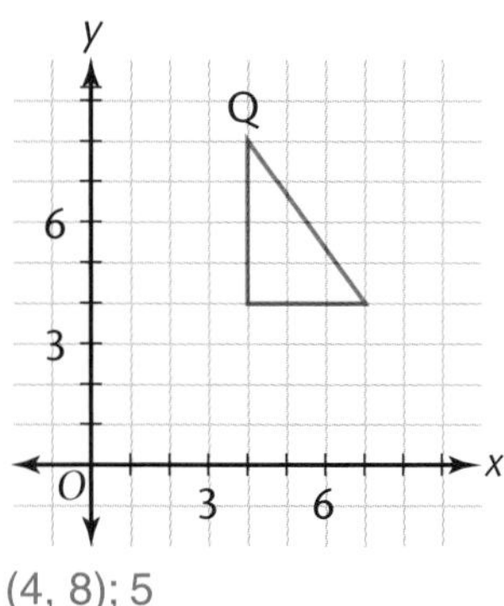

(4, 8); 5

7.

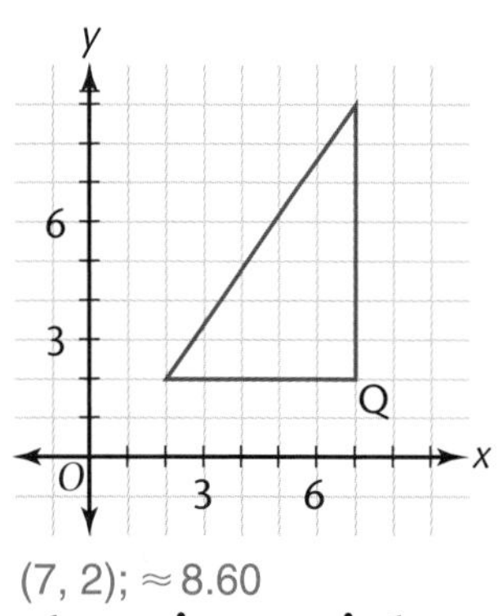

(7, 2); ≈ 8.60

8.

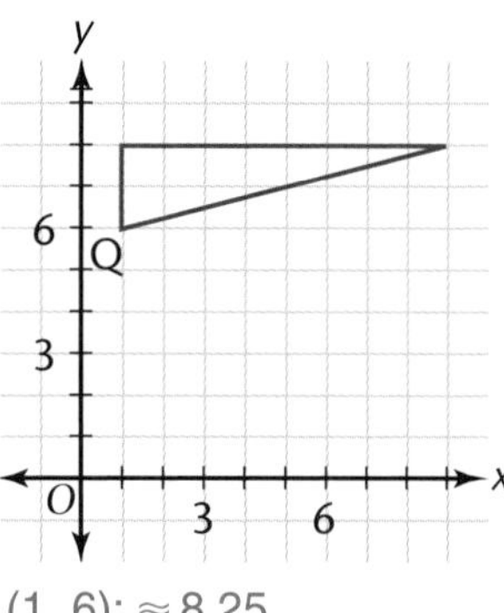

(1, 6); ≈ 8.25

Find the distance between the two given points.

9. $A(4, 7)$, $B(1, 3)$ 5

10. $P(5, 6)$, $Q(17, 11)$ 13

11. $R(-5, -2)$, $S(-9, 3)$ ≈ 6.40

Plot △*PQR* on graph paper. Decide which description(s) apply to △*PQR*: scalene (no sides equal), isosceles (2 sides equal), equilateral (3 sides equal), or right triangle. Justify your answer.

12. $P(-1, 3)$, $Q(4, 6)$, $R(4, 0)$

13. $P(-2, 2)$, $Q(2, 5)$, $R(8, -3)$

14. Rescue Service A Coast Guard communications center is located at (0, 0). A rescue helicopter located 3 miles west and 2 miles north of the center must respond to a ship located 5 miles east and 7 miles south of the center. What distance must the helicopter travel to reach the ship? About 12.04 mi

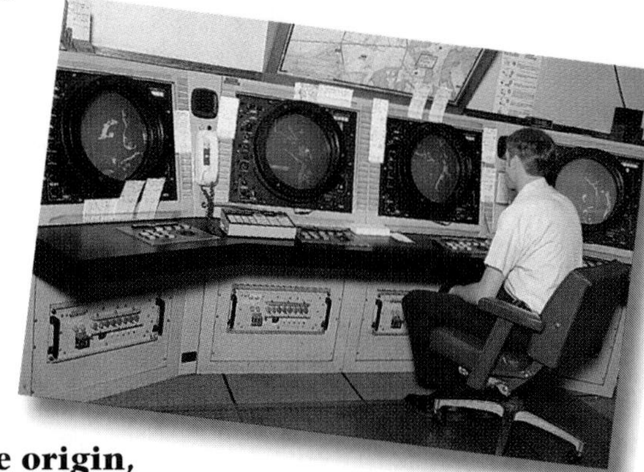

Plot each segment in a coordinate plane, and find the coordinates of the midpoint of the segment.

15. $\overline{AB}$, with $A(2, 5)$ and $B(9, 5)$

16. $\overline{EF}$, with $E(-4, -1)$ and $F(-2, -7)$

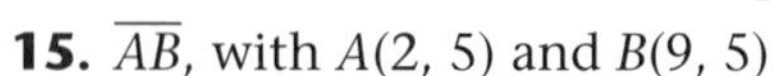

17. $\overline{GH}$, with $G(-5, 6)$ and $H(-1, -2)$

18. $\overline{IJ}$, with $I(4, 3)$ and $J(9, -4)$

The locations of two points are given relative to the origin, (0, 0). Find the coordinates of the midpoint of the segment connecting the two points.

19. *A* is 2 units east and 3 units north of 0. *B* is 4 units west and 5 units north of 0. (− 1, 4)

20. *C* is 2 units west and 5 units south of 0. *D* is 6 units west and 1 unit south of 0. (− 4, − 3)

21. Use the method from Example 2 on page 444 to find the coordinates of the midpoint of $\overline{PQ}$, with $P(10, 7)$ and $Q(2, 3)$. (6, 5)

22. If the coordinates of *P* are (a, b) and the coordinates of *Q* are (c, d), what are the coordinates of the midpoint of $\overline{PQ}$? $\left(\frac{a+c}{2}, \frac{b+d}{2}\right)$

The midpoint of $\overline{PQ}$ is *M*. Find the missing coordinates.

23. $P(2, -5)$, $Q(6, 7)$, $M(\underline{?}, \underline{?})$ (4, 1)

24. $P(4, 8)$, $Q(\underline{?}, \underline{?})$, $M(10, 7)$ (16, 6)

25. $P(\underline{?}, \underline{?})$, $Q(6, -2)$, $M(9, 4)$ (12, 10)

26. $P(3, \underline{?})$, $Q(\underline{?}, 5)$, $M(2, 8)$ (3, 11); (1, 5)

Complete Exercises 27–32.

27. Plot $\overline{RS}$, with $R(2, 1)$ and $S(6, 3)$.

28. Find the coordinates of M, the midpoint of $\overline{RS}$, and plot M. $M(4, 2)$

29. Find the slope of $\overline{RS}$. $\frac{1}{2}$

30. What is the slope of any line perpendicular to $\overline{RS}$? -2

31. Find an equation for k, the line that is perpendicular to $\overline{RS}$ and contains M. Draw k on your graph. The line k is called the *perpendicular bisector* of $\overline{RS}$ because it is perpendicular to $\overline{RS}$ and contains the midpoint of $\overline{RS}$.

32. Choose any two points on k other than M. Name them A and B. Compute the distance from A to R and the distance from A to S. Compute the distance from B to R and the distance from B to S.

33. Use right triangles and the "Pythagorean" Right-Triangle Theorem to show that (4, 7) is the midpoint between (1, 3) and (7, 11).

34. **Portfolio Activity** Complete the problem in the portfolio activity on page 411. Round your answers to the nearest hundredth.

Look Back

Factor each trinomial. [Lesson 7.6]

35. $y^2 + 35y + 300$
$(y + 20)(y + 15)$

36. $x^2 + 30x + 216$
$(x + 12)(x + 18)$

Technology Find the vertex of each parabola. Use graphing technology to assist you. **[Lesson 8.1]**

37. $y = (x - 2)^2 + 3$ $(2, 3)$

38. $y = (x + 3)^2 - 1$ $(-3, -1)$

39. $y = -2(x - 1)^2 - 4$ $(1, -4)$

40. $y = -(x + 4)^2 + 1$ $(-4, 1)$

41. Use an algebraic technique to find the x-intercept(s) of the graph of the function $y = x^2 - 6x - 10$. Give an exact solution(s). **[Lesson 8.4]**
$3 \pm \sqrt{19}$

Use properties of radicals to simplify each expression. Use a calculator to check your answers. [Lesson 9.2]

42. $\sqrt{3}(\sqrt{12} - \sqrt{75})$
-9

43. $\frac{\sqrt{75}}{\sqrt{3}}$
5

44. $\frac{\sqrt{36} + \sqrt{81}}{\sqrt{9}}$
5

Look Beyond

Technology Use a calculator to evaluate each expression.
NOTE: To evaluate $9^{\frac{1}{2}}$, use the keystrokes of 9 [y^x] [(] 1 [÷] 2 [)] or 9 [^] [(] 1 [÷] 2 [)].

45. $9^{\frac{1}{2}}$ 3

46. $16^{\frac{1}{2}}$ 4

47. $39^{\frac{1}{2}}$ ≈ 6.24

48. $2^{\frac{1}{2}}$ ≈ 1.41

49. What is calculated if you do not use parentheses around the exponent $\frac{1}{2}$?

50. Make a conjecture about the meaning of the exponent $\frac{1}{2}$.

LESSON 9.6 Exploring Geometric Properties

Why *In coordinate geometry, algebra is used to find relationships between the line segments that form geometric figures. The distance formula and the midpoint formula are useful in connecting geometry with algebra. This connection can help you solve problems.*

The sandstone blocks of Stonehenge, an ancient monument in England, once formed a circle with a 160-foot radius.

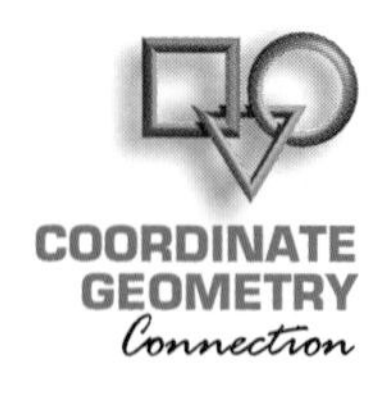

A *circle* is the set of all points in a plane that are the same distance from a given point called the *center*. You can use the distance formula to find the equation of the circle with its *center* at (0, 0) and a *radius* of 160.

If (x, y) is on the circle, then its distance from (0, 0) must be 160.

$$d = \sqrt{(x-0)^2 + (y-0)^2} = 160$$
$$\sqrt{x^2 + y^2} = 160$$

Square both sides and simplify.

$$x^2 + y^2 = 160^2 \text{ or}$$
$$x^2 + y^2 = 25{,}600$$

The equation $x^2 + y^2 = 160^2$ represents all possible points that form the circle.

Exploration 1 The Equation of a Circle

1. The first circle is given to the student. The second circle is shown on the same axes at the right.

1. Construct a circle with center (0, 0) and radius 3. Then repeat with a radius of 1.
2. Find the equation of each circle.
3. What is the equation of a circle with center (0, 0) and radius r?
4. What is the equation of a circle with center (1, 1) and radius 1? ❖

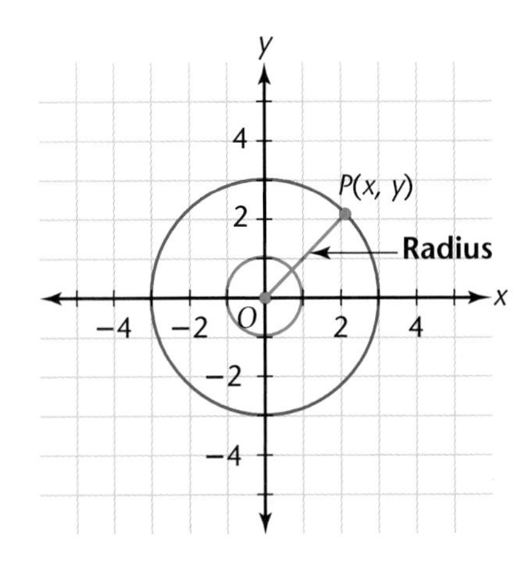

2. $x^2 + y^2 = 9$; $x^2 + y^2 = 1$
3. $x^2 + y^2 = r^2$
4. $(x - 1)^2 + (y - 1)^2 = 1$

Exploration 2 The Diagonals of a Rectangle

1. Use the distance formula to find the length of each diagonal of rectangle $ABCO$. 5, 5

2. Answers may vary.

2. Construct two other rectangles, and find the lengths of the diagonals for each rectangle.

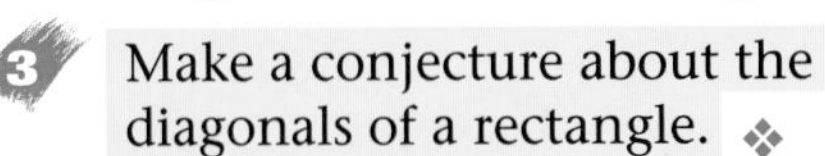

3. Make a conjecture about the diagonals of a rectangle. ❖

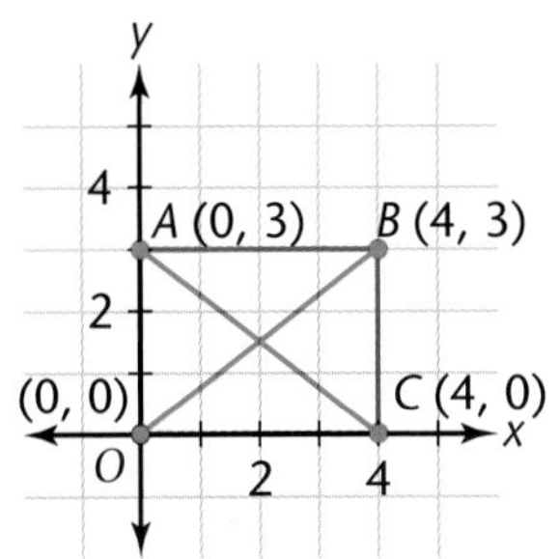

The diagonals of a rectangle are equal in length.

When you change from numbers to letters, the conjecture becomes more general. It can then represent whatever numbers are specified for the letters.

Exploration 3 Finding the Formula

1. length of a diagonal = $\sqrt{a^2 + b^2}$

1. Use 0, a, and b from the diagram to write a formula for finding the length of a diagonal of a rectangle.

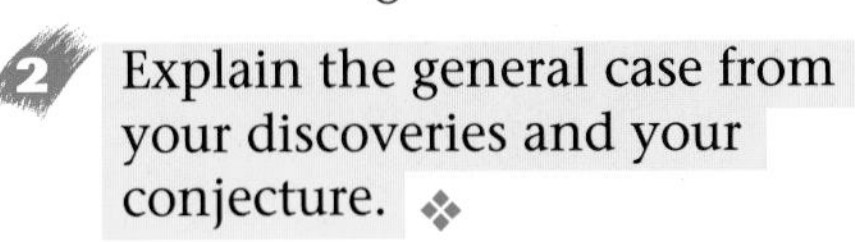

2. Explain the general case from your discoveries and your conjecture. ❖

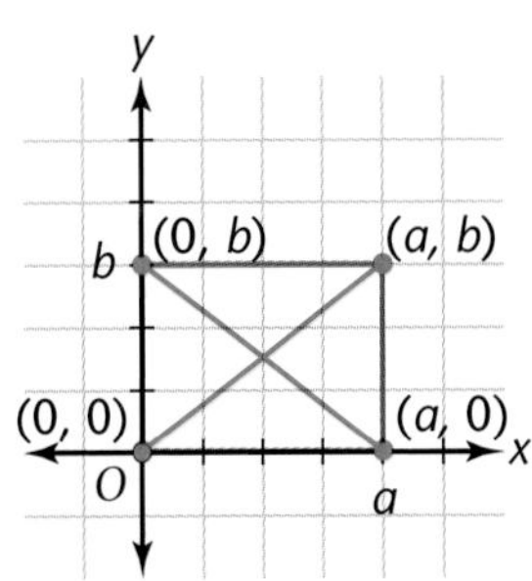

The midpoint formula is also helpful for showing geometric relationships algebraically.

Exploration 4 Segments and Midpoints

1. Construct three triangles on a coordinate plane. For each triangle, find the midpoints of two sides, and draw a segment to connect those midpoints.

   What do you notice about the length of segment AB at the right and the length of the side parallel to that segment?

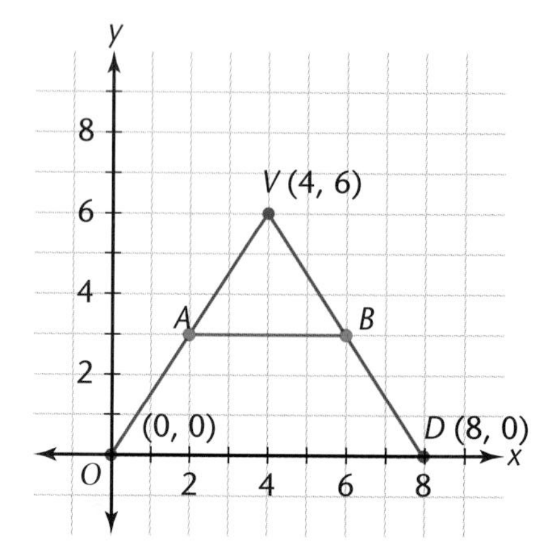

2. Make a conjecture that describes this relationship.

3. Test the other triangles to see if your conjecture is true for them.

4. Use coordinate geometry with letters to represent the nonzero coordinates, and show that your conjecture is true. ❖

One of the interesting properties of a parabola is that it represents the set of all points the same distance from a given line and a given point. This is illustrated in the following application.

APPLICATION

Surveying Engineers plan to locate a maintenance facility at a point that is the same distance from a TV tower and a road. The TV tower is located 2 miles from the road. Point P is the same distance from the TV tower, T, and the road, X. Is there an equation that describes the possible placement of the maintenance facility?

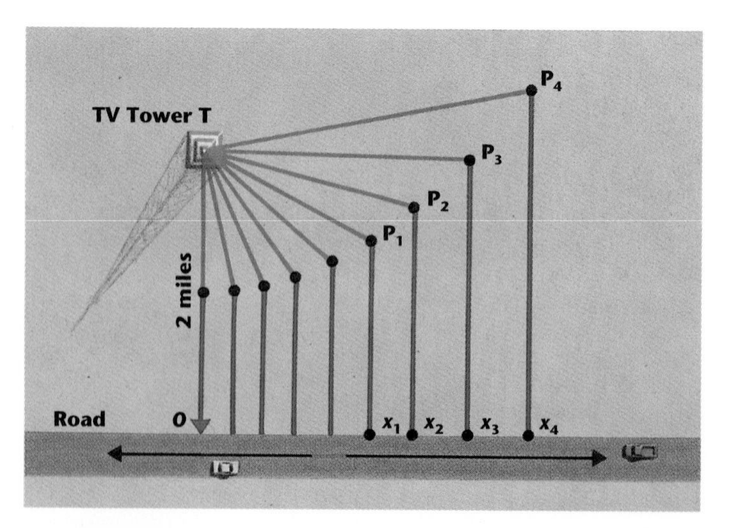

A coordinate grid can be used to model the problem. Let $T(0, 2)$ be the TV tower and $X(x, 0)$ be a point on the road x units away from the origin. Let the point $P(x, y)$ be any point equidistant from T and X. You can use coordinate geometry to find the equation of all points that are equidistant from the x-axis ($y = 0$) and the point $(0, 2)$.

Let $P(x, y)$ be any point that is equidistant from $(x, 0)$ and $(0, 2)$.

$$d_1 = \sqrt{(x - x)^2 + (y - 0)^2} = \sqrt{y^2} = y$$
$$d_2 = \sqrt{(x - 0)^2 + (y - 2)^2} = \sqrt{x^2 + (y - 2)^2}$$

Since d_1 and d_2 are equal distances, the two expressions are equal. Substitute the values, square both sides of the equation, and simplify. This is the function that will produce the graph.

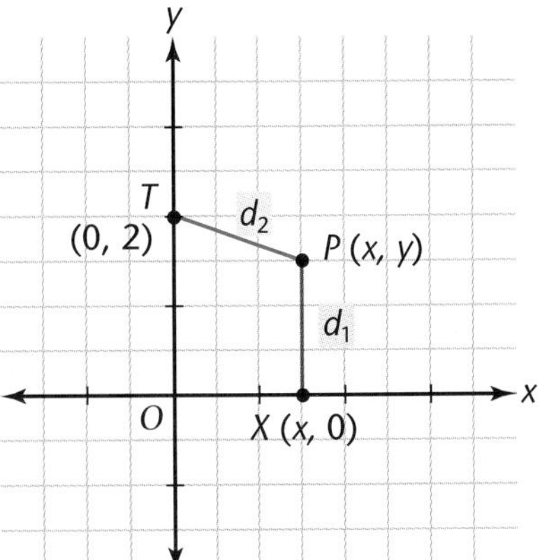

$d_1 = d_2$	Given
$y = \sqrt{x^2 + (y - 2)^2}$	Substitute the values.
$y^2 = x^2 + (y - 2)^2$	Square both sides.
$y^2 = x^2 + y^2 - 4y + 4$	Simplify.
$4y = x^2 + 4$	Addition Property of Equality
$y = \frac{1}{4}x^2 + 1$	Division Property of Equality

The equation that describes the possible placements of the maintenance facility is $y = \frac{1}{4}x^2 + 1$. ❖

What is the name for the graph of $y = \frac{1}{4}x^2 + 1$? What is the vertex? What is the line of symmetry? The equation is for a parabola. The vertex is (0, 1). The axis of symmetry is the y-axis.

Cultural Connection: Africa In ancient Egypt, African surveyors used geometry to solve algebraic problems. They noticed that by using the diagonal of a 1-cubit square (Square B) as the side of a larger square (Square C), the area of the new square would be double the original square. Recall that $(\sqrt{2})^2 = 2$. They made the diagonal of a 1-cubit square part of their system of measurement. It was called a *double-remen* and was equal to $\sqrt{2}$ cubits.

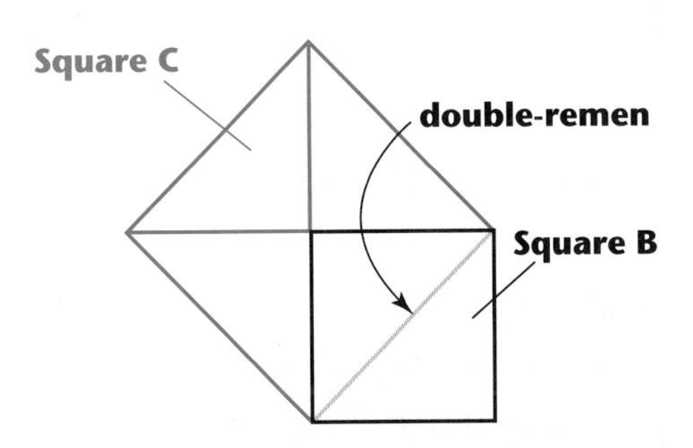

1. Square B is 1 square cubit in area. How can you determine that square C has an area of 2 square cubits?
2. Since the area of square C is 2 square cubits, how would you explain why the side of square C has length $\sqrt{2}$ cubits?

EXERCISES & PROBLEMS

Communicate

1. Discuss the ways that geometry can help solve problems in algebra.
2. What is meant by the set of all points in a plane equidistant from a given point?
3. Describe how to use the distance formula.
4. Explain how the distance formula is used to express the set of all points equidistant from a given point.
5. How do you find the length of a diagonal of a rectangle by using the distance formula?

Practice & Apply

Find an equation of a circle with its center at the origin and with the given radius.

6. radius = 3
   $x^2 + y^2 = 9$
7. radius = 7
   $x^2 + y^2 = 49$
8. radius = 15
   $x^2 + y^2 = 225$
9. radius = r
   $x^2 + y^2 = r^2$

10. Copy and complete the table of ordered pairs that satisfy the equation $x^2 + y^2 = 100$.

								± 10	± 8	± 8	± 6	± 6	0	0
x	0	6	−6	8	−8	10	−10	?	?	?	?	?	?	?
y	?	?	?	?	?	?	?	0	6	−6	8	−8	10	−10
	± 10	± 8	± 8	± 6	± 6	0	0							

11. Plot the ordered pairs on graph paper.
12. What are the x-intercepts? 10, −10
13. What are the y-intercepts? 10, −10
14. What shape is the graph? circle

Use graph paper to sketch each circle below. Find an equation of a circle with

15. center (2, 3) and radius = 5.
16. center (−2, 3) and radius = 5.
17. center at (2, −3) and radius = 5.
18. center (−2, −3) and radius = 5.
19. Rewrite the equations for Exercises 15–18 with a radius of 7.
20. Generalize. Find an equation of a circle with center (h, k) and radius, r.
    $(x - h)^2 + (y - k)^2 = r^2$

Use your generalization to identify the center and radius for each circle.

21. $(x - 1)^2 + (y - 5)^2 = 9$ (1, 5); 3
22. $(x + 3)^2 + (y - 4)^2 = 16$ (−3, 4); 4
23. $(x - 2)^2 + (y + 4)^2 = 50$ (2, −4); $5\sqrt{2}$
24. $(x + 5)^2 + (y + 10)^2 = 10$ (−5, −10); $\sqrt{10}$

Geometry Plot triangle PQR on graph paper, with $P(2, 1)$, $Q(4, 7)$, and $R(12, 3)$.

25. Find the coordinates of M, the midpoint of $\overline{PQ}$, and plot M. (3, 4)

26. Find the coordinates of N, the midpoint of $\overline{PR}$, and plot N. (7, 2)

27. Draw $\overline{MN}$ and find its length (simplify your answer). $2\sqrt{5}$ or 4.47

28. Find the length of $\overline{QR}$. $4\sqrt{5}$

29. Compare the lengths of $\overline{MN}$ and $\overline{QR}$. $QR = 2MN$

30. Find the slopes of $\overline{MN}$ and $\overline{QR}$. $-\frac{1}{2}; -\frac{1}{2}$

31. What does the slope information tell you about $\overline{MN}$ and $\overline{QR}$? parallel

32. Use the distance formula to find an equation for all points $P(x, y)$ such that the sum of the distances from P to $A(-8, 0)$ and from P to $B(8, 0)$ is 20.

33. Find an equation for all points $P(x, y)$ such that the sum of the distances from P to $A(0, -4)$ and from P to $B(0, 4)$ is 10.

32. $\sqrt{(x + 8)^2 + y^2} + \sqrt{(x - 8)^2 + y^2} = 20$

33. $\sqrt{x^2 + (y + 4)^2} + \sqrt{x^2 + (y - 4)^2} = 10$

Look Back

34. Identify which of the following expressions correctly represents the distance between 2 and 5 on a number line. **[Lesson 1.6]** d, f, h

a. $2 + 5$ **b.** $2 - 5$ **c.** $5 + 2$ **d.** $5 - 2$

e. $|2 + 5|$ **f.** $|2 - 5|$ **g.** $|5 + 2|$ **h.** $|5 - 2|$

35. Evaluate $\frac{3^0 \cdot 3^2}{3^2} - 3^2$. **[Lesson 6.3]** -8

Write each number in scientific notation. [Lesson 6.4]

36. 2,300,000 2.3×10^6

37. 0.00000125 1.25×10^{-6}

Simplify each polynomial. [Lesson 7.2]

38. $(x^2 + 3x + 5) + (7x^2 - 5x - 10)$ $8x^2 - 2x - 5$

39. $(8b^2 - 15) - (2b^2 + b + 1)$ $6b^2 - b - 16$

Find the value of the discriminant for each equation. [Lesson 8.5]

40. $3x^2 - 6x + 3 = 0$ 0

41. $2x^2 + 3x - 2 = 0$ 25

Graph each quadratic inequality. [Lesson 8.6]

42. $y < 2x^2 - 3x + 4$

43. $y > -4x^2 + 6x - 5$

Look Beyond

Geometry In geometry, you can prove several theorems by using coordinates. Prove each of the following.

44. The diagonals of a rectangle bisect each other.

45. The diagonals of an isosceles trapezoid have the same length.

LESSON 9.7 Tangent Function

why *The tangent function is sometimes used for solving problems by indirect measure. If you know the measurement of an acute angle and one leg of a right triangle, you can determine the length of the other leg without measuring it. Also, you have a way to determine the acute angle of the triangle if you know the lengths of the two legs.*

Dale works with a milling machine in a machine shop. A blueprint calls for a part with the dimensions as shown. What should be the measure of angle *B* where Dale is to make a cut?

Consider the triangle that is to be removed. Examine the following ratio with respect to angle *B*:

$$\frac{\text{length of the leg opposite the angle}}{\text{length of the leg adjacent to the angle}} = \frac{AC}{BC} = \frac{2}{3}$$

In a right triangle, this ratio is called the **tangent** of angle *B*, or tan *B*. The goal is to find an angle whose tangent is $\frac{2}{3}$. "The angle whose tangent is $\frac{2}{3}$" is written $\tan^{-1}\left(\frac{2}{3}\right)$, and is also read "the inverse tangent of *x*."

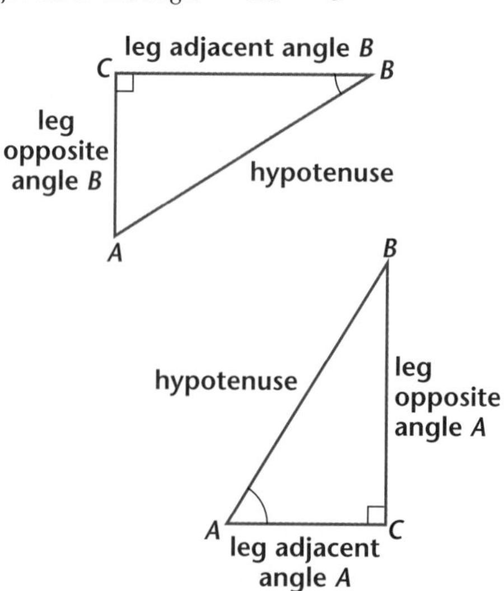

Exploration Angles and Tangents

In the figure at the right, $\overline{CA}$ is the base of all 7 triangles. For example, one triangle is CAB_1. As angle A increases in increments of 10°, the length of the side that is opposite angle A for each triangle also increases.

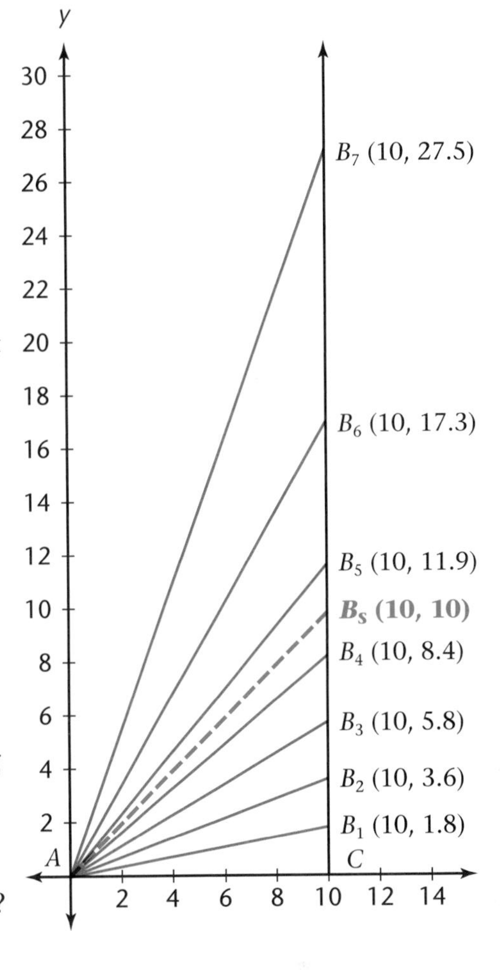

1 Find the tangent ratio for each given angle measure. Round the ratio to the nearest tenth.

a. 10° (angle CAB_1) 0.2

b. 20° (angle CAB_2) 0.4

c. 30° (angle CAB_3) 0.6

d. 40° (angle CAB_4) 0.8

e. 50° (angle CAB_5) 1.2

f. 60° (angle CAB_6) 1.7

g. 70° (angle CAB_7) 2.7

Refer to your answers for **a–g** for Steps 2–5.

2 How does the value of the tangent ratio increase as the measure of the angle doubles? (Compare, for example, tan 10° and tan 20°, and tan 30° and tan 60°.)

3 For which of the angles in Step 1 is the tangent ratio less than 1? 10°; 20°; 30°; 40°

4 The measure of angle CAB_S is 45°. What can you say about the length of the side opposite this angle and the length of the side adjacent to this angle? They have the same length.

5 How can you find the length of the side opposite angle A if you know that the measure of angle A is 45° and you know the length of the side adjacent to A? Explain. ❖

2. As the number of degrees doubles, the value of the tangent ratio increases.
5. The angle is 45°. The tangent ratio is 1 and equal to the ratio of the length of the side opposite the angle A to the length of the side adjacent to angle A. Therefore, by solving for the length of the adjacent side, the result is the length of the adjacent side is equal to the length of the opposite side.

The function $f(x) = \tan x$ is an example of a trigonometric function. The domain of f is a set of angle measures. Thus, $f(45°) = \tan 45° = 1$.

You can use a calculator to find the tangent of an angle using the TAN key. Since you are using degrees to measure the angles, be sure the calculator mode is set for degrees. For example, for $f(45) = \tan 45$, press TAN 45 ENTER, or =, and the calculator will display the value 1.

You can also find the measure of an acute angle if you know the tangent value. To find the angle x, where x represents the degree measure of the acute angle, use the TAN⁻¹ key. For example, if $\tan x = 1$, press TAN⁻¹ 1 ENTER, and the calculator will display 45.

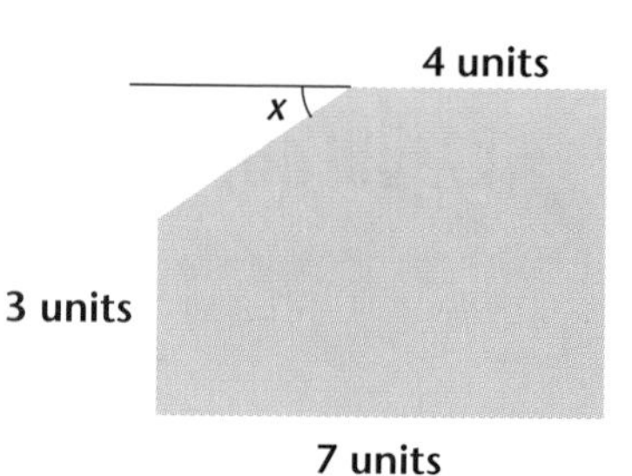

Calculator

Now you can solve Dale's problem. That is, find the angle whose tangent is $\frac{2}{3}$, or approximately 0.67.

TAN⁻¹ (2 ÷ 3) ≈ 34

Dale should cut the metal at a 34° angle.

EXAMPLE 1

Find the tangent of each acute angle of each triangle.

A

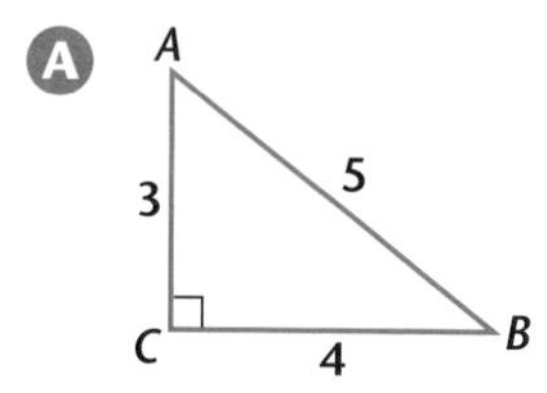

B

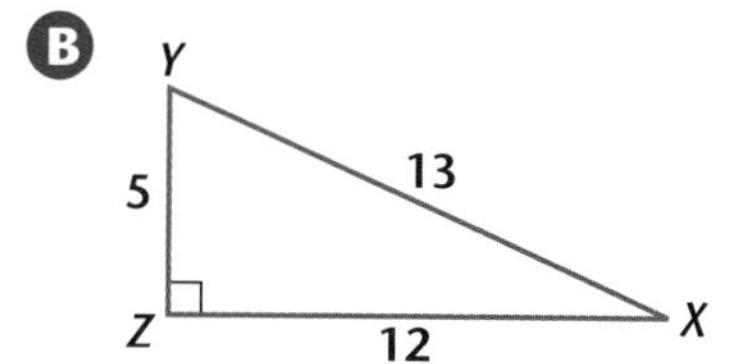

Solution ➤

A $\tan A = \frac{\text{opposite}}{\text{adjacent}} = \frac{4}{3}$, $\tan B = \frac{\text{opposite}}{\text{adjacent}} = \frac{3}{4}$

B $\tan Y = \frac{\text{opposite}}{\text{adjacent}} = \frac{12}{5}$, $\tan X = \frac{\text{opposite}}{\text{adjacent}} = \frac{5}{12}$ ❖

EXAMPLE 2

An airplane climbs to 30,000 feet at a steady rate. If the altitude is reached after the airplane has covered a distance of 35 miles on the ground, what is its angle of elevation?

Solution ➤

Draw a diagram to model the problem. Since 5280 feet = 1 mile, change 35 miles to feet by multiplying by 5280.

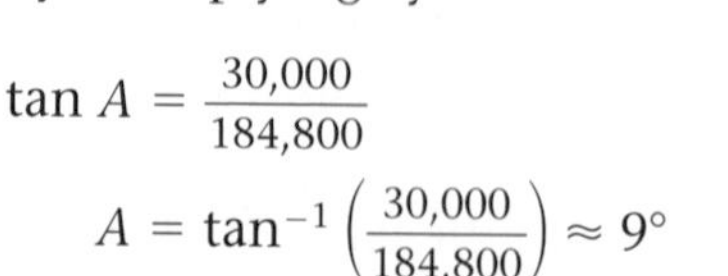

Calculator

$$\tan A = \frac{30{,}000}{184{,}800}$$

$$A = \tan^{-1}\left(\frac{30{,}000}{184{,}800}\right) \approx 9^\circ$$

The airplane is climbing at an angle of about 9°. The angle of elevation is approximately 9°. ❖

EXAMPLE 3

A group of skateboarders wants to build a ramp with an angle of 12°. What should be the rise for each 10 meters of run in order to achieve a 12° angle?

Solution ➤

Draw a diagram, and write the tangent ratio using the information that you have.

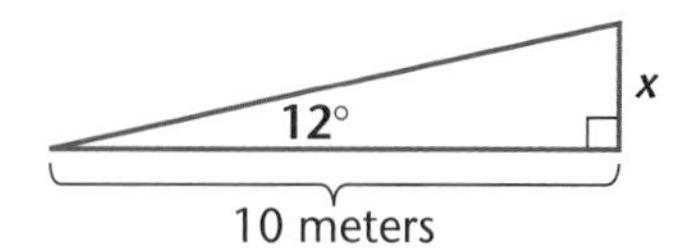

Solve for x to find the rise.

$$\tan 12^\circ = \frac{x}{10}$$

$$0.2126 \approx \frac{x}{10}$$

$$(10)(0.2126) \approx x$$

$$2.126 \approx x$$

The ramp should rise about 2.1 meters for each 10 meters of run. ❖

EXERCISES & PROBLEMS

Communicate

1. Explain how the tangent function is related to the slope of a line.
2. How can you find the length of the side opposite an acute angle, A, in a right triangle when you know the measure of angle A and the length of the other leg?
3. How can you find the measure of an acute angle in a right triangle when you know the lengths of the legs opposite of and adjacent to that angle?

Practice & Apply

Technology Use a calculator to find the tangent of each angle to the nearest ten-thousandth.

4. 45° 1.0000 **5.** 30° 0.5774 **6.** 60° 1.7321 **7.** 0° 0.0000

8. 38° 0.7813 **9.** 57° 1.5399 **10.** 89° 57.2900 **11.** 89.9° 572.9572

What is the approximate measure of the angle for each approximate tangent value?

12. 1.7321 60° **13.** 1.1918 50° **14.** 0.2679 15° **15.** 3.7321 75°

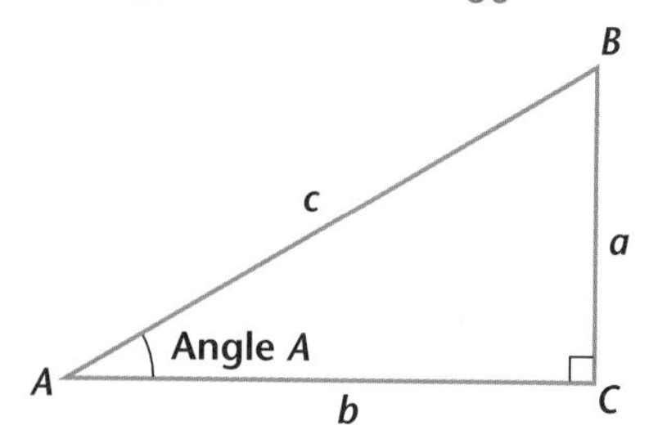

The lower case letters represent the side lengths, and the capital letters represent the angles of a general right triangle. Use the given information to find the indicated side length to the nearest hundredth.

16. If m$\angle A = 30°$ and $b = 12$ meters, find a. 6.93 m

17. If m$\angle B = 60°$ and $a = 10$ feet, find b. 17.32 ft

18. If a is 3 centimeters and tan A is 0.75, what is b? 4 cm

19. If b is 23 feet long and tan A is 1.0, what is a? 23 ft

20. If a is 6 inches and m$\angle A \approx 31°$, what is b? $\approx$ 10 in.

21. If m$\angle A \approx 35°$ and b is 20 millimeters, what is a? 14 mm

22. Construction Mr. Fernandez wants the roof of his house to have an angle of 20°. The slope of the roof is the same on both sides and has the same length on either side of the center. How high does the roof rise if the length of the roof is 14 meters? About 2.5 m

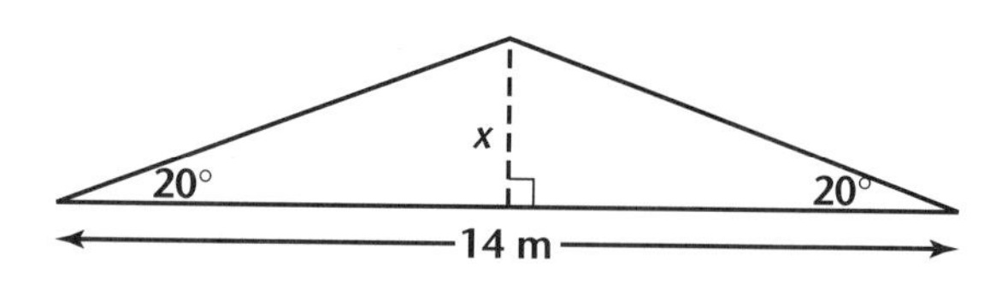

23. A water tower is 600 feet from the observer. Sighting the top of the water reservoir forms an angle of 28°. A sighting at the bottom of the reservoir has an angle of 26°. How tall is the reservoir part of the water tower? About 26.4 ft

24. Transportation If a road has a 7% grade, its slope is $\frac{7}{100}$. What is the measure of the angle of the road's incline? About 4°

25. Hobbies A model airplane and a car begin at the same place. After 15 seconds the car has traveled $\frac{1}{4}$ mile, and the elevation angle from the starting point to the airplane is 30°. If the model airplane is directly above the car, what is the altitude of the airplane? 0.14 mi, or about 739 ft

Model airplane
?
30°
Car
Starting point
$\frac{1}{4}$ mile

Look Back

26. Solve the equation $4x - 5 = 7(x - 3) + 2$ for x. **[Lesson 1.4]** $4\frac{2}{3}$

27. What is the probability that after 1 roll of 2 number cubes, the sum of the dots equals 7? **[Lesson 4.6]** $\frac{1}{6}$

28. Determine whether the quadratic equation $2x^2 + x - 15 = 0$ has a real solution. If it does, use the quadratic formula to determine the solution(s). **[Lesson 8.5]** yes; $2\frac{1}{2}$, -3

29. Solve $x^2 - 5x + 6 \geq 0$. **[Lesson 8.4]** $x \geq 3$ or $x \leq 2$

30. Graph $x^2 - 16 < 0$. **[Lesson 8.6]**

31. **Geometry** The 45-45-90 triangle and the 30-60-90 triangle have sides with special relationships.

Use the "Pythagorean" Right-Triangle Theorem to show that the sides actually have the relationship shown. **[Lesson 9.4]**

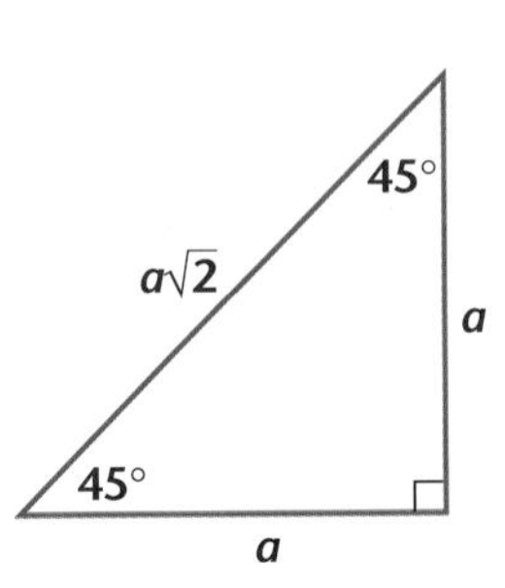

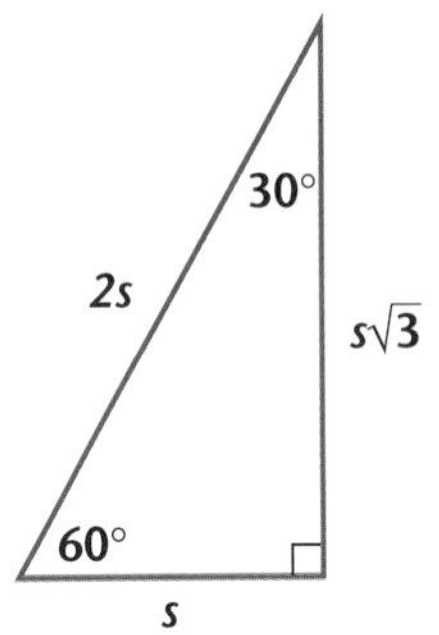

32. Points $P(4, 9)$ and $Q(10, 5)$ are given on a coordinate plane. Find the coordinates of M, the midpoint of segment PQ. **[Lesson 9.5]** (7, 7)

Using the information given in Exercise 32, find the length of each segment. [Lesson 9.5]

33. PQ $2\sqrt{13}$ **34.** PM $\sqrt{13}$ **35.** MQ $\sqrt{13}$

Look Beyond

Besides the tangent (tan), the sine (sin) and cosine (cos) are two other basic trigonometric functions. They provide additional tools for solving problems by indirect measure.

$$f(x) = \tan(x) = \frac{b}{a}$$

$$g(x) = \sin(x) = \frac{b}{c}$$

$$h(x) = \cos(x) = \frac{a}{c}$$

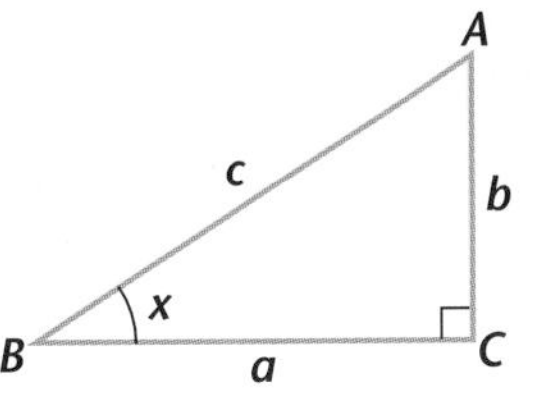

36. If the tangent of angle x is $\frac{3}{4}$, what are the sine and the cosine of x? $\frac{3}{5}, \frac{4}{5}$

HINT: Use the "Pythagorean" Right-Triangle Theorem to find c.

37. Explain why $(\sin x)^2 + (\cos x)^2 = 1$ is true for any right triangle.

LESSON 9.8 Sine and Cosine Functions

why *In addition to the tangent function, the sine and cosine functions are also used to solve problems by indirect measurement.*

As Paul rides the log ride, he feels as if he is dropping straight down. Architects who designed the water slide used trigonometry to plan the angle and height of the slide.

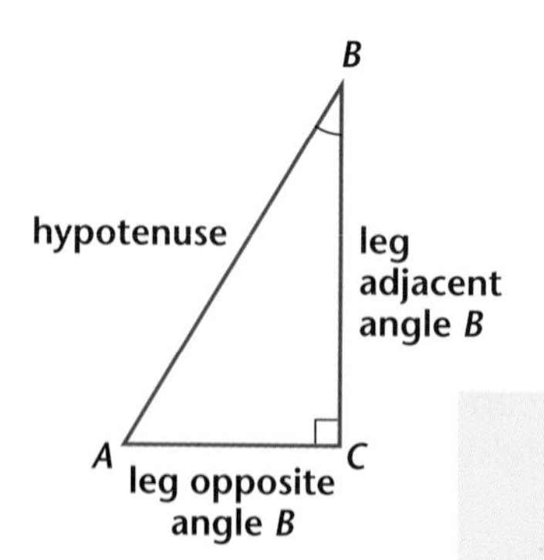

Recall the tangent ratio from Lesson 9.7.

$$\tan B = \frac{\text{length of the leg opposite } \angle B}{\text{length of the leg adjacent to } \angle B}$$

There are two other ratios, *sine* and *cosine,* that are based on the right triangle.

SINE AND COSINE

$$\text{sine of } \angle B \text{, or } \sin B = \frac{\text{length of leg opposite } \angle B}{\text{length of hypotenuse}}$$

$$\text{cosine of } \angle B \text{, or } \cos B = \frac{\text{length of leg adjacent to } \angle B}{\text{length of hypotenuse}}$$

Exploration 1 *Trigonometric Ratios*

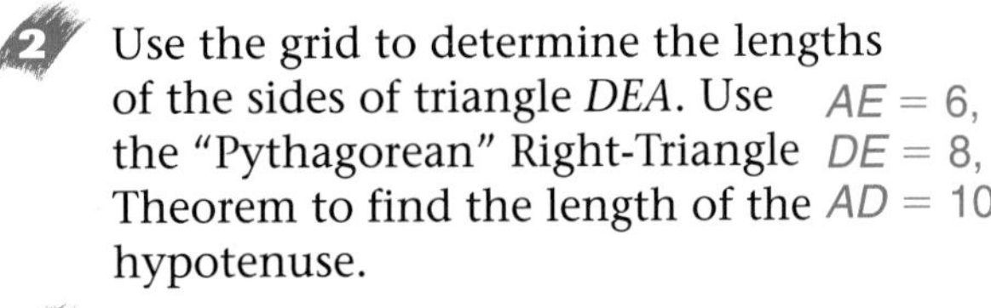

1. Triangle BCA is a right triangle in this figure. Name the two other right triangles with A as a vertex. $\triangle DEA$, $\triangle FGA$

2. Use the grid to determine the lengths of the sides of triangle DEA. Use the "Pythagorean" Right-Triangle Theorem to find the length of the hypotenuse. $AE = 6$, $DE = 8$, $AD = 10$

3. Use the grid to determine the lengths of the sides of triangle FGA. Use the "Pythagorean" Right-Triangle Theorem to find the length of the hypotenuse. $AG = 9$, $FG = 12$, $AF = 15$

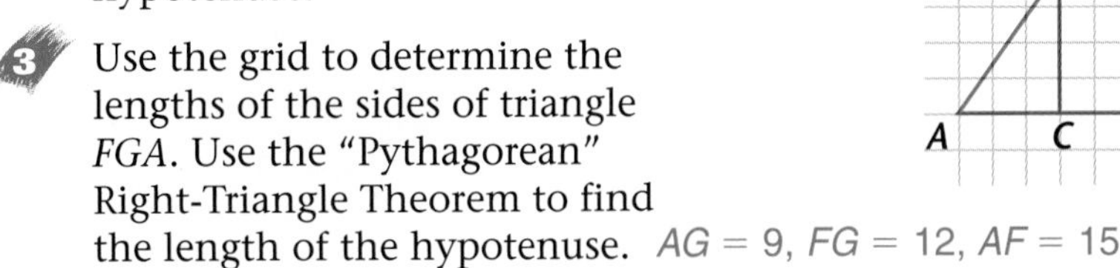

Use the grid to complete the following table:

	Length of leg opposite $\angle A$	Length of leg adjacent $\angle A$	Length of hypotenuse	tan A	sin A	cos A
$\triangle BCA$	4	3	$\sqrt{4^2 + 3^2} = 5$	$\frac{4}{3} \approx 1.\overline{3}$	$\frac{4}{5} = 0.8$	$\frac{3}{5} = 0.6$
4. $\triangle DEA$	?	?	?	?	?	?
5. $\triangle FGA$	?	?	?	?	?	?

6. As the lengths of the legs increase, how does the length of the hypotenuse change? The hypotenuse increases by multiples of 5.

7. The lengths of the sides of each triangle change, but the measure of $\angle A$ is the same. Is the ratio for sin A the same or different in each triangle? Is the ratio for cos A the same or different in each triangle? Is the ratio for tan A the same or different in each triangle?

8. Using the information in the table, make a conjecture about the trigonometric ratios for any given angle measure. ❖

7. The ratios for sin A, cos A, and tan A are the same for each triangle.
8. Similar triangles have the same trigonometric ratios.

Use triangle BCA shown on the grid in the exploration above. Compare sin A and cos A to sin B and cos B. Which values are equal? Explain why these values are equal.

You can use your calculator to find the sine or cosine of an angle. Because the angle is given in degrees, your calculator must be in the degree mode. To determine sin 10°, use the SIN key.

SIN 10° ≈ 0.174

cos 65° ≈ 0.4226 tan 40° ≈ 0.8391

Use the COS and TAN keys on your calculator to find cos 65° and tan 40°.

EXAMPLE 1

Recreation A parasailer is towed behind a boat. The rope holding the parasailer is 35 feet long. The parasailer is at an angle of 30° with the surface of the water. Approximately how high above the water is the parasailer? (The rope is connected to the boat 1 foot above the surface of the water.)

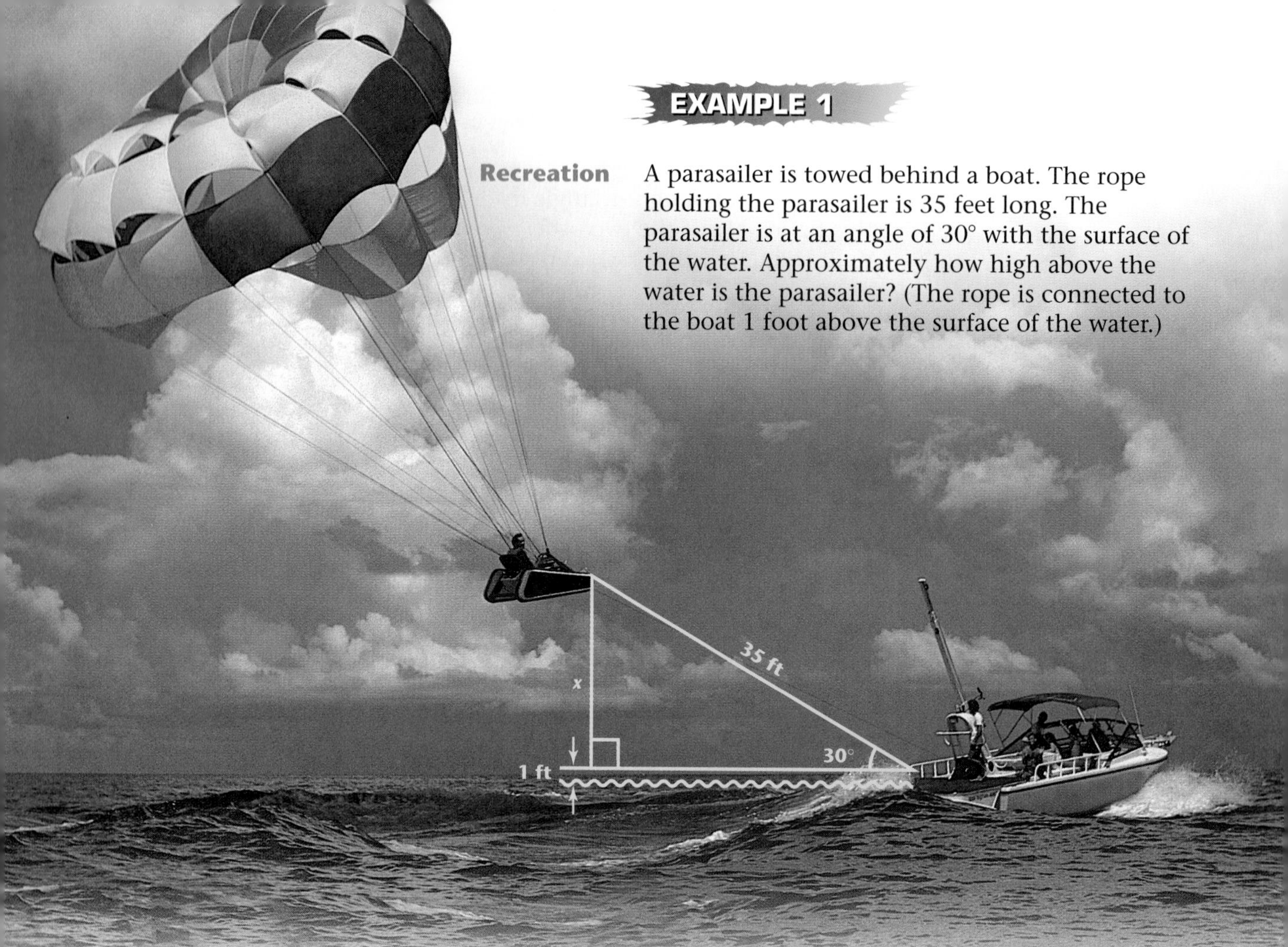

Solution

Use the sine ratio to find the length of b, the side opposite the 30° angle.

$$\sin 30^\circ = \frac{\text{opposite}}{\text{hypotenuse}}$$

$$\sin 30^\circ = \frac{b}{35}$$

$$0.5 = \frac{b}{35} \qquad \text{Evaluate } \sin 30^\circ.$$

$$(0.5)(35) = \left(\frac{b}{35}\right)(35) \qquad \text{Multiplication Property of Equality}$$

$$17.5 = b$$

Since the rope is connected to the boat 1 foot above the water, add 1 to 17.5. Thus, the parasailer is 18.5 feet above the surface of the water. ❖

Try This Use cosine to find a. Use sine to find b. Round your answers to the nearest hundredth.

$\cos 42^\circ = \frac{a}{45}$; $a \approx 33.44$

$\sin 42^\circ = \frac{b}{45}$; $b \approx 30.11$

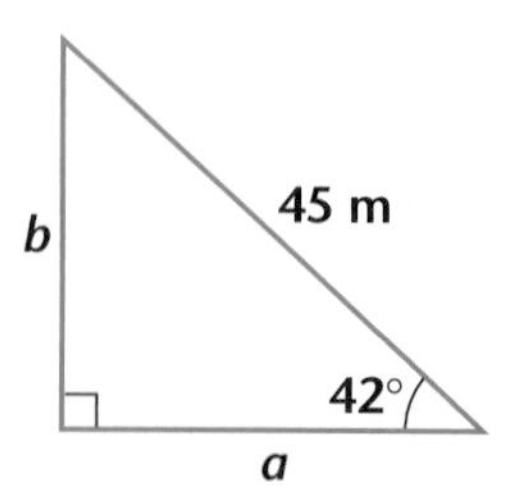

EXAMPLE 2

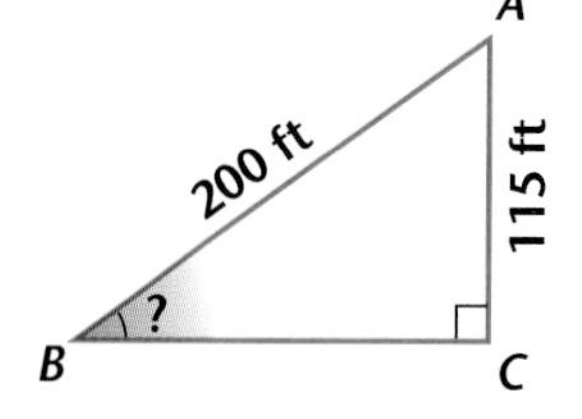

Recreation Suppose that a waterslide is 200 feet long and rises to a height of 115 feet. What is the angle of the slide with the ground?

Solution ➤

Scientific Calculator

Since the length of the hypotenuse and the length of the side opposite $\angle B$ are known, use the sine ratio to find $\angle B$.

$$\sin B = \frac{115}{200}$$

$$\sin B = 0.575$$

$$B = \sin^{-1}(0.575)$$

Use the SIN⁻¹ key. $B \approx 35°$

The angle of the slide with the ground is about 35°. ❖

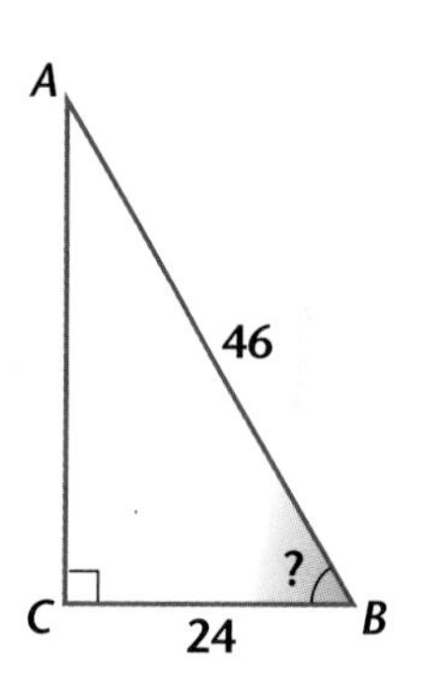

Try This Find the measure of $\angle B$ in the diagram at the right. $B \approx 59°$

EXERCISES & PROBLEMS

Communicate

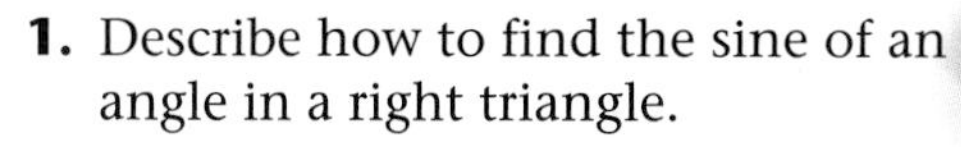

1. Describe how to find the sine of an angle in a right triangle.
2. Describe how to find the cosine of an angle in a right triangle.
3. Explain how to find the length that the ramp should be in order to create a 4° incline on a 3-inch curb.
4. Is the cosine ratio in a right triangle always less than 1? Is the sine ratio in a right triangle always less than 1? Explain.

Practice & Apply

Evaluate each expression to the nearest ten-thousandth.

5. $\sin 24°$ 0.4067 **6.** $\cos 32°$ 0.8480 **7.** $\cos 88°$ 0.0349 **8.** $\sin 18°$ 0.3090

9. $\sin 66.8°$ 0.9191 **10.** $\cos 84.2°$ 0.1011 **11.** $\cos 27.7°$ 0.8854 **12.** $\sin 48.6°$ 0.7501

For Exercises 13–22, use right triangle ABC at the right. Find the indicated length or angle measure. 14. ≈37.3 16. ≈13.8

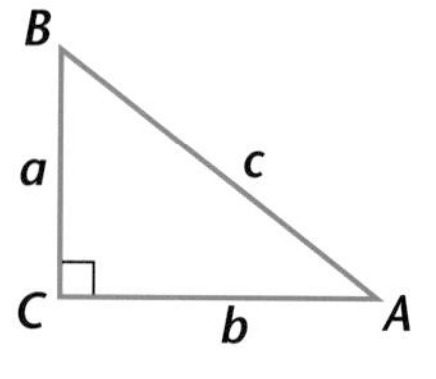

13. $b = 7, c = 10, m\angle B = \underline{?}$ ≈44.4° **14.** $m\angle B = 35°, c = 65, b = \underline{?}$

15. $b = 6, c = 13, m\angle A = \underline{?}$ ≈62.5° **16.** $m\angle B = 51°, c = 22, a = \underline{?}$

17. $b = 15, a = 25, m\angle B = \underline{?}$ ≈31.0° **18.** $m\angle B = 41°, c = 6, b = \underline{?}$

19. $a = 27, c = 36, m\angle B = \underline{?}$ **20.** $m\angle B = 30°, b = 11, c = \underline{?}$ 22

21. $a = 7, c = 14, m\angle B = \underline{?}$ 60° **22.** $m\angle A = 45°, c = 4, b = \underline{?}$ ≈2.8

18. ≈3.9 19. ≈41.4°

What acute angle measure has each approximate sine value? Give your answer to the nearest tenth of a degree.

23. 0.56 34.1° **24.** 0.892 63.1° **25.** 0.129 7.4° **26.** 0.759 49.4°

27. 0.5 30.0° **28.** 0.707 45.0° **29.** 0.563 34.3° **30.** 0.445 26.4°

31. 0.26 15.1° **32.** 0.972 76.4° **33.** 0.323 18.8° **34.** 0.75 48.6°

What acute angle measure has each approximate cosine value? Give your answer to the nearest tenth of a degree.

35. 0.126 82.8° **36.** 0.5 60.0° **37.** 0.886 27.6° **38.** 0.707 45.0°

39. 0.99 8.1° **40.** 0.81 35.9° **41.** 0.54 57.3° **42.** 0.78 38.7°

43. 0.612 52.3° **44.** 0.643 50.0° **45.** 0.199 78.5° **46.** 0.333 70.5°

Advertising To advertise its grand opening, a pet store rents a giant parrot balloon to place on top of the building. The balloon is attached with guy wires to the roof. The guy wires are 30 feet long.

47. If the guy wires are attached to the roof 20 feet away from the balloon ($b = 20$), what angle do the guy wires make with the roof? ≈48°

48. If the guy wires are attached to the balloon at a height of 25 feet above the roof ($a = 25$), what angle do the guy wires make with the roof? ≈56°

49. In order to make a 45° angle with the roof, how far away from the balloon should the guy wires be attached to the roof? ≈21 feet

50. At what height above the roof should the guy wires be attached to the balloon in order to make a 55° angle with the roof? ≈25 feet

51. If the guy wires are attached to the roof 15 feet away from the balloon, what angle do the guy wires make with the roof? 60°

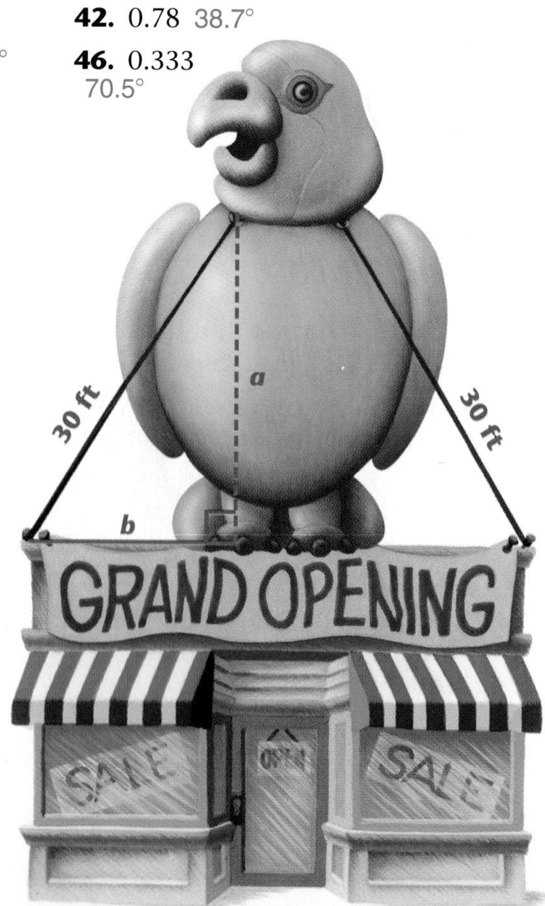

Construction A wheelchair ramp is to have an angle of 4.5° with the ground. A porch is 18 inches high. Use this information for Exercises 52 and 53.

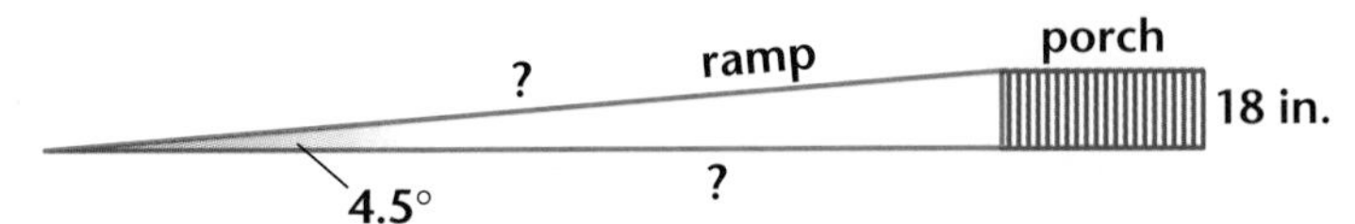

52. How long will the ramp be? ≈229.4 inches or about 19.1 feet

53. How far from the porch should the ramp begin? ≈228.7 inches or about 19.1 feet

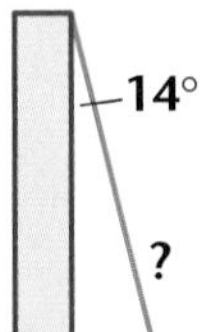

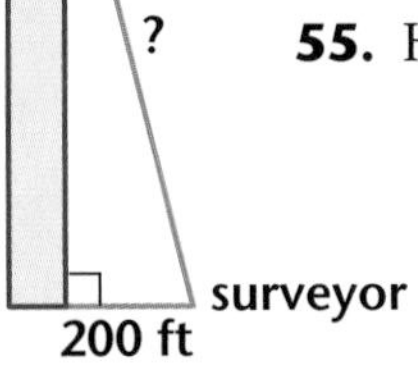

Surveying A surveyor is 200 feet from the base of tower. Using a tool called a *transit,* she finds that the angle from the top of the tower is 14°.

54. How far is the surveyor from the top of the tower? ≈826.7 feet

55. How tall is the tower? ≈802.2 feet

56. A 12-foot ladder is leaning against a building. The ladder makes a 75° angle with the ground. How far away from the base of the building is the base of the ladder? About 3.1 feet

Look Back

Solve each equation. Round answers to the nearest hundredth. [Lesson 8.2]

57. $x^2 = 36$ 6, −6

58. $x^2 = 144$ 12, −12

59. $(x + 4)^2 - 36 = 0$ 2, −10

60. $(x - 1)^2 = 11$

60. $1 \pm \sqrt{11}$, or ≈4.32 and ≈−2.32

Complete the square. [Lesson 8.3]

61. $x^2 + 10x$ $+25$, $(x + 5)^2$

62. $x^2 + 40x$ $+400$, $(x + 20)^2$

63. $x^2 - 5x$ $+\frac{25}{4}$, $\left(x - \frac{5}{2}\right)^2$

Use the quadratic formula to solve each equation. Give answers to the nearest hundredth. [Lesson 8.5]

64. $t^2 + 6t - 22 = 0$ $-3 \pm \sqrt{31}$, or ≈2.57 and ≈−8.57

65. $a^2 + 9a = 0$ 0, −9

66. $h^2 + 6h + 9 = 0$ −3

Look Beyond

Geometry Use trigonometry to find the area of each figure.

67. $A \approx 19.3$

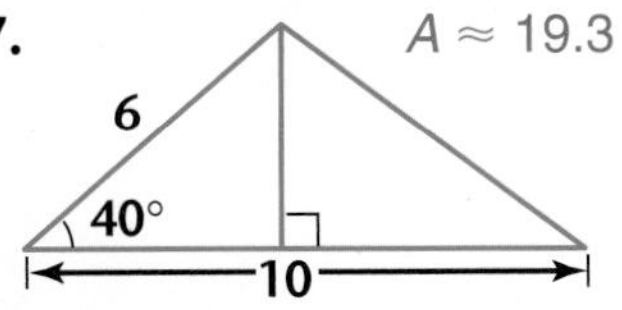

68. $A \approx 60.23$

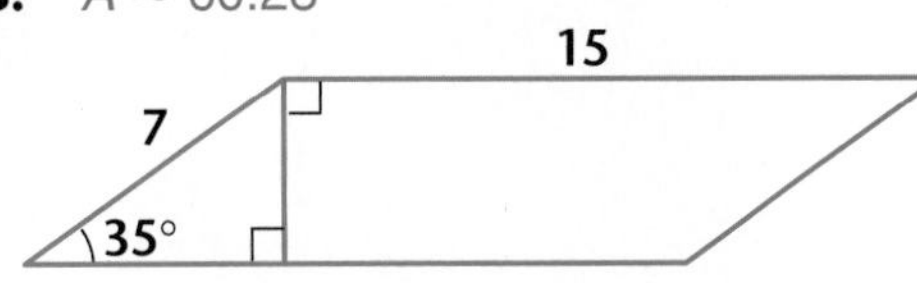

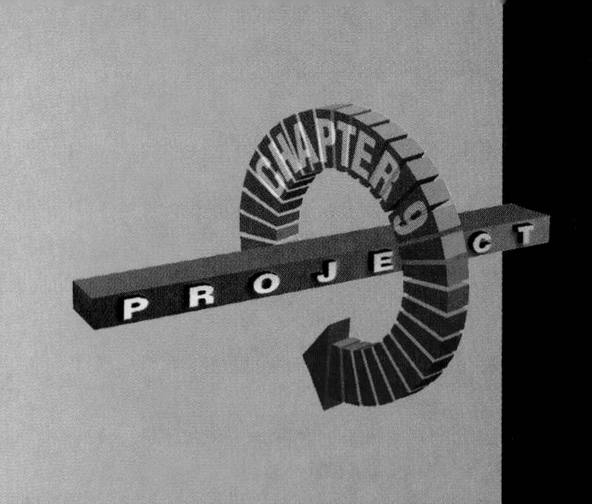

RADICAL RECTANGLES

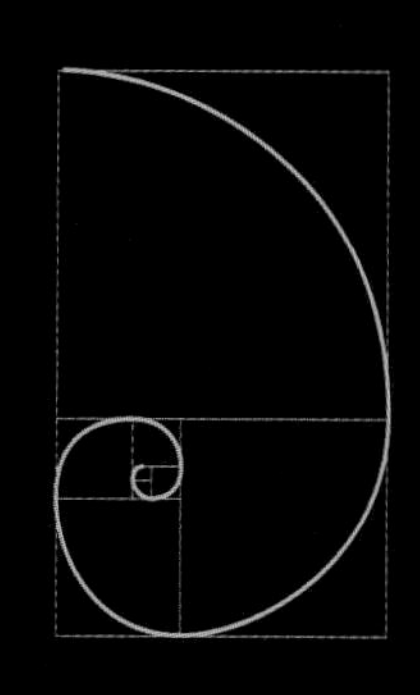

DID YOU KNOW THAT SOME RECTANGLES ARE considered more beautiful than others? Greek architects of the fifth century B.C.E. believed that the golden rectangle was the most beautiful rectangle possible. Great artists all over the world have used this rectangle in their creations.

What is a golden rectangle? It is a rectangle whose length and width form a special ratio. If a is the longer side of the rectangle and b is the shorter side, then the golden ratio, m, is the proportion

$$\frac{b}{a} = \frac{a}{a+b}.$$

This proportion can be rewritten as $a^2 = b(a + b)$ and can be solved by the quadratic formula. The ratio is

$$m = \frac{\sqrt{5}-1}{2} \approx 0.618.$$

Masaccio, an Italian Renaissance artist, used properties of the golden rectangle in his Fresco *The Tribute Money*.

Photo: Scala/ Art Resource, NY

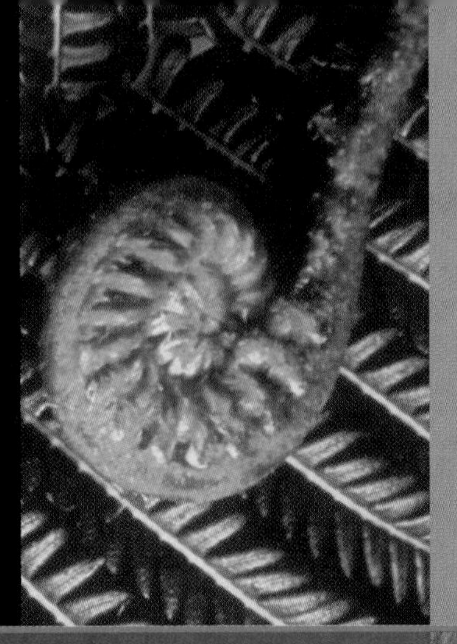

A golden rectangle will produce a similar rectangle when a square is cut from the end of the original rectangle. A similar rectangle is one that has an identical shape, but has a different size.

Activity 1

Construct an approximate golden rectangle with a 3-by-5 inch note card. Since the golden ratio has the ratio of b to a or 0.61803, this is approximately $\frac{6}{10}$ or $\frac{3}{5}$.

You can see that the approximation is close. If you fold one side and draw a vertical line to form the square portion, the remaining rectangle will have a ratio of $\frac{2}{3} \approx 0.667$ which is close to 0.61803.

Activity 2

There is another way you can create a golden rectangle that is more accurate. Begin with a square. Find the midpoint of the base. Use a compass to measure the distance from, the midpoint, M, to the vertex, V, of the opposite side. Extend the base of the square, and draw an arc to the point of the intersection with the base, B. Use this to complete the rectangle.

Activity 3

Construct a large golden rectangle. Fold a corner of the rectangle, and draw the line to create a smaller rectangle. Repeat the process with each of the resulting rectangles. Continue this process several times. You can produce a spiral by connecting the corresponding vertex of each square.

Examine the decreasing rectangles. Explore and describe the mathematical relationships that result. For example, consider the length of the side of the smallest square as 1. Use the "Pythagorean" Right-Triangle Theorem to determine the lengths of the diagonals.

The spiral pattern appears frequently in nature. It is shown here in the flower.

Chapter 9 Review

Vocabulary

converse	443	irrational numbers	413	radical sign	414
cosine	460	midpoint formula	444	radicand	420
dense set	414	principal square root	414	rational numbers	413
distance formula	442	"Pythagorean" Right-Triangle Theorem	436	sine	460
hypotenuse	436			tangent	454

Key Skills & Exercises

Lesson 9.1

➤ Key Skills

Find square roots.

When the square root is irrational, find the value to the nearest hundredth.

$\sqrt{16} = 4$ $\quad\sqrt{\frac{9}{25}} = \frac{3}{5}$ $\quad\sqrt{18} \approx 4.24$ $\quad-\sqrt{0.012} \approx -0.11$

➤ Exercises

Find each square root. If the square root is irrational, find the value to the nearest hundredth.

1. $\sqrt{20}$ 4.47 **2.** $\sqrt{115}$ 10.72 **3.** $\sqrt{134}$ 11.58 **4.** $-\sqrt{67}$ -8.19 **5.** $\sqrt{\frac{9}{16}}$ $\frac{3}{4}$ **6.** $-\sqrt{0.09}$ -0.3

Lesson 9.2

➤ Key Skills

Simplify radical expressions.

a. $\sqrt{32} = \sqrt{16 \cdot 2} = \sqrt{16} \cdot \sqrt{2} = 4\sqrt{2}$

b. $\sqrt{\frac{4}{3}} = \frac{\sqrt{4}}{\sqrt{3}} = \frac{2}{\sqrt{3}} \cdot \frac{\sqrt{3}}{\sqrt{3}} = \frac{2\sqrt{3}}{3}$

c. $6\sqrt{3} + 5\sqrt{2} - 3\sqrt{3} + \sqrt{2} = (6-3)\sqrt{3} + (5+1)\sqrt{2} = 3\sqrt{3} + 6\sqrt{2}$

d. $(4\sqrt{5})^2 = (4\sqrt{5})(4\sqrt{5}) = (4 \cdot 4)(\sqrt{5} \cdot \sqrt{5}) = 16 \cdot 5 = 80$

➤ Exercises

Simplify.

7. $\sqrt{150}$ $5\sqrt{6}$ **8.** $\sqrt{\frac{16}{27}}$ $\frac{4\sqrt{3}}{9}$ **9.** $\sqrt{2} + 3\sqrt{7} - 3\sqrt{2}$ $3\sqrt{7} - 2\sqrt{2}$ **10.** $(2\sqrt{3})^2$ 12 **11.** $\sqrt{3}(2 - \sqrt{12})$ $2\sqrt{3} - 6$

Lesson 9.3

➤ Key Skills

Solve radical equations. Use radicals to solve equations.

$\sqrt{x+2} = 5 \longrightarrow (\sqrt{x+2})^2 = 5^2 \longrightarrow x + 2 = 25 \longrightarrow x = 23$

$2x^2 = 36 \longrightarrow x^2 = 18 \longrightarrow x = \pm\sqrt{18} \longrightarrow x = \pm\sqrt{9 \cdot 2} = \pm 3\sqrt{2}$

➤ *Exercises*

Solve each equation. Check your solution.

12. $\sqrt{x-7}=2$ 11

13. $\sqrt{3x+4}=1$ −1

14. $\sqrt{x+6}=x$ 3

15. $\sqrt{x^2-2x+1}=x-5$ No solution

16. $x^2=40$ $\pm 2\sqrt{10}$, or $\approx \pm 6.325$

17. $2x^2-32=0$ 4, −4

Lesson 9.4

➤ *Key Skills*

Use the "Pythagorean" Right-Triangle Theorem to find the length of a missing side of a right triangle.

Find the length of the hypotenuse of the triangle.

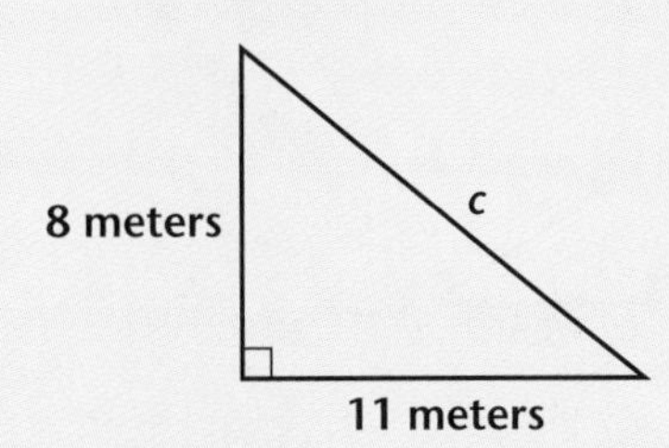

$$\begin{aligned} a^2+b^2&=c^2\\ 8^2+11^2&=c^2\\ 64+121&=c^2\\ 185&=c^2\\ c&=\pm\sqrt{185}\approx\pm 13.6 \end{aligned}$$

Since c represents the length of a segment, it must be positive. Therefore, $c \approx 13.6$ meters.

➤ *Exercises*

Use the "Pythagorean" Right-Triangle Theorem to find the length of the missing side of each right triangle.

18.

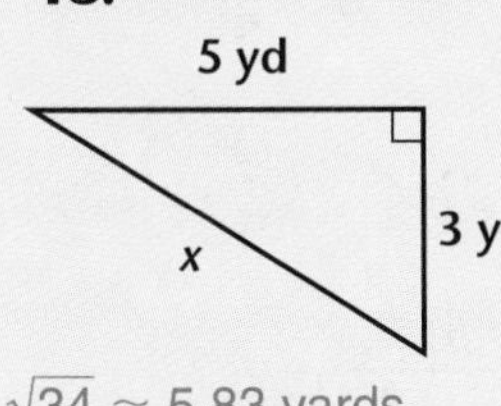

$\sqrt{34} \approx 5.83$ yards

19.

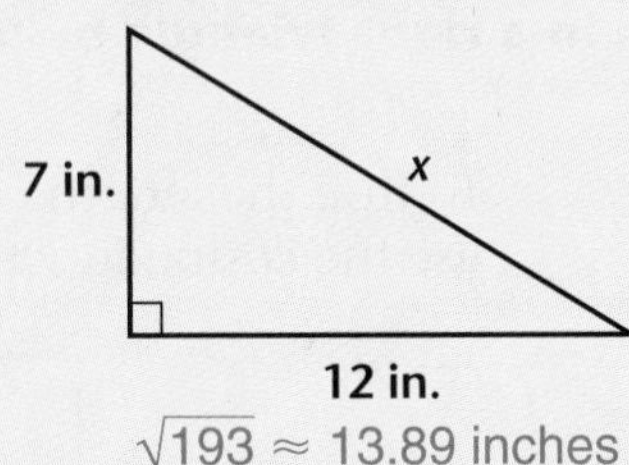

$\sqrt{193} \approx 13.89$ inches

20.

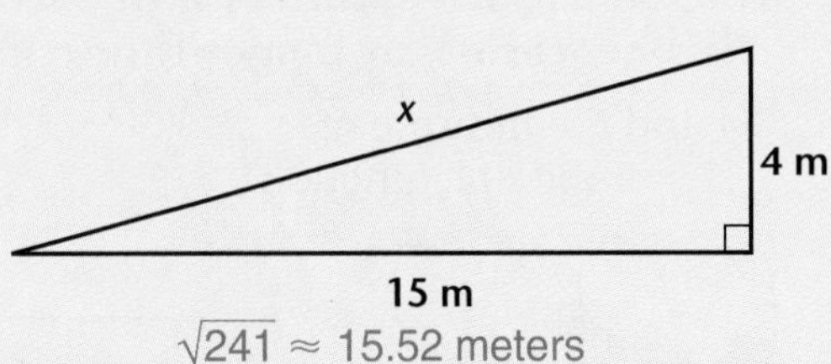

$\sqrt{241} \approx 15.52$ meters

Lesson 9.5

➤ *Key Skills*

Use the distance formula to find the distance between two points.

Use the distance formula to find the distance between (2, 3) and (−4, −1).

$$\begin{aligned} d&=\sqrt{(x_2-x_1)^2+(y_2-y_1)^2}\\ d&=\sqrt{(-4-2)^2+(-1-3)^2}\\ d&=\sqrt{36+16}\\ d&=\sqrt{52}=2\sqrt{13}\approx 7.2 \end{aligned}$$

Use the midpoint formula to find the midpoint of a segment.

Use the midpoint formula to find the midpoint of the segment whose endpoints are (−2, 6) and (3, 4).

$$\left(\frac{x_1+x_2}{2}, \frac{y_1+y_2}{2}\right)=\left(\frac{-2+3}{2}, \frac{6+4}{2}\right)=\left(\frac{1}{2}, 5\right)$$

➤ *Exercises*

25. $\left(-5, \frac{3}{2}\right)$ 26. $\left(\frac{7}{2}, 1\right)$

Find the distance between the two given points.

21. $A(0, 3)$, $B(2, 8)$ $\sqrt{29} \approx 5.39$

22. $X(-1, 5)$, $Y(3, -8)$ $\sqrt{185} \approx 13.60$

23. $G(7, 2)$, $F(-6, -1)$ $\sqrt{178} \approx 13.34$

Find the midpoint of each segment whose endpoints are given.

24. $M(5, 2)$, $N(3, 6)$ (4, 4)

25. $P(-3, 4)$, $Q(-7, -1)$

26. $R(8, -2)$, $S(-1, 4)$

Lesson 9.6

➤ Key Skills

Write the equation of a circle given the center and the radius.

The equation of a circle with center at the origin and a radius of 4 is $x^2 + y^2 = 4^2$.

The equation of a circle with center $(2, -3)$ and a radius of 5 is $(x - 2)^2 + (y + 3)^2 = 5^2$.

➤ Exercises

Write the equation of each circle with the given center and radius.

27. center $(0, 0)$, radius 2 $x^2 + y^2 = 4$

28. center $(0, 2)$, radius 6 $x^2 + (y - 2)^2 = 36$

29. center $(-2, 5)$, radius 3 $(x + 2)^2 + (y - 5)^2 = 9$

Lesson 9.7

➤ Key Skills

Find the tangent of an acute angle of a right triangle.

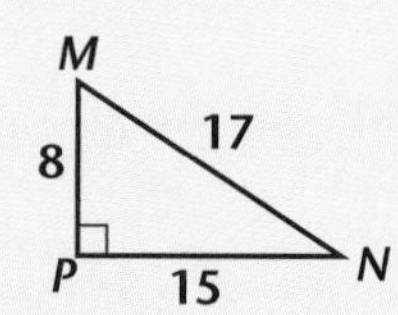

$\tan M = \frac{15}{8}$ $\tan N = \frac{8}{15}$

➤ Exercises

Find the tangent for the given angle of triangle *RST*.

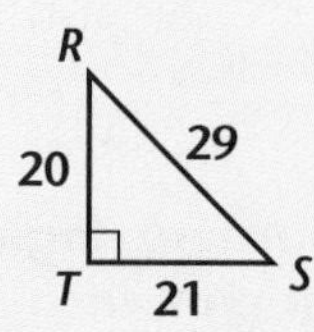

30. $\tan R$ $\frac{21}{20}$

31. $\tan S$ $\frac{20}{21}$

Lesson 9.8

➤ Key Skills

Find an angle measure or a side length in a right triangle by using the sine, cosine, or tangent function.

To find the measure of $\angle A$, use the sine function.

$$\sin A = \frac{5}{8}$$
$$\sin A = 0.625$$
$$A = \sin^{-1}(0.625)$$
$$A \approx 38.7°$$

To find the length of side b, use the cosine function.

$$\cos 28° = \frac{b}{10}$$
$$0.8829 \approx \frac{b}{10}$$
$$8.8 \approx b$$

➤ Exercises

Find the indicated length or angle measure to the nearest whole number.

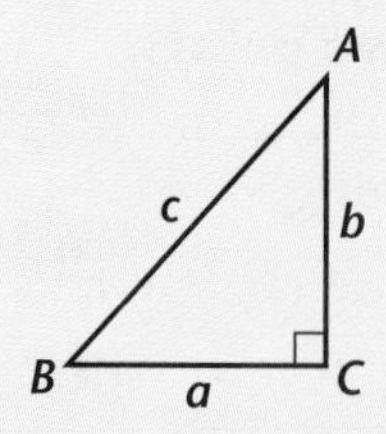

32. $a = 4$, $b = 9$, $m\angle A = \underline{\ ?\ }$ 24°

33. $m\angle A = 35°$, $c = 12$, $a = \underline{\ ?\ }$ 7

34. $b = 6$, $c = 10$, $m\angle A = \underline{\ ?\ }$ 53°

Applications

35. The bottom of a circular swimming pool has an area of 80 square feet. Find the radius to the nearest tenth. Use 3.14 for π. About 5.0 feet

36. Physics The time that it takes an object to fall d feet is given by the formula $t = \sqrt{\frac{d}{16}}$, where t is time in seconds. About how long, to the nearest second, does it take for an object dropped from the top of a 50-foot building to hit the ground? 2 seconds

Chapter 9 Assessment

Simplify.

1. $\sqrt{50} \cdot \sqrt{3}$ $5\sqrt{6}$ **2.** $\sqrt{\frac{72}{10}}$ $\frac{6\sqrt{5}}{5}$ **3.** $(\sqrt{10} - 3)^2$ $19 - 6\sqrt{10}$

4. $6\sqrt{6} + 5\sqrt{6} - \sqrt{3}$ $11\sqrt{6} - \sqrt{3}$ **5.** $\sqrt{\frac{m^8}{n^{10}}}$, where $m \geq 0$ and $n > 0$ $\frac{m^4}{n^5}$

Solve each equation. Check your solution.

6. $\sqrt{4x + 2} = 8$ $15\frac{1}{2}$ **7.** $\sqrt{x + 2} = x + 2$ $-1, -2$ **8.** $4x^2 = 48$ $\pm 2\sqrt{3}$, or $\approx \pm 3.5$

Find the length, to the nearest tenth, of the missing side(s) of each right triangle.

9.

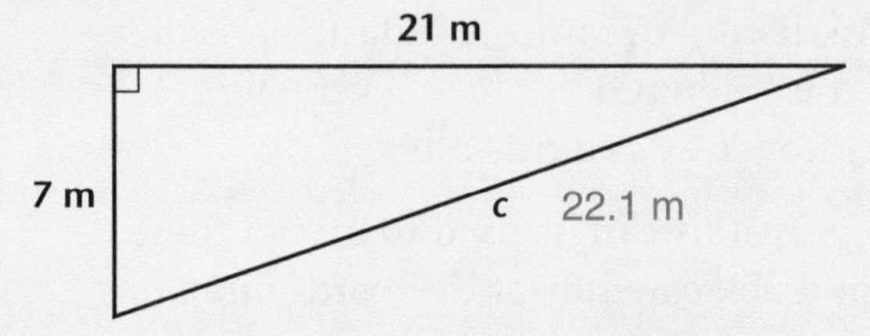

10.

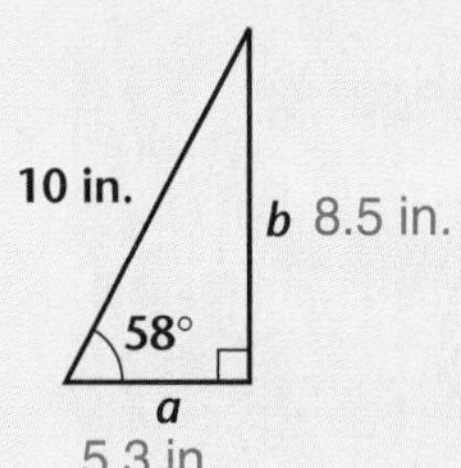

11. One boat travels 9 miles south. Another boat travels 6 miles west and meets the first boat. How far apart were the two boats when they started? $\approx$ 10.82 miles

13. $\frac{\sqrt{2}}{2}$ **14.** $\frac{12}{13}$ **15.** $\frac{12}{5}$ **16.** $\frac{5\sqrt{41}}{41}$ **17.** $\frac{5\sqrt{41}}{41}$

Find the indicated trigonometric ratio for the right triangles.

12. tan M 1
13. sin K
14. cos A
15. tan B
16. cos R
17. sin T

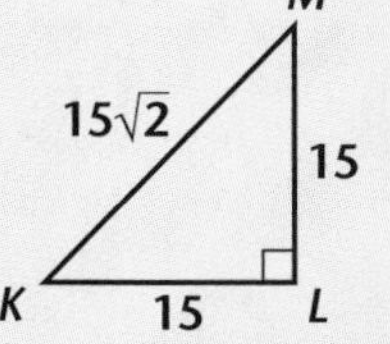

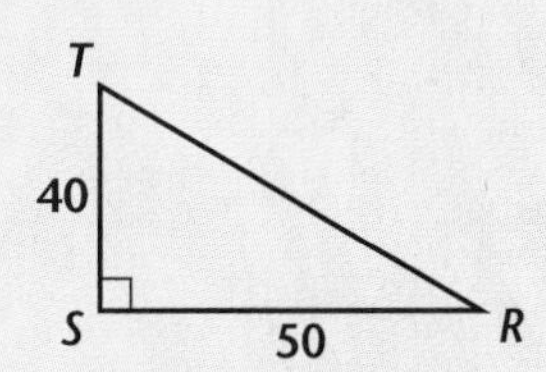

18. A ladder is leaning against a house. It forms an angle of 65° with the ground. If the ladder touches the house at a point 12 feet high, how far is the bottom of the ladder from the bottom of the house? How long is the ladder? About 5.6 ft; about 13.2 ft

Write the equation of each circle with the given center and radius.

19. $x^2 + y^2 = 81$

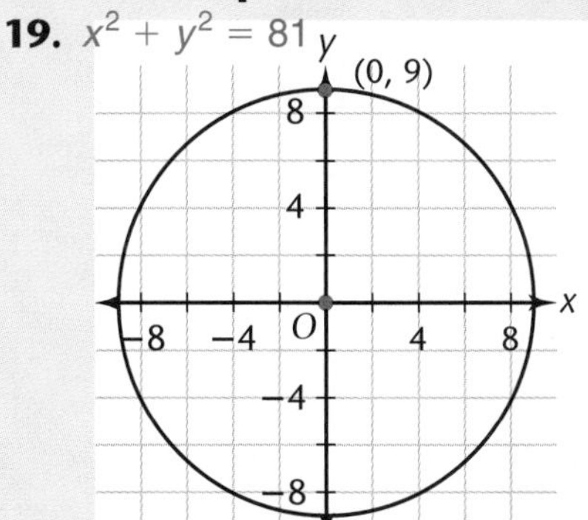

20. $(x - 3)^2 + (y + 2)^2 = 49$

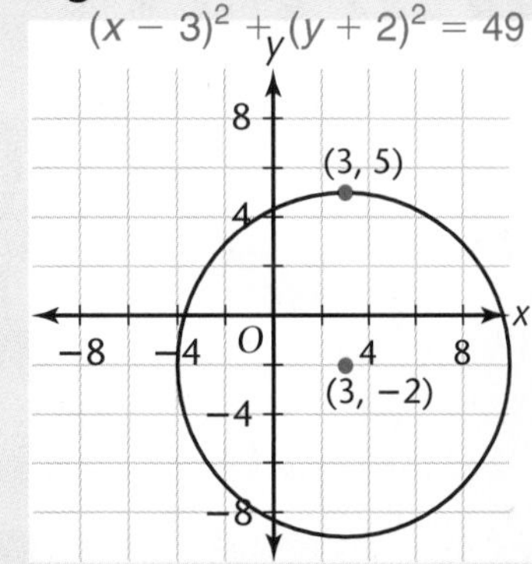

CHAPTER 10

Rational Functions

LESSONS

Have you ever noticed that it takes you longer to row upstream than downstream? Have you ever used a lever to lift a heavy object? Do you know what a gear ratio is?

The rate of the current affects your time spent rowing. The distance from the balance point determines the amount of lift for a lever. The gear ratio affects how fast you ride a bicycle.

Rational expressions can be used to model these relationships. Look carefully at the words *rational* and *irrational*. The core of each word is *ratio*. A **ratio** compares two quantities, and a **rational expression** is a fraction that compares two algebraic expressions.

Portfolio Activity

A bicycle controls the speed and effort of pedaling by the relationship of the teeth on the pedal gears to the teeth on the wheel gears. For example, if the pedal gear has 54 teeth and the wheel gear has 16 teeth, the gear ratio (wheel to pedal) is $\frac{16}{54} = 0.296$. The reciprocal is $\frac{54}{16} = 3.375$, or about 3.4. This represents the number of turns of the wheel for every full turn of the pedal. Assuming that the circumference of a bicycle tire is approximately 2.19 meters, how far will the rider travel during one full turn of the pedal?

a. The wheel gears on a bicycle might have 13, 14, 15, 16, 18, 21, and 24 teeth and the pedal gear, 49 or 54 teeth. Find all the possible gear ratios. Then find the reciprocal for each gear ratio.

b. Assume that a bicycle rider turns the pedal 60 full turns per minute. Choose 2 gear combinations and calculate the miles per hour for each.
NOTE: 1 kilometer per hour is approximately 0.62 miles per hour.

LESSON 10.1 Rational Expressions

The art institute charges a \$55 membership fee and \$4.50 for each pottery lesson.

why *People often compare data. Baseball players compare their batting averages. Business owners compare profit to total income. Students figure their average expenses. Rational expressions can be used to model comparisons such as these.*

When Saul's *total* expenses for membership and pottery lessons are considered, the following formula represents the average cost per lesson.

$$\text{Average cost} = \frac{\text{total expenses}}{\text{number of lessons}}$$

If Saul takes only 1 lesson, his cost will be \$55 + \$4.50, or \$59.50. If Saul takes 2 lessons and then stops, his average cost per lesson will be $\frac{\$55 + \$4.50(2)}{2}$, or \$32.00.

If x is the number of lessons taken, then $55 + 4.50x$ is the total expenses for all lessons plus membership. So, the averge cost per lesson can be written as:

$$\text{Average cost} = \frac{55 + 4.50x}{x}.$$

To compare the average cost per lesson for 20 lessons with the average cost for 30 lessons, evaluate the following expression for 20 lessons and 30 lessons:

$$\frac{55 + 4.50x}{x}$$

A spreadsheet or graphics calculator can be used to quickly evaluate the expression for any value of x.

A	B	C
x	55 + 4.50x	(55 + 4.50x)/x
0	55.00	error
1	59.50	59.50
2	64.00	32.00
3	68.50	22.83
4	73.00	18.25
5	77.50	15.50
6	82.00	13.67
20	145.00	**7.25**
30	190.00	**6.33**

The average cost per lesson for 20 lessons is $7.25. If Saul takes 30 lessons, his average cost per lesson is $6.33. Why does the entry in row 2, column C, read "error"? The error is caused by trying to divide by zero.

The expression $\frac{55 + 4.50x}{x}$ is an example of a *rational expression.*

RATIONAL EXPRESSION

If P and Q are polynomials and $Q \neq 0$, then an expression in the form $\frac{P}{Q}$ is a rational expression.

Since $55 + 4.50x$ and x are polynomials and $x \neq 0$, $\frac{55 + 4.50x}{x}$ is a rational expression.

A function of the form $y = \frac{P}{Q}$, or $f(x) = \frac{P(x)}{Q(x)}$, where $\frac{P}{Q}$ is a rational expression is called a **rational function**. Thus, $g(x) = \frac{55 + 4.50x}{x}$ is a rational function. Since you cannot divide by zero, g is *undefined* when x is 0. A rational expression is undefined when its denominator is equal to zero.

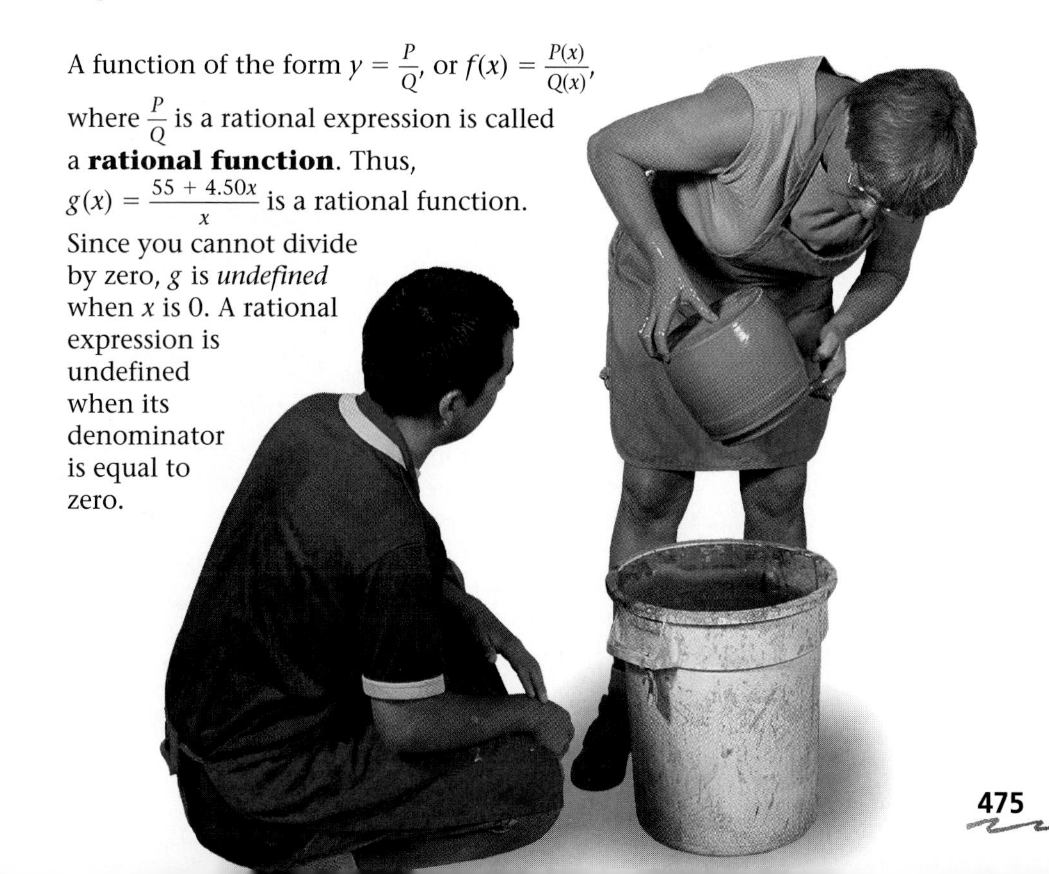

EXAMPLE 1

For what value or values is each rational expression undefined?

A $\frac{1}{x}$ **B** $\frac{a^2 - b^2}{a - b}$ **C** $\frac{m + n}{m^2 - 4m + 3}$

Solution ➤

A rational expression is undefined when its denominator is equal to zero.

A $\frac{1}{x}$ is undefined when x is 0.

B $\frac{a^2 - b^2}{a - b}$ is undefined when $a - b = 0$. That is, when $a = b$.

C $\frac{m + n}{m^2 - 4m + 3}$ is undefined when $m^2 - 4m + 3 = 0$. Since $m^2 - 4m + 3$ can be factored as $(m - 1)(m - 3)$, the rational expression $\frac{m + n}{m^2 - 4m + 3}$ is undefined when m is 1 or m is 3. ❖

Try This For what value or values is each rational expression undefined?

a. $\frac{y - 2}{y^2 - 5y + 6}$ **b.** $\frac{3x + 6x}{x - y}$

$y = 2$ or $y = 3$ $y = x$

Recall that the parent function for rational functions is $f(x) = \frac{1}{x}$. Recall from Chapter 5 that the graph of $f(x) = \frac{1}{x}$ is a hyperbola with a vertical asymptote of $x = 0$.

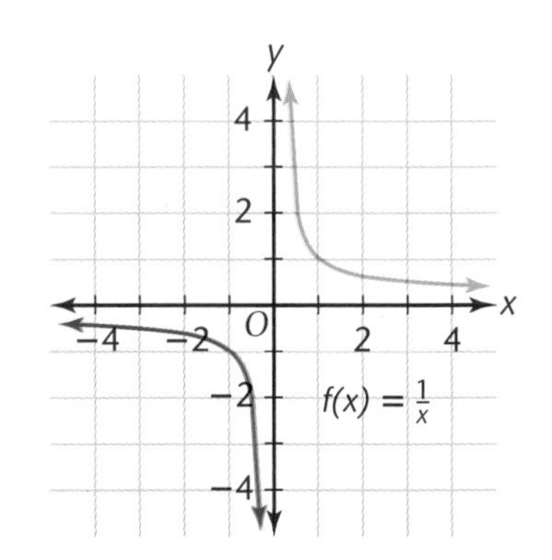

EXAMPLE 2

A Describe the transformations applied to the parent function $f(x) = \frac{1}{x}$ to produce the graph of $g(x) = \frac{55 + 4.50x}{x}$.

B Use the graph from part **A** to find the number of pottery lessons that can be taken for an average cost of \$5.60 per lesson.

Solution ➤

A The parent function is $f(x) = \frac{1}{x}$. To compare g with f, first rewrite g by using the Distributive Property, and simplify.

$$g(x) = \frac{55 + 4.50x}{x} = \frac{55}{x} + \frac{4.50x}{x} = \frac{55}{x} + 4.5$$

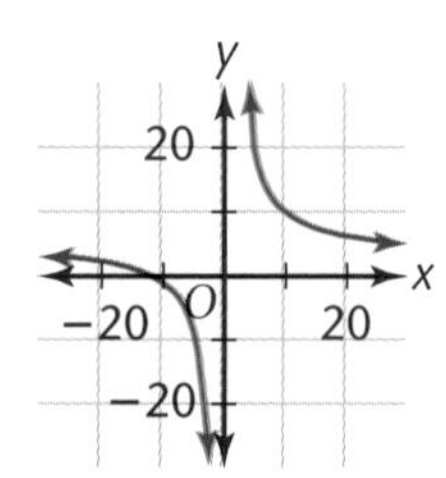

The graph of the function $g(x) = \frac{55}{x} + 4.5$ shows that the parent function has been transformed by a vertical stretch of 55 units and a translation of 4.5 units upward.

B Use the graph to find the x-value for which $f(x) = 5.60$.

The x-value, 50, represents 50 lessons.

Therefore, the average cost per lesson for 50 lessons is $5.60. ❖

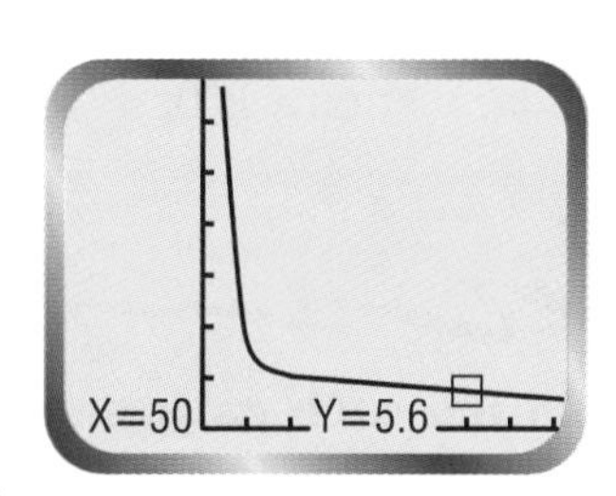

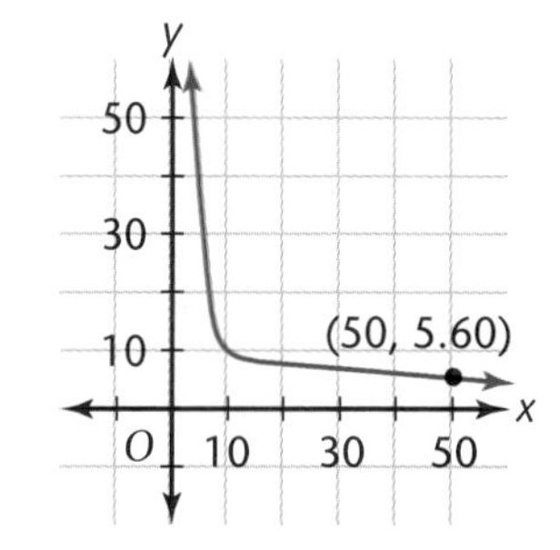

EXAMPLE 3

Describe the transformation applied to the parent function $f(x) = \frac{1}{x}$ to produce the graph of $g(x) = \frac{5}{x - 3} + 2$.

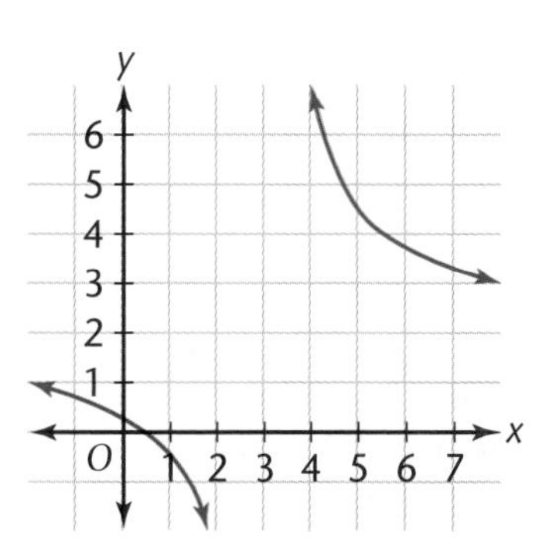

Solution ➤

The information from the function $g(x) = \frac{5}{x - 3} + 2$ shows that the parent function has been transformed by a stretch by a factor of 5 units, a shift of 3 units to the right, and a shift of 2 units upward. ❖

CT– Changing *a* results in a stretch; changing *h* results in a shift to the right or left; and changing *k* results in a shift up or down.

Try This Describe the transformations applied to the parent function $f(x) = \frac{1}{x}$ to produce the graph of $g(x) = \frac{3}{2x + 3} - 5$.

Shift to the left 1.5 units, down 5 units, and stretch by a factor of $\frac{3}{2}$.

Begin with the function $f(x) = a\left(\frac{1}{x - h}\right) + k$. First substitute different values for a, then h, then k, while keeping the other two values constant, and graph the results. Describe the effects of varying a, h, and k on the graph.

Communicate

1. Explain how a rational expression is defined.
2. What condition causes a rational expression to be undefined?

Graph the function $h(x) = \frac{12x - 7}{x}$.

3. Describe all the transformations applied to $g(x) = \frac{1}{x}$ in order produce the graph of $h(x)$.
4. Discuss how to express $h(x)$ in a form which shows that it is a transformation of the parent function $g(x) = \frac{1}{x}$.
5. Describe the behavior of $h(x)$ when x is 0.

Practice & Apply

Identify the values for which each rational expression is undefined.

6. $\frac{3x + 1}{x}$ 0 **7.** $\frac{6}{y - 2}$ 2 **8.** $\frac{2m - 5}{6m^2 - 3m}$ $0, \frac{1}{2}$ **9.** $\frac{x^2 + 2x - 3}{x^2 + 4x - 5}$ $1, -5$

Evaluate each rational expression for $x = 1$ and $x = -2$. Write "undefined" if appropriate.

10. $\frac{5x - 1}{x}$ **11.** $\frac{4x}{x - 1}$ **12.** $\frac{x^2 - 1}{x^2 - 4}$ **13.** $\frac{x^2 + 2x}{x^2 + x + 2}$

Describe all the transformations applied to $f(x) = \frac{1}{x}$ in order to produce the graph of each rational function.

14. $g(x) = \frac{1}{x - 5}$ **15.** $g(x) = \frac{2}{x}$ **16.** $g(x) = \frac{-1}{x} + 3$

17. $g(x) = \frac{1}{x + 3}$ **18.** $g(x) = \frac{4}{x + 2} - 5$ **19.** $g(x) = \frac{-2}{x - 1} + 4$

Graph each rational function. List the value for which the function is undefined.

20. $h(x) = \frac{1}{x} + 1$ **21.** $h(x) = \frac{1}{x + 2}$ **22.** $h(x) = \frac{-1}{x} + 3$

23. $h(x) = \frac{3}{x - 1}$ **24.** $h(x) = \frac{2}{x - 3} + 4$ **25.** $h(x) = \frac{-2}{x - 4} + 3$

Technology The air pressure, P, in pounds per square inch, in the tires of a car can be modeled by the rational equation $P = \frac{(1.1)10^5}{V}$, where V is volume in cubic inches.

26. Use graphing technology to graph the rational function P.
27. The owner's manual calls for a pressure of 32 pounds per square inch. What volume of air does a tire hold at this pressure?

Look Back

28. Your class orders 500 mugs with all the names of the class members. In the first week, 356 mugs were sold. What percent of the mugs was sold the first week? **[Lesson 1.2]** 71.2%

29. Write an equation for the line that passes through the points (2, 1) and (−3, 5). **[Lesson 2.2]** $y = -\frac{4}{5}x + \frac{13}{5}$

30. Find the inverse of $\begin{bmatrix} 0.3 & 0.4 \\ 0.5 & 0.6 \end{bmatrix}$. **[Lesson 3.4]** $\begin{bmatrix} -30 & 20 \\ 25 & -15 \end{bmatrix}$

31. How many different 4-digit numbers are possible if the digits 1–9 can be used with repetition? **[Lesson 4.5]** 6561

32. What transformation is applied to $f(x) = x^2$ to obtain the graph of $y = (x - 5)^2$? **[Lesson 5.5]** Shifts 5 units to the right

33. Solve $x^2 - 20 = -x$. **[Lesson 8.2]** 4, −5

34. Cultural Connection: Americas Benjamin Banneker, the first African American astronomer, is most famous for the almanacs he created from 1791 to 1797. Banneker also enjoyed solving mathematical problems, such as this one from his journal. **[Lesson 9.4]** 102.65 feet

If the length of each ladder is 60 feet, find the width of the street that runs between the buildings to the nearest hundredth.

Look Beyond

35. Describe the transformations that produce $y = \frac{5}{x - 1} + 2$ from $y = \frac{1}{x}$.

36. Sketch the graph of $y = \frac{5}{x - 1} + 2$.

LESSON 10.2 Inverse Variation

A volunteer organization has 5000 trees to plant for an ecology project.

why

Variation describes how a change in one quantity changes another quantity. Recall from Chapter 2 that direct variation describes a linear relationship between two variables. Inverse variation describes a nonlinear relationship.

If 1 person plants all 5000 trees, it will take 500 hours; if 2 people plant, it will take $\frac{500}{2}$, or 250, hours; if 3 people plant, it will take $\frac{500}{3}$, or $166\frac{2}{3}$, hours, and so on. This relationship is an example of *inverse variation*.

INVERSE VARIATION

In an **inverse variation**, y varies inversely as x. That is, $xy = k$, or $y = \frac{k}{x}$, where k is the **constant of variation**, $k \neq 0$, and $x \neq 0$.

EXAMPLE 1

Ecology How many hours are needed to plant all 5000 of the trees when 50 people plant them?

Solution ➤

Since the number of hours required to plant all the trees varies inversely as the number of people planting trees, write an inverse variation equation, $xy = k$ or $y = \frac{k}{x}$.

Let x represent the number of people, and let y represent the number of hours. First find the constant of variation, k.

$$\begin{aligned} xy &= k \\ 1 \cdot 500 &= k \\ 500 &= k \end{aligned} \qquad \begin{aligned} xy &= k \\ 2 \cdot 250 &= k \\ 500 &= k \end{aligned} \qquad \begin{aligned} xy &= k \\ 3 \cdot 166\tfrac{2}{3} &= k \\ 500 &= k \end{aligned}$$

The constant is 500, so substitute 500 for k. To find the number of hours required for 50 people, substitute 50 for x.

$$\begin{aligned} xy &= k \\ 50 \cdot y &= 500 \\ y &= 10 \end{aligned}$$

Thus, it would take 10 hours for 50 people to plant all 5000 trees. ❖

Try This If y is 4 when x is 12, write an equation that shows how y is related to x. Assume that y varies inversely as x. $xy = 48$ or $y = \frac{48}{x}$

In Example 1, k is the constant of variation. What physical quantity does k represent? The constant of variation, k, represents the total number of work hours needed to plant 5000 trees.

Cultural Connection: Africa One of the earliest references to variation comes from an ancient Egyptian papyrus that describes the commodity exchange system used in some Egyptian cities. Bread was used as a form of currency, and its value was based on the amount of wheat that was used to make each loaf. A rating system was created to value the bread.

The table shows the value of 1000 loaves of bread at various ratings.

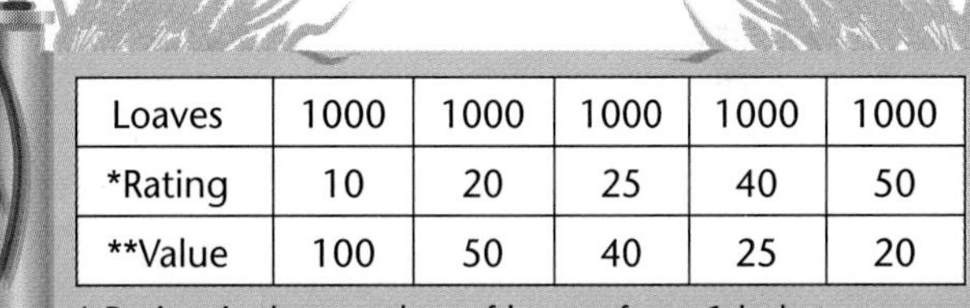

Loaves	1000	1000	1000	1000	1000
*Rating	10	20	25	40	50
**Value	100	50	40	25	20

* Rating is the number of loaves from 1 hekat.
** Value is the total number of hekats used.

Notice that as the *rating* for each 1000 loaves *increases* the *value* for each 1000 loaves *decreases*. This is an example of inverse variation.

Algebraically, the variables of rating and value are inversely related.

$$\begin{aligned} x \cdot y &= k \\ \text{rating} \cdot \text{value} &= \text{loaves} \\ 20\,\frac{\text{loaves}}{\text{hekat}} \cdot 50 \text{ hekats} &= 1000 \text{ loaves} \end{aligned}$$

The bread's rating is said to vary inversely as its value. ❖

Physics

A 60-gram mass on one end of a fulcrum is balanced by placing an appropriate weight on the other end. This procedure is repeated for fulcrum distances of 10, 15, 20, and 25 centimeters from the end. The table below gives the findings of a group of students who performed the experiment.

Fulcrum distance	5	10	15	20	25
Mass	60	30	20	15	12

Examine the products of these corresponding distance and mass pairs. Is there an inverse variation between these variables? Explain. Yes; the product is constant when the variables are multiplied together.

EXAMPLE 2

Analyze the experiment just described.

A Write a sentence that describes the variation between the variables.

B Write an equation that relates the variables.

C Graph your equation, and analyze the type of function that the graph represents.

Solution

This experiment illustrates inverse variation because the product of each distance from the balance point and the corresponding mass is constant.

A The balance mass varies inversely as the distance from the end to the fulcrum.

B The inverse variation relationship means that

$$\text{mass} \cdot \text{fulcrum distance} = k.$$

For the values of 15 grams and 20 centimeters,

$$15 \cdot 20 = 300.$$

The constant of variation, k, is 300.

C The equation for this inverse variation is $m \cdot d = 300$.

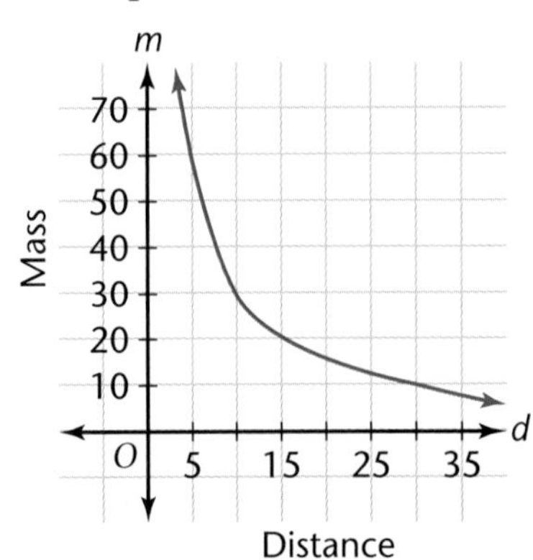

If $md = 300$, then $m = \frac{300}{d}$. The equation represents a rational function that is a transformation of the parent function $y = \frac{1}{x}$ by a vertical stretch of 300. ❖

Investments The amount of time, t, it takes to double your money at *compound* interest can be approximated by the function $t = \frac{72}{r}$, where r is the annual interest rate written as a whole number. In business this function is usually referred to as the **Rule of 72**.

CT– The graph of $t(r) = \frac{72}{r}$ has been stretched by a factor of 72. The reasonable domain and range for the Rule of 72 function are all positive real numbers. The constant of variation stretches the graph by a factor of 72.

EXAMPLE 3

A How long does it take to double your money at 6% compound interest?

B Graph the Rule of 72 function.

Solution ➤

A To use the Rule of 72, substitute 6 for r; then $t = \frac{72}{6} = 12$. It takes about 12 years to double your money at 6%.

B Choose interest rates for r. Make a table and sketch a graph from the points.

r%	t
4	18
6	12
8	9
12	6

❖

Describe the graph of the Rule of 72 as a transformation of the parent graph, $y = \frac{1}{x}$. What are the reasonable domain and range of the function for the Rule of 72? What effect does the constant of variation, 72, have on the function?

EXERCISES & PROBLEMS

Communicate

1. Give two equivalent equations expressing inverse variation if y varies inversely as x.
2. Explain what inverse variation between rate and time means. Give a real-world example of inverse variation.
3. If y is 3 when x is 8 and y varies inversely as x, explain how to find y when x is 2.
4. Explain how to use the Rule of 72 to find how long it would take to double your money at 4% interest compounded yearly.
5. Describe the parent function for inverse variation functions.

Practice & Apply

Determine whether each equation is an equation of inverse variation.

6. $rt = 400$ Yes **7.** $y = \frac{-28}{x}$ Yes **8.** $x - y = 10$ No **9.** $\frac{n}{5} = \frac{3}{m}$ Yes

10. $\frac{x}{y} = \frac{1}{2}$ No **11.** $x = 10y$ No **12.** $a = \frac{42}{b}$ Yes **13.** $r = t$ No

For Exercise 14–19, y varies inversely as x. If y is

14. 8 when x is 6, find x when y is 12. 4

15. 9 when x is 12, find y when x is 36. 3

16. 3 when x is 32, find x when y is 4. 24

17. 3 when x is -8, find x when y is -4. 6

18. $\frac{3}{5}$ when x is -60, find y when x is 2. -18

19. $\frac{3}{4}$ when x is 12, find y when x is 27. $\frac{1}{3}$

20. 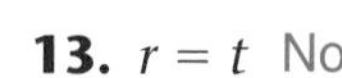 **Geometry** If the area is constant, the base of a triangle varies inversely as the height. When the base is 22 centimeters, the height is 36 centimeters. Find the length of the base when the height is 24 centimeters.
33 centimeters

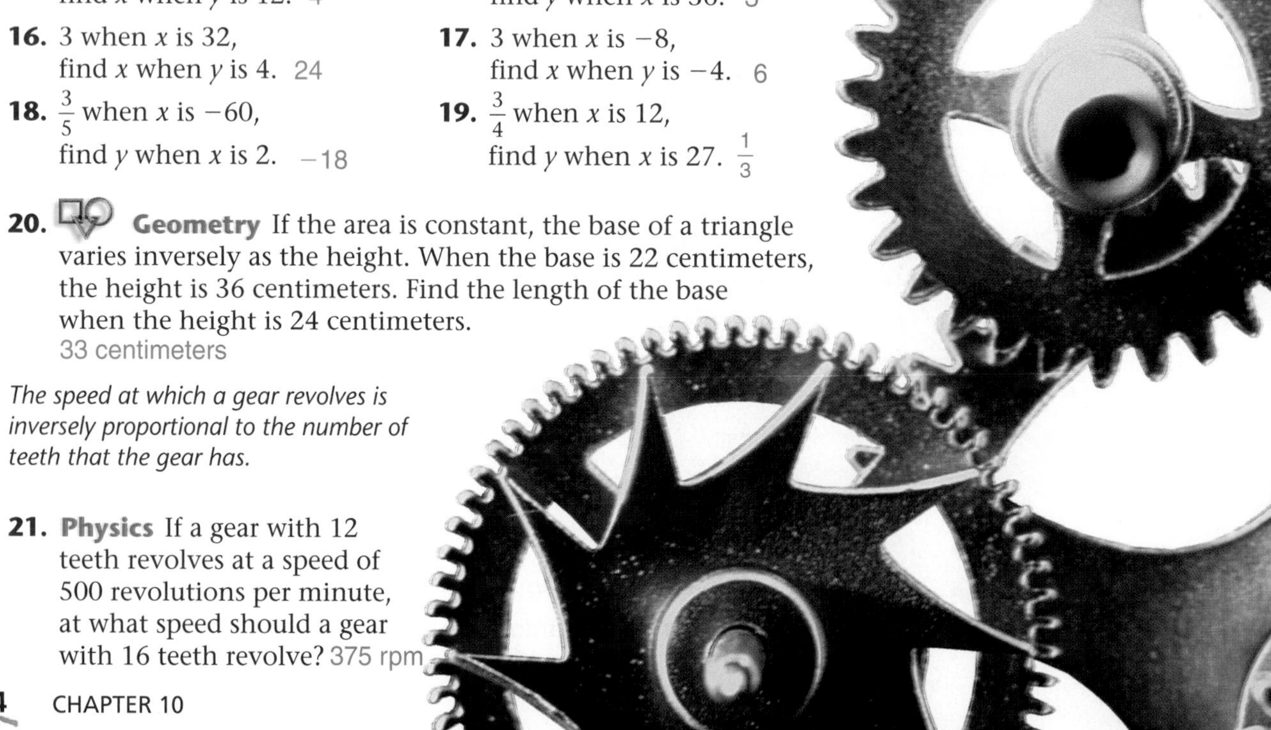

The speed at which a gear revolves is inversely proportional to the number of teeth that the gear has.

21. **Physics** If a gear with 12 teeth revolves at a speed of 500 revolutions per minute, at what speed should a gear with 16 teeth revolve? 375 rpm

22. Music A harp string vibrates to produce sound. The number of these vibrations varies inversely as its length. If a string 28 centimeters long vibrates 510 times per second, how long is a string that vibrates 340 times per second? 42 centimeters

23. Time varies inversely as the average rate of speed over a given distance. When Mike travels 6 hours, his average rate is 80 kilometers per hour. Find Mike's time when his rate is 90 kilometers per hour. $5\frac{1}{3}$ hr, or 5 hr, 20 min

24. The frequency of a radio wave varies inversely as its wavelength. If a 200-meter wave has a frequency of 3000 kilocycles, what is the wavelength of a wave that has a frequency of 2000 kilocycles? 300 meters

25. Travel Time varies inversely as the average speed over a given distance. A jet takes about 2.7 hours to fly from Boston to Paris, averaging 2200 kilometers per hour. How long would it take a plane traveling at an average speed of 1760 kilometers per hour to make the same trip? 3.375 hours, or 3 hr, 22.5 min

26. Investments If Harold invests $500 of his 4-H prize money at 5% interest compounded yearly, how long will it take him to double his money? Approximately 15 years, since interest is compounded at the end of the year.

27. Geometry The area of a rectangle is 36 square centimeters. What is the length of a rectangle with the same area that has twice the original width? Explain why the formula for the area of a rectangle is a model for inverse variation. Half the original length; because the dimensions of a rectangle with a constant area vary inversely. Let area = k; then $lw = k$.

Look Back

28. The first three terms of a sequence are 1, 5, and 12. The second difference is a constant 3. Find the next three terms of the sequence. **[Course 1]** 22, 35, 51

29. The 5-inch-by-7-inch picture of your award winning cat, Kat, is surrounded by a frame with a width of $3x$ inches. Find the dimensions of the outside of the frame. **[Lesson 1.2]** $5 + 6x$ in. by $7 + 6x$ in.

30. The total cost for your order of computer discs, including the $1.95 handling charge, is $57.60. The discs are sold in boxes of 10 and were on sale for $7.95 a box. How many discs were in the order? **[Lesson 1.4]** 70 discs

31. What are the coordinates of the vertex of $y = -x^2 - 4x$? **[Lesson 8.1]** $(-2, 4)$

Look Beyond

32. Subtract the rational expression. $\frac{1}{x-3} - \frac{x}{x^2-9}$ $\frac{3}{x^2-9}$

33. The intensity, I, of light from a light source varies inversely as the square of the distance, d^2, from the light source. If I is 30 units when d is 6 meters, find I when d is 3 meters. 120 units

LESSON 10.3 Simplifying Rational Expressions

why *In arithmetic, a fraction is simplified when it is expressed in lowest terms. In algebra, rational expressions are simplified in much the same way. Algebraic operations on rational expressions follow the same mathematical steps as those used on fractions.*

Recall the steps to reduce $\frac{18}{24}$ to lowest terms.

$\frac{18}{24} = \frac{2 \cdot 3 \cdot 3}{2 \cdot 2 \cdot 2 \cdot 3}$ Write both the numerator and denominator as a product of prime factors.

$= \frac{3}{2 \cdot 2} \cdot \frac{\mathbf{2}}{\mathbf{2}} \cdot \frac{\mathbf{3}}{\mathbf{3}}$ Rewrite to show fractions equal to one, $\frac{\mathbf{2}}{\mathbf{2}} = 1$ and $\frac{\mathbf{3}}{\mathbf{3}} = 1$.

$= \frac{3}{4}$ Simplify.

The same procedure is used to simplify or reduce a rational expression to lowest terms. A rational expression is in simplest form when the numerator and denominator have no common factors other than 1 or -1.

EXAMPLE 1

Simplify $\frac{5a + 10}{5a}$. Write any restrictions on the variable.

Solution ➤

$\frac{5a + 10}{5a} = \frac{5(a + 2)}{5a}$ Factor the numerator and denominator.

$= \frac{5}{5} \cdot \frac{a + 2}{a}$ Rewrite to show fractions equal to 1: $\frac{5}{5} = 1$.

$= \frac{a + 2}{a}$, where $a \neq 0$ Simplify.

One way to check your work is to replace the variable by a convenient value. For example, substitute 10 for *a*, and evaluate.

$$\frac{5a + 10}{5a} = \frac{a + 2}{a}$$

$$\frac{5(10) + 10}{5(10)} \stackrel{?}{=} \frac{10 + 2}{10}$$

$$\frac{6}{5} = \frac{6}{5} \quad \text{True}$$ ❖

Why is it important that the check for the original fraction and the check for the reduced fraction be the same?

STEPS TO REDUCE RATIONAL EXPRESSIONS

1. Factor the numerator and denominator.
2. Express, or think of, each common-factor pair as 1.
3. Simplify.

CT– The expression $\frac{x}{x^2 + x}$ can be reduced by factoring the denominator. $\frac{x}{x^2 + x} = \frac{x}{x(x + 1)} = \frac{1}{x + 1}$

Be sure that only common factors are eliminated when you simplify. Notice the common factors in the examples below.

Expression	Common factor	Reduced form
$\frac{x(a + 3)}{2x}$, where $x \neq 0$	x	$\frac{a + 3}{2}$
$\frac{x + 4}{x}$, where $x \neq 0$	none	cannot be simplified further
$\frac{2x(x + 1)}{4x^2}$, where $x \neq 0$	$2x$	$\frac{x + 1}{2x}$, where $x \neq 0$
$\frac{2x^2 + 3}{x}$, where $x \neq 0$	none	cannot be simplified further

Explain why $\frac{x}{x^2 + x} = \frac{1}{x + 1}$, where $x \neq 0$ and $x \neq -1$, is a true statement.

Sometimes a rational expression can be simplified by first factoring the numerator and denominator. Remember to check the denominator for restrictions on the value of the variable.

EXAMPLE 2

Simplify $\frac{3x - 6}{x^2 + x - 6}$. Write any restrictions on the variable.

Solution ➤

$\frac{3x - 6}{x^2 + x - 6} = \frac{3(x - 2)}{(x + 3)(x - 2)}$ — Factor numerator and denominator.

$= \frac{3}{x + 3} \cdot \frac{\mathbf{x - 2}}{\mathbf{x - 2}}$ — Rewrite to show fractions equal to 1.

$= \frac{3}{x + 3}$, where $x \neq 2, -3$ — Simplify. Write restrictions from the original expression. ❖

Try This Simplify $\frac{x^2 + 3x - 4}{x - 1}$. Write any restrictions on the variable. $x + 4, x \neq 1$

Simplifying rational expressions may require many steps. Examine each expression carefully to see if it can be factored.

EXAMPLE 3

Simplify $\frac{x^2 + x - 20}{16 - x^2}$. Write any restrictions on the variable.

Solution ➤

Rewrite the denominator with the variable first. In this case, factor -1 from the denominator.

$\frac{x^2 + x - 20}{16 - x^2} = \frac{x^2 + x - 20}{(-1)(x^2 - 16)}$ — Factor -1 from the denominator.

$= \frac{(x + 5)\mathbf{(x - 4)}}{(-1)(x + 4)\mathbf{(x - 4)}}$ — Factor the numerator and denominator.

$= \frac{x + 5}{(-1)(x + 4)}$ — The common factor is $x - 4$. Simplify.

$= \frac{x + 5}{-x - 4}$, where $x \neq 4, -4$ — Multiply $x + 4$ in the denominator by -1. ❖

CT– You must add the restriction $x \neq 1$ because if $x = 1$, $\frac{x - 1}{1 - x} = \frac{0}{0}$, which is undefined.

Explain why the opposite of $(x - 4)$ can be written as $(4 - x)$.
The opposite of $(x - 4)$ is $-(x - 4)$. By the Distributive Property, $-(x - 4) = -x + 4$; and by the Commutative Property, $-x + 4 = 4 - x$.

Try This Simplify $\frac{5x^3 - 20x}{12 - 3x^2}$. Write any restrictions on the variable. $-\frac{5x}{3}, x \neq 2, -2$

Explain why you cannot write $\frac{x - 1}{1 - x} = -1$ without adding restrictions. What restrictions must be added?

EXAMPLE 4

Find the ratio of the volume of a right circular cylinder with radius r and height h to its surface area.

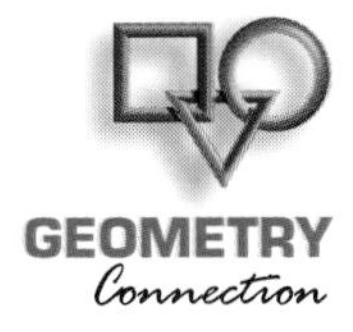

Solution ➤

$$\frac{\text{Volume}}{\text{Surface area}} = \frac{\pi r^2 h}{2\pi r^2 + 2\pi rh} \quad \text{Given}$$

$$= \frac{\pi r rh}{2\pi r(r + h)} \quad \text{Factor.}$$

$$= \frac{rh}{2(r + h)}, \text{ where } r \neq -h \quad \text{Simplify.}$$

Therefore, the ratio $\frac{\text{Volume}}{\text{Surface area}}$ is $\frac{rh}{2(r + h)}$. ❖

Try This Find the ratio of the surface area of a cube to its volume.

$\frac{6}{x}$, where x is the length of a side.

EXERCISES & PROBLEMS

Communicate

1. Explain what makes a rational expression undefined.
2. Discuss how to find the restrictions on the variable in a rational expression.
3. Explain what is meant by a common factor.
4. Describe the process that reduces $\frac{x + 1}{x^2 + 2x + 1}$ to lowest terms.
5. Explain how the expression $x - 7$ is related to $7 - x$.

Practice & Apply

For what values of the variable is each rational expression undefined?

6. $\frac{10x}{x-3}$ 3

7. $\frac{10}{5-y}$ 5

8. $\frac{r-6}{r}$ 0

9. $\frac{7p}{p-4}$ 4

10. $\frac{k-3}{3-k}$ 3

11. $\frac{(a+3)(a+4)}{(a-3)(a+4)}$ 3, −4

12. $\frac{c-4}{2c-10}$ 5

13. $\frac{3}{y(y^2-5y+6)}$ 0, 2, 3

Write the common factors.

14. $\frac{9}{12}$ 3

15. $\frac{3(x+4)}{6x}$ 3

16. $\frac{x-y}{(x+y)(x-y)}$ $x - y$

17. $\frac{r+3}{r^2+5r+6}$ $r + 3$

Simplify. Write any restrictions on the variable.

18. $\frac{16(x+1)}{30(x+2)}$

19. $\frac{3(a+b)}{6(a-b)}$

20. $\frac{4(c+2)}{10(2+c)}$

21. $\frac{3(x+y)(x-y)}{6(x+y)}$

22. $\frac{6m+9}{6}$

23. $\frac{7t+21}{t+3}$

24. $\frac{12+8x}{4x}$

25. $\frac{3d^2+2d}{3d+1}$

26. $\frac{b+2}{b^2-4}$

27. $\frac{x-2}{x^2+2x-8}$

28. $\frac{-(a+1)}{a^2+8a+7}$

29. $\frac{4-k}{k^2-k-12}$

30. $\frac{c^2-9}{3c+9}$

31. $\frac{3n-12}{n^2-7n+12}$

32. $\frac{y^2+2y-3}{y^2+7y+12}$

33. $\frac{a^2-b^2}{(a+b)^2}$

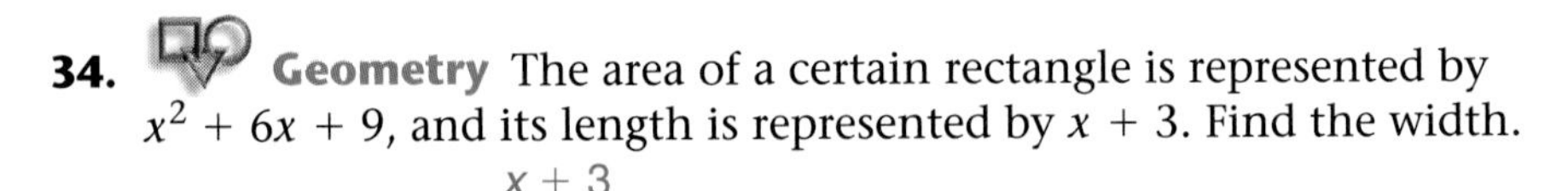

34. **Geometry** The area of a certain rectangle is represented by $x^2 + 6x + 9$, and its length is represented by $x + 3$. Find the width. $x + 3$

35. **Biology** Two equal populations of animals, one a predator and the other its prey, were released into the wild. Five years later a survey showed that the predator population had multiplied by 3, while the prey population had multiplied by 8. Then 12 predators and 32 prey were released into the area. What is the numerical ratio of predator to prey after the release? 3:8

Geometry Exercises 36 and 37 refer to the cube below.

36. To find the volume of a cube, multiply the length by the width by the height. Write an expression to represent the volume of this cube. x^3

37. To find the total surface area of a cube, multiply the area of one face by 6. Write an expression to represent the total surface area of this cube. $6x^2$

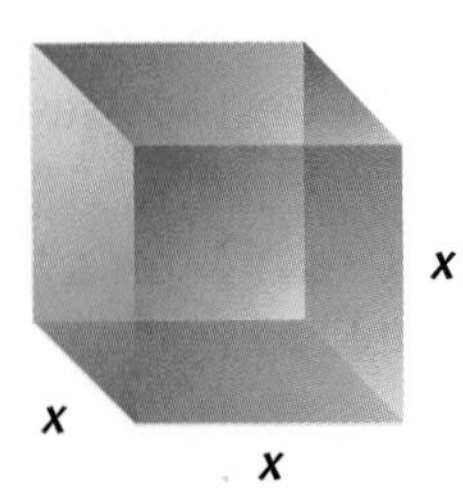

Look Back

38. Savings How much does Heather have to deposit in the bank at 4.5% interest, compounded yearly, to have $1500 for the stereo on her 16th birthday? **[Lesson 1.2]** $1373.60

In 2 years, Heather wants to buy a stereo for her 16th birthday.

39. Chin-Lyn buys a new sweater at a 20% discount and pays $23.60, before tax. What is the sweater's original price? **[Lesson 1.3]** $29.50

40. Find the equation of the line connecting the points $\left(\frac{2}{3}, -2\right)$ and $\left(-\frac{5}{6}, 3\frac{1}{3}\right)$. **[Lesson 2.2]** $y = -\frac{32}{9}x + \frac{10}{27}$, or $96x + 27y = 10$

41. Multiply $\begin{bmatrix} -3 & 2 \\ -1 & 4 \end{bmatrix} \cdot \begin{bmatrix} 5 & -3 \\ 3 & -6 \end{bmatrix}$. **[Lesson 3.3]** $\begin{bmatrix} -9 & -3 \\ 7 & -21 \end{bmatrix}$

42. Probability How many different 3-letter arrangements can be formed from the letters in the word PHONE? **[Lesson 4.5]** 60

43. If $f(x) = 3^x$, then explain how $g(x) = 3^{x+2} - 5$ compares with $f(x)$. **[Lesson 6.6]**

44. Multiply $(x + 2)(x - 2)$. **[Lesson 7.4]** $x^2 - 4$

45. Write the equation of the parabola that passes through the point $(0, -1)$ and whose vertex is $(2, 3)$. **[Lesson 8.1]** $y = -(x - 2)^2 + 3$

46. Which of the parabolas open downward? **[Lesson 8.1]** *b, c*

a. $y = 3x^2 - 8$ **b.** $y = 9 - x^2$ **c.** $y = -x^2 + 4$ **d.** $y = \frac{3}{4}x^2 - 6$

47. Simplify $\sqrt{\frac{2}{3}} - \sqrt{96}$. **[Lesson 9.2]**

$\frac{-11\sqrt{6}}{3}$

Look Beyond

48. Eight squares are the same size, but each is a different color. The squares are stacked on top of each other as shown. Identify the order of the squares numerically from top to bottom. The top square is number 1.

LESSON 10.4 Operations With Rational Expressions

why *Operations with rational expressions are like operations with ordinary fractions. You will need your arithmetic skills in order to add, subtract, multiply, and divide rational expressions.*

Anna's rowing rate is 4 times the rate of the river's current.

Each year a local club sponsors the River Challenge Day at the riverfront in order to support safety and recreational activities. One of the most popular activities is the Current Challenge. The current challenge requires a participant to paddle a kayak up and down a 1-mile portion of the river. Is there a simple rational expression that models Anna's round-trip time?

Write a formula for the total time it takes to paddle upstream and downstream.

$$\underbrace{\text{time upstream}}_{\text{against current}} + \underbrace{\text{time downstream}}_{\text{with current}} = \text{total time}$$

Write the formula for time in terms of distance and rate.

If distance = rate · time, then time $= \frac{\text{distance}}{\text{rate}}$.

Thus, $\frac{\text{distance}}{\text{rate against current}} + \frac{\text{distance}}{\text{rate with current}} = \text{total time}$.

The distance downstream is 1 mile. Let x represent the rate of the current. From the caption on page 492, Anna's rate is $4x$.

$$\frac{1 \text{ mile}}{4x \text{ mph} - x \text{ mph}} + \frac{1 \text{ mile}}{4x \text{ mph} + x \text{ mph}} = \text{total time}$$

The equation that models the relationship is

$$\frac{1}{3x} + \frac{1}{5x} = \text{total time, where } x \neq 0.$$

To find a simple rational form for $\frac{1}{3x} + \frac{1}{5x}$, add the rational expressions in the same way that you add rational numbers with unlike denominators. Compare the methods for adding $\frac{1}{30} + \frac{1}{50}$ and $\frac{1}{3x} + \frac{1}{5x}$.

Adding Rational Numbers	Adding Rational Expressions	
$\frac{1}{30} + \frac{1}{50}$	$\frac{1}{3x} + \frac{1}{5x}$	Given
$\frac{5}{5} \cdot \frac{1}{30} + \frac{3}{3} \cdot \frac{1}{50}$	$\frac{5}{5} \cdot \frac{1}{3x} + \frac{3}{3} \cdot \frac{1}{5x}$	Find a common denominator.
$\frac{5}{150} + \frac{3}{150}$	$\frac{5}{15x} + \frac{3}{15x}$	Change to common denominators.
$\frac{8}{150}$	$\frac{8}{15x}$	Simplify.

Anna's time in terms of the rate of the current is $\frac{8}{15x}$ where $x \neq 0$.

In working with rational expressions, why is it important to remember to check the denominator for restrictions on the variable in the original expression?

When adding the rational expressions on page 493, it is necessary to find a common denominator by multiplying: $\frac{5}{5} \cdot \frac{1}{3x} = \frac{5}{15x}$ and $\frac{3}{3} \cdot \frac{1}{5x} = \frac{3}{15x}$. This illustrates the method used for multiplying any two rational expressions.

EXAMPLE 1

Multiply $\frac{x-2}{x+3} \cdot \frac{x+3}{x-5}$. Write any restrictions on the variable.

Solution ➤

$\frac{x-2}{x+3} \cdot \frac{x+3}{x-5} = \frac{(x-2)(x+3)}{(x+3)(x-5)}$	Multiply the numerators and denominators.
$= \frac{x-2}{x-5} \cdot \frac{x+3}{x+3}$	Find fractions that equal 1.
$= \frac{x-2}{x-5}$, where $x \neq 5, -3$	Simplify, and write any restrictions on the variable. ❖

Try This Multiply $\frac{7}{x} \cdot \frac{x^2}{14} \cdot \frac{5}{x}$. Write any restrictions on the variable. $\frac{5}{2}$, $x \neq 0$

One way to add or subtract two rational expressions such as $\frac{1}{x+2}$ and $\frac{1}{x-1}$ is to rewrite the expressions so that they have a common denominator. This can be done by multiplying each expression by the appropriate equivalent for 1.

$$\frac{x}{x+2} + \frac{x}{x-1} = \frac{\mathbf{x-1}}{\mathbf{x-1}} \cdot \frac{x}{x+2} + \frac{\mathbf{x+2}}{\mathbf{x+2}} \cdot \frac{x}{x-1}$$

Since the expressions now have common denominators, the numerators can be added.

$$\frac{(x-1)\,x}{(x-1)(x+2)} + \frac{(x+2)\,x}{(x+2)(x-1)} = \frac{(x^2-x)+(x^2+2x)}{(x-1)(x+2)}$$

$$= \frac{2x^2+x}{(x-1)(x+2)} \text{ or } \frac{2x^2+x}{x^2+x-2}, \text{ where } x \neq 1, -2$$

How do you choose the appropriate equivalents for 1?

EXAMPLE 2

Simplify $\frac{a}{a^2-4} + \frac{2}{a+2}$. Write any restrictions on the variable.

Solution ➤

Since $a^2 - 4 = (a+2)(a-2)$, multiply the second expression by $\frac{a-2}{a-2}$, and add.

$$\frac{a}{a^2-4} + \frac{2}{a+2} = \frac{a}{(a+2)(a-2)} + \frac{2(a-2)}{(a+2)(a-2)}$$

$$= \frac{a+2a-4}{(a+2)(a-2)}$$

$$= \frac{3a-4}{a^2-4}, \text{ where } a \neq 2, -2 \text{ ❖}$$

EXAMPLE 3

Simplify $\frac{x-4}{3x-3} - \frac{x-3}{2x-2}$. Write any restrictions on the variable.

Solution ➤

To find the common denominator, factor the denominators.

$$\frac{x-4}{3(x-1)} - \frac{x-3}{2(x-1)}$$

A common denominator is $3 \cdot 2(x-1) = 6(x-1)$, where $x \neq 1$.

CT– Factoring out −1 is sometimes helpful because it allows opposites to be expressed in a form in which a common factor can be identified, and then the expression can be reduced by that factor.

$\frac{x-4}{3(x-1)} - \frac{x-3}{2(x-1)} = \frac{2(x-4)}{6(x-1)} - \frac{3(x-3)}{6(x-1)}$	Change to common denominators.
$= \frac{2x-8-3x+9}{6(x-1)}$	Subtract the numerators.
$= \frac{-x+1}{6(x-1)}$	Simplify.
$= \frac{-1(x-1)}{6(x-1)}$	Factor −1 from the numerator.
$= -\frac{1}{6}$, where $x \neq 1$	Simplify. ❖

Explain when to factor −1 from an expression. How does factoring −1 simplify a rational expression?

EXERCISES & PROBLEMS

Communicate

Refer to the expressions $\frac{x}{x+1}$ and $\frac{3}{x}$ for Exercises 1–5.

1. Explain how a common denominator may be found for the two rational expressions.
2. Describe how to find the sum of these rational expressions.
3. Describe how to find the difference of these rational expressions.
4. Describe how to find the product of these rational expressions.
5. How do you think you can find the quotient of the first expression divided by the second expression?

Practice & Apply

Perform the indicated operations. Simplify, and write any restrictions on the variables.

6. $\frac{5}{3x}+\frac{2}{3x}$ [$\frac{7}{3x}$, $x \neq 0$]

7. $\frac{8}{x+1}-\frac{5}{x+1}$ [$\frac{3}{x+1}$, $x \neq -1$]

8. $\frac{2x}{y+4}-\frac{5x}{y+4}$ [$-\frac{3x}{y+4}$, $y \neq -4$]

9. $\frac{5x+4}{a-c}-\frac{7+3x}{a-c}$ [$\frac{2x-3}{a-c}$, $a \neq c$]

10. $\frac{2}{a}+\frac{3}{b}$ [$\frac{3a+2b}{ab}$, $a \neq 0, b \neq 0$]

11. $\frac{7}{3t}-\frac{8}{2t}$ [$-\frac{5}{3t}$, $t \neq 0$]

12. $\frac{t}{2rs}+\frac{3s}{rt}$ [$\frac{6s^2+t^2}{2rst}$, $r \neq 0$, $s \neq 0$, $t \neq 0$]

13. $\frac{5}{pq}-\frac{m}{p^2q}$ [$\frac{5p-m}{p^2q}$, $p \neq 0$, $q \neq 0$]

14. $\frac{-2}{x+1}+\frac{3}{2(x+1)}$ [$\frac{-1}{2(x+1)}$, $x \neq -1$]

15. $x-\frac{x-4}{x+4}$ [$\frac{x^2+3x+4}{x+4}$, $x \neq -4$]

16. $\frac{-3-d}{d-1}+2$ [$\frac{d-5}{d-1}$, $d \neq 1$]

17. $\frac{a}{b}+\frac{c}{d}$ [$\frac{ad+bc}{bd}$, $b \neq 0$, $d \neq 0$]

18. $\frac{5+m}{3+n}\cdot\frac{3+n}{a+b}$ [$\frac{5+m}{a+b}$, $n \neq -3$, $a \neq -b$]

19. $\frac{x+2}{x(x+1)}\cdot\frac{x^2}{(x+2)(x+3)}$

20. $\frac{q^2-1}{q^2}\cdot\frac{q}{q+1}$ [$\frac{q-1}{q}$, $q \neq 0, -1$]

21. $\frac{1}{x+1}\cdot\frac{2}{x}$

22. $\frac{3}{x-1}\cdot\frac{5}{x}$

23. $\frac{y}{y-4}-\frac{1}{y}$

24. $\frac{-2}{x+1}+\frac{3}{x}$

25. $\frac{a-2}{a+1}+\frac{5}{a+3}$

26. $\frac{m+5}{m+2}\cdot\frac{m-3}{m-1}$

27. $\frac{2r}{(r+5)(r+1)}-\frac{r}{r+5}$

28. $\frac{3}{b^2+b-6}+\frac{5}{b-2}$

29. $\frac{4}{y-2}+\frac{5y}{y^2-4y+4}$

30. $\frac{2}{y-5}-\frac{5y-3}{y^2+y-30}$

31. $\frac{x}{1-x}+\frac{1}{x-1}$

32. $\frac{2}{x^2-9}-\frac{1}{2x+6}$

[19. $\frac{x}{x^2+4x+3}$, $x \neq 0, -1, -2, -3$]

Ecology Students at an aquarium are creating an exhibit of a bay ecosystem. They determine that the water in the bay contains 4.0% salt by mass. The water in the aquarium tanks is currently 6.0% salt by mass. The students plan to dilute 750 kilograms of the water in the aquarium by adding enough pure water to get the 4.0% salt concentration of the bay. The concentration can be determined by the following formula:

$$4\%\text{ salt concentration} = \frac{\text{aquarium-water salt concentration} \cdot \text{mass of aquarium water}}{\text{mass of aquarium water} + \text{mass of pure water}}$$

33. Write a rational function for this problem to represent the resulting concentration when a mass of pure water, w, is added.

34. Graph the function.

35. How much pure water needs to be added to reach a concentration of 4.0% salt by mass?

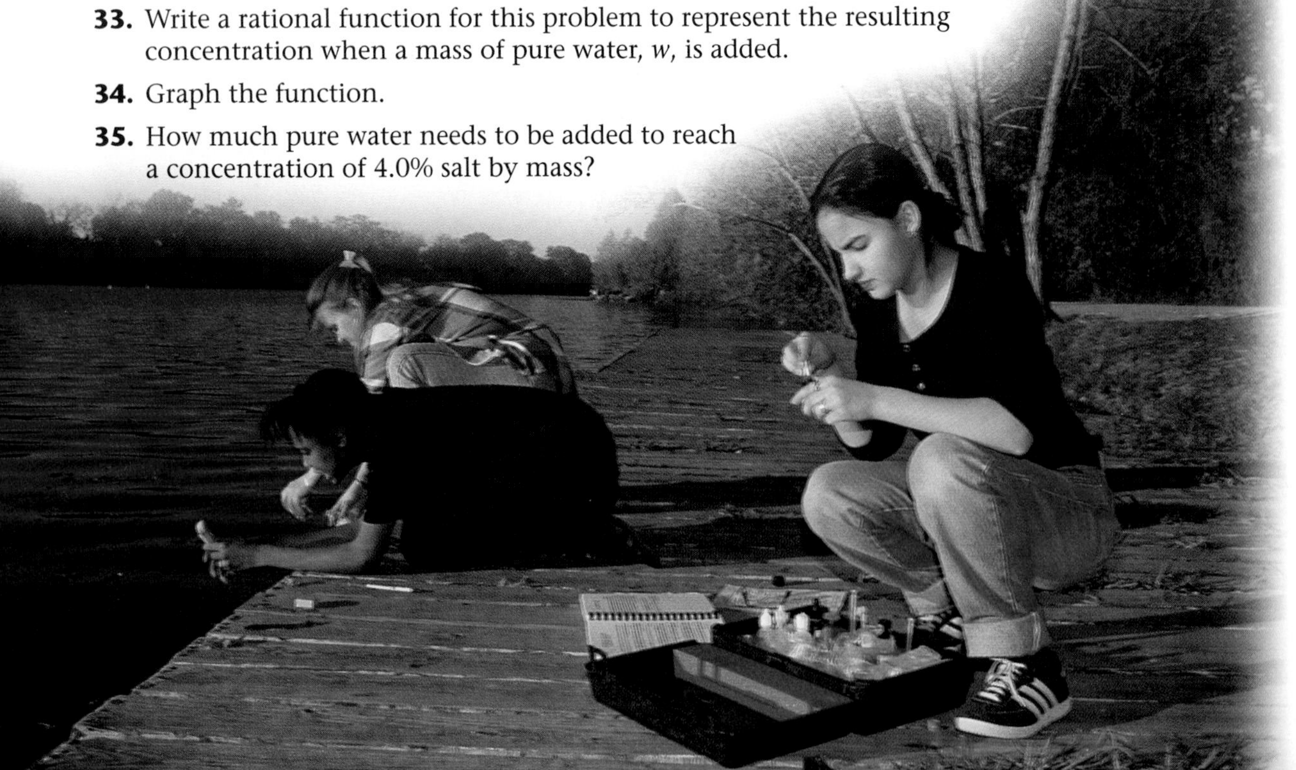

Look Back

36. State the pattern and find the next three terms of this sequence.
$-1, 6, 15, 26, \underline{?}, \underline{?}, \underline{?}$ **[Course 1]** The 2nd differences are constant: +2; terms: 39, 54, 71.

37. Solve $0.3(x - 90) = 0.7(2x - 70)$. **[Lesson 1.4]** 20

38. Graph the solution to $y < -2x + 6$ *and* $y \leq 3x - 4$. **[Lesson 2.7]**

39. Add $\begin{bmatrix} -3 & 7 \\ 5 & 11 \end{bmatrix} + \begin{bmatrix} 7 & -13 \\ 22 & 2 \end{bmatrix}$. **[Lesson 3.2]** $\begin{bmatrix} 4 & -6 \\ 27 & 13 \end{bmatrix}$

Cultural Connection: Americas The Blackfoot people of Montana played a game of chance in which they threw 4 decorated bones made from buffalo ribs. Three were blank on one side, but the fourth, called the chief, was engraved on both sides like a coin with heads and tails. The highest score was 6 points, for a throw of 3 blanks and the chief landing heads up. For a throw of 3 blanks and the chief landing heads down, the score was 3 points.

40. What is the probability of throwing 3 blanks and the chief landing heads up? **[Lesson 4.1]** $\frac{1}{16}$

41. What is the probability of throwing 3 blanks and the chief landing heads down? **[Lesson 4.1]** $\frac{1}{16}$

42. **Probability** If a bag contains 4 red marbles, 3 black marbles, and 1 white marble, what is the probability that you will select a white marble? **[Lesson 4.1]** $\frac{1}{8}$

43. Express 25,000,000,000 in scientific notation. **[Lesson 6.5]** 2.5×10^{10}

44. Graph $y = (x - 1)^2 - 4$. **[Lesson 8.1]**

Look Beyond

A proportion is the equality of two ratios such as $\frac{5}{10} = \frac{2}{4}$; both ratios equal $\frac{1}{2}$. This reduced fraction is called the **constant of the proportion**. Find the missing term in each proportion, and state the constant of the proportion.

45. $\frac{26}{39} = \frac{x}{9}$ 6; $\frac{2}{3}$ **46.** $\frac{14}{x} = \frac{49}{28}$ 8; $\frac{7}{4}$ **47.** $\frac{15}{40} = \frac{27}{x}$ 72; $\frac{3}{8}$

How Worried Should You Be?

CENTRAL HIGH HEALTH FAIR

Now, the testing question

Concerns about iron's role in heart disease, if ultimately confirmed, surely will touch off a debate over iron tests. Who should be tested? And should testing be a matter of public-health policy? . . .

The virtue of such broad screening would be to help ferret out many of the estimated 32 million Americans who carry a faulty gene that prompts their body to harbor too much iron. There is no test for the gene itself, but anyone with a high blood level of iron is a suspect. Siblings and children would be candidates for testing as well, since the condition is inherited . . .

Two widely used blood tests for iron generally cost from \$25 to \$75 each. The first measures ferritin, a key iron-storing protein . . .

The results should always be discussed with a doctor, since some physicians take recent studies seriously enough to worry about even moderately elevated ferritin levels. Moreover, the test might bear repeating. Erroneously high readings are relatively common.

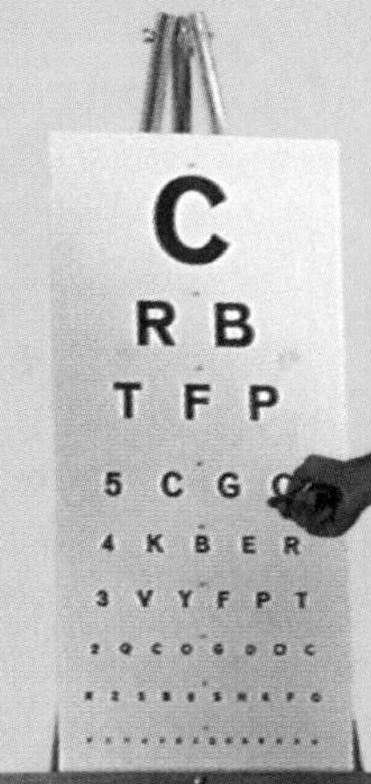

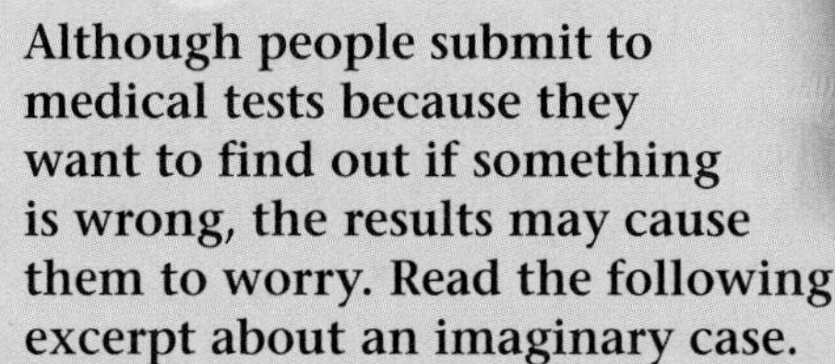

Although people submit to medical tests because they want to find out if something is wrong, the results may cause them to worry. Read the following excerpt about an imaginary case.

> Assume that there is a test for cancer which is 98 percent accurate; i.e., if someone has cancer, the test will be positive 98 percent of the time, and if one doesn't have it, the test will be negative 98 percent of the time. Assume further that 0.5 percent—one out of two hundred people—actually have cancer. Now imagine that you've taken the test and that you've tested positive. The question is: How depressed should you be?

To find out whether your reaction would be justified, you need to look at the numbers more closely.

Cooperative Learning

Activity 1 Use the data in the excerpt to complete **a–g**. Explain each answer.

a. Of the 10,000 people tested, how many would you expect to have cancer?

b. How many of those people that have cancer would you expect to test positive? (We'll call such cases *true positives.*)

c. Of the 10,000 people tested, how many would you expect to not have cancer?

d. How many of those people that do not have cancer would you expect to test positive? (We'll call such cases *false positives.*)

e. Add your results in **b** and **d** to find the total number of people expected to test positive.

f. What is the probability (P) that if you tested positive, you actually have cancer?

g. How does your answer in part **f** compare with your response to the question, "How concerned should you be?"

Remember:

$$\text{Probability} = \frac{\text{considered outcome}}{\text{total number of outcomes}}$$

$$P = \frac{\text{number of true positives}}{\text{total number expected to test positive}}$$

Activity 2 The data you used in Activity 1 are made up. To explore what happens with different sets of data, you can find a general formula for P, the probability that if you tested positive, you actually have cancer.

Let N = the number of people tested,
a = the accuracy of the test (if test is 98% accurate, then $a = 0.98$), and
r = the portion of people tested who actually have what's being tested for (in the example, $r = 0.005$).

a. Write a formula for the number of true positives. HINT: Look back at **a** and **b** of Activity 1.

b. Write a formula for the number of false positives. HINT: Look back at **c** and **d** of Activity 1.

c. Write a formula for P. HINT: Look back at **e** and **f** of Activity 1.

LESSON 10.5 Solving Rational Equations

Many relationships such as time, work, and average costs are modeled by rational expressions in equations. Graphing and algebraic methods can be used to solve rational equations.

Sam rode his bike 6 miles before his chain broke, and he had to walk home.

Sports During the walk home, Sam estimated that he bikes 5 times as fast as he walks. If the entire trip took him 2 hours and 24 minutes, find his biking rate.

The time for each part of the trip can be represented using the formula $\frac{\text{distance}}{\text{rate}} = \text{time}$. The total time can be represented by the formula:

$$\frac{\text{distance biked}}{\text{rate}} + \frac{\text{distance walked}}{\text{rate}} = \text{total time}$$

If x is Sam's walking rate, then $5x$ is his biking rate.

$$\frac{6 \text{ miles}}{5x \text{ mph}} + \frac{6 \text{ miles}}{x \text{ mph}} = 2\frac{2}{5} \text{ hours}$$

Therefore, the equation that models the problem is $\frac{6}{5x} + \frac{6}{x} = \frac{12}{5}$, where $x \neq 0$. It can be solved by two methods.

Common Denominator Method Multiply the expressions on each side of the original equation by a common denominator. This will clear the equation of fractions. Using the least common denominator saves extra steps in the solution.

$\frac{6}{5x} + \frac{6}{x} = \frac{12}{5}$	Given
$5x\left(\frac{6}{5x} + \frac{6}{x}\right) = 5x\left(\frac{12}{5}\right)$	Multiply both sides by $5x$, the least common denominator.
$6 + 30 = 12x$	Simplify.
$12x = 36$	Simplify.
$x = 3$	Division Property of Equality

Graphics Calculator

Graphing Method Graph $\mathbf{Y_1} = \frac{6}{5x} + \frac{6}{x}$ and $\mathbf{Y_2} = \frac{12}{5}$. You find that their intersection is at $x = 3$.

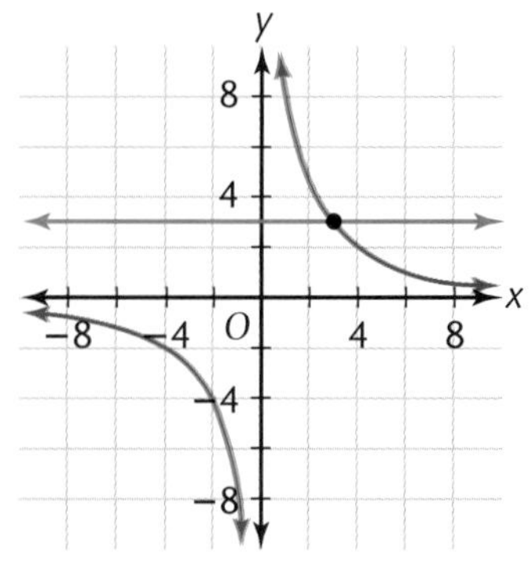

Sam walks at 3 miles per hour. His biking rate is 5 times that, or 15 miles per hour.

What is the reasonable domain for the rational function $\mathbf{Y_1}$ above? What is the reasonable range for $\mathbf{Y_1}$?

The x-values represent walking rates, in miles per hour. The y-values represent time, in hours. Therefore, the reasonable domain and reasonable range are both all positive real numbers.

EXAMPLE 1

Solve $\frac{6}{x-1} + 2 = \frac{12}{x^2-1}$ using the common-denominator method and the graphing method.

Solution A ➤

Common Denominator Method

Factor. Notice that this equation is undefined for $x = 1$ and $x = -1$.

$$\frac{6}{x-1} + 2 = \frac{12}{(x+1)(x-1)}$$

Multiply both sides by the lowest common denominator $(x + 1)(x - 1)$.

$$(x+1)(x-1)\left[\frac{6}{x-1} + 2\right] = (x+1)(x-1)\left[\frac{12}{(x+1)(x-1)}\right]$$

$6(x+1) + 2(x+1)(x-1) = 12$	Divide out common factors.
$6x + 6 + 2x^2 - 2 = 12$	Simplify.
$2x^2 + 6x + 4 = 12$	Simplify.
$2x^2 + 6x - 8 = 0$	Standard quadratic form
$2(x+4)(x-1) = 0$	Factor.
$x + 4 = 0$ or $x - 1 = 0$	Zero Product Property
$x = -4$ or $x = 1$	Solve.

Reject 1 as a solution because it makes the *original* equation undefined. Thus, −4 is the only solution. Always check all possible solutions to be sure that the original equation is defined.

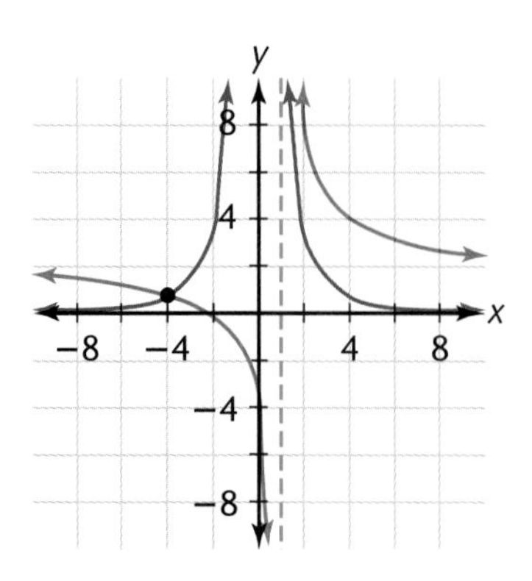

Solution B ➤

Graphing Method

Graph $Y_1 = \frac{6}{x-1} + 2$ and $Y_2 = \frac{12}{x^2-1}$. The two functions have a common asymptote at $x = 1$ and a clear point of intersection at $x = -4$.

The solution is $x = -4$. ❖

Explain the effect of an undefined function value on the graph of a rational function.

The undefined values become asymptotes of the graph.

EXAMPLE 2

Juanita is starting a lawn-mowing business in her neighborhood. The startup cost for the business is $400.

Juanita has expenses of $6 per lawn that include a helper and the cost of gasoline. How many lawns must she mow before the average cost per lawn is $20?

Solution ➤

Let x represent the number of lawns mowed. Substitute the known values.

$$\text{average cost per lawn} = \frac{\text{start-up cost} + \left(\text{cost per lawn}\right) \times \left(\text{number of lawns}\right)}{\text{number of lawns}}$$

$$20 = \frac{400 + 6x}{x}, \text{ where } x \neq 0$$

Common-denominator method

$$\frac{20}{1} = \frac{400 + 6x}{x}$$

$$20x = 400 + 6x$$

$$14x = 400$$

$$x = 28\frac{4}{7}$$

Graphing method

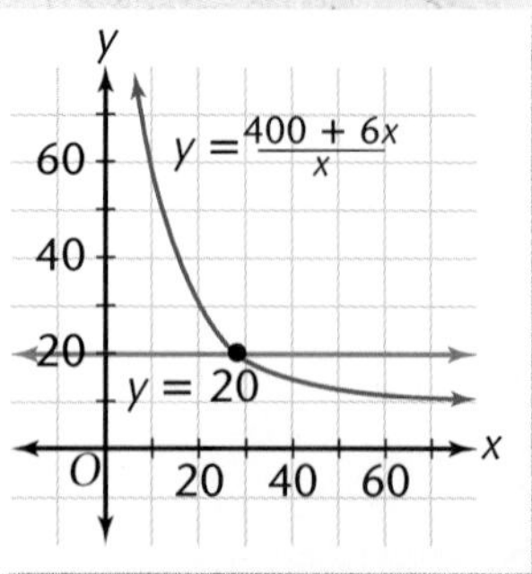

Juanita would have to mow 29 lawns before the average cost per lawn is $20. ❖

Try This Solve this rational equation by finding the least common denominator. Verify the solution by graphing. $\frac{20}{x^2 - 4} + 6 = \frac{10}{x - 2}$

$x = 3$ or $x = -\frac{4}{3}$

EXAMPLE 3

Max and his younger brother Carl are raising extra money by delivering papers. If the boys work together, how long will it take them to finish this job?

Max can deliver his papers in 80 minutes, but it takes Carl 2 hours to deliver the same number of papers.

Solution ➤

Let t represent the number of minutes the boys work together. The completed job is equal to the sum of all the fractional parts and is represented by 1 complete job.

fractional part of the job done by Max	+	fractional part of the job done by Carl	=	the complete job, or 1 whole unit
(rate of Max's work) · (time worked)	+	(rate of Carl's work) · (time worked)	=	1
$\frac{1}{80} \cdot t$	+	$\frac{1}{120} \cdot t$	=	1
$\frac{t}{80}$	+	$\frac{t}{120}$	=	1

Continue to solve $\frac{t}{80} + \frac{t}{120} = 1$. Multiply each term by the lowest common denominator of 80 and 120.

$$240 \cdot \frac{t}{80} + 240 \cdot \frac{t}{120} = 240 \cdot 1$$
$$3t + 2t = 240$$
$$5t = 240$$
$$t = 48$$

Working together, it would take Max and Carl 48 minutes to deliver all the papers.

In 48 minutes, Max does $\frac{48}{80}$ of the job and Carl does $\frac{48}{120}$ of the job.

$\frac{48}{80} + \frac{48}{120} = \frac{6}{10} + \frac{4}{10} = \frac{10}{10} = 1$ complete job ❖

EXERCISES & PROBLEMS

Communicate

1. Discuss what "undefined variable values of rational equations" means.
2. How is the "realistic domain" of a rational equation determined in applications?

Refer to the equation $\frac{3}{c} - \frac{2}{c-1} = \frac{6}{c}$ for Exercises 3–5.

3. Describe the steps needed to find the lowest common denominator.
4. Explain how to solve the rational equation by graphing.
5. How would you find the undefined values for the equation? What are they?
6. Explain how to write the rate of work on a particular job as a fraction.

Practice & Apply

Solve each rational equation by finding the least common denominator.

7. $\frac{x-1}{x} + \frac{7}{3x} = \frac{9}{4x}$ $\frac{11}{12}$
8. $\frac{10}{2x} + \frac{4}{x-5} = 4$ $6\frac{1}{4}, 1$
9. $\frac{x-3}{x-4} = \frac{x-5}{4+x}$ $3\frac{1}{5}$
10. $\frac{x+3}{x} + \frac{6}{5x} = \frac{7}{2x}$ $-\frac{7}{10}$
11. $\frac{x}{x-3} = 5 + \frac{x}{x-3}$ None
12. $\frac{x+3}{x^2-9} - \frac{6}{x-3} = 5$ 2
13. $\frac{4}{y-2} + \frac{5}{y+1} = \frac{1}{y+1}$ $\frac{1}{2}$
14. $\frac{10}{x+3} - \frac{3}{5} = \frac{10x+1}{3x+9}$ 2
15. $\frac{3}{x-2} - \frac{6}{x^2-2x} = 1$ 3

Solve each rational equation by graphing.

16. $\frac{12x-7}{x} = \frac{17}{x}$ 2
17. $\frac{3}{7x+x^2} = \frac{9}{x^2+7x}$ None
18. $\frac{20-x}{x} = x$ $-5, 4$
19. $\frac{2}{x} + \frac{1}{3} = \frac{4}{x}$ 6
20. $\frac{1}{2} + \frac{1}{x} = \frac{1}{2x}$ -1
21. $\frac{x-3}{x-4} = \frac{x-5}{x+4}$ 3.2
22. $\frac{3}{x-2} - \frac{6}{x^2-2x} = 1$ 3
23. $\frac{1}{x-2} + \frac{16}{x^2+x-6} = -3$ $-\frac{1}{3}, -1$
24. $\frac{1}{x-2} + 3 = \frac{-16}{x^2+x-6}$

24. $-\frac{1}{3}, -1$

25. The sum of two numbers is 56. If the larger number is divided by the smaller, the quotient is 1 with a remainder of 16. Find the two numbers. 20, 36

Marisa pedals downstream at a still-water rate of 20 meters per minute for 500 meters before turning around to pedal upstream to where she started.

Recreation If Marisa maintains the same still-water rate, she will only pedal 300 meters upstream in the same time it takes her to pedal 500 meters downstream.

26. Write rational expressions that represent the time it takes Marisa to pedal downstream and upstream in the current.

27. What is the rate of the current?
5 meters per minute

28. **Government** Jeff, Phil, and Allison decide to support their favorite candidate by stuffing envelopes for a candidate's political campaign. They find that they each work at different rates. Phil takes 2 hours longer than Jeff to stuff 1000 envelopes, and Allison takes 3 hours less than Phil to do the same. One day they work together for 4 hours. Write rational expressions that represent the rate at which each works.

Small Business In their shop, it takes Tom about 5 hours to paint a car working alone. It takes Shelley about 7 hours to paint a car.

29. Write rational expressions that represent the fractional parts of the work that Tom and Shelley complete when painting a car together.

30. Write and solve a rational equation to find the time it would take them to paint a car if they work together.
$\frac{x}{5} + \frac{x}{7} = 1$; $x = 2\frac{11}{12}$ hours or 2 hours 55 minutes

Look Back

31. Solve $x - 4 = \frac{-3}{4}(x + 2)$. **[Lesson 1.4]** $\frac{10}{7}$ or $1\frac{3}{7}$

32. Solve $|2x - 7| \leq 7$. **[Lesson 1.6]**

33. Solve $x + 5y = 9$ and $3x - 2y = 10$ by graphing. **[Lesson 2.3]**

34. A club has 20 members. In how many ways can 3 officers be selected if all members are eligible for these positions? **[Lesson 4.5]** 6840

35. Multiply $(3x - 8)(5x + 2)$. **[Lesson 7.4]** $15x^2 - 34x - 16$

36. If $f(x) = \frac{x^2 - 3x - 10}{(x - 4)^2}$, find $f(-2)$. **[Lesson 10.1]** 0

Look Beyond

37. 250 seconds

37. As a game, Antonio climbs up 6 steps in 3 seconds and hops back 4 steps in 2 seconds. How long will it take him to reach the 100th step?

LESSON 10.6 Exploring Proportions

Why *You have used ratios and proportions to solve percent problems. Because a ratio containing a variable is a rational expression, proportions can be solved by using the methods in the previous lesson.*

In the first round of a free-throw contest, Al made 28 out of 40 shots. Alicia made 21 out of 30. Who wins the first round?

There are several methods to determine the winner. One way is to use fractions.

Al	**Alicia**
$\frac{28}{40} = \frac{7}{10}$	$\frac{21}{30} = \frac{7}{10}$

Since both fractions reduce to $\frac{7}{10}$, $\frac{28}{40}$ is equal to $\frac{21}{30}$, and the contest ends in a tie. The equation $\frac{28}{40} = \frac{21}{30}$ is called a proportion.

A **proportion** is an equation which states that two ratios are equal.

A proportion $\frac{a}{b} = \frac{c}{d}$ is sometimes written $a : b = c : d$.

Exploration 1 *Solving Proportions*

1 For each statement, identify the extremes and the means.

a. $\frac{1}{2} = \frac{5}{10}$ **b.** $\frac{3}{4} = \frac{9}{12}$ **c.** $\frac{4}{6} = \frac{6}{9}$ **d.** $\frac{6}{2\sqrt{2}} = \frac{3\sqrt{2}}{2}$

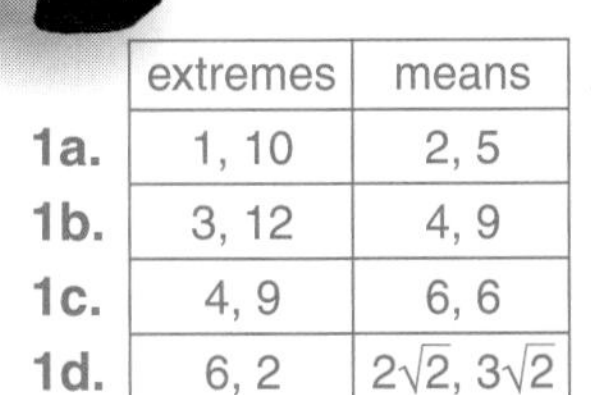

	extremes	means
1a.	1, 10	2, 5
1b.	3, 12	4, 9
1c.	4, 9	6, 6
1d.	6, 2	$2\sqrt{2}$, $3\sqrt{2}$

2 For each statement, find the product of the extremes and the product of the means.

a. $\frac{1}{2} = \frac{5}{10}$ **b.** $\frac{3}{4} = \frac{9}{12}$ **c.** $\frac{4}{6} = \frac{6}{9}$ **d.** $\frac{6}{2\sqrt{2}} = \frac{3\sqrt{2}}{2}$

	extremes	means
2a.	10	10
2b.	36	36
2c.	36	36
2d.	12	12

3 Make a conjecture about the relationship between the product of the extremes and the product of the means in a proportion.
The two products are equal.

4 In $\frac{a}{b} = \frac{c}{d}$, ad and bc are often called the *cross products*. Restate your conjecture from Step 3 using cross products. Use your conjecture to tell which of the equations below are proportions.

a. $\frac{6}{10} = \frac{20}{25}$ **b.** $\frac{2 + 4\sqrt{3}}{-4} = \frac{11}{2 - 4\sqrt{3}}$ **c.** $\frac{24}{15} = \frac{160}{100}$

4. The product of the extremes equals the product of the means in a proportion. Thus, the cross products are equal. The equations in 4**b** and 4**c** are proportions.

5 Explain how to use your conjecture to solve the proportion $\frac{x}{4} = \frac{42}{24}$.
Cross multiply: $x \cdot 24 = 4 \cdot 42$. Then solve: $24x = 168$ and $x = 7$.

6 Use your conjecture to solve each of the following proportions:

a. $\frac{4}{b} = \frac{6}{b + 3}$, where $b \neq -3, 0$ 6 **b.** $\frac{c - 6}{7} = \frac{1}{c}$, where $c \neq 0$ 7, −1

7 How are the proportions in Step 6 alike? different? ❖

Exploration 2 *True or False?*

Suppose $\frac{a}{b} = \frac{c}{d}$. Substitute values for a, b, c, and d that make $\frac{a}{b} = \frac{c}{d}$ true. Find the cross products for each proportion below, using the same values for a, b, c, and d.

a. $\frac{b}{a} = \frac{d}{c}$ **b.** $\frac{a+b}{c} = \frac{a-b}{c}$ **c.** $\frac{a+b}{b} = \frac{c+d}{d}$

d. $\frac{a}{c} = \frac{b}{d}$ **e.** $\frac{a-b}{b} = \frac{c-d}{d}$ **f.** $\frac{a}{b} = \frac{a+c}{b+d}$

Which of the proportions in Step 1 seem to be true for any substitutions you make?

State a rule for the relationship between the original proportion, $\frac{a}{b} = \frac{c}{d}$, and each proportion in Step 1 that seems to be true. ❖

APPLICATION

Sports Competing for the best free-throw percentage, Al made 28 out of 40 free throws, for a 70% average. To win the contest, Al needs to shoot 76%. How many more consecutive free throws must Al make to have a 76% average?

Let x represent how many more successful free throws Al must make. Write a proportion for Al's new percentage.

$$70\% = \frac{28}{40} = \frac{\text{successes}}{\text{number attempted}} \qquad 76\% = \frac{\text{successes} + x}{\text{number made} + x}$$

$\frac{76}{100} = \frac{28 + x}{40 + x}$, where $x \neq -40$	Given
$76(40 + x) = 100(28 + x)$	Cross products are equal.
$3040 + 76x = 2800 + 100x$	Distributive Property
$240 = 24x$	Simplify.
$10 = x$	Division Property of Equality

Al must make 10 more consecutive free throws to shoot 76%. ❖

APPLICATION

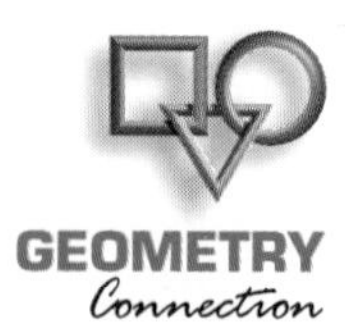

The measures of the three angles of a triangle are in the ratio 1:2:3. Find the measure of each angle.

Draw a picture. Let the angle measures be x, $2x$, and $3x$. Recall that the sum of the degree measures of the angles of a triangle is 180°.

$$x + 2x + 3x = 180$$
$$6x = 180$$
$$x = 30 \qquad 2x = 60 \qquad 3x = 90$$

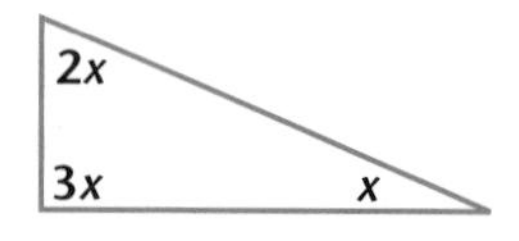

The angle measures are 30°, 60°, and 90°. ❖

A theorem in geometry states that if a line is parallel to one side of a triangle and intersects the other two sides, then it divides the two sides proportionately. Illustrate the theorem with a proportion.

$\frac{a}{a+c} = \frac{b}{b+d}$

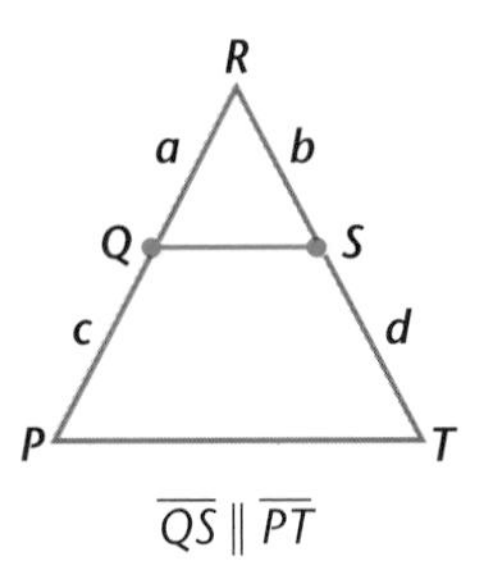

EXERCISES & PROBLEMS

Communicate

1. What is a proportion? Give an example.
2. Explain how to find the means and the extremes of a proportion.
3. Make a statement relating the cross products of the means and of the extremes of a proportion.
4. Discuss how to solve for x in the proportion $\frac{1}{2} = \frac{2x+1}{4}$.
5. Explain how to write a proportion to solve the following: A car travels 320 kilometers on 40 liters of gas. How far can the car travel on 75 liters?

Practice & Apply

Identify the means and extremes of each proportion.

6. $\frac{x}{3} = \frac{7}{4}$ **7.** $\frac{x}{10} = \frac{2}{5}$ **8.** $\frac{3}{4} = \frac{y}{5}$ **9.** $\frac{17}{2} = \frac{8.5}{x}$

10. $\frac{10}{m-2} = \frac{m}{7}$ **11.** $\frac{2w-3}{5} = \frac{w}{3}$ **12.** $\frac{3}{n+6} = \frac{2}{n}$ **13.** $\frac{c+3}{5} = \frac{c-2}{4}$

18. 5, $m \neq 3$, $m \neq 0$ **21.** 2, $y \neq 3$, $y \neq 5$ **25.** 12, -14, $r \neq -2$ **29.** -6, 4, $n \neq -2$

Solve for the variable. Write any restrictions on the variable.

14. $\frac{8}{14} = \frac{4}{x}$ 7, $x \neq 0$ **15.** $\frac{33}{n} = \frac{3}{1}$ 11, $n \neq 0$ **16.** $\frac{p}{4} = \frac{4}{16}$ 1 **17.** $\frac{65}{13} = \frac{x}{5}$ 25

18. $\frac{2}{m-3} = \frac{5}{m}$ **19.** $\frac{w}{3} = \frac{w+4}{7}$ 3 **20.** $\frac{x-1}{3} = \frac{x+1}{5}$ 4 **21.** $\frac{1}{y-3} = \frac{3}{y-5}$

22. $\frac{p}{4} = \frac{10}{p-3}$ 8, -5, $p \neq 3$ **23.** $\frac{12}{c} = \frac{c+4}{1}$ 2, -6, $c \neq 0$ **24.** $\frac{x}{2} = \frac{30}{x-4}$ -6, 10, $x \neq 4$ **25.** $\frac{r}{3} = \frac{56}{r+2}$

26. $\frac{2}{x} = \frac{x}{9}$ $\pm 3\sqrt{2}$, $x \neq 0$ **27.** $\frac{m-5}{4} = \frac{m+3}{3}$ -27 **28.** $\frac{x}{3} = \frac{6}{2x}$ 3, -3, $x \neq 0$ **29.** $\frac{2n}{3} = \frac{16}{n+2}$

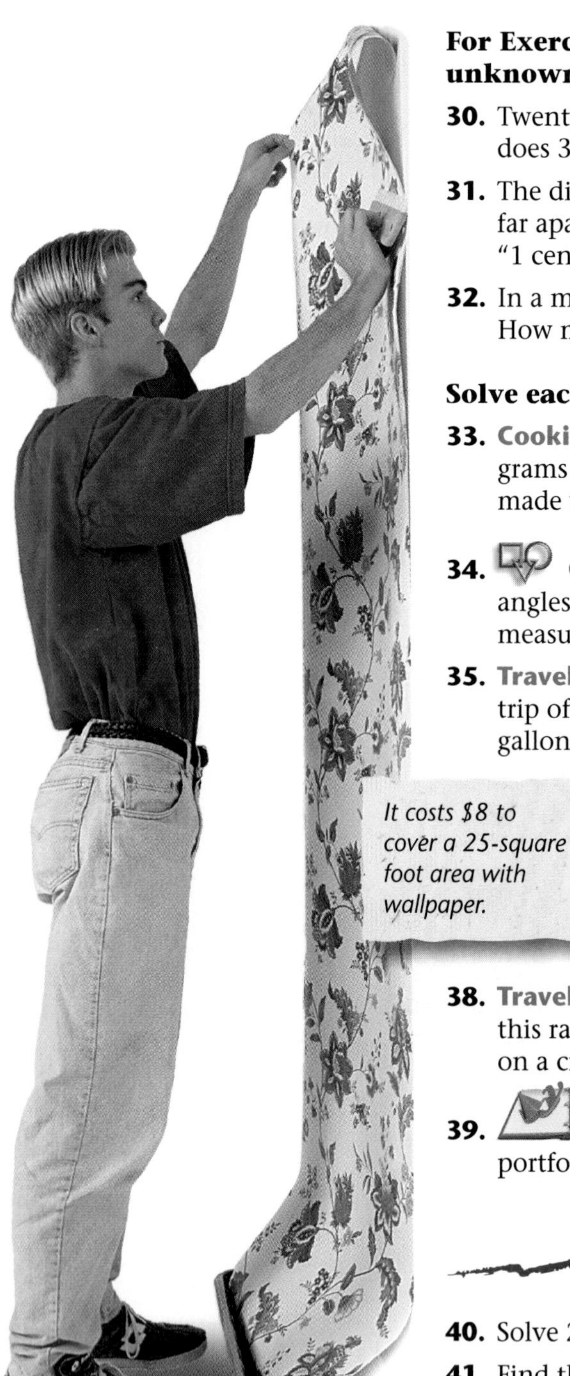

It costs $8 to cover a 25-square foot area with wallpaper.

For Exercises 30–32, use a variable to represent the unknown. Write a proportion to solve each problem.

30. Twenty-five feet of copper wire weighs 1 pound. How much does 325 feet of copper wire weigh? $\frac{25}{1} = \frac{325}{x}$; 13 pounds

31. The distance between two cities is 750 kilometers. Find how far apart the cities are on a map with a scale that reads "1 centimeter = 250 kilometers." $\frac{750}{x} = \frac{250}{1}$; 3 cm

32. In a mixture of concrete, the ratio of sand to cement is 1 : 4. How many bags of cement are needed for 100 bags of sand? $\frac{1}{4} = \frac{100}{x}$; 400 bags

Solve each problem.

33. Cooking A recipe for making blueberry muffins requires 1600 grams of flour to make 40 muffins. How many muffins can be made with 1000 grams of flour? 25 muffins

34. Geometry Find the measures of two supplementary angles if their measures are in a ratio of 3 : 6. The sum of the measures of supplementary angles is 180°. 60°, 120°

35. Travel Mrs. Sanchez used 15 gallons of gasoline to drive on a trip of 450 miles. How far can she drive on a full tank of 20 gallons? 600 miles

36. $52.80

36. Interior Design How much will it cost to cover a wall with an area of 165 square feet?

37. A recent poll found that 4 out of 6 people nationwide use Klean toothpaste. How many people can be expected to use this toothpaste in a city of 30,000? 20,000 people

38. Travel A plane has a cruising speed of 650 miles per hour. At this rate, how long does it take the plane to travel 3250 miles on a cross-country flight? 5 hours

39. Portfolio Activity Complete the problem in the portfolio activity on page 473.

Look Back

40. Solve $2(x - 3) = 6x$. **[Lesson 1.4]** $-\frac{3}{2}$ or $-1\frac{1}{2}$

41. Find the slope of the line parallel to $6x - 3y = 24$. **[Lessons 2.1, 2.2]** 2

42. Simplify $\begin{bmatrix} 2 & -1 \\ 3 & 0 \end{bmatrix}\begin{bmatrix} 1 & 4 \\ -3 & 2 \end{bmatrix}$. **[Lesson 3.3]** $\begin{bmatrix} 5 & 6 \\ 3 & 12 \end{bmatrix}$

Look Beyond

43. $\frac{9}{5}$ inches or $1\frac{4}{5}$ inches

43. Interior Design A room is 12 feet long by 9 feet wide. Jay uses a scale of 1 inch = 5 feet to draw a blueprint. What is the width of the room on the blueprint?

LESSON 10.7 Proof in Algebra

why *Throughout this book statements have been made that are assumed to be true. Many of these statements can be proven by a logical argument.*

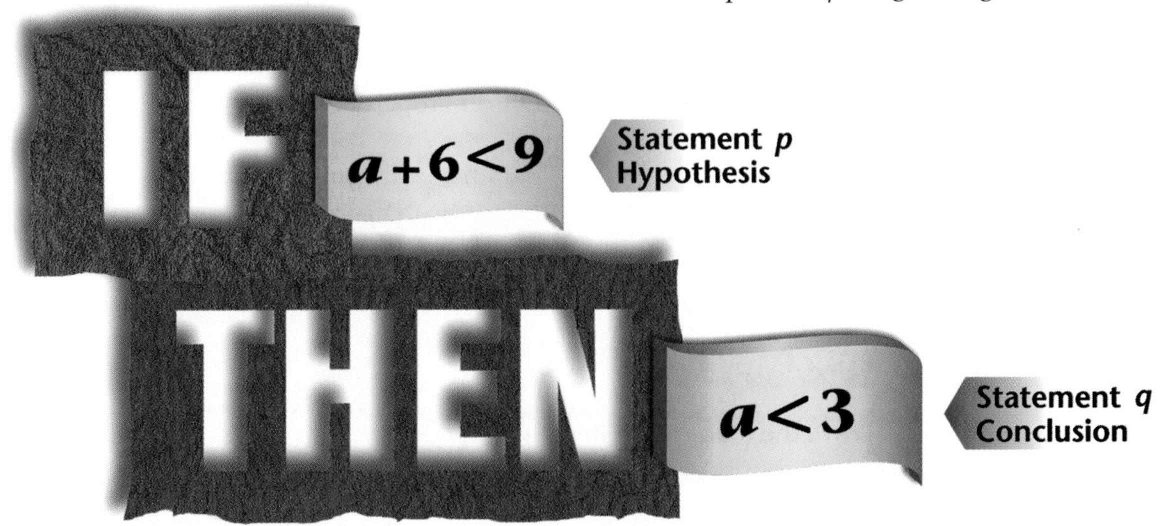

A statement that has been proven is called a **theorem**. In the *if p then q* form of a theorem, p is called the **hypothesis** and q is called the **conclusion**. A **proof** provides a logical justification that q is true whenever p is true.

In a proof, the hypothesis is assumed to be true. Each step in a proof must be based on established definitions, on axioms or postulates (statements assumed to be true), or on previously proven theorems. Study the following proof.

Prove: *If* $a + 6 < 9$, *then* $a < 3$.

$a + 6 < 9$	Hypothesis or given
$a + 6 - 6 < 9 - 6$	Subtraction Property of Inequality
$a < 3$	Simplify.

Therefore, if $a + 6 < 9$, then $a < 3$.

The conclusion follows logically from the hypothesis, and the theorem is proven.

If you interchange the hypothesis and the conclusion, the new statement is called the **converse** of the original statement. "If q, then p" is the converse of "if p, then q." What is the converse of the following statement: If $a + 6 < 9$, *then* $a < 3$? Is the converse true? If $a < 3$, then $a + 6 < 9$. True

In Lesson 10.6 on proportions, one conjecture made is

$$\text{if } \frac{a}{b} = \frac{c}{d}, \text{ then } \frac{a+b}{b} = \frac{c+d}{d}.$$

Hypothesis: $\frac{a}{b} = \frac{c}{d}$ **Conclusion:** $\frac{a+b}{b} = \frac{c+d}{d}$

The conjecture can be proven by adding 1 to each side of the original proportion in the hypothesis.

$\frac{a}{b} = \frac{c}{d}$	Hypothesis
$\frac{a}{b} + 1 = \frac{c}{d} + 1$	Addition Property of Equality
$\frac{a}{b} + \frac{b}{b} = \frac{c}{d} + \frac{d}{d}$	$\frac{b}{b} = 1$ and $\frac{d}{d} = 1$; substitution
$\frac{a+b}{b} = \frac{c+d}{d}$	Definition of Addition for Rational Numbers

EXAMPLE 1

Prove the following statement:

$$\text{If } \frac{a}{b} = \frac{c}{d}, \text{ then } ad = bc.$$

Solution ➤

The hypothesis is $\frac{a}{b} = \frac{c}{d}$, and the conclusion is $ad = bc$.

$\frac{a}{b} = \frac{c}{d}$	Hypothesis
$bd\left(\frac{a}{b}\right) = bd\left(\frac{c}{d}\right)$	Multiplication Property of Equality
$da = bc$	Simplify
$ad = bc$	Commutative Property

Therefore, if $\frac{a}{b} = \frac{c}{d}$, then $ad = bc$. ❖

Try This Prove: If $a + c = b + c$, then $a = b$.

The following definitions can be used in proofs.

DEFINITION OF EVEN NUMBERS

An even number is of the form $2n$, where n is an integer. For example, 16 is even because $16 = 2(8)$.

DEFINITION OF ODD NUMBERS

An odd number is of the form $2n + 1$, where n is an integer. For example, 17 is odd because $17 = 2(8) + 1$.

EXAMPLE 2

Prove the following statement:

$2k + 2$ is an even number if k is an integer.

Solution ➤

k is an integer.	Hypothesis
$k + 1$ is an integer.	Definition of Integer
Let $n = k + 1$. Then n is an integer.	Substitution (of n for $k + 1$)
$2n$ is an even number.	Definition of Even Numbers
$2(k + 1)$ is an even number.	Substitution (of $k + 1$ for n)
$2k + 2$ is an even number.	Distributive Property

Therefore, $2k + 2$ is an even number if k is an integer. ❖

Try This Prove that $2k + 3$ is an odd number if k is an integer.

EXAMPLE 3

Prove the following statement:

If two numbers are odd, then their sum is even.

Solution ➤

Hypothesis: ***a* and *b* are odd.** **Conclusion:** ***a* + *b* is even.**

If two numbers are odd, **then** their sum is even.

Proof:

a and b are two odd numbers.	Hypothesis or given
Let $a = 2k + 1$ and $b = 2p + 1$.	Definition of Odd Numbers (k and p are integers.)
$a + b = (2k + 1) + (2p + 1)$	Addition of Real Numbers
$= 2k + 2p + 2$	Simplify.
$= 2(k + p + 1)$	Distributive Property

$2(k + p + 1)$ is in the form $2n$, where n is equal to $k + p + 1$ and $k + p + 1$ is an integer. Thus, $2(k + p + 1)$ is an even number. ❖

CT– The converse statement would be "If I live in the United States, then I live in Texas." The converse statement is not always true. Some Americans live in Texas, but many live in other states.

Suppose you are given the following statement: If I live in Texas, then I live in the United States. Is the converse true?

EXERCISES & PROBLEMS

Communicate

1. What does a proof in algebra provide?
2. Name three types of reasons that can be used in proofs of theorems.
3. Identify the part of an if-then statement that is assumed to be true. What part has to be proven?
4. Explain what is meant by the converse of a statement.
5. Explain how to prove that if $7x + 9 = -5$, then $x = -2$. Give a reason for each step.

Practice & Apply

Give a reason for each step to prove that if $3m - 4 = -19$, then $m = -5$.

	Proof	
	$3m - 4 = -19$	**Hypothesis or given**
6.	$3m - 4 + 4 = -19 + 4$	Addition Property of Equality
7.	$3m + 0 = -15$	Definition of Opposites
8.	$3m = -15$	Identity Property of Addition
9.	$\frac{3m}{3} = \frac{-15}{3}$	Division Property of Equality
10.	$1 \cdot m = -5$	Reciprocal Property
11.	$m = -5$	Identity Property of Multiplication

Give a reason for each step in the proof that $(a + b) + (-a) = b$.

	Proof	
12.	$(a + b) + (-a)$	Given
13.	$(a + b) + (-a) = (b + a) + (-a)$	Commutative Property of Addition
14.	$(a + b) + (-a) = b + [a + (-a)]$	Associative Property of Addition
15.	$(a + b) + (-a) = b + 0$	Definition of Opposites
16.	$(a + b) + (-a) = b$	Identity Property for Addition

Prove each statement. Give a reason for each step.

17. For all y, $2y + 3y = 5y$.
18. For all x, $5x - 3x = 2x$.
19. For all a, $(-3a)(-2a) = 6a^2$.
20. If $7n - 3n = 32$, then $n = 8$.
21. If $3(x + 2) = -15$, then $x = -7$.
22. If $\frac{2}{3}x + 5 = -9$, then $x = -21$.
23. If $\frac{3}{4}m + 8 = -1$, then $m = -12$.

Write a proof for each statement. Let all variables represent real numbers. Complete Exercises 24–32.

24. If $a + (b + c)$, then $(a + b) + c$.

25. If $a = b$, then $a + c = b + c$.

26. $a(b - c) = ab - ac$

27. $(ax + b) + ay = a(x + y) + b$

28. The square of an even number is even.

29. The square of an odd number is odd.

30. For all real numbers a and b, $a - b$ and $b - a$ are opposites.
HINT: Prove that $(a - b) + (b - a) = 0$ is true. Then refer to the Definition of Opposites.

31. If $x \neq 0$, then $(xy)\frac{1}{x} = y$.

32. Let a, b, c, and d be integers with b and $d \neq 0$.
Prove the arithmetic sum: $\frac{a}{b} + \frac{c}{d} = \frac{da + bc}{bd}$.

Look Back

33. Solve $x = 16x + 45$. **[Lesson 1.3]** -3

34. Write the equation of the line perpendicular to $2x - 3y = 9$ and passing through the point $(-6, 12)$. **[Lesson 2.2]** $y = -\frac{3}{2}x + 3$

35. Solve for x and y. $\begin{bmatrix} 1 & 3 \\ 2 & 1 \end{bmatrix} \cdot \begin{bmatrix} x \\ y \end{bmatrix} = \begin{bmatrix} 5 \\ 0 \end{bmatrix}$ **[Lesson 3.5]** $-1, 2$

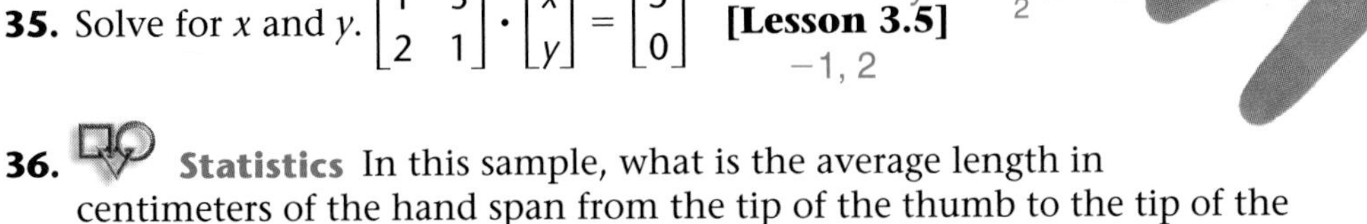

36. **Statistics** In this sample, what is the average length in centimeters of the hand span from the tip of the thumb to the tip of the little finger: 16.2, 18.5, 23.0, 21.2, 22.9, 21.1, 20.6, 19.2? **[Lesson 4.3]** 20.3 cm

37. Describe the transformations applied to $f(x) = x^2$ for $g(x) = -2(x - 1)^2 + 3$. **[Lesson 5.2]**

38. $x^2 - 10x + 24$ square units

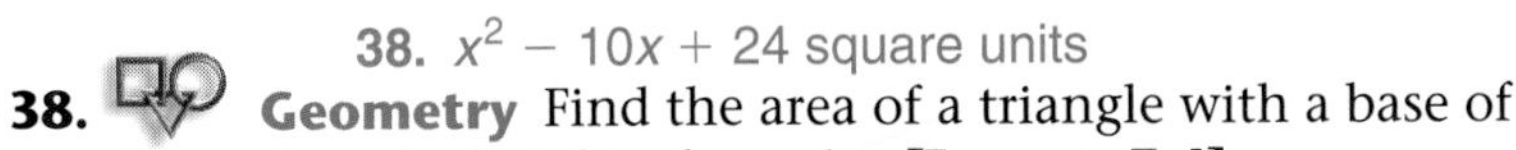

38. **Geometry** Find the area of a triangle with a base of $2(x - 4)$ and a height of $x - 6$. **[Lesson 7.4]**

39. Solve $x^2 - 8 = 2x$. **[Lesson 8.5]** $-2, 4$

40. Solve $\sqrt{x - 5} = -3$. **[Lesson 9.3]** No solution

41. **Geometry** Find the midpoint of the segment joining the points $(-2.5, 4)$ and $(2.5, 6)$. **[Lesson 9.5]** $(0, 5)$

42. Simplify $\frac{x^2 - 9}{x + 3}$. **[Lesson 10.3]** $x - 3, x \neq -3$

Look Beyond

43. A bug was trying to crawl to the top of an 18-foot drain pipe. Each day it climbed 4 feet, but each night it slipped back 3 feet, how many days did it take to crawl out of the pipe? 15 days

A Different Dimension

DIMENSION

When you apply mathematics to solve problems, the numbers are only one part of the problem to consider. Basic dimensions such as mass, length, force, time, and temperature contribute essential information about the problem. Some dimensions are derived from powers or combinations of dimensions. The measurement of these dimensions includes various units such as feet or centimeters to identify relative size.

Part of dimensional analysis is knowing how to convert measurements from one unit to another.

1 mile	= 5280 feet
1 hour	= 60 minutes
1 minute	= 60 seconds

A test car travels on the track at 200 *miles per hour.* What is the equivalent speed in feet per second? Express the original speed in terms of its dimensions.

$\frac{200 \text{ miles}}{1 \text{ hour}}$ The *per* indicates that hour is in the denominator.

Express the equivalent dimensions as a fraction, and multiply.

$$\frac{1 \text{ hour}}{60 \text{ minutes}} \cdot \frac{200 \text{ miles}}{1 \text{ hour}}$$

Since units can be treated as factors, the expression can be simplified by dividing identical expressions.

$$\frac{1 \cancel{\text{hour}}}{60 \text{ minutes}} \cdot \frac{200 \text{ miles}}{1 \cancel{\text{hour}}} = \frac{200 \text{ miles}}{60 \text{ minutes}}$$

Now multiply by the next equivalent expression.

$$\frac{200 \text{ miles}}{60 \cancel{\text{minutes}}} \cdot \frac{1 \cancel{\text{minute}}}{60 \text{ seconds}} = \frac{200 \text{ miles}}{3600 \text{ seconds}} = \frac{1 \text{ mile}}{18 \text{ seconds}}$$

Finally, multiply the result by the equivalent fraction to change from miles to feet.

$$\frac{1 \cancel{\text{mile}}}{18 \text{ seconds}} \cdot \frac{5280 \text{ feet}}{1 \cancel{\text{mile}}} = \frac{5280 \text{ feet}}{18 \text{ seconds}} \approx \frac{293.3 \text{ feet}}{1 \text{ second}}$$

Thus, 200 miles per hour is approximately equivalent to about 293.3 feet per second.

Dimensional Analysis

Some derived dimensions are powers of a basic dimension. The centimeter is a unit for measuring length. Length squared is area. The units of area are square centimeters. Length cubed is volume. The units of volume are cubic centimeters.

You can use dimensional analysis to find the number of square feet in 16 square yards. Begin with the basic conversion 3 feet = 1 yard. This can be written $\frac{3 \text{ ft}}{1 \text{ yd}} = 1$.

$$\frac{16 \text{ yd}^2}{1} \cdot \left(\frac{3 \text{ ft}}{1 \text{ yd}}\right)^2 = \frac{16 \cancel{\text{yd}^2}}{1} \cdot \frac{9 \text{ ft}^2}{1 \cancel{\text{yd}^2}} = 144 \text{ square feet}$$

Since 1 inch is approximately 2.54 centimeters, you can use dimensional analysis to relate cubic centimeters and liters to cubic inches.

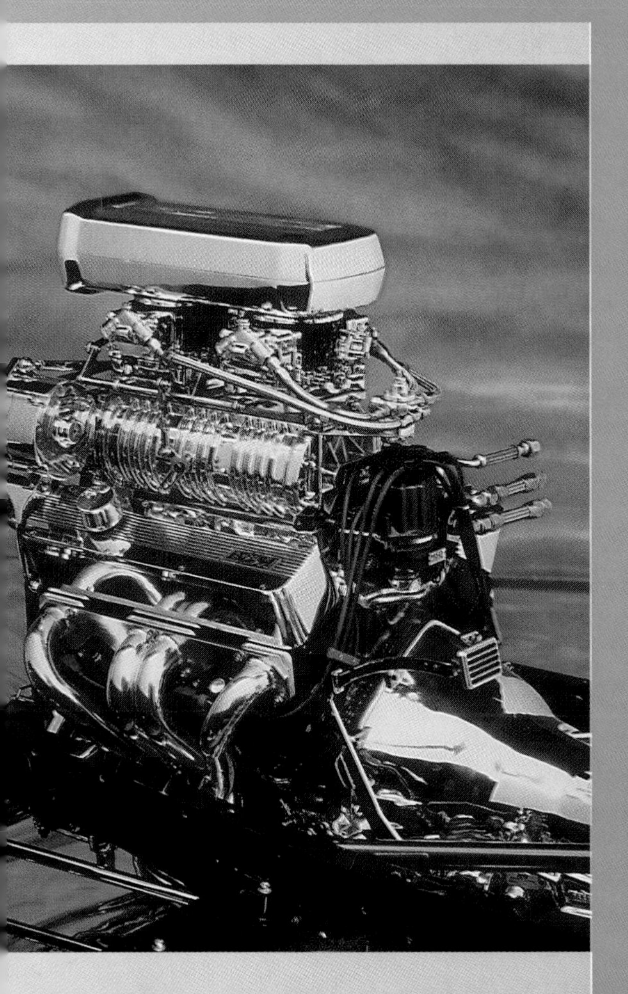

Activity 1

Use dimensional analysis to convert 220 kilometers per hour into meters per second. Find at least three other ways to express 220 kilometers in different units. Show the steps for converting to those units.

Activity 2

A racing automobile has a 7.1 liter engine. Use dimensional analysis to determine the cubic inch displacement of the engine. Consider a liter as 1000 cubic centimeters. Find the engine size of three different kinds of cars and compare the displacement of these engines with the size of a racing engine.

Activity 3

Find two change-of-dimension problems that you might encounter in your activities, and show how to change the dimensions.

Chapter 10 Review

Vocabulary

conclusion	511	inverse variation	480	proof	511
extremes	506	means	506	rational expression	475
hypothesis	511	proportion	506	rational function	475

Key Skills & Exercises

Lesson 10.1

➤ Key Skills

Identify transformations applied to $f(x) = \frac{1}{x}$. List the value(s) for which the function is undefined.

The transformation of $g(x) = \frac{-3}{x-1} - 6$ is a reflection over the x-axis, a vertical stretch by 3 units, a shift of 1 unit to the right, and a shift of 6 units downward. The vertical asymptote is $x = 1$, because the rational expression $\frac{-3}{x-1} - 6$ is undefined when x is 1.

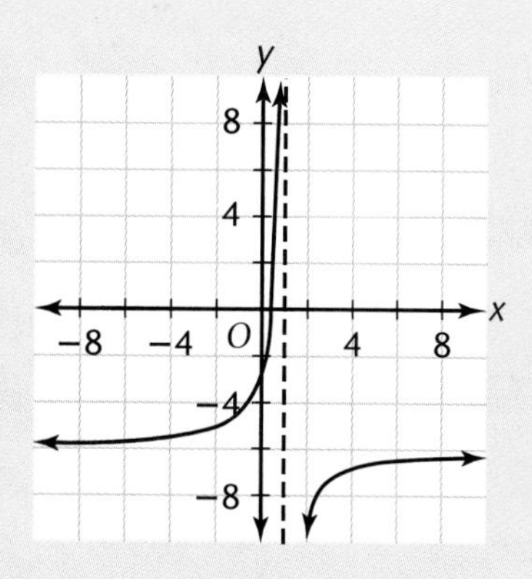

➤ Exercises

Describe the transformations applied to $f(x) = \frac{1}{x}$. List any values for which the function is undefined.

1. $g(x) = \frac{-2}{x} + 4$ **2.** $h(x) = \frac{1}{x-3} + 2$ **3.** $g(x) = \frac{-3}{x+4} - 1$

Lesson 10.2

➤ Key Skills

Identify and use inverse variation.

If y varies inversely as x and y is 10 when x is 3, solve the equation $10 = \frac{k}{3}$ to find the constant of variation, k.

$$xy = k \longrightarrow (10)(3) = k \longrightarrow 30 = k$$

Use the constant of variation, 30, to find x when y is 5.

$$xy = k \longrightarrow x(5) = 30 \longrightarrow x = 6$$

➤ Exercises

4. If x varies inversely as y and x is 75 when y is 20, find y when x is 25.

5. If p varies inversely as q and q is 36 when p is 15, find p when q is 24.

6. If m varies inversely as n and m is 0.5 when n is 5, find m when n is 50.

4. 60 **5.** 22.5 **6.** 0.05

Lesson 10.3

➤ Key Skills

Simplify rational expressions.

$$\frac{-x^2 + x + 2}{x^2 + 3x + 2} = \frac{(-1)(x + 1)(x - 2)}{(x + 1)(x + 2)} = \frac{(-1)(x - 2)}{x + 2} = \frac{-x + 2}{x + 2}, \text{ where } x \neq -1, -2$$

➤ Exercises

Simplify. Write any restrictions on the variable.

7. $\frac{8(x + 1)}{10(x + 1)^2}$ $\frac{4}{5(x + 1)}$, $x \neq -1$

8. $\frac{2x^2 + 6x}{6x - 30}$ $\frac{x^2 + 3x}{3x - 15}$, $x \neq 5$

9. $\frac{2x^2 - x - 1}{1 - x^2}$ $\frac{2x + 1}{-x - 1}$, $x \neq 1, -1$

Lesson 10.4

➤ Key Skills

Add, subtract, and multiply rational expressions.

a.
$$\frac{5}{x - 3} + \frac{2}{x - 5} = \frac{5(x - 5)}{(x - 3)(x - 5)} + \frac{2(x - 3)}{(x - 5)(x - 3)}$$
$$= \frac{5x - 25 + 2x - 6}{x^2 - 8x + 15} = \frac{7x - 31}{x^2 - 8x + 15}, \text{ where } x \neq 3, 5$$

b.
$$\frac{3x + 6}{x + 5} \cdot \frac{x^2 - 25}{x + 2} = \frac{3(x + 2) \cdot (x + 5)(x - 5)}{(x + 5)(x + 2)}$$
$$= 3 \cdot \frac{x + 2}{x + 2} \cdot \frac{x + 5}{x + 5} \cdot (x - 5) = 3(x - 5), \text{ where } x \neq -5, -2$$

➤ Exercises

Perform the indicated operations, and simplify. Write any restrictions on the variable.

10. $\frac{2}{5x} + \frac{5}{10y}$

11. $\frac{16}{y^2 - 16} - \frac{2}{y + 4}$

12. $\frac{t^2 - 9}{6} \cdot \frac{9}{3 - t}$

10. $\frac{4y + 5x}{10xy}$, $x \neq 0, y \neq 0$ **11.** $\frac{24 - 2y}{y^2 - 16}$, $y \neq \pm 4$ **12.** $\frac{-3t - 9}{2}$, $t \neq 3$

Lesson 10.5

➤ Key Skills

Solve rational equations. $\frac{1}{x + 3} + \frac{3}{4x} = \frac{2}{x}$, where $x \neq -3, 0$

$$4x(x + 3)\left(\frac{1}{x + 3} + \frac{3}{4x}\right) = 4x(x + 3)\left(\frac{2}{x}\right)$$
$$4x + (x + 3)3 = 8(x + 3)$$
$$4x + 3x + 9 = 8x + 24$$
$$7x + 9 = 8x + 24$$
$$-x = 15$$
$$x = -15$$

➤ Exercises

Solve.

13. $\frac{z+2}{2z} + \frac{z+3}{z} = 5$, where $z \neq 0$ 13. $\frac{8}{7}$

14. $\frac{6-x}{6x} = \frac{1}{x+1}$, where $x \neq -1, 0$ 2, −3

15. $\frac{5}{w+6} + \frac{2}{w} = \frac{9w+6}{w^2+6w}$, where $w \neq -6, 0$ 3

Lesson 10.6

➤ Key Skills

Solve proportions.

$$\frac{x}{8} = \frac{-2}{x+10}, \text{ where } x \neq -10$$

$$\begin{aligned} x(x+10) &= 8(-2) \\ x^2 + 10x &= -16 \\ x^2 + 10x + 16 &= 0 \\ (x+8)(x+2) &= 0 \\ x + 8 = 0 \quad &\text{or} \quad x + 2 = 0 \\ x = -8 \quad & \quad\quad x = -2 \end{aligned}$$

➤ Exercises

Solve.

16. $\frac{x}{3} = \frac{3}{9}$ 1

17. $\frac{x+1}{2} = \frac{x-1}{3}$ −5

18. $\frac{7}{x+3} = \frac{x}{4}$, where $x \neq -3$ 4, −7

Lesson 10.7

➤ Key Skills

Prove algebraic statements.

Prove that $a + 2(b + a) = 3a + 2b$.

Proof

$a + 2(b + a)$	Hypothesis
$a + 2b + 2a$	Distributive Property
$3a + 2b$	Addition of monomials

➤ Exercises

Prove each algebraic statement. Give a reason for each step.

19. If $2a + 3 = 5$, then $a = 1$.

20. For all a and b, $(a - b) - (b - a) = 2a - 2b$.

Applications

21. **Health** Millie joins a health club. There is a $300 fee to join and then a monthly charge of $20. Find a rational expression to represent Millie's average cost per month for belonging to the health club. $\frac{300 + 20x}{x}, x \neq 0$

22. **Geometry** The area of a certain triangle is represented by $x^2 - 2x - 15$, and its height is $x - 5$. Find the expression for the length of the base. $2(x + 3)$

23. **Travel** A trip that is 189 miles long can take 1 hour less if Gerald increases the average speed of his car by 12 miles per hour. Find the original average speed of his car. 42 mph

Chapter 10 Assessment

1. For what value is $f(x) = \frac{1}{x - 2}$ undefined? 2
2. Describe all the transformations applied to $f(x) = \frac{1}{x}$ in order to graph $g(x) = \frac{-4}{x - 5} + 1$. Write the restriction for the variable.

If *x* varies inversely as *y* and *x* is 0.2 when *y* is 7, find

3. y when x is 4. 0.35
4. x when y is 10. 0.14

5. If 6 feet of wire weighs 0.7 kilograms, how much does 50 feet of wire weigh? About 5.83 kilograms
6. Write the common denominator for $\frac{3}{x - 2}$ and $\frac{2}{x + 4}$. $(x - 2)(x + 4)$

Simplify.

7. $\frac{4x^2(x - 1)}{6x}$
8. $\frac{x - y}{x^2 - y^2}$
9. $\frac{5}{6y} + \frac{7}{4y}$
10. $\frac{6}{x - 8} - \frac{7}{8 - x}$
11. $\frac{4 - x^2}{6x} \cdot \frac{3x^2 - x}{5x + 10}$
12. $\frac{a}{a - 3} - \frac{2a}{a + 4}$

13. The area of a certain rectangle is represented by $6x^2 + x - 35$, and its length is $2x + 5$. Find the width. $3x - 7$

Solve.

14. $\frac{5}{2y} - \frac{3}{10} = \frac{1}{y}$ 5
15. $\frac{x^2 + 5}{3} - \frac{6x}{3} = 0$ 1, 5
16. $\frac{2(y - 1)}{y + 2} - \frac{1}{y + 2} = 1$, where $y \neq -2$ 5
17. $\frac{4}{7} = \frac{x}{49}$ 28
18. $\frac{x + 4}{2} = \frac{7}{4}$ $-\frac{1}{2}$
19. $\frac{x}{3} = \frac{4}{x - 1}$, where $x \neq 1$ 4, -3

20. A large truck has a capacity of 3 tons more than a smaller truck. If the ratio of their capacities is 5 to 2, find the capacity of both trucks. 2 tons and 5 tons
21. The ratio of a number to 5 more than the same number is 2 to 5. Find the number. $\frac{10}{3}$ or $3\frac{1}{3}$
22. A blueprint for a house shows a room as a square that has sides of 3 centimeters. The scale indicates that the actual length of a side of the square room is 12 meters. The actual length of another square room is 16 meters. What is the length of this room in centimeters on the blueprint? 4 cm
23. Sean and Sam are bricklayers. Sean can build a wall with bricks in 5 hours, and Sam can build a wall of the same size in 4 hours. How long will it take them to build the wall if they work together? $2\frac{2}{9}$ hours
24. A boat can travel 90 kilometers downstream in the same time that it can travel 60 kilometers upstream. If the boat travels 15 kilometers per hour in still water, find the rate of the current. 3 km per hr
25. Prove that $3b - (b - a) = 2b + a$. Give a reason for each step.

Chapters 1 – 10 Cumulative Assessment

College Entrance Exam Practice

Quantitative Comparison For Questions 1–4, write
A if the quantity in Column A is greater than the quantity in Column B;
B if the quantity in Column B is greater than the quantity in Column A;
C if the two quantities are equal; or
D if the relationship cannot be determined from the information given.

	Column A	Column B	Answers
1. B	The d_{12} entry of $D = \begin{bmatrix} 4 & 5 \\ 6 & 7 \end{bmatrix}$	The d_{21} entry of $D = \begin{bmatrix} 4 & 5 \\ 6 & 7 \end{bmatrix}$	(A) (B) (C) (D) **[Lesson 3.1]**
2. A	$\frac{5 \times 10^{-2}}{7 \times 10^{-5}}$	$\frac{5 \times 10^{2}}{7 \times 10^{5}}$	(A) (B) (C) (D) **[Lesson 6.3]**
3. C	$\frac{2\sqrt{2}}{\sqrt{5}}$	$\frac{2\sqrt{10}}{5}$	(A) (B) (C) (D) **[Lesson 9.2]**
4. C	The solution to the equation $\frac{8}{3x} + \frac{9}{x} = \frac{7}{3}$	5	(A) (B) (C) (D) **[Lesson 10.5]**

5. For the function $g(x) = \frac{-3}{x-4} - 2$, which transformation has *not* been applied to the parent function $f(x) = \frac{1}{x}$? **[Lessons 5.6, 10.1]** d

a. a shift of 2 units downward **b.** a reflection over the x-axis
c. a vertical stretch of 3 units **d.** a shift of 4 units to the left

6. Which is the product $(6x - 2)(6x - 2)$? **[Lesson 7.4]** d

a. $36x^2 - 4$ **b.** $36x^2 + 4$
c. $36x^2 - 12x + 4$ **d.** $36x^2 - 24x + 4$

7. Which is the solution to the equation $x^2 + 6x - 40 = 0$? **[Lesson 8.4]** c

a. -4 and 10 **b.** 4 and 10
c. 4 and -10 **d.** -4 and -10

8. Which is the complete solution to the equation $x^2 = 144$? **[Lesson 9.3]** c

a. 12 **b.** 72 **c.** 12 and -12 **d.** 72 and -72

9. In the figure, which trigonometric function describes the ratio $\frac{4}{5}$? **[Lessons 9.7, 9.8]** b

a. $\sin B$ **b.** $\sin A$
c. $\tan B$ **d.** $\cos A$

B
4 5
C 3 A

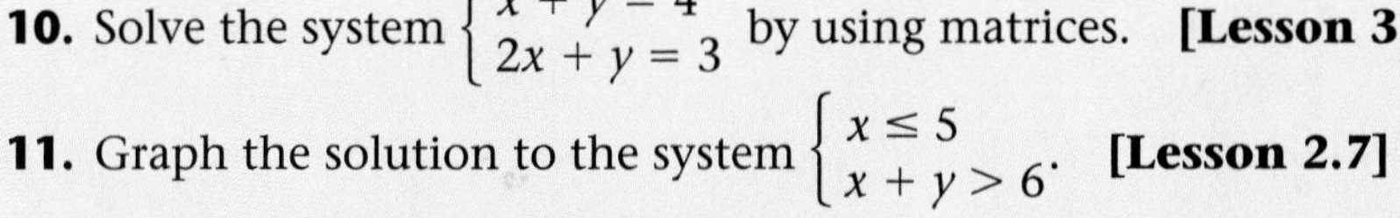

10. Solve the system $\begin{cases} x + y = 4 \\ 2x + y = 3 \end{cases}$ by using matrices. **[Lesson 3.5]** $(-1, 5)$

11. Graph the solution to the system $\begin{cases} x \leq 5 \\ x + y > 6 \end{cases}$. **[Lesson 2.7]**

12. Identify the parent function of $y = 5|x|$. Then tell what transformation was applied to the parent function. **[Lessons 5.2–5.5]**

13. Write the polynomial $3x^2 + 16x + 5$ as the product of two binomial factors. **[Lesson 7.7]** $(3x + 1)(x + 5)$

14. Find $-\sqrt{29}$ to the nearest hundredth. **[Lesson 9.1]** -5.39

15. A right triangle has legs of 16 inches and 18 inches. Find the length of the hypotenuse to the nearest tenth of an inch. **[Lesson 9.4]** 24.1 in.

16. Find the distance between the points $P(-2, 3)$ and $Q(4, 7)$. Give your answer as a radical in simplest form. **[Lesson 9.5]** $2\sqrt{13}$

17. What is an equation of a circle with its center at (0, 0) and a radius of 7 units? **[Lesson 9.6]** $x^2 + y^2 = 49$

18. The number x varies inversely as the number y. When x is 0.6, y is 50. Find the value of y when x is 3. **[Lesson 10.2]** 10

19. Perform the indicated operation and simplify: $\frac{4}{3x - 6} + \frac{5}{3x + 3}$. **[Lessons 10.3, 10.4]** $\frac{3x - 2}{x^2 - x - 2}, x \neq -1, 2$

20. Prove: $a + 3(a + b) = 4a + 3b$. **[Lesson 10.7]**

21. Complete the square for $x^2 - 10x$. **[Lesson 8.3]** $x^2 - 10x + 25$

22. Multiply $(8.2 \times 10^7)(2.1 \times 10^9)$. Write your answer in scientific notation. **[Lesson 6.4]** 1.722×10^{17}

Free-Response Grid The following questions may be answered using a free-response grid commonly used by standardized test services.

23. Solve $9(x - 5) + 20 = 6(1 - x) - 26$ for x. **[Lesson 1.4]** $\frac{1}{3}$

24. In a group of people, 18 like chamber music, 25 like jazz, and 7 like both. How many people are in the group? **[Lesson 4.4]** 36

25. Evaluate $g(-2)$ for the function $g(x) = |3 - 2x| - x$. **[Lesson 5.1]** 9

26. What is the greater of the two solutions to $(x - 4)^2 - 9 = 0$? **[Lesson 8.2]** 7

27. Solve $\frac{x - 12}{6} = \frac{x - 10}{5}$ for x. **[Lesson 10.6]** 0

28. Find the value of the discriminant for the quadratic equation $2a^2 - 8a = 8$. **[Lesson 8.5]** 128

Info Bank

EXTRA PRACTICE

Lesson 1.1

Solve and check each equation.

1. $x + 4 = 11$ 7 **2.** $y - 5 = 9$ 14 **3.** $x + 14 = -15$ −29

4. $y - 10 = 5$ 15 **5.** $x - 4 = -7$ −3 **6.** $6 = x - 13$ 19

7. $-8 = y - 12$ 4 **8.** $y - 18 = -2$ 16 **9.** $-13 = y - 1$ −12

10. $x + 5 = -20$ −25 **11.** $21 = x - 3$ −24 **12.** $x - 3.2 = 4.5$ 7.7

13. $y + 8.9 = -1.7$ −10.6 **14.** $x - 7.8 = -2.4$ 5.4 **15.** $y - 8 = -2$ 6

16. $-3 = y - 2$ −1 **17.** $y - 1 = 7.6$ 8.6 **18.** $-1 = x + 15$ −16

19. $x - 25 = 35$ 60 **20.** $x - 5.8 = 3.7$ 9.5 **21.** $x + 1.3 = -0.8$ −2.1

22. $y - 7 = -14$ −7 **23.** $-7.5 = x + 4.8$ −12.3 **24.** $-0.9 = y - 1$ 0.1

25. $12 = y - 19$ 31 **26.** $-9 = y - 14$ 5 **27.** $3.8 = x - 4.9$ 8.7

28. $x - 14.5 = -2.6$ 11.9 **29.** $y - 12 = -2$ 10 **30.** $x - 6 = 1.8$ 7.8

Lesson 1.2

4. −0.1 **8.** 17 **12.** −72 **16.** −0.13 **20.** 14 **24.** −2.4 **28.** −4 **32.** 36 **36.** −3 **40.** −1.25 **44.** 0.3125

Solve and check each equation.

1. $7y = -49$ −7 **2.** $-8x = 32$ −4 **3.** $-0.1z = 9$ −90 **4.** $8w = -0.8$

5. $-5w = -25$ 5 **6.** $-1.35 = -9z$ 0.15 **7.** $4r = -96$ −24 **8.** $-170 = -10t$

9. $\frac{w}{3} = 12$ 36 **10.** $\frac{r}{-7} = 8$ −56 **11.** $\frac{t}{-9} = -9$ 81 **12.** $\frac{h}{6} = -12$

13. $7y = -42$ −6 **14.** $-5a = -12.5$ 2.5 **15.** $-0.9q = -63$ 70 **16.** $-1.69 = 13m$

17. $\frac{3}{4}p = 36$ 48 **18.** $\frac{5}{8}g = 15$ 24 **19.** $\frac{2}{5}t = 18$ 45 **20.** $\frac{6}{7}y = 12$

21. $-2z = -48$ 24 **22.** $-1.5 = 6q$ −0.25 **23.** $7.6 = -3.8y$ −2 **24.** $-6 = 2.5r$

25. $14x = -7$ −0.5 **26.** $-9r = -45$ 5 **27.** $100h = -10$ −0.1 **28.** $-50t = 200$

29. $\frac{j}{15} = 5.5$ 82.5 **30.** $-\frac{m}{12} = 3.8$ −45.6 **31.** $\frac{x}{-9} = -10.7$ 96.3 **32.** $\frac{4}{9}v = -16$

33. $\frac{11}{12}g = -121$ −132 **34.** $\frac{9}{10}h = -10$ $-11\frac{1}{9}$ **35.** $\frac{-2}{5}f = -50$ 125 **36.** $\frac{6}{-15}k = 1.2$

37. $-5 = 3.5r$ $-1\frac{3}{7}$ **38.** $-9x = -8.1$ 0.9 **39.** $12y = -4.8$ −0.4 **40.** $-6 = 4.8w$

41. $\frac{-7}{8}q = -5$ $5\frac{5}{7}$ **42.** $\frac{9}{-2} = 3j$ −1.5 **43.** $-5y = \frac{2}{7}$ $-\frac{2}{35}$ **44.** $\frac{5}{-8} = -2g$

Solve each proportion. **48.** −3.5 **52.** $6\frac{2}{3}$ **56.** −1.25

45. $\frac{x}{-3} = \frac{5}{18}$ $-\frac{5}{6}$ **46.** $\frac{9}{25} = \frac{z}{5}$ 1.8 **47.** $\frac{w}{8} = \frac{-2}{5}$ −3.2 **48.** $\frac{t}{-6} = \frac{7}{12}$

49. $\frac{5}{8} = \frac{m}{-10}$ −6.25 **50.** $\frac{r}{-7} = \frac{14}{21}$ $-4\frac{2}{3}$ **51.** $\frac{b}{14} = \frac{-8}{28}$ −4 **52.** $\frac{4}{-9} = \frac{r}{-15}$

53. $\frac{h}{16} = \frac{-4}{9}$ $-7\frac{1}{9}$ **54.** $\frac{-16}{24} = \frac{t}{32}$ $-21\frac{1}{3}$ **55.** $\frac{-18}{20} = \frac{-r}{6}$ 5.4 **56.** $\frac{-q}{5} = \frac{1}{4}$

Lesson 1.3

Solve and check each equation.

1. $4x + 1 = 9$ 2
2. $6m - 5 = 13$ 3
3. $8r - 3 = 29$ 4
4. $8t - 7 = 11$ 2.25
5. $3g - 10 = 2$ 4
6. $15x - 5 = 40$ 3
7. $40s - 9 = 1$ 0.25
8. $3p - 9 = -3$ 2
9. $20r + 3 = -17$ −1
10. $8q - 5 = -17$ −1.5
11. $4w - 9 = -17$ −2
12. $3t + 25 = 10$ −5
13. $30z - 10.5 = 4.5$ 0.5
14. $-10y + 6 = -52$ 5.8
15. $-16x - 5 = -13$ 0.5
16. $16 + 5x = 13.5$ −0.5
17. $-17.1 = 3t - 18$ 0.3
18. $3 - 8r = -0.2$ 0.4
19. $0.9w + 6.1 = 7$ 1
20. $1.2r - 7 = -1$ 5
21. $34.4 = 7t - 0.6$ 5

Lesson 1.4

Solve and check each equation.

1. $3x + 12 = x + 16$ 2
2. $7w - 13 = 5w - 7$ 3
3. $6t + 4 = 3t + 2$ $-\frac{2}{3}$
4. $10 - 5z = 7z + 22$ −1
5. $6x - 4 = -4x + 1$ 0.5
6. $-5x + 3 = 1 - 10x$ −0.4
7. $7 - 4g = 8g + 4$ 0.25
8. $3x - 5 = x - 4.8$ 0.1
9. $3(x + 1) = 2x + 7$ 4
10. $3w - 1 - 4w = 4 - 2w$ 5
11. $8x - 5 = 4x + 4 - 2x$ 1.5
12. $15 - 3y = y + 13 + y$ 0.4
13. $2(8y - 7) = 2(3 + 8y)$ No solution
14. $2(2x + 3) = 8x + 5$ 0.25
15. $3 - 2(x - 1) = 2 + 4x$ 0.5
16. $3(2r - 1) + 5 = 5(r + 1)$ 3
17. $5(x + 1) + 2 = 2x + 11$ $1\frac{1}{3}$
18. $4z - (z + 6) = 3z - 4$ No solution

Lesson 1.5

Solve each inequality. Graph your solution on a number line.

1. $x + 2 \leq -3$
2. $c - 2 < 7$
3. $4 > z - 5$
4. $8y \leq 24$
5. $-2t \geq 18$
6. $3t > 5$
7. $\frac{m}{3} < 5$
8. $\frac{w}{-2} \geq 9$
9. $\frac{5}{6}a \leq 10$
10. $2x - 1 < 6$
11. $4y - 5 \geq -1$
12. $5t + 2 \leq -17$
13. $4y - 2 > 11.6$
14. $8 \geq 5 - w$
15. $18 \leq 15 + 3q$
16. $1.2x - 8.9 > -2.9$
17. $3.5r - 2.5 < -0.75$
18. $-11.6 > 3x - 14.3$
19. $\frac{2}{3}x - 4 < 12$
20. $-\frac{5}{8}g - 10 \geq -25$
21. $\frac{-x}{7} + 9 < -8$
22. $\frac{-4}{5}w - 2 > -10$
23. $\frac{y}{-8} - 5 \geq -5.5$
24. $\frac{-3}{4}v - 7 \leq -16$

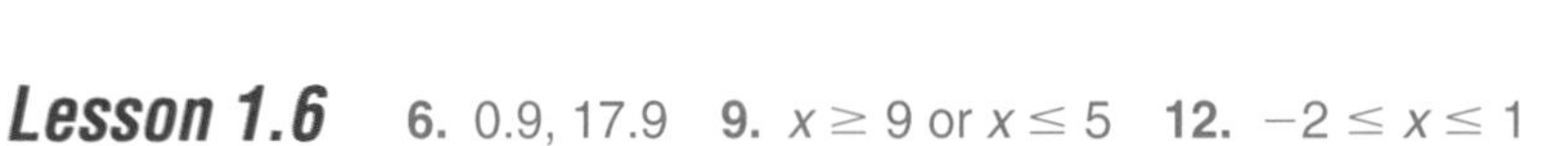

Lesson 1.6 6. 0.9, 17.9 9. $x \geq 9$ or $x \leq 5$ 12. $-2 \leq x \leq 1$

Solve each equation or inequality. Check each solution.

1. $|x + 6| = 3$ −9, −3
2. $|t - 2| = 4$ −2, 6
3. $|s + 5| = 9$ −14, 4
4. $|y - 2.5| = 4.5$ −2, 7
5. $|j + 4.9| = 7.8$ −12.7, 2.9
6. $|h - 9.4| = 8.5$
7. $|x + 10| \leq 9$ $-19 \leq x \leq -1$
8. $|x + 3| < 5$ $-8 < x < 2$
9. $|x - 7| \geq 2$
10. $|2x - 2| < 2$ $x > 2$ or $x < 0$
11. $|3w - 1| \leq 5$ $-1\frac{1}{3} \leq w \leq 2$
12. $|4v + 2| \leq 6$
13. $|4y + 4.4| < 6.25$ $-2.6625 < y < 0.4625$
14. $|3t - 5.8| \leq 4.55$ $0.416 \leq t \leq 3.45$
15. $|0.05f + 1| \leq 2.9$ $-78 \leq f \leq 38$

Write and solve an inequality for each situation. Graph the solution on a number line.

16. The distance between a number and 4 is less than 3.
17. The distance between a number and −6 is greater than 5.
18. The distance between a number and 10 is less than 1.5.
19. The distance between a number and 12 is at least 2.2.

Lesson 2.1

Determine the slope of each line.

1.

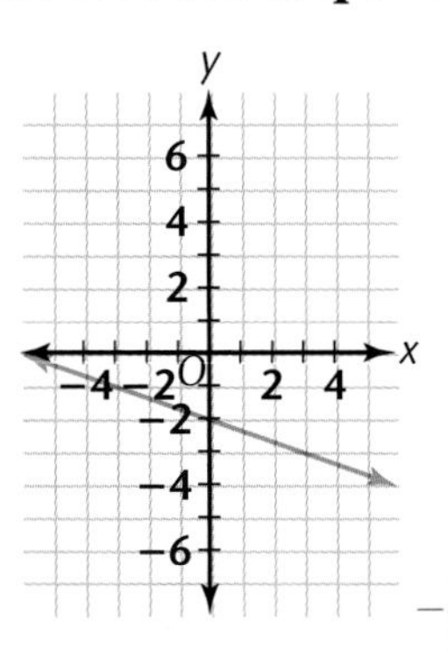

$-\frac{1}{3}$

2.

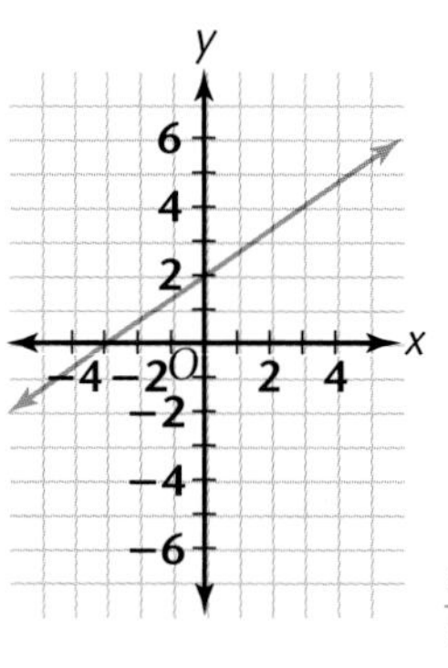

$\frac{2}{3}$

3.

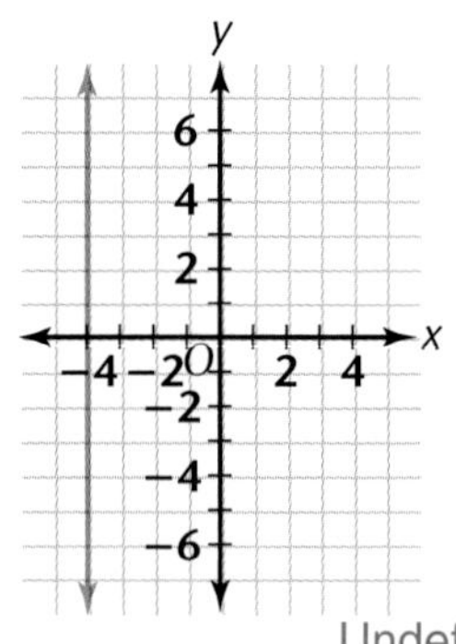

Undefined

Determine the slope of the line through the two given points.

4. $(-7, 3)$ and $(4, 2)$ $-\frac{1}{11}$
5. $(6, -1)$ and $(-6, 2)$ $-\frac{1}{4}$
6. $(5, 3)$ and $(-4, -1)$
7. $(-2.5, 0.5)$ and $(3.5, -0.5)$ $-\frac{1}{6}$
8. $(-7, 12)$ and $(7, 12)$ 0
9. $(6, 5)$ and $(5, -6)$

Graph each line described below. 6. $\frac{4}{9}$ 9. 11

10. Slope −2 through the point (1, 2)
11. Slope $\frac{1}{2}$ through the point (4, −1)
12. Slope $-\frac{3}{4}$ through the point (3, −6)
13. Through the point (5, 2) with 0 slope

Lesson 2.2

Use x- and y-intercepts to draw the graph of each line.

1. $2x + 3y = 6$ **2.** $4x + y = 2$ **3.** $x - 2y = 4$

4. $3x - y = 6$ **5.** $1.5x - 3y = 3.6$ **6.** $3x - 1.5y = 2$

Convert the equation to slope-intercept form. Then state the slope and graph the equation.

7. $2x - y = -5$ **8.** $x - 2y = -4$ **9.** $2x + 3y = 3$

10. $x - 5y = 2$ **11.** $3x - 3y = 5$ **12.** $4x - y = 8$

13. $2x - 5y = 3$ **14.** $2.4x + 6y = 3$ **15.** $6.5x - 2.5y = 3$

Write the equation of each line described below in standard form.

16. Slope $\frac{1}{4}$ with y-intercept -2 $x - 4y = 8$

17. Slope -1 with y-intercept 5

18. Through $(-3, 1)$ with slope $\frac{4}{3}$ $4x - 3y = -15$

19. Through $(4, -5)$ with slope 0

20. Through $(-5, -4)$ with undefined slope $x = -5$

21. Slope $-\frac{2}{5}$ through $(-4, 0)$

22. Slope $\frac{2}{7}$ with y-intercept $\frac{1}{2}$ $4x - 14y = -7$

23. Slope $-\frac{2}{3}$ with y-intercept -4

24. Through $(-6, 3)$ and $(3, 4)$ $x - 9y = -33$

25. Through $(3, 10)$ and $(-1, 9)$

26. Through $(0, 2)$ and $(-4, 5)$ $3x + 4y = 8$

27. Through $(7, -5)$ and $(2, -1)$

17. $x + y = 5$ **19.** $y = -5$ **21.** $2x + 5y = -8$ **23.** $2x + 3y = -12$
25. $x - 4y = -37$ **27.** $4x + 5y = 3$

Lesson 2.3

Solve by graphing. Round approximate solutions to the nearest tenth. Check algebraically.

1. $\begin{cases} 3x + 4y = 2 \\ x - y = 1 \end{cases}$ (0.9, −0.1) **2.** $\begin{cases} 2y = 5x - 1 \\ x = y + 2 \end{cases}$ (−1, −3) **3.** $\begin{cases} -3y = 2x - 6 \\ x + 4y = 8 \end{cases}$ (0, 2)

4. $\begin{cases} x = 300 - y \\ x - 200 = y \end{cases}$ (250, 50) **5.** $\begin{cases} 4x + 3y = 18 \\ 5x - 6y = 15 \end{cases}$ (3.9, 0.8) **6.** $\begin{cases} x = 5 \\ 3y = 2x - 1 \end{cases}$ (5, 3)

7. $\begin{cases} y = 4 \\ 2x - y = 1 \end{cases}$ (2.5, 4) **8.** $\begin{cases} 4x - 2y = 8 \\ 3y = x + 9 \end{cases}$ (4.2, 4.4) **9.** $\begin{cases} y - 5x = 10 \\ x - 5y = 10 \end{cases}$ (−2.5, −2.5)

10. $\begin{cases} x = -4 \\ y = -2 \end{cases}$ (−4, −2) **11.** $\begin{cases} 15x + y = 50 \\ y = 35x - 40 \end{cases}$ (1.8, 23) **12.** $\begin{cases} 7x - 5y = 1 \\ 3x + 2y = 2 \end{cases}$ (0.4, 0.4)

Algebraically determine whether the point (3, 8) is a solution for each pair of equations.

13. $\begin{cases} 2x + y = 14 \\ x + y = 11 \end{cases}$ Yes **14.** $\begin{cases} y = -x - 5 \\ y = x + 5 \end{cases}$ No **15.** $\begin{cases} 4x - y = -4 \\ 3x - 2y = 7 \end{cases}$ No

Algebraically determine whether the point (−4, 6) is a solution for each pair of equations.

16. $\begin{cases} 2x + y = 2 \\ x - y = 10 \end{cases}$ No **17.** $\begin{cases} y = -x + 2 \\ y = x + 10 \end{cases}$ Yes **18.** $\begin{cases} 6x - y = -30 \\ 3x - 2y = 0 \end{cases}$ No

Lesson 2.4

Solve and check each system by substitution.

1. $\begin{cases} 2x = y - 1 \\ y = 3x \end{cases}$ (1, 3)
2. $\begin{cases} x = -y \\ 2x + y = -2 \end{cases}$ (−2, 2)
3. $\begin{cases} 2x + 2y = 4 \\ x = 10 - 3y \end{cases}$ (−2, 4)
4. $\begin{cases} y = 7 - x \\ 2x + 3y = -1 \end{cases}$ (22, −15)
5. $\begin{cases} x + y = -1 \\ 2x + y = 3 \end{cases}$ (4, −5)
6. $\begin{cases} 2x + 3y = -8 \\ y = 9 - x \end{cases}$ (35, −26)
7. $\begin{cases} y + 3x = 1 \\ 2y + 5x = 5 \end{cases}$ (−3, 10)
8. $\begin{cases} 5 = 2y + x \\ x = 20y \end{cases}$ $\left(4\frac{6}{11}, \frac{5}{22}\right)$
9. $\begin{cases} 4x + 4y = 1 \\ -x = 2y \end{cases}$ (0.5, −0.25)

First graph each system and estimate the solution. Then use the substitution method to get an exact solution.

10. $\begin{cases} x + 4y = 5 \\ 8y - x = -8 \end{cases}$ (6, −0.25)
11. $\begin{cases} 2y + x = 4 \\ 4y = x - 1 \end{cases}$ (3, 0.5)
12. $\begin{cases} x = y \\ 3x + 6y = 6 \end{cases}$ $\left(\frac{2}{3}, \frac{2}{3}\right)$
13. $\begin{cases} x = -y \\ 16x + 4y = 9 \end{cases}$ (0.75, −0.75)
14. $\begin{cases} 5x - y = 3 \\ 10x = y \end{cases}$ (−0.6, −6)
15. $\begin{cases} 3x = y - 4 \\ 2y - 6x = 12 \end{cases}$ No solution
16. $\begin{cases} x + y = 8 \\ 2x - y = 7 \end{cases}$ (5, 3)
17. $\begin{cases} 2x + y = -5 \\ x + 2y = -2 \end{cases}$ $\left(-2\frac{2}{3}, \frac{1}{3}\right)$
18. $\begin{cases} 2x + 2y = 8 \\ y - x = 24 \end{cases}$ (−10, 14)

Lesson 2.5

Solve each system of equations by elimination and check.

1. $\begin{cases} x - y = 9 \\ x + y = 7 \end{cases}$ (8, −1)
2. $\begin{cases} x + 3y = 7 \\ 2x - 3y = -4 \end{cases}$ (1, 2)
3. $\begin{cases} 5x + 4y = 12 \\ 3x - 4y = 4 \end{cases}$ (2, 0.5)
4. $\begin{cases} x + y = 1 \\ x - 2y = 2 \end{cases}$ $\left(1\frac{1}{3}, -\frac{1}{3}\right)$
5. $\begin{cases} 3x + 3y = 6 \\ 2x - y = 1 \end{cases}$ (1, 1)
6. $\begin{cases} x + 8y = 3 \\ 4x - 2y = 7 \end{cases}$ $\left(1\frac{14}{17}, \frac{5}{34}\right)$
7. $\begin{cases} 4x - y = 4 \\ x + 2y = 3 \end{cases}$ $\left(1\frac{2}{9}, \frac{8}{9}\right)$
8. $\begin{cases} 3x - 5y = -13 \\ 4x + 3y = 2 \end{cases}$ (−1, 2)
9. $\begin{cases} 2x + 3y = 8 \\ 3x + 2y = 17 \end{cases}$ (7, −2)

Solve each system of equations using any method.

10. $\begin{cases} x - y = 1 \\ 3x - y = 3 \end{cases}$ (1, 0)
11. $\begin{cases} y = 3x - 1 \\ 3x + 4y = 16 \end{cases}$ $\left(1\frac{1}{3}, 3\right)$
12. $\begin{cases} 4x + 2y = -8 \\ x = 2y - 7 \end{cases}$ (−3, 2)
13. $\begin{cases} x + y = 15 \\ \frac{1}{6}x = \frac{1}{9}y \end{cases}$ (6, 9)
14. $\begin{cases} 3x + \frac{1}{3}y = 10 \\ 2x - 5 = \frac{1}{3}y \end{cases}$ (3, 3)
15. $\begin{cases} 2x = y + 36 \\ 3x = \frac{1}{2}y + 26 \end{cases}$ (4, −28)
16. $\begin{cases} y = -2x \\ 5x + 3y = 1 \end{cases}$ (−1, 2)
17. $\begin{cases} x + y = 6 \\ -2x + y = -3 \end{cases}$ (3, 3)
18. $\begin{cases} 0.2m - 0.3n = 0 \\ 0.4m - 0.2n = 0.2 \end{cases}$ (0.75, 0.5)
19. $\begin{cases} 3a + 7b = 5 \\ 3b = -7 - 2a \end{cases}$ (−12.8, 6.2)
20. $\begin{cases} 2x + y = 45 \\ 3x - y = 5 \end{cases}$ (10, 25)
21. $\begin{cases} 4x + 2y = -8 \\ \frac{1}{2}x - y = -\frac{7}{2} \end{cases}$ (−3, 2)
22. $\begin{cases} 4x + y = 5 \\ -4x + \frac{3}{2}y = 5 \end{cases}$ (0.25, 4)
23. $\begin{cases} 3x + 6 + 8y = 0 \\ 4x - 2y = 11 \end{cases}$ (2, −1.5)
24. $\begin{cases} x + y = 6 \\ x - y = 4.5 \end{cases}$ (5.25, 0.75)

Lesson 2.6

Solve each system using any method.

1. $\begin{cases} x - 6y = 2 \\ 5x - y = 3 \end{cases}$ $\left(\frac{16}{29}, -\frac{7}{29}\right)$

2. $\begin{cases} x - 5y = -42 \\ x - y = -5 \end{cases}$ (4.25, 9.25)

3. $\begin{cases} 4x + 5y = 14 \\ -x - y = -10 \end{cases}$ (36, −26)

4. $\begin{cases} 2y - 3 = x \\ 2x + 2y = 7 \end{cases}$ $\left(1\frac{1}{3}, 2\frac{1}{6}\right)$

5. $\begin{cases} 5x + y = 3 \\ 2y = 5x \end{cases}$ (0.4, 1)

6. $\begin{cases} 3x + y = 13 \\ 2x - y = 2 \end{cases}$ (3, 4)

7. $\begin{cases} 2x - 3y = 5 \\ x - 5y = 0 \end{cases}$ $\left(3\frac{4}{7}, \frac{5}{7}\right)$

8. $\begin{cases} y - 4x = 11 \\ 2y + x = 6 \end{cases}$ $\left(-1\frac{7}{9}, 3\frac{8}{9}\right)$

9. $\begin{cases} x + 2y = 3 \\ 5x - 3y = 2 \end{cases}$ (1, 1)

10. $\begin{cases} y = -3x \\ x = 38 + 6y \end{cases}$ (2, −6)

11. $\begin{cases} 2x + 4y = 8 \\ x + 2y = 4 \end{cases}$ All real numbers

12. $\begin{cases} 3x - 8y = 4 \\ 6x = 42 + 16y \end{cases}$ No solution

Determine which of the following systems are dependent, independent, or inconsistent.

13. $\begin{cases} x + 2y = 6 \\ x + 2y = 8 \end{cases}$

14. $\begin{cases} 2x + 3y = 4 \\ 8 - 6y = 4x \end{cases}$

15. $\begin{cases} y = 3x - 4 \\ 6x + 2y = -8 \end{cases}$

16. $\begin{cases} x + y = 3 \\ x + y = 4 \end{cases}$

17. $\begin{cases} 9x - 2 = 4y \\ x = 6 - y \end{cases}$

18. $\begin{cases} 3x + 6 = 7y \\ x = 11 - 2y \end{cases}$

19. $\begin{cases} 8x + 3y = 48 \\ 4x = 24 - 1.5y \end{cases}$

20. $\begin{cases} 4x - 3y = 12 \\ 12x + 9y = -36 \end{cases}$

21. $\begin{cases} 2x + y = 1 \\ x = 1 - 0.5y \end{cases}$

13. Inconsistent **14.** Dependent **15.** Independent **16.** Inconsistent
17. Independent **18.** Independent **19.** Dependent **20.** Independent
21. Inconsistent

Lesson 2.7

Graph the common solution of each system. Choose a point from the solution. Check both inequalities.

1. $\begin{cases} x > 3 \\ y < 6 \end{cases}$

2. $\begin{cases} y > 2 \\ y > -x + 2 \end{cases}$

3. $\begin{cases} y \geq 2x + 1 \\ y \leq -x + 1 \end{cases}$

4. $\begin{cases} y \geq 3x \\ 3y \leq 5x \end{cases}$

5. $\begin{cases} y + 3 \geq x \\ x + y \geq -1 \end{cases}$

6. $\begin{cases} y > x + 1 \\ y < x + 3 \end{cases}$

7. $\begin{cases} y - x < 1 \\ y - x > 3 \end{cases}$

8. $\begin{cases} 2y + x < 4 \\ 3x - y > 6 \end{cases}$

9. $\begin{cases} y + 2 < x \\ 2y - 3 > 2x \end{cases}$

Match the system of inequalities to the graph that represents the solution.

10. $\begin{cases} y \leq 3x + 6 \\ y > -\frac{1}{2}x + 1 \end{cases}$ b

11. $\begin{cases} y > 3x + 6 \\ y \leq -\frac{1}{2}x + 1 \end{cases}$ a

a.

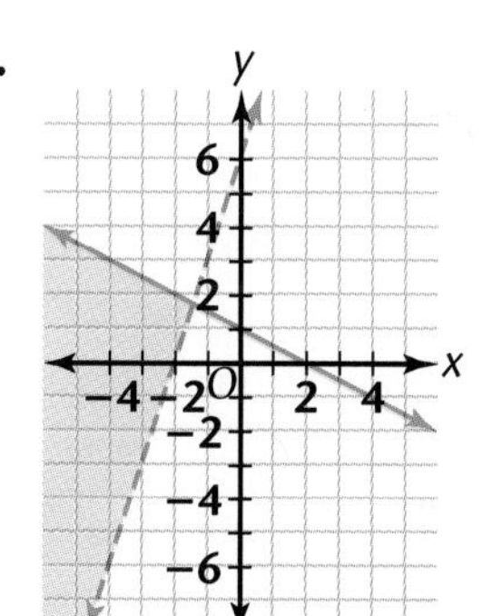

b.

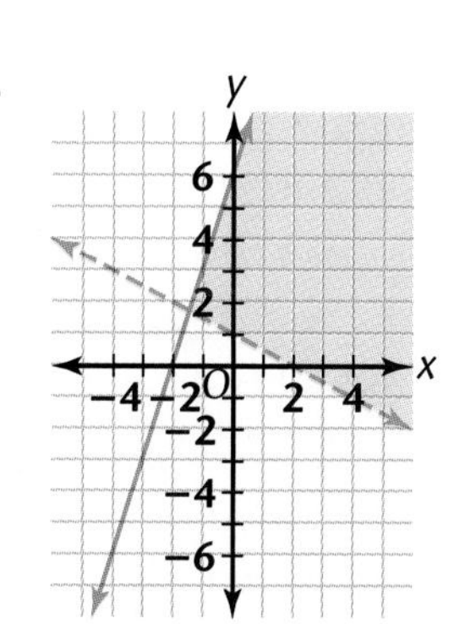

Lesson 2.8

Solve each problem.

1. Jenny has 83 coins in nickels and dimes. She has a total of $6.95. How many of each coin does she have? 56 dimes, 27 nickels
2. How many ounces of a 25% acid solution should be mixed with a 50% solution to produce 100 ounces of a 45% acid solution?
3. A plane leaves Chicago and heads for Columbus, which is 300 miles away. The plane, flying with the wind, takes 40 minutes. The plane returns to Chicago, traveling against the wind, in 45 minutes. Find the rate of the wind and the rate of the plane with no wind.
4. A two-digit number is equal to 7 times the units digit. If 18 is added to the number, its digits are reversed. Find the original two-digit number. 35
5. Tia is 25 years younger than her father. The sum of their ages is 75. How old is each now? Her father is 50. Tia is 25.

2. 20 ounces of the 25% solution, 80 ounces of the 50% solution
3. wind: 25 mph; plane: 425 mph

Lesson 3.1

Use the matrix below for Exercises 1–7.

Road Mileages Between U.S. Cities

	Boston, MA	Dallas, TX	Chicago, IL	Denver, CO
Boston, MA	0	1815	983	1991
Dallas, TX	1815	0	931	801
Chicago, IL	983	931	0	1050
Denver, CO	1991	801	1050	0

$= R$

1. What do the entries in matrix R represent? Distance between given cities
2. What are the row and column dimensions of R? 4×4
3. How many miles is it from Chicago to Denver? What is the matrix address of this entry? 1050; r_{34}, r_{43}
4. How many miles is it from Boston to Dallas? What is the matrix address of this entry? 1815; r_{12}, r_{21}
5. How many miles is it from Dallas to Denver? What is the matrix address of this entry? 801; r_{24}, r_{42}
6. What does the entry r_{23} represent? Distance from Chicago to Dallas
7. What does the entry r_{41} represent? Distance from Denver to Boston

Lesson 3.2

Perform each of the matrix operations. If an operation is not possible, explain why.

1. $\begin{bmatrix}-6 & 15\\ 7 & -14\end{bmatrix} - \begin{bmatrix}0 & -12\\ 8 & 6\end{bmatrix}$ $\begin{bmatrix}-6 & 27\\ -1 & -20\end{bmatrix}$

2. $\begin{bmatrix}8 & 5\\ 6 & -23\end{bmatrix} + \begin{bmatrix}14 & 0\\ 1 & 2\end{bmatrix} - \begin{bmatrix}2 & 3 & -11\\ -6 & 2 & 4\end{bmatrix}$

3. $\begin{bmatrix}5 & 7 & -9\\ -8 & 23 & -4\end{bmatrix} - \begin{bmatrix}-2 & 17 & 23\\ -40 & 21 & -6\end{bmatrix}$

4. $\begin{bmatrix}2.1 & 3.5\\ 7.8 & 1.9\\ 0.6 & 3.3\end{bmatrix} + \begin{bmatrix}3.4 & 7.9 & 1.4\\ 7.1 & 6.4 & 0.6\end{bmatrix}$

5. $\begin{bmatrix}4.2 & 2.4\\ 1.9 & 2.5\\ 3.7 & 5.4\end{bmatrix} + \begin{bmatrix}2.1 & 4.9\\ 7.2 & 0.6\\ 4.2 & 4.8\end{bmatrix}$ $\begin{bmatrix}6.3 & 7.3\\ 9.1 & 3.1\\ 7.9 & 10.2\end{bmatrix}$

6. $\begin{bmatrix}-\frac{1}{2} & \frac{2}{5}\\ \frac{1}{4} & \frac{3}{4}\end{bmatrix} + \begin{bmatrix}\frac{1}{2} & \frac{4}{5}\\ \frac{3}{4} & 2\end{bmatrix}$ $\begin{bmatrix}0 & 1\frac{1}{5}\\ 1 & 2\frac{3}{4}\end{bmatrix}$

7. $\begin{bmatrix}0.6 & 7.3 & -6.5\\ -0.8 & 18.1 & -9.7\end{bmatrix} - \begin{bmatrix}4.8 & -10.4 & 7\\ -2.6 & 5.9 & 8.4\end{bmatrix}$

8. $\begin{bmatrix}15 & 8.5 & 1\\ 0 & 7 & 9.25\\ 6 & 12 & 6\end{bmatrix} + \begin{bmatrix}-2 & 5 & 6.5\\ -1 & -2 & 0.75\\ 15 & -3 & 2\end{bmatrix}$

2. Not possible because the last two matrices do not have the same dimensions.
4. Not possible because the two matrices do not have the same dimensions.

3. $\begin{bmatrix}7 & -10 & -32\\ 32 & 2 & 2\end{bmatrix}$ 7. $\begin{bmatrix}-4.2 & 17.7 & -13.5\\ 1.8 & 12.2 & -18.1\end{bmatrix}$ 8. $\begin{bmatrix}13 & 13.5 & 7.5\\ -1 & 5 & 10\\ 21 & 9 & 8\end{bmatrix}$

Lesson 3.3

The dimensions of two matrices are given. If the two matrices can be multiplied, indicate the dimensions of the product matrix, *AB*. If they cannot, write *not possible.*

2. Not possible

	A	*B*	*AB*			*A*	*B*	*AB*
1.	3 by 2	2 by 3	?	3 × 3	**2.**	1 by 2	1 by 3	?
3.	4 by 3	4 by 5	?	Not possible	**4.**	1 by 3	3 by 2 1 × 2	?
5.	4 by 2	2 by 2	?	4 × 2	**6.**	3 by 5	5 by 2 3 × 2	?
7.	2 by 1	1 by 6	?	2 × 6	**8.**	3 by 4	2 by 3	?

8. Not possible

Use the given matrices to evaluate each operation. Write *not possible* when appropriate.

$M = \begin{bmatrix}2 & 4\\ -1 & 5\end{bmatrix}$ $N = \begin{bmatrix}-4 & 6 & 1\\ 3 & -6 & 4\end{bmatrix}$ $P = \begin{bmatrix}-3 & -2\\ 6 & 0\\ 5 & 7\end{bmatrix}$ $Q = \begin{bmatrix}2 & 4 & 3\\ 0 & 9 & -3\end{bmatrix}$

9. $N + Q$ **10.** $3P$ **11.** $-4M$ **12.** $N + P$ **13.** $M - N$

14. $0.5Q$ **15.** MN **16.** $2Q - N$ **17.** PN **18.** PQ

Find the missing element(s) in each product matrix.

19. $-2\begin{bmatrix}-2 & 4\\ 3 & -1\end{bmatrix} = \begin{bmatrix}4 & ?\\ -6 & 2\end{bmatrix}$ −8

20. $\begin{bmatrix}-2 & 7\\ 4 & 1\end{bmatrix}\begin{bmatrix}-4 & 2\\ 3 & 5\end{bmatrix} = \begin{bmatrix}29 & ?\\ -13 & ?\end{bmatrix}$ 31; 13

Lesson 3.4

Find the inverse matrix for each matrix.

1. $\begin{bmatrix} 6 & 4 \\ -7 & -1 \end{bmatrix}$ **2.** $\begin{bmatrix} -1 & 12 \\ 3 & -5 \end{bmatrix}$ **3.** $\begin{bmatrix} 2 & 0.5 \\ 4 & 0.2 \end{bmatrix}$

4. $\begin{bmatrix} -2 & 0 \\ 3 & 0 \end{bmatrix}$ **5.** $\begin{bmatrix} -3.5 & 6.8 \\ 1.4 & -4.2 \end{bmatrix}$ **6.** $\begin{bmatrix} -6 & 15 \\ -14 & 10 \end{bmatrix}$

Determine whether each matrix has an inverse.

7. $\begin{bmatrix} -1 & 17 \\ 3 & 9 \end{bmatrix}$ Yes **8.** $\begin{bmatrix} 7 \\ -2 \\ 3 \end{bmatrix}$ No **9.** $\begin{bmatrix} 0.2 & 5 \\ 6 & 0.1 \end{bmatrix}$ Yes

10. $\begin{bmatrix} \frac{1}{6} & 7 \\ 6 & \frac{1}{8} \end{bmatrix}$ Yes **11.** $\begin{bmatrix} 8 & -5 & 10 \end{bmatrix}$ No **12.** $\begin{bmatrix} 9 & -1 & 2 \\ 4 & -9 & 6 \end{bmatrix}$ No

Lesson 3.5

Represent each system of equation in matrix form, *AX* = *B*.

1. $\begin{cases} 3y = 2x - 2 \\ 4x = 3y + 12 \end{cases}$ **2.** $\begin{cases} 4x - y = 2 \\ x = 9y + 1 \end{cases}$ **3.** $\begin{cases} 7 - 2y = 5x \\ y = 2x + 9 \end{cases}$

4. $\begin{cases} 7a - b + 2c = -1 \\ 5a + b + c = 5 \\ -3a + 4b - c = -2 \end{cases}$ **5.** $\begin{cases} 3x + y - z = 1 \\ x + y - 6z = 2 \\ 4x + 2y + 9z = -2 \end{cases}$ **6.** $\begin{cases} 5m - n + p = 0 \\ 2m + 2n + 2p = 5 \\ m - n - p = 7 \end{cases}$

Solve each system of equations using matrices if possible.

7. $\begin{cases} 3x + y = 11 \\ x + y = 7 \end{cases}$ (2, 5) **8.** $\begin{cases} 3x = 9 \\ 2x + y = 4 \end{cases}$ (3, −2) **9.** $\begin{cases} 5x + 2y = 8 \\ 3x - y = 7 \end{cases}$ (2, −1)

10. $\begin{cases} 6x + 2y = 11 \\ 3x - 8y = 1 \end{cases}$ $\left(1\frac{2}{3}, \frac{1}{2}\right)$ **11.** $\begin{cases} x + 8y = -3 \\ 2x - 6y = -17 \end{cases}$ (−7, 0.5) **12.** $\begin{cases} x + 2y = 9 \\ 9y - 2x = -5 \end{cases}$ (7, 1)

13. Which of the following ordered pairs represents the solution to

$\begin{bmatrix} 4 & 8 \\ 2 & -3 \end{bmatrix}\begin{bmatrix} x \\ y \end{bmatrix} = \begin{bmatrix} 7 \\ 0 \end{bmatrix}$? a

a. $\left(\frac{3}{4}, \frac{1}{2}\right)$ **b.** (3, 4) **c.** $\left(\frac{4}{3}, 2\right)$ **d.** (28, 14)

14. Which of the following ordered pairs represents the solution to

$\begin{bmatrix} 5 & 3 \\ 7 & 5 \end{bmatrix}\begin{bmatrix} x \\ y \end{bmatrix} = \begin{bmatrix} -5 \\ -11 \end{bmatrix}$? b

a. (−2, −5) **b.** (2, −5) **c.** (2, 5) **d.** (−2, 5)

Lesson 4.1

The results of spinning a spinner with regions I, II, III, and IV 100 times are given in the table below. Use this table for Exercises 1–3.

Region	I	II	III	IV
Frequency	34	29	22	15

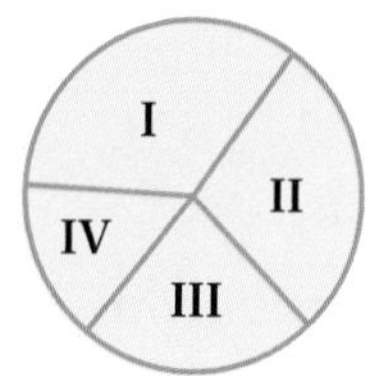

1. Is it equally likely for the spinner to land in any region? Explain.
2. Find the experimental probabilities for landing in each region.
3. Find the sum of the experimental probabilities for landing in Regions I and IV. Then find the sum of the experimental probabilities for landing in Regions II and III. Are these sums almost the same? Why or why not?

The results of 20 trials of the experiment in which a coin is tossed 3 times as shown below. The experiment simulates a 50% probability of rain on 3 consecutive days. In the experiment, H represents rain and T represents no rain.

HHH	TTH	HHT	THH	HTH	TTT	TTH	THT	TTT	HHT
THH	HTH	TTH	HHH	THT	THT	HHH	TTH	THT	TTH

Use the table to find the experimental probability of rain on

4. all three days. $\frac{3}{20}$ or 15%
5. at least one day. $\frac{9}{10}$ or 90%
6. the first day. $\frac{7}{20}$ or 35%
7. the second day. $\frac{11}{20}$ or 55%
8. the first and third days. $\frac{1}{4}$ or 25%
9. none of the days. $\frac{1}{10}$ or 10%
10. Find the sum of the experimental probabilities from Exercises 4 and 9. Is the sum 1? Explain. No, there are other probabilities besides getting all rain or no rain.

Describe the output of each expression.

11. INT(RAND*3) {0,1,2}
12. INT(RAND*5)+1 {1,2,3,4,5}
13. INT(RAND*100)+5 {5,6,7,...,104}

Write the command to generate the random numbers described.

14. integer values from 0 to 5 INT(RAND*6)
15. integer values from 10 to 15 INT(RAND*6)+10

Lesson 4.2

Describe a way to simulate the random selection.

1. a day of a 30-day month
2. 1 student out of 28 students
3. which of 6 classes gets to eat lunch first
4. 1 team out of 10 teams
5. 1 book from a collection of 15 books
6. 1 number from 50 different numbers

Lesson 4.3

Find the mean, median, mode, and range for each set of data.

1. 36, 24, 18, 35, 42 31; 35; no mode; 24
2. 13.5, 15.1, 16.4, 15.1, 16.0, 14.8
3. 165, 137, 184, 175, 186, 165, 184, 190
4. 68.5, 98.3, 37.4, 92.6, 38.5, 40.9, 70.6
5. If the range of scores is 36, the median score 90, and the highest score is 95, find the lowest score. 59

2. 15.15; 15.1; 15.1; 2.9 3. 173.25; 179.5; 165 and 184; 53 4. ≈63.8; 68.5; no mode; 60.9

Lesson 4.4

List the integers from 20 to 40, inclusive, which are

1. even AND odd.
2. even OR odd.
3. even AND multiples of 3.
4. even OR multiples of 3.
5. odd OR multiples of 3.
6. even AND multiples of 5.
7. even OR multiples of 5.
8. odd AND multiples of 5.
9. odd OR multiples of 5.
10. multiples of 3 AND multiples of 5.

In a class of 25 students, 14 have brown hair and 19 have blue eyes. This includes 8 who have both.

11. Show this in a Venn diagram. Include all 25 students.
12. How many students have either brown hair OR blue eyes? 25
13. How many students have brown hair, but not blue eyes? 6
14. How many students have blue eyes, but not brown hair? 11
15. How many students have neither brown hair or blue eyes? 0

A survey about favorite items on the lunch menu had the results shown in the table.

	Pizza	Salad	Tacos
Boys	194	102	148
Girls	91	136	129

16. How many students are girls? 356
17. How many students like pizza the best? 285
18. How many students like salad the best? 238
19. How many students are girls AND like salad the best? 136
20. How many students are girls OR like salad the best? 458
21. How many students are girls AND like pizza the best? 91
22. How many students are girls OR like pizza the best? 550
23. How many students are boys? 444
24. How many students like tacos the best? 277
25. How many students are boys AND like tacos the best? 148
26. How many students are boys OR like tacos the best? 573
27. How many students are boys AND like salad the best? 102
28. How many students are boys OR like salad the best? 580

Lesson 4.5

A menu contains 3 appetizers, 4 salads, 6 main courses, and 5 desserts.

1. How many ways are there to choose one of each? 360
2. How many combinations include a piece of apple pie, one of the 5 desserts? 72
3. How many combinations include a shrimp cocktail, one of the 3 appetizers? 120

A regular deck of 52 playing cards contains 26 red cards and 26 black cards. The deck contains 4 queens, 2 black and 2 red.

One card is drawn. Find the number of ways to draw each.

4. a black queen 2
5. a queen 4
7. a red card OR a queen 28
6. a red card 26

Two cards are drawn. The first card is replaced before drawing the second card. Find the number of ways to draw each.

8. a red card, then a queen 30
9. a queen, then a black card 30
10. a queen, then a red card 30

How many ways are there to arrange the letters in each word?

11. dog 6
12. mail 24
13. heart 120
14. computer 40,320

Lesson 4.6

An integer from 50 to 100 is drawn at random. Find the probability it is

1. even AND a multiple of 10. $\frac{6}{51}$
2. even OR a multiple of 10. $\frac{26}{51}$
3. odd AND a multiple of 10. 0
4. odd OR a multiple of 10. $\frac{31}{51}$
5. even AND a multiple of 9. $\frac{3}{51}$ or $\frac{1}{17}$
6. even OR a multiple of 9. $\frac{29}{51}$
7. odd AND a multiple of 9. $\frac{3}{51}$ or $\frac{1}{17}$
8. odd OR a multiple of 9. $\frac{28}{51}$
9. even AND odd. 0
10. a multiple of 10 OR a multiple of 9. $\frac{11}{51}$

A letter of the alphabet is selected at random. Find the probability that it is the following:

11. *q* $\frac{1}{26}$
12. *a* or *z* $\frac{2}{26}$ or $\frac{1}{13}$
13. *r*, *s* or *t* $\frac{3}{26}$
14. one of the letters of the word *math.* $\frac{4}{26}$ or $\frac{2}{13}$
15. a letter from *a* to *m*, inclusive. $\frac{13}{26}$ or $\frac{1}{2}$
16. one of the letters of the word *games.* $\frac{5}{26}$

Lesson 4.7

Nine cards numbered 1 to 9 are in a box. A card is drawn and replaced. Then a second card is drawn. Find the probability that

1. both are prime. $\frac{16}{81}$
2. the first one is even and the second is odd. $\frac{20}{81}$
3. one is even and the other is odd. $\frac{40}{81}$
4. one is prime and the other is even. $\frac{32}{81}$

A box contains 5 quarters, 8 dimes, 2 nickels, and 6 pennies. One coin is drawn and replaced. Then a second coin is drawn. Find the probability that

5. both are quarters. $\frac{25}{441}$
6. both are dimes. $\frac{64}{441}$
7. both are nickels. $\frac{4}{441}$
8. both are pennies. $\frac{36}{441}$
9. one is a quarter and one is a penny.
10. at least one is a penny or a dime.
11. the first one is a nickel and the other is a quarter.

9. $\frac{60}{441}$ 10. $\frac{392}{441}$ 11. $\frac{10}{441}$

Lesson 5.1

Which of the following relations are functions? Explain.

1. $\{(2, 3), (3, 5), (2, 8)\}$
2. $\{(6, -2), (5, -2), (8, -2)\}$
3. $\{(0, 3), (2, 5), (-2, -5)\}$

4.

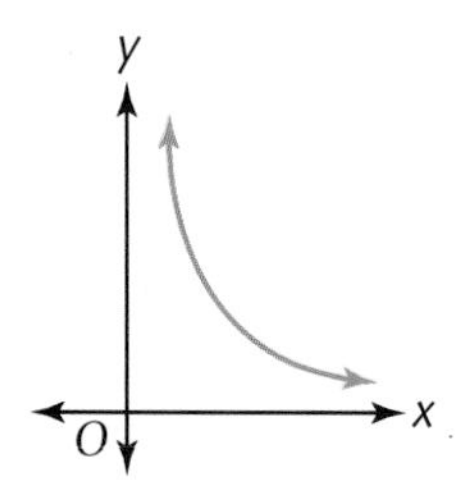

5.

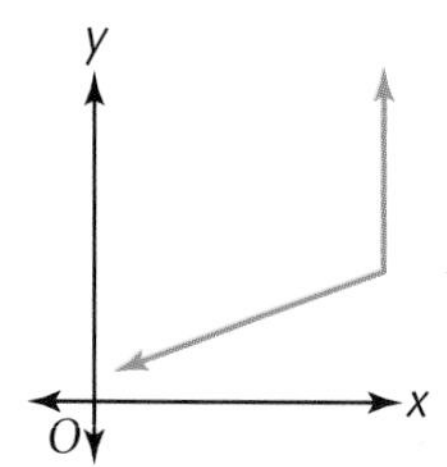

6.

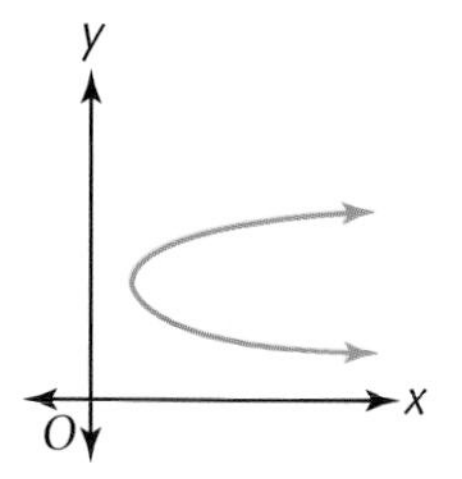

If $f(x) = 4x$, find the following.

7. $f(3)$ 12
8. $f(0)$ 0
9. $f(6)$ 24
10. $f(-1)$ −4
11. $f(-5)$ −20
12. $f(0.5)$ 2

Evaluate each function for $x = 2$.

13. $f(x) = x^2$ 4
14. $g(x) = 3x$ 6
15. $k(x) = |x|$ 2
16. $h(x) = \frac{1}{x}$ $\frac{1}{2}$
17. $f(x) = x^2 - 5x$ −6
18. $h(x) = -2|x|$ −4
19. $h(x) = \frac{-4}{x}$ −2
20. $k(x) = 3x^2 + 9$ 21
21. $g(x) = -x^2 + x - 7$ −9
22. $k(x) = -5x^2 - 3x + 1$ −25
23. $h(x) = |x - 6|$ 4

Which of these ordered pairs satisfy $|x| + |y| = 15$?

24. $(-8, -7)$ Yes
25. $(6, -9)$ Yes
26. $(10, -5)$ Yes
27. $(14, 1)$ Yes
28. $(16, -1)$ No
29. $(-2, 17)$ No
30. $(20, -5)$ No
31. $(-13, 2)$ Yes

Lesson 5.2

Identify the parent function for each graph. Then describe the transformation using the terms *stretch, reflect,* and *shift.*

1.

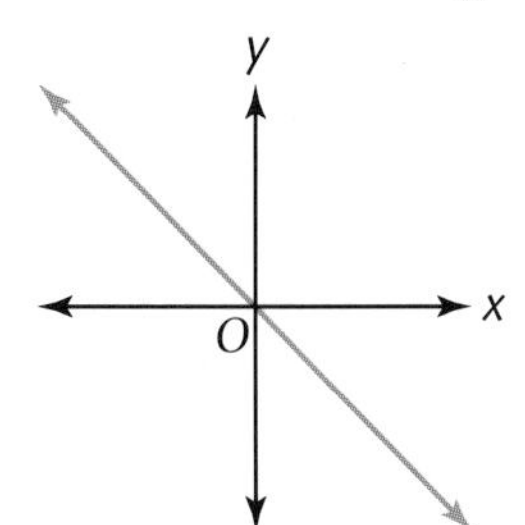

2.

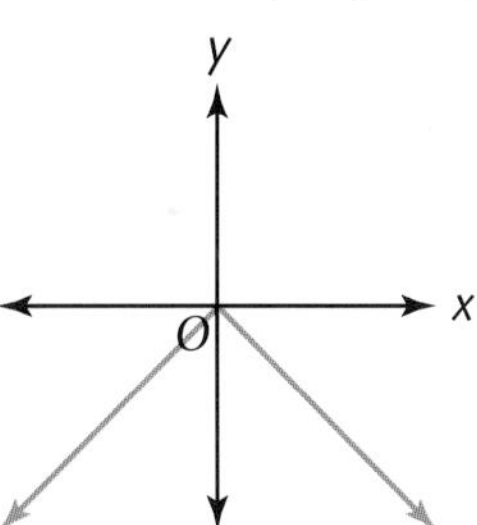

3.

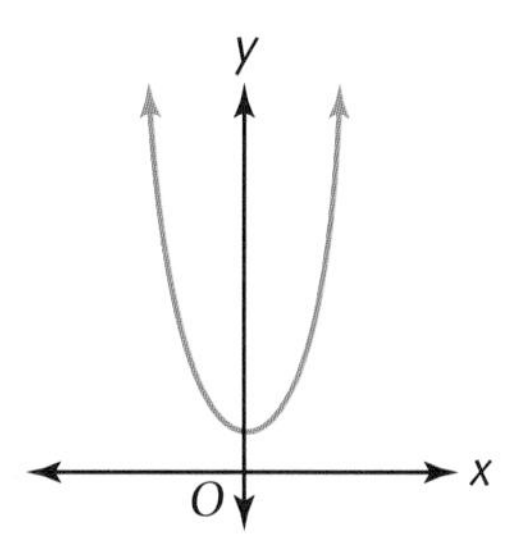

Graph each function. For each function, describe the transformation on the graph of the parent function, $y = |x|$.

4. $y = 2|x|$ **5.** $y = 2 + |x|$ **6.** $y = -2|x|$ **7.** $y = |x - 2|$

Graph each function. For each function, describe the transformation on the graph of the parent function, $y = x^2$.

8. $y = x^2 - 3$ **9.** $y = 3x^2$ **10.** $y = -3x^2$ **11.** $y = (x + 3)^2$

Lesson 5.3

Given the function $y = 3x^2$,

1. identify the parent function.

2. graph the function and parent function on the same coordinate plane.

3. tell how the graph of the parent function was changed by the 3.

Given the function $y = \frac{4}{x}$,

4. identify the parent function.

5. graph the function and parent function on the same coordinate plane.

6. tell how the graph of the parent function was changed by the 4.

Given the function $y = 5|x|$,

7. identify the parent function.

8. graph the function and parent function on the same coordinate plane.

9. tell how the graph of the parent function was changed by the 5.

Determine whether each function has a graph that is a stretch of the parent function.

10. $y = x^2 + 4$ **11.** $y = 3|x|$ **12.** $y = \frac{x^2}{4}$ **13.** $y = \frac{1}{x} + 5$

Lesson 5.4

Graph each function.

1. $y = -2x^2$ **2.** $y = -|x|$ **3.** $y = 4x^2$ **4.** $y = \frac{4}{x}$

Determine whether each function has a graph that is a vertical reflection of the parent function.

5. $y = -x^2$
Vertical reflection

6. $y = |x| - 1$
Not a vertical reflection

7. $y = \frac{-1}{x}$
Vertical reflection

8. $y = -x$
Vertical reflection

Determine whether each function has a graph that is a horizontal reflection of the parent function.

9. $y = (-x)^2$
Horizontal reflection

10. $y = |-x|$
Horizontal reflection

11. $y = x - 1$
Not a horizontal reflection

12. $y = -|x|$
Not a horizontal reflection

Find $f(4)$, and use this information to tell which kind of reflection (horizontal or vertical) is applied to $y = x^2 + 3$.

13. $y = -(x^2 + 3)$
$f(4) = -19$; vertical

14. $y = (-x)^2 + 3$
$f(4) = 19$; horizontal

Lesson 5.5

For each function, identify the parent function. Tell what type of transformation is applied to each, and then graph it.

1. $y = -2x^2$ **2.** $y = x^2 - 1$ **3.** $y = (x - 1)^2$ **4.** $y = 2(-x)^2$

5. $y = -3|x|$ **6.** $y = |x| + 2$ **7.** $y = |x + 2| + 2$ **8.** $y = |-3x|$

Consider the parent function $f(x) = x^2$. Note that $f(2) = 4$, so the point (2, 4) is on the graph. In each of the following, use the given fact to describe how the parent function is translated.

9. $f(x) = x^2 + 2$ contains the point (2, 6).

10. $f(x) = x^2 - 2$ contains the point (2, 2).

11. $f(x) = (x + 2)^2$ contains the point (2, 16).

12. $f(x) = (x - 2)^2$ contains the point (2, 0).

The point (6, 9) is on the graph of $f(x)$. Tell what happens to (6, 9) when each of the following transformations are applied to the function.

13. vertical translation by 5

14. vertical translation by -3

15. vertical translation by -9

16. vertical stretch by 4

17. horizontal translation by 2

18. horizontal translation by -1

19. horizontal translation by -10

20. vertical stretch by 5

Lesson 5.6

Sketch the graph of each function.

1. $k(x) = 3(x + 1)^2$ **2.** $n(x) = 3x^2 + 1$ **3.** $h(x) = |x + 2| - 1$

4. $q(x) = 5|x - 2|$ **5.** $f(x) = 2^x - 2$ **6.** $m(x) = \frac{2}{x} - 1$

Tell what happens to the point (4, 6) on $y = \frac{3}{2}x$ when each transformation is applied. HINT: If you are not sure, draw a sketch.

7. a vertical stretch by 2, followed by a vertical translation by 2

8. a vertical reflection, followed by a vertical stretch by 3

9. a vertical reflection, followed by a vertical translation by −2

10. a vertical stretch by −2, followed by a horizontal translation by 4

Tell what happens to the point (4, 8) on $y = 0.5x^2$ when each transformation is applied. HINT: If you are not sure, draw a sketch.

11. a vertical stretch by 2, followed by a vertical translation by 2

12. a vertical reflection, followed by a vertical stretch by 3

13. a vertical reflection, followed by a vertical translation by −2

14. a vertical stretch by −2, followed by a horizontal translation by 4

24. 0.11
32. a^{18}

37. 10^8
38. $10^6 + 10^2$
39. 10^4
40. $10^6 - 10^2$
41. 10^{12}
42. 10^8

Lesson 6.1

Change each of the following to customary notation.

1. 2^3 8 **2.** 2^2 4 **3.** 2^5 32 **4.** 2^4 16

5. 10^3 1000 **6.** 10^5 100,000 **7.** 10^2 100 **8.** 6^2 36

9. 5^2 25 **10.** 3^3 27 **11.** 4^2 16 **12.** 9^2 81

Evaluate each of the following. 16. 1,000,000,000,000 20. 100,000

13. $(3^2)^3$ 729 **14.** $(2^4)^2$ 256 **15.** $(5^2)^2$ 625 **16.** $(10^3)^4$

17. $\frac{5^4}{5^3}$ 5 **18.** $\frac{3^5}{3^2}$ 27 **19.** $\frac{4^7}{4^5}$ 16 **20.** $\frac{10^8}{10^3}$

21. $\frac{2^2 \cdot 2^4}{2^3}$ 8 **22.** $\frac{5^4 \cdot 5^3}{5^6}$ 5 **23.** $\frac{10^3 \cdot 10^6}{10^7}$ 100 **24.** $\frac{10^4 + 10^5}{10^6}$

25. $(y^3)^4$ y^{12} **26.** $x^3 \cdot x^5$ x^8 **27.** $(w^2)^5$ w^{10} **28.** $q^6 \cdot q^4$ q^{10}

29. $z^3 \cdot (z^2)^4$ z^{11} **30.** $(r^3)^3 \cdot r^3$ r^{12} **31.** $(m^2)^3 \cdot (m^4)^2$ m^{14} **32.** $(a^2)^4 \cdot (a^5)^2$

33. $\frac{d^4}{d^2}$ d^2 **34.** $\frac{f^4(f^3)}{f^5}$ f^2 **35.** $\frac{(w^3)^2}{w^4}$ w^2 **36.** $\frac{3^2 + 3^3}{3^4}$ $\frac{4}{9}$

Write each expression as a single power of 10. Then evaluate.

37. $10^6 \cdot 10^2$ 100,000,000 **38.** $10^6 + 10^2$ 1,000,100 **39.** $10^6 \div 10^2$ 10,000

40. $10^6 - 10^2$ 999,900 **41.** $(10^6)^2$ 1,000,000,000,000 **42.** 10^{6+2} 100,000,000

Lesson 6.2 7. $-12a^4b^3c^4$ 8. $0.36n^5m^{17}$ 9. $-0.42a^7b^{10}$

Find each product.

1. $(6x^2)(5x^4)$ $30x^6$
2. $(20d^6)(3d^2)$ $60d^8$
3. $(-3f^4)(3f^3)$ $-9f^7$
4. $(-p^3)(12p^2)$ $-12p^5$
5. $(-2c^3)(-3c^5)$ $6c^8$
6. $(-x^3y^4)(-x^4y^2)$ x^7y^6
7. $(24a^3b^3)(-0.5ac^4)$
8. $(3.6n^2m^9)(0.1n^3m^8)$
9. $(2.1a^5b^7)(-0.2a^2b^3)$
10. $(3x^3y^2)^3(-4xyz^4)$ $-108x^{10}y^7z^4$
11. $(-6p^3qr^5)^2(2pq)^2$ $144p^8q^4r^{10}$
12. $(2.4c^2d^2)(-0.5b^3cd^3)^2$ $0.6b^6c^4d^8$

Find each quotient.

13. $\frac{9a^5}{3a^3}$ $3a^2$
14. $\frac{80d^3}{8d^2}$ $10d$
15. $\frac{-3x^6}{6x^3}$ $-0.5x^3$
16. $\frac{-q^7}{5q^4}$ $-0.2q^3$
17. $\frac{ab^5}{-4b^4}$ $-\frac{ab}{4}$
18. $\frac{-2c^2d^3}{-4cd}$ $0.5cd^2$
19. $\frac{r^5t^6}{6r^4t^2}$ $\frac{rt^4}{6}$
20. $\frac{-6pq^5}{-2pq^2}$ $3q^3$
21. $\frac{-x^3y^5}{-x^2y^2}$ xy^3
22. $\frac{32a^7b^3}{-1.6a^2b}$ $-20a^5b^2$
23. $\frac{0.9g^9}{0.003g^4}$ $300g^5$
24. $\frac{8x^4y^5}{2x^3y^2}$ $4xy^3$
25. $\frac{24mn^4}{6n^2}$ $4mn^2$
26. $\frac{15c^4d^5}{0.3c^2d^2}$ $50c^2d^3$
27. $\frac{8u^2v^6w^3}{0.2v^4w^2}$ $40u^2v^2w$
28. $\frac{16r^4s^3t}{0.04rs^2}$ $400r^3st$

Simplify each expression. 33. $-27a^6b^6$ 34. $16x^2y^8z^4$ 44. $4p^7q^{18}$

29. $(4x)^2$ $16x^2$
30. $(4y^3)^2$ $16y^6$
31. $(2b^3)^4$ $16b^{12}$
32. $(-3t^2)^4$ $81t^8$
33. $(-3a^2b^2)^3$
34. $(-4xy^4z^2)^2$
35. $4(2b^4)^4$ $64b^{16}$
36. $10(-3p^5)^2$ $90p^{10}$
37. $(5n^3p)^2$ $25n^6p^2$
38. $(8j^4)^2$ $64j^8$
39. $(-3t^3)^4$ $81t^{12}$
40. $-5(2pq^3)^3$ $-40p^3q^9$
41. $(ab^2)(a^4)^2$ a^9b^2
42. $(x^2y^3)^3(y^4)^2$ x^6y^{17}
43. $(m^4n^2)^5(m^3)^2$ $m^{26}n^{10}$
44. $(2p^2q^3)^2(pq^4)^3$
45. $\left(\frac{m}{n}\right)^4$ $\frac{m^4}{n^4}$
46. $\left(\frac{2w}{x^2}\right)^3$ $\frac{8w^3}{x^6}$
47. $\left(\frac{-6b^2}{c^3}\right)^2$ $\frac{36b^4}{c^6}$
48. $\left(\frac{3v^5}{2w}\right)^2$ $\frac{9v^{10}}{4w^2}$

Lesson 6.3

Evaluate each expression.

1. 4^{-2} $\frac{1}{16}$
2. $(-4)^2$ 16
3. -4^2 -16
4. -4^{-2} $-\frac{1}{16}$
5. $(-4)^{-2}$ $\frac{1}{16}$
6. 2^{-4} $\frac{1}{16}$
7. 10^{-3} 0.001
8. $(-10)^3$ -1000
9. -6^2 -36
10. $(-6)^2$ 36
11. 5^{-2} $\frac{1}{25}$
12. -5^2 -25

Write each expression without a negative or zero exponent.

13. r^2s^{-3} $\frac{r^2}{s^3}$
14. $x^{-2}y^4$ $\frac{y^4}{x^2}$
15. c^0d^4 d^4
16. $r^{-3}t^5$ $\frac{t^5}{r^3}$
17. ab^0c^{-4} $\frac{a}{c^4}$
18. $g^{-7}h^{-1}k$ $\frac{k}{g^7h}$
19. $-y^3z^{-5}$ $-\frac{y^3}{z^5}$
20. $(w^2z^{-3})^0$ 1
21. $-w^{-3}w^{-2}$ $-\frac{1}{w^5}$
22. n^6n^{-2} n^4
23. j^4j^{-6} $\frac{1}{j^2}$
24. $-x^3x^{-3}$ -1
25. $\frac{n^4}{n^{-2}}$ n^6
26. $\frac{t^{-5}}{t^2}$ $\frac{1}{t^7}$
27. $\frac{g^{-4}}{g^{-3}}$ $\frac{1}{g}$
28. $\frac{rs^{-6}}{r^{-3}}$ $\frac{r^4}{s^6}$
29. $\frac{4d^{-5}}{d^4}$ $\frac{4}{d^9}$
30. $\frac{-3a^{-3}}{2a^2b}$ $-\frac{3}{2a^5b}$
31. $\frac{2r^4}{r^{-3}s}$ $\frac{2r^7}{s}$
32. $\frac{10w^{-3}}{w^2v^{-2}}$ $\frac{10v^2}{w^5}$
33. $\frac{(3a^2)(5a^3)}{a^{-1}}$ $15a^6$
34. $\frac{(2n^4)(10n^2)}{5n^{-3}}$ $4n^9$
35. $\frac{(-4b^{-2})(-2b^4)}{2b}$ $4b$
36. $\frac{(6d^{-5})(-3d^{-4})}{2d^{-3}}$ $-\frac{9}{d^6}$
37. $\frac{c^{-3}c^2}{c^{-4}c^3}$ 1
38. $\frac{a^{-3}a^{-6}}{b^{-5}b^{-2}}$ $\frac{b^7}{a^9}$
39. $\frac{-t^{-7}t^2}{r^{-5}r^2}$ $-\frac{r^3}{t^5}$
40. $\frac{-w^2w^{-5}}{-2w^3}$ $\frac{1}{2w^6}$

Lesson 6.4

2. 7×10^9 4. 7.3×10^5 8. 9.4×10^3 11. 3.5×10^{-7} 12. 9.8×10^{-4}

Write each number in scientific notation.

1. 4,000,000 4×10^6 **2.** 7,000,000,000 **3.** 2,500,000 2.5×10^6 **4.** 730,000

5. 67,000 6.7×10^4 **6.** 490,000,000 4.9×10^8 **7.** 8,800,000 8.8×10^6 **8.** 9400

9. 0.0005 5×10^{-4} **10.** 0.0000004 4×10^{-7} **11.** 0.00000035 **12.** 0.00098

13. 0.000044 4.4×10^{-5} **14.** 0.0028 2.8×10^{-3} **15.** 0.00056 5.6×10^{-4} **16.** 0.00000021 2.1×10^{-7}

Write each number in customary notation. 23. 0.0000002 24. 0.000000056

17. 4×10^3 4,000 **18.** 8×10^5 800,000 **19.** 7×10^7 70,000,000 **20.** 8.2×10^2 820

21. 7×10^{-3} 0.007 **22.** 9×10^{-4} 0.0009 **23.** 2×10^{-7} **24.** 5.6×10^{-8}

25. 4.5×10^6 **26.** 7.9×10^5 790,000 **27.** 8.45×10^8 **28.** 9.78×10^7

29. 1.7×10^{-3} **30.** 6.65×10^{-9} **31.** -9.7×10^5 **32.** -8.033×10^{-7}

25. 4,500,000 27. 845,000,000 28. 97,800,000 29. 0.0017 30. 0.00000000665

Evaluate. Answer using scientific notation. 31. −970,000 32. −0.0000008033

33. $(3 \times 10^3)(4 \times 10^6)$ 1.2×10^{10} **34.** $(7 \times 10^2)(1 \times 10^8)$ 7×10^{10} **35.** $(6 \times 10^4) + (4 \times 10^4)$

36. $(8 \times 10^9) + (9 \times 10^9)$ **37.** $(7 \times 10^5) - (6 \times 10^5)$ 1×10^5 **38.** $(8 \times 10^2) - (3 \times 10^2)$

39. $(3.1 \times 10^4)(5.5 \times 10^4)$ **40.** $(5.8 \times 10^6)(6.2 \times 10^{10})$ **41.** $(3 \times 10^2)(4.9 \times 10^{12})$

42. $\frac{8 \times 10^4}{4 \times 10^2}$ 2×10^2 **43.** $\frac{6 \times 10^8}{2 \times 10^3}$ 3×10^5 **44.** $\frac{7 \times 10^{12}}{2 \times 10^5}$ 3.5×10^7

45. $\frac{(3 \times 10^5)(6 \times 10^2)}{9 \times 10^4}$ 2×10^3 **46.** $\frac{(5 \times 10^{10})(6 \times 10^2)}{5 \times 10^9}$ 6×10^3 **47.** $\frac{(4 \times 10^3)(6 \times 10^8)}{8 \times 10^5}$ 3×10^6

48. $(3 \times 10^8)(6 \times 10^6)(3 \times 10^{-4})$ **49.** $(5 \times 10^{-4})(6 \times 10^3)(7 \times 10^{-5})$

35. 10×10^4 36. 17×10^9 38. 5×10^2 39. 1.705×10^9 40. 3.596×10^{17}
41. 1.47×10^{15} 48. 5.4×10^{11} 49. 2.1×10^{-4}

Lesson 6.5

Graph each function.

1. $y = 3^x$ **2.** $y = 6^x$ **3.** $y = 8^x$ **4.** $y = 0.2^x$

Graph each function. In each case describe the effect of the 2 or the 4 on the graph of the parent function, $y = 2^x$.

5. $y = 2^x + 2$ **6.** $y = 2 \cdot 2^x$ **7.** $y = 2^{x+4}$ **8.** $y = 2^{4x}$

Let $f(x) = 3^x$. Evaluate.

9. $f(2)$ 9 **10.** $f(0)$ 1 **11.** $f(1)$ 3 **12.** $f(3)$ 27 **13.** $f(-2)$ $\frac{1}{9}$ **14.** $f(-1)$ $\frac{1}{3}$

The population of a city was about 300,000 in 1994 and was growing at a rate of about 0.4% per year.

15. What multiplier is used to find the new population each year? 1.004

16. Use this information to estimate the population for the year 2000. 307,272

Lesson 6.6

Answers may vary for Exercises 1–5. 6. About 500 years

Use the method of carbon-14 dating to estimate the age of an object which has the given percent of its carbon-14 remaining.

1. 30% About 12,000 years
2. 2% About 35,000 years
3. 50% About 5700 years
4. 15% About 16,000 years
5. 80% About 2900 years
6. 90%

An investment is growing at a rate of 7% per year and now has a value of $7600. Find the value of the investment

7. in 2 years. $8,701.24
8. in 8 years. $13,058.21
9. 3 years ago. $6,203.86
10. 5 years ago. $5,418.69

An investment is losing money at a rate of 1% per year and now has a value of $48,000. Find the value of the investment

11. in 5 years. $45,647.52
12. in 10 years. $43,410.34
13. 2 years ago. $48,974.59
14. 8 years ago. $52,018.72

The population of a city in the United States was 164,998 in 1990 and was decreasing at a rate of about 1.2% per year. In how many years will the city's population fall below

15. 150,000? About 8 years
16. 125,000? About 23 years
17. 110,000? About 34 years
18. 50,000? About 99 years

Lesson 7.1

Give surface area and volume functions for each geometric solid.

1. A cube with an edge length, x
2. A rectangular solid with a 5 in.-by-6 in. base and a height of x in.
3. A rectangular solid with a 4.2 cm-by-9.5 cm base and a height of x cm
4. A cylinder with height of 6 inches and a radius of x inches
5. A cone with height of 9 feet and a radius of x feet

Show that each equation is true by substituting the x-values -1, 0, and 1.

6. $x^2 + 4x + 3 = (x + 1)(x + 3)$
7. $x^2 - 7x + 10 = (x - 5)(x - 2)$
8. $x^2 - 7x + 12 = (x - 4)(x - 3)$
9. $x^2 + 7x + 6 = (x + 6)(x + 1)$
10. $(x + 4)(x - 3) = x^2 + x - 12$
11. $(x - 4)(x - 5) = x^2 - 9x + 20$
12. $x^2 - 16 = (x + 4)(x - 4)$
13. $(x - 5)(x + 5) = x^2 - 25$
14. $(x + 4)(x - 6) = x^2 - 2x - 24$
15. $x^2 + 9x + 14 = (x + 7)(x + 2)$
16. $x^2 - 100 = (x - 10)(x + 10)$
17. $(x - 4)(x + 3) = x^2 - x - 12$

Lesson 7.2

Rewrite each polynomial in standard form.

1. $9x^3 - 4x - 2 + x^2$ $9x^3 + x^2 - 4x - 2$
2. $y - 4y^4 + 3 - 2y^2$ $-4y^4 - 2y^2 + y + 3$
3. $r - r^3$ $-r^3 + r$

Write the degree of each polynomial.

4. $6t^2 - t + 1$ 2
5. $3x - 2$ 1
6. $9x^3 + 8x^4 - 4$ 4
7. $p^5 - p^2 + p$ 5

Use vertical form to add.

8. $(4x^4 + 3x^2 - 3) + (6x^4 + 2x^2 + 5)$
9. $(7w^3 - 4w + 3) + (5w^3 - 3w - 3)$
10. $(8z^4 - 4z^2 - z + 4) + (z^2 - z)$
11. $(x^2 + 3x + 4) + (3x + 7)$ $x^2 + 6x + 11$

8. $10x^4 + 5x^2 + 2$ 9. $12w^3 - 7w$ 10. $8z^4 - 3z^2 - 2z + 4$

Use horizontal form to add.

12. $(x^3 + 4x - 1) + (x^2 - 5)$ $x^3 + x^2 + 4x - 6$
13. $(w^4 + 3w^2 - 5) + (w^2 - 5w)$
14. $(3z^3 - 4z + 5) + (8z^3 - 5z^2 - 1)$
15. $(7r^4 - 5r^3 + r) + (8r^2 - 6r - 8)$

13. $w^4 + 4w^2 - 5w - 5$ 14. $11z^3 - 5z^2 - 4z + 4$ 15. $7r^4 - 5r^3 + 8r^2 - 5r - 8$

Use vertical form to subtract.

16. $2x^3 + x^2 - 1 - (x^3 - x^2)$ $x^3 + 2x^2 - 1$
17. $3w^3 - 5w + 4 - (6w + 2)$ $3w^3 - 11w + 2$
18. $9x^4 - 5x^2 + 7 - (8x^4 + 2)$ $x^4 - 5x^2 + 5$
19. $3z^5 + z^3 - 8 - (z^5 - z^3 + 8)$ $2z^5 + 2z^3 - 16$

Use horizontal form to subtract.

20. $y^3 - y^2 - 1 - (4y + 3)$ $y^3 - y^2 - 4y - 4$
21. $5x^2 - 3x - 9 - (3x^2 - 5x + 5)$
22. $6y^3 - 4y^2 + 3 - (5y^3 + 3y^2 + y - 1)$
23. $8y^2 - 7 - (5y^2 + 3y - 6)$ $3y^2 - 3y - 1$

21. $2x^2 + 2x - 14$ 22. $y^3 - 7y^2 - y + 4$

Simplify. Express all answers in standard form.

24. $(2 - x - x^4) + (x + 4.5x^4 + 5)$ $3.5x^4 + 7$
25. $6 + 4x - 3.6x^2 - (10.6x - 65 + 6.8x^2)$ $-10.4x^2 - 6.6x + 71$
26. $9.9x + 4.8x^2 - 9.5 + (6.6x^2 + 7 - 4.75x)$ $11.4x^2 + 5.15x - 2.5$

Lesson 7.3

Use the Distributive Property to find each product.

1. $5(x + 4)$ $5x + 20$
2. $6(2x + 6)$ $12x + 36$
3. $7(x + 11)$ $7x + 77$
4. $4(x - 7)$
5. $3(2x - 6)$ $6x - 18$
6. $-4(x + 2)$ $-4x - 8$
7. $-3(x - 2)$ $-3x + 6$
8. $5(3x - 4)$
9. $x(2 - x)$ $-x^2 + 2x$
10. $x(3 + 3x)$ $3x^2 + 3x$
11. $-x(4 + x)$ $-x^2 - 4x$
12. $-x(x - 4)$
13. $5x(x + 3)$ $5x^2 + 15x$
14. $2x(4 - x)$ $-2x^2 + 8x$
15. $x(-x + 6)$ $-x^2 + 6x$
16. $-2x(x - 4)$

4. $4x - 28$ 8. $15x - 20$ 12. $-x^2 + 4x$ 16. $-2x^2 + 8x$

Identify the zeroes for each function.

17. $y = (x + 4)(x + 2)$ $-4, -2$
18. $y = (x - 2)(x + 5)$ $-5, 2$
19. $y = (x + 10)(x + 12)$ $-12, -10$
20. $y = (x - 3)(x + 3)$ $-3, 3$
21. $y = (x + 14)(x - 2)$ $-14, 2$
22. $y = (x + 8)(x + 8)$ -8
23. $y = (3x + 3)(3x - 3)$ $-1, 1$
24. $y = (2x + 4)(x - 4)$ $-2, 4$

Lesson 7.4

Use the Distributive Property to find each product.

1. $(x + 2)(x + 3)$ $x^2 + 5x + 6$ **2.** $(z - 5)(z - 1)$ $z^2 - 6z + 5$ **3.** $(a + 2)(a - 4)$

4. $(b - 2)(b - 3)$ $b^2 - 5b + 6$ **5.** $(d + 3)(d + 4)$ $d^2 + 7d + 12$ **6.** $(y - 2)(y + 5)$

7. $(c + 6)(c + 6)$ $c^2 + 12c + 36$ **8.** $(w + 4)(w - 3)$ $w^2 + w - 12$ **9.** $(m - 2)(m - 4)$

3. $a^2 - 2a - 8$ 6. $y^2 + 3y - 10$ 9. $m^2 - 6m + 8$

Use the FOIL method to find each product.

10. $(x + 3)(x + 6)$ $x^2 + 9x + 18$ **11.** $(w + 10)(w + 2)$ $w^2 + 12w + 20$ **12.** $(c - 5)(c - 5)$

13. $(y - 3)(y - 9)$ $y^2 - 12y + 27$ **14.** $(r - 4)(r + 7)$ $r^2 + 3r - 28$ **15.** $(w + 8)(w - 3)$

16. $(3z - 1)(2z + 1)$ **17.** $(4x + 3)(x - 2)$ $4x^2 - 5x - 6$ **18.** $(2t - 1)(2t + 3)$

19. $(5x - 2)(3x + 2)$ **20.** $(8s - 1)(s + 4)$ $8s^2 + 31s - 4$ **21.** $(4w - 5)(3w - 1)$

22. $\left(y - \frac{1}{2}\right)\left(y + \frac{3}{4}\right)$ $y^2 + \frac{y}{4} - \frac{3}{8}$ **23.** $\left(x - \frac{1}{3}\right)\left(x + \frac{1}{3}\right)$ $x^2 - \frac{1}{9}$ **24.** $\left(z + \frac{2}{5}\right)\left(z + \frac{1}{5}\right)$

25. $(1.3m + 4)(0.7m - 2)$ **26.** $(5.6r - 1)(0.4r + 2)$ **27.** $(0.9g - 15)(1.1g + 5)$

12. $c^2 - 10c + 25$ 15. $w^2 + 5w - 24$ 16. $6z^2 + z - 1$ 18. $4t^2 + 4t - 3$

19. $15x^2 + 4x - 4$ 21. $12w^2 - 19w + 5$ 24. $z^2 + \frac{3}{5}z + \frac{2}{25}$ 25. $0.91m^2 + 0.2m - 8$

26. $2.24r^2 + 10.8r - 2$ 27. $0.99g^2 - 12g - 75$

Lesson 7.5

Identify each polynomial as prime or composite.

1. $5x^2 - 15$ **2.** $t^2 + 6$ **3.** $w^2 - 4$ **4.** $4m^2 + 9$

5. $7t^2 - 8t$ **6.** $25c - 25d$ **7.** $16x^2 - 1$ **8.** $w^3 - w^2$

9. $3x - 2$ **10.** $8x^3 - 9$ **11.** $7x^2 + x$ **12.** $16c - 8$

Factor each polynomial by removing the GCF.

13. $4x^2 + 16$ **14.** $7w^2 + 21$ **15.** $8a^2 - 4a$

16. $5n^3 - 10n^2 + 15$ **17.** $80 - 75d^3 - 50d^2$ **18.** $6x^2y - 14xy + 2x^2y^2$

19. $3s^2t^3 + 15st^3 - 12s^2t^2$ **20.** $4z^3w - 16z^2w^3 + 24z^4w^2$ **21.** $25ab^4 + 20a^3b^2 - 15ab^3$

Write each as the product of two binomials.

22. $x(x - 1) + 2(x - 1)$ **23.** $7(y + 4) - x(y + 4)$ **24.** $b(c + d) + a(c + d)$

25. $4(w - 3) - v(w - 3)$ **26.** $x(y - z) + 4(y - z)$ **27.** $s(x + 5) - t(x + 5)$

28. $q(4 - s) - 5(4 - s)$ **29.** $3w(x + y)^2 - 4(x + y)^2$ **30.** $mn(p - q)^3 + ab(p - q)^3$

Factor.

31. $15a - 3ay + 4y - 20$ **32.** $5x + 3xy - 21y - 35$ **33.** $2ax + 6xc + ba + 3bc$

34. $4p + 4a + ay + yp$ **35.** $am - ab + 3my - 3by$ **36.** $a^2 - 2ab + a - 2b$

37. $xr + 6w + rw + 6x$ **38.** $x^2 + 5xy + ax + 5ay$ **39.** $5ab - 14ab + 10a^2 - 7b^2$

40. $15c - 9d + 10c^2 - 6cd$ **41.** $7t^2 - 4ts + 16s^2 - 28ts$ **42.** $7a + 7c + ab + bc$

43. $4xy + 10x^2 + 15x + 6y$ **44.** $15m - 3mn - 20p + 4np$ **45.** $15 - 24xy - 5x + 8x^2y$

Lesson 7.6

Use the generalization of a perfect-square trinomial or the difference of two squares to find each product.

1. $(x + 2)^2$ $x^2 + 4x + 4$ **2.** $(2x + 1)^2$ $4x^2 + 4x + 1$ **3.** $(m - 4)^2$ $m^2 - 8m + 16$
4. $(a - 6)^2$ $a^2 - 12a + 36$ **5.** $(4y - 5)^2$ $16y^2 - 40y + 25$ **6.** $(9q + 1)^2$ $81q^2 + 18q + 1$
7. $(x + 3y)(x - 3y)$ $x^2 - 9y^2$ **8.** $(7a - b)(7a + b)$ $49a^2 - b^2$ **9.** $(5a - 3)(5a + 3)$ $25a^2 - 9$

Find the missing terms in each perfect-square trinomial.

10. $a^2 - 10a + \underline{?}$ 25 **11.** $25z^2 + \underline{?} + 4$ $20z$ **12.** $9b^2 - 12b + \underline{?}$ 4
13. $\underline{?} - 4y + 1$ $4y^2$ **14.** $9r^2 + 30r + \underline{?}$ 25 **15.** $\underline{?} + 42x + 49$ $9x^2$

Factor each polynomial completely.

16. $y^2 - 9$ $(y + 3)(y - 3)$ **17.** $9t^2 - 1$ $(3t + 1)(3t - 1)$ **18.** $z^2 - 144$ $(z + 12)(z - 12)$
19. $x^2 - 4x + 4$ $(x - 2)^2$ **20.** $y^2 + 16y + 64$ $(y + 8)^2$ **21.** $25w^2 - 16$ $(5w + 4)(5w - 4)$
22. $1 - 81q^2$ $(1 + 9q)(1 - 9q)$ **23.** $x^2 - y^2$ $(x + y)(x - y)$ **24.** $x^2 + 12x + 36$ $(x + 6)^2$
25. $q^2 - 20q + 100$ $(q - 10)^2$ **26.** $49m^2 - 14m + 1$ **27.** $25r^2 + 20r + 4$ $(5r + 2)^2$
28. $64t^2 + 16t + 1$ $(8t + 1)^2$ **29.** $1 - 10b + 25b^2$ $(1 - 5b)^2$ **30.** $w^2 + 22w + 121$ $(w + 11)^2$
26. $(7m - 1)^2$

Lesson 7.7

Write each trinomial in factored form. Are the signs of the factors the same or opposite?

1. $x^2 - 3x - 10$ **2.** $x^2 - 3x - 28$ **3.** $x^2 + 9x + 20$
4. $x^2 - 9x + 8$ **5.** $x^2 - 12x - 45$ **6.** $x^2 + 10x + 16$
21. $(y + 3x)(b + 4m)$ 26. $(a - 4c)(a + b)$ 29. $6(n^2 - 4n + 1)$

Write each trinomial as a product of its factors.

7. $b^2 + 12b + 27$ $(b + 9)(b + 3)$ **8.** $x^2 - 8x + 15$ $(x - 5)(x - 3)$ **9.** $z^2 + 2z - 3$ $(z + 3)(z - 1)$
10. $s^2 + 22s + 21$ $(s + 21)(s + 1)$ **11.** $n^2 - n - 20$ $(n - 5)(n + 4)$ **12.** $h^2 + 15h + 26$ $(h + 13)(h + 2)$
13. $y^2 - 49$ $(y + 7)(y - 7)$ **14.** $w^2 - 16w + 64$ $(w - 8)^2$ **15.** $z^2 + 14z + 24$ $(z + 12)(z + 2)$
16. $y^2 - 8y + 7$ $(y - 7)(y - 1)$ **17.** $28n^2 + 18n$ $2n(14n + 9)$ **18.** $5a^2 + 20b^2$ $5(a^2 + 4b^2)$
19. $s^2 - 12s - 13$ $(s - 13)(s + 1)$ **20.** $r^2 - r - 72$ $(r - 9)(r + 8)$ **21.** $by + 4my + 3xb + 12mx$
22. $3x^2 - 12$ $3(x + 2)(x - 2)$ **23.** $t^2 - t - 65$ Prime **24.** $m^2 + 10m - 39$ $(m + 13)(m - 3)$
25. $y^2 + 7y + 12$ $(y + 4)(y + 3)$ **26.** $a^2 - 4ac + ab - 4bc$ **27.** $4y^2 + 2y - 6$ $2(2y + 3)(y - 1)$
28. $4t^2 - 6t - 40$ $2(t - 4)(2t + 5)$ **29.** $6n^2 - 24n + 6$ **30.** $a^2 + 18a + 51$ Prime
31. $2y^2 - 128$ $2(y + 8)(y - 8)$ **32.** $16 - 9w^2$ $(4 + 3w)(4 - 3w)$ **33.** $27y^2 - 36y + 12$ $3(3y - 2)^2$

Lesson 8.1

Describe how the graph of each function is a transformation of the graph of $y = x^2$.

1. $y = (x - 3)^2 + 1$ **2.** $y = (x + 2)^2 - 4$ **3.** $y = 2(x + 1)^2 - 2$
4. $y = (x + 3)^2$ **5.** $y = -(x - 4)^2$ **6.** $y = 3(x - 6)^2$
7. $y = 4(x - 1)^2 - 1$ **8.** $y = -2(x - 5)^2 + 2$ **9.** $y = -5(x + 2)^2 - 5$
10. $y = \frac{1}{2}(x - 1)^2 + 4$ **11.** $y = -\frac{1}{3}(x + 9)^2 - 3$ **12.** $y = \frac{2}{5}(x - 4)^2 - 5$

Determine the vertex and axis of symmetry for each function.

13. $y = -2(x + 1)^2 - 2$ **14.** $y = 3(x - 2)^2 + 3$ **15.** $y = (x - 3)^2 + 4$
16. $y = \frac{1}{2}(x + 2)^2 + 1$ **17.** $y = -3(x + 3)^2$ **18.** $y = -5(x + 1)^2 + 4$

Find the zeros of each polynomial function by graphing.

19. $y = x^2 + 7x + 12$ −4, −3 **20.** $y = x^2 + 4x + 4$ −2 **21.** $y = x^2 - x - 30$ −5, 6
22. $y = x^2 + 9x + 8$ −8, −1 **23.** $y = x^2 + 2x - 15$ −5, 3 **24.** $y = x^2 + x - 20$ −5, 4
25. $y = x^2 - 2x + 1$ 1 **26.** $y = x^2 + 7x - 8$ −8, 1 **27.** $y = x^2 + 10x + 9$ −9, −1

Graph each function. Then find the zeros, the axis of symmetry, and the vertex.

28. $y = x^2 + 10x + 25$ **29.** $y = x^2 + 4x + 3$ **30.** $y = x^2 - 2x - 8$
31. $y = x^2 + 8x + 15$ **32.** $y = x^2 - 6x - 7$ **33.** $y = x^2 - 2x - 24$

Lesson 8.2

Find each square root. If necessary, round to the nearest hundredth.

1. $\sqrt{100}$ 10 **2.** $\sqrt{169}$ 13 **3.** $\sqrt{484}$ 22 **4.** $\sqrt{49}$ 7 **5.** $\sqrt{50}$ 7.07
6. $\sqrt{89}$ 9.43 **7.** $\sqrt{28}$ 5.29 **8.** $\sqrt{14}$ 3.74 **9.** $\sqrt{99}$ 9.95 **10.** $\sqrt{110}$ 10.49
11. $\sqrt{200}$ 14.14 **12.** $\sqrt{65}$ 8.06 **13.** $\sqrt{95}$ 9.75 **14.** $\sqrt{250}$ 15.81 **15.** $\sqrt{135}$ 11.62

22. −8, 0 23. −3, −1 24. −3, 9 26. $1 \pm 2\sqrt{3}$, or −2.46 and 4.46
27. $-5 \pm \sqrt{35}$, or −10.92 and 0.92

Solve each equation to the nearest hundredth.

16. $x^2 = 36$ ±6 **17.** $x^2 = 121$ ±11 **18.** $x^2 = 64$ ±8 **19.** $x^2 = 400$ ±20
20. $x^2 = \frac{16}{25}$ $\pm\frac{4}{5}$ **21.** $x^2 = \frac{64}{121}$ $\pm\frac{8}{11}$ **22.** $(x + 4)^2 - 16 = 0$ **23.** $(x + 2)^2 - 1 = 0$
24. $(x - 3)^2 - 36 = 0$ **25.** $(x + 3)^2 = 64$ −11, 5 **26.** $(x - 1)^2 = 12$ **27.** $(x + 5)^2 = 35$

Find the vertex, axis of symmetry, and zeros of each function.

28. $f(x) = (x - 9)^2 - 16$ **29.** $g(x) = (x + 3)^2 - 4$ **30.** $h(x) = (x - 4)^2 - 2$

Lesson 8.3

Complete the square.

1. $x^2 + 12x$ $+ 36$
2. $x^2 + 10x$ $+ 25$
3. $x^2 + 4x$ $+ 4$
4. $x^2 + 20x$ $+ 100$
5. $x^2 - 30x$ $+ 225$
6. $x^2 + 11x$ $+ 30.25$
7. $x^2 + 30x$ $+ 225$
8. $x^2 - 21x$ $+ 110.25$

Find the minimum value for each quadratic function.

9. $f(x) = x^2 + 5$ $(0, 5)$
10. $f(x) = x^2 - 9$ $(0, -9)$
11. $f(x) = x^2 + 1$ $(0, 1)$
12. $f(x) = x^2 - 2$ $(0, -2)$

Rewrite each function in the form $y = (x - h)^2 + k$. Find the vertex.

13. $y = x^2 - 6x$
14. $y = x^2 - 12x$
15. $y = x^2 + 2x$
16. $y = x^2 + 6x + 5$
17. $y = x^2 - 2x - 3$
18. $y = x^2 - 8x + 12$
19. $y = x^2 - 10x + 21$
20. $y = x^2 - 12x + 32$
21. $y = x^2 + 5x + 4$
22. $y = x^2 + 6x - 7$
23. $y = x^2 + 7x + 12$
24. $y = x^2 - 4x - 24$
25. $y = x^2 + \frac{1}{2}x - 2$
26. $y = x^2 - \frac{1}{4}x + 6$
27. $y = x^2 - \frac{1}{3}x + 5$

26. $\left(x - \frac{1}{8}\right)^2 + 5\frac{63}{64}; \left(\frac{1}{8}, 5\frac{63}{64}\right)$ 27. $\left(x - \frac{1}{6}\right)^2 + 4\frac{35}{36}; \left(\frac{1}{6}, 4\frac{35}{36}\right)$

Lesson 8.4

Find the zeros of each function.

1. $y = x^2 - 2x - 3$ $-1, 3$
2. $y = x^2 + 5x + 4$ $-4, -1$
3. $y = x^2 - 8x + 12$ $2, 6$
4. $y = x^2 - 2x - 35$ $-5, 7$
5. $y = x^2 - 6x + 8$ $2, 4$
6. $y = x^2 - 6x + 5$ $1, 5$

Solve each equation by factoring or by completing the square.

7. $x^2 - x - 2 = 0$ $-1, 2$
8. $x^2 + 5x + 4 = 0$ $-4, -1$
9. $x^2 - x - 6 = 0$ $-2, 3$
10. $x^2 + 3x - 10 = 0$ $-5, 2$
11. $x^2 + 3x - 28 = 0$ $-7, 4$
12. $x^2 + 7x - 30 = 0$ $-10, 3$
13. $x^2 + 6x - 7 = 0$ $-7, 1$
14. $x^2 + 2x - 8 = 0$ $-4, 2$
15. $x^2 + 4x - 45 = 0$ $-9, 5$
16. $x^2 + 3x - 10 = 0$ $-5, 2$
17. $x^2 - x - 12 = 0$ $-3, 4$
18. $x^2 - 2x - 3 = 0$ $-1, 3$
19. $x^2 - 12x + 20 = 0$ $2, 10$
20. $x^2 - 3x - 4 = 0$ $-1, 4$
21. $x^2 + 7x + 12 = 0$ $-4, -3$
22. $x^2 + 2x - 15 = 0$ $-5, 3$
23. $x^2 - 4x = 0$ $0, 4$
24. $x^2 + 5x - 24 = 0$ $-8, 3$
25. $x^2 - x - 2 = 0$ $-1, 2$
26. $x^2 - 9x + 20 = 0$ $4, 5$
27. $x^2 + 5x - 6 = 0$ $-6, 1$

Find the point or points where the graphs intersect. Graph to check.

28. $\begin{cases} y = 4 \\ y = x^2 \end{cases}$ $(2, 4)$, $(-2, 4)$
29. $\begin{cases} y = x - 2 \\ y = x^2 - 4x + 4 \end{cases}$ $(3, 1)$, $(2, 0)$
30. $\begin{cases} y = x + 4 \\ y = x^2 + 3x - 4 \end{cases}$ $(-4, 0)$, $(2, 6)$
31. $\begin{cases} y = x - 5 \\ y = x^2 - 3x - 10 \end{cases}$ $(5, 0)$, $(-1, 6)$
32. $\begin{cases} y = 5 \\ y = x^2 - 2x - 3 \end{cases}$ $(-2, 5)$, $(4, 5)$
33. $\begin{cases} y = x + 2 \\ y = x^2 - 2x - 8 \end{cases}$ $(-2, 0)$, $(5, 7)$

Lesson 8.5

Identify *a*, *b*, and *c* for each quadratic equation.

1. $x^2 - 4x + 5 = 0$ 1, −4, 5
2. $x^2 - 6x + 10 = 0$ 1, −6, 10
3. $8x^2 - 5x + 2 = 0$ 8, −5, 2
4. $3x^2 - 6 + 3x = 0$ 3, 3, −6
5. $-x^2 + 7x - 6 = 0$ −1, 7, −6
6. $-5x + x^2 - 9 = 0$ 1, −5, −9

Find the value of the discriminant for each quadratic equation.

7. $x^2 + 3x - 4 = 0$ 25
8. $x^2 + 4x + 1 = 0$ 12
9. $x^2 + 6x - 2 = 0$ 44
10. $x^2 - 6x + 1 = 0$ 32
11. $x^2 - 5x + 9 = 0$ −11
12. $x^2 + 6x - 10 = 0$ 76
13. $4x^2 + 8x + 3 = 0$ 16
14. $5x^2 - 125 = 0$ 2500
15. $x^2 + x - 12 = 0$ 49
16. $3x^2 + 14x - 5 = 0$ 256
17. $-4x^2 + 8x + 3 = 0$ 112
18. $x^2 - 7x - 8 = 0$ 81

Use the quadratic formula to solve each equation. Give answers to the nearest hundredth when necessary. Check by substitution.

19. $x^2 + 7x + 6 = 0$ −6, −1
20. $y^2 + 8y + 15 = 0$ −5, −3
21. $x^2 + 4x + 3 = 0$ −3, −1
22. $w^2 - 6w = 0$ 0, 6
23. $x^2 - 16 = 0$ ±4
24. $y^2 + 10y = 0$ −10, 0
25. $-4x^2 + 16x + 13 = 0$
26. $8x^2 + 10x + 3 = 0$
27. $2x^2 + 7x - 15 = 0$ $-5, \frac{3}{2}$
28. $4x^2 - 8x + 1 = 0$
29. $-x^2 + 3x - 1 = 0$
30. $3x^2 + 6x + 2 = 0$

25. $\frac{4 \pm \sqrt{29}}{2}$, or −0.69 and 4.69 26. $-\frac{3}{4}$ and $-\frac{1}{2}$ 28. $\frac{2 \pm \sqrt{3}}{2}$, or 0.13 and 1.87

Solve each equation to the nearest hundredth.

31. $x^2 - 9x + 20 = 0$ 4, 5
32. $x^2 + 10x - 2 = 0$
33. $x^2 + 13x = -42$ −7, −6
34. $3x^2 - 5x - 2 = 0$ −0.33, 2
35. $2x^2 + x = 5$ −1.85, 1.35
36. $4x^2 - 7x - 2 = 0$ −0.25, 2
37. $7x^2 - 3x - 1 = 0$ 0.65, −0.22
38. $3x^2 - 5x + 2 = 0$ 0.67, 1
39. $11x^2 - x - 3 = 0$
40. $x^2 - 12x + 27 = 0$ 3, 9
41. $9x^2 - 20x + 11 = 0$ 1, 1.22
42. $3x^2 - 7x - 2 = 0$ −0.26, 2.59

32. 0.20, −10.20 39. −0.48, 0.57

29. $\frac{3 \pm \sqrt{5}}{2}$, or 0.38 and 2.62 30. $\frac{-3 \pm \sqrt{3}}{3}$, or −1.58 and −0.42

Lesson 8.6

Solve each quadratic inequality using the Zero Product Property. Graph the solution on a number line.

1. $x^2 - x - 12 > 0$ $x > 4$ or $x < -3$
2. $x^2 + 3x - 10 < 0$ $-5 < x < 2$
3. $x^2 - 4 \geq 0$ $x \geq 2$ or $x \leq -2$
4. $x^2 + 7x + 12 \leq 0$ $-4 \leq x \leq -3$
5. $x^2 - 10x + 21 < 0$ $3 < x < 7$
6. $x^2 + 4x - 12 \leq 0$
7. $x^2 - 25 \leq 0$ $-5 \leq x \leq 5$
8. $x^2 + 7x + 12 > 0$
9. $x^2 + 8x + 15 < 0$
10. $x^2 - 6x + 9 \leq 0$ 3
11. $x^2 + 14x + 49 \geq 0$
12. $x^2 - 10x + 25 > 0$

6. $-6 \leq x \leq 2$ 8. $x > -3$ or $x < -4$ 9. $-5 < x < -3$ 11. All real numbers

Graph each quadratic inequality. Shade the solution region. 12. All real numbers

13. $y < x^2$
14. $y < x^2 + 2$
15. $y > -x^2$
16. $y \geq x^2$
17. $y < x^2 - x$
18. $y < 3 - x^2$
19. $y \geq x^2 - 4x$
20. $y < x^2 - x - 12$
21. $y > x^2 - 4x - 5$

Lesson 9.1

Find each square root to the nearest hundredth.

1. $\sqrt{625}$ 25 **2.** $-\sqrt{144}$ −12 **3.** $\sqrt{12}$ 3.46 **4.** $-\sqrt{50}$ −7.07 **5.** $-\sqrt{18}$ −4.24 **6.** $\sqrt{100}$ 10

7. $-\sqrt{0.09}$ −0.3 **8.** $\sqrt{0.078}$ 0.28 **9.** $\sqrt{\frac{9}{16}}$ 0.75 **10.** $-\sqrt{\frac{25}{49}}$ −0.71 **11.** $-\sqrt{0.0085}$ −0.09 **12.** $\sqrt{0.0072}$ 0.08

Describe how the graph of each function is a transformation of the graph of $y = \sqrt{x}$.

13. $y = \sqrt{x} - 2$ **14.** $y = 3\sqrt{x}$ **15.** $y = \sqrt{x + 1}$

16. $y = -\sqrt{x - 9}$ **17.** $y = 3\sqrt{x + 6}$ **18.** $y = \sqrt{x - 5} + 6$

19. $y = 2 - \sqrt{x - 1}$ **20.** $y = 4.50\sqrt{x + 1.25}$ **21.** $y = 4.75 + 3.00\sqrt{x - 1.50}$

Write the equation of the function graphed.

22.

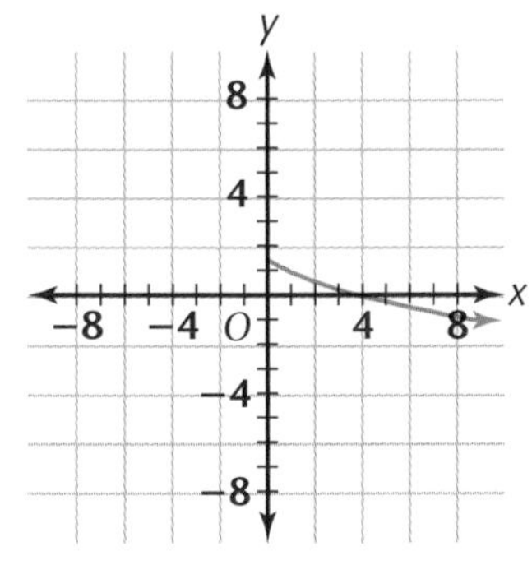

$y = -\sqrt{x} + 2$

23.

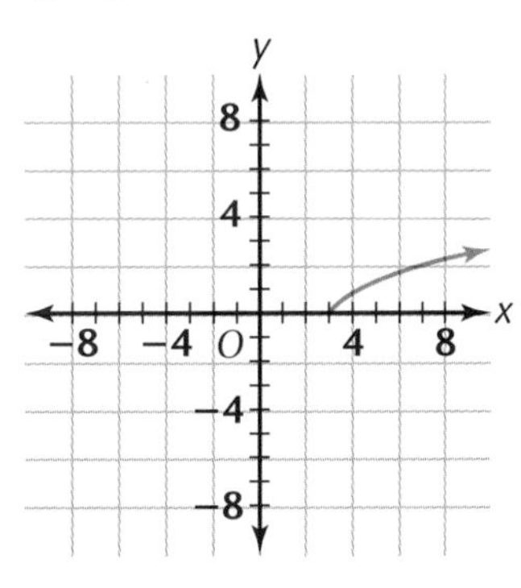

$y = \sqrt{x - 3}$

24.

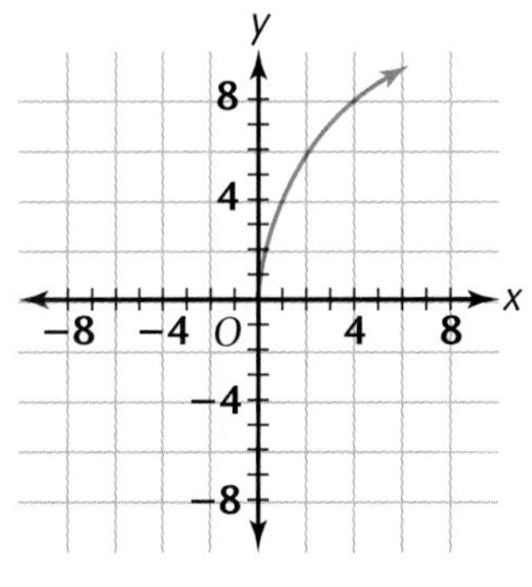

$y = 4\sqrt{x}$

Lesson 9.2

Simplify each radical by factoring.

1. $\sqrt{36}$ 6 **2.** $\sqrt{225}$ 15 **3.** $\sqrt{324}$ 18 **4.** $\sqrt{2500}$ 50

5. $\sqrt{108}$ $6\sqrt{3}$ **6.** $\sqrt{80}$ $4\sqrt{5}$ **7.** $\sqrt{192}$ $8\sqrt{3}$ **8.** $\sqrt{1440}$ $12\sqrt{10}$

9. $\sqrt{605}$ $11\sqrt{5}$ **10.** $\sqrt{784}$ 28 **11.** $\sqrt{1944}$ $18\sqrt{6}$ **12.** $\sqrt{768}$ $16\sqrt{3}$

26. $6\sqrt{3} - 24$ **27.** $3\sqrt{6} - 6\sqrt{3}$ **28.** $7\sqrt{3} + 3\sqrt{5}$ **30.** $4\sqrt{7} + 11$ **31.** $3\sqrt{3} - 7$ **32.** $6\sqrt{2} - 2\sqrt{10}$

Simplify.

13. $\sqrt{5}\sqrt{20}$ 10 **14.** $\sqrt{2}\sqrt{18}$ 6 **15.** $\sqrt{21}\sqrt{98}$ $7\sqrt{42}$ **16.** $\sqrt{45}\sqrt{5}$ 15

17. $\sqrt{\frac{100}{25}}$ 2 **18.** $\sqrt{\frac{72}{3}}$ $2\sqrt{6}$ **19.** $\frac{\sqrt{60}}{\sqrt{6}}$ $\sqrt{10}$ **20.** $\frac{\sqrt{216}}{\sqrt{6}}$ 6

21. $(4\sqrt{3})^2$ 48 **22.** $(6\sqrt{25})^2$ 900 **23.** $(8\sqrt{2})^2$ 128 **24.** $(5\sqrt{15})^2$ 375

25. $3(\sqrt{6} + 4)$ $3\sqrt{6} + 12$ **26.** $3(\sqrt{12} - 8)$ **27.** $\sqrt{6}(3 - \sqrt{18})$ **28.** $\sqrt{3}(7 + \sqrt{15})$

29. $(\sqrt{6} - 4)(\sqrt{6} + 4)$ −10 **30.** $(\sqrt{7} + 2)^2$ **31.** $(\sqrt{3} - 2)(\sqrt{3} + 5)$ **32.** $\sqrt{2}(\sqrt{5} - 1)^2$

Lesson 9.3

18. No solution

Solve each equation if possible. If not possible, explain why.

1. $\sqrt{x-3} = 1$ 4
2. $\sqrt{x+5} = 3$ 4
3. $\sqrt{x-9} = 8$ 73
4. $\sqrt{3x} = 6$ 12
5. $\sqrt{5x} = 10$ 20
6. $\sqrt{x-4} = 2$ 8
7. $\sqrt{8-x} = 2$ 4
8. $\sqrt{17-x} = 4$ 1
9. $\sqrt{3x+1} = 5$ 8
10. $\sqrt{x+2} = x-4$ 7
11. $\sqrt{1-2x} = x+1$ 0
12. $\sqrt{x-2} = x-4$ 6
13. $\sqrt{x+3} = x-3$ 6
14. $\sqrt{3x-5} = x-5$ 10
15. $\sqrt{3x-14} = 6-x$ 5
16. $\sqrt{x^2-3x+10} = x+1$ 1.8
17. $\sqrt{x^2+x-8} = x-2$ 2.4
18. $\sqrt{x^2+6x-4} = x-5$
19. $x^2 = 80$ $\pm 4\sqrt{5}$
20. $x^2 = 27$ $\pm 3\sqrt{3}$
21. $x^2 = 98$ $\pm 7\sqrt{2}$
22. $9x^2 = 10$ $\pm \frac{\sqrt{10}}{3}$
23. $16x^2 = 5$ $\pm \frac{\sqrt{5}}{4}$
24. $25x^2 = 1$ $\pm \frac{1}{5}$
25. $3x^2 = 27$ ± 3
26. $5x^2 = 75$ $\pm \sqrt{15}$
27. $6x^2 = 72$ $\pm 2\sqrt{3}$
28. $x^2 - 6x + 9 = 0$ 3
29. $x^2 + 4x + 2 = 0$ $-2 \pm \sqrt{2}$
30. $x^2 - 8x + 16 = 0$ 4
31. $x^2 = 16$ ± 4
32. $\sqrt{x} = 16$ 256
33. $|x| = 16$ ± 16
34. $x^2 = -16$
35. $\sqrt{x} = -16$
36. $|x| = -16$

34. Not possible; square cannot be negative.
35. Not possible; principal square root cannot be negative.
36. Not possible; absolute value cannot be negative.

Lesson 9.4

Copy and complete the table. Use a calculator and round the answer to the nearest tenth.

	Leg	Leg	Hypotenuse	
1.	2	6	?	6.3
3.	6	?	10	8
5.	7	15	?	16.6
7.	?	9	20	17.9
9.	15	18	?	23.4
11.	10	24	?	26
13.	?	40	41	9

	Leg	Leg	Hypotenuse	
2.	8	?	17	15
4.	?	12	13	5
6.	7	12	?	13.9
8.	6	7	?	9.2
10.	?	12	14	7.2
12.	8	?	30	28.9
14.	1.5	2	?	2.5

Solve for *x*.

15.

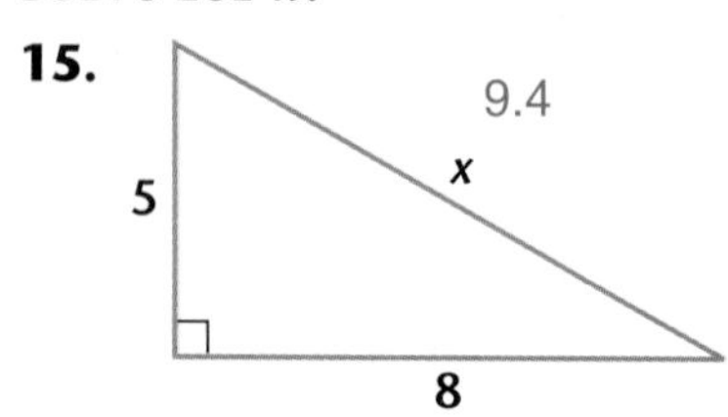

16.

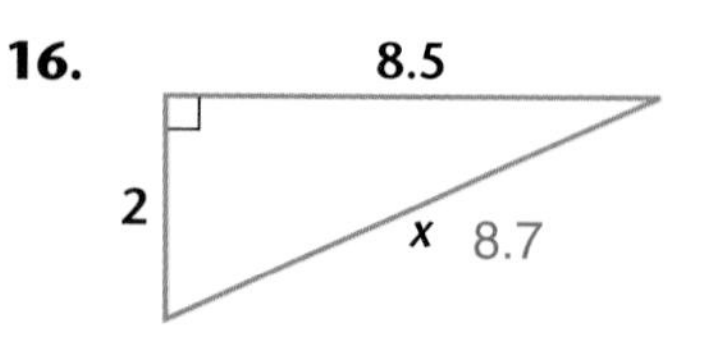

17.

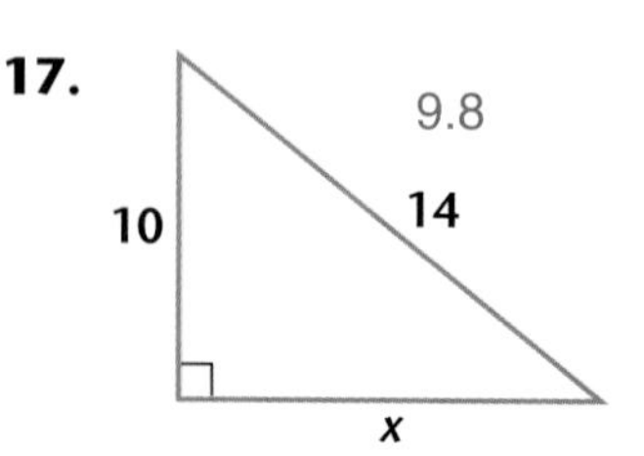

Lesson 9.5

For each figure, find the coordinates of Q and the lengths of the three sides of the triangle.

1.

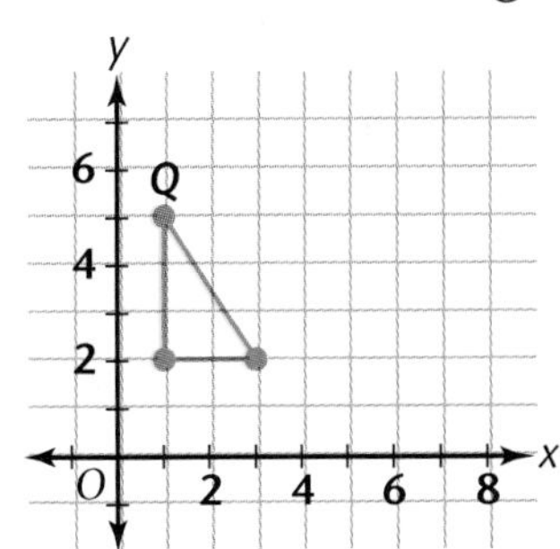

$Q(1, 5)$; 2, 3, $\sqrt{13}$

2.

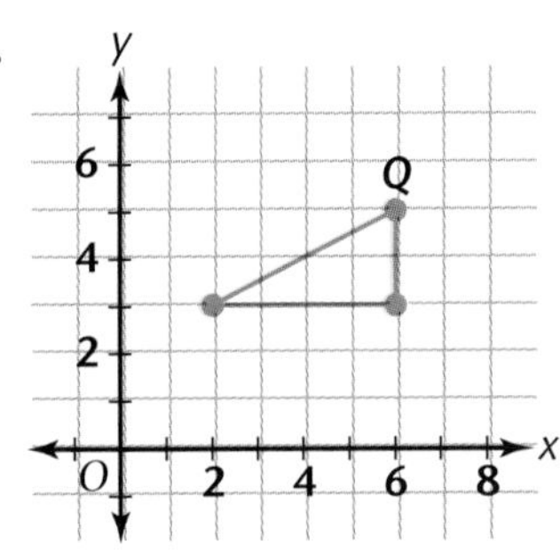

$Q(6, 5)$; 2, 4, $2\sqrt{5}$

3.

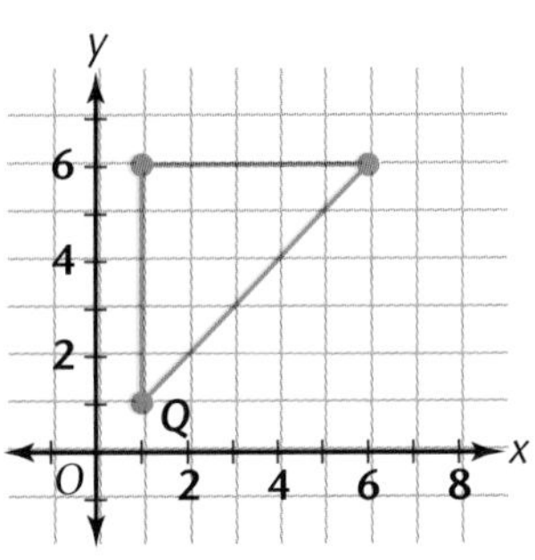

$Q(1, 1)$; 5, 5, $5\sqrt{2}$

Find the distance between the two given points.

4. $A(3, 6)$, $B(1, 2)$ 4.5
5. $M(3, 7)$, $N(12, 19)$ 15
6. $X(-5, -2)$, $Y(-8, 3)$ 5.8
7. $D(6, 7)$, $E(-1, 1)$ 9.2
8. $P(-5, 3)$, $Q(5, 5)$ 10.2
9. $R(3, 4)$, $S(-4, -1)$ 8.6

Plot each segment in a coordinate plane, and find the coordinates of the midpoint of the segment.

10. $\overline{AB}$, with $A(4, 6)$ and $B(8, 8)$ (6, 7)
11. $\overline{MN}$, with $M(-4, 4)$ and $N(0, -8)$ (−2, −2)
12. $\overline{XY}$, with $X(-6, 5)$ and $Y(-2, 3)$ (−4, 4)
13. $\overline{CD}$, with $C(7, 8)$ and $D(4, 1)$ (5.5, 4.5)

The midpoint of $\overline{PQ}$ is M. Find the missing coordinates.

14. $P(5, 8)$, $Q(1, 2)$, $M(\underline{?}, \underline{?})$ (3, 5)
15. $P(6, 2)$, $Q(\underline{?}, \underline{?})$, $M(8, 6)$ (10, 10)
16. $P(-4, 6)$, $Q(8, 9)$, $M(\underline{?}, \underline{?})$ (2, 7.5)
17. $P(15, -9)$, $Q(\underline{?}, \underline{?})$, $M(5, -3)$ (−5, 3)

Lesson 9.6

Find an equation of a circle with center at the origin and having the given radius.

1. radius = 3
$x^2 + y^2 = 9$
2. radius = 5
$x^2 + y^2 = 25$
3. radius = 10
$x^2 + y^2 = 100$
4. radius = 16
$x^2 + y^2 = 256$

Use graph paper to sketch each circle below. Find an equation of a circle with

5. center (1, 2) and radius = 2.
6. center (−1, 2) and radius = 2.
7. center (1, −2) and radius = 2.
8. center (−1, −2) and radius = 2.

Identify the center and radius for each circle.

9. $(x - 1)^2 + (y - 2)^2 = 16$ (1, 2); $r = 4$
10. $(x - 4)^2 + (y - 5)^2 = 1$ (4, 5); $r = 1$
11. $(x + 3)^2 + (y - 4)^2 = 25$ (−3, 4); $r = 5$
12. $(x + 2)^2 + (y - 6)^2 = 9$ (−2, 6); $r = 3$
13. $(x + 7)^2 + (y - 1)^2 = 3$ (−7, 1); $r = \sqrt{3}$
14. $(x + 10)^2 + (y - 9)^2 = 7$ (−10, 9); $r = \sqrt{7}$

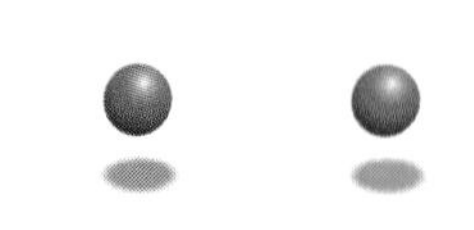

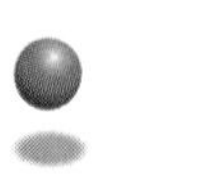

Lesson 9.7

Find the tangent of each acute angle of each triangle.

1. X 24 Y 7 25 Z

2.

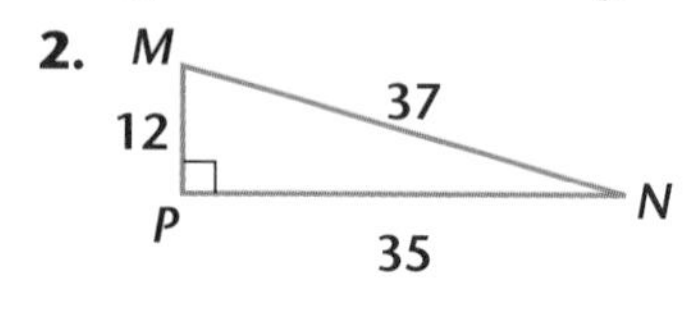

3.

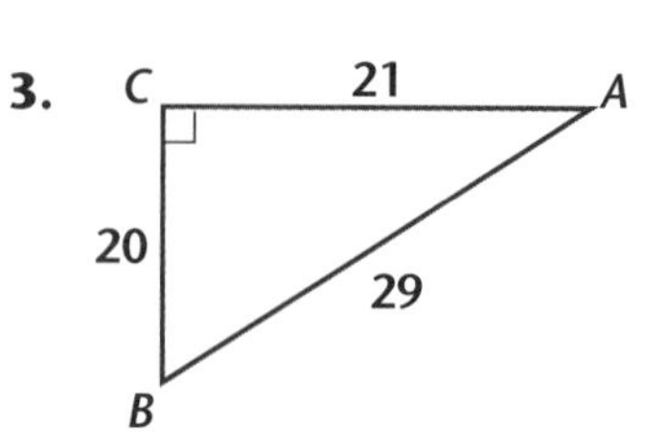

Use a calculator to find the tangents of the following angles to the nearest ten-thousandth.

4. 60° 1.7321 **5.** 35° 0.7002 **6.** 2° 0.0349 **7.** 49° 1.1504 **8.** 52° 1.2799

9. 5° 0.0875 **10.** 73° 3.2709 **11.** 44.9° 0.9965 **12.** 40.5° 0.8541 **13.** 29.9° 0.5750

Find the angle measure with the approximate tangent value. Give your answer to the nearest tenth of a degree.

14. 0.4663 25.0° **15.** 5.1446 79.0° **16.** 0.1228 7.0° **17.** 0.2493 14.0°

18. 0.6745 34.0° **19.** 14.3007 86.0° **20.** 1.2800 52.0° **21.** 0.9004 42.0°

Lesson 9.8

Evaluate each expression to the nearest ten-thousandth.

1. sin 45° 0.7071 **2.** cos 34° 0.8290 **3.** sin 83° 0.9925 **4.** cos 62° 0.4695

5. sin 2° 0.0349 **6.** cos 58.5° 0.5225 **7.** cos 10.9° 0.9820 **8.** sin 26.4° 0.4446

Find the indicated angle measure to the nearest tenth.

9. $b = 6$, $c = 9$, $m\angle B = \underline{?}$ 41.8°

10. $a = 5$, $b = 12$, $m\angle B = \underline{?}$ 67.4°

11. $b = 2$, $c = 7$, $m\angle A = \underline{?}$ 73.4°

12. $a = 9$, $c = 12$, $m\angle B = \underline{?}$ 41.4°

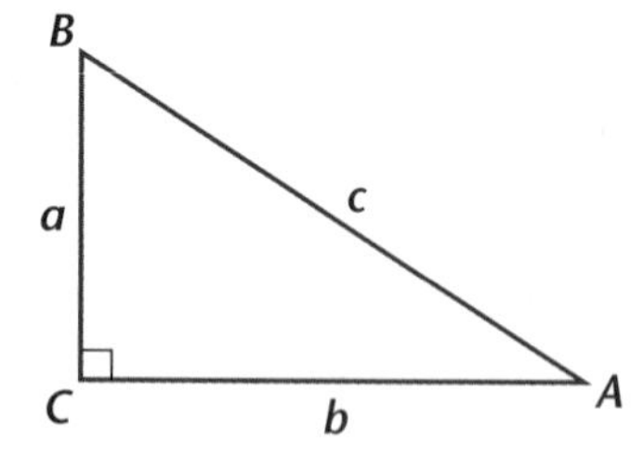

Find the angle measure with the approximate sine value. Give your answer to the nearest tenth of a degree.

13. 0.2924 17.0° **14.** 0.9781 78.0° **15.** 0.8090 54.0° **16.** 0.0523 3.0°

Find the angle measure with the approximate cosine value. Give your answer to the nearest tenth of a degree.

17. 0.9994 2.0° **18.** 0.0523 87.0° **19.** 0.5878 54.0° **20.** 0.7547 41.0°

Lesson 10.1

For each rational expression, determine the restrictions on the variable.

1. $\frac{4x + 1}{x}$ $x \neq 0$
2. $\frac{5}{y - 1}$ $y \neq 1$
3. $\frac{2m - 1}{3m + 1}$ $m \neq -\frac{1}{3}$
4. $\frac{3v - 2}{4v^2 - 2v}$ $v \neq 0, \frac{1}{2}$
5. $\frac{x^2 - 2x - 3}{x^2 + 4x + 3}$ $x \neq -3, -1$
6. $\frac{5b + 3}{b^2 + 4b + 4}$ $x \neq -2$
7. $\frac{w^2 - 4}{w^2 - 5w + 6}$ $x \neq 3, 2$
8. $\frac{q^2 + 6}{q^2 - 16}$ $q \neq \pm 4$
9. $\frac{x^2 - 4x}{x^2 + 3x + 2}$ $x \neq -2, -1$
10. $\frac{y^2 - 1}{y^2 + 6y + 9}$ $y \neq -3$
11. $\frac{7h + 14}{15h^2 - 25h}$ $h \neq 0, \frac{5}{3}$
12. $\frac{8d^2 + 4d}{6d - 3}$ $d \neq \frac{1}{2}$

Describe all the transformations to $f(x) = \frac{1}{x}$ needed to graph each of the following rational functions.

13. $f(x) = \frac{1}{x - 3}$
14. $f(x) = \frac{3}{x}$
15. $f(x) = \frac{1}{x} - 2$
16. $f(x) = \frac{-1}{x} + 4$
17. $f(x) = \frac{4}{x - 2}$
18. $f(x) = \frac{4}{x - 1} + 3$
19. $f(x) = \frac{-2}{x + 1} - 1$
20. $f(x) = \frac{5}{x} + 3$
21. $f(x) = \frac{-3}{x - 3} + 6$

Graph each rational function. List the restrictions on the domain of the function.

22. $h(x) = \frac{1}{x} + 2$
23. $h(x) = \frac{1}{x - 1}$
24. $h(x) = \frac{-2}{x}$
25. $h(x) = \frac{4}{x + 1}$
26. $h(x) = \frac{3}{x + 2} - 1$
27. $h(x) = \frac{-2}{x - 2} + 4$

Lesson 10.2

Determine whether each equation is an inverse variation equation.

1. $xy = 200$ Yes
2. $x = \frac{25}{y}$ Yes
3. $x + y = 36$ No
4. $\frac{a}{3} = \frac{5}{b}$ Yes
5. $\frac{c}{d} = \frac{2}{3}$ No
6. $r = 50t$ No
7. $m = \frac{n}{75}$ No
8. $w = 7t$ No
9. $\frac{-12}{x} = y$ Yes
10. $p = \frac{1}{q}$ Yes
11. $bh = 14$ Yes
12. $\frac{1}{10}m = n$ No

For Exercises 13–26, y varies inversely as x. If y is

13. 24 when x is 8, find y when x is 4. 48
14. 3 when x is 12, find x when y is 4. 9
15. 6 when x is 2, find x when y is 4. 3
16. 12 when x is 15, find x when y is 18. 10
17. 15 when x is 21, find x when y is 27. $11\frac{2}{3}$
18. -6 when x is -2, find y when x is 5. 2.4
19. -8 when x is 2, find x when y is 7. $-2\frac{2}{7}$
20. 99 when x is 11, find x when y is 11. 99
21. 2 when x is 5, find y when x is 20. 0.5
22. 6.9 when x is 1.7, find y when x is 5.1. 2.3
23. $\frac{1}{3}$ when x is 5, find x when y is $6\frac{2}{3}$. $\frac{1}{4}$
24. 7 when x is $\frac{2}{3}$, find x when y is $\frac{2}{3}$. 7
25. 5.6 when x is 2.8, find x when y is 4.48. 3.5
26. 8.1 when x is 2.7, find y when x is 3.6. 6.075

Lesson 10.3

For what values of the variable is each rational expression undefined?

1. $\frac{5x}{x + 15}$ -15 **2.** $\frac{12}{7 - y}$ 7 **3.** $\frac{8 - w}{w}$ 0 **4.** $\frac{9q}{q - 4}$ 4

5. $\frac{t - 6}{6 - t}$ 6 **6.** $\frac{(a + 1)(a - 3)}{(a - 1)(a - 3)}$ $1, 3$ **7.** $\frac{s - 5}{3s - 6}$ 2 **8.** $\frac{4}{d(d^2 - 2d - 8)}$ $-2, 0, 4$

Write the common factors.

9. $\frac{8}{14}$ 2 **10.** $\frac{16}{36}$ 4 **11.** $\frac{4(x - 3)}{8x}$ 4 **12.** $\frac{5(x + 4)}{20}$ 5

13. $\frac{x + y}{(x - y)(x + y)}$ $x + y$ **14.** $\frac{6w - 12}{3}$ 3 **15.** $\frac{x^2 - 9}{x + 3}$ $x + 3$ **16.** $\frac{r + 4}{r^2 + 8r + 16}$ $r + 4$

17. $\frac{x^2 - y^2}{(x - y)^2}$ $x - y$ **18.** $\frac{y^2 + 3y - 4}{y - 1}$ $y - 1$ **19.** $\frac{(x + 2)^2}{x^2 - x - 8}$ 1 **20.** $\frac{2 - f}{f^2 + 5f - 14}$ $f - 2$

Simplify.

21. $\frac{12(x + 3)}{24(x - 1)}$ $\frac{x + 3}{2(x - 1)}$ **22.** $\frac{4(x - y)}{16(x + y)}$ $\frac{x - y}{4(x + y)}$ **23.** $\frac{8(d + 1)}{28(1 + d)}$ $\frac{2}{7}$ **24.** $\frac{5(c + 9)(c - 9)}{10(c - 9)}$

25. $\frac{7w - 14}{7}$ $w - 2$ **26.** $\frac{6x - 18}{x - 3}$ 6 **27.** $\frac{10a + 15b}{5a}$ $\frac{2a + 3b}{a}$ **28.** $\frac{4x^2 - 3x}{4x - 3}$ x

29. $\frac{w^2 - 25}{w + 5}$ $w - 5$ **30.** $\frac{y + 4}{y^2 + 3y - 4}$ $\frac{1}{y - 1}$ **31.** $\frac{z^2 + 10z + 9}{-(z + 9)}$ $-(z + 1)$ **32.** $\frac{6 - t}{t^2 - 3t - 18}$

33. $\frac{4f + 8}{f^2 - 4}$ $\frac{4}{f - 2}$ **34.** $\frac{m^2 - 2m - 4}{3m - 12}$ **35.** $\frac{g^2 - 9}{g^2 + 6g - 27}$ $\frac{g + 3}{g + 9}$ **36.** $\frac{n^2 - n - 20}{n^2 + 9n + 20}$

24. $\frac{c + 9}{2}$ **32.** $\frac{-1}{t + 3}$ **34.** Cannot be simplified **36.** $\frac{n - 5}{n + 5}$

Lesson 10.4

Perform the indicated operations. Simplify and state restrictions for variables.

1. $\frac{4}{3y} + \frac{5}{3y}$ **2.** $\frac{7}{s - 2} - \frac{3}{s - 2}$ **3.** $\frac{4m}{n + 3} - \frac{9m}{n + 3}$

4. $\frac{7w - 3}{x - y} - \frac{4 + 5w}{x - y}$ **5.** $\frac{x}{x + 1} + \frac{1}{x + 1}$ **6.** $\frac{z}{2} + \frac{z - 6}{2}$

7. $\frac{5}{m} + \frac{2}{n}$ **8.** $\frac{5}{4t} - \frac{3}{t}$ **9.** $\frac{x}{3zw} + \frac{4z}{wx}$

10. $\frac{6}{ab} - \frac{c}{ab^2}$ **11.** $\frac{-3}{w - 2} + \frac{4}{5(w - 2)}$ **12.** $\frac{m + n}{m - 2} + \frac{m - n}{2 - m}$

13. $a - \frac{a - 5}{a + 5}$ **14.** $\frac{-4 - g}{g - 1} + 3$ **15.** $\frac{a}{b} - \frac{c}{d}$

16. $\frac{7 - x}{4 + y} \cdot \frac{4 + y}{m + n}$ **17.** $\frac{a^2}{(a + 4)(a + 1)} \cdot \frac{a + 1}{a(a + 4)}$ **18.** $\frac{x}{x + 3} \cdot \frac{x^2 - 9}{x^2}$

19. $\frac{4}{y + 2} \cdot \frac{1}{y}$ **20.** $\frac{-6}{r - 3} \cdot \frac{2}{r}$ **21.** $\frac{t^2 - t}{3} \cdot \frac{6}{t - 1}$

22. $\frac{5}{(x - 1)(x + 2)} + \frac{4}{x - 1}$ **23.** $\frac{3}{c + 5} - \frac{4}{(c + 2)(c + 5)}$ **24.** $\frac{v - 2}{v + 2} + \frac{3}{v - 3}$

25. $\frac{3x}{x^2 + x - 12} + \frac{x}{x - 3}$ **26.** $\frac{4}{x - 6} - \frac{3}{x^2 - 3x - 18}$ **27.** $\frac{7}{n - 3} + \frac{2m}{n^2 - 9}$

28. $\frac{1}{y - 1} + \frac{y}{1 - y}$ **29.** $\frac{4}{h + 2} - \frac{3h - 1}{h^2 + 6h + 8}$ **30.** $\frac{2}{x^2 - 4} + \frac{3}{4x + 8}$

Lesson 10.5

Solve each rational equation.

1. $\frac{11}{2x} - \frac{2}{3x} = \frac{1}{6}$ 29
2. $\frac{1}{x} + \frac{5x}{x+1} = 5$ 0.25
3. $\frac{x-1}{x+3} = \frac{x+4}{3-x}$ No solution
4. $\frac{1}{4} + \frac{2}{x} = \frac{11}{12}$ 3
5. $\frac{2w}{3} = 2 + \frac{w+3}{6}$ 5
6. $a + \frac{a}{a-1} = \frac{4a-3}{a-1}$ 3
7. $1 + \frac{3}{z-1} = \frac{4}{3}$ 10
8. $\frac{1}{2x} = \frac{1}{x^2} - \frac{1}{9}$ −6, 1.5
9. $\frac{4y}{3y-2} + \frac{2y}{3y+2} = 2$ −2
10. $\frac{5}{p-1} + \frac{p}{p+1} = 1$ −1.5
11. $\frac{x+3}{x} + \frac{x-12}{x} = 5$ −3
12. $\frac{w-1}{w+1} + 1 = \frac{2w}{w-1}$ 0
13. $\frac{2q}{q-1} + \frac{q-5}{q^2-1} = 1$ −4
14. $\frac{14}{y-6} - \frac{1}{2} = \frac{6}{y-8}$ 10, 20
15. $\frac{a-2}{a} = \frac{1}{a} + \frac{a-3}{a-6}$ 3
16. $\frac{c-1}{3-c} = c - \frac{2}{c-3}$ −1
17. $\frac{x-5}{x^2-1} + \frac{2x}{x-1} = 1$ −4
18. $\frac{m}{3m+6} = \frac{2}{5} + \frac{m}{5m+10}$ −3

Solve the following rational equations by the graphing method.

19. $\frac{16}{x+8} = \frac{4}{9}$ 28
20. $\frac{2}{x+1} = \frac{3}{x+2}$ 1
21. $\frac{2x}{x-4} - \frac{3}{5} = 5$ $6.\overline{2}$

Lesson 10.6

Identify the means and extremes of each proportion.

1. $\frac{x}{6} = \frac{4}{9}$
2. $\frac{7}{1.5} = \frac{z}{14}$
3. $\frac{19}{c} = \frac{4+c}{3}$
4. $\frac{m+1}{4} = \frac{m-2}{3}$

Solve for the variable indicated.

5. $\frac{9}{15} = \frac{r}{20}$ 12
6. $\frac{12}{21} = \frac{20}{q}$ 35
7. $\frac{y}{5} = \frac{8}{-1}$ −40
8. $\frac{16}{h+8} = \frac{4}{9}$ 28
9. $\frac{7}{10} = \frac{3x-2}{5}$ $\frac{11}{6}$
10. $\frac{4w+3}{9} = \frac{1}{2}$ $\frac{3}{8}$
11. $\frac{2x}{10} = \frac{4-x}{5}$ 2
12. $\frac{3}{m+10} = \frac{4}{16}$ 2
13. $\frac{8}{5t+1} = \frac{-1}{3}$ −5
14. $\frac{2c}{12} = \frac{4c-15}{6}$ 5
15. $\frac{7}{f-3} = \frac{f+4}{f-3}$ No solution
16. $\frac{9}{m+5} = \frac{6}{m-3}$ 19
17. $\frac{7}{2} = \frac{21}{2x-1}$ 3.5
18. $\frac{2+d}{8} = \frac{d-5}{9}$ −58
19. $\frac{x}{3} = \frac{2}{x+5}$ −6, 1
20. $\frac{2m+4}{2} = \frac{m-3}{3}$ −4.5
21. $\frac{x+5}{6} = \frac{1}{x}$ −6, 1
22. $\frac{1}{m} = \frac{m+1}{6}$ −3, 2
23. $\frac{x}{x^2-3} = \frac{5}{x+4}$ $-\frac{3}{2}, \frac{5}{2}$
24. $\frac{12}{y} = \frac{8-y}{1}$ 2, 6

Lesson 10.7

Prove each of the following. Give a reason for each step.

1. For all x, $3x + 4x = 7x$.
2. For all y, $7y - 6y = y$.
3. For all m, $(-4m)(-3m) = 12m^2$.
4. If $8x - 3x = 35$, then $x = 7$.
5. If $3c - 1 = 11$, then $c = 4$.
6. If $4(y + 2) = -24$, then $y = -8$.
7. If $\frac{3}{4}x + 5 = 2$, then $x = -4$.
8. If $\frac{3}{5}n - 6 = 0$, then $n = 10$.
9. If $\frac{d}{3} - 7 = -4$, then $d = 9$.
10. If $5 - \frac{q}{4} = 8$, then $q = -12$.

Graphing Calculator Keystroke Guide

Essential keystroke sequences (using the TI-82 or TI-83 graphing calculator) are presented below for the most significant activities and examples found in the chapter lessons that require or suggest the use of a graphing calculator.

internetconnect

Keystrokes for other graphing calculator models are on the HRW Web site. Visit the HRW Web site at go.hrw.com and enter the keyword MJ1 CALC.

LESSON 1.1

Page 11

```
WINDOW
 Xmin=-10
 Xmax=10
 Xscl=1
 Ymin=-12
 Ymax=12
 Yscl=1
 Xres=1
```

Graph the equations $Y_1 = X + 2$ and $Y_2 = 6$ on your graphing calculator and find the point of intersection of the two lines. Use viewing window $[-10, 10]$ by $[-12, 12]$.

[Y=] **(Y1=)** [X,T,θ,n] [+] **2** [ENTER] **(Y2 =) 6**

Use the intersect feature in the CALCULATE menu to find the point of intersection.

[2nd] [TRACE] (CALC) [5:intersect] [ENTER] **(First curve?)** [ENTER] **(Second curve?)** [ENTER] **(Guess?)** [ENTER]

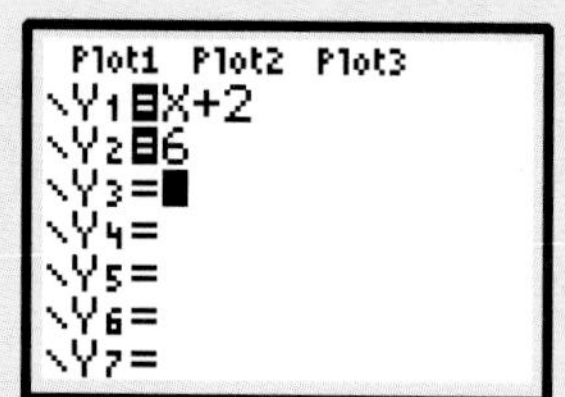

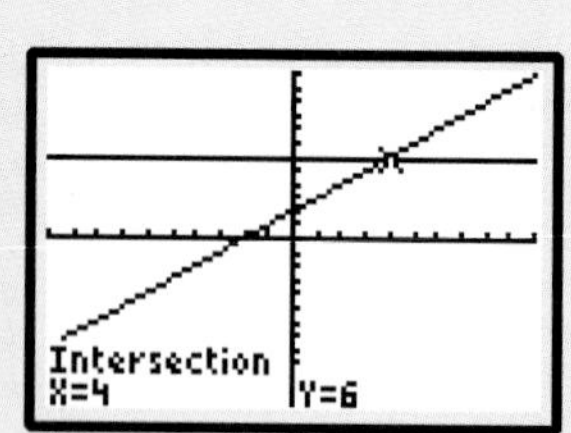

Page 25

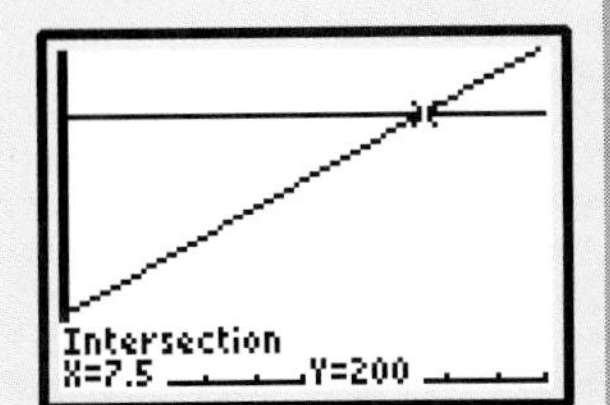

X	Y_1	Y_2
6	170	200
6.5	180	200
7	190	200
7.5	200	200
8	210	200
8.5	220	200
9	230	200

X=7.5

Graph the equations $Y_1 = 20X + 50$ and $Y_2 = 200$ on your graphing calculator. The x-coordinate of the point of intersection is the solution. Use viewing window [0, 10] by [0, 248].

Y= **(Y1=) 20** X,T,θ,n + **50** ENTER **(Y2=) 200** GRAPH

Use the intersect feature in the CALCULATE menu to find the point of intersection.

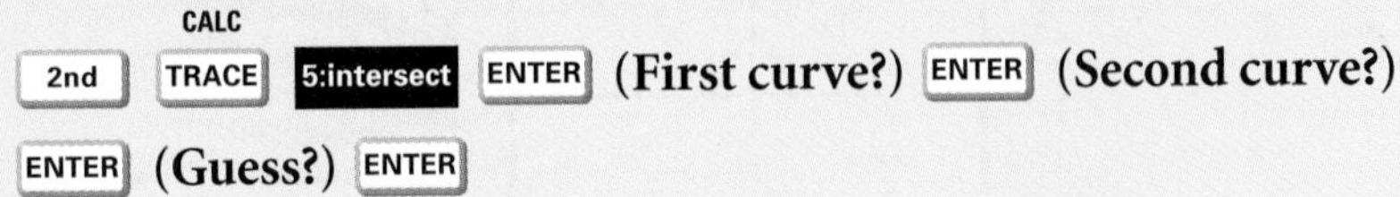

CALC
2nd TRACE 5:intersect ENTER **(First curve?)** ENTER **(Second curve?)**

ENTER **(Guess?)** ENTER

Use the table feature to solve the problem.

TBLSET
2nd WINDOW **(Tblstart=) 6** ENTER **(ΔTbl=) .5** ENTER
⇑ TI-82-TblMin=

TABLE
(Indpnt:) AUTO ▼ **(Depend:)** AUTO 2nd GRAPH

LESSON 1.4

Page 30

Graph both equations $Y_1 = 0.10X + 80$ and $Y_2 = 0.18X + 30$, find the x-coordinate of the point of intersection. Use viewing window [0, 960] Xscl = 80 by [0, 280] Yscl = 20.

Y= **(Y1=) .10** X,T,θ,n + **80** ENTER **(Y2=) .18** X,T,θ,n **+ 30** GRAPH

Use the intersect feature in the CALCULATE menu to find the point of intersection.

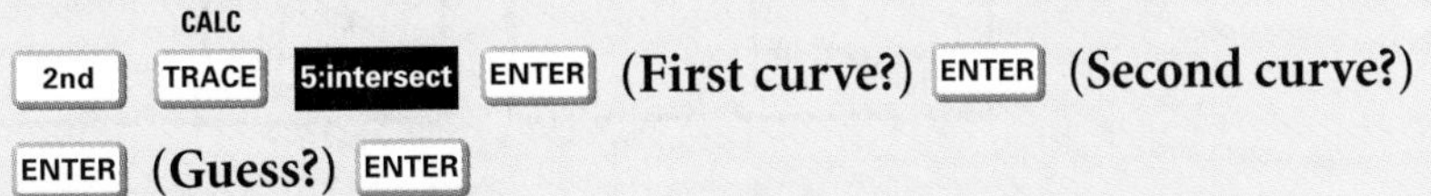

CALC
2nd TRACE 5:intersect ENTER **(First curve?)** ENTER **(Second curve?)**

ENTER **(Guess?)** ENTER

Use the table feature to find the value of x for which both y-values are equal.

TBLSET
2nd WINDOW **(TblStart=) 622** ENTER **(ΔTbl=) 1** ENTER **(Indpnt:)** AUTO
⇑ TI-82: TblMin=

TABLE
▼ **(Depend:)** AUTO 2nd GRAPH

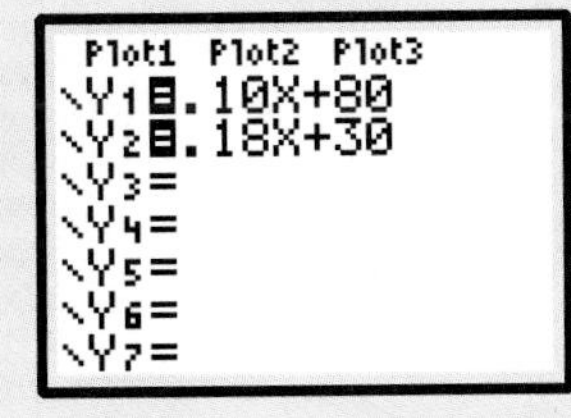

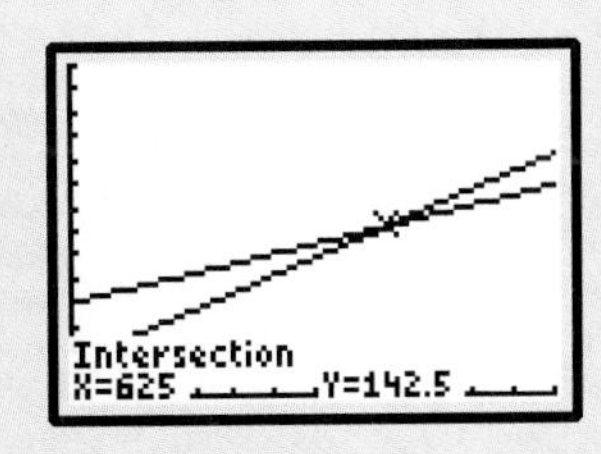

X	Y_1	Y_2
622	142.2	141.96
623	142.3	142.14
624	142.4	142.32
625	142.5	142.5
626	142.6	142.68
627	142.7	142.86
628	142.8	143.04

X=625

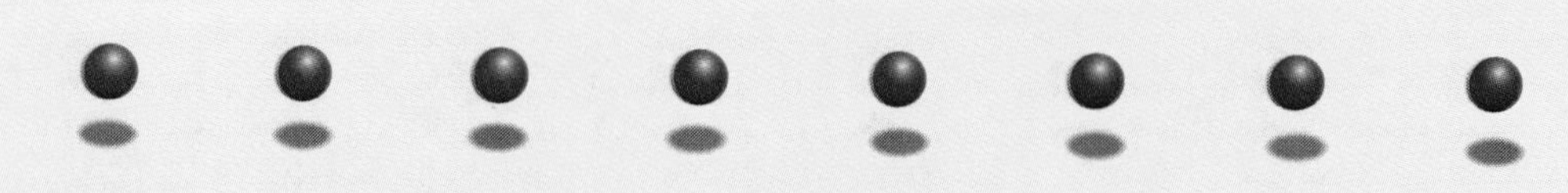

Page 32

Graph the equations $Y_1 = 4X + 10$ and $Y_2 = 4X + 12$ on the same calculator display to illustrate that the equation $4w + 10 = 4w + 12$ has no solution. Use viewing window $[-5, 5]$ by $[-4, 4]$ with scales of 1.

Use a keystroke sequence similar to that used in the previous example.

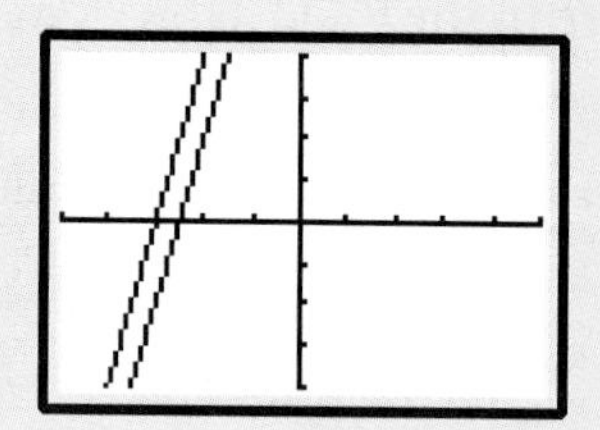

LESSON 1.6

Page 44

Solve the inequality $|x - 2| \geq 4$ by graphing the functions $Y_1 = |x - 2|$ and $Y_2 = 4$ on the same calculator display. Use the standard viewing window.

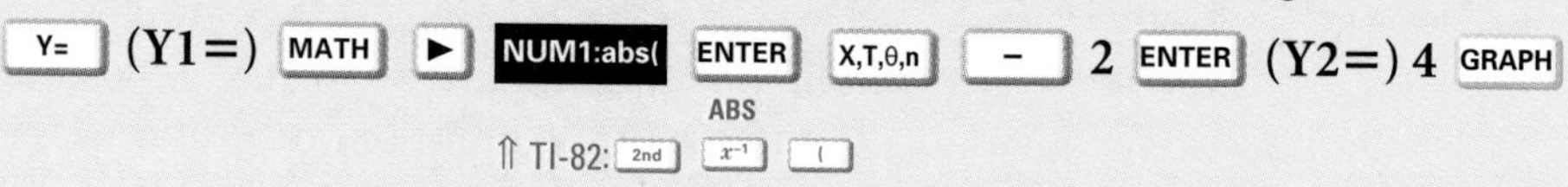

Y=
(Y1=)
MATH
▶
NUM1:abs(
ENTER
X,T,θ,n
–
2
ENTER
(Y2=) 4
GRAPH
ABS
⇑ TI-82: 2nd x⁻¹ (

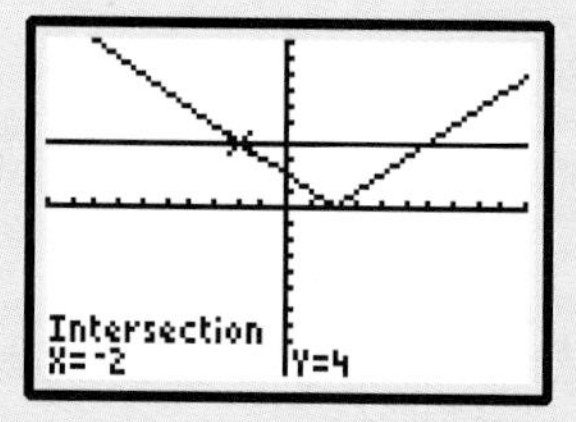

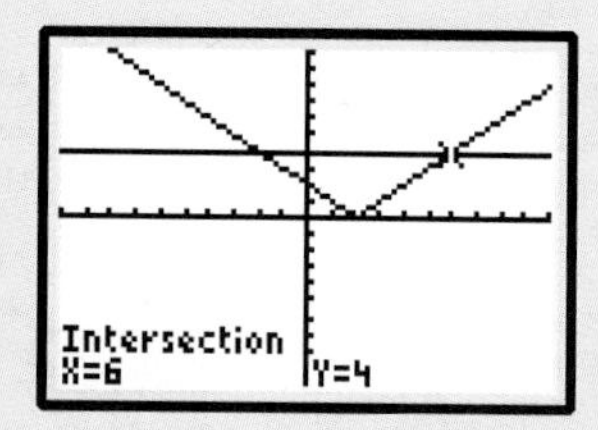

LESSON 2.2

Page 69

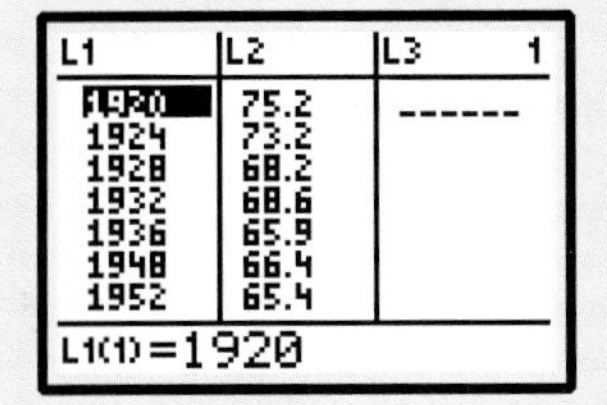

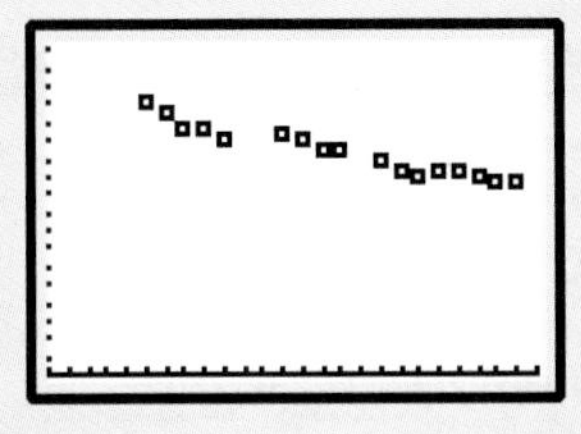

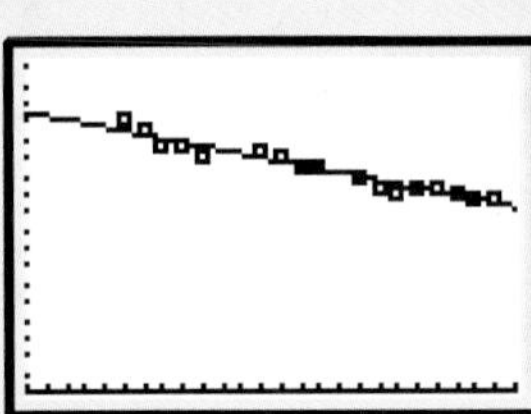

Enter the data into the statistics list of your graphing calculator and graph the points. Use viewing window [1900, 2000] Xscl = 4 by [0, 95] Yscl = 5.

STAT EDIT 1:Edit ENTER Under the heading **L1,** enter the data values that represent the year.

1920 ENTER **1924** ENTER **1928** ENTER **1932** ENTER **1936** ENTER **1948** ENTER

1952 ENTER **1956** ENTER **1960** ENTER **1968** ENTER **1972** ENTER **1976** ENTER

1980 ENTER **1984** ENTER **1988** ENTER **1992** ENTER **1996** ENTER

Under **L2,** enter the data values that represent the time in seconds.

75.2 ENTER **73.2** ENTER **68.2** ENTER **68.6** ENTER **65.9** ENTER **66.4** ENTER

65.4 ENTER **62.2** ENTER **61.9** ENTER **58.7** ENTER **56.58** ENTER **55.49** ENTER

56.53 ENTER **55.79** ENTER **55.05** ENTER **53.98** ENTER **54.10**

(Xlist:) 2nd 1 (L1) ENTER **(Ylist:)** 2nd 2 (L2) ENTER **(Mark:)** ▫ ENTER GRAPH

Graph the regression equation.

STAT ► CALC 4:LinReg (ax+b) ENTER

⇑ TI-82 5:LinReg (ax+b)

Y= VARS VARS 5:Statistics ENTER ► ► EQ 1:RegEQ ENTER GRAPH

⇑ TI-82 7:Reg EQ

LESSON 2.3

EXAMPLE 1

Page 74

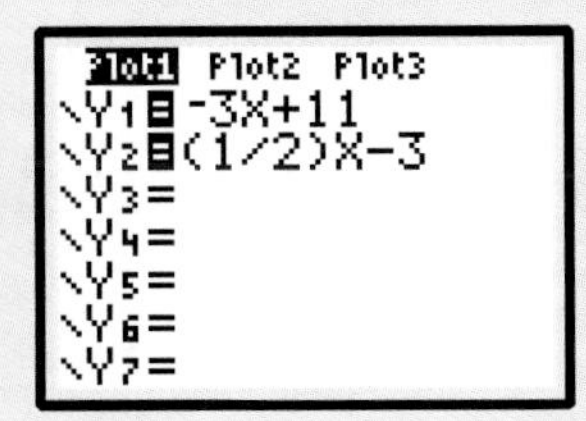

Graph $y = -3x + 11$ and $y = \frac{1}{2}x - 3$ on the same calculator display. To find the common solution, find the coordinates of the point of intersection. Use viewing window $[-2, 6]$ by $[-5, 3]$ with scales of 1.

Y= **−3** X,T,θ,n + **11** ENTER **(Y2=)** (**1** ÷ **2)** X,T,θ,n **− 3** 2nd TRACE (CALC) ▼ 5:intersect ENTER **(First curve?)** ENTER **(Second curve?)** ENTER **(Guess?)** ENTER

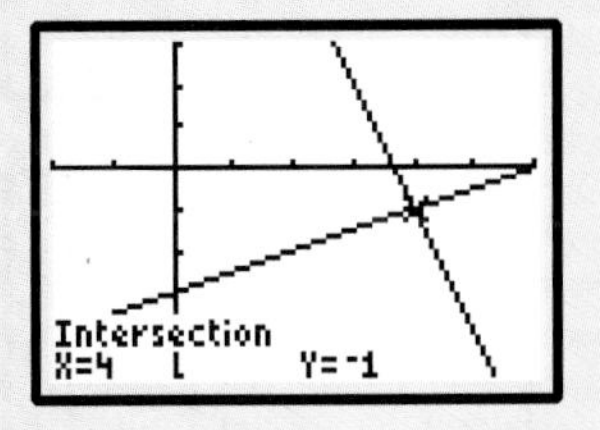

EXAMPLE 3

Page 76

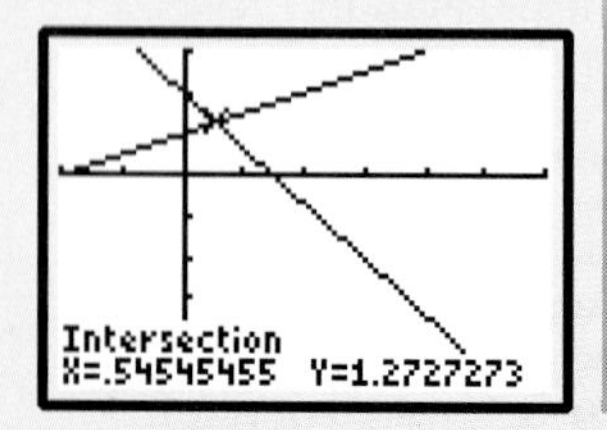

Graph $y = -\frac{4}{3}x + 2$ and $y = \frac{1}{2}x + 1$ on the same calculator display. Use viewing window [−2, 6] by [−5, 3] with scales of 1.

Y= (Y1=) ((−) 4 ÷ 3) X,T,θ,n + 2 ENTER **(Y2=)**

(1 ÷ 2) X,T,θ,n + 2 GRAPH 2nd TRACE (CALC) 5:intersect ENTER

(First curve?) ENTER (Second curve?) ENTER (Guess?) ENTER

LESSON 3.5

EXAMPLE 1

Page 153

[A]
[[6 -1]
[2 -7]]
[B]
[[3]
[1]]

[A]^-1
[[.175 -.025]
[.05 -.15]]
Ans*[B]
[[.5]
[0]]

Use matrices to solve this system of equations. $\begin{cases} 6x - y = 3 \\ 2x - 7y = 1 \end{cases}$

MATRX ► ► EDIT 1: [A] ENTER (MATRIX[A]) 2 ENTER 2 ENTER 6 ENTER

− 1 ENTER 2 ENTER − 7 ENTER 2nd MODE (QUIT) MATRX Names 1: [A] ENTER ENTER

MATRX ► ► EDIT 2: [B] ENTER (MATRIX[B]) 2 ENTER 1 ENTER 3 ENTER

1 ENTER 2nd MODE (QUIT) MATRX Names 2: [B] ENTER ENTER

MATRX Names 1: [A] ENTER x^{-1} ENTER

X MATRX Names 2: [B] ENTER ENTER

EXAMPLE 2

Page 154

[A]
[[14 12]
[10 7]]
[B]
[[49.5]
[33]]

[B]
[[49.5]
[33]]
[A]^-1[B]
[[2.25]
[1.5]]

Use a graphing calculator to solve the matrix equation. $\begin{cases} 14h + 12d = 49.50 \\ 10h + 7d = 33.00 \end{cases}$

MATRX ► ► EDIT 1: [A] ENTER (MATRIX[A]) 2 ENTER 2 ENTER 14 ENTER

12 ENTER 10 ENTER 7 ENTER 2nd MODE (QUIT) MATRX Names 1: [A] ENTER ENTER

MATRX ► ► EDIT 2: [B] ENTER (MATRIX[B]) 2 ENTER 1 ENTER

49.50 ENTER 33.00 ENTER 2nd MODE (QUIT) MATRX Names 2: [B] ENTER ENTER MATRX

Names 1: [A] ENTER x^{-1} MATRX Names 2: [B] ENTER ENTER

LESSON 4.3

EXAMPLE 2

Page 179

Make a scatter plot of the batting averages for each year for the National League. Find the correlation coefficient and the line of best fit. Use viewing window [1987, 1995] Xscl = 1 by [0.3, 0.4] Yscl = .01.

STAT | EDIT | 1:Edit | ENTER

L1	L2	L3 2
1989	.339	------
1990	.329	
1991	.342	
1992	.343	
1993	.363	
------	------	

L2(1)=.339

Under the heading **L1,** enter the data values that represent the year.

1989 ENTER **1990** ENTER **1991** ENTER **1992** ENTER **1993** ENTER

Under **L2,** enter the data values that represent the average.

.339 ENTER **.329** ENTER **.342** ENTER **.343** ENTER **.363** ENTER

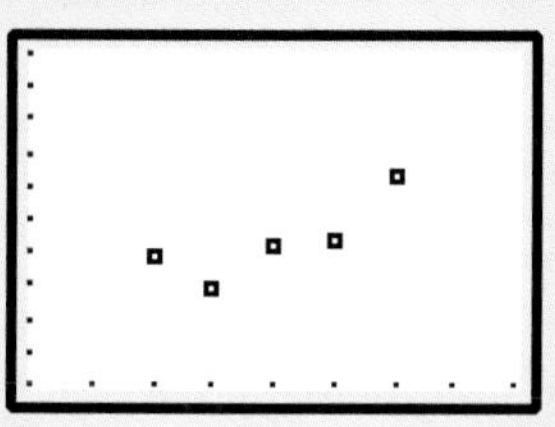

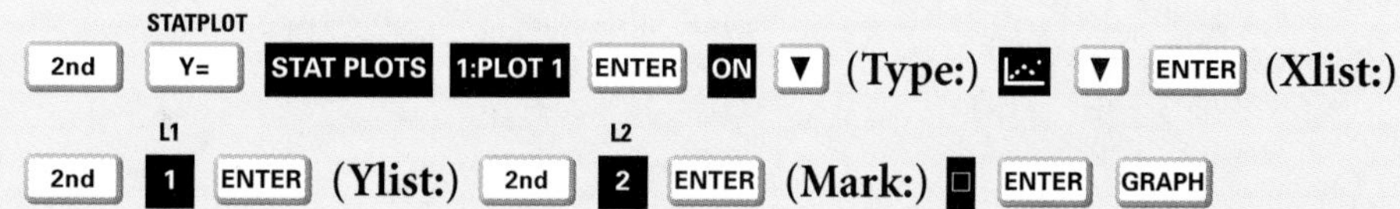

Find the correlation coefficient.

⇑ TI-82- 5:LinReg (ax+b)

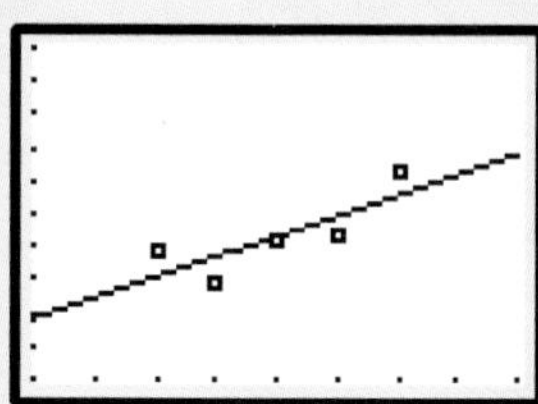

Graph the line of best fit.

⇑ TI-82- 7:RegEQ

LESSON 5.2

Page 229

For step 1, graph the parent function $y = |x|$. Use the standard viewing window.

Y= (Y1=) MATH ► NUM 1:abs ENTER X,T,θ,n GRAPH

⇑ TI-82- 2nd ABS x^{-1} (

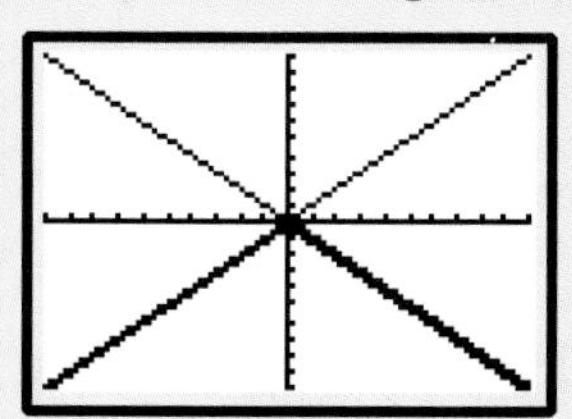

For Step 2a, graph $y = -|x|$ on the same calculator display.

Y= ▼ (Y2=) (–) MATH ► NUM 1:abs ENTER X,T,θ,n

⇑ TI-82- 2nd ABS x^{-1} (

For Steps 2b–2f and Steps 6 and 8, use a keystroke sequence similar to that used in step 1.

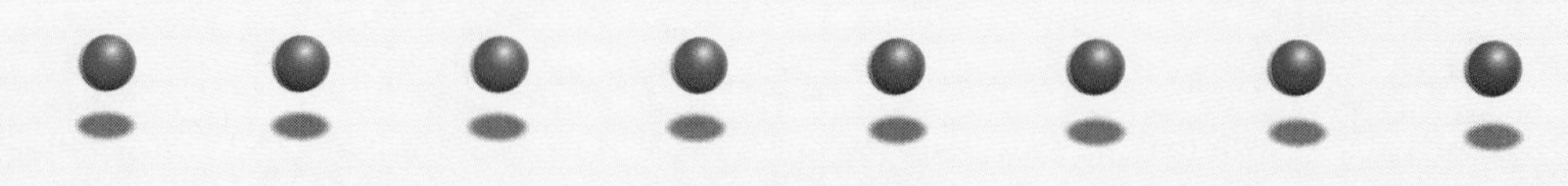

Page 230

Graph the equation $y = (x + 2)^2$ and compare the graph with the parent function $y = x^2$. Use viewing window $[-5, 5]$ by $[0, 7]$ with scales of 1.

Y= (Y1=) X,T,θ,n x^2 ENTER ▼ (Y2=) (X,T,θ,n + 2) x^2 GRAPH

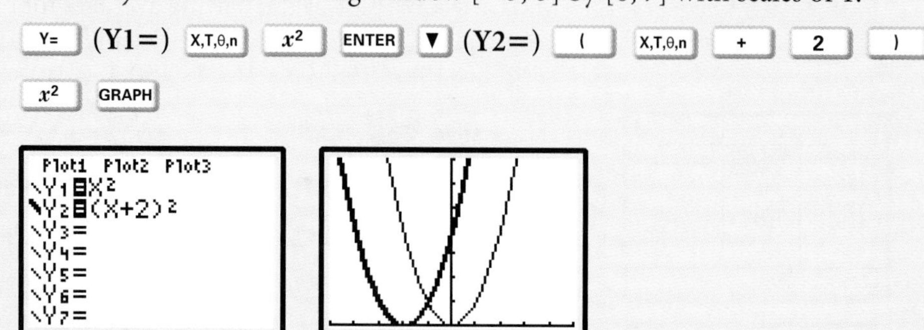

EXAMPLE 3

Page 237

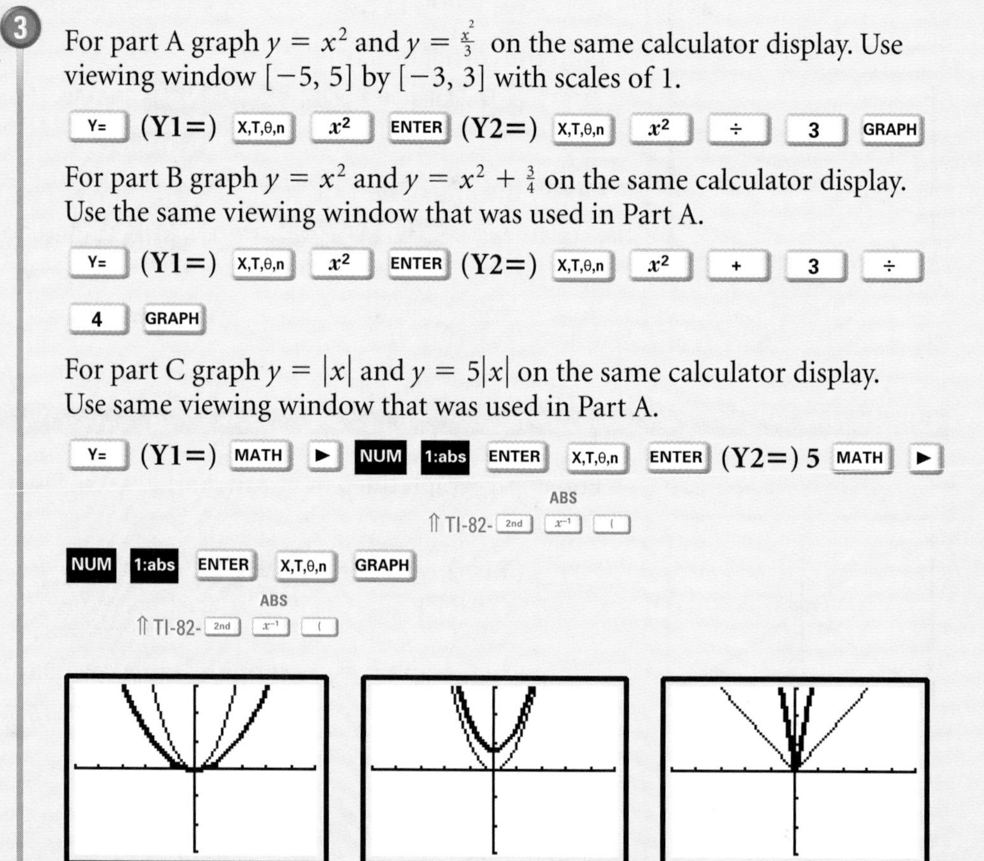

For part A graph $y = x^2$ and $y = \frac{x^2}{3}$ on the same calculator display. Use viewing window $[-5, 5]$ by $[-3, 3]$ with scales of 1.

Y= (Y1=) X,T,θ,n x^2 ENTER (Y2=) X,T,θ,n x^2 ÷ 3 GRAPH

For part B graph $y = x^2$ and $y = x^2 + \frac{3}{4}$ on the same calculator display. Use the same viewing window that was used in Part A.

Y= (Y1=) X,T,θ,n x^2 ENTER (Y2=) X,T,θ,n x^2 + 3 ÷ 4 GRAPH

For part C graph $y = |x|$ and $y = 5|x|$ on the same calculator display. Use same viewing window that was used in Part A.

Y= (Y1=) MATH ► NUM 1:abs ENTER X,T,θ,n ENTER (Y2=) 5 MATH ►

ABS
⇑ TI-82- 2nd x^{-1} (

NUM 1:abs ENTER X,T,θ,n GRAPH

ABS
⇑ TI-82- 2nd x^{-1} (

EXAMPLE 1

Page 255

Graph $y = |x|$ and $y = 2|x + 4|$ on the same calculator display. Use viewing window $[-9.4, 9.4]$ by $[-6.2, 6.2]$ with scales of 1.

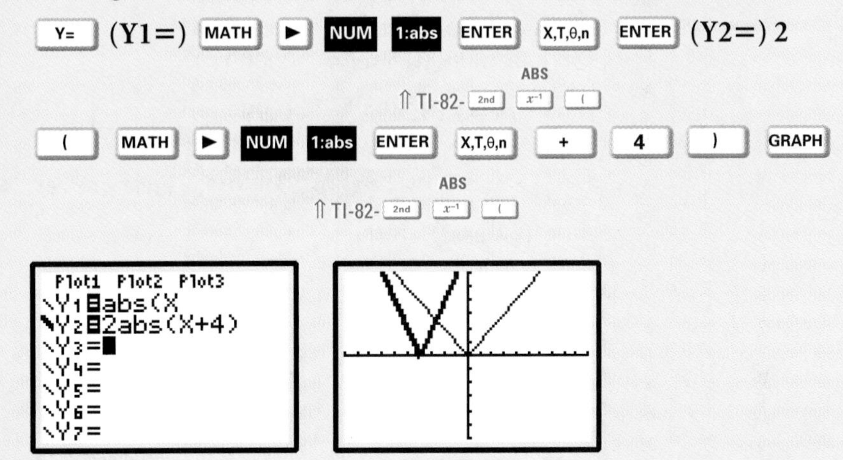

LESSON 6.5

EXAMPLE 4

Page 296

Graph $y = 3^x$ and $y = 4^x$ on the same calculator display. Use viewing window $[-2, 2]$ by $[-2, 5]$ with scales of 1.

Y= (Y1=) 3 ^ X,T,θ,n ENTER (Y2=) 4 ^ X,T,θ,n GRAPH

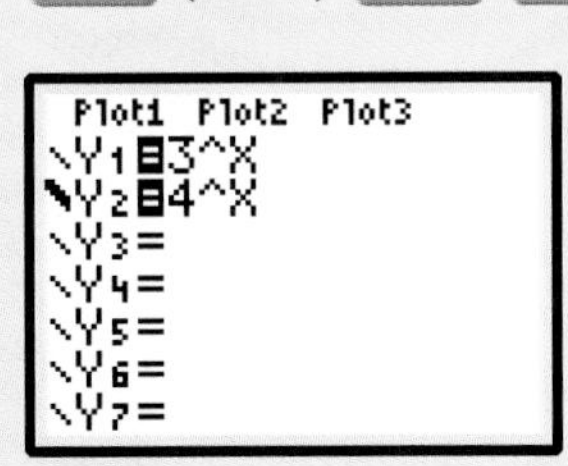

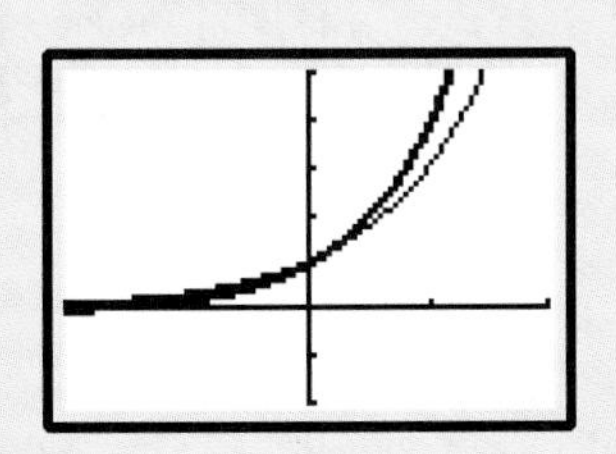

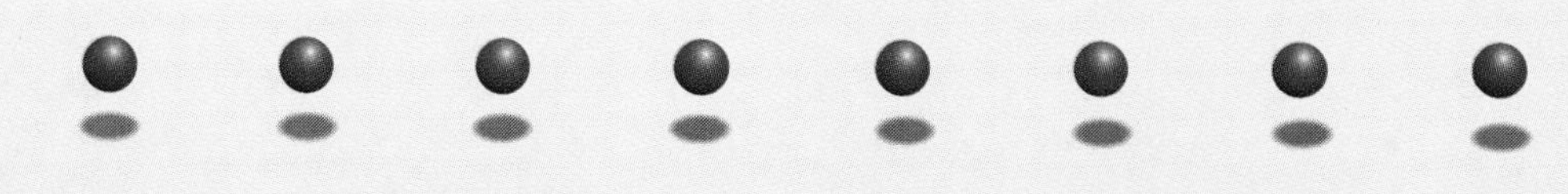

EXAMPLE 5

Page 302

Graph $Y_1 = 152{,}494(0.99)^x$ and $Y_2 = 140{,}000$ on the same calculator display and find the point of intersection. Use viewing window [0, 15] Xscl = 1 by [130000, 150000] Yscl = 100000.

GRAPH 2nd CALC TRACE 5:intersect ENTER **(First curve?)** ENTER **(Second curve?)** ENTER **(Guess?)** ENTER

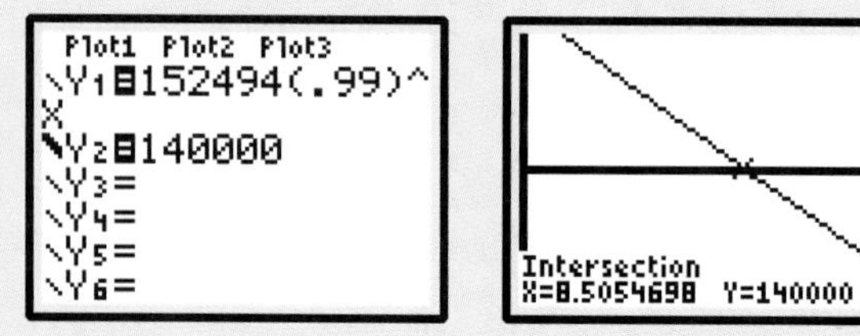

Use the table feature to solve the problem.

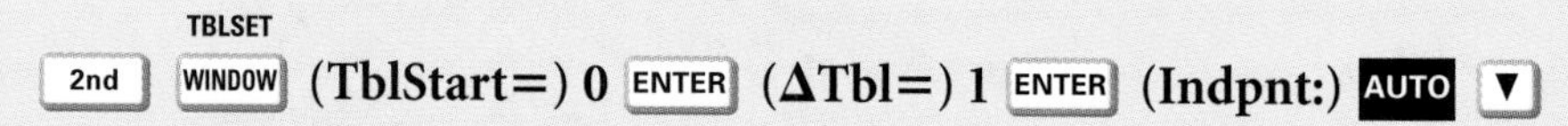

(Depend:) AUTO 2nd TABLE GRAPH

X	Y1	Y2
7	142134	140000
8	140713	140000
9	139306	140000
10	137913	140000
11	136534	140000
12	135168	140000
13	133817	140000

X=7

EXAMPLE 6

Page 303

For Part A graph $y = 2^x$ and $2^x - 3$ on the same calculator display. Use viewing window $[-5, 5]$ by $[-3, 10]$ with scales of 1.

[Y=] **(Y1=)** [2] [^] [X,T,θ,n] [ENTER] **(Y2=)** [2] [^] [X,T,θ,n] [–] [3] [GRAPH]

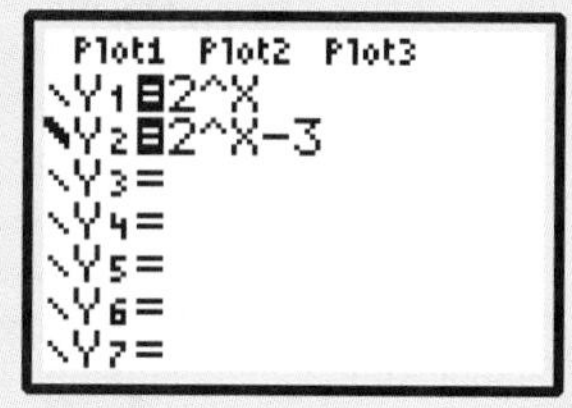

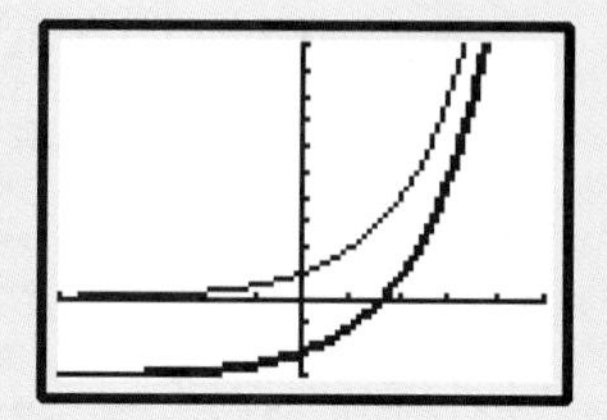

For Part B graph $y = 2^x$ and $y = 2^{x-3}$ on the same calculator display. Use viewing window $[-4, 10]$ by $[-3, 10]$ with scales of 1.

[Y=] **(Y1=)** [2] [^] [X,T,θ,n] [ENTER] **(Y2=)** [2] [^] [(] [X,T,θ,n] [–] [3] [)] [GRAPH]

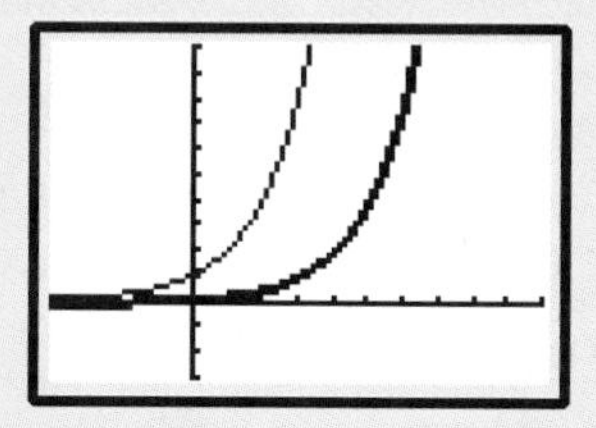

Page 318

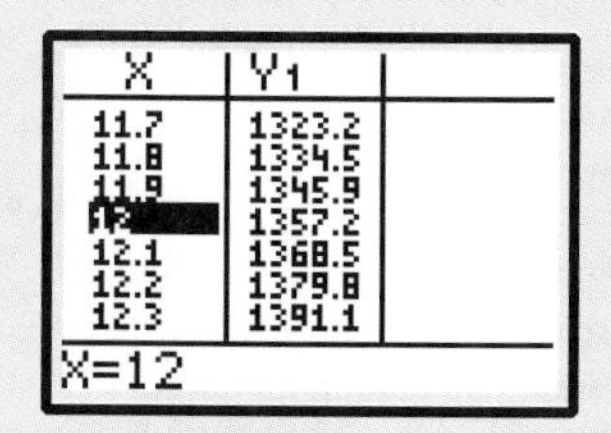

X	Y1
11.7	1323.2
11.8	1334.5
11.9	1345.9
12	1357.2
12.1	1368.5
12.2	1379.8
12.3	1391.1

X=12

Use a graphing calculator to make a table for the function $V(h) = 36\pi h$ where h is the height of the can. Substitute X for h, and enter $Y = 36\pi X$ into your graphing calculator.

[Y=] **(Y1=) 36** [2nd] [^] (π) [X,T,θ,n] [2nd] [WINDOW] (TBLSET) **(TblStart=) 0** [ENTER]

⇑ TI-82: TblMin=

(ΔTbl=) .1 [ENTER] **(Indpnt:)** [AUTO] [▼] **(Depend:)** [AUTO] [2nd] [GRAPH] (TABLE)

Press the down arrow key until the volume of 1357.2 occurs in the table.

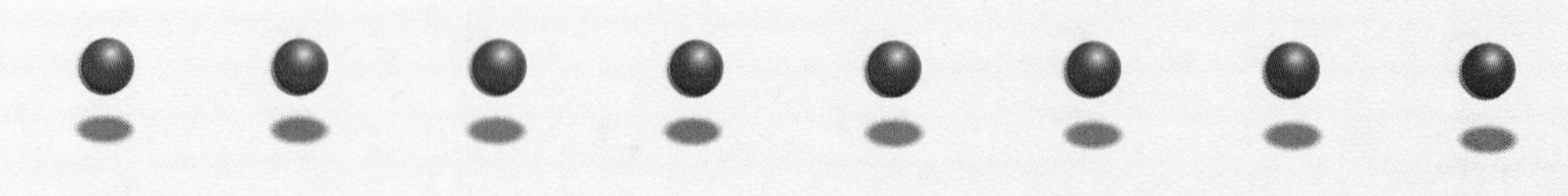

Page 319

Build a table of values for the equations. Enter $Y_1 = (X + 2)^3$ and $Y_2 = X^3 + 6X^2 + 12X + 8$, and compare the y-values in the table.

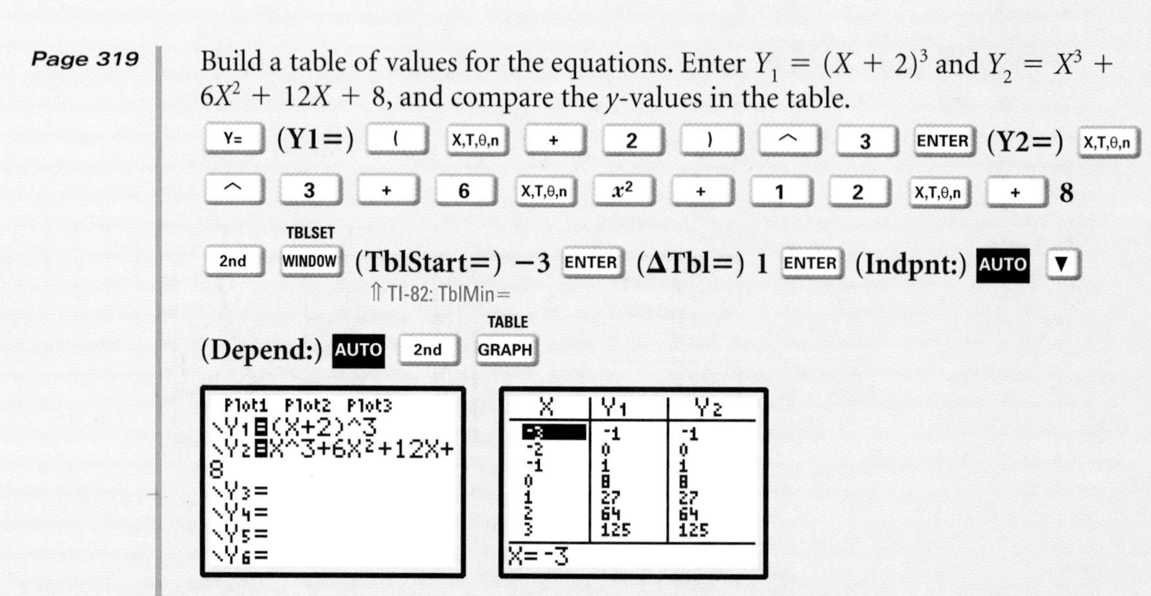

LESSON 7.3

Page 331

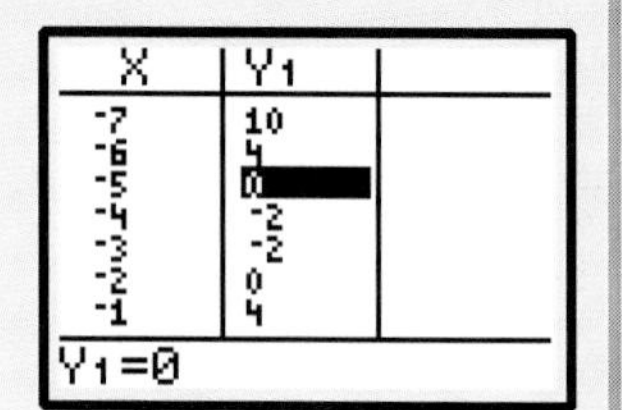

Use your graphing calculator to find the zeroes of the function $y = (x + 5)(x + 2)$. Build a table of values for the function, and look for the x-values that give a y-value of zero.

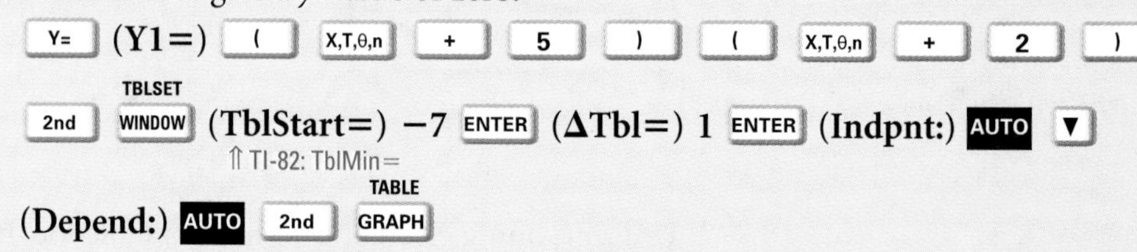

Y= (Y1=) (X,T,θ,n + 5) (X,T,θ,n + 2)
TBLSET
2nd WINDOW (TblStart=) −7 ENTER (ΔTbl=) 1 ENTER (Indpnt:) AUTO ▼
⇑ TI-82: TblMin=
TABLE
(Depend:) AUTO 2nd GRAPH

EXAMPLE 5

Page 348

Graph $y = x^2 - 10x + 25$. Use viewing window [0, 10] by [−1, 5] with scales of 1. Explain how the graph of a perfect-square trinomial can give you the factors of the expression. Use your graphing calculator to factor $x^2 - 16x + 64$.

Y= (Y1=) X,T,θ,n x^2 − 10 X,T,θ,n + 25 GRAPH

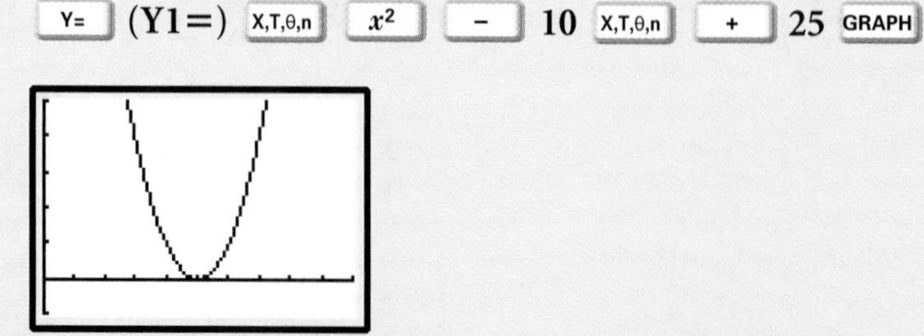

LESSON 8.2

Page 369

Graph the function $h(t) = -16t^2 + 185$. Find the point where $y \approx 0$. Use viewing window $[-5, 5]$ by $[-60, 186]$ with scales of 1.

Y= (Y1=) −16 X,T,θ,n x^2 + 185 ENTER (Y2=) 0 GRAPH 2nd

CALC
TRACE 5:intersect ENTER (First curve?) ENTER (Second curve?) ENTER ►

(Guess?) ENTER

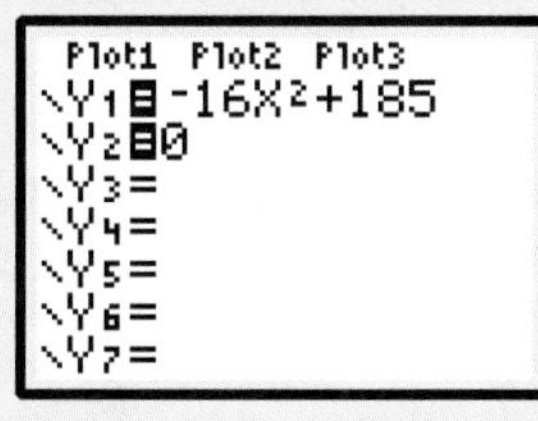

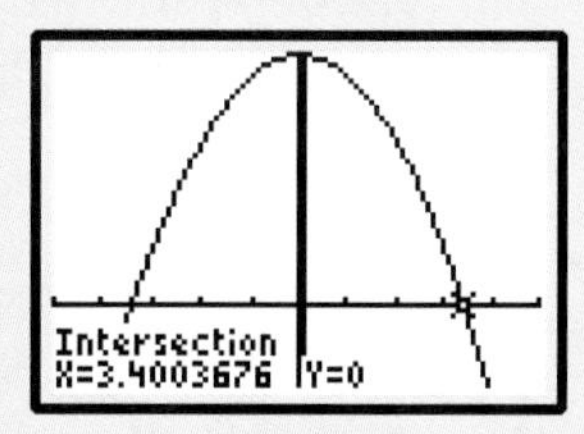

Use a similar keystroke sequence to find the other point of intersection.

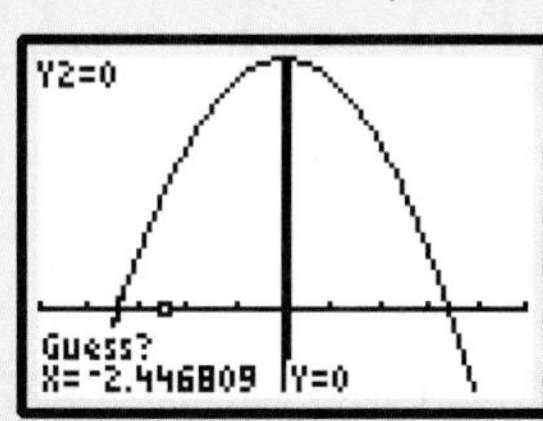

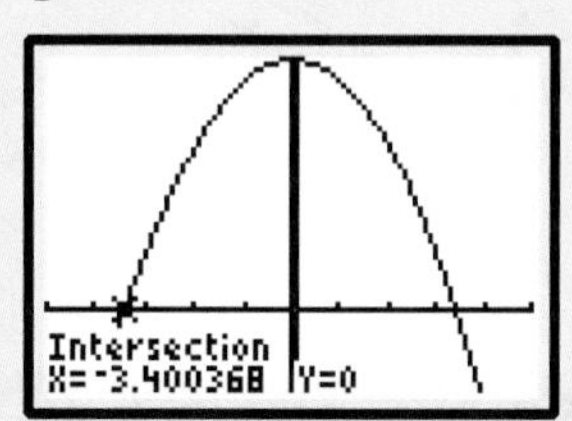

LESSON 8.4

EXAMPLE 3

Page 384

Graph $y = x^2 + 3x - 15$ and $y = -5$ on the same calculator display and find the point(s) of intersection. Use viewing window $[-16, 16]$ by $[-20, 20]$.

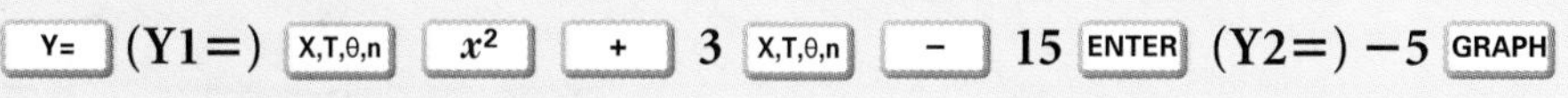

Y= (Y1=) X,T,θ,n x^2 + 3 X,T,θ,n − 15 ENTER (Y2=) −5 GRAPH

CALC
2nd TRACE 5:intersect ENTER (First curve?) ENTER (Second curve?)

ENTER ► (Guess?) ENTER

Use a similar keystroke sequence to find the other point of intersection.

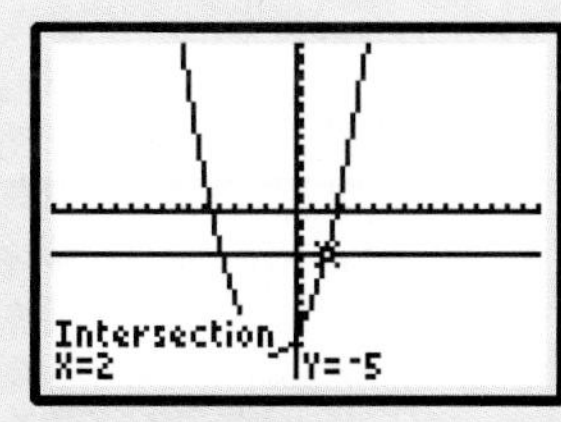

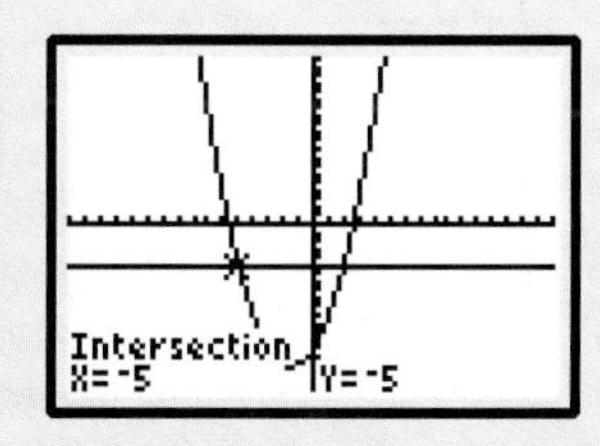

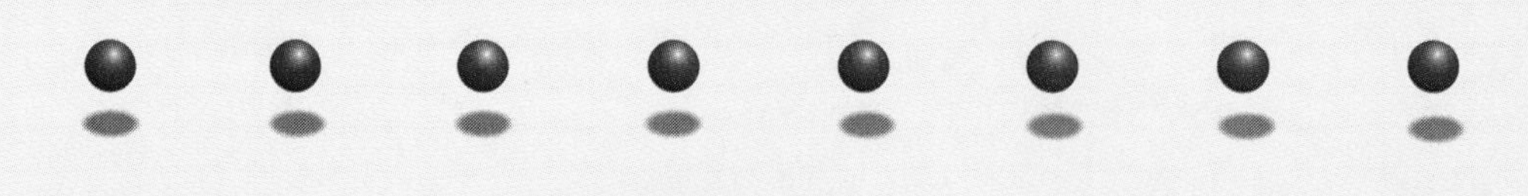

EXAMPLE 5
Page 386

Find the points where the graph of $y = x - 3$ intersects the graph of $y = x^2 - 10x + 21$. Use viewing window $[-16, 16]$ by $[-20, 20]$ with scales of 1.

Y= (Y1=) X,T,θ,n x^2 − 10 X,T,θ,n + 21 ENTER (Y2=) X,T,θ,n − 3 GRAPH 2nd TRACE (CALC) ▼ 5:intersect ENTER (First curve?) ENTER (Second curve?) ENTER Move the cursor to the left side of the calculator screen near the point of intersection ◄ (Guess?) ENTER

Use a similar keystroke sequence to find the other point of intersection.

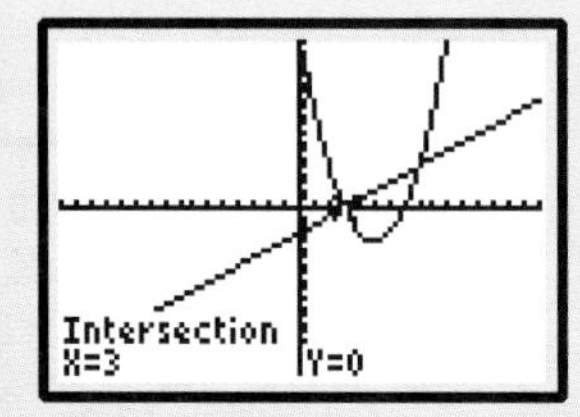

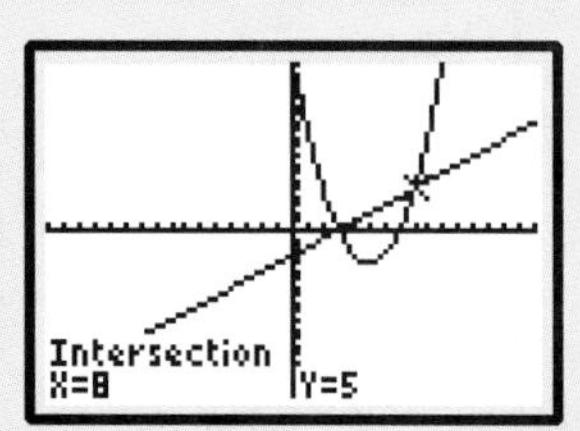

LESSON 9.3

EXAMPLE 2
Page 429

Enter the left side of the equation in the graphing calculator as $Y_1 = \sqrt{x + 2}$. Then enter the right side of the equation as $Y_2 = 3$. Find the point(s) of intersection. Use the viewing window $[-10, 10]$ by $[-5, 5]$ with scales of 1.

On the TI-82 the right parentheses is not automatically placed on the screen. It must be entered.

Y= (Y1=) 2nd x^2 (√) X,T,θ,n + 2 ENTER (Y2=) 3 GRAPH 2nd TRACE (CALC) 5:intersect ENTER (First curve?) ENTER (Second curve?) ENTER (Guess?) ENTER

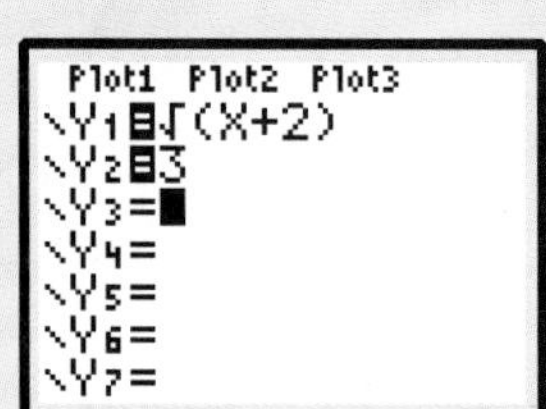

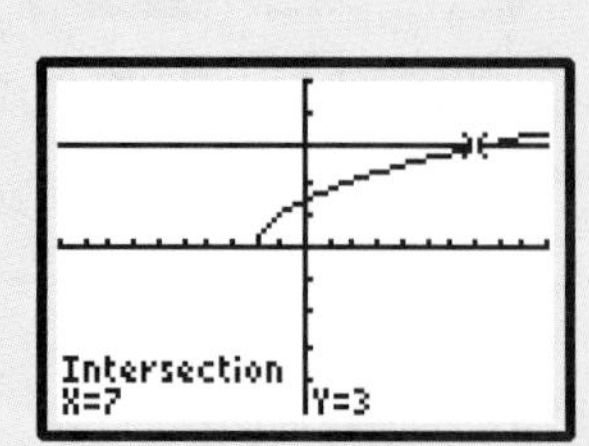

LESSON 10.1

EXAMPLE 2
Page 477

Graph $y = \frac{55 + 4.50x}{x}$ and $y = 5.60$ on the same calculator display and find the points of intersection. Use viewing window $[-2, 75]$ by $[-2, 10]$ with scales of 1.

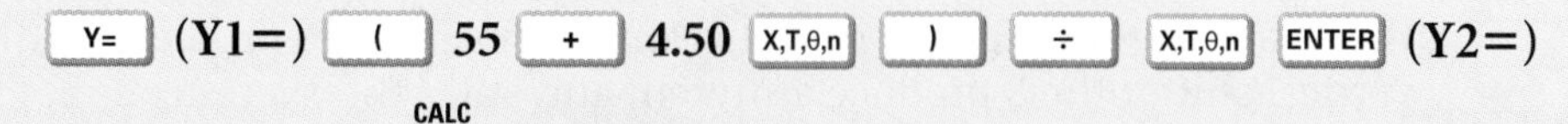

[Y=] (Y1=) [(] 55 [+] 4.50 [X,T,θ,n] [)] [÷] [X,T,θ,n] [ENTER] (Y2=)

5.6 [GRAPH] [2nd] [TRACE] (CALC) [5:intersect] [ENTER] (First curve?) [ENTER] (Second curve?)

[ENTER] (Guess?) [ENTER]

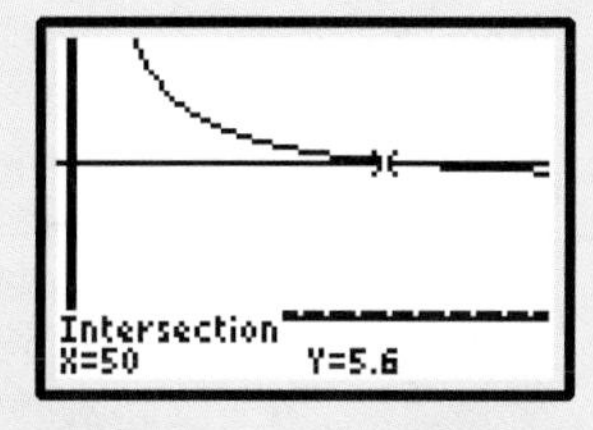

LESSON 10.5

EXAMPLE 1
Page 501

Graph $y = \frac{6}{5x} + \frac{6}{x}$ and $y = \frac{12}{5}$ on the same calculator display and find the point(s) of intersection. Use viewing window $[-7, 8]$ by $[-8, 8]$ with scales of 1.

[Y=] (Y1=) [(] 6 [÷] [(] 5 [X,T,θ,n] [)] [)] [+] [(] 6

[÷] [X,T,θ,n] [)] [ENTER] (Y2=) 12 [÷] 5 [GRAPH] [2nd] [TRACE] (CALC) [2nd]

[5:intersect] [ENTER] (First curve?) [ENTER] (Second curve?) [ENTER] (Guess?) [ENTER]

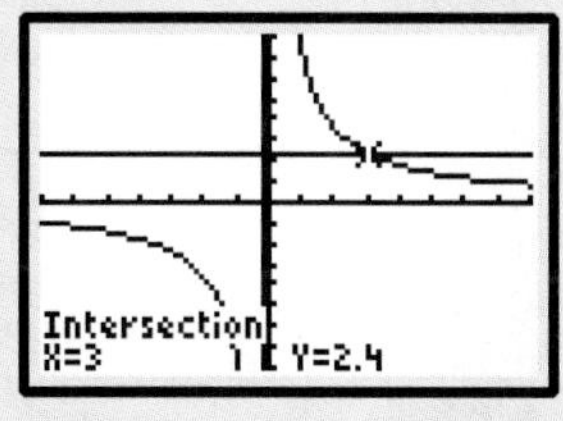

Functions and Their Graphs

Throughout the text a variety of functions were studied—linear, exponential, absolute value, quadratic, and reciprocal. The simplest form of any function is called the parent function. Each parent function has a distinctive graph. Changes made to the parent function will alter the graph but retain the distinctive features of the parent function. Some changes produce a horizontal or vertical shift, others produce a stretch, and still others a reflection. Pages 734–737 summarize some parent functions and the changes that transform them.

Linear Function

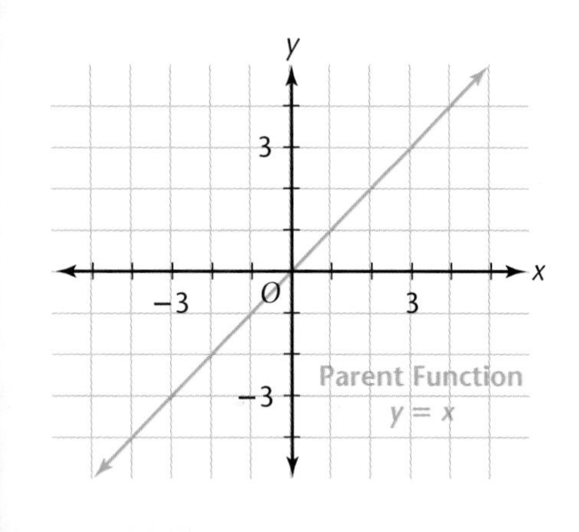

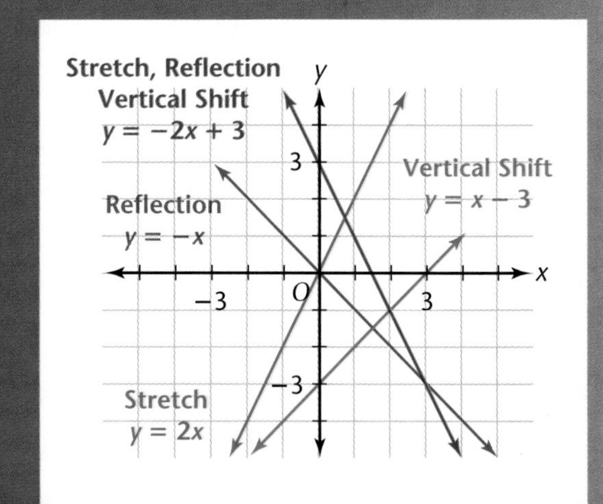

Exponential Function

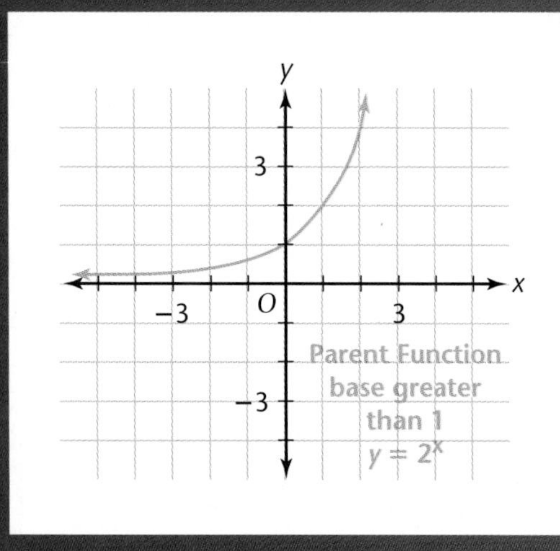

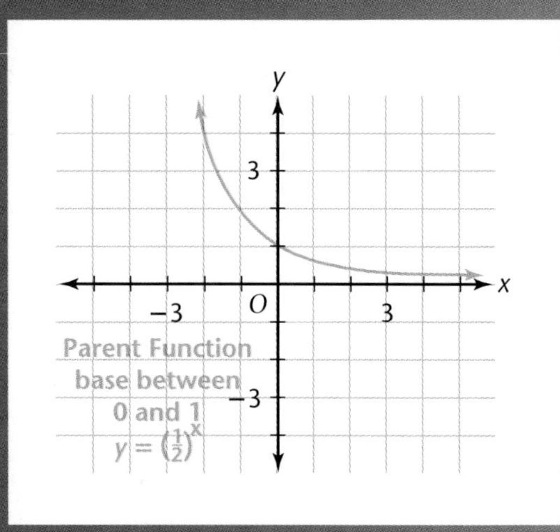

Absolute Value Function

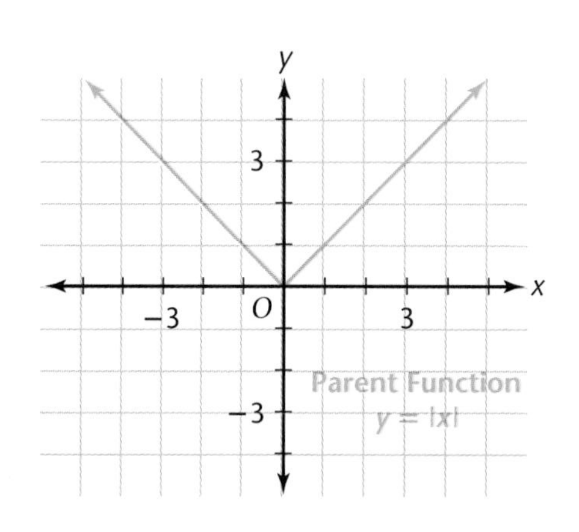

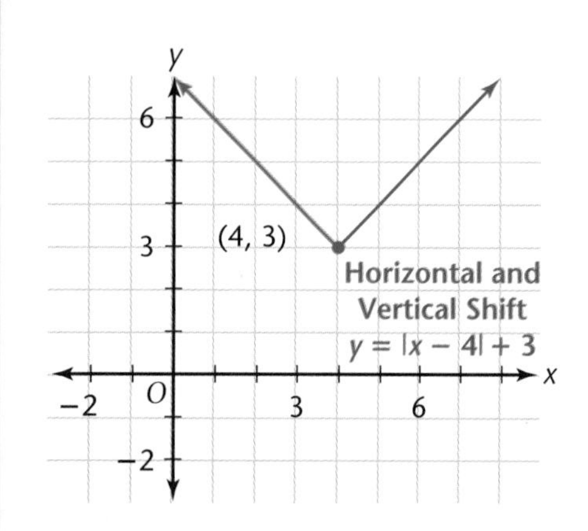

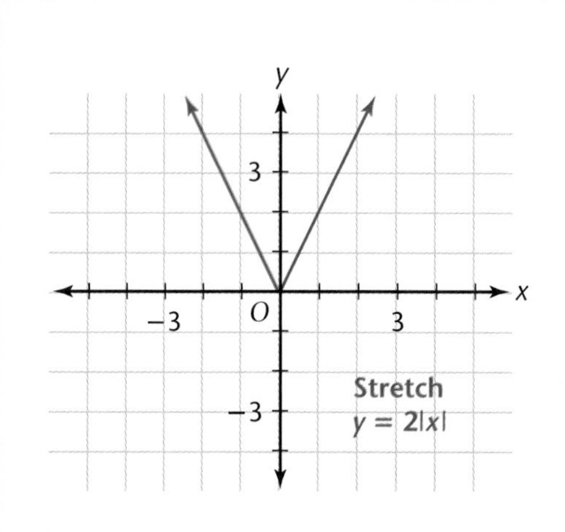

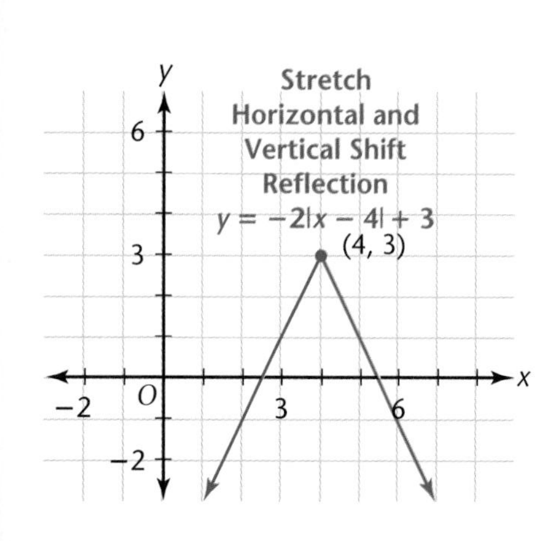

Quadratic Function

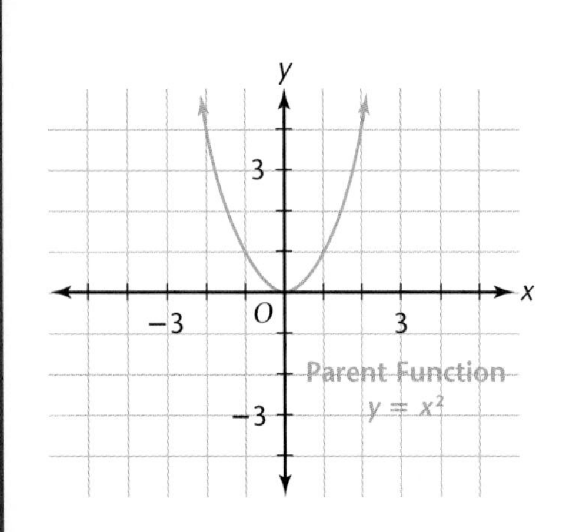

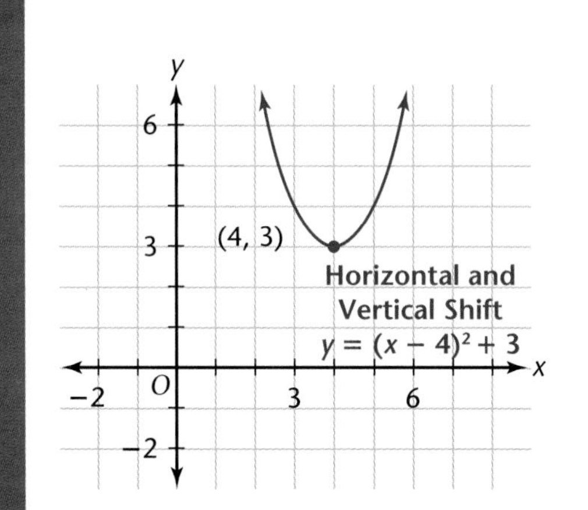

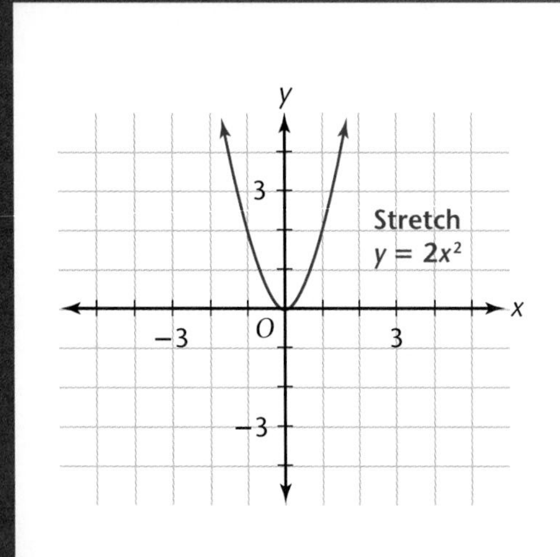

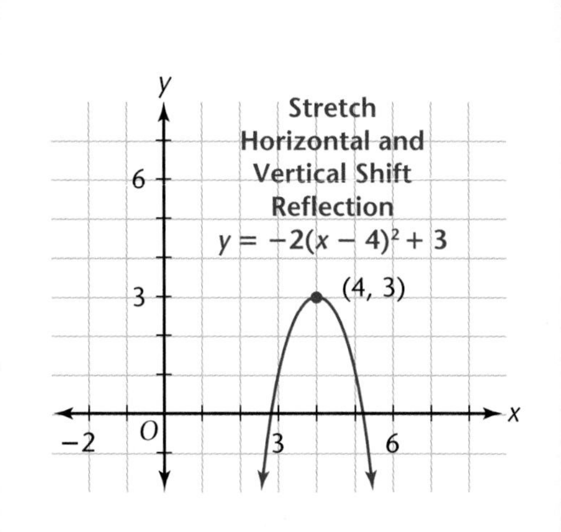

Reciprocal Function

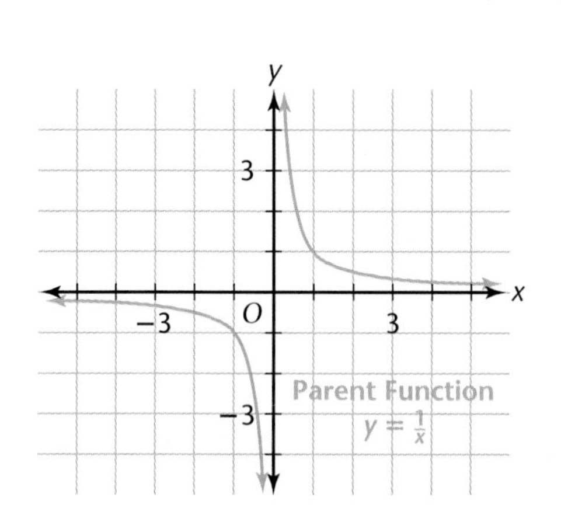

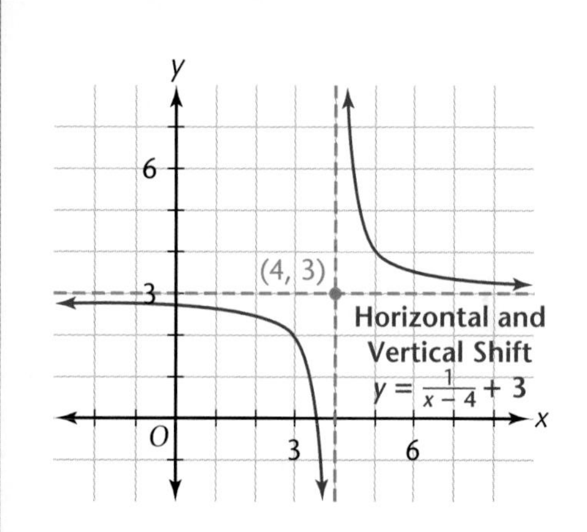

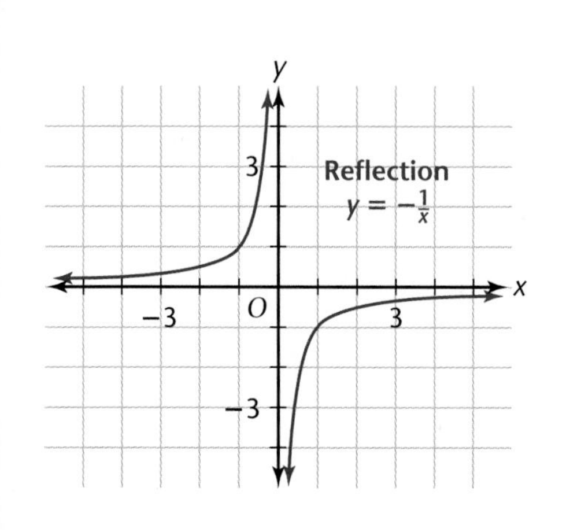

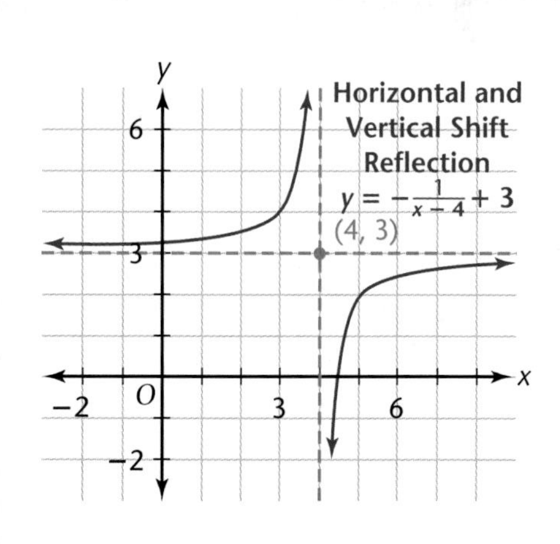

Table of Squares, Cubes, and Roots

No.	Squares	Cubes	Square Roots	Cube Roots
1	1	1	1.000	1.000
2	4	8	1.414	1.260
3	9	27	1.732	1.442
4	16	64	2.000	1.587
5	25	125	2.236	1.710
6	36	216	2.449	1.817
7	49	343	2.646	1.913
8	64	512	2.828	2.000
9	81	729	3.000	2.080
10	100	1,000	3.162	2.154
11	121	1,331	3.317	2.224
12	144	1,728	3.464	2.289
13	169	2,197	3.606	2.351
14	196	2,744	3.742	2.410
15	225	3,375	3.873	2.466
16	256	4,096	4.000	2.520
17	289	4,913	4.123	2.571
18	324	5,832	4.243	2.621
19	361	6,859	4.359	2.668
20	400	8,000	4.472	2.714
21	441	9,261	4.583	2.759
22	484	10,648	4.690	2.802
23	529	12,167	4.796	2.844
24	576	13,824	4.899	2.884
25	625	15,625	5.000	2.924
26	676	17,576	5.099	2.962
27	729	19,683	5.196	3.000
28	784	21,952	5.292	3.037
29	841	24,389	5.385	3.072
30	900	27,000	5.477	3.107
31	961	29,791	5.568	3.141
32	1,024	32,768	5.657	3.175
33	1,089	35,937	5.745	3.208
34	1,156	39,304	5.831	3.240
35	1,225	42,875	5.916	3.271
36	1,296	46,656	6.000	3.302
37	1,369	50,653	6.083	3.332
38	1,444	54,872	6.164	3.362
39	1,521	59,319	6.245	3.391
40	1,600	64,000	6.325	3.420
41	1,681	68,921	6.403	3.448
42	1,764	74,088	6.481	3.476
43	1,849	79,507	6.557	3.503
44	1,936	85,184	6.633	3.530
45	2,025	91,125	6.708	3.557
46	2,116	97,336	6.782	3.583
47	2,209	103,823	6.856	3.609
48	2.304	110,592	6.928	3.634
49	2,401	117,649	7.000	3.659
50	2,500	125,000	7.071	3.684
51	2,601	132,651	7.141	3.708
52	2,704	140,608	7.211	3.733
53	2,809	148,877	7.280	3.756
54	2,916	157,464	7.348	3.780
55	3,025	166,375	7.416	3.803
56	3,136	175,616	7.483	3.826
57	3,249	185,193	7.550	3.849
58	3,364	195,112	7.616	3.871
59	3,481	205,379	7.681	3.893
60	3,600	216,000	7.746	3.915
61	3,721	226,981	7.810	3.936
62	3,844	238,328	7.874	3.958
63	3,969	250,047	7.937	3.979
64	4,096	262,144	8.000	4.000
65	4,225	274,625	8.062	4.021
66	4,356	287,496	8.124	4.041
67	4,489	300,763	8.185	4.062
68	4,624	314,432	8.246	4.082
69	4,761	328,509	8.307	4.102
70	4,900	343,000	8.367	4.121
71	5,041	357,911	8.426	4.141
72	5,184	373,248	8.485	4.160
73	5,329	389,017	8.544	4.179
74	5,476	405,224	8.602	4.198
75	5,625	421,875	8.660	4.217
76	5,776	438,976	8.718	4.236
77	5,929	456,533	8.775	4.254
78	6,084	474,552	8.832	4.273
79	6,241	493,039	8.888	4.291
80	6,400	512,000	8.944	4.309
81	6,561	531,441	9.000	4.327
82	6,724	551,368	9.055	4.344
83	6,889	571,787	9.110	4.362
84	7,056	592,704	9.165	4.380
85	7,225	614,125	9.220	4.397
86	7,396	636,056	9.274	4.414
87	7,569	658,503	9.327	4.431
88	7,744	681,472	9.381	4.448
89	7,921	704,969	9.434	4.465
90	8,100	729,000	9.487	4.481
91	8,281	753,571	9.539	4.498
92	8,464	778,688	9.592	4.514
93	8,649	804,357	9.644	4.531
94	8,836	830,584	9.695	4.547
95	9,025	857,375	9.747	4.563
96	9,216	884,736	9.798	4.579
97	9,409	912,673	9.849	4.595
98	9,604	941,192	9.899	4.610
99	9,801	970,299	9.950	4.626
100	10,000	1,000,000	10.000	4.642

Table of Random Digits

Line / Col	(1)	(2)	(3)	(4)	(5)	(6)	(7)	(8)	(9)	(10)	(11)	(12)	(13)	(14)
1	10480	15011	01536	02011	81647	91646	69179	14194	62590	36207	20969	99570	91291	90700
2	22368	46573	25595	85393	30995	89198	27982	53402	93965	34095	52666	19174	39615	99505
3	24130	48360	22527	97265	76393	64809	15179	24830	49340	32081	30680	19655	63348	58629
4	42167	93093	06243	61680	07856	16376	39440	53537	71341	57004	00849	74917	97758	16379
5	31570	39975	81837	16656	06121	91782	60468	81305	49684	60672	14110	06927	01263	54613
6	77921	06907	11008	42751	27756	53498	18602	70659	90655	15053	21916	81825	44394	42880
7	99562	72905	56420	69994	98872	31016	71194	18738	44013	48840	63213	21069	10634	12952
8	96301	91977	05463	07972	18876	20922	94595	56869	69014	60045	18425	84903	42508	32307
9	89579	14342	63661	10281	17453	18103	57740	84378	25331	12566	58678	44947	05585	56941
10	85475	36857	53342	53988	53060	59533	38867	62300	08158	17983	16439	11458	18593	64952
11	28918	69578	88231	33276	70997	79936	56865	05859	90106	31595	01547	85590	91610	78188
12	63553	40961	48235	03427	49626	69445	18663	72695	52180	20847	12234	90511	33703	90322
13	09429	93969	52636	92737	88974	33488	36320	17617	30015	08272	84115	27156	30613	74952
14	10365	61129	87529	85689	48237	52267	67689	93394	01511	26358	85104	20285	29975	89868
15	07119	97336	71048	08178	77233	13916	47564	81056	97735	85977	29372	74461	28551	90707
16	51085	12765	51821	51259	77452	16308	60756	92144	49442	53900	70960	63990	75601	40719
17	02368	21382	52404	60268	89368	19885	55322	44819	01188	65225	64835	44919	05944	55157
18	01011	54092	33362	94904	31273	04146	18594	29852	71585	85030	51132	01915	92747	64951
19	52162	53916	46369	58586	23216	14513	83149	98736	23495	64350	94738	17752	35156	35749
20	07056	97628	33787	09998	42698	06691	76988	13602	51851	46104	88916	19509	25625	58104
21	48663	91245	85828	14346	09172	30168	90229	04734	59193	22178	30421	61666	99904	32812
22	54164	58492	22421	74103	47070	25306	76468	26384	58151	06646	21524	15227	96909	44592
23	32639	32363	05597	24200	13363	38005	94342	28728	35806	06912	17012	64161	18296	22851
24	29334	27001	87637	87308	58731	00256	45834	15398	46557	41135	10367	07684	36188	18510
25	02488	33062	28834	07351	19731	92420	60952	61280	50001	67658	32586	86679	50720	94953
26	81525	72295	04839	96423	24878	82651	66566	14778	76797	14780	13300	87074	79666	95725
27	29676	20591	68086	26432	46901	20849	89768	81536	86645	12659	92259	57102	80428	25280
28	00742	57392	39064	66432	84673	40027	32832	61362	98947	96067	64760	64584	96096	98253
29	05366	04213	25669	26422	44407	44048	37937	63904	45766	66134	75470	66520	34693	90449
30	91921	26418	64117	94305	26766	25940	39972	22209	71500	64568	91402	42416	07844	69618
31	00582	04711	87917	77341	42206	35126	74087	99547	81817	42607	43808	76655	62028	76630
32	00725	69884	62797	56170	86324	88072	76222	36086	84637	93161	76038	65855	77919	88006
33	69011	65795	95876	55293	18988	27354	26575	08625	40801	59920	29841	80150	12777	48501
34	25976	57948	29888	88604	67917	48708	18912	82271	65424	69774	33611	54262	85963	03547
35	09763	83473	73577	12908	30883	18317	28290	35797	05998	41688	34952	37888	38917	88050
36	91567	42595	27958	30134	04024	86385	29880	99730	55536	84855	29080	09250	79656	73211
37	17955	56349	90999	49127	20044	59931	06115	20542	18059	02008	73708	83517	36103	42791
38	46503	18584	18845	49618	02304	51038	20655	58727	28168	15475	56942	53389	20562	87338
39	92157	89634	94824	78171	84610	82834	09922	25417	44137	48413	25555	21246	35509	20468
40	14577	62765	35605	81263	39667	47358	56873	56307	61607	49518	89656	20103	77490	18062
41	98427	07523	33362	64270	01638	92477	66969	98420	04880	45585	46565	04102	46880	45709
42	34914	63976	88720	82765	34476	17032	87589	40836	32427	70002	70663	88863	77775	69348
43	70060	28277	39475	46473	23219	53416	94970	25832	69975	94884	19661	72828	00102	66794
44	53976	54914	06990	67245	68350	82948	11398	42878	80287	88267	47363	46634	06541	97809
45	76072	29515	40980	07391	58745	25774	22987	80059	39911	96189	41151	14222	60697	59583
46	90725	52210	83974	29992	65831	38857	50490	83765	55657	14361	31720	57375	56228	41546
47	64364	67412	33339	31926	14883	24413	59744	92351	97473	89286	35931	04110	23726	51900
48	08962	00358	31662	25388	61642	34072	81249	35648	56891	69352	48373	45578	78547	81788
49	95012	68379	93526	70765	10592	04542	76463	54328	02349	17247	28865	14777	62730	92277
50	15664	10493	20492	38391	91132	21999	59516	81652	27195	48223	46751	22923	32261	85653

Source: Interstate Commerce Commission

GLOSSARY

absolute value For any number x, if x is a positive integer or zero, the the absolute value of x is x, or $|x| = x$; and if x is a negative integer, the the absolute value of x is the opposite of x, or $|x| = -x$. (42)

Addition Principle of Counting Suppose there are m ways to make a first choice, n ways to make a second choice, and t ways that have been counted twice. Then there are $m + n - t$ ways to make the first choice OR the second choice. (188)

Addition Property of Equality If equal amounts are added to the expressions on each side of an equation, the expressions remain equal. (12)

Addition Property of Inequality If equal amounts are added to the expressions on each side of an inequality, the resulting inequality is still true. (36)

address Each entry in a matrix can be located by its matrix address. (123)

approximate solution A reasonable estimate for a point of intersection for a system of equations. (76)

area The number of nonoverlapping unit squares that will cover the interior of a figure. (232)

ascending order The order of the terms of a polynomial when they are ordered from left to right, from the least to the greatest degree of the variable. (323)

Associative Property for Addition For all numbers a, b, and c, $(a + b) + c = a + (b + c)$. (132)

asymptotes The lines that a hyperbola will approach, but not touch. (237)

axis of symmetry The line along which a figure can be folded so that the two halves of the figure match exactly. (228, 364)

base The number raised to an exponent. In an expression of the form x^a, x is the base. (266)

binomial The sum or difference of two monomials. (323)

boundary line A line that divides a coordinate plane into two half-planes. (96)

coefficient matrix A matrix containing the coefficients of a system of equations. (152)

common binomial factor A binomial factor that is common to all terms of a polynomial. (342)

common monomial factor A monomial factor that is common to all terms of a polynomial. (341)

common solution A solution that is the same for two or more equations. (73)

Commutative Property for Addition For any numbers a and b, $a + b = b + a$. (132)

complement The number of ways that an event does not occur. (207)

completing the square A method for solving quadratic equations. (374)

composite number A whole number greater than 1 that is not a prime number. (341)

conclusion In a theorem written in if-then form, the conclusion is the *then* part, and must be proven to be true. (511)

consistent system A system of equations that has one or more solutions. (91)

constant A number that represents a fixed amount. (5)

constant matrix A matrix containing the constant terms of a system of equations. (152)

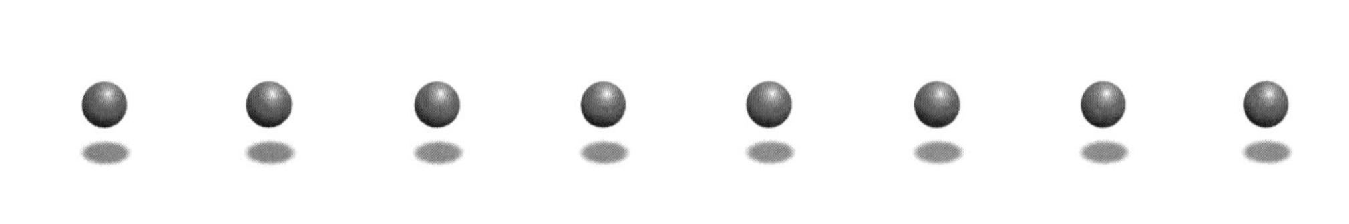

converse "If q, then p" is the converse of "if p then q." (443, 511)

converse of the "Pythagorean" Right-Triangle Theorem If a, b, and c are the lengths of the sides of a triangle with $a^2 + b^2 = c^2$, $c > a$ and $c > b$, then the triangle is a right triangle. (443)

correlation An indication of the proportion of the relationship between two sets of data. (619)

correlation coefficient The measure of how closely a set of data falls along a line. (69, 179)

cosine function A ratio based on a right triangle such that the cosine of an angle is the ratio of the length of the leg adjacent to the angle to the length of the hypotenuse. (460)

cross products In $\frac{a}{b} = \frac{c}{d}$, ad and bc are called the cross products. (507)

degree of a polynomial The degree of a polynomial in one variable is the exponent with the greatest value of any of the polynomial's terms. (323)

dependent system A system of equations with an infinite number of solutions. The graphs of the two equations are the same line. This also called a consistent system. (91)

dependent variable A variable whose value depends on the value of another variable. (5, 223)

descending order The order of the terms of a polynomial when they are ordered from left to right, from the greatest to the least degree of the variable. (323)

difference of two squares A polynomial of the form $a^2 - b^2$ that can be written as the product of two factors, $a^2 - b^2 = (a - b)(a + b)$. (347)

direct variation If y varies directly as x, then $\frac{y}{k} = k$ or $y = kx$. The k is called the constant of variation. (60)

discriminant The expression $b^2 - 4ac$ in the quadratic formula. (393)

disjoint Sets with no members in common. (186)

distance formula The distance, d, between two points $A(x_1, y_1)$ and $B(x_2, y_2)$ is given by $d = \sqrt{(x_2 - x_1)^2 + (y_2 - y_1)^2}$. (442)

Distributive Property For all numbers a, b, and c, $a(b + c) = ab + ac$ and $(b + c)a = ba + ca$. (30)

Division Property of Equality If the expressions on each side of an equation are divided by equal nonzero amounts, the expressions remain equal. (17)

Division Property of Inequality If the expressions on each side of an inequality are divided by the same positive number, the resulting inequality is still true. If the expressions on each side of an inequality are divided by the same negative number and the inequality sign is reversed, the resulting inequality is still true. (36)

Division Property of Square Roots For all numbers $a \geq 0$ and $b > 0$, $\sqrt{\frac{a}{b}} = \frac{\sqrt{a}}{\sqrt{b}}$. (422)

domain The set of first coordinates, or x-values, in an ordered pair of a function. These values are represented on the horizontal axis in a coordinate plane. (7, 223)

elimination method A method used to solve a system of equations in which one variable is eliminated by adding opposites. (84)

entry The information that appears in each position of a matrix. (123)

equally likely outcomes Outcomes of an

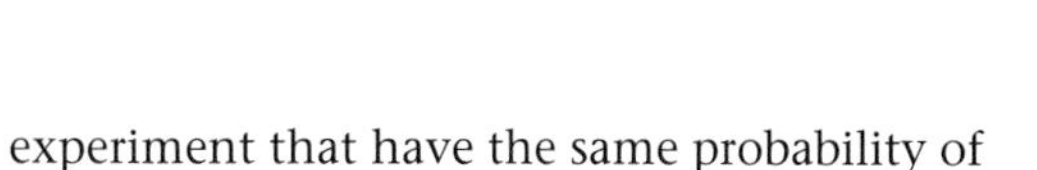

experiment that have the same probability of happening. (199)

equation Two equivalent expressions separated by an equal sign. (2)

equation method A method for solving problems involving percent in which the parts of the problem are translated into an algebraic equation. (366)

equivalent ratios Ratios that have the same value. (19)

even numbers An even number is of the form $2n$, where n is an integer. (512)

experimental probability Let t be the number of trials in an experiment. Let f be the number of times that a successful event occurs. The experimental probability, P, of the event is given by $P = \frac{f}{t}$. (167)

exponent The number that tells how many times a number is used as a factor. In an expression of the form x^a, a is the exponent. (267)

exponential decay A situation in which a number is repeatedly multiplied by a number between 0 and 1. (302)

exponential growth A situation in which a number is repeatedly multiplied by a number greater than 1. (302)

extremes In the proportion $\frac{a}{b} = \frac{c}{d}$, a and d are the extremes. (506)

factor Numbers or polynomials that are multiplied to form a product. (341)

FOIL method A method for multiplying two binomials.

1. Multiply **F**irst terms.
2. Multiply **O**utside terms.
3. Multiply **I**nside terms. Add products from 2 and 3.
4. Multiply **L**ast terms. (336)

frequency table A representation used to summarize statistical data. (180)

function A set of ordered pairs for which no two pairs have the same first coordinate. (5, 221)

general growth formula Let P be the amount after t years growing at a yearly rate of growth, r, expressed as a decimal. If the original amount is A, then $P = A(1 + r)^t$. (295)

greatest common factor (GCF) The greatest factor that is common to all terms of a polynomial. (341)

horizontal translation The graph of $y = f(x - h)$ is translated horizontally h units from the graph of $y = f(x)$. (248)

hyperbola The graph of the reciprocal function. (237)

hypotenuse The side opposite the right angle in a right triangle. (436)

hypothesis In an if-then form of a theorem, the hypothesis is the *if* part, and is assumed to be true. (511)

identity An equation in which the expressions on both sides of the equal sign are equivalent expressions. (319)

identity matrix for addition The sum of a given matrix and the identity matrix is the given matrix. The identity matrix for addition is the zero matrix. (131)

identity matrix for multiplication The product of a given matrix and the identity matrix is the given matrix. (145)

inconsistent system A system of equations with no solution. The graphs of the equations in an inconsistent system are parallel lines. (91)

independent events Two events are independent if the occurrence of the first event does not affect the probability of the second event occurring. (206)

independent system A system of equations that has one solution. The graphs of the equations are lines that intersect at one point. This is also called a consistent system. (92)

independent variable A variable whose value does not depend on the value of another variable. (5, 223)

intersection The data that is the same in different sets. The word AND is used to denote an intersection. (186)

inverse matrix The matrix A has an inverse matrix, A^{-1}, if $AA^{-1} = I = A^{-1}A$, where I is the identity matrix for multiplication. (145)

inverse variation If y varies inversely as x, then $xy = k$, or $y = \frac{k}{x}$, where k is the constant of variation, $k \neq 0$, $x \neq 0$. (480)

irrational numbers Numbers whose decimal part never terminates or repeats. Irrational numbers cannot be represented by the ratio of two integers. (413)

legs of a right triangle The sides that form the right angle of a right triangle. (435)

line of best fit Represents an approximation of the data in a scatter plot. (68, 179)

linear inequality An inequality with a boundary line that can be expressed in the form $y = mx + b$. The solution to a linear inequality is the set of all ordered pairs that make the inequality true. (478)

matrix Data arranged in a table of rows and columns and enclosed by brackets [ ]. The plural of *matrix* is *matrices*. (123)

matrix equality Two matrices are equal when their dimensions are the same and their corresponding entries are equal. (125)

matrix equation An equation for a system of linear equations of the form $AX = B$, where A is the coefficient matrix, X is the variable matrix, and B is the constant matrix. (152)

maximum value The y-value of the vertex of a parabola that opens down. (376)

mean A measure of central tendency in which all pieces of data are added and the sum is divided by the number of pieces of data. (178)

means In the proportion $\frac{a}{b} = \frac{c}{d}$, b and c are the means. (506)

median The middle number in a set of data arranged in order. If there are two middle numbers, the median is the mean of the two numbers. (179)

midpoint formula The midpoint, M, of PQ where $P(x_1, y_1)$ and $Q(x_2, y_2)$ is $M = \left(\frac{x_1 + x_2}{2}, \frac{y_1 + y_2}{2}\right)$. (444)

minimum value The y-value of the vertex of a parabola that opens up. (376)

mode The piece of data that occurs most often. There can be more than one mode for a set of data. (179)

monomial An algebraic expression that is either a constant, a variable, or a product of a constant and one or more variables. (274)

Multiplication Principle of Counting If there are m ways to make a first choice and n ways to make a second choice, then there are $m \cdot n$ ways to make a first choice AND a second choice. (193)

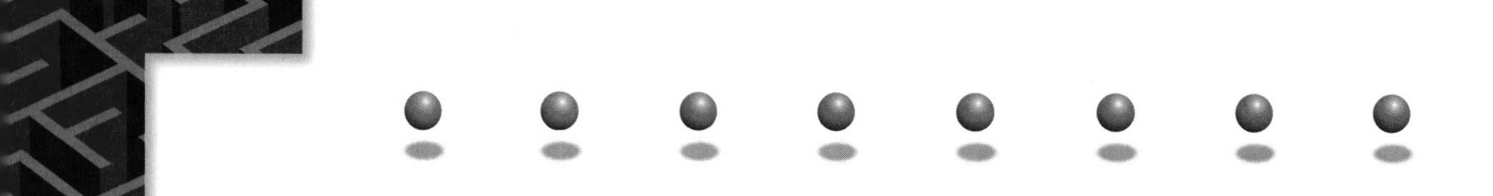

Multiplication Property of Equality If the expressions on each side of an equation are multiplied by equal nonzero amounts, the expressions remain equal. (17)

Multiplication Property of Inequality If the expressions on each side of an inequality are multiplied by the same positive number, the resulting inequality is still true. If the expressions on each side of an inequality are multiplied by the same negative number and the inequality sign is reversed, the resulting inequality is still true. (36)

Multiplication Property of Square Roots For all numbers a and b, such that a and $b \geq 0$, $\sqrt{ab} = \sqrt{a}\sqrt{b}$. (422)

negative exponent If x is any number except zero and n is a positive integer, then $x^{-n} = \frac{1}{x^n}$. (280)

neutral pair A pair of algebra tiles with opposite signs. (12)

odd numbers An odd number is of the form $2n + 1$, where n is an integer. (512)

outcome The result of an experiment. (198)

parabola The graph of a quadratic function. (364)

parent function The most basic of a family of functions. (228)

perfect square A number whose square root is a positive integer. (412)

perfect-square trinomial A trinomial of the form $a^2 + 2ab + b^2$ or $a^2 - 2ab + b^2$. The factored form of $a^2 + 2ab + b^2$ is $(a + b)(a + b) = (a + b)^2$, and of $a^2 - 2ab + b^2$ is $(a - b)(a - b) = (a - b)^2$. (345)

perimeter The distance around a geometric figure that is contained in a plane. (232)

point-slope form The form $y - y_1 = m(x_2 - x_1)$ is the point-slope form for the equation of a line. The coordinates x_1 and y_1 are taken from a given point (x_1, y_1), and the slope is m. (235)

polynomial A monomial or the sum or difference of two or more monomials. (322)

polynomial function A function that consists of one or more monomials. (318)

power An expression of the form x^a, where x is the base and a is the exponent. (266)

Power-of-a-Power Property If x is any number and a and b are any positive integers, then $(x^a)^b = x^{ab}$. (274)

Power-of-a-Product Property If x and y are any numbers and n is a positive integer, then $(xy)^n = x^n y^n$. (275)

prediction The hypothesized outcome of a probability experiment. (186)

prime number A whole number greater than 1 whose only factors are itself and 1. (341)

prime polynomial A polynomial that has no polynomial factors with integral coefficients except itself and 1. (341)

principal square root The positive square root of a number that is indicated by the radical sign. (414)

probability The ratio of the number of successful outcomes to the number of possible outcomes. (166)

product matrix The result of multiplying two matrices. (139)

Product-of-Powers Property If x is any number and a and b are any positive integers, then $x^a \cdot x^b = x^{a+b}$. (273)

proof A logical justification that shows that the conculsion to an if-then statement is true. (511)

proportion An equation containing two or more equivalent ratios. (19, 506)

"Pythagorean" Right-Triangle Theorem Given a right triangle with legs of length a and b and hypotenuse of length c, $c^2 = a^2 + b^2$ or $a^2 + b^2 = c^2$. (436)

quadratic formula The solutions of the quadratic equation, $ax^2 + bx + c = 0$, are $x = \frac{-b \pm \sqrt{b^2 - 4ac}}{2a}$. (391)

quadratic inequality An inequality that contains one or more quadratic expressions. (397)

Quotient-of-Powers Property If x is any number except 0 and a and b are any positive integers, with $a > b$, then $\frac{x^a}{x^b} = x^{a-b}$. (276)

radical expression An expression that contains a square root. (421)

radical sign The sign used to denote the square root. (414)

radicand The number under a radical sign. (420)

RAND A command used by many calculators and computers to generate random numbers. (166)

range The difference between the highest and lowest numbers in a set of data. (179) The set of second coordinates, or y-values, in an ordered pair of a function. These values are represented on the vertical axis in a coordinate plane. (7, 223)

rate of change The amount of increase or decrease in the range values of a function relative to the increase or decrease in the domain values. (56, 59)

ratio The comparison of two quantities. (472)

rational expression If P and Q are polynomials and $Q \neq 0$, then an expression in the form $\frac{P}{Q}$ is a rational expression. (475)

rational function A function of the form $y = \frac{P}{Q}$, or $f(x) = \frac{P(x)}{Q(x)}$. (475)

rational numbers Numbers that can be represented by a terminating or repeating decimal. (413)

rationalizing the denominator The process of simplifying an expression so that no radical expression exists in the denominator. (423)

ratios Expressions that compare two quantities. (19)

real numbers The combination of all rational and irrational numbers. (414)

reasonable domain Meaningful x-values of a function that models a real-world situation. (7)

Reciprocal Property For any nonzero number, r, there is a number, $\frac{1}{r}$, such that $r \cdot \frac{1}{r} = 1$. (17)

reflection A transformation that flips a figure over a given line. (241)

regression line The line of best fit on a graphics calculator. (69)

relation A relation pairs elements from one set with elements of another set. (220)

rise The vertical change in a line. (56)

run The horizontal change in a line. (56)

scalar The number by which each entry in a matrix is multiplied. (138)

scalar multiplication Multiplication of a matrix by a scalar. (138)

scale factor The number by which a parent function is multiplied to create a vertical stretch. (235)

scatter plot A display of data that has been organized into ordered pairs and graphed on the coordinate plane. (179)

scientific notation A number written in scientific notation is written with two factos, a number from 1 to 10, but not including 10, and a power of 10. (286)

simplest radical form A radical expression is in simplest form if the radicand contains no perfect square factors greater than 1, contains no fractions, and if there is no radical sign in the denominator of a fraction. (421)

simulation An experiment with mathematical characteristics that are similar to the actual event. (172)

sine function A ratio based on a right triangle such that the sine of an angle is the ratio of the length of the leg opposite the angle to the length of the hypotenuse. (460)

slope Measure of the steepness of a line. (56)

slope formula Given two points with coordinates (x_1, y_1) and (x_2, y_2), the slope formula is $m = \frac{\text{change in } y}{\text{change in } x} = \frac{y_2 - y_1}{x_2 - x_1}$. (66)

slope-intercept form The slope-intercept formula or form for a line with slope m and y-intercept b is $y = mx + b$. (65)

square matrix A matrix with equal row and column dimensions. (124)

square root If k is greater than zero and $x^2 = k$, then $x = \sqrt{k}$ or $x = -\sqrt{k}$, also written $\pm\sqrt{k}$. For $k = 0$. (369, 420)

standard form of a linear equation An equation of the form $Ax + By = C$ is in standard form when A, B, and C are integers, A and B are not both zero, and A is not negative. (66)

standard form of a polynomial A polynomial is in standard form when the terms of the polynomial are ordered from left to right, from the greatest to the least degree of the variable. (323)

statistics The science of collecting, analyzing, describing, and interpreting data. (178, 584)

stretch A transformation in which the graph of a parent function is stretched by the amount of the scale factor. (234)

substitution method A method used to solve a system of equations in which variables are replaced with known values or algebraic expressions. (79)

Subtraction Property of Equality If equal amounts are subtracted from the expressions on each side of an equation, the expressions remain equal. (11)

Subtraction Property of Inequality If equal amounts are subtracted from the expressions on each side of an inequality, the resulting inequality is still true. (36)

successful event An event that produces the desired outcome. (167)

symmetric A figure is symmetric with respect to a line called the axis of symmetry if the figure can be folded along this axis to produce the left and right halves of the figure that match exactly. (228)

system of equations Two or more equations in two or more variables. (73)

system of linear inequalities Two or more inequalities in two or more variables. The

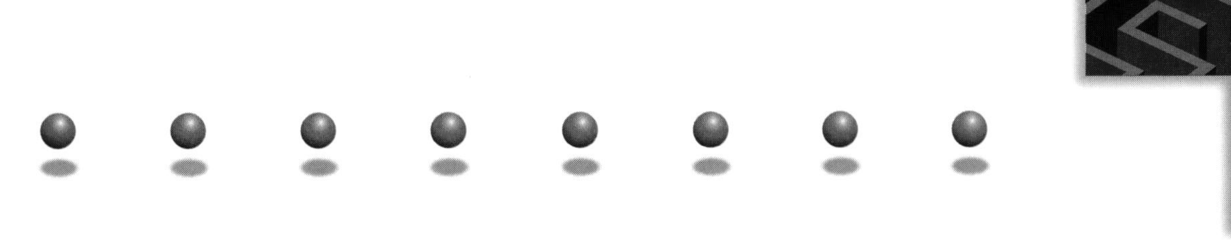

solution for a system of linear inequalities is the intersection of the solution for each inequality. (97)

tangent function In a right triangle, the ratio of the length of the leg opposite an acute acute of the triangle to the length of the leg adjacent to the acute angle. (454)

theorem A statement that has been proven. (511)

theoretical probability Let n be the number of equally likely outcomes in an event. Let s be the number of successful outcomes in the same event. Then the theoretical probability that the event will occur is $P = \frac{s}{n}$. (199)

transformation A variation such as a stretch, reflection, or shift of a parent function. (228)

translation A transformation that shifts the graph of a function horizontally or vertically. (246)

tree diagram A diagram that is used to find all the possible selections in a situation in which one choice AND other choices need to be made. (192)

trial The simulation of one event in an experiment. (167)

trinomial The sum or difference of three monomials. (323)

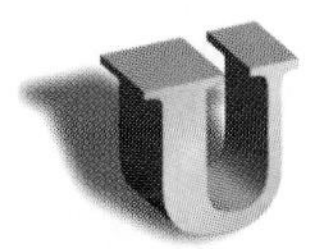

union The combined data from different sets of data. The word OR is used to denote the union of sets. (186)

variable A number that represents an amount that is not fixed. (5)

variable matrix A matrix containing the variables of a system of equations. (152)

Venn diagram A diagram used to represent the relationships among different sets of data. (186)

vertex The point where a parabola or the graph of an absolute value function changes direction. (228, 364)

vertical-line test A test used to determine whether the graph of a relation is a function. A relation is a function if any vertical line intersects the graph of the relation no more than once. (222)

vertical translation The graph of $y = f(x) + k$ is translated vertically by k units from the graph of $y = f(x)$. (247)

***y*-intercept** The point where a line crosses the y-axis. (64)

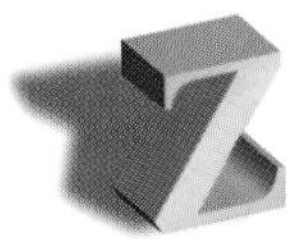

zero exponent If x is any number except zero, $x^0 = 1$. (280)

zero matrix A matrix that is filled with zeros. The zero matrix is the identity matrix for matrix addition. (131)

zero of a function The x-value of the point where a function crosses the x-axis. (365)

Zero Product Property If a and b are real numbers such that $ab = 0$, then $a = 0$ or $b = 0$. (383)

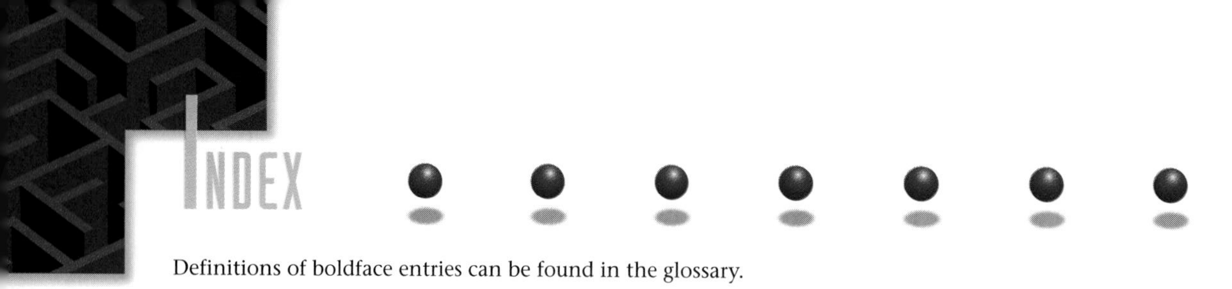

INDEX

Definitions of boldface entries can be found in the glossary.

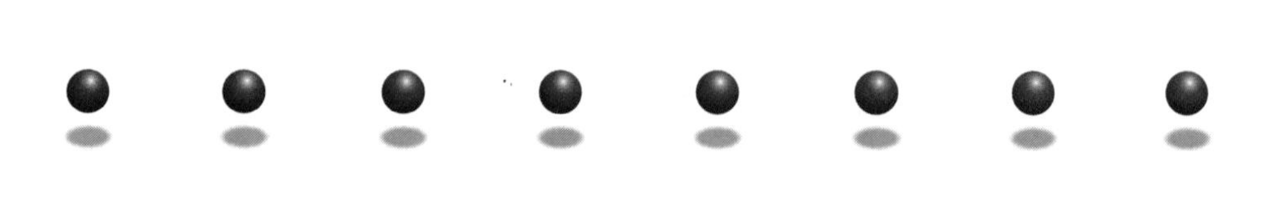

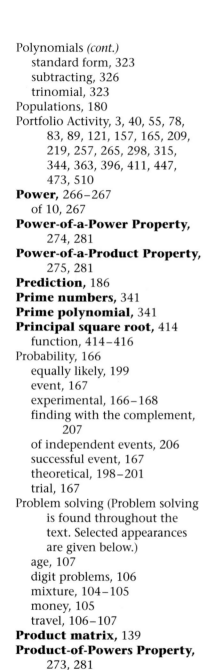

CREDITS

Photos

Abbreviations used: (t) top, (c) center, (b) bottom, (l) left, (r) right, (bckgd) background.

FRONT COVER: (l), Terry Vine/Tony Stone Images; (c), Glen Allison/Tony Stone Images; (r), Rex Ziak/Tony Stone Images.

TABLE OF CONTENTS: Page vi (tr), Rick Rickman/Duomo Photography; vii (cr), John Langford/HRW Photo; viii (bl), Scott Van Osdol/HRW Photo; ix (b), David Phillips/HRW Photo; (cr), Shinichi Kanno/FPG International; 1 (b), John Langford/HRW Photo. **CHAPTER ONE:** Page 2 (c), 3 (c), Sam Dudgeon/HRW Photo; 3 (bkgd), Peter Van Steen/HRW Photo; 3 (b), (inset), Sam Dudgeon/HRW Photo; 4 (all), 5 (tr), John Langford/HRW Photo; 6 (b), Sam Dudgeon/HRW Photo; 7 (r), 8-9 (b), John Langford/HRW Photo; 10-11 (b), 10 (t), 13 (tr), Sam Dudgeon/HRW Photo; 14 (tr), Randall Alhadeff/HRW Photo; 15 (tr), source photo courtesy Austin History Center; 16 (tr), Sam Dudgeon/HRW Photo; 18 (br), Randall Alhadeff/HRW Photo; 20 (b), Sam Dudgeon/HRW Photo; 22 (c), 23 (l), Peter Van Steen/HRW Photo; 24 (c), Randall Alhadeff/HRW Photo; 27 (tr), Peter Van Steen/HRW Photo; 28 (r), Steven Ferry/HRW Photo; 32 (br), Randall Alhadeff/HRW Photo; 35 (t), Peter Van Steen/HRW Photo; 39 (bc), Sam Dudgeon/HRW Photo; 39 (br), Bob Daemmrich Photography; 41 (c), John Langford/HRW Photo; location access courtesy of Sea World of Texas; 42-43 (c), John Langford/HRW Photo; 42 (tr), Sam Dudgeon/HRW Photo; 45 (tr), Peter Van Steen/HRW Photo; 47 (tr), Steven Ferry/HRW Photo. **CHAPTER TWO:** Page 54-55 (bkgd), 54 (bc), Superstock; 61 (tr), Peter Van Steen/HRW Photo; 62 (bl), Robert Wolf; 64 (t), John Langford/HRW Photo; 67 (t), Steven Ferry/HRW Photo; 68 (tl), The Bettman Archive, UPI/Corbis-Bettmann; 68 (tr), Rick Rickman/Duomo Photography; 68 (c), Sam Dudgeon/HRW Photo; 71 (tl), Tony Stone Images; 71 (tr), Pascal Rondeau/Allsport USA; 72 (t), Steven Ferry/HRW Photo; 73 (c), Jeff Zaruba/The Stock Market; 74 (br), Michelle Bridwell/HRW Photo; 76-77 (bkgd), A. Upitis/Image Bank; 79 (tr), Superstock; 80-81 (bc), David Phillips/HRW Photo; 82 (tr), Dennis Fagan/HRW Photo; 82 (bl), Robert Reiff/FPG International; 84 (t), 88 (cl), David Phillips/HRW Photo; 88 (br), Sam Dudgeon/HRW Photo; 90 (t), 91 (br), Dennis Fagan/HRW Photo; 96 (t), Superstock; 99 (tl), (tr), Michelle Bridwell/HRW Photo; 101 (tr), David Phillips/HRW Photo; 103 (tr), Michelle Bridwell/HRW Photo; 104 (tr), (cl), Ron Kimball; 104 (bc), Dennis Fagan/HRW Photo; 105 (bl), Sam Dudgeon/HRW Photo; 107 (bl), Nawrocki Stock Photo; 108 (bc), Michelle Bridwell/HRW Photo; 109 (tl), Joe Towers/The Stock Market; 109 (c), A. Edgeworth/The Stock Market; 110-111 (bkgd), Superstock; 110 (tl), Sam Dudgeon/HRW Photo; 110 (br), Michelle Bridwell/HRW Photo. **CHAPTER THREE:** Page 120-121 (bkgd), 120 (bl), © Natioal Geographic Society; 120 (cr), Erich Lessing/Art Resource, N.Y.; 120 (br), Werner Forman/Art Resource, N.Y.; 121(tl), Erich Lessing/Art Resource, N.Y.; 121 (c), Scala/Art Resource, N.Y.; 121 (tr), Art Resource, N.Y.; 122 (tc), Michelle Bridwell/HRW Photo; 123 (tc), Sam Dudgeon/HRW Photo; 124 (t), Tom Van Sant/The Stock Market; 126 (br), Sam Dudgeon/HRW Photo; 129 (r), David Phillips/HRW Photo; 130 (bl), Sam Dudgeon/HRW Photo; 132 (c), Dennis Fagan/HRW Photo; 133 (br), Michelle Bridwell/HRW Photo; 134 (tr), Jonathan Daniel/Allsport USA; 135 (tr), Michelle Bridwell/HRW Photo; 136-137 (bkgd), Sam Dudgeon/HRW Photo; 136 (tr), Mike Anich/Adventure Photo; 136 (bl), D. R. Fernandez & M. L. Peck/Adventure Photo; 138 (t), David Phillips/HRW Photo; 139 (br), Michelle Bridwell/HRW Photo; 148 (br), 150 (tl), Dennis Fagan/HRW Photo; 151 (t), Sam Dudgeon/HRW Photo; 152 (tc), The Bettman Archive; 154 (t), 155 (c), Steven Ferry/HRW Photo; 157 (c), Michelle Bridwell/HRW Photo; 159 (br), FPG International. **CHAPTER FOUR:** Page 164-165 (bkgd), Sam Dudgeon/HRW Photo; 164 (c), (bc), 165 (tr), (cl), John Langford/HRW Photo; 165 (inset), Sam Dudgeon/HRW Photo; 166 (l), Steven Ferry/HRW Photo; 168 (bl), Sam Dudgeon/HRW Photo; TROUBLE® is a trademark of Hasbro, Inc. © 1997 Hasbro, Inc.; All Rights Reserved; used with permission; 170 (r), Peter Van Steen/HRW Photo; 170 (inset), Nawrocki Stock Photo; 172 (tr), (l), David Phillips/HRW Photo; 173 (b), Sam Dudgeon/HRW Photo; 174 (l), 175 (tr), David Phillips/HRW Photo; 176 (tr), David Phillips/HRW Photo; 178 (r), R. Stewart/Allsport USA; 182 (tr), Warren Morgan/Westlight; 182 (c), Dick Reed/The Stock Market; 182 (br), Superstock; 184 (c), Annie Griffiths Belt/Westlight; 185 (r), David Phillips/HRW Photo; 186 (tc), Ken Lax/HRW Photo; 189 (tr), David Phillips/HRW Photo; 190 (c), Mel Digiacomo/Image Bank; 191 (all), Michelle Bridwell/HRW Photo; 192 (tl), David Phillips/HRW Photo; 192 (tc), Ken Karp/HRW Photo; 192 (tr), Ken Lax/HRW Photo; 193 (tl), (tc), Sam Dudgeon/HRW Photo; 193 (tr), Rodney Jones/HRW Photo; 194 (tl), 195 (br), 196 (r), Sam Dudgeon/HRW Photo; 196 (br), Michelle Bridwell/HRW Photo; 197 (tr), Sam Dudgeon/HRW Photo; 198 (c), Mary Kate Denny/PhotoEdit; 199 (tr), (bl), 200 (cl), (c), Sam Dudgeon/HRW Photo; 201 (tl), Michelle Bridwell/HRW Photo; 201 (br), David Phillips/HRW Photo; 202-203 (c), Science Photo Library/Photo Researchers; 202 (tl), Superstock; 204 (bl), David Phillips/HRW Photo; 206 (tr), 208-209 (c), Sam Dudgeon/HRW Photo; 210 (c), Michelle Bridwell/HRW Photo; 211 (r), Sam Dudgeon/HRW Photo. **CHAPTER FIVE:** Page 218-219 (t), 218 (b), 219 (l), Scott Van Osdol/HRW Photo; 220 (t), Tibor Bognar/The Stock Market; 221 (t), Jonathan A. Meyers/FPG International; 225 (br), David Phillips/HRW Photo; 228 (tr), (inset), Michelle Bridwell/HRW Photo; 228 (cl), Sam Dudgeon/HRW Photo; 231 (br), David Phillips/HRW Photo; 234 (r), Ron Thomas/FPG International; 238 (tr), David Phillips/HRW Photo; 240 (br), 241 (cl), (t), Sam Dudgeon/HRW Photo; 243 (br), Scott Van Osdol/HRW Photo; 244 (bl), Dennis Cox/FPG International; 245 (tc), Sam Dudgeon/HRW Photo; 245 (tr), Tom Ives/The Stock Market; 248 (t), Sam Dudgeon/HRW Photo; 250 (tl), Douglas Dawson Gallery, c. 1920, Earthenware wine vessel, Bamileke culture, Cameroon grasslands; 250 (tl), Douglas Dawson Gallery, c. 1200 A.D., Earthenware slip decorated vessel, D'jenne culture, Mali; 250 (br), Michelle Bridwell/HRW Photo; 250 (inset), Sam Dudgeon/HRW Photo, Painting by Janet Brooks; 252 (cl), (bl), Sam Dudgeon/HRW Photo; 253 (t), 254 (br), 256 (tr), David Phillips/HRW Photo; 258-259 (r), 259 (tl), 260 (bl), Scott Van Osdol/HRW Photo. **CHAPTER SIX:** Page 264-265 (bkgd), Lee Kuhn/FPG International; 264 (tr), Thomas Craig/FPG International; 264 (bc), Stan Osolinski/FPG International; 264 (br), Frithfoto/Bruce Coleman; 265 (tr), James King-Holmes/Science Photo Library/Photo Researchers; 265 (c), Joe McDonald/Bruce Coleman; 265 (b), Douglas Faulkner/Photo Researchers; 266 (bl), (bc), Scott Van Osdol/HRW Photo; 269 (tl), Visuals Unlimited; 269 (c), Sam Dudgeon/HRW Photo; 276 (br), David Phillips/HRW Photo; 278 (tr), Sam Dudgeon/HRW Photo; 279 (tc), Telegraph Colour Library/FPG International; 282 (br), David Phillips/HRW Photo; 283 (br), Van Nostrand Reinhold Company, The VNR Concise Encyclopedia of Mathematics; 284-285 (bkgd), Sam Dudgeon/HRW Photo; 284 (tr), Rick Friedman/The New York Times; 285 (tl), Scott Van Osdol/HRW Photo; 285 (tr), Rick Friedman/The New York Times; 286-287 (t), Superstock; 286 (cl), Scott Van Osdol/HRW Photo; 286 (c), John Sanford/Photo Researchers, Science Photo Library; 288 (bkgd), Dr. E. R. Degginger/Color-Pic, Inc.; 290 (tr), Scott Van Osdol/HRW Photo; 291 (r), NASA; 293 (t), May Polycarpe/HRW Photo; 293 (b), Jack Zehrt/FPG International; 294 (b), Telegraph Colour Library/FPG International; 295 (b), NRSC LTD/Photo Researchers, Science Photo Library; 297 (tr), Michelle Bridwell/HRW Photo; 298 (tl), Alese & Mort Pechter/The Stock Market; 298 (cl), Ulf Sjostedt/FPG International; 298 (t), New York Public Library; 299 (cl), James King Holmes/Photo Researchers, Science Photo Library; 299 (t), Ken Reid/FPG International; 301 (t), Matt Bradley/Bruce Coleman; 303 (br), Michelle Bridwell/HRW Photo; 304-305 (b), Richard Stockton; 306-307 (bkgd), Andrew Leonard/The Stock Market; 307 (bl), Michelle Bridwell/HRW Photo. **CHAPTER SEVEN:** Page 316 (tr), 317 (br), Sam Dudgeon/HRW Photo; 318 (tr), 319 (br), Peter Van Steen/HRW Photo; 320 (b), Sam Dudgeon/HRW Photo; 321 (tr), Peter Van Steen/HRW Photo; 322 (c), 326 (br), Scott Van Osdol/HRW Photo; 329 (c), Peter Van Steen/HRW Photo; 335 (t), Scott Van Osdol/HRW Photo; 337 (cl), 340 (tr), Michelle Bridwell/HRW Photo; 343 (tr), Scott Van Osdol/HRW Photo; 344 (r), Sam Dudgeon/HRW Photo; 348 (c), Dennis Fagan/HRW Photo; 349 (c), David Phillips/HRW Photo; 350 (c), John Langford/HRW Photo; 351 (t), 352 (b), Michelle Bridwell/HRW Photo; 355 (c), Robert Wolf; 356 (bc), (br), Ron Kimball; 357 (b), Michelle Bridwell/HRW Photo; 358-359 (t), Robert Burch/Bruce Coleman; 360 (tr), David Phillips/HRW Photo; 361 (br), Sam Dudgeon/HRW Photo. **CHAPTER EIGHT:** Page 362 (bl), Shinichi Kanno/FPG International; 364-365 (bkgd), Michael Brohm/Nawrocki Stock Photo; 364 (br), Superstock; 365 (tr), David Phillips/HRW Photo; 365 (c), Nawrocki Stock Photo; 365 (bl), (t), (c), Ron Kimball; 367 (l), Carlos V. Causo/Bruce Coleman; 368 (cl), (c), (bl), (br), Sam Dudgeon/HRW Photo; 372 (bl), Michelle Bridwell/HRW Photo; 373 (tr), David Phillips/HRW Photo; 374-375 (t), Superstock; 374 (c), Nancy Engebretson/Phoenix Gazette; 376 (br), Michelle Bridwell/HRW Photo; 378 (bc), Sam Dudgeon/HRW Photo; 379 (r), David Phillips/HRW Photo; 381 (tr), 382 (b), Michelle Bridwell/HRW Photo; 383 (tc), Robert Frerck/Tony Stone Images; 383 (r), Peter Brandt; 384 (cl), (c), 386 (tr), 387 (tl), David Madison; 389 (tr), Michelle Bridwell/HRW Photo; 391 (tc), Bernarn Kappelmeyer/FPG International; 391 (cl), Spencer Jones/FPG International; 395 (r), Dave Gleiter/FPG International. **CHAPTER NINE:** Page 410 (tr), W. Warren/Westlight; 411 (tl), 411 (tc), (tr), Michelle Bridwell/HRW Photo; 411 (c), Superstock; 412 (tc), (br), Michelle Bridwell/HRW Photo; 416 (cl), (bl), Rhonda Bishop/Contact Press/Woodfin Camp; 419 (br), Michelle Bridwell/HRW Photo; 420 (t), 423 (c), Dennis Fagan/HRW Photo; 425 (tr), Michelle Bridwell/HRW Photo; 427 (br), (t), Dennis Fagan/HRW Photo; 428 (br), John Langford/HRW Photo; 431 (bl), Superstock; 432 (br), John Langford/HRW Photo; 433 (br), M. & C. Werner/Comstock; 435 (tr), Henry Georgi/Comstock; 435 (c), Sam Dudgeon/HRW Photo; 435 (bl), Dave Barnuff/Nawrocki Stock Photo; 437 (c), Tony Stone Images; 439 (c), Superstock; 440 (c), David Phillips/HRW Photo; 441 (tr), Matthew McVay/Tony Stone Images; 442 (tl), Nawrocki Stock Photo; 443 (tl), Michael J. Howell/Superstock; 445 (br), Dennis Fagan/HRW Photo; 446 (c), Index Stock Photography; 448 (c), Georg Gerster/Comstock; 448 (t), Index Stock Photography; 449 (bkgd), Bill Ross/Tony Stone Images; 451 (cl), Alan Bolesta/Index Stock Photography; 452 (tr), David Phillips/HRW Photo; 454 (tl), Michelle Bridwell/HRW Photo; 457 (tl), Superstock; 457 (br), Michelle Bridwell/HRW Photo; 458 (cl), Superstock; 459 (bl), Dennis Fagan/HRW Photo; 460 (t), John Langford/HRW Photo, Location access courtesy Sea World of Texas; 462 (t), David Lawrence/The Stock Market; 463 (br), Sam Dudgeon/HRW Photo; 466 (c), Scala/Art Resource, NY; 467 (tl), Paul Chelsey/Tony Stone Images; 467 (c), Dennis Fagan/HRW Photo; 467 (br), Ron Rovtar/Photonica. **CHAPTER TEN:** Page 472 (bc), Nawrocki Stock Photo; 473 (tc), David Phillips/HRW Photo; 474 (t), 475 (br), 476 (br), Michelle Bridwell/HRW Photo; 480 (t), David Phillips/HRW Photo; 482 (t), Michelle

Bridwell/HRW Photo; 484-485 (bc), Superstock; 484 (bl), Robert A. Lubeck/Animals Animals; 486 (c), Dennis Fagan/HRW Photo; 489 (tl), Index Stock Photography; 489 (br), Dennis Fagan/HRW Photo; 491 (tr), David Phillips/HRW Photo; 492-493 (t), 495 (br), Michelle Bridwell/HRW Photo; 496 (cl), David Barnes/Tony Stone Images; 496 (b), Michelle Bridwell/Frontera Fotos; 497 (c), Museum of the Rockies, Montana State University; 498-499 (bkgd), Superstock; 498 (tr), (bl), David Phillips/HRW Photo; 500 (tr), Dennis Fagan/HRW Photo; 501 (b), Sam Dudgeon/HRW Photo; 502 (c), 503 (tr), 504 (tr), David Phillips/HRW Photo; 505 (tr), Robert Frerck/Tony Stone Images; 506-507 (all), David Phillips/HRW Photo; 509 (c), Michelle Bridwell/HRW Photo; 510 (l), Dennis Fagan/HRW Photo; 510 (t), *Hot Rod Magazine*; 514 (tr), 515 (all), Sam Dudgeon/HRW Photo; 517 (bl), Ron Kimball.

Illustrations

Abbreviated as follows: (t) top; (b) bottom; (l) left; (r) right; (c) center.
All art, unless otherwise noted, by Holt, Rinehart and Winston.

TABLE OF CONTENTS: Page v., Nishi Kumar; vi., Michael Morrow. **CHAPTER 1:** Page 7, Michael Morrow; 9, Michael Morrow; 15, Uhl Studios; 19, Boston Graphics, Inc.; 28, Michael Morrow; 33, Lori Osiecki; 42, Boston Graphics, Inc.; 46, Philip Knowles. **CHAPTER 2:** Page 56-57, Scott Johnston; 56 (bl) Michael Morrow; 57(tr) Michael Morrow; 63, Michael Morrow; 65, Boston Graphics, Inc.; 68, Boston Graphics, Inc.; 71, Stephen Durke; 72, Boston Graphics, Inc.; 78(cl), Boston Graphics, Inc.; 78(tl), Michael Morrow; 79, Michael Morrow; 81, Michael Morrow; 83, Michael Morrow; 84, Boston Graphics, Inc.; 87, Michael Morrow; 89, Michael Morrow; 90, Michael Morrow; 93, Michael Morrow; 95, Michael Morrow; 102, Boston Graphics, Inc.; 106, Michael Morrow; 107, Michael Morrow. **CHAPTER 3:** Page 122, Nishi Kumar; 126, Boston Graphics, Inc.; 127, Boston Graphics, Inc.; 128, Michael Morrow; 133, Michael Morrow; 141, Nishi Kumar; 143, Nishi Kumar; 144, Boston Graphics, Inc.; 145, Nishi Kumar; 150, Nishi Kumar; 151, Michael Morrow; 152, Michael Morrow; 156(t), Boston Graphics, Inc.; 156(b), Nishi Kumar. **CHAPTER 4:** Page 167, Stephen Durke; 169, Michael Morrow; 178, Michael Morrow; 181, Michael Morrow; 183(t), Michael Morrow; 183(b), Boston Graphics, Inc.; 198, David Fischer; 202, Michael Morrow; 204, Nishi Kumar; 208, Michael Morrow. **CHAPTER 5:** Page 220, Michael Morrow; 221, David Fischer; 223, Michael Morrow; 224, Michael Morrow; 232, Michael Morrow; 239, Michael Morrow; 244, Nishi Kumar; 249, Nishi Kumar; 251, David Fischer; 258, Michael Morrow. **CHAPTER 6:** Page 266, Michael Morrow; 272, Michael Morrow; 273, Michael Morrow; 278, David Fischer; 279, David Fischer; 286, David Fischer; 288, Michael Morrow; 293(t), Nishi Kumar; 293(b), David Fischer; 294-295, David Fischer, 300, David Fischer. **CHAPTER 7:** Page 314-315, Nishi Kumar; 331, Michael Morrow; 338, David Fischer; 345, Nishi Kumar. **CHAPTER 8:** Page 546-547, Nishi Kumar; 368, David Fischer; 379, Nishi Kumar; 380, Ophelia Wong; 397, Nishi Kumar; 400, Michael Morrow. **CHAPTER 9:** Page 414, Nishi Kumar; 416, David Fischer; 428, Michael Morrow; 438, David Fischer; 441, David Fischer; 444, Michael Morrow; 450, David Fischer; 454, David Fischer; 464, Philip Knowles; 465, Michael Morrow. **CHAPTER 10:** Page 479, David Fischer; 481, Boston Graphics, Inc.; 483, Michael Morrow.

Design

CHAPTER OPENERS: Nishi Kumar; 2–3, 54–55, 120–121, 164–165, 218–219, 264–265, 314–315, 362–363, 410–411, 472–473. **CHAPTER PROJECTS**: Paradigm Design; 48–49, 104–105, 158–159, 210–211, 258–259, 306–307, 356–357, 402–403, 466–467, 516–517.

Permissions

Grateful acknowledgment is made to the following sources for permission to reprint copyrighted material.

Hill and Wang, a division of Farrar, Strauss & Giroux, Inc.: From *Innumercy* by John Allen Paulos. Copyright © 1988 by John Allen Paulos.

The New York Times Company: From graph, "Getting Lost in the Shuffle," and from "In Shuffling Cards, 7 is Winning Number" by Gina Kolata from The New York Times, January 9, 1990. Copyright © 1990 by The New York Times Company.

Penguin Books, Ltd.: From page 274, of *The Crest of the Peacock: Non-European Roots of Mathematics* by George Gheverghese Joseph. Copyright © 1991 by George Gheverghese Joseph. Published by Penguin Books, 1992. First published by I.B. Tauris.

Time Inc.: From "A Miraculous Sky Rescue" from Time, May 4, 1987. Copyright © 1987 by Time Inc.

U S. News & World Report: From "Now, the Testing Question," from *U. S. News & World Report,* September 21, 1992. Copyright © 1992 by U. S. News & World Report.

The Wall Street Journal: From "Counting Big Bears" from *The Wall Street Journal,* May 2, 1990. Copyright © 1990 by Dow Jones & Company, Inc. All Rights Reserved Worldwide.

ADDITIONAL ANSWERS

Lesson 1.0

Exercises and Problems, **pages 8–9**

19. \$70; To find the answer, find 7 on the horizontal axis. From that point move vertically to the graph of the function. The corresponding y-value represents the wages earned.

20. His hourly wage is \$10. To find the answer, find the slope of the graph; this is the hourly wage. Or use the graph to find that Lyle worked 4 hours for \$40. Divide 40 by 4.

21. When Lyle works more than 8 hours in one day, his hourly wage increases. The slope increases and the graph gets steeper.

22. A reasonable domain is all real numbers from 0 to 24. It is impossible to work a negative number of hours, but a fraction of an hour is possible. Also, since these are daily wages the domain cannot go past 24 hours.

24. Multiply the number of mums by 15 to get the amount of money raised.

25. Extend the graph, and draw a line containing the points shown on the graph. Find 10 on the horizontal axis. From that point move vertically to the graph of the function. The corresponding y-value is the amount of money raised.

26. Draw a line containing the points. Find \$75 on the vertical axis. From that point move horizontally to the graph of the function. The corresponding x-value is the number of mums that must be sold.

27. A reasonable domain is all positive integers.

Lesson 1.1

Communicate, **page 14**

1. Isolate the x-tile by adding 2 positive 1-tiles to each side. Remove neutral pairs and determine the value of x by adding the remaining tiles; $x = 1$.

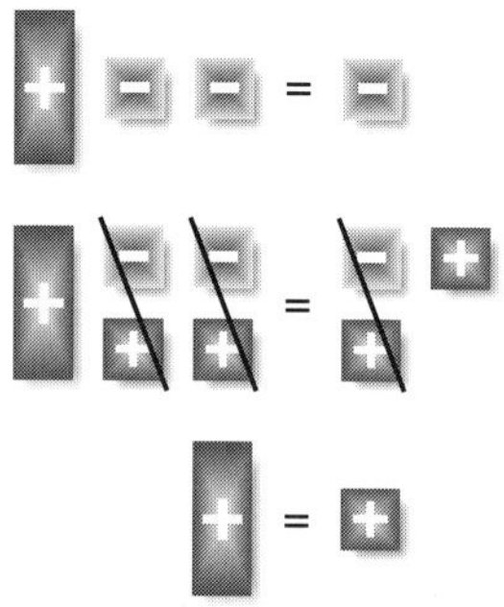

2. Substitute your solution for the variable in the equation; simplify. If the resulting statement is true, then the solution is correct; if the resulting statement is false, an error occurred in solving the problem.

3. To solve an addition equation, use the Subtraction Property of Equality; to solve a subtraction equation, use the Addition Property of Equality.

4. Answers may vary. Possible example: Jon's father gave him \$3 for washing the family car. If Jon adds the \$3 to the money he has already saved and now has \$12, how much money did Jon have before he washed the car?

Look Beyond, **page 15**

57. Let x = a number. $x + 5 = 6$

58. Let g = Gayle's age. Let j = Joel's age. $g = 2j$

59. Let w = length of white rope. Let y = length of yellow rope. $w = \frac{2}{3}y$

Lesson 1.2

Critical Thinking, **page 20**

Given the proportion $\frac{a}{b} = \frac{c}{d}$, use the

Multiplication Property of Equality to multiply each side by bd.

$\frac{a}{b} = \frac{b}{c}$ Given

$bd\left(\frac{a}{b}\right) = bd\left(\frac{b}{c}\right)$ Mult Prop

$da = bc$ Simplify.

$ad = bc$ Comm Prop

Communicate, **page 21**

1. To solve $\frac{x}{3} = 19$, multiply each side of the equation by 3; $x = 57$.
2. Answers may vary. Possible situation: 80% or the retail price is the price marked.
3. For any nonzero number, r, there is a number, $\frac{1}{r}$, such that $r \cdot \frac{1}{r} = 1$. When a number is multiplied by its reciprocal, the answer is 1; $8 \cdot \frac{1}{8} = \frac{8}{8} = 1$.
4. Multiply each side of the equation by $\frac{7}{2}$ to have a coefficient of 1 for x.

$$\frac{2z}{7} = \frac{4}{9}$$
$$\frac{7}{2} \cdot \frac{2z}{7} = \frac{4}{9} \cdot \frac{7}{2}$$
$$\frac{14z}{14} = \frac{28}{18}$$
$$z = 1\frac{5}{9}$$

5. Divide each side of the equation by 5.

$$5f = 28.5$$
$$\frac{5f}{5} = \frac{28.5}{5}$$
$$f = 5.7$$

Look Back, **page 22**

80.

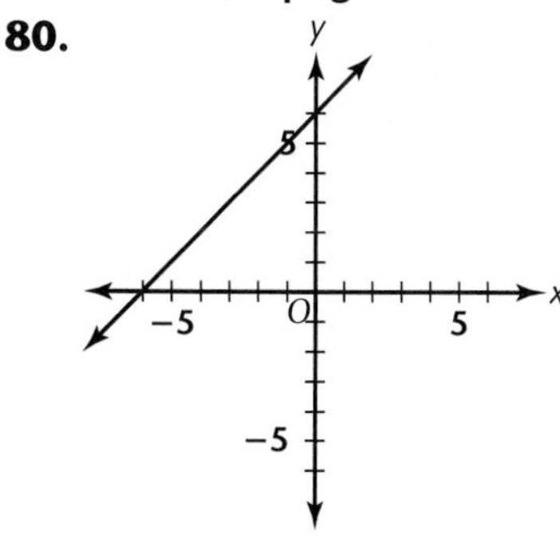

81.

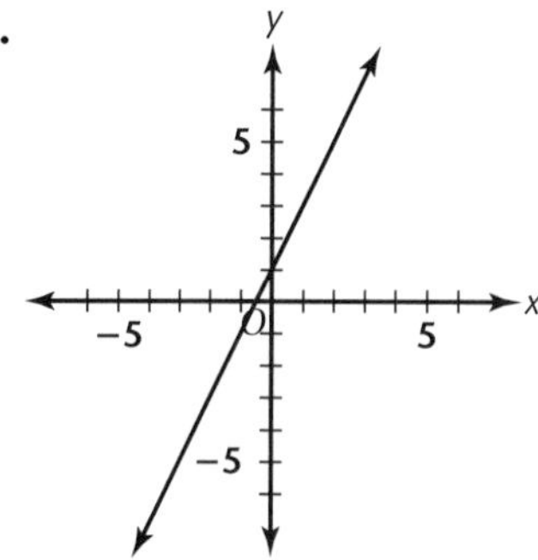

82.

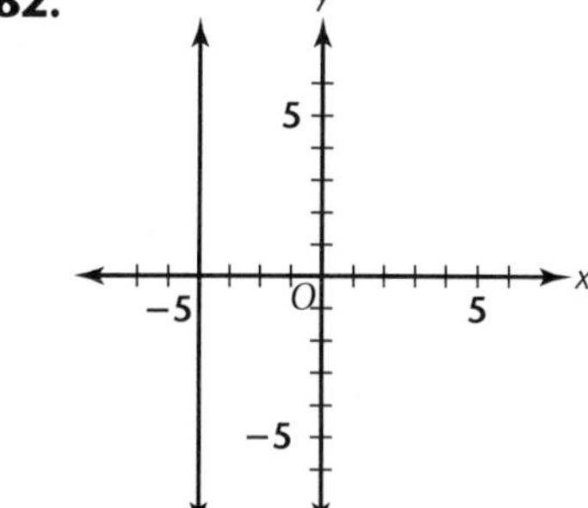

Lesson 1.3

Exploration, **page 24**

4.

$43 = 4s - 5$	Given
$43 + 5 = 4s - 5 + 5$	Add Prop
$48 = 4s$	Simplify.
$\frac{48}{4} = \frac{4s}{4}$	Div Prop
$12 = s$	Simplify.

12 units is the side of the first square.

Check: $43 = 4s - 5$

$43 \stackrel{?}{=} 4(12) - 5$

$43 = 43$ True

5.

$60 = 4s + 8$

$52 = 4s$

$13 = s$

The side of square A is 13 units in length.

Check: $60 = 4s + 8$

$60 \stackrel{?}{=} 4(13) + 8$

$60 = 60$ True

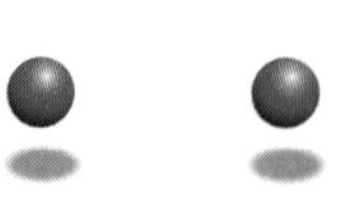
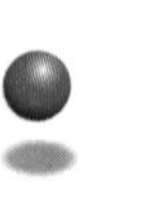

Communicate, page 26

1. To solve the equation $2k - 6 = 8$, add 6 to each side of the equation, simplify, and divide each side of the equation by 2, then simplify; $k = 7$.
2. Substitution, solving the equation, and graphing are three methods of determining whether the solution to $3m - 4 = 11$ is 5.
3. Answers may vary. Possible answer: The school store ordered stationery for \$2.50 a box, plus a \$25 shipping charge, for a total of \$150. How many boxes did they order?
4. To solve $\frac{x-9}{3} = -7$, multiply each side of the equation by 3, which is the reciprocal of $\frac{1}{3}$. Simplify, and then add 9 to each side of the equation, and simplify again; $x = -12$.

Lesson 1.4

Communicate, page 32

For Exercises 1–3, the equations may be solved using algebra tiles, the algebraic method, graphing, or tables.

1. Methods may vary; use algebra tiles or graphing.
   Tile method:

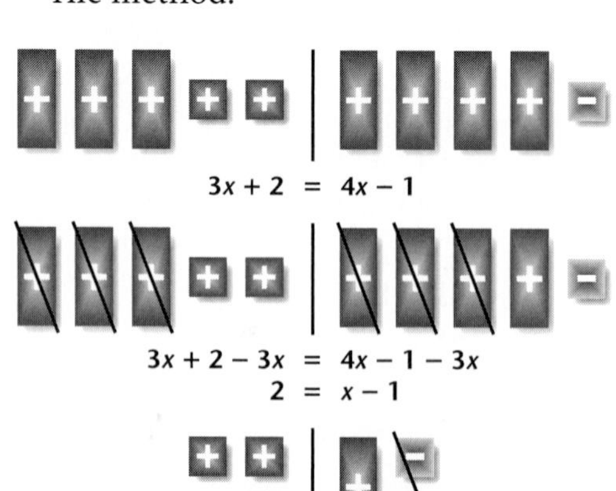

Graphing method:
Graph $y = 3x + 2$ and $y = 4x - 1$; the point of intersection is the solution; $x = 3$.

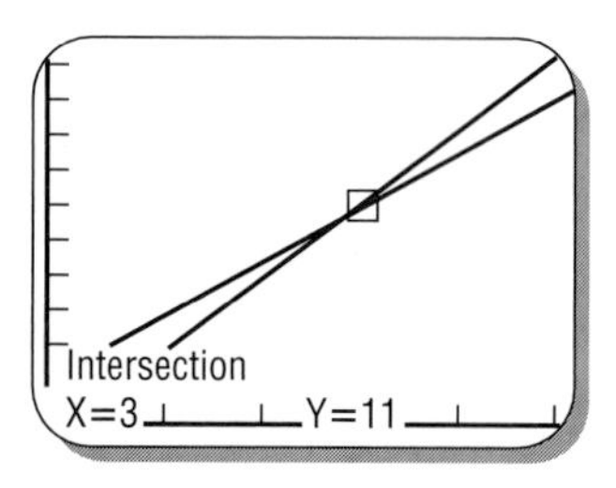

2. Methods may vary. Use the algebraic method or tables.
   Algebraic method:

$$7x - 5 = 2x - 3$$
$$7x - 5 - 2x = 2x - 3 - 2x$$
$$5x - 5 = -3$$
$$5x - 5 + 5 = -3 + 5$$
$$5x = 2$$
$$\frac{5x}{5} = \frac{2}{5}$$
$$x = \frac{2}{5} = 0.4$$

Table method:
Use a table to find the value of x for which both y-values are equal; $x = 0.4$ or $\frac{2}{5}$. Enter $Y_1 = 7x - 5$ and $Y_2 = 2x - 3$ in a calculator.

X	Y_1	Y_2
0	−5	−3
.2	−3.6	−2.6
.4	−2.2	−2.2
.6	−8	−1.8
.8	.6	−1.4
1	2	−1
1.2	3.4	−.6
X=.4		

3. Methods may vary. Use the algebraic method or algebra tiles.
   Algebraic method:

$$13 + 2x = 5x + 1$$
$$13 + 2x - 2x = 5x + 1 - 2x$$
$$13 = 3x + 1$$

$$13 - 1 = 3x + 1 - 1$$
$$12 = 3x$$
$$\frac{12}{3} = \frac{3x}{3}$$
$$4 = x$$

Tile method:

$13 + 2x = 5x + 1$

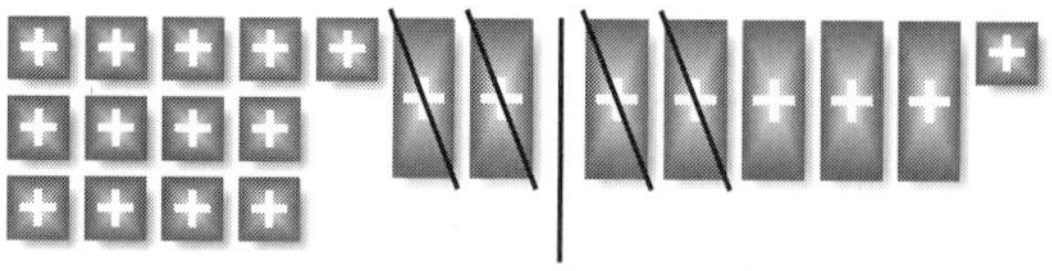

$13 + 2x - 2x = 5x + 1 - 2x$
$13 = 3x + 1$

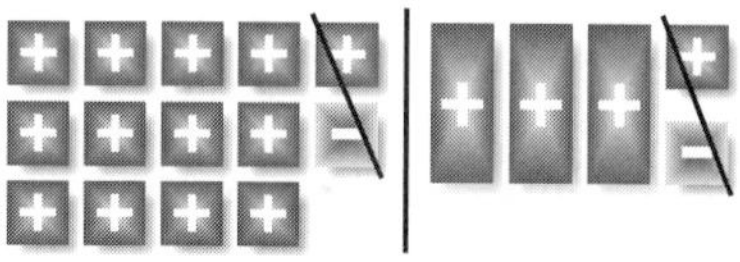

$13 - 1 = 3x + 1 - 1$
$12 = 3x$

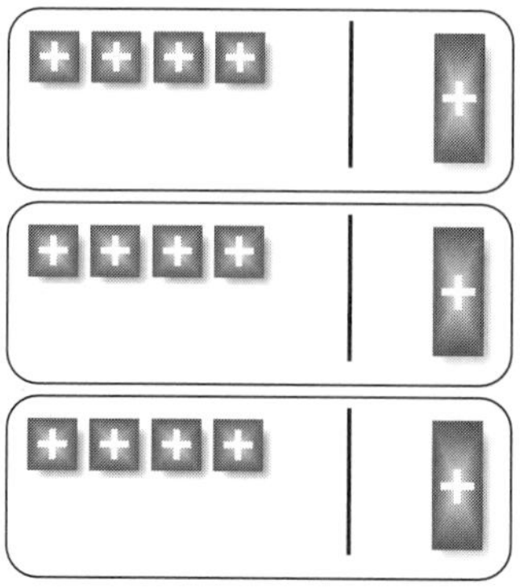

$\frac{12}{3} = \frac{3x}{3}$

$4 = x$

4. An equation has no solution when the variables cancel out and the resulting statement is false.

5. An equation has a solution of all real numbers when the variables cancel out and the resulting equality is true.

Lesson 1.5

Try This, **page 37**

Let t = number of T-shirts.

$$2t + 28{,}000 \leq 40{,}000$$
$$2t + 28{,}000 - 28{,}000 \leq 40{,}000 - 28{,}000$$
$$2t \leq 12{,}000$$
$$t \leq 6000$$

The maximum number of T-shirts that can be given away is 6000.

Communicate, **page 38**

1. To solve the inequality $2x - 5 < 3$, add 5 to each side of the inequality, simplify, then divide each side by 2 and simplify.

$$2x - 5 < 3$$
$$2x - 5 + 5 < 3 + 5$$
$$2x < 8$$
$$\frac{2x}{2} < \frac{8}{2}$$
$$x < 4$$

To graph $x < 4$ on a number line, circle 4, and shade all points to the left of 4.

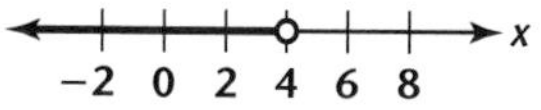

2. The properties of inequality allow you to solve an inequality the same way the properties of equality allow you to solve an equation. The only difference in the properties is that you must remember to reverse the inequality sign when multiplying or dividing each side by a negative number.

3. When dividing each side by a negative number, you must remember to reverse the inequality sign; $x > -2$.

4. To solve the inequality $-\frac{x}{2} \geq 12$, multiply each side of the inequality by -2 and reverse the inequality sign.

$$-2 \cdot \frac{-x}{2} \leq 12 \cdot -2$$
$$x \leq -24$$

5. Addition Property of Inequality

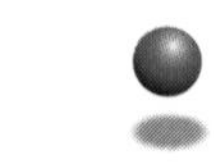

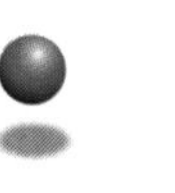

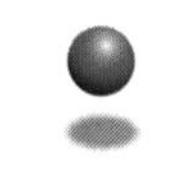

6. Division Property of Inequality
7. Subtraction Property of Inequality and Multiplication Property of Inequality
8. Addition Property of Inequality, Subtraction Property of Inequality, and Division Property of Inequality

Practice and Apply, **pages 38–40**

9. $x \leq -5$

−6 −5 −4 −3 −2 −1 0 x

10. $t < -1\frac{1}{4}$

$-1\frac{1}{4}$

−3 −2 −1 0 t

11. $r \geq -21$

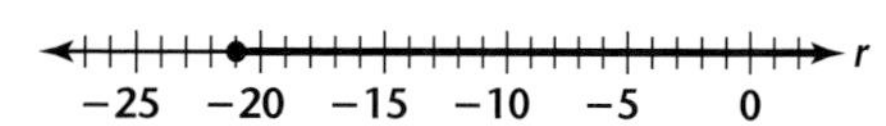

12. $x > 8\frac{1}{3}$

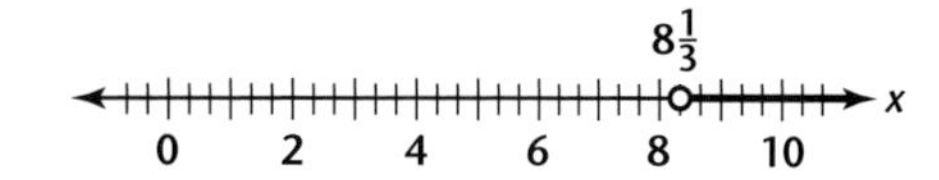

13. $b \geq -2\frac{4}{5}$

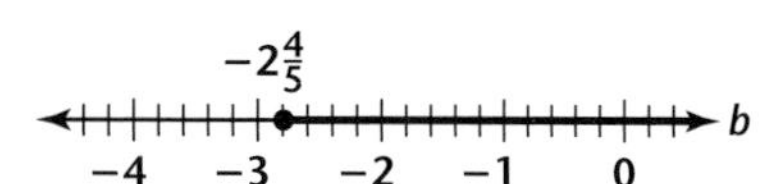

14. $y \geq 5\frac{1}{4}$

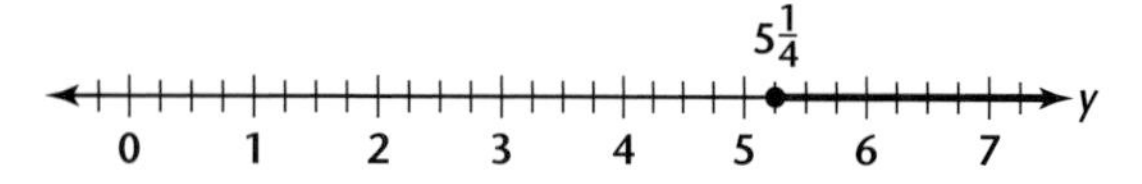

15. $y < -3.1$

−3.5 −3.4 −3.3 −3.2 −3.1 −3.0 y

16. $x < 13$

17. $z \geq 4.875$

4.875

3 4 5 6 z

18. $11.05 \leq x$

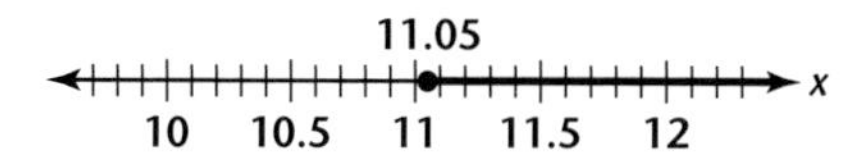

19. $-4.375 < q$

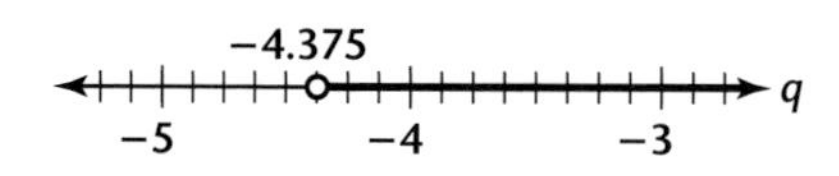

20. $x \leq 60$

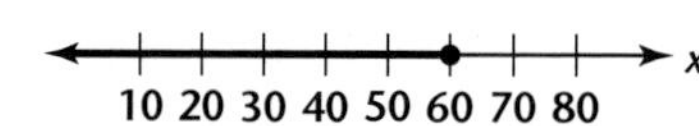

21. $x \leq -21$

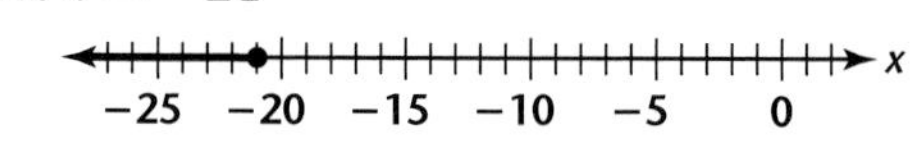

22. $m \geq -56$

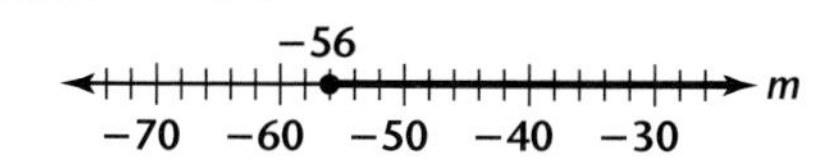

23. $a < 12.5$

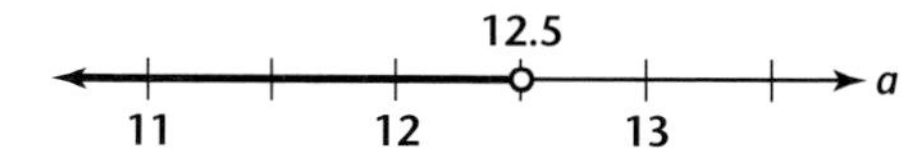

27. $x \leq -4$

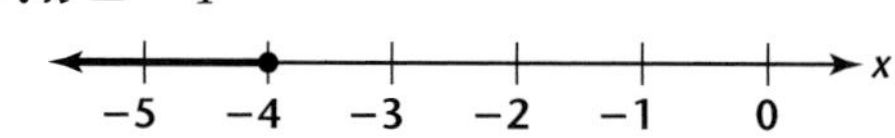

28. $t < -2\frac{2}{3}$

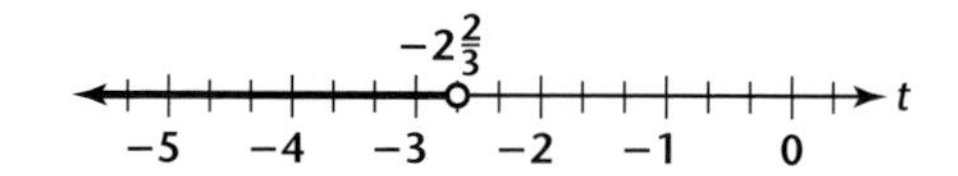

29. $b \leq -3$

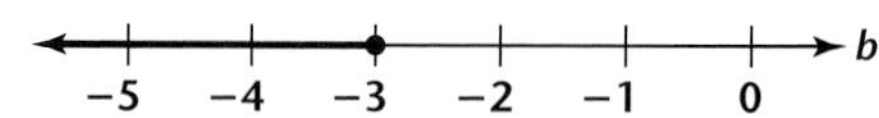

30. $x > 5\frac{4}{5}$

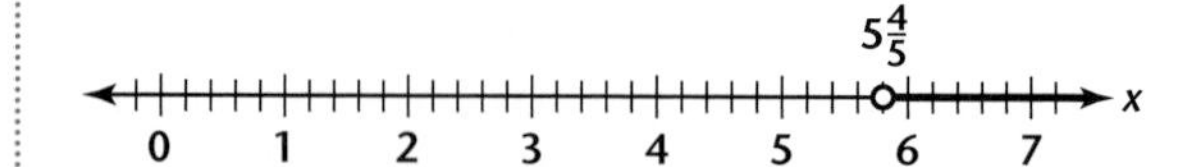

31. $b \geq -4\frac{2}{3}$

$-4\frac{2}{3}$
−5 −4 −3 −2 −1 0 b

32. $y \geq 7$

0 2 4 6 8 10 12 14 y

33. $y < -2\frac{1}{6}$

$-2\frac{1}{6}$
−3 −2 −1 0 y

34. $x < 6$

0 1 2 3 4 5 6 7 x

35. $z \leq 1.125$

1.125
0 1 2 z

36. $5.2 \leq x$

5.2
5 5.5 6 6.5 7 x

37. $q < 7.375$

7.375
6 7 8 q

38. $h \leq 4$

0 1 2 3 4 5 h

39. $n > -13.5$

−13.5
−14 −12 −10 −8 −6 −4 −2 0 n

40. $d > 9.5$

9.5
0 2 4 6 8 10 12 14 d

41. $t \leq 3\frac{1}{3}$

$3\frac{1}{3}$
0 1 2 3 4 5 t

Extra Practice, **page 527**

1. $x \leq -5$

−9 −8 −7 −6 −5 −4 −3 −2 −1 0 1 2 3 4 5 6 7 8 9 x

2. $c < 9$

−4 −2 0 2 4 6 8 10 12 c

3. $z < 9$

−4 −2 0 2 4 6 8 10 12 z

4. $y \leq 3$

−4 −2 0 2 4 6 8 10 12 y

5. $t \leq -9$

−12 −10 −8 −6 −4 −2 0 2 4 6

6. $t > 1\frac{2}{3}$

$1\frac{2}{3}$
−2 −1 0 1 2 t

7. $m < 15$

0 2 4 6 8 10 12 14 16 m

8. $w \leq -18$

−18 −14 −10 −6 −2 0 2 w

9. $a \leq 12$

−2 0 2 4 6 8 10 12 14 a

10. $x < 3.5$

3.5
−2 0 2 4 x

11. $y \geq 1$

−2 0 2 4 y

12. $t \leq -3.8$

−3.8
−4 −2 0 2 t

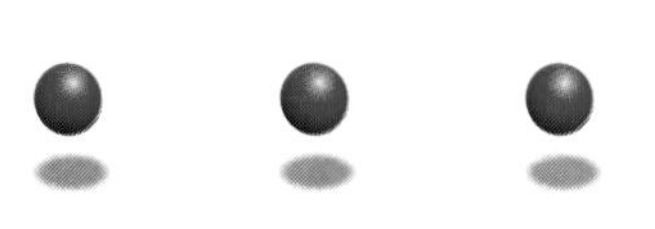

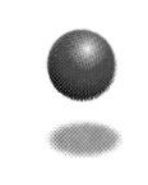

13. $y > 3.4$

14. $w \geq -3$

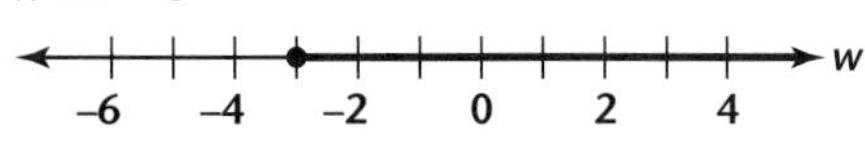

15. $q \geq 1$

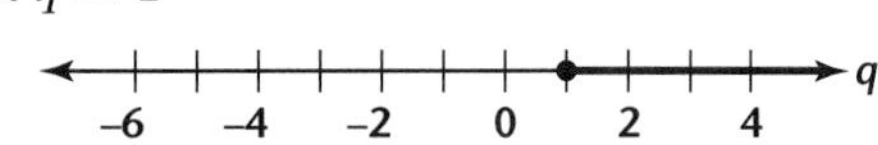

16. $x > 5$

17. $r < 0.5$

18. $x < 0.9$

19. $x < 24$

20. $q \leq 24$

21. $x > 119$

22. $w < 10$

23. $y \leq 4$

24. $v \geq 12$

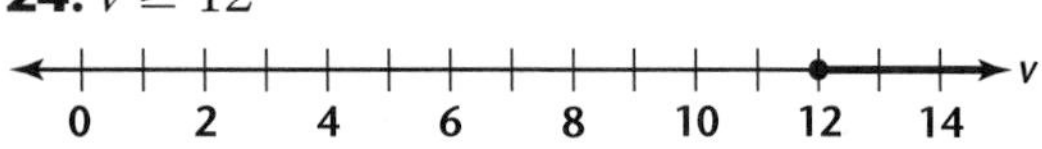

Lesson 1.6

Communicate, **page 45**

1. Absolute value is defined as being the distance a number is from the origin on a number line. In defining absolute value, if $a \geq 0$, then the absolute value of a is a. If $a < 0$, then the absolute value of a is $-a$ (the opposite of a).
2. When defining absolute value on a number line, opposite sides from the origin are taken into consideration; therefore, there are two possible solutions.
3. Let x = guessed number of marbles. The amount guessed minus the actual amount (1500) needs to be ± 10 of the actual 1500 marbles; $|x - 1500| \leq 10$. The guessed amount of marbles needs to be ± 10 of the actual amount of 1500. Since the amount guessed can be up to 10 more than 1500 or up to 10 less than 1500, we write an inequality using absolute value.
4. Answers may vary. One example: $|-8| = -(-8) = 8$; The distance from 0 to -8 is 8.

Practice and Apply, **pages 45–47**

14. $w = 1\frac{1}{6}$ or $w = -5\frac{5}{6}$

15. $m = 1.6$ or $m = 1.28$

16. $t = 2$ or $t = -0.4$

17. $x \geq 2$ or $x \leq 0$

18. $-6 \leq x \leq 1$

19. $-37 < h < 3$

20. $j < -9$ or $j > 35$

21. $d < -\frac{25}{27}$ or $d > 1\frac{20}{27}$

22. $a < -5.4$ or $a > -5.4$

23. $c \geq \frac{7}{17}$ or $c \leq -\frac{9}{17}$

24. $y \geq 3\frac{1}{3}$ or $y \leq 2\frac{2}{3}$

25. $-\frac{1}{7} < v < \frac{5}{7}$

26. $y = 6.9$ or $y = 0.1$

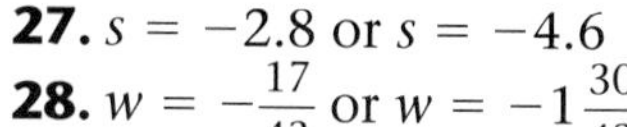

27. $s = -2.8$ or $s = -4.6$

28. $w = -\frac{17}{43}$ or $w = -1\frac{30}{43}$

29. $m = \frac{5}{13}$ or $m = \frac{3}{13}$

30. $t = 15$ or $t = -3$

31. $x \geq 7.8$ or $x \leq -0.6$

32. $-5.7 \leq x \leq -1.7$

33. $-4.02 < h < -1.98$

34. $j > 45$ or $j < -7$

35. $d > 2$ or $d < -1\frac{5}{7}$

36. $-\frac{6}{11} < a < -\frac{2}{11}$

37. $c \geq \frac{50}{403}$ or $c \leq -1\frac{47}{403}$

38. $y \geq 4.15$ or $y \leq 2.05$

39. $x = 1.1$ or $x = -11.5$

40. $\frac{21}{115} < v < \frac{49}{115}$

41. $|x - 5| = 2$; $x = 7$ or $x = 3$

42. $|x - 3| > 3$; $x > 6$ or $x < 0$

43. $|x + 2| < 4$; $-6 < x < 2$

44. $|x + 5| < 2.3$; $-7.3 < x < -2.7$

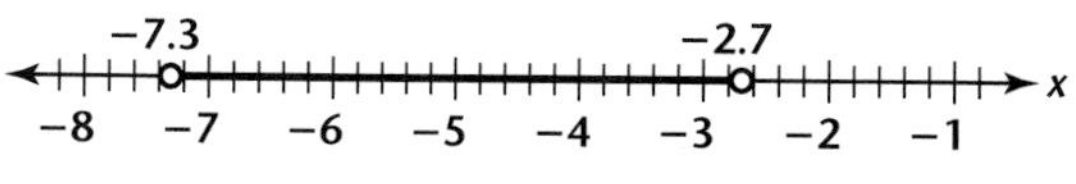

45. $|x - 15| \geq 1.5$; $x \leq 13.5$ or $x \geq 16.5$

46. $x \geq 7$ or $x \leq -1$

47. $-9 \leq x \leq 3$

48. $-17 < h < 3$

49. $j > 11$ or $j < 7$

50. $d > 2$ or $d < -1$

51. $-2.4 < a < 1.6$

52. $c \geq 1$ or $c \leq -2$

53. $y \geq 4$ or $y \leq 2$

54. $0.85 < x < 1.15$

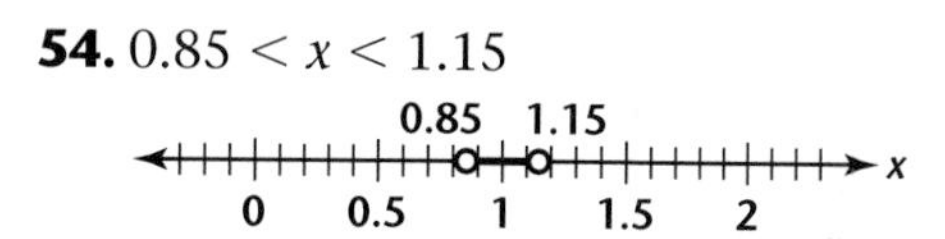

55. $|14 - x| \leq 0.2$; $13.8 \leq x \leq 14.2$

56.

57.

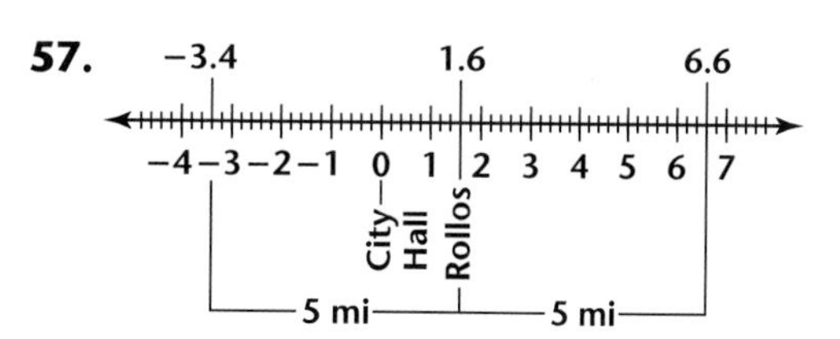

58. $|c - 1.6| \leq 5$; $-3.4 \leq c \leq 6.6$

59. The distance that Rollo's delivery customers are from City Hall

60. $|1.5 - x| \leq 0.03$; $1.47 \leq x \leq 1.53$

61.

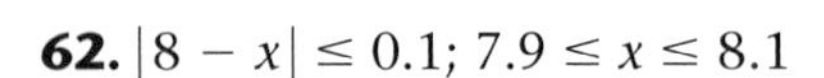

62. $|8 - x| \leq 0.1$; $7.9 \leq x \leq 8.1$

63.

Extra Practice, **page 528**

16. $|n - 4| < 3$; $1 < n < 7$

17. $|n - (-6)| > 5$; $n < -11$ or $n > -1$

18. $|n - 10| < 1.5$; $8.5 < n < 11.5$

19. $|n - 12| \geq 2.2$; $n \leq 9.8$ or $n \geq 14.2$

Chapter 1 Project

Activity, **page 49**

Answers given were calculated using 3.14 for π.

3. Answers may vary.

Lane	Outer Radius
1	≈29.261205 meters
2	≈30.511205 meters
3	≈31.761205 meters
4	≈33.011205 meters
5	≈34.261205 meters
6	≈35.511205 meters

4. Answers may vary.

Lane	Perimeter
1	400 meters
2	≈417.850007 meters
3	≈435.700007 meters
4	≈453.550007 meters
5	≈471.400007 meters
6	≈489.250007 meters

Stagger the starting points of each runner by 17.9 meters.

Chapter 1 Review, pages 50–52

19. $x > -8$

20. $x \geq 2$

21. $y > 5$

22. $t \geq 3\frac{1}{5}$

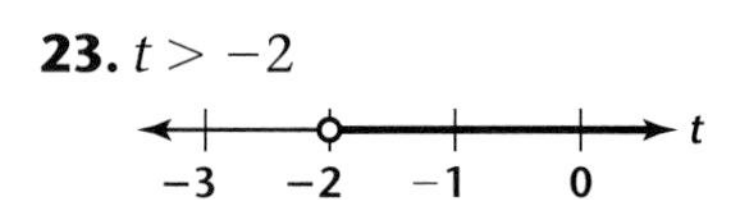

23. $t > -2$

24. $x \leq 4$

25. $x = 3$ or $x = 9$

26. $t < -11$ or $t > 1$

27. $y \leq -4$ or $y \geq 3$

28. $-5 \leq z \leq 9$

Chapter 1 Assessment, page 53

18. $x > 7$

19. $x \geq -11\frac{2}{3}$

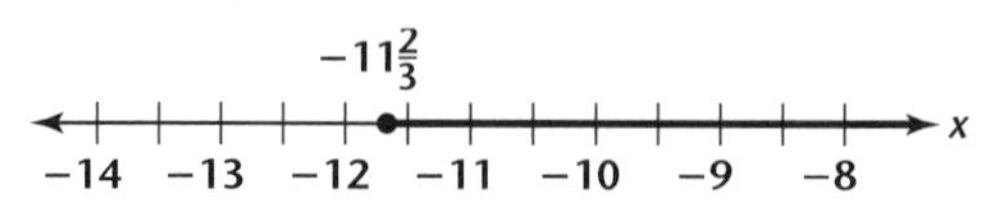

20. $x > -24$

−28 −27 −26 −25 −24 −23 −22

30. $|35 - x| \leq 0.2$; $34.8 \leq x \leq 35.2$

31.

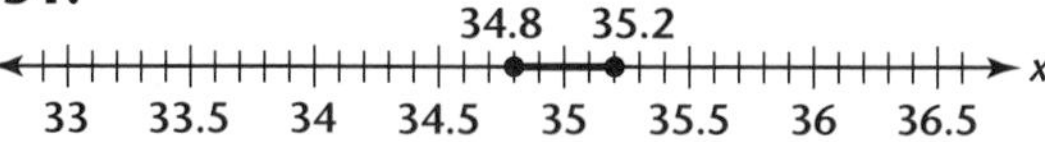

Lesson 2.1

Exploration 2, Part 1, page 58

1.

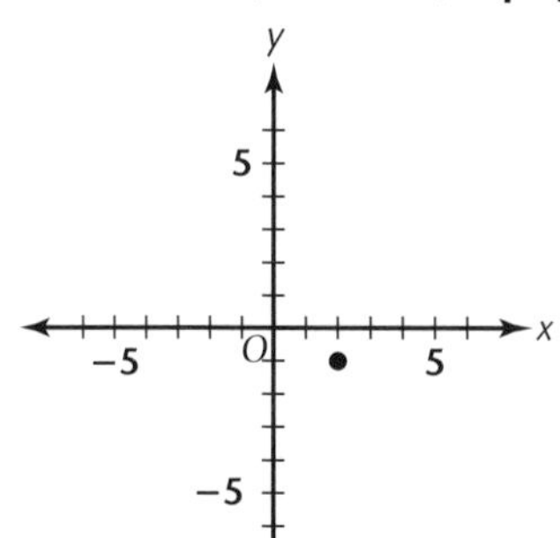

4.

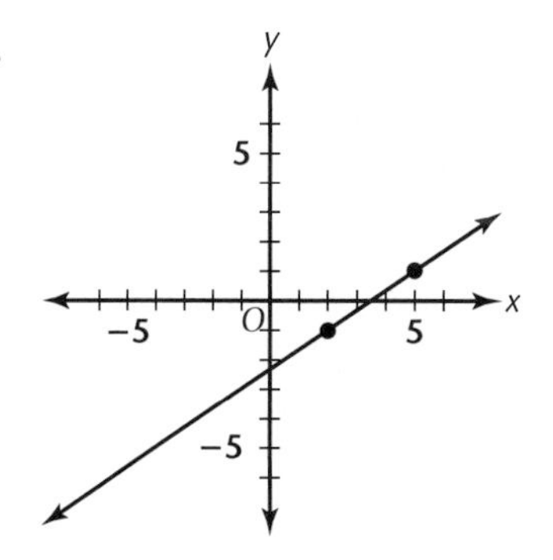

5. If the slope is $\frac{a}{b}$, $a, b > 0$, and the point (x, y) is on the line, start at (x, y), go up a units and to the right b units. Plot this point because this point is also on the line. Connect the two points with a line to draw the graph.

Exploration 2, Part 2, page 58

1.

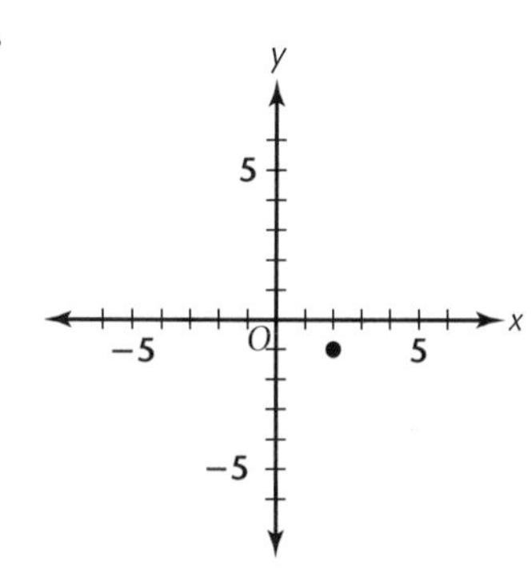

4.

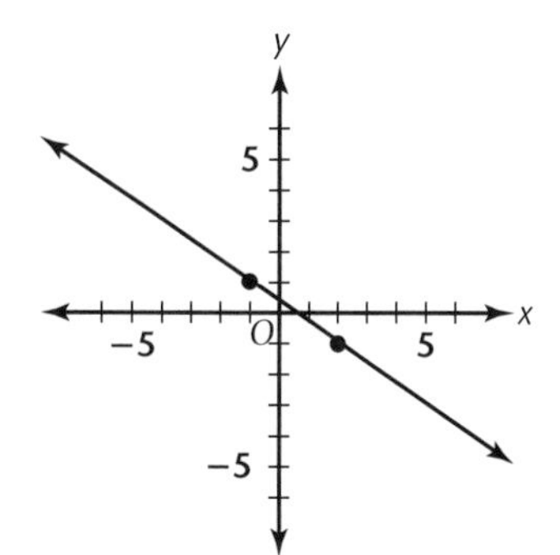

5. If the slope is $-\frac{a}{b}$, $a, b > 0$, and the point (x, y) is on the line, start at (x, y), go down a units and to the right b units or up a units and to the left b units. Plot this point because this point is also on the line. Connect the two points to draw the graph of the line.

Exploration 3, Part 1, pages 58–59

4. Answers may vary. One example is given.

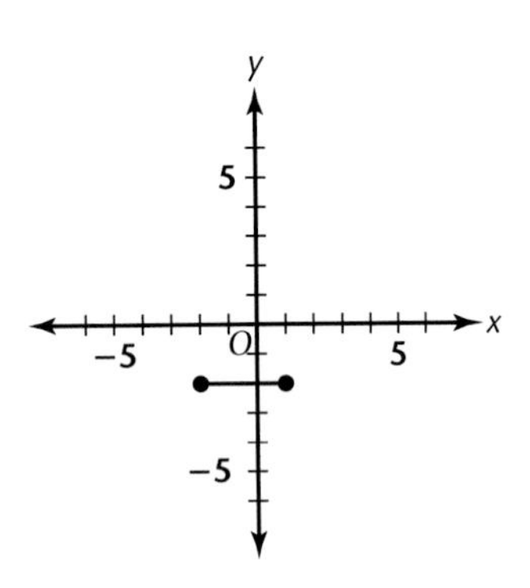

rise = 0, run = 3; $\frac{\text{rise}}{\text{run}} = \frac{0}{3} = 0$; Slopes of horizontal lines are equal to zero.

Exploration 3, Part 2, page 59

5. Answers may vary. One example is given.

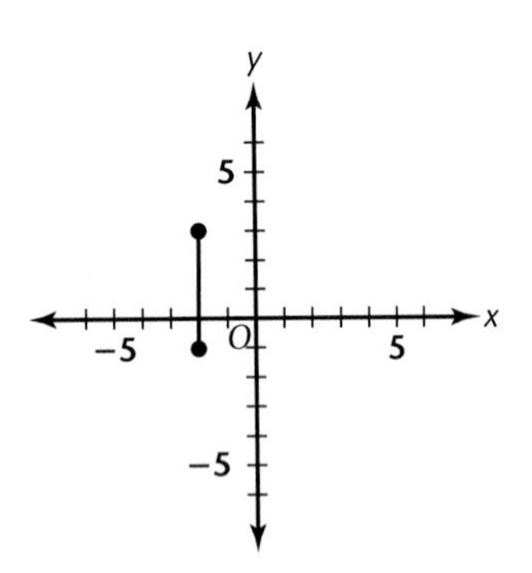

rise = 3, run = 0; $\frac{\text{rise}}{\text{run}} = \frac{3}{0}$ = undefined; Slopes of vertical lines are undefined.

Exploration 4, page 59

1.

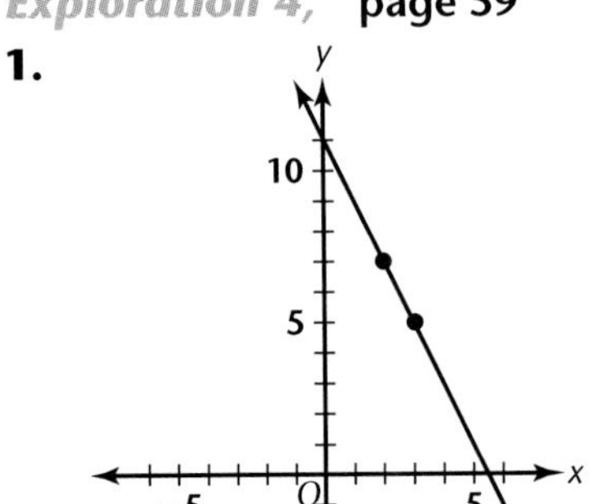

Try This, page 60

$y = 8x$

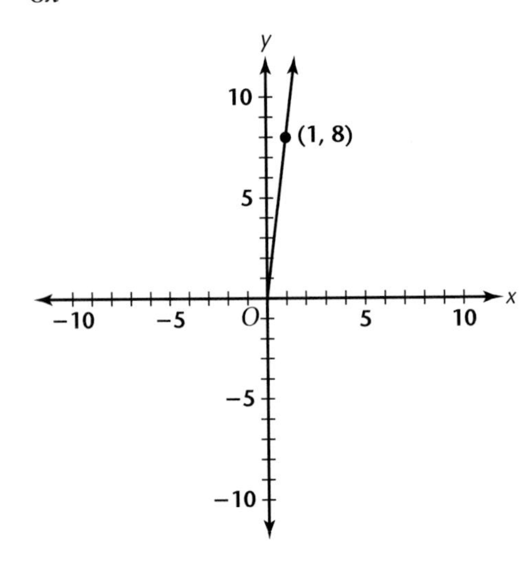

Critical Thinking, page 60

Slope $= \frac{y_1 - y_2}{x_1 - x_2}$ and slope $= \frac{y_2 - y_1}{x_2 - x_1}$; slope $\neq \frac{y_2 - y_1}{x_1 - x_2}$ and slope $\neq \frac{y_1 - y_2}{x_2 - x_1}$. The ratios $\frac{y_1 - y_2}{x_1 - x_2}$ and $\frac{y_2 - y_1}{x_2 - x_1}$ measure the $\frac{\text{rise}}{\text{run}}$ from $P_1(x_1, y_1)$ to $P_2(x_2, y_2)$. The other ratios do not. To find the slope or $\frac{\text{rise}}{\text{run}}$ between two points, the order of subtracting the coordinates must be consistent.

Communicate, **page 61**

2. Choose two points on the line, and calculate the $\frac{\text{rise}}{\text{run}}$ from one point to another.

5. The bicyclist travels at a steady rate for 7 minutes, stops for about 3 minutes, then continues for another 5 minutes. The total distance the bicyclist traveled was 50 yards.

6. Only the first quadrant is used because distance is positive.

Practice and Apply, **pages 61–63**

31.

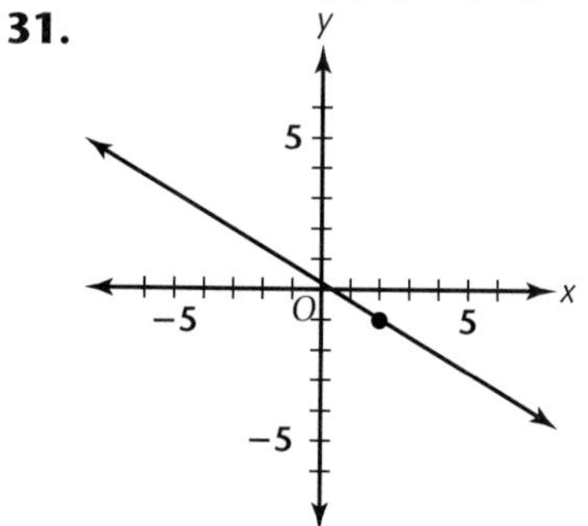

32.

33.

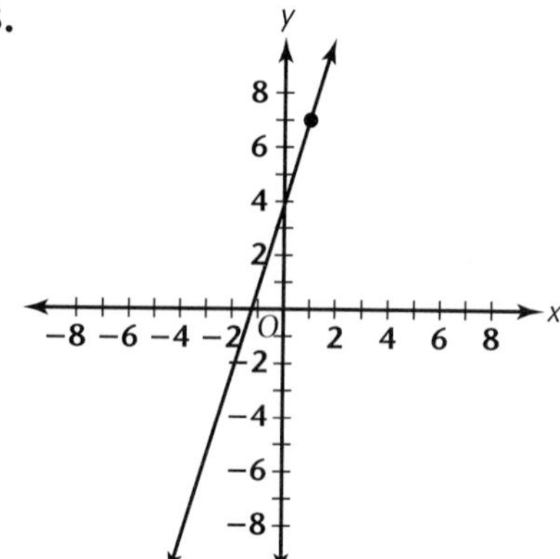

34.

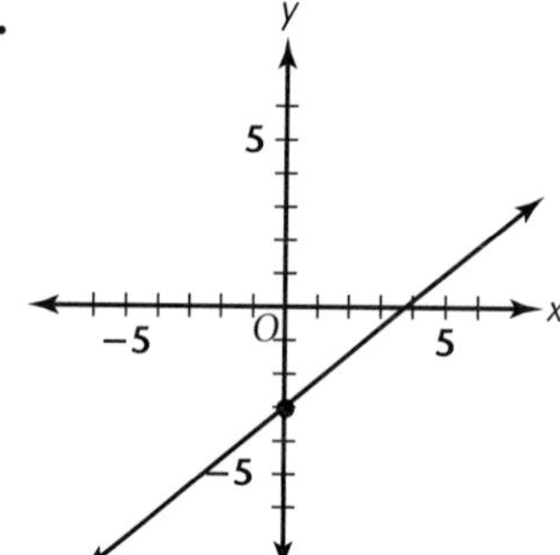

35.

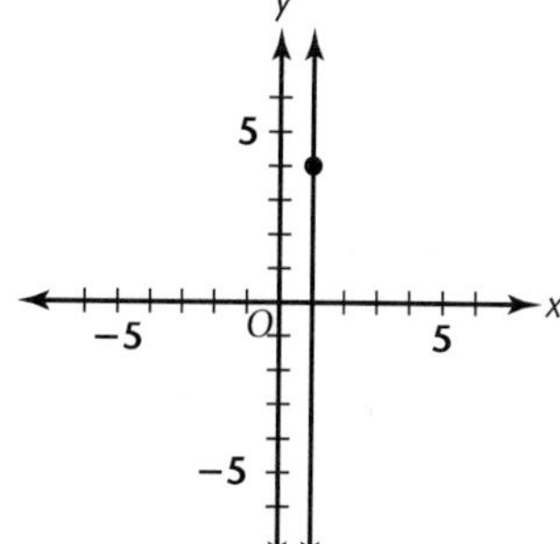

36.

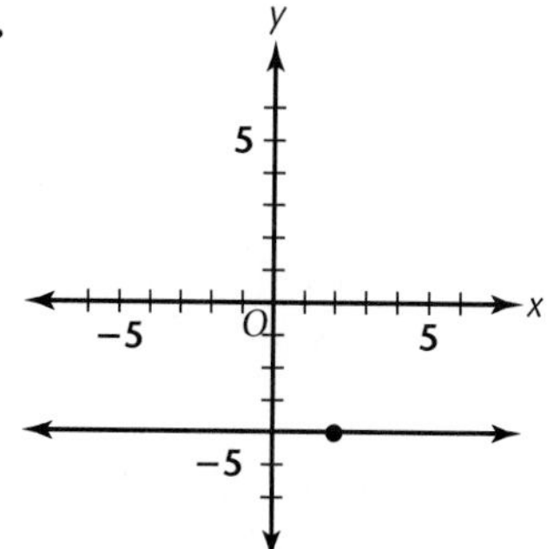

37.

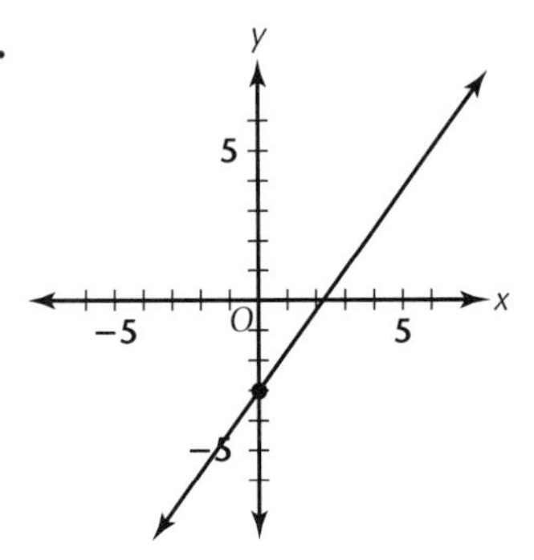

38.

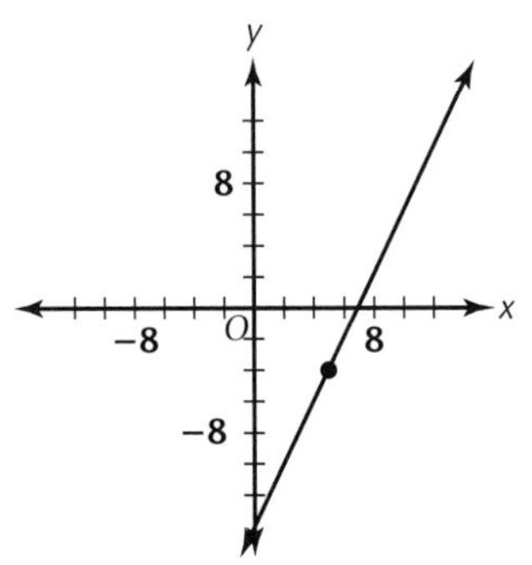

39.

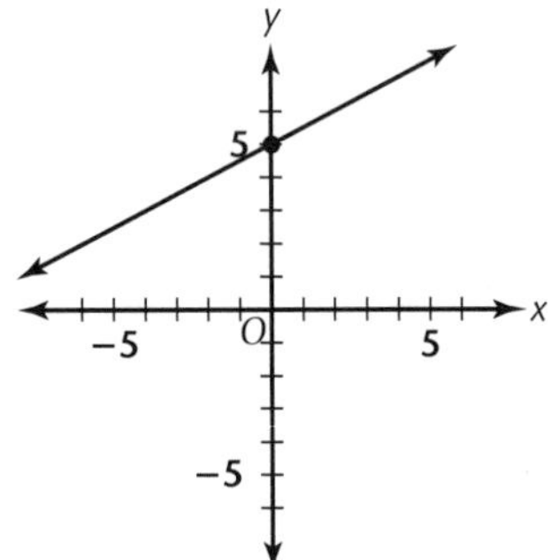

40.

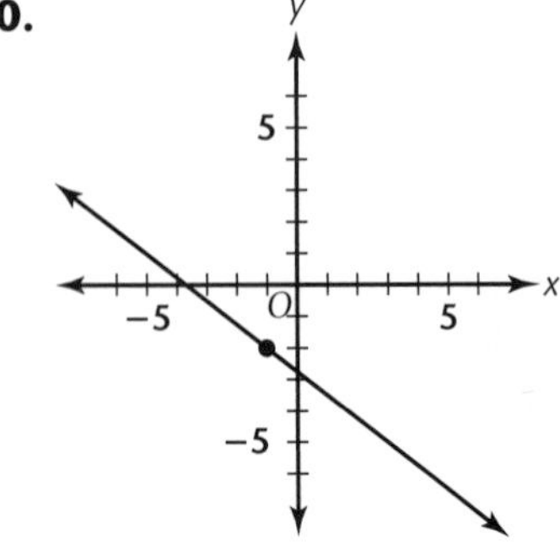

41.

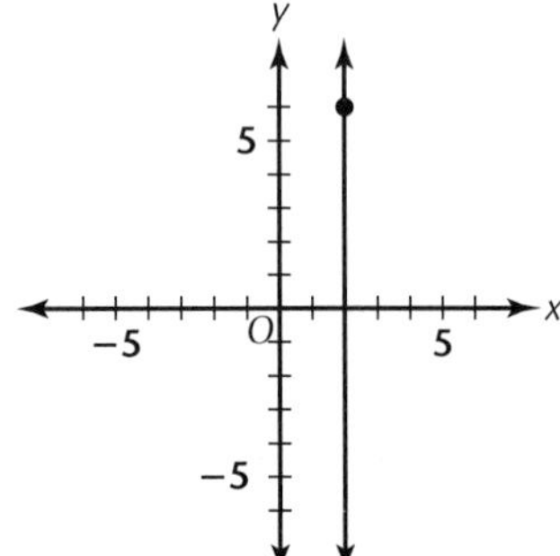

42.

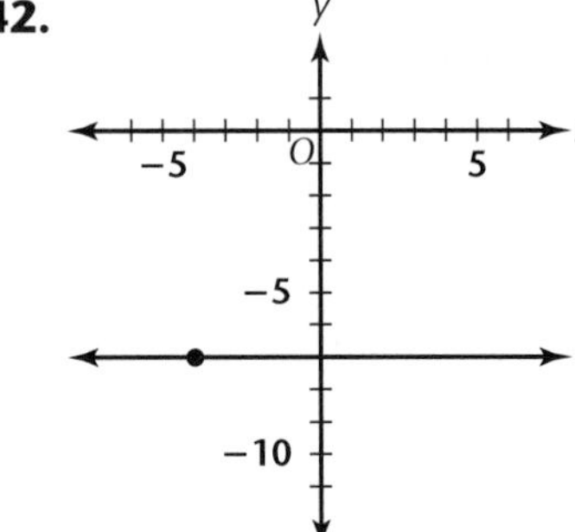

43–44. All the points lie on a straight line.

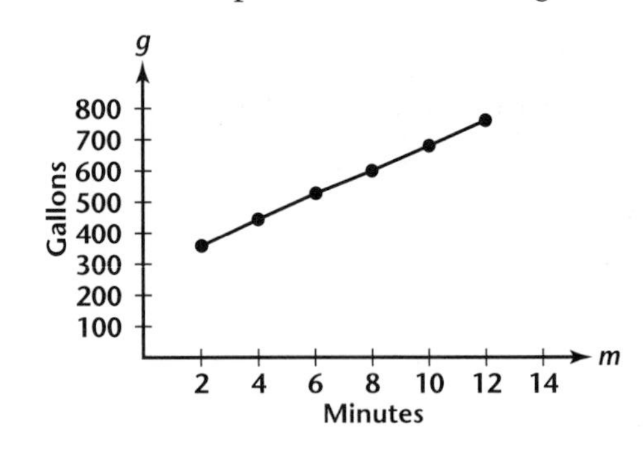

45. Slope = 40; Yes; the slope is the same regardless of which point is used.

46. Gallons of water filling the tank per minute

47. (0, 280); the number of gallons of water already in the tank before pouring the water to fill the tank

48. No, this would represent negative time.

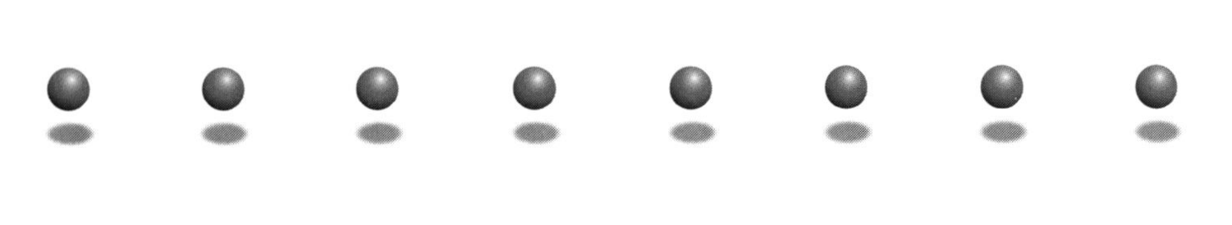

49.

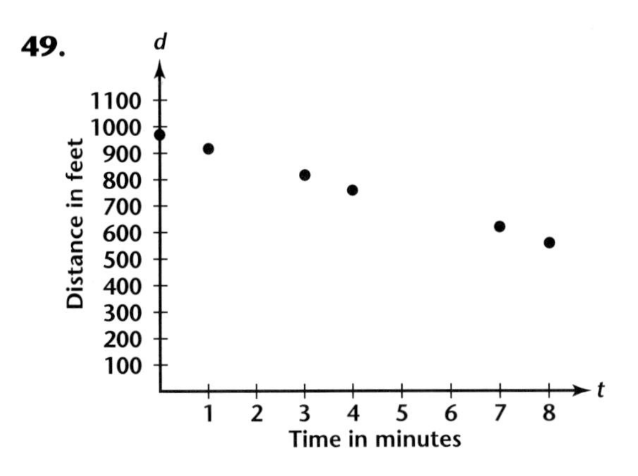

The points are in the first quadrant because time and distance are positive values.

50.

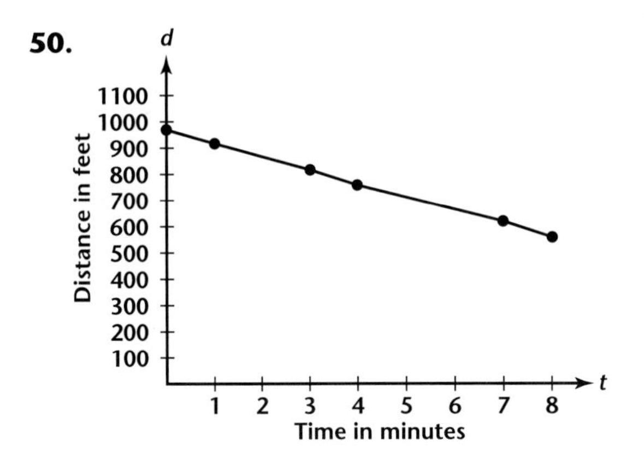

Yes, it appears that Nick was walking at a constant rate.

51. Slope = -50

52. The slope is negative since he is getting closer to Mike's house, which means the distance is decreasing over time.

53. Let d = distance in feet above the ground and t = the amount of time in minutes; $d = 2000t$.

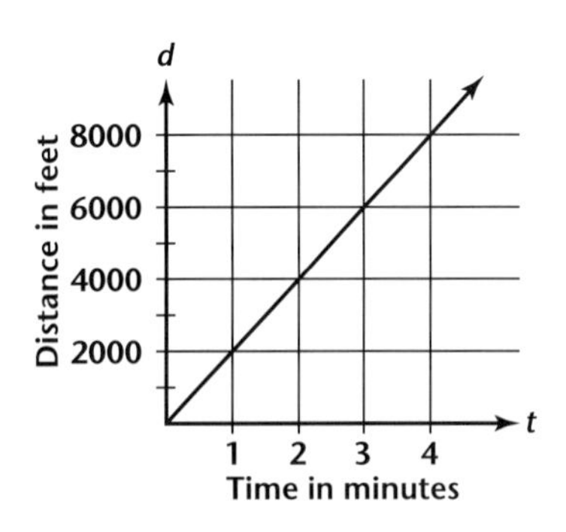

Look Beyond, **page 63**

62. Yes, the increase in cost is constant; the equation is $y = 10x + 5$.

63. No, the increase is not constant.

64. No, the increase is not constant.

65. Yes, the speed is a constant 50 miles per hour.

Extra Practice, **page 528**

10.

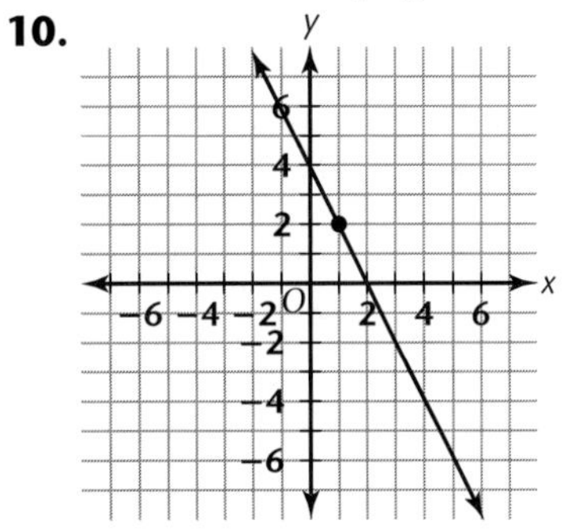

11.

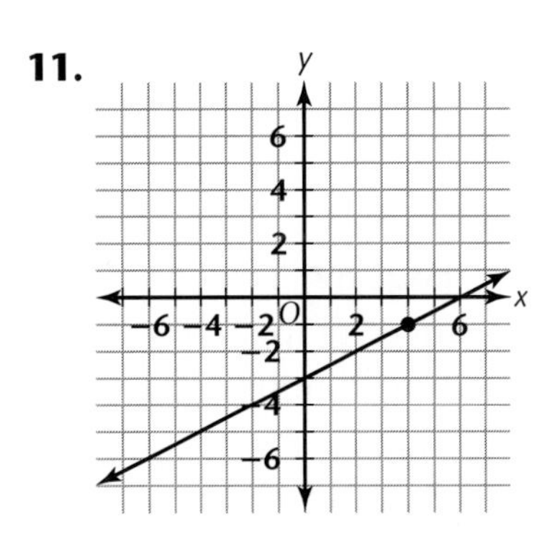

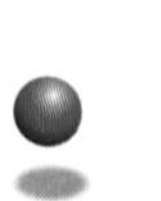

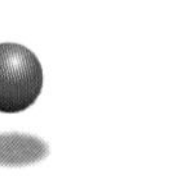

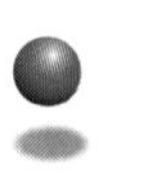

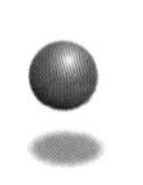

12.

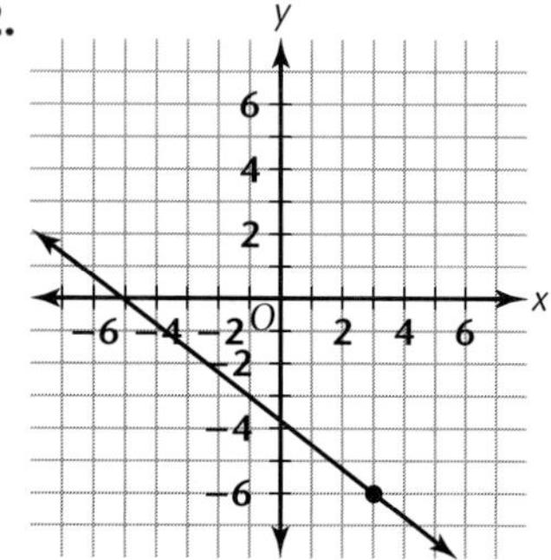

13.

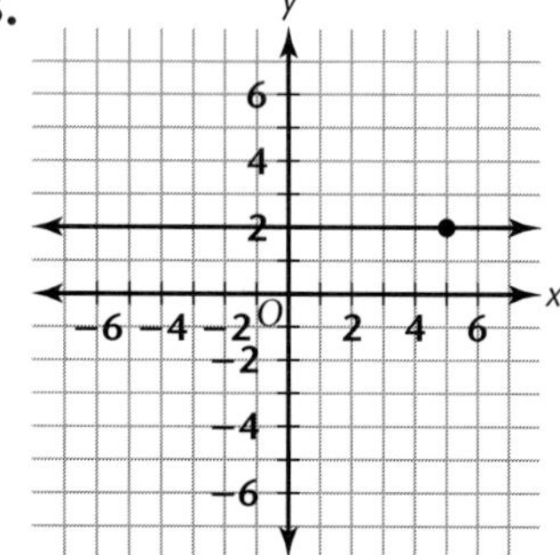

Lesson 2.2

Try This, **page 65**

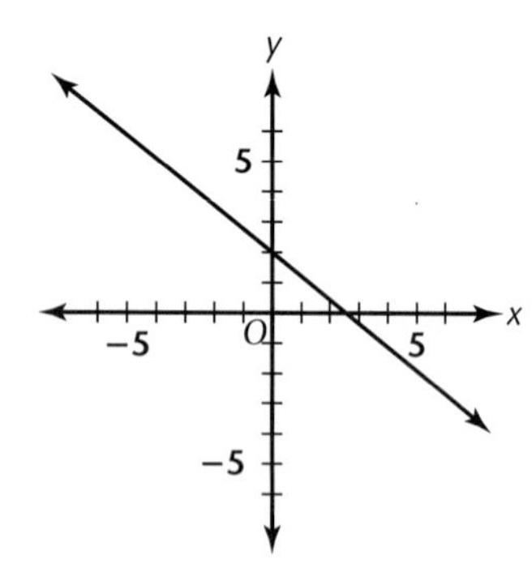

Communicate, **page 69**

1. Start at the y-intercept (0, 2). Use the slope $\frac{3}{4}$ to move up 3 units and to the right 4 units to point (4, 5). Connect the two points and extend the line.

2. First, let $x = 0$ and solve for y.

$$2x - 3y = 12$$
$$2(0) - 3y = 12$$
$$-3y = 12$$
$$\frac{-3y}{-3} = \frac{12}{-3}$$
$$y = -4$$

The y-intercept is $(0, -4)$.
Second, let $y = 0$ and solve for x.

$$2x - 3y = 12$$
$$2x - 3(0) = 12$$
$$2x = 12$$
$$\frac{2x}{2} = \frac{12}{2}$$
$$x = 6$$

The x-intercept is (6, 0). Plot $(0, -4)$ and (6, 0), connect the points, and extend the line.

3. A regression equation is a good fit if the points follow a linear pattern and the line follows the points almost exactly.

4. Let (x_1, y_1) and (x_2, y_2) be the given points. Find the slope of the line connecting the points, $m = \frac{y_2 - y_1}{x_2 - x_1}$. Substitute the slope and one of the points into the equation $y = mx + b$ to solve for b. Use the slope and the value of b to write $y = mx + b$.

5. Solve for y; $y = \frac{3}{4}x - 2$. Use this equation to enter in the calculator.

Practice and Apply, **pages 70–71**

6.

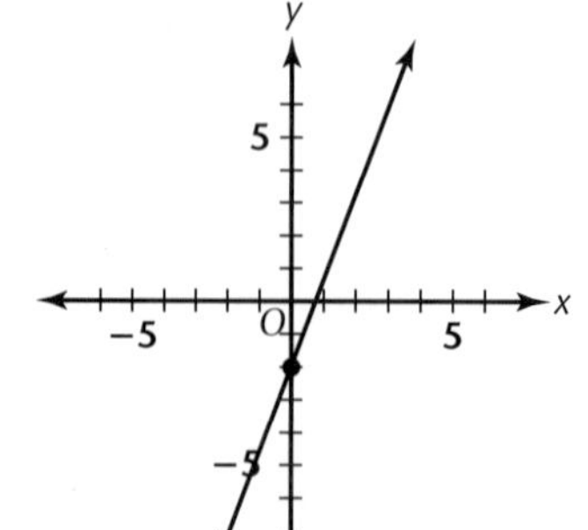

7.

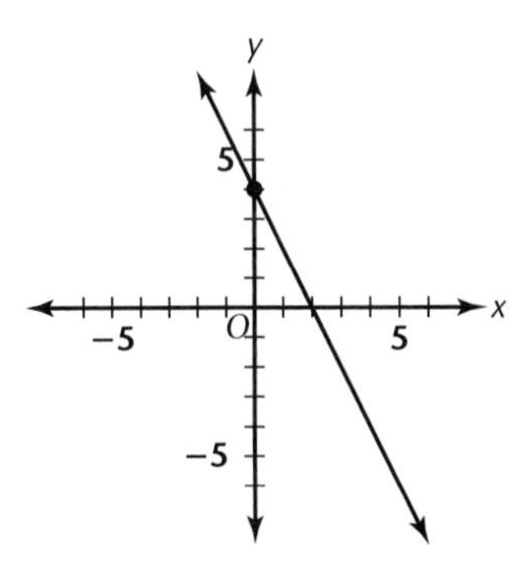

8.

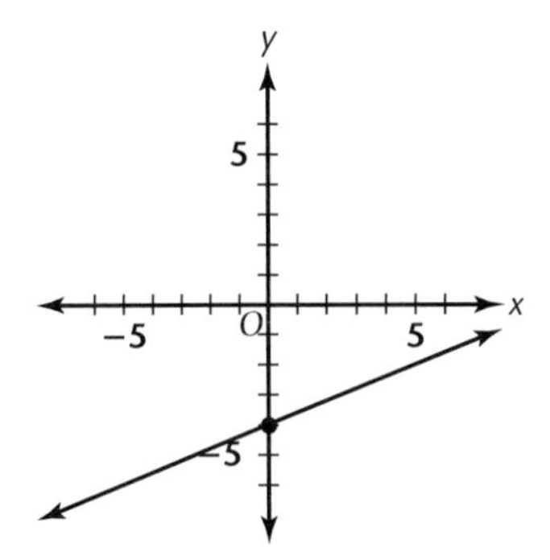

9.

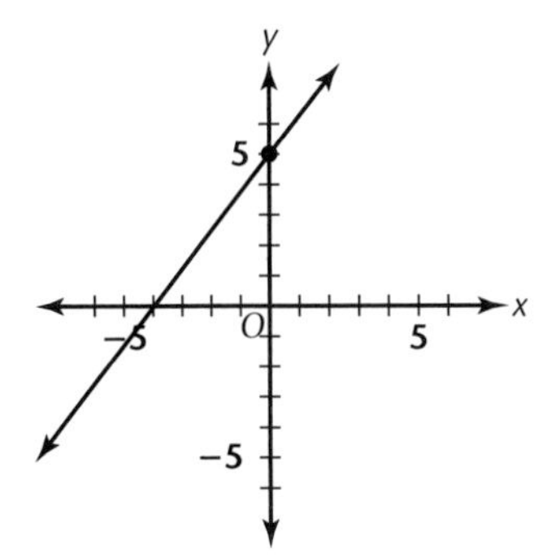

10.

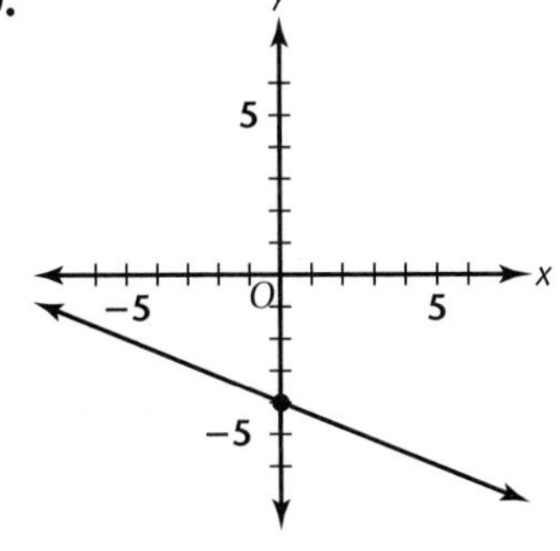

11.

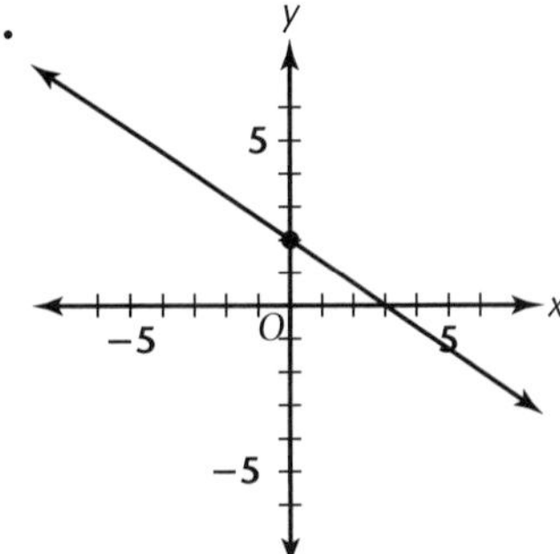

12.

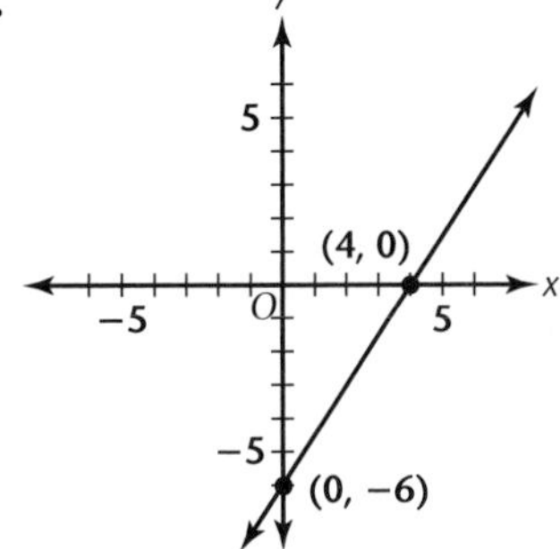

13.

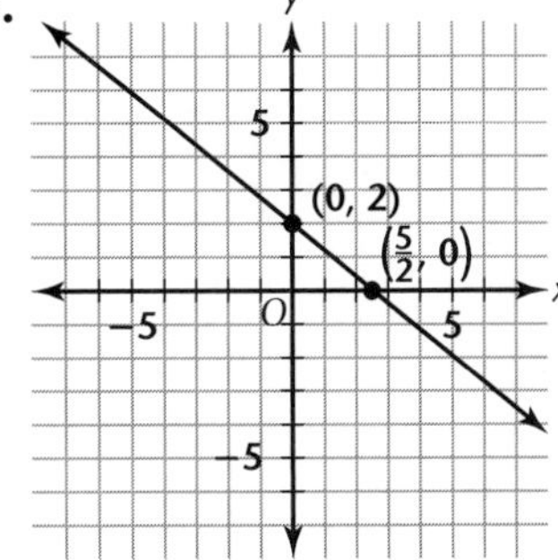

14.

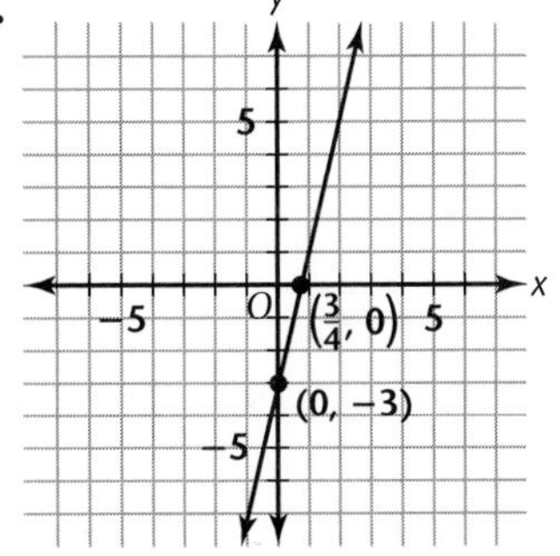

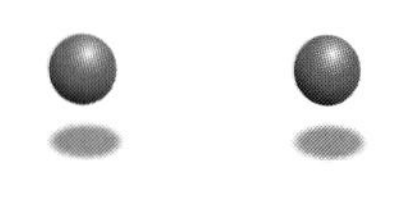

15.

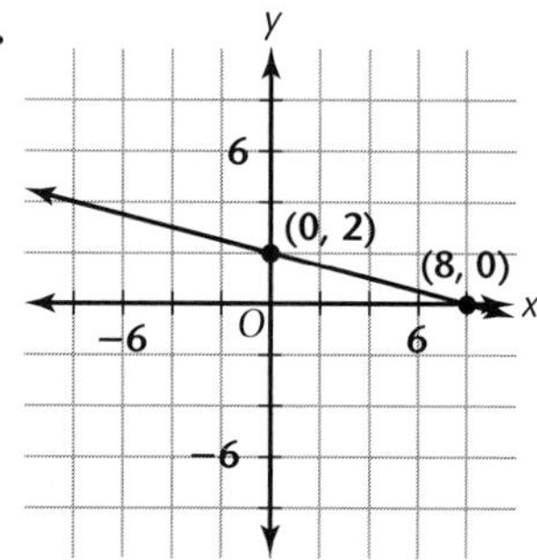

16.

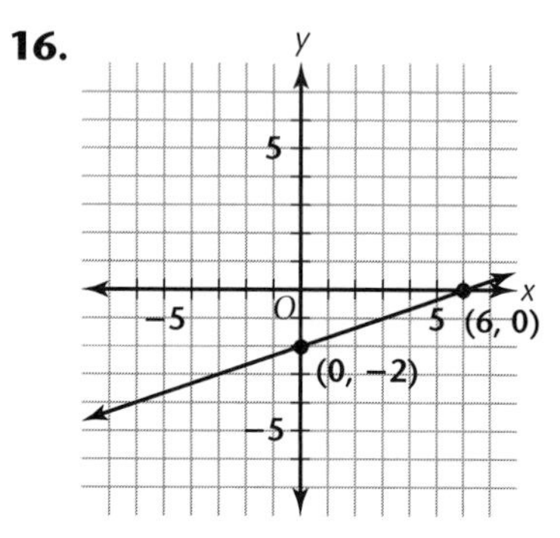

17.

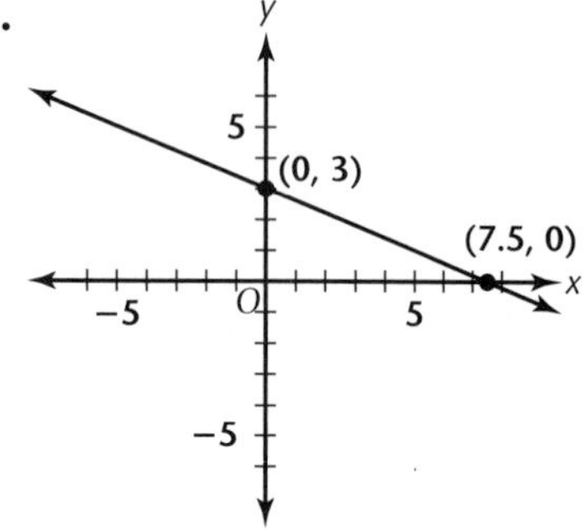

18. $y = -\frac{4}{5}x + 3$; $m = -\frac{4}{5}$

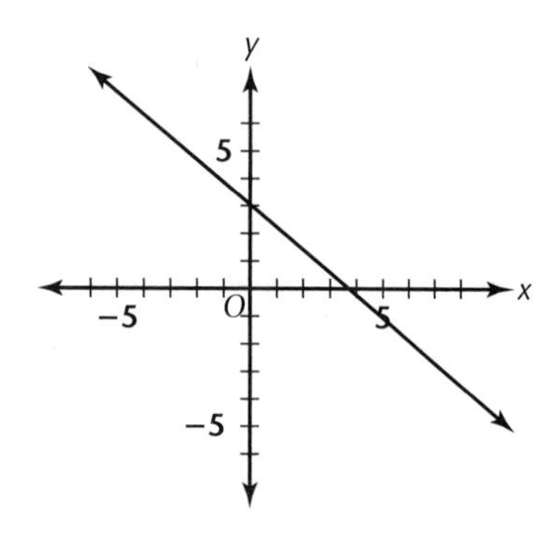

19. $y = 3x - 24$; $m = 3$

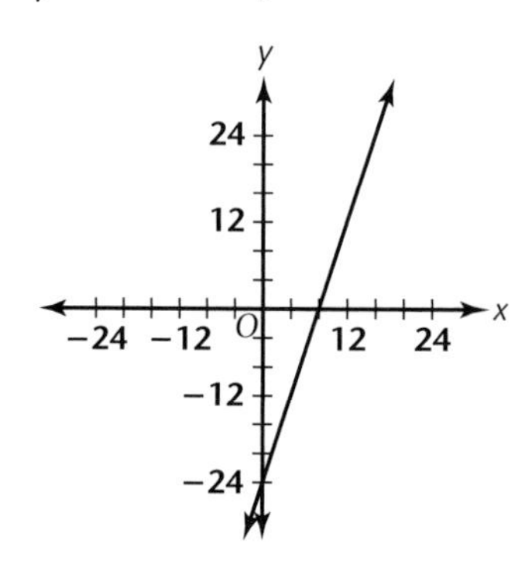

20. $y = -0.4x + 3.75$; $m = -0.4$

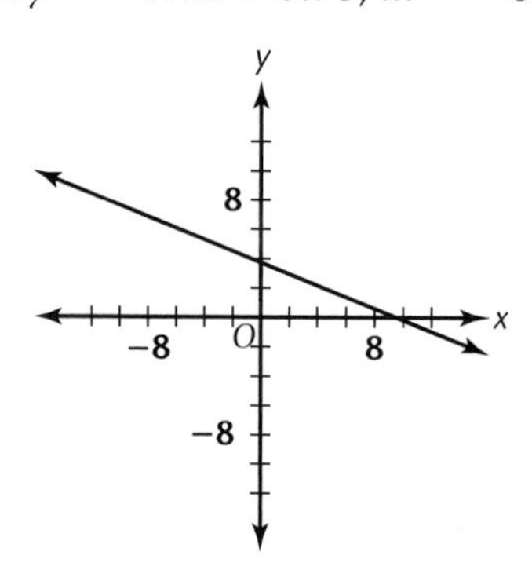

21. $y = 1.4x - 7.5$; $m = 1.4$

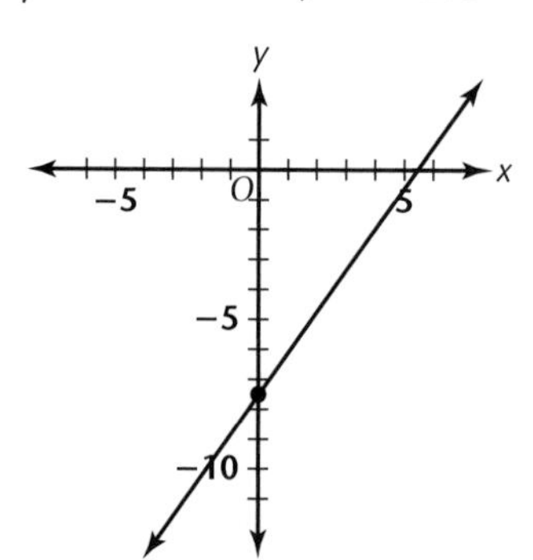

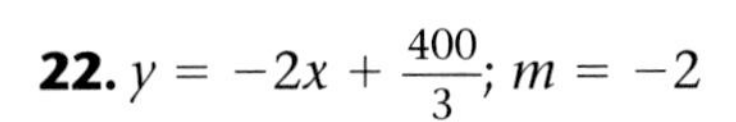

22. $y = -2x + \frac{400}{3}$; $m = -2$

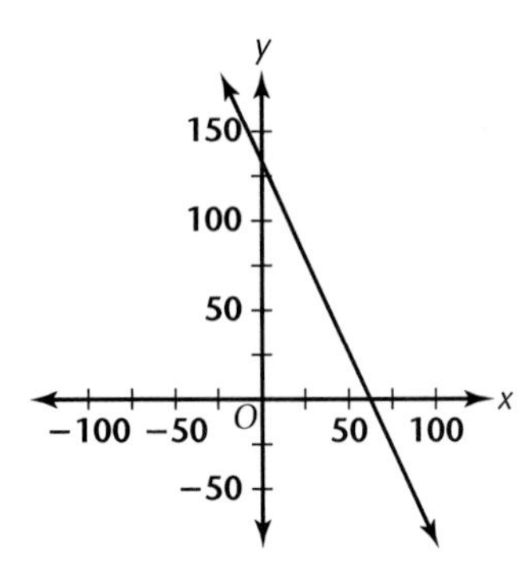

23. $y = -3.7x + 5.2$; $m = -3.7$

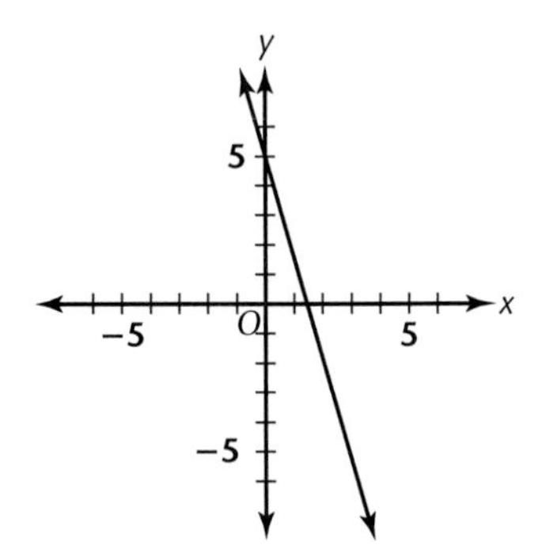

34.

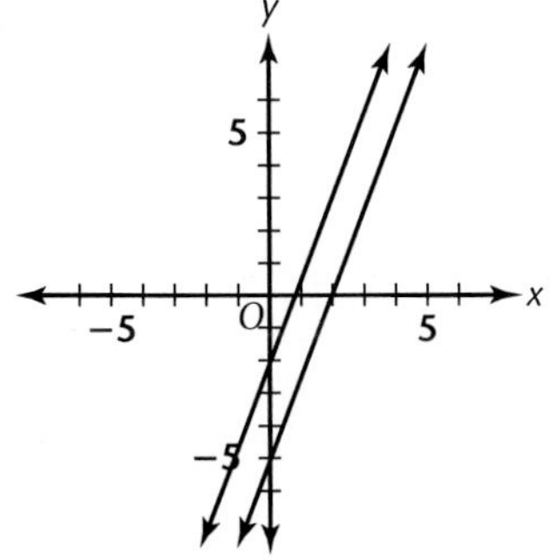

They have the same slope and are parallel.

42.

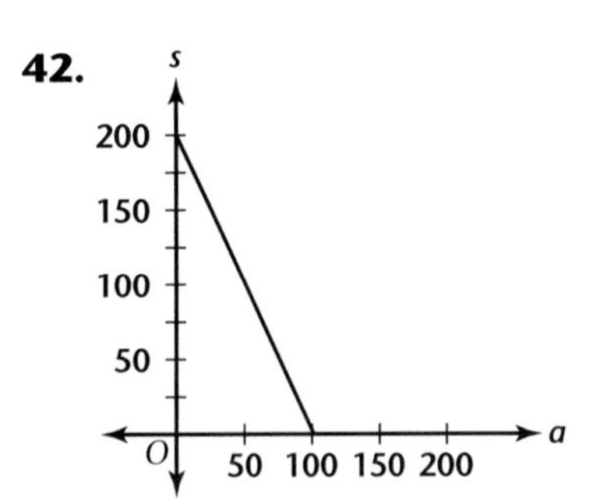

43. Answers may vary. One example is given.

a	*s*
0	200
12	176
24	152
36	128
48	104
60	80
72	56

Look Beyond, **page 71**

52.

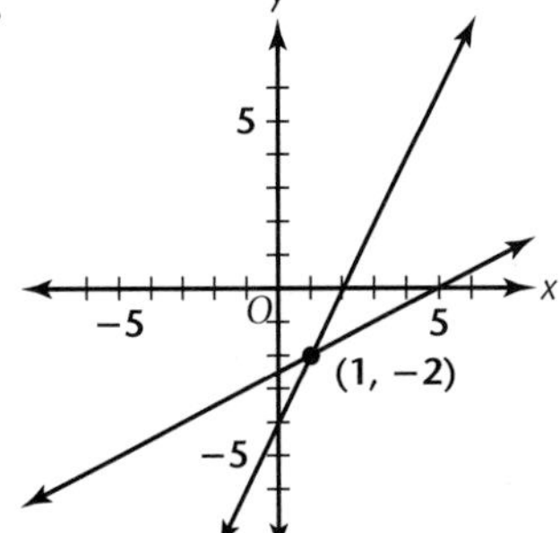

Extra Practice, **page 529**

1.

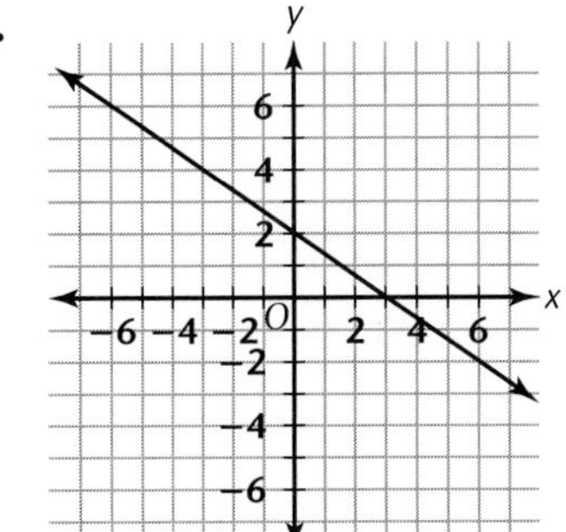

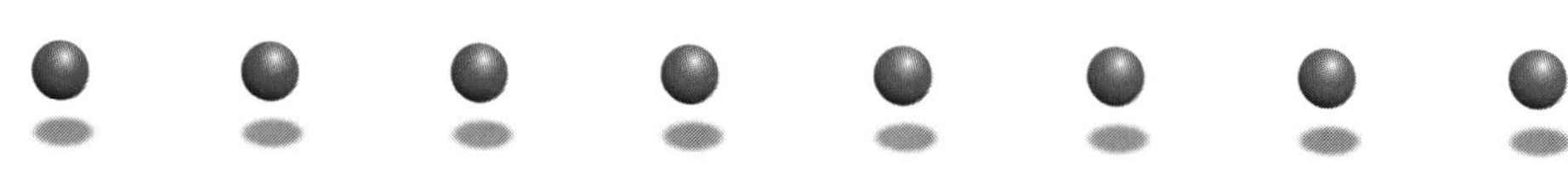

2.

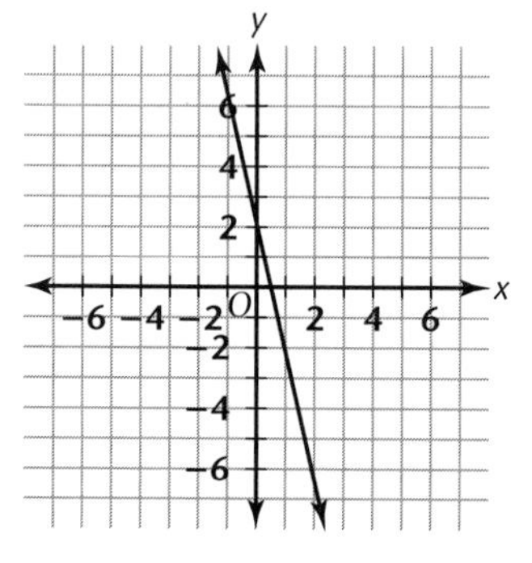

3.

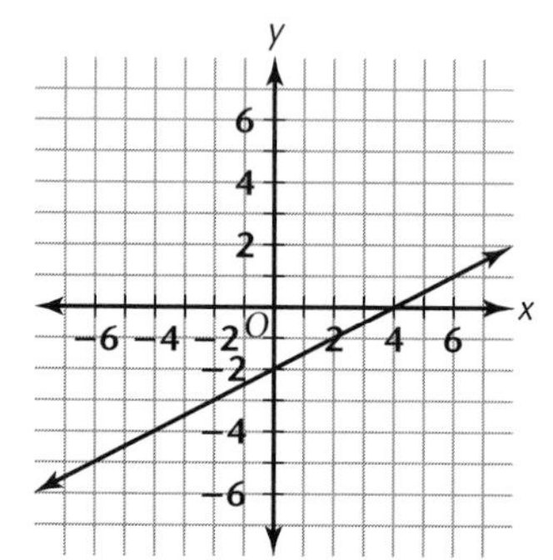

4.

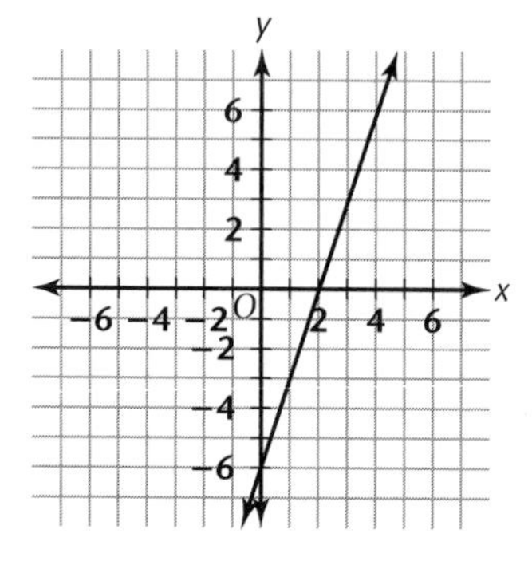

5.

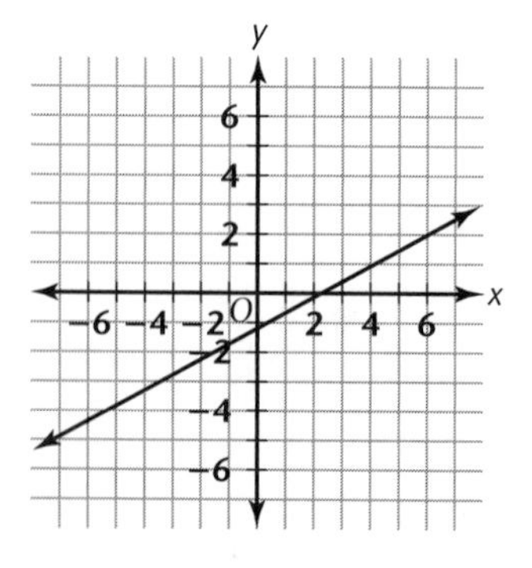

6.

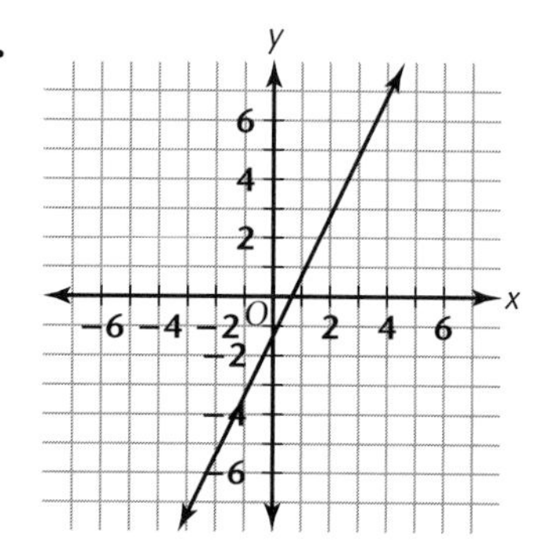

7. $y = 2x + 5; m = 2$

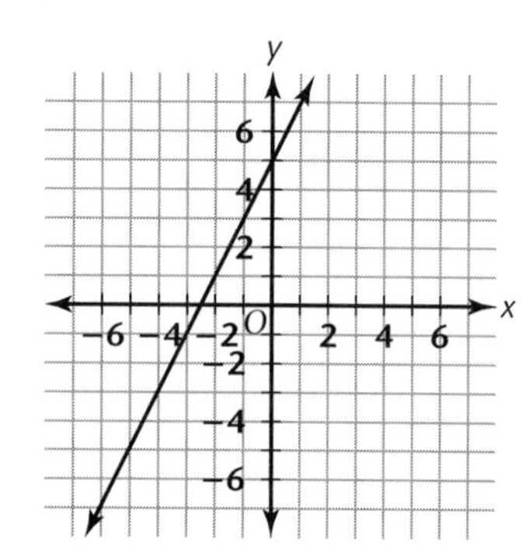

8. $y = \frac{1}{2}x + 2; m = \frac{1}{2}$

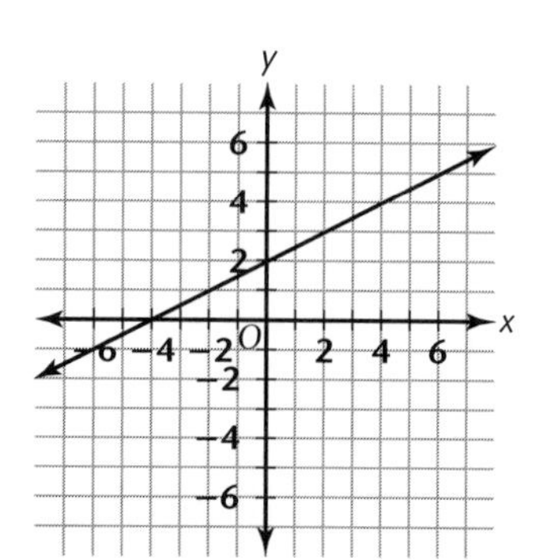

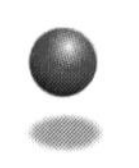       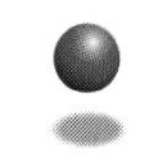

9. $y = -\frac{2}{3}x + 1;\ m = -\frac{2}{3}$

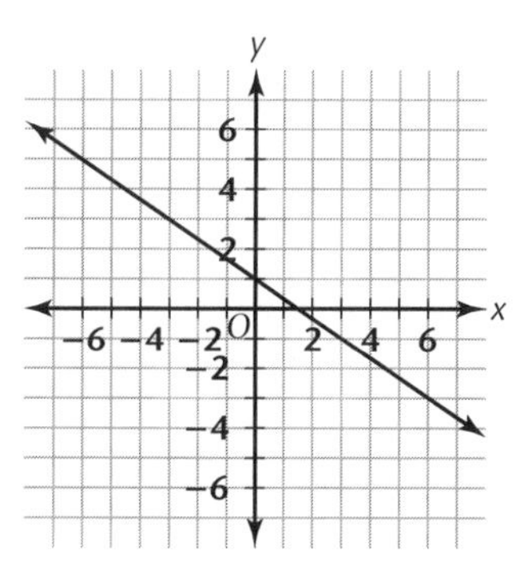

10. $y = \frac{1}{5}x - \frac{2}{5};\ m = \frac{1}{5}$

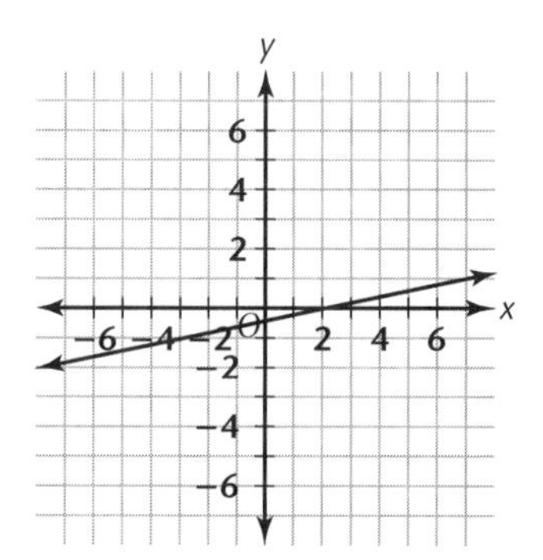

11. $y = x - \frac{5}{3};\ m = 1$

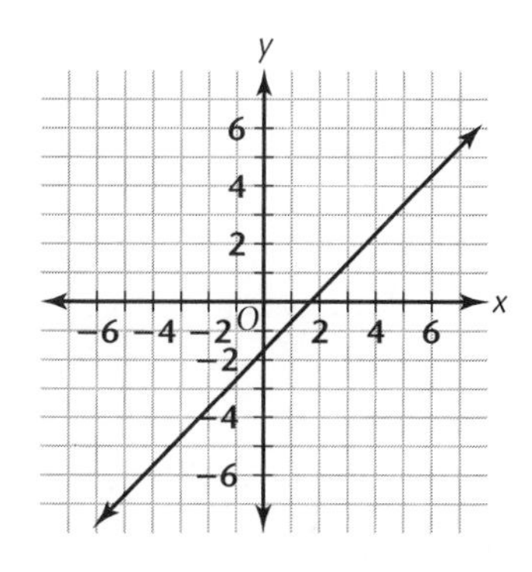

12. $y = 4x - 8;\ m = 4$

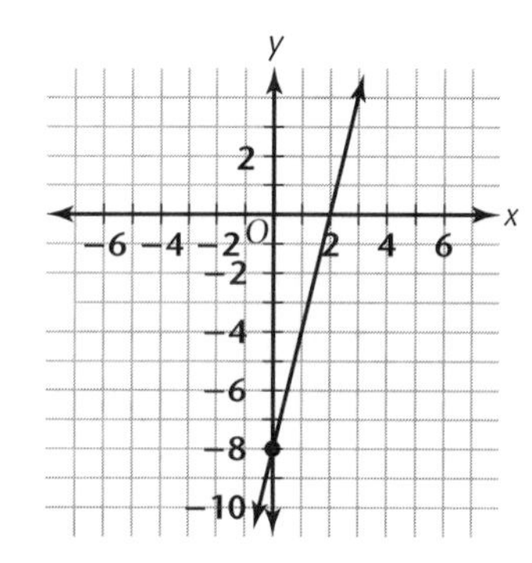

13. $y = \frac{2}{5}x - \frac{3}{5};\ m = \frac{2}{5}$

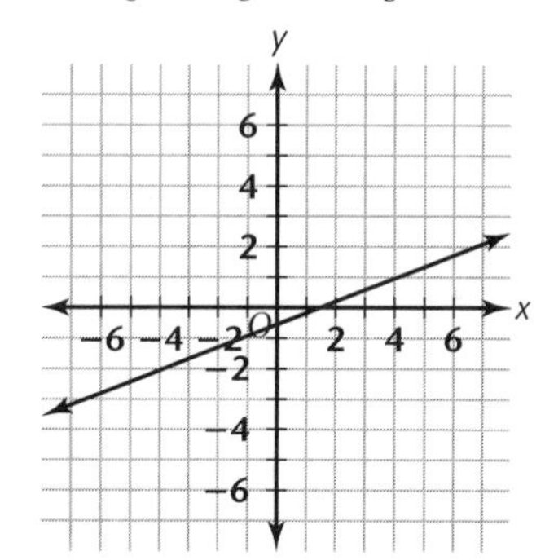

14. $y = -0.4x + 0.5;\ m = -0.4$

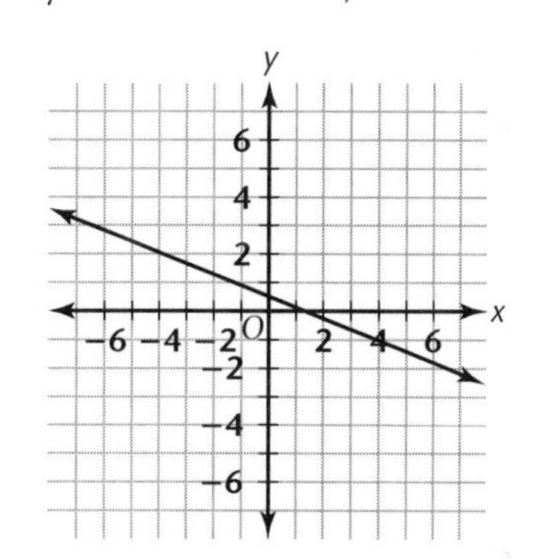

15. $y = 2.6x - 1.2$; $m = 2.6$

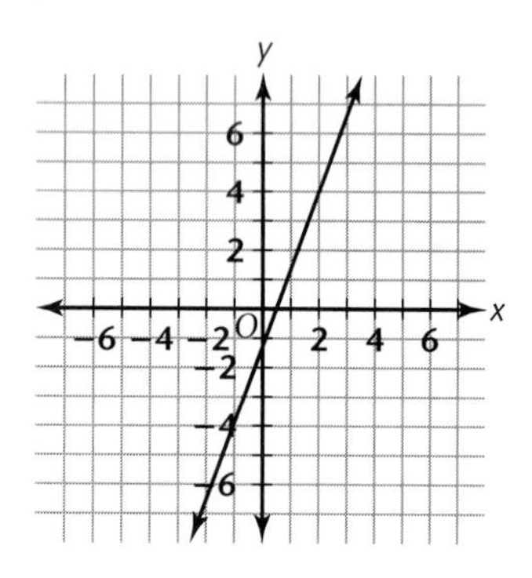

Lesson 2.3

Try This, **page 74**

$(-1, 0)$

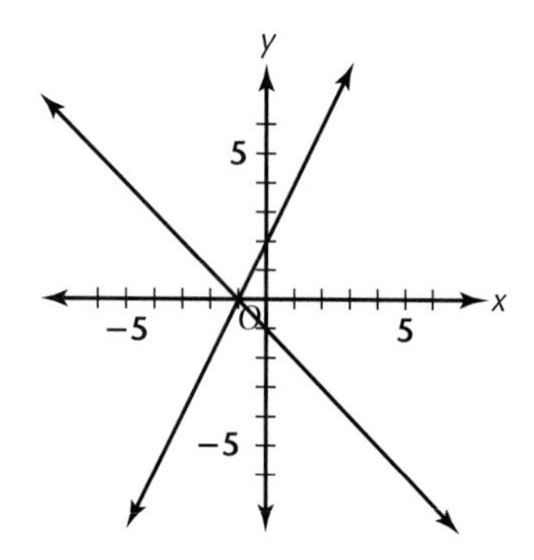

Communicate, **pages 76–77**

4. Write the equations in slope-intercept form. Graph the equations by substituting values for x and finding values for y to plot points and graph lines. Find the point where the two lines intersect. A good estimate of the solution is $\left(\frac{7}{2}, -\frac{1}{2}\right)$.

Practice and Apply, **pages 77–78**

18. No

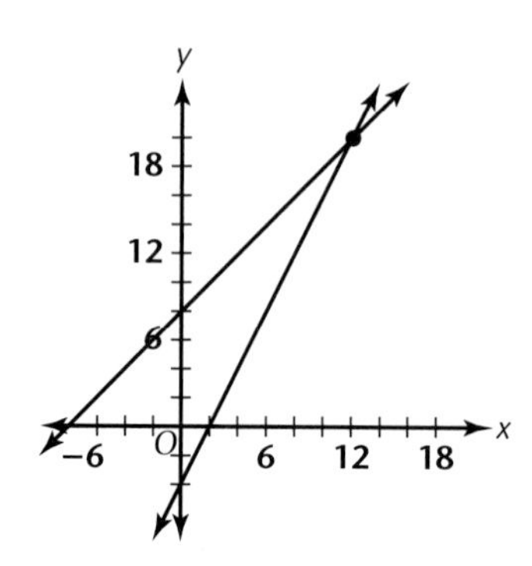

19. (2, 10); Yes

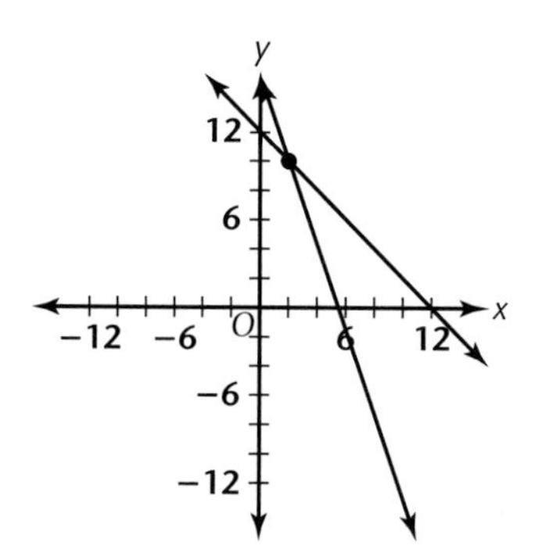

20. (2, 10); Yes

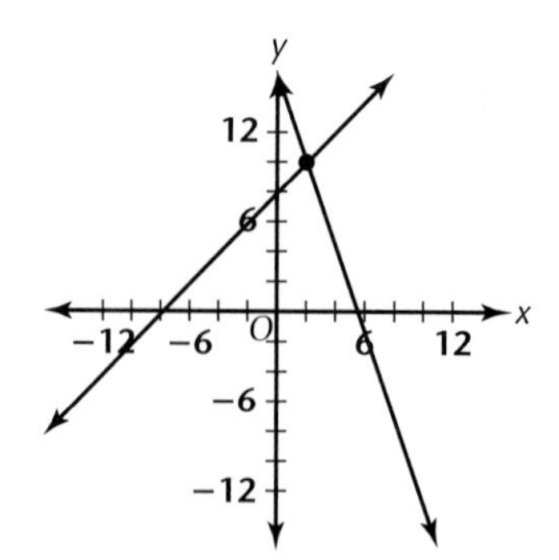

21. No intersection

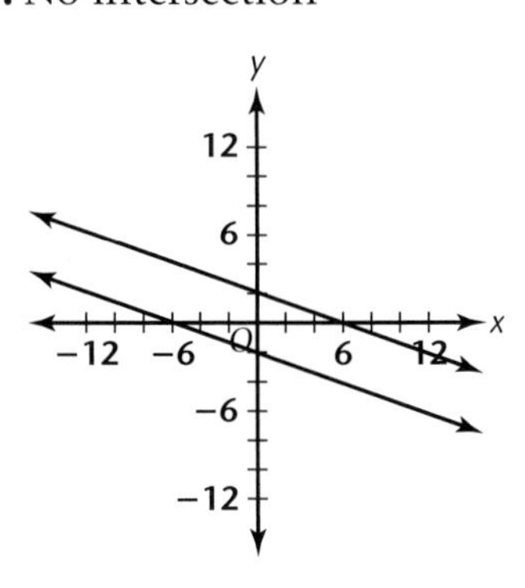

25. $y = -6x + 23$
26. $y = -\frac{1}{3}x + 6$

27.

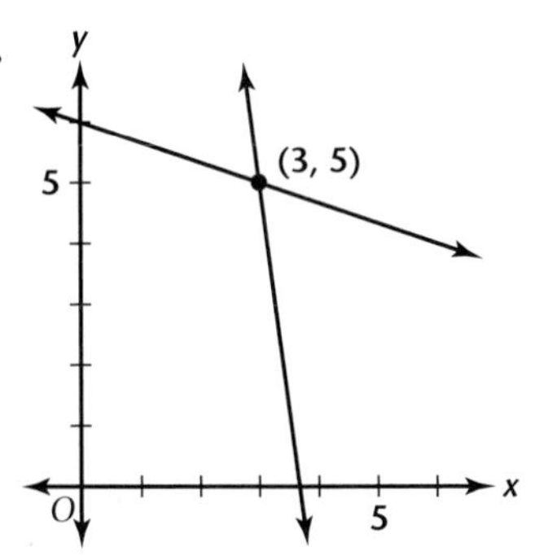

28. (3, 5) satisfies both equations.

29–30.

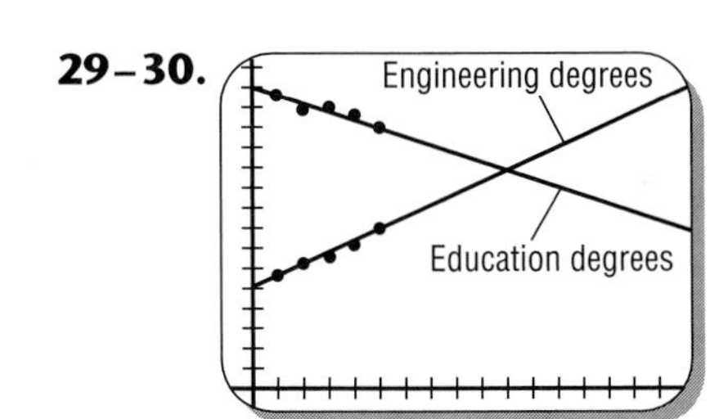

32. Answers may vary. For example, choose engineering because the graph shows an increase in degrees for engineers. Engineering is a growing field. Or choose education because fewer degrees might mean more demand for people in education.

Look Back, **page 78**

39. $-3 < x \leq 5$; all real numbers between -3 and 5, including 5 but not including -3.

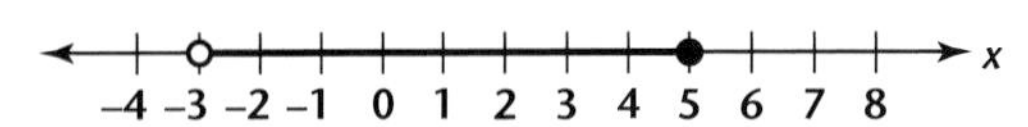

Lesson 2.4

Exploration 2, **page 80**

3. Both methods yield the same solution. The method in Step 1 involved substituting an expression for x which yielded a solution for y. The method in Step 2 involved substituting an expression for y which yielded a solution for x.

4. Solve for either variable. Substitute the value of one variable in terms of the other variable (as an expression) into the other equation. Solve, and then substitute the known variable into the equation to find the value of the other variable.

Exploration 3, **page 80**

4. Solve for the variable whose coefficient is 1. Substitute this expression into the other equation to solve for the value of the unknown variable. Substitute this value into the equation to determine the value of the other variable.

Communicate, **page 82**

1. Substitute 42 for y and solve for x; $x = 17$.

2. Solve the first equation for y, because the coefficient of y is 1; solve for x and y. The solution is (2, 10).

3. Solve the first equation for x, because the coefficient of x is 1; solve for x and y. The solution is (10, 1).

Practice and Apply, **pages 82–83**

Estimates may vary for Exercises 12–19.

12. $\left(\frac{1}{4}, \frac{1}{4}\right)$ or (0.25, 0.25)

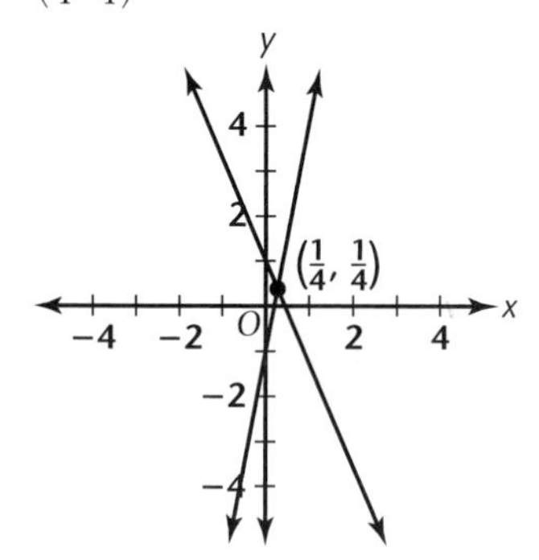

13. $\left(\frac{1}{3}, \frac{1}{3}\right)$ or approximately (0.3, 0.3)

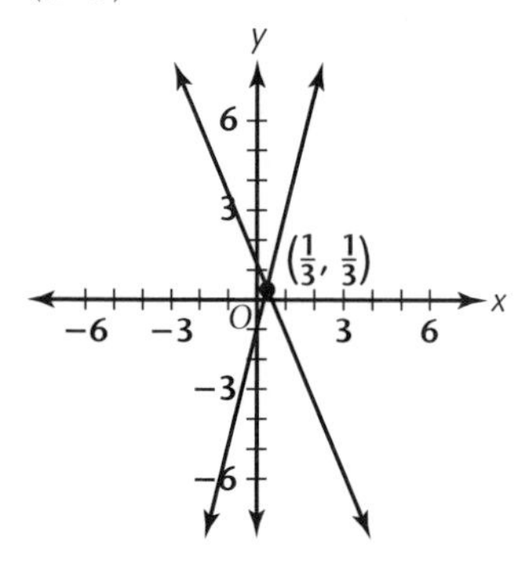

14. (6, 6)

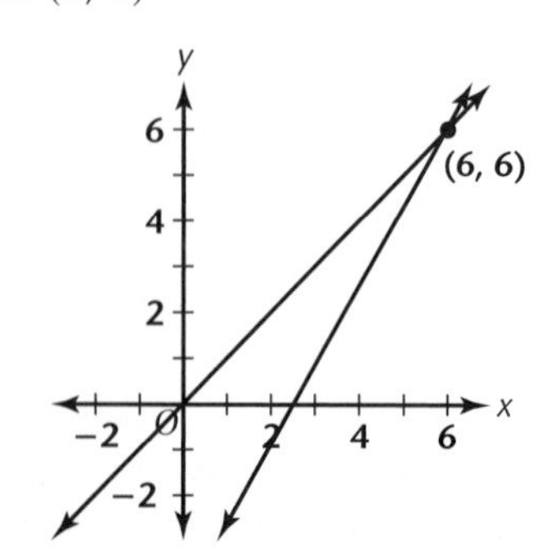

15. (−18, −28)

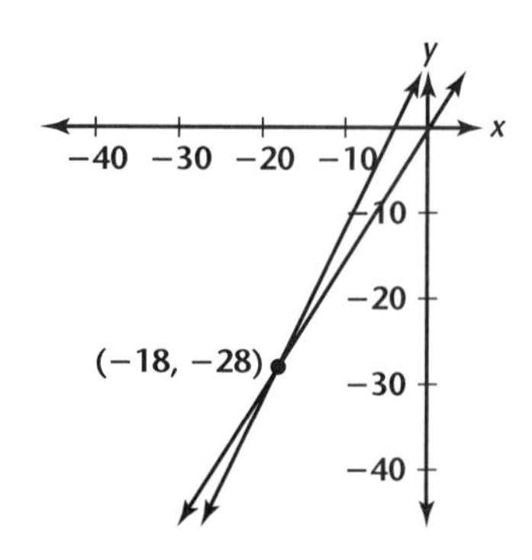

16. $\left(\frac{1}{5}, \frac{11}{5}\right)$ or (0.2, 2.2)

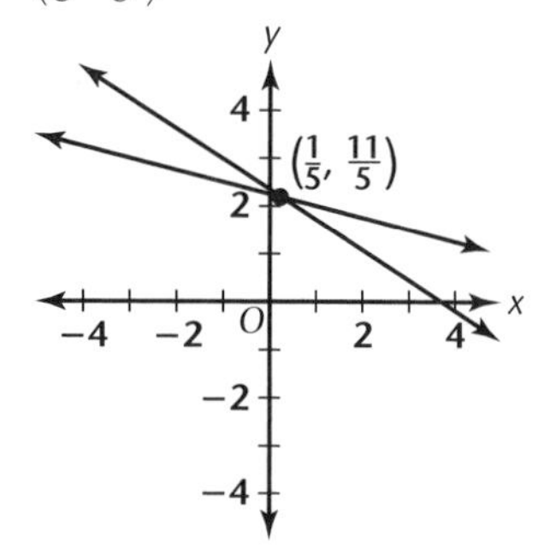

17. $\left(\frac{57}{10}, \frac{78}{10}\right)$ or (5.7, 7.8)

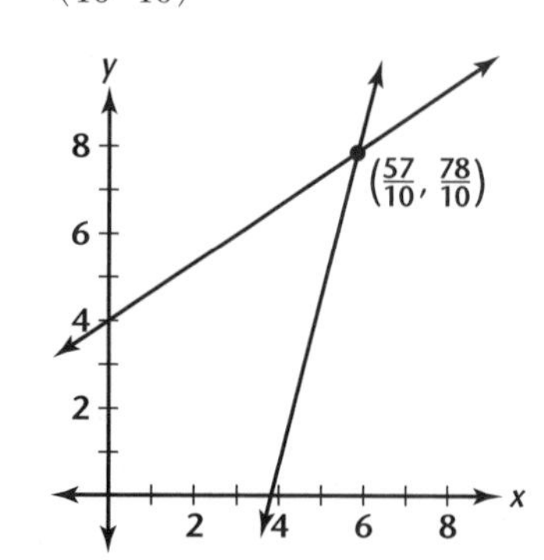

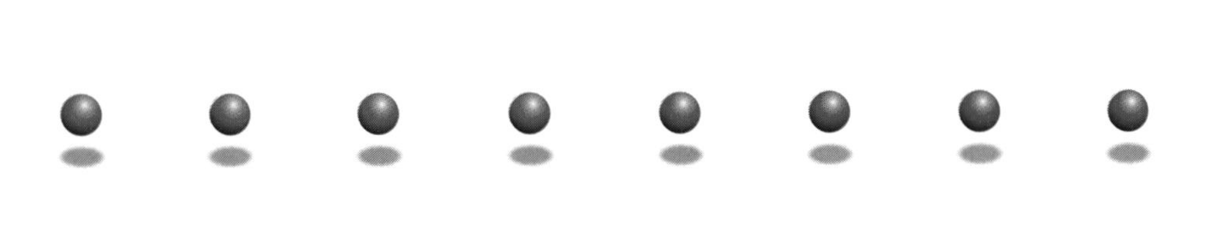

18. (6, −1)

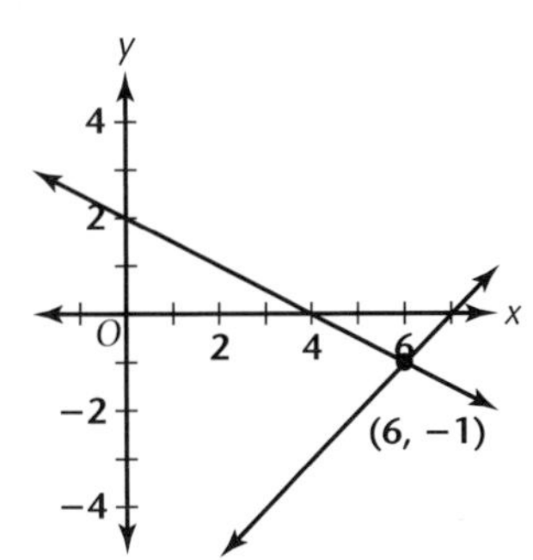

19. (4, 2)

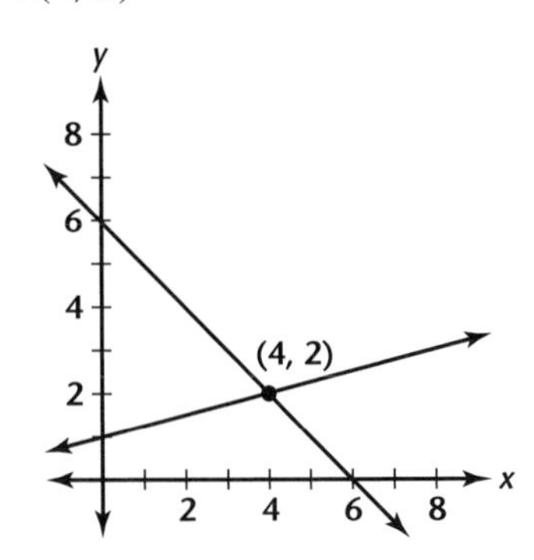

20. $\begin{cases} y = 7452 - 30x \\ y = 754 + 4x \end{cases}$; 197 seconds

21. $\begin{cases} x + y = 27 \\ y = 3 + x \end{cases}$; 15 and 12

22. $\begin{cases} x = 3y - 4 \\ 3 + 2x - 2y = 11 \end{cases}$; 8 and 4

23. $\begin{cases} 2l + 2w = 210 \\ l = 2w \end{cases}$; About 35 meters by 70 meters

Lesson 2.5

Communicate, **page 87**

1. y and $-y$ are opposites. Use the Addition Property of Equality to solve for x, then y. (9, 4)
2. $-3y$ and $3y$ are opposites. Use the Addition Property of Equality. (4, 0)
3. $2a$ and $-2a$ are opposites. Use the Addition Property of Equality. $\left(\frac{13}{2}, -7\right)$
4. Multiply each side of $x - 2y = 13$ by -4.
5. Multiply each side of $a - 2b = 7$ by -2.
6. Multiply each side of $3m - 5n = 11$ by 3 and $2m - 3n = 1$ by 5.
7. Multiply each side of equation 1 by -2. Solve for x, then y. $\left(11, -\frac{13}{3}\right)$
8. Multiply each side of equation 1 by -3 and each side of equation 2 by 2. Solve for x and y. $(-2, -1)$
9. Multiply each side of equation 1 by -3. Solve for x and y. $\left(\frac{1}{3}, -\frac{1}{2}\right)$

Practice and Apply, **pages 87–89**

17. 1 pizza, \$6.99; 1 soda, \$0.79

27. $\begin{cases} b + s = 6000 \\ 0.05b + 0.09s = 380 \end{cases}$; $b = \$4000$, $s = \$2000$

28. $\begin{cases} c + s = 25 \\ 6.99c + 10.99s = 230.75 \end{cases}$; $c = 14$, $t = 11$

29. $\begin{cases} 2l + 2w = 24 \\ l = 3w \end{cases}$; $l = 9$ cm, $w = 3$ cm

30. $\begin{cases} 2n + 4m = 205 \\ 3n + 8m = 342.50 \end{cases}$; $m = \$17.50$, $n = \$67.50$

31. (−2, −4)

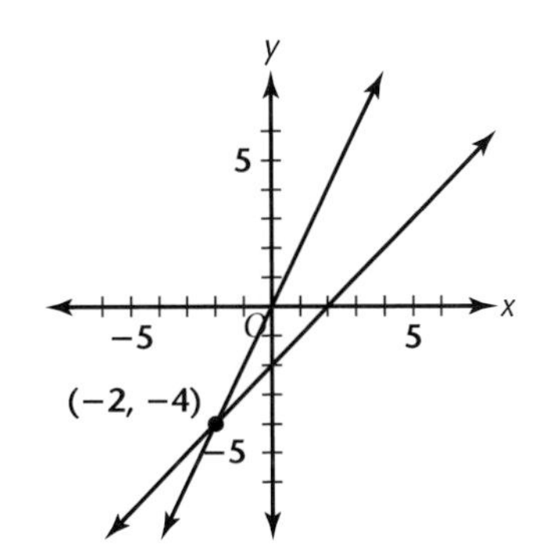

32. (−2, −4)

33. Answers may vary. For example, use either substitution or elimination.

Look Beyond, **page 89**

45. The lines have the same slope and are parallel.

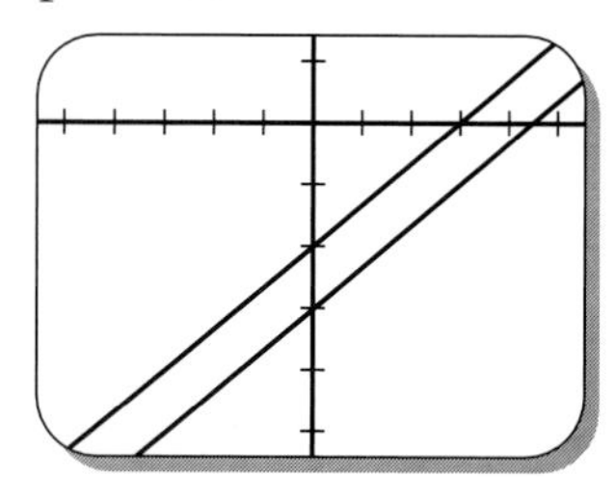

Lesson 2.6

Ongoing Assessment, **page 92**

Independent systems have lines with different slopes. Both dependent and inconsistent systems have lines with the same slope. Independent systems have lines that intersect at one point, dependent systems have only one line, and inconsistent systems have parallel lines. Independent systems have one solution, dependent systems have an infinite number of solutions, and inconsistent systems have no solutions. Independent systems have the same or different y-intercepts, dependent systems have the same y-intercept, and inconsistent systems have different y-intercepts.

Try This, **page 93**

First equation Second equation

First Equation **Second Equation**

Communicate, **pages 93–94**

1. Since both the slopes and y-intercepts of both lines are the same, the system is dependent.

2. Since the slopes of these two lines are not the same but their y-intercepts are, the system is independent.

3. Since the slopes of both equations are the same but their y-intercepts are not, the system is inconsistent.

4. Multiply both sides of the equation by the same number. The slope and the y-intercept remain the same.

5. Use the slope of the given equation (-3) and choose a different y-intercept.

6. If two lines intersect at only one point, the lines have different slopes. The y-intercepts may or may not be the same.

22. $\overline{AB}$: $y = -\frac{3}{2}x + 10$; $\overline{CD}$: $y = -\frac{3}{2}x + 3$

23. Inconsistent; the lines are parallel; they have the same slope.

24. Answers may vary. For example, $y = \frac{2}{3}x + \frac{4}{3}$. Graph of any line that contains point (4, 4) other than that of $\overline{AB}$.

25. Answers may vary. For example, $2y = -3x + 6$. Graph should be the same as that of $\overline{CD}$.

Look Back, **page 95**

40.

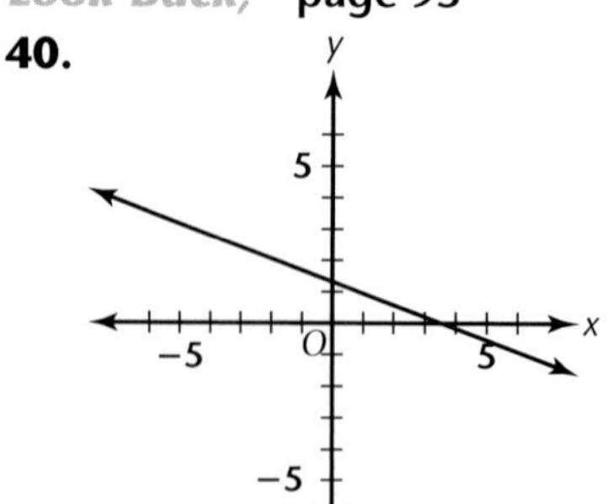

Lesson 2.7

Exploration, **page 96**

1.

Critical Thinking, **page 100**

0 hours of lawn mowing and 18 hours of library work: \$126; 30 hours of library work and 0 hours of lawn mowing: \$210

Communicate, **page 101**

1. If the inequality symbol includes the equal sign, then use a solid line, otherwise use a dashed line.
2. Graph the boundary line, $y = -\frac{1}{2}x + 1$. Substitute a point that is not on the line. If it satisfies the inequality, shade the side of the line containing the point. If the point does not satisfy the inequality, shade the other side of the line. Answers may vary. Try (0, 0), because 0 is a simple number to check.
3. Graph the boundary lines using dashed lines. Locate a point that satisfies each inequality, and shade the intersection of the half-planes that contains each point.
4. Graph the boundary lines using dashed lines. Locate a point that satisfies each inequality, and shade the intersection of the half-planes that contains each point.
5. Graph the boundary lines using solid lines. Locate a point that satisfies each inequality, and shade the intersection of the half-planes that contains each point.
6. Graph $x + y \leq -2$ using a solid line, and graph $x + y > -2$ using a dashed line. Locate a point that satisfies each inequality, and shade the intersection of the half-planes that contains each point.
7. They plan for *no more* than 20 perennials, so $p \leq 20$. Their budget has room for *at least* \$30, so $5p + 1.5a \geq 30$. Use a solid line to graph the boundary line, $p = 20$. Use a solid line to graph $5p + 1.5a = 30$. Locate a point that satisfies each inequality, and shade the intersection of the half-planes that contains that point. Only positive integers in the overlapping region make sense.
$6 \leq p \leq 20$, $a \geq 6$

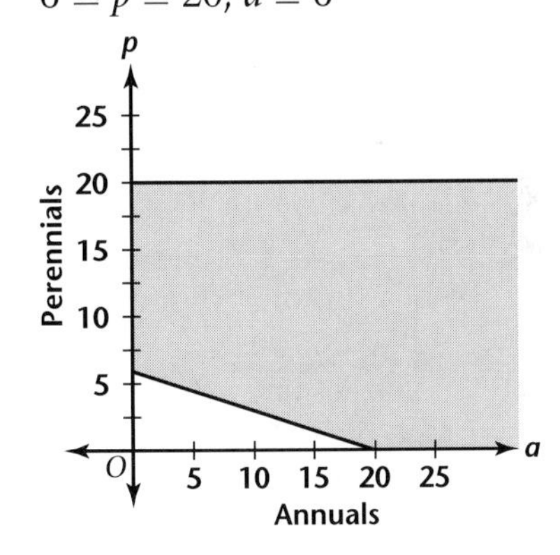

Practice and Apply, **pages 101–103**

8.

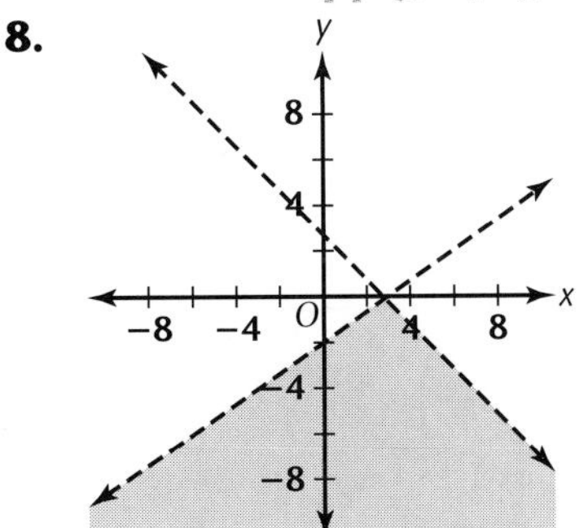

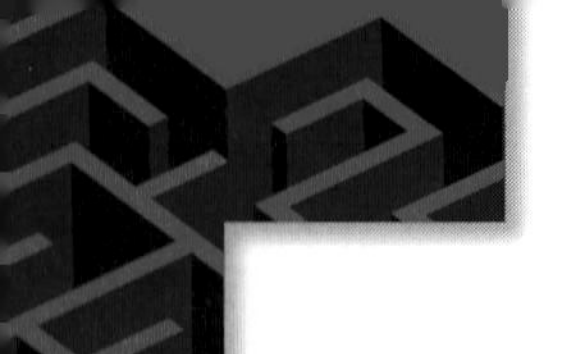        

9.

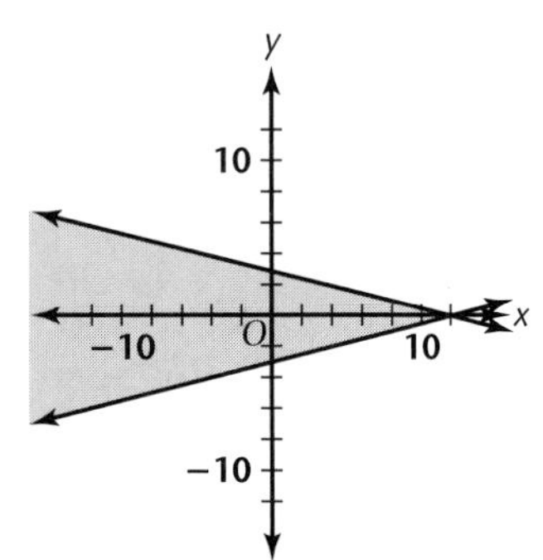

10.

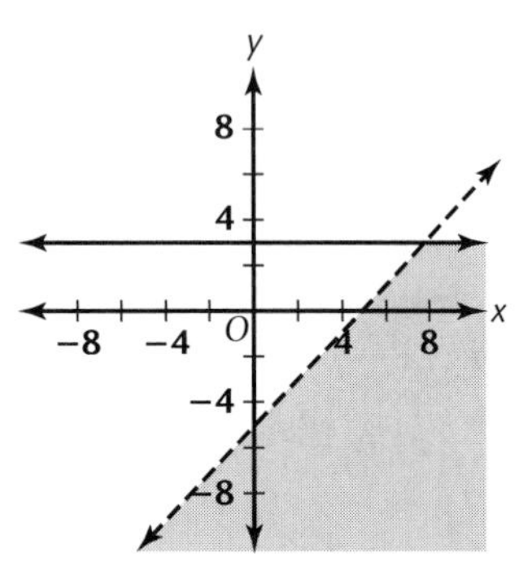

11.

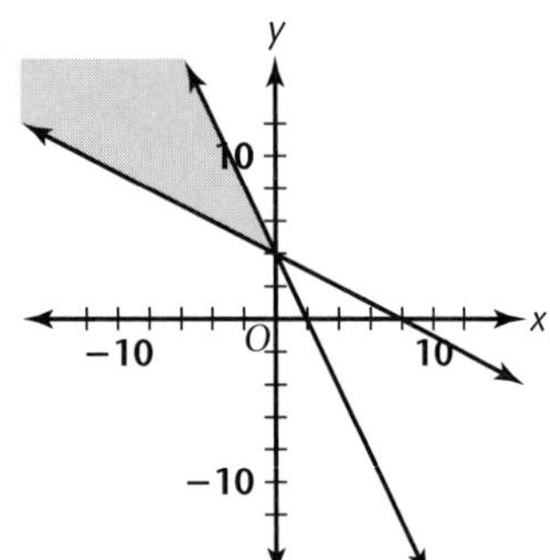

12.

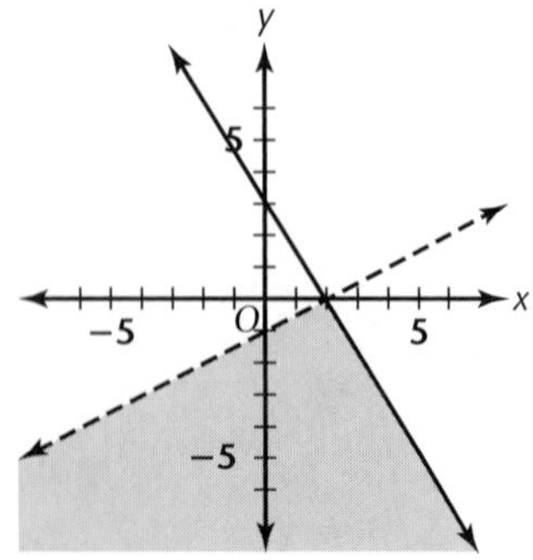

13.

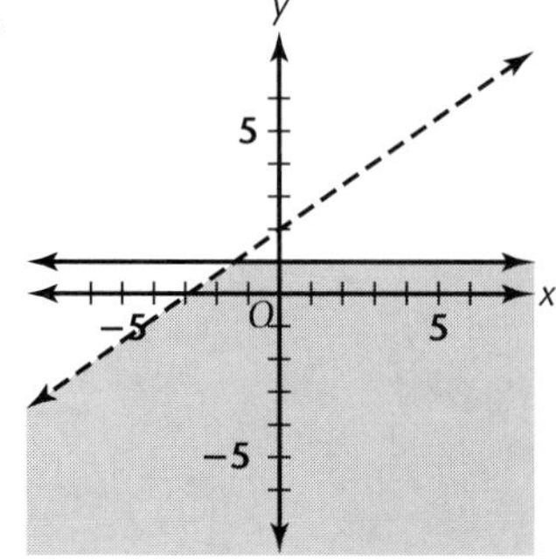

14.

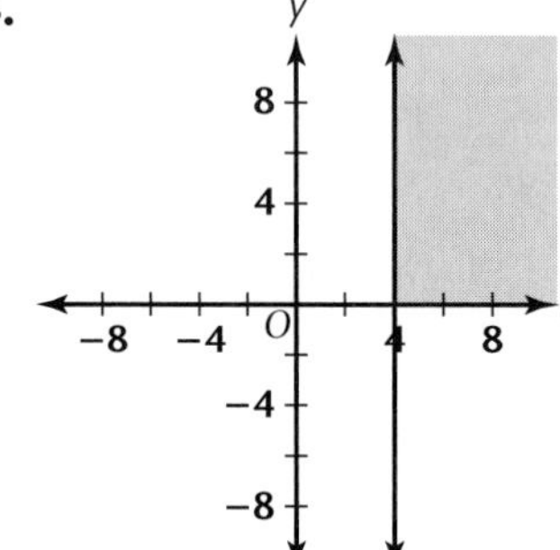

15.

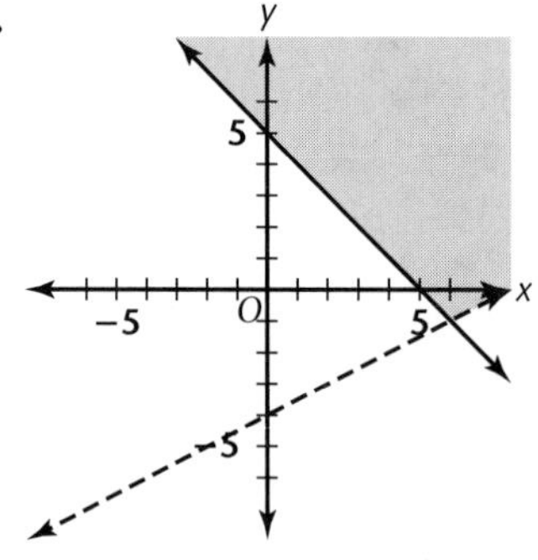

16.

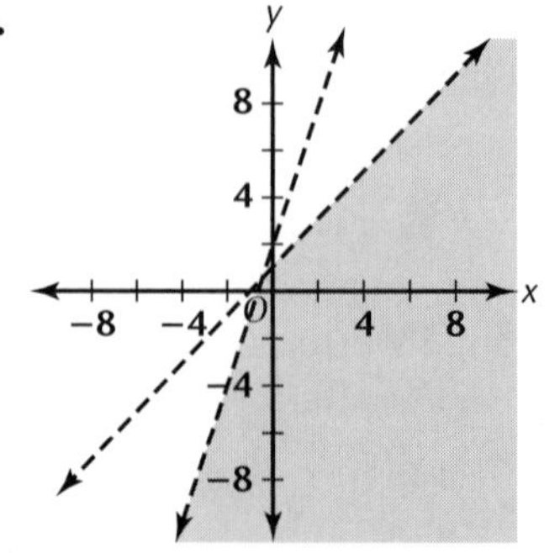

17.

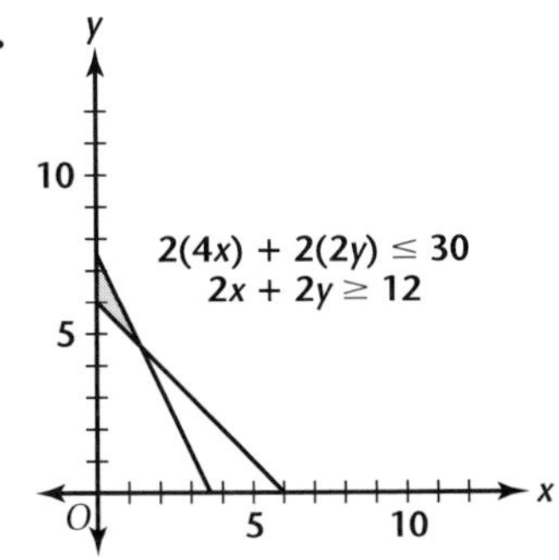

21. $\begin{cases} y \geq 2 \\ y < 3x - 1 \\ y < -\frac{1}{2}x + 9.5 \\ y \geq 3x - 22 \end{cases}$

22. $y \geq 2x$ and $y \leq 2x + 6$

23.

Baskets	0	1	2	3	4	5	6	7	8
Free throws	16	14	12	10	8	6	4	2	0

27.

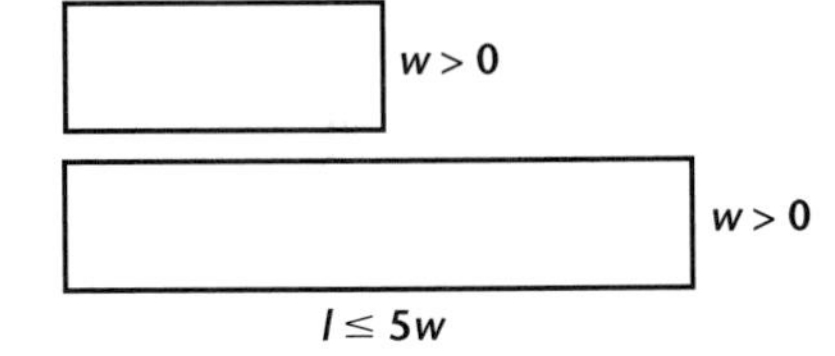

Extra Practice, **page 531**

1.

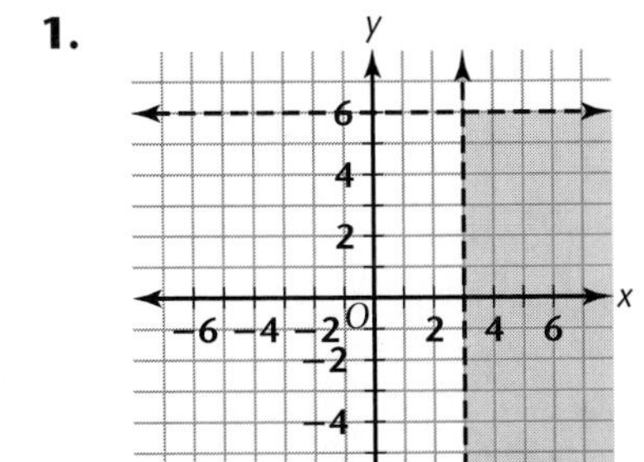

2.

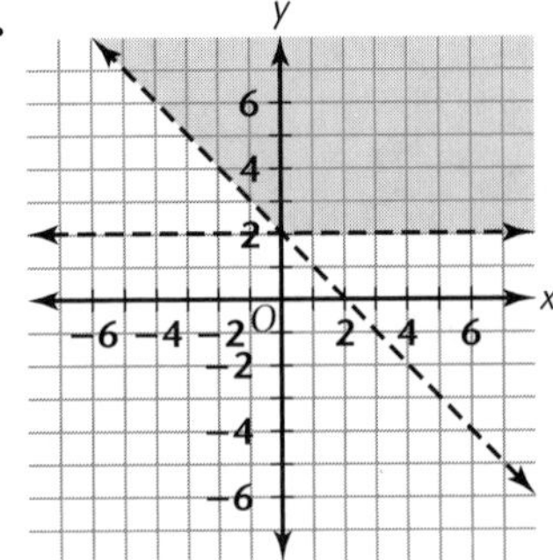

3.

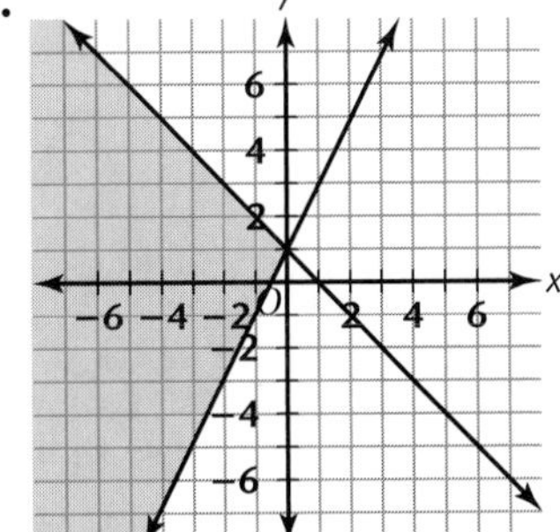

4.

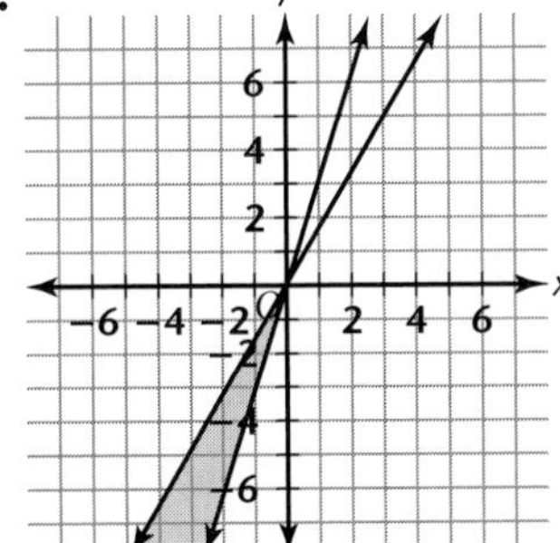

5.

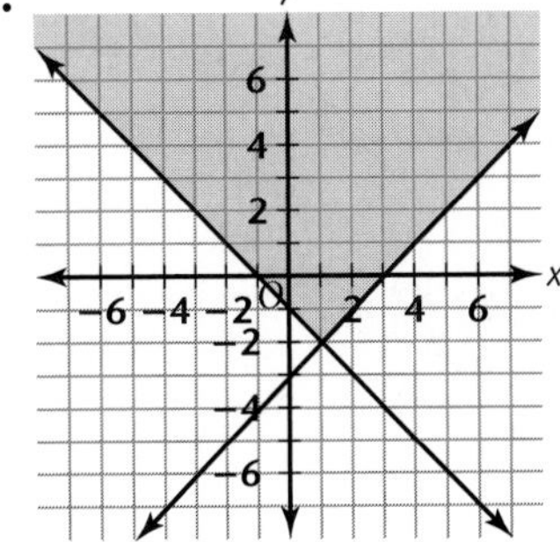

 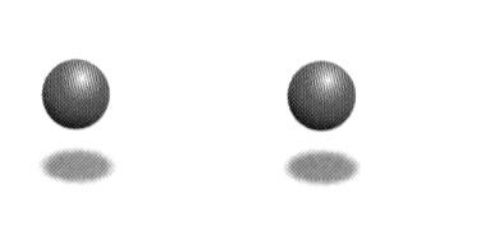    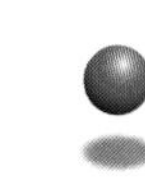 

6.

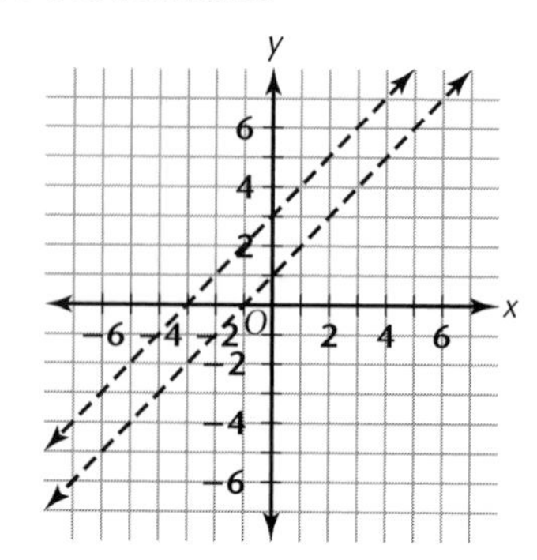

7. No solution

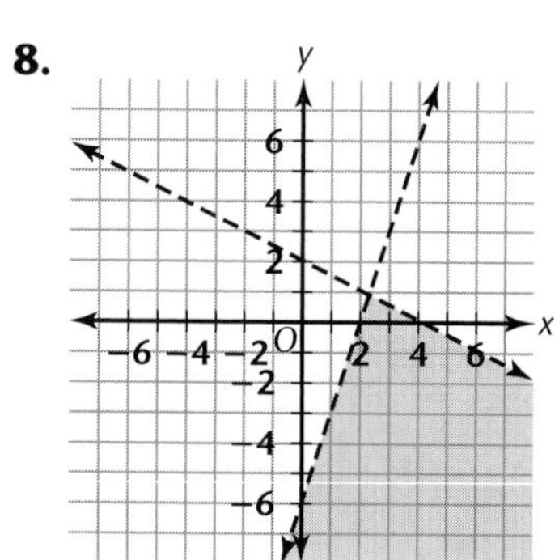

8.

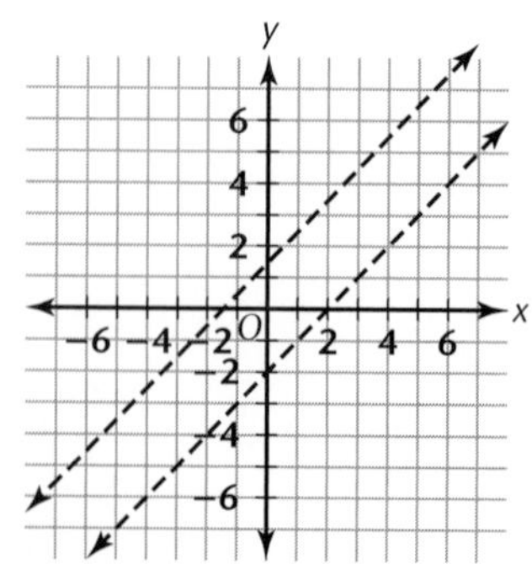

9. No solution

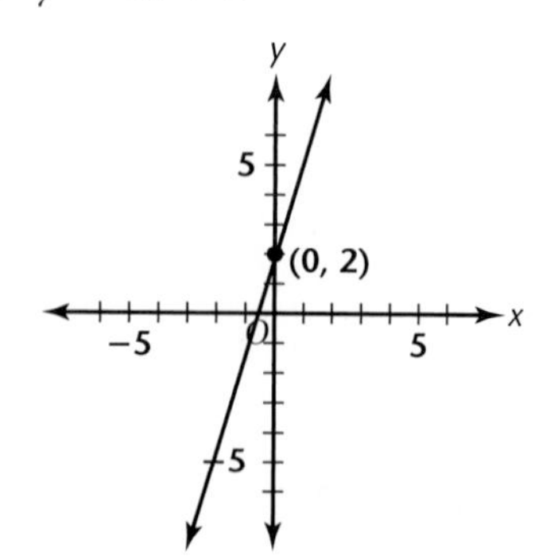

Lesson 2.8

Communicate, **page 108**

1. There are 20 nickels and quarters, so $q + n = 20$. The money mixture totaled \$2.60, so $0.25q + 0.05n = 2.60$. 8 quarters

2. There are 20 liters of the final solution, so $x + y = 20$. There are 2.6 liters of glucose in the final solution, so $0.25x + 0.05y = 2.6$. 8 liters

3–4. Let x = the speed of the bird, and let y = the speed of the wind. Then $x + y$ = the speed of the bird flying with the wind, and $x - y$ = the speed of the bird flying against the wind.

5. The bird's speed is 3 times faster flying with the wind than flying against the wind. $(x + y) = 3(x - y)$

6. Let t represent the tens digit and u represent the units digit. The sum of the digits is 8, so $t + u = 8$. The number is 62. To check the answer, add 16 to 62 which gives 78. Multiply 26 by 3 which also gives 78.

Chapter 2 Project

Activity, **page 111**

Mountain Crawler: 20 models; Dune Crawler: 45 models

Chapter 2 Review, pages 112–116

4. $y = 4x + 2$

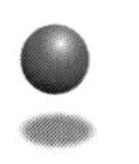

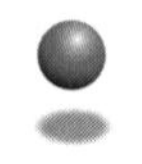

5. $x + y = 6$

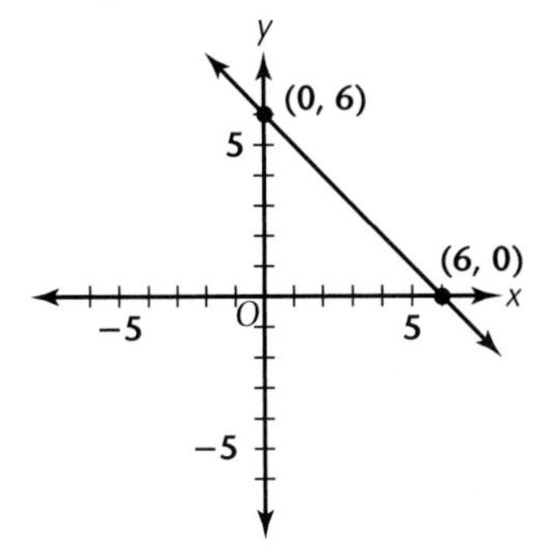

6. $3x - 4y = 12$

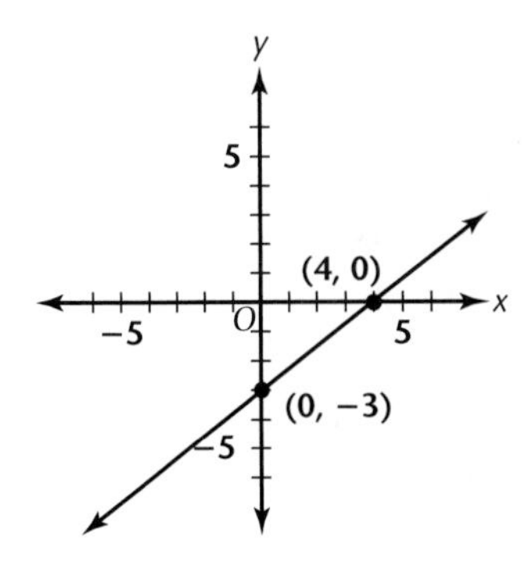

7. $y = -3x + 1$

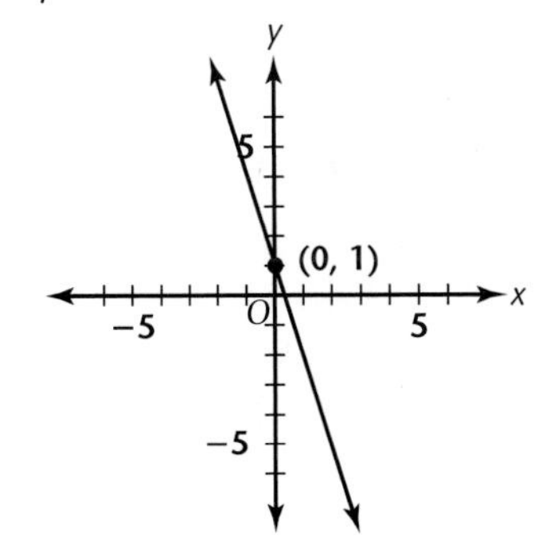

8. $(1, -5)$

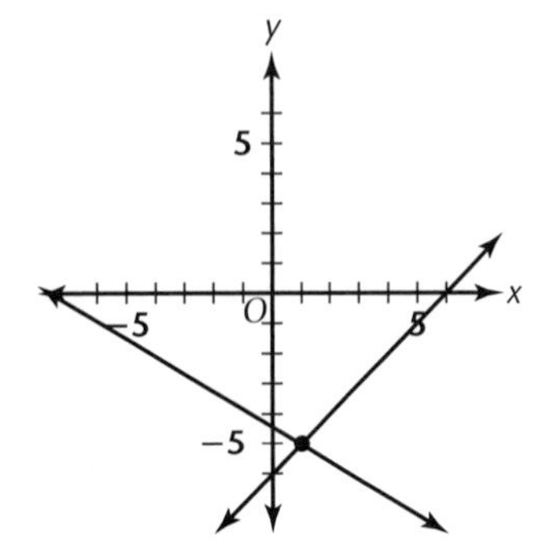

9. $(0, 0)$

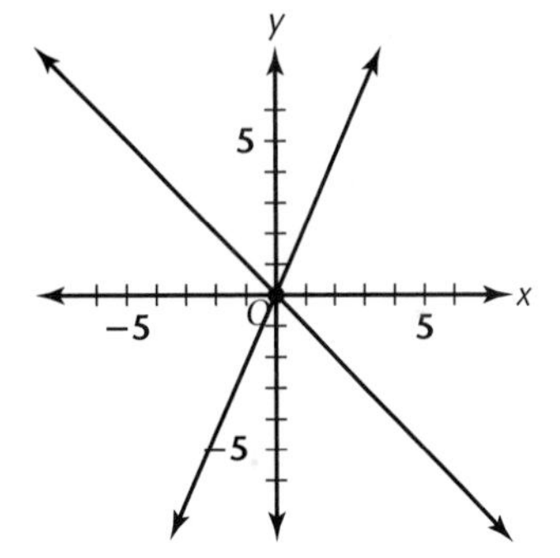

10. $(-3, 1)$

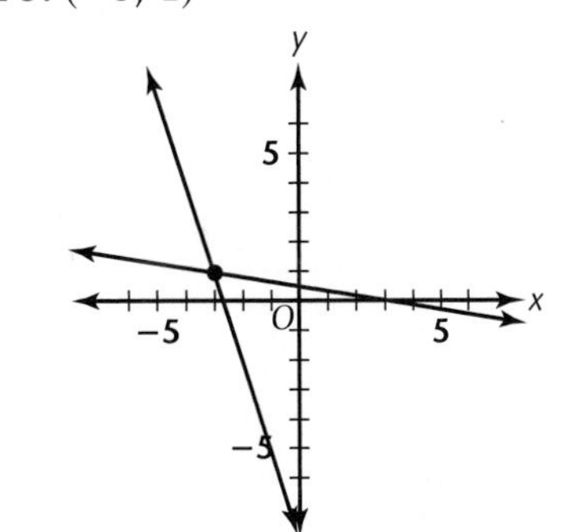

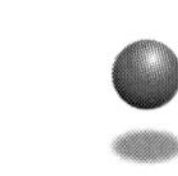

11. (2, 0)

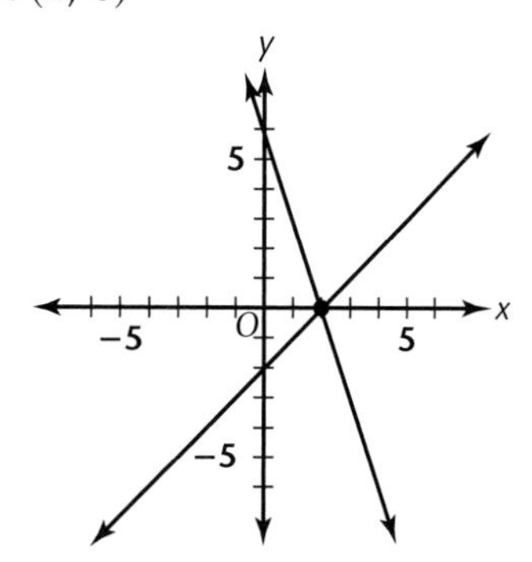

24.

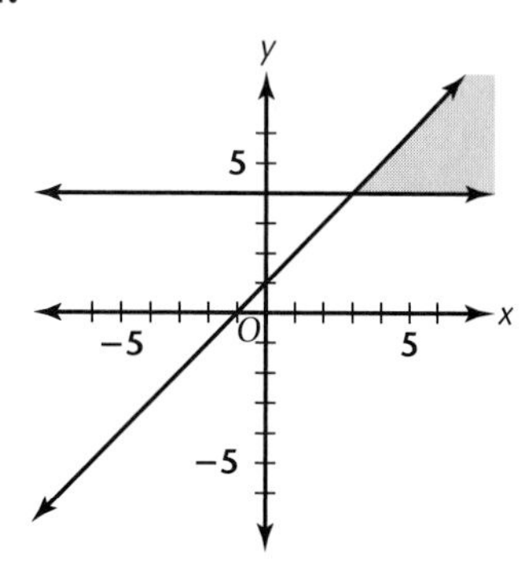

25.

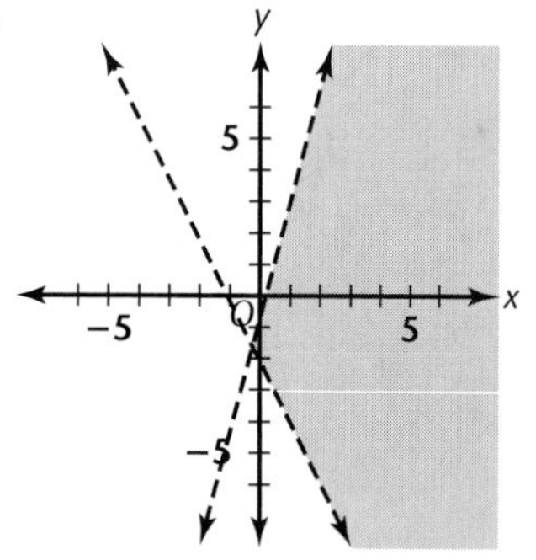

26.

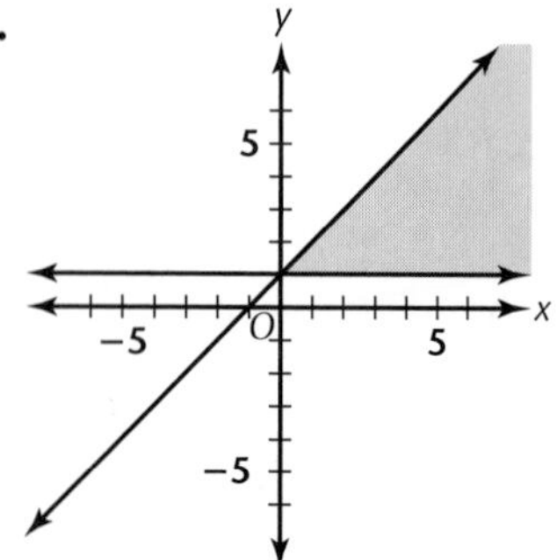

27.

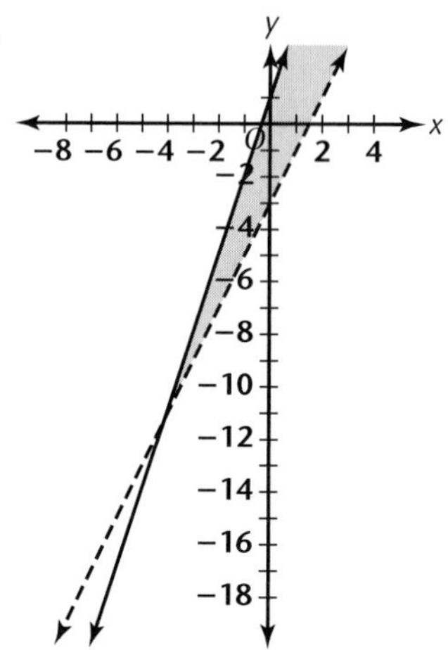

Chapter 2 Assessment, page 117

1.

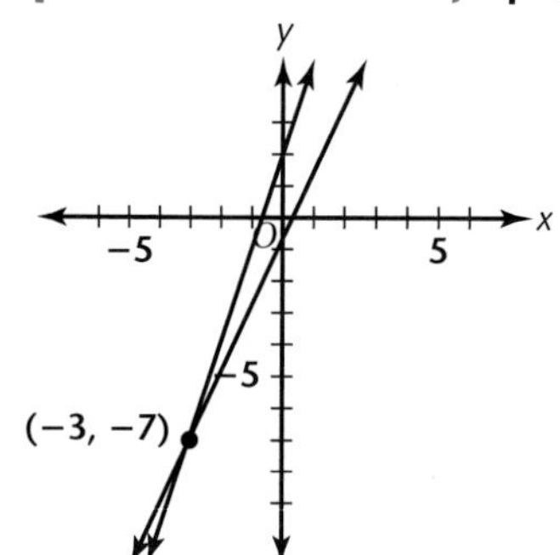

2.

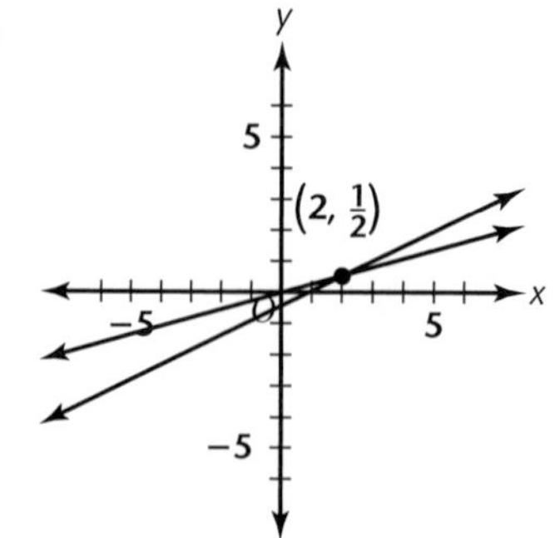

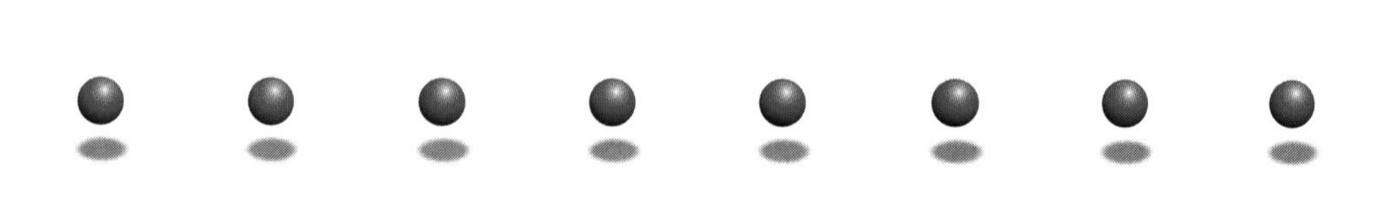

3.

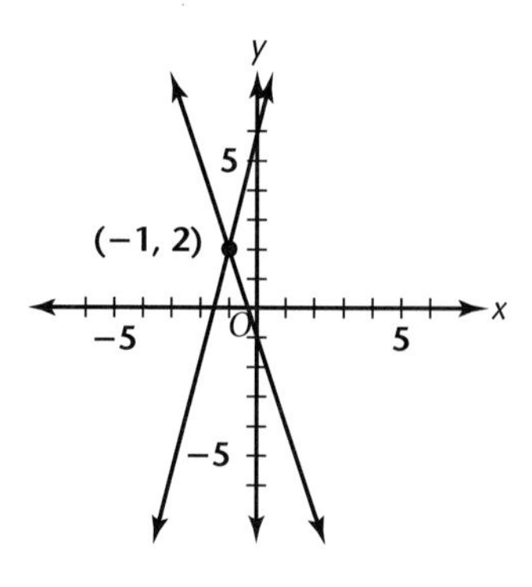

7.

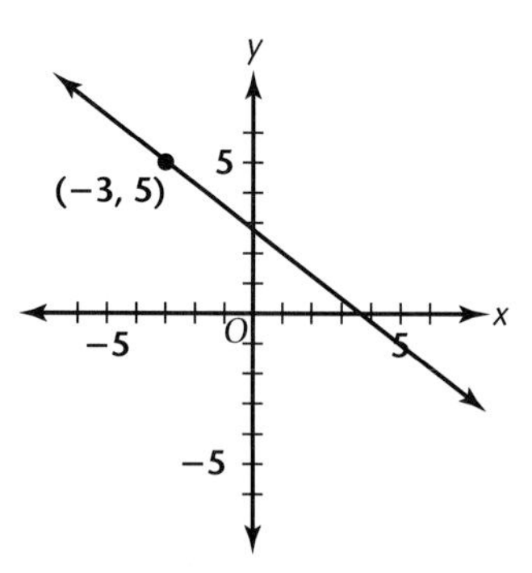

16.

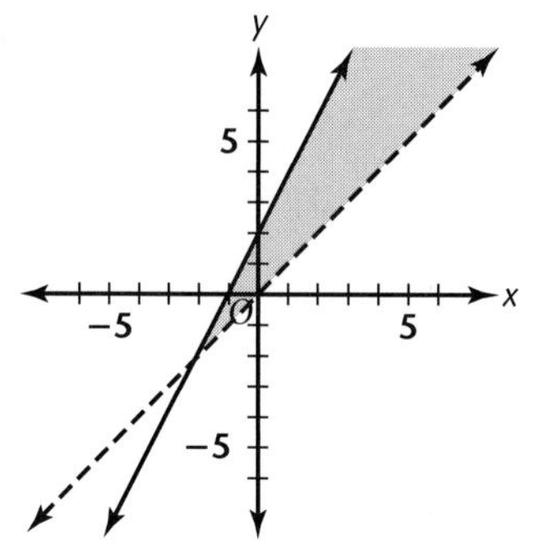

17.

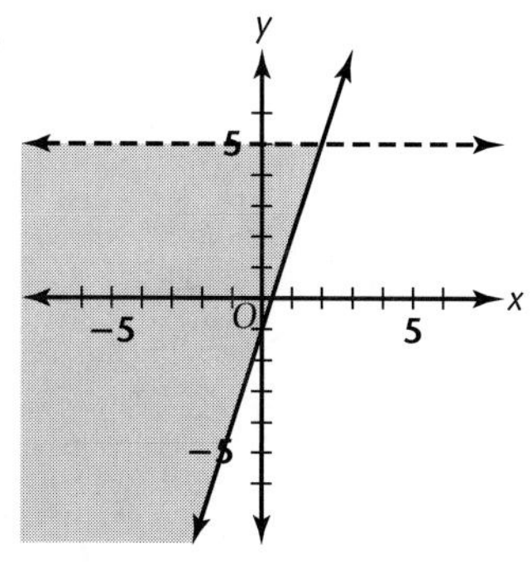

18.

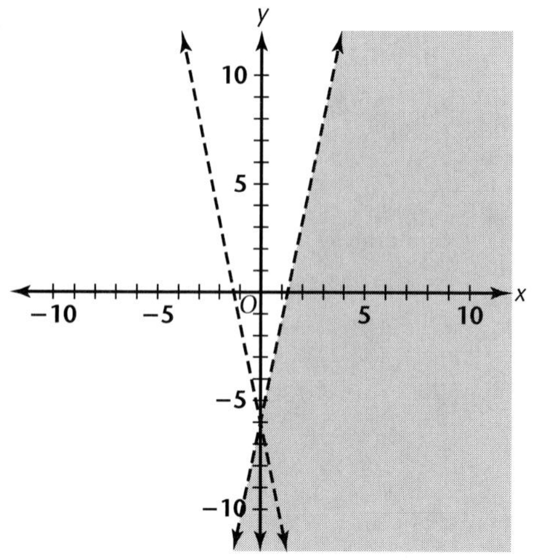

19. $y = x - 6; m = 1$

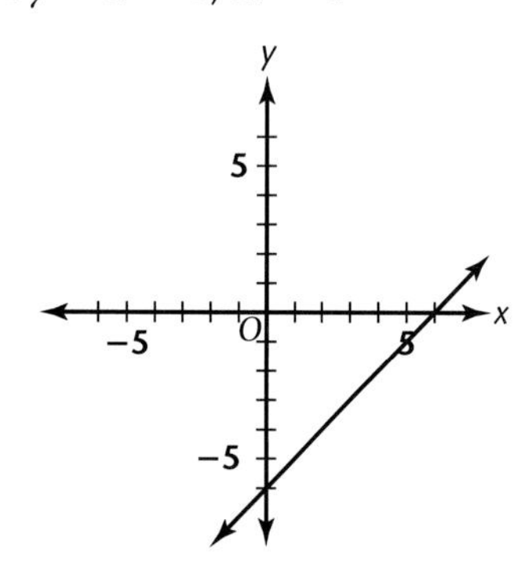

20. $y = -\frac{1}{3}x - 3; m = -\frac{1}{3}$

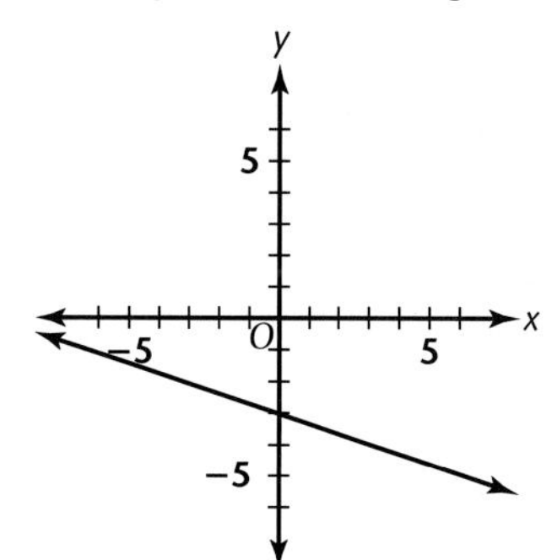

21. $y = 0.6x - 8$; $m = 0.6$

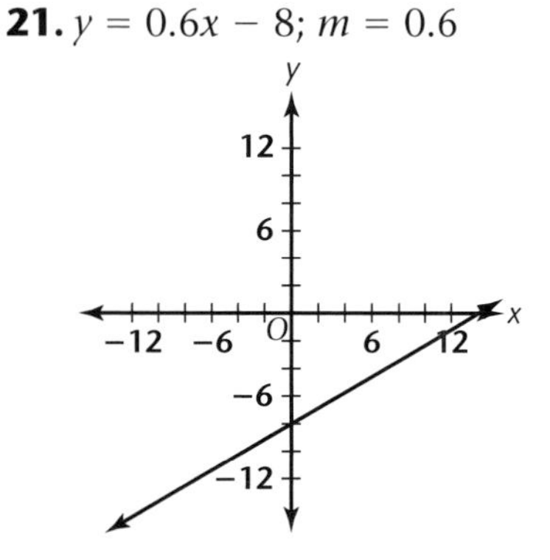

Chapters 1–2 Cumulative Assessment, pages 118–119

19.

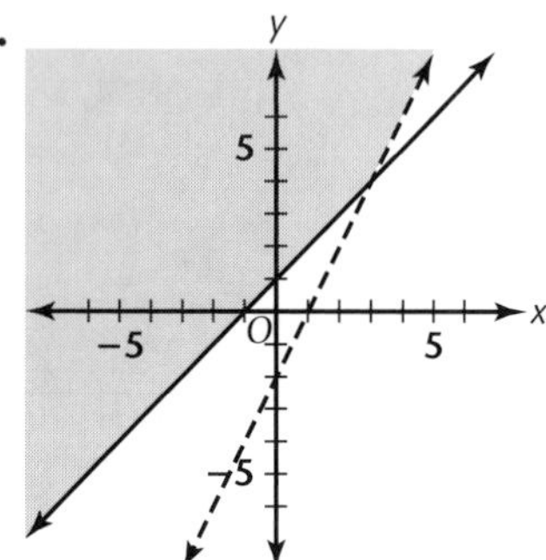

Lesson 3.1

Communicate, **page 126**

1. Place all of the data representing sales of men's clothing items in one matrix and the data representing the sale of women's clothing items in the other matrix.

Men's Clothing Sales (in thousands)

	1985	1989	1993
Suits	14.6	10.8	11.5
Shirts	16.7	16.2	17.3
Jeans	242.7	210.5	186.9

Women's Clothing Sales (in thousands)

	1985	1989	1993
Suits	17.4	12.3	8.6
Shirts	25.6	21.3	16.2
Jeans	98.2	90.1	80.3

Practice and Apply, **pages 126–128**

10. $L =$

	Atlantis	Columbia	Discovery
1988	1	0	1
1989	2	1	2
1990	2	2	2
1991	3	1	3

16.

	Chula Vista	Abilene
1980	85,000	99,000
1981	89,500	100,000
1982	94,000	101,000
1983	98,500	102,000
1984	103,000	103,000
1985	107,500	104,000
1986	112,000	105,000
1987	116,500	106,000
1988	121,000	107,000
1989	125,500	108,000
1990	130,000	109,000

Look Back, **page 128**

27. $x \leq -4$ and $x \geq -4$

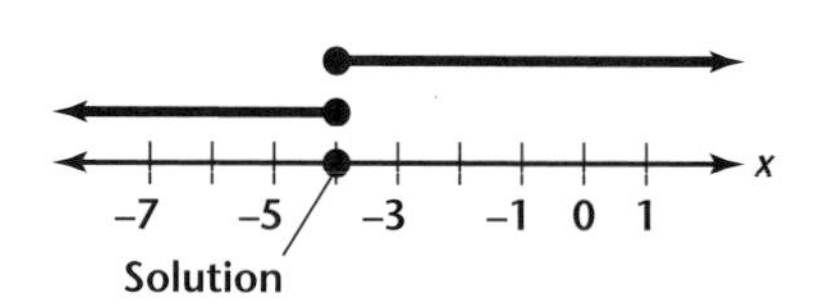

28.

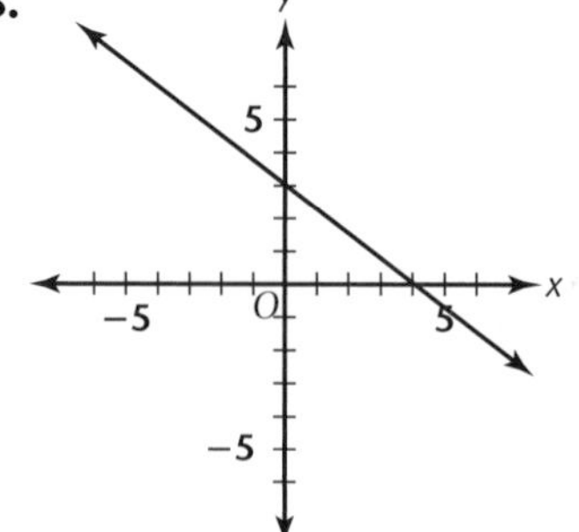

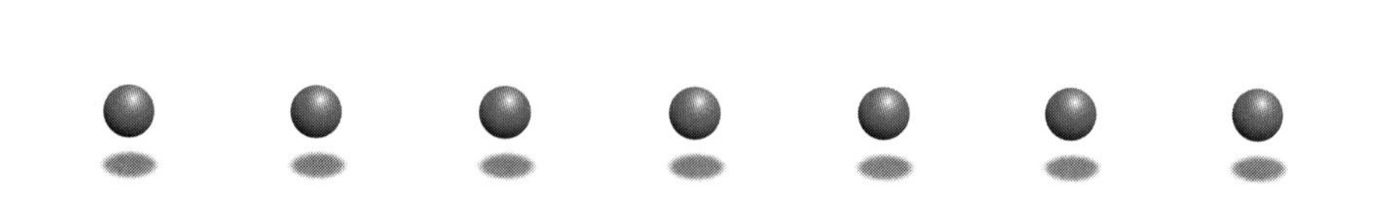

Lesson 3.2

Exploration, **page 131**

4. In order to subtract matrices, the matrices must have the same dimensions. The subtractions in **a** and **d** can be performed because the matrices have the same dimensions. The results are the same except that the signs of the corresponding elements are opposite because the order of the matrices being subtracted is reversed. The subtraction in **b** involves only one matrix so the dimensions are the same. The result is a matrix of all 0s because every element is subtracted from itself. The subtractions in **c** and **e** can be performed because the matrices have the same dimensions. The subtraction in **f** cannot be performed because the matrices have different dimensions.

Ongoing Assessment, **page 131**

The zero matrix for C has 2 rows and 2 columns of zeros. The zero matrix for D has 2 rows and 3 columns of zeros.

Communicate, **page 132**

1. Add the corresponding entries of each matrix. $\begin{bmatrix} 9 & 12 \\ 13 & -12 \end{bmatrix}$

2. Subtract the corresponding entries of each matrix. $\begin{bmatrix} -9 & -4 \\ 29 & -4 \\ 13 & -6 \end{bmatrix}$

4. Add the corresponding entries of each matrix. $\begin{bmatrix} 1 & \frac{1}{2} & -1 \\ \frac{4}{3} & \frac{8}{5} & \frac{1}{2} \end{bmatrix}$

Practice and Apply, **pages 132–134**

12. $\begin{bmatrix} 4 & -5 \\ 5.1 & -4.8 \end{bmatrix}$

13. $\begin{bmatrix} 1.1 & -11.4 & 7.1 \\ -8 & -8.1 & 7.7 \\ -1.3 & 8.4 & 5.84 \end{bmatrix}$

14. $\begin{bmatrix} -\frac{19}{15} & \frac{32}{15} \\ \frac{7}{15} & \frac{1}{2} \end{bmatrix} = \begin{bmatrix} -1.267 & 2.133 \\ 0.467 & 0.5 \end{bmatrix}$

15. $\begin{bmatrix} -4.3 & 1.1 & 7.8 \\ -4.7 & 16.7 & -1.8 \end{bmatrix}$

16. $\begin{bmatrix} -3 & 3 \\ -10 & 9 \\ -5 & 0 \end{bmatrix}$

21.

	Mu	Ar	Te	He
Enrollment	[1837	1219	1822	1592]

30. $-A = \begin{bmatrix} -12 & -8 \\ -7 & 3 \end{bmatrix}$

$-B = \begin{bmatrix} 13 & 1 \\ 8 & 6 \end{bmatrix}$

$-C = \begin{bmatrix} 1 & -7 \\ 1 & 9 \end{bmatrix}$

Look Back, **page 135**

36. $x > -3$ or $x \leq -5$

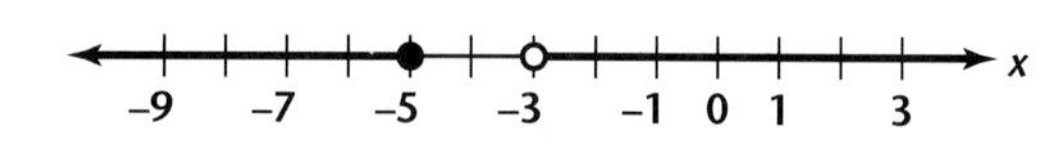

38.

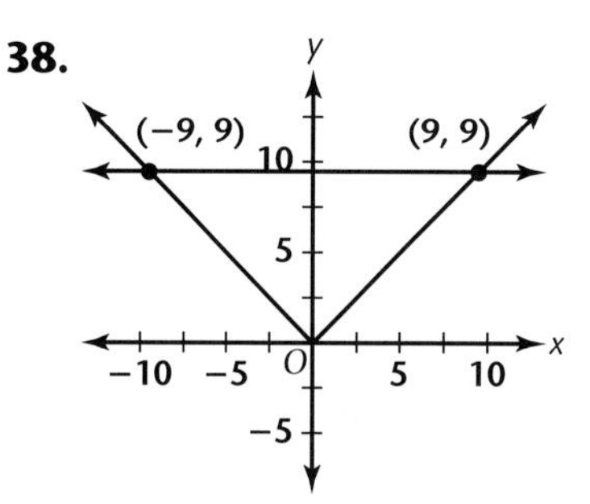

39. Strong positive

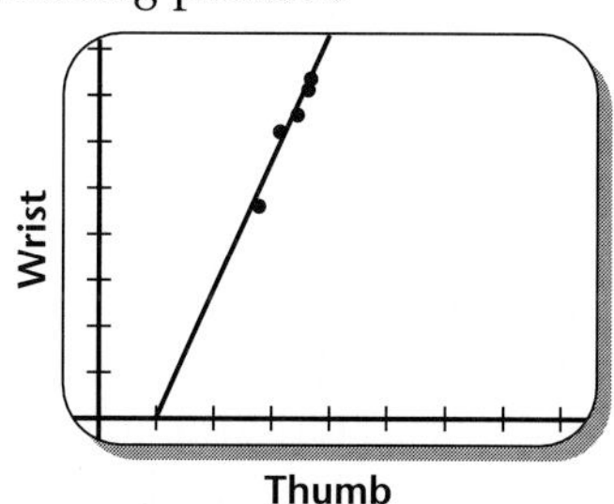

As the thumb size increases, the wrist size increases.

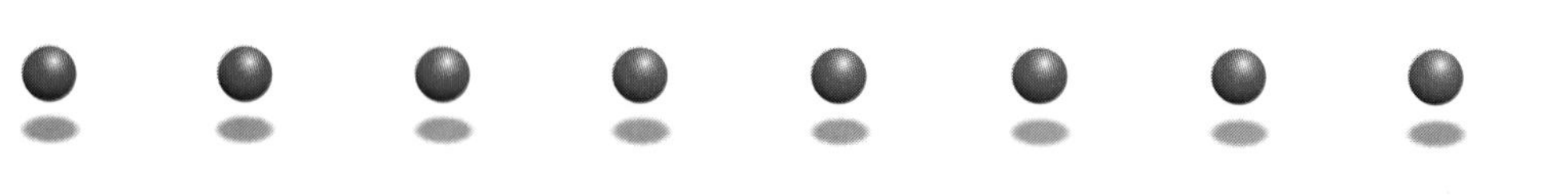

Look Beyond, **page 135**

41. $\begin{bmatrix} -3 & 5 \\ 0 & 0 \\ 3 & 8 \end{bmatrix}$

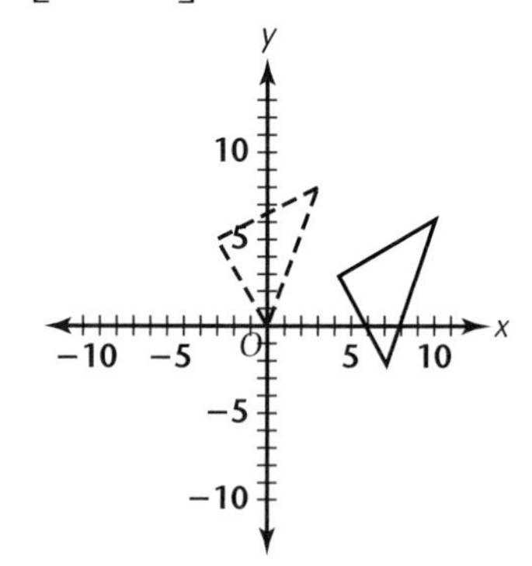

Eye Witness Math, **page 137**

1. a–c. Answers may vary. The following show the results from a batch of 450 objects with 50 marked.

Sample	Total number of objects in sample	Number of marked objects in sample
1	30	2
2	30	4
3	30	3
4	30	4

1e. Sample calculation based on table above:

$M_s = \frac{2 + 4 + 3 + 4}{4} = \frac{13}{4} = 3.25$

$N_s = 30$

$M_b = 50$

$\frac{N_b}{N_s} = \frac{M_b}{M_s} \rightarrow \frac{N_b}{30} = \frac{50}{3.25}$

$N_b = 462$

1f. Answer will depend on estimate and actual number of objects.

2. Answers may vary. Capture a number of animals you are studying, tag or mark them in some way, and return them to the wild. After some time passes, capture a sample of the animals and count how many in the sample are marked. Set up the equation used in the activity and solve for N_b to get an estimate of the number of animals in that whole population.

3. Answers may vary. No, because it would be too hard to capture enough grizzlies without harming them.

Lesson 3.3

Communicate, **page 142**

5. Since the number of columns in the first matrix is equal to the number of rows in the second matrix, matrix multiplication can be performed. To multiply the matrices: Multiply (2)(5) and (−4)(6). Place the sum of the products (−14), in address p_{11} in the product matrix. Multiply (2)(−1) and (−4)(5). Place the sum of the products (−22), in address p_{12} in the product matrix. Multiply (1)(5) and (0)(6). Place the sum of the products, (5), in address p_{21} in the product matrix. Multiply (1)(−1) and (0)(5). Place the sum of the products, (−1), in address p_{22} in the product matrix. The product matrix is $\begin{bmatrix} -14 & -22 \\ 5 & -1 \end{bmatrix}$.

6. Since the number of columns in the first matrix is equal to the number of rows in the second matrix, matrix multiplication can be performed. To multiply the matrices: Multiply (1)(4) and (2)(2). Place the sum of the products, (8), in address p_{11} in the product matrix. Multiply (1)(1) and (2)(1). Place the sum of the products, (3) in address p_{12} in the product matrix. Multiply (1)(3) and (2)(6). Place the sum of the products, (15), in address p_{13} in the product matrix. Multiply (1)(4) and (−3)(2). Place the sum of the products, (−2), in address p_{21} in the

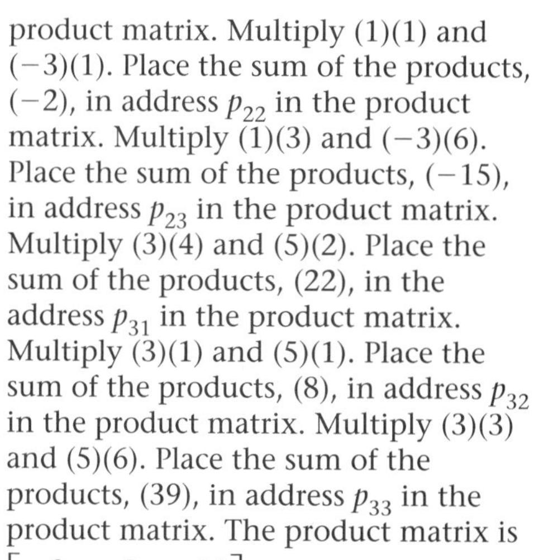

product matrix. Multiply (1)(1) and (−3)(1). Place the sum of the products, (−2), in address p_{22} in the product matrix. Multiply (1)(3) and (−3)(6). Place the sum of the products, (−15), in address p_{23} in the product matrix. Multiply (3)(4) and (5)(2). Place the sum of the products, (22), in the address p_{31} in the product matrix. Multiply (3)(1) and (5)(1). Place the sum of the products, (8), in address p_{32} in the product matrix. Multiply (3)(3) and (5)(6). Place the sum of the products, (39), in address p_{33} in the product matrix. The product matrix is

$$\begin{bmatrix} 8 & 3 & 15 \\ -2 & -2 & -15 \\ 22 & 8 & 39 \end{bmatrix}.$$

Practice and Apply, **pages 142–144**

27. The Associative Property of Multiplication is true for numbers and matrices. One example is

$$\left([4 \quad 3]\begin{bmatrix} 2 & 1 \\ 5 & 3 \end{bmatrix}\right)\begin{bmatrix} 6 \\ -1 \end{bmatrix} = [125] \text{ and}$$

$$[4 \quad 3]\left(\begin{bmatrix} 2 & 1 \\ 5 & 3 \end{bmatrix}\begin{bmatrix} 6 \\ -1 \end{bmatrix}\right) = [125]$$

31. $[22]; \begin{bmatrix} 3 & 6 & 9 \\ 2 & 4 & 6 \\ 5 & 10 & 15 \end{bmatrix}$

37.

	Cost	Sale price
Freshman	163.75	450.00
Sophomore	145.00	400.00
Junior	90.00	250.00
Senior	237.50	650.00

38.

	Profit
Freshman	286.25
Sophomore	255.00
Junior	160.00
Senior	412.50

Look Back, **page 144**

47.

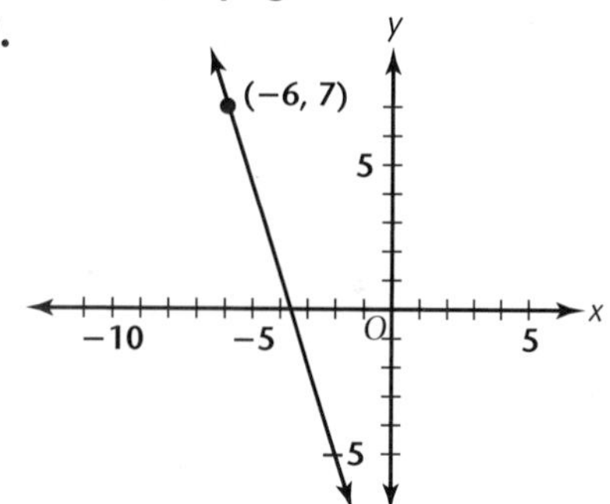

Extra Practice, **page 533**

9. $\begin{bmatrix} -2 & 10 & 4 \\ 3 & 3 & 1 \end{bmatrix}$

10. $\begin{bmatrix} -9 & -6 \\ 18 & 0 \\ 15 & 21 \end{bmatrix}$

11. $\begin{bmatrix} -8 & -16 \\ 4 & -20 \end{bmatrix}$

12. Not Possible

13. Not Possible

14. $\begin{bmatrix} 1 & 2 & 1.5 \\ 0 & 4.5 & -1.5 \end{bmatrix}$

15. $\begin{bmatrix} 4 & -12 & 18 \\ 19 & -36 & 19 \end{bmatrix}$

16. $\begin{bmatrix} 8 & 2 & 5 \\ -3 & 24 & -10 \end{bmatrix}$

17. $\begin{bmatrix} 6 & -6 & -11 \\ -24 & 36 & 6 \\ 1 & -12 & 33 \end{bmatrix}$

18. $\begin{bmatrix} -6 & -30 & -3 \\ 12 & 24 & 18 \\ 10 & 83 & -6 \end{bmatrix}$

Lesson 3.4

Communicate, **page 148**

1. $\begin{bmatrix} -1 & 6 \\ 0 & 3 \end{bmatrix}\begin{bmatrix} a & b \\ c & d \end{bmatrix} = \begin{bmatrix} 1 & 0 \\ 0 & 1 \end{bmatrix}$

$-a + 6c = 1 \qquad -b + 6d = 0$

$0a + 3c = 0 \qquad 0b + 3d = 1$

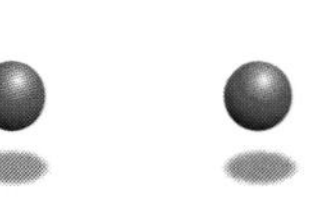

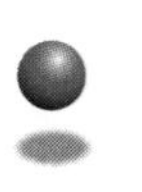

Solve for a, b, c, and d. $\begin{bmatrix} -1 & 2 \\ 0 & \frac{1}{3} \end{bmatrix}$

Practice and Apply, **pages 148–149**

27.

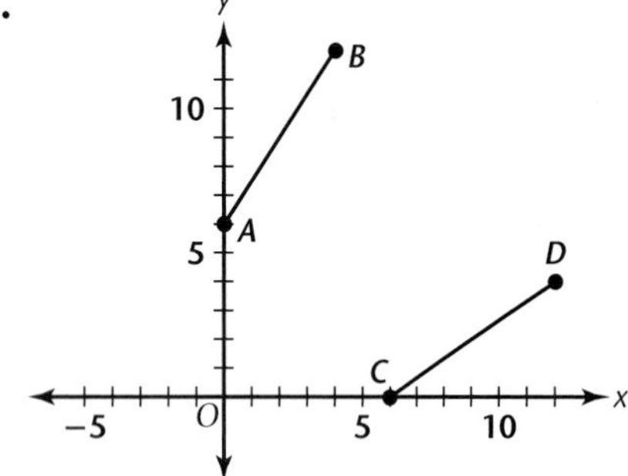

They are reciprocals.

Extra Practice, **page 534**

1. $\begin{bmatrix} -\frac{1}{22} & -\frac{2}{11} \\ \frac{7}{22} & \frac{3}{11} \end{bmatrix}$

2. $\begin{bmatrix} \frac{5}{31} & \frac{12}{31} \\ \frac{3}{31} & \frac{1}{31} \end{bmatrix}$

3. $\begin{bmatrix} -\frac{1}{8} & \frac{5}{16} \\ \frac{5}{2} & -\frac{5}{4} \end{bmatrix}$

4. There is no inverse.

5. $\begin{bmatrix} -\frac{30}{37} & -\frac{340}{259} \\ -\frac{10}{37} & -\frac{25}{37} \end{bmatrix}$

6. $\begin{bmatrix} \frac{1}{15} & -\frac{1}{10} \\ \frac{7}{75} & -\frac{1}{25} \end{bmatrix}$

Lesson 3.5

Communicate, **page 155**

3. Find the inverse of the coefficient matrix: $\begin{bmatrix} 0 & 1 \\ -1 & -2 \end{bmatrix}$. Multiply the inverse matrix on the left of each side of the matrix equation:

$\begin{bmatrix} 0 & 1 \\ -1 & -2 \end{bmatrix}\begin{bmatrix} -2 & -1 \\ 1 & 0 \end{bmatrix}\begin{bmatrix} x \\ y \end{bmatrix} =$

$\begin{bmatrix} 0 & 1 \\ -1 & -2 \end{bmatrix}\begin{bmatrix} -9 \\ -9 \end{bmatrix}$. The result is

$\begin{bmatrix} x \\ y \end{bmatrix} = \begin{bmatrix} -9 \\ 27 \end{bmatrix}$, so the solution is $(-9, 27)$.

4. Find the inverse of the coefficient matrix: $\begin{bmatrix} 0.5 & 0.5 \\ 0.5 & -0.5 \end{bmatrix}$. Multiply it on the left of each side of the matrix equation:

$\begin{bmatrix} 0.5 & 0.5 \\ 0.5 & -0.5 \end{bmatrix}\begin{bmatrix} 1 & 1 \\ 1 & -1 \end{bmatrix}\begin{bmatrix} x \\ y \end{bmatrix} =$

$\begin{bmatrix} 0.5 & 0.5 \\ 0.5 & -0.5 \end{bmatrix}\begin{bmatrix} 6 \\ 2 \end{bmatrix}$. The result is $\begin{bmatrix} x \\ y \end{bmatrix} = \begin{bmatrix} 4 \\ 2 \end{bmatrix}$,

so the solution is $(4, 2)$.

Practice and Apply, **pages 155–157**

5. $\begin{bmatrix} -6 & 4 \\ 5 & -2 \end{bmatrix}\begin{bmatrix} x \\ y \end{bmatrix} = \begin{bmatrix} -12 \\ 10 \end{bmatrix}$

6. $\begin{bmatrix} 6 & -1 & -3 \\ -3 & 1 & -3 \\ -2 & 3 & 1 \end{bmatrix}\begin{bmatrix} a \\ b \\ c \end{bmatrix} = \begin{bmatrix} 2 \\ 1 \\ -6 \end{bmatrix}$

7. $\begin{bmatrix} 2 & 3 & 4 \\ 6 & 12 & 16 \\ 4 & 9 & 8 \end{bmatrix}\begin{bmatrix} x \\ y \\ z \end{bmatrix} = \begin{bmatrix} 8 \\ 31 \\ 17 \end{bmatrix}$

Look Beyond, **page 157**

30. If you graph the equation for the relationship $y = -0.15x^2 + 2x + 2$, the vertex is $\left(\frac{20}{3}, \frac{26}{3}\right)$ or about $(6.67, 8.67)$. Yes, the third player can if he/she can jump high enough so that his/her hands reach 8.67 ft which is the maximum height that the ball will reach.

31.

Sequence	0	2	6	12	20
1st Differences		2	4	6	8
2nd Differences			2	2	2

Since the 2nd differences are constant, the function is quadratic.

Extra Practice, **page 534**

1. $\begin{bmatrix} 2 & -3 \\ 4 & -3 \end{bmatrix}\begin{bmatrix} x \\ y \end{bmatrix} = \begin{bmatrix} 2 \\ 12 \end{bmatrix}$

2. $\begin{bmatrix} 4 & -1 \\ 1 & -9 \end{bmatrix}\begin{bmatrix} x \\ y \end{bmatrix} = \begin{bmatrix} 2 \\ 1 \end{bmatrix}$

3. $\begin{bmatrix} 5 & 2 \\ 2 & -1 \end{bmatrix}\begin{bmatrix} x \\ y \end{bmatrix} = \begin{bmatrix} 7 \\ -9 \end{bmatrix}$

4. $\begin{bmatrix} 7 & -1 & 2 \\ 5 & 1 & 1 \\ -3 & 4 & -1 \end{bmatrix}\begin{bmatrix} a \\ b \\ c \end{bmatrix} = \begin{bmatrix} -1 \\ 5 \\ -2 \end{bmatrix}$

5. $\begin{bmatrix} 3 & 1 & -1 \\ 1 & 1 & -6 \\ 4 & 2 & 9 \end{bmatrix}\begin{bmatrix} x \\ y \\ z \end{bmatrix} = \begin{bmatrix} 1 \\ 2 \\ -2 \end{bmatrix}$

6. $\begin{bmatrix} 5 & -1 & 1 \\ 2 & 2 & 2 \\ 1 & -1 & -1 \end{bmatrix}\begin{bmatrix} m \\ n \\ p \end{bmatrix} = \begin{bmatrix} 0 \\ 5 \\ 7 \end{bmatrix}$

Chapter 3 Project, pages 158–159

Activity 1

HAVE*FUN

Activity 2

Coded Message $= \begin{bmatrix} \text{G J G B I D O X G N I} \\ \text{Q J C T R K T S F R I} \end{bmatrix}$

Decoder $= \begin{bmatrix} 2 & -1.5 \\ 1 & -0.5 \end{bmatrix}$

Activity 3

Answers may vary.

Lesson 4.1

Exploration 1, **page 167**

1. Answers may vary. A sample answer is given.

HEADS	HEADS
TAILS	TAILS
HEADS	TAILS
TAILS	HEADS
TAILS	TAILS

2. Answers may vary. This sample answer references data from Step 1.
$P(\text{heads}) = \frac{4 \text{ heads}}{10 \text{ trials}} = \frac{4}{10} = \frac{2}{5} = 0.40$, or 40%
$P(\text{tails}) = \frac{6 \text{ tails}}{10 \text{ trials}} = \frac{6}{10} = \frac{3}{5} = 0.60$, or 60%

3. Answers may vary. A sample answer is given. There are 25 students in my class. Given 10 trials per student, there were 250 trials in my class.
$P(\text{heads}) = \frac{136 \text{ heads}}{250 \text{ trials}} = \frac{136}{250} = 0.544$, or 54.4%
$P(\text{tails}) = \frac{114 \text{ tails}}{250 \text{ trials}} = \frac{114}{250} = 0.456$, or 45.6%

4. $P(\text{heads}) + P(\text{tails}) = 40\% + 60\% = 100\%$ from experiment in Step 1. $P(\text{heads}) + P(\text{tails}) = 54.4\% + 45.6\% = 100\%$ from class data in Step 3. The two sums are equal, because the number of trials = number of heads + number of tails so that $P(\text{heads}) + P(\text{tails}) = \frac{\text{heads}}{\text{trials}} + \frac{\text{tails}}{\text{trials}} = \frac{\text{heads + tails}}{\text{trials}} = \frac{\text{trials}}{\text{trials}} = 1$, or 100%. The sum of the probabilities of all possible events should always equal 1, or 100%. If not, then there are some events not included in the list of possible events.

5. Answers may vary. A sample answer is given. In my experiment, the probability for heads was less than the probability for tails. However, the class as a whole experimentally determined the probability for heads to be more than the probability for tails. Also, the increased number of trials produced a more precise value for the probabilities of heads and tails.

Critical Thinking, **page 167**

Using the probability of heads as determined from the class' data, $P(\text{heads}) =$

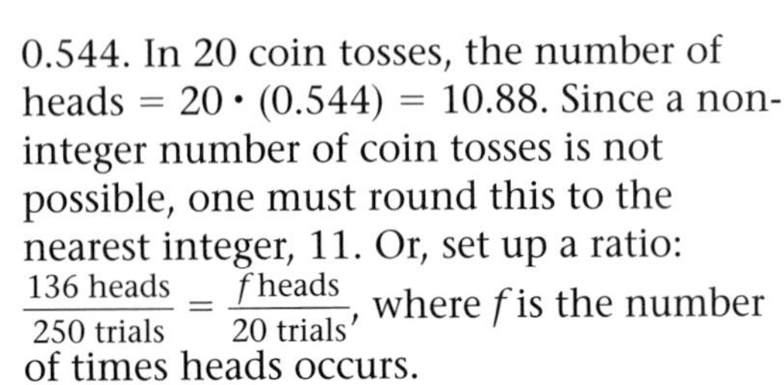

0.544. In 20 coin tosses, the number of heads = 20 • (0.544) = 10.88. Since a non-integer number of coin tosses is not possible, one must round this to the nearest integer, 11. Or, set up a ratio: $\frac{136 \text{ heads}}{250 \text{ trials}} = \frac{f \text{ heads}}{20 \text{ trials}}$, where f is the number of times heads occurs.

$$136 \times 20 = 250 \cdot f$$
$$\frac{136 \times 20}{250} = \frac{250}{250} \cdot f$$
$$\frac{2720}{250} = 1 \cdot f$$
$$10.88 = f$$

Again, round 10.88 to the nearest integer, 11. Note that great precision does not always make conclusions easier to determine.

Exploration 2, page 168

1. Answers may vary as students may guess. Since you have a 1 in 6 chance of rolling a 4, $P(4) = \frac{1}{6}$.
$$\frac{1}{6} = \frac{f}{12}$$
$$12\left(\frac{1}{6}\right) = 12\left(\frac{f}{12}\right)$$
$$2 = f$$
So heads should appear 2 times.
2. INT(RAND*6) + 1. Generate random numbers from 1 to 6 beginning with 1. Answers may vary. 12 applications of this operation might generate 5, 1, 5, 6, 1, 2, 4, 2, 1, 6, 2, 2. "4" occurs one time in this sample result.
4. From the answer to Step 2, $P(4) = \frac{1 \text{ success}}{12 \text{ trials}} = \frac{1}{12}$, or ≈8%.

Communicate, page 169

1. Answers may vary. A sample answer is given. Experimental probability involves performing an experiment and using the results of that experiment to determine a number that serves as a measurement of the likelihood of an event occurring. Examples include counting the number of students in my class wearing blue jeans to class on a given day. Another example is counting the number of people who put ketchup on their French fries to determine the likelihood of someone putting ketchup on their French fries.
2. Since the $P(\text{heads}) + P(\text{tails}) = 100\%$, then if you know the probability of tails, you subtract that number from 100% in order to find the probability of heads. Since $P(\text{heads}) + P(\text{tails}) = 100\%$, then $P(\text{heads}) = 100\% - P(\text{tails})$.
3. Yes. Since probability is a ratio of successes to trials, both Janis and Juanita could observe the same ratio. For example, $\frac{5}{10} = \frac{6}{12}$.

Practice and Apply, pages 169–171

5. No. Given that the regions are of different sizes, the probability of landing in each region will be different.
9. 21% + 26% = 47%. 47% is somewhat close to 53%. These experimental probabilities are close since the combined area of regions I and II is the same as the area of region III.

10.

Group	P(heads)
1	70%
2	60%
3	40%
4	20%
5	30%

12.

Group	P(heads)
1	30%
2	40%
3	60%
4	80%
5	70%

14. 100%; The sum of the probabilities for all possible events must equal 100%.
15. 100%; The sum of the probabilities for all possible events must equal 100%.
21. 25% + 15% = 40%; No; other outcomes occurred besides rain on both days and no rain on either day.

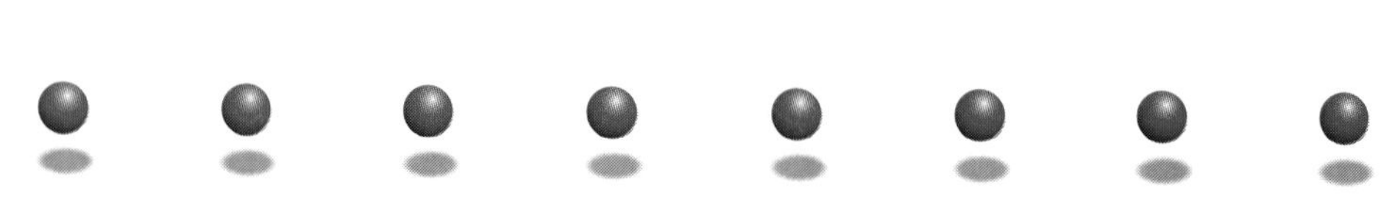
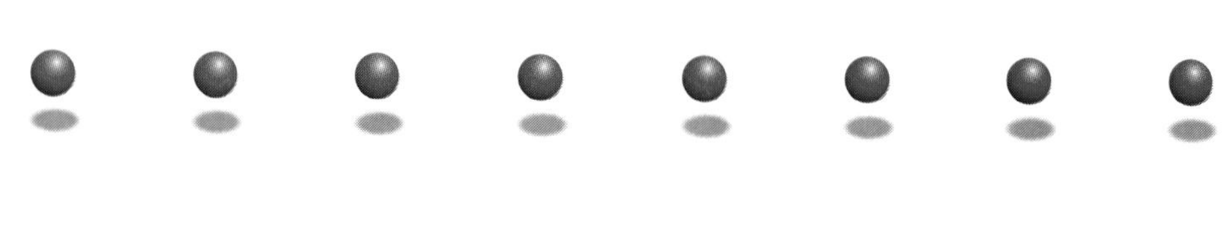

22. $P = \frac{17}{20} + \frac{3}{20} = 1$; Yes; the event of rain on at least one day includes HH, TH, and HT. The event of no rain is TT. Combining these two lists includes all possible outcomes; therefore, the sum of the experimental probabilities for these outcomes must equal 1.

23. A random integer x such that $0 < x < 1$. Therefore, all outputs will be 0s.

24. A random integer x such that $0 \leq x \leq 4$; in other words, 0, 1, 2, 3, or 4

25. A random integer x such that $0 \leq x \leq 9$; in other words, 0, 1, 2, 3, 4, 5, 6, 7, 8, or 9

26. A random integer x such that $1 \leq x \leq 10$; in other words, 1, 2, 3, 4, 5, 6, 7, 8, 9, 10

27. A random integer x such that $1 \leq x \leq 12$; in other words, 1, 2, 3, 4, 5, 6, 7, 8, 9, 10, 11, or 12

28. A random integer x such that $1 \leq x \leq 365$

29. INT(RAND*10) + 1; Answers may vary. A sample answer is provided.

Pairs generated	Symbols
5, 3	T H
6, 10	T T
4, 1	H H
6, 4	T H
9, 9	T T
10, 7	T T
7, 5	T T
9, 6	T T
3, 6	H T
3, 10	H T

30. Answers may vary. Using the sample results from Exercise 29, $P(\text{HH}) = \frac{1}{10} = 0.10 = 10\%$.

31. Answers may vary. Using the sample results from Exercise 29, $P(\text{HT or TH or HH}) = \frac{5}{10} = 0.50 = 50\%$.

32. Answers may vary. Using the sample results from Exercise 29, $P(\text{TT}) = \frac{5}{10} = 0.50 = 50\%$.

33. 50% + 50% = 100%; Yes; all possible events are included in this sum.

34. Answers may vary. Using the sample results from Exercise 29, 10% + 50% = 60%. No; this sum excludes the nonzero probability of rain occurring on just one day.

35. INT(RAND*10) + 1; Answers may vary. A sample answer is provided.

3,5	H H
7, 1	T H
5, 9	H T
10, 8	T T
9, 10	T T
3, 6	H H
5, 6	H H
6, 3	H H
8, 6	T H
2, 5	H H

36. Answers may vary. Using the sample results from Exercise 35, $P(\text{HH}) = \frac{5}{10} = 0.50 = 50\%$.

37. Answers may vary. Using the sample results from Exercise 35, $P(\text{HT or TH or HH}) = \frac{8}{10} = 0.80 = 80\%$.

38. Answers may vary. Using the sample results from Exercise 35, $P(\text{TT}) = \frac{2}{10} = 0.20 = 20\%$.

39. Answers may vary. 50% + 20% = 70%; No; this sum does not include the nonzero probability of rain occurring on just one day.

Look Beyond, **page 171**

51. Possible Outcomes
BBBB
BBBG
BBGB
BBGG
BGBB
BGBG
BGGB
BGGG
GBBB
GBBG
GBGB
GBGG
GGBB
GGBG
GGGB
GGGG

Extra Practice, **page 535**

1. No, the areas are different sizes.

2. $\frac{17}{50}$ or 34%; 29%, 22%, 15%

3. $\frac{49}{100}$; $\frac{51}{100}$; The sums are almost the same because the areas of the combined regions are almost the same.

Lesson 4.2

Communicate, **page 176**

1a. Choose a way to generate random numbers and describe what each result represents.

1b. Decide how to simulate one trial of the experiment.

1c. Carry out a large number of trials and record the results.

2. A coin toss; heads is a shot made, tails is a shot missed. A random number generator; 1 is a shot made, 0 is a shot missed. A number cube; an even number is a shot made, an odd number is a shot missed.

3. Let the numbers 1, 2, and 3 represent the occurrence of rain, and the numbers 4, 5, 6, 7, 8, 9, and 10 represent no rain.

4. Answers may vary. For example: hitting a nail on the head with a hammer, kicking a football through the goal post, and getting an answer on the phone when you call

5. Use 5 different-colored slips of paper, where each color represents a different answer.

6. Generate random positive integers from 1 to 10 on a calculator. Let the number 1 represent a late flight, and the numbers 2 to 10 represent a flight on time. Generate 3 numbers. Let this represent 1 trial. Repeat for 20 trials. Count the number of trials where all 3 numbers are greater than 1. Divide this number by 20 to find the experimental probability.

Practice and Apply, **pages 176–177**

11. Let the numbers 1 to 8 represent the number of occurrences of rain, and the numbers 9 and 10 represent no rain. Generate 2 random numbers, let the results represent the weather on day 1 and day 2, respectively. Repeat for 10 trials. Divide by 10 the number of trials where rain occurred on at least one day. The quotient is the experimental probability.

12. Let the numbers 1 to 4 represent the number of occurrences of rain, let 5 represent no rain, and let 6 mean "roll the number cube again." Toss the number cube twice with each toss representing 1 day. Record the result. Repeat for 10 trials. Divide by 10 the number of trials where rain occurred on at least one day. The quotient is the experimental probability.

For Exercises 13–18, answers may vary.

13. Use a calculator to generate a number from 1 to 7; use INT(RAND*7) + 1.

14. Use a calculator to generate a number from 1 to 365; use INT(RAND*365) + 1.

15. Use a calculator to generate a number from 1 to 24; use INT(RAND*24) + 1.

16. Use a calculator to generate a number from 1 to 8; use INT(RAND*8) + 1.

17. Generate 100 pairs of random numbers from 1 to 100. Count the number of pairs in which both numbers are less than or equal to 20. Let that number equal n. Then the experimental probability is $\frac{n}{100}$.

18. Generate 20 sets of 5 random numbers from 1 to 10. Then count the number of sets that have exactly 2 numbers less than or equal to 3. Let that number equal n. Then the experimental probability is $\frac{n}{20}$.

19. Answers may vary. Use a calculator to generate the numbers 1 and 2. Let 1 represent a boy, and 2 represent a girl. Generate 4 numbers; record the results. Repeat for 10 trials. Divide by 10 the number of trials where there are 2 boys and 2 girls.

20. Answers may vary. Use a calculator to generate the numbers 0 and 1. Let 0 represent an incorrect response, and 1 represent a correct response. Generate 3 numbers; record the results. Repeat for 10 trials. Divide by 10 the number of trials where all three responses are correct.

21. Answers may vary. Use a calculator to generate the numbers 1 to 5. Let 1 represent a correct response on a question, and 2 to 5 represent an incorrect response. Generate 4 numbers; record the results. Repeat for 100 trials. Divide by 100 the number of trials where 3 or 4 correct responses were given. This is the experimental probability, approximately $\frac{17}{625}$.

22. Answers may vary. Use a number cube to generate the numbers from 1 to 6. Let 1 represent a correct response, and 2 to 5 represent an incorrect response, and 6 means roll again. Generate 6 numbers; record the results. Repeat for 100 trials. Divide by 100 the number of trials where 5 or 6 correct responses were given. This is the experimental probability, approximately $\frac{1}{625}$.

Look Back, **page 177**

23.

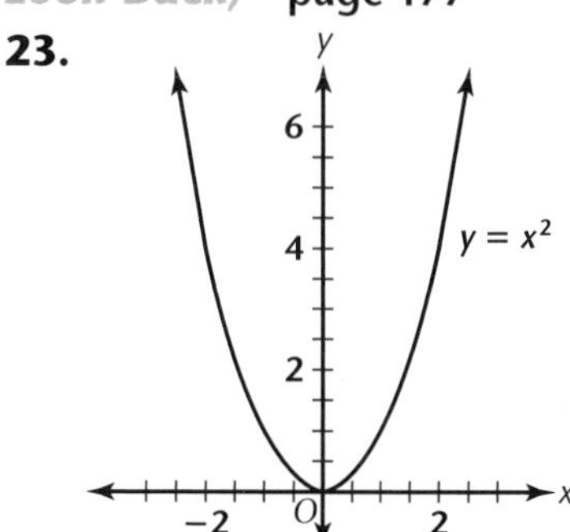

The graph of a quadratic function is a parabola.

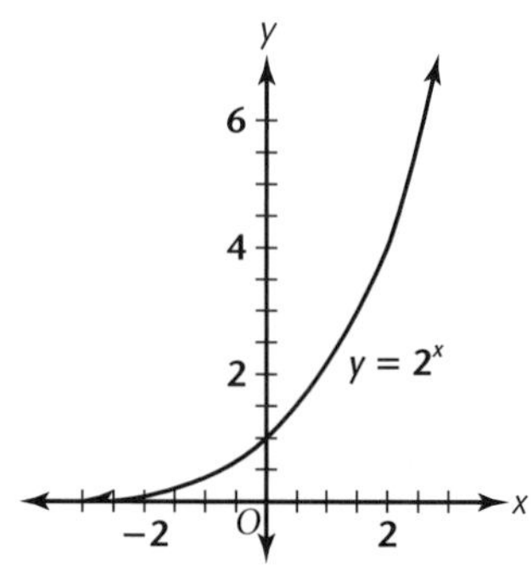

The graph of an exponential function is a curve that increases very rapidly.

28.

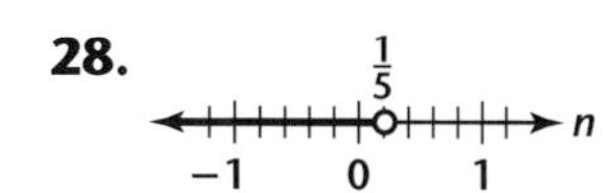

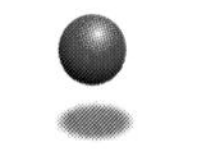   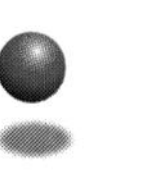     

Extra Practice, **page 535**

Answers for Exercises 1–6 may vary. Sample answers are given.

1. Use a calculator or computer to generate random integers from 1 to 30. Each number represents a different day of the month.
2. Use a calculator or computer to generate random integers from 1 to 28. Each number represents a different student.
3. Use a calculator or computer to generate random integers from 1 to 6. Each number represents a different class.
4. Use a calculator or computer to generate random integers from 1 to 10. Each number represents a different team.
5. Use a calculator or computer to generate random integers from 1 to 15. Each number represents a different book.
6. Use a calculator or computer to generate random integers from 1 to 50. Each number represents a different number.

Lesson 4.3

Exploration, Part 1 **pages 180–181**

1. Answers may vary. A sample answer is given. Place the categories horizontally along the base of a vertical scale. Draw rectangles of equal width but with varying height to indicate the frequency of each response within the rating categories.
2. 3 is the mode. It is a useful measure in this survey because it allows you to see the most popular opinion about country music.
3. *Spatially,* tolerate (3) falls halfway between strongly dislike and strongly like. Since there are approximately an even number of responses on both sides of this halfway point and numerous people answered 3, it must be the median.
4. 1, 1, 2, 2, 2, 2, 3, 3, 3, 3, 3, 3, 3, 3, 3, 4, 4, 4, 4, 4, 5, 5; mean ≈ 3.05
5. Tolerate: $\frac{9}{22}$, or 41%; strongly like or strongly dislike: $\frac{2}{22} + \frac{2}{22} = \frac{2}{11}$, or ≈18%

Exploration, Part 2, **page 181**

1. 4, like; The median for rap music is higher than the median for country music.
2. There are two modes, like and strongly like, 4 and 5. The modes for rap music are both higher than the mode for country music.
3. Mean: 3.23 ⟹ between tolerate and like. The mean for rap music is higher than the mean for country music, but both means are between tolerate and like.
4. 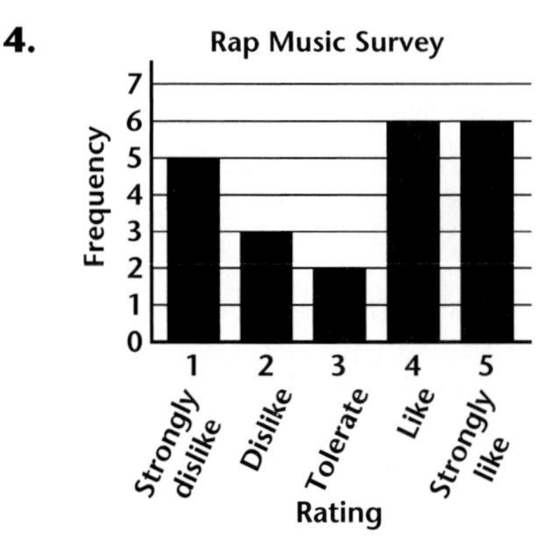

The bar graph for rap music is higher on the ends and lower in the middle. The bar graph for country music is higher in the middle and low at both ends. These surveys seem to indicate that people feel strongly about rap music—they either love it or hate it. On the other hand, people tend to be

middle-of-the-road about country music.

5. Tolerate: $\frac{2}{22}$, or ≈9%; strongly like or strongly dislike: $\frac{6}{22} + \frac{5}{22} = \frac{11}{22}$, or 50%

6. The surveys for country music and rap music both report that 23% of the students "like" the music. However, 50% of the students felt strongly about rap music compared with 18% for country music. Also, 41% of the students tolerate country music, while only 9% tolerate rap music. None of the measures of central tendency indicate this major difference between the surveys.

Communicate, **pages 181–182**

1. The mean is the sum of all the data divided by the number of items. The median is the number in the middle of the ordered data. The mode is the most frequently occurring item of data.

2. The 5 numbers were placed in order from the least to the greatest; 0.342 is the middle value.

3. The student was wrong. When arranged in an ascending order (5, 8, 10, 17, 20), the middle number is 10 and not 8.

4. Arrange the data from the least to the greatest and find the mean or average of the 2 middle numbers.

5. Plot the frequency on the y-axis from zero to fifty; plot the ratings on the x-axis. Above each rating, go up to the corresponding frequency and create a bar from that height.

6. To find the median, place all the ratings in ascending order. Since there are an even number of ratings, the median is the mean of the two middle values, 2 and 2. So the median is 2. To find the mode, notice which rating was chosen more than any other. 2 is the mode. To find the mean, list all the ratings, add them up, and divide by 100. The mean is $\frac{(1)(15) + (2)(41) + (3)(34) + (4)(7) + (5)(3)}{100} =$ 2.42, basically midway between often and sometimes.

Practice and Apply, **pages 182–184**

17. Arrange in order, then find the mean or average of the 2 middle values.

21. Impossible; 18 must occur more than once (since it is the mode), and 20 must be the third number (since it is the median of the 5 pieces of data). Therefore, data must be {18, 18, 20, x, y}. The last 2 pieces of data can be no less than 20 (since 20 is the median). They also cannot equal 20 (since 18 is the mode), and $x \neq y$ (since there would then be two modes, 18 and x). Therefore, the smallest possible values for x and y are 21 and 22. The mean of {18, 18, 20, 21, 22} is 19.8. Since the smallest possible x and y were used, the mean can never be less than 19.8.

22.

City	Range
San Francisco, CA	15.8
Washington, DC	45.4
Miami, FL	15.6

The temperature fluctuation during one year is much greater in Washington, DC than in San Francisco or Miami. While San Francisco and Miami have very similar ranges of temperatures throughout the year, the temperature in Miami is consistently higher than the temperature in San Francisco during any given month. Despite vastly different ranges in temperature, San Francisco and Washington, DC have similar average temperatures.

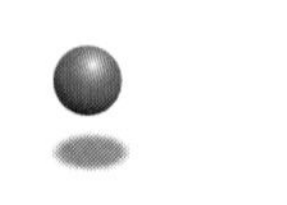
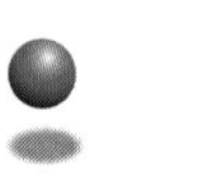

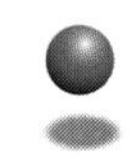

25.

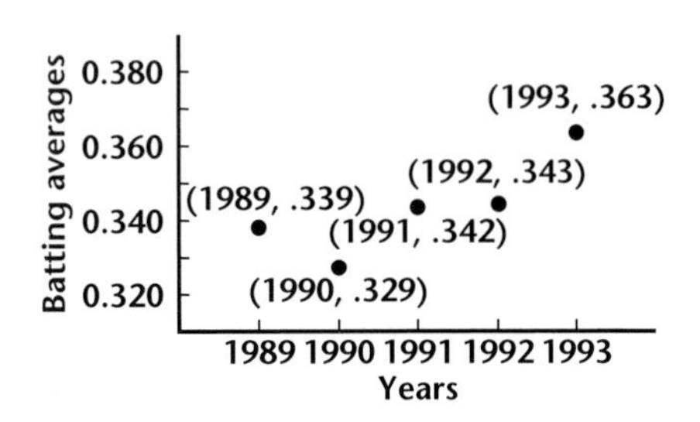

26. The averages are generally getting better. The greatest change in one year happened from 1992 and 1993, an increase of 0.02 points.

28. The correlation coefficient for the National League is 0.52, much lower than the American League. This means that a line better fits the data for the American League than for the National League.

Look Back, **page 184**

37.

2 Cities	Absolute value of difference in temperature
Juneau/Duluth	17°F
Juneau/Houston	15°F
Duluth/Houston	32°F

40. The points form an isosceles triangle.

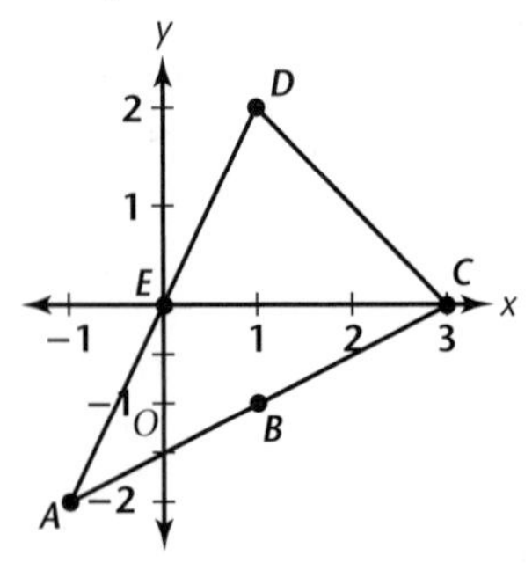

The line connecting the points A and D is $y = 2x$.

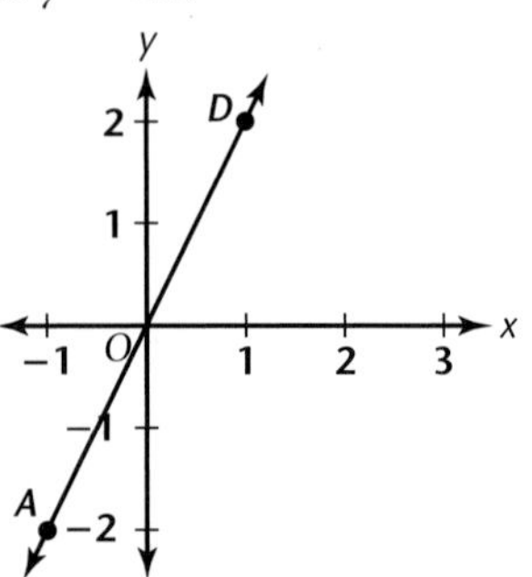

41. The slope of any line parallel to the line connecting the points C and D is -1.

Look Beyond, **page 184**

45. If there are n arrangements of a set of m letters, then adding a letter would produce $n(m + 1)$.

Lesson 4.4

Exploration 1, **page 187**

1.

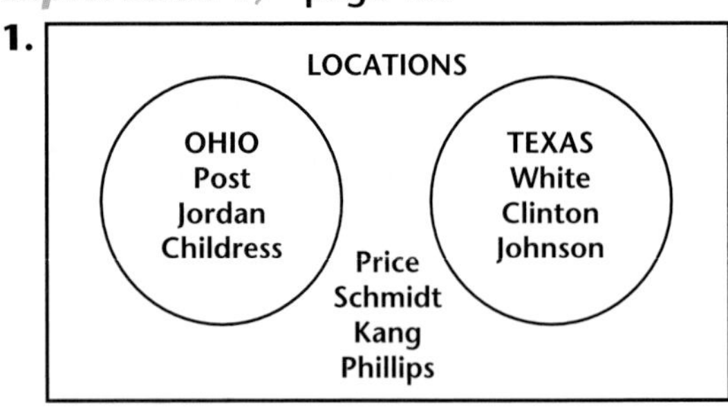

Exploration 2, **page 187**

1.

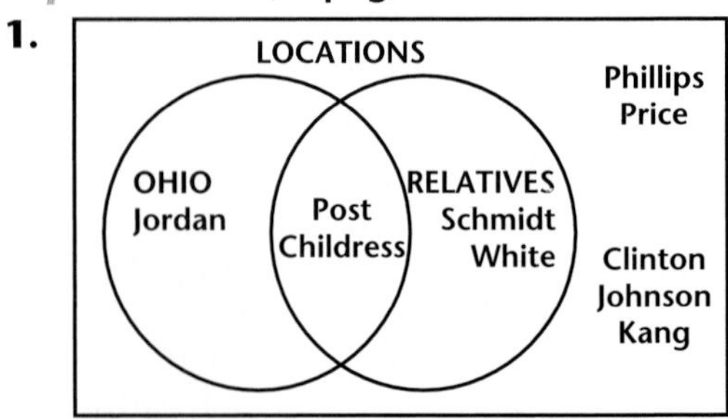

Practice and Apply, **pages 189–190**

28.

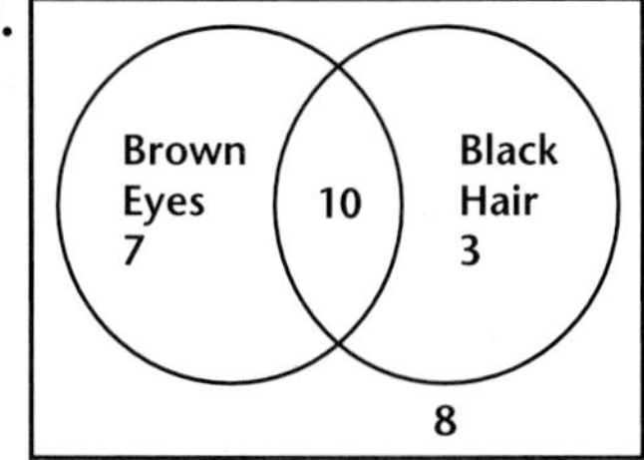

Look Back, **page 191**

46. Answers may vary. Use a calculator to randomly generate the numbers 1 to 10. Let the numbers 1, 2, and 3 represent a hit; let the numbers 4, 5, 6, 7, 8, 9, and 10 represent a miss. Generate 4 numbers; record the results. Repeat for 10 trials. Divide the number of trials where two hits are made by 10. This is the experimental probability.

Extra Practice, **page 536**

1. None

2. 20, 21, 22, . . . , 40

3. 24, 30, 36

4. 20, 21, 22, 24, 26, 27, 28, 30, 32, 33, 34, 36, 38, 39, 40

5. 21, 23, 24, 25, 27, 29, 30, 31, 33, 35, 36, 37, 39

6. 20, 30, 40

7. 20, 22, 24, 25, 26, 28, 30, 32, 34, 35, 36, 38, 40

8. 25, 35

9. 20, 21, 23, 25, 27, 29, 30, 31, 33, 35, 37, 39, 40

10. 30

11.

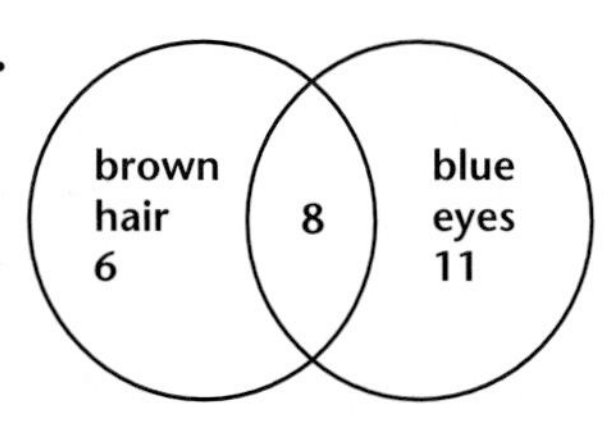

Lesson 4.5

Critical Thinking, **page 195**

Answers may vary. Sample answer: An advantage of a tree diagram is that it shows all the possible outcomes. An advantage of the Multiplication Principle of Counting is that it is shorter and faster to complete, especially when there are many possible outcomes.

Communicate, **page 196**

4. There are 2 volleyball teams, so there are initially 2 branches. Each team has 8 players, so each branch has 8 branches. The total number of branches is 2×8 or 16.

5. There are 4 ways to draw a three. There are 26 ways to draw a red card. $4 \cdot 26 = 104$

Lesson 4.6

Critical Thinking, **page 200**

Events will vary. For 0, the event can never occur. For 1, the event must always occur. The range of values for theoretical probability is from 0 to 1, inclusive.

Communicate, **page 201**

1. experimental = $\frac{\text{number of times event occurs}}{\text{number of trials}}$; theoretical = $\frac{\text{number of successful outcomes}}{\text{number of equally likely outcomes}}$

2. No, he has a 50-50 chance on one question, but with each subsequent question, he decreases the probability of getting all of the questions correct by $\frac{1}{2}$.

3. Experimental; this probability is based on the actual results of an experiment.

4. No; the probability of all heads or all tails is only $\frac{1}{16} + \frac{1}{16}$ or $\frac{1}{8}$.

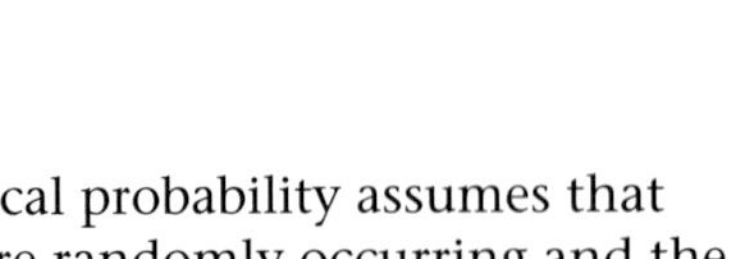

5. Theoretical probability assumes that events are randomly occurring and the situations are fair, that is, not biased toward one outcome.

Look Back, **page 203**

26. $\frac{3.18}{53.00} = \frac{x}{100}$

$$x = \frac{3.18}{53.00}(100)$$
$$= 6$$

The sales tax is 6%.

28. $-2 \leq n < 14$

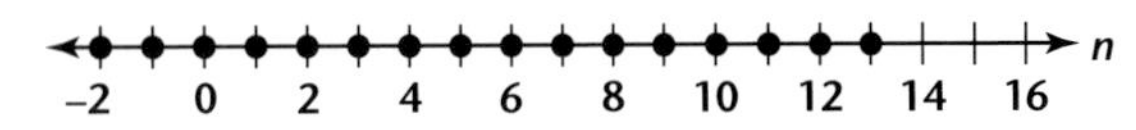

29.

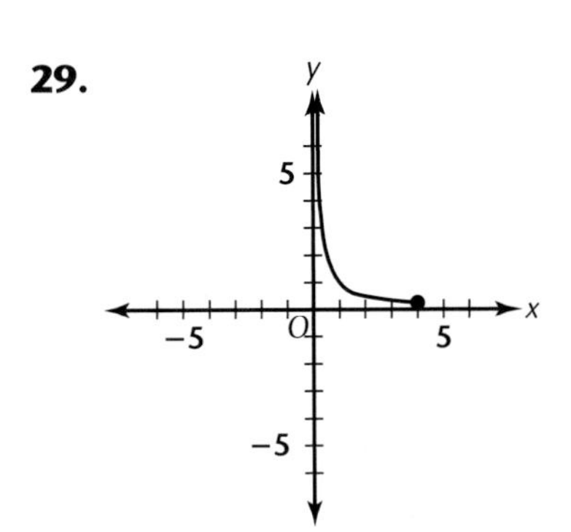

31. $y \approx 0.54x + 1.81$

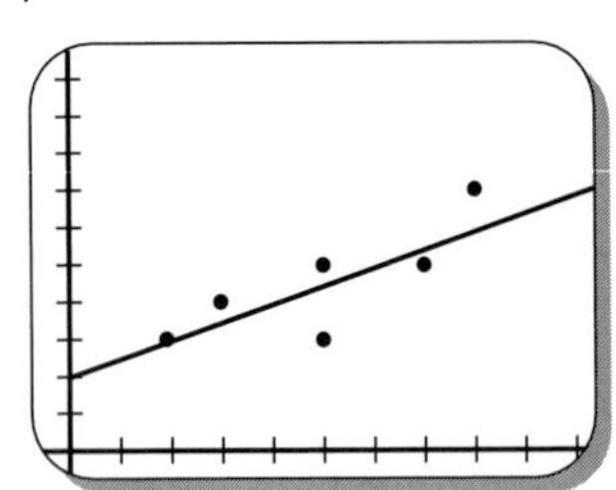

Lesson 4.7

Try This, **page 207**

The probability that the sum is less than 3 is $\frac{1}{36}$. So the probability that the sum is at least 3 is $1 - \frac{1}{36}$, or $\frac{35}{36}$.

Communicate, **pages 207–208**

5. Subtract the number of ways the event could not occur from the total number of possible outcomes. The difference represents the number of ways the event could occur. Divide the difference by the total number of possible outcomes to find the probability of an event occurring.

6. Count the squares on the left side of the vertical line. Divide by 100; 60%

7. Count the squares above the horizontal line. Divide by 100; 80%.

8. Count the number of squares that are to the left of the vertical line **and** are also above the horizontal line. Divide by 100; 48%.

9. Count the number of squares to the left of the vertical line. Add this number to the number of squares above the horizontal line. Subtract the number of squares found in Exercise 8 from this total. Divide by 100; 92%.

Practice and Apply, **pages 208–209**

18. Independent, because it does not matter what the results of the first coin toss are.

19. Independent, because it does not matter what the results of the first coin toss are.

20. Dependent, because to have two heads in a row, the first coin toss must be heads.

29. Answers may vary for experimental probability.
Experimental Probability
Randomly generate 0 or 1. Let 0 represent an incorrect response and 1 represent a correct response. Generate 5 numbers, and record the results. This represents 1 trial. Repeat for 30 trials. Divide the number of trials where 4 or

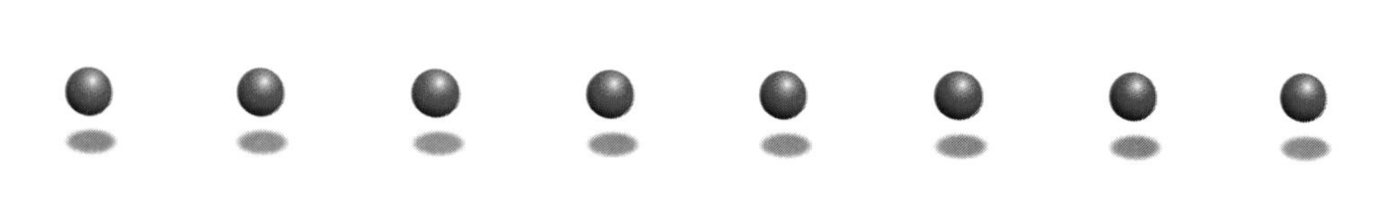

5 questions were answered correctly by 30. This is the experimental probability. The probability should be about 19%.
Theoretical Probability
Theoretical probability is 18.75%.

Look Back, page 209

30. Reciprocal function

x	−2	−1	0	1	2
y	−2	−4	undefined	4	2

31. Quadratic function

x	−2	−1	0	1	2
y	16	4	0	4	16

32. Absolute-value function

x	−2	−1	0	1	2
y	8	4	0	4	8

33. Linear function

x	−2	−1	0	1	2
y	8	4	0	−4	−8

Look Beyond, page 209

39. You have $\frac{1}{2000}$ or 0.0005 chance of winning \$25 from the second station. You have a $\frac{1}{100,000}$ or 0.00001 chance of winning \$1000 from the first station. If you call the second station 100,000 times, you would have a chance of winning 50 times, for a total of \$1250. If you call the first station 100,000 times, you would have a chance of winning once, or \$1000. It is more lucrative financially to call the second station.

Chapter 4 Project

Activity 1, page 211

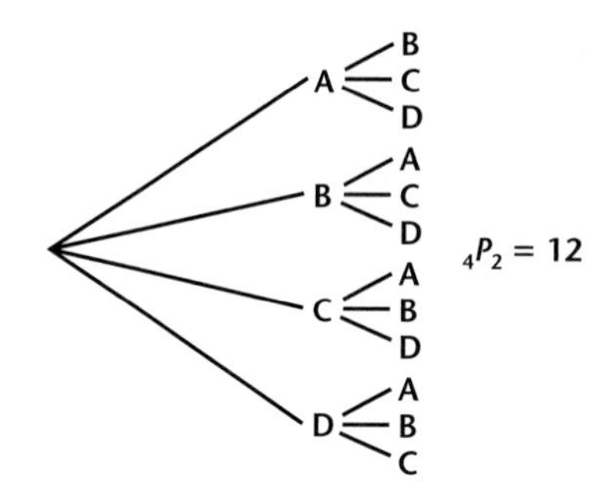

Activity 2, page 211

$$\begin{aligned} {}_5P_1 &= 5 \\ {}_5P_2 &= 20 \\ {}_5P_3 &= 60 \\ {}_5P_4 &= 120 \\ + {}_5P_5 &= 120 \\ \hline & 325 \end{aligned}$$

TPO [TOP] [TIP] TNP
TPI TOI TIO TNO
TPN [TON] [TIN] TNI
POI [PIN] PIO PNI OIP ONT OPI OTN
PON [PIT] PNO PTI OIN ONP OPN OTI
[POT] PNT PTO PTN OIT OTP [OPT] ONI
INT ITP IPO [ION] NTP NPO [NOT] [NIT]
INP ITN IPT IOT NTO NPT NOP [NIP]
INO ITO IPN IOP NTI NPI NOI NIO

12 English words; $_5P_3 = 60$
The probability is $\frac{12}{60} = 20\%$.

Chapter 4 Review, pages 212–214

5. Perform 10 trials using INT(RAND*2). Let 0 indicate false and 1 indicate true.

6. Flip a coin 10 times and let heads represent true and tails represent false.

7. The median score is 3, and it is slightly smaller than the mean score of 3.35.

9.

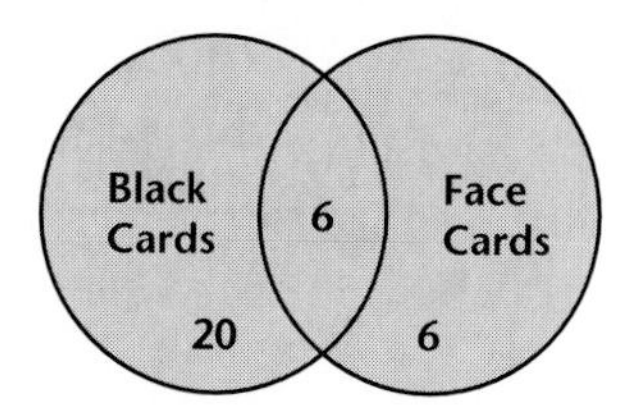

Chapter 4 Assessment, page 215

1. Generates a rational number between 0 and 5.

2. Generates an integer from the set {0, 1, 2, 3, 4}.

3. Generates an integer from the set {4, 5, 6, 7, 8}.

5. Use the expression INT(RAND*6) + 1 to generate a random integer from 1 to 6, and then use this number to pick a phone number.

14. The probability of each outcome is equal. The formula for determining probability is based on equally likely outcomes. $P = \frac{s}{n}$, where n is the number of equally likely outcomes.

25. Answers may vary. A sample answer is provided. Independent events have no relationship to each other, whereas dependent events do affect one another.

Chapters 1–4 Cumulative Assessment, pages 216–217

11. $x > -4$

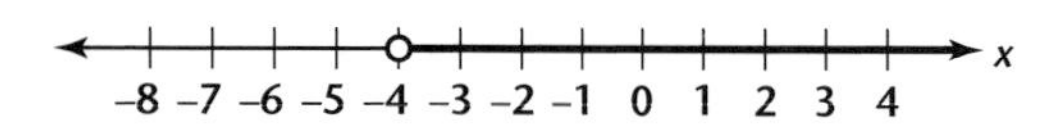

12. $y = -\frac{2}{5}x + \frac{1}{2}$; The slope is $-\frac{2}{5}$.

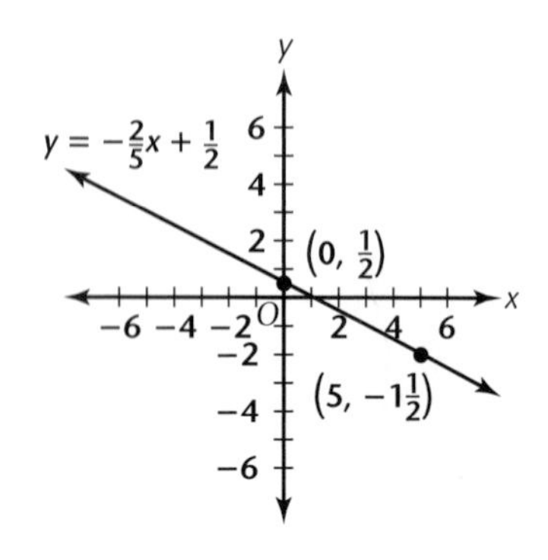

Lesson 5.1

Ongoing Assessment, page 223

The absolute value of any number is always non-negative because absolute value represents distance, which can never be less than zero.

Communicate, page 225

1. A relation pairs elements from one set with elements of a second set. A function is a set of ordered pairs for which there is exactly one second coordinate for each first coordinate.

2. The possible values of the independent variable are the domain, and the possible values of the dependent variable are the range.

3. No vertical line will intersect the graph of a function more than once since a function has only one y-value (or dependent variable) for each x-value (independent variable).

4. Substitute and evaluate $2x^2 - 5x$ for each of the integers in the domain, -2, -1, 0, 1, and 2; $f(-2) = 18$, $f(-1) = 7$, $f(0) = 0$, $f(1) = -3$, $f(2) = -2$. The range is $-3, -2, 0, 7, 18$.

5. The first element (the x-coordinate), 6, is from the domain and the second element (the y-coordinate), -46, is from the range.

6. The -24 is the x-coordinate (the first element), and the 16 is the y-coordinate (the second element). Therefore, the ordered pair is $(-24, 16)$.

Practice and Apply, pages 226–227

22. Arches, Bryce Canyon, Canyonlands, Capitol Reef, and Zion

23. Grand Canyon, Mesa Verde, Rocky Mountains, Chaco Culture, Arches, Bryce Canyon, Canyonlands, Capitol Reef, and Zion
24. Petrified Forest, Carlsbad Caverns, Rocky Mountains, Chaco Culture, Arches, Canyonlands, and Capitol Reef
26. No; the first coordinate, 9, has more than one second coordinate.
27. Yes; there is one y-coordinate for each x-coordinate.
28. Yes; there is one y-coordinate for each x-coordinate; it passes the vertical-line test.
29. No; there is more than one y-coordinate for each x-coordinate; it does not pass the vertical-line test.
42. x, $f(x)$, D: {all real numbers}, R: {all real numbers}
43. x, $f(x)$, D: {all real numbers}, R: $\{y \geq 0\}$
44. x, $f(x)$, D: {all real numbers}, R: $\{y \geq 0\}$
45. x, $f(x)$, D: {all real numbers}, R: $\{y \geq 0\}$
46. x, $f(x)$, D: {all real numbers}, R: $\{y \leq 0\}$
47. x, $f(x)$, D: {all real numbers}, R: $\{y \leq 0\}$
48. x, $f(x)$, D: {all real numbers}, R: $\{y \geq 0\}$
49. x, $f(x)$, D: {all real numbers}, R: $\{y \geq 0\}$
50. Yes; yes, if C represents temperature in degrees Celsius and F represents temperature in degrees Fahrenheit, then $F = \frac{9}{5}C + 32$. In function notation, $F(C) = \frac{9}{5}C + 32$, so Fahrenheit temperature is a function of Celsius temperature.
51. Yes; yes, since $C = \frac{5}{9}(F - 32)$, where F represents degrees Fahrenheit and C represents degrees Celsius, then $C(F) = \frac{5}{9}(F - 32)$, so Celsius temperature is a function of Fahrenheit temperature.

Look Back, page 227

55. $y = 2x - 12$

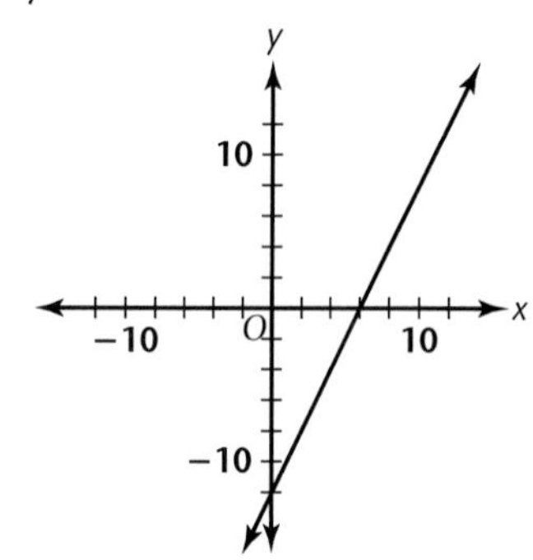

Look Beyond, page 227

62. Let $Y_1 = \sqrt{x}$, and $Y_2 = -\sqrt{x}$.

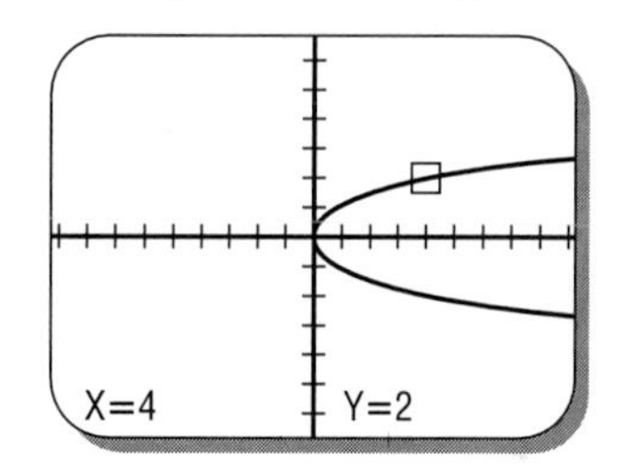

63. Let $Y_1 = \sqrt{25 - x^2}$, and $Y_2 = -\sqrt{25 - x^2}$.

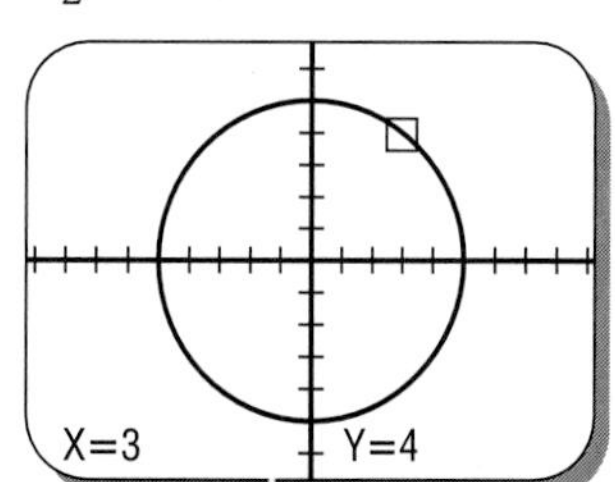

Extra Practice page 538

1. No, the x-value of 2 is paired with two y-values, 3 and 8.
2. Yes, each x-value is paired with exactly one y-value.
3. Yes, each x-value is paired with exactly one y-value.
4. Yes, each x-value is paired with exactly one y-value.

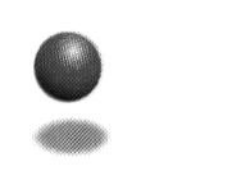   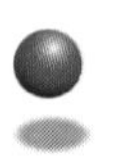 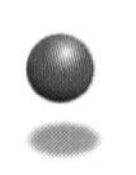  

5. No, there are x-values paired with more than one y-value.
6. No, there are x-values paired with more than one y-value.

Lesson 5.2

Exploration 1, **page 229**

1.

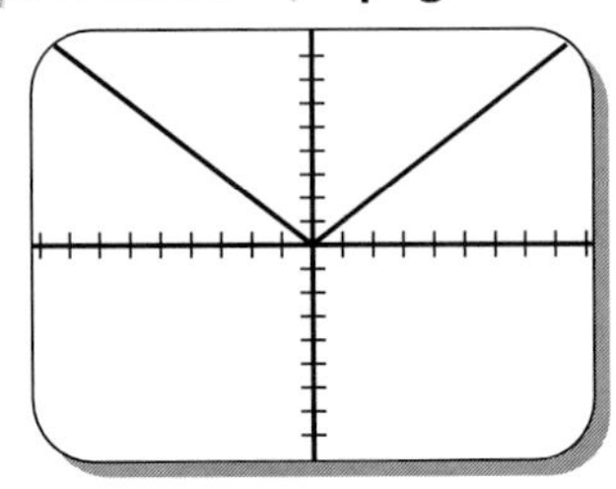

2a. Reflected through the x-axis.

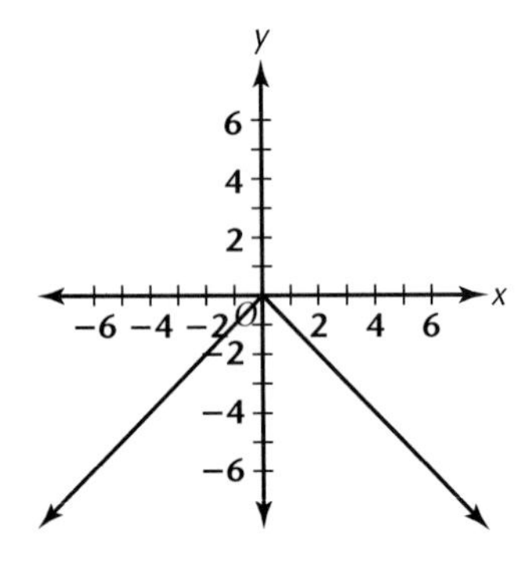

2b. Stretched by a factor of 2

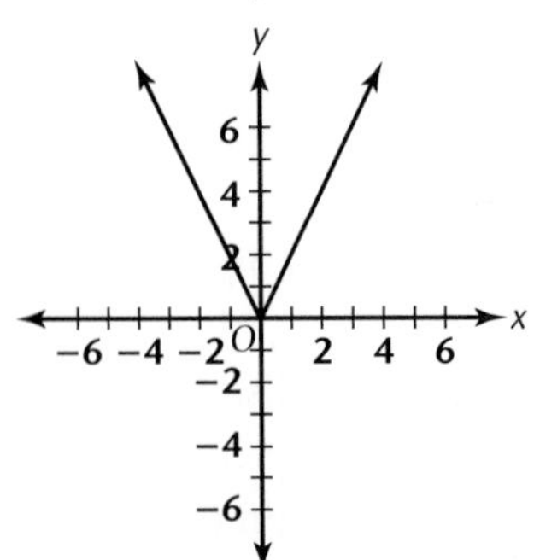

2c. Shifted vertically by −2 units

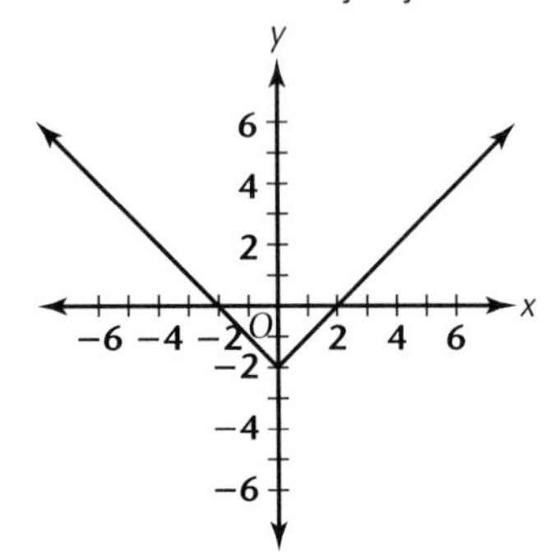

2d. Reflected through the x-axis and shifted vertically by 4 units

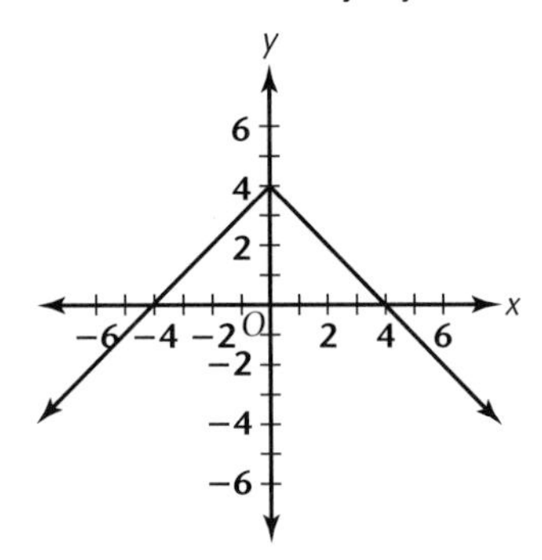

2e. Stretched by a factor of 2 and shifted vertically by −4 units

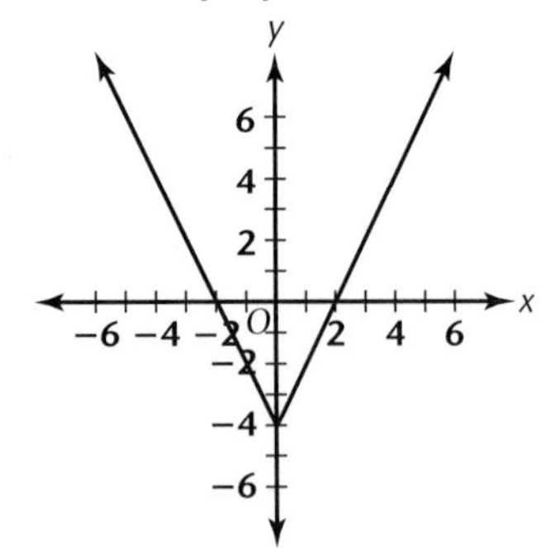

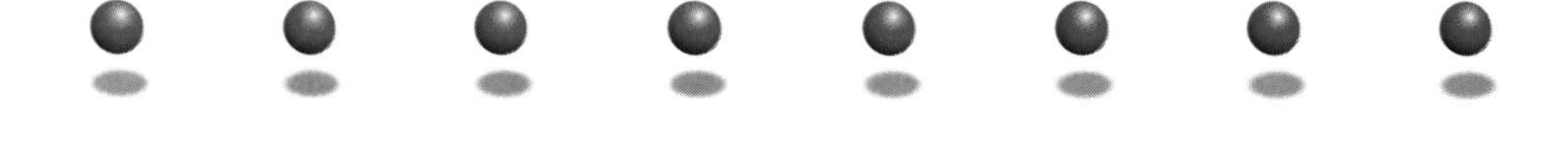

2f. Stretched by a factor of 3, reflected through the x-axis, and shifted horizontally by -2 units

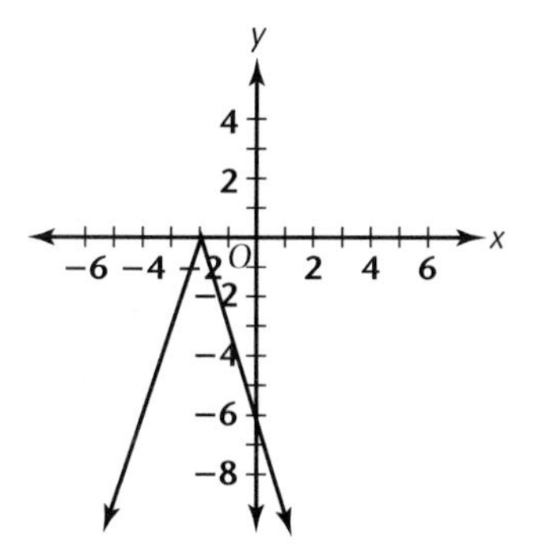

3. The function in 2a matches graph **III**; 2b matches graph **I**; 2c matches graph **II**; 2d matches **V**; 2e matches **IV**; and 2f matches graph **VI**.

4. To stretch the graph, multiply it by a number, the larger the number the narrower (closer to the y-axis) the graph appears. To reflect $y = |x|$, multiply $|x|$ by -1. To shift the graph vertically, add or subtract a number to or from x after its absolute value has been taken. Adding shifts the graph up and subtracting shifts it down. To shift the graph horizontally, add or subtract a number from x and then take the absolute value of this result. Adding shifts the graph to the left and subtracting shifts it to the right.

5. If it has been stretched, then the sides of the graph would be nearer to the y-axis than the lines $y = x$ and $y = -x$. If it has been reflected, then it would open downward. If it has been shifted, then the vertex of the graph would not be at (0, 0).

6a. Stretch the graph by a factor of 3.

6b. Stretch the graph by a factor of 5 and reflect it through the x-axis.

6c. Shift the graph vertically by 2 units.

7. The graph will be shifted horizontally to the left by 1 unit.

8a.

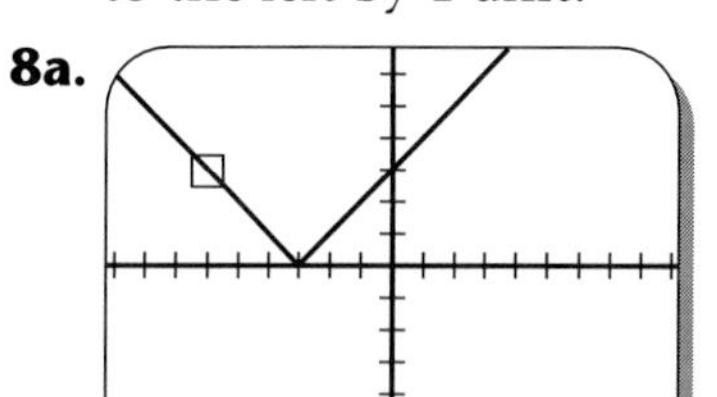

8b.

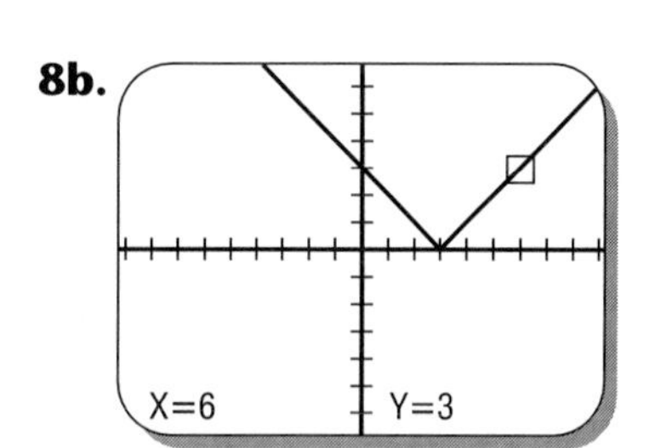

Exploration 2, **page 230**

1.

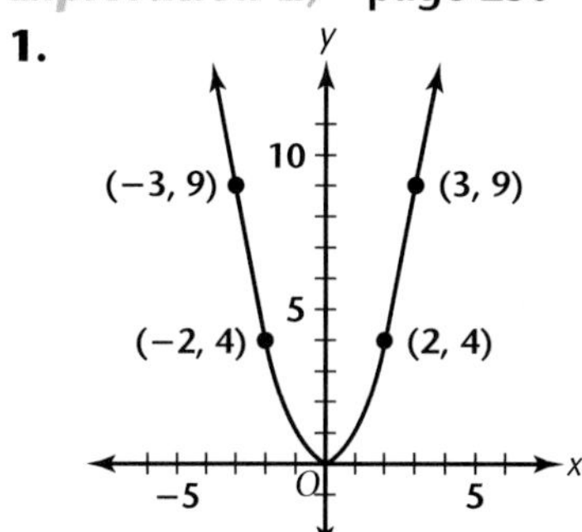

2a. Reflected through the x-axis

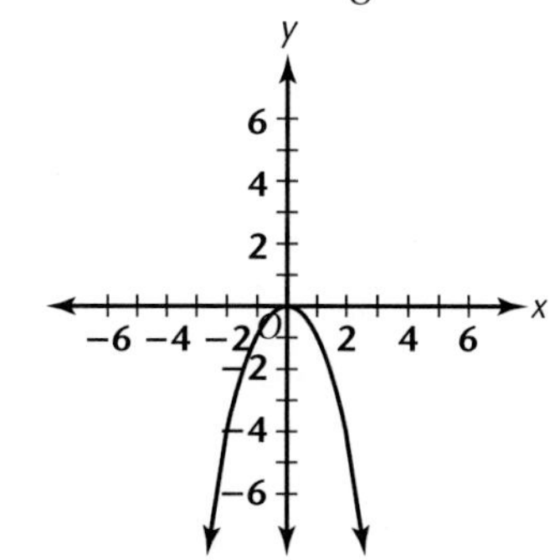

2b. Stretched by a factor of 2

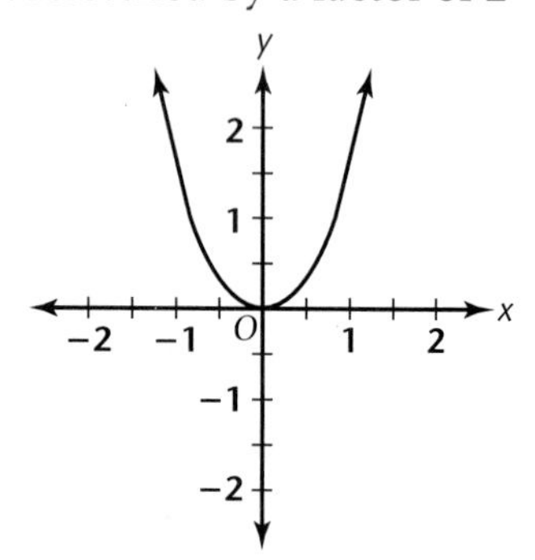

2c. Shifted vertically by −2 units

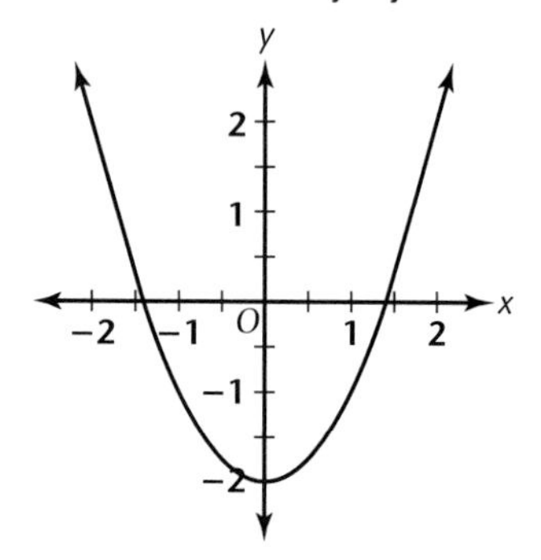

2d. Shifted horizontally to the left by 4 units

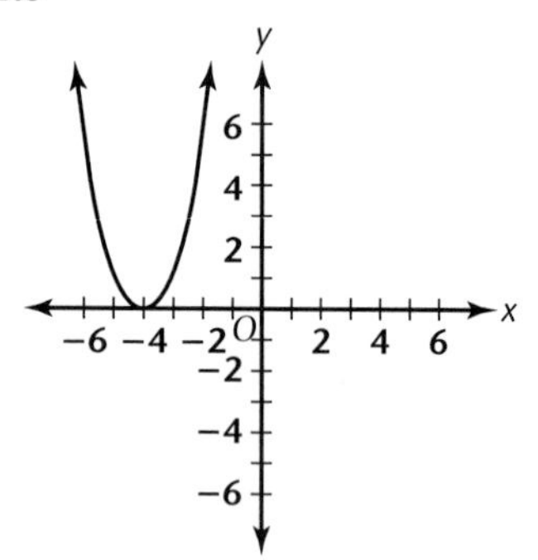

3. Graphs **2a** of Explorations 1 and 2 are reflected through the x-axis and open downward. Graphs **2b** of Explorations 1 and 2 open upward and are stretched by a factor of 2. Graphs **2c** of Explorations 1 and 2 are shifted down by 2 units. Graph **2d** in Exploration 2 is shifted left 4 units while graph **2d** in Exploration 1 is translated up 4 units.

4. To stretch the graph, you multiply it by a number. To reflect the graph, multiply x^2 by -1. To shift the graph vertically, add or subtract a number to or from x^2. Adding shifts the graph up and subtracting shifts it down. To shift the graph horizontally, add or subtract a number to or from x and then square this result. Adding shifts the graph to the left and subtracting shifts it to the right.

5. Stretch the function by a factor of 2, reflect it through the x-axis, and shift it vertically by 3 units.

Practice and Apply, **pages 232–233**

10. Stretched by a factor of 4

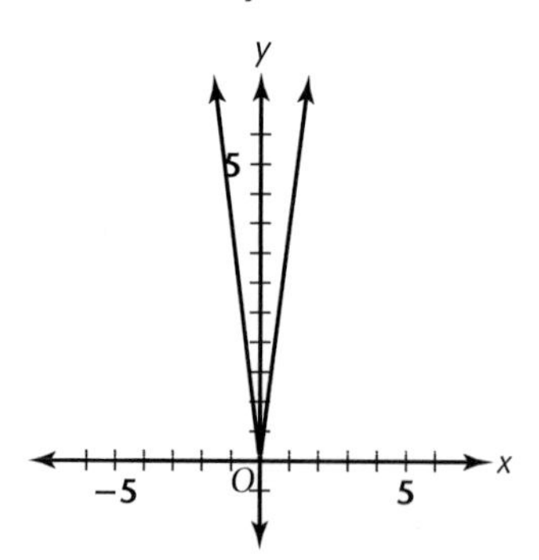

11. Shifted vertically by 4 units

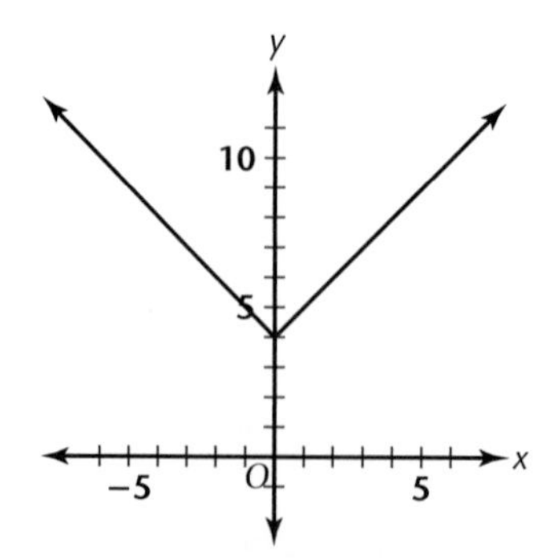

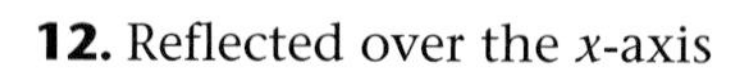

12. Reflected over the x-axis

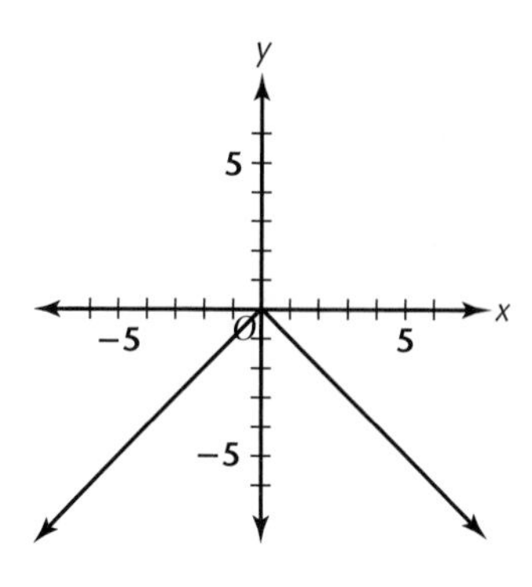

13. Stretched by a factor of 0.5

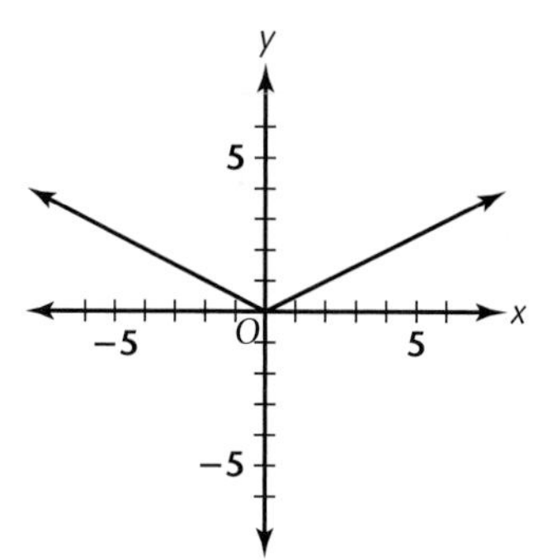

14. Shifted horizontally by 5 units

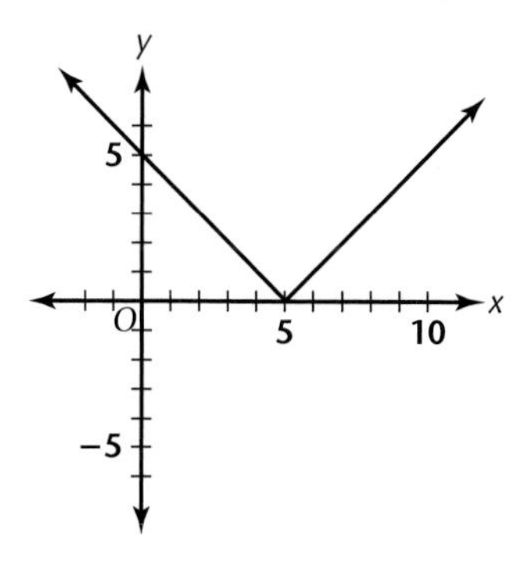

15. Shifted vertically by −3 units

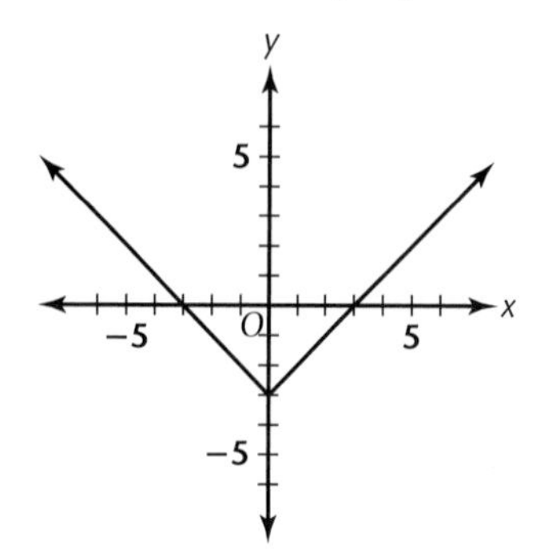

16. Shifted vertically by 5 units

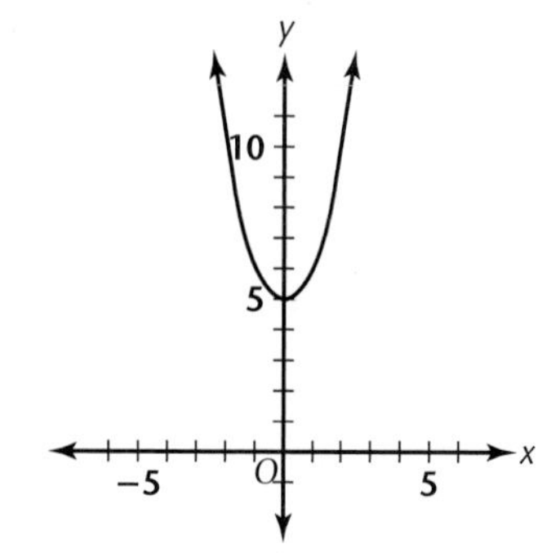

17. Stretched by a factor of 4

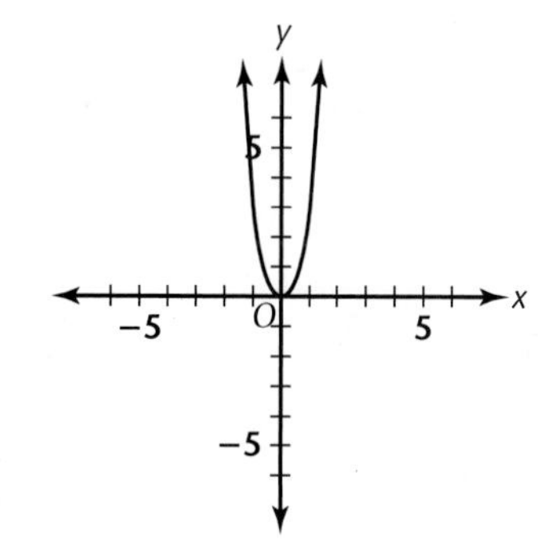

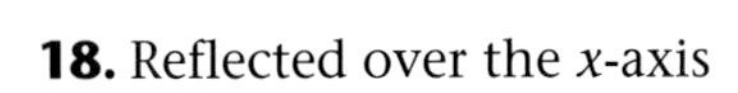

18. Reflected over the x-axis

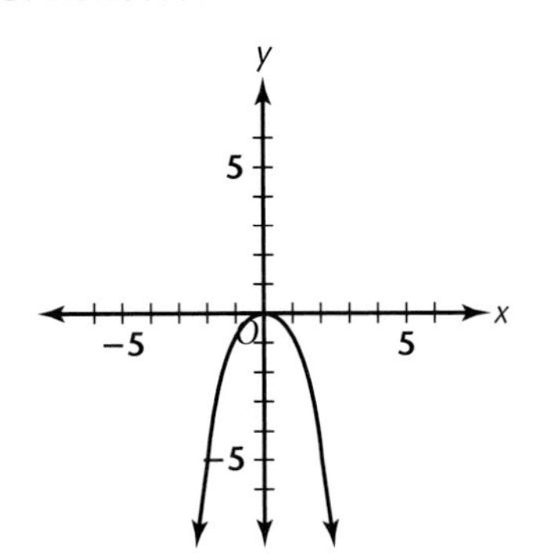

19. Shifted vertically by $\frac{1}{3}$ unit

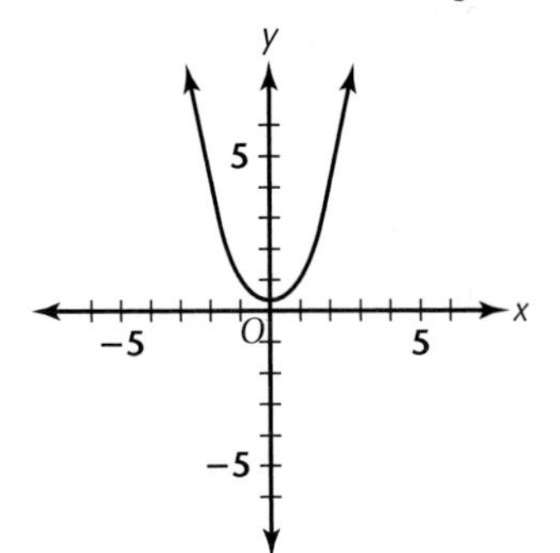

20. Shifted horizontally by -2 units

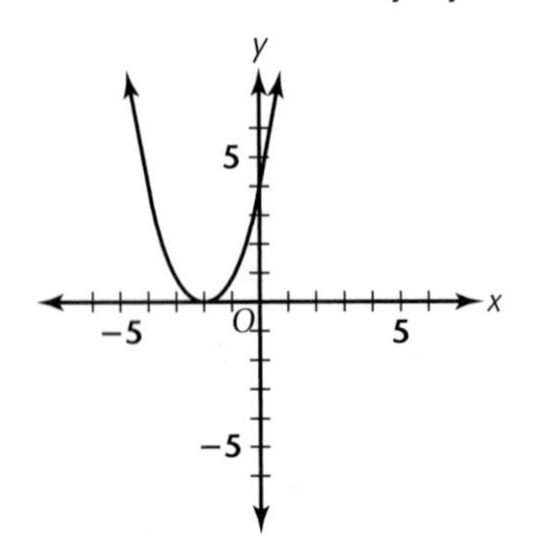

21. Shifted vertically by -3 units

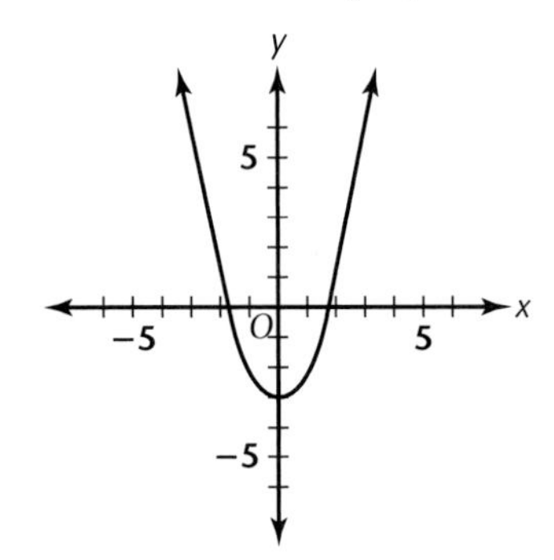

22. $y = x^2$; stretched by a factor of 5

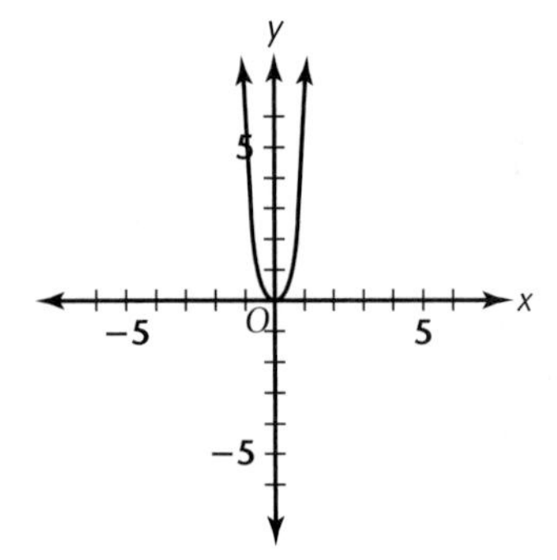

23. $y = x$; stretched by a factor of 10

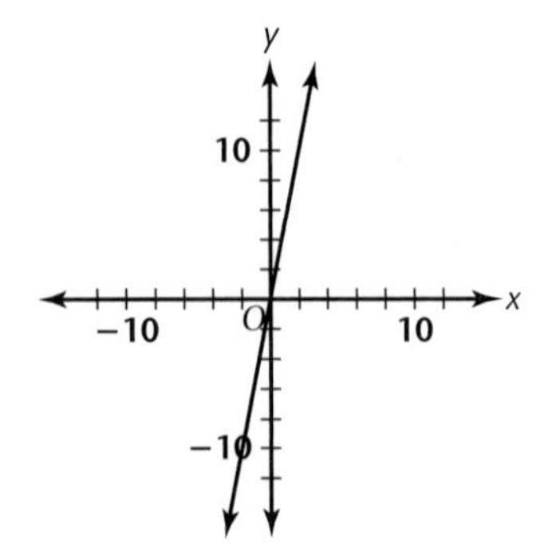

24. $y = \frac{1}{x}$; shifted horizontally by -2 units

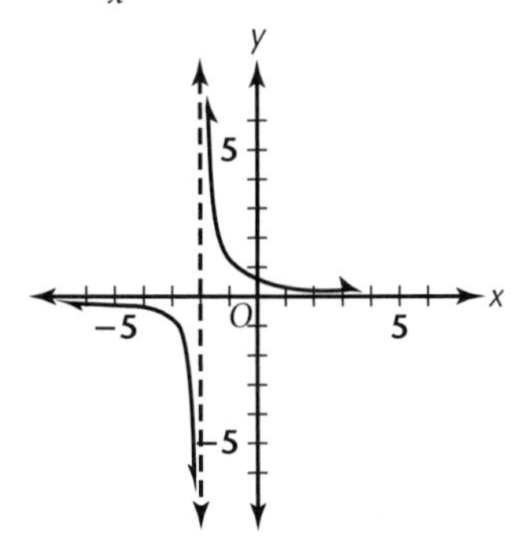

25. $y = |x|$; shifted horizontally by 2 units

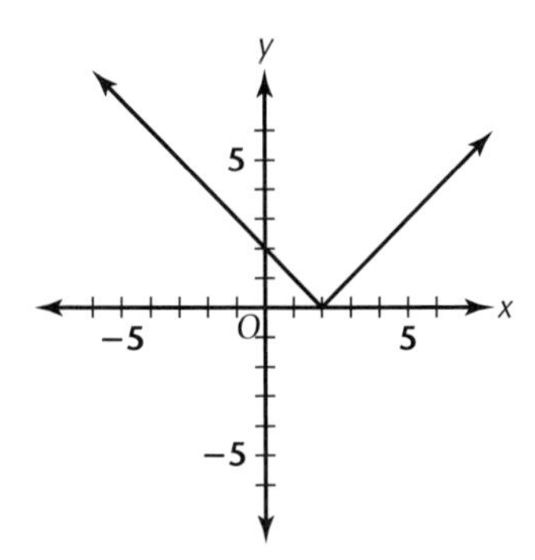

26.

x	$A = x^2$
2	4
1	1
0	0
−1	1
−2	4

27.

x	$A = (x + 4)^2$
0	16
−1	9
−2	4
−3	1
−4	0
−5	1
−6	4

28.

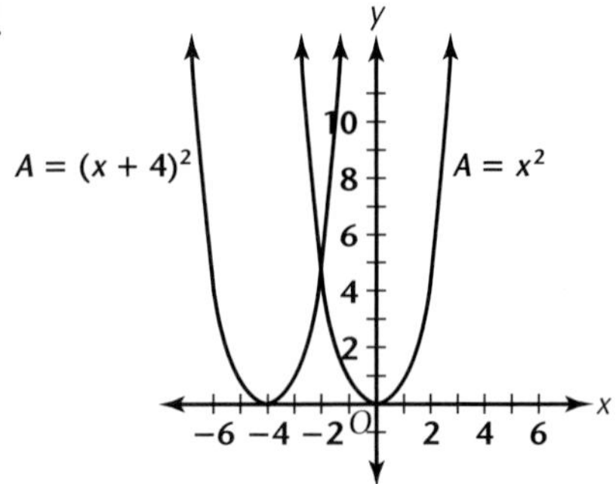

Look Back, **page 233**

38.

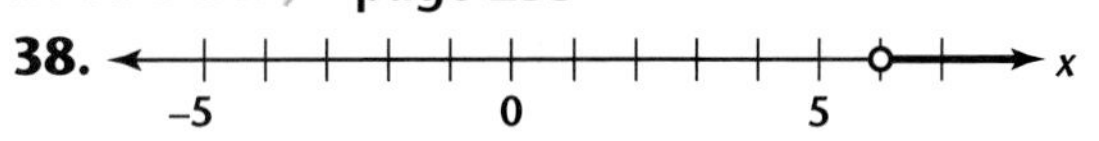

39.

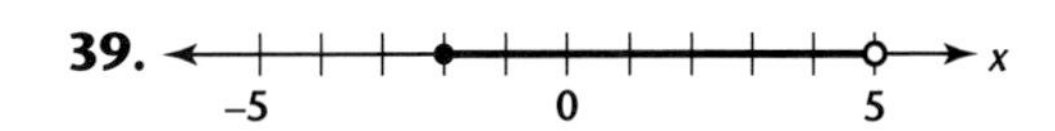

Extra Practice, **page 539**

1. $y = x$; The graph of $y = x$ is reflected over the y-axis.
2. $y = |x|$; The graph of $y = |x|$ is reflected over the x-axis.
3. $y = x^2$; The graph of $y = x^2$ is shifted up.
4. Stretched by a factor of 2

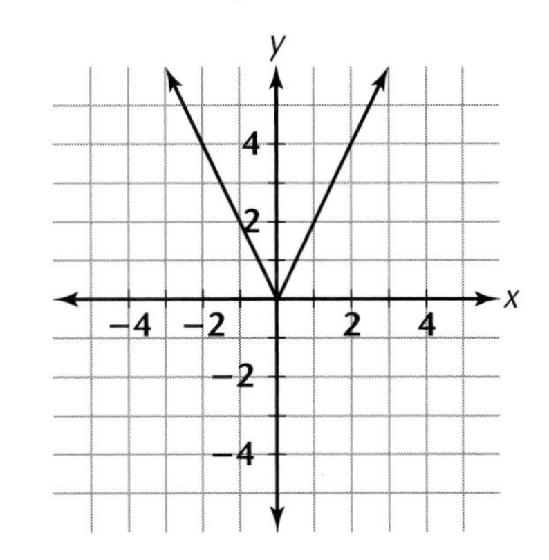

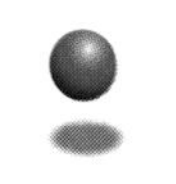

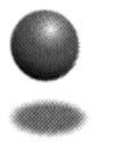

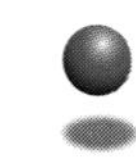

5. Shifted vertically by 2 units

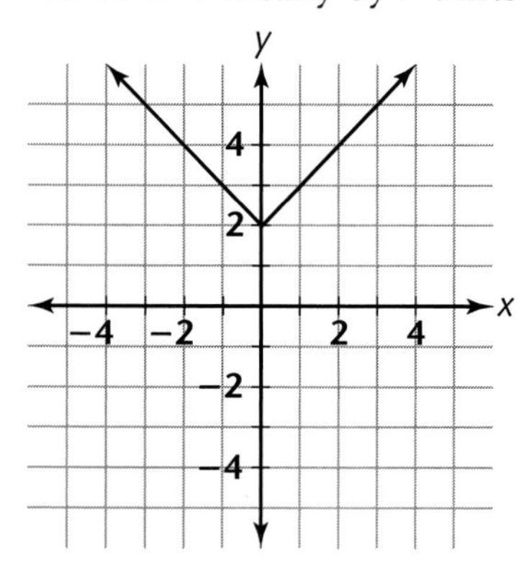

6. Stretched by a factor of 2 and reflected over the x-axis

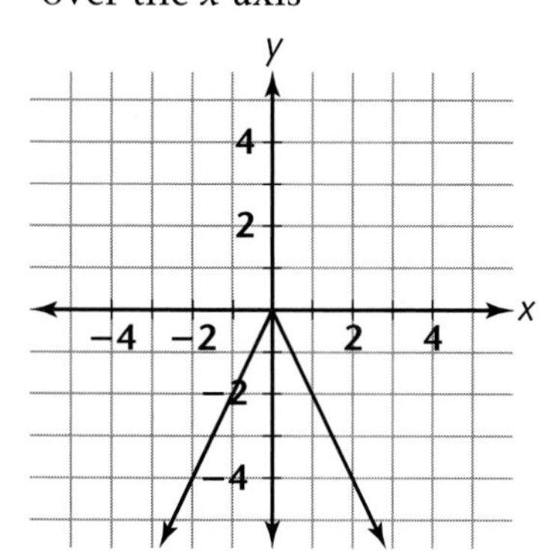

7. Shifted horizontally by 2 units

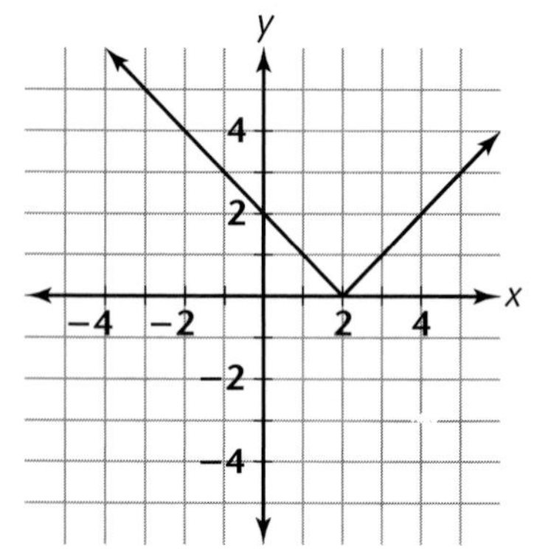

8. Shifted vertically by −3 units

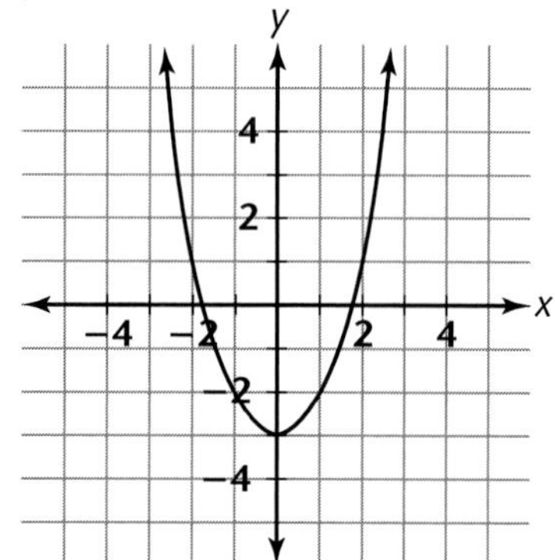

9. Stretched by a factor of 3

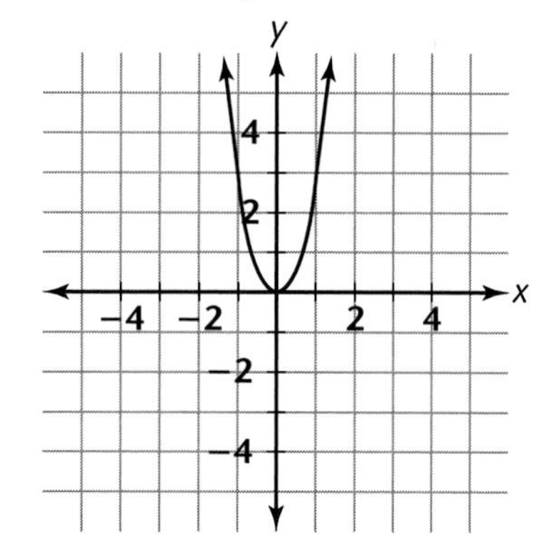

10. Stretched by a factor of 3 and reflected over the x-axis

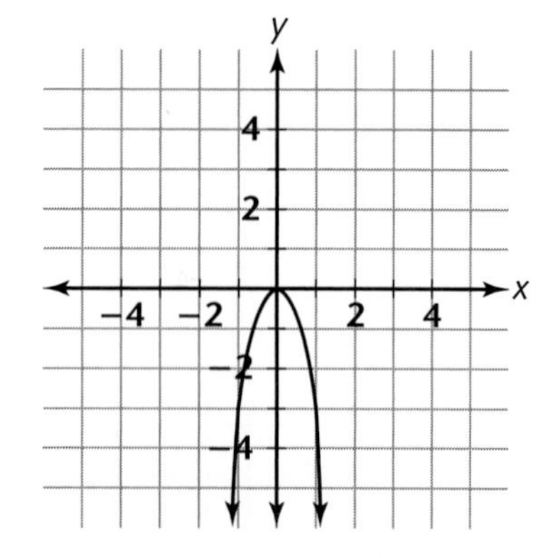

11. Shifted horizontally by -3 units

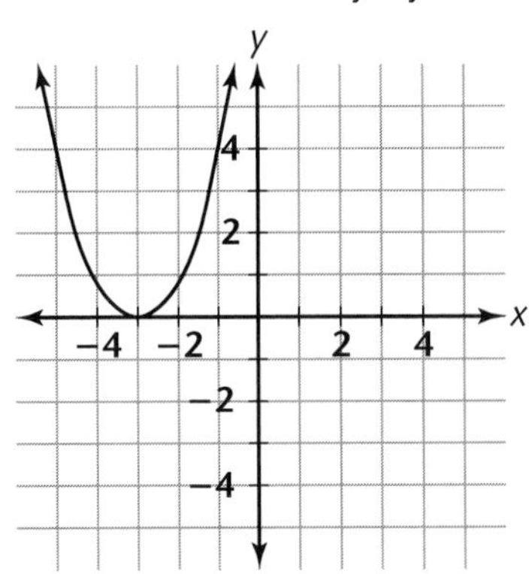

Lesson 5.3

Try This, **page 236**

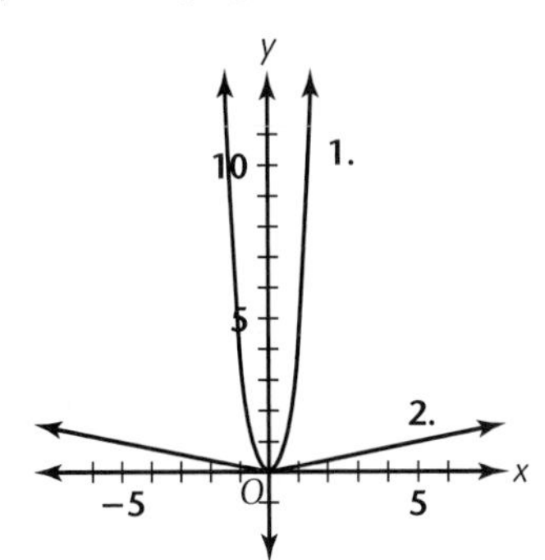

1. $f(x) = 4x^2$
2. $g(x) = \frac{|x|}{4}$

Communicate, **page 238**

1. When a is 1, the graph is the parent function $y = x^2$. As the value of a increases, the parabola becomes narrower (closer to the y-axis).
2. As a gets smaller and takes on values between 1 and 0, the parabola becomes wider (moves away from the y-axis).
3. When compared with the parent function, the y-values are changed by the stretch factor a.
4. When a is 1, the graph is the parent function $y = \frac{1}{x}$. As the value of a increases, the curves of the hyperbola move away from the origin.
5. As a gets smaller and takes on values between 1 and 0, the curves of the hyperbola move closer to the origin because the denominators of the y-values are becoming larger.

Practice and Apply, **pages 238–240**

8.

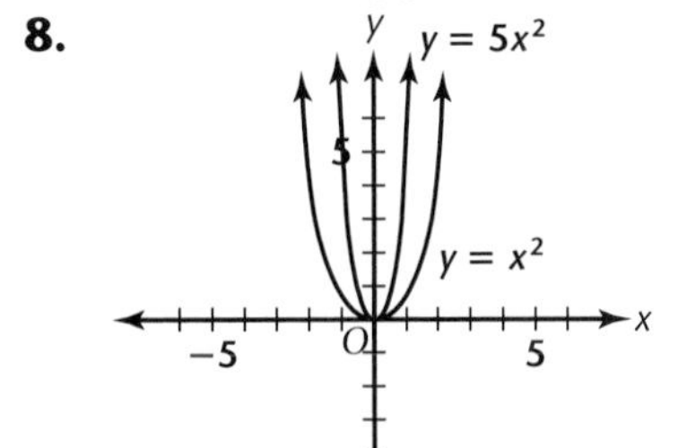

10.

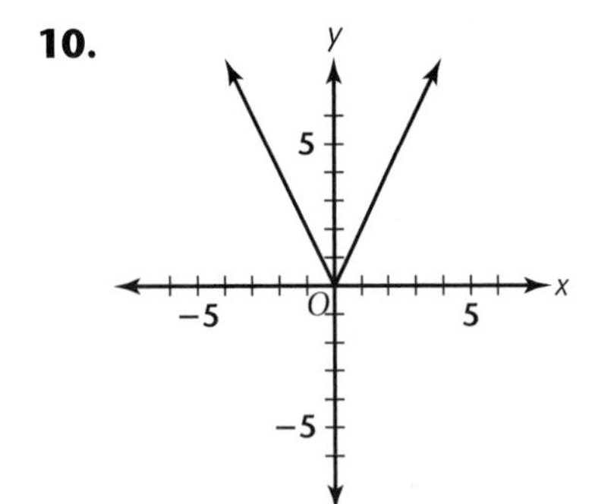

11.

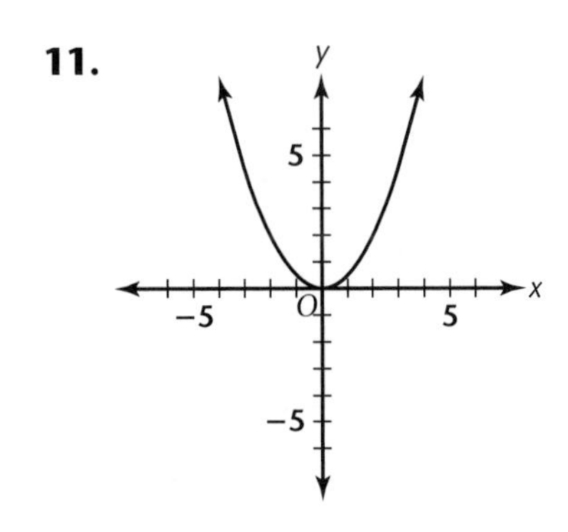

12.

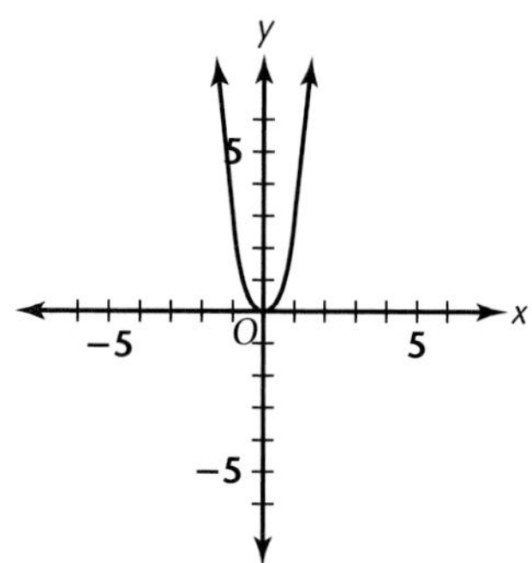

16.

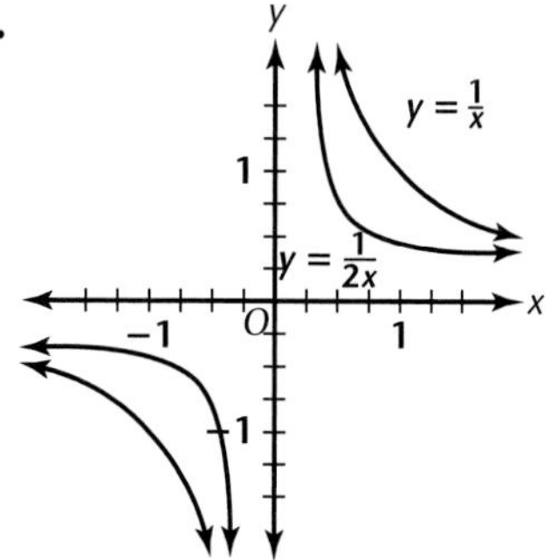

17. The scale factor of $\frac{1}{2}$ in $y = \frac{1}{2x}$ stretched the graph of $y = \frac{1}{x}$ by a factor of $\frac{1}{2}$. The graph of $y = \frac{1}{2x}$ is closer to the origin because the y-values are $\frac{1}{2}$ as large.

26. $y = |x|$; stretch by a factor of 3

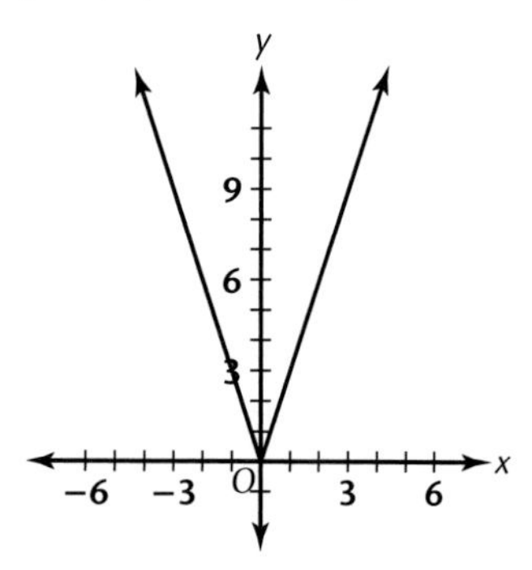

27. $y = x^2$; shift up by 3 units

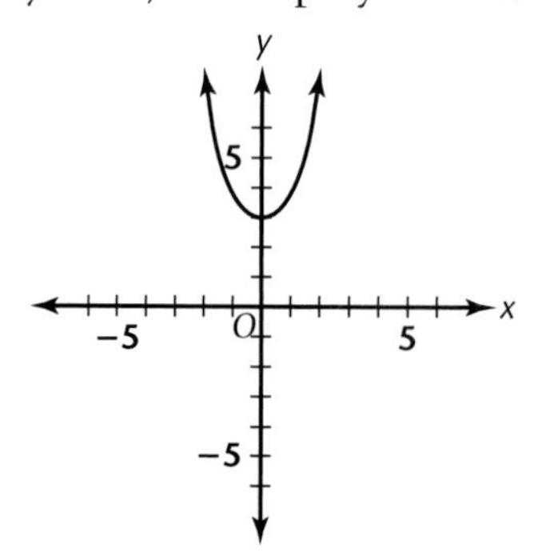

28. $y = x^2$; stretch by a factor of $\frac{1}{9}$

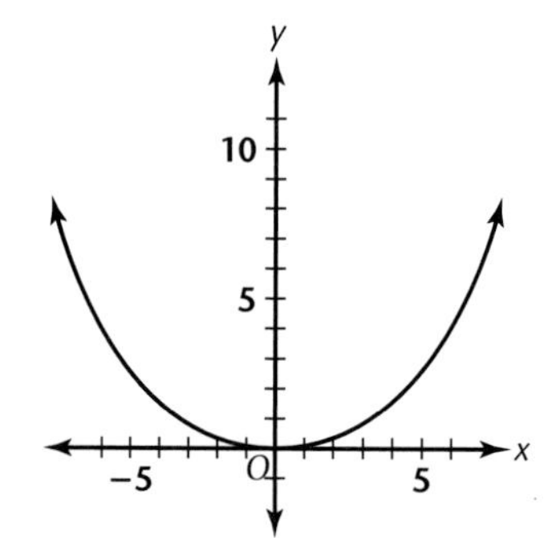

29.

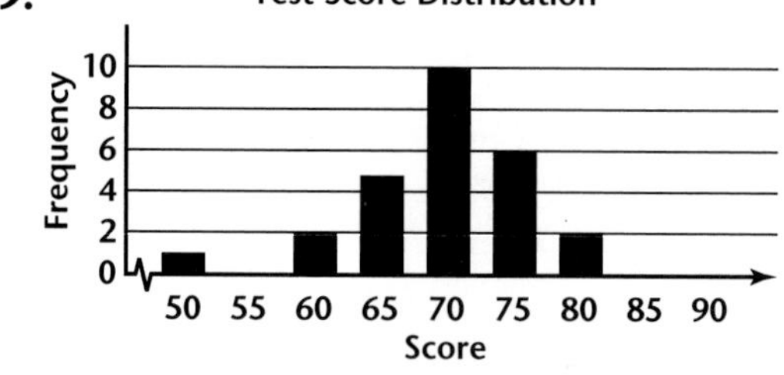

30.

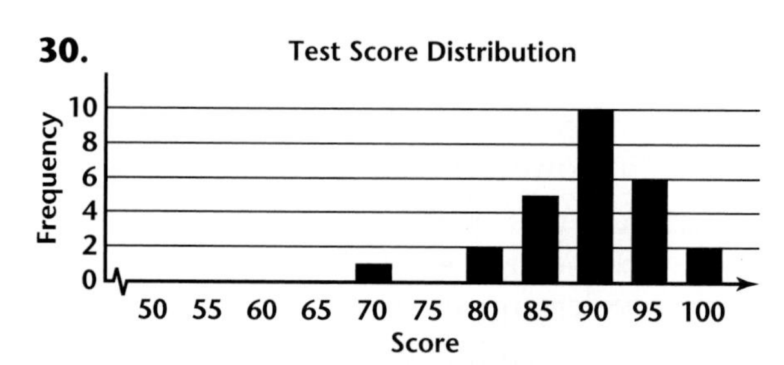

Horizontal shift of 20 points to the right.

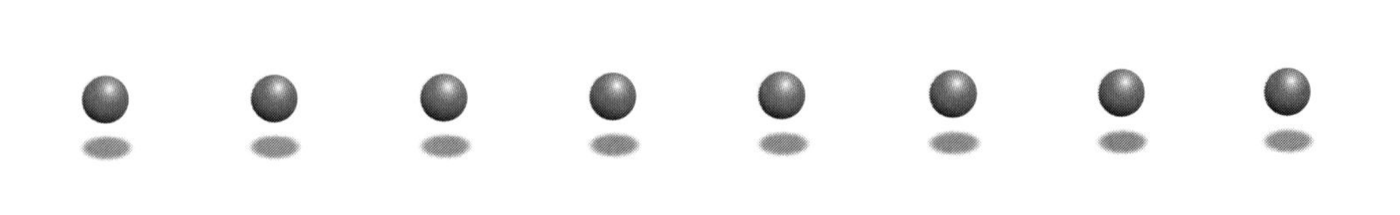

31.

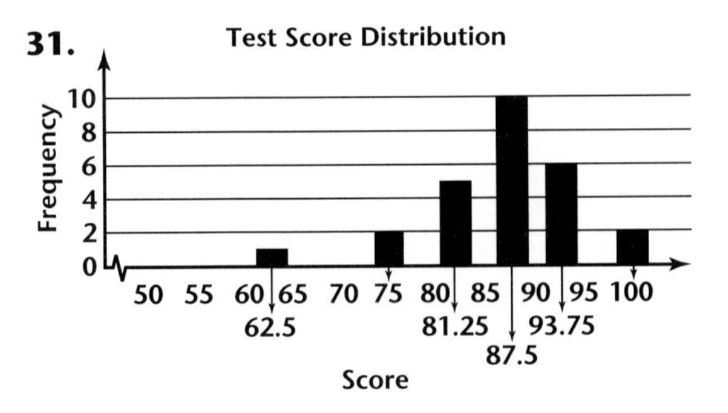

Moves each bar $\frac{5}{4}$ points to the right.

Extra Practice, **page 539**

1. $y = x^2$

2.

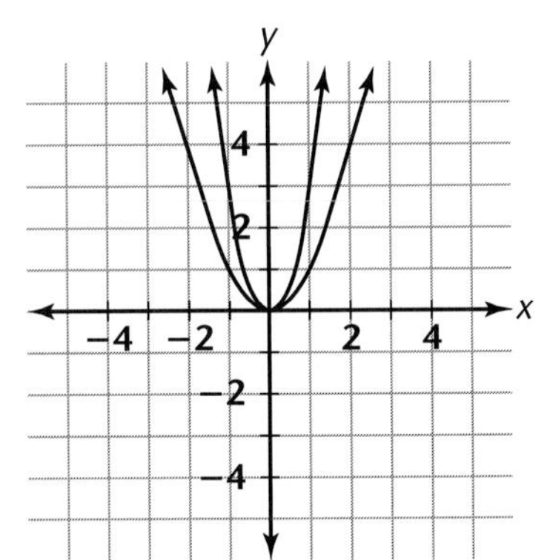

3. The graph of $y = x^2$ is stretched by a factor of 3 to become the graph of $y = 3x^2$.

4. $y = \frac{1}{x}$

5.

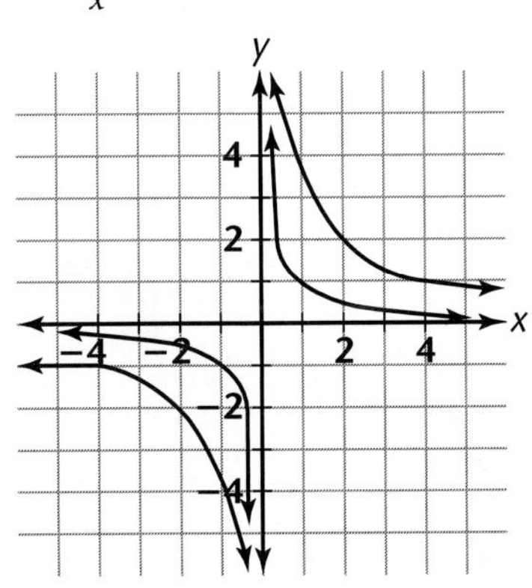

6. The graph of $y = \frac{1}{x}$ is stretched by a factor of 4 to become the graph of $y = \frac{4}{x}$.

7. $y = |x|$

8.

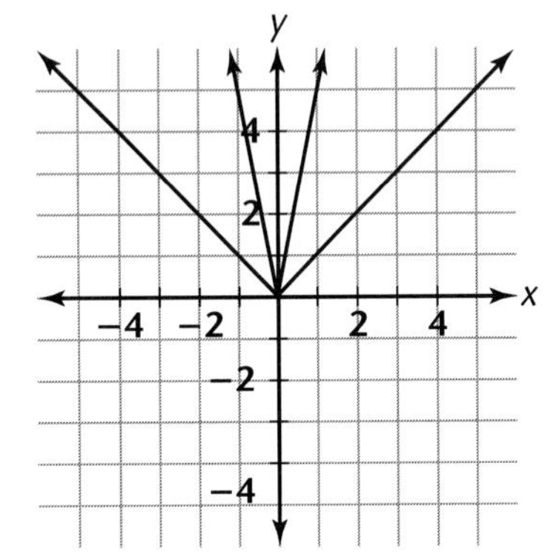

9. The graph of $y = |x|$ is stretched by a factor of 5 to become the graph of $y = 5|x|$.

10. Not a stretch

11. Stretch

12. Stretch

13. Not a stretch

Lesson 5.4

Exploration, **page 241**

1.

x	−2	−1	0	1	2
y	4	1	0	1	4

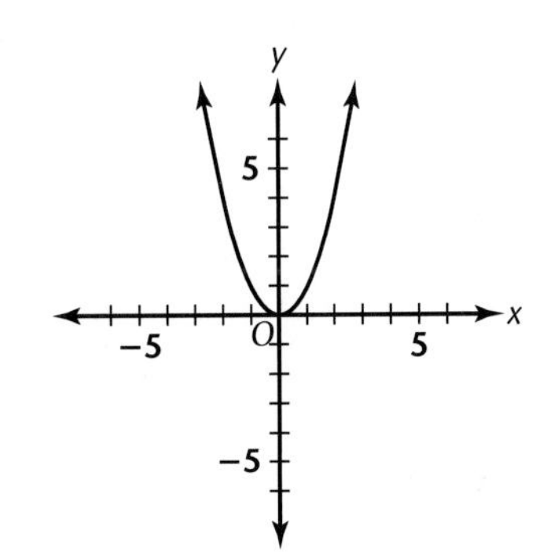

2.

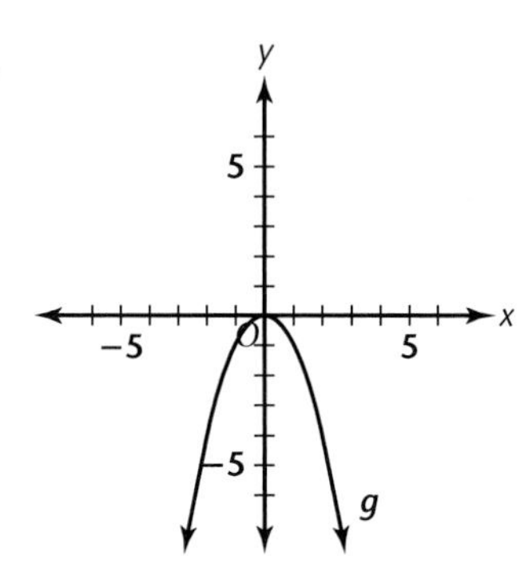

3.

x	−2	−1	0	1	2
y	−4	−1	0	−1	−4

6a. $f(x) = |x|$

x	−2	−1	0	1	2
y	2	1	0	1	2

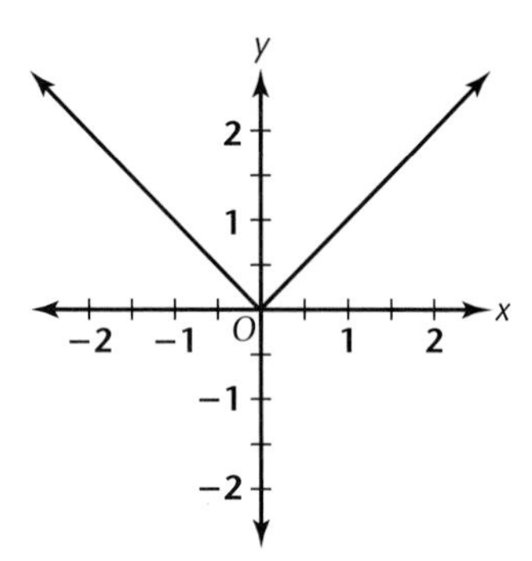

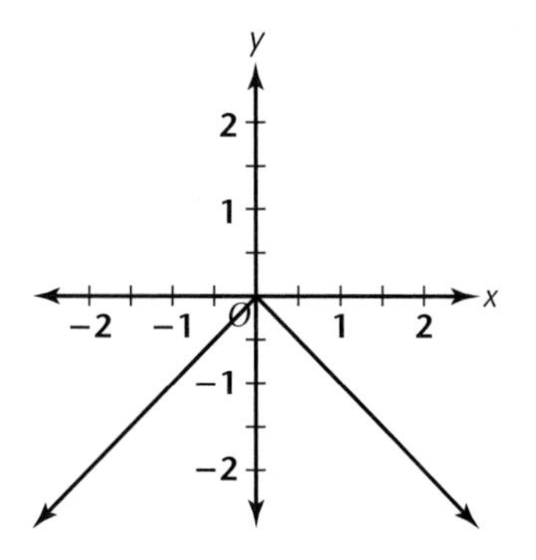

x	−2	−1	0	1	2
y	−2	−1	0	−1	−2

6b. $f(x) = \frac{2}{x}$

x	−2	−1	0	1	2
y	−1	−2	Undef	2	1

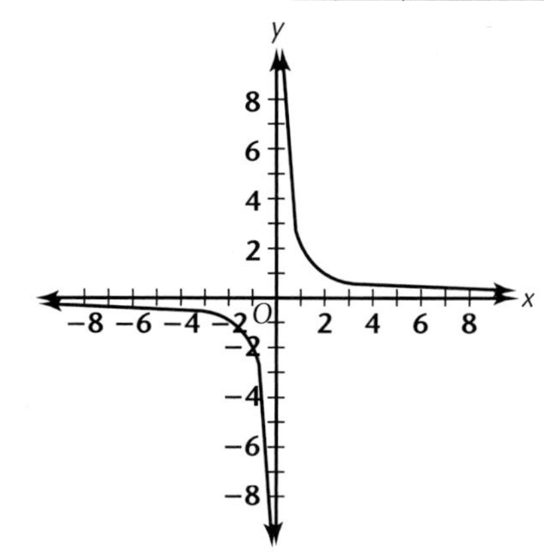

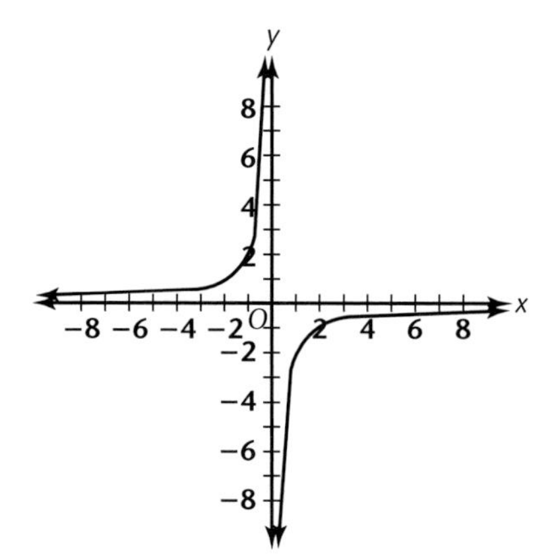

x	−2	−1	0	1	2
y	1	2	Undef	−1	−2

6c. $f(x) = -2x^2$

x	−2	−1	0	1	2
y	−8	−2	0	−2	−32

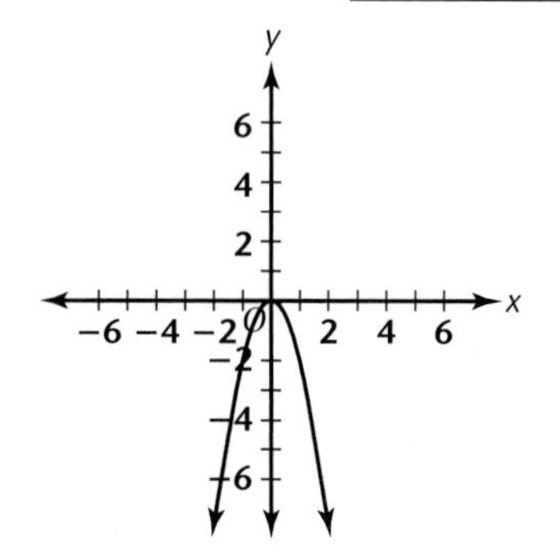

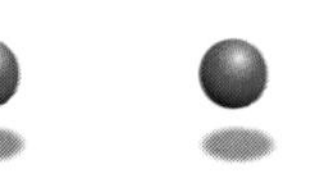

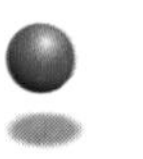

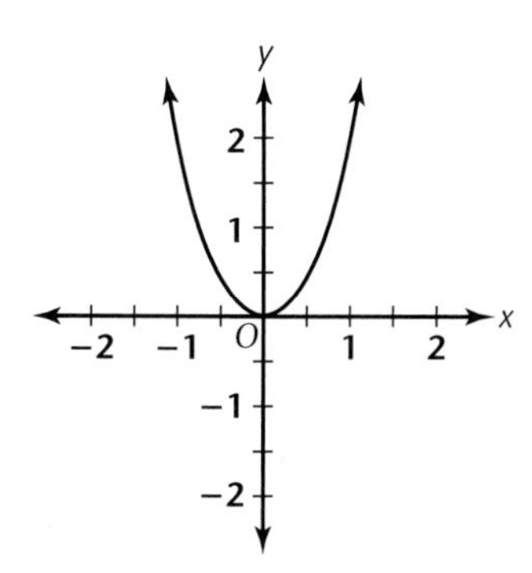

x	−2	−1	0	1	2
y	8	2	0	2	32

To generate the graph reflected through the x-axis, multiply the y-values of the parent function by -1.

Communicate, **page 243**

2. The original function is multiplied by -1 giving $g(x) = -a\left(\frac{1}{x}\right)$.

3. Select points on the graph of $f(x) = 2x$ and multiply their y-values by -1.

4. The new function is the opposite of the original function.

5. The graph remains the same. For example, if $f(x) = x^2$, then $g(x) = (-x)^2$ is the reflection of $f(x)$ through the y-axis.

6. $y = -(x)^2$ is a vertical reflection of $y = x^2$. $y = (-x)^2$ is a horizontal reflection of $y = x^2$, which is the same graph as that of $y = x^2$. If the negative sign is outside the parentheses, then the square of the x-value is taken *before* the resulting value is multiplied by -1. If the negative sign is inside the parentheses, then the x-values are squared *after* the x-value is multiplied by -1.

Practice and Apply, **page 244**

11.

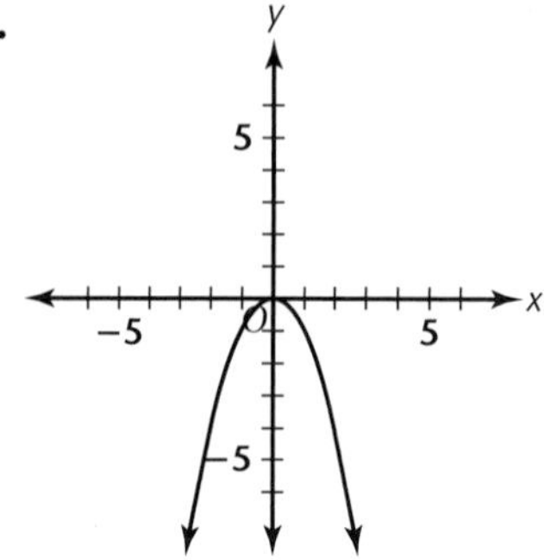

12.

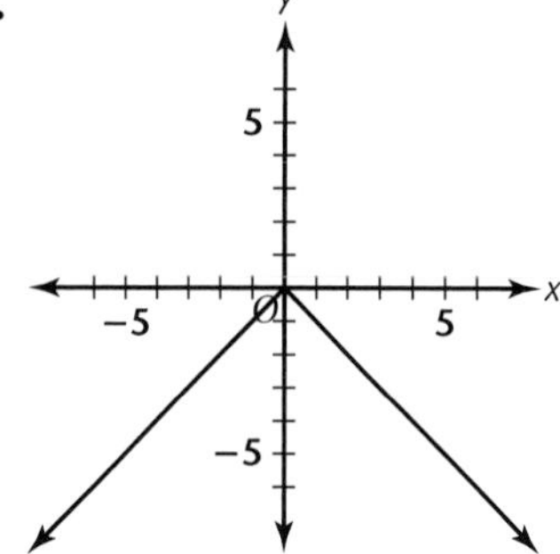

13.

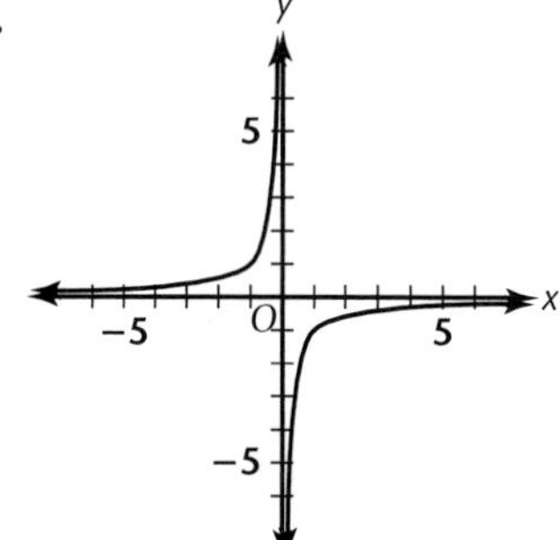

14. The graphs are the same. They both have a vertex at (0, 0), have a stretch of $\frac{1}{2}$ of the parent function $y = x^2$, and are reflected through the x-axis.

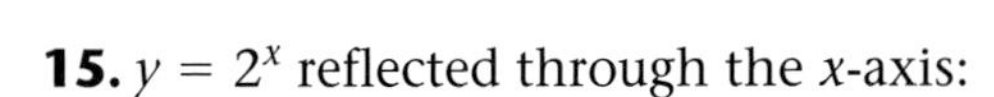

15. $y = 2^x$ reflected through the x-axis:

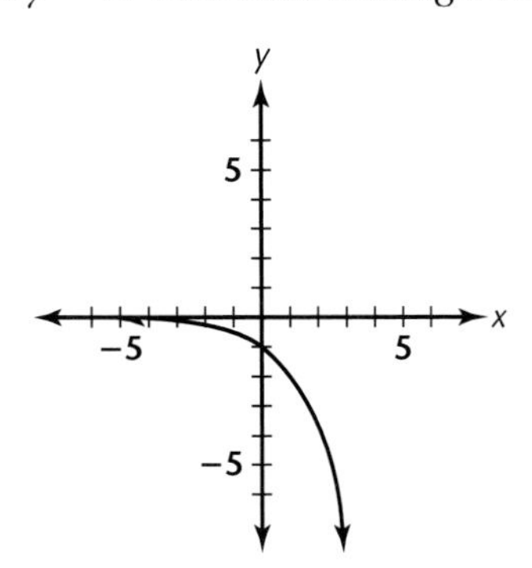

$y = 2^x$ reflected through the y-axis:

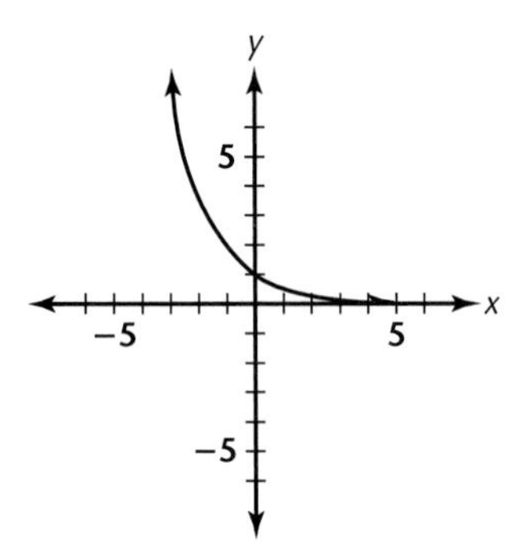

16.

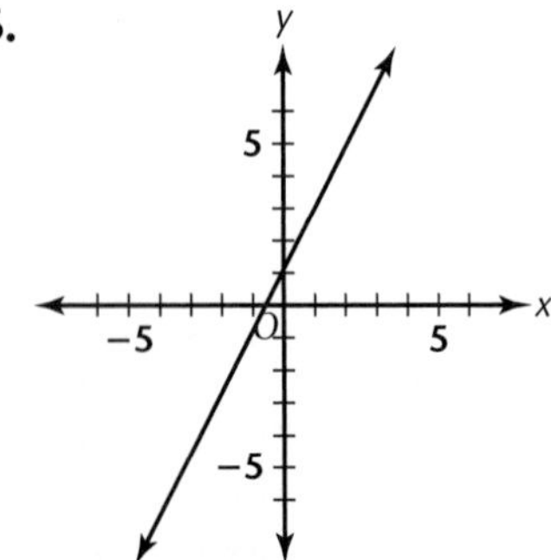

20.

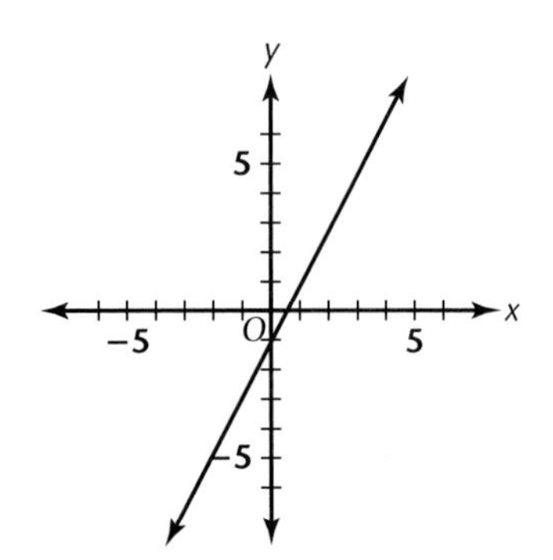

21. The graphs of the functions in Exercises 18 and 19 are parallel; they have the same slope. They have different y-intercepts.

22.

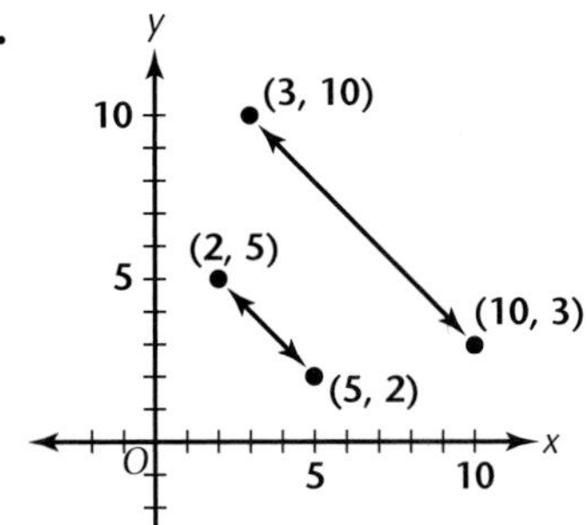

23.

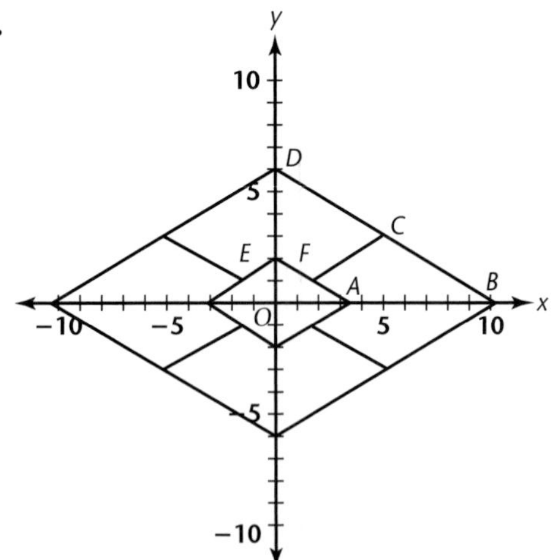

24. Since the y-values are the same for a given x on either side of the y-axis, plot a y-value for a positive x and use the same y-value for negative x.

Look Back, **page 245**

30.

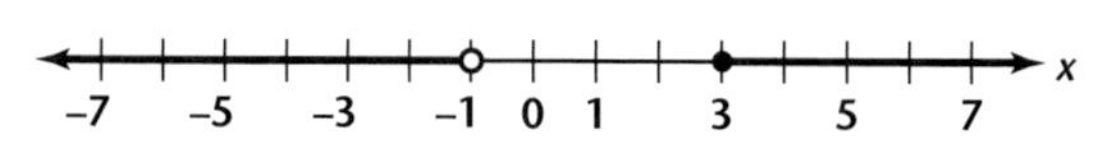

Look Beyond, **page 245**

37.

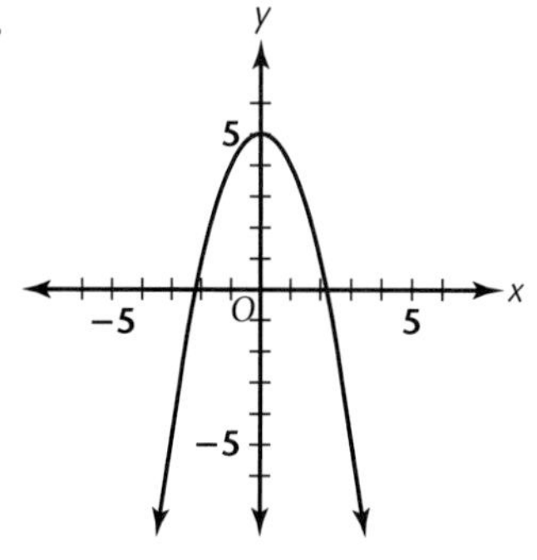

Extra Practice, **page 540**

1.

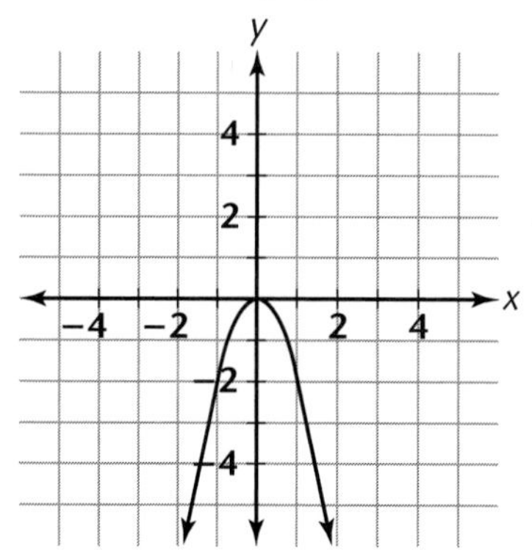

2.

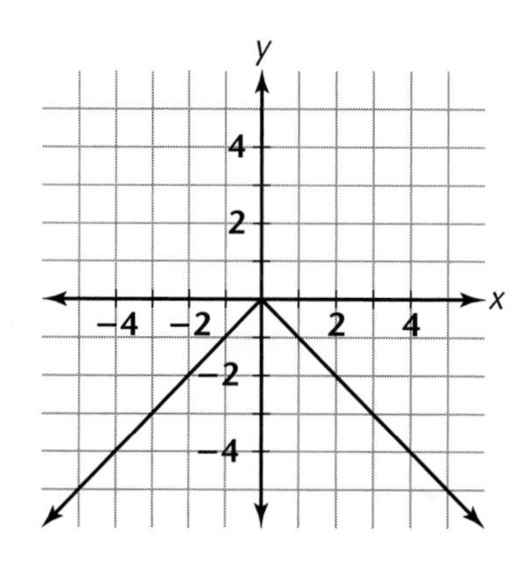

3.

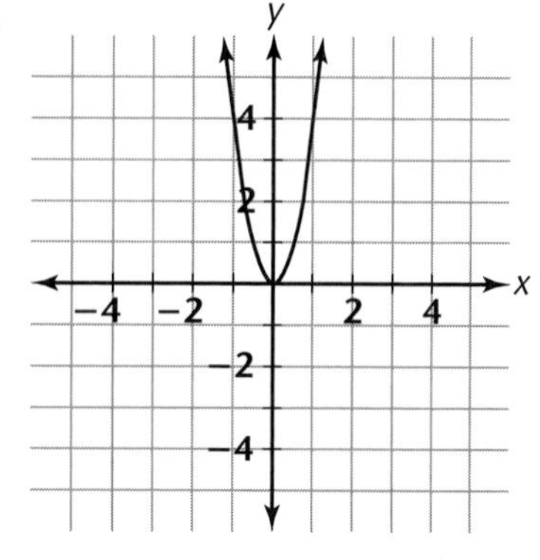

4.

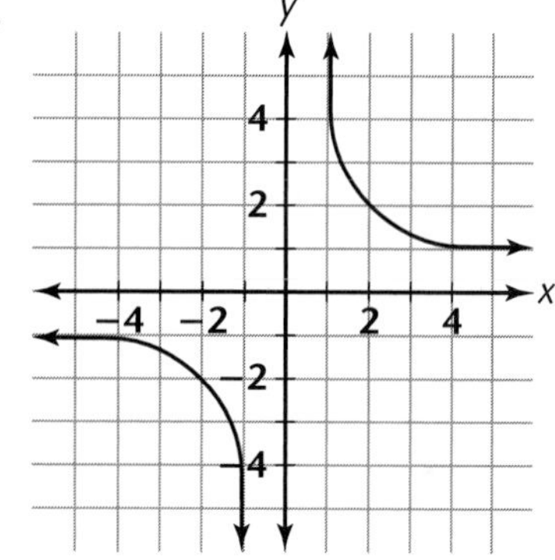

Lesson 5.5

Try This, **page 247**

The parent function, $y = |x|$, is translated up by 4 units.

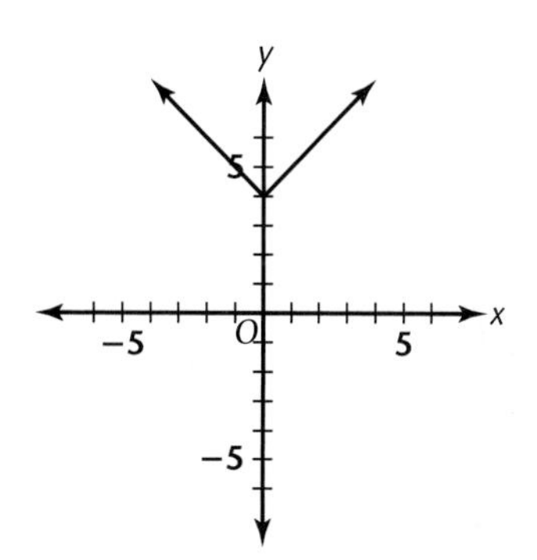

Try This, **page 248**

The parent function, $y = |x|$, is shifted right by 6 units.

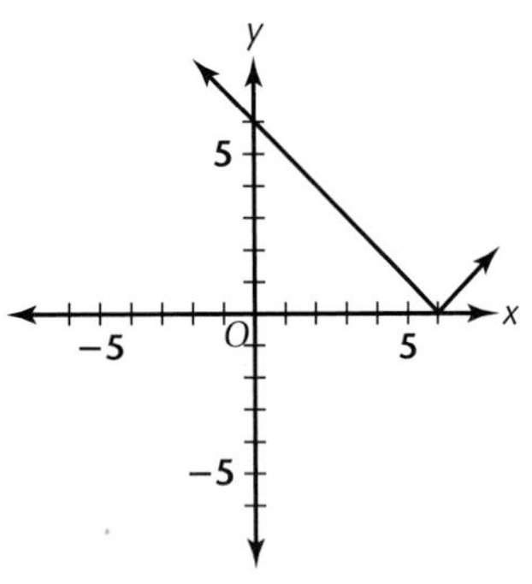

Communicate, **page 250**

1. A translation shifts a graph horizontally or vertically without changing its shape.

4. h increases or decreases the x-value at each point.

5. k increases or decreases the y-value at each point.

6. The graph of the parent function, $y = x^2$, is moved right 4 units and up 3 units. The translated vertex is (4, 3).

Practice and Apply, **pages 251–252**

7. $y = x^2$; reflection through the x-axis

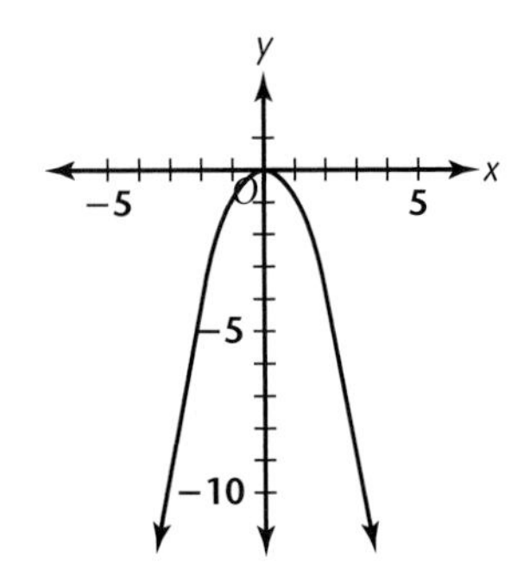

8. $y = x^2$; reflection through the y-axis

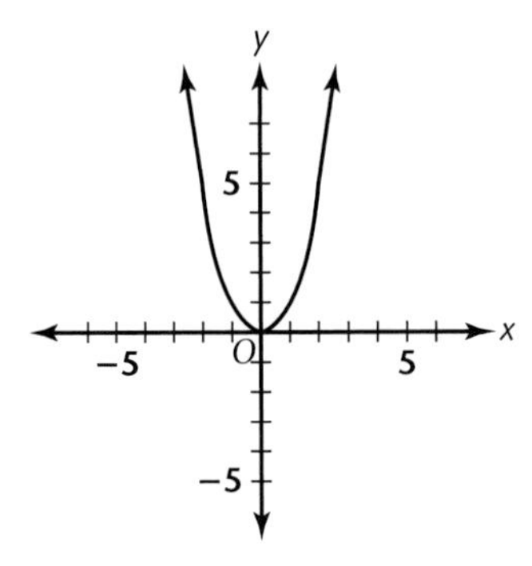

9. $y = x^2$; vertical translation 3 units up

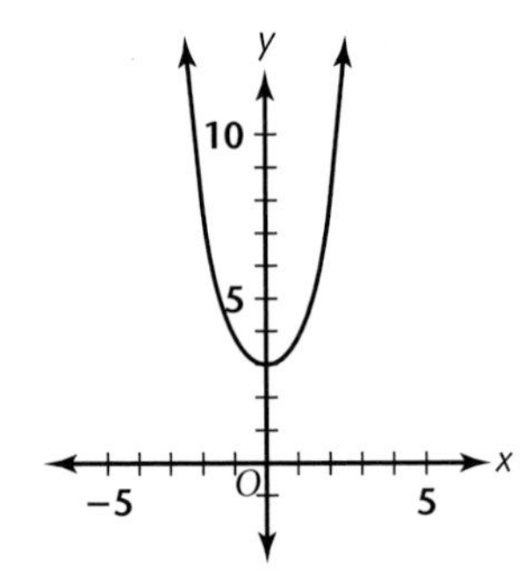

10. $y = x^2$; horizontal translation 3 units to the left

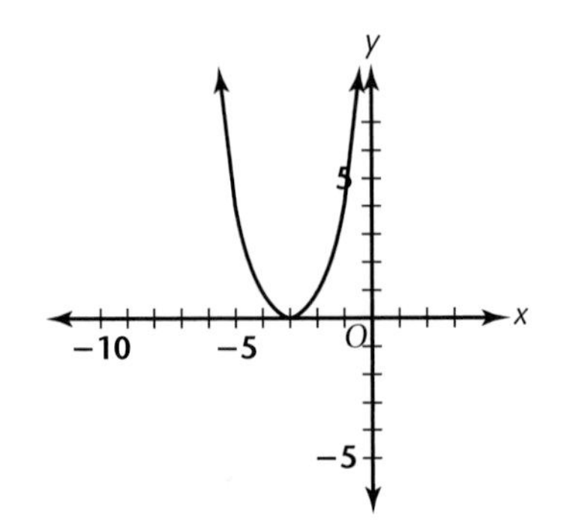

11. $y = x^2$; horizontal translation 6 units to the right and vertical translation 1 unit down

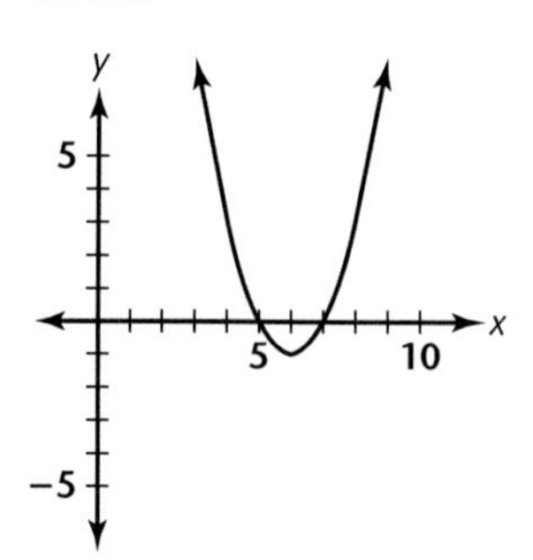

12. $y = x^2$; reflection through the x-axis and vertical translation 6 units down

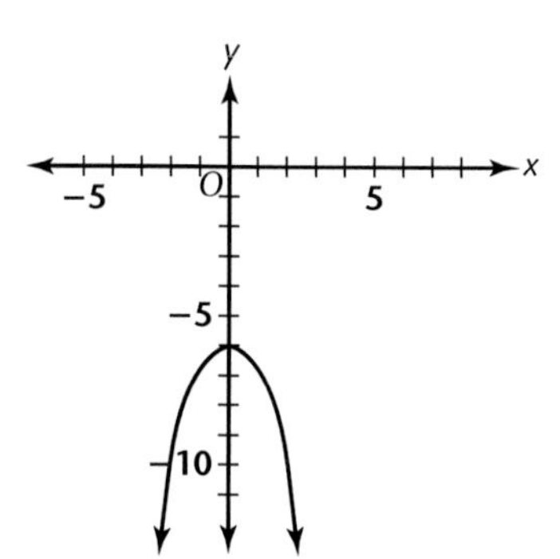

13. $y = |x|$; vertical translation 1 unit down

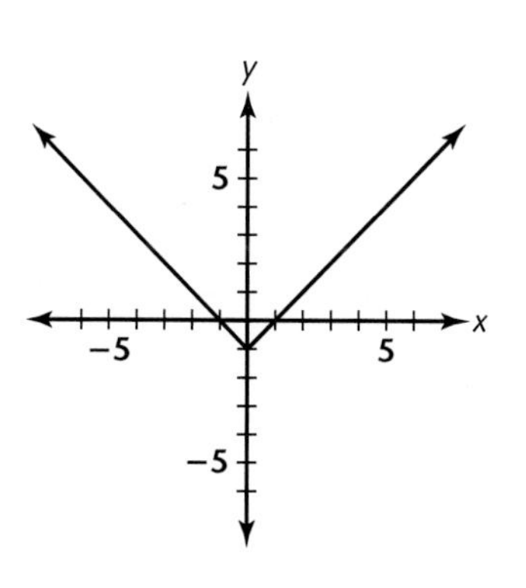

14. $y = |x|$; horizontal translation 1 unit to the right

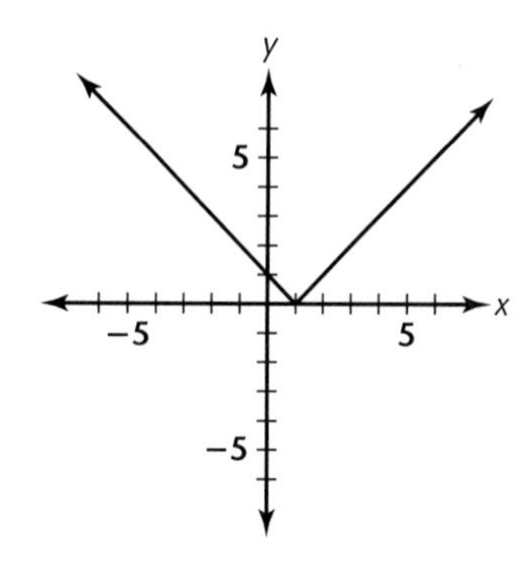

Look Back, **page 252**

36.

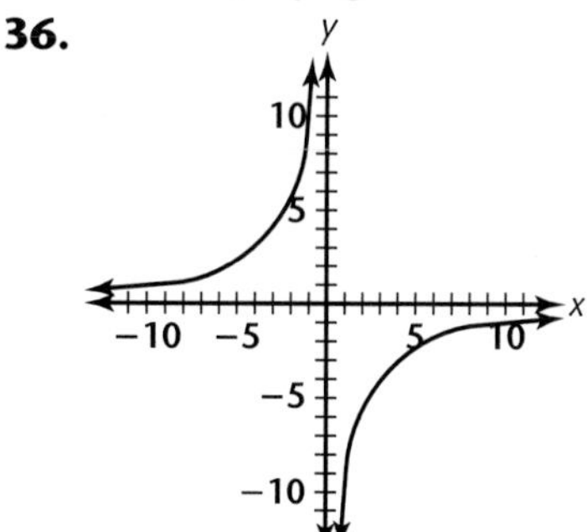

37. $y = 3x^2$; Answers may vary. For example, by examining the first two points, y could be $3x$. The third point, however, does not work with this rule. So, divide both the third and fourth y-values by 3: 4 and 9. These numbers are both perfect squares: $4 = 2^2$ and $9 = 3^2$. Also, with the first and second points, $0 = 3(0)^2$ and $3 = 3(1)^2$. Test this rule on the fifth, sixth, seventh, and eighth points:

$48 = 3(16) = 3(4)^2 \Rightarrow y = 3x^2$
$75 = 3(25) = 3(5)^2 \Rightarrow y = 3x^2$
$108 = 3(36) = 3(6)^2 \Rightarrow y = 3x^2$
$147 = 3(49) = 3(7)^2 \Rightarrow y = 3x^2$

Therefore, the rule is $y = 3x^2$.

38. 

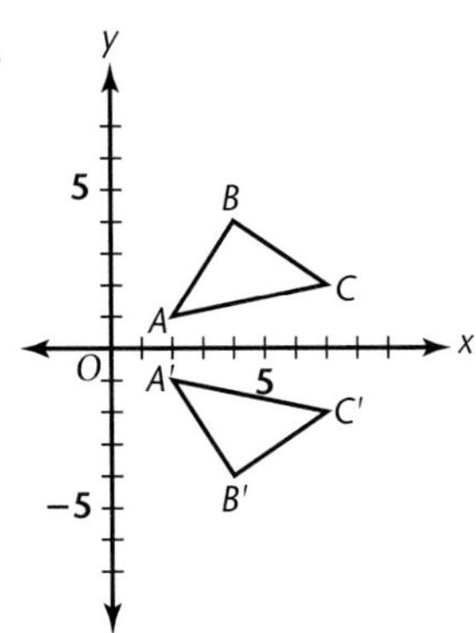

(2, −1), (4, −4), and (7, −2)

39.

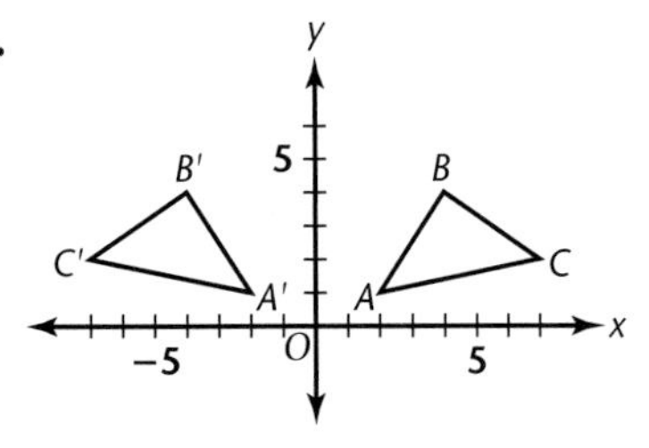

(−2, 1), (−4, 4), and (−7, 2)

Look Beyond, **page 252**

40–41.

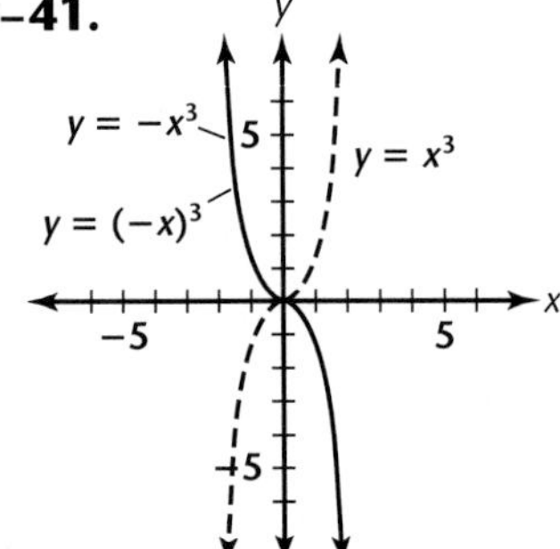

$y = -x^3$ is a reflection of $y = x^3$ over the x-axis and $y = (-x)^3$ is a reflection of $y = x^3$ over the y-axis; however, the graphs of $y = -x^3$ and $y = (-x)^3$ are identical.

Extra Practice, **page 540**

1. $y = x^2$; A vertical reflection over the x-axis and a stretch by a factor of 2

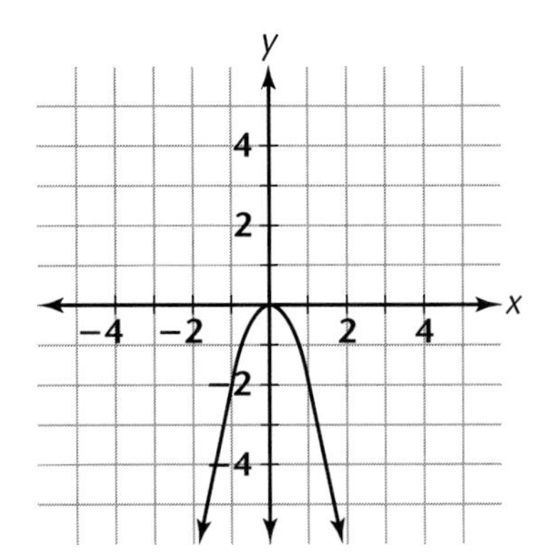

2. $y = x^2$; A shift 1 unit down

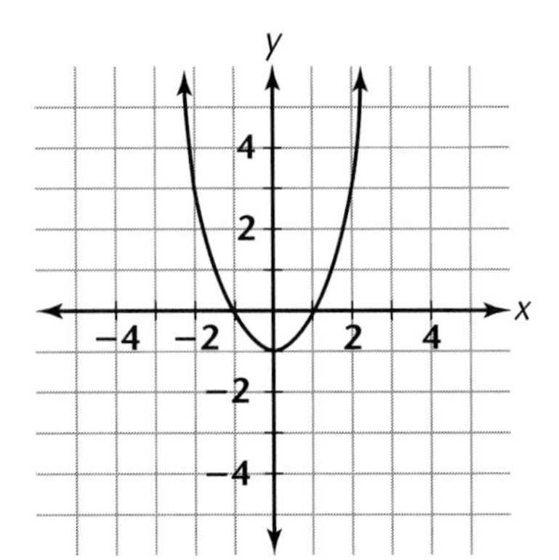

3. $y = x^2$; A shift 1 unit to the right

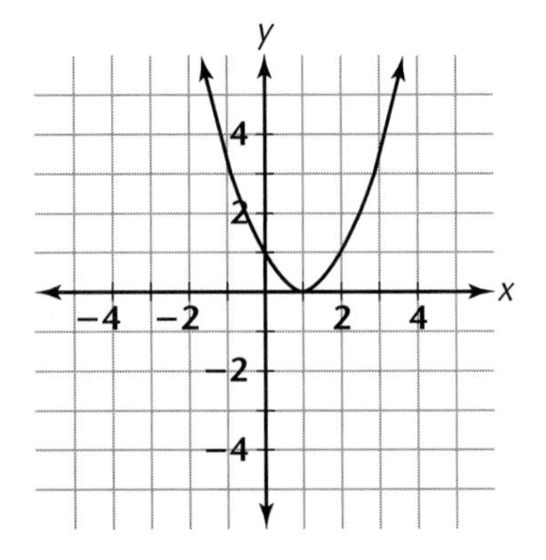

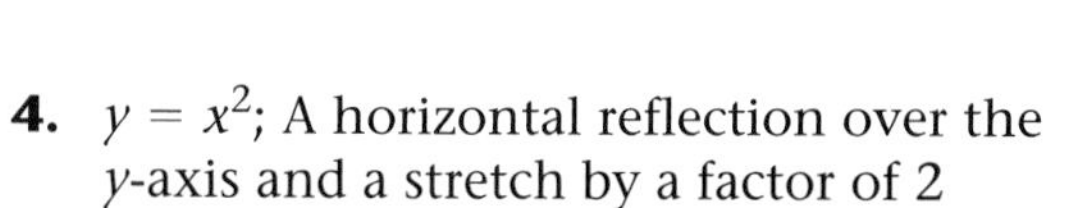

4. $y = x^2$; A horizontal reflection over the y-axis and a stretch by a factor of 2

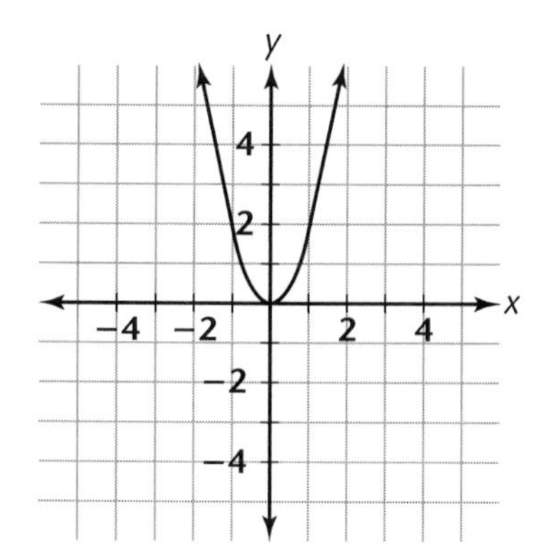

5. $y = |x|$; A vertical reflection over the x-axis and a stretch by a factor of 3

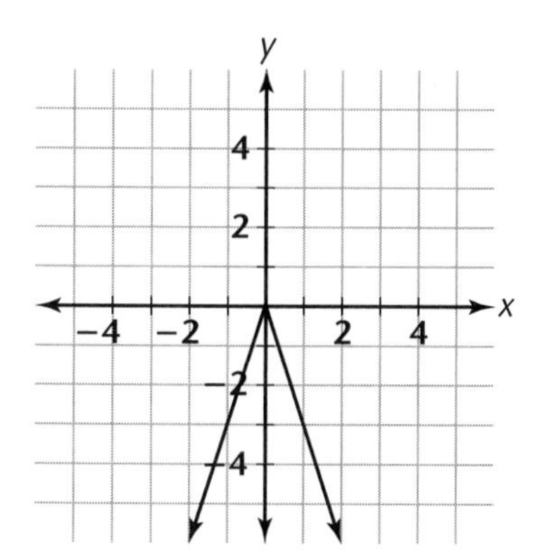

6. $y = |x|$; A shift up by 2 units

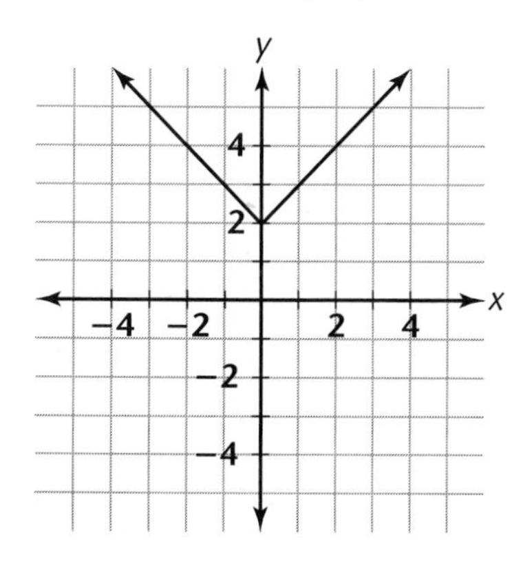

7. $y = |x|$; A shift to the left by 2 units and shift up 2 units

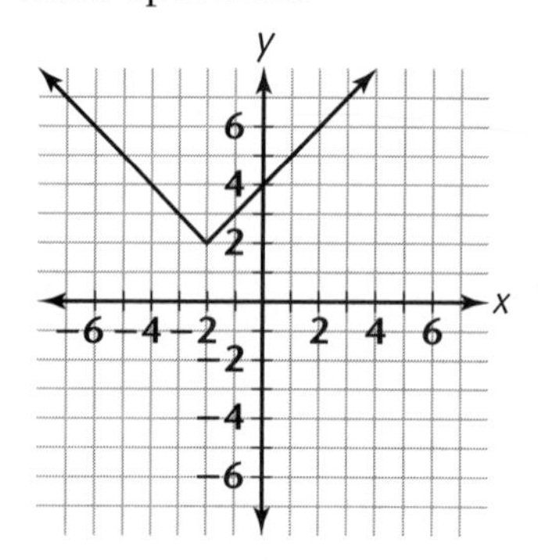

8. $y = |x|$; A stretch by a factor of 3

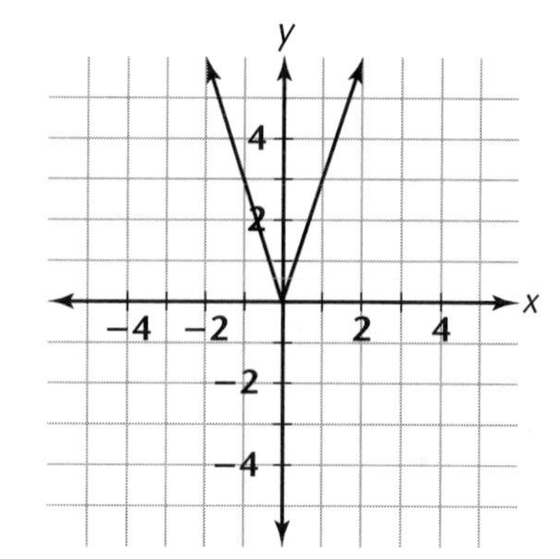

9. Shift 2 units up

10. Shift 2 units down

11. Shift 2 units to the left

12. Shift 2 units to the right

13. (6, 9) is translated to (6, 14).

14. (6, 9) is translated to (6, 6).

15. (6, 9) is translated to (6, 0).

16. (6, 9) is translated to (6, 36).

17. (6, 9) is translated to (8, 9).

18. (6, 9) is translated to (5, 9).

19. (6, 9) is translated to (−4, 9).

20. (6, 9) is translated to (6, 45).

Lesson 5.6

Communicate, **page 256**

1. Translations occur when a constant is added or subtracted. Thus, the number

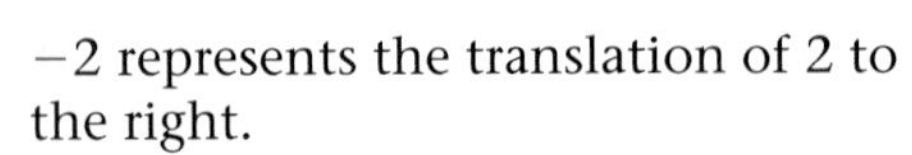

-2 represents the translation of 2 to the right.

2. A scale factor stretches a graph. Since $R = \frac{100}{h}$ can be written as $R = 100\left(\frac{1}{h}\right)$, the scale factor 100 represents a vertical stretch of 100.

3. Answers may vary. For example, for the parent absolute value function, the graph is a V with the vertex at (0, 0). The graph opens upward and contains the portions of the lines $y = x$ and $y = -x$ that lie above the x-axis.

4. Since it is a horizontal shift, apply the translation to the x-value of the parent function $y = \frac{1}{x}$. It is a shift to the left, so add 4 to x giving $x + 4$. The translated function is $y = \frac{1}{x + 4}$.

5. The function is reflected by the negative sign preceding a, stretched by a, shifted horizontally by $-h$, and shifted vertically by $+k$.

6. The graph of the parent function, $y = |x|$, has a vertex at (0, 0). Since the graph of $y = |x + 7| - 3$ is the graph of $y = |x|$ shifted to the left by 7 units and down by 3 units, plot the vertex of the transformed graph at $(-7, -3)$. Then draw the V-shaped graph.

Practice and Apply, pages 256–257

7.

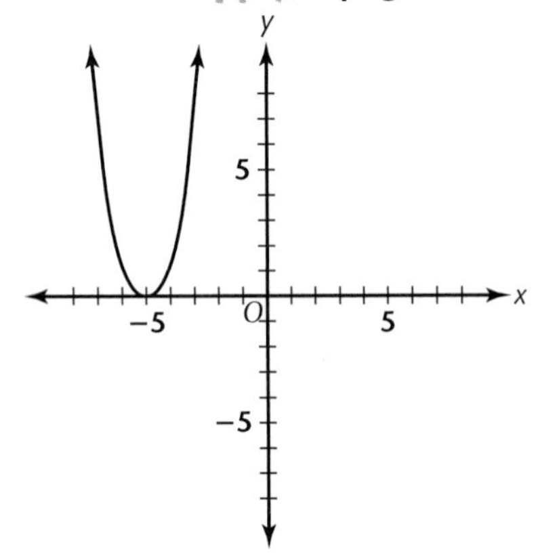

8.

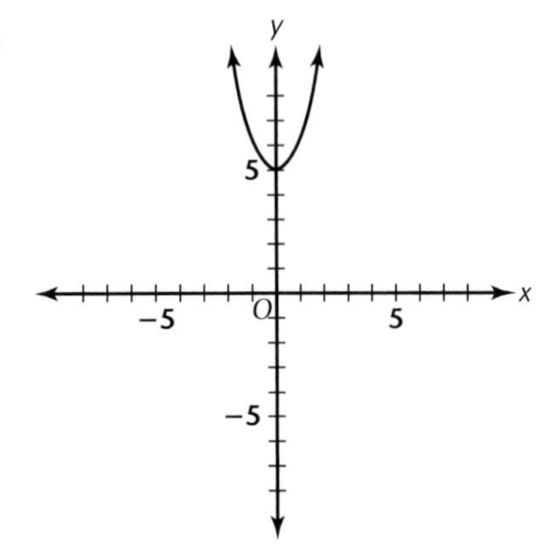

9.

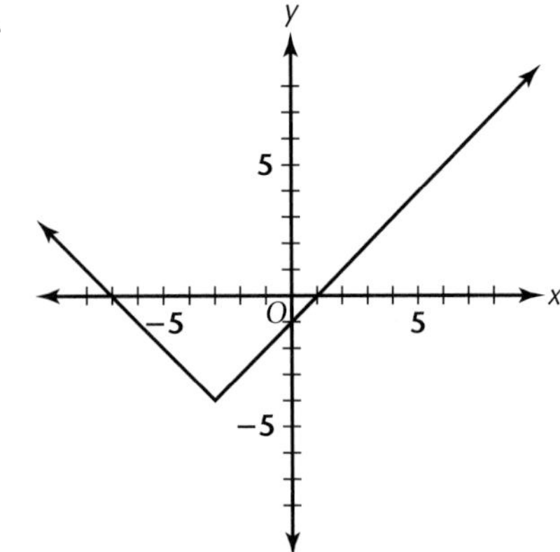

10.

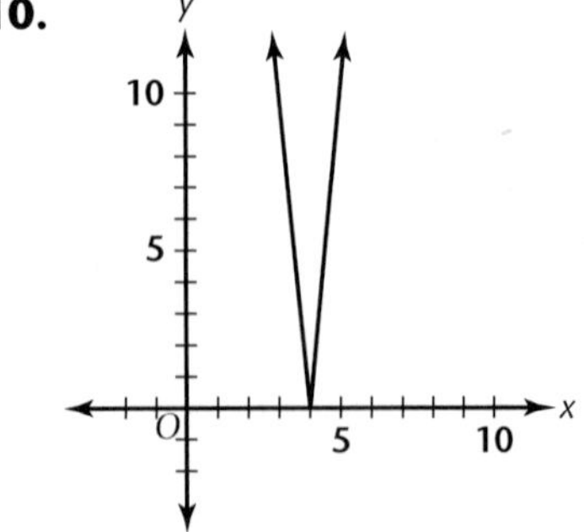

11.

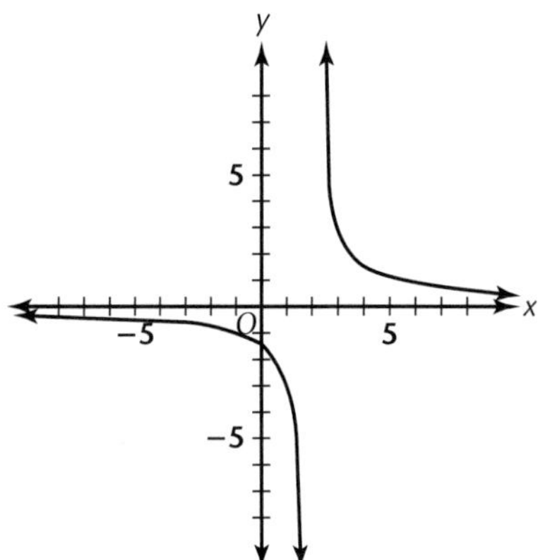

12.

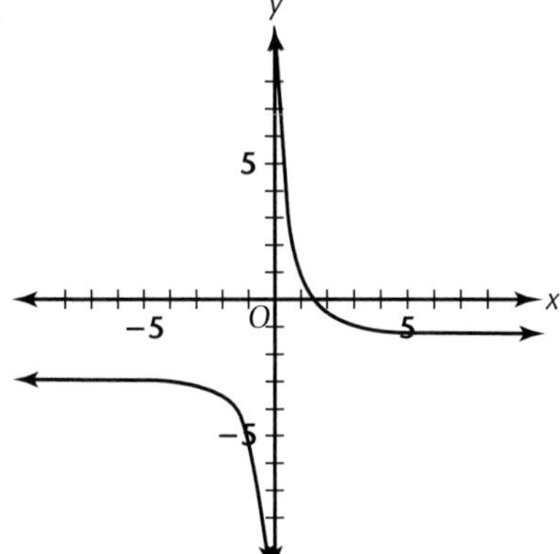

13.

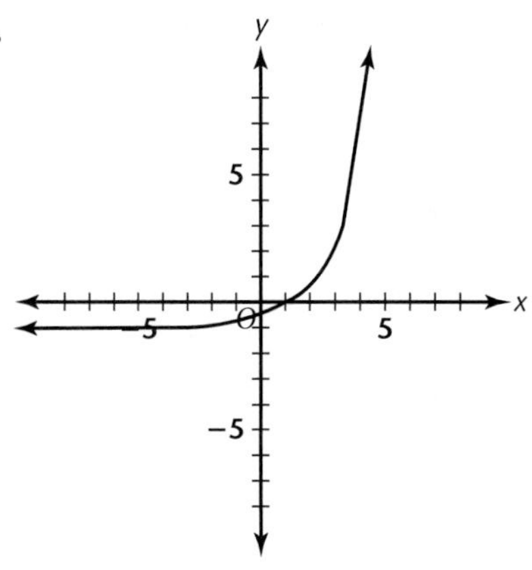

14.

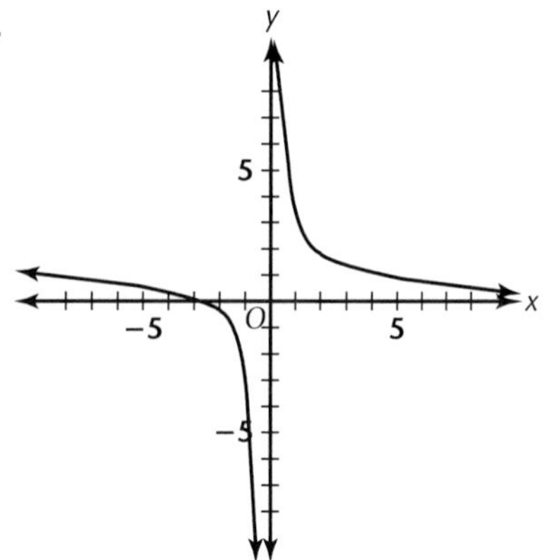

15.

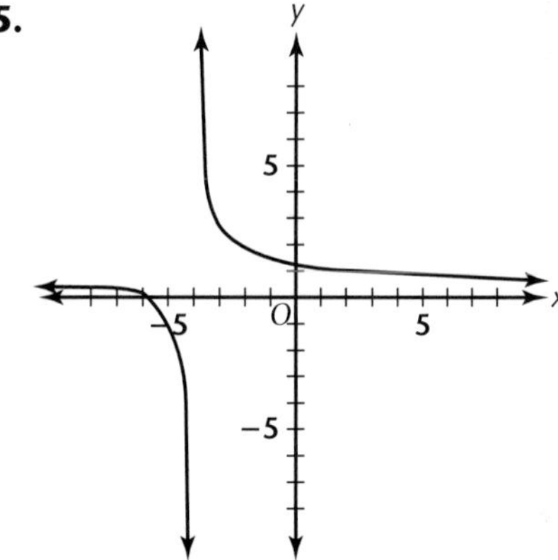

22. $A'(5, -3)$; $A''(-4, -3)$; yes;
$(x, y)' = (x, -y)$; $(x, y)'' = (x - 9, -y)$

Look Back, **page 257**

26. $-12 \leq x \leq 22$

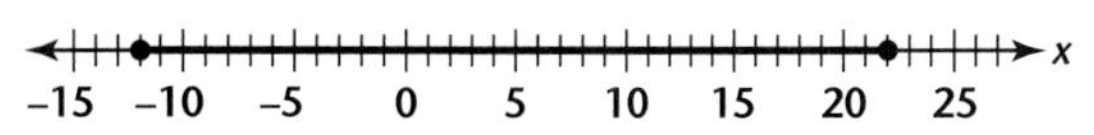

27.

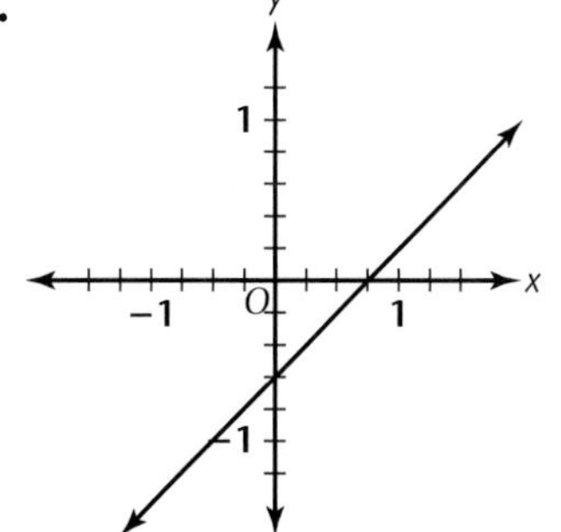

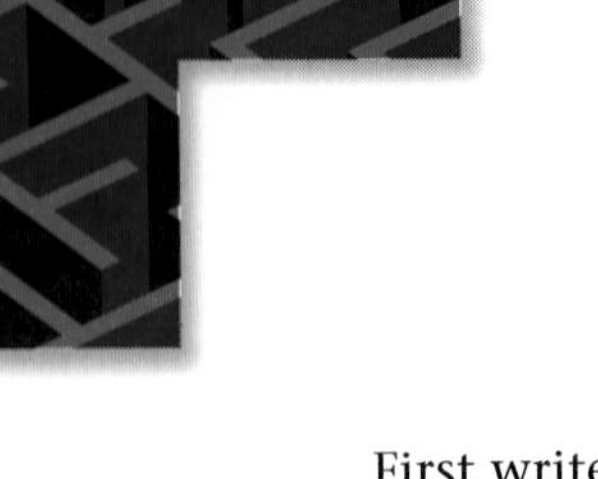

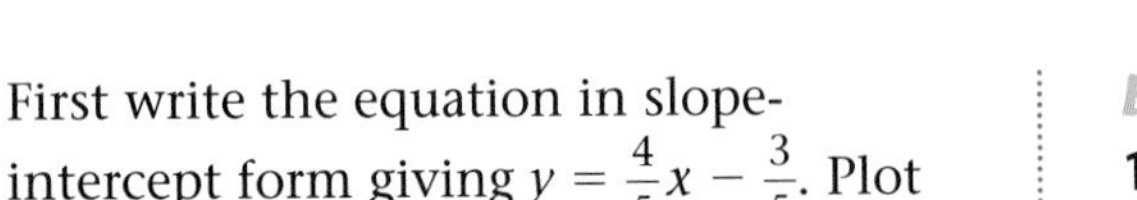

First write the equation in slope-intercept form giving $y = \frac{4}{5}x - \frac{3}{5}$. Plot the y-intercept of $\left(0, -\frac{3}{5}\right)$. Apply the slope of $\frac{4}{5}$ (up 4 units and right 5 units) to this point to generate a second point $\left(1, \frac{1}{5}\right)$. Draw a line through these two points.

Look Beyond, **page 257**

33.

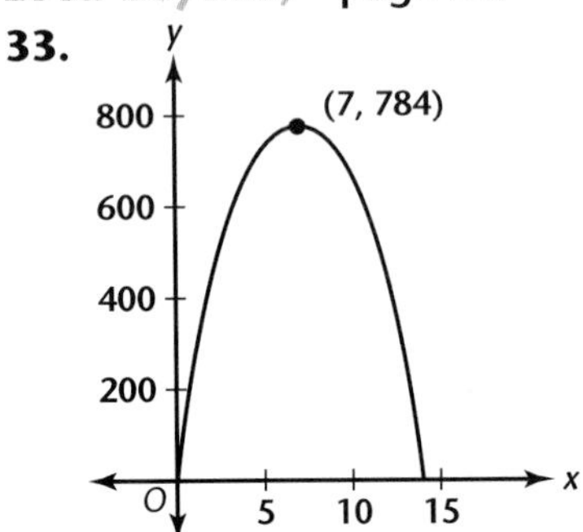

34. The parent function is $y = x^2$. The negative sign in front of the 16 denotes a vertical reflection over the x-axis. The 16 represents a stretching of the y-values by a factor of 16. The -7 denotes a shift to the right by 7 units. The 784 is a shift up of 784 units.

35.

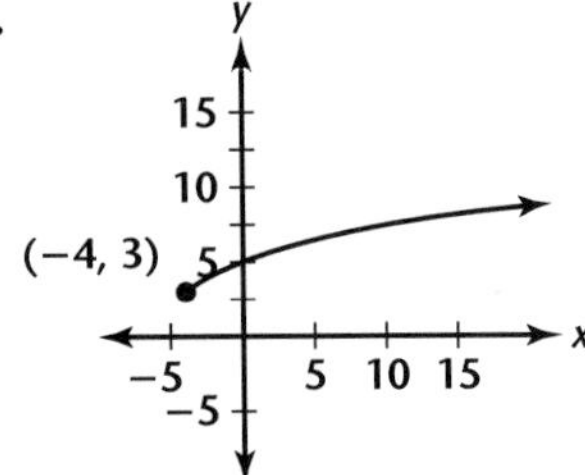

36.

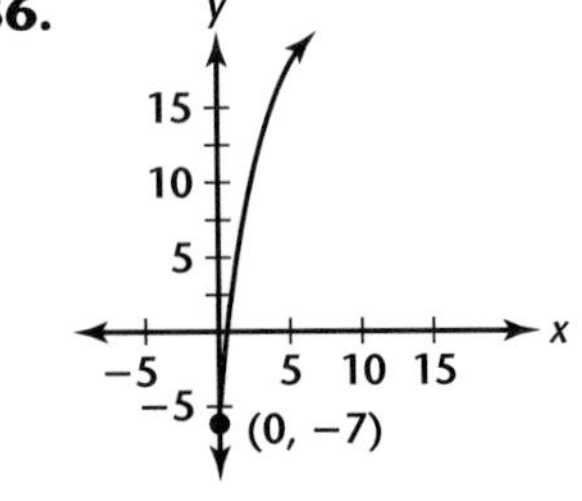

Extra Practice, **page 541**

1.

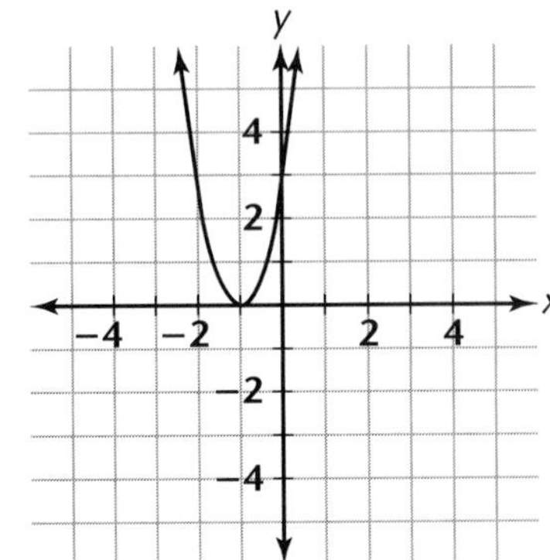

2.

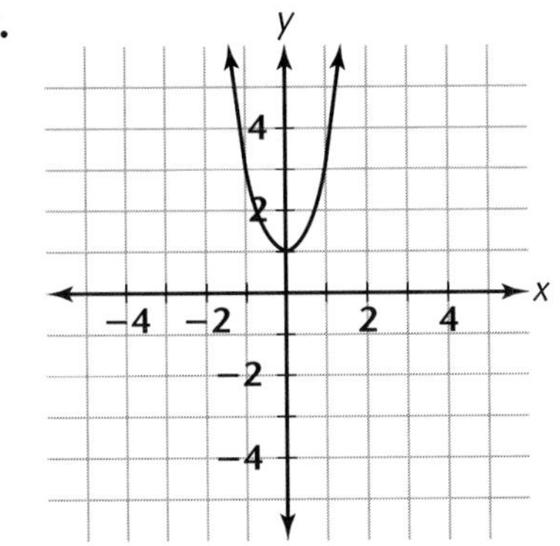

3.

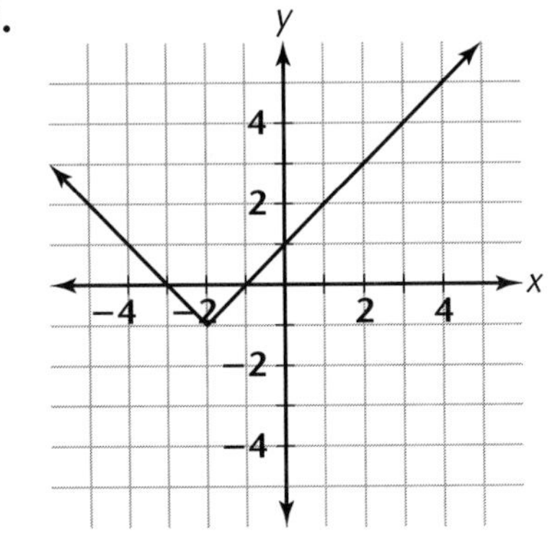

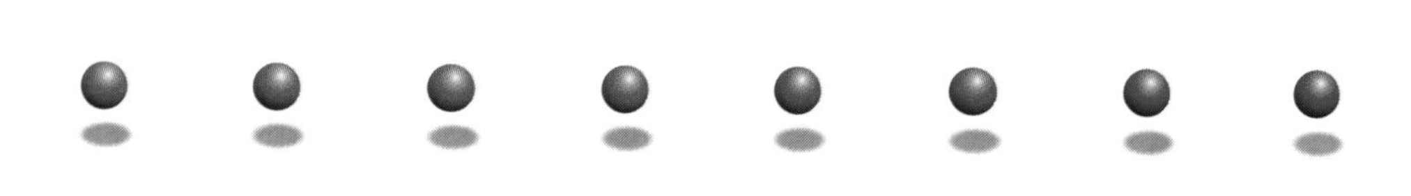

4.

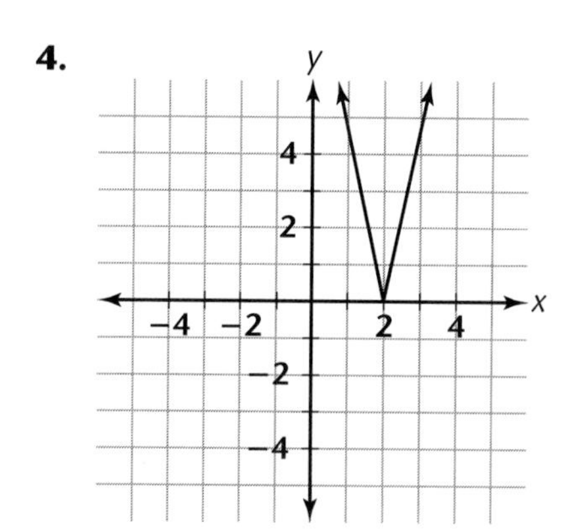

5.

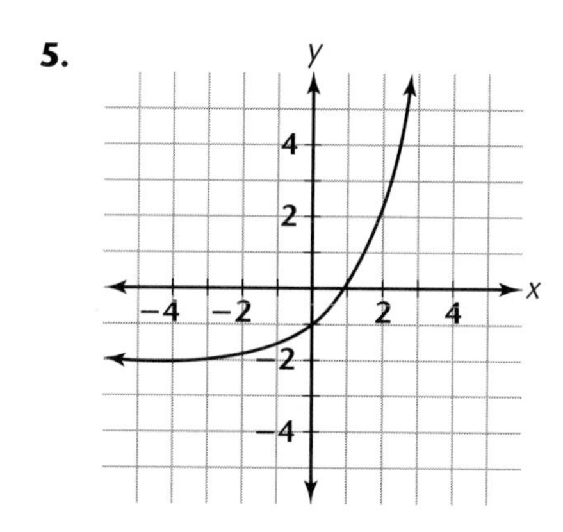

6.

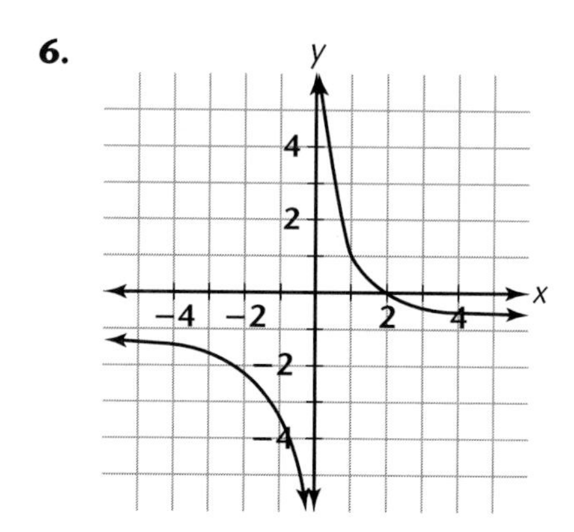

7. (4, 6) is translated to (4, 14).

8. (4, 6) is translated to (4, −18).

9. (4, 6) is translated to (4, −8).

10. (4, 6) is translated to (8, −12).

11. (4, 8) is translated to (4, 18).

12. (4, 8) is translated to (4, −24).

13. (4, 8) is translated to (4, −10).

14. (4, 8) is translated to (8, −16).

Chapter 5 Project

Activity 1, **page 259**

Answers may vary.

Activity 2, **page 259**

Answers may vary.

Chapter 5 Review, pages 260–262

10.

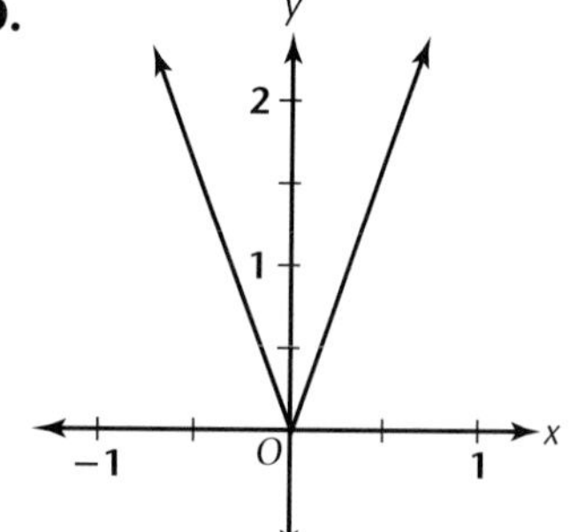

11.

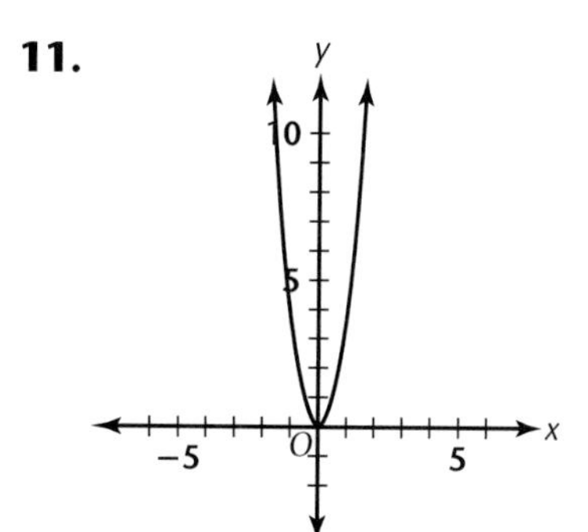

12.

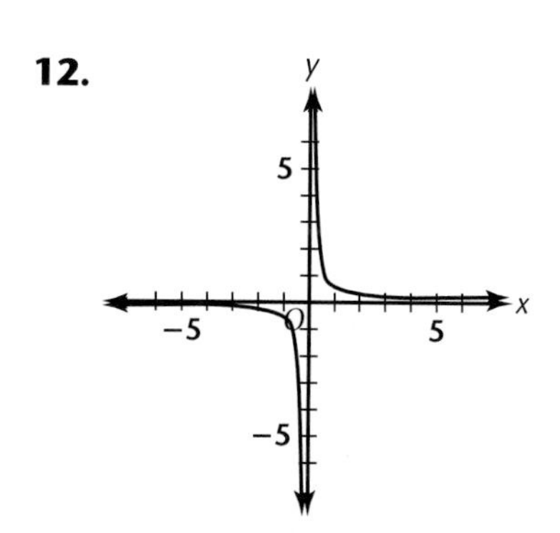

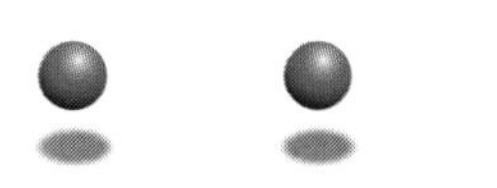

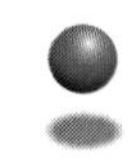

13.

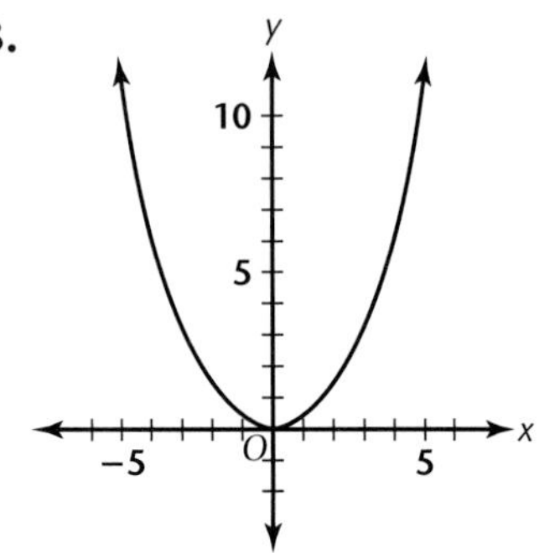

14.

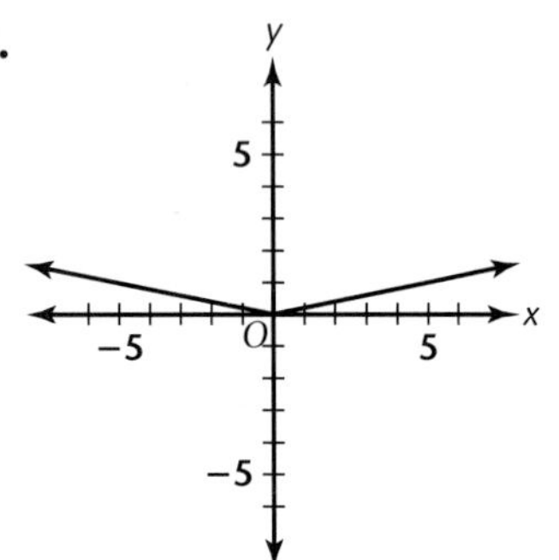

15.

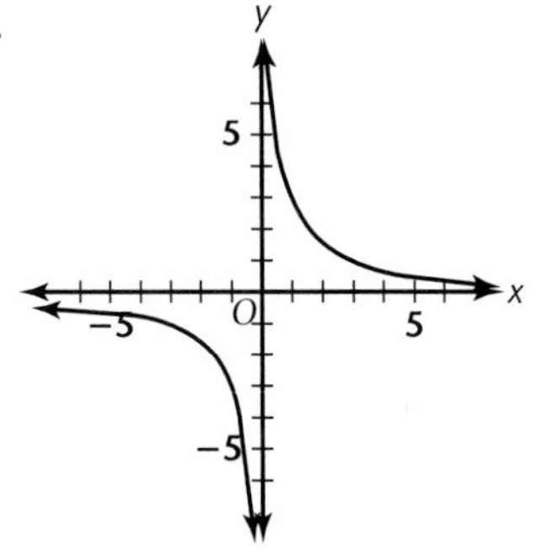

16.

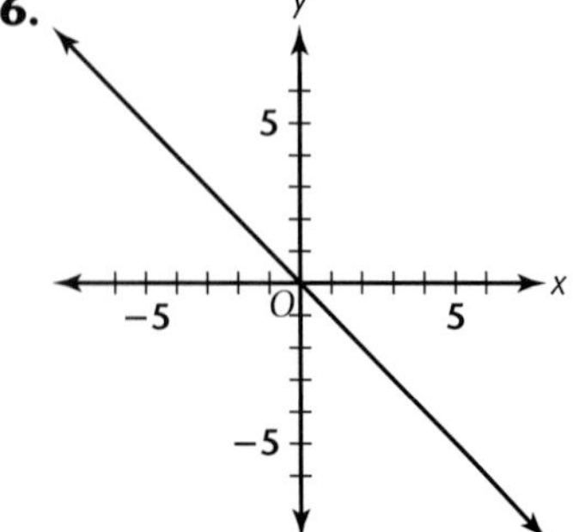

17.

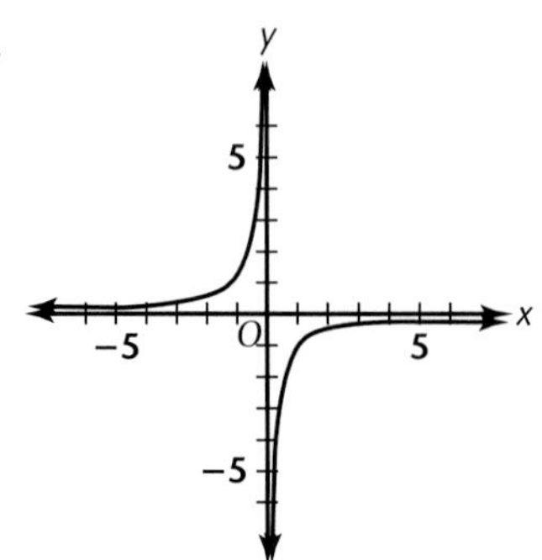

18.

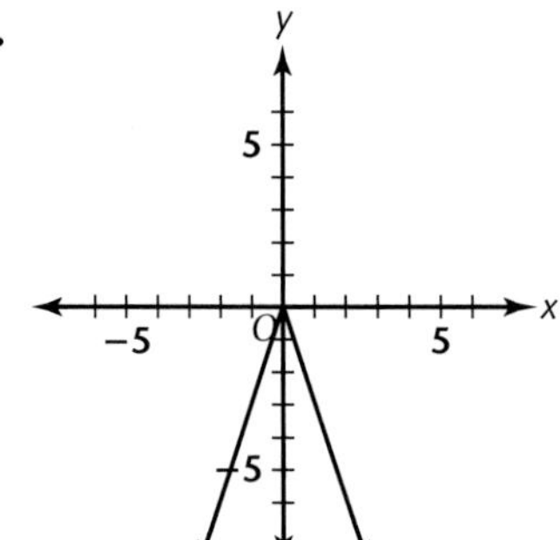

19.

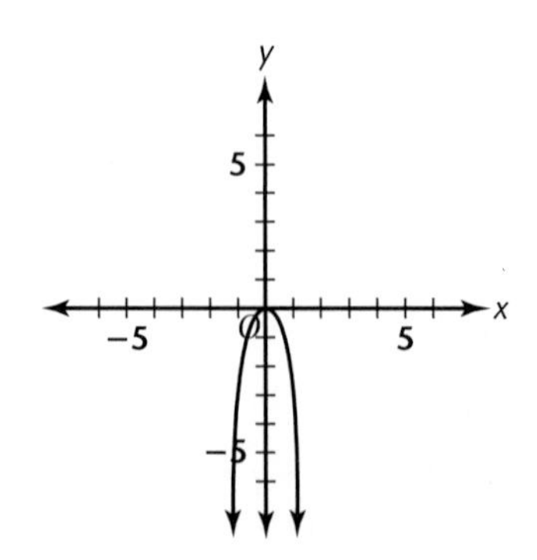

20. $y = x^2$; shift 1 unit up

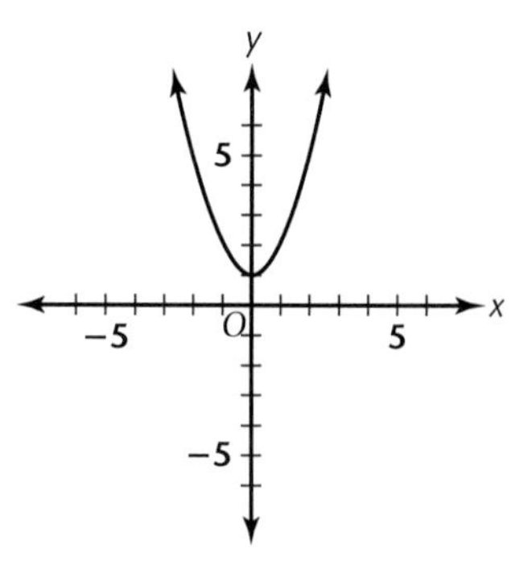

21. $y = x^2$; shift 1 unit left

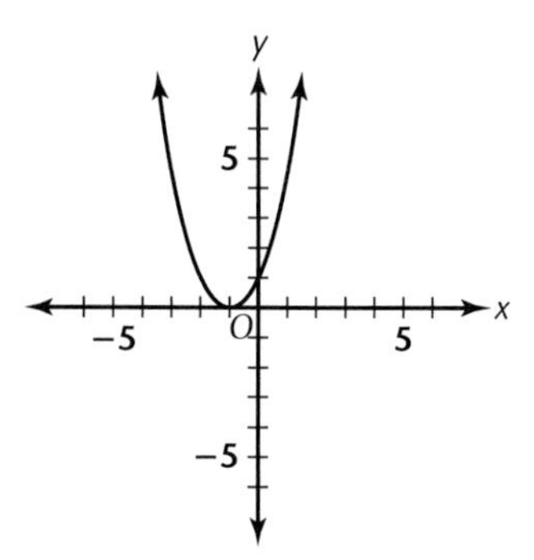

22. $y = |x|$; shift down 4 units

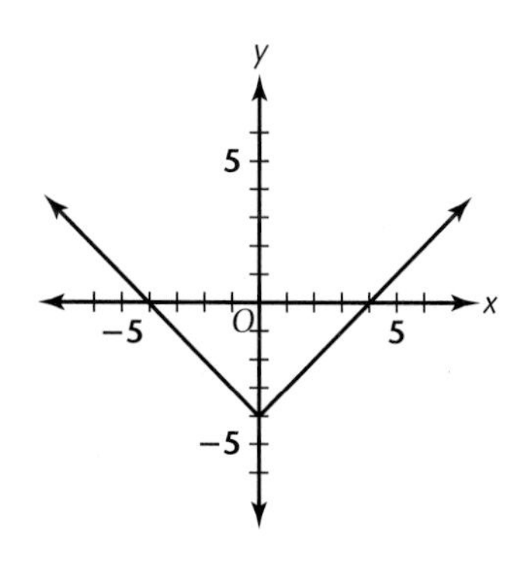

23. $y = |x|$; shift right 4 units

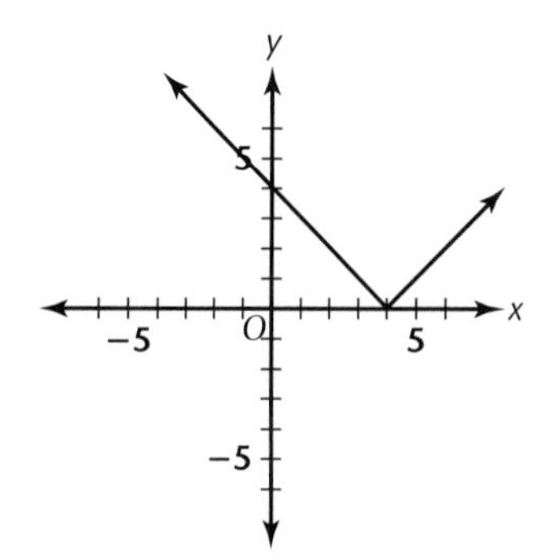

24.

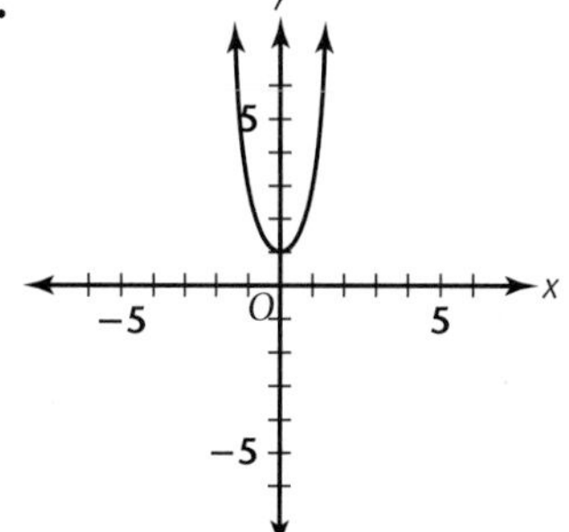

25.

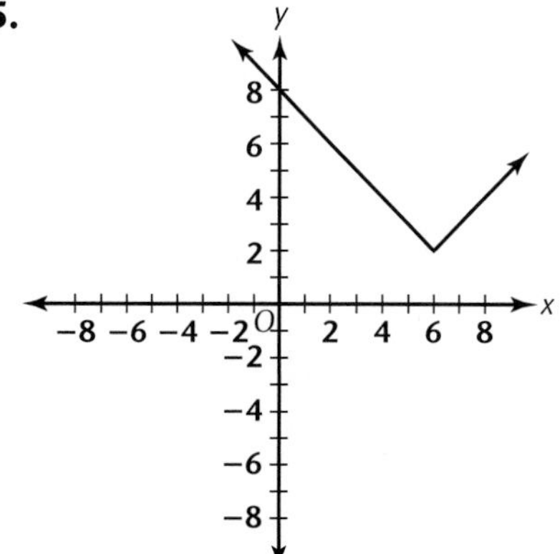

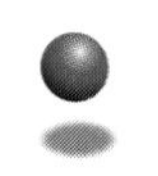

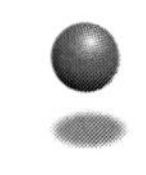

26.

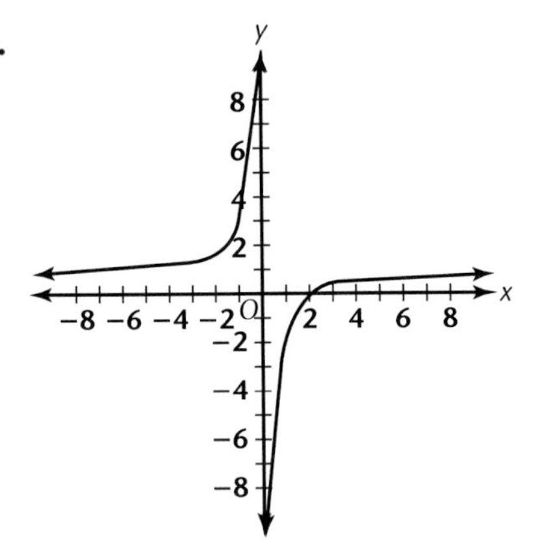

Chapter 5 Assessment, page 263

6. Stretch by a factor of 3

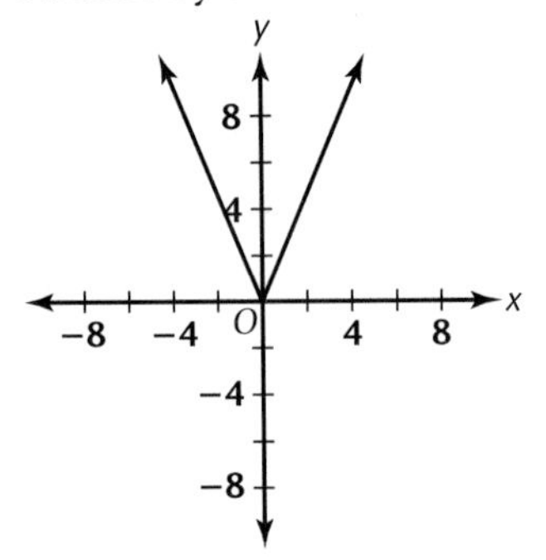

7. Vertical shift by −5 units

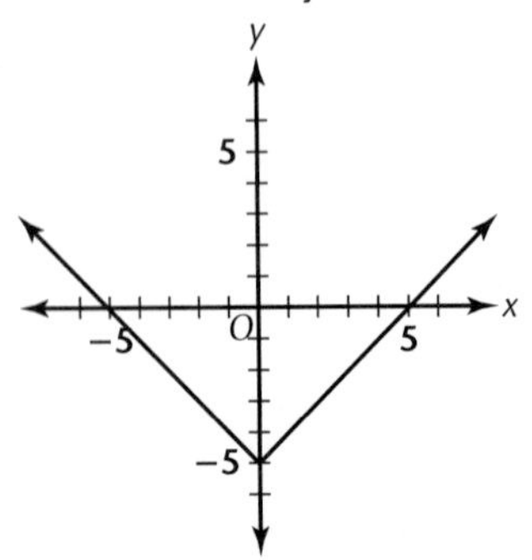

8. Reflected across the x-axis

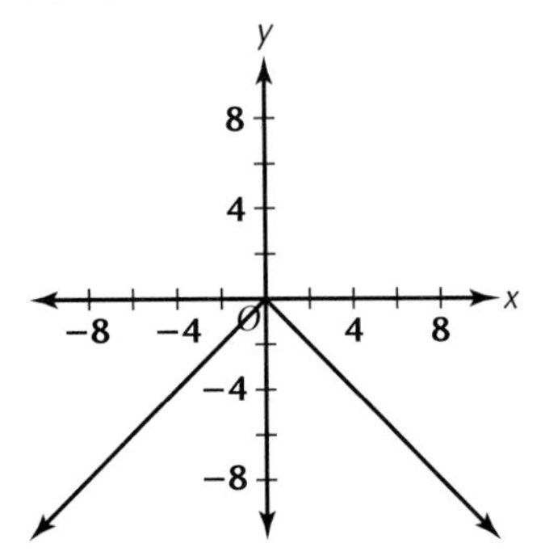

9. Horizontal shift by −4 units

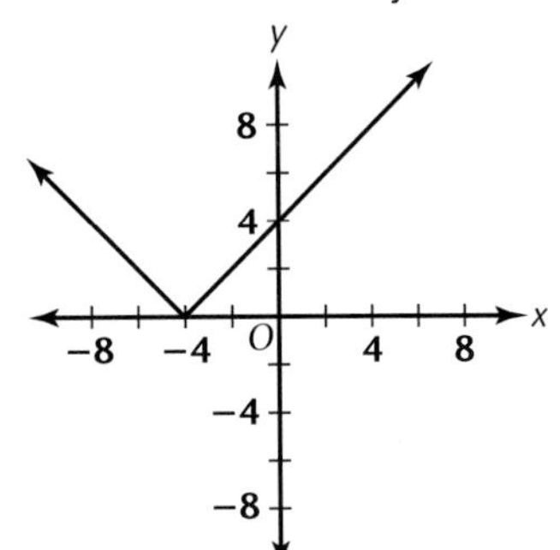

11.

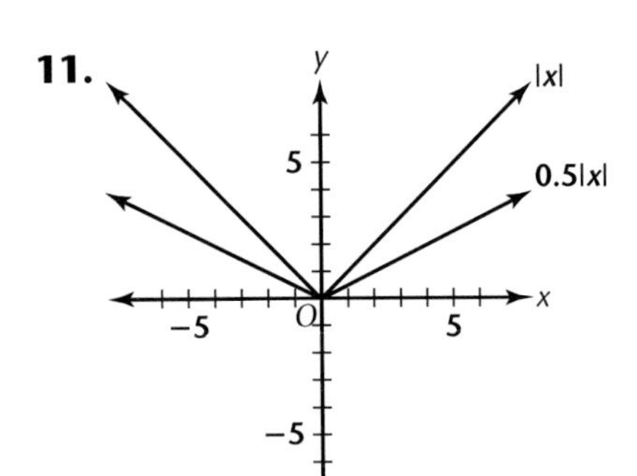

19.

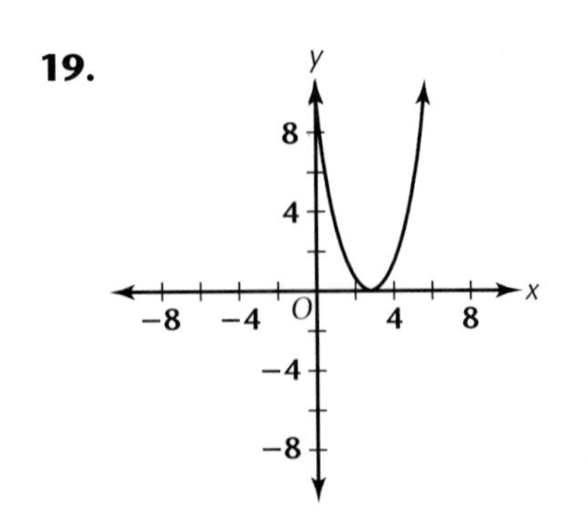

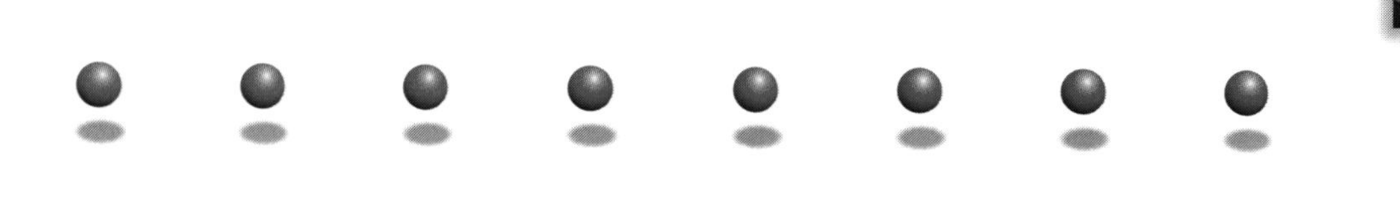

20.

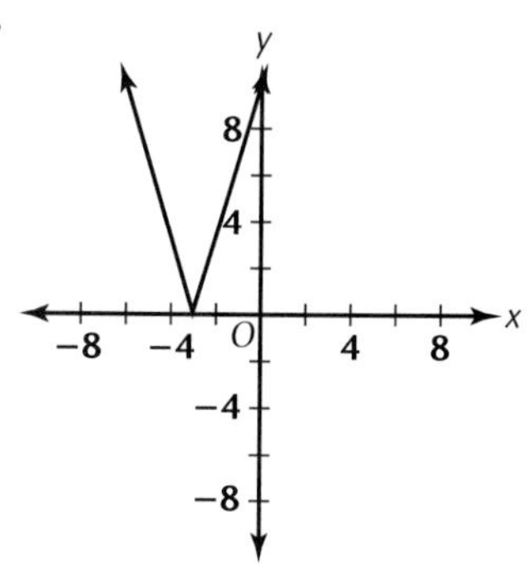

21.

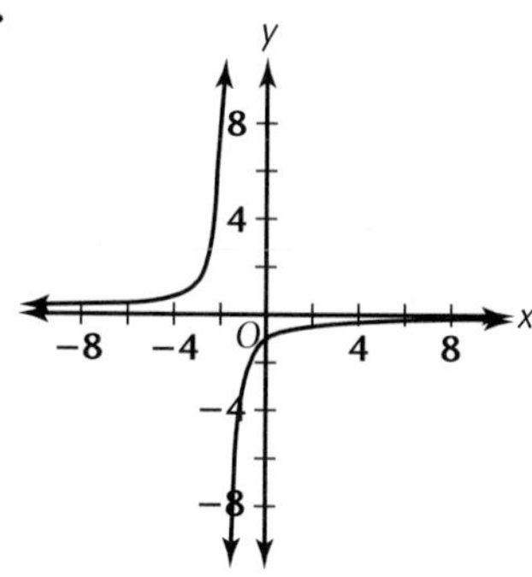

23. When a quadratic is written in the form $y = a(x - h)^2 + k$, and $a > 0$, the minimum point is (h, k). For the function $y = 2(x + 4)^2 - 1$, the minimum point is $(-4, -1)$.

24. Graph $y = \frac{4}{x}$ and $y = 3 + x$. Find their points of intersection OR graph $y = \frac{4}{x} - (3 + x)$ and determine where it crosses the x-axis. The solutions are -4 and 1.

25. Graph $y = |1 - 3x|$ and $y = |x| + 5$. Find their points of intersection OR graph $y = |1 - 3x| - (|x| + 5)$ and determine where it crosses the x-axis. The solutions are 3 and -2.

Lesson 6.1

Exploration 1, page 267

	Decimal form	Exponent form
3a.	1000 • 1,000,000 = 1,000,000,000	$10^3 \cdot 10^6 = 10^9$
3b.	10 • 1000 = 10,000	$10^1 \cdot 10^3 = 10^4$
3c.	4 • 16 = 64	$2^2 \cdot 2^4 = 2^6$
3d.	9 • 9 = 81	$3^2 \cdot 3^2 = 3^4$

4. Decimal form:

$$\underbrace{a \cdot a \cdot a}_{m \text{ times}} \cdot \ . \ . \ . \ \cdot \underbrace{a \cdot a \cdot a}_{n \text{ times}} = \underbrace{a \cdot a \cdot a \cdot a \cdot \cdot \cdot}_{m + n \text{ times}}$$

Exponent form: $a^m \cdot a^n = a^{m+n}$

Exploration 2, page 268

2a.

	Decimal	Exponent
Numerator	100,000	10^5
Denominator	100	10^2
Quotient	1000	10^3

2b.

	Decimal	Exponent
Numerator	64	2^6
Denominator	4	2^2
Quotient	16	2^4

2c.

	Decimal	Exponent
Numerator	81	3^4
Denominator	27	3^3
Quotient	3	3^1

3. 10^4

	Decimal	Exponent
Numerator	100,000	10^5
Denominator	10	10^1
Quotient	10,000	10^4

Communicate, page 270

1. The exponent indicates the number of zeros in customary notation.
2. The exponent indicates how many times the base is used as a factor.
3. Addition; because you are expressing the total number of times the base is used as a factor
4. Subtraction; because you are expressing the number of times the base occurs as a factor in the numerator and the denominator
5. Multiplication; because you are finding the number of times the base is used as a factor
6. 8^{17}; $8^{17} > 8^{16}$ because the exponent $17 > 16$ and $8^{17} > 5^{17}$ because the base $8 > 5$.
7. You are multiplying powers of the same base.
8. You are dividing powers of the same base.
9. You are finding a power of a power.

Practice and Apply, pages 270–272

10. Decimal form: $1{,}000{,}000 \cdot 10{,}000 = 10{,}000{,}000{,}000$
Exponent form: $10^6 \cdot 10^4 = 10^{10}$
11. Decimal form: $1000 \cdot 100{,}000 = 100{,}000{,}000$
Exponent form: $10^3 \cdot 10^5 = 10^8$
12. Decimal form: $8 \cdot 16 = 128$
Exponent form: $2^3 \cdot 2^4 = 2^7$
13. Decimal form: $81 \cdot 9 = 729$
Exponent form: $3^4 \cdot 3^2 = 3^6$

14.

	Decimal	Exponent
Numerator	1000	10^3
Denominator	100	10^2
Quotient	10	10^1

15.

	Decimal	Exponent
Numerator	1,000,000	10^6
Denominator	1000	10^3
Quotient	1000	10^3

16.

	Decimal	Exponent
Numerator	64	2^6
Denominator	8	2^3
Quotient	8	2^3

17.

	Decimal	Exponent
Numerator	243	3^5
Denominator	81	3^4
Quotient	3	3^1

54.

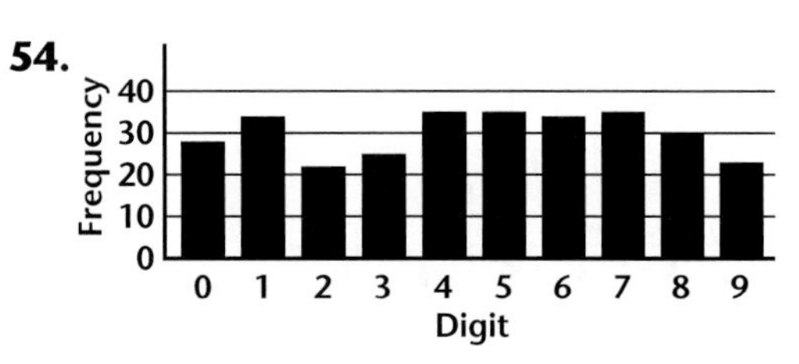

The distribution is fairly flat.

Look Back, page 272

61. $\frac{2}{3}$ or $66\frac{2}{3}\%$; experimental; Joshua's daily practice is the experiment on which the probability is based.

Lesson 6.2

Communicate, page 276

1. $3x^2$ is a monomial because it is the product of a constant and a variable with a positive integer exponent. $3x^{-2}$ is not a monomial because the variable has a negative exponent.
2. The Product-of-Powers Property states that to multiply powers of the same base you add the exponents. The Quotient-of-Powers Property states that to divide powers of the same base you subtract the exponents.

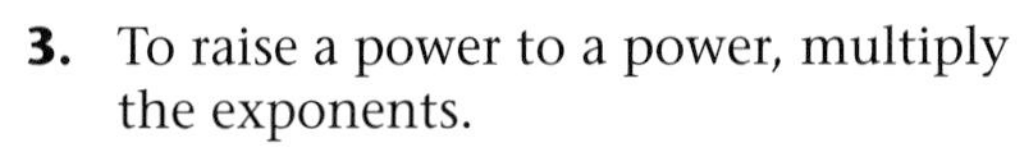

3. To raise a power to a power, multiply the exponents.
4. Multiplication is commutative.
5. To multiply powers of the same base, add the exponents.

Practice and Apply, **page 277**

57. $(10^2)^3 = 100^3$ or $(10^2)^3 = 10^{2 \cdot 3} = 10^6$, not 10^8, 10^5, or 100^3.

Look Beyond, **page 278**

67. $1^2 + 8^2 = 65$; $4^2 + 7^2 = 65$; $14^2 + 87^2 = 7765$; $41^2 + 78^2 = 7765$; 14 and 87 are each composed of two digits, one from the base numbers in the two members (sides) of the equation $1^2 + 8^2 = 4^2 + 7^2$; the same is true for 41 and 78.

68. $17^2 + 84^2 = 7345$; $71^2 + 48^2 = 7345$; Yes, 17 and 71 are composed of the same two digits reversed, as are 84 and 48.

69. $0^2 + 5^2 = 25$; $3^2 + 4^2 = 25$; $03^2 + 54^2 = 30^2 + 45^2 = 2925$: $x = 30$, $y = 45$

70. $x = 40$, $y = 35$; $4^2 + 6^2 = 52$; $3^2 + 7^2 = 58$: so $4^2 + 6^2 \neq 3^2 + 7^2$; no.

Lesson 6.3

Ongoing Assessment, **page 279**

On most calculators, 10^{-1} can be evaluated in two ways.

a. 10 [x^{-1}] [ENTER]
b. 10 [y^x] − 1 [=]

Communicate, **page 281**

2. $5a^{-2} = 5 \cdot \frac{a^3}{a^5} = 5 \cdot \frac{a \cdot a \cdot a}{a \cdot a \cdot a \cdot a \cdot a} = 5 \cdot \frac{1}{a^2} = \frac{5}{a^2}$

3. $\frac{2^5}{2^5} = \frac{2 \cdot 2 \cdot 2 \cdot 2 \cdot 2}{2 \cdot 2 \cdot 2 \cdot 2 \cdot 2} = 2^{5-5} = 2^0 = 1$; true for any positive base.

4. No, since 5^2 and 4^{-3} do not have the same base, $5^2 \cdot 4^{-3}$ cannot be simplified using the properties of exponents.

5. The parentheses are around both 3 and a; $(3a)^{-2} = \frac{1}{(3a)^2} = \frac{1}{9a^2}$

Practice and Apply, **pages 282–283**

29. Mentally: subtract exponents to obtain 2.56^1 or 2.56.

30. Mentally: except for the final division. $2.56^{-1} = \frac{1}{2.56} = 0.391$.

31. Mentally: any nonzero power of 0 is 0, so $0^7 = 0$.

32. Mentally: the exponent 0 gives the value 1, so $7^0 = 1$.

33. Mentally: the 0 exponent makes the value of the product 1.

34. Mentally: anything multiplied by 0 is zero, any nonzero power of 0 is 0.

35. Answers may vary. On some calculators, an error code appears because the value is too large. On other calculators, the answer is given in scientific notation.

36. An error code appears because the value is undefined.

37. $0^0 = 0^{1-1} = \frac{0^1}{0^1} \neq 1$; division by zero is undefined.

38. An error message occurs; because 0 raised to the zero power is undefined.

Look Back, **page 283**

40. $y = -2$

41. $y = \frac{-c}{ab - b}$; If $a = 5$, $b = 3$, and $c = 24$, then $y = \frac{-c}{ab - b} = -2$.

42.

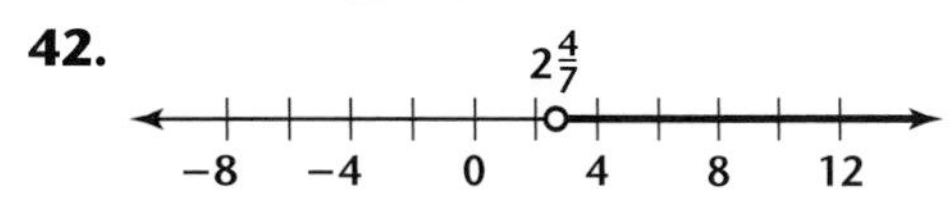

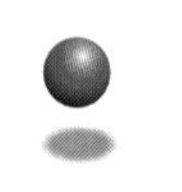

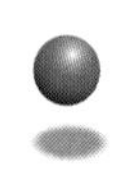

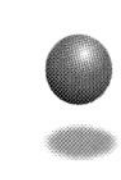

The solution set is all values greater than $2\frac{4}{7}$.

44. $y = -2(x + 3)^2 - 4$

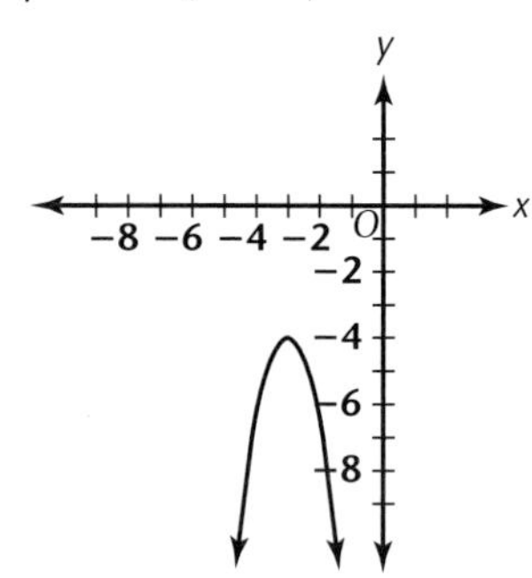

Eyewitness Math, **pages 284–285**

1. About 6 or 7 times greater (although other estimates may be reasonable, as well); The probability of the card still being first is about 0.07 and the probability of it being thirty-fifth is about 0.01.

2a. In a perfect shuffle, the cards from the two halves of the deck alternate: you get one card from one half, one from the other hand, then one from the first half, and so on. In an ordinary shuffle, you may get two cards from one half, then three from the other, then one from the first, and so on.

2b. No. The first four cards alternate, but then the 3 of hearts should come before the 9 of hearts. Also, the 4 and 5 of hearts are together; they should be separated by a card from the right half of the deck.

3a. 2

3b. 3

3c. 4

4a. 5, because 32 is 2^5, which continues the pattern in activity 3

4b. It does take 5 perfect shuffles for the deck of 32 cards to return to its original order.

5. No, because the machine wouldn't mix the cards randomly, and, with a certain number of shuffles, it might not mix them at all.

Lesson 6.4

Critical Thinking, **page 288**

$12 \times 10^{-2} = 1.2 \times 10^1 \times 10^{-2} = 1.2 \times 10^{-1}$; $0.12 \times 10^{-2} = 1.2 \times 10^{-1} \times 10^{-2} = 1.2 \times 10^{-3}$

Communicate, **page 290**

2. The value is multiplied by 0.00000001; 8.

3. The value is multiplied by 1,000,000,000,000; −12.

4. It is 93,000,000 miles from Earth to the sun. Move the decimal point 7 places to the left: 9.3×10^7 miles.

5. Write each number in scientific notation first, then multiply the non-power factors and multiply the powers of 10. Check that the final answer is expressed in scientific notation.

$240{,}000 \times 0.006 = 2.4 \times 10^5 \times 6 \times 10^{-3} = 2.4 \times 6 \times 10^{5+(-3)}$

$= 14.4 \times 10^2 = 1.44 \times 10^1 \times 10^2 = 1.44 \times 10^3$

Look Back, **page 292**

69. $16 < r$

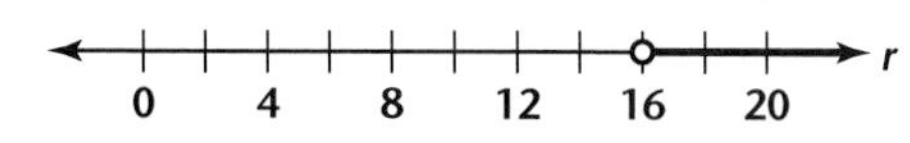

70. $g \leq -1$

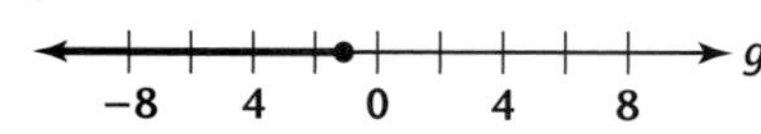

71. $x \geq 2$ or $x \leq -2\frac{1}{4}$

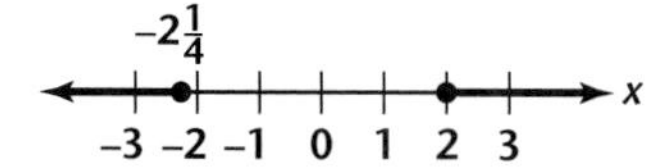

72. $-4 < t < -2$

$-6\ -5\ -4\ -3\ -2\ -1$

74.

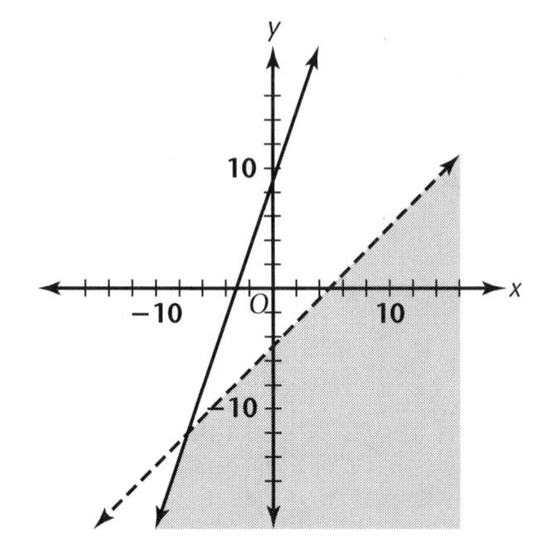

The solution is all the points below the line $y = x - 5$ and to the right of $y = 3x + 9$.

Lesson 6.5

Communicate, page 297

1. Add 100% to the yearly rate of increase: 101.9% or 1.019.
2. Graphs illustrate how quickly the population grows.
3. P = final amount, A = original amount, r = rate of increase or growth, t = time in years

Practice and Apply, page 297

7.

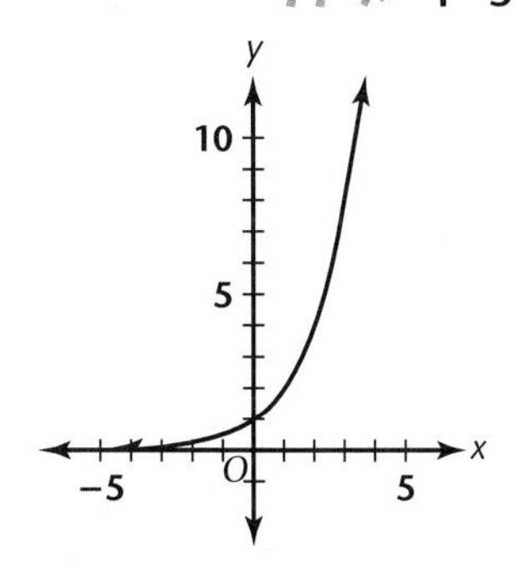

8.

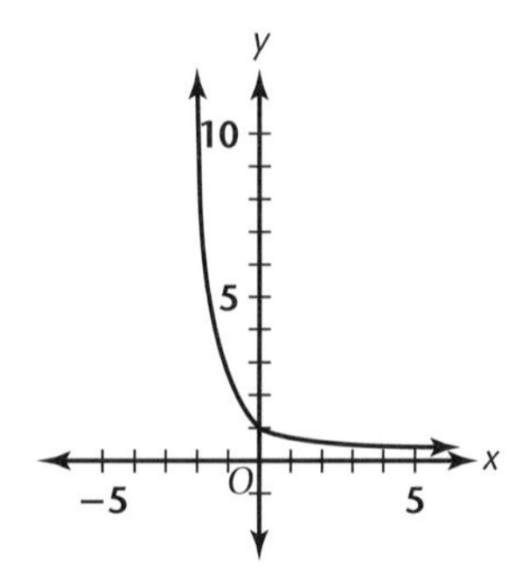

9.

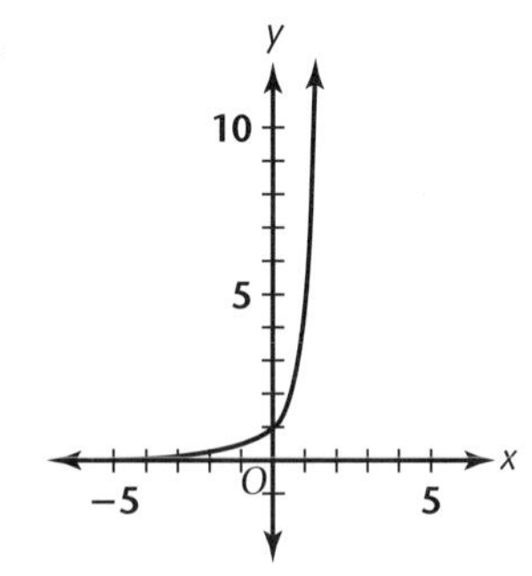

10.

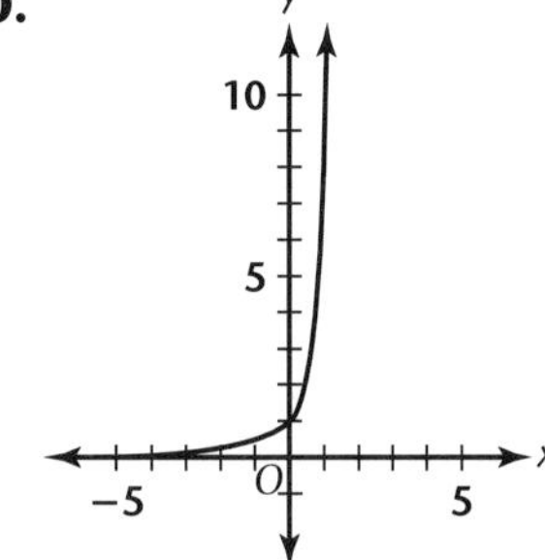

11.

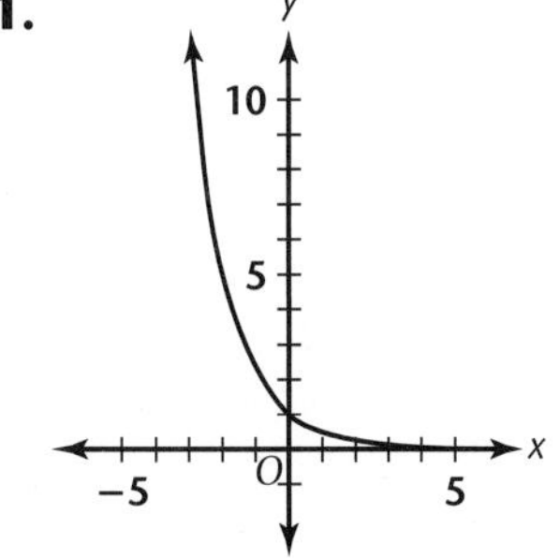

12.

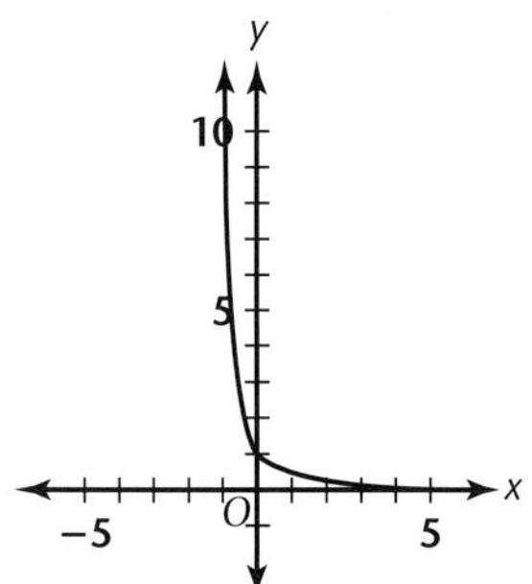

15. $y = 2^x + 3$ shifts the parent graph $y = 2^x$ vertically up 3 units.

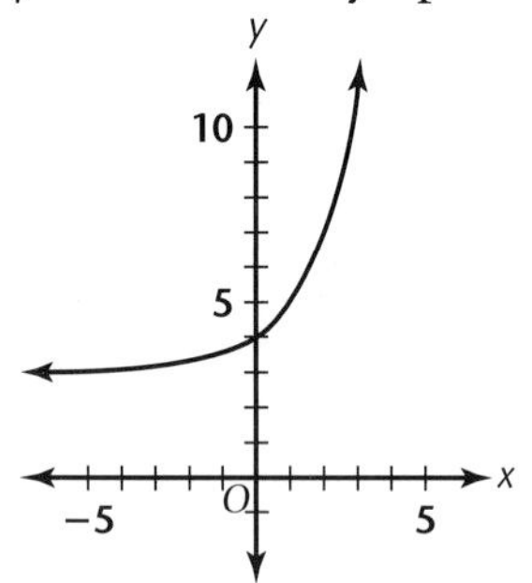

16. $y = 2^{(x+3)}$ shifts the parent graph horizontally 3 units to the left.

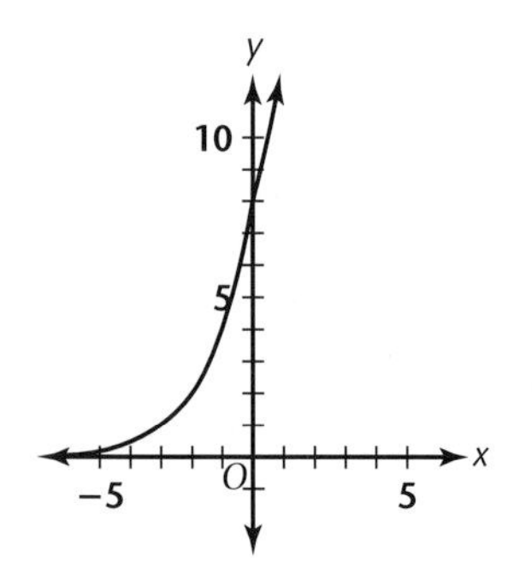

17. $y = 3 \cdot 2^x$ shifts the parent graph vertically by a factor of 3.

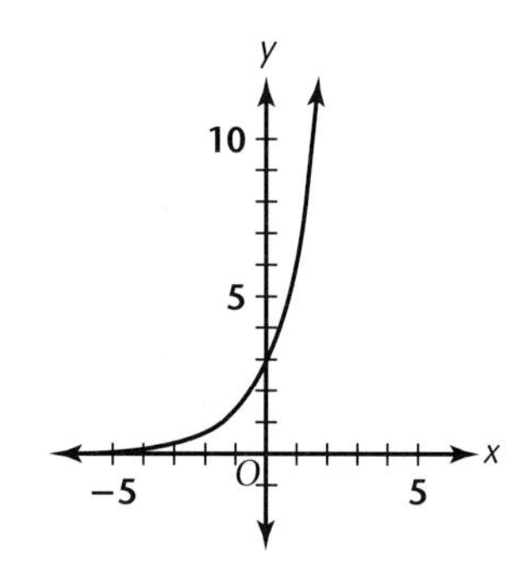

18. $y = 2^{3x}$ stretches the parent graph by a power of 3.

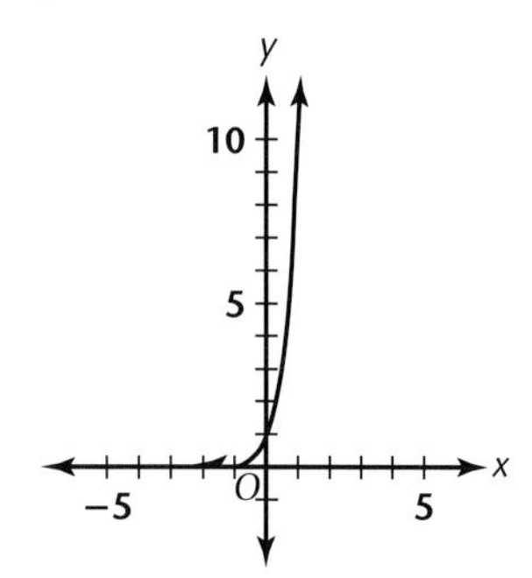

Look Beyond, **page 298**

44–45.

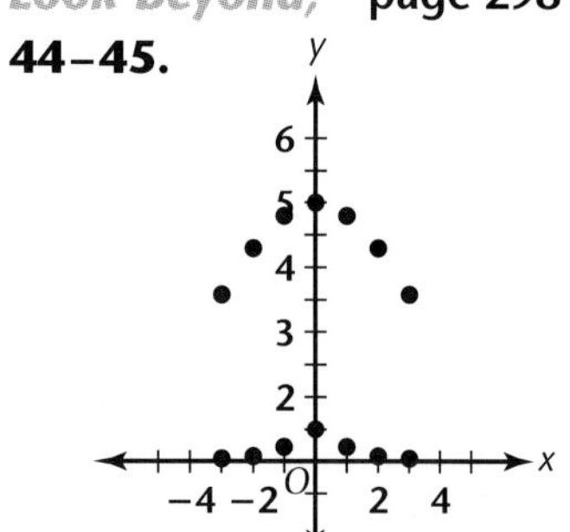

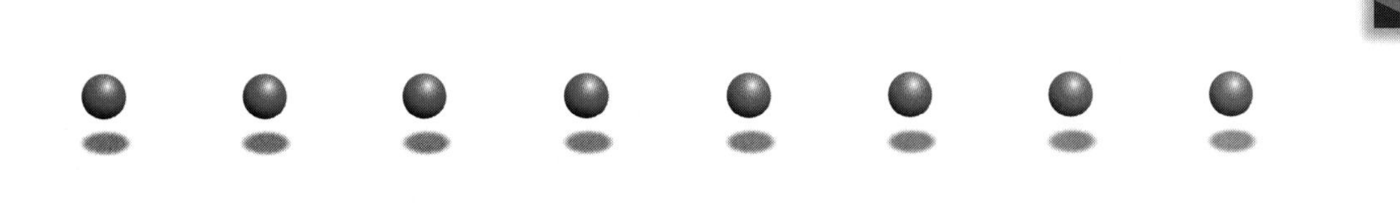

Extra Practice, **page 543**

1.

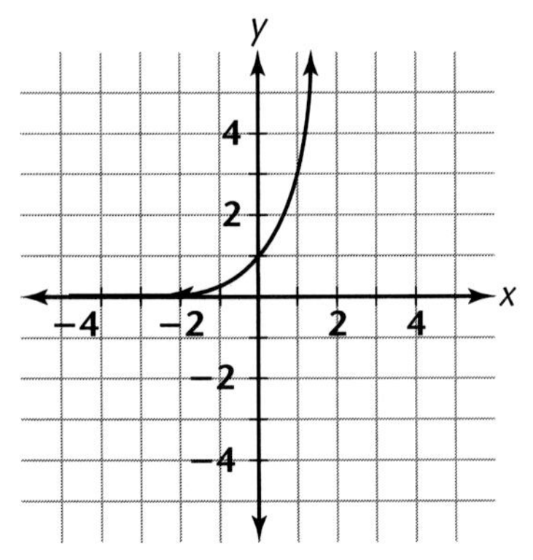

2.

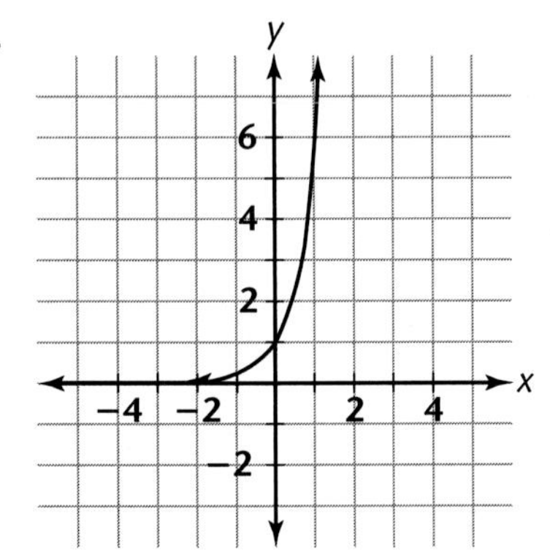

3.

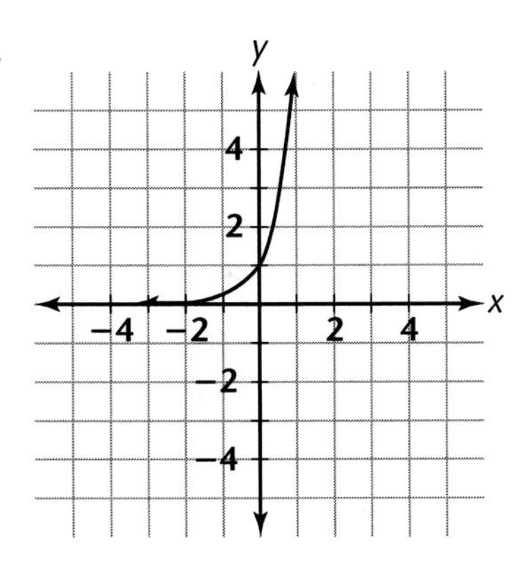

4.

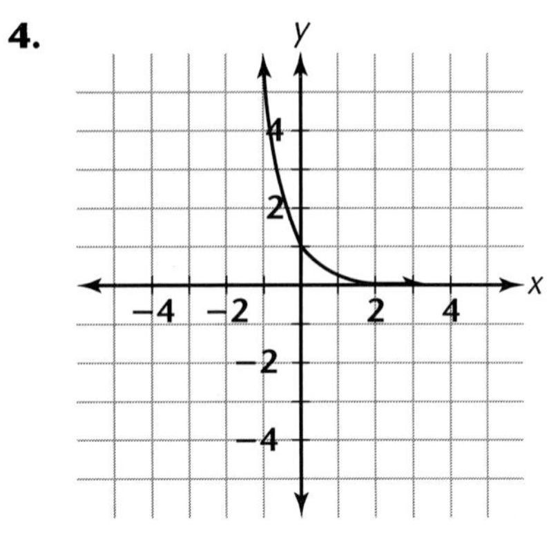

5.

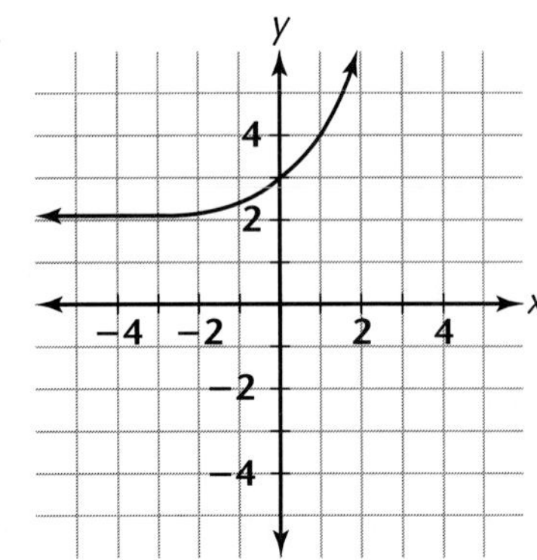

Vertical shift 2 units up

6.

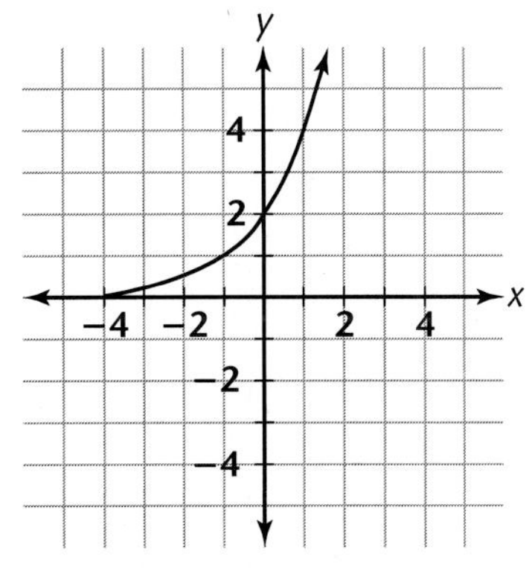

Stretch by a factor of 2

7. 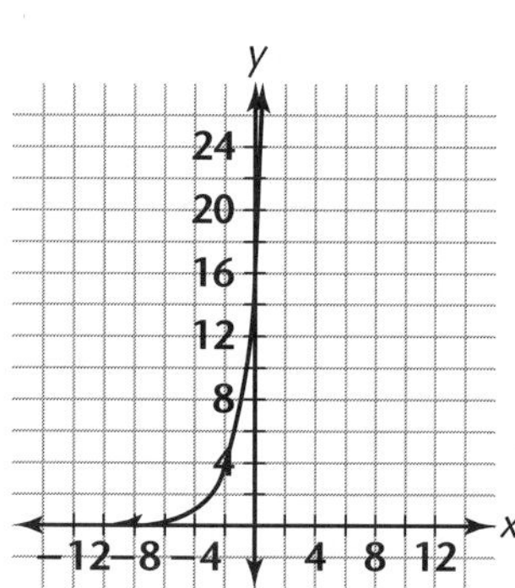

Horizontal shift by 4 units to the left

8. 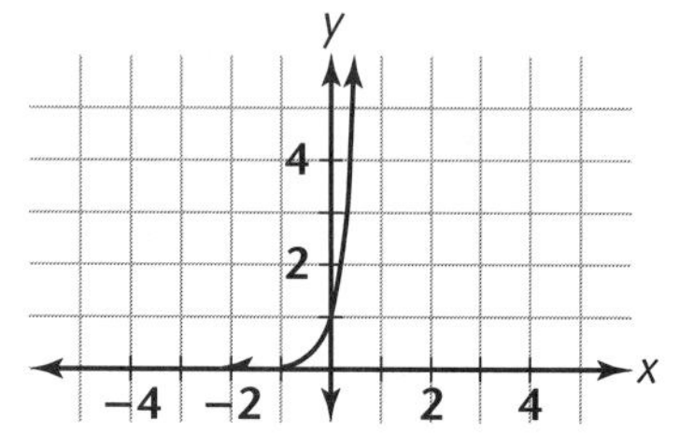

Stretch by a factor of 4

Lesson 6.6

Communicate, page 303

1. Examples may vary; for example, interest paid on loans, growth of bacteria.
2. The remaining amount, P, equals the original amount, A, multiplied by the constant multiplier, $1 - r$, raised to the *exponent* for the number of years of decay, t.
3. Decay is a decreasing measure.
4. For a horizontal translation, a value, h, is subtracted from the *exponent* of the parent function, $y = b^{x-h}$, $h \neq 0$.
5. For vertical translation, a value, k, is added to the parent function, $y = b^x + k$, $k \neq 0$.
6. Answers may vary; for example: $3^1, 3^2, 3^3, 3^4, 3^5, 3^6, \ldots = 3, 9, 27, 81, 243, 729, \ldots$. First differences: 6, 18, 54, 162, 486, . . . ; Each difference is 3 times the previous difference.

Practice and Apply, pages 304–305

9. 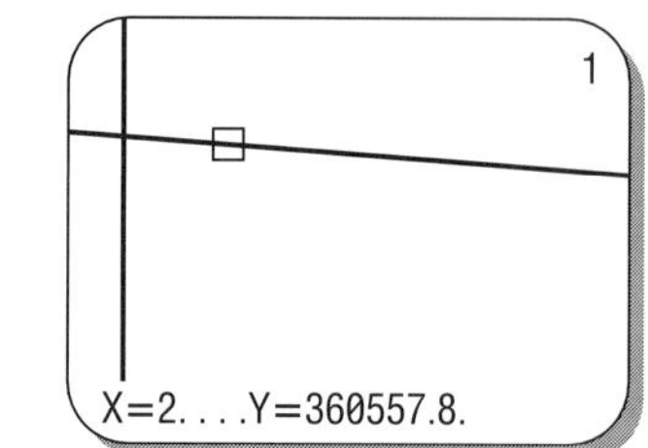

31. Shifts parent graph up 5 units

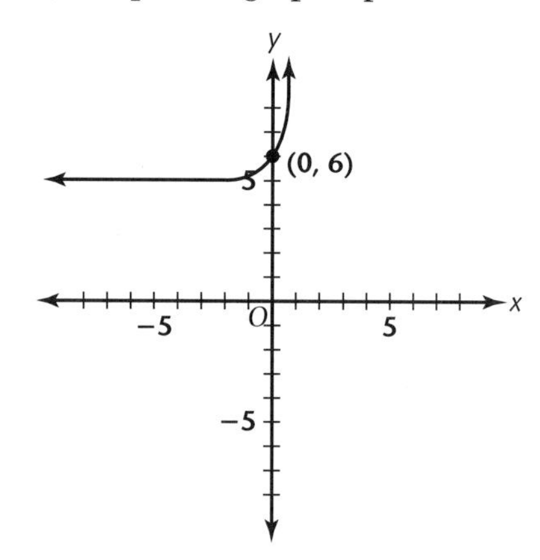

32. Shifts parent graph 5 units to the left

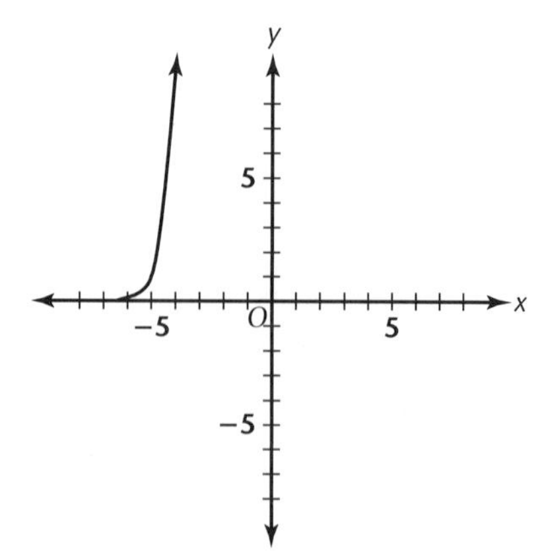

33. Stretched by a factor of 5

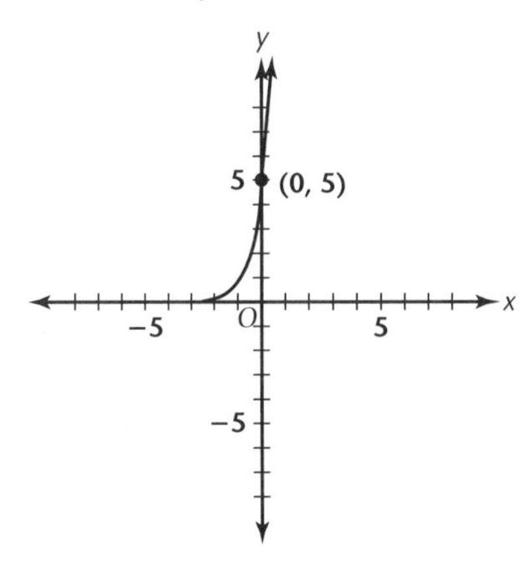

34. Stretched by a power of 5; graph includes the points $(-1, 0.00001)$, $(0, 1)$, and $(1, 100{,}000)$. No choice of axes will depict a reasonable graph.

Look Back, **page 305**

40. Yes, because the result of the first toss does not affect the outcome of the next toss.

Look Beyond, **page 305**

42. ≈ 2.7182818

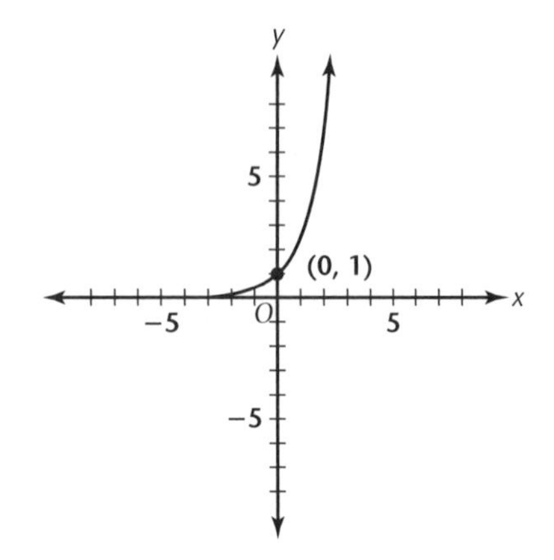

Chapter 6 Project

Activity, **page 307**

Answers may vary.

Chapters 1–6 Cumulative Assessment, pages 312–313

14. The parent function is $y = |x|$; $y = -2|x|$ is a reflection and a stretch.

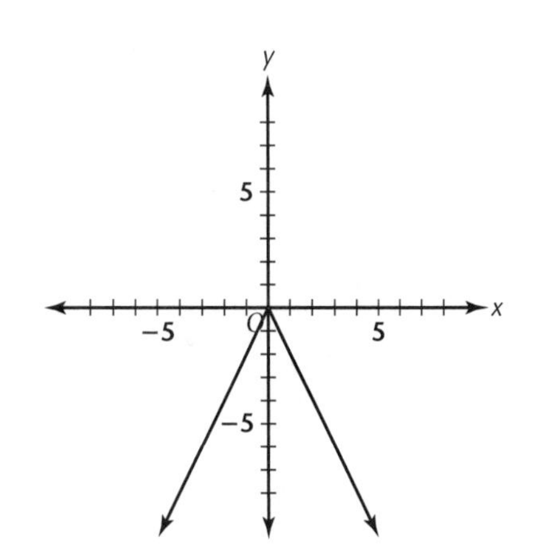

15.

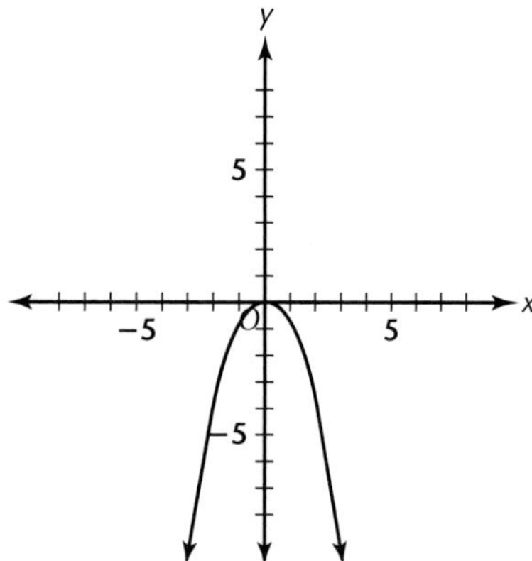

16.

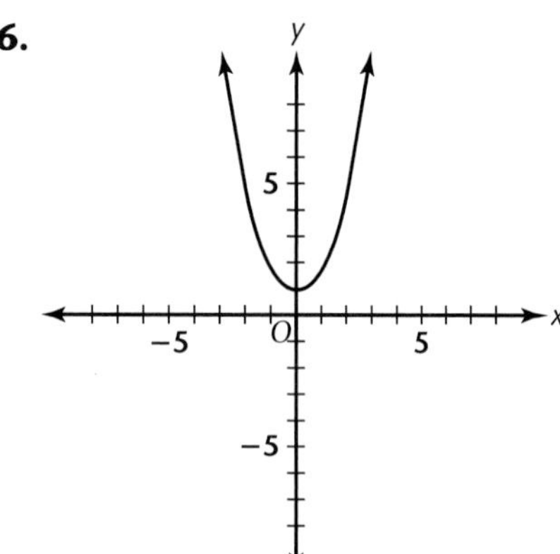

17.

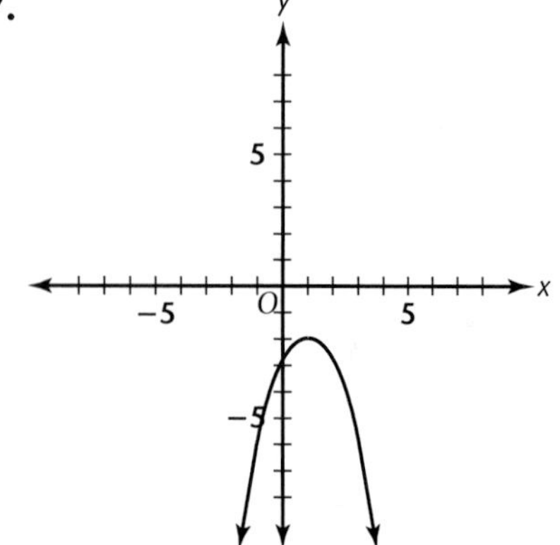

Lesson 7.1

Communicate, **page 319**

1. A polynomial function is a function that consists of one or more monomials. A monomial is an algebraic expression that consists of either a constant, a variable, or a product of a constant and one or more variables. Answers may vary. One possible example: $P(x) = 3x^2 + 4$.
2. $V = x^3$; $S = 6x^2$
3. Enter $Y_1 = x^2 - 4$ and $Y_2 = (x + 2)(x - 2)$ into the calculator. Show a table of values for integer x-values from -3 to 3, and compare the y-values for the two functions. If these y-values are equal, then the equation is true for integer x-values from -3 to 3.

Practice & Apply, **pages 320–321**

8.

x	$x^2 + 5x + 6$	$(x + 2)(x + 3)$
−1	2	2
0	6	6
1	12	12

9.

x	$(x - 4)(x - 2)$	$x^2 - 6x + 8$
−1	15	15
0	8	8
1	3	3

10.

x	$x^2 - 5x + 6$	$(x - 2)(x - 3)$
−1	12	12
0	6	6
1	2	2

11.

x	$(x + 7)(x - 1)$	$x^2 + 6x - 7$
−1	−12	−12
0	−7	−7
1	0	0

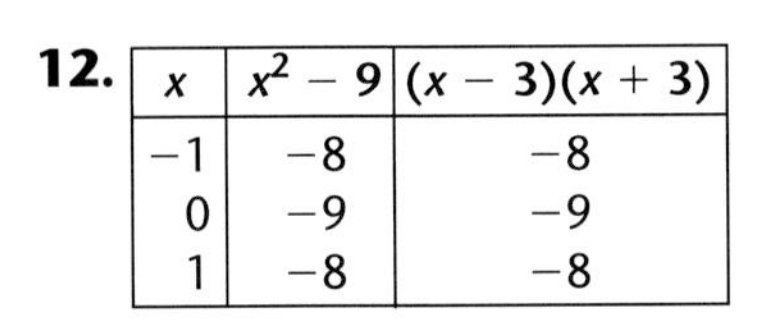

12.

x	$x^2 - 9$	$(x - 3)(x + 3)$
−1	−8	−8
0	−9	−9
1	−8	−8

13.

x	$(x - 6)(x + 4)$	$x^2 - 2x - 24$
−1	−21	−21
0	−24	−24
1	−25	−25

14.

x	$x^3 - 8$	$(x - 2)(x^2 + 2x + 4)$
−1	−9	−9
0	−8	−8
1	−7	−7

15.

x	$(x + 5)(x + 3)$	$x^2 + 8x + 15$
−1	8	8
0	15	15
1	24	24

16.

x	$x^3 + 9x^2 + 27x + 27$	$(x + 3)^3$
−1	8	8
0	27	27
1	64	64

17.

x	$(x - 6)(x - 7)$	$x^2 - 13x + 42$
−1	56	56
0	42	42
1	30	30

18.

x	$(x - 3)^2$	$x^2 - 6x + 9$
−1	16	16
0	9	9
1	4	4

19.

x	$(x + 5)^2$	$x^2 + 10x + 25$
−1	16	16
0	25	25
1	36	36

20. $C = 36\pi$ inches, or ≈ 113.1 inches; $S = 1296\pi$ square inches, or ≈ 4071.5 square inches; $V = 7776\pi$ cubic inches, or $\approx 24{,}429$ cubic inches

21. $C = 16\pi$ inches, or ≈ 50.3 inches; $S = 256\pi$ square inches, or ≈ 804.2 square inches; $V = \frac{2048}{3}\pi$ cubic inches, or ≈ 2144.7 cubic inches

22. $C = 10\pi$ inches, or ≈ 31.4 inches; $S = 100\pi$ square inches, or ≈ 314.2 square inches; $V = \frac{500}{3}\pi$ cubic inches, or ≈ 523.6 cubic inches

23. $C = 4\pi$ inches, or ≈ 12.6 inches; $S = 16\pi$ square inches, or ≈ 50.3 square inches; $V = \frac{32}{3}\pi$ cubic inches, or ≈ 33.5 cubic inches

24. $C = 2\pi$ inches, or ≈ 6.3 inches; $S = 4\pi$ square inches, or ≈ 12.6 square inches; $V = \frac{4}{3}\pi$ cubic inches, or ≈ 4.2 cubic inches

25. $C = \pi$ inches, or ≈ 3.1 inches; $S = \pi$ square inches, or ≈ 3.1 square inches; $V = \frac{1}{6}\pi$ cubic inches, or ≈ 0.5 cubic inches

26. $V = 666\frac{2}{3}\pi$ cubic feet, or ≈ 2094.4 cubic feet

27. $S = 300\pi$ square feet, or 942.5 square feet

28. $V(x) = 20\pi x^2 + \frac{4}{3}\pi x^3$

29. $S(x) = 40\pi x + 4\pi x^2$

30.

Radius	S (in sq ft)	V (in cu ft)
1	≈138.2	≈67
2	≈301.6	≈284.8
3	≈490.1	≈678.6
4	≈703.7	≈1273.4
5	≈942.5	≈2094.4
6	≈1206.4	≈3166.7

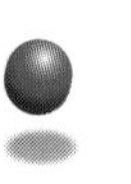

Look Beyond, **page 321**

40. Answers may vary. The larger the odd power, the steeper the graph. All of these functions have rotational symmetry about the origin.

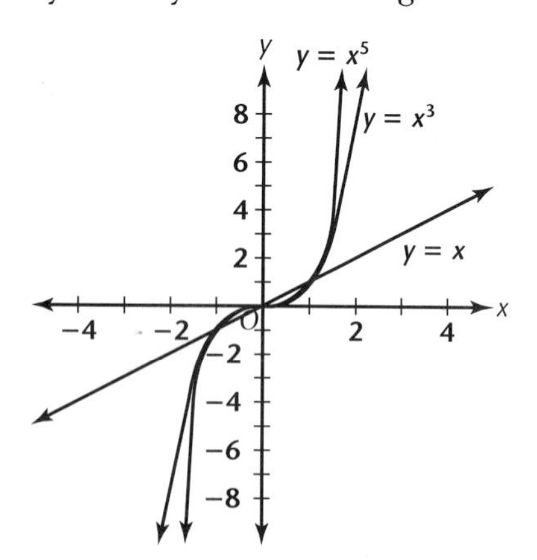

41. Answers may vary. The larger the even power, the steeper the graph. All of these functions have reflectional symmetry with respect to the y-axis.

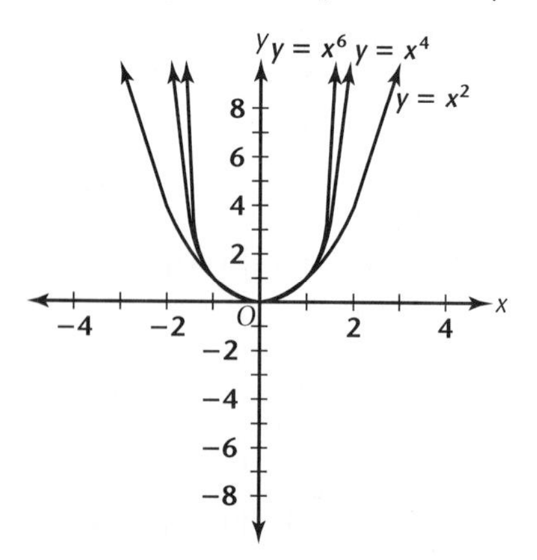

42. Answers may vary. $y = x^2 - 2$ and $y = x^4 - 2x^2$ are symmetrical with respect to the y-axis; $y = x^3 - 2x$ is symmetrical with respect to the origin. All three functions have the same x-intercepts at $\sqrt{2}$ and $-\sqrt{2}$. $y = x^4 - 2x^2$ and $y = x^3 - 2x$ have an x-intercept at 0.

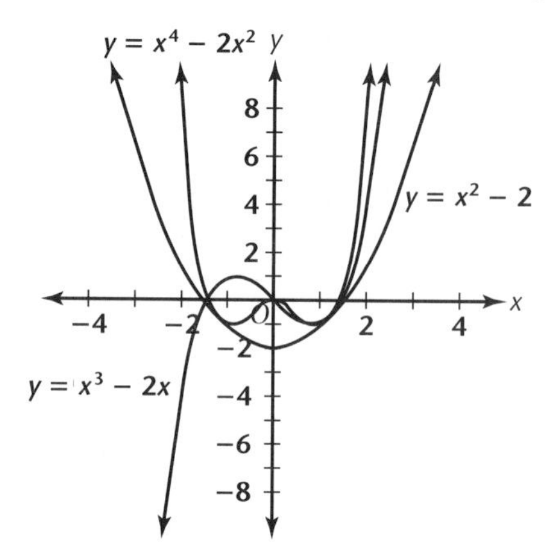

Extra Practice, **page 544**

1. $S(x) = 6 \cdot x^2$ square units;
$V(x) = x^3$ cubic units

2. $S(x) = 22x + 60$ square inches;
$V(x) = 30x$ cubic inches

3. $S(x) = 27.4x + 79.8$ square centimeters;
$V(x) = 39.9x$ cubic centimeters

4. $S(x) = 2\pi x^2 + 12\pi x$ square inches;
$V(x) = 6\pi x^2$ cubic inches

5. $S(x) = \pi x^2 + \pi x\sqrt{x^2 + 81}$ square feet;
$V(x) = 3\pi x^2$ cubic feet

6.

x	$x^2 + 4x + 3$	$(x + 1)(x + 3)$
−1	0	0
0	3	3
1	8	8

7.

x	$x^2 - 7x + 10$	$(x - 5)(x - 2)$
−1	18	18
0	10	10
1	4	4

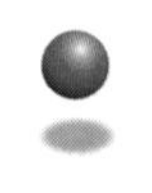
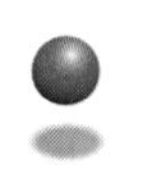

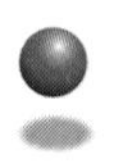

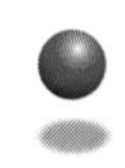

8.

x	$x^2 - 7x + 12$	$(x - 4)(x - 3)$
-1	20	20
0	12	12
1	6	6

9.

x	$x^2 + 7x + 6$	$(x + 6)(x + 1)$
-1	0	0
0	6	6
1	14	14

10.

x	$(x + 4)(x - 3)$	$x^2 + x - 12$
-1	-12	-12
0	-12	-12
1	-10	-10

11.

x	$(x - 4)(x - 5)$	$x^2 - 9x + 20$
-1	30	30
0	20	20
1	12	12

12.

x	$x^2 - 16$	$(x + 4)(x - 4)$
-1	-15	-15
0	-16	-16
1	-15	-15

13.

x	$(x - 5)(x + 5)$	$x^2 - 25$
-1	-24	-24
0	-25	-25
1	-24	-24

14.

x	$(x + 4)(x - 6)$	$x^2 - 2x - 24$
-1	-21	-21
0	-24	-24
1	-25	-25

15.

x	$x^2 + 9x + 14$	$(x + 7)(x + 2)$
-1	6	6
0	14	14
1	24	24

16.

x	$x^2 - 100$	$(x - 10)(x + 10)$
-1	-99	-99
0	-100	-100
1	-99	-99

17.

x	$(x - 4)(x + 3)$	$x^2 - x - 12$
-1	-10	-10
0	-12	-12
1	-12	-12

Lesson 7.2

Communicate, **page 326**

1. Use five x^2-tiles for $5x^2$, two negative x-tiles for $-2x$, and three 1-tiles for 3.

2. Determine the greatest exponent of all of the terms in the polynomial.

3. Trinomial; cubic

4.

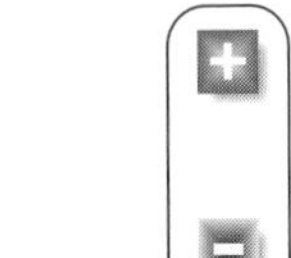

$3x^2 + 2$

5. Change the signs of the terms in the trinomial that is to be subtracted to their opposites. Then group like terms and simplify to $2b^3 - 3b + 4$.

6. Arrange the terms in descending order according to their exponents. The polynomial in standard form is $3x^4 + 5x^2 - 2x - 6$.

Practice & Apply, **pages 327–328**

7.

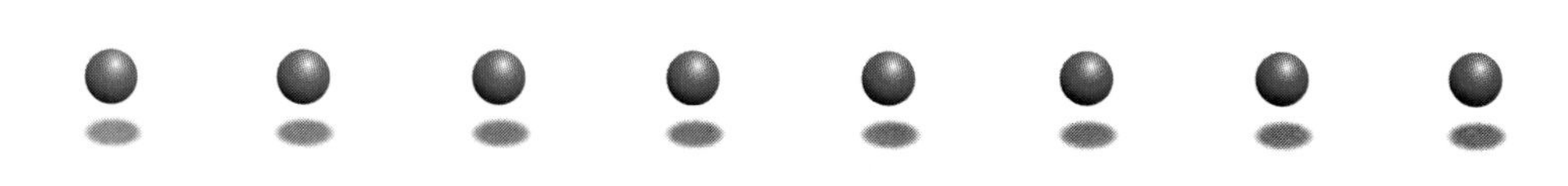

8.

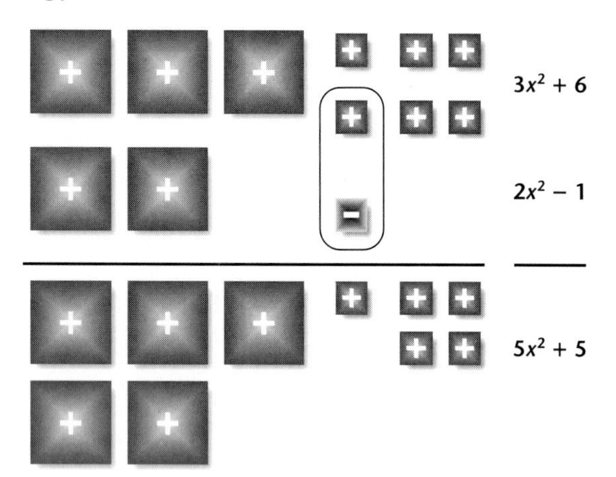

Lesson 7.3

Exploration 1, **pages 329–330**

6.
$$(x + 2)(2x + 3) = 2x^2 + 7x + 6$$
$$(10 + 2)[2(10) + 3] = 2(10^2) + 7(10) + 6$$
$$(12)(23) = 200 + 70 + 6$$
$$276 = 276 \quad \text{True}$$

7. Use the factors $2x + 2$ and $x + 4$. Arrange the product rectangle so that $x(x + 4) = x^2 + 4x$, $x(x + 4) = x^2 + 4x$ and $2(x + 4) = 2x + 8$; the product is $2x^2 + 10x + 8$.

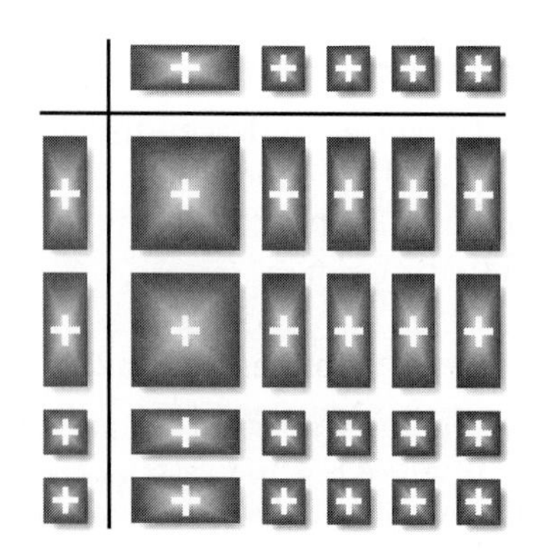

Check by substituting a value for x to prove the equality is true;
$$(2x + 2)(x + 4) = 2x^2 + 10x + 8$$
$$[2(1) + 2](1 + 4) = 2(1)^2 + 10(1) + 8$$
$$(4)(5) = 2 + 10 + 8$$
$$20 = 20 \quad \text{True}$$

Exploration 2, **pages 330–331**

4.
$$(10 + 2)(10 - 3) = 10^2 - 10 - 6$$
$$(12)(7) = 100 - 10 - 6$$
$$84 = 84 \quad \text{True}$$

5. $x^2 - 6x + 8$

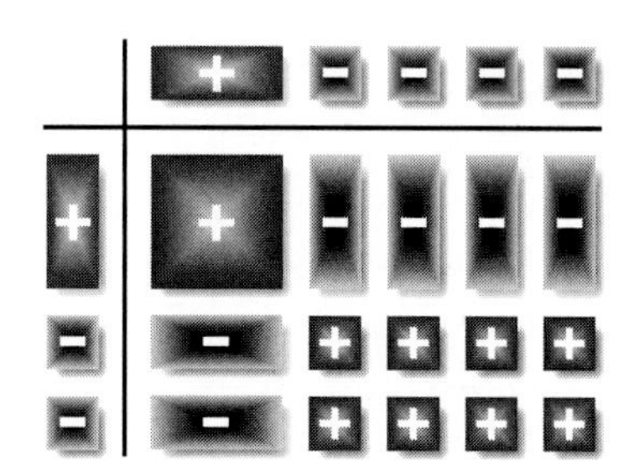

$$(x - 2)(x - 4) = x^2 - 6x + 8$$
$$(10 - 2)(10 - 4) = 10^2 - 6(10) + 8$$
$$(8)(6) = 100 - 60 + 8$$
$$48 = 48 \quad \text{True}$$

6. 0; 0; when these numbers are substituted for the x-value, they cause one of the factors and the product to be zero.

Exploration 3, Part I, **page 331**

3.

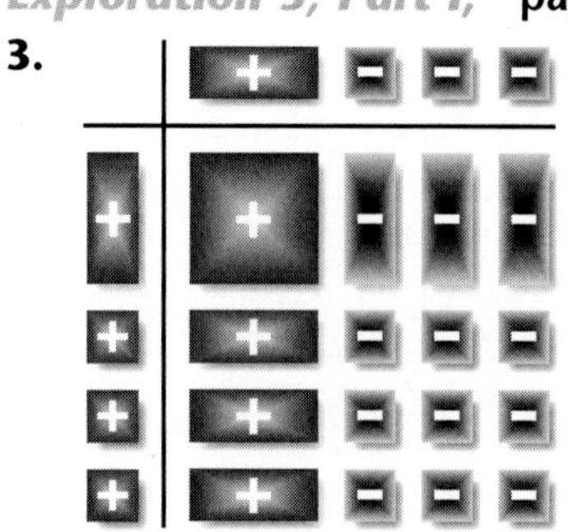

$x^2 - 9$; the x-terms cancel.

4. Square the first term and subtract the square of the second term.

Exploration 3, Part II, page 332

2. $x^2 + 6x + 9$

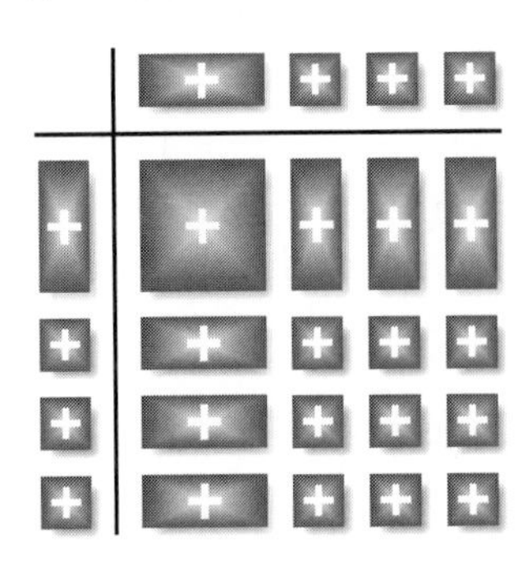

3. $x^2 + 4x + 4$ is similar to $x^2 + 6x + 9$ because the first terms are the products of the first terms of the factors, the second terms are the sum of the product of the outer terms plus the product of the inner terms of the factors, and the last terms are the products of the last terms of the factors.
4. Multiply the first terms of the factors; x^2. Add the product of the outer terms and the product of the inner terms; $5x + 5x = 10x$. Multiply the last terms; $(5)(5) = 25$; $x^2 + 10x + 25$.

Communicate, page 332

1. $x^2 + 5x + 6$

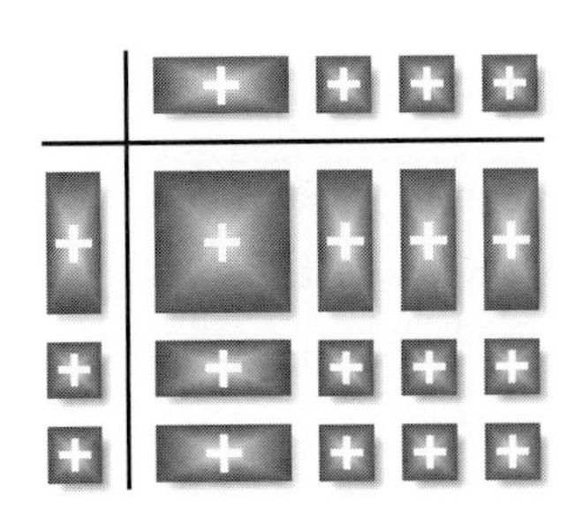

$x^2 - 8x + 16$

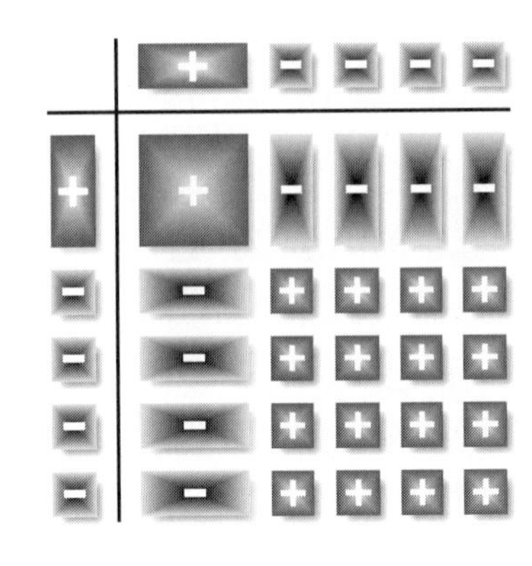

$x^2 + 12x + 36$

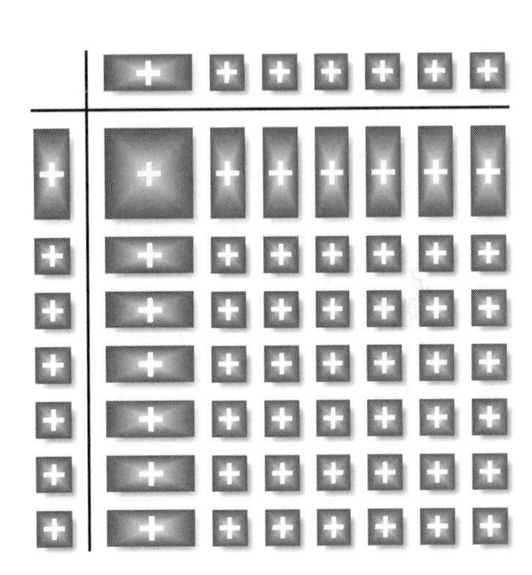

2. Answers may vary; one example is $(x - 2)(x + 1) = x^2 - x - 2$.
3. Square the first terms of the factors and subtract the square of the last terms of the factors.

4. $(x + 4)^2 = (x + 4)(x + 4)$
$= x^2 + 8x + 16$

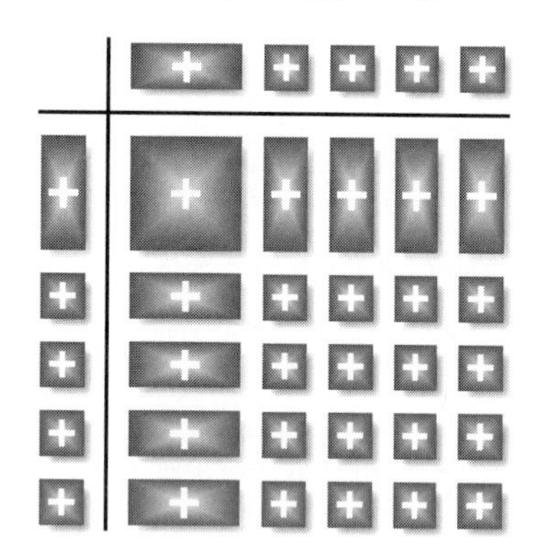

When the exponent, 2, is on the outside of the parentheses, the whole quantity inside the parentheses is squared.

Practice & Apply, **pages 332–333**

29. $x^2 + 7x + 10$

30. $x^2 + x - 2$

31. $2x^2 + 3x + 1$

32. $x^2 + 3x - 10$

33. $4x^2 - 4$

34. $x^2 + 3x + 2$

35. $3x^2 - 16x - 35$

36. $25x^2 - 1$

37. $x^2 - 4$

38. $x^2 + 2x + 1$

39. $4x^2 - 11x + 6$

40. $x^2 + 12x + 36$

41. $x^2 - 16$

42. $x^2 - 2x - 35$

43. $x^2 - 12x + 36$

56. The diagram indicates multiplying the 6 by the 10s place and then multiplying the 6 by the 1s place of the number 23. To solve you add 120 and 18 together.

$$6(20 + 3) = 120 + 18 = 138$$

57. The diagram indicates multiplying each of the numbers on the left by each of the numbers at the top of the columns. The numbers at the top represent $12 = 10 + 2$, and the numbers at the side represent $14 = 10 + 4$. When each of the rows and columns are multiplied and then added together,

$$10 \cdot 10 + 10 \cdot 2 + 4 \cdot 10 + 4 \cdot 2 = 168.$$

58. By multiplying each term of $(x + 1)(x - 1)$ and adding the results, the product is $x^2 - 1$.

Look Back, **page 334**

62.

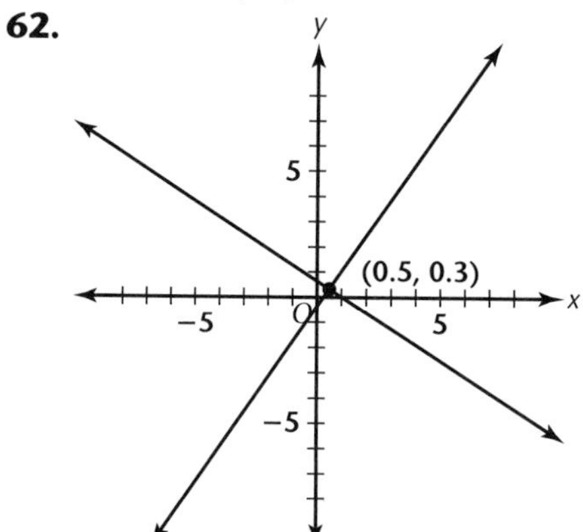

69. The events are independent if he replaces the card and dependent if he keeps the card before drawing another.

70. Shifted down 4 units

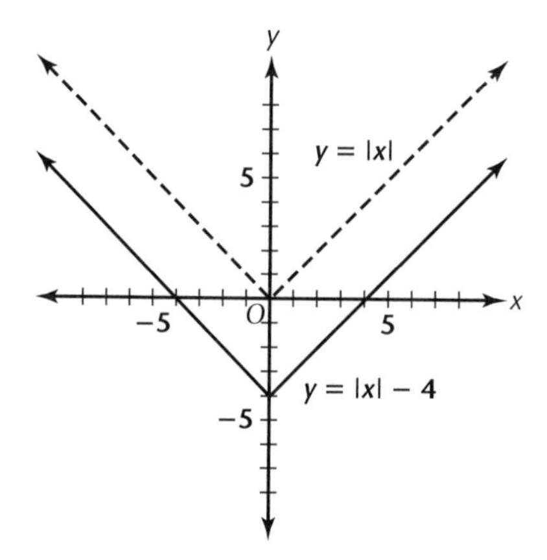

71. Shifted to the right by 4 units; at $x = 4$ the function is undefined at the vertical asymptote.

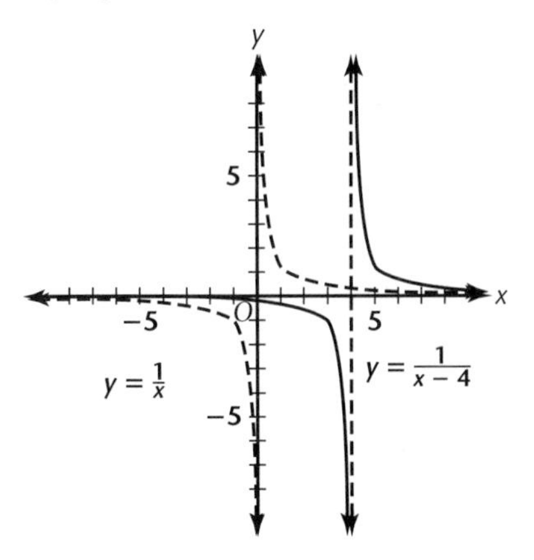

Look Beyond, **page 334**

78. The zeros are the value of x where the graph crosses the x-axis and the y-value of the coordinate is 0.

Lesson 7.4

Exploration 1, **page 335**

1. $x + 2$

$x + 3$

Ongoing Assessment, **page 335**

$$(x + 4)(x + 5) = x(x + 4) + 5(x + 4)$$
$$= x^2 + 4x + 5x + 20$$
$$= x^2 + 9x + 20$$

Critical Thinking, **page 337**

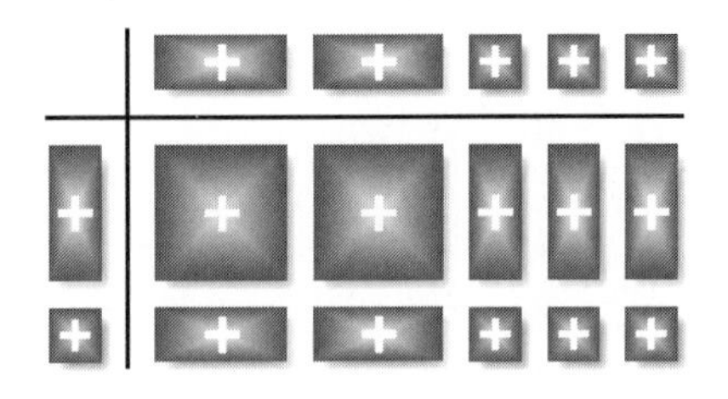

Distributive Property:

$$(2x + 3)(x + 1) = 2x(x + 1) + 3(x + 1)$$
$$= 2x^2 + 2x + 3x + 3$$
$$= 2x^2 + 5x + 3$$

FOIL method:

$$(2x + 3)(x + 1)$$
$$= (2x)(x) + (2x)(1) + (3)x + 3(1)$$
$$= 2x^2 + 5x + 3$$

Communicate, **page 338**

1a. $x^2 + 3x + 2$

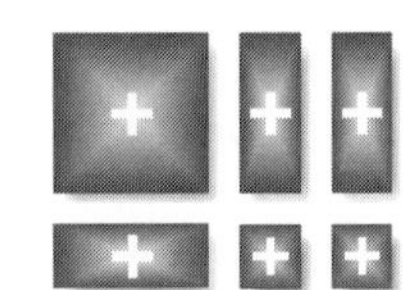

1b. $x^2 - 3x + 2$

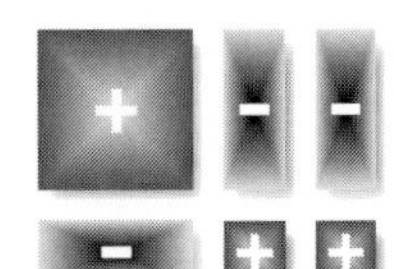

2a. $(x + 1)(x + 2) = x(x + 2) + 1(x + 2) = x^2 + 2x + x + 2 = x^2 + 3x + 2$

2b. $(x - 1)(x - 2) = x(x - 2) - 1(x - 2) = x^2 - 2x - x + 2 = x^2 - 3x + 2$

3. $(2x + 3)(x - 4) = 2x^2 + 3x - 8x - 12 = 2x^2 - 5x - 12$

4. $(x + 2)^2$ has 4 x-tiles that $x^2 + 2^2$ does not.

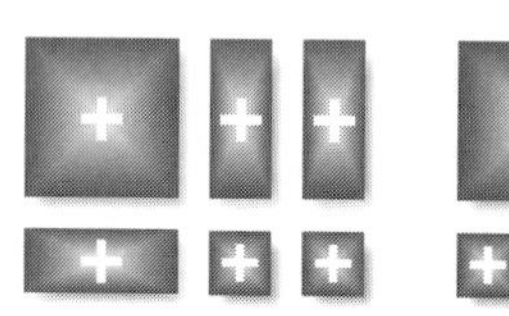

Practice & Apply, **pages 338–339**

11. $y^2 + 8y + 15$

12. $w^2 + 10w + 9$

13. $b^2 - 4b - 21$

14. $3y^2 - 5y + 2$

15. $5p^2 + 8p + 3$

16. $4q^2 - 1$

17. $4x^2 + 4x - 15$

18. $20m^2 - 7m - 3$

19. $6w^2 - 43w + 72$

20. $9x^2 - 30x + 25$

21. $14s^2 - 17s - 6$

22. $y^2 - \frac{1}{4}$

23. $y^2 + \frac{2}{9}y - \frac{5}{27}$

24. $c^4 + 3c^2 + 2$

25. $4a^4 + 12a^2 + 9$

26. $a^2 + 3ac + 2c^2$

27. $p^2 + 2pq + q^2$

28. $a^4 - b^2$

29. $x^3 + x^2y + xy + y^2$

30. $2c^2 + 3cd + d^2$

31. $0.96m^2 - 0.8m - 20$

32. $(x + 8)$ by $(x - 4)$

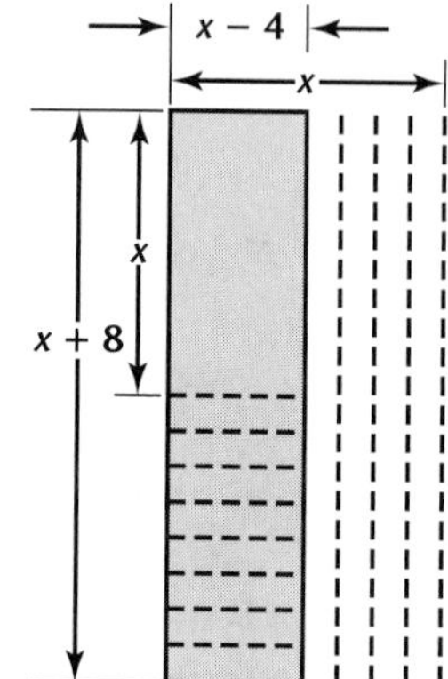

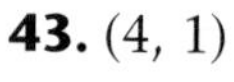

Look Back, **page 339**

43. (4, 1)

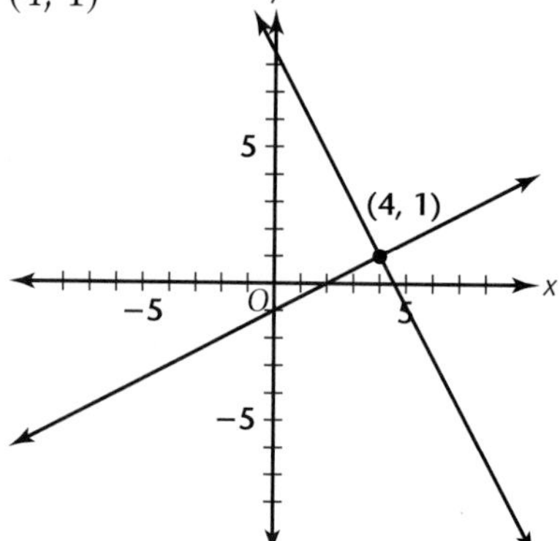

44. There is no solution because the lines are parallel.

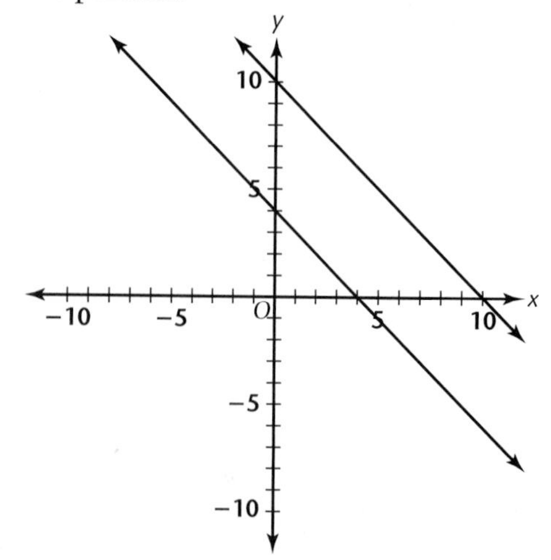

Lesson 7.5

Exploration, pages 340–341

4. 1 has only one distinct factor and prime numbers have exactly two distinct factors; 1 can be factored into any number of 1s, but they are not distinct.

5. A prime number is a number which has exactly two distinct factors, 1 and itself.

Communicate, page 343

1. A prime polynomial is one that has no polynomial factors with integral coefficients except itself and 1.

2. Distributive Property

3. Divide each term in the polynomial by the greatest common factor (GCF).

4. 30; Find the largest number that divides 60 and 150 evenly.

5. x^3y^2; Factor out variables that are common to both terms with the highest possible degree.

6. $(x + y)$; First check for a possible GCF between 25 and 39. There is no GCF, so the coefficients do not change. But both expressions have $(x + y)$ in common. Therefore, $(x + y)$ is the GCF.

7. Group terms that have common factors. The result is $(y^2 + 2y) + (3y + 6)$.

Practice & Apply, pages 343–344

23. $(x + 2)(x + 1)$

24. $(5 - x)(y + 3)$

25. $(a + b)(x + y)$

26. $(3q - 4)(4 + p)$

27. $(x + 2)(x - 1)$

28. $(r + t)(x - 4)$

29. $(5a + 4)(a - 3)$

30. $(2w - 3)(w + 4)$

31. $(x - 2)(2 - x)$ or $-(x - 2)^2$

32. $(8 - x)(y - 1)$

33. $(2r - 3)(r - s)^2$

34. $(ax + bz)(u - v)^n$

35. $(2 + a)(x + y)$

36. $(n + 3)(u + v)$

37. $(4b - 5)(3a - 2)$

38. $(a + 12)(x + y)$

39. $15(x + y)$

40. $(x + 4)(x + 3)$

41. $2(n + 7)(n - 3)$

42. $2p(7p - q)$ or $-2p(q - 7p)$

43. $5(x - 2)(2d + 3)^3$

Look Back, page 344

51.

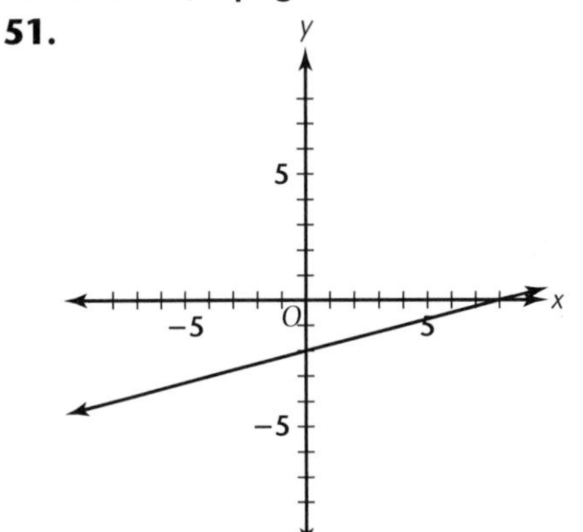

Look Beyond, page 344

57. 5 and 15

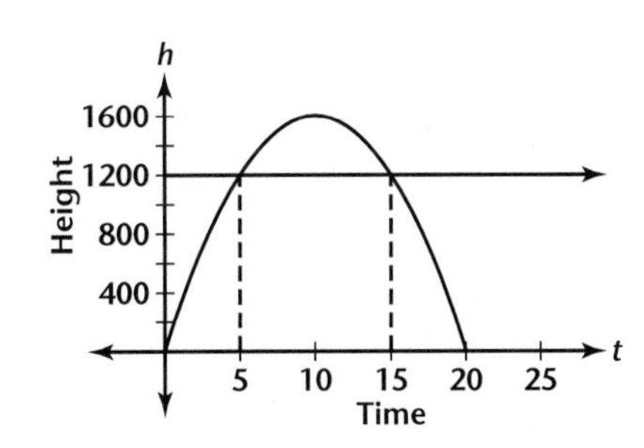

Extra Practice, **page 546**

1. Composite
2. Prime
3. Composite
4. Prime
5. Composite
6. Composite
7. Composite
8. Composite
9. Prime
10. Prime
11. Composite
12. Composite
13. $4(x^2 + 4)$
14. $7(w^2 + 3)$
15. $4a(2a - 1)$
16. $5(n^3 - 2n^2 + 3)$
17. $5(16 - 15d^3 - 10d^2)$
18. $2xy(3x - 7 + xy)$
19. $3st^2(st + 5t - 4s)$
20. $4z^2w(z - 4w^2 + 6z^2w)$
21. $5ab^2(5b^2 + 4a^2 - 3b)$
22. $(x - 1)(x + 2)$
23. $(y + 4)(7 - x)$
24. $(c + d)(a + b)$
25. $(w - 3)(4 - v)$
26. $(y - z)(x + 4)$
27. $(x + 5)(s - t)$
28. $(4 - s)(q - 5)$
29. $(x + y)^2(3w - 4)$
30. $(p - q)^3(mn + ab)$
31. $(y - 5)(4 - 3a)$
32. $(3y + 5)(x - 7)$
33. $(a + 3c)(2x + b)$
34. $(p + a)(4 + y)$
35. $(m - b)(a + 3y)$
36. $(a - 2b)(a + 1)$
37. $(x + w)(r + 6)$
38. $(x + 5y)(x + a)$
39. $(5a - 7b)(2a + b)$
40. $(5c - 3d)(3 + 2c)$
41. $(7t - 4s)(t - 4s)$
42. $(a + c)(7 + b)$
43. $(5x + 2y)(2x + 3)$
44. $(5 - n)(3m - 4p)$
45. $(5 - 8xy)(3 - x)$

Lesson 7.6

Ongoing Assessment, **page 345**

$$(2x + 6)^2 = (2x)^2 + (2)(2x)(6) + 6^2 = 4x^2 + 24x + 36$$
$$(2x - 6)^2 = (2x)^2 + (2)(2x)(-6) + (-6)^2 = 4x^2 - 24x + 36$$

Middle terms have opposite signs.

Ongoing Assessment, **page 348**

When $y = 0$, $x = 5$; therefore, $0 = (x - 5)(x - 5)$ and $x = 5$.

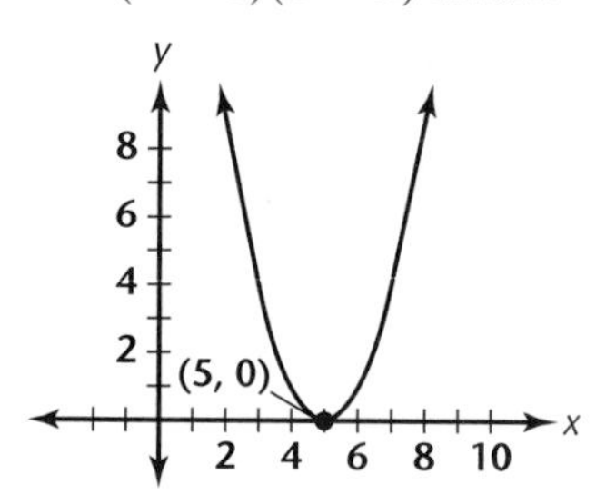

When $y = 0$, $x = 8$; $x^2 - 16x + 64 = (x - 8)(x - 8)$

Communicate, **page 348**

1. Determine the 2 equal factors of the first term, x^2, namely, x and x. Determine the 2 equal factors of the third term, 100, namely, 10 and 10. Since the trinomial is a perfect square the factors are $(x + 10)$ and $(x + 10)$.

2. Since the first and third terms are perfect squares, determine their square roots, $2x$ and 3. Check to see if the middle term of the trinomial is equal to twice the product of the square roots of $4x^2$ and 9. Since the values are the same, the factors are $(2x - 3)$ and $(2x - 3)$.

3. Find the positive square root of p^2 and 121. The square roots are p and 11, respectively. Since the binomial is a difference of two squares, its factors are the sum and the difference of the square roots, $(p + 11)$ and $(p - 11)$.

Practice & Apply, **pages 348–350**

37. $(x + 2)(x - 2)(5 + x)(5 - x)$

38. $(x - 1)^3$

39. $(3x + 5)(x + 2)(x - 2)$

40. $(x - y)(x + y)^3$

41. The rectangle formed has dimensions $a + b$ and $a - b$; therefore, by comparing areas, $a^2 - b^2 = (a + b)(a - b)$.

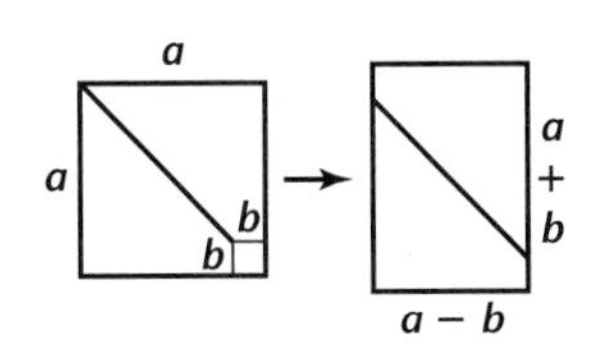

46. Size of rectangle may vary.

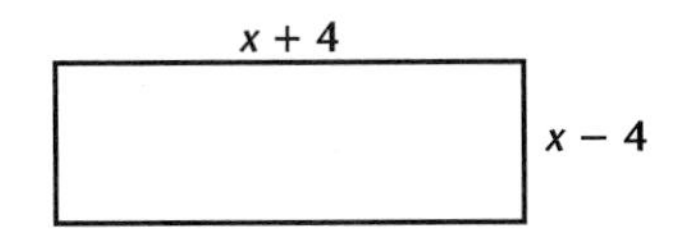

47. The area of the rectangle is $(x + 4)(x - 4)$ or $x^2 - 16$, which is the area of the original shaded area.

Look Back, **page 350**

54.

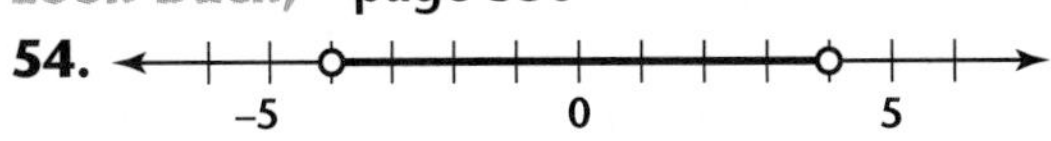

57.

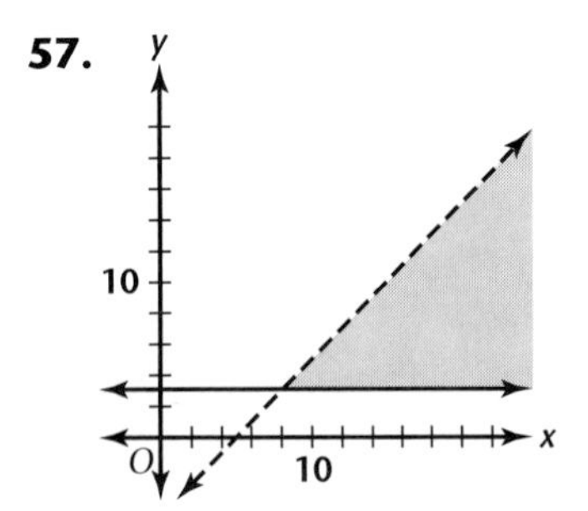

Look Beyond, **page 350**

64. 1 and −42, −1 and 42, 2 and −21, −2 and 21, 3 and −14, −3 and 14, 6 and −7, −6 and 7; $-6 + 7 = 1$; the possible signs of the factors.

Lesson 7.7

Try This, **page 352**

a. To make a rectangle, you have to add neutral pairs of x-tiles. Then the rectangle has dimensions of $x - 4$ and $x + 2$. Check: $(x - 4)(x + 2) = x^2 - 2x - 8$

b. To make a rectangle, arrange the tiles so the dimensions are $x - 2$ and $x - 4$. Since $(-)(-) = (+)$, the units are positive. Check: $(x - 2)(x - 4) = x^2 - 6x + 8$

Ongoing Assessment, **page 353**

The other pairs of factors (1, 12 and 2, 6) will not work because their sum is not 1. Also, using −4 and 3 will not work because that sum is −1.

Communicate, **page 354**

1. Arrange tiles in a rectangle; $(x - 1)(x - 4)$.

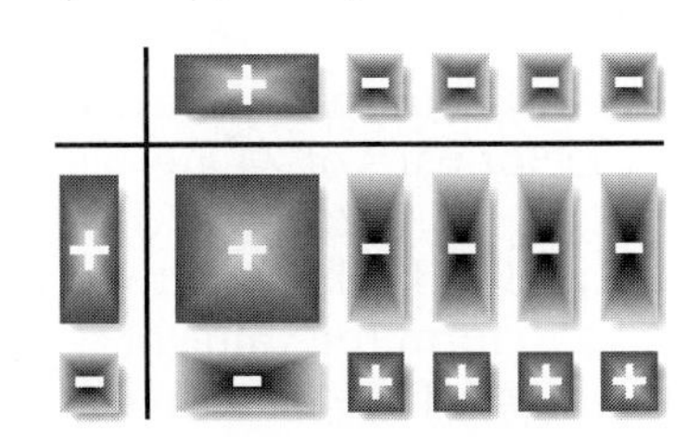

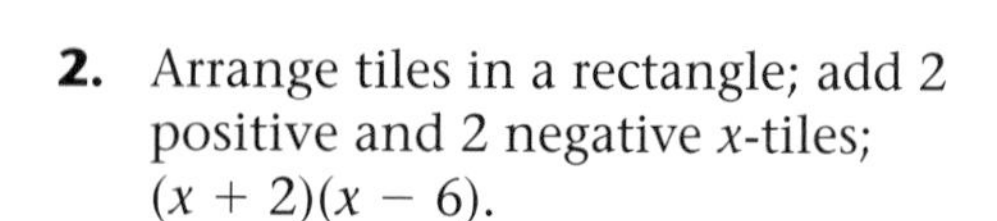

2. Arrange tiles in a rectangle; add 2 positive and 2 negative x-tiles; $(x + 2)(x - 6)$.

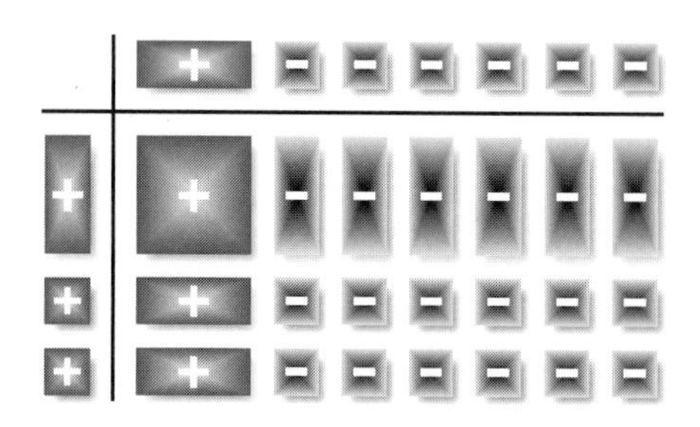

3. Arrange tiles in a rectangle; $(x + 3)(x + 3)$.

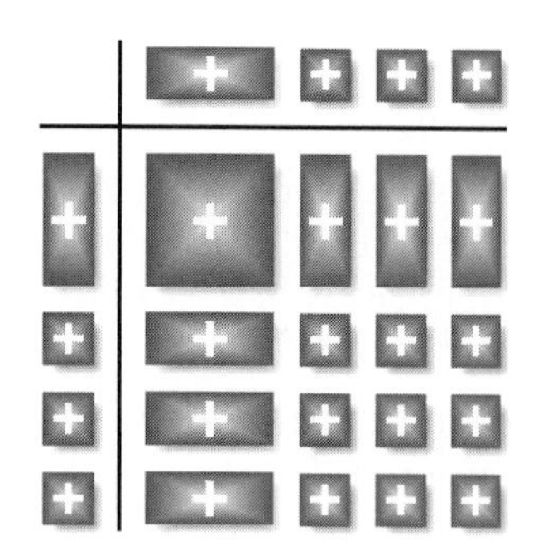

4. $(x + 3)(x - 2)$; the signs are opposite because the third term of the trinomial is negative.
5. $(x - 2)(x - 5)$; the signs are the same because the third term of the trinomial is positive.
6. $(x + 5)(x - 3)$; the signs are opposite because the third term of the trinomial is negative.
7. Determine the factor pairs of the third term. The factor pairs of 24 are $-1, 24$; $1, -24$; $-2, 12$; $2, -12$; $-3, 8$; $3, -8$; $-4, 6$; $4, -6$. Check which factors can be arranged to form the value of -5, the middle term of the trinomial.

Practice & Apply, **pages 354–355**

9. $(x + 2)(x + 3)$;

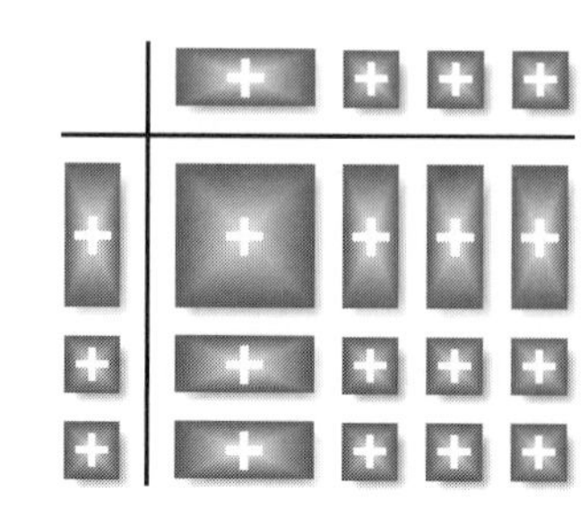

14. $1, -36$; $-1, 36$; $2, -18$; $-2, 18$; $3, -12$; $-3, 12$; $4, -9$; $-4, 9$; $6, -6$; $-6, 6$
15. $1, 54$; $-1, -54$; $2, 27$; $-2, -27$; $3, 18$; $-3, -18$; $6, 9$; $-6, -9$
16. $1, 48$; $-1, -48$; $2, 24$; $-2, -24$; $3, 16$; $-3, -16$; $4, 12$; $-4, -12$; $6, 8$; $-6, -8$
17. $-1, 144$; $1, -144$; $-2, 72$; $2, -72$; $-3, 48$; $3, -48$; $-4, 36$; $4, -36$; $-6, 24$; $6, -24$; $-8, 18$; $8, -18$; $-9, 16$; $9, -16$; $-12, 12$; $12, -12$
18. $(a - 7)(a + 5)$
19. $(p + 6)(p - 2)$
20. $(y - 3)(y - 2)$
21. $(b - 8)(b + 3)$
22. $(n - 2)(n - 9)$
23. $(z - 4)(z + 5)$
24. $(x - 7)(x + 4)$
25. $(s - 21)(s - 3)$
26. $4xy(x - 1)(x - 4)$
27. $6y(y - 1)(y - 2)$
28. $(x - 9)^2$
29. $(a + 6)(a - 1)(a + 3)$
30. $5x(x - 1)(x - 9)$
31. $(x + 6)(x - 6)(x + 2)$
32. $-(x - 2)(x + 2)(x^2 + 2)$
33. $16(2p^2 + 1)(2p^2 - 1)$
34. $(z - 9)(z + 4)$

35. $(x - 1)^2$

36. $5x^2y(5 - x)(5 + x)$

37. $(a + b)(2x + y)$

Look Back, **page 355**

44.

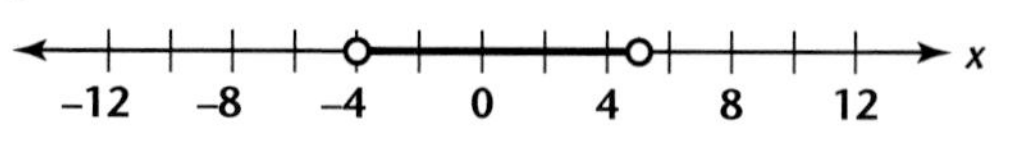

45. $\begin{bmatrix} 3 & 1 \\ 2 & -1 \end{bmatrix}\begin{bmatrix} x \\ y \end{bmatrix} = \begin{bmatrix} 7 \\ 3 \end{bmatrix}$; (2, 1)

Look Beyond, **page 355**

48. $x = 0$

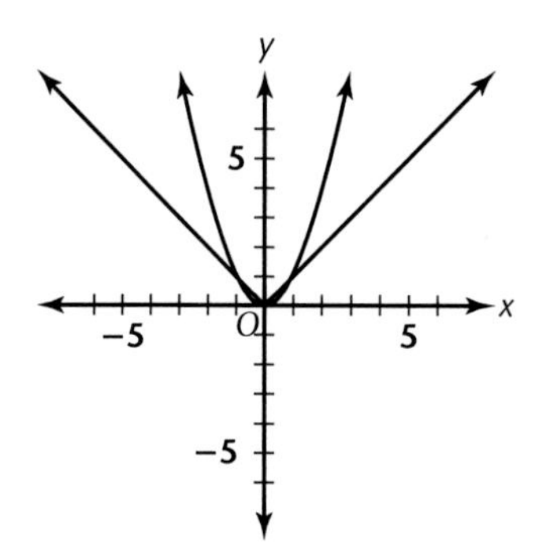

Extra Practice, **page 547**

1. $(x - 5)(x + 2)$; opposite
2. $(x - 7)(x + 4)$; opposite
3. $(x + 4)(x + 5)$; same
4. $(x - 8)(x - 1)$; same
5. $(x - 15)(x + 3)$; opposite
6. $(x + 8)(x + 2)$; same

Chapter 7 Project

Activity 1, **page 356**

1. $a^3 + 3a^2b + 3ab^2 + b^3$
2. $a^4 + 4a^3b + 6a^2b^2 + 4ab^3 + b^4$

   $a^5 + 5a^4b + 10a^3b^2 + 10a^2b^3 + 5ab^4 + b^5$
3. The number of terms is one more than the exponent of the binomial.
4. Eight

5.

1 6 15 20 15 6 1

1 7 21 35 35 21 7 1

1 8 28 56 70 56 28 8 1

1 9 36 84 126 126 84 36 9 1

1 10 45 120 210 252 210 120 45 10 1

6. 1, 7, 21, 35, 35, 21, 7, 1
7. The exponents of a decrease by 1 as you move from left to right. The exponents of b increase by 1 as you move from left to right.
8. $a^7 + 7a^6b + 21a^5b^2 + 35a^4b^3 + 35a^3b^4 + 21a^2b^5 + 7ab^6 + b^7$

Activity 2, **page 357**

Answers may vary.

Activity 3, **page 357**

2. 1
3. 1
4. 2
5. 2
6. $P(\text{3 heads}) = \frac{1}{8}$; $P(\text{3 tails}) = \frac{1}{8}$; $P(\text{2 heads, 1 tail}) = \frac{3}{8}$; $P(\text{2 tails, 1 heads}) = \frac{3}{8}$

Activity 4, **page 357**

1. $(h + t)^3 = h^3 + 3h^2t + 3ht^2 + t^3$
2. $h^3 = \left(\frac{1}{2}\right)^3 = \frac{1}{8}$; $3h^2t = 3\left(\frac{1}{2}\right)^2\left(\frac{1}{2}\right) = \frac{3}{8}$; $3ht^2 = 3\left(\frac{1}{2}\right)\left(\frac{1}{2}\right)^2 = \frac{3}{8}$; $t^3 = \left(\frac{1}{2}\right)^3 = \frac{1}{8}$

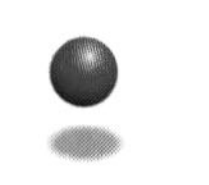

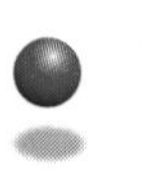

3a. The value of h^3 is the same as the probability of getting 3 heads. The value of $3h^2t$ is the same as the probability of getting 2 heads and 1 tail. The value of $3ht^2$ is the same as the probability of getting 1 head and 2 tails. The value of t^3 is the same as the probability of getting 3 tails.

3b. Answers may vary.

4. $(h + t)^5 = h^5 + 5h^4t + 10h^3t^2 + 10h^2t^3 + 5ht^4 + t^5$; $P(5 \text{ heads}) = \left(\frac{1}{2}\right)^5 = \frac{1}{32}$; $P(4 \text{ heads, } 1 \text{ tail}) = 5\left(\frac{1}{2}\right)^4\left(\frac{1}{2}\right) = \frac{5}{32}$

Chapter 7 Review, pages 358–360

2.

x	y_1	y_2
−3	5	5
−2	0	0
−1	−3	−3
0	−4	−4
1	−3	−3
2	0	0
3	5	5

27. $8x^2(2x + 1)$

28. $3y(3y^6 + 2y^2 + 1)$

29. $b^2(b^4 + 15b - 30)$

30. $8m^3(3m^6 - 2m + 1)$

31. $10a^2(6a^2 + 2a + 1)$

32. $25p(4p^7 - 2p^5 - 1)$

33. $(f + 1)(d + h)$

34. $(y - 3)(3y - 4)$

35. $(x - y)(5 + x)$

36. $(z - 4)(x + y)$

37. $(2 - t)(5 - t)$

38. $(c^2 + 1)(6 + c)$

Chapter 7 Assessment, page 361

12. To multiply $(x + 4)(x - 8)$ using FOIL method, first multiply the **F**irst terms together: $x \cdot x = x^2$. Then multiply the **O**uter terms together: $x \cdot 8 = 8x$. Then multiply the **I**nner terms: $4 \cdot x = 4x$. Then multiply the **L**ast terms: $4 \cdot (-8) = -32$. Add the products: $x^2 + 8x + 4x - 32$. Then simplify: $x^2 + 12x - 32$.

Lesson 8.1

Exploration 1, page 364

4.

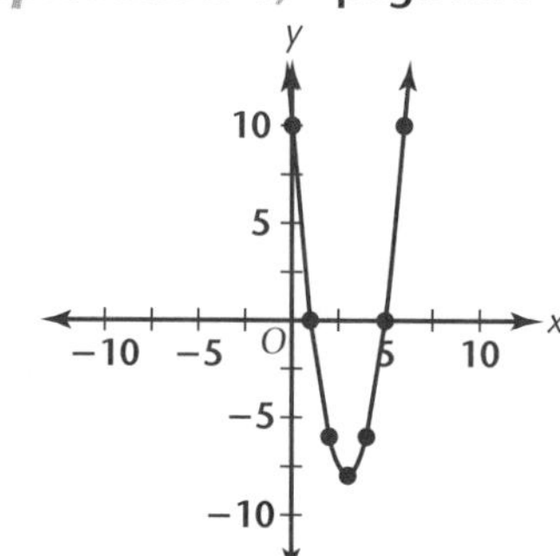

Exploration 2, page 365

3.

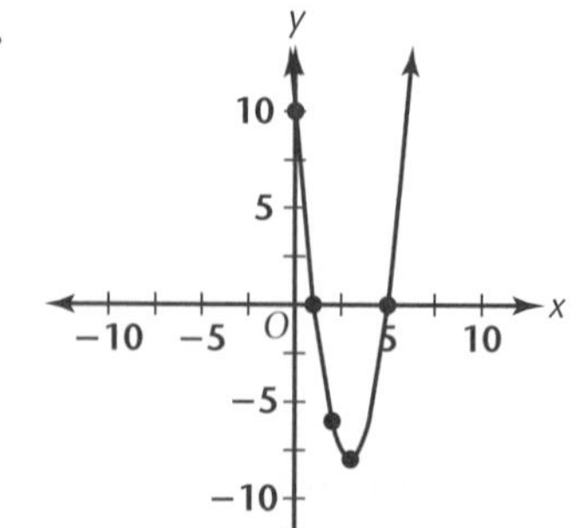

Exploration 3, **page 365**

1.

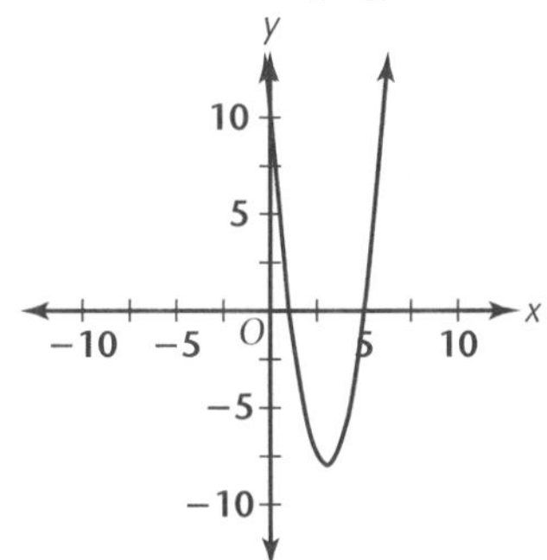

Critical Thinking, **page 365**

The zeros of the function are the beginning and ending times of the rocket's flight.

Communicate, **page 366**

1. The second differences are constant.
2. The parent graph has been stretched by a factor of 2, shifted horizontally 3 units to the right, and shifted vertically 8 units down.
3. The vertex (h, k) is $(3, -4)$.
4. The axis of symmetry, $x = h$, is $x = 3$.
5. Factor the quadratic expression. Then set each linear factor equal to zero and solve these equations. $(x - 4)(x - 4) = 0$; $x = 4$, $x = 4$
6. Find the average of the two zeros. The axis of symmetry is $x = 4$.

Practice and Apply, **pages 366–367**

7. Yes; the second differences are constant.
8. Shift right 2 units, up 3 units
9. Shift right 5 units, down 2 units; stretch by a factor of 3
10. Shift right 2 units, up 1 unit, and reflect with respect to the x-axis
11. Shift left 1 unit
12. Shift right 2 units, up 1 unit, and reflect with respect to the x-axis; stretch by a factor of 3
13. Shift right 2 units, up 3 units; stretch by a factor of $\frac{1}{2}$
14. vertex: $(-4, -3)$; $x = -4$
15. vertex: $(2, 3)$; $x = 2$
16. vertex: $(3, -7)$; $x = 3$
17. vertex: $(5, 2)$; $x = 5$
18. vertex: $(-3, -2)$; $x = -3$
19. vertex: $(-5, 7)$; $x = -5$

29.

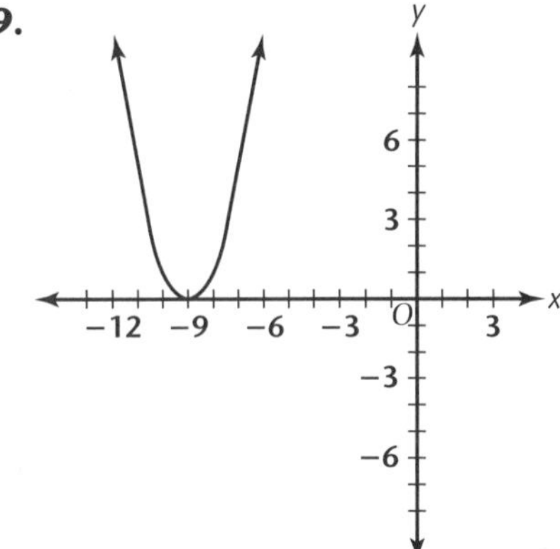

Zero: -9; axis of symmetry: $x = -9$; min.: 0

30.

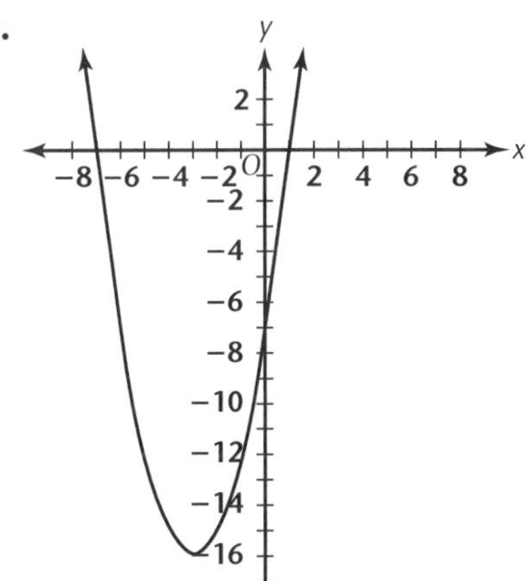

Zeros: 1 and -7; axis of symmetry: $x = -3$; min.: -16

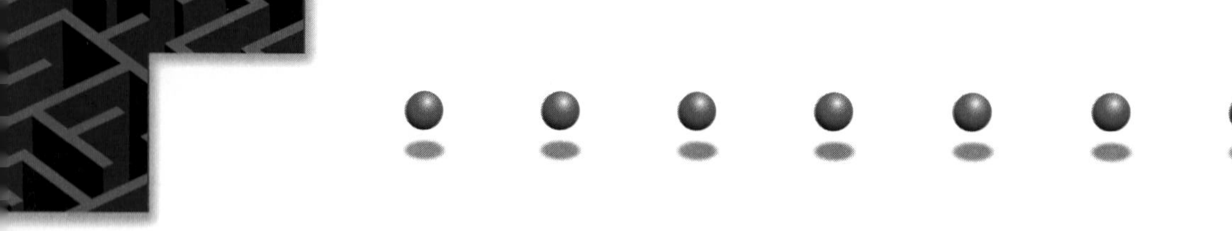

31.

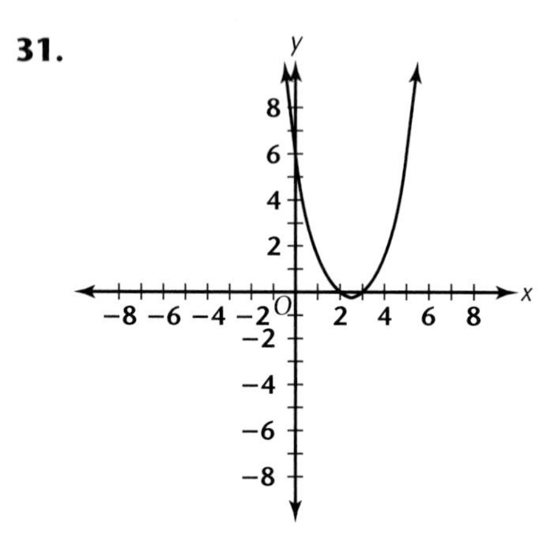

Zeros: 2 and 3; axis of symmetry: $x = 2.5$; min.: -0.25

Look Back, **page 367**

38.

Look Beyond, **page 367**

46. ≈43.9 cubic meters, the minimum volume is the area of the base times the height of the 100,000 cans. $V = \pi r^2 h$, r is the radius of the base of the can and h is the height.

39.

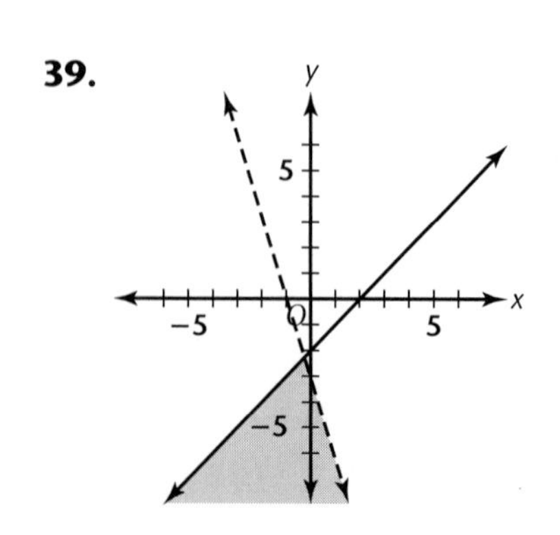

Extra Practice, **page 548**

1. Horizontal shift by 3 units to the right, vertical shift by 1 unit up
2. Horizontal shift by 2 units to the left, vertical shift by 4 units down
3. Horizontal shift by 1 unit to the left, vertical shift by 2 units down, stretch by a factor of 2
4. Horizontal shift by 3 units to the left
5. Horizontal shift by 4 units to the right, opens downward
6. Horizontal shift by 6 units to the right, stretch by a factor of 3
7. Horizontal shift by 1 unit to the right, vertical shift by 1 unit down, stretch by a factor of 2
8. Horizontal shift by 5 units to the right, vertical shift by 2 units up, stretch by a factor of 2, opens downward
9. Horizontal shift by 2 units to the left, vertical shift by 5 units down, stretch by a factor of 5, opens downward
10. Horizontal shift by 1 unit to the right, vertical shift by 4 units up, stretch by a factor of $\frac{1}{2}$
11. Horizontal shift by 9 units to the left, vertical shift by 3 units down, stretch by a factor of $\frac{1}{3}$, opens downward
12. Horizontal shift by 4 units to the right, vertical shift by 5 units down, stretch by a factor of $\frac{2}{5}$

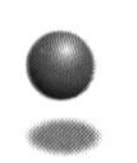
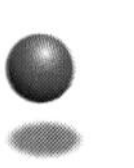

13. $(-1, -2)$, $x = -1$

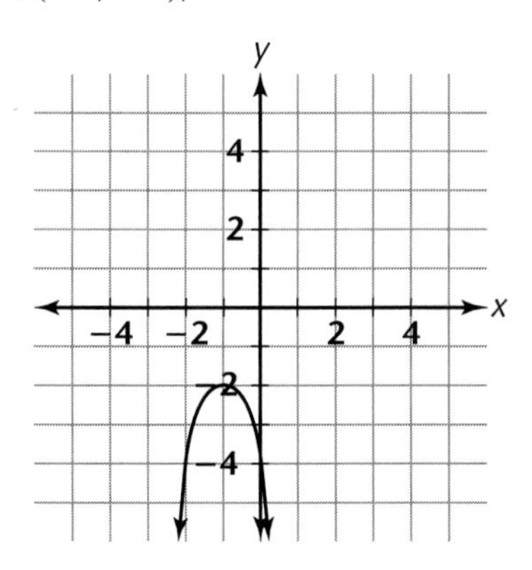

14. $(2, 3)$, $x = 2$

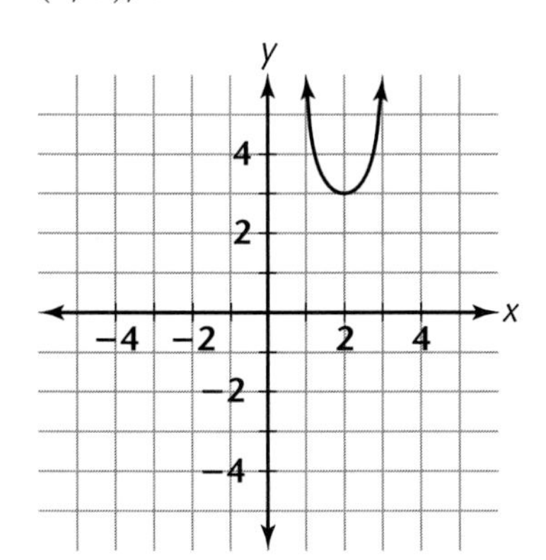

15. $(3, 4)$, $x = 3$

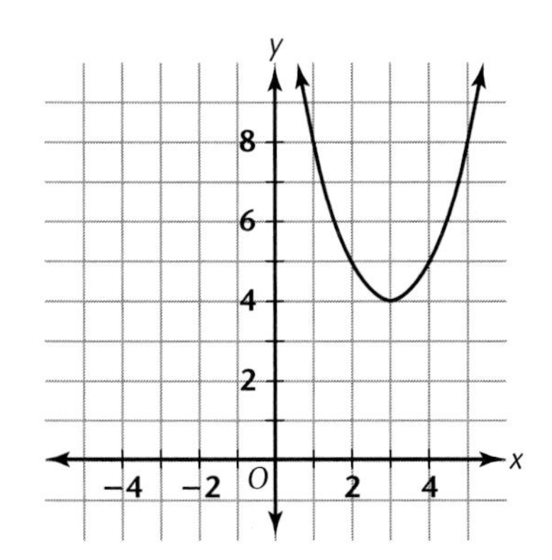

16. $(-2, 1)$, $x = -2$

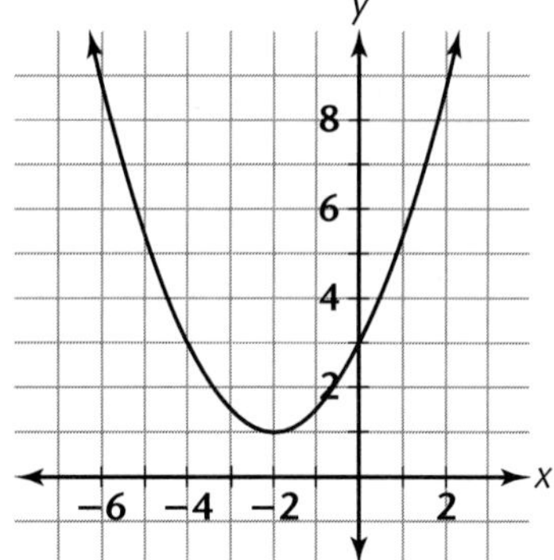

17. $(-3, 0)$, $x = -3$

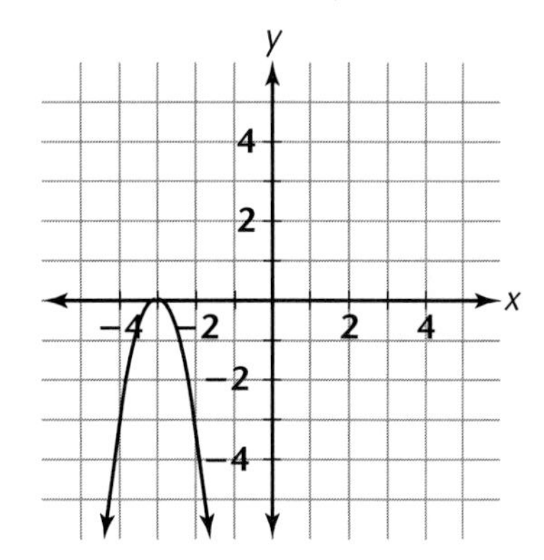

18. $(-1, 4)$, $x = -1$

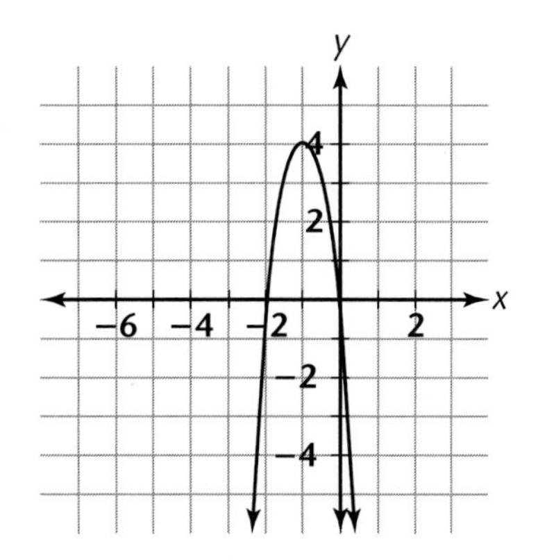

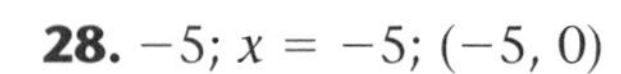

28. -5; $x = -5$; $(-5, 0)$

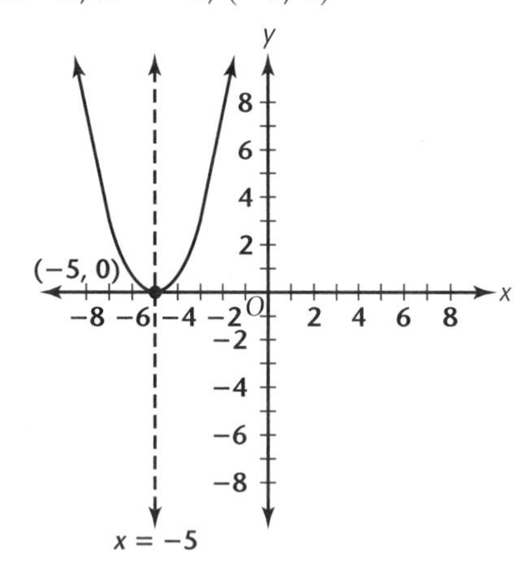

29. $-3, -1$; $x = -2$; $(-2, -1)$

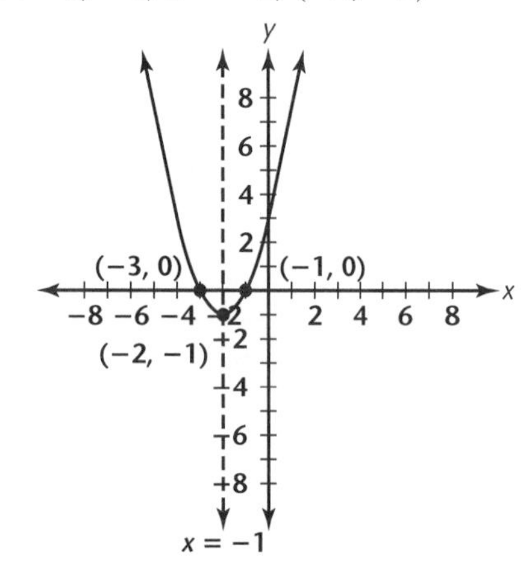

30. $-2, 4$; $x = 1$; $(1, -9)$

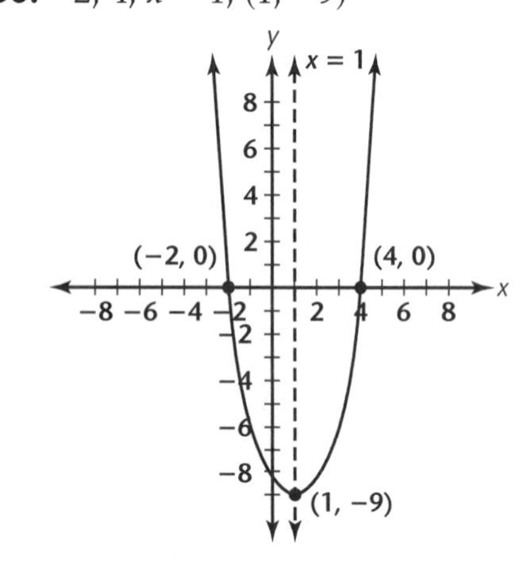

31. $-5, -3$; $x = -4$; $(-4, -1)$

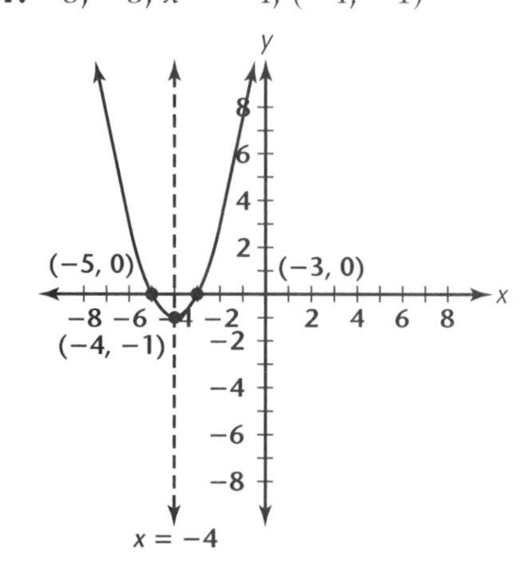

32. $-1, 7$; $x = 3$; $(3, -16)$

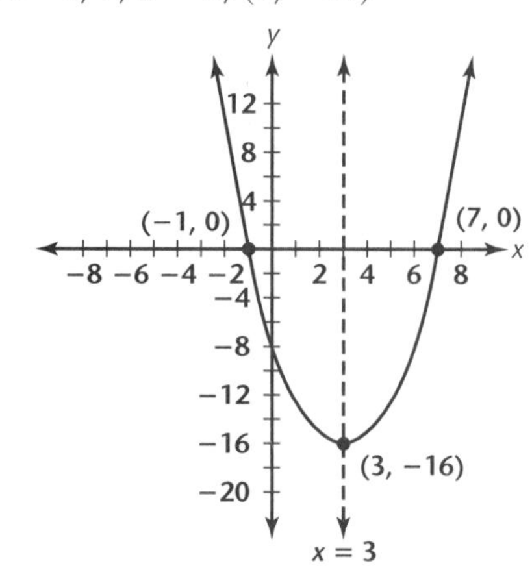

33. $-4, 6$; $x = 1$; $(1, -25)$

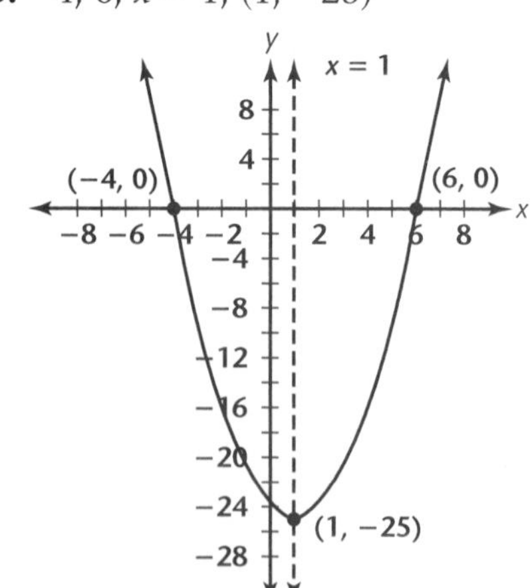

Lesson 8.2

Ongoing Assessment, page 368

Make another table that uses 3.40 to 3.50 for values of t.

Communicate, page 372

1. The value for x for which y is 0 is the point at which the function crosses the x-axis. If the function represents a real relationship such as between time and height, negative values may have no meaning.
2. Find the numbers which when squared give 100; positive or negative.
3. Since $x^2 = 64$ and $64 \geq 0$, $x = \pm 8$.
4. Since $x^2 = 8$ and $8 \geq 0$, $x = \pm\sqrt{8}$.
5. Since $x^2 = \frac{16}{100}$ and $\frac{16}{100} \geq 0$, $x = \pm\frac{4}{10}$.
6. Add 25 to both sides of the equation. Solve the resulting equation.
   $(x + 3)^2 = 25$
   $x + 3 = \pm 5$
   $x = 2$ or $x = -8$
7. $(x - 8)^2 = 2$
   $x - 8 = \pm\sqrt{2}$
   $x = 8 \pm \sqrt{2}$
8. Plot the vertex at $(-4, -5)$. Solve the equation $(x + 4)^2 - 5 = 0$ to find the zeros. Mark these two points on the x-axis. Sketch the graph by drawing a parabola passing through the three points plotted and symmetrical about the axis of symmetry, $x = -4$.

Practice and Apply, pages 372–373

9.

t	y
5	−50
4	76
3	174
2	244
1	286
0	300

t	y
−1	286
−2	244
−3	174
−4	76
−5	−50

The graph crosses the x-axis at about 4.5 and −4.5.

46. vertex = $(4, -9)$; $x = 4$; zeros: 1, 7

47. vertex = $(-2, -1)$; $x = -2$; zeros: −3, −1

48. vertex = $(4, -3)$; $x = 4$; zeros: ≈2.27, ≈5.73

52.

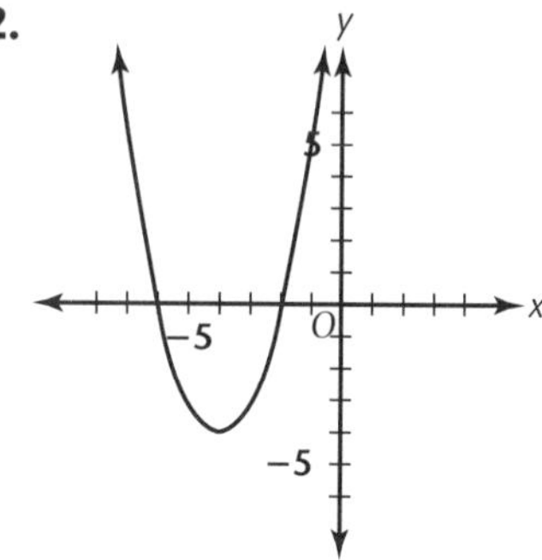

53.

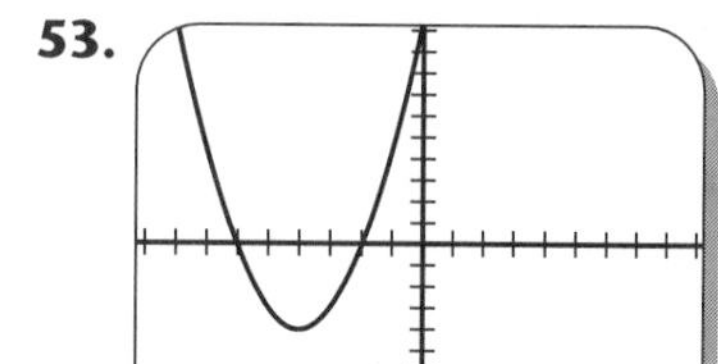

Look Back, page 373

65. $(a + b)(a - b)$; $(a^2 + b^2)(a + b)(a - b)$; $(a^4 + b^4)(a^2 + b^2)(a + b)(a - b)$; 7; 64 is 2^6, so $(a^{64} - b^{64})$ has 6 + 1 factors.

Extra Practice, page 548

28. $(9, -16)$; $x = 9$; 5, 13

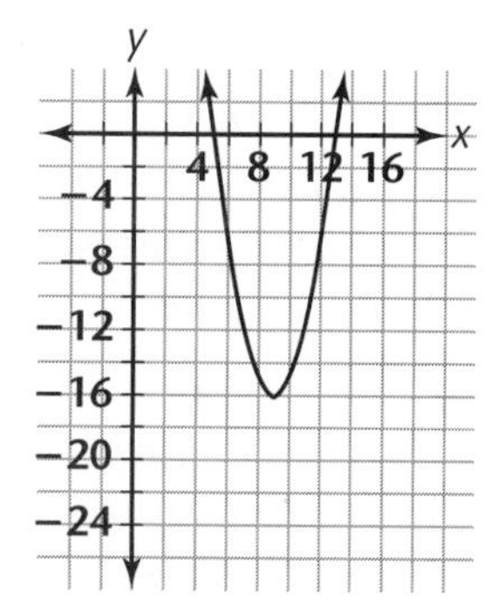

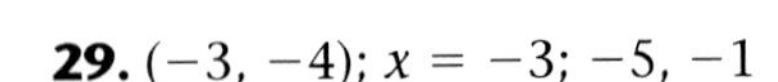

29. $(-3, -4)$; $x = -3$; -5, -1

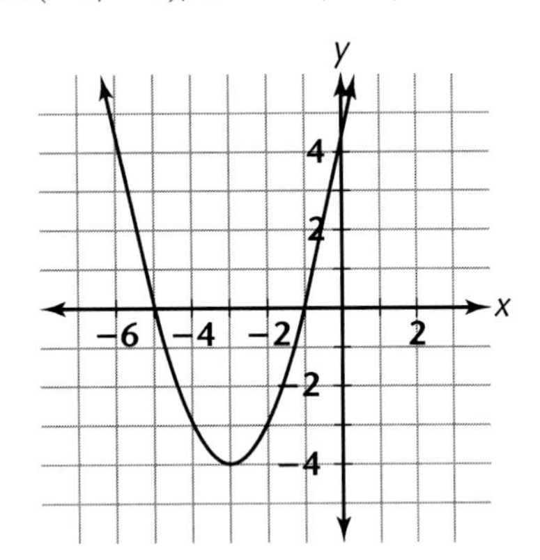

30. $(4, -2)$; $x = 4$; 2.59, 5.41

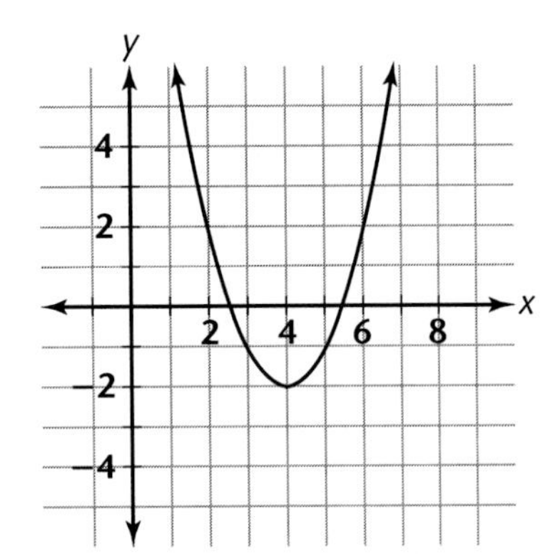

Lesson 8.3

Communicate, **page 377**

1. Start with one x^2-tile, arrange the four x-tiles in two groups of 2. Four 1-tiles are needed to complete the square.
2. Find half of the coefficient of x and then add its square to the expression. Factor the trinomial into a perfect square. $x^2 - 7x + \frac{49}{4} = \left(x - \frac{7}{2}\right)^2$
3. The first three terms are a perfect square trinomial, so the equation can be written as $y = (x + 5)^2 - 25$.
4. Write $y = x^2 + 2$ in the form $y = (x - h)^2 + k$. h is the x-value and k is the y-value of the vertex. Since $y = (x - 0)^2 + 2$; the minimum value is 2.
5. Complete the square to find the vertex. Since $y = \left(x - \frac{1}{2}\right)^2 - \frac{1}{4}$, the vertex is $\left(\frac{1}{2}, -\frac{1}{4}\right)$, and the minimum value is $-\frac{1}{4}$.
6. Write $y = x^2 - 3$ in $y = (x - h)^2 + k$ form. Since $y = (x - 0)^2 - 3$, the vertex is $(0, -3)$, and the minimum value is -3.
7. $y = x^2 - 10x + 11$
   $= x^2 - 10x + 25 - 25 + 11$
   $= (x - 5)^2 - 14$
   The vertex is $(5, -14)$; $y = (x - 5)^2 - 14$ is in the form $y = (x - h)^2 + k$; h is the x-value of the vertex and k is the y-value of the vertex.

Practice and Apply, **pages 378–379**

8. $x^2 + 14x + 49$; $(x + 7)^2$
9. $x^2 - 14x + 49$; $(x - 7)^2$
10. $x^2 + 8x + 16$; $(x + 4)^2$
11. $x^2 - 8x + 16$; $(x - 4)^2$
12. $x^2 + 6x + 9$; $(x + 3)^2$
13. $x^2 - 2x + 1$; $(x - 1)^2$
14. $x^2 + 12x + 36$; $(x + 6)^2$
15. $x^2 - 12x + 36$; $(x - 6)^2$
16. $x^2 + 7x + \frac{49}{4}$; $\left(x + \frac{7}{2}\right)^2$
17. $x^2 - 10x + 25$; $(x - 5)^2$
18. $x^2 + 15x + \frac{225}{4}$; $\left(x + \frac{15}{2}\right)^2$
19. $x^2 - 5x + \frac{25}{4}$; $\left(x - \frac{5}{2}\right)^2$
20. $x^2 + 16x + 64$; $(x + 8)^2$
21. $x^2 + 20x + 100$; $(x + 10)^2$
22. $x^2 - 9x + \frac{81}{4}$; $\left(x - \frac{9}{2}\right)^2$
23. $x^2 + 40x + 400$; $(x + 20)^2$
24. $y = (x + 4)^2 - 16$
25. $y = (x - 2)^2 - 4$

26. $y = (x - 5)^2 - 25$

27. $y = (x + 7)^2 - 49$

28. $y = (x - 8)^2 - 64$

29. $y = (x + 10)^2 - 100$

45.

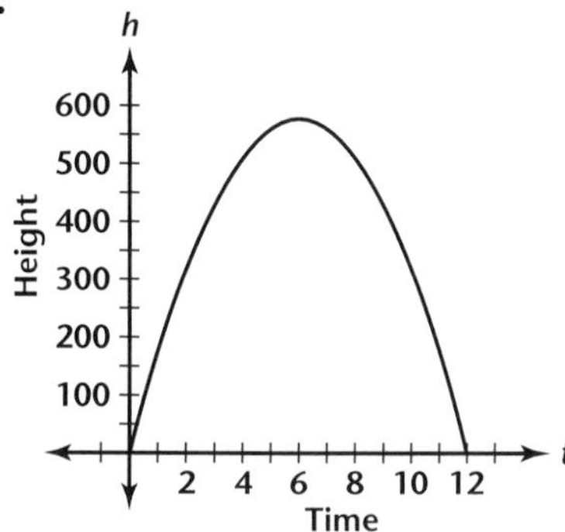

Look Back, **page 379**

53.

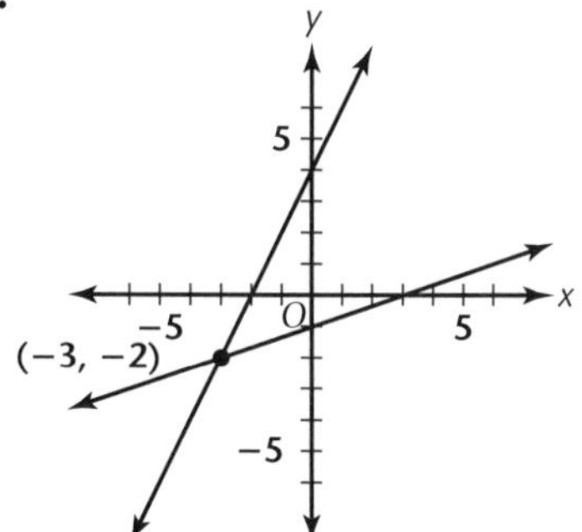

54.

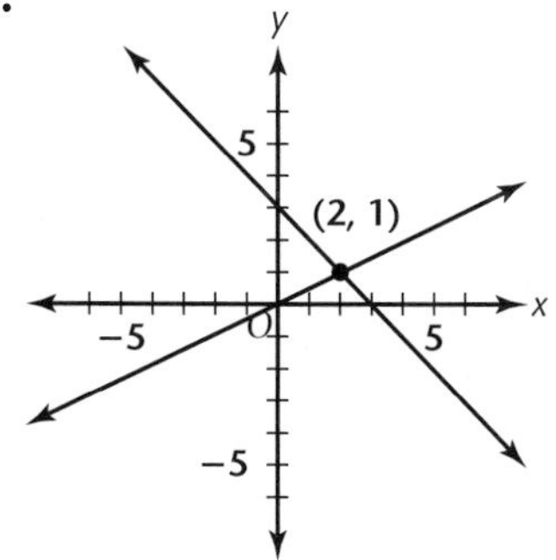

Look Beyond, **page 379**

61.

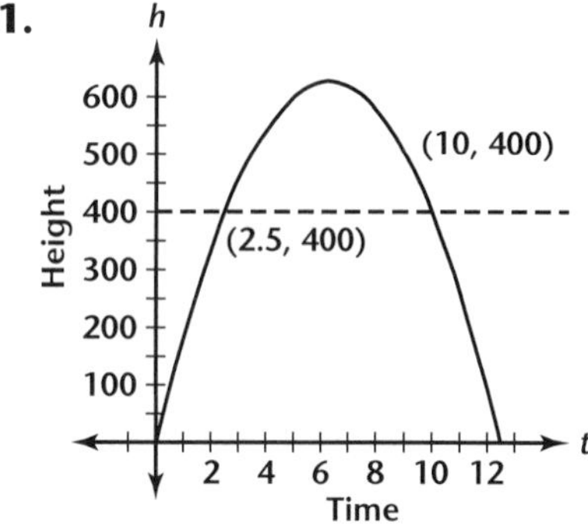

Eyewitness Math, **pages 380–381**

1a. 4 seconds

1b. 1500 feet

1c. No, 1500 ft/4 s = 375 ft/s or 256 mph

2a. 135 mph = 198 ft/s; time to impact at 3500 ft: (3500 ft)/(198 ft/sec) or 17.7 sec; time to impact at 2000 ft: (2000 ft)/(198 ft/sec) or 10.1 sec

2b. 17.7 sec − 10.1 sec = 7.5 sec (or about 7 to 8 seconds)

3. $t = 200/(v_r - v_w)$

4a. $t = 200/(v_r - v_w) = 200/(206 - 198) = 200/8 = 25$; 25 seconds

4b. 3350 feet

5. Answers will vary, but numbers used should be consistent with figures given in the lesson and with each other. For example, speeds, times, and distances should satisfy basic equations of motion.

Extra Practice, **page 549**

13. $y = (x - 3)^2 - 9$; $(3, -9)$

14. $y = (x - 6)^2 - 36$; $(6, -36)$

15. $y = (x + 1)^2 - 1$; $(-1, -1)$

16. $y = (x + 3)^2 - 4$; $(-3, -4)$

17. $y = (x - 1)^2 - 4$; $(1, -4)$

18. $y = (x - 4)^2 - 4$; $(4, -4)$

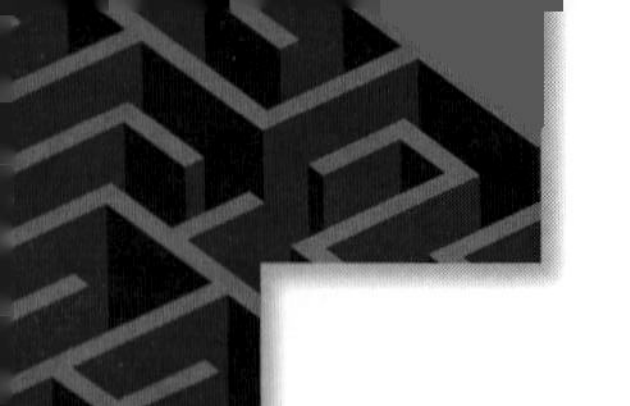

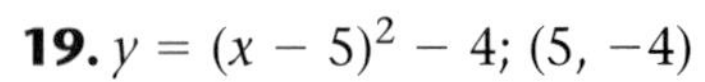

19. $y = (x - 5)^2 - 4; (5, -4)$
20. $y = (x - 6)^2 - 4; (6, -4)$
21. $y = (x + 2.5)^2 - 2.25; (-2.5, -2.25)$
22. $y = (x + 3)^2 - 16; (-3, -16)$
23. $y = (x + 3.5)^2 - 0.25; (-3.5, -0.25)$
24. $y = (x - 2)^2 - 28; (2, -28)$
25. $y = \left(x + \frac{1}{4}\right)^2 - 2\frac{1}{16}; \left(-\frac{1}{4}, -2\frac{1}{16}\right)$
26. $y = \left(x - \frac{1}{8}\right)^2 + 5\frac{63}{64}; \left(\frac{1}{8}, 5\frac{63}{64}\right)$
25. $y = \left(x - \frac{1}{6}\right)^2 + 4\frac{35}{36}; \left(-\frac{1}{6}, -4\frac{35}{36}\right)$

Lesson 8.4

Communicate, page 387

1. They have the same value.
2. Complete the square on the first two terms of the trinomial. If $f(x) = 0$, $0 = (x - 3)^2 - 1$. $\pm\sqrt{1} = x - 3$; $x = 2$ or $x = 4$
3. Take half the coefficient of x and square it. Add this number to both sides of the equation. Express the perfect square on the left side as a square of a linear factor and simplify the integers on the right side. Solve the resulting equation; $x = 5$ or 3
4. $$\begin{aligned} x^2 + 10x &= 24 \\ x^2 + 10x - 24 &= 0 \\ (x + 12)(x - 2) &= 0 \\ x = -12 \text{ or } x &= 2 \end{aligned}$$
5. Factoring, because this is a perfect square trinomial.
6. Factoring, because this trinomial has two linear factors.
7. Solve the equation $2 = x^2 - 7x + 12$. Solve by factoring or find the intersection of the graphs of $f(x) = x^2 - 7x + 12$ and $f(x) = 2$.
8. Graph the line $y = x - 1$ and the parabola $y = x^2 - 3x + 2$ and observe where they intersect.

Practice and Apply, pages 387–389

30. 0, −10
31. 4, 6
32. $\approx 2.30, \approx -1.30$
33. 2, −6
34. $\approx 1.19, \approx -4.19$
35. $\approx 3.30, \approx -0.30$
36. $\approx 5.54, \approx -0.54$
37. 0, 1

Look Back, page 389

52. $\begin{bmatrix} 1 & -2 \\ 4 & 2 \end{bmatrix} \begin{bmatrix} x \\ y \end{bmatrix} = \begin{bmatrix} 1 \\ -1 \end{bmatrix}$

53. $\begin{bmatrix} 5 & -2 \\ 3 & 5 \end{bmatrix} \begin{bmatrix} x \\ y \end{bmatrix} = \begin{bmatrix} 11 \\ 19 \end{bmatrix}$

Look Beyond, page 389

66. $i^1 = i, i^2 = -1, i^3 = -i, i^4 = 1, i^5 = i, i^6 = -1$; the pattern has a cycle of 4. Since $92 \div 4 = 23$ without a remainder (divides evenly), i^{92} will have the same value as i^4; $i^{92} = 1$.

Lesson 8.5

Exploration, page 393

2. If $b^2 - 4ac > 0$, there are two solutions. If $b^2 - 4ac = 0$, there is one solution. If $b^2 - 4ac < 0$, there are no real solutions (the graph does not cross the x-axis).

Communicate, **page 395**

1. When the equation is written in the form $2n^2 + 3n - 7 = 0$, the coefficient of n^2 is a, the coefficient of n is b, and the constant term is c.
2. In $b^2 - 4ac$, substitute $a = 2$, $b = 3$, and $c = -7$.
3. Find the value of $b^2 - 4ac$ when $a = 2$, $b = -1$, and $c = -2$. Since this value is greater than zero (17), the equation has two solutions.
4. Find the value of $b^2 - 4ac$ when $a = 1$, $b = -2$, and $c = 7$. Since this value is less than zero (−24), the equation has no solutions.
5. Rewrite the equation so that it equals zero. Use $a = 4$, $b = 12$, and $c = -7$ in the quadratic formula.
6. The factors are $\left(x + \frac{1}{2}\right)\left(x + \frac{5}{2}\right)$. Multiply each expression by 2 to eliminate the fractions. The equation is $(2x + 1)(2x + 5) = 0$.

Practice and Apply, **pages 395–396**

7. $a = 1, b = -4, c = -5$
8. $a = 1, b = -5, c = -4$
9. $a = 5, b = -2, c = -1$
10. $a = -1, b = -3, c = 7$
11. $a = 3, b = 0, c = -45$
12. $a = -1, b = -3, c = 7$
13. 13; 2 real solutions
14. 36; 2 real solutions
15. 12; 2 real solutions
16. 100; 2 real solutions
17. 128; 2 real solutions
18. 64; 2 real solutions
19. 89; 2 real solutions
20. −47; no real solution
21. 0; 1 real solution

Look Back, **page 396**

54.

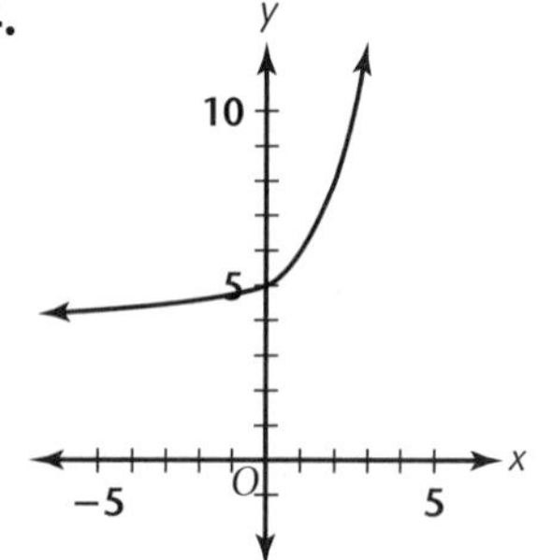

55.

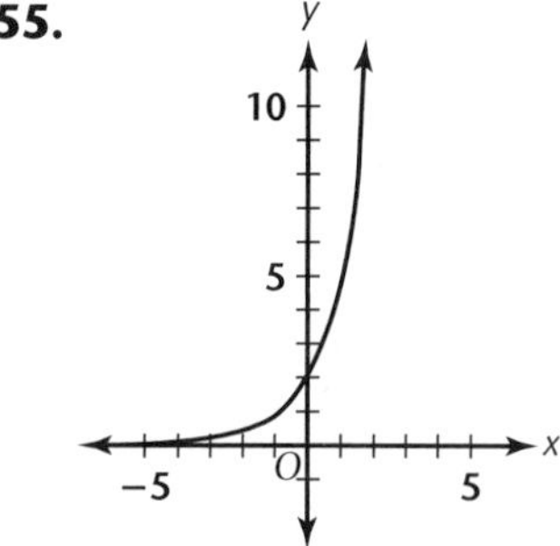

56.

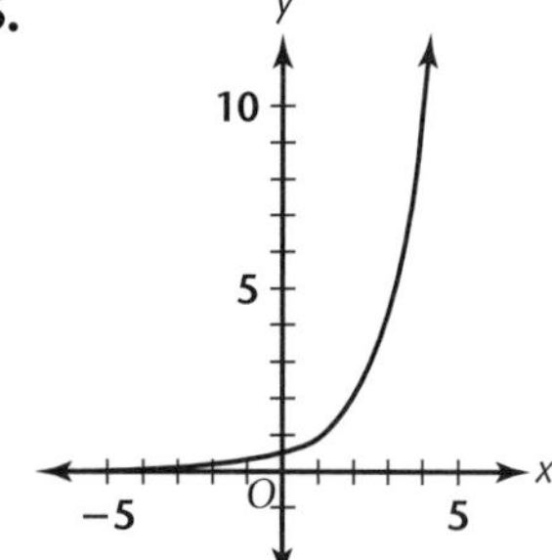

Look Beyond, **page 396**

60. First graph the parabola $f(x) = x^2 + 3x + 2$; use a dashed line for the parabola and choose test points to determine where to shade.

Lesson 8.6

Try This, **page 398**

$x \le -4$ or $x \ge 3$

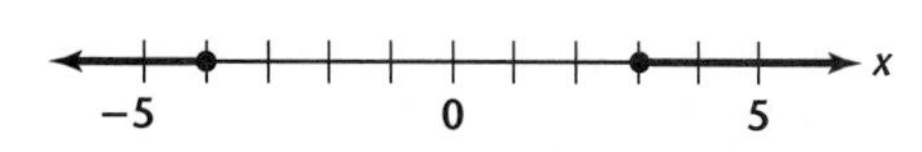

Communicate, **page 400**

1. Factor; solve $x^2 - 9x + 18 = 0$ for x: $x = 3$ or $x = 6$. Plot 3 and 6 with open circles on the number line. Test each interval. Graph $x < 3$ or $x > 6$.
2. Since the inequality does not include *equal to,* circle the two solutions found from Exercise 1. Test values for x in $x^2 - 9x + 18 > 0$ from each of the three intervals created on the number line to see which range of values makes the inequality true.
3. The boundary is dashed since the inequality does not include *equal to.*
4. The boundary is solid because the inequality includes *equal to.*
5. The boundary is solid because the inequality includes *equal to.*
6. First sketch the parabola $y = x^2 - 3$ with a broken line. Test one point between the zeros of the function. If the resulting inequality is true, shade that side of the parabola. If the test is false, shade the other side.
7. Solve Exercise 7 using the same steps as described in Exercise 6, except graph the parabola using solid lines.
8. Solve Exercise 8 using the same steps as described in Exercise 6, except graph the parabola using solid lines.

Practice and Apply, **page 401**

9. $x < 2$ or $x > 4$

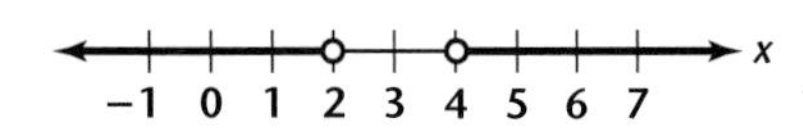

10. $-2 < x < -1$

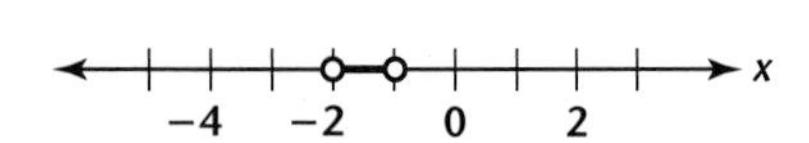

11. $2 \le x \le 7$

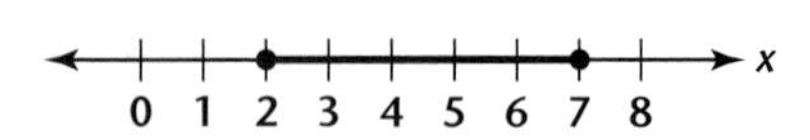

12. $x < -5$ or $x > 3$

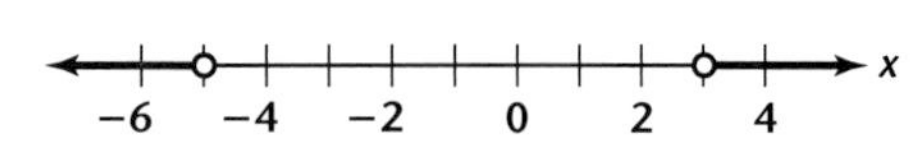

13. $1 < x < 6$

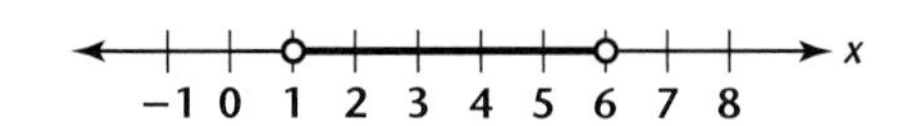

14. $-2 < x < 10$

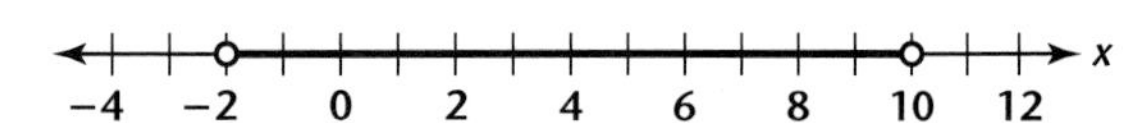

15. $3 \le x \le 5$

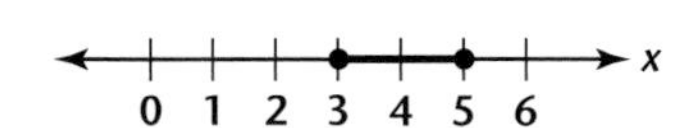

16. $-3 \le x \le 6$

17. $x \le 5$ or $x \ge 6$

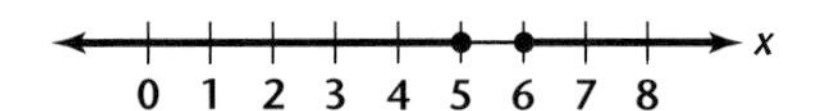

18. $x \ge 6$ or $x \le -2$

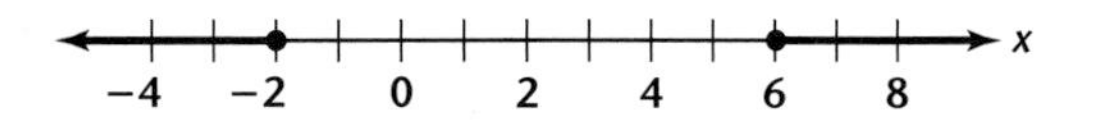

19. $x < -7$ or $x > 2$

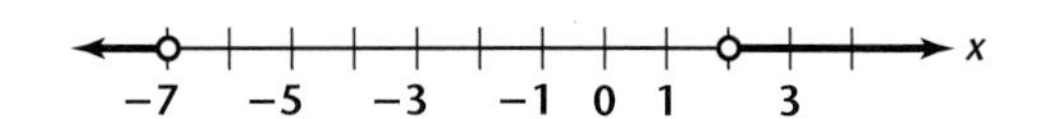

20. $x \le -6$ or $x \ge -4$

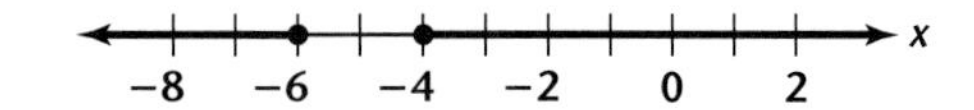

21.

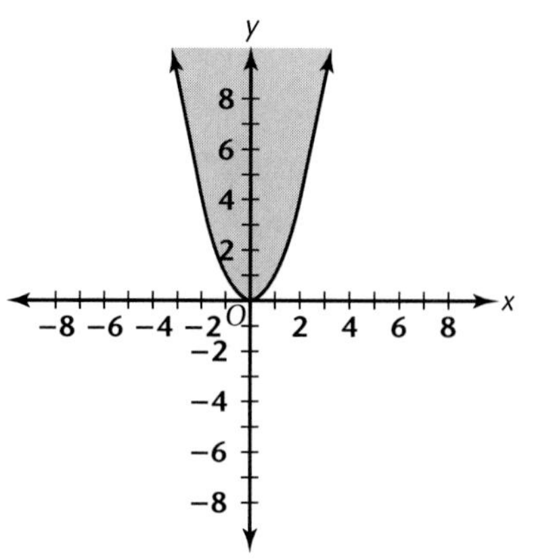

22.

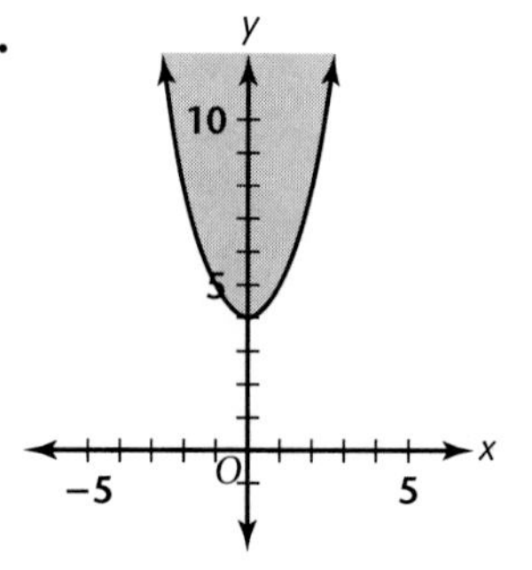

23.

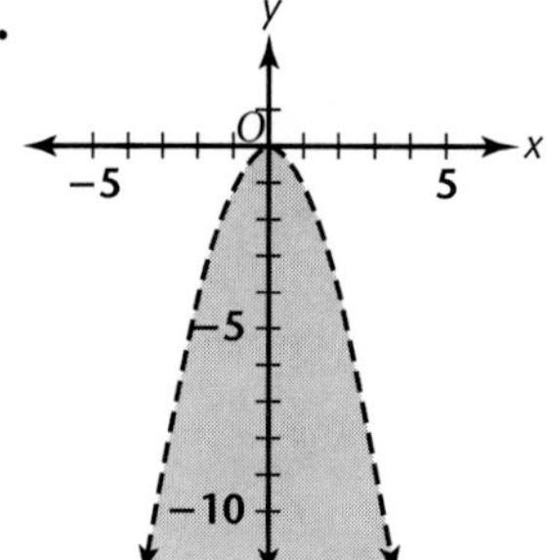

24.

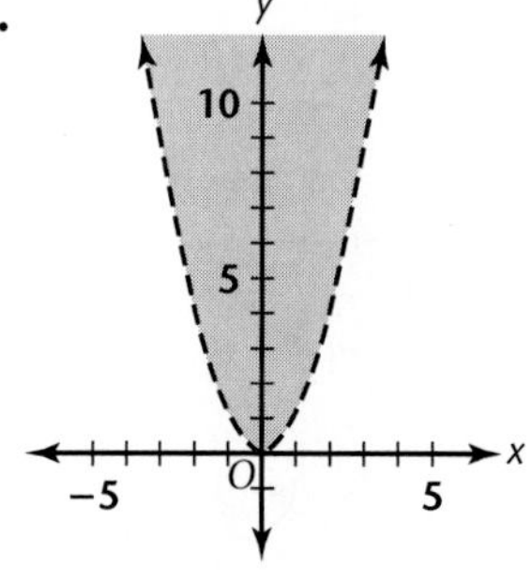

25.

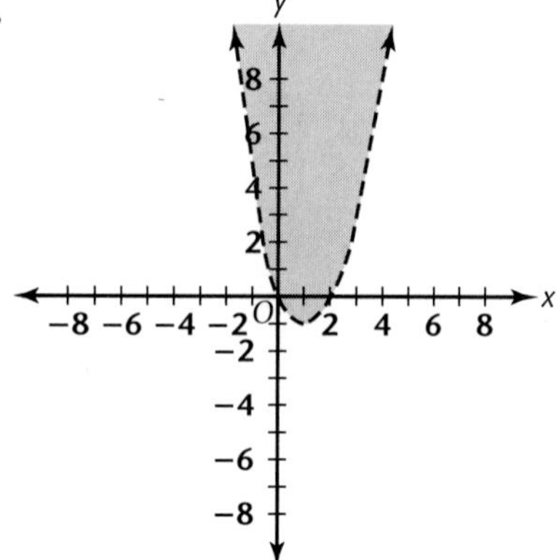

26.

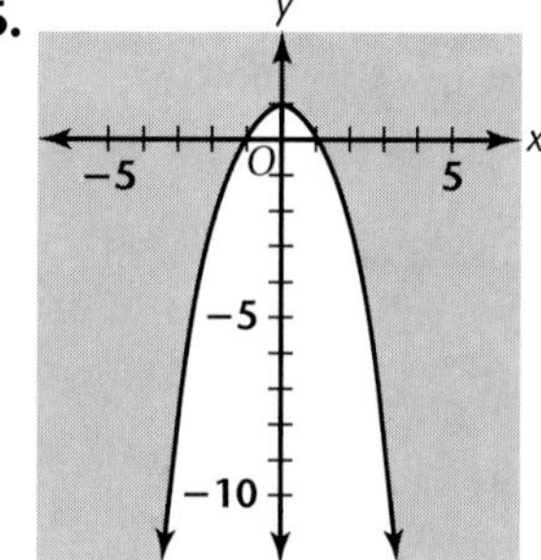

27.

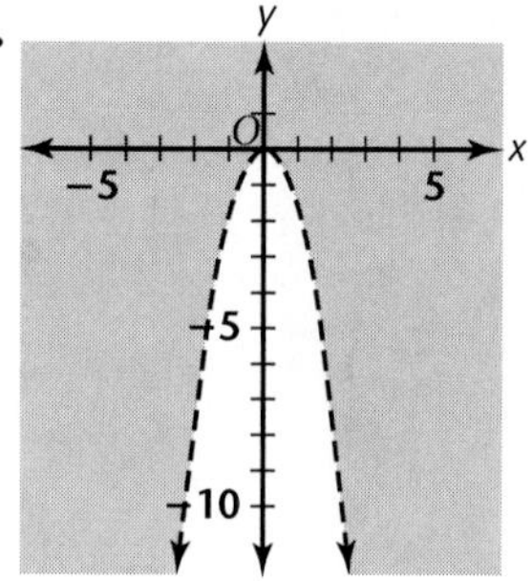

28.

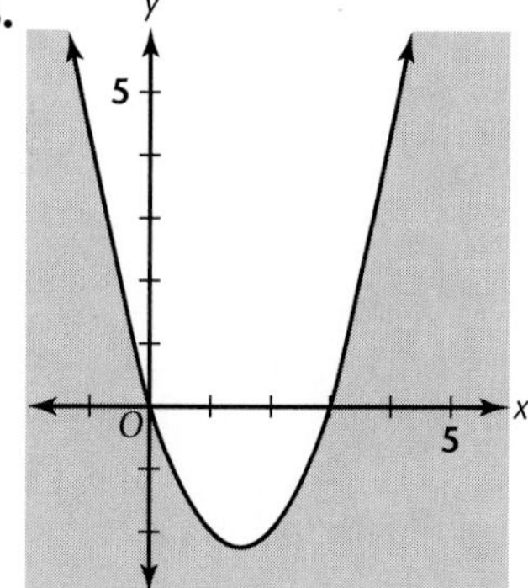

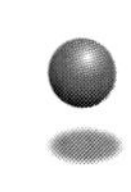

29.

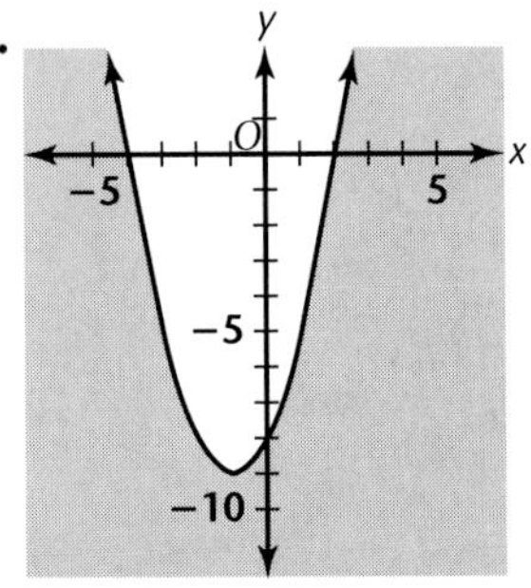

30.

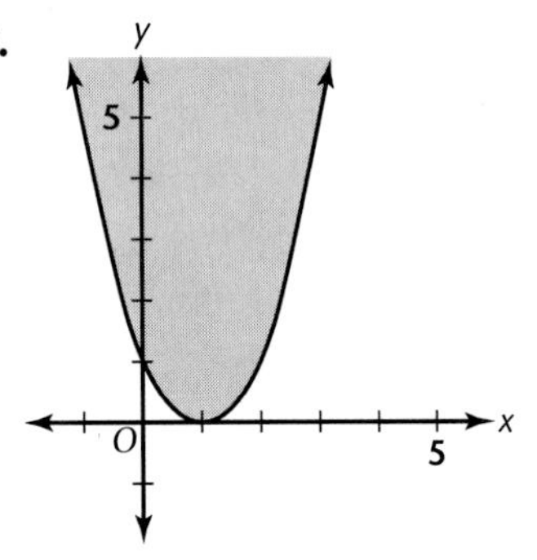

31.

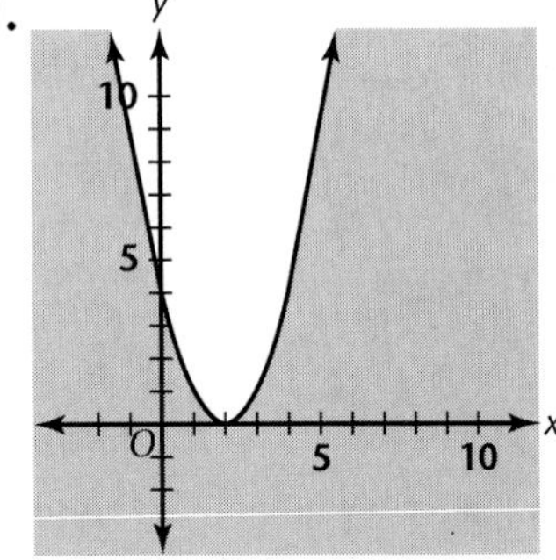

32.

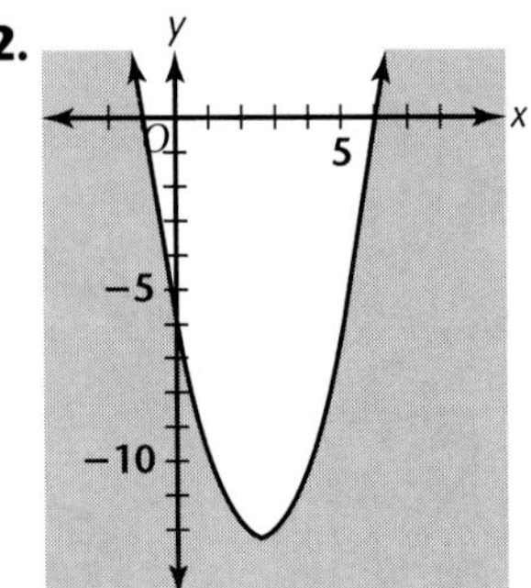

33. Less than 4 seconds and greater than 16 seconds

34. Between 4 and 16 seconds

35.

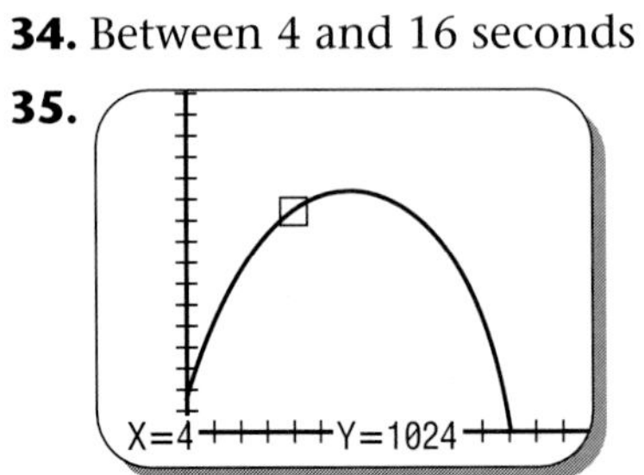

Extra Practice, **page 550**

13.

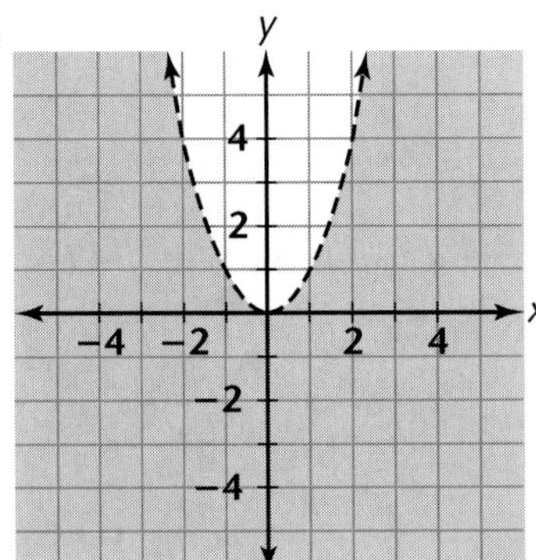

14.

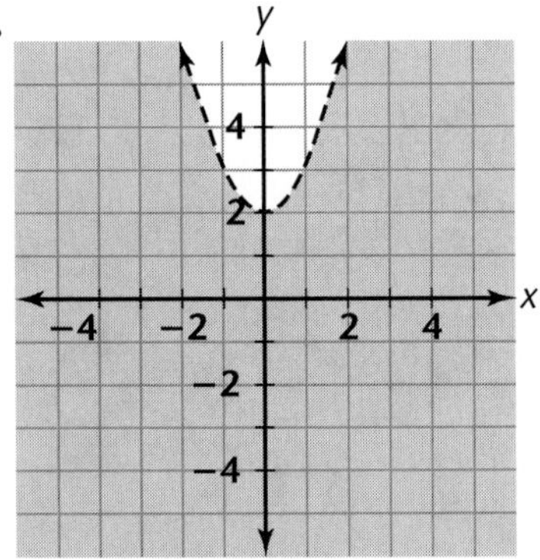

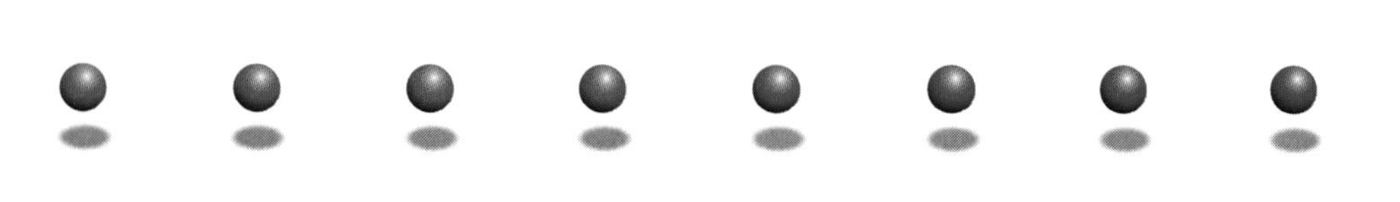

15.

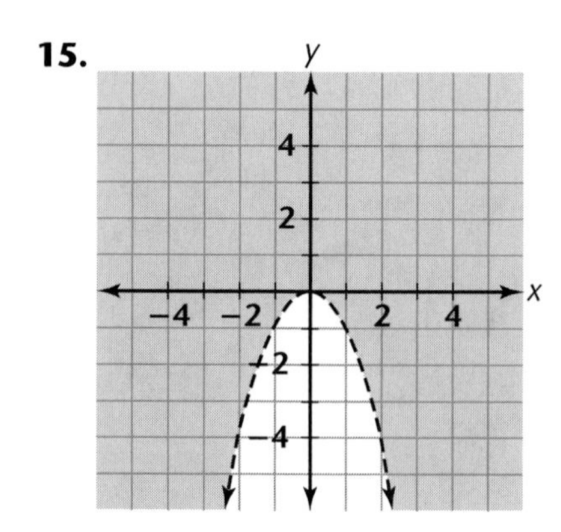

16.

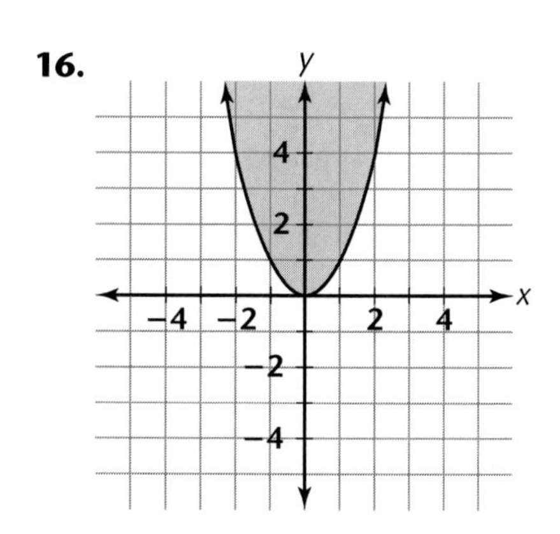

17.

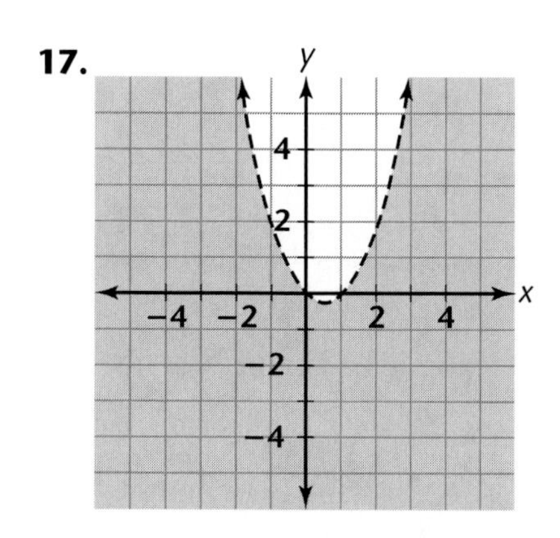

18.

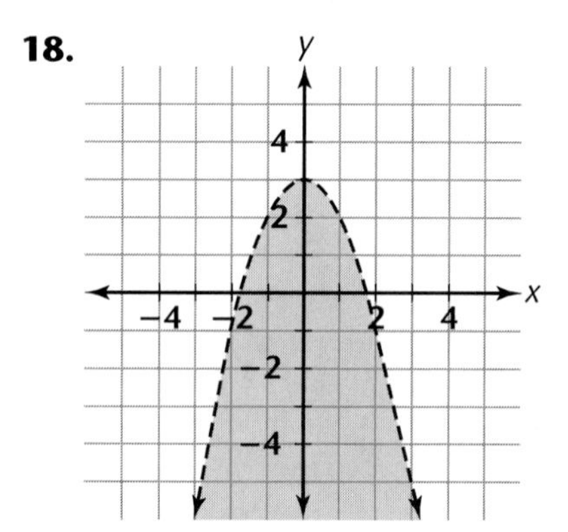

19.

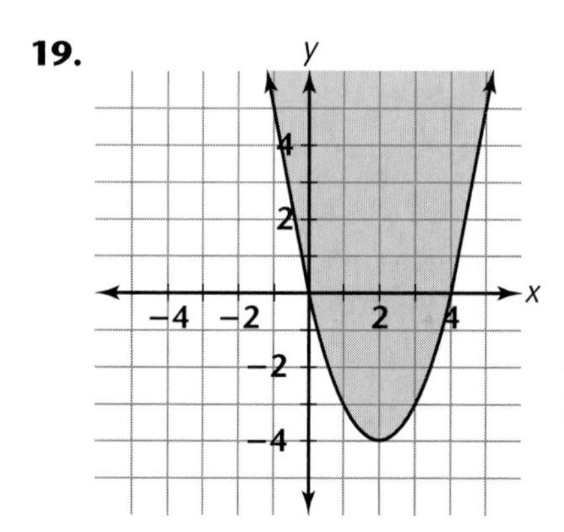

20.

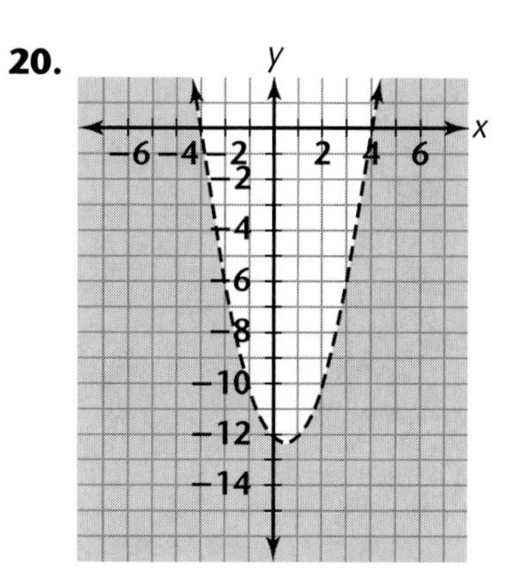

21.

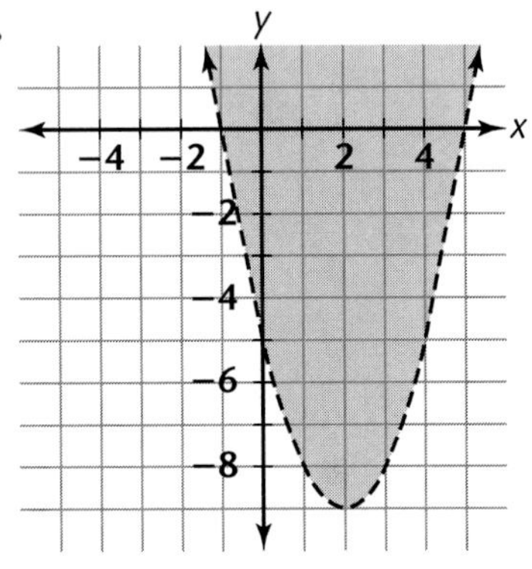

Chapter 8 Project

Activity, **page 403**

Answers may vary.

Chapter 8 Review, pages 404–406

1. Vertex: $(-1, -4)$; axis of symmetry: $x = -1$; zeros: 1, -3

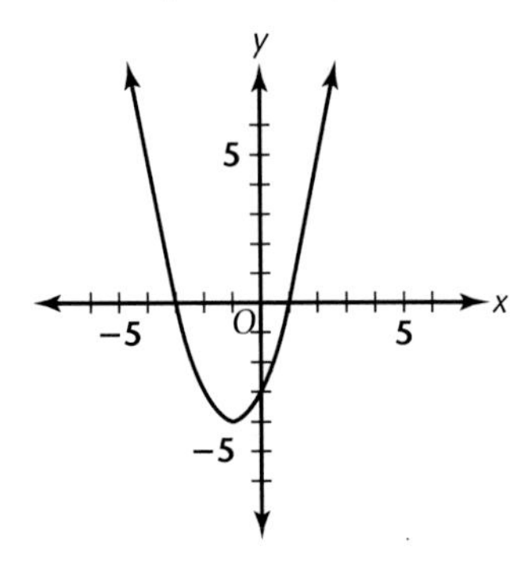

2. Vertex: $(5, -1)$; axis of symmetry: $x = 5$; zeros: 4, 6

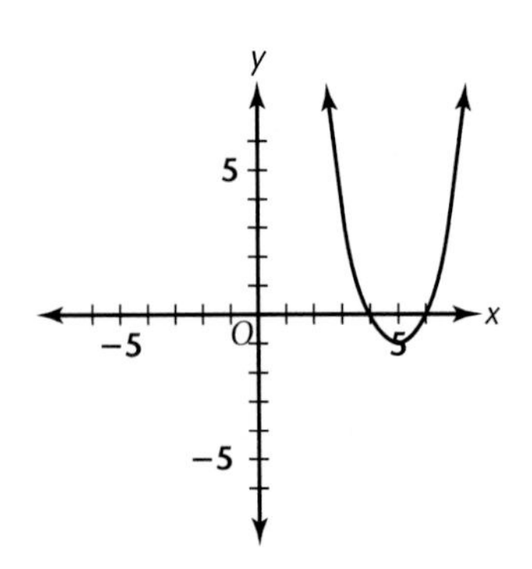

3. Vertex: $(3, -2)$; axis of symmetry: $x = 3$; zeros: 2, 4

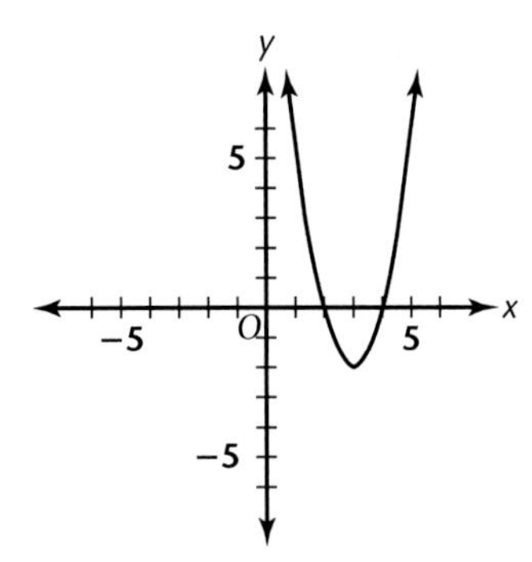

20. -32; no real solution

21. 89; two real solutions: ≈ 1.80, -0.55

22. 0; one real solution: 0.5

23. 9; two real solutions: 3, 6

24. 121; two real solutions: 1.5, ≈ -0.33

25. 36; two real solutions: 0.25, -0.5

26. $x \leq -4$ or $x \geq -2$

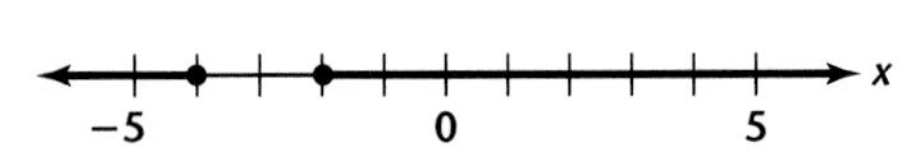

27. $-2 < x < 5$

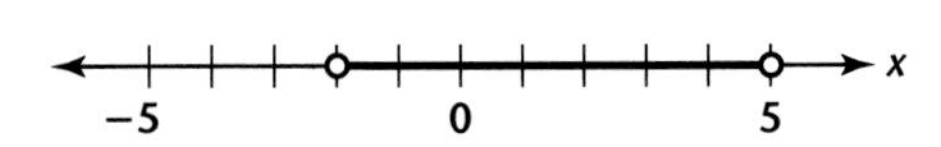

28. $x < -6$ or $x > -3$

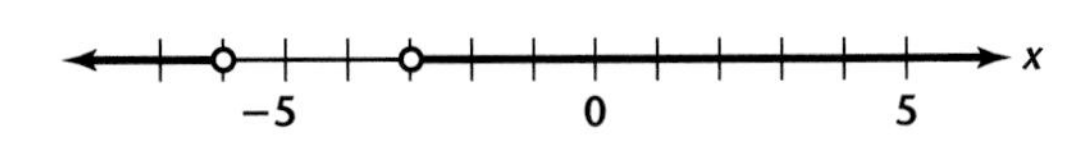

29.

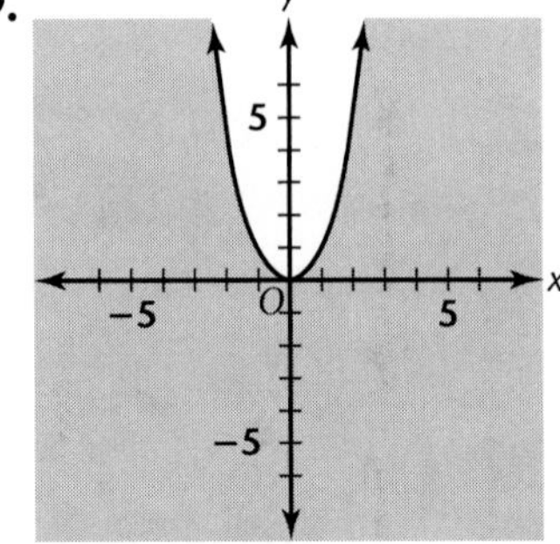

30.

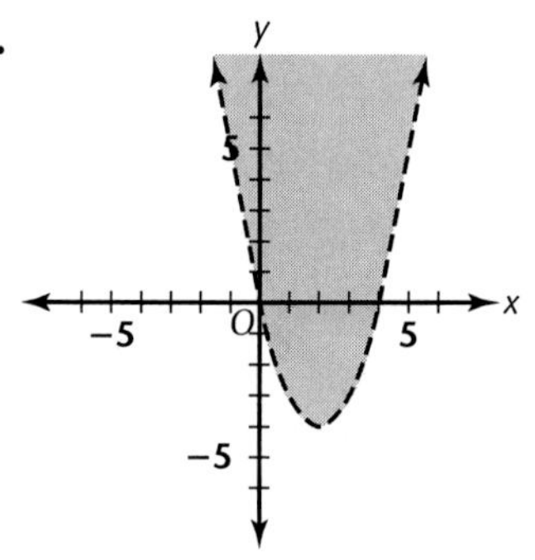

31.

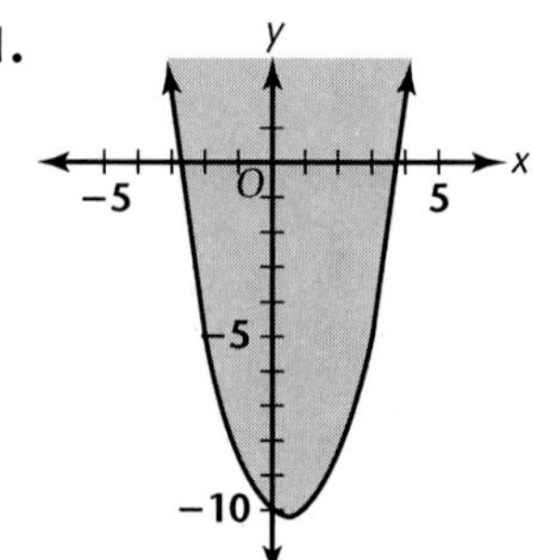

Chapter 8 Assessment, page 407

2. Vertex: $(6, -4)$; axis of symmetry: $x = 6$; zeros: 4, 8

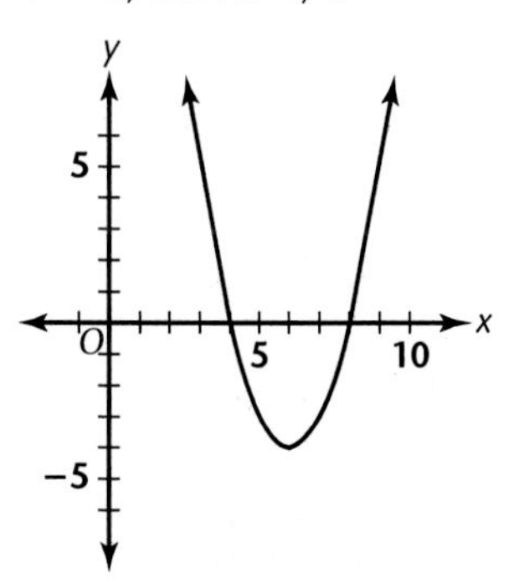

3. Vertex: $(-3, -1)$; axis of symmetry: $x = -3$; zeros: -2, -4

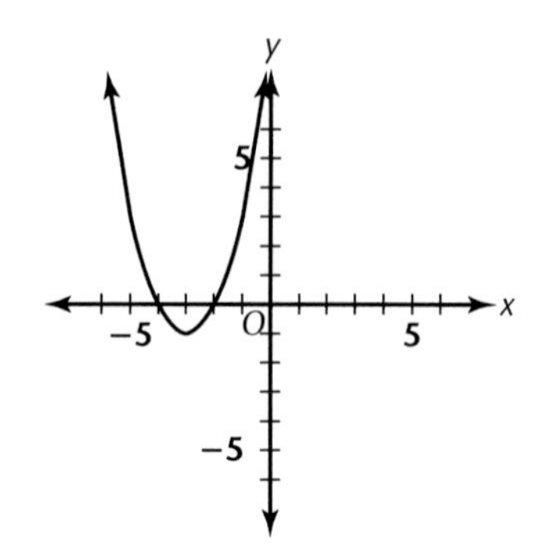

22. $x < -5$ or $x > 1$

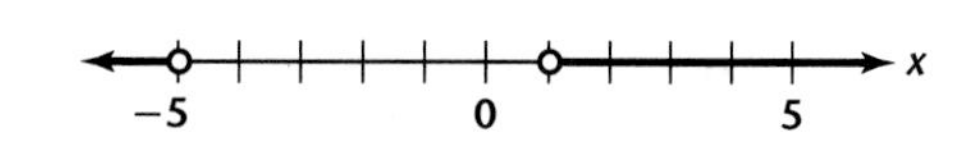

23. $-7 \leq x \leq -3$

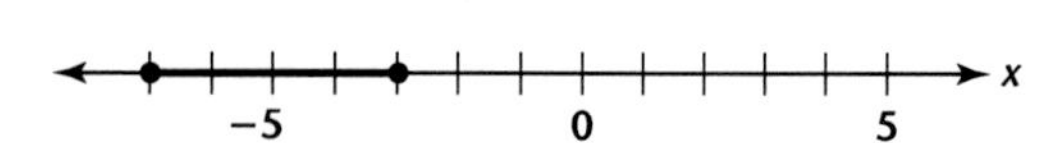

24.

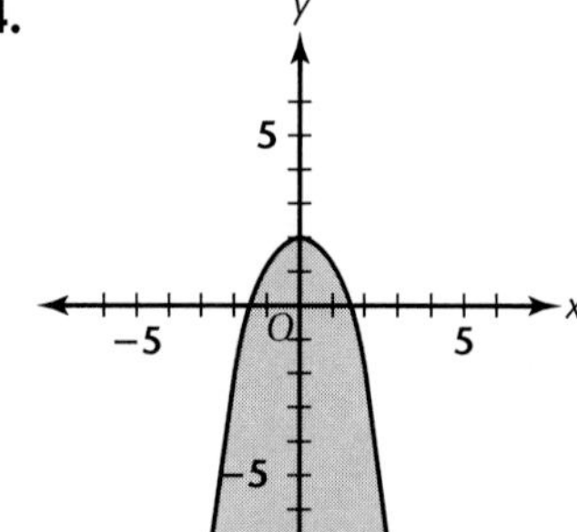

25.

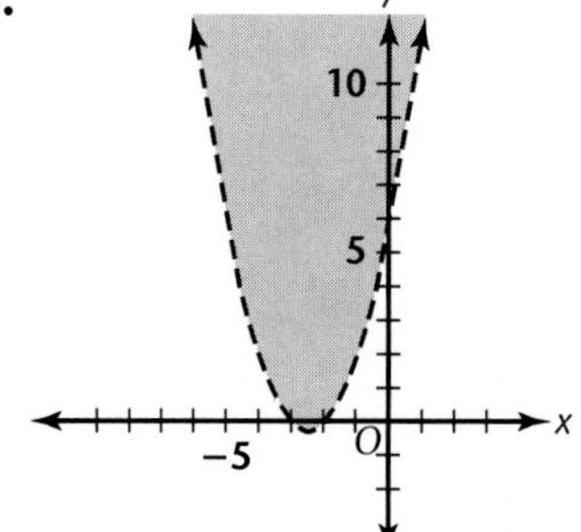

Chapters 1–8 Cumulative Assessment, pages 408–409

14. Multiply the terms in the following order, then add.
$(x)(2x) + (x)(-5) + (4)(2x) + (4)(-5) =$
$2x^2 - 5x + 8x - 20 = 2x^2 + 3x - 20$

19.

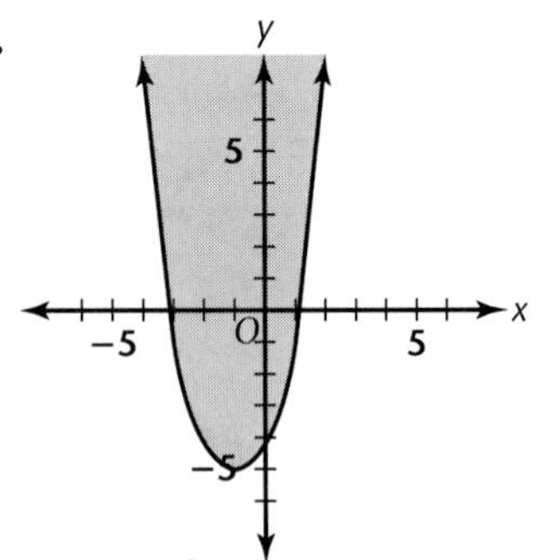

Lesson 9.1

Exploration 2, **page 415**

1.

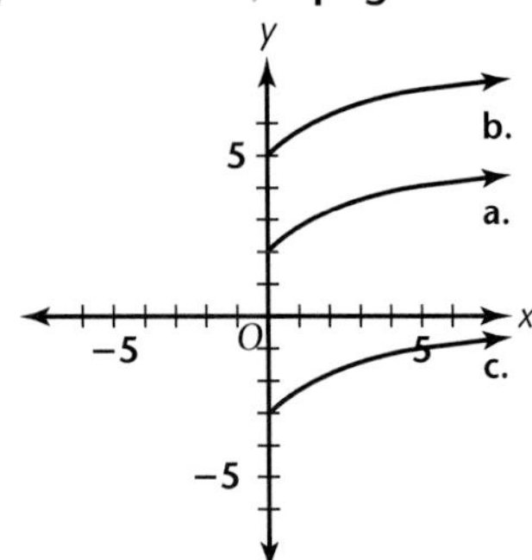

2. **a.** $g(x)$ is the graph of $f(x)$ shifted vertically up by 2 units.
b. $h(x)$ is the graph of $f(x)$ shifted vertically up by 5 units.
c. $p(x)$ is the graph of $f(x)$ shifted vertically down by 3 units. When k is added to $f(x) = \sqrt{x}$, the graph of $f(x)$ is shifted vertically by k units. A positive k-value results in an upward shift, a negative k-value results in a downward shift.

3.

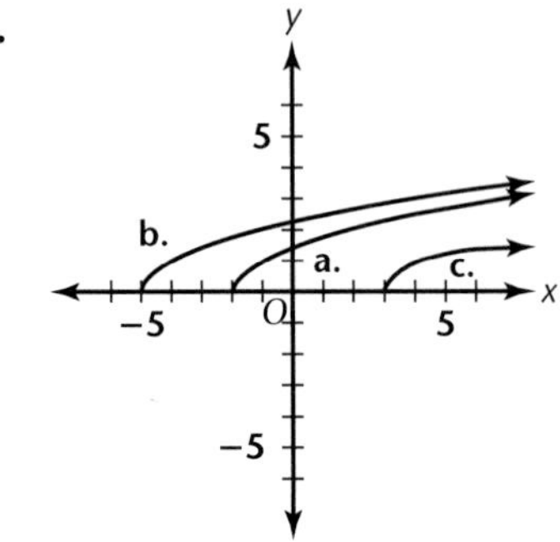

4. **a.** $g(x)$ is the graph of $f(x)$ shifted horizontally 2 units to the left.
b. $h(x)$ is the graph of $f(x)$ shifted horizontally 5 units to the left.
c. $p(x)$ is the graph of $f(x)$ shifted horizontally 3 units to the right. When h is subtracted from x, the graph of $f(x)$ is shifted horizontally by h units. A positive h-value results in a shift to the right; when $h = 2$, the function $y = \sqrt{x - 2}$ is a shift to the right. A negative h-value results in a shift to the left; when $h = -3$, the function $y = \sqrt{x - (-3)}$ is a shift to the left.

5.

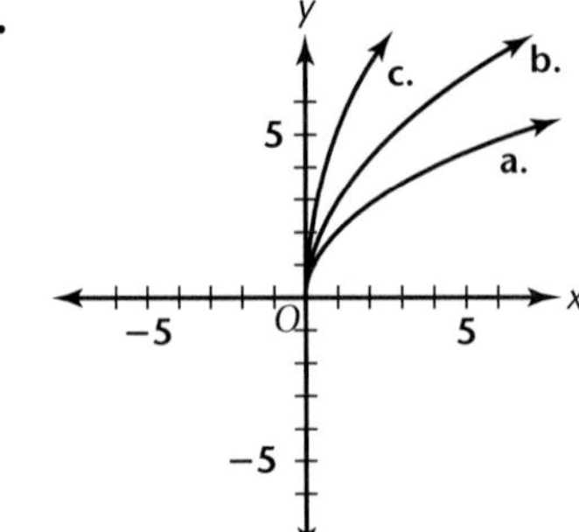

6. **a.** $g(x)$ is the graph of $f(x)$ vertically stretched by a factor of 2.
b. $h(x)$ is the graph of $f(x)$ vertically stretched by a factor of 3.

c. $p(x)$ is the graph of $f(x)$ stretched by a factor of 5. When $f(x) = \sqrt{x}$ is multiplied by a scale factor, a, the graph of $f(x)$ is stretched from the vertex (0, 0). When $a > 1$, then the graph of $f(x)$ is stretched upward. When $0 < a < 1$, the graph is flattened.

7.

8. **a.** $g(x)$ is the graph of $f(x)$ reflected through the x-axis and stretched by a factor of 2.

b. $h(x)$ is the graph of $f(x)$ reflected through the x-axis and stretched by a factor of 3.

c. $p(x)$ is the graph of $f(x)$ reflected through the x-axis and stretched by a factor of 5. When $f(x)$ is multiplied by a, where $a < 0$, the graph of $f(x)$ is reflected through the x-axis and then stretched by the scale factor a.

9. The graph of $y = \sqrt{x} + a$ is a vertical shift up or down of the graph of $y = \sqrt{x}$ by a or $-a$ units, respectively. The graph of $y = \sqrt{x + a}$ is a horizontal shift left or right of the graph of $y = \sqrt{x}$ by a or $-a$ units, respectively. The graph of $y = a\sqrt{x}$ is a stretch of the graph of $y = \sqrt{x}$ by a scale factor of a, and a flattening stretch toward the x-axis, when $0 < a < 1$. The graph of $y = -a\sqrt{x}$ is a reflection through the x-axis of the graph of $y = \sqrt{x}$ followed by a stretch by a scale factor of a.

Ongoing Assessment, page 415

g: Stretch by a factor of 2, vertical shift of 4, horizontal shift of 3 to the right; h: Stretch by a factor of −2, vertical shift of −3, horizontal shift of 4 to the left

Communicate, page 417

1. Draw a square that contains the number of graph paper squares equal to the perfect square number.

2. Draw two squares, one inside the other, whose areas are equal to the perfect square numbers just less than and just greater than the number that is not a perfect square. Determine the length of the side of each square. The square root of the non-square number is between these two values. Make an estimate for the square root of the non-square number. Use the guess-and-check method to find a value such that when it is squared, it is approximately the original number.

3. An irrational number is a number that cannot be represented by a repeating or terminating decimal. Answers may vary. For example, π, 6.10100111 . . . , $\sqrt{19}$.

4. Answers may vary. For example, use the square root key on a calculator; use the guess-and-check method with square numbers.

5. To move the square-root function 2 to the right, subtract 2 from x before taking the square root. To move the function up 3 units, add 3 to $\sqrt{x - 2}$. The result is $f(x) = \sqrt{x - 2} + 3$

6. A negative number multiplied by itself gives a positive result. The same is true for a positive number.

Practice and Apply, **pages 417–419**

22. Answers may vary. For example:

+number	$(+\text{number})^2$	$\sqrt{(+\text{number})^2}$
52	2704	52
54	2916	54
60	3600	60
75	5625	75
90	8100	90

The square root of a positive number squared is that positive number.

23. Answers may vary. For example:

−number	$(-\text{number})^2$	$\sqrt{(-\text{number})^2}$
−52	2704	52
−54	2916	54
−60	3600	60
−75	5625	75
−90	8100	90

The square root of a negative number squared is the positive value of that number.

24. Vertical shift up by 6 units

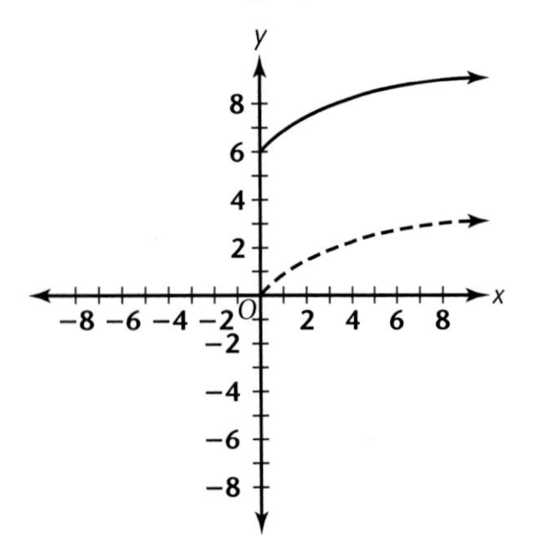

25. Stretch by a factor of 4

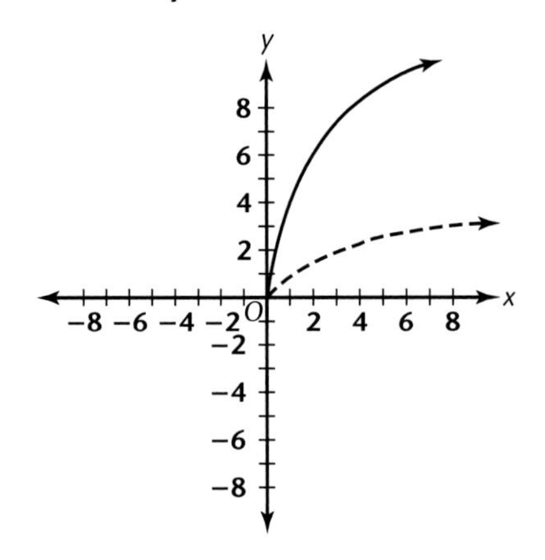

26. Horizontal shift to the right by 2 units

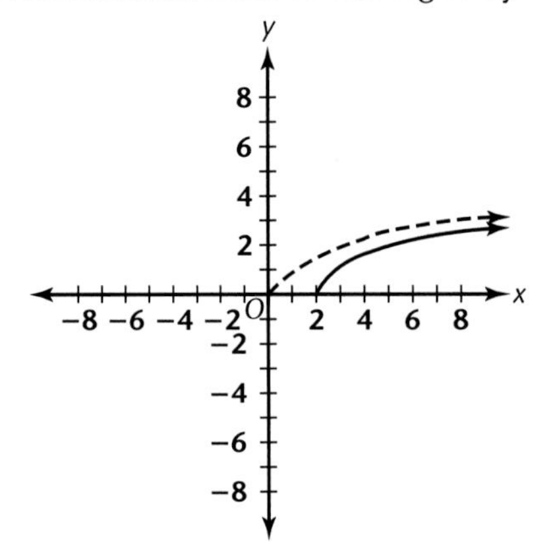

27. Reflected through the *x*-axis; vertical shift up by 1 unit

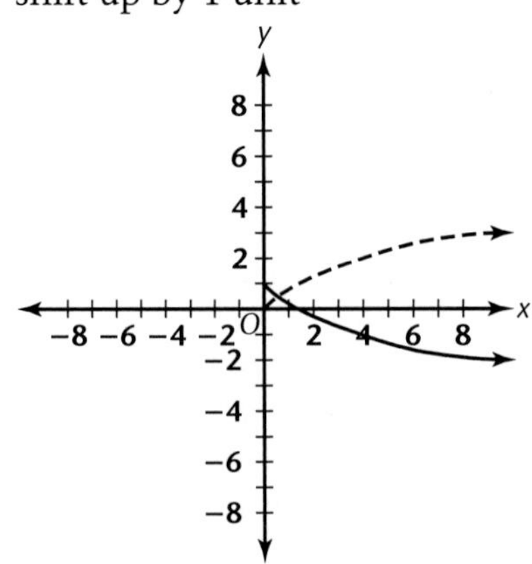

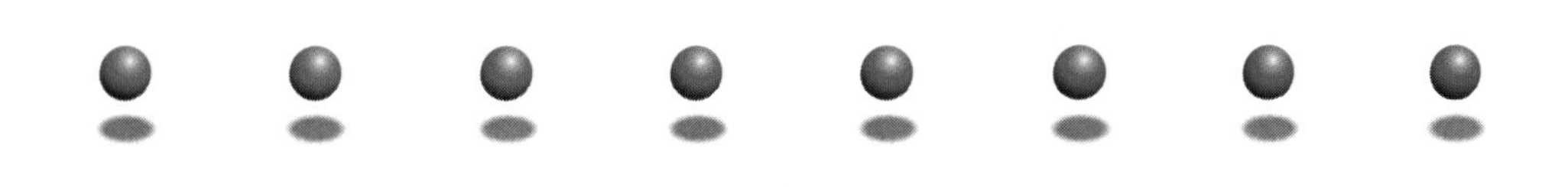

28. Stretched by a factor of 2; horizontal shift by 1 unit left; vertical shift down by 3 units

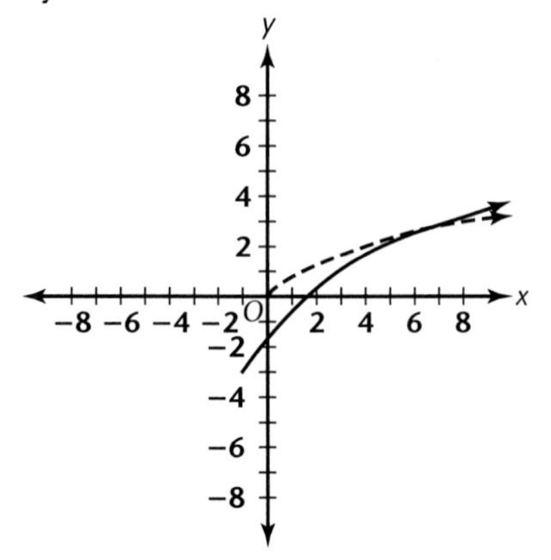

29. Reflected through the x-axis; stretched by a factor of 2; horizontal shift right by 1 unit; vertical shift up by 6 units

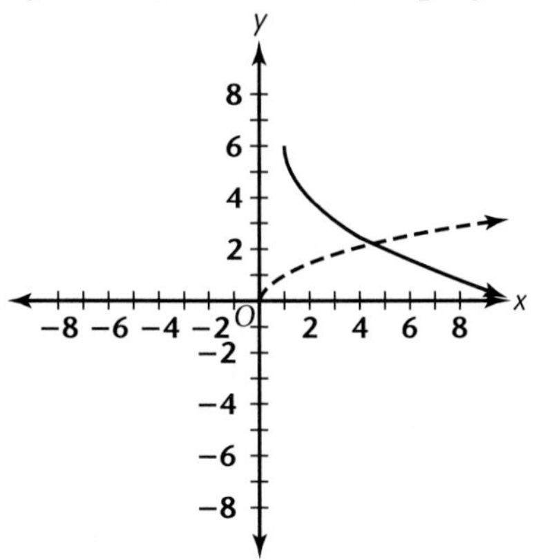

36. The graph of the parent function is shifted horizontally to the right by 4 units, and then reflected through the x-axis.

37. The graph of the parent function is shifted horizontally to the left by 3 units, then shifted vertically down 4 units, and stretched by a scale factor of 5.

38. The graph of the parent function is shifted horizontally to the left by 3.75 units, stretched by a scale factor of 3.5, reflected through the x-axis then shifted upward by 2 units.

42.

x	0	1	2	3	4	5	6	7	8	9	10
$\sqrt{x}$	0	1	1.4	1.7	2	2.2	2.4	2.6	2.8	3	3.2

43.

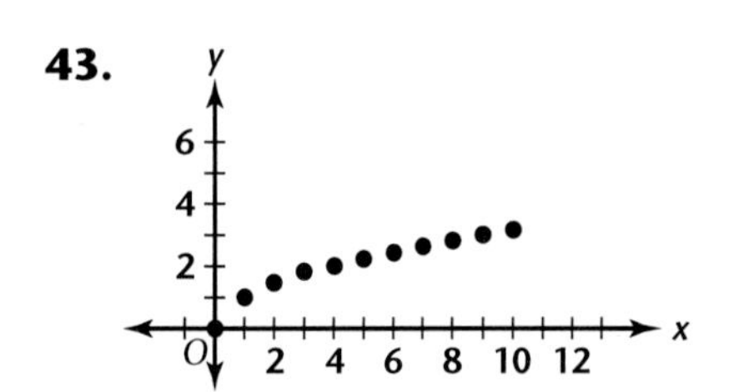

47. Yes

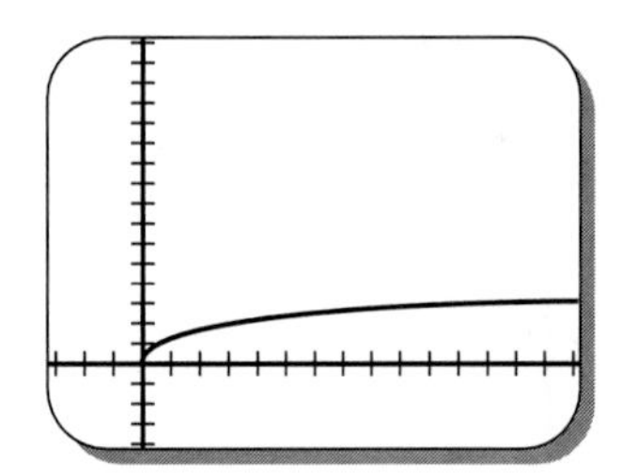

Look Back, **page 419**

48.

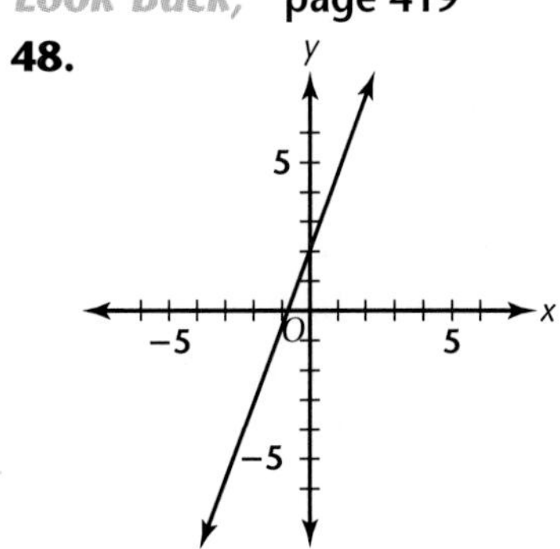

49.

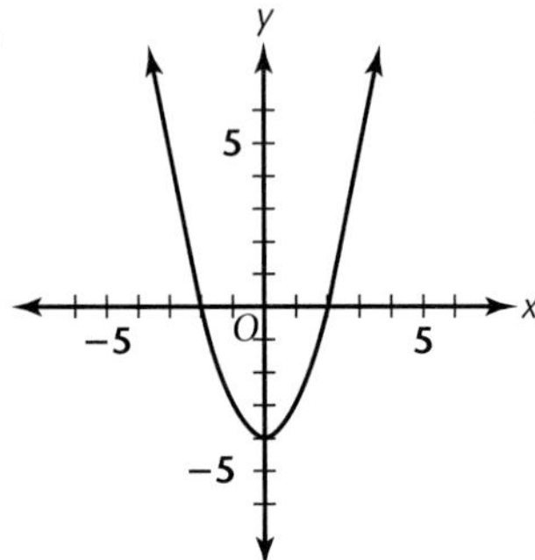

50.

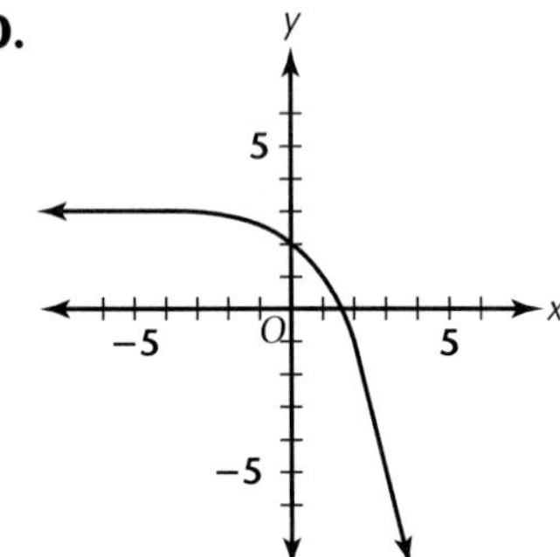

51.

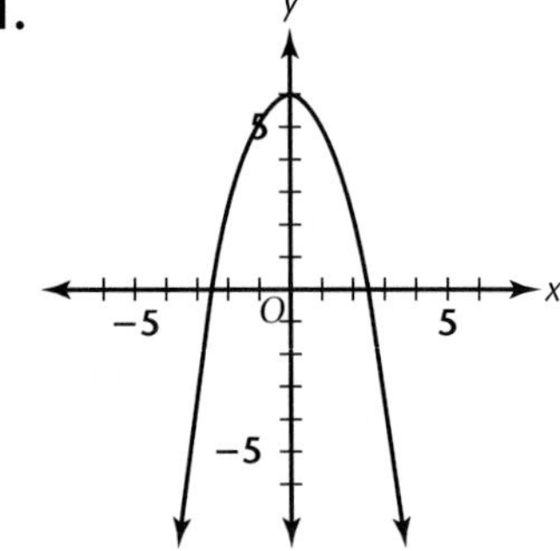

Extra Practice, **page 551**

13. Vertical shift by 2 units down

14. Stretch by a factor 3

15. Horizontal shift by 1 unit to the left

16. Horizontal shift by 9 units to the right, reflection over the x-axis

17. Horizontal shift by 6 units to the left, stretch by a factor of 3

18. Horizontal shift by 5 units to the right, vertical shift by 6 units up

19. Horizontal shift by 1 unit to the right, vertical shift by 2 units up, reflection through the x-axis

20. Horizontal shift by 1.25 units to the left, stretch by a factor of 4.5 units

21. Horizontal shift by 1.5 units to the right, vertical shift by 4.75 units up, stretch by a factor of 3

Lesson 9.2

Communicate, **page 425**

1. When $a \geq 0$

2. When simplifying an expression, perfect square factors are factored out of the radical; an expression is in simplified form when there are no perfect square factors under the radical sign, and there are no radicals in the denominator.

3. $a \leq 3$

4. Answers may vary. For example, the Multiplication Property of Square Roots states that the square root of a product is the same as the product of square roots of its factors; the Division Property of Square Roots states that the square root of a quotient is the same as the quotient of the square roots of the dividend and divisor.

5. When the expressions under the radical signs are equal.

6. Answers may vary. For example, use the FOIL method. $(\sqrt{3} + 2)(\sqrt{2} + 3)$
$= (\sqrt{3})(\sqrt{2}) + 3\sqrt{3} + 2\sqrt{2} + (2)(3)$
$= \sqrt{6} + 3\sqrt{3} + 2\sqrt{2} + 6$

Lesson 9.3

Communicate, **page 432**

1. When both positive and negative quantities are squared, the result is positive.

2. Square both sides of the equation, and then isolate and solve for the variable.
3. No; the principal square root of a real number can never be negative.
4. For $a^2 = b^2 + c^2$, when you solve for a, the two values are $\sqrt{b + c}$ and $-\sqrt{b + c}$. For $a^2 = (b + c)^2$, when you solve for a, the two values are $(b + c)$ and $-(b + c)$, or $-b - c$.
5. Express the left side as a perfect square binomial: $(x + b)^2 = c^2$. Thus, $x = c - b$ or $x = -c - b$. Answers may vary. For example, let $b = 2$ and $c = 3$. The equation becomes $x^2 + 4x + 4 = 9$. Factor both sides: $(x + 2)^2 = 3^2$. Thus, $x + 2 = \pm 3$. Solving for x, $x = 1$ or $x = -5$.
6. Radicals are used to identify the variable x in the equation $ax^2 + bx + c = 0$.

Practice and Apply, **pages 433–434**

22. $x = 4$

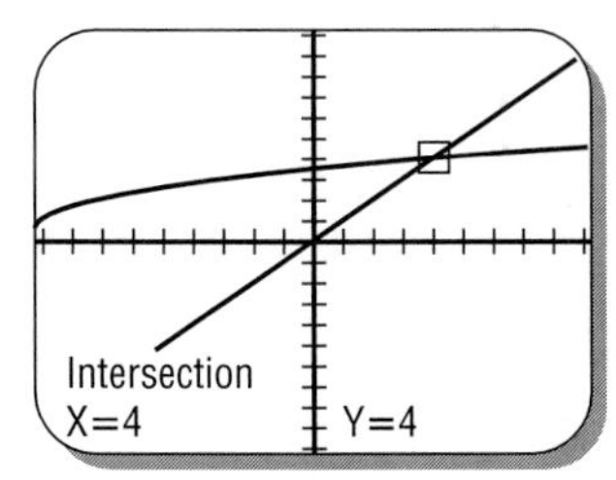

23. No solution

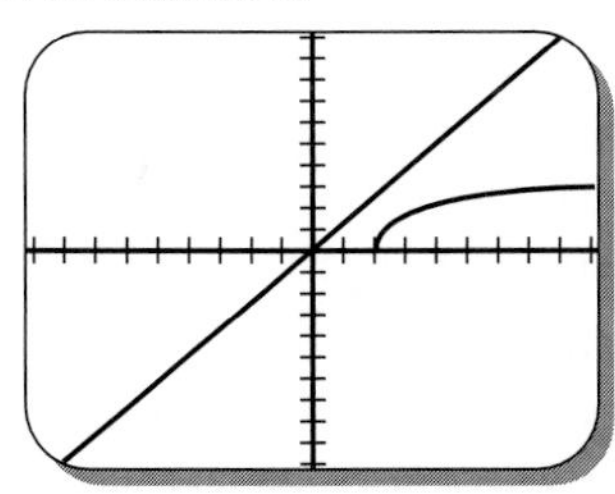

24. Two solutions, $x = 1$ or $x = 4$

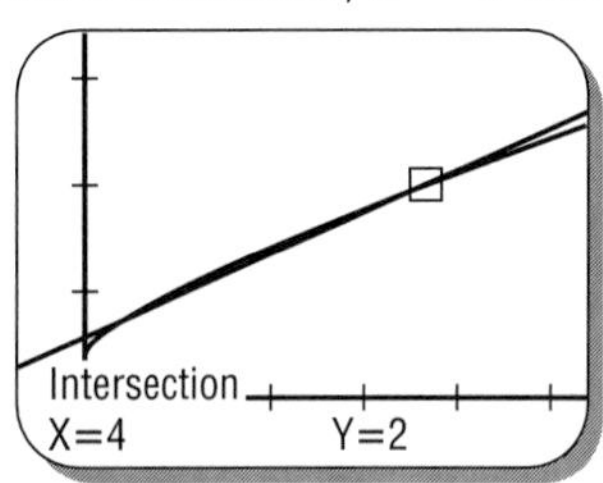

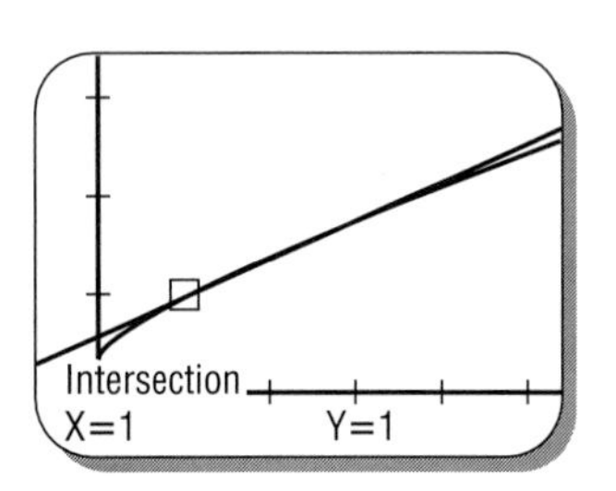

40. No solution; the square root of a negative number is not a real number.

41. No solution; the principal square root of a number cannot be negative.

42. No solution; the absolute value of a number cannot be negative.

Look Back, **page 434**

48. The lines are parallel. The second graph is the first graph shifted to the right by 2 units.

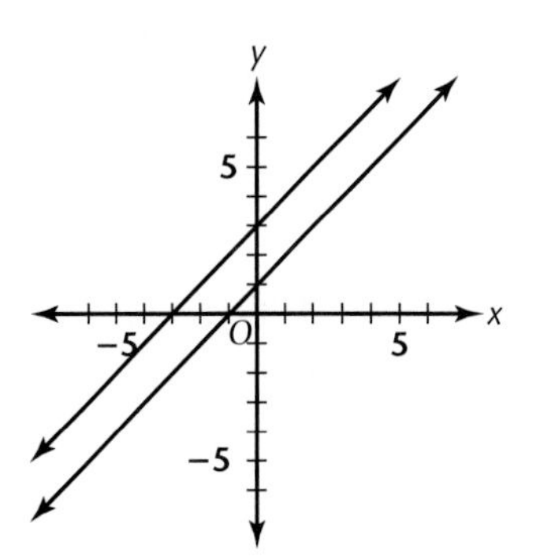

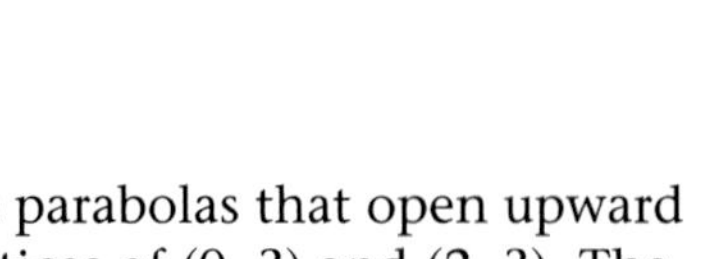

49. Both are parabolas that open upward with vertices of (0, 3) and (2, 3). The graphs intersect at (1, 4). The second graph is shifted to the right by 2 units.

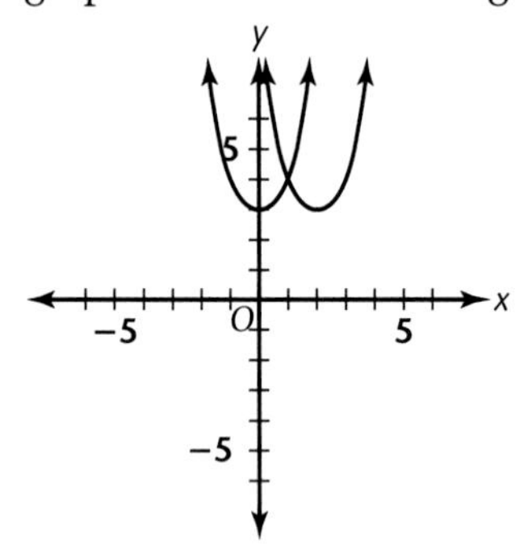

50. Both are V-shaped graphs that open upward; one has a vertex of (0, 3), and the other has a vertex of (2, 3). They both contain the point (1, 4). The second graph is shifted to the right by 2 units.

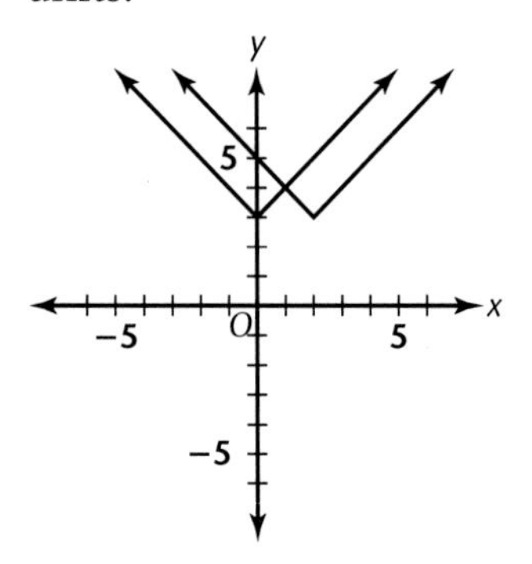

53. $x < -3$ or $x > 4$

−4 −3 −2 −1 0 1 2 3 4 x

54. $-3 < y < -2$

−4 −3 −2 −1 0 1 2 3 4 y

55. $-2 < a < 8$

−5 −4 −3 −2 −1 0 1 2 3 4 5 6 7 8 a

Look Beyond, **page 434**

56. $y = \pm\sqrt{-x^2 + 4}$; (0, 0)

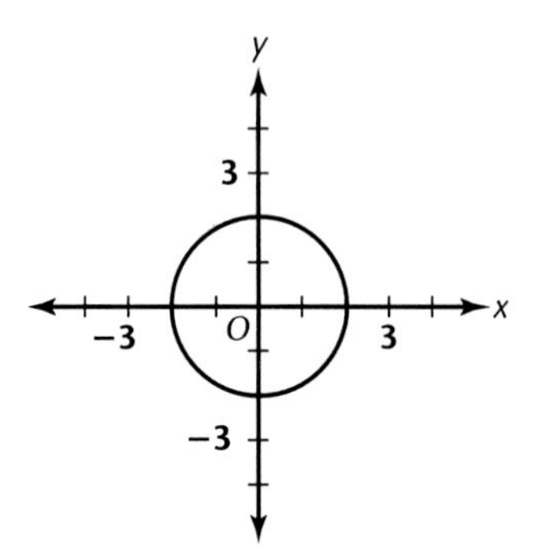

The center of the circle is the origin.

Lesson 9.4

Ongoing Assessment, **page 435**

Side a	Side b	Side c	a^2	b^2	c^2	$a^2 + b^2$
13	4	5	9	16	25	25
5	12	13	25	144	169	169
16	30	34	256	900	1156	1156
13	84	85	169	7056	7225	7225

The last two columns of the table have equal entries, so $c^2 = a^2 + b^2$ for these values.

Ongoing Assessment, **page 437**

All triangles with sides that are multiples of 3, 4, and 5, respectively, are similar, thus all are right triangles.

Communicate, **page 438**

1. The side opposite the right angle is the hypotenuse. The other sides are the legs.
2. In a right triangle, the sum of the squares on the legs is equal to the square on the hypotenuse.
3. Square the lengths of the legs, add the squares together and then determine the square root of the sum.

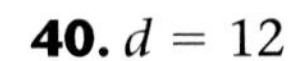

4. Subtract the square of the known leg from the square of the hypotenuse. Then find the square root of the difference.
5. Answers may vary. For example, draw a right triangle. Measure the length of its sides, square the lengths of the legs and then add the squares. Show that the sum is equal to the length of the hypotenuse.
6. Let the length of the rectangle be one leg of the right triangle, and the width be the other leg. The diagonal of the rectangle will be the hypotenuse of the right triangle. Apply the "Pythagorean" Right-Triangle Theorem to determine the length of the diagonal.

40. $d = 12$

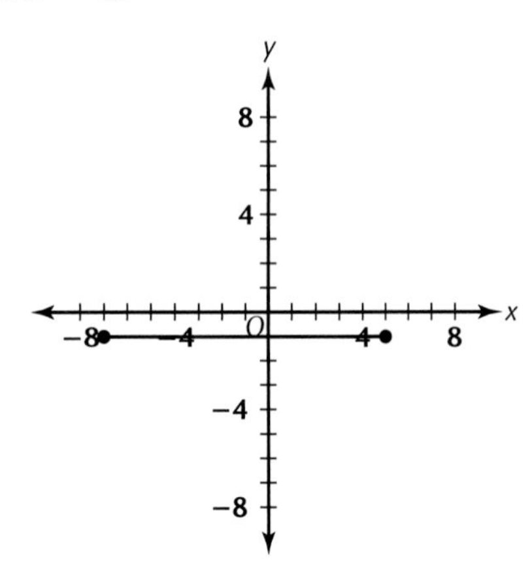

Practice & Apply, **pages 439–440**

23.

Equilateral Triangles								
Side	4	6	8	10	12	20	n	34
Half of a Side	2	3	4	5	6	10	$\frac{n}{2}$	17
Altitude	$2\sqrt{3}$	$3\sqrt{3}$	$4\sqrt{3}$	$5\sqrt{3}$	$6\sqrt{3}$	$10\sqrt{3}$	$\frac{n}{2}\sqrt{3}$	$17\sqrt{3}$

Look Beyond, **page 440**

For Exercises 39–42, measure the number of units between the points on the grid.

39. $d = 7$

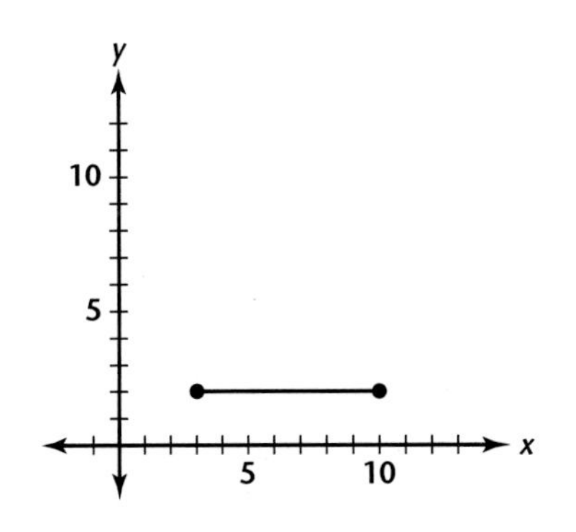

41. $d = 7$

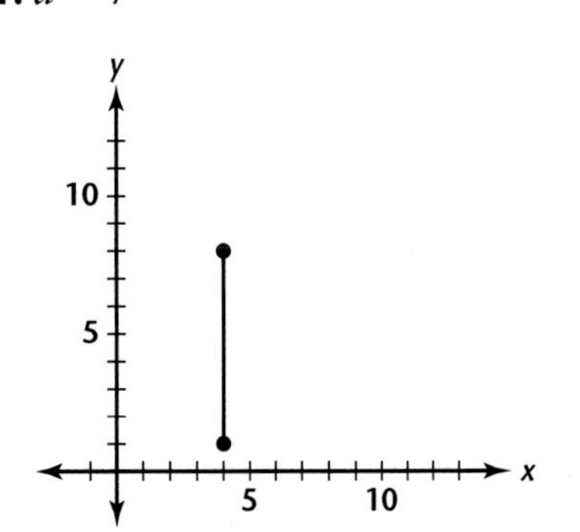

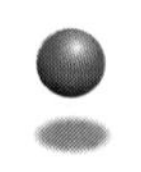
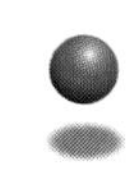

42. $d = 3$

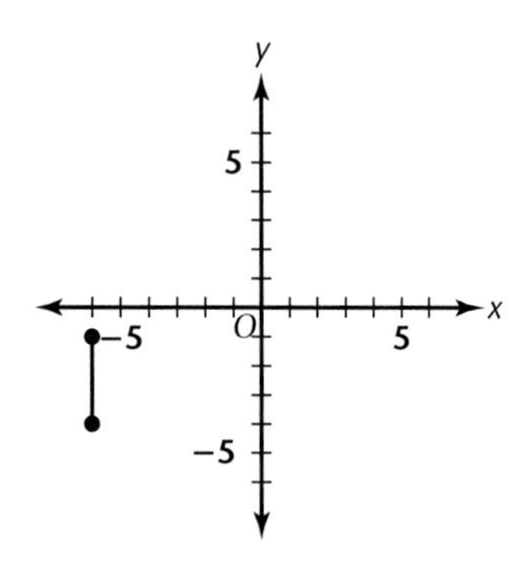

Lesson 9.5

Exploration 2, page 444

1.

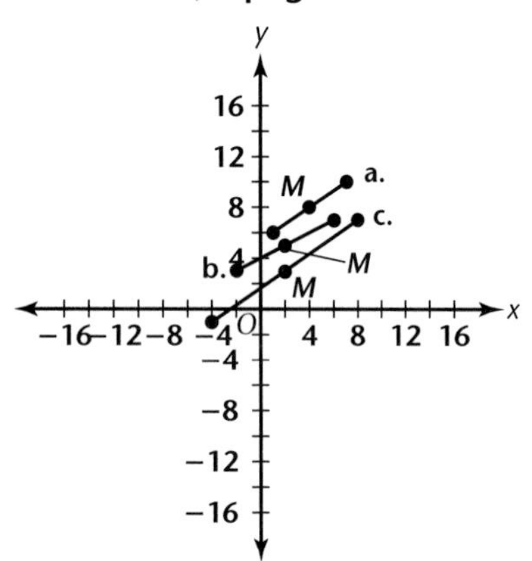

a. (4, 8)
b. (2, 5)
c. (2, 3)

2. **a.** The lengths of line segments *PM* and *MQ* are both $\sqrt{13}$.
b. The lengths of line segments *PM* and *MQ* are both $2\sqrt{5}$.
c. The lengths of line segments *PM* and *MQ* are both $2\sqrt{13}$.

3. Average the *x*-coordinates.

4. Average the *y*-coordinates.

5. (27.5, 41.5)

6. 26.50 = 26.50; $AM = MB$

Communicate, page 445

1. A point is a position on a line. A coordinate is a number which is used to locate a point.

2. The "Pythagorean" Right-Triangle Theorem can be used to find the distance between the endpoints of two perpendicular line segments.

3. Square the difference between the two *x*-coordinates. Repeat for the *y*-coordinates. Add these two values and take the square root of the sum.

4. The *x*-coordinate of the midpoint is found by taking the average of the *x*-coordinates. The *y*-coordinate is found by the same method.

5. Interchange the words *if* and *then* in the statement.

Practice and Apply, pages 446–447

12. Isosceles; the length of both $\overline{PQ}$ and $\overline{PR}$ is $\sqrt{34}$ units.

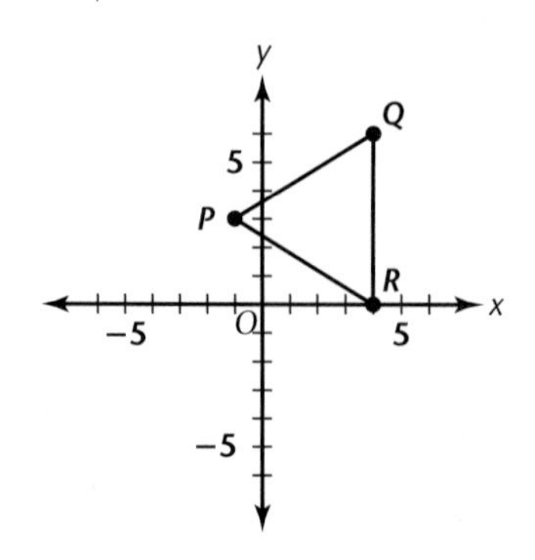

13. Right scalene; the lengths of $\overline{PQ}$, $\overline{QR}$, and $\overline{RP}$ are 5, 10, and $5\sqrt{5}$, respectively, and $PQ^2 + QR^2 = RP^2$.

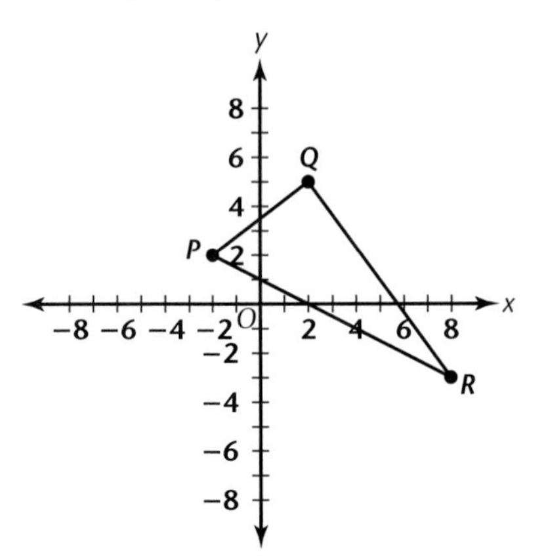

15. (5.5, 5)

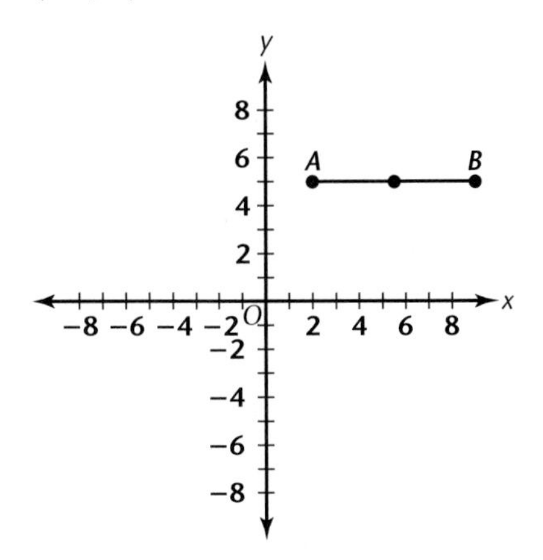

16. (−3, −4)

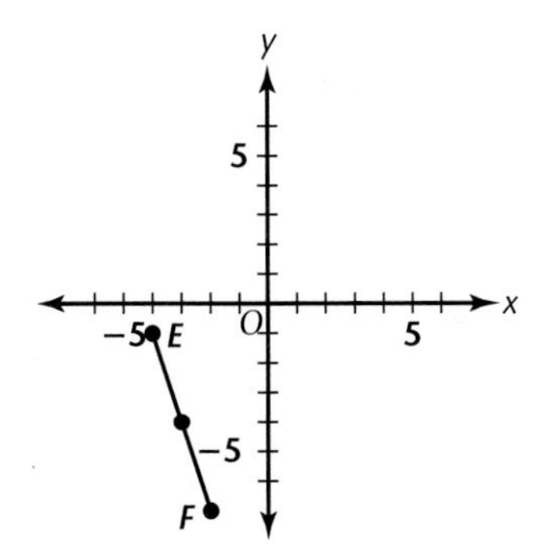

17. (−3, 2)

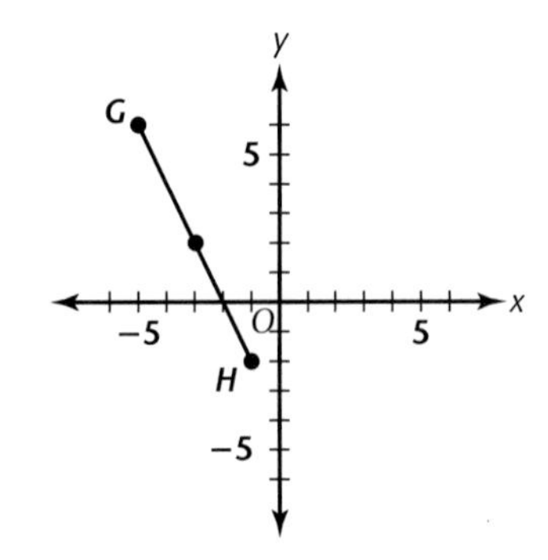

18. (6.5, −0.5)

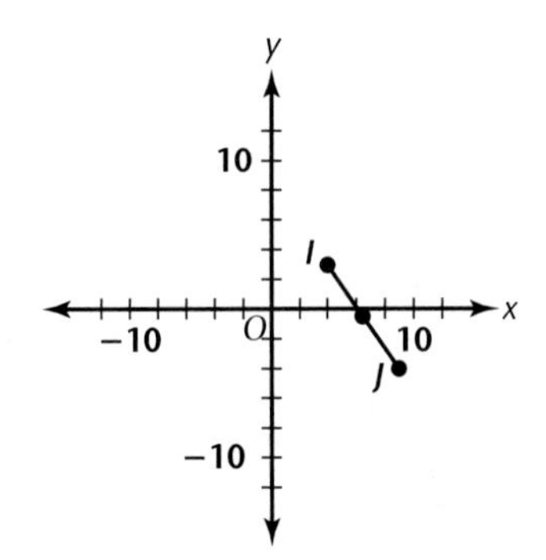

27.

28. $M(4, 2)$

31. $y = -2x + 10$

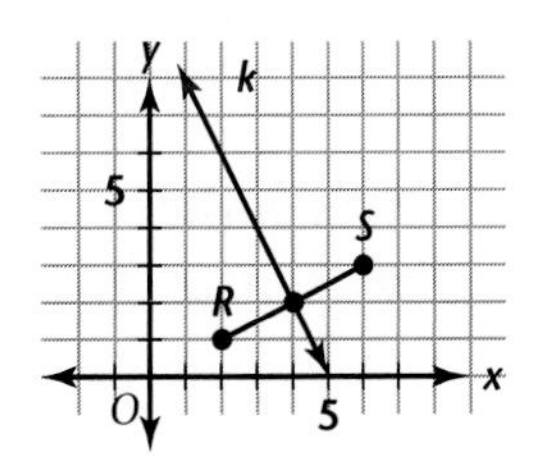

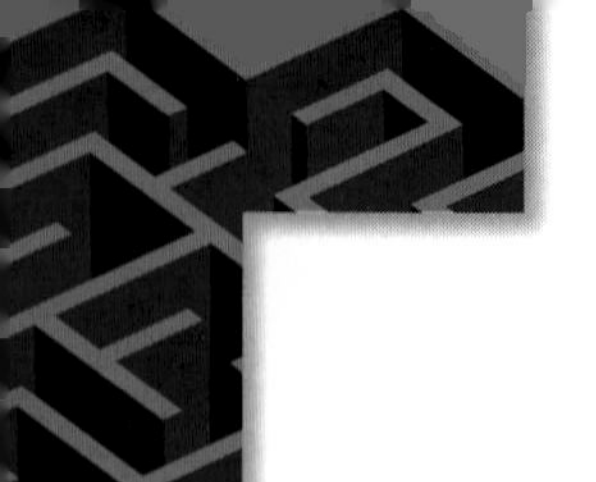
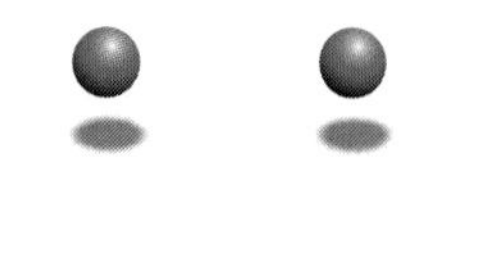

32. Answers may vary. Any point on the perpendicular bisector of a segment is equidistant from the endpoints of the segment.

33.

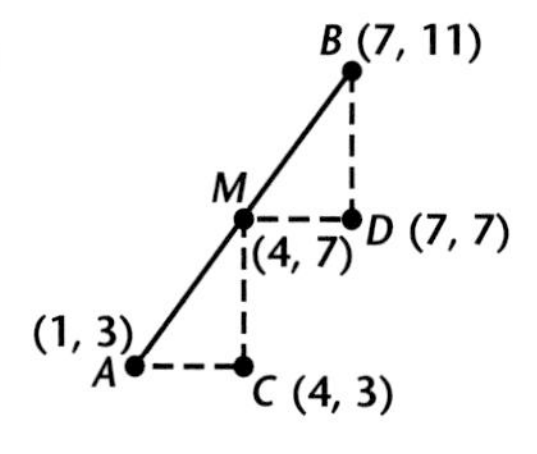

$$(AC)^2 + (CM)^2 = (AM)^2$$
$$3^2 + 4^2 = (AM)^2$$
$$9 + 16 = (AM)^2$$
$$\sqrt{25} = AM$$
$$5 = AM$$
$$(MD)^2 + (BD)^2 = (MB)^2$$
$$3^2 + 4^2 = (MB)^2$$
$$9 + 16 = (MB)^2$$
$$\sqrt{25} = MB$$
$$5 = MB$$

Since $AM = MB$, M is the midpoint of $\overline{AB}$.

34. a. $S = 4.26\sqrt{d}$

b. (1) 42.6 mph; (2) Approx. 166.69 feet

c. $S = 25.56$ mph; technically the officer was correct. However, speedometers do not usually measure speed to this precision, and the passenger would be considered correct.

Look Beyond, **page 447**

49. The base number is raised to the first power, then that answer is divided by two.

50. Raising a number to the power of $\frac{1}{2}$ is the same as taking the square root of a number.

Lesson 9.6

Exploration 3, **page 449**

2. To determine the length of the diagonal of a rectangle, first square the lengths of both sides then add the squares together, and take the square root of the sum. Both diagonals will have the same length.

Exploration 4, **page 450**

1. Students' three triangles may vary. The length of $\overline{AB}$ is $\frac{1}{2}$ the length of the side to which it is parallel.

2. In a triangle, the segment connecting the midpoints of two sides is $\frac{1}{2}$ the length of the third side.

3. It should also be true.

4. Triangles may vary. Label the vertices of the triangle as follows:

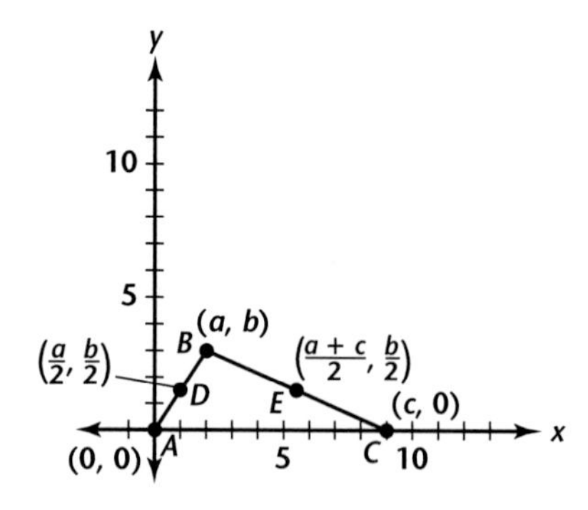

Let D be the midpoint of $\overline{AB}$, then the coordinates of D are $\left(\frac{a}{2}, \frac{b}{2}\right)$. Let E be the midpoint of $\overline{BC}$, then the coordinates are $\left(\frac{a+c}{2}, \frac{b}{2}\right)$. The length of $\overline{DE}$ and $\overline{AC}$ can be found by applying the distance formula.

Length of $\overline{DE}$

$= \sqrt{\left(\frac{a}{2} - \frac{a+c}{2}\right)^2 + \left(\frac{b}{2} - \frac{b}{2}\right)^2}$

$= \sqrt{\left(\frac{a}{2} - \frac{a}{2} - \frac{c}{2}\right)^2 + (0)^2}$

$= \sqrt{\frac{c^2}{4}}$

$= \frac{c}{2}$

Length of $\overline{AC} = \sqrt{(c-0)^2 + (0-0)^2}$
$= \sqrt{c^2}$
$= c$

This shows that the length of the line segment connecting the midpoints of the two sides of a triangle is half the length of the third side.

Ongoing Assessment, page 451

1. It is made of four triangles, each with an area of $\frac{1}{2}$ square cubit.
2. $A = s^2$
   $2 = s^2$
   $\sqrt{2} = \sqrt{s^2}$
   $\sqrt{2} = s$

Communicate, page 452

1. Answers may vary. For example, by using geometric figures to model a problem you can solve problems in algebra.
2. This is the definition of a circle.
3. Subtract the x-coordinates and square the difference. Repeat for the y-coordinates. Add these two values and take the square root of the sum.
4. Let r be the common distance, (x, y) the given point and (x_n, y_n) be any of the equidistant points. The distance between the given point and any of the equidistant points can be expressed as $r = \sqrt{(x - x_n)^2 + (y - y_n)^2}$. Square both sides of the equation for $r^2 = (x - x_n)^2 + (y - y_n)^2$, which is the equation of a circle.
5. Find the distance between any two opposite corners of the triangle.

Practice and Apply, pages 452–453

11.

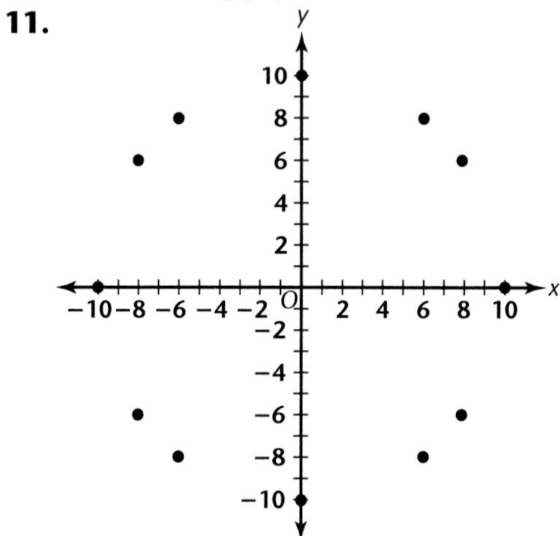

15. $(x - 2)^2 + (y - 3)^2 = 25$
16. $(x + 2)^2 + (y - 3)^2 = 25$
17. $(x - 2)^2 + (y + 3)^2 = 25$
18. $(x + 2)^2 + (y + 3)^2 = 25$
19. $(x - 2)^2 + (y - 3)^2 = 49;$
    $(x + 2)^2 + (y - 3)^2 = 49;$
    $(x - 2)^2 + (y + 3)^2 = 49;$
    $(x + 2)^2 + (y + 3)^2 = 49$

Look Back, page 453

42.

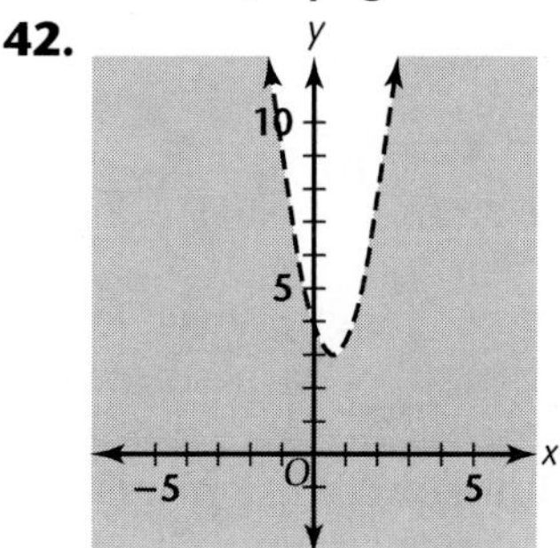

43.

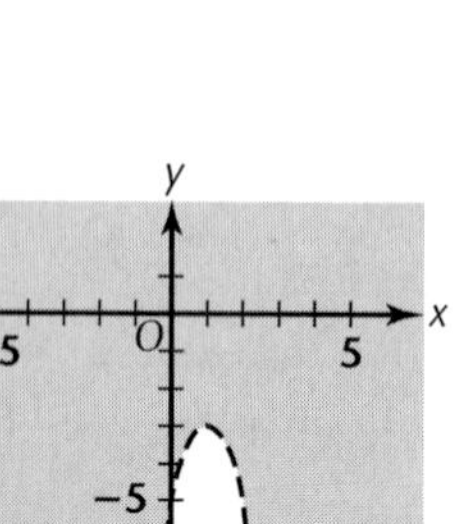

Look Beyond, page 453

44. Label the vertices of the rectangle as shown.

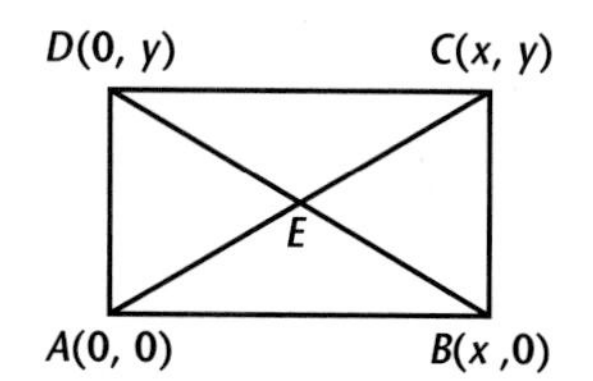

The coordinates of the diagonals' midpoint E are $\left(\frac{x}{2}, \frac{y}{2}\right)$. To verify that E is the midpoint of $\overline{AC}$ and $\overline{BD}$, apply the distance formula for line segments AE and EC, and for the line segments BE and ED. The result is $AE = EC = BE = ED = \frac{\sqrt{x^2 + y^2}}{2}$.

45. Label the vertices of the isosceles trapezoid as shown.

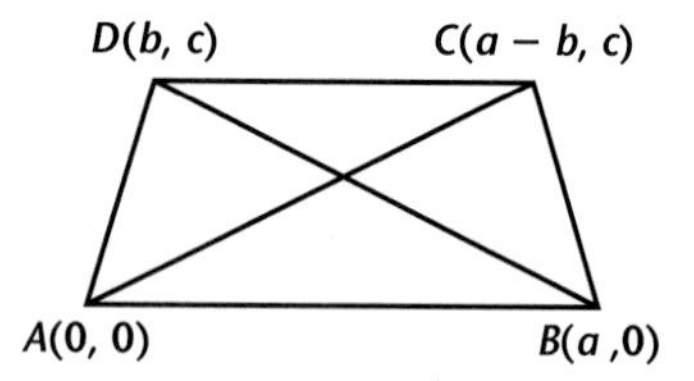

Let a represent the length of the longer side of the trapezoid or its base, and b represent the x-coordinate of point D, which is the horizontal distance D is from origin. Since the trapezoid is isosceles, then the horizontal distance of C from the origin is $a - b$. Apply the distance formula to find the length of the diagonals. The result is the length of $AC = BD = \sqrt{(a + b)^2 + c^2}$.

Extra Practice, page 553

5. $(x - 1)^2 + (y - 2)^2 = 4$

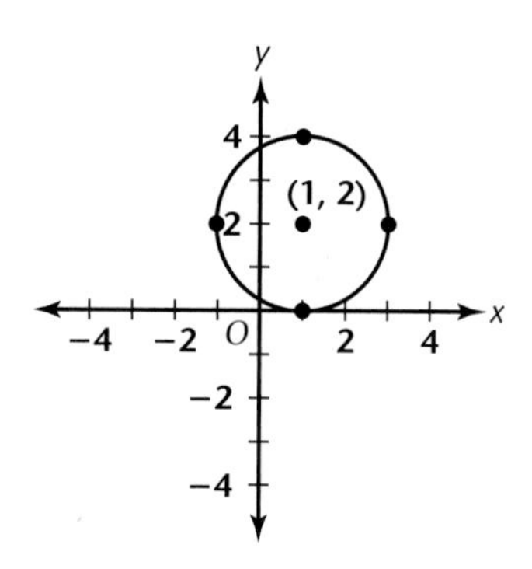

6. $(x + 1)^2 + (y - 2)^2 = 4$

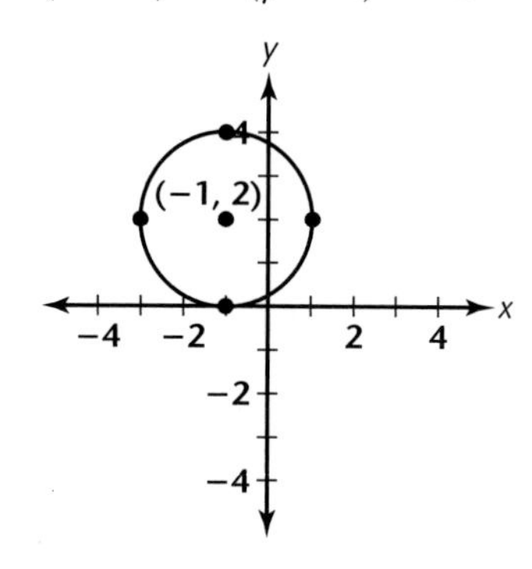

7. $(x - 1)^2 + (y + 2)^2 = 4$

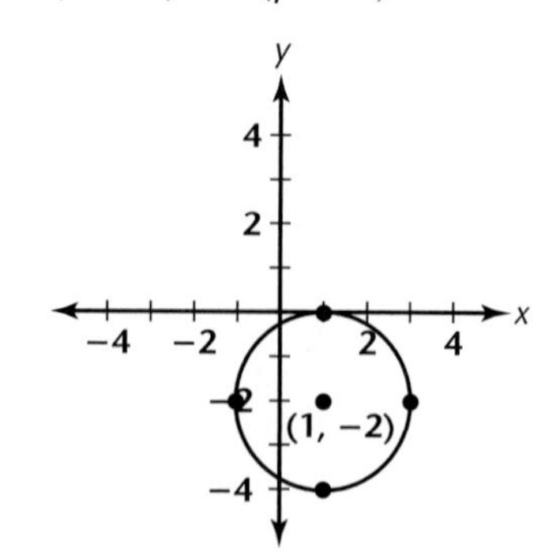

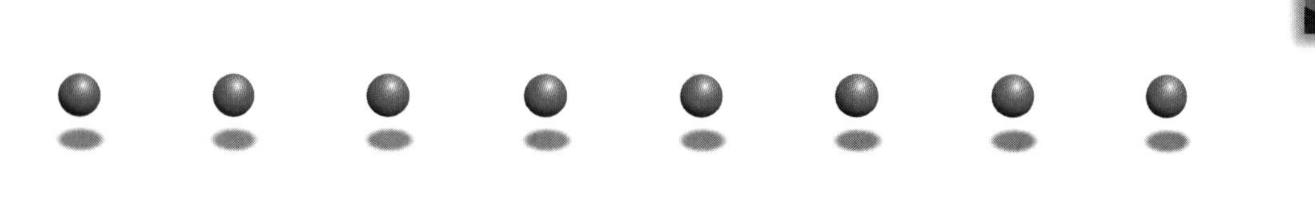

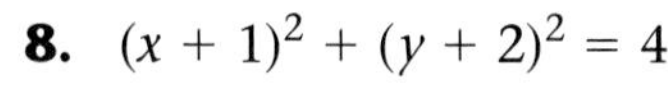

8. $(x + 1)^2 + (y + 2)^2 = 4$

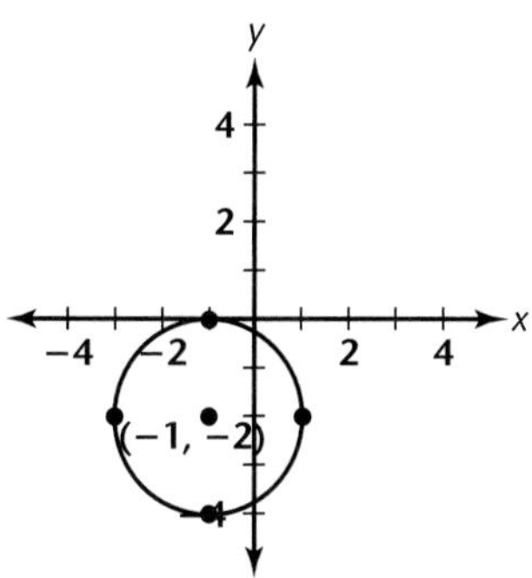

Lesson 9.7

Communicate, page 457

1. The slope of a line is the ratio of the change in y to the change in x. The tangent function is the ratio of the length of the side opposite an angle (the change in y) to the length of the side adjacent an angle (the change in x).

2. $\tan A = \dfrac{\text{length of opposite leg}}{\text{length of adjacent leg}}$
Solve for the length of the opposite leg = tan A (length of the adjacent leg). Use a calculator to determine the value of tan A and calculate the length of the opposite leg.

3. Use a calculator to apply the inverse tangent function to the ratio.

Look Back, page 459

30. $x > -4$ and $x < 4$

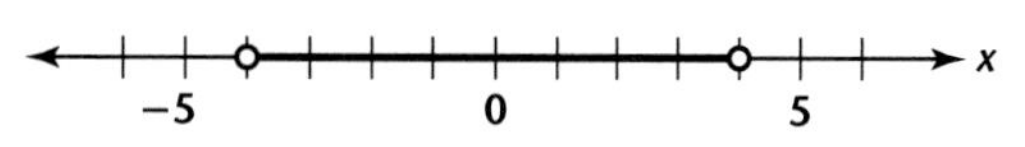

31. 45-45-90 triangle:

$$(a\sqrt{2})^2 = a^2 + a^2$$
$$2a^2 = 2a^2$$

30-60-90 triangle:

$$s^2 + (s\sqrt{3})^2 = (2s)^2$$
$$s^2 + 3s^2 = 4s^2$$
$$4s^2 = 4s^2$$

Look Beyond, page 459

37. $\cos(x) = \dfrac{a}{c}$ Solving for a, gives $a = c\cos(x)$.

$\sin(x) = \dfrac{b}{c}$ Solving for b, gives $b = c\cos(x)$.

The "Pythagorean" Right-Triangle Theorem gives $c^2 = a^2 + b^2$, or $c^2 = c^2(\cos(x))^2 + c^2(\sin(x))^2$. After dividing each term by c^2, $(\cos(x))^2 + (\sin(x))^2 = 1$. This is true for any right triangle because the hypotenuse value, c, is never equal to zero and does not affect the equation.

Extra Practice, page 554

1. $\tan Y = \dfrac{7}{24}$; $\tan Z = \dfrac{24}{7}$

2. $\tan M = \dfrac{35}{12}$; $\tan N = \dfrac{12}{35}$

3. $\tan A = \dfrac{20}{21}$; $\tan B = \dfrac{21}{20}$

Lesson 9.8

Exploration 1, page 461

		Length of leg opposite ∠A	Length of leg adjacent ∠A	Length of hypotenuse	tan A	sin A	cos A
	△BCA	4	3	$\sqrt{4^2 + 3^2} = 5$	$\frac{4}{3} \approx 1.\overline{3}$	$\frac{4}{5} = 0.8$	$\frac{3}{5} = 0.6$
4.	△DEA	8	6	$\sqrt{8^2 + 6^2} = 10$	$\frac{8}{6} \approx 1.\overline{3}$	$\frac{8}{10} = 0.8$	$\frac{6}{10} = 0.6$
5.	△FGA	12	9	$\sqrt{12^2 + 9^2} = 15$	$\frac{12}{9} \approx 1.\overline{3}$	$\frac{12}{15} = 0.8$	$\frac{9}{15} = 0.6$

Critical Thinking, **page 461**

$\sin A = 0.8$, $\sin B = 0.6$, $\cos A = 0.6$, $\cos B = 0.8$; $\sin A = \cos B$ and $\cos A = \sin B$. These values are equal because the side opposite A is adjacent to B, and the side adjacent to A is opposite B. The hypotenuse remains the same.

Communicate, **page 463**

1. The sine of an angle is equal to the ratio of the length of the opposite side over the length of the hypotenuse.
2. The cosine of an angle is equal to the ratio of the length of the adjacent side over the length of the hypotenuse.
3. Draw a right triangle reflecting the given information: Use trigonometric ratios with the given information. $\sin 4° = \frac{3}{r}$. Solve for r; $r \approx 43$ inches.
4. Yes, the cosine and sine ratio are always less than 1 because both the opposite and adjacent legs are smaller than the hypotenuse. If the larger number is the denominator, the ratio must be less than 1.

Chapter 9 Project

Activities 1–3, **page 467**

Answers will vary due to accuracy of measurements.

Lesson 10.1

Communicate, **page 478**

1. A rational expression is an expression that can be expressed in the form $\frac{P}{Q}$, where P and Q are both polynomials in the same variable and $Q \neq 0$.
2. A rational expression $\frac{P}{Q}$ is undefined whenever the value of Q is zero.
3. The graph of $h(x)$ is obtained from the parent function $g(x) = \frac{1}{x}$ by a stretch factor of 7, a reflection in the x-axis, and a vertical shift of 12 units.

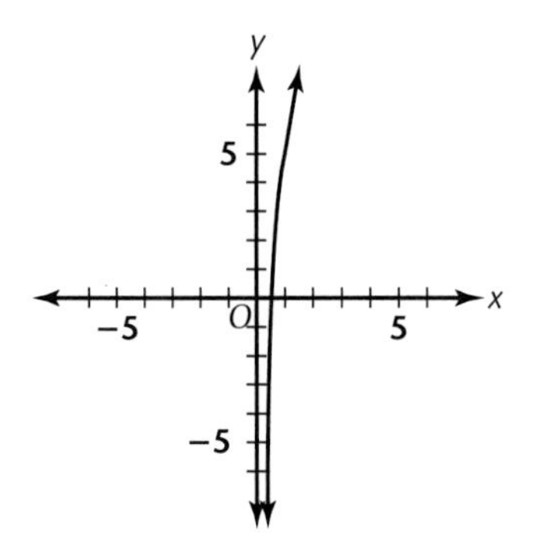

4. Express the function as the sum of two terms, $\frac{a}{x}$ and b, where a and b are constants. Since $h(x) = \frac{12x - 7}{x} = 12 - \frac{7}{x} = -\frac{7}{x} + 12$, then for h, a is -7 and b is 12. Now, the parent rational function, $g(x) = \frac{1}{x}$, is stretched by the a factor, -7, and vertically shifted by the b factor, 12.
5. When x is 0, $h(x)$ is undefined. The line $x = 0$ is a vertical asymptote of the graph.

Practice and Apply, **pages 478–479**

10. 4; 5.5

11. Undefined; $\frac{8}{3}$

12. 0; undefined

13. 0.75; 0

14. Shift to the right 5 units

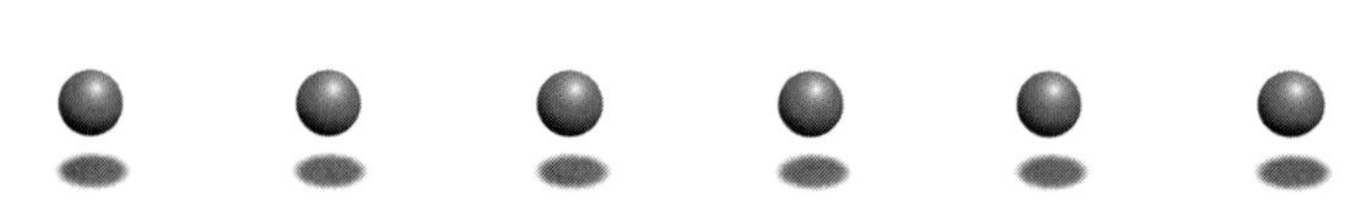

15. Stretch by a factor of 2

16. Reflection over the x-axis and shift up 3 units

17. Shift to the left by 3 units

18. Stretch by a factor of 4, shift to the left 2 units, and shift down 5 units

19. Reflection over the x-axis, stretch by a factor of 2, shift to the right 1 unit and up 4 units

20. $x \neq 0$

21. $x \neq -2$

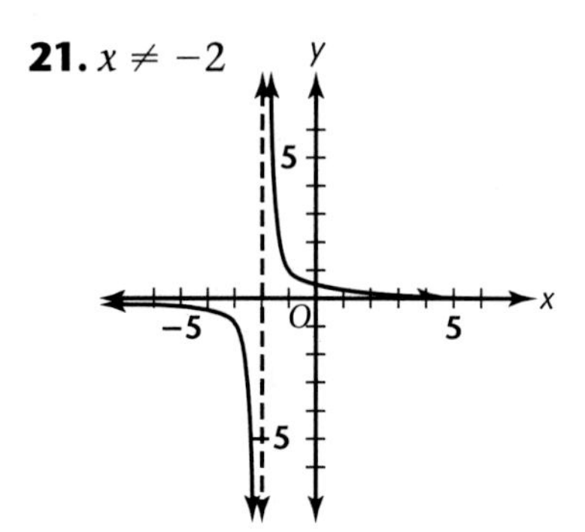

22. $x \neq 0$

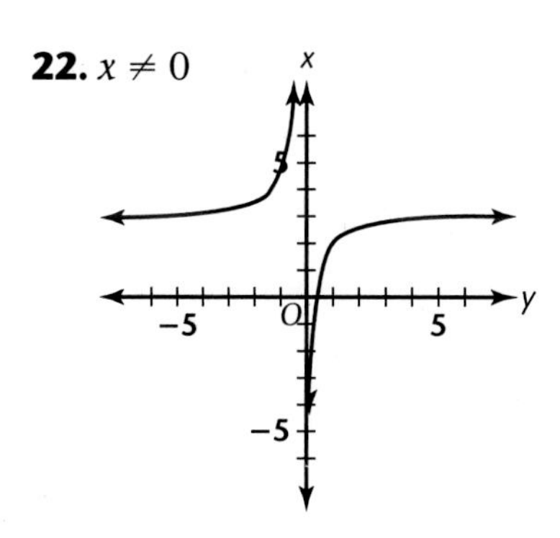

23. $x \neq 1$

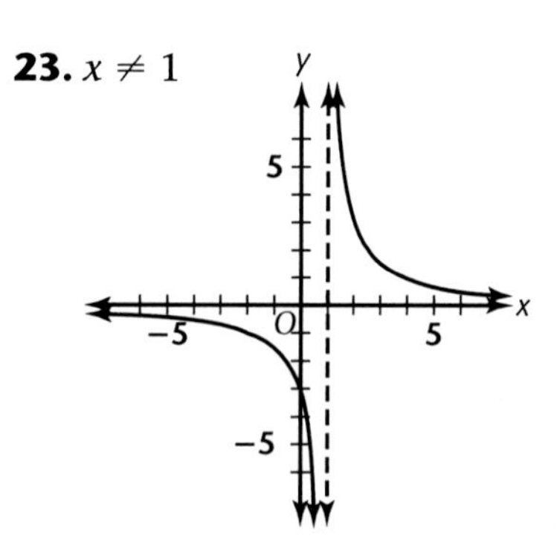

24. $x \neq 3$

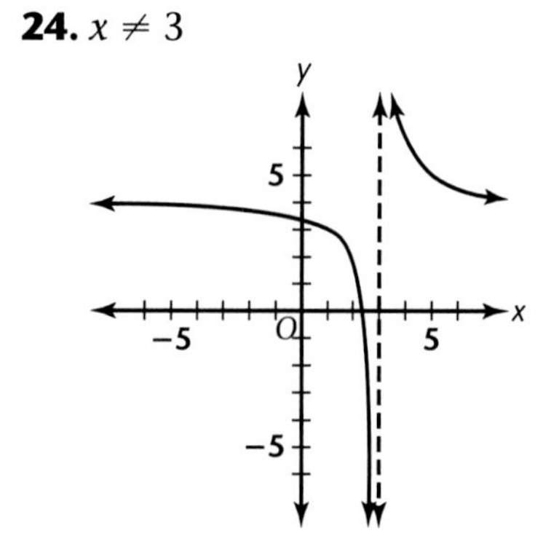

25. $x \neq 4$

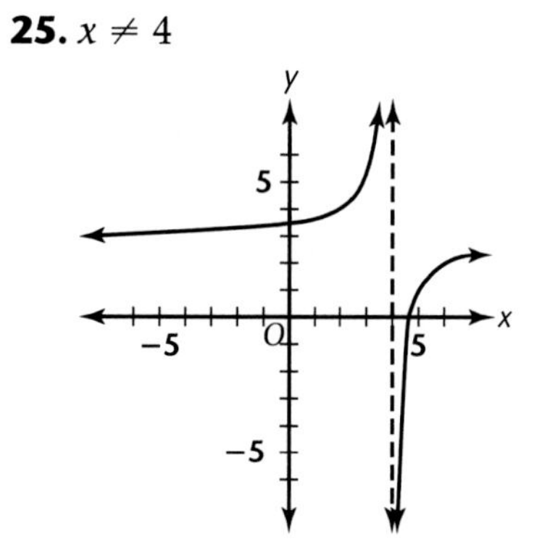

26.

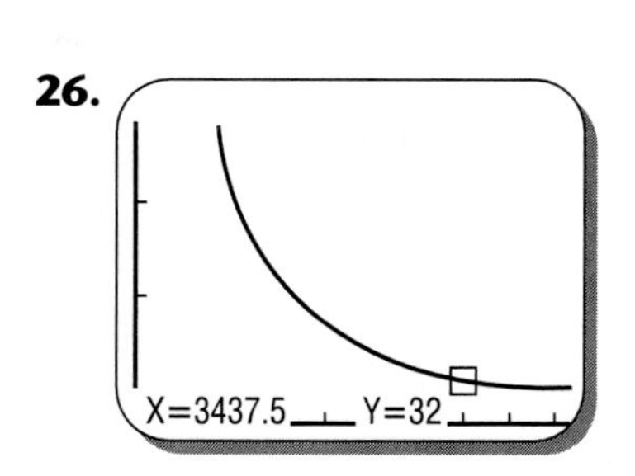

27. 3437.5 cubic inches

Look Beyond, **page 479**

35. Vertical shift up by 2, horizontal shift right by 1, stretched by a factor of 5

36.

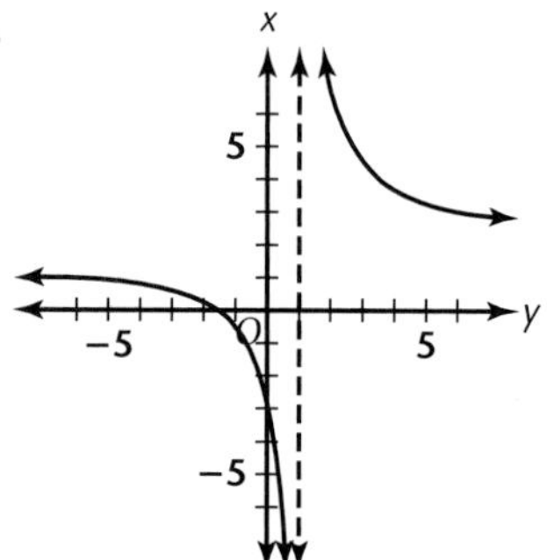

Extra Practice, **page 555**

13. Shift 3 units right

14. Vertical stretch by 3

15. Shift 2 units down

16. Reflection over the x-axis, shift 4 units up

17. Stretch by a factor of 4, shift 2 units right

18. Stretch by a factor of 4, shift 1 unit right and 3 units up

19. Reflection over the x-axis, stretch by a factor of 2, shift 1 unit left and 1 unit down

20. Stretch by a factor of 5, shift 3 units up

21. Reflection over the x-axis, stretch by a factor of 3 units, shift 3 units right and 6 units up

22. $x \neq 0$

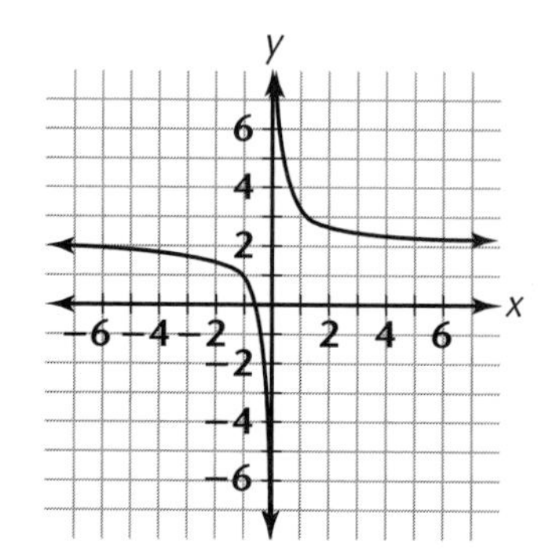

23. $x \neq 1$

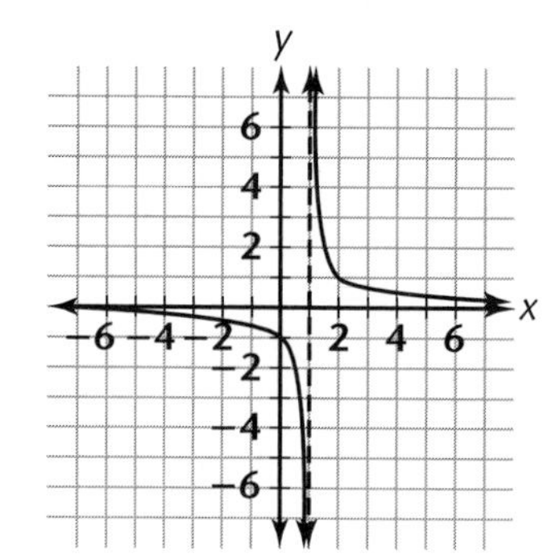

24. $x \neq 0$

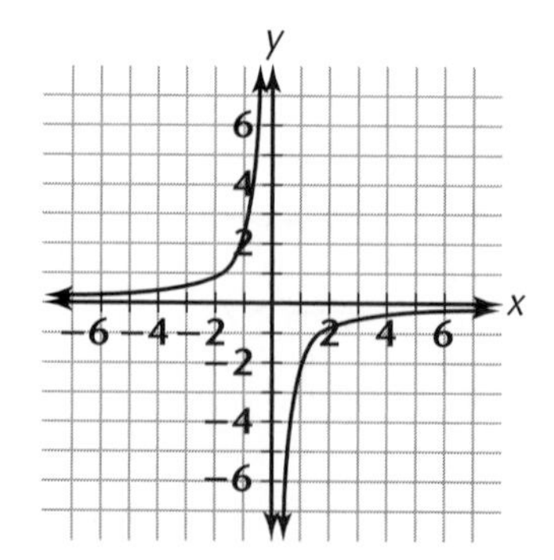

25. $x \neq -1$

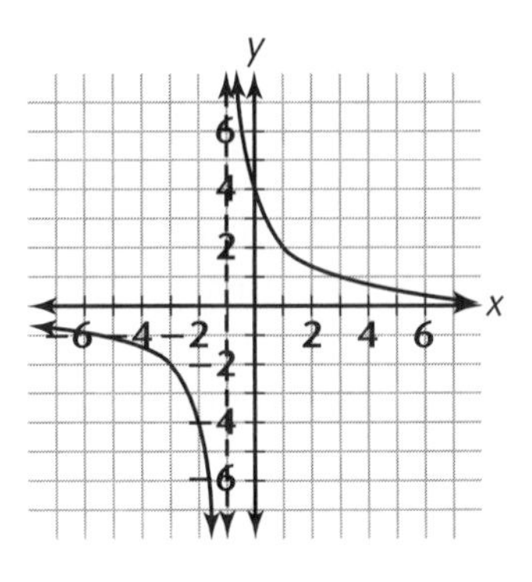

26. $x \neq -2$

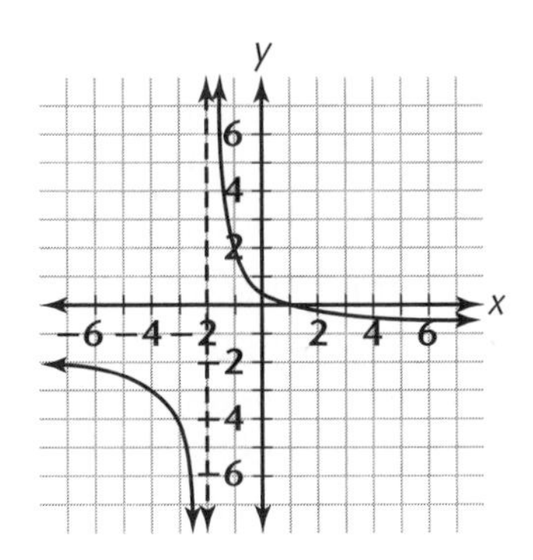

27. $x \neq 2$

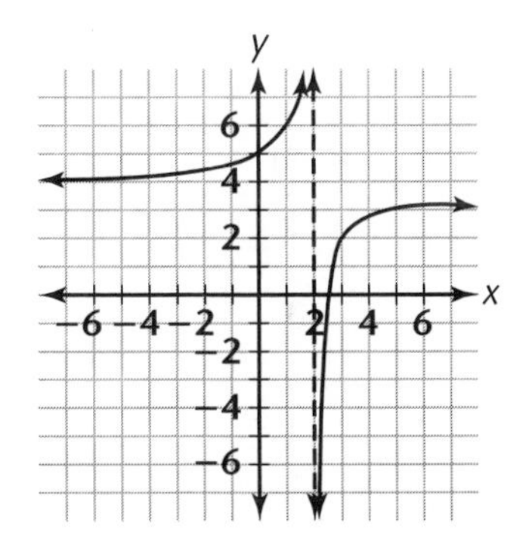

Lesson 10.2

Communicate, page 484

1. $xy = k$; $y = \frac{k}{x}$
2. An inverse variation between rate and time means that as the rate increases, the time decreases. For example, the faster you drive, the less time it takes to drive a certain distance.
3. If y varies inversely as x, then $y = \frac{k}{x}$, where k is the constant of variation. Substitute $y = 3$ and $x = 8$ to find the value of k. Then use the equation of variation to find the value of y when $x = 2$.
4. The Rule of 72 is $t = \frac{72}{r}$. To find how long it would take money to double at 4%, divide 72 by 4.
5. $y = \frac{k}{x}$, where k is the constant of variation and $x \neq 0$.

Lesson 10.3

Ongoing Assessment, page 487

If they are different, an error has been made. It is important that the check be the same for the original fraction and its reduced form because you are trying to check that the two expressions are equivalent.

Communicate, page 489

1. Any value of the variable that makes the value of the denominator 0 results in an undefined expression.
2. Set the expression in the denominator equal to zero and solve the resulting equation to find the value of the variable that would make the expression undefined.
3. A common factor divides evenly into each of two or more numbers or expressions.

4. Factor the denominator to obtain $\frac{x+1}{(x+1)(x+1)}$. Then remove the common factors from the numerator and the denominator. State any restrictions to avoid division by zero. So the expression simplifies to $\frac{1}{x+1}$, $x \neq -1$.

5. $x - 7$ is the opposite of $7 - x$ because $-1(x - 7) = -x + 7 = 7 - x$.

Practice and Apply, page 490

18. $\frac{8(x+1)}{15(x+2)}$, $x \neq -2$
19. $\frac{a+b}{2(a-b)}$, $a \neq b$
20. $\frac{2}{5}$, $c \neq -2$
21. $\frac{x-y}{2}$, $x \neq -y$
22. $\frac{2m+3}{2}$
23. 7, $t \neq -3$
24. $\frac{3+2x}{x}$, $x \neq 0$
25. $\frac{d(3d+2)}{3d+1}$, $d \neq -\frac{1}{3}$
26. $\frac{1}{b-2}$, $b \neq 2, -2$
27. $\frac{1}{x+4}$, $x \neq 2, -4$
28. $\frac{-1}{a+7}$, $a \neq -1, -7$
29. $\frac{-1}{k+3}$, $k \neq 4, -3$
30. $\frac{c-3}{3}$, $c \neq -3$
31. $\frac{3}{n-3}$, $n \neq 3, 4$
32. $\frac{y-1}{y+4}$, $y \neq -4, -3$
33. $\frac{a-b}{a+b}$, $a \neq -b$

Look Back, page 491

43. $g(x)$ is obtained from $f(x)$ by a shift of 2 units to the left and 5 units down.

Look Beyond, page 491

48. The top left square is number 2. The squares number counterclockwise, ending at number 8 at the top right.

Lesson 10.4

Ongoing Assessment, page 493

Restrictions on the denominator in the original expression are values that cause the denominator to equal zero and cause the rational expression to be undefined. Solutions cannot include these values.

Ongoing Assessment, page 494

Decide what factor you have to multiply the denominator of the fraction by to get the least common denominator. Then multiply the fraction by that factor divided by itself.

Communicate, page 495

1. Multiply $x + 1$ by x or x by $x + 1$.
2. After the numerators are rewritten so the fractions have common denominators, add the numerators.
3. Follow the steps in 2, but subtract.
4. Multiply the numerators, then multiply the denominators. Simplify.
5. Multiply the first expression by the reciprocal of the second expression. Simplify.

Practice and Apply, page 496

21. $\frac{2}{x^2+x}$, $x \neq 0, -1$
22. $\frac{15}{x^2-x}$, $x \neq 0, 1$
23. $\frac{y^2-y+4}{y^2-4y}$, $y \neq 0, 4$
24. $\frac{x+3}{x^2+x}$, $x \neq 0, -1$
25. $\frac{a^2+6a-1}{a^2+4a+3}$, $a \neq -3, -1$
26. $\frac{m^2+2m-15}{m^2+m-2}$, $m \neq 1, -2$
27. $\frac{r-r^2}{r^2+6r+5}$, $r \neq -1, -5$
28. $\frac{5b+18}{b^2+b-6}$, $b \neq 2, -3$
29. $\frac{9y-8}{y^2-4y+4}$, $y \neq 2$

30. $\frac{-3}{y+6}$, $y \neq 5, -6$

31. -1, $x \neq 1, 2$

32. $\frac{7-x}{2x^2-18}$, $x \neq 3, -3$

33. $C = \frac{45}{750+w}$, where w represents the number of kilograms of pure water added

34.

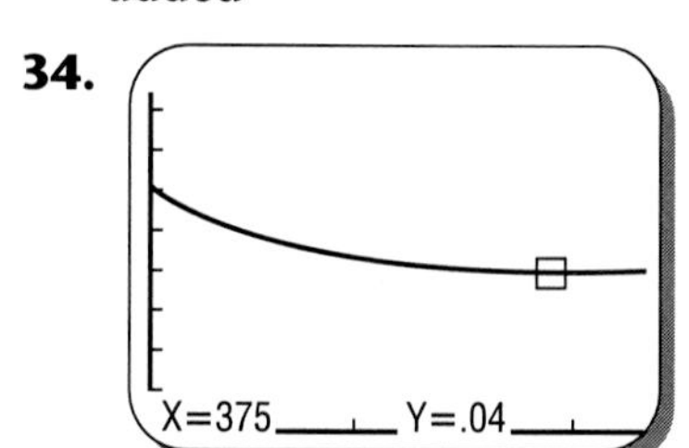

35. 375 kilograms

Look Back, page 497

38.

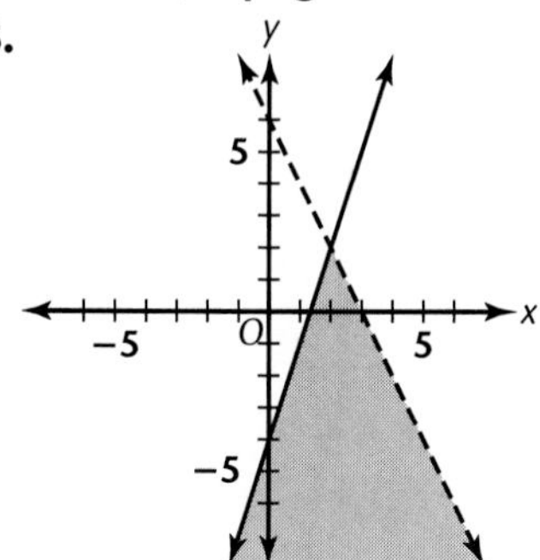

44.

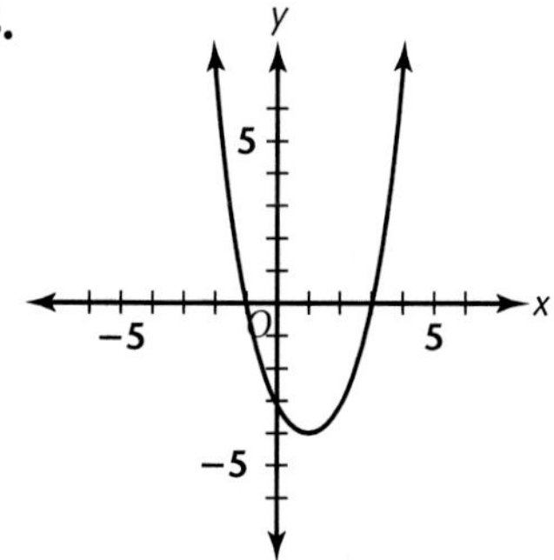

Extra Practice, page 556

1. $\frac{3}{y}$, $y \neq 0$
2. $\frac{4}{s-2}$, $s \neq 2$
3. $\frac{-5m}{n+3}$, $n \neq -3$
4. $\frac{2w-7}{x-y}$, $x \neq y$
5. 1, $x \neq -1$
6. $z - 3$
7. $\frac{5n+2m}{mn}$, $m \neq 0, n \neq 0$
8. $\frac{-7}{4t}$, $t \neq 0$
9. $\frac{x^2+12z^2}{3wxz}$, $w \neq 0, x \neq 0, z \neq 0$
10. $\frac{6b-c}{ab^2}$, $a \neq 0, b \neq 0$
11. $\frac{-11}{5w-10}$, $w \neq 2$
12. $\frac{2n}{m-2}$, $m \neq 2$
13. $\frac{a^2+4a+5}{a+5}$, $a \neq -5$
14. $\frac{2g-7}{g-1}$, $g \neq 1$
15. $\frac{ad-bc}{bd}$, $b \neq 0, d \neq 0$
16. $\frac{7-x}{m+n}$, $m \neq -n, y \neq -4$
17. $\frac{a}{a^2+8a+16}$, $a \neq -4, -1, 0$
18. $\frac{x-3}{x}$, $x \neq -3, 0$
19. $\frac{4}{y^2+2y}$, $y \neq -2, 0$
20. $\frac{-12}{r^2-3r}$, $r \neq 0, 3$
21. $2t$, $t \neq 1$
22. $\frac{4x+13}{x^2+x-2}$, $x \neq -2, 1$
23. $\frac{3c+2}{c^2+7c+10}$, $c \neq -5, -2$
24. $\frac{v^2-2v+12}{v^2-v-6}$, $v \neq -2, 3$
25. $\frac{x^2+7x}{x^2+x-12}$, $x \neq -4, 3$
26. $\frac{4x+9}{x^2-3x-18}$, $x \neq -3, 6$
27. $\frac{7n+21+2m}{n^2-9}$, $n \neq \pm 3$
28. -1, $y \neq 1$

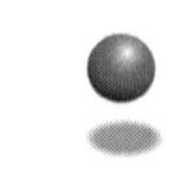

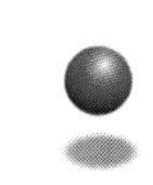

29. $\frac{h+17}{h^2+6h+8}$; $h \neq -4, -2$

30. $\frac{3x+2}{4x^2-16}$, $x \neq \pm 2$

Eyewitness Math, page 499

1a. 0.5% of 10,000 = (0.005)(10,000) = 50

1b. 98% of 50 = (0.98)(50) = 49

1c. 9950; Since 0.5% of the 10,000 are assumed to have cancer, then the remaining 99.5% can be assumed not to: (0.995)(10,000) = 9950. Or, since 50 out of 10,000 have it, 10,000 − 50, or 9950, do not.

1d. 199; The test gives correct results 98% of the time and therefore gives incorrect results 2% of the time. So, (9950)(0.02) = 199.

1e. 248; 49 true positives plus 199 false positives

1f. $P = 49/248$, or about 0.2. You would have about a 20% chance of actually having cancer.

1g. Answers may vary. For those who originally figured there would be a 98% chance of having cancer, the 20% figure should be surprising. Though worried still, they should be much less worried with a 20% chance than with a 98% chance.

2a. The number of true positives are arN.

2b. The number of false positives are $(1-a)(N-rN)$ or $(1-a)(1-r)N$.

2c. $P = \frac{arN}{(1-a)(1-r)N + arN} = \frac{ar}{(1-a)(1-r) + ar}$

Lesson 10.5

Communicate, page 504

1. The values of the variable that make the denominator of any rational expressions in the equation have value 0.
2. The realistic domain usually contains positive numbers because x is representing some type of measurement.
3. Find the lowest common multiple of the three denominators. The common denominator is $c(c-1)$.
4. Graph two functions, Y_1 = the left side of the equation and Y_2 = the right side of the equation, and find their point(s) of intersection. The x-value(s) at the point(s) of intersection is the solution provided this value does not make the denominator of any of the terms in the equation have value 0.
5. Look at the denominator of each rational expression in the equation to find any values of the variable that make them have value 0. This equation is undefined for $c = 0$ and $c = 1$.
6. $\text{Rate} = \frac{1}{\text{number of hours worked}}$

Practice and Apply, pages 504–505

26. Upstream: $\frac{300}{20-x}$, downstream: $\frac{500}{20+x}$, where x represents the rate of the current

28. Jeff $\frac{1}{x}$, Phil $\frac{1}{x+2}$, Allison $\frac{1}{x-1}$, where x represents the time for Jeff to complete the job alone

29. Tom $\frac{x}{5}$, Shelley $\frac{x}{7}$, where x represents the time to paint one car together

Look Back, page 505

32.
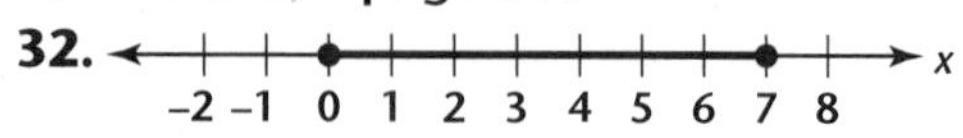

33. $x = 4, y = 1$

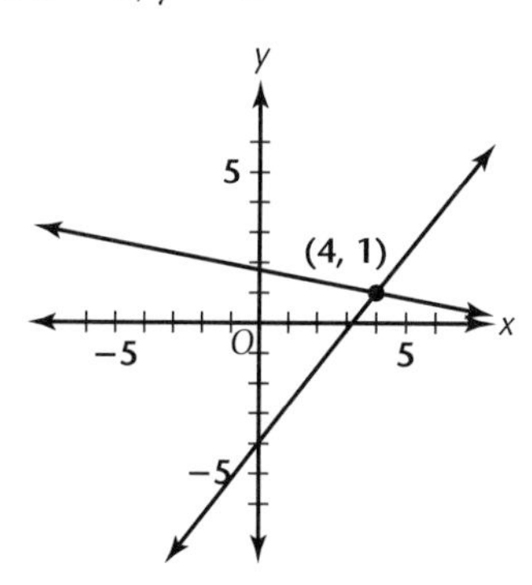

transposed, the resulting proportion is a true proportion; **e.** The differences of the numerator and the denominator of the original proportion divided by the denominators of the original proportion are equal to each other and are true proportions; **f.** The left side of the original proportion is equal to the sum of the numerators of the original proportion divided by the sum of the denominators of the original proportion, and is a true proportion.

Lesson 10.6

Exploration 1, page 507

7. Both proportions involve one variable and can be solved by using the cross product. The cross product in **a** results in a linear equation with one solution, while the cross product in **b** results in a quadratic equation with two solutions.

Exploration 2, page 508

1. Answers may vary. Examples are given for $a = 2$, $b = 4$, $c = 3$, and $d = 6$.

1a. $bc = 12$; $ad = 12$

1b. $c(a + b) = 18$; $c(a - b) = -6$

1c. $d(a + b) = 36$; $d(a + b) = 36$

1d. $ad = 12$; $bc = 12$

1e. $d(a - b) = -12$; $b(c - d) = -12$

1f. $a(b + d) = 20$; $b(a + c) = 20$

2. a, c, d, e, f

3. **a.** The reciprocals of the original proportion are equal to each other and are true proportions; **c.** The sums of the numerator and the denominator of the original proportion divided by the denominators of each side of the original proportion are equal to each other and are true proportions; **d.** If the means of the original proportion are

Communicate, page 509

1. A proportion is an equation that states that two ratios are equal. For example, $\frac{2}{3} = \frac{4}{6}$.

2. The extremes of a proportion are the numerator of the first ratio and the denominator of the second ratio. The means are the denominator of the first ratio and the numerator of the second ratio.

3. The product of the means is equal to the product of the extremes.

4. Find the cross product and solve the resulting linear equation: $4 = 4x + 2$, $x = 0.5$.

5. Let d represent the distance traveled on 75 liters. Write a proportion that involves two equal ratios comparing the number of liters used to the distance traveled: $\frac{40}{320} = \frac{75}{d}$. The car can travel 600 kilometers.

Practice and Apply, pages 509–510

6. Means: 3, 7; extremes: x, 4

7. Means: 10, 2; extremes: 5, x

8. Means: 4, y; extremes: 3, 5

9. Means: 8.5, 2; extremes: 17, x

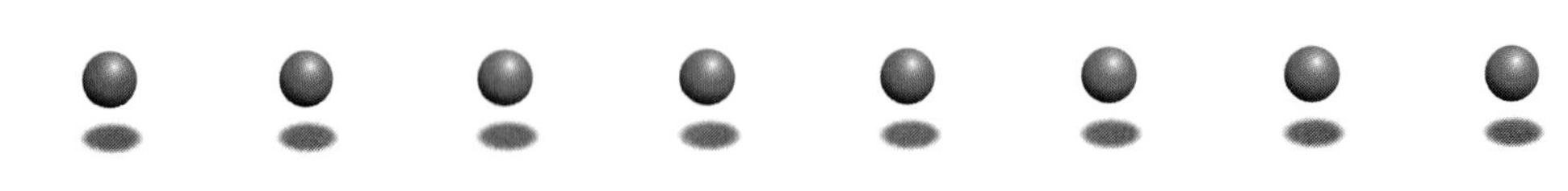

10. Means: $m - 2$, m; extremes: 10, 7

11. Means: 5, w; extremes: 3, $2w - 3$

12. Means: 2, $n + 6$; extremes: 3, n

13. Means: 5, $c - 2$; extremes: 4, $c + 3$

39. 7.45 m

39a. Gear ratio $= \frac{\text{wheel teeth}}{\text{pedal teeth}}$; $\frac{13}{49}, \frac{14}{49}, \frac{15}{49}, \frac{16}{49}, \frac{18}{49}, \frac{21}{49}, \frac{24}{49}, \frac{13}{54}, \frac{14}{54}, \frac{15}{54}, \frac{16}{54}, \frac{18}{54}, \frac{21}{54}, \frac{24}{54}$.
Reciprocal $= \frac{\text{pedal teeth}}{\text{wheel teeth}}$; 3.77, 3.5, 3.27, 3.06, 2.72, 2.33, 2.04; 4.15, 3.86, 3.6, 3.38, 3, 2.57, 2.25

39b. Answers may vary.

Extra Practice, **page 557**

1. Means: 6, 4; extremes: 9, x
2. Means: 1.5, z; extremes: 7, 14
3. Means: c, $4 + c$; extremes: 19, 3
4. Means: 4, $m - 2$; extremes: $m + 1$, 3

Lesson 10.7

Try This, **page 512**

$a + c = b + c$	Given
$a + c - c = b + c - c$	Subtraction Property of Equality
$a = b$	Simplify.

Try This, **page 513**

k is an integer.	Hypothesis
$k + 1$ is an integer.	Definition of an Integer
Let $n = k + 1$. Then n is an integer.	Substitution of n for $k + 1$
$2n + 1$ is an odd number.	Definition of Odd Numbers
$2(k + 1) + 1$ is an odd number.	Substitution of $k + 1$ for n
$2k + 3$ is an odd number.	Distributive Property; Simplify.

Communicate, **page 514**

1. A proof in algebra gives a logical justification to show that a conclusion is true.
2. Definitions, postulates, and already proved theorems
3. The hypothesis, the statement that follows the word "if," is assumed to be true. The conclusion, the statement that follows the word "then," is the part that has to be proved.
4. In the converse of a statement, the conclusion and the hypotheses are interchanged.
5. Start with the given equation $7x + 9 = -5$. Add -9 to both sides using the Addition Property of Equality: $7x + 9 - 9 = -5 - 9$. Since 9 and -9 are opposites, their sum is 0, so the equation simplifies to $7x = -14$. Divide both sides by 7 using the Division Property of Equality: $x = -2$.

Practice and Apply, **pages 514–515**

17–32. Proofs may vary.

17.

$2y + 3y = 5y$	Given
$y(2 + 3) = 5y$	Distributive Property
$y(5) = 5y$	Addition of Integers
$5y = 5y$	Commutative Property

18.

$5x - 3x = 2x$	Given
$x(5 - 3) = 2x$	Distributive Property
$x(2) = 2x$	Subtraction of Integers
$2x = 2x$	Commutative Property

19.

$(-3a)(-2a) = 6a^2$	Given
$(-3)a \cdot a(-2) = 6a^2$	Commutative Property
$(-3)a^2(-2) = 6a^2$	Definition of a Power
$(-3)(-2)a^2 = 6a^2$	Commutative Property
$6a^2 = 6a^2$	Multiplication of Integers

20. $7n - 3n = 32$ Given
$n(7 - 3) = 32$ Distributive Property
$n(4) = 32$ Subtraction of Integers
$\frac{4n}{4} = \frac{32}{4}$ Commutative Property, Division Property of Equality
$n = 8$ Division of Integers

21. $3(x + 2) = -15$ Given
$3x + 6 = -15$ Distributive Property
$3x + 6 - 6 = -15 - 6$ Subtraction Property of Equality
$3x + 0 = -21$ Property of Opposites
$3x = -21$ Identity for Addition
$\frac{3x}{3} = \frac{-21}{3}$ Division Property of Equality
$x = -7$ Division of Integers

22. $\frac{2}{3}x + 5 = -9$ Given
$\frac{2}{3}x + 5 - 5 = -9 - 5$ Subtraction Property of Equality
$\frac{2}{3}x + 0 = -14$ Property of Opposites
$\frac{2}{3}x = -14$ Identity for Addition
$\left(\frac{3}{2}\right)\left(\frac{2}{3}x\right) = \left(\frac{3}{2}\right)(-14)$ Multiplication Property of Equality
$x = -\frac{42}{2}$ Reciprocal Property
$x = -21$ Division of Integers

23. $\frac{3}{4}m + 8 = -1$ Given
$\frac{3}{4}m + 8 - 8 = -1 - 8$ Subtraction Property of Equality
$\frac{3}{4}m + 0 = -9$ Property of Opposites
$\left(\frac{4}{3}\right)\left(\frac{3}{4}m\right) = \left(\frac{4}{3}\right)(-9)$ Multiplication Property of Equality
$m = -\frac{36}{3}$ Reciprocal Property
$m = -12$ Division of Integers

24. $a + (b + c)$ Given
$= (a + b) + c$ Associative Property

25. $a = b$ Given
$a + c = b + c$ Addition Property of Equality

26. $a(b - c)$ Given
$= ab - ac$ Distributive Property

27. $(ax + b) + ay$ Given
$= (ax + ay) + b$ Commutative and Associative Properties
$= a(x + y) + b$ Distributive Property

28. If a number is even, then it has 2 as a prime factor and is divisible by 2. When the number is squared the number 2 will appear at least twice and the number will be divisible by 2. If n is even and k is an integer, then $n = 2k$. $(n)^2 = (2k)^2$ and since $4k^2$ is divisible by 2, n^2 must be divisible by 2.

29. If n is odd, then $n = k + 1$, where k is an even number. Then $n^2 = (k + 1)(k + 1) = k^2 + 2k + 1$. From Exercise 28, if k is even, then k^2 is also even. Since $2k$ is also always even, the sum $k^2 + 2k + 1$ must be odd.

30. If $(a - b) + (b - a) = 0$, then $a - b$ and $b - a$ are opposites. $(a - b) + (b - a)$
$= a + (-b + b) + (-a)$ Associative Property
$= a + 0 + (-a)$ Property of Opposites
$= a + (-a)$ Identity for Addition
$= 0$ Property of Opposites
Therefore, $a - b$ and $b - a$ are opposites.

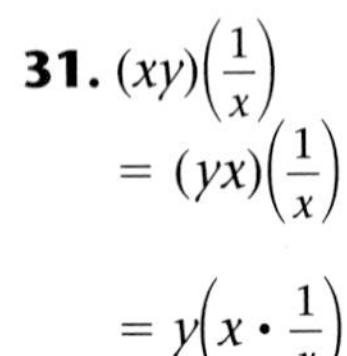

31. $(xy)\left(\frac{1}{x}\right)$ Given

$= (yx)\left(\frac{1}{x}\right)$ Commutative Property

$= y\left(x \cdot \frac{1}{x}\right)$ Associative Property

$= y(1)$ Reciprocal Property

$= y$ Identity for Multiplication

32. $\frac{a}{b} + \frac{c}{d}$ Given

$= 1\left(\frac{a}{b}\right) + 1\left(\frac{c}{d}\right)$ Identity for Multiplication

$= \left(\frac{d}{d}\right)\left(\frac{a}{b}\right) + \left(\frac{b}{b}\right)\left(\frac{c}{d}\right)$ Substitution Property

$= \frac{da}{db} + \frac{bc}{bd}$ Multiplication of Rational Numbers

$= \frac{ad}{bd} + \frac{bc}{bd}$ Commutative Property

$= \frac{ad + bc}{bd}$ Addition of Rational Numbers

Look Back, **page 515**

37. Stretch by a factor of 2, reflect through the x-axis, and shift 1 unit to the right and 3 units up.

Extra Practice, **page 557**

1. $3x + 4x$ Given
$3x + 4x = 7x$ Addition of Monomials

2. $7y - 6y$ Given
$7y - 6y = y$ Subtraction of Monomials

3. $(-4m)(-3m)$ Given
$(-4m)(-3m) = 12m^2$ Multiplication of Monomials

4. $8x - 3x = 35$ Given
$5x = 35$ Subtraction of Monomials
$\frac{5x}{5} = \frac{35}{5}$ Division Property of Equality
$x = 7$ Division of Integers

5. $3c - 1 = 11$ Given
$3c - 1 + 1 = 11 + 1$ Addition Property of Equality
$3c = 12$ Property of Opposites
$\frac{3c}{3} = \frac{12}{3}$ Division Property of Equality
$c = 4$ Division of Integers

6. $4(y + 2) = -24$ Given
$4y + 8 = -24$ Distributive Property
$4y + 8 - 8 = -24 - 8$ Subtraction Property of Equality
$4y = -32$ Property of Opposites
$\frac{4y}{4} = \frac{-32}{4}$ Division Property of Equality
$y = -8$ Division of Integers

7. $\frac{3}{4}x + 5 = 2$ Given
$\frac{3}{4}x + 5 - 5 = 2 - 5$ Subtraction Property of Equality
$\frac{3}{4}x = -3$ Property of Opposites
$\frac{4}{3} \cdot \frac{3}{4}x = \frac{4}{3} \cdot -3$ Multiplication Property of Equality
$x = -4$ Reciprocal Property

8. $\frac{3}{5}n - 6 = 0$ Given
$\frac{3}{5}n - 6 + 6 = 0 + 6$ Addition Property of Equality
$\frac{3}{5}n = 6$ Property of Opposites

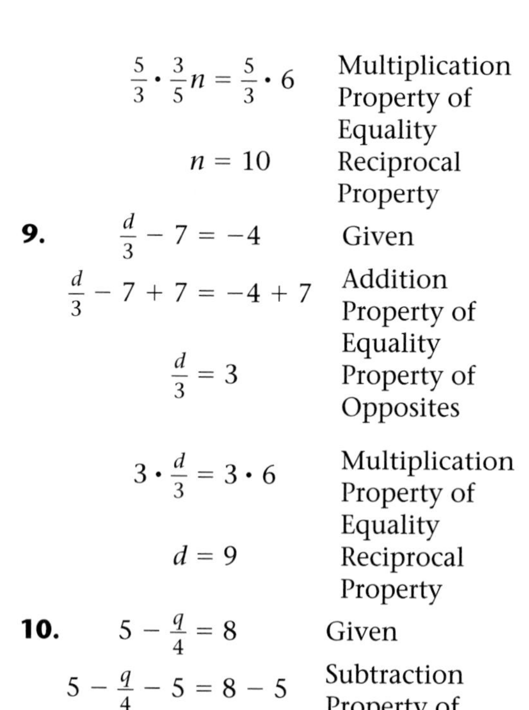

$\frac{5}{3} \cdot \frac{3}{5}n = \frac{5}{3} \cdot 6$	Multiplication Property of Equality
$n = 10$	Reciprocal Property

9.

$\frac{d}{3} - 7 = -4$	Given
$\frac{d}{3} - 7 + 7 = -4 + 7$	Addition Property of Equality
$\frac{d}{3} = 3$	Property of Opposites
$3 \cdot \frac{d}{3} = 3 \cdot 6$	Multiplication Property of Equality
$d = 9$	Reciprocal Property

10.

$5 - \frac{q}{4} = 8$	Given
$5 - \frac{q}{4} - 5 = 8 - 5$	Subtraction Property of Equality
$-\frac{q}{4} = 3$	Property of Opposites
$-4 \cdot -\frac{q}{4} = -4 \cdot 3$	Multiplication Property of Equality
$q = -12$	Reciprocal Property

Chapter 10 Project

Activity 1, page 517

61.1 meters per second

Activity 2, page 517

433.3 cubic inches

Activity 3, page 517

Answers may vary.

Chapter 10 Review, pages 518–520

1. Stretch by a factor of 2, reflect through the x-axis, and shift up 4 units; $x \neq 0$.

2. Shift to the right 3 units and up 2 units; $x \neq 3$.

3. Stretch by a factor of 3, reflect through the x-axis, shift 4 units left and 1 unit down; $x \neq -4$.

19–20. Proofs may vary.

19.

$2a + 3 = 5$	Given
$2a + 3 - 3 = 5 - 3$	Subtraction Property of Equality
$2a + 0 = 2$	Property of Opposites; Simplify.
$2a = 2$	Identity for Addition
$\frac{2a}{2} = \frac{2}{2}$	Division Property of Equality
$a = 1$	Reciprocal Property

20.

$(a - b) - (b - a)$	Given
$= a - b - b + a$	Distributive Property
$= a + (-b) + (-b) + a$	Definition of Subtraction
$= a + (-2b) + a$	Addition of Monomials
$= a + a + (-2b)$	Commutative Property of Addition
$= 2a + (-2b)$	Addition of Monomials
$= 2a - 2b$	Definition of Subtraction

Chapter 10 Assessment, page 521

2. Stretch by a factor of 4, reflect over the x-axis, shift 5 units to the right and 1 unit up; $x \neq 5$.

7. $\frac{2x(x-1)}{3}, x \neq 0$

8. $\frac{1}{x+y}, x \neq \pm y$

9. $\frac{31}{12y}, y \neq 0$

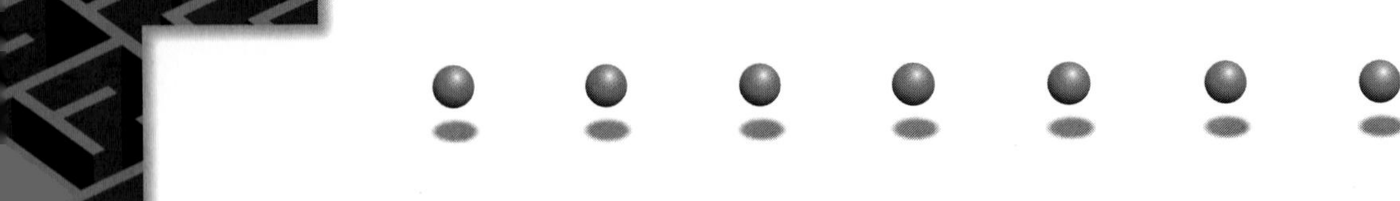

10. $\frac{13}{x-8}, x \neq 8$

11. $\frac{-3x^2+7x-2}{30}, x \neq 0, -2$

12. $\frac{10a-a^2}{a^2+a-12}, a \neq 3, -4$

25. $3b - (b - a) = 3b - b + a$ Distributive Property
$= (3b - b) + a$ Associative Property for Multiplication
$= 2b + a$ Subtraction of Monomials

Chapters 1–10 Cumulative Assessment, pages 522–523

11.

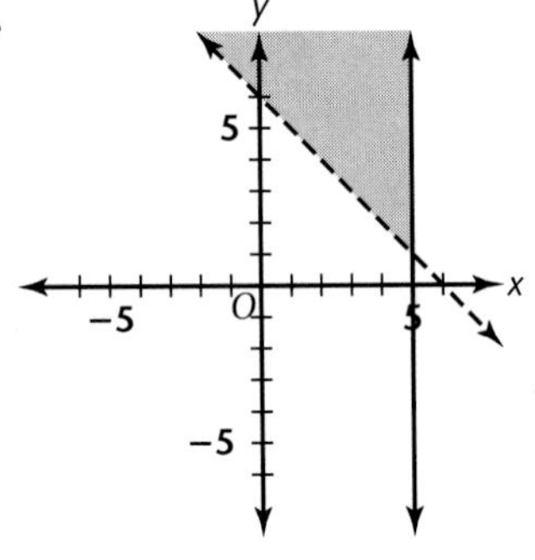

12. $y = |x|$, stretched by a factor of 5

20. Proofs may vary.

$a + 3(a + b) = 4a + 3b$	Given
$a + 3a + 3b = 4a + 3b$	Distributive Property
$4a + 3b = 4a + 3b$	Addition of Monomials